Elementary Differential Equations with Boundary Value Problems

ELEMENTARY DIFFERENTIAL EQUATIONS WITH BOUNDARY VALUE PROBLEMS

FOURTH EDITION

WILLIAM R. DERRICK

University of Montana

STANLEY I. GROSSMAN

University College London

ADDISON-WESLEY

An imprint of Addison Wesley Longman, Inc.

Reading, Massachusetts • Menlo Park, California • New York • Harlow, England
Don Mills, Ontario • Sydney • Mexico City • Madrid • Amsterdam

Sponsoring Editor: Kevin Connors/Laurie Rosatone
Project Coordination: Elm Street Publishing Services, Inc.
Art Development Editor: Vita Jay
Design Administrator: Jess Schaal
Text and Cover Design: Ellen Pettengell
Cover Photo: Hank Morgan/Photo Researchers, Inc.
Photo Research: Nina Page/Judy Ladendorf
Production Administrator: Randee Wire
Compositor: Interactive Composition Corporation
Printer and Binder: R.R. Donnelley & Sons Company
Cover Printer: Phoenix Color Corporation

Elementary Differential Equations with Boundary Value Problems, Fourth Edition

Library of Congress Cataloging-in-Publication Data

Derrick, William R.
Elementary differential equations with boundary value problems / William R. Derrick, Stanley I. Grossman. — 4th ed.
p. cm.
Rev. ed. of: Introduction to differential equations with boundary value problems. 3rd ed. c1987.
Includes index.
ISBN 0-673-98555-5
1. Differential equations. 2. Boundary value problems. I. Grossman, Stanley I. II. Derrick, William R. Introduction to differential equations with boundary value problems. 3rd ed. III. Title.
QA371.D432 1996
515′.35—dc20 96–13215
CIP

ISBN 0-673-98555-5

1 2 3 4 5 6 7 8 9—DOW—99 98 97 96

To the memory of Marvin A. Derrick
and Benjamin and Regina Grossman

CONTENTS

PREFACE

Since the time of Newton, differential equations have been of fundamental importance in the application of mathematics to the physical sciences. Lately, differential equations have gained increasing importance in the biological and social sciences. New information concerning the theory of differential equations and its application to all sciences appears at an increasing rate, leading to a continued interest in its study by students in all scientific disciplines.

Today, the subject of differential equations is well understood. There is a large and mathematically elegant body of theory that enables us to study differential equations in a very systematic way. This provides the writers of textbooks in the subject area with a challenge: to provide the theoretical tools needed to understand fully how various topics fit together while not losing sight of the important applications that led to the development of the theory.

In this book, the theory is not slighted. Every theorem that can be proved using the mathematics available to the student who has studied calculus is proved in the text. Occasionally, more difficult tools—the notion of uniform convergence, for example—are used to prove results that are central to the study of differential equations. However, the primary goal of this book is to teach students how to use differential equations in applied areas.

Here are some features that illustrate how our dual goals have been met:

Examples

As students, we learned how to solve differential equations by reading examples and solving problems. *Elementary Differentiai Equations with Boundary Value Problems* contains approximately 400 examples—many more than are commonly found in standard differential equations texts. Most examples contain all the algebraic and calculus steps needed to complete the solution. It is infuriating to be told, "It easily follows that . . . ;" when it is not at all obvious. Students have a right to see the "whole hand," so to speak, so that they always know how to get from A to B. In many instances, explanations are highlighted in color to make a step easier to follow.

Exercises

The text contains more than 2,500 exercises, including both drill and applied problems, as well as exercises designed to be solved using a computer algebra system.

Exercises provide the most important learning tool in any undergraduate mathematics texbook. We stress to our students that no matter how well they think they understand our lectures or the textbook, they do not really know the material until they have worked problems. A vast difference exists between understanding someone else's solution and solving a problem by yourself. Learning mathematics without doing problems is about as easy as learning to ski without going to the slopes.

Chapter Summaries

At the end of each chapter (except Chapter 4) a detailed review of the important results of that chapter appears and page references are included. These reviews should prove to be very helpful in studying for exams.

Chapter Review Exercises

At the end of each chapter we have provided a collection of review exercises. Any student who can do these exercises can feel confident that he or she understands the material in the chapter.

Self-Quiz Problems

Almost every problem set that doesn't consist solely of applications begins with multiple-choice and true-false questions that require little or no computation. Answers to these problems appear at the end of the self-quiz problems. These problems test whether the student understands the basic ideas in the section, and they should be done before tackling the more standard problems that follow. The Self-Quiz Problems are new to this edition.

Captions

Captions have been added to almost every example and figure in the text. They make it easier for students to determine at a glance exactly what is being done in each example and will be very helpful when tackling the problem sets. The captions on figures clearly describe exactly what each figure illustrates.

Applications

Differential equations is *applied* mathematics. Consequently, this book includes many applied examples and exercises. The majority of the examples are drawn from the physical sciences. However, a wide range of applications in other areas is provided as well. Here is a partial list of applications in just the first four chapters:

Computation of escape velocity and reentry into the atmosphere (Section 2.1)
Electrical circuits (Sections 2.7, 4.4)
Harmonic and forced harmonic motion (Sections 4.1–4.3)
Newton's law of cooling (Section 1.1)
The collapse of the Tacoma Narrows Bridge (Section 4.3)
Chemical mixture problems (Section 2.6)
A model of epidemics (Section 2.3)
Quality control in business (Section 3.9)
Foucault's Pendulum (Section 4.1)
The Apollo Project (Section 2.1)
The Hubble telescope and lithotripsy (Section 2.5)
Logistic growth (Section 2.2)
Magnetic induction (Section 2.7)
Bifurcation and chaos (Section 2.9)

Perspectives

In nine sections of the book, we digress into short "perspectives" that explore an important topic from a different point of view. For example, every differential equations text discusses separable differential equations. We go further: after solving a number of differential equations in Section 2.1 by separating the variables we ask (and answer) the question, "When is a differential equation separable?" Some of the other perspectives are:

Singular Solutions (Section 2.2)
Green's Functions (3.5)
The Taylor Series Method (5.1)
Oscillatory Solutions of Linear Second-Order Differential Equations with Variable Coefficients (5.6)
Numerical Instability Caused by Propagation of Round-Off Error (8.6)
Stiff Equations (8.7)
The Matrix Exponential (9.7)
Lyapunov's Method (10.4)

Difference Equations

Differential equations serve as models for phenomena that are changing continuously. Many phenomena, however, change at discrete time intervals. Models in this case take the form of difference equations. Difference and differential equations are closely related. An introduction to first- and second-order difference equations is given in Sections 2.9 and 3.9. In those sections we demonstrate the similarity in the pattern of solutions for those two types of equations (the continuous and the discrete).

Mathematical Models

Most textbooks present mathematical models of a physical situation without questioning their efficacy. We believe that students should always question the reasonableness of such models. Consequently, in three early sections (1.1, 2.3, and 4.3), we have asked students to think about a mathematical model. In sections 2.3 and 4.3, students are also asked to consider whether a linear model is imposed because it is the best model of reality or, rather, because it's easier to solve than a nonlinear one.

Technology

Today, a majority of students studying differential equations have access to computers with software having symbolic manipulation capability. Commonly used software programs include Mathematica®, Maple®, and Derive®. In addition, many students are familiar with programming languages like FORTRAN or C. We believe that students should be encouraged to use computers in this course; many share this viewpoint. There are many text supplements on the market with titles like "Differential Equations with Mathematica® (or Maple®, Derive®, etc.)."

We help students and their instructors to incorporate computer calculations into this course in three ways:

Each chapter in the book ends with a section on using a symbolic manipulation program to solve problems that arise in that chapter. These sections are *generic;* that is, they are designed to be used with any appropriate software material currently available. In each section a set of commands that are not program specific is given, together with appropriate examples and exercises. The student is left to implement these commands on whatever symbolic manipulator he or she has available.

To supplement these sections, a manual is available that covers the use of Mathematica®, Maple®, and Derive®. This manual is designed to help the student solve many of the exercises in the text using each of these sophisticated computer algebra systems.

In some of the numerical sections in the text, we have added FORTRAN and C problems that can be used to carry out certain computations.

While using a computer program can certainly help in solving differential equation problems, **the availability of a computer is not a requirement for using this book.** Computer material can be ignored, at the instructor's discretion, without loss of continuity.

Numerical Methods

Most differential equations cannot be solved in terms of elementary functions. For this reason, it is important that a differential equations textbook include a discussion of numerical methods for obtaining solutions. We provide this material in three ways:

In Section 1.4, before a student has solved many differential equations, we discuss Euler's method. This provides the reader with a simple way to approximate the solution at a given point to a first-order initial-value problem.

Later, in Chapter 8, we provide a number of techniques for solving a variety of differential equations.

Also in Chapter 8, we discuss the error in numerical approximation. Many books omit this important topic. To describe numerical techniques without discussing how errors can propagate is a bit like learning how to drive without learning the rules of the road.

Existence and Uniqueness Theory

Proofs of the existence and uniqueness of solutions involve mathematical tools beyond those taught in a first calculus course. For that reason, we break our discussion of this topic into two parts. First, we discuss successive approximations in Section 2.8. Second, we prove both local and global existence-uniqueness theorems for scalar equations and systems of equations in Appendix 3. In that appendix, we discuss uniform convergence and related topics.

Systems of Differential Equations

In order to study systems of differential equations in the most general setting, it is useful to know something about matrices and their eigenvalues and eigenvectors. Many students do not have this background. Therefore, in order to make this topic accessible to all readers, we provide three approaches.

The **method of elimination** and its allied **method of determinants** (Section 7.3) provide efficient and elementary procedures for solving systems of two first-order linear differential equations in two unknown functions.

The **method of Laplace transforms** (Section 7.4) converts a system of n linear first-order differential equations into a system of n linear equations, which may then be solved by the techniques of linear algebra.

Chapter 9 provides a discussion of **matrix methods** for the solution of systems of linear differential equations. A brief review of matrices is given in Section 9.1 and a complete introduction to eigenvalues and eigenvectors is given in Section 9.4.

First-Order Partial Differential Equations

Most elementary textbooks give a very cursory introduction to first-order partial differential equations. We delve deeper into this important topic. In Sections 12.1 and 12.2, we discuss **characteristics** and use them to solve both linear and certain nonlinear equations.

Determinants and Complex Numbers

In certain sections of the book, students need to know something about determinants and complex numbers. Everything they need to know is contained in Appendices 4 and 5.

Integral Tables

In the real world, most nontrivial integrals are computed with the use of an integral table. The table in Appendix 1 of this book contains 220 entries. Every integral that comes up in this text can be found there.

Biographical Sketches

Mathematics becomes more interesting if one knows something about its historical development. We chose seven mathematicians who contributed greatly to the understanding of differential equations. Extended biographies of Euler, Bernoulli, Legendre, Laplace, Frobenius, Fourier and Wronski appear on pages 29, 60, 111, 223, 243, 259, 267, and 545.

Numbering and Notation in the Text

Numbering in the book is fairly standard. *Within each section,* examples, problems, theorems, and equations are numbered consecutively, starting with 1. Reference to an example, problem, theorem, or equation *outside the section in which it appears* is by chapter, section, and number. Thus, Example 4 in Section 2.3 is called, simply, Example 4 in that section, but outside the section it is referred to as Example 2.3.4. Finally, the ends of examples or groups of examples are marked with a ◀ and the ends of proofs are marked with a ♦.

Chapter Interdependence

The following diagram indicates interdependence—that is, which chapters depend on earlier material.

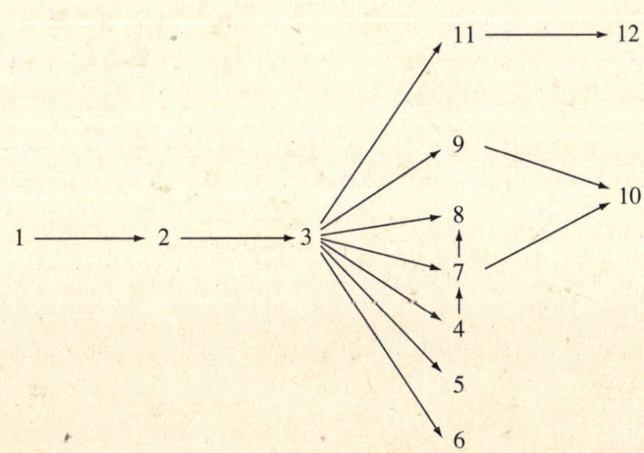

ACKNOWLEDGMENTS

We are grateful to the Harcourt Brace College Publishers Publishing Company for permission to use, in Chapter 7, material from the book *Elementary Linear Algebra, Fifth Edition* (1995), by Stanley Grossman. Some of the material for the biographical sketches came from the excellent book *An Introduction to the History of Mathematics, Fifth Edition* (1983), by Howard Eves. We are grateful to W. B. Saunders Company for permission to use this information.

Some of the examples and exercises in this book first appeared in *Mathematics for the Biological Sciences* by Stanley Grossman and James E. Turner, originally published by Macmillan in 1974. We are grateful to Professor Turner for permission to use this material.

A number of reviewers provided useful suggestions for improving the quality of this text. In particular, we would like to thank the following:

Richard Bagby
New Mexico State University

Mihaly Bakonyi
Georgia State University

Barbara Bertram
Michigan Technical University

William Bray
University of Maine

Patricia Clark
Rochester Institute of Technology

Gregory J. Davis
University of Wisconsin—Green Bay

David Ellis
San Francisco State University

Garrett Etgen
University of Houston

Juan Gatica
University of Iowa

G. Elton Graves
Rose Hulman Institute of Technology

Alfred Gray
University of Maryland

Heini Halberstam
University of Illinois

Noel Harbertson
California State University

Bernard J. Harris
Northern Illinois University

James Herod
Georgia Institute of Technology

Allan M. Krall
Pennsylvania State University

Roger Marty
Cleveland State University

Kevin McLeod
University of Wisconsin—Milwaukee

Peter Morris
Pennsylvania State University

Gregory B. Passty
Southwest Texas State University

Steven Pruess
Colorado School of Mines

Louis Rall
University of Wisconsin—Madison

Murali Rao
University of Florida

Jerry Schuur
Michigan State University

Ted Suffridge
University of Kentucky

Betty Tang
Arizona State University

Mo Tavakoli
Chaffey College

H. Clare Wiser
Washington State University

Elizabeth Yanik
Emporia State University

John A. Ziegler
Southern Technical Institute

We would also like to thank Jeremy Haefner, University of Colorado–Colorado Springs; Glenn Ledder, University of Nebraska; Robert Maynard, Tidewater Community College; and Bruno Welfert, Arizona State University, for assistance with the computer algebra systems manual, which was written jointly by the first author and Andrew G. Keck, University of Nevada—Las Vegas.

Professor Leon Gerber, at St. John's University in New York City, prepared the Student's and Instructor's manuals; he made many useful suggestions for the improvement of the problem sets. This book is a better teaching tool because of him.

Putting together a book like this is always a difficult process. Our task was made easier by the highly professional production work done by Michele Heinz and the staff at Elm Street Publishing Services, Inc. in Hinsdale, Illinois. We are grateful for her efforts that resulted not only in a nicely produced book, but also a nicely produced book that came out on schedule.

Finally, we especially want to thank our editors, George Duda and Kevin Connors, whose timely suggestions and attention to detail helped significantly in improving this book.

William R. Derrick
Stanley I. Grossman

1 INTRODUCTION TO DIFFERENTIAL EQUATIONS

INTRODUCTION

Mathematical models for many of the basic laws of the physical sciences and, more recently, of the biological and social sciences are formulated in terms of mathematical equations involving certain known and unknown quantities and their derivatives. Such equations are called **differential equations.**

Differential equations arise in situations where there is a derivative lurking around; that is, whenever something is changing. Before the twentieth century, much of mathematics was developed in order to provide solutions to physical problems. The history of the study of differential equations coincides, in part, with the history of physics. In fact, it is fair to say that differential equations constitute the most important part of what we now call *applied mathematics.*

The study of differential equations is a vast subject. This book forms merely an introduction to the subject. Many topics in differential equations are today the object of advanced research.

In this first chapter we will provide a preliminary answer to four basic questions: How do differential equations arise in a practical situation? What do we mean by a *solution* to a differential equation and under what circumstances does such a solution exist? How can we obtain a graph of a solution without finding that solution in closed form? How can we find solutions *numerically,* such as, on a computer?

Each of the four sections in this chapter provides a partial answer to one of these questions. The remainder of this book will furnish more detailed answers.

1.1 REVIEW OF SOME ELEMENTARY DIFFERENTIAL EQUATIONS

In this section, we show how some differential equations arise and how their solutions are obtained. The equations in this section were discussed in your calculus course.

Before citing any examples, however, we should emphasize that in the study of differential equations, the most difficult problem often is to describe a real situation quantitatively. To do this, it is usually necessary to make simplifying assumptions that can be expressed in mathematical terms. Thus, for example, we initially describe the motion of a mass in space by assuming that (a) it is a point and (b) there is no friction or air resistance. These assumptions are not realistic, but the scientist can often glean valuable information from even highly idealized models that, once understood, can be modified to take other observable factors into account.

Our first example is a differential equation that can be solved directly by integration.

EXAMPLE 1 ▶ **Determining distance when the velocity is given**

A ball is dropped at rest from a certain height. Its velocity after t seconds due to the earth's gravitational field is given by

$$v(t) = 32.2t \text{ ft/sec} \tag{1}$$

(when air resistance is ignored). How far has the ball fallen after t seconds?

Solution Let $s(t)$ denote the distance the ball has fallen after t seconds. Then, as velocity is the derivative of position, we have

$$\frac{ds}{dt} = v(t)$$

or

$$\frac{ds}{dt} = 32.2t. \tag{2}$$

Equation **(2)** is a called a **differential equation** because it is an equation involving a derivative. We can solve it directly by integration:

$$\begin{aligned} s &= \int v(t)\, dt = \int 32.2t\, dt \\ &= 32.2\frac{t^2}{2} + C = 16.1t^2 + C. \end{aligned}$$

But $s(0)$ = distance fallen after 0 seconds = *initial* distance fallen = 0 and

$$s(0) = 16.1 \cdot 0^2 + C = 0 + C = C.$$

Thus $C = 0$ and, after t seconds the ball has fallen

$$s(t) = 16.1t^2 \text{ ft}.$$

We note that the differential equation $ds/dt = 32.2t$ together with the implicitly given **initial condition** $s(0) = 0$ is called an **initial-value problem.** We shall say more about initial-value problems later in this section. ◀

Example 1 is a special case of the equations of motion discussed in calculus and physics courses. In our second example we derive these equations under some simplifying assumptions.

EXAMPLE 2 ▶ **A model of free fall**

Newton's law of gravitation states that the magnitude of the gravitational force exerted by the earth on a body of mass m is proportional to its mass and inversely proportional to the square of its distance $r = r(t)$ from the center of the earth. Thus, if k denotes the constant of proportionality, then

$$F = \frac{km}{r^2}.$$

By Newton's second law of motion

$$F = ma = m\frac{d^2r}{dt^2},$$

where $a = a(r)$ is the acceleration of the body when the distance to the center of the earth is r. Hence, equating both forces and dividing by m, we get

$$\frac{d^2r}{dt^2} = \frac{k}{r^2}. \tag{3}$$

Equation **(3)** is called a **second-order differential equation** because it involves a second derivative (and no derivative of order greater than 2).

If the object is falling, the velocity $v = dr/dt$ is negative because the distance to the center of the earth, r, is decreasing when t is increasing. Moreover, if the speed at which the object falls is increasing, the acceleration $a = dv/dt = d^2r/dt^2$ is also negative because v is decreasing, that is, becoming more and more negative. Thus, the constant k in Equation **(3)** is negative.

Let R be the mean radius of the earth ($R \approx 6378$ km ≈ 3963 miles) (see Figure 1.1) and denote the acceleration of gravity at the earth's surface by $a(R) = -g$. Then

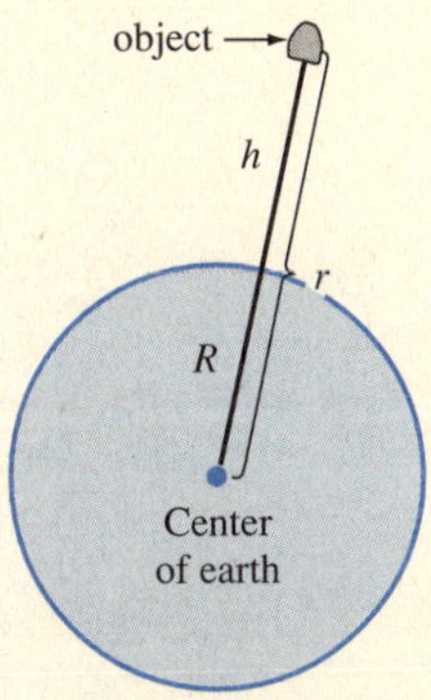

FIGURE 1.1
An object is h units above the surface of the earth and r units from the center of the earth

$$-g = a(R) = \frac{k}{R^2},$$

so that $k = -gR^2$. Hence we obtain the equation of motion

$$\frac{d^2r}{dt^2} = -g\frac{R^2}{r^2}, \tag{4}$$

where g is approximately 9.81 m/sec^2 (= 32.2 ft/sec^2). If we substitute $r = R + h$, where $h = h(t)$ is the height of the body from the surface of the earth, then $dr/dt = dh/dt$ and Equation **(4)** becomes

$$\frac{d^2h}{dt^2} = -g\frac{R^2}{(R + h)^2}.$$

When the height h is very small in comparison to the radius of the earth R, the ratio $R/(R + h)$ is very close to 1, so the differential equation is well approximated by the usual equation found in calculus books:

$$\frac{d^2h}{dt^2} = -g. \tag{5}$$

Note

Equation **(5)** is an equation of constant acceleration, a reasonable model for free fall only if h is very small compared to R. For example, if $h = 1000$ m $= 1$ km, then

$$\frac{h}{R} = \frac{1}{6378} \approx 0.000157 = 0.0157\%.$$

Here h is a very small percentage of R. However, if h is, say, 2000 km, then

$$\frac{h}{R} = \frac{2000}{6378} \approx 0.3136 = 31.36\%$$

which is not a small percentage. In this case, the model of constant acceleration would not represent reality very well.

Determining the height, h, at which Equation **(5)** ceases to represent a meaningful model of reality is a very difficult problem. We will not address that problem in this book. It must be left for a more advanced course in differential equations.

We now solve Equation **(5)** by integration: Integrating both sides of Equation **(5)** with respect to t, we obtain

$$h'(t) = -gt + C_1.$$

The constant C_1 can be determined by setting $t = 0$. We obtain $C_1 = h'(0)$, the initial velocity. Thus the velocity of the body at any time t is given by the differential equation.

$$h'(t) = -gt + h'(0). \tag{6}$$

Integrating once more with respect to t, we have

$$h(t) = -\frac{gt^2}{2} + h'(0)t + C_2.$$

The constant C_2 can also be found by setting $t = 0$. We obtain $C_2 = h(0)$, the initial height. Thus the height of the body at any time t is

$$h(t) = -\frac{gt^2}{2} + h'(0)t + h(0). \tag{7}$$

Equations **(6)** and **(7)** approximate the velocity and altitude of an object subject to free fall. As an example, suppose a ball is dropped at rest from the top of a building 45 meters high. Then the initial velocity is $h'(0) = 0$ and the initial height is $h(0) = 4500$ centimeters. If we wish to find the time it takes for the ball to strike the ground, we substitute these values in Equation **(7)** to obtain

$$0 = h(t) = -\frac{981t^2}{2} + 4500,$$

since the height at impact is zero. Solving for t, we have

$$490.5t^2 = 4500 \quad \text{or} \quad t^2 \approx 9.17.$$

Thus $t \approx \pm 3.03$. Since $t = -3.03$ has no physical significance, the answer is $t \approx 3.03$ seconds.

Note

The equation

$$\frac{d^2h}{dt^2} = -g$$

together with the given values $h(0)$ and $h'(0)$ is called a **second-order initial-value problem.** ◀

The differential equations in Examples 1 and 2 were solved directly by integration. If it were always possible to do this, then differential equations would be a direct application of integral calculus and there would be no need for separate books on differential equations. However, most solvable differential equations can only be solved by other techniques. One type of differential equation that we can solve is often discussed in a beginning calculus class.

Let $y = f(x)$ represent some physical quantity such as the volume of a substance, the population of a certain species, the mass of a decaying radioactive substance, the number of dollars invested in bonds, and so on. Then the rate of growth or decline of $f(x)$ is given by its derivative dy/dx. Thus, if $f(x)$ is growing at a constant rate, then $dy/dx = k$, and $y = kx + C$, that is, $y = f(x)$ is a straight-line function.

It is sometimes more interesting and more appropriate to consider the **relative rate of growth,** defined by

$$\text{relative rate of growth} = \frac{\text{actual rate of growth}}{\text{size of } f(x)} = \frac{f'(x)}{f(x)} = \frac{\frac{dy}{dx}}{y}. \tag{8}$$

The relative rate of growth indicates the fraction increase or decrease in f. For example, an increase of 100 individuals for a species with a population size of 500 would probably have a significant impact, being an increase of 20%. On the other hand, if the population were 1,000,000, then the addition of 100 would hardly be noticed, being an increase of only 0.01%.

In many applications, we are told that the relative rate of growth of the given physical quantity is constant. That is,

$$\frac{\frac{dy}{dx}}{y} = \alpha \tag{9}$$

or

$$\frac{dy}{dx} = \alpha y \tag{10}$$

where α is the constant percentage increase or decrease in the quantity.

Another way to view Equation **(10)** is that it tells us that the function is changing at a rate proportional to itself. If the constant of proportionality α is greater than 0, the quantity is increasing; while if $\alpha < 0$, it is decreasing.

Equation **(10)** is called a **first-order** differential equation. It is different from Equations **(2)** and **(5)** because now the unknown function y appears on *both sides* of the equation. A **solution** to Equation **(10)** is a differentiable function y for which $y' = \alpha y$. If we tried to integrate both sides of Equation **(10)** with respect to x, we would obtain

$$\int \frac{dy}{dx}\, dx = \alpha \int y(x)\, dx + C$$

or

$$y = \alpha \int y(x)\, dx + C.$$

But this doesn't help at all because we don't yet know what the function y is.

To solve Equation **(10)**, we carry out the following steps: We rewrite Equation **(10)** as

$$\frac{dy}{y} = \alpha\, dx \tag{11}$$

where y is a function of x, and then integrate to obtain

$$\int \frac{dy}{y} = \int \alpha\, dx \tag{12}$$

or

$$\ln|y| = \alpha x + C.$$

Hence

(If $\ln a = b$, then $a = e^b$.)

$$|y| = e^{\alpha x + C} = e^{\alpha x} e^{C},$$

or

$$y = ce^{\alpha x}, \qquad \text{where } c = \pm e^{C}. \tag{13}$$

Solution to the differential equation of constant relative growth

$$y = ce^{\alpha x} \tag{14}$$

where c can be any real number. To check this, we simply differentiate:

$$\frac{dy}{dx} = \frac{d}{dx} ce^{\alpha x} = c \frac{d}{dx} e^{\alpha x} = c\alpha e^{\alpha x} = \alpha(ce^{\alpha x}) = \alpha y,$$

so that $y = ce^{\alpha x}$ does satisfy Equation **(10).** We therefore have proven our first theorem.

THEOREM 1 ♦

If α is any real number, then there are an infinite number of solutions to the differential equation $y' = \alpha y$. They take the form $y = ce^{\alpha x}$ for any real number c. ♦

If $\alpha > 0$, we say that the quantity described by $f(x)$ is **growing exponentially.** If $\alpha < 0$, it is **decaying exponentially** (see Figure 1.2). Of course, if $\alpha = 0$, then there is no growth and $f(x)$ remains constant.

For a physical problem, we often know more than the rate of growth. We may also know one particular value of y, say $y(x_0) = y_0$. This value is called an **initial condition,** and it will give us a unique solution to the problem. The differential Equation **(10)** together with an initial condition is called an **initial-value problem.** We will find the unique solutions to a number of initial-value problems in this section.

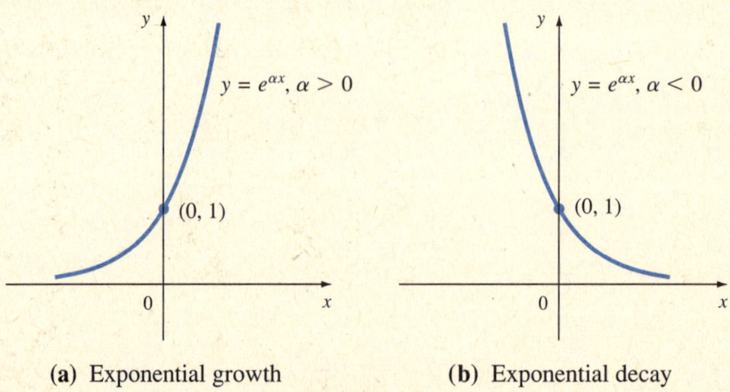

FIGURE 1.2
Curves showing exponential growth and decay

EXAMPLE 3 ▶ **Solving an initial-value problem**

(a) Find all solutions to $dy/dx = 3y$.
(b) Find the solution that satisfies the initial condition $y(0) = 2$.

Solution **(a)** Since $\alpha = 3$, all solutions are of the form $y = ce^{3x}$.
(b) $2 = y(0) = ce^{3 \cdot 0} = c \cdot 1 = c$, so $c = 2$ and the unique solution is $y = 2e^{3x}$. ◀

EXAMPLE 4 ▶ A model of population growth

A bacterial population is growing continuously at a rate equal to 10% of its present population. Its initial size is 10,000 organisms. How many bacteria are present after 10 days? After 30 days?

Solution Since the percentage growth of the population is $10\% = 0.1$, we have

$$\frac{\frac{dP}{dt}}{P} = 0.1 \qquad \text{or} \qquad \frac{dP}{dt} = 0.1P; \text{ where } t \text{ is time.}$$

Here $\alpha = 0.1$, and all solutions have the form

$$P(t) = ce^{0.1t}$$

where t is measured in days. Since $P(0) = 10{,}000$, we have

$$ce^{(0.1)(0)} = c = 10{,}000 \qquad \text{and} \qquad P(t) = 10{,}000e^{0.1t}.$$

This is the unique solution to the initial-value problem. After 10 days $P(10) = 10{,}000e^{(0.1)(10)} = 10{,}000e \approx 27{,}183$, and after 30 days $P(30) = 10{,}000e^{0.1(30)} = 10{,}000e^3 \approx 200{,}855$ bacteria. ◀

HISTORICAL NOTE

When it applies to human populations, exponential growth is often referred to as **Malthusian growth.** In 1798, the English economist Thomas Robert Malthus (1766–1834) claimed that increases in population tend to exceed increases in the means of subsistence and that therefore sexual restraint should be exercised.* More precisely, he claimed that populations grow exponentially (like the function $e^{\alpha t}$) while food and other necessities grow only geometrically (like the function t^{α}). His theories are still debated. A more realistic model of population growth (logistic growth) is discussed in Example 2.2.3.

Before giving further examples, we summarize the principal result of this section.

The equation of exponential growth or decay

If $P(t)$ is growing or declining continuously at a constant relative rate (so that $P'(t)/P(t) = \pm\alpha$) with initial size $P(0)$, then $P(t)$ is given by (with $\alpha > 0$)

$$P(t) = P(0)e^{\alpha t} \qquad \text{or} \qquad P(t) = P(0)e^{-\alpha t}. \tag{15}$$

The term $P(0)e^{\alpha t}$ represents exponential growth and the term $P(0)e^{-\alpha t}$ represents exponential decay.

EXAMPLE 5 ▶ Population growth in India

The population of India was estimated to be 574,220,000 in 1974 and 746,388,000 in 1984. Assume that the relative growth rate remains constant and that growth is continuous.

---

* T. R. Malthus, "An Essay on the Principle of Population As It Affects the Future Improvement of Society," 1798.

(a) Estimate the population in 1994.
(b) When will the population reach 1.5 billion?

Solution **(a)** From Equation (15), we have

$$P(t) = P(0)e^{\alpha t}.$$

We have $\alpha > 0$ because the population is increasing.

Treat the year 1974 as year zero. Then 1984 = year 10. We are told that

$$P(0) = 574{,}220{,}000 \quad \text{and} \quad P(10) = 746{,}388{,}000.$$

Thus

$$\begin{aligned}
P(t) &= 574{,}220{,}000e^{\alpha t} \\
P(10) &= 746{,}388{,}000 = 574{,}220{,}000e^{10\alpha} \\
e^{10\alpha} &= \frac{746{,}388{,}000}{574{,}220{,}000} \approx 1.3 \\
\ln e^{10\alpha} &\approx \ln 1.3 \\
10\alpha &\approx \ln 1.3 \qquad \ln e^x = x \\
\alpha &\approx \frac{\ln 1.3}{10}.
\end{aligned}$$

The year 1994 is year 20. Thus

$$\begin{aligned}
P(20) &= 574{,}220{,}000e^{20\alpha} = 574{,}220{,}000e^{2\ln 1.3} \\
&= 574{,}220{,}000(1.3)^2 \approx 970{,}432{,}000.
\end{aligned}$$

(b) We seek a number t such that $P(t) = 1{,}500{,}000{,}000$. That is,

$$\begin{aligned}
P(t) &= 1{,}500{,}000{,}000 = 574{,}220{,}000e^{\alpha t} \\
e^{\alpha t} &= \frac{1{,}500{,}000{,}000}{574{,}220{,}000} \approx 2.61 \\
\ln e^{\alpha t} &= \ln 2.61 \\
\alpha t &= \ln 2.61 \\
t &= \frac{\ln 2.61}{\alpha} \\
&= \frac{\ln 2.61}{\frac{\ln 1.3}{10}} = \frac{10 \ln 2.61}{\ln 1.3} \\
&\approx 36.6 \text{ years.}
\end{aligned}$$

Then

$$\text{Year } 36.6 = 1974 + 36.6 = 2010.6.$$

We conclude that the population of India would reach 1.5 billion sometime in the year 2010 if its population growth rate continued at the same rate as it was between 1974 and 1984. ◀

In this problem we found that

$$\alpha \approx \frac{\ln 1.3}{10} \approx 0.0262.$$

This means that the population is growing at a rate of about 2.62% a year. We stress that all the calculations in this problem were made under the assumption that this percentage did not vary and will not vary at all from 1974 until now and into the foreseeable future. This is not a reasonable assumption. Thus our answers are not reasonable.

In this example, we created a mathematical model to solve a problem based on certain assumptions about population growth. Solving the problem was much easier than checking the validity of our assumptions. You should keep this in mind whenever you use a mathematical model. Put more concisely, we stress that while a mathematical model may be a good (or not so good) model of reality,

A MATHEMATICAL MODEL IS NOT REALITY

EXAMPLE 6 ▶ **Newton's law of cooling: a model of heating and cooling**

Newton's law of cooling states that the rate of change of the temperature difference between an object and its surrounding medium is proportional to the temperature difference. If $D(t)$ denotes this temperature difference at time t and if α denotes the constant of proportionality, then we obtain

$$\frac{dD}{dt} = -\alpha D.$$

The minus sign indicates that this difference decreases. (If the object is cooler than the surrounding medium—usually air—it will warm up; if it is hotter, it will cool.) The solution to this differential equation is

$$D(t) = ce^{-\alpha t}.$$

If we denote the initial ($t = 0$) temperature difference by D_0, then

$$D(t) = D_0 e^{-\alpha t} \tag{16}$$

is the formula for the temperature difference for any $t > 0$. Notice that for t large $e^{-\alpha t}$ is very small, so that, as we have all observed, temperature differences tend to die out rather quickly.

We can rewrite Equation **(16)** in a form that is often more useful. Let $T(t)$ denote the temperature of the object at time t, let T_s denote the temperature of the surroundings (which is assumed to be constant throughout), and let T_0 denote the initial temperature of the object. Then $D(t) = T(t) - T_s$ and $D_0 = T_0 - T_s$, so Equation **(16)** becomes

$$T(t) - T_s = (T_0 - T_s)e^{-\alpha t}$$

or

$$T(t) = T_s + (T_0 - T_s)e^{-\alpha t}. \tag{17}$$

We now may ask: In terms of the constant α, how long does it take for the temperature difference to decrease to half its original value?

Solution The original value is D_0. We are therefore looking for a value of t for which $D(t) = \frac{1}{2}D_0$. That is, $\frac{1}{2}D_0 = D_0 e^{-\alpha t}$, or $e^{-\alpha t} = \frac{1}{2}$. Taking natural logarithms, we obtain

$$-\alpha t = \ln\frac{1}{2} = -\ln 2 = -0.6931 \qquad \text{and} \qquad t \approx \frac{0.6931}{\alpha}.$$

Notice that this value of t does *not* depend on the initial temperature difference D_0. ◀

EXAMPLE 7 ▶ **Using Newton's law of cooling**

With the air temperature equal to 30°C, an object with an initial temperature of 10°C warmed to 14°C in 1 hour.

(a) What was its temperature after 2 hours?
(b) After how many hours was its temperature 25°C?

Solution Here $T_s = 30$, $T_0 = 10$, and $T(1) = 14$. From Equation **(17),** we have

$$T(t) = 30 - 20e^{-\alpha t}.$$

But

$$\begin{aligned} 14 = T(1) &= 30 - 20e^{-\alpha} \\ 20e^{-\alpha} &= 16 \\ e^{-\alpha} &= \frac{4}{5} = 0.8. \end{aligned}$$

Thus,

$$T(t) = 30 - 20e^{-\alpha t} = 30 - 20(e^{-\alpha})^t = 30 - 20(0.8)^t.$$

We can now answer the two questions.

(a) $T(2) = 30 - 20(0.8)^2 = 30 - 20(0.64) = 17.2$°C.
(b) We need to find t such that $T(t) = 25$. That is,

$$25 = 30 - 20(0.8)^t, \quad \text{or} \quad (0.8)^t = \frac{1}{4}, \quad \text{or} \quad t \ln(0.8) = -\ln 4,$$

and

$$t = \frac{-\ln 4}{\ln(0.8)} \approx \frac{1.3863}{0.2231} \approx 6.2 \text{ hr} = 6 \text{ hr } 12 \text{ min.}$$ ◀

EXAMPLE 8 ▶ **Carbon dating: a model for estimating the age of an artifact**

Carbon dating is a technique used by archaeologists, geologists, and others who want to estimate the ages of certain artifacts and fossils they uncover. The technique is based on certain properties of the carbon atom. In its natural state the nucleus of the carbon atom ^{12}C has 6 protons and 6 neutrons. The **isotope** carbon-14, ^{14}C, is produced through cosmic-ray bombardment of nitrogen in the atmosphere. Carbon-14 has 6 protons and 8 neutrons and is **radioactive.** It decays by beta emission. That is, when an atom of ^{14}C decays, it gives up an electron to form a stable nitrogen atom ^{14}N. We make the assumption that the ratio of ^{14}C to ^{12}C in the atmosphere is constant. This assumption has been shown experimentally to be approximately valid, for although ^{14}C is being constantly lost through **radioactive decay** (as this process is often termed), new ^{14}C is constantly being produced by the cosmic bombardment of nitrogen in the upper atmosphere. Living plants and animals do not distinguish between ^{12}C and ^{14}C, so at the time of death the ratio of ^{12}C to ^{14}C in an organism is the same as the ratio in the atmosphere. However, this ratio changes after death since ^{14}C is converted to ^{14}N but no further ^{14}C is taken in.

It has been observed that ^{14}C decays at a rate proportional to its mass and that its **half-life** is approximately 5730 years.* That is, if a substance starts with 1 g of ^{14}C, then 5730 years later it will have $\frac{1}{2}$ g of ^{14}C, the other $\frac{1}{2}$ g having been converted to ^{14}N.

We may now pose a question typically asked by an archaeologist. A fossil is unearthed and it is determined that the ratio of ^{14}C present is 40% of what it would be for a similarly sized living organism. What is the approximate age of the fossil?

Solution Let $M(t)$ denote the mass of ^{14}C present in the fossil. Then since ^{14}C decays at a rate proportional to its mass, we have

$$\frac{dM}{dt} = -\alpha M,$$

where α is the constant of proportionality. Then $M(t) = ce^{-\alpha t}$, where $c = M_0$, the initial amount of ^{14}C present. When $t = 0$, $M(0) = M_0$; when $t = 5730$ years, $M(5730) = \frac{1}{2}M_0$, since half the original amount of ^{14}C has been converted to ^{12}C. We can use this fact to solve for α since we have

$$\frac{1}{2}M_0 = M_0 e^{-\alpha \cdot 5730} \quad \text{or} \quad e^{-5730\alpha} = \frac{1}{2}.$$

Thus,

$$(e^{-\alpha})^{5730} = \frac{1}{2}, \quad \text{or} \quad e^{-\alpha} = \left(\frac{1}{2}\right)^{1/5730}, \quad \text{and} \quad e^{-\alpha t} = \left(\frac{1}{2}\right)^{t/5730},$$

so

$$M(t) = M_0\left(\frac{1}{2}\right)^{t/5730}.$$

Now we are told that after t years (from the death of the fossilized organism to the present) $M(t) = 0.4M_0$, and we are asked to determine t. Then

$$0.4M_0 = M_0\left(\frac{1}{2}\right)^{t/5730},$$

and taking natural logarithms (after dividing by M_0), we obtain

$$\ln 0.4 = \frac{t}{5730}\ln\left(\frac{1}{2}\right) \quad \text{or} \quad t = \frac{5730\ln(0.4)}{\ln\left(\frac{1}{2}\right)} \approx 7575 \text{ years.}$$

The carbon-dating method has been used successfully on numerous occasions. It was this technique that established that the Dead Sea scrolls were prepared and buried about two thousand years ago. ◀

---

* This number was first determined in 1949 by the American chemist W. S. Libby, who based his calculations on the wood from sequoia trees, whose ages were determined by rings marking years of growth. Libby's method has come to be regarded as the archaeologist's absolute measuring scale. But in truth, this scale is flawed. Libby used the assumption that the atmosphere had at all times a constant amount of ^{14}C. However, the American chemist C. W. Ferguson of the University of Arizona deduced from his study of tree rings in 4000-year-old American giant trees that before 1500 B.C. the radiocarbon content of the atmosphere was considerably higher than it was later. This result implied that objects from the pre-1500 B.C. era were much older than previously believed, because Libby's "clock" allowed for a smaller amount of ^{14}C than actually was present. For example, a find dated at 1800 B.C. was in fact from 2500 B.C. This has had a considerable impact on the study of prehistoric times. For a fascinating discussion of this subject, see Gerhard Herm, *The Celts* (New York: St. Martin's Press, 1975), pages 90–92.

In this section we solved a very simple differential equation. Other, more elaborate differential equations can be solved by a variety of methods. We will show you how beginning in Chapter 2. However, it should be apparent from these examples and the problems that follow that a large variety of applied problems can be solved by considering a simple differential equation. It is the purpose of this book to show you how to deal with other, more complicated differential equations arising in physical applications.

Thinking about a mathematical model

In many of the examples given in this section, we assumed that the relative rate of growth was constant so that

$$y(x) = y(0)e^{kx}.$$

Suppose that $k > 0$. Let's give it a value, say $k = 0.1$. This means that something—a population or investment, for example, is growing continuously at a rate of 10% per time period. Did you stop to consider the fact that this rate of growth *cannot possibly continue* for an indefinite period of time?

To illustrate this fact, suppose that a certain insect population grows at 10% a month and starts with 1000 individuals. Then the population after t months is

$$P(t) = 1000e^{0.1t}.$$

TABLE 1.1

t	$P(t) = 1000e^{0.1t}$
10	2718
20	7389
30	20,086
60	403,429
120	162,754,791
240	2.649×10^{13}
480	7.017×10^{23}

The population after t months is given in Table 1.1. We see that after 240 months (20 years) there would be over 26 trillion insects. After 40 years there would be over 7×10^{23} insects. That is an unimaginably large number. To give you some idea, let us do some calculations.

1. The radius of the earth is approximately 4000 miles.
2. The surface area of a sphere is given by $S = 4\pi r^2$ so surface area of the earth $\approx 4\pi(4000)^2 \approx 200{,}000{,}000$ square miles.
3. There are 5280^2 square feet in a square mile so the surface area of the earth in square feet $\approx (200{,}000{,}000)(5280^2) \approx 5.6 \times 10^{15}$ square feet.
4. We divide:

$$\frac{7 \times 10^{23} \text{ insects}}{5.6 \times 10^{15} \text{ sq ft}} \approx 125{,}000{,}000 \text{ insects per square foot.}$$

   That is, 125 million insects occupying every square foot on earth! Where would they fit? What would they eat?

The situation is ridiculous. This should come as no surprise because as $t \to \infty$, $e^{kt} \to \infty$ if $k > 0$, so any population (or anything else) grows without bound if it is growing exponentially.

Since nothing on earth can possibly grow without bound, it seems that the model of this section is useless. This is not true. Many quantities grow exponentially—like continuously compounded interest, for example. But nothing can grow exponentially indefinitely. There are always limits to growth. Any population growing exponentially will eventually run out of space or food and suffer other effects from overcrowding. Then nature will force the growth rate to change.

But unlimited growth is not the only problem with this model. Is the assumption that growth continues at a constant relative rate a reasonable one? If a bond pays 7% compounded continuously for 20 years, then you are certain that, at least for 20 years, your money will grow exponentially at the constant rate of 7%.

However, for another kind of problem, the answer is likely to be no. For example, the population of the U.S. grew at an average annual rate of 0.93% from

1980 to 1990. Changes in immigration law could increase or decrease the number of immigrants arriving in the U.S. each year. Better education in the schools might lead to fewer children born to unwed teenagers. Better health care could lead to drops in infant mortality and a corresponding increase in the population growth rate. You can undoubtably think of other factors.

So the assumption of indefinite growth at a constant rate is unrealistic for many different types of problems. Models based on this assumption may be very useful. But they are often useful only for limited periods of time. As with all mathematical models, you must question the validity of the assumptions inherent in the model before you make predictions based on the model.

We stress again that

A MATHEMATICAL MODEL IS NOT REALITY.

SELF-QUIZ

I. Suppose y depends smoothly on x (i.e., $dy/dx = y'$ exists). Also suppose that its relative rate of growth is constant. Then y satisfies a differential equation of the form ________ (where α is a nonzero constant).

(a) $yy' = \alpha$ **(c)** $y' + \frac{1}{y} = \alpha$

(b) $\frac{y'}{y} = \alpha$ **(d)** $y' = \frac{\alpha}{y}$

II. The general solution to $y'/y = 10$ is ________.

(a) $\ln|y| = 10x + C$ **(c)** $x = K + e^{10y}$

(b) $y = K \cdot e^{10x}$ **(d)** $y = 5y^2 + K$

III. Suppose y is a smooth function of x such that $y'/y = -3$; then $\lim_{x\to\infty} y$ ________.

(a) $= 1$ **(b)** $= -\frac{3}{2}$ **(c)** $= \infty$

(d) $= 0$ **(e)** doesn't exist

(f) can't be found from this information alone

IV. Suppose y is a smooth function of x such that $y'/y = -3$; then $\lim_{x\to 0} y$ ________.

(a) $= 1$ **(b)** $= -\frac{3}{2}$ **(c)** $= \infty$ **(d)** $= 0$

(e) doesn't exist

(f) can't be determined from this information

Answers to Self-Quiz

I. b **II.** a, b (a implies b) **III.** d **IV.** f [The limit is $y(0)$.]

PROBLEMS 1.1*

In Problems 1–4, use Newton's law of cooling to determine how long to bake a cake at the given oven temperature, assuming that it takes exactly 30 minutes to change 70°F dough into a 170°F cake in a 350° oven.

1. 250°F **2.** 400°F **3.** 300°F **4.** 200°F

5. The growth rate of a bacteria population is proportional to its size. Initially the population is 10,000, and after 10 days it is 25,000. What is the population size after 20 days? After 30 days?

6. Suppose that is Problem 5 the population after 10 days is 6000. What is the population after 20 days? After 30 days?

7. The population of a certain city grows 6% a year. If the population in 1970 is 250,000, what is the population in 1980? In 2000?

8. When the air temperature is 70°F, an object cools from 170°F to 140°F in 0.5 hour.

a. What is the temperature after 1 hour?

b. When does the temperature reach 90°F? [*Hint*: Use Newton's law of cooling.]

9. A hot coal (temperature 150°C) is immersed in ice water (temperature 0°C). After 30 seconds the temperature of the coal is 60°C. Assume that the ice water is kept at 0°C.

(a) What is the temperature of the coal after 2 minutes?

(b) When does the temperature of the coal reach 10°C?

---

*To complete most of these problems, you will need to use a hand calculator with ln and e^x function keys (INV ln gives e^x on some calculators).

10. A ball is thrown upward with an initial velocity v_0 meters per second from the top of a building h_0 meters high. Find how high the ball travels and determine when it hits the ground for the following choices of v_0 and h_0 (neglect air resistance and let $g = 9.8 \text{ m/s}^2$)
(a) $v_0 = 49$ m/s, $h_0 = 539$ m
(b) $v_0 = 14$ m/s, $h_0 = 21$ m
(c) $v_0 = 21$ m/s, $h_0 = 175$ m
(d) $v_0 = 7$ m/s, $h_0 = 56$ m
(e) $v_0 = 7.7$ m/s, $h_0 = 42$ m

11. A fossilized leaf contains 70% of a "normal" amount of ^{14}C. How old is the fossil?

12. Forty percent of a radioactive substance disappears in 100 years.
(a) What is its half-life?
(b) After how many years will 90% be gone?

13. Salt decomposes in water into sodium $[Na^+]$ and chloride $[Cl^-]$ ions at a rate proportional to its mass. Suppose there are 25 kilograms of salt initially and 15 kilograms after 10 hours.
(a) How much salt is left after one day?
(b) After how many hours is there less than 0.5 kilograms of salt left?

14. X rays are absorbed into a uniform, partially opaque body as a function not of time but of penetration distance. The rate of change of the intensity $I(x)$ of the X ray is proportional to the intensity. Here x measures the distance of penetration. The more the X ray penetrates, the lower the intensity is. The constant of proportionality is the density D of the medium being penetrated.
(a) Formulate a differential equation describing this phenomenon.
(b) Solve for $I(x)$ in terms of x, D, and the initial (surface) intensity $I(0)$.

15. Radioactive beryllium is sometimes used to date fossils found in deep-sea sediment. The decay of beryllium satisfies the equation

$$\frac{dA}{dt} = -\alpha A, \qquad \text{where } \alpha = 1.5 \times 10^{-7},$$

where t is measured in years. What is the half-life of beryllium?

***16.** In a certain medical treatment a tracer dye is injected into the pancreas to measure its function rate. A normally active pancreas secretes 4% of the dye each minute. A physician injects 0.3 gram of the dye, and 30 minutes later 0.1 gram remains. How much dye would remain if the pancreas were functioning normally?

17. The estimated world population in 1986 was 4,845,000,000. Assume that the population grows at a constant rate of 1.9%. When will the world population reach 8 billion?

18. In Problem 17, at what constant rate would the population grow if it reached 6 billion in the year 2000?

19. In 1900 the world's population was 1.571 billion while in 1950 it was 2.517 billion. Assuming the Malthusian law of population growth, predict the world's population in the year 2000.

20. The population of the United States was 76,212,168 in 1900 and 92,228,496 in 1910. If population had grown at a constant percentage until 1990, what would the population have been in 1980 and 1990?

21. The population of Australia was approximately 13,400,000 in 1974 and 16,643,000 in 1990. Assume constant relative growth.
(a) Predict the population in the year 2000.
(b) When will the population be 20 million?

22. The population of New York State was 17,558,165 in 1980 and 17,990,455 in 1990. Assume a constant relative growth in population.
(a) Predict the population in 1995.
(b) When will the population reach 18.5 million?

23. Atmospheric pressure is a function of altitude above sea level and is given by $dP/da = \beta P$, where β is a constant. The pressure is measured in millibars (mbar). At sea level ($a = 0$), $P(0)$ is 1013.25 mbar which means that the atmosphere at sea level will support a column of mercury 1013.25 millimeters high at a standard temperature of 15°C. At an altitude of $a = 1500$ meters, the pressure is 845.6 mbar.
(a) What is the pressure at $a = 4000$ meters?
(b) What is the pressure at 10 kilometers?
(c) In California the highest and lowest points are Mount Whitney (4418 meters) and Death Valley (86 meters below sea level). What is the difference in their atmospheric pressures?
(d) What is the atmospheric pressure at the top of Mount Everest (elevation 8848 meters)?
(e) At what elevation is the atmospheric pressure equal to 1 mbar?

* This and similar mathematical models in medicine are discussed by J. S. Rustagi, "Mathematical Models in Medicine," *International Journal of Mathematical Education in Science and Technology* 2 (1971): 193–203.

24. A bacteria population is known to grow exponentially. The following data are collected:

Number of Days	*Number of Bacteria*
5	936
10	2,190
20	11,993

(a) What is the initial population?
(b) If the present growth rate continues, what is the population after 60 days?

1.2 FUNDAMENTAL THEORY

A great variety of types of differential equations can arise in the study of familiar phenomena. It is clearly necessary (and expedient) to study, independently, more restricted classes of these equations.

The most obvious classification is based on the nature of the derivative(s) in the equation. A differential equation involving only ordinary derivatives (derivatives of functions of one variable) is called an **ordinary differential equation,** whereas one containing partial derivatives is called a **partial differential equation.** We postpone the further classification of partial differential equations until the last chapter.

Definition Order

The **order** of a differential equation is defined as the order of the highest derivative appearing in the equation.

EXAMPLE 1 ▶ **The orders of three differential equations**

The following are examples of differential equations with indicated orders.

(a) $dy/dx = ay$. (first order)
(b) $x''(t) - 3x'(t) + x(t) = \cos t$. (second order)
(c) $(y^{(4)})^{3/5} - 2y'' = \cos x$. (fourth order) ◀

Much of this book is concerned with the solutions of differential equations. Thus we need to explain what we mean by a *solution*. First we note that any nth-order differential equation can be written in the form

$$F(x, y, y', \dots, y^{(n)}) = 0. \tag{1}$$

For example, $y' = f(x, y)$ can be written $y' - f(x, y) = 0$. Here $F(x, y, y') = y' - f(x, y)$. The second-order equation $y'' + a(x)y' + b(x)y = f(x)$ can be written $F(x, y, y', y'') = y'' + a(x)y' + b(x)y - f(x) = 0$.

Definition Solution to an nth-order differential equation on an interval

A **solution** to an nth-order differential equation on an interval (a, b) is a function that is n times differentiable on that interval and that satisfies the differential equation.

Note

A solution to a differential equation does not contain any derivatives. Symbolically, this means that a solution of differential Equation **(1)** is a function $y(x)$ whose derivatives $y'(x), y''(x), \ldots, y^{(n)}(x)$ exist and that satisfies the equation

$$F(x, y(x), y'(x), \ldots, y^{(n)}(x)) = 0$$

for all values of the independent variable x in some interval where $F(x, y(x), y'(x), \ldots, y^{(n)}(x))$ is defined.

As we saw in Section 1.1, we are often interested in solving a first-order differential equation

$$\frac{dy}{dx} = f(x, y)$$

subject to the condition that $y = y_0$ when $x = x_0$, or

$$y(x_0) = y_0.$$

This is an example of an **initial-value problem.** The condition $y(x_0) = y_0$ is called an **initial condition** and x_0 is called the **initial point.** More generally, we have the following.

Definition Initial-value problem

An **initial-value problem** consists of an nth-order differential equation (of any order) together with n initial conditions of the form $y(x_0) = a_0$, $y'(x_0) = a_1, \ldots, y^{(n-1)}(x_0) = a_{n-1}$, that must be satisfied by the solution of the differential equation and its derivatives at the initial point x_0.

EXAMPLE 2 ▶ **Two initial-value problems**

The following are examples of initial-value problems.

(a) $dy/dx = 2y - 3x$, $y(0) = 2$. (Here $x = 0$ is the initial point.)

(b) $x''(t) + 5x'(t) + \sin tx(t) = 0$, $x(1) = 0$, $x'(1) = 7$. (Here $t = 1$ is the initial point.) ◀

Definition Solution to an initial-value problem

We define a **solution** to an nth-order initial-value problem as a function that is n times differentiable on an interval (a, b), satisfies the given differential equation on that interval, and satisfies the n given initial conditions.*

EXAMPLE 3 ▶ **Solution of an initial-value problem**

The function $y(x) = 2e^{3x}$ is a solution of the initial-value problem

$$\frac{dy}{dx} = 3y, \qquad y(0) = 2,$$

because $y(0) = 2e^{3 \cdot 0} = 2e^0 = 2$ and

$$\frac{dy}{dx} = 2\frac{d}{dx}(e^{3x}) = 6e^{3x} = 3y. \quad ◀$$

* The interval (a, b) could be infinite; that is, $a = -\infty$, or $b = \infty$, or both.

> **Definition** Boundary value problem
>
> A **boundary value problem** consists of a differential equation and a set of conditions that must be satisfied by the solution of the differential equation or its derivatives at no fewer than two different points.

EXAMPLE 4 ▶ **Two boundary value problems**

The following are examples of boundary value problems:

(a) $d^2y/dx^2 + 5xy = \cos x$, $y(0) = 0$, $y'(1) = 2$.
(b) $dy/dx + 5xy = 0$, $y(0)y(1) = 2$. ◀

In each of Examples 1.1.1–1.1.4, we see that once initial conditions are satisfied, the resulting initial-value problem has a unique solution (note that two initial conditions were required in Example 1.1.2, the initial height $h(0)$ and the initial velocity $h'(0)$. It is reasonable to ask whether every initial-value problem has a unique solution. Essentially we are asking two questions:

1. Is there a solution to the problem?
2. If there is a solution, is it the only one?

As we see in the next two examples, the answer may be *no* to each question.

EXAMPLE 5 ▶ **An initial-value problem with no solution**

The initial-value problem

$$\left(\frac{dy}{dx}\right)^2 + y^2 + 1 = 0, \qquad y(0) = 1$$

has no real-valued solutions, since the left-hand side is always positive for real-valued functions. ◀

EXAMPLE 6 ▶ **An initial-value problem with more than one solution**

The initial-value problem

$$\frac{dy}{dx} = xy^{1/3}, \qquad y(0) = 0 \tag{2}$$

has at least two solutions in the interval $-\infty < x < \infty$. Note that the functions

$$y = 0 \qquad \text{and} \qquad y = \frac{1}{3\sqrt{3}}x^3$$

both satisfy the initial condition and the differential equation in **(2)**, trivially for the first, and by the following computation for the second:

$$y' = \frac{1}{\sqrt{3}}x^2 = x\left(\frac{x}{\sqrt{3}}\right) = xy^{1/3}. \qquad y^{1/3} = \left(\frac{x^3}{3\sqrt{3}}\right)^{1/3} = \frac{(x^3)^{1/3}}{(3^{3/2})^{1/3}} = \frac{x}{3^{1/2}}$$ ◀

The physical examples given in Section 1.1 are such that in each case we know that a solution exists. However, there is an inherent danger of confusing physical reality with the mathematical model given by the differential equation we use to represent the real problem. It may well be that our modeling is faulty, in which case the equations obtained may bear no connection with reality. Then solutions to the equations need not exist.

Examples 5 and 6 illustrate that it is important to obtain theorems that guarantee the existence of a unique solution. The following result, originally developed by

Liouville and proved in its most general form by Picard, is very useful because of the ease with which its hypotheses are checked.

THEOREM 1 ♦ **Existence-uniqueness theorem**

Let f and $\partial f/\partial y$ be continuous in a rectangle R given by $a < x < b$, $c < y < d$ that contains the point (x_0, y_0) (see Figure 1.3). Then there exists a number $h > 0$ such that the following holds: The interval $x_0 - h < x < x_0 + h$ is contained in $a < x < b$ and, in that interval, there is a unique solution $y = y(x)$ of the initial-value problem

$$\frac{dy}{dx} = f(x, y), \qquad y(x_0) = y_0. \quad ♦$$

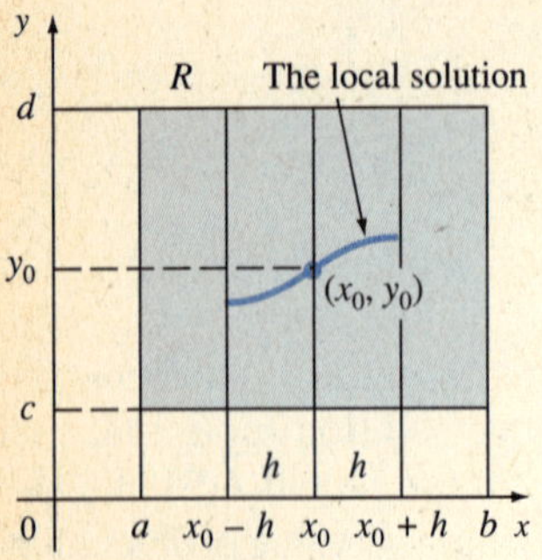

FIGURE 1.3
A solution exists in an open interval containing x_0

Theorem 1 is called an **existence-uniqueness theorem,** because it provides criteria guaranteeing the existence of a unique solution. It requires that we check that f and $\partial f/\partial y$ are continuous functions in a rectangle R containing the initial point (in many of our examples, f and $\partial f/\partial y$ are continuous everywhere). When these criteria hold, the theorem guarantees that a unique solution exists in some region containing the initial point (see Figure 1.3). Since the guarantee is only for a (possibly) small region around the initial point, we call such a theorem a **local** existence-uniqueness theorem. We now illustrate how to use Theorem 1, but we defer its proof to Section 2.8 and Appendix 3, where we also prove a **global** existence-uniqueness theorem for linear initial-value problems.

EXAMPLE 7 ▶ **Verifying uniqueness**

Does the initial-value problem

$$\frac{dy}{dx} = x^2 + y^2, \qquad y(0) = 1$$

have a unique solution in a region around its initial point (0, 1)?

Solution Since $f(x, y) = x^2 + y^3$ and $\partial f/\partial y = 3y^2$ are continuous everywhere, they are certainly continuous in any rectangle R containing the initial point (0, 1). Hence a unique local solution to the initial-value problem exists. ◀

EXAMPLE 8 ▶ **An initial-value problem where Theorem I does not apply**

If we look again at Equation **(2)** in Example 6 $[dy/dx = xy^{1/3}, y(0) = 0]$, we see that $\partial f/\partial y = x/3y^{2/3}$, which is not defined (and therefore not continuous) at the initial point (0, 0). Thus Theorem 1 does not apply to Equation **(2).** ◀

Theorem 1 is only one of many known existence-uniqueness theorems. One can relax its conditions while still retaining its conclusions. We will refer to a variety of existence-uniqueness theorems in subsequent sections.

SELF-QUIZ

Answer True *or* False.

I. $y' = 2xy$, $y(0) = y(1)$ is an initial-value problem.

II. $y' = 2xy$, $y(1) + 5 = 0$ is an initial-value problem.

III. $y' = \sqrt{y}$, $y(0) = 0$ must have a unique solution.

IV. $y' = xy$, $y(0) = 1$ must have a unique solution.

Answers to Self-Quiz

I. False (It is a boundary value problem) **II.** True **III.** False ($y \equiv 0$ and $y = x^2/4$ are both solutions. $\partial f/\partial t = 1/2\sqrt{y}$ does not exist at $y = 0$.) **IV.** True

PROBLEMS 1.2

In Problems 1–7 state the order of each of the following differential equations.

1. $y' + ay = \sin^2 x$ **2.** $\left(\frac{d^2x}{dt^2}\right) - 3x\frac{dx}{dt} = 4\cos t$ **3.** $s'''(t) - s''(t) = 0$ **4.** $\frac{d^5y}{dx^5} = 0$

5. $y'' + y = 0$ **6.** $\left(\frac{dx}{dt}\right)^3 = x^5$ **7.** $x' - x^2 = 3x'''$

In Problems 8–13 state whether each of the following differential equations is an initial-value problem or a boundary value problem.

8. $y'' + \omega^2 y = 0$, $y(0) = 0$, $y(1) = 1$ (here ω^2 is a constant)

9. $y'' + \omega^2 y = 0$, $y'(0) = 0$, $y'(0) = 1$ (ω^2 constant)

10. $y'' + \omega^2 y = 0$, $y'(0) = 0$, $y'(1) = 1$ (ω^2 constant)

11. $\left(\frac{dx}{dt}\right)^3 - 4x^2 = \sin t$, $x(0) = 3$

12. $y''' + 3y'' - (y')^2 + e^y = \sin x$, $y(0) = 0$, $y'(0) = 3$, $y''(0) = 5$

13. $y''' + 3y'' - (y')^2 + e^y = \sin x$, $y(0) = y(1) = 0$, $y'(0) = 2$

In Problems 14–18 verify that the given function or functions are solutions to the given differential equation.

14. $y'' + y = 0$; $y_1 = 2\sin x$, $y_2 = -5\cos x$

15. $y''' - y'' + y' - y = x$; $y_1 = e^x - x - 1$, $y_2 = 3\cos x - x - 1$, $y_3 = \cos x + \sin x + e^x - x - 1$

16. $x^2y'' - 2xy' + 2y = 0$; $y_1 = x$, $y_2 = x^2$, $y_3 = 2x - 3x^2$

17. $y'' - y = e^x$; $y_1 = \frac{x}{2}e^x$, $y_2 = \left(4 + \frac{x}{2}\right)e^x + 3e^{-x}$

18. $x^2y'' + 5xy' + 4y = 0$; $y_1 = \frac{4\ln x}{x^2}$, $y_2 = \frac{-6}{x^2}$ $(x > 0)$

---

19. By "guessing" that there is a solution to the equation

$$y'' - 4y' + 5y = 0$$

of the form $y = e^{ax}\cos bx$, find this solution. Try to "guess" a second solution.

20. By "guessing" that there is a solution to

$$y'' - 3y' - 4y = 0$$

of the form $y = e^{ax}$ for some constant a, find two solutions of the equation.

21. Given that $y_1(x)$ and $y_2(x)$ are two solutions of the equation in Problem 20, check to see that $y_3(x) = c_1y_1(x) + c_2y_2(x)$ is also a solution, where c_1 and c_2 are arbitrary constants, by substituting y_3 into the differential equation.

22. Determine $\phi(x)$ so that the functions $\sin \ln x$ and $\cos \ln x (x > 0)$ are solutions of the differential equation

$$[\phi(x)y']' + \frac{y}{x} = 0.$$

23. Show that $\sin(1/x)$ and $\cos(1/x)$ are solutions of the differential equation

$$\frac{d}{dx}\left(x^2\frac{dy}{dx}\right) + \frac{y}{x^2} = 0.$$

24. Verify that $y_1 = \sinh x$ and $y_2 = \cosh x$ are solutions of the differential equation $y'' - y = 0$. [*Hint:* $\cosh x = \frac{1}{2}(e^x + e^{-x})$; $\sinh x = \frac{1}{2}(e^x - e^{-x})$.]

25. Suppose that $\phi(x)$ is a solution of the initial-value problem $y'' + yy' = x^3$, with $y(-1) = 1$, $y'(-1) = 2$. Find $\phi''(-1)$ and $\phi'''(-1)$.

26. Let $\phi(x)$ be a solution to $y' = x^2 + y^2$ with $y(1) = 2$. Find $\phi'(1)$, $\phi''(1)$, and $\phi'''(1)$.

In Problems 27–30 determine the region in the xy-plane for which the existence of a unique solution through one of its points is guaranteed by the existence-uniqueness theorem (Theorem 1).

27. $y' = \frac{x - y}{x + y}$ **28.** $y' = x - y^2$

29. $y' = \sqrt{1 - y^2}$ **30.** $y' = \frac{1}{(x - y)^2}$

31. Show that $y = (1 - x)^{-1}$ is a solution of the initial-value problem

$$y' = y^2, \qquad y(0) = 1.$$

Where is this solution valid? [This illustrates why Theorem 1 is a local result.]

32. Show that the initial-value problem $y' = y^{1/5}$, $y(0) = 0$ has three solutions: $y \equiv 0$, $y = (\frac{4}{5}x)^{5/4}$, and $y = -(\frac{4}{5}x)^{5/4}$. Does this contradict the existence-uniqueness theorem? Explain.

1.3 DIRECTION FIELDS

In Section 1.1 we saw how to solve one kind of first-order differential equation. A first-order differential equation is **linear** if it can be written in the form $y' + a(x)y = f(x)$ where a and f are functions of x only. Otherwise it is **nonlinear.** Many (actually, most) nonlinear first-order equations cannot be solved by any of the methods discussed in Section 1.1 and Chapter 2. For example, it is impossible to find a closed form solution to the initial-value problem

$$\frac{dy}{dx} = (1 + x)\sin y - x, \qquad y(0) = 1$$

although the basic existence-uniqueness Theorem 1.2.1 guarantees that such a solution exists.

There are many ways to deal with this vexing problem. One is to look for numerical solutions; in other words, rather than looking for a function $y(x)$ that solves the problem for every value of the independent variable x, we seek approximate solutions for particular values of x. We will take this approach when we describe Euler's method in Section 1.4. We will discuss numerical method more fully in Chapter 8.

Another approach is to try to describe, without solving the equation, how solutions "behave." Typical questions that can be asked in such a *qualitative* approach are

1. Do solutions grow without bound as x increases?
2. Do solutions tend to zero or some other value?
3. Do solutions oscillate between certain values?

Much of the modern research on differential equations centers on these questions. In the rest of this section, we describe one relatively simple way to obtain information about the solution to a differential equation.

Consider the first-order differential equation

$$y' = f(x, y). \tag{1}$$

Equation **(1)** contains a great deal of information. Under certain conditions, the differential equation has a unique solution if we specify an initial condition; that is, there is a unique function $y(x)$ satisfying $y'(x) = f(x, y)$ and $y(x_0) = y_0$ for arbitrarily chosen numbers x_0 and y_0. The function $y(x)$ is a curve in the xy-plane. Even though we may not be able to find the curve $y(x)$, *we do know its slope, y', at every point.* If the solution $y(x)$, passes through the point (x, y), since $y' = f(x, y)$, we may conclude that

the slope of the tangent line to the curve $y(x)$ at the point (x, y) is given by $f(x, y)$.

This fact can be very useful. We know that the solution to the initial-value problem $y' = f(x, y)$, $y(x_0) = y_0$, passes through the point (x_0, y_0) in the xy-plane.

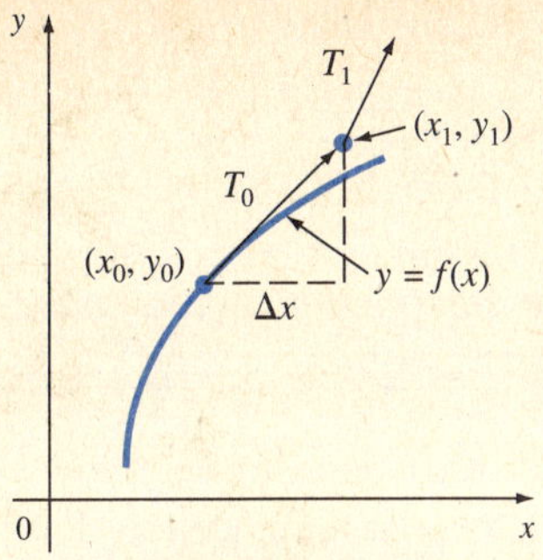

FIGURE 1.4
The tangent line to the solution of the differential equation $y' = f(x, y)$ at the point (x_0, y_0) is the line passing through (x_0, y_0) with slope $f(x_0, y_0)$.

At that point the tangent line has its slope equal to $f(x_0, y_0)$ (since $dy/dx = f(x, y)$). In Figure 1.4 we draw the (unknown) solution curve (in blue). We don't know the solution but we can draw the tangent line, T_0, to the curve at (x_0, y_0). If Δx is small, then the tangent line gives a good approximation to the solution curve near (x_0, y_0). We follow the tangent line a short distance to a new point which we label (x_1, y_1) and draw a new tangent line there. In Figure 1.4 it is labelled T_1. The slope of T_1 is equal to $f(x_1, y_1)$. Continuing in this manner we can get a rough idea of the shape of the solution curve. The smaller we take the step size Δx, the better will be the approximation to the true solution curve—seen as a collection of sometimes connected tangent lines (such a collection is called a **piecewise linear function.**) Thus we know the direction of the solution curve $y(x)$ through any point in the xy-plane. The set of all these directions in the plane is called the **direction field** (or **slope field**) of the differential equation $y' = f(x, y)$. In many cases, we can use the direction field to sketch the solution to a differential equation without actually solving the differential equation.

EXAMPLE 1 ▶ **Sketching a direction field**

Consider the initial-value problem

$$y' = 2xy, \quad y(0) = 1. \tag{2}$$

This is a linear equation which can be solved either by separation of variables discussed in Section 2.1 or by the methods of Section 2.3. The unique solution is $y = e^{x^2}$. Forgetting for the moment that we do know the solution, we will sketch its direction field to illustrate this very useful technique.

Since $y'(x) = 2xy$, it is apparent that

$$\begin{aligned} y'(x) &> 0, \quad \text{if } xy > 0 \\ y'(x) &< 0, \quad \text{if } xy < 0. \end{aligned}$$

Thus $y'(x) > 0$ in the first and third quadrants and $y'(x) < 0$ in the second and fourth quadrants. The direction field is sketched in Figure 1.5. This figure consists of arrows with ends at a grid of points (x, y) and slopes $f(x, y) = 2xy$, where x and y are the coordinates of each grid point. Note that since x and y are positive in the first quadrant, the slopes of the tangent lines to any solution curve are positive, so the solution curves increase and become steeper as x and y become larger. Along

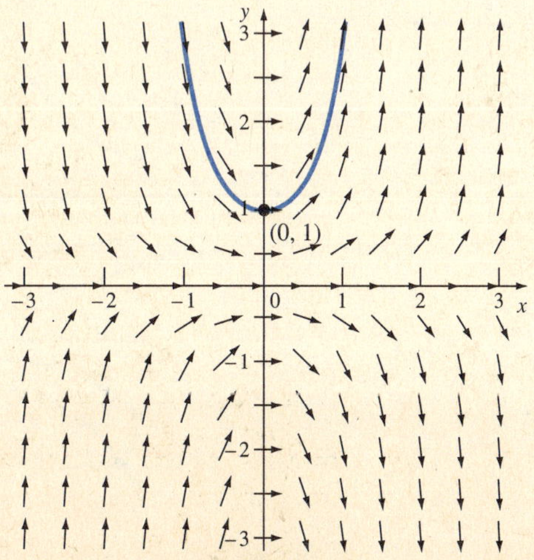

FIGURE 1.5
Direction field of $y' = 2xy$

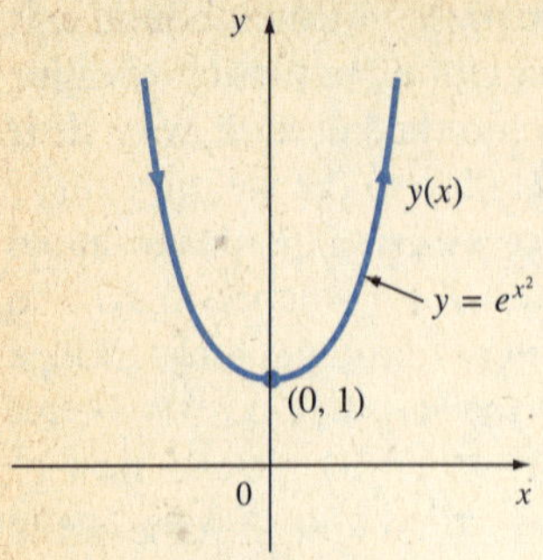

FIGURE 1.6
Graph of the solution to the initial-value problem
$y' = 2xy, \quad y(0) = 1$

the axes, the solution is flat (has a horizontal tangent) because the derivative y' is zero. In the second quadrant, the slopes of the tangent lines are negative, since $y' < 0$. Similar conditions apply in the third and fourth quadrants.

For the particular initial-value problem we are considering in Equation **(2)**, we know that the solution curve must satisfy the initial condition $y(0) = 1$. Thus the solution curve must pass through the point $(0, 1)$. Since that point is on the y-axis, the curve is initially flat and moves into the first quadrant with increasing values of x. As x increases, the solution curve begins to rise as $y'(x) > 0$ in the first quadrant. Hence $y > 1$ for all values of $x > 0$. On the other hand, if we allow x to be negative, the solution curve extends into the second quadrant. Since the slope is negative in this quadrant, the solution curve decreases as the curve moves to the right. Because xy becomes larger in absolute value as we move away from the y-axis, the curve becomes steeper. Plotting this information, we obtain the curve in Figure 1.6, which is also superimposed on Figure 1.5. This curve is, of course, the graph of $y = e^{x^2}$. ◀

EXAMPLE 2 ▶ Sketching a direction field

Consider the differential equation

$$y'(x) = y(y + 1). \tag{3}$$

The following facts are evident:

$$\begin{aligned} y' &> 0, \quad \text{if} \quad y < -1, \\ y' &= 0, \quad \text{if} \quad y = -1, \\ y' &< 0, \quad \text{if} \quad -1 < y < 0, \\ y' &= 0, \quad \text{if} \quad y = 0, \\ y' &> 0, \quad \text{if} \quad y > 0. \end{aligned}$$

Since there is no x-term in the right-hand side of Equation **(3)**, the direction field depends only on the values of y. The direction field, together with eight possible solutions (or **trajectories**) for eight different initial values $(0, y_0)$, is given in Figure 1.7. Note that the curve $y(x) \equiv 0$ is the unique solution for the initial value $(0, 0)$, and the curve $y(x) \equiv -1$ is the unique solution passing through the initial

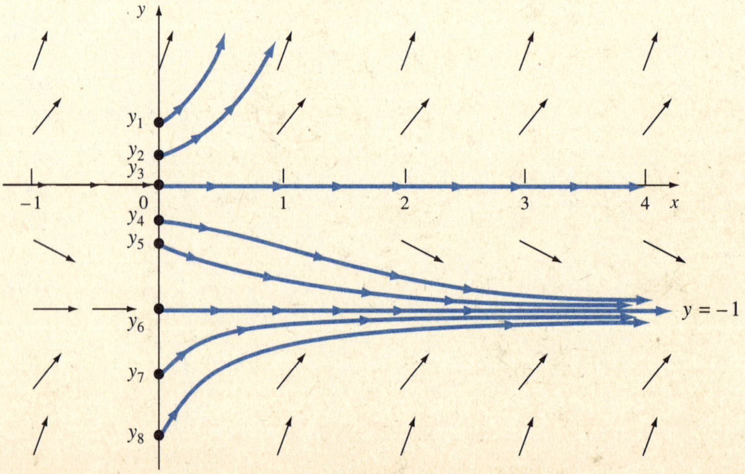

FIGURE 1.7
Direction field and trajectories for $y' = y(y + 1)$

value $(0, -1)$. Note, too, that if $y(0) > 0$, then $y(x)$ increases without bound as x increases. On the other hand, if $y(0) < 0$, then all solutions approach the line $y(x) \equiv -1$. This is a substantial amount of information obtained with very little work.

Equation **(3)** can be solved by separation of variables (Section 2.1). You should verify that

$$y = \frac{1}{Ce^{-x} - 1} \tag{4}$$

is a solution for every constant C. Another solution is $y \equiv 0$ [this satisfies the initial condition $y(0) = 0$].

If $C < 1$, then the denominator in Equation **(4)** is always negative and tends to -1 as $x \to \infty$. This shows that if $y(0) < 0$, the solution curve approaches -1 as x becomes arbitrarily large. If $C > 1$, so that $y(0) > 0$, there is a positive value for x such that $Ce^{-x} = 1$ and the denominator of Equation **(4)** is zero. This occurs when $x = \ln C$, and since $y(0) = 1/(C - 1)$ by Equation **(4)**, $C = 1/y(0) + 1 = (y(0) + 1)/y(0)$. Thus the solution curve tends to infinity as x approaches $\ln[y(0) + 1)/y(0)]$. This confirms the shape of our rough sketches in Figure 1.7. ◀

Of course, if we can solve the differential equation directly, we need not plot the direction field. Nevertheless, direction fields provide a quick and useful, if crude, tool for getting an idea of the shape of the solution. And if a solution is not readily obtainable, direction fields provide an important first step in analyzing the behavior of solutions.

EXAMPLE 3 ▶ **Generating a direction field on a computer**

Using MATHEMATICA, MAPLE, or DERIVE, it is possible to generate direction fields for almost any first-order differential equations. In Figure 1.8(a) we provide direction fields for the equation $y' = (1 + x)\sin y - x$. In Figure 1.8(b) and (c), the tangent lines are drawn with smaller and smaller values for Δx. The following graphs were produced on MATHEMATICA. ◀

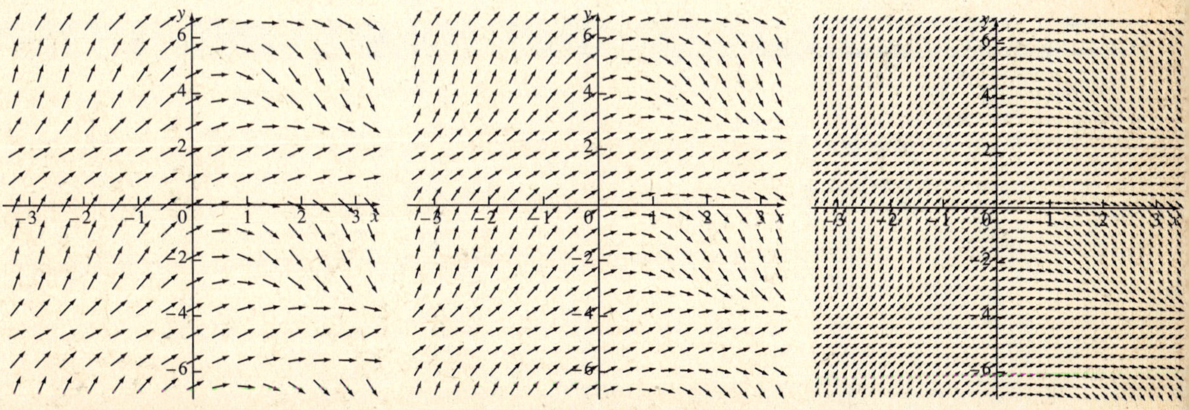

(a) $\Delta s = 1$ (b) $\Delta s = \frac{1}{2}$ (c) $\Delta s = 0.2$

FIGURE 1.8
Direction fields for $y' = (1 + x)\sin y - x$, $-3 \le x \le 3$, and $-2\pi \le y \le 2\pi$

SELF-QUIZ

Answer True *or* False.

I. The direction field consists of tangent vectors to the solution curves.

II. The tangent vectors in a direction field always point to the right.

Answers to Self-Quiz

I. True **II.** True

PROBLEMS 1.3

1. (a) Plot the direction field for the differential equation

$$y' = y^{4/5}.$$

(b) Plot the solution that satisfies $y(0) = 2$.
(c) Plot the solution that satisfies $y(0) = -1$.

2. (a) Plot the direction field for the equation

$$y' = y^{3/5}.$$

(b) Plot the solution that satisfies $y(0) = 3$.
(c) Plot the solution that satisfies $y(0) = -2$.

3. (a) Plot the direction field for the equation

$$y' = 2y(3 - y).$$

(b) Plot the solution that satisfies $y(0) = -1$.
(c) Plot the solution that satisfies $y(0) = 2$.
(d) Plot the solution that satisfies $y(0) = 4$.

4. (a) Plot the direction field for the equation

$$y' = (y^2 - 4)(y - 4).$$

(b) Plot the solution that satisfies $y(0) = -3$.
(c) Plot the solution that satisfies $y(0) = -1$.
(d) Plot the solution that satisfies $y(0) = 1$.
(e) Plot the solution that satisfies $y(0) = 3$.
(f) Plot the solution that satisfies $y(0) = 5$.

5. (a) Plot the direction field for the equation

$$y' = x^2 + y^2.$$

(b) Plot the solution that satisfies $y(1) = 2$.

6. (a) Plot the direction field of the equation

$$y' = \frac{2xy}{1 + y^2}.$$

(b) Plot the solution that satisfies $y(1) = 1$.
(c) Plot the solution that satisfies $y(1) = -1$.

7. Plot the direction field of the equation

$$y' = \frac{4y - 5x}{y + x}.$$

8. (a) Plot the direction field of the equation

$$y' = 1 + x + y.$$

(b) Plot the solution that satisfies $y(0) = 1$.

9. (a) Plot the direction field of the equation

$$y' = e^{xy} - 1.$$

(b) Discuss the solution that passes through the origin.
(c) Compare the solution through $(0, 1)$ to that through $(0, -1)$.

COMPUTER EXERCISES

In Problems 10–15 use computer software to generate direction fields for the given differential equations in the given interval.

10. $\dfrac{dy}{dx} = 2y; \quad -2 \le x \le 2, 0 \le y \le 6$

11. $\dfrac{dy}{dx} = 3y(1 - y); \quad -2 \le x \le 3, -3 \le y \le 4$

12. $\dfrac{dy}{dx} = xy^2 - yx^2; \quad -2 \le x \le 2, \quad -2 \le y \le 2$

13. $\dfrac{dy}{dx} = |x|y; \quad -4 \le x \le 4, -3 \le y \le 3$

14. $\dfrac{dy}{dx} = -2y \cos y - x; \quad -4 \le x \le 4, -2\pi \le y \le 2\pi$

15. $\dfrac{dy}{dx} = e^x \ln|y|; \quad -2 \le x \le 2, -8 \le y \le 8$

1.4 SOLVING DIFFERENTIAL EQUATIONS NUMERICALLY: EULER'S METHOD

In Section 1.1 we solved the differential equation of exponential growth and decay. In Section 1.2 we discussed conditions under which first-order initial-value problems have unique solutions. Finding these solutions, however, is sometimes difficult, or even impossible. Nevertheless, it is usually possible to approximate a solution to great accuracy and, for most applications, this is really all one needs. In this section we present an elementary numerical technique for approximating the unique solution to a first-order, initial-value problem.

Before presenting this numerical technique, it is useful to discuss the situations in which numerical methods could or should be employed. Such methods are used frequently when other methods are not applicable. Even when other methods do apply, there may be an advantage in having a numerical solution; solutions in terms of more exotic special functions are sometimes difficult to interpret. There may also be computational advantages: the exact solution may be extremely tedious to obtain. Finally, there are situations (see, for example, Section 2.2) for which we can solve a differential equation but for which the solution is given implicitly; in these cases a numerical method can tell us more about the solution.

On the other hand, care must always be exercised in using any numerical scheme, as the accuracy of the solution depends, not only on the "correctness" of the numerical method being used, but also on the precision of the device (hand calculator or computer) used for the computations.

We assume that the initial-value problem

$$\frac{dy}{dx} = f(x, y), \qquad y(x_0) = y_0 \tag{1}$$

has a unique solution $y(x)$ on some interval containing x_0. The technique we describe below approximates this solution $y(x)$ only at a finite number of points

$$x_0, \qquad x_1 = x_0 + h, \qquad x_2 = x_0 + 2h, \ldots, \qquad x_n = x_0 + nh,$$

where h is some (nonzero) real number. The method provides a value y_k that is an approximation of the exact value $y(x_k)$ for $k = 0, 1, \ldots, n$.

The Euler method

This crude but very simple procedure was first used by Leonhard Euler, a mathematical genius whose biography appears on page 29. The idea is to approximate $y(x_1) = y_1$ by assuming that $f(x, y)$ varies so little on the interval $x_0 \le x \le x_1$ that only a very small error is made by replacing it by the constant value $f(x_0, y_0)$. Integrating

$$\frac{dy}{dx} = f(x, y)$$

from x_0 to x_1, we obtain

$$\begin{aligned} y_1 - y_0 = y(x_1) - y(x_0) &= \int_{x_0}^{x_1} \frac{dy}{dx}\,dx \\ &= \int_{x_0}^{x_1} f(x, y)\,dx \approx f(x_0, y_0)(x_1 - x_0). \end{aligned} \tag{2}$$

Since $h = x_1 - x_0$, $y_1 = y_0 + hf(x_0, y_0)$.

Repeating the process with (x_1, y_1) to obtain y_2, and so on, we obtain the **difference equation**

Definition Euler's method

$$y_{n+1} = y_n + hf(x_n, y_n), \tag{3}$$

where y_n is an approximation to $y(x_0 + nh)$, $n = 0, 1, 2, \ldots$. We solve Equation **(3)** iteratively, that is, by first finding y_1, then using it to find y_2, and so on.

The geometric meaning of Equation **(3)** is easily seen by considering the direction field of differential Equation **(1)**: we are simply following the tangent to the solution curve passing through (x_n, y_n) for a small horizontal distance. Looking at Figure 1.9, where the smooth curve is the unknown exact solution to the initial-value problem (1), we see how Equation **(3)** approximates the exact solution. Since $f(x_0, y_0)$ is the slope of the exact solution at (x_0, y_0), we follow this line to the point (x_1, y_1). Some solution to the differential equation passes through this point. We follow its tangent line at this point to reach (x_2, y_2), and so on. The differences Δ_k are errors at the kth stage in the process.

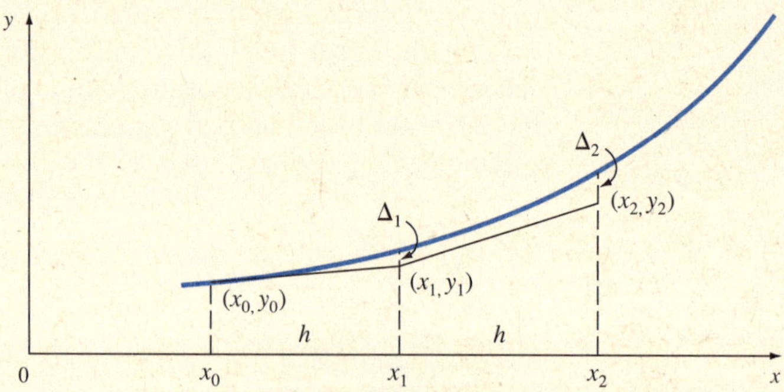

FIGURE 1.9
The true solution is in blue. The Euler approximation is in black.

EXAMPLE 1 ▶ **Using Euler's method with different values of *h***

Estimate $y(1)$ where $y(x)$ satisfies the initial-value problem

$$\frac{dy}{dx} = y, \qquad y(0) = 1.$$

Solution From Section 1.1 we know that the solution to this problem is $y(x) = e^x$. Thus $y(1) = e^1 = e \approx 2.71828$. Let us apply the Euler method with $h = \frac{1}{5}$ to this familiar problem to see if this answer is obtained.

We begin by dividing the interval $[0, 1]$ into five subintervals. Then $h = \frac{1}{5} = 0.2$ and $f(x_n, y_n) = y_n$. Thus we obtain, successively,

$$\begin{aligned}
y_0 &= y(0) = 1\\
y_1 &= y_0 + 0.2y_0 = (1 + 0.2)y_0 = (1.2)y_0 = 1.2\\
y_2 &= y_1 + 0.2y_1 = (1.2)y_1 = (1.2)^2\\
y_3 &= y_2 + 0.2y_2 = (1.2)y_2 = (1.2)^3\\
y_4 &= y_3 + 0.2y_3 = (1.2)y_3 = (1.2)^4\\
y_5 &= y_4 + 0.2y_4 = (1.2)y_4 = (1.2)^5 \approx 2.48832.
\end{aligned}$$

Thus, with $h = 0.2$, the Euler method yields $y(1) \approx 2.48832$. Since we know that $y(1) = e \approx 2.71828$, our error at $x = 1$ is given by

$$\text{error} = 2.71828 - 2.48832 = 0.22996.$$

What happens if we double the number of subintervals? If $n = 10$, then $h = 0.1$, and, computing as before, we obtain $y(1) \approx y_{10} = (1.1)^{10} \approx 2.59374$.

The error is given by

$$\text{error} = 2.71828 - 2.59374 = 0.12454.$$

In general, if $h = 1/n$, then

$$y_1 = y_0 + hy_0 = (1 + h)y_0 = 1 + h = 1 + \frac{1}{n}$$

$$y_2 = y_1\left(1 + \frac{1}{n}\right) = \left(1 + \frac{1}{n}\right)^2$$

$$y_3 = y_2\left(1 + \frac{1}{n}\right) = \left(1 + \frac{1}{n}\right)^3$$

$$\vdots$$

$$y_n = \left(1 + \frac{1}{n}\right)^n.$$

TABLE 1.2

n	$\left(1 + \frac{1}{n}\right)^n$
1	2
2	2.25
5	2.48832
10	2.59374
100	2.70481
1,000	2.71692
10,000	2.71815
100,000	2.71827
1,000,000	2.71828

Thus

$$y(1) \approx \left(1 + \frac{1}{n}\right)^n.$$

Different values of $(1 + 1/n)^n$ are given in Table 1.2 (to five decimal places of accuracy).

The numbers in Table 1.2 should not be surprising. In fact, in elementary calculus the number e is *defined* by

$$e = \lim_{n\to\infty}\left(1 + \frac{1}{n}\right)^n.$$

Thus $e = \lim_{n\to\infty} y_n$, where y_n is the nth iterate in Euler's method with $h = 1/n$; that is, y_n approximates $y(1)$ with better and better accuracy as n increases. ◀

EXAMPLE 2 ▶ **Approximating the solution of an initial-value problem using Euler's method**

Find an approximate value for $y(1)$ if $y(x)$ satisfies the initial-value problem

$$\frac{dy}{dx} = y + x^2, \qquad y(0) = 1. \tag{4}$$

Use five subintervals in your approximation.

Solution Here $h = 1/n = 1/5 = 0.2$ and we wish to find $y(1)$ by approximating the solution at $x = 0.0, 0.2, 0.4, 0.6, 0.8$, and 1.0. We see that $f(x_n, y_n) = y_n + x_n^2$, and the Euler method [Equation (3)] yields

$$y_{n+1} = y_n + h \cdot f(x_n, y_n) = y_n + h(y_n + x_n^2).$$

Since $y_0 = y(0) = 1$, we obtain

$$y_1 = y_0 + h \cdot (y_0 + x_0^2) = 1 + 0.2(1 + 0^2) = 1.2$$

$$y_2 = y_1 + h \cdot (y_1 + x_1^2) = 1.2 + 0.2[1.2 + (0.2)^2] = 1.448 \approx 1.45$$

$$y_3 = y_2 + h(y_2 + x_2^2) = 1.45 + 0.2[1.45 + (0.4)^2] \approx 1.77$$

$$y_4 = y_3 + h(y_3 + x_3^2) = 1.77 + 0.2[1.77 + (0.6)^2] \approx 2.20$$

$$y_5 = y_4 + h(y_4 + x_4^2) = 2.20 + 0.2[2.20 + (0.8)^2] \approx 2.77.$$

We arrange our work as shown in Table 1.3.

TABLE 1.3

x_n	y_n	$f(x_n, y_n) = y_n + x_n^2$	$y_{n+1} = y_n + hf(x_n, y_n)$
0.0	1.00	1.00	1.20
0.2	1.20	1.24	1.45
0.4	1.45	1.61	1.77
0.6	1.77	2.13	2.20
0.8	2.20	2.84	2.77
1.0	2.77		

The value $y_5 = 2.77$, corresponding to $x_5 = 1.0$, is our approximate value for $y(1)$. Equation **(4)** has the exact solution $y = 3e^x - x^2 - 2x - 2$ (check this), so that $y(1) = 3e - 5 \approx 3.154$. Thus the Euler method estimate was off by about 12%.* This is not surprising, because we treated the derivative as a constant over intervals of length 0.2 unit. The error that arises in this way is called **discretization error,** because the "discrete" function $f(x_n, y_n)$ is substituted for the "continuously valued" function $f(x, y)$. It is usually true that if we reduce the step size h, we can improve the accuracy of our answer, since then the "discretized" function $f(x_n, y_n)$ is closer to the true value of $f(x, y)$ over the interval $[0, 1]$. This is illustrated in Figure 1.10 with $h = 0.2$ and $h = 0.1$. Indeed, carrying out similar calculations with $h = 0.1$ yields an approximation of $y(1)$ of 2.94, which is a good deal more accurate (an error of less than 7%).† ◀

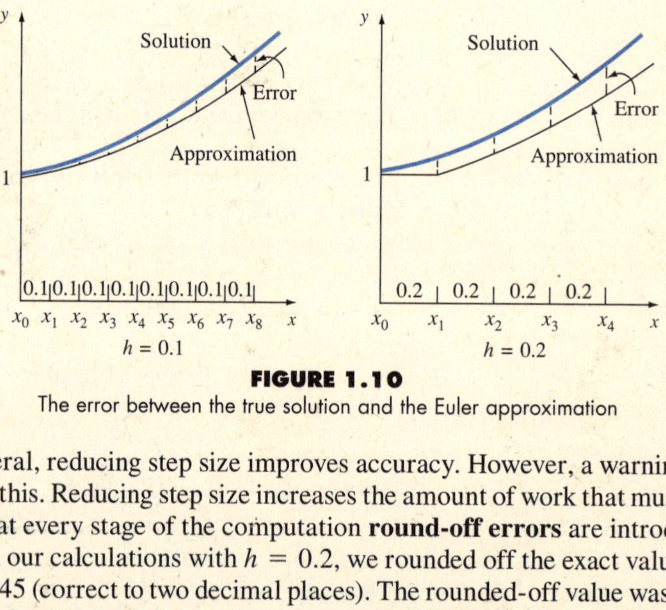

FIGURE 1.10
The error between the true solution and the Euler approximation

In general, reducing step size improves accuracy. However, a warning must be attached to this. Reducing step size increases the amount of work that must be done. Moreover, at every stage of the computation **round-off errors** are introduced. For example, in our calculations with $h = 0.2$, we rounded off the exact value 1.448 to the value 1.45 (correct to two decimal places). The rounded-off value was then used to calculate further values of y_n. It is not unusual for a computer solution of a more complicated differential equation to take several thousand individual computations, thus having several thousand round-off errors. In some problems the accumulated round-off error can be so large that the resulting computed solution is

*$y(1) - y_5 \approx 3.15 - 2.77 = 0.38$ and $\frac{0.38}{3.15} \approx 0.12 = 12\%$.
†$y(1) - y_{10} \approx 3.15 - 2.94 \approx 0.21$ and $\frac{0.21}{3.15} \approx 0.067 \approx 7\%$.

sufficiently inaccurate to invalidate the result. The statement that reducing step size improves accuracy is made under the assumption (often true) that the average of the round-off errors is zero. In any event, it should be clear that reducing the step size, thereby increasing the number of computations, is a procedure that should be carried out carefully. In general, each problem has an optimal step size, and a smaller than optimal step size yields a greater error due to accumulated round-off errors.

LEONHARD EULER (1707–1783)

Leonhard Euler

Leonhard Euler (pronounced "Oiler") was born in Basel, Switzerland. His father, a clergyman, planned that his son follow him into the ministry. The father was gifted at mathematics and, together with Johann Bernoulli, instructed young Leonhard in that subject as well as in theology, astronomy, physics, medicine, and several Eastern languages.

In the early eighteenth century, Catherine I of Russia, wife of Peter the Great, founded the St. Petersburg Academy. In 1727 Euler applied and was accepted for a chair in the faculty of medicine and physiology at the academy. However, Catherine died the day Euler arrived in Russia and the academy was plunged into turmoil. By 1730 Euler found himself in the chair of natural philosophy, from which he pursued his mathematical career. In 1741 he went to Berlin to head the Prussian Academy, accepting an invitation from Frederick the Great. Twenty-five years later he returned to St. Petersburg, where he died in 1783 at the age of 76.

Euler was the most prolific writer in the history of mathematics. He found results in virtually every branch of pure and applied mathematics. Although German was his native language, he wrote mostly in Latin, and occasionally in French. His amazing productivity did not decline even when, in 1766, he became totally blind. During his lifetime Euler published 530 books and papers. When he died, he left so many unpublished manuscripts that the St. Petersburg Academy was still publishing his work in its *Proceedings* almost half a century later. Euler enriched such diverse areas as hydraulics, celestial mechanics, lunar theory, and the theory of music, as well as mathematics.

Euler had a phenomenal memory. As a young man he memorized the entire Aeneid by Virgil (in Latin), and many years later he could still recite the entire work. He was able to solve astonishingly complex mathematical problems in his head, including, it is said, problems in astronomy that stymied Newton. The French academician François Arago once commented that Euler could calculate without effort "just as men breathe, as eagles sustain themselves in the air."

Euler wrote in a mathematical language that is largely in use today. The following symbols, among many others, were first used by him:

$f(x)$	for functional notation
e	for the base of the natural logarithm
Σ	for the summation sign
i	to denote the imaginary unit

Euler's textbooks are models of clarity. His texts include *Introductio in analysis infinitorum* (1748), *Institutiones calculi differentialis* (1755), and the three-volume *Institutiones calculi integralis* (1768–1774). These and others of his works served as models for many of today's mathematics textbooks.

It is said that Euler did for mathematical analysis what Euclid did for geometry. It is no wonder that so many mathematicians who followed express the debt they owe him.

SELF-QUIZ

Answer True *or* False.

I. The only error that arises using Euler's method is discretization error.

II. Euler's method always underestimates the exact solution.

III. Euler's method follows the direction field a horizontal distance of h.

Answers to Self-Quiz

I. False (round-off also occurs) **II.** False (not if the solution is concave down) **III.** True

PROBLEMS 1.4

Solve Problems 1–10 by using the Euler method and the indicated value of h.

1. $\frac{dy}{dx} = x + y,\ y(0) = 1.$

Approximate $y(1)$ with $h = 0.2$.

2. $\frac{dy}{dx} = x - y,\ y(1) = 2.$

Approximate $y(3)$ with $h = 0.4$.

3. $\frac{dy}{dx} = \frac{x - y}{x + y},\ y(2) = 1.$

Approximate $y(1)$ with $h = -0.2$.

4. $\frac{dy}{dx} = \frac{y}{x} + \left(\frac{y}{x}\right)^2,\ y(1) = 1.$

Approximate $y(2)$ with $h = 0.2$.

5. $\frac{dy}{dx} = x\sqrt{1 + y^2},\ y(1) = 0.$

Approximate $y(3)$ with $h = 0.4$.

6. $\frac{dy}{dx} = x\sqrt{1 - y^2},\ y(1) = 0.$

Approximate $y(2)$ with $h = 0.125$.

7. $\frac{dy}{dx} = \frac{y}{x} - \frac{5}{2}x^2y^3,\ y(1) = \frac{1}{\sqrt{2}}.$

Approximate $y(2)$ with $h = 0.125$.

8. $\frac{dy}{dx} = \frac{-y}{x} + x^2y^2,\ y(1) = \frac{2}{9}.$

Approximate $y(3)$ with $h = \frac{1}{3}$.

9. $\frac{dy}{dx} = ye^x,\ y(0) = 2.$

Approximate $y(2)$ with $h = 0.2$.

10. $\frac{dy}{dx} = xe^y,\ y(0) = 0.$

Approximate $y(1)$ with $h = 0.1$.

In Problems 11–20, use the Euler method to graph approximately the solution of the given initial-value problem by plotting the points (x_k, y_k) over the indicated range, where $x_k = x_0 + kh$.

11. $y' = xy^2 + y^3,\ y(0) = 1,\ h = 0.02,\ 0 \le x \le 0.1$

12. $y' = x + \sin(\pi y),\ y(1) = 0,\ h = 0.2,\ 1 \le x \le 2$

13. $y' = x + \cos(\pi y),\ y(0) = 0,\ h = 0.4,\ 0 \le x \le 2$

14. $y' = \cos(xy),\ y(0) = 0,\ h = \pi/4,\ 0 \le x \le \pi$

15. $y' = \sin(xy),\ y(0) = 1,\ h = \pi/4,\ 0 \le x \le 2\pi$

16. $y' = \sqrt{x^2 + y^2},\ y(0) = 1,\ h = 0.5,\ 0 \le x \le 5$

17. $y' = \sqrt{y^2 - x^2},\ y(0) = 1,\ h = 0.1,\ 0 \le x \le 1$

18. $y' = \sqrt{x + y^2},\ y(0) = 1,\ h = 0.2,\ 0 \le x \le 1$

19. $y' = \sqrt{x + y^2},\ y(1) = 2,\ h = -0.2,\ 0 \le x \le 1$

20. $y' = \sqrt{x^2 + y^2},\ y(1) = 5,\ h = -0.2,\ 0 \le x \le 1$

1.5 USING SYMBOLIC MANIPULATING PROGRAMS

We shall show in a sequence of short notes at the end of each chapter how symbolic manipulating programs, such as MAPLE, MATHEMATICA, DERIVE, MATLAB, or MATHCAD, can be used in solving and graphing differential equations. In this section we will show their use in computing approximations to first-order differential equations with initial conditions, using the Euler method that was discussed in Section 1.4. The solution we will get is a numerical approximation to the solution of the initial-value problem.

Properties of symbolic manipulators

Programs such as MATHEMATICA, MAPLE, DERIVE, MATLAB, and MATHCAD are computer programs that can do many diverse mathematical tasks. In addition to the numerical approximation of differential equations, they can also integrate or differentiate a function and perform algebraic operations, using the symbols we use to define the function. You may have noticed while studying the Euler method that the method is done repeatedly to obtain approximate values of the dependent variable at consecutive steps of the independent variable. This is an example of an *iteration*, a procedure that has many applications in mathematics.

Since all of the programs for symbolic manipulation are different, the command to do any particular task will change from program to program. Consequently, we shall develop a set of *generic commands*, which we shall use to explain each process. Once we have described the generic algorithm for attacking a particular type of differential equation, each symbolic manipulating program requires you to define a "script" of command to execute sequentially the steps in the algorithm.

We shall use the letter x to denote the independent variable in this section, but any other letter could have been used instead. The following basic generic commands will be used to construct the script for the Euler method:

`ITERATE(f(v),v,v₀,n)` = {do n iterations of the function $v_{k+1} = f(v_k)$ beginning at v_0 and save the values $v_0, v_1, \ldots, v_n$ in a vector. The symbol v can be either a variable or a vector of variables.}.

`SUBST(f(x,y) [=c], x=x₀, y=y₀)` = {substitute $x = x_0$ and $y = y_0$ in the function $f(x, y)$ [or in the equation $f(x, y) = c$]}.

`LIM(f(x),x,x₀)` = {limit of $f(x)$ as $x \to x_0$}.

Euler's method

We saw in Section 1.4 that Euler's method for approximating the solution to the initial-value problem

$$\frac{dy}{dx} = f(x, y), \qquad y(x_0) = y_0, \tag{1}$$

is given by the two equations

$$\begin{aligned} x_{k+1} &= x_k + h, \\ y_{k+1} &= y_k + hf(x_k, y_k), \end{aligned} \tag{2}$$

where h is the step size and x_0, y_0 are the initial values.

Consider the vector $v_k = [x_k, y_k]$, and observe that if we put the equations in **(2)** in vector form, then

$$v_{k+1} = v_k + h[1, f(v_k)]$$

where the h term multiplies each of the entries of the following vector.

Using the generic commands listed above, we can describe the Euler method by the following "script":

```
EULER(x,y,f(x,y),x₀,y₀,h,n):=
                    ITERATE(v+h*[1,SUBST(f,[x, y]=v)],v,[x₀,y₀],n)
```

or, equivalently by

```
EULER(x,y,f(x,y),x0,y0,h,n):=
                    ITERATE(v+h*[1,LIM(f,[x, y],v)],v,[x0,y0],n).
```

We shall illustrate how this script works by an example.

EXAMPLE 1 ▶

Approximate the solution of the initial-value problem

$$\frac{dy}{dx} = y + x^2, \qquad y(0) = 1,$$

using Euler's method, with $h = 0.2$ and n = number of steps = 5.

Solution This is Example 1.4.2. Using our generic commands, we would obtain the numerical solution by typing:

```
EULER(x, y, y+x^2, 0, 1, 0.2, 5)
```

The program would then execute the following steps:

```
ITERATE(v+h*[1,SUBST(f,[x,y]=v)],v,[x0,y0],n)=
ITERATE(v+0.2*[1,SUBST(y+x^2,[x,y]=v)],v,[0,1],5)=[[0,1],
[0,1]+0.2*[1,1]=[0.2,1.2]
[0.2,1.2]+0.2*[1,1.24]=[0.4,1.448]
[0.4,1.448]+0.2*[1,1.608]=[0.6,1.7696]
[0.6,1.7696]+0.2*[1,2.1296]=[0.8,2.19552]
[0.8,2.19552]+0.2*[1,2.83552]=[1,2.762624]]
```

so we would obtain the vector

```
[[0,1],[0.2,1.2], [0.4,1.448],[0.6,1.7696],
[0.8,2.19552], [1, 2.762624]].
```

Using this vector, it would be possible to graph the numerical solution to the problem. All of the symbolic manipulating programs have ways in which they can graph the numerical solution. Since the commands for plotting and entering the data are different, you should consult the manual of your program. ◀

PROBLEMS 1.5

Solve each of the following initial-value problems by Euler's method using one of the symbolic manipulating programs: MAPLE, MATHEMATICA, DERIVE, MATLAB, or MATCAD, and plot the graph of the solution. Print the graph of your solution.

1. $\dfrac{dy}{dx} = xy^2 + y^3$, $y(0) = 1$, $h = 0.02$, $n = 20$

2. $\dfrac{ds}{dt} = s + \cos(\pi t)$, $s(1) = 0$, $h = 0.05$, $n = 20$

3. $y' = \sqrt{x^2 + y^2}$, $y(0) = 1$, $h = 0.05$, $n = 20$

4. $y' = \sqrt{x + y^2}$, $y(1) = 2$, $h = 0.05$, $n = 20$

5. $y' = \dfrac{x + y^2}{x + 2y^2}$, $y(1) = 2$, $h = 0.1$, $n = 20$

SUMMARY OUTLINE OF CHAPTER 1

Free Fall: $h(t) = -\frac{g}{2}t^2 + v_0 t + h_0$, $v(t) = -gt + v_0$, $a(t) = -g$. p. 2

Relative Rate of Growth: Let $y = f(x)$.

$$\text{relative rate of growth} = \frac{\text{actual rate of growth}}{\text{size of } f(x)} = \frac{f'(x)}{f(x)} = \frac{dy/dx}{y} \quad \text{p. 5}$$

Exponential Growth and Decay: The differential equation

$$\frac{dy}{dx} = \alpha y \ (*)$$

is called the differential equation of exponential growth or decay (growth if $\alpha > 0$ and decay if $\alpha < 0$). All solutions to (*) are given by

$$y = ce^{\alpha x}$$

where c is an arbitrary real number. p. 6

The Initial-Value Problem of Exponential Growth or Decay: If $P(t)$ is growing or declining continuously at a constant relative rate (so that $P'(t)/P(t) = \pm\alpha$) with initial size $P(0)$, then $P(t)$ is given by (with $\alpha > 0$)

$$P(t) = P(0)e^{\alpha t} \quad \text{or} \quad P(t) = P(0)e^{-\alpha t}. \quad \text{p. 7}$$

Newton's Law of Cooling: $T' = \alpha(T - T_s)$ where T is the temperature of an object and T_s is the temperature of the surrounding medium. p. 9

Order of a Differential Equation: The **order** of a differential equation is defined as the order of the highest derivative appearing in the equation. p. 15

Solution to an *n*th-order Differential Equation: A **solution** to an nth-order differential equation is a function that is n times differentiable and that satisfies the differential equation. p. 15

Initial-Value Problem: An **initial-value problem** consists of an nth-order differential equation (of any order) together with n initial conditions that must be satisfied by the solution of the differential equation and its derivatives at the initial point. p. 16

Solution to an Initial-Value Problem: We define a **solution** to an nth-order initial-value problem as a function that is n times differentiable, satisfies the given differential equation, and satisfies the n given initial conditions. p. 16

Boundary Value Problem: A **boundary value problem** consists of a differential equation and a collection of values that must be satisfied by the solution of the differential equation or its derivatives at no fewer than two different points. p. 17

Existence-Uniqueness Theorem: Let $y' = f(x, y)$, $y(x_0) = y_0$, and suppose f and $\partial f/\partial y$ are continuous in the region $a < x < b$, $c < y < d$ containing (x_0, y_0). Then, a unique solution exists in some subinterval $|x - x_0| < h$ contained in $a < x < b$. p. 18

Direction Field: Set of directions in the plane satisfying the differential equation $y' = f(x, y)$. Any solution tracks the direction field. p. 21

Euler's Method: $y_{n+1} = y_n + hf(x_n, y_n)$, $x_{n+1} = x_n + h$, where $y' = f(x, y)$, $y(0) = y_0$. p. 25

REVIEW EXERCISES FOR CHAPTER 1

In Exercises 1–4 find all solutions to the given differential equation. When an initial condition is given, find the particular solution that satisfies that condition.

1. $\frac{dy}{dx} = 3x$

2. $\frac{dx}{dt} = -2t$, $x(0) = 4$

3. $\frac{dx}{dt} = \frac{x}{2}$, $x(0) = -3$

4. $\frac{dy}{dx} = 100y$

5. The relative annual rate of growth of a population is 15%. If the initial population is 10,000, what is the population after 5 years? After 10 years?
6. In Exercise 5, how long does it take for the population to double?
7. When a loaf of bread is taken out of the oven, its temperature is 125°C. Room temperature is 23°C. The temperature of the bread is 80°C after 10 minutes.
   (a) What is its temperature after 20 minutes?
   (b) How long does the bread take to cool to 25°C?
8. A fossil contains 35% of the normal amount of ^{14}C. What is its approximate age?
9. What is the half-life of an exponentially decaying substance that loses 20% of its mass in one week?
10. How long does it take the substance in Exercise 9 to lose 75% of its mass? 95% of its mass?
11. **(a)** Plot the direction field for the differential equation $y' = y^{2/5}$.
    (b) Plot the solution that satisfies $y(0) = 5$.
    (c) Plot the solution that satisfies $y(0) = -2$.
12. **(a)** Plot the direction field for the differential equation $y' = -4xy/(1 + y^2)$.
    (b) Plot the solution that satisfies $y(2) = -1$.
    (c) Plot the solution that satisfies $y(2) = 3$.

In Exercises 13–18 use the Euler method and the given value of h to obtain an approximate solution at the indicated value of x.

13. $\dfrac{dy}{dx} = \dfrac{e^x}{y}$, $y(0) = 2$.
    Approximate $y(3)$ with $h = \dfrac{1}{2}$.
14. $\dfrac{dy}{dx} = \dfrac{e^y}{x}$, $y(1) = 0$.
    Approximate $y(\frac{1}{2})$ with $h = -0.1$.
15. $\dfrac{dy}{dx} = \dfrac{y}{\sqrt{1 + x^2}}$, $y(0) = 1$.
    Approximate $y(3)$ with $h = \dfrac{1}{2}$.
16. $xy\dfrac{dy}{dx} = y^2 - x^2$, $y(1) = 2$.
    Approximate $y(3)$ with $h = \dfrac{1}{2}$.
17. $\dfrac{dy}{dx} = y - xy^3$, $y(0) = 1$.
    Approximate $y(3)$ with $h = \dfrac{1}{2}$.
18. $\dfrac{dy}{dx} = \dfrac{2xy}{3x^2 - y^2}$, $y\left(-\dfrac{3}{8}\right) = -\dfrac{3}{4}$.
    Approximate $y(6)$ with $h = \dfrac{3}{8}$.
19. **(a)** Apply the Euler method with $h = 0.2$ and estimate $y(1)$ for the initial-value problem
    $$y' = y^3, \qquad y(0) = 1.$$
    (b) Show that $y = (1 - 2x)^{-1/2}$ is a local solution to the initial-value problem in part (a).
    (c) Explain the apparent contradiction.

2 First-Order Equations

In this chapter we provide a variety of techniques for solving first-order ordinary differential equations. In Sections 2.6 and 2.7 we give some interesting applications using these methods. For each method we assume that f and $\partial f/\partial y$ are continuous in some rectangle containing the initial point (x_0, y_0). According to Theorem 1.2.1 (p. 18) the differential equation has a unique solution in the interval $(x_0 - h, x_0 + h)$ for some number $h > 0$.

2.1 Separation of Variables

In Section 1.1 we solved the differential equation $dy/dx = \alpha y$. There are, however, several ways to solve this equation. Here is the technique we used in Section 1.1:

$$\frac{dy}{dx} = \alpha y.$$

Step 1 Treating the differentials dx and dy algebraically, put all the x's on one side of the equation and all the y's on the other. That is, *separate the variables:*

$$\frac{dy}{y} = \alpha dx.$$

Step 2 Integrate:

$$\int \frac{dy}{y} = \int \alpha dx.$$

and

$$\ln|y| = \alpha x + C.$$

Step 3 Solve for y:

$$|y| = e^{\alpha x + C} = e^{\alpha x}e^{C}$$

$$y = \pm e^{C}e^{\alpha x} = C_1 e^{\alpha x} \quad \text{where } C_1 = \pm e^{C} \text{ is an arbitrary constant.}$$

The technique used above can be applied to a variety of differential equations.

Consider the differential equation

$$\frac{dy}{dx} = f(x, y) \tag{1}$$

and suppose that the function $f(x, y)$ can be factored into a product,

$$f(x, y) = g(x)h(y), \tag{2}$$

where $g(x)$ and $h(y)$ are each functions of only one variable. When this occurs, Equation **(1)** can be solved by the method of **separation of variables.** To solve the equation, we substitute the product **(2)** into **(1)** to obtain

$$\frac{dy}{dx} = g(x)h(y),$$

or

$$\frac{1}{h(y)}\frac{dy}{dx} = g(x).$$

Here we are assuming that $h(y) \neq 0$. Integrating both sides of this equation with respect to x, we have

$$\int \frac{1}{h(y)}\frac{dy}{dx}\,dx = \int g(x)\,dx + C, \tag{3}$$

which can be rewritten as

$$\int \frac{1}{h(y)}\,dy = \int g(x)\,dx + C,^{*} \tag{4}$$

since the left side of Equation **(3)** is precisely equal to the left side of Equation **(4)** under a change of variables. Observe that this procedure, in effect, allows us to treat the derivative dy/dx as if it were a fraction: We simply separate all the terms involving x on one side and those involving y on the other. This gives the method its name. If both integrals in Equation **(4)** can be evaluated, a solution to the differential Equation **(1)** is obtained. We illustrate this method with several examples.

EXAMPLE 1 ▶ **Solving by separating variables**

Solve the differential equation

$$\frac{dy}{dx} = 2xy.$$

Solution Let $g(x) = 2x$ and $h(y) = y$ and separate variables obtaining

$$\frac{dy}{y} = 2x\,dx \qquad \text{so} \qquad \int \frac{dy}{y} = 2\int x\,dx + C.$$

Integration yields

$$\ln|y| = x^2 + C \qquad \text{or} \qquad |y| = e^{x^2+C} = e^C e^{x^2},$$

since $e^{\ln a} = a$. The term e^C is an arbitrary *positive* constant. Since we want to solve for y, we need to remove the absolute value. This is done by setting

$$y = \pm e^C e^{x^2}.$$

---

* There is no need to use two constants of integration,

$$\int \frac{dy}{h(y)} + C_1 = \int g(x)\,dx + C_2,$$

since a single arbitrary constant $C = C_2 - C_1$ serves the same purpose.

But $\pm e^C$ can be any nonzero arbitrary real number, so we replace this term by the letter k, yielding

$$y = ke^{x^2}. \tag{5}$$

Since k is any nonzero arbitrary real number, it follows that the original differential equation for Example 1 has infinitely many solutions, one for each value of k. Observe that we also have a solution if $k = 0$, then $y \equiv 0$ and $dy/dx \equiv 0$, satisfying the differential equation. We call Equation **(5)** the *general solution* of this differential equation.

It should not surprise us that a differential equation has infinitely many solutions: this is simply a consequence of the integration process, since each integral involves an arbitrary constant.

In every case where we wish to single out a specific solution, we require additional information. Often that additional information is given as an *initial condition,* such as $y(1) = 2$. Had that condition been given in Example 1, we would set $x = 1$ and $y = 2$ in Equation **(5)** obtaining

$$2 = y(1) = ke^{1^2} = ke$$

or $k = \frac{2}{e}$. Then, the *particular solution* for that initial-value problem is

$$y(x) = \frac{2}{e}e^{x^2} = 2e^{x^2-1}. \quad ◀$$

Warning

Do not do the following:

There is a common error that students make when they try to separate variables. Consider the following problem, *done incorrectly:*

$$\frac{dy}{dx} = y + x$$

$$dy = y\,dx + x\,dx$$

$$\int dy = \int y\,dx + \int x\,dx$$

$$y = \frac{y^2}{2} + \frac{x^2}{2} + C$$

What is wrong? *It is not correct* that

$$\int y\,dx = \frac{y^2}{2},$$

because y is a function of x. For example, if $y = \cos x$, then

$$\int y\,dx = \int \cos x\,dx = \sin x + C \neq \frac{y^2}{2} + C = \frac{\cos^2 x}{2} + C.$$

Of course

$$\int y\,dx = \frac{y^2}{2} + C.$$

So, be careful. (Using the techniques of Section 2.3, we can show that the solutions to $dy/dx = y + x$ are $y = -x - 1 + Ce^x$ for every real number C.)

EXAMPLE 2 ▶ **Separation of variables**

Solve the initial-value problem

$$\frac{dy}{dx} = 3x^2e^{-y}, \qquad y(0) = 2. \tag{6}$$

Solution Since $e^{-y} = 1/e^y$, when we separate variables we get

$$e^y\,dy = 3x^2\,dx \qquad \text{or} \qquad \int e^y\,dy = \int 3x^2\,dx + C.$$

Integrating, we have

$$e^y = x^3 + C \qquad \text{or} \qquad y = \ln(x^3 + C).$$

To determine C we use the initial condition. Set $x = 0$; then

$$2 = y(0) = \ln(0^3 + C) = \ln C \qquad \text{or} \qquad C = e^2.$$

Thus the particular solution to the initial-value problem is $y = \ln(x^3 + e^2)$.

Observe that this solution is defined only in the domain where $x^3 + e^2 > 0$, or $x > -e^{2/3}$, since the natural logarithm is only defined for positive numbers. Thus a solution to a differential equation need not be defined for all values of the independent variable x. ◀

Note

In Example 1, $f(x, y) = 2xy$ and both f and $\partial f/\partial y$ are continuous everywhere in R^2. Here the solution $y = 2e^{x^2-1}$ exists everywhere. In Example 2, $f = 3x^2e^{-y}$ and both f and $\partial f/\partial y$ are continuous everywhere in R^2. However, the solution $y = \ln(x^3 + e^2)$ to the initial-value problem is defined only for $x > -e^{2/3}$. This illustrates the fact that the basic existence-uniqueness theorem (Theorem 1.2.1) ensures a solution only in some interval containing x_0, not necessarily for all real values of x_0 (even if f and $\partial f/\partial y$ are everywhere continuous).

EXAMPLE 3 ▶ **Separation of variables**

Solve the differential equation

$$\frac{dy}{dx} = 2x(y^2 + 1).$$

Solution Separating variables, we have

$$\frac{dy}{1 + y^2} = 2x\,dx \qquad \text{or} \qquad \int \frac{dy}{1 + y^2} = 2\int x\,dx + C,$$

or $\tan^{-1} y = x^2 + C$. Thus, $y = \tan(x^2 + C)$. Note that the solution is only defined for values of x satisfying $|x^2 + C| < \frac{\pi}{2}$, because that is the range of the arc tangent function. This suggests that any solution to this differential equation quickly becomes unbounded. In Figure 2.1 we illustrate the solution that satisfies the initial condition $y(0) = -1$. ◀

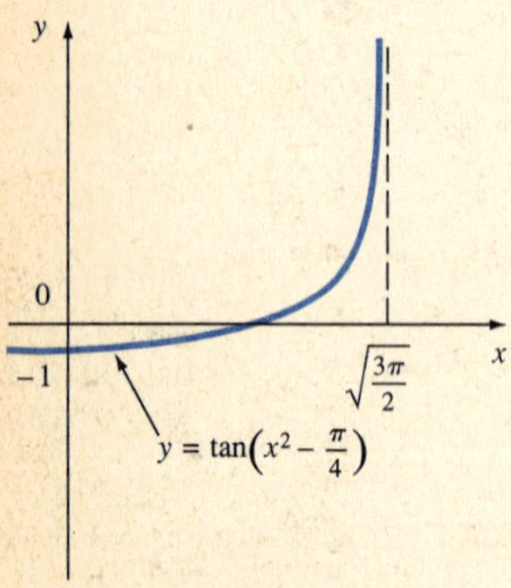

FIGURE 2.1
Solution is unbounded at $x = \sqrt{3\pi/2}$.

EXAMPLE 4 ▶ **Escape velocity**

In Chapter 1 (p. 2) we studied the motion of a body falling freely subject to the gravitational force of the earth. In that example we assumed that the distance the body fell was small in comparison to the radius R of the earth.* However, if we wish to study the equation of motion for a communications satellite or an interplanetary vehicle, the distance r of the object from the center of the earth may be considerably larger than R. In such a case the approximation we made in obtaining Equation (1.1.3) (p. 3) is no longer valid. Returning to Equation (1.1.4)

$$\frac{d^2r}{dt^2} = -g\frac{R^2}{r^2}, \tag{7}$$

and setting $v = dr/dt$, we see by the chain rule that

$$\frac{d^2r}{dt^2} = \frac{dv}{dt} = \frac{dv}{dr}\frac{dr}{dt} = \frac{v\,dv}{dr}. \tag{8}$$

Hence Equation **(7)** can be rewritten as

$$\frac{v\,dv}{dr} = -g\frac{R^2}{r^2},$$

where $g(\approx 9.81\text{ m/s}^2)$ and R are constant. Separating variables and integrating, we have

$$\int v\,dv = -gR^2\int\frac{dr}{r^2} + C,$$

or

$$\frac{1}{2}v^2 = \frac{gR^2}{r} + C,$$

Assuming that the object is at the surface of the earth when $t = 0$, we get

$$\frac{1}{2}v(0)^2 = g\frac{R^2}{R} + C,$$

or

$$C = \frac{1}{2}v(0)^2 - gR.$$

Thus

$$v^2 = 2g\frac{R^2}{r} + v(0)^2 - 2gR. \tag{9}$$

For the object to escape the gravitational force of the earth, it is necessary that $v > 0$ for all time t. If we select $v(0) = \sqrt{2gR}$, the last two terms in Equation **(9)** cancel, so that $v^2 > 0$ for all r. Observe that any smaller choice for $v(0)$ allows the right side of Equation **(9)** to be zero for some sufficiently large value of r. Thus $v(0) = \sqrt{2gR} \approx 11.2$ kilometers per second is the initial velocity an object needs to escape the gravitational attraction of the earth. This is called the **escape velocity.** ◀

---

* The radius of the earth is approximately 6378 kilometers (3963 miles) at the equator and 6357 kilometers (3950 miles) at the poles.

The substitution we used in Equation **(8)** can always be used to reduce an equation involving a second derivative to an equation involving only first derivatives, provided that the independent variable does not appear explicitly in the equation.

EXAMPLE 5 ▶ **Reentry into atmosphere**

One of the problems facing space vehicles is their reentry into the earth's atmosphere. In particular, we need to know the maximum deceleration of the vehicle because the greater the deceleration, the greater the heat of the vehicle upon reentry. For simplicity, assume that gravity has a negligible effect in determining the maximum deceleration during reentry of a vehicle heading toward (a flat) earth at a constant angle α and speed V. Let $s(t)$ denote the distance to impact of the reentry vehicle (see Figure 2.2). Then ds/dt is its velocity and d^2s/dt^2 its deceleration. Assume that the density of the earth's atmosphere at height z is

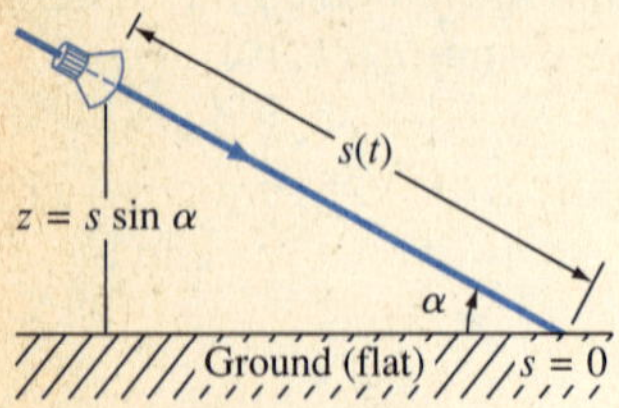

FIGURE 2.2
Reentry into atmosphere

$$\rho(z) = \rho_0 e^{-z/k_0},$$

where ρ_0 and k_0 are constants, and that the **drag force** of the atomsphere is proportional to the product of the air density and the square of the velocity of the vehicle. By Newton's second law of motion,

$$m\frac{d^2s}{dt^2} = \beta\rho_0 e^{-z/k_0}\left(\frac{ds}{dt}\right)^2,$$

where m is the mass of the vehicle and β is the constant of proportionality. Dividing both sides by m and substituting $N = \beta\rho_0/m$ and $k = k_0/\sin\alpha$, we get the equation of motion $[-z/k_0 = -s\sin\alpha/k_0 = -s/(k_0/\sin\alpha) = -s/k]$

$$\frac{d^2s}{dt^2} = Ne^{-s/k}\left(\frac{ds}{dt}\right)^2. \qquad \textbf{(10)}$$

Letting $v = ds/dt$ and using the same technique as in Equation **(8)**, we have

$$\frac{d^2s}{dt^2} = \frac{dv}{dt} = \frac{dv}{ds}\frac{ds}{dt} = v\frac{dv}{ds},$$

or

$$v\frac{dv}{ds} = Ne^{-s/k}v^2.$$

Separating variables, we get

$$\int\frac{dv}{v} = \int Ne^{-s/k}\,ds + C_1,$$

or

$$\ln|v| = -kNe^{-s/k} + C_1,$$

so

$$v = Ce^{-kNe^{-s/k}}.$$

The arbitrary constant C can now be determined by noting that the vehicle is traveling at velocity V when it is very far from earth. Thus, letting s tend to ∞, we have

$$V = Ce^0 = C,$$

or

$$\frac{ds}{dt} = v(s) = Ve^{-kNe^{-s/k}}. \tag{11}$$

Combining Equations **(10)** and **(11)**, we get

$$\frac{d^2s}{dt^2} = Ne^{-s/k}(Ve^{-kNe^{-s/k}})^2. \tag{12}$$

To find the *maximum* deceleration, we must determine the value of s that maximizes the right side $f(s)$ of Equation **(12)**. This can be found by differentiating the right side of Equation **(12)** with respect to s by logarithmic differentiation and using the first derivative test of calculus:

$$f'(s) = f(s)\left(-\frac{1}{k} + 2Ne^{-s/k}\right) = 0.$$

Solving for s, we get $e^{-s/k} = 1/2kN$, or $s = k\ln(2kN)$. Here we are assuming that $2kN > 1$ so that $\ln 2kN > 0$ and $s > 0$. This value is easily seen to be a maximum, since $f > 0$ and the term in parentheses is decreasing. Substituting $s_{\max} = k\ln(2kN)$ into Equation **(12)**, we have

$$\left.\left(\frac{d^2s}{dt^2}\right)\right|_{s_{\max}} = Ne^{-\ln 2kN}(Ve^{-kNe^{-\ln 2kN}})^2$$

$$= \frac{N}{2kN}(Ve^{-(kN/2kN)})^2 = \frac{V^2}{2ek} \overset{k = k_0/\sin\alpha}{=} \frac{V^2\sin\alpha}{2ek_0},$$

which is independent of the drag coefficient $N = \beta\rho_0/m$. ◀

Apollo project

On July 16, 1969, NASA launched one of its most ambitious projects: Apollo 11, the fist manned lunar landing. Aboard Columbia on that day were astronauts Neil Armstrong, Buzz Aldrin, and Michael Collins. Blast-off occurred at 9:32 A.M. (EDT), and after reaching orbital height, the engines fired at 12:16 P.M. to put the craft on a translunar path. Columbia entered a lunar orbit at 1:22 P.M. on July 19, and Eagle, the lunar lander, touched down at 4:17 P.M. on July 20. Armstrong stepped on the moon at 10:56 P.M. and was joined by Aldrin 18 minutes later. After conducting some experiments and collecting samples, Eagle blasted off at 1:56 P.M. on July 21 to rejoin the mother ship. Columbia left lunar orbit at 12:56 A.M. on July 22, and in the early afternoon of July 24 began the crucial step of entering the Earth's atmosphere. The drag force of the atmosphere quickly raised the temperature of the craft's outer shell to thousands of degrees. This was one of the many problems that NASA had solved in designing the space program: How to protect the space craft from being vaporized on reentry. Special ceramic tiles coated the side of the craft on which the drag force was exerted. Splashdown occurred at 12:50 P.M.

Figure 2.3 is a view of Eagle from Columbia, and Figure 2.4 is a shot of Columbia from Eagle. Figure 2.5 shows an artist's rendition of the Apollo 11 mission.

FIGURE 2.3
Eagle (the lunar lander) as seen from Columbia
Source: NASA.

FIGURE 2.4
Columbia as seen from Eagle
Source: NASA.

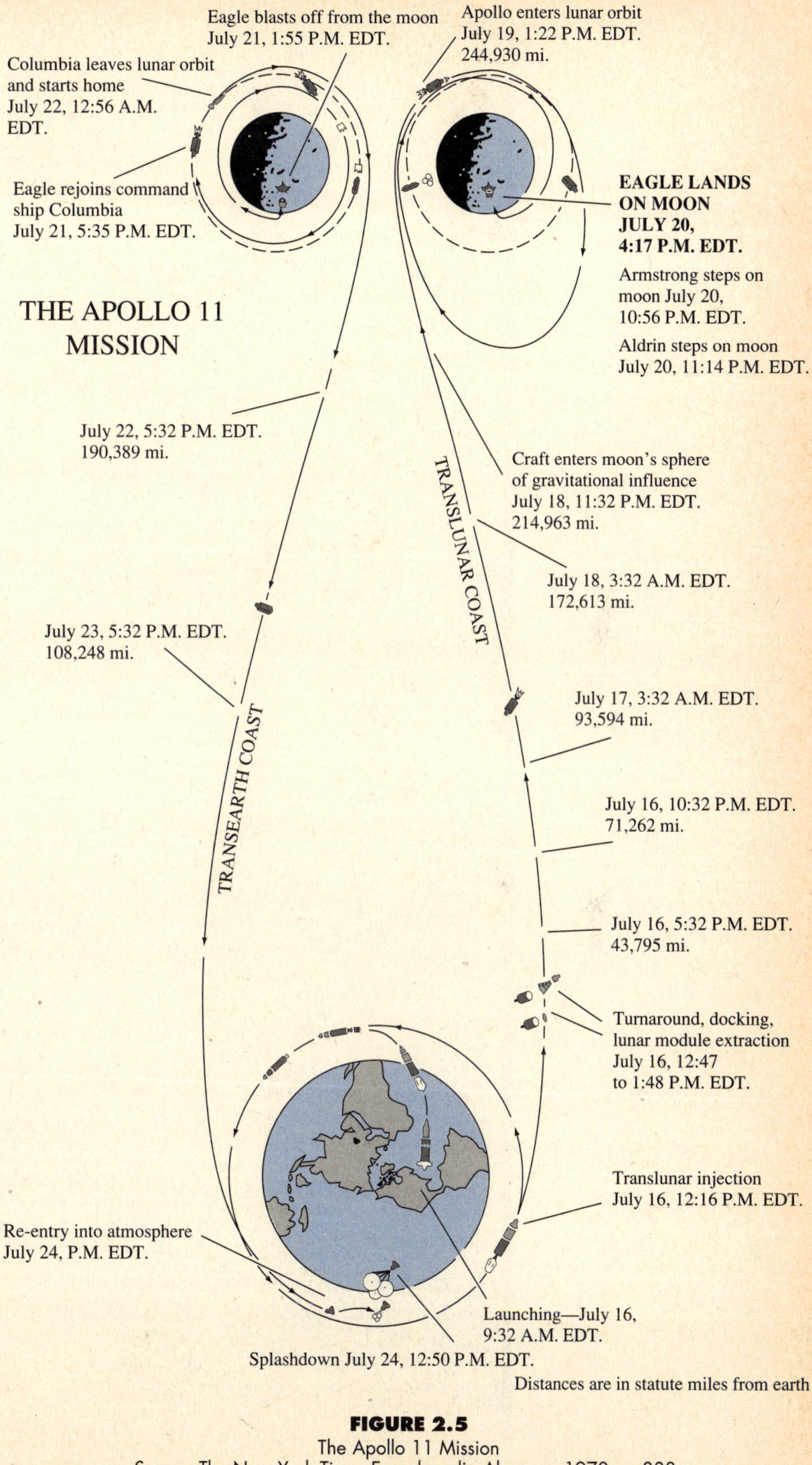

FIGURE 2.5
The Apollo 11 Mission
Source: The New York Times Encyclopedia Almanac, 1970, p. 333.

Perspective When is a differential equation separable?

In this section we saw how to solve a first-order differential equation when the equation was separable. However, it is not always clear that an equation is separable. For example, it is obvious that $f(x, y) = e^x \cos y$ is separable, but it is not so obvious that $f(x, y) = 2x^2 + y - x^2y + xy - 2x - 2$ is separable.* In this perspective, we give conditions that ensure that an equation is separable.†

Theorem 1
Suppose that $f(x, y) = g(x)h(y)$, where both g and h are differentiable. Then

$$f(x, y)f_{xy}(x, y) = f_x(x, y)f_y(x, y). \tag{13}$$

Here the subscripts denote partial derivatives.

Proof
Observe that

$$\begin{aligned} f_x(x, y) &= g'(x)h(y) \\ f_y(x, y) &= g(x)h'(y) \\ f_{xy}(x, y) &= g'(x)h'(y). \end{aligned}$$

Hence

$$\begin{aligned} f(x, y)f_{xy}(x, y) &= g(x)h(y)g'(x)h'(y) = [g'(x)h(y)][g(x)h'(y)] \\ &= f_x(x, y)f_y(x, y). \quad \blacklozenge \end{aligned}$$

It turns out that, under further conditions, if Equation **(13)** holds, then $f(x, y)$ is separable. For what follows, D denotes an open disk in the xy-plane; that is, $D = \{(x, y): (x - a)^2 + (y - b)^2 < r^2\}$, where a, b, and r are real numbers and $r > 0$.

Theorem 2
Suppose that in D, f, f_x, f_y, and f_{xy} exist and are continuous, $f(x, y) \neq 0$, and Equation **(13)** holds. Then there are continuously differentiable functions $g(x)$ and $h(y)$ such that, for every $(x, y) \in D$,

$$f(x, y) = g(x)h(y). \tag{14}$$

Proof
Since $f(x, y) \neq 0$ and f is continuous on D, f has the same sign on D. Assume $f(x, y) > 0$ for $(x, y) \in D$. A similar proof works if $f(x, y) < 0$ if f is replaced by $-f$. Now, from the quotient rule of differentiation,

$$\frac{\partial}{\partial y}\frac{f_x(x, y)}{f(x, y)} = \frac{f(x, y)f_{xy}(x, y) - f_x(x, y)f_y(x, y)}{f^2(x, y)} \overset{\text{Equation (13)}}{=} 0$$

When the partial derivative with respect to y of a function of x and y is zero, then that function must be a function of x only. Thus there is a

* $2x^2 + y - x^2y + xy - 2x - 2 = (1 + x - x^2)(y - 2)$.

† The results here are based on the paper "When is an Ordinary Differential Equation Separable?" by David Scott in *American Mathematical Monthly* 92 (1982): 422–423.

function $\alpha(x)$ such that

$$\frac{f_x(x, y)}{f(x, y)} = \alpha(x).$$

Also, since $f(x, y) > 0$, $\ln f(x, y)$ is defined and

$$\frac{\partial}{\partial x} \ln f(x, y) = \frac{f_x(x, y)}{f(x, y)} = \alpha(x).$$

The function $\alpha(x)$ is continuous in D because it is the quotient of continuous functions and the function in the denominator is nonzero. Let $\beta(x) = \int \alpha(x)\, dx$. Then

$$\ln f(x, y) = \int \left[\frac{\partial}{\partial x} \ln f(x, y)\right] dx = \int \alpha(x)\, dx = \beta(x) + \gamma(y),$$

where γ is a function of y only. (The partial derivative with respect to x of a function of y only is zero. Thus $\gamma(y)$ represents the most general constant of integration.) Finally, let $g(x) = e^{\beta(x)}$ and $h(y) = e^{\gamma(y)}$. Then

$$f(x, y) = e^{\ln f(x, y)} = e^{\beta(x) + \gamma(y)} = e^{\beta(x)} e^{\gamma(y)} = g(x)h(y). \quad \blacklozenge$$

Example 1
Let $f(x, y) = 2x^2 + y - x^2y + xy - 2x - 2$. Then

$$\begin{aligned}
f_x(x, y) &= 4x - 2xy + y - 2 \\
f_y(x, y) &= 1 - x^2 + x \\
f_{xy}(x, y) &= -2x + 1 \\
f(x, y) f_{xy}(x, y) &= (2x^2 + y - x^2y + xy - 2x - 2)(-2x + 1) \\
&= -4x^3 - xy + 2x^3y - 3x^2y + 6x^2 + 2x + y - 2
\end{aligned}$$

and

$$\begin{aligned}
f_x(x, y) f_y(x, y) &= (4x - 2xy + y - 2)(1 - x^2 + x) \\
&= 2x - xy + y - 2 - 4x^3 + 2x^3y - 3x^2y + 6x^2.
\end{aligned}$$

Since the last two expressions are equal, we conclude by Theorem 2 that $f(x, y)$ is separable. ◀

Example 2
Let $f(x, y) = 1 + xy$. Then

$$\begin{aligned}
f_x(x, y) &= y \\
f_y(x, y) &= x \\
f_{xy}(x, y) &= 1 \\
f(x, y) f_{xy}(x, y) &= 1 + xy
\end{aligned}$$

and

$$f_x(x, y) f_y(x, y) = xy.$$

Since the last two expressions are unequal, we conclude that $f(x, y)$ is not separable. ◀

SELF-QUIZ

Answer True *or* False.

I. $\frac{dy}{dx} = x^2 + y$ is separable.

II. $\frac{dy}{dx} = x^2(y + \ln y)$ is separable.

III. $\frac{dy}{dx} = 3x + xy^2$ is separable.

IV. $\frac{dy}{dx} = xy + y - x - 1$ is separable.

V. $\frac{dy}{dx} = xy + y - x$ is separable.

Answers to Self-Quiz

I. False **II.** True **III.** True **IV.** True **V.** False

PROBLEMS 2.1

In Problems 1–25 find the general solution by separating variables. If a specific condition is given, find the particular solution that satisfies that condition and sketch its graph using either a graphing calculator or a computer algebra system.

1. $\frac{dy}{dx} = x\,e^y$

2. $xy' = 3y,\ y(2) = 5$

3. $\frac{dP}{dQ} = 2\sqrt{P}$

4. $\frac{dy}{dx} = 2x\sqrt{y}$

5. $\frac{dx}{dt} = t\,x\,e^{t^2}$

6. $\frac{dr}{ds} = s\,r\,e^{s^2},\ r(0) = 1$

7. $\frac{dy}{dt} + 4y = y(e^{-t} + 4)$

8. $\frac{dz}{dx} = \frac{\sqrt{x(z^2 + 4)}}{z},\ z(1) = 1$

9. $\frac{dy}{dx} = \frac{e^y x}{e^y + x^2 e^y}$

10. $\frac{dx}{dy} = x \cos y,\ x\left(\frac{\pi}{2}\right) = 1$

11. $\frac{dz}{dr} = r^2(1 + z^2)$

12. $\frac{dy}{dx} + y = y(xe^{x^2} + 1),\ y(0) = 1$

13. $\frac{dP}{dQ} = P(\cos Q + \sin Q)$

14. $\frac{dy}{dx} = y^2(1 + x^2),\ y(0) = 1$

15. $\frac{ds}{dt} + 2s = st^2,\ s(0) = 1$

16. $\frac{dx}{dt} + (\cos t)e^x = 0$

17. $\cot x \frac{dy}{dx} + y + 3 = 0$

18. $\frac{dx}{dt} = x(1 - \sin t),\ x(0) = 1$

19. $x^2 \frac{dy}{dx} + y^2 = 0,\ y(1) = 3$

20. $x^2y' + e^y = 0$

21. $yy' = e^x$

22. $y' + y = y(xe^x + 1)$

23. $e^x\left(\frac{dx}{dt} + 1\right) = 1,\ x(0) = 1$

24. $e^x(x' + 1) = e^x,\ x(0) = 1$

25. $e^x(x' - 1) = e^x,\ x(0) = 1$

26. **Obsidian dating*** is a technique used by archaeologists which allows the dating of certain artifacts well beyond the reliable 40,000-year range of carbon dating. Obsidian, a glassy volcanic rock, absorbs water from the atmosphere. The water forms a hydration layer—a compound of water molecules and obsidian molecules. The depth of the hydration layer $x(t)$ beneath the surface of an obsidian artifact is a function of the time t that has elapsed since the manufacture of that artifact. The velocity at which the hydration layer grows is inversely proportional to its depth. Find the depth of the layer for all time t.

27. A rocket is launched from an initial position (x_0, y_0) with an initial speed v_0 and with an angle θ $(0 \le \theta \le \frac{\pi}{2})$. Find its horizontal and vertical coordinates $x(t)$ and $y(t)$ as functions of time. Assume that there is no air resistance, and that the force of gravity g is constant.

28. The half-life of a radioactive substance is defined as the time required to decompose 50% of the substance. If $r(t)$ denotes the amount of the radioactive substance present after t years, $r(0) = r_0$, and the half-life is H years, what is a differential equation for $r(t)$ taking all side conditions into account?

---

* We wish to thank H. W. Vayo for bringing to our attention the article by I. Friedman and R. L. Smith, "A New Dating Method Using Obsidian," *American Antiquity* 25 (1960): 476–522.

29. A bacteria population is known to double every 3 hours. If the population initially consists of 1000 bacteria, how long does it take for the population to reach 10,000?

30. The economist Vilfredo Pareto (1848–1923) discovered that the rate of decrease of the number of people y in a stable economy having an income of at least x dollars is directly proportional to the number of such people and inversely proportional to their income. Obtain an expression (**Pareto's law**) for y in terms of x.

31. It is the last lap on a speedway, the finish line is straight ahead 2 miles away, and you lead your nearest opponent by 3 miles. Both of you are going at the same speed, but you just ran out of gasoline. Your car decelerates at a rate proportional to the square of your instantaneous speed. At the end of a mile, your speed is exactly half of what it was when you ran out of gas. Will you win the race?

32. On a certain day it begins to snow early in the morning, and the snow continues to fall at a constant rate. The velocity at which a snowplow is able to clear a road is inversely proportional to the height of the accumulated snow. The snowplow starts at 11 A.M. and clears 4 miles by 2 P.M. By 5 P.M. it clears another 2 miles. When did it start snowing?*

33. Snowplow chase.† If a second snowplow starts at noon along the same path as the snowplow in Problem 32, when does it catch up to the first snowplow?

34. Snowplow collision.‡ Suppose three identical snowplows start clearing the same road at 10 A.M., 11 A.M., and noon. If all three collide sometime after noon, when did it start snowing?

2.2 PRECAUTIONS WHEN SEPARATING VARIABLES

You should always test your answer carefully when you use the method of separation of variables. You may have noticed in the previous section that some of the solutions were valid only for limited values of the independent variable. For example, the solution to Example 2.1.2 was only valid for values of $x > -e^{2/3}$. Frequently, in the process of separating variables, we may divide by a term that becomes zero for some value of the variables, or introduce extraneous solutions by some algebraic step. It is important always to analyze the resulting solution to be sure that it does, indeed, solve the problem. In this section we shall explore three common pitfalls, and indicate some of the precautions one must take when using this method.

EXAMPLE 1 ▶ **A differential equation which has solutions only for certain initial conditions**

Solve the differential equation

$$\frac{dx}{dt} = t\sqrt{1 - x^2}, \qquad -1 \leq x \leq 1$$

with the initial conditions: **(a)** $x(0) = 0$, **(b)** $x(0) = \frac{1}{2}$, **(c)** $x(0) = -\sqrt{2}/2$, and **(d)** $x(0) = 2$.

Solution Separating variables yields

$$\int \frac{dx}{\sqrt{1 - x^2}} = \int t\,dt + C.$$

* Based on problem E275 of the Otto Dunkel Memorial Problem Book, *American Mathematical Monthly* 64 (1957): 54.

† This problem was first proposed by Fred Wan, *Applied Mathematics Notes,* (January 1975): 6–11.

‡ This problem was first proposed by M. S. Klamkin, *American Mathematical Monthly* 59 (1952): 42 (problem E963).

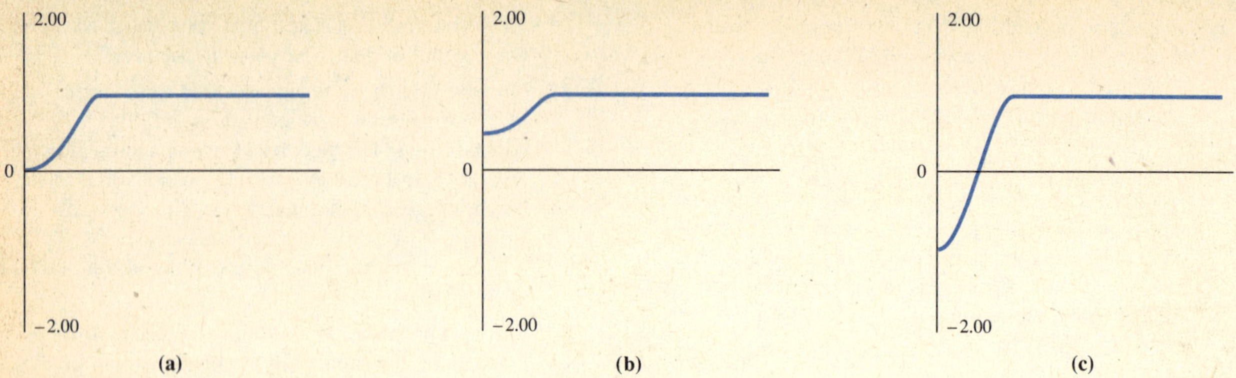

FIGURE 2.6
Graphs of $x = \sin(t^2/2 + C)$ for three different values of $C = \sin^{-1}(x(0))$. (a) $x(0) = 0$; (b) $x(0) = 1/2$; (c) $x(0) = -\sqrt{2}/2$

Note that the function being integrated is defined only for $-1 < x < 1$. By entry 86 of the Table of Integrals,

$$\sin^{-1} x = \frac{1}{2}t^2 + C. \tag{1}$$

(a) If $x(0) = 0$, setting $t = 0$ in Equation **(1)** yields $C = \sin^{-1} 0 = 0$. Thus, $x = \sin \frac{1}{2}t^2$ in the interval $|t| < \sqrt{\pi}$, since the range of the arc sine function is the interval $[-\frac{1}{2}\pi, \frac{1}{2}\pi]$. But what happens if t is outside this interval? The answer is shown in Figure 2.6(a), where we see that for $t \geq \sqrt{\pi}$ the solution is $x \equiv 1$. This could also have been inferred by looking at the original differential equation: When $x = 1$, the slope $dx/dt = 0$, so the solution stays at that value. Note that the solution is not simply $x = \sin \frac{1}{2}t^2$. Instead,

$$x(t) = \begin{cases} \sin \dfrac{1}{2}t^2, & 0 \leq t \leq \sqrt{\pi}, \\ 1, & t \geq \sqrt{\pi}. \end{cases}$$

(b) For $x(0) = \frac{1}{2}$, setting $t = 0$ gives $\sin^{-1}(\frac{1}{2}) = C$, or $C = \frac{\pi}{6}$. The graph is shown in Figure 2.6(b). Note, again, that once $x(t)$ reaches the value 1, the solution remains at that value.

(c) When $x(0) = -\sqrt{2}/2$, we have $\sin^{-1}(-\sqrt{2}/2) = C$, so $C = -\frac{\pi}{4}$. The solution is shown in Figure 2.6(c) and remains at $x = 1$ after it reaches that value.

(d) If $x(0) = 2$, there is **no solution**, since 2 is not in the domain of the arc sine. This, too, could have been seen from the differential equation, as the term inside the root is negative. Summarizing, we see that the separation of variables solution $x = \sin(\frac{1}{2}t^2 + C)$ is valid only for certain initial conditions, and then only for certain ranges of t. ◀

EXAMPLE 2 ▶ Separation of variables

Solve the differential equation

$$\frac{dy}{dx} + \sqrt{\frac{1 - y^2}{1 - x^2}} = 0. \tag{2}$$

Solution Before we begin, we investigate where the term inside the root is nonnegative. Observe that it is nonzero in the five shaded regions in Figure 2.7; those are the only regions where a solution can exist. If we are trying to solve an initial-value problem consisting of Equation **(2)** and an initial condition, only initial conditions in the screened areas will have solutions. Furthermore, the solutions may terminate abruptly if we reach the boundary of the screened regions.

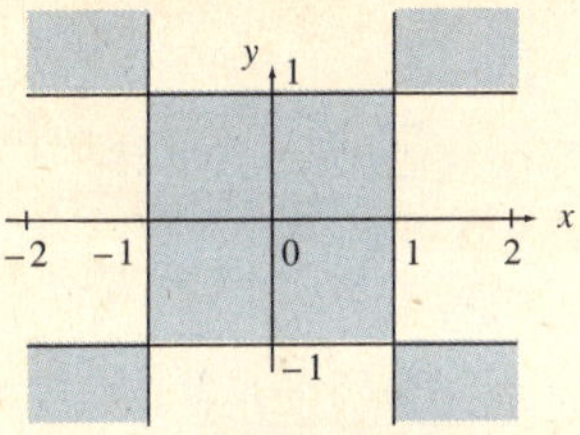

FIGURE 2.7
Domains of solutions

Separating variables, we obtain when $|x| < 1$ and $|y| < 1$

$$\int \frac{dy}{\sqrt{1-y^2}} = -\int \frac{dx}{\sqrt{1-x^2}} + C, \tag{3}$$

and integrating, we have

$$\sin^{-1} y = C - \sin^{-1} x, \tag{4}$$

Then, $y = \sin(C - \sin^{-1} x) = \sin C \cos(\sin^{-1} x) - \cos C \sin(\sin^{-1} x)$, and since the cosine is positive over the range of the arc sine, $\cos\alpha = \sqrt{1-\sin^2\alpha}$, and $\sin^2(\sin^{-1} x) = [\sin(\sin^{-1} x)]^2 = x^2$, we see that

$$y = (\sin C)\sqrt{1-x^2} - x\cos C. \tag{5}$$

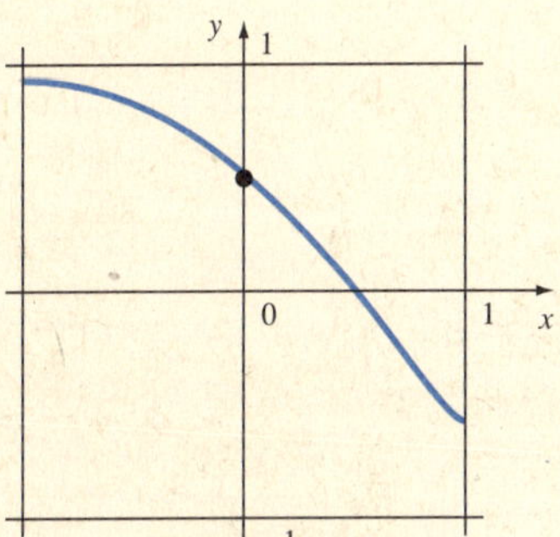

FIGURE 2.8
Solution satisfying $y(0) = 0.5$

It is not hard to check that Equation **(5)** is the solution of the differential equation in the set $\{(x, y) : |x| < 1, |y| < 1\}$ (see Problem 1). Figure 2.8 illustrates the solution for the initial conditions $y(0) = 0.5$.

If $|x| > 1$ and $|y| > 1$, we rewrite Equation **(3)** as

$$\int \frac{dy}{\sqrt{y^2-1}} = -\int \frac{dx}{\sqrt{x^2-1}} + C, \tag{6}$$

and the solution is now $y = x\cosh C - (\sinh C)\sqrt{x^2-1}$ (check!). ◀

EXAMPLE 3 ▶ **Logistic growth**

Let $P(t)$ denote the population of a species at time t. The **growth rate** per individual of the population is defined as the growth of the population divided by the size of the population; for example, if the birth rate is 3.2 per hundred and the death rate is 1.8 per hundred, then the growth rate is $3.2 - 1.8 = 1.4$ per hundred $= \frac{1.4}{100} = 0.014$. We then write $dP/dt = 0.014P$.

Suppose that in a given population the average birth rate is a positive constant β. It is reasonable to assume that the average death rate is proportional to the number of individuals in the population. Greater populations mean greater crowding and more competition for food and territory. We call this constant of proportionality δ (which is greater than zero). Since dP/dt is the growth rate of the population, the growth rate per individual of the population is

$$\frac{1}{P}\frac{dP}{dt}.$$

The differential equation that governs the growth of this population is

$$\frac{1}{P}\frac{dP}{dt} = \beta \quad \delta P.$$

Multiplying both sides of this equation by P, we have

$$\frac{dP}{dt} = P(\beta - \delta P), \tag{7}$$

which is called the **logistic equation**.* The growth shown by this equation is called **logistic growth**.

Separating the variables, we have

$$\int \frac{dP}{P(\beta - \delta P)} = \int dt + C. \tag{8}$$

Using partial fractions,† we set

$$\frac{1}{P(\beta - \delta P)} = \frac{A}{P} + \frac{B}{\beta - \delta P} = \frac{A(\beta - \delta P) + BP}{P(\beta - \delta P)} = \frac{(B - A\delta)P + \beta A}{P(\beta - \delta P)}.$$

Equating coefficients of P and 1, we obtain

$$\begin{aligned} B - A\delta &= 0 \\ \beta A &= 1 \end{aligned}$$

from which it follows that $A = \frac{1}{\beta}$ and $B = \frac{\delta}{\beta}$.

Thus,

$$\frac{1}{P(\beta - \delta P)} = \frac{1}{\beta P} + \frac{\delta}{\beta(\beta - \delta P)}.$$

Substituting the right-hand side of this equation into Equation **(8)** and integrating, we obtain

$$\frac{1}{\beta}\ln|P| - \frac{1}{\beta}\ln|\beta - \delta P| = t + C,$$

or

$$\ln|P| - \ln|\beta - \delta P| = \beta t + \beta C \overset{C_1 = \beta C}{=} \beta t + C_1$$

$$\ln\left|\frac{P}{\beta - \delta P}\right| = \beta t + C_1 \qquad \ln x - \ln y = \ln\tfrac{x}{y}$$

$$\left|\frac{P}{\beta - \delta P}\right| = e^{\beta t + C_1} \qquad \text{if } \ln x = y, \text{ then } x = e^y$$

$$\frac{P(t)}{\beta - \delta P(t)} = \pm e^{\beta t + C1} \overset{e^{x+y} = e^x e^y}{=} \pm e^{C_1}e^{\beta t} \overset{C_2 = \pm e^{C_1}}{=} C_2 e^{\beta t} \tag{9}$$

Warning

We will frequently make such changes of constant without further notice.

Setting $t = 0$, we find that

$$\frac{P(0)}{\beta - \delta P(0)} = C_2,$$

and substituting this value of C_2 into Equation **(9)** yields

$$\frac{P(t)}{\beta - \delta P(t)} = \frac{P(0)}{\beta - \delta P(0)} e^{\beta t}.$$

* The logistic equation is sometimes referred to as **Verhulst's equation** in honor of P. F. Verhulst (1804–1849), a Belgian mathematician who proposed this model for human population growth in 1838.

† A review of partial fractions is given in Section 6.2.

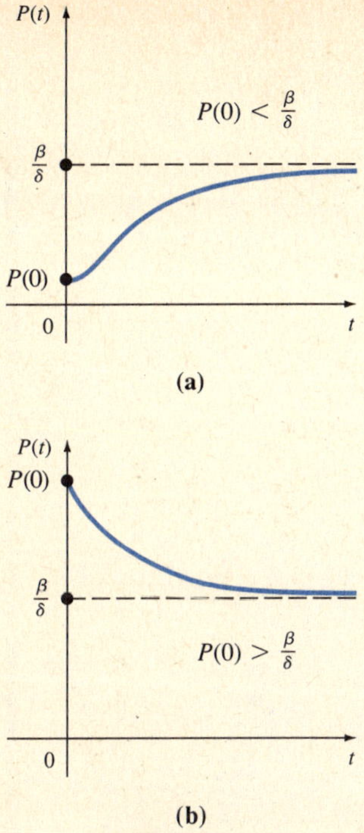

FIGURE 2.9 Graphs of

$$P(t) = \frac{\beta}{\delta + \left[\frac{\beta}{P(0)} - \delta\right]e^{-\beta t}}$$

(logistic growth)

Cross-multiplying and solving for $P(t)$, we obtain

$$P(t)[\beta - \delta P(0)] = P(0)[\beta - \delta P(t)]\, e^{\beta t},$$

or

$$\beta P(t) - \delta P(t)P(0) = \beta P(0)e^{\beta t} - \delta P(0)P(t)e^{\beta t}.$$

Thus

$$P(t)[\beta - \delta P(0) + \delta P(0)e^{\beta t}] = \beta P(0)e^{\beta t},$$

and

$$P(t) = \frac{\beta P(0)e^{\beta t}}{\beta - \delta P(0) + \delta P(0)e^{\beta t}}$$

Finally, dividing both the numerator and denominator by $P(0)e^{\beta t}$, we obtain

$$P(t) = \frac{\beta}{\delta + \left(\frac{\beta}{P(0)} - \delta\right)e^{-\beta t}}. \tag{10}$$

Observe that since $\beta > 0$, the term $e^{-\beta t}$ approaches zero as t increases. Thus the population approaches a limiting value of $\frac{\beta}{\delta}$, since setting $P = \frac{\beta}{\delta}$ in Equation **(7)** yields $dP/dt = 0$. In Figure 2.9, we sketch the curves of logistic growth in the cases $P(0) < \frac{\beta}{\delta}$ and $P(0) > \frac{\beta}{\delta}$. ◀

Perspective Singular solutions

In describing the solution of a first-order differential equation, we have used the term **general solution** for a *family* of solutions of the differential equation containing one arbitrary constant. The term **particular solution** has been used for a solution that is free of arbitrary constants, usually as the result of requiring that the solution satisfy some initial condition. In most of the examples that we have considered, the particular solution was obtained by choosing a specific value for the arbitrary constant in the general solution.

Sometimes a differential equation has a particular solution that cannot be obtained by selecting a specific value for the arbitrary constant in the general solution. Consider the initial-value problem

$$\frac{dy}{dx} = xy^{1/3}, \qquad y(0) = 0. \tag{11}$$

Separating variables and integrating, we have

$$\int y^{-1/3}\,dy = \int x\,dx + C$$

and

$$\frac{3}{2}y^{2/3} = \frac{x^2}{2} + C,$$

with the general solution

$$y = \pm\left(\frac{x^2}{3} + C\right)^{3/2}.$$

Thus *two* particular solutions of the initial-value problem **(11)** (obtained by setting $C = 0$) are $y = \pm x^3/3\sqrt{3}$. However, $y \equiv 0$ is *also* a solution of the initial-value problem, and this solution cannot be obtained from the general solution, no matter what value we assign to C. Solutions that cannot be obtained from the general solution are called **singular solutions.**

The reason the solution $y \equiv 0$ is lost is that the initial step of separating variables involves dividing by $y^{1/3}$, and division by zero is not allowed.

One final point of interest: These are not all of the solutions of the initial-value problem **(11)**. It possesses infinitely many more solutions, defined by the piecewise functions (graphed in Figure 2.10):

$$y = \begin{cases} 0, & x < a, \\ \pm\left(\dfrac{x^2 - a^2}{3}\right)^{3/2}, & x \geq a, \end{cases}$$

for any choice of the parameter $a \geq 0$. (Check.)

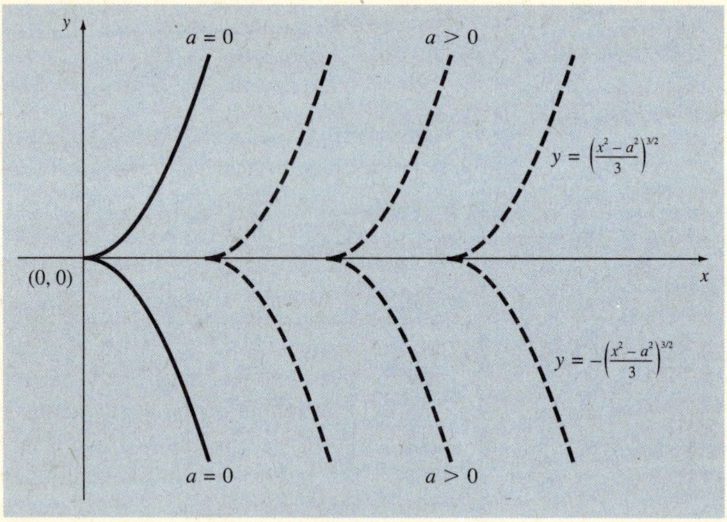

FIGURE 2.10
Infinitely many solutions to $y' = xy^{1/3}$, $y(0) = 0$

SELF-QUIZ

Indicate where a solution to the following differential equations might exist.

I. $\dfrac{dy}{dx} = \dfrac{y}{\sqrt{1 - x^2}}$ **II.** $\dfrac{dy}{dx} = \dfrac{x}{\sqrt{1 - y^2}}$ **III.** $\dfrac{dy}{dx} = \dfrac{\sqrt{1 - x^2}}{y}$

Answers to Self-Quiz

I. $|x| < 1$ **II.** $|y| < 1$ **III.** $|x| \leq 1, \quad y \neq 0$

PROBLEMS 2.2

1. Show that Equation **(5)** is the solution to the differential equation **(2)** in the set $\{(x, y) : |x| < 1, |y| < 1\}$.

In Problems 2–18, find a general solution for the given differential equation or a particular solution for the given initial-value problem. Indicate where the solution is defined. For particular solutions sketch a graph.

2. $\dfrac{dy}{dx} = \dfrac{e^x}{2y}$

3. $(1 + x)\dfrac{dy}{dx} = -3y,\ y(6) = 7$

4. $\dfrac{dr}{ds} = s(9 - r^2),\ r(0) = 2$

5. $\dfrac{dy}{dt} = 2\,t\,y(4 - y^2)$

6. $\dfrac{dy}{dx} = \dfrac{y^3 + 2y}{x^2 + 3x},\ y(1) = 1$

7. $\dfrac{dy}{dx} = \dfrac{x^2 - xy - x + y}{xy - y^2}$

8. $\dfrac{ds}{dr} = \dfrac{s^2 + s - 2}{r^2 - 2r - 8},\ s(0) = 0$

9. $\dfrac{dx}{dt} = t\,x\sqrt{x - 4}$

10. $\dfrac{dy}{dx} = \sqrt{1 - y^2}$

11. $\dfrac{dz}{dx} = z\dfrac{\sqrt{z^2 - 1}}{x},\ z(1) = 2$

12. $\dfrac{dP}{dQ} = \sqrt{\dfrac{P^2 - 4}{Q^2 - 1}}$

13. $\dfrac{dy}{dx} = \sqrt{\dfrac{4 - y^2}{x^2 - 1}}$

14. $\dfrac{dy}{dx} = \sqrt{x^3(1 - y^2)}$

15. $\dfrac{dy}{dx} + \sqrt{\dfrac{9 - y^2}{1 - x^2}} = 0$

16. $(\tan y)\dfrac{dy}{dx} - \tan x = 0,\ y(0) = 0$

17. $\dfrac{dy}{dx} = xy\sqrt{y^2 - 4}$

18. $\dfrac{dy}{dx} = x^2\sqrt{4 - y^2}$

19. A pond is shaped like a cone, with radius r and depth d. Water flows into the pond at a constant rate i and is lost through evaporation at a rate proportional to the surface area. These are the only sources of inflow and outflow.

(a) Show that the volume $V(t)$ of water at time t satisfies the differential equation

$$V' = i - k\pi\left(\frac{3rV}{\pi d}\right)^{2/3}.$$

(b) Solve this equation.

(c) What condition must be satisfied so that the pond will not overflow?

20. A large open cistern filled with water has the shape of a hemisphere with radius 25 feet. The bowl has a circular hole of radius 1 foot in the bottom. By **Torricelli's law,*** water flows out of the hole with the same speed it would attain in falling freely from the level of the water to the hole. How long does it take for all the water to flow from the cistern?

21. In Problem 20, find the shape of the cistern that ensures that the water level drops at a constant rate.†

22. The table below shows data for the growth of yeast in a culture. Use Equation **(10)**, with $\beta = 0.55$ and $\delta = 8.3 \times 10^{-4}$, to calculate the predicted growth, and find the percentage error between observed and predicted values at $t = 0$, 9, and 18 hours.

Time in Hours	*Yeast Biomass*	*Time in Hours*	*Yeast Biomass*
0	9.6	10	513.3
1	18.3	11	559.7
2	29.0	12	594.8
3	47.2	13	629.4
4	71.1	14	640.8
5	119.1	15	651.1
6	174.6	16	655.9
7	257.3	17	659.6
8	350.7	18	661.8
9	441.0		

Data from R. Pearl, "The Growth of Population," *Quarterly Review of Biology* 2 (1927): 532–548.

2.3 LINEAR FIRST-ORDER DIFFERENTIAL EQUATIONS

A nth-order differential equation is **linear** if it can be written in the form

$$\frac{d^n y}{dx^n} + a_{n-1}(x)\frac{d^{n-1}y}{dx^{n-1}} + \ldots + a_1(x)\frac{dy}{dx} + a_0(x)y = f(x).$$

Hence a first-order linear equation has the form

$$\frac{dy}{dx} + a(x)y = f(x), \qquad (1)$$

* Evangelista Torricelli (1608–1647) was an Italian physicist.

† The ancient Egyptians (ca 1380 B.C.) used water clocks based on this principle to tell time.

and a second-order linear equation can be written as

$$\frac{d^2y}{dx^2} + a(x)\frac{dy}{dx} + b(x)y = f(x).$$

The notation indicates that $a(x)$, $b(x)$, $f(x)$ and so on are functions of x alone. In most of this book we will assume that they are continuous functions on some interval of the real line.

homogeneous equation
nonhomogeneous equation

If the function $f(x)$ is the zero function, the linear differential equation is said to be **homogeneous.** Otherwise, we say that the linear differential equation is **nonhomogeneous.** Any differential equation that cannot be written in the form above is said to be **nonlinear.** For example,

$$\frac{dy}{dx} = y^2$$

is nonlinear.

Before dealing with the nonhomogeneous first-order linear equation **(1),** it is important to discuss the solution of the homogeneous equation

$$\frac{dy}{dx} + a(x)y = 0, \tag{2}$$

or

$$\frac{dy}{dx} = -a(x)y.$$

Separating variables, we have

$$\int \frac{dy}{y} = -\int a(x)\,dx + C.$$

Integrating, we have

$$\ln|y| = -\int a(x)\,dx + C$$

and

$$y = C_1 e^{-\int a(x)\,dx}. \tag{3}$$

If this notation bothers you, simply think of $\int a(x)\,dx$ as an antiderivative of a. This is the general solution to Equation **(2);** it indicates that a solution is obtainable whenever the antiderivative can be found. We illustrate this situation with two examples.

EXAMPLE 1 ▶ **Solving a homogeneous differential equation**

Solve the homogeneous differential equation

$$y' + 3y = 0.$$

Solution Rewriting the equation as

$$y' = -3y$$

and separating variables, we have

$$\int \frac{dy}{y} = -3\int dx + C$$

$$\ln|y| = -3x + C$$

$$y = C_1 e^{-3x}. \quad ◀$$

EXAMPLE 2 ▶ **Solving a homogeneous initial-value problem**

Solve the equation

$$\frac{dy}{dx} = -2xy, \qquad y(1) = 1 \tag{4}$$

Solution Separating variables, we have

$$\int \frac{dy}{y} = -2\int x\,dx + C$$
$$\ln|y| = -x^2 + C$$
$$y = C_1 e^{-x^2}.$$

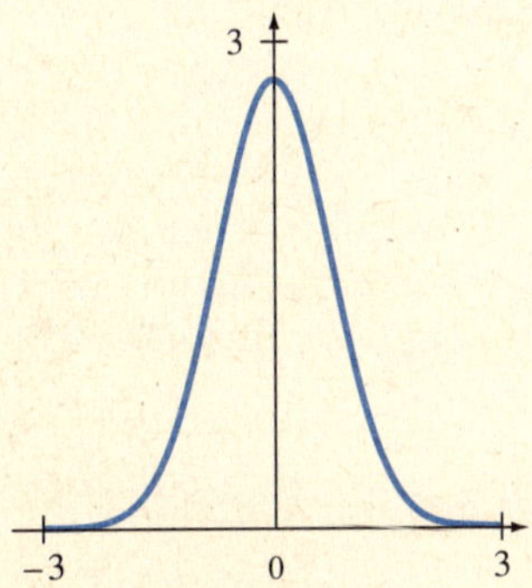

FIGURE 2.11
Graph of $y = e^{1-x^2}$

Since $y(1) = 1 = C_1 e^{-1}$, it follows that $C_1 = e$; the particular solution to the initial-value problem **(4)** is therefore

$$y = e^{1-x^2}.$$

A graph of this solution is given in Figure 2.11. ◀

We see that Equation **(3)** can be written in the form

$$ye^{\int a(x)\,dx} = C \tag{5}$$

by multiplying both sides by $e^{\int a(x)\,dx}$. If we differentiate both sides of Equation **(5),** we obtain

$$y'e^{\int a(x)\,dx} + a(x)ye^{\int a(x)\,dx} = [y' + a(x)y]e^{\int a(x)\,dx} = 0, \tag{6}$$

$$\left(\tfrac{d}{dx}e^{\int a(x)\,dx} = e^{\int a(x)\,dx}\tfrac{d}{dx}\int a(x)\,dx = a(x)e^{\int a(x)\,dx}\right)$$

since the derivative of an indefinite integral is the integrand. Notice that the expression in brackets is the left side of the original differential equation **(2).** We call the exponential

$$e^{\int a(x)\,dx} \tag{7}$$

integrating factor

an **integrating factor** for the linear Equation **(2),** because when we multiply Equation **(2)** by Equation **(7),** the left side of the result can be integrated.

We can use an integrating factor to obtain the solution of the nonhomogeneous linear equation

$$\frac{dy}{dx} + a(x)y = f(x). \tag{8}$$

We multiply both sides of Equation **(8)** by the integrating factor $e^{\int a(x)\,dx}$ to obtain

$$[y' + a(x)y]e^{\int a(x)\,dx} = f(x)e^{\int a(x)\,dx}. \tag{9}$$

But as we have seen in going from Equation **(5)** to Equation **(6),** the left side of Equation **(9)** is the derivative of $ye^{\int a(x)\,dx}$. Thus

$$\frac{d}{dx}\left[ye^{\int a(x)\,dx}\right] = f(x)e^{\int a(x)\,dx}. \tag{10}$$

Integrating both sides of Equation **(10)** with respect to x, we have

$$\int \frac{d}{dx}\left[ye^{\int a(x)\,dx}\right]dx = \int f(x)e^{\int a(x)\,dx}\,dx + C,$$

or

$$ye^{\int a(x)\,dx} = \int f(x)e^{\int a(x)\,dx}\,dx + C.$$

The solution of a linear first-order differential equation (8)

$$y = \left[\int f(x)e^{\int a(x)\,dx}\,dx + C\right]e^{-\int a(x)\,dx}. \tag{11}$$

Equation **(11)** provides an expression for the general solution of the first-order nonhomogeneous linear differential Equation **(8)**. **It is usually better to go through the process of multiplying both sides of (8) by the integrating factor to obtain the solution than to try to memorize Equation (11).** We illustrate this process with several examples.

EXAMPLE 3 ▶ **Solving a nonhomogeneous initial-value problem**

Solve the nonhomogeneous linear equation

$$y' = y + x^2, \qquad y(0) = 1. \tag{12}$$

Solution Rewriting Equation **(12)** in the form

$$y' - y = x^2, \tag{13}$$

we see that $a(x) = -1$; thus the integrating factor is

$$e^{-\int dx} = e^{-x}.$$

Multiplying both sides of Equation **(13)** by e^{-x}, we get

$$e^{-x}(y' - y) = x^2e^{-x},$$

or

$$(ye^{-x})' = x^2e^{-x}.$$

Integrating both sides, we have

$$ye^{-x} = \int x^2e^{-x}\,dx + C$$

$$= C - (x^2 + 2x + 2)e^{-x},$$

↑ integrate by parts twice

so

$$y = Ce^x - (x^2 + 2x + 2).$$

Finally, setting $x = 0$, we get

$$1 = y(0) = C - 2,$$

so $C = 3$ and the solution of the initial-value problem is

$$y = 3e^x - (x^2 + 2x + 2).$$

Note the solution can be written as the **superposition**, or sum, of two functions: $y = 3e^x$, which is a solution to the homogeneous equation $y' = y$ and $y = -(x^2 + 2x + 2)$ which is a solution to the nonhomogeneous equation $y' = y + x^2$. A graph of this solution appears in Figure 2.12. ◀

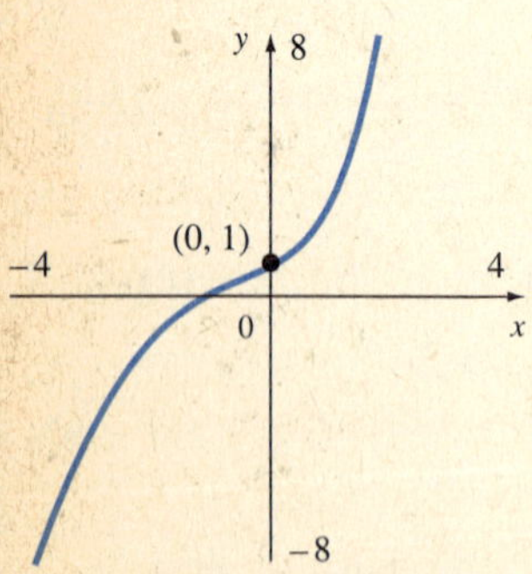

FIGURE 2.12
Graph of
$y = 3e^x - (x^2 + 2x + 2)$

EXAMPLE 4 ▶ **Solving a nonhomogeneous initial-value problem**

Consider the equation $dy/dx = x^3 - 2xy$, where $y = 1$ when $x = 1$. Rewriting the equation as $dy/dx + 2xy = x^3$, we see that $a(x) = 2x$ and the integrating factor is $e^{\int a(x)\,dx} = e^{x^2}$. Multiplying both sides by e^{x^2} and integrating we have

$$e^{x^2}y = \int x^3 e^{x^2}\,dx + C,$$

so

$$y = e^{-x^2}\left(\int x^3 e^{x^2}\,dx + C\right).$$

We can integrate by parts as follows:

$$\int x^3 e^{x^2}\,dx = \int x^2(xe^{x^2})\,dx = \frac{x^2 e^{x^2}}{2} - \int xe^{x^2}\,dx = e^{x^2}\left(\frac{x^2-1}{2}\right).$$

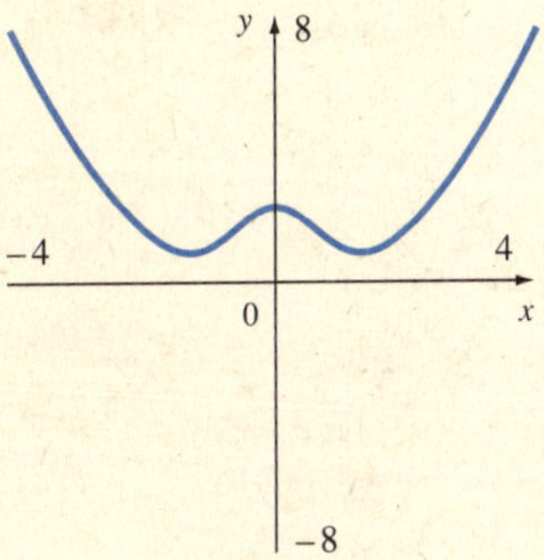

FIGURE 2.13
Graph of
$y = \frac{1}{2}(x^2 - 1) + e^{1-x^2}$

Replacing this term for the integral above, we have

$$y = e^{-x^2}\left[e^{x^2}\left(\frac{x^2-1}{2}\right) + C\right] = \frac{x^2-1}{2} + Ce^{-x^2}.$$

Setting $x = 1$ yields

$$1 = y(1) = Ce^{-1}.$$

Thus $C = e$ and the solution to the problem is

$$y = \frac{1}{2}(x^2 - 1) + e^{1-x^2}.$$

A graph of the solution is given in Figure 2.13. ◀

EXAMPLE 5 ▶ **Determining temperature as a function of input volume***

A closed tank containing V gallons of water at temperature T_0 is stirred constantly. It has a inlet pipe which supplies water at temperature T_W. Suppose that water is drained from the tank at the same rate at which it is put in, so that the volume of the water in the tank remains constant at V gallons. Let T denote the temperature of the water in the closed tank after the passage of p gallons through the system. Find T as a function of p.

Solution The situation is depicted in Figure 2.14.

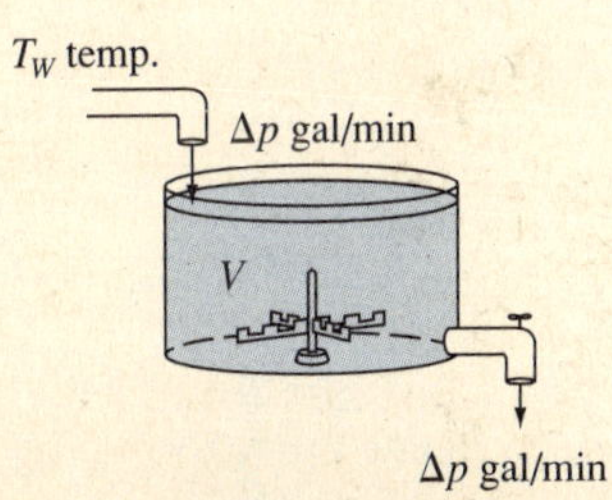

FIGURE 2.14
The volume V in the tank remains constant.

Let the total heat in the system at time t be $Q(t)$ calories. (One calorie is the amount of heat necessary to raise 1 cubic centimeter of water 1°C.) If the temperature of the water in the tank changes by the amount ΔT, then the change in the total heat in a system of V gallons is

$$\Delta Q = kV\,\Delta T,$$

where k is the number of cubic centimeters in a gallon. Similarly, if Δp gallons run through the system at a temperature T, then since water is entering the system at a temperature T_W, the net change in heat in the system is given by

$$\Delta Q = k(T_w - T)\,\Delta_p.$$

* This problem is adapted from one that appeared in *The American Mathematical Monthly* in March, 1944.

Equating the expressions for ΔQ yields

$$V\,\Delta T = (T_W - T)\,\Delta p,$$

or dividing by $V\,\Delta p$ and rewriting

$$\frac{\Delta T}{\Delta p} = -\frac{T}{V} + \frac{T_W}{V}.$$

To obtain the instantaneous rate of change of T with respect to p, we take the limit as $\Delta p \to 0$ to obtain

$$\frac{dT}{dp} = \frac{-T}{V} + \frac{T_W}{V}$$

or

$$\frac{dT}{dp} + \frac{1}{V}T = \frac{T_W}{V}. \tag{14}$$

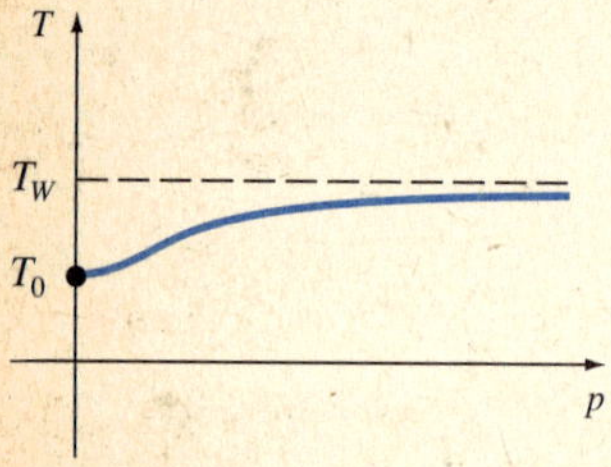

(a) $T_0 < T_W$

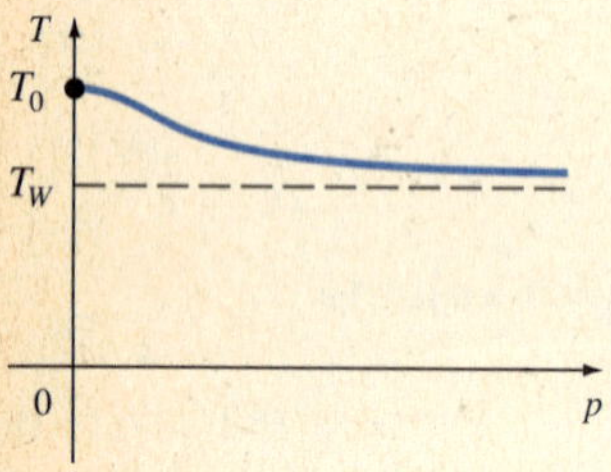

(b) $T_0 > T_W$

FIGURE 2.15
Graphs of
$T = T_W + (T_0 - T_W)e^{-p/V}$.

An integrating factor for Equation **(14)** is $e^{p/V}$. Thus,

$$\frac{d}{dp}(Te^{p/V}) = \frac{T_W}{V}e^{p/V}$$

$$Te^{p/V} = T_We^{p/V} + C$$

$$T(p) = T_W + Ce^{-p/V}.$$

But, $T(0) = T_0$, the temperature of the water before any water enters or flows out. So

$$T(0) = T_W + C = T_0$$

and

$$C = T_0 - T_W.$$

So

$$T(p) = T_W + (T_0 - T_W)e^{-p/V}.$$

Note the unsurprising result that as $p \to \infty$, $T(p) \to T_W$, the temperature of the incoming water. Graphs of the solution for $T_0 > T_W$ and $T_0 < T_W$ are given in Figure 2.15. ◀

Bernoulli's equation*

Certain nonlinear first-order equations can be reduced to linear equations by a suitable change of variables. The equation

$$\frac{dy}{dx} + a(x)y = f(x)y^n, \tag{15}$$

which is known as **Bernoulli's equation**, is of this type. Set $z = y^{1-n}$. Then $z' = (1-n)y^{-n}y'$, so if we multiply both sides of Equation **(15)** by $(1-n)y^{-n}$, we obtain

$$(1-n)y^{-n}y' + (1-n)a(x)y^{1-n} = (1-n)f(x),$$

* See the biographical sketch on page 60.

or

$$\frac{dz}{dx} + (1 - n)a(x)z = (1 - n)f(x).$$

The equation is now linear and may be solved as before.

EXAMPLE 6 ▶ **Solving Bernoulli's equation when $n = 3$**

Solve

$$\frac{dy}{dx} - \frac{y}{x} = -\frac{5}{2}x^2y^3. \tag{16}$$

Solution Here $n = 3$, so we let $z = y^{-2}$. Then $z' = -2y^{-3}y'$, and multiply both sides of Equation **(16)** by $-2y^{-3}$ to obtain

$$-2y^{-3}y' + \frac{2}{x}y^{-2} = 5x^2,$$

or

$$z' + \frac{2z}{x} = 5x^2. \tag{17}$$

An integrating factor for this linear equation is

$$e^{2\int dx/x} = e^{2\ln x} = e^{\ln x^2} = x^2.$$

Multiplying both sides of Equation **(17)** by x^2, we have

$$x^2z' + 2xz = 5x^4$$
$$(x^2z)' = 5x^4.$$

Thus

$$x^2z = 5\int x^4\,dx + C = x^5 + C.$$

Hence

$$y^{-2} = z = x^3 + Cx^{-2},$$

or

$$y = (x^3 + Cx^{-2})^{-1/2}. \quad ◀$$

A similar procedure can be used to solve

$$\frac{dy}{dx} + a(x)y = f(x)y\ln y. \tag{18}$$

Let $z = \ln y$. Then $z' = y'/y$, so dividing Equation **(18)** by y, we obtain the linear equation

$$\frac{dz}{dx} + a(x) = f(x)z.$$

or, in standard form,

$$\frac{dz}{dx} - f(x)z = -a(x).$$

Linear versus nonlinear differential equations

As we have seen in this section, all linear first-order initial-value problems can be solved—if not exactly, then to any desired precision by using a numerical integration technique. On the other hand, the only nonlinear equations we can solve in closed form are those that are separable, exact, homogeneous of degree zero (see Sections 2.4 and 2.5), or those that have some special form like the Bernoulli equations just discussed.

In Chapters 3, 4, 5, 6, and 7, we will discuss linear, second and higher-order differential equations, and systems. We do not even mention nonlinear equations in those chapters because, except in a few very carefully chosen situations, it is impossible to find closed-form solutions.

For this reason, scientists over the last three centuries have attempted to model physical, biological, or economic phenomena by using linear differential equations when a nonlinear equation might give a much better reflection of reality. We have seen this already. In Section 1.1, we showed that, by modeling population growth using the linear, Malthusian equation $dP/dt = \alpha P$, we end up with the absurd result of a population becoming infinitely large. On the other hand, if we use the nonlinear logistic model described in Section 2.2, then we obtain the much more reasonable result of a population approaching a stable equilibrium.

This is not to say that all linear models lead to absurd results. We have seen, and will continue to see, many important phenomena modeled accurately by linear equations. However, when you encounter a linear "application," you should always be at least a little bit skeptical. You should ask, is this linear equation a true representation of reality or is it there because some scientist finds it easy to solve? In the latter case, it is necessary to discard the model and find a nonlinear equation that may be impossible to solve exactly, but which *does* better mirror reality.

JAKOB BERNOULLI (1654–1705)

Jakob Bernoulli
(Offentliche Bibliothek der Universitat, Basel, Switzerland)

One of the most distinguished families in the history of mathematics and science are the Bernoullis of Switzerland. The family record starts in the seventeenth century with two brothers, Jakob and Johann. These two men gave up earlier vocational interests and became mathematicians when Leibniz's papers began to appear in the *Acta eruditorum.* Among the first mathematicians to realize the surprising power of the calculus, they applied the tool to a great diversity of problems. From 1687 until his death, Jakob occupied the mathematics chair at the University of Basel. The two brothers, often bitter rivals, maintained an almost constant exchange of ideas with Leibniz and with each other.

Among Jakob Bernoulli's contributions to mathematics are the early use of polar coordinates, the derivation in both rectangular and polar coordinates of the formula for the radius of curvature of a plane curve, the study of the catenary curve with extensions to strings of variable density and strings under the action of a central force, the study of a number of other higher-plane curves, the discovery of the so-called **isochrone**—or curve along which a body will fall with uniform vertical velocity (it turned out to be a semicubical parabola with a vertical cusptangent), the determination of the form taken by an elastic rod fixed at one end and carrying a weight at the other, the form assumed by a flexible rectangular sheet having two opposite edges held horizontally fixed at the same height and loaded with a heavy liquid, and the shape of a rectangular sail filled with wind. He also proposed and discussed the problem of isoperimetric figures (planar closed paths of given species and fixed perimeter which include a maximum area) and was one of the first mathematicians to work in the calculus of variations. He was also one of the early students of mathematical probability; his book in this field, the *Ars conjectandi*, was published posthumously in 1713. There are several mathematical entities that now bear Jakob Bernoulli's name; among them are the *Bernoulli distribution* and *Bernoulli theorem* of statistics and probability theory, the *Bernoulli equation* met by every student of a first course in differential equations, the *Bernoulli numbers* and *Bernoulli polynomials* of number-theory interest, and the *lemniscate of Bernoulli* encountered in a first course in the calculus. In Jakob Bernoulli's solution to the problem of the isochrone curve, which was published in the *Acta eruditorum* in 1690, we meet for the first time the word *integral* in a calculus sense.

Jakob Bernoulli was struck by the way the equiangular spiral reproduces itself under a variety of transformations and asked, in imitation of Archimedes, that such a spiral be engraved on his tombstone along with the inscription "Eadem mutata resurgo" (I shall arise the same, though changed).

SELF-QUIZ

I. The general solution to $dx/dt + 3x = 0$ is ______.
(a) $y = e^{-3x} + C$ **(b)** $y = Ce^{-3x}$
(c) $x = e^{-3t} + C$ **(d)** $x = Ce^{-3t}$

II. e^{3x} is an integrating factor for ______.
(a) $\dfrac{dy}{dx} - 3y = x^2$ **(b)** $\dfrac{dy}{dx} + \dfrac{y}{3} = x - 3$
(c) $\dfrac{dy}{dx} + 3y = x + 4$ **(d)** $3\dfrac{dy}{dx} + y = \dfrac{x}{5}$

III. The general solution to $dy/dx + 3y = 9x$ is ______.
(a) $y = e^{-3x} + C(3x - 1)$
(b) $y = (3x - 1) + Ce^{-3x}$
(c) $y = C(e^{-3x} + (3x - 1))$
(d) $y = e^{-3x} + (3x - 1) + C$

IV. ______ is an integrating factor for $dy/dx + 2xy = -2xe^{-x^2}$.
(a) e^{-x^2} **(b)** e^{x^2}
(c) $e^{x^2/2}$ **(d)** $-2xe^{-x^2}$

V. $\sin x$ is an integrating factor for ______.
(a) $\dfrac{dy}{dx} + (\cot x)y = x$ **(b)** $\dfrac{dy}{dx} + (\cos x)y = -x$
(c) $\dfrac{dy}{dx} + y = -\cos x$ **(d)** $\dfrac{dy}{dx} + \dfrac{y}{x} = \cot x$

VI. Suppose $f(x)$ is a solution of $dy/dx + y = 3$. Answer *True* or *False* to each of the following:
(a) $\lim_{x\to\infty} f(x)$ does not exist.
(b) $\lim_{x\to\infty} f(x) = 0$.
(c) $\lim_{x\to\infty} f(x) = 3$.
(d) $\lim_{x\to\infty} f'(x)$ does not exist.
(e) $\lim_{x\to\infty} f'(x) = 3$.
(f) $\lim_{x\to\infty} f'(x) = 0$.
(g) If $f(0) > 3$, then f always decreases.
(h) If $f(5) > \pi$, then f always decreases.
(i) If $f(0) < \pi$, then $f'(x) > 0$ for all x.
(j) If $f(-5) < 3$, then $f'(x) > 0$ for all x.

Answers to Self-Quiz

I. d **II.** c **III.** b **IV.** b **V.** a **VI.** **(a)** False **(b)** False **(c)** True **(d)** False **(e)** False **(f)** True **(g)** True **(h)** True **(i)** False **(j)** True

PROBLEMS 2.3

In Problems 1–21 find the general solution for each equation. When an initial condition is given, find the particular solution that satisfies the condition and sketch its graph.

1. $\dfrac{dx}{dt} = 3x$

2. $\dfrac{dy}{dx} + 22y = 0,\ y(1) = 2$

3. $\dfrac{dx}{dt} = x - 1,\ x(0) = 1$

4. $\dfrac{dy}{dt} = 2y + 1$

5. $\dfrac{dy}{dx} - 7y = x$

6. $\dfrac{dx}{dt} + 3x = t,\ x(0) = -1$

7. $\dfrac{dz}{dx} + 5z = x + 1$

8. $\dfrac{dy}{dx} - 5y = x - 1,\ x(1) = 1$

9. $\dfrac{dy}{dx} - 5y = 1 - x,\ x(1) = -1$

10. $\frac{dy}{dx} + y = x^2$

11. $\frac{dy}{dx} - y = x^2 + 2$

12. $\frac{dy}{dx} - xy = e^{\frac{1}{2}x^2}$

13. $\frac{dx}{dt} = tx + t,\ x(0) = 1$

14. $\frac{dy}{dx} + y = \sin x,\ y(0) = 0$

15. $\frac{dx}{dy} - x \ln y = y^y$

16. $\frac{dy}{dx} + y = \frac{1}{1 + e^{2x}}$

17. $\frac{dy}{dx} - \frac{3}{x}y = x^3,\ y(1) = 4$

18. $\frac{dx}{dt} + x \cot t = 2t \csc t$

19. $x' - 2x = t^2 e^{2t}$

20. $y' + \frac{2}{x}y = \frac{\cos x}{x^2},\ y(\pi) = 0$

21. $\frac{ds}{du} + s = ue^{-u} + 1$

22. Solve the equation

$$y - x\frac{dy}{dx} = \frac{dy}{dx}y^2e^y$$

by reversing the roles of x and y (that is, treat x as the dependent variable).

23. Use the method shown in Problem 22 to solve

$$\frac{dy}{dx} = \frac{1}{e^{-y} - x}.$$

24. Find the solution of $dy/dx = 2(2x - y)$ that passes through the point $(0, -1)$.

25. In a study* on the rate at which education is being forgotten or made obsolete, the following linear first-order differential equation was used:

$$x' = 1 - kx,$$

where $x(t)$ denotes the education of an individual at time t and k is a constant given by the rate at which that education is being lost. Obtain an equation for x at time t.

26. Data collected in a botanical experiment† led to the differential equation

$$\frac{dI}{dw} = 0.088(2.4 - I).$$

Find the value of I as $w \to \infty$.

27. Assume that there exists an upper bound B for the size y of a crop in a given field. E. A. Mitscherlich proposed in 1939 the use of the linear differential equation

$$\frac{dy}{dt} = k(B - y)$$

as a model for agricultural growth. Find the general solution of this equation.

28. Suppose a population is growing at a rate proportional to its size and, in addition, individuals are immigrating into the population at a constant rate.

(a) Find the linear differential equation governing this situation.

(b) Find its general solution.

29. Use the method we have developed in this section for the solution of Bernoulli's Equation (15) to solve the logistic equation (see Example 2.2.3)

$$\frac{dP}{dt} = P(\beta - \delta P).$$

‡**30.** Let $N(t)$ be the biomass§ of a fish species in a given area of the ocean and suppose the rate of change of the biomass is governed by the logistic equation

$$\frac{dN}{dt} = rN\left(1 - \frac{N}{K}\right),$$

where the net proportional growth rate r is a constant and K is the carrying capacity for that species in that area. Assume that the rate at which fish are caught depends on the biomass. If E is the constant effort expended to harvest that species, then

(a) find the resulting growth rate of the biomass;**

(b) solve the differential equation in part (a) using the Bernoulli method.

[*Note:* Effort can be measured in man-hours per year, tonnage of the fishing fleet, or volume seined by the fleet's nets.]

31. Suppose fish are harvested at a constant rate h independent of their biomass.** Answer parts (a) and (b) of Problem 30 for this situation.

32. Find the effort E that maximizes the sustainable yield in Problem 30. [This is the limit of the yield as $t \to \infty$.]

* L. Southwick and S. Zionts, "An Optimal-Control-Theory Approach to the Education Investment Decision." *Operations Research* 22 (1974): 1156–1174.

† R. L. Specht, "Dark Island Heath," *Australian Journal of Botany* 5 (1957): 137–172.

‡ A number of other examples of models of renewable resources are given in C. W. Clark, *Mathematical Bioeconomics* (New York: Wiley-Interscience, 1976).

§ The **biomass** of a species is defined as the total number of living organisms in a given area, expressed in terms of living or dry weight per given area.

This is called the **Schaefer model, after the biologist M. B. Schaefer.

33. Show that if $h > rK/4$, in Problem 31, the species will become extinct regardless of the initial size of the biomass.

34. The differential equation governing the velocity v of an object of mass m subject to air resistance proportional to the instantaneous velocity is

$$m\frac{dv}{dt} = -mg - kv.$$

Solve the equation and determine the limiting velocity of the object as $t \to \infty$.

35. Repeat Problem 34 if air resistance is proportional to the square of the instantaneous velocity.

36. An infectious disease is introduced into a large population. The proportion of people exposed to the disease increases with time. Suppose that $P(t)$ is the proportion of people exposed to the disease within t years of its introduction. If $P'(t) = [1 - P(t)]/3$ and $P(0) = 0$, after how many years does the proportion increase to 90%?

In Problems 37–42 find the general solution for each equation and a particular solution when an initial condition is given.

37. $\dfrac{dy}{dx} = -\dfrac{(6y^2 - x - 1)y}{2x}$

38. $y' = -y^3xe^{-2x} + y$

39. $x\dfrac{dy}{dx} + y = x^4y^3,\ y(1) = 1$

40. $tx^2\dfrac{dx}{dt} + x^3 = t\cos t$

41. $\dfrac{dy}{dx} + \dfrac{3}{x}y = x^2y^2,\ y(1) = 2$

42. $xyy' - y^2 + x^2 = 0$

43. We have seen that the equation

$$y' + f(x)y = 0$$

has the general solution

$$y = ce^{-\int f(x)\,dx}.$$

This fact prompted J. L. Lagrange (1736–1813) to seek a solution of the equation

$$y' + f(x)y = g(x) \qquad \textbf{(i)}$$

of the form

$$y = c(x)e^{-\int f(x)\,dx},$$

where $c = c(x)$ is a function of x.*

(a) Show that $c'(x) = g(x)e^{\int f(x)\,dx}$.

(b) Integrate part (a) to obtain the general solution of Equation **(i)**.

44. Use the method in Problem 43 to solve the equation

$$y' + \frac{1}{x}y = e^x.$$

45. Consider the second-order linear equation

$$y'' + 5y' + 6y = 0. \qquad \textbf{(ii)}$$

(a) Let $z = y' + 2y$. Show that Equation **(ii)** reduces to

$$z' + 3z = 0. \qquad \textbf{(iii)}$$

(b) Solve Equation **(iii)**, substitute z in the equation $y' + 2y = z$, and use the methods given in this section to obtain the solution to Equation **(ii)**.

46. Use the procedure outlined in Problem 45 to find the general solution to

$$y'' + (a + b)y' + aby = 0,$$

where a and b are constants and $a \neq b$. (*Hint:* Let $z = y' + ay$.)

47. Use the method given in Problem 45 to find the general solution to

$$y'' + 2ay' + a^2y = 0,$$

where a is a constant.

In Problems 48 and 49 the function f *is given by*

$$f(x) = \begin{cases} 1, & \text{if } 0 \le x \le 1, \\ 0, & \text{if } x > 1. \end{cases}$$

48. Solve the initial-value problem

$$y' + y = f(x), \qquad y(0) = 0.$$

49. Solve the initial-value problem

$$y' + f(x)y = 0, \qquad y(0) = 1.$$

[*Hint:* Solve separately on $0 \le x \le 1$ and $x > 1$ and match at $x = 1$.]

†50. Show that the general solution of the differential equation

$$y' - 2xy = x^2$$

is of the form $y = f(x) + ce^{x^2}$, where

$$\left| f(x) + \frac{x}{2} \right| \le \frac{1}{4x}, \text{ for } x \ge 2.$$

---

* This technique, called the method of **variation of constants,** or **variation of parameters,** can be extended to equations of higher order (see Section 3.5).

† R. C. Buck, "On 'Solving' Differential Equations," *American Mathematical Monthly* 63 (1956): 414.

*51. Show that if the family of solutions of

$$y' + a(x)y = f(x), \qquad a(x)f(x) \neq 0,$$

all cross the line $x = c$, the tangent lines to the solution curves at the points of intersection are **concurrent** (have a point in common).

52. Daniel Bernoulli obtained the following differential equation (in 1760),

$$S' = -pS + \frac{S}{N}N' + \frac{pS^2}{mN},$$

in studying the effects of smallpox. Here $N(t)$ is the number of individuals that survive at age t, $S(t)$ is the number that are susceptible to smallpox at age t (the disease imparts lifetime immunity if survived), p is the probability of a susceptible individual getting the disease, and $\frac{1}{m}$ is the proportion of those who die from the disease. Let $y = \frac{N}{S}$ and find a solution of the resulting equation.

2.4 EXACT DIFFERENTIAL EQUATIONS

The **total differential** dg of a function of two variables $g(x, y)$ is defined by

$$dg = \frac{\partial g}{\partial x}\,dx + \frac{\partial g}{\partial y}\,dy.$$

EXAMPLE 1 ▶ Computing a total differential

Let $g(x, y) = x^2y^3 + e^{4x}\sin y$. Compute the total differential dg.

Solution

$$\frac{\partial g}{\partial x} = 2xy^3 + 4e^{4x}\sin y,$$

and

$$\frac{\partial g}{\partial y} = 3x^2y^2 + e^{4x}\cos y.$$

Hence

$$dg = (2xy^3 + 4e^{4x}\sin y)\,dx + (3x^2y^2 + e^{4x}\cos y)\,dy. \quad ◀$$

We now use partial derivatives to solve ordinary differential equations. Suppose that we take the total differential of both sides of the equation $g(x, y) = c$:

$$dg = \frac{\partial g}{\partial x}\,dx + \frac{\partial g}{\partial y}\,dy = 0. \tag{1}$$

For example, the equation $xy = c$ has the total differential $y\,dx + x\,dy = 0$, which may be rewritten as the differential equation $y' = -y/x$. Reversing the situation, suppose that we start with the differential equation

$$M(x, y)\,dx + N(x, y)\,dy = 0. \tag{2}$$

If we can find a differentiable function $g(x, y)$ such that

$$\frac{\partial g}{\partial x} = M \quad \text{and} \quad \frac{\partial g}{\partial y} = N,$$

then Equation **(2)** becomes $dg = 0$, so that $g(x, y) = c$ is the general solution of Equation **(2)**. In this case $M\,dx + N\,dy$ is said to be an **exact differential,** and Equation **(2)** is called an **exact differential equation.**

---

* Problem 3 of the William Lowell Putnam Mathematical Competition, *American Mathematical Monthly* 61 (1954): 545.

It is very easy to determine whether a differential equation is exact by using the **cross-derivative test:**

Test for exactness

Suppose that $\partial M/\partial y$ and $\partial N/\partial x$ exist.

The equation $M(x, y)\,dx + N(x, y)\,dy = 0$ is exact if and only if

$$\frac{\partial M}{\partial y} = \frac{\partial N}{\partial x} \tag{3}$$

If $M\,dx + N\,dy$ is exact, then we can find a relation linking x and y by finding the function g given above. The procedure for doing this is illustrated in Examples 2 and 3.

EXAMPLE 2 ▶ **Solving an exact differential equation**

Solve the equation

$$(1 - \sin x \tan y)\,dx + (\cos x \sec^2 y)\,dy = 0.$$

Solution Letting $M(x, y) = 1 - \sin x \tan y$ and $N(x, y) = \cos x \sec^2 y$, we have

$$\frac{\partial M}{\partial y} = -\sin x \sec^2 y = \frac{\partial N}{\partial x},$$

so the equation is exact. We now seek a function g of two variables such that $\partial g/\partial x = M$ and $\partial g/\partial y = N$. But if $\partial g/\partial x = M$, then

$$g(x, y) = \int M\,dx = \int (1 - \sin x \tan y)\,dx$$

$$= x + \cos x \tan y + h(y). \tag{4}$$

The *constant of integration* $h(y)$ occurring in Equation **(4)** is an arbitrary function of y since we must introduce the most general term that vanishes under partial differentiation with respect to x. But

$$\cos x \sec^2 y = N(x, y) = \frac{\partial g}{\partial y} = \cos x \sec^2 y + h'(y).$$

differentiating (4) with respect to y

This means that $h'(y) = 0$, so $h(y) = k$, a constant. Thus the most general relation linking x and y is

$$g(x, y) = x + \cos x \tan y + k = C, \quad \text{another constant}$$

or

$$x + \cos x \tan y = C_1. \quad ◀$$

EXAMPLE 3 ▶ **Solving an exact equation**

Find a solution to the equation

$$(y^2 + 2xy + 1)\,dx + (2xy + x^2 + 2)\,dy = 0 \tag{5}$$

Solution Here $M(x, y) = y^2 + 2xy + 1$ and $N(x, y) = 2xy + x^2 + 2$. Equation **(5)** is exact because

$$\frac{\partial M}{\partial y} = 2y + 2x = \frac{\partial N}{\partial x},$$

so we calculate

$$g = \int M\,dx = xy^2 + x^2y + x + h(y).$$

To find $h(y)$, we take the partial derivative of g with respect to y:

$$2\,xy + x^2 + 2 = N = \frac{\partial g}{\partial y} = 2xy + x^2 + h'(y),$$

implying that $h'(y) = 2$. Thus, $h(y) = 2y + c$, and the most general relation linking x and y in Equation **(5)** is

$$x\,y^2 + x^2\,y + x + 2\,y = C. \blacktriangleleft$$

Multiplying by an integrating factor to make an equation exact

It should be apparent that exact equations are comparatively rare, since the condition in Equation **(3)** requires a precise balance of the functions M and N. For example,

$$(3x + 2y)\,dx + x\,dy = 0$$

is not exact. However, if we multiply the equation by x, the new equation

$$(3x^2 + 2\,xy)\,dx + x^2\,dy = 0$$

is exact.

The problem reduces to finding an integrating factor $\mu(x, y)$ so that even if $Mdx + Ndy$ is not exact,

$$\mu M dx + \mu N\,dy = 0 \tag{6}$$

is exact. Finding integrating factors is generally very difficult because one must often proceed by trial and error. A procedure that is sometimes useful is to note, by Equation **(6)**, that

$$\mu\frac{\partial M}{\partial y} + M\frac{\partial \mu}{\partial y} = \frac{\partial}{\partial y}(\mu M) = \frac{\partial}{\partial x}(\mu N) = \mu\frac{\partial N}{\partial x} + N\frac{\partial \mu}{dx};$$

therefore

$$\frac{1}{\mu}\left(N\frac{\partial \mu}{\partial x} - M\frac{\partial \mu}{\partial y}\right) = \frac{\partial M}{\partial y} - \frac{\partial N}{\partial x}. \tag{7}$$

In case the integrating factor μ depends only on x, Equation **(7)** becomes

$$\frac{1}{\mu}\frac{d\mu}{dx} = \frac{\dfrac{\partial M}{\partial y} - \dfrac{\partial N}{\partial x}}{N} = k(x, y). \tag{8}$$

Since the left-hand side of Equation **(8)** consists only of functions of x, k *must* also be a function of x only. If this is indeed true, then μ can be found by separating the variables: $\mu(x) = e^{\int k(x)\,dx}$. A similar result holds if μ is a function of y alone, in which case

$$K = \frac{\dfrac{\partial M}{\partial y} - \dfrac{\partial N}{\partial x}}{-M}$$

is also a function of y. In this case, $\mu(y) = e^{\int K(y)\,dy}$ is the integrating factor.

EXAMPLE 4 ▶ **Making a differential equation exact by multiplying by an integrating factor**

Solve the equation

$$(3x^2 - y^2)\,dy - 2xy\,dx = 0.$$

Solution In this problem, $M = -2xy$ and $N = 3x^2 - y^2$, so

$$\frac{\partial M}{\partial y} = -2x \qquad \text{and} \qquad \frac{\partial N}{\partial x} = 6x.$$

Then

$$K = \frac{\dfrac{\partial M}{\partial y} - \dfrac{\partial N}{\partial x}}{-M} = \frac{-4}{y}$$

and

$$\mu = e^{-4\int y^{-1}\,dy} = e^{-4\ln y} = y^{-4}.$$

Multiplying the differential equation by y^{-4}, we obtain the exact equation

$$-\frac{2x}{y^3}\,dx + \left(\frac{3x^2 - y^2}{y^4}\right)dy = 0.$$

Then

$$g = \int M\,dx = -\frac{x^2}{y^3} + h(y),$$

so that

$$\frac{3x^2}{y^4} - \frac{1}{y^2} = N = \frac{\partial g}{\partial y} = \frac{3x^2}{y^4} + h'(y),$$

or, $h'(y) = -y^{-2}$. Hence, $h(y) = y^{-1} + c$, and a relation linking x and y is

$$g(x, y) = \frac{1}{y} - \frac{x^2}{y^3} = c,$$

or

$$cy^3 - y^2 + x^2 = 0. \quad ◀$$

SELF-QUIZ

Answer True *or* False *in Problems I to IV.*

I. $(4y^2 - 4x^2)\,dx + (8xy - \ln y)\,dy$ is exact.

II. $(4x^2 - 4y^2)\,dx + (8xy - \ln y)\,dy$ is exact.

III. $2x \cos y\,dx + x^2 \sin y\,dy$ is exact.

IV. $2x \cos y\,dx - x^2 \sin y\,dy$ is exact.

V. Which of the following is an integrating factor for $2y^2\,dx + 3xy\,dy$?
(a) x **(b)** y **(c)** xy **(d)** x^2 **(e)** y^2

VI. Which of the following is an integrating factor for $2 \cos 2x\,dx + 3 \sin 2x\,dy$?
(a) e^{2y} **(b)** e^{3y} **(c)** $\sin 3x$ **(d)** $\cos 3y$ **(e)** $\frac{2}{3}$

Answers to Self-Quiz

I. True **II.** False **III.** False **IV.** True **V.** (c) **VI.** (b)

PROBLEMS 2.4

In Problems 1–19 verify that each given differential equation is exact and find the general solution. Find a particular solution when an initial condition is given and sketch its graph using a graphing calculator or a computer algebra system.

1. $2xy\,dx + (x^2 + 1)\,dy = 0$
2. $3x^2\,dx + (x^3 + 1)\,dy = 0$
3. $y(y + 2x)\,dx + x(2y + x)\,dy = 0$
4. $y\cos(xy)\,dx + x\cos(xy)\,dy = 0$
5. $(x - y\cos x)\,dx - \sin x\,dy = 0,\ y\left(\frac{\pi}{2}\right) = 1$
6. $\cosh 2x \cosh 2y\,dx + \sinh 2x \sinh 2y\,dy = 0$
7. $(ye^{xy} + 4y^3)\,dx + (xe^{xy} + 12xy^2 - 2y)\,dy = 0,$ $y(0) = 2$
8. $(3x^2 \ln x + x^2 - y)\,dx - x\,dy = 0,\ y(1) = 5$
9. $(2xy + e^y)\,dx + (x^2 + xe^y)\,dy = 0$
10. $(x^2 + y^2)\,dx + 2xy\,dy = 0,\ y(1) = 1$
11. $\left(\frac{1}{x} - \frac{y}{x^2 + y^2}\right)dx + \left(\frac{x}{x^2 + y^2} - \frac{1}{y}\right)dy = 0$
12. $[x\cos(x + y) + \sin(x + y)]\,dx +$ $x\cos(x + y)\,dy = 0,\ y(1) = \frac{\pi}{2} - 1$
13. $\left(4x^3y^3 + \frac{1}{x}\right)dx + \left(3x^4y^2 - \frac{1}{y}\right)dy = 0,\ x(e) = 1$
14. $\left(\frac{\ln(\ln y)}{x} + \frac{2}{3}xy^3\right)dx + \left(\frac{\ln x}{y\ln y} + x^2y^2\right)dy = 0$

In Problems 15–24 find an integrating factor for each differential equation and obtain the general solution.

15. $y\,dx + (y - x)\,dy = 0$
16. $(x^2 + y^2 + x)\,dx + y\,dy = 0$
17. $2y^2\,dx + (2x + 3xy)\,dy = 0$
18. $(x^2 + 2y)\,dx - x\,dy = 0$
19. $(x^2 + y^2)\,dx + (3xy)\,dy = 0$
20. $x\,y\,dx - x^2\,dy = 0$
21. $(x^2 + 2xy^2 - y)\,dx + (y^2 + 2x^2y - x)\,dy = 0$
22. $y(2y + 3x)\,dx + x(3y + 2x)\,dy = 0$
23. $(4y^2 + \frac{3}{x}\sin y)\,dx + (2xy + \cos y)\,dy = 0$
24. $[y^2(x^2 + y) + 2x]\,dx + [2y(x^2 + y) + 1]\,dy = 0$
25. Solve $xy\,dx + (x^2 + 2y^2 + 2)\,dy = 0$.
26. Let $M = yF(xy)$ and $N = xG(xy)$. Show that $1/(xM - yN)$ is an integrating factor for
$$M\,dx + N\,dy = 0.$$
27. Use the result of Problem 26 to solve the equation
$$2x^2y^3\,dx + x^3y^2\,dy = 0.$$
28. Solve $(x^2 + y^2 + 1)\,dx - (xy + y)\,dy = 0$. [*Hint:* Try an integrating factor of the form $\mu(x, y) = (x + 1)^n$.]

2.5 HOMOGENEOUS EQUATIONS OF DEGREE ZERO AND SUBSTITUTION TECHNIQUES

Consider the differential equation

$$\frac{dy}{dx} = f(x, y). \tag{1}$$

If for every real number t for which $f(tx, ty)$ is defined, $f(tx, ty) = f(x, y)$, then the function f is said to be **homogeneous of degree zero.** Note that the word "homogeneous" used in this section has nothing to do with the word "homogeneous" when applied to a linear differential equation (as in Section 2.3).

EXAMPLE 1 ▶ **Homogeneous function of degree zero**

The function $f(x, y) = (x + y)/(x - y)$ is homogeneous of degree zero because

$$f(tx, ty) = \frac{tx + ty}{tx - ty} = \frac{t(x + y)}{t(x - y)} = \frac{x + y}{x - y} = f(x, y). \blacktriangleleft$$

Now if we substitute $t = \frac{1}{x}$, it follows that if f is homogeneous of degree zero, then

$$f(x, y) = f(tx, ty) = f\left(1, \frac{y}{x}\right) = F\left(\frac{y}{x}\right),$$

is a function of the variable $\frac{y}{x}$. Equation **(1)** now becomes

$$\frac{dy}{dx} = F\left(\frac{y}{x}\right) \tag{2}$$

and the right-hand side of the equation can be written as a function of the variable $\frac{y}{x}$. It is then natural to try the substitution $z = \frac{y}{x}$. Since the function y depends on x, so does the function z. Differentiating $y = xz$ with respect to x, we have

$$\frac{dy}{dx} = z + x\frac{dz}{dx}. \tag{3}$$

Substituting $z = \frac{y}{x}$ and replacing the left-hand side of Equation **(2)** by the right-hand side of Equation **(3)**, we obtain

$$z + x\frac{dz}{dx} = F(z).$$

The variables in this equation can be separated, since

$$x\frac{dz}{dx} = F(z) - z,$$

and so

$$\frac{dz}{F(z) - z} = \frac{dx}{x}.$$

A solution can now be obtained by integrating both sides of this equation and replacing each z by $\frac{y}{x}$.

Note

We stress that the substitution $z = \frac{y}{x}$ always transforms a homogeneous differential equation of degree zero into a separable differential equation.

We illustrate this procedure in the next two examples.

EXAMPLE 2 ▶ **Solving a homogeneous equation of degree zero**

Solve the equation

$$\frac{dy}{dx} = \frac{x - y}{x + y}.$$

Solution If $f(x, y) = (x - y)/(x + y)$, then

$$f(tx, ty) = \frac{tx - ty}{tx + ty} = \frac{t(x - y)}{t(x + y)} = \frac{x - y}{x + y} = f(x, y);$$

so f is homogeneous of degree zero.

Dividing the numerator and denominator of the right-hand side of this equation by x yields

$$\frac{dy}{dx} = \frac{1 - \dfrac{y}{x}}{1 + \dfrac{y}{x}} = \frac{1 - z}{1 + z} = F(z).$$

Replacing the left-hand side by $z + x(dz/dx)$ and separating variables, we have (after some algebra)

$$\left(\frac{1+z}{1-2z-z^2}\right) dz = \frac{dx}{x}.$$

After integrating, we obtain

$$\ln(1 - 2z - z^2) = -2 \ln x + c = \ln(cx^{-2});$$

exponentiating and replacing z by $\frac{y}{x}$ leads to the implicit solution

$$1 - \frac{2y}{x} - \frac{y^2}{x^2} = \frac{c}{x^2}.$$

Finally, multiplying both sides by x^2 yields $x^2 - 2xy - y^2 = c$. ◀

EXAMPLE 3 ▶ **Determining the shape of a mirror**

Find the shape of a curved mirror such that light from a source at the origin is reflected in a beam parallel to the x-axis.

Solution By symmetry, the mirror has the shape of a surface of revolution obtained by revolving a curve about the x-axis. Let (x, y) be any point on the cross section in the xy-plane (see Figure 2.16). The law of reflection states that the angle of incidence α must equal the angle of reflection β; thus $\alpha = \beta = \theta$. Since the interior angles of a triangle add up to 180°, we have $\psi = \alpha + \theta = 2\theta$. What is important about the angles θ and ψ is that

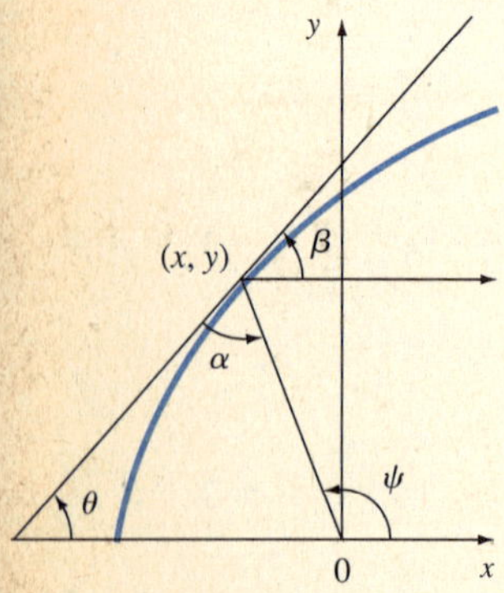

FIGURE 2.16
Curved mirror that reflects parallel beams

$$y' = \tan\theta \quad \text{and} \quad \frac{y}{x} = \tan\psi.$$

Using the trigonometric formula for the tangent of a double angle, we have

$$\frac{y}{x} = \tan\psi = \tan 2\theta = \frac{2\tan\theta}{1 - \tan^2\theta} = \frac{2y'}{1-(y')^2}.$$

Solving for y', we obtain the quadratic equation

$$y(y')^2 + 2xy' - y = 0,$$

and by the quadratic formula we find that

$$y' = \frac{-x \pm \sqrt{x^2+y^2}}{y}. \qquad \textbf{(4)}$$

Since $y' > 0$, we need only consider the + sign of the ± in Equation **(4),** because $\sqrt{x^2+y^2} > |x|$, yielding the nonlinear differential equation

$$y' = \frac{-x + \sqrt{x^2+y^2}}{y}. \qquad \textbf{(5)}$$

Dividing the numerator and denominator of Equation **(5)** by x, and remembering that $\sqrt{x^2+y^2}/x = \sqrt{(x^2+y^2)/x^2}$, we obtain

$$\frac{dy}{dx} = \frac{-1 + \sqrt{1 + \left(\frac{y}{x}\right)^2}}{\frac{y}{x}} \qquad \textbf{(6)}$$

Substituting $z = \frac{y}{x}$ into Equation **(6)** and using Equation **(3)**, we have

$$z + x\frac{dz}{dx} = \frac{-1 + \sqrt{1 + z^2}}{z},$$

or, subtracting z from both sides and simplifying the right-hand side, we have

$$x\frac{dz}{dx} = \frac{\sqrt{1 + z^2} - (1 + z^2)}{z}.$$

Separating variables, we obtain

$$\frac{z\,dz}{\sqrt{1 + z^2}(1 - \sqrt{1 + z^2})} = \frac{dx}{x}. \tag{7}$$

To integrate the left-hand side of Equation **(7)** we substitute

$$u = 1 - \sqrt{1 + z^2} \quad \text{and} \quad du = -\frac{z\,dz}{\sqrt{1 + z^2}},$$

obtaining

$$-\int \frac{du}{u} = \int \frac{dx}{x},$$

or $-\ln|u| = \ln|x| + c$. Exponentiating both sides, we have

$$\frac{1}{u} = cx;$$

inverting both sides leads to (since c is arbitrary)

$$1 - \sqrt{1 + z^2} = -\frac{c}{x}.$$

We add the minus sign to the arbitrary constant to make the answers appear more compact.

Replacing z by $\frac{y}{x}$ and simplifying, we have

$$1 + \frac{c}{x} = \sqrt{1 + \left(\frac{y}{x}\right)^2},$$

which may be squared to obtain

$$1 + \frac{2c}{x} + \frac{c^2}{x^2} = 1 + \frac{y^2}{x^2}. \tag{8}$$

Canceling the ones and multiplying both sides of Equation **(8)** by x^2, we get

$$y^2 = 2cx + c^2,$$

which is the equation of the family of all parabolas with foci at the origin that are symmetric with respect to the x-axis. In Figure 2.17 we sketch four of these parabolas. ◀

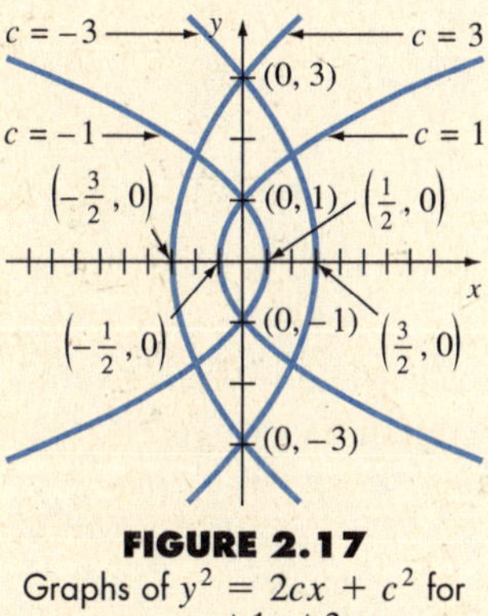

FIGURE 2.17
Graphs of $y^2 = 2cx + c^2$ for $c = \pm 1, \pm 3$

The Hubble telescope and lithotripsy

Example 3 is not just an odd application of mathematics. The idea of designing a mirror that reflects all beams parallel to an axis has many valuable applications: flashlights, searchlights, and automobile headlights to name a few. In all of these uses, electricity generates light at the focus, and the parabolic shape of the mirror reflects the light parallel to the axis. This concept is also used in reverse for telescopes, T.V. satellite dishes and microwave relay stations, radar detectors, and so on. A particular use of these concepts was made in the design of the Hubble Space Telescope. The telescope was put in space so that it would not be subject to atmospheric interference and would be able to detect objects undetected by other telescopes. One of the initial disappointments with the Hubble telescope has been that the mirror's shape is incorrect. It suffers from spherical aberration, which makes its images fuzzy. The telescope was equipped with actuators that can adjust the mirror's shape, but because the possibility that the mirror would be ground to an incorrect shape was not considered, the actuators are in the wrong place and are not strong enough to compensate for the defect. Apparently NASA tried to save money by not performing a rigorous performance test before launching the telescope, and was forced to make costly repairs in space to fix the problem. Figure 2.18 shows the Hubble telescope in space.

A related application is in the field of lithotripsy. Ellipses reflect waves from one focus to the other focus. This geometric fact has been used in medicine. A lithotriptor is a machine that produces an ultrasound wave at one focus of a metallic frame shaped as half of a spherical ellipsoid. Sound waves reflecting off this frame are used as shock waves to blast kidney and bile stones located at the other focus. The patient is placed so that the kidney stone is located correctly using X rays. Figure 2.19 shows a lithotriptor.

Another useful technique applies to equations of the form

$$\frac{dy}{dx} = F(ax + by + c), \tag{9}$$

FIGURE 2.18
The Hubble Telescope
Source: NASA.

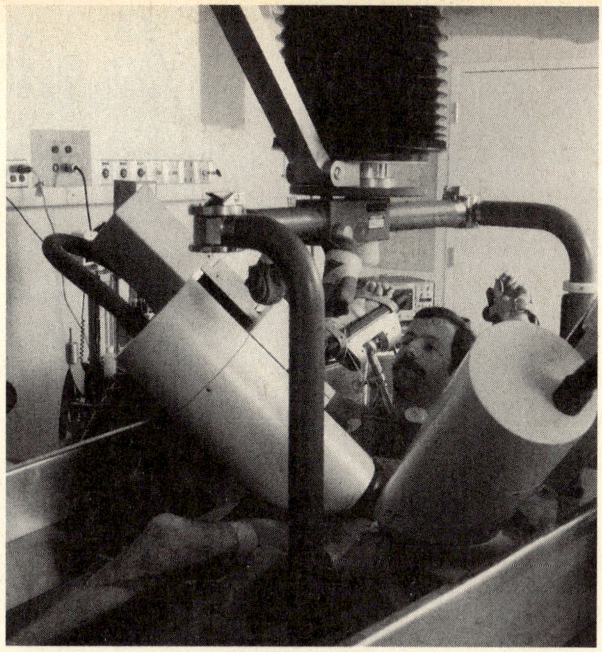

FIGURE 2.19
A lithotriptor
Source: Hank Morgan/Rainbow.

where a, b, and c are real constants. If the substitution $z = ax + by + c$ is made in Equation **(9)**, we obtain

$$\frac{z' - a}{b} = F(z),$$

since $z' = a + by'$. This equation has separable variables.

EXAMPLE 4 ▶ **Solving a differential equation by making an appropriate substitution**

Solve

$$\frac{dy}{dx} = (x + y + 1)^2 - 2.$$

Solution Letting $z = x + y + 1$, we have $z' = 1 + y'$. Thus

$$z' - 1 = y' = (x + y + 1)^2 - 2 = z^2 - 2,$$

or

$$z' = z^2 - 1.$$

We now separate variables to obtain

$$\int \frac{dz}{z^2 - 1} = \int dx. \quad \textbf{(10)}$$

We can integrate the left-hand side of Equation **(10)** by partial fractions, using

$$\frac{1}{z^2 - 1} = \frac{1}{2}\left(\frac{1}{z - 1} - \frac{1}{z + 1}\right),$$

so that Equation (**10**) yields

$$\frac{1}{2}[\ln|z-1| - \ln|z+1|] = x + c,$$

or

$$\ln\left|\frac{z-1}{z+1}\right| = 2x + c.$$

Exponentiating both sides, we obtain

$$\frac{z-1}{z+1} = ce^{2x}, \qquad \text{or} \qquad (z-1) = ce^{2x}(z+1).$$

Gathering all terms involving z on one side and the remaining terms on the other, we get

$$z = \frac{1+ce^{2x}}{1-ce^{2x}}.$$

Finally, we replace z by $x + y + 1$, which yields

$$y = \frac{1+ce^{2x}}{1-ce^{2x}} - x - 1. \quad ◀$$

As these examples indicate, we may make *any* substitution we want in trying to solve a differential equation. Of course, *there is no guarantee of success.* Moreover, there may be more than one substitution that yields the answer, as illustrated in the following example.

EXAMPLE 5 ▶ **Making a substitution to solve the mirror problem of Example 3**

Here we reconsider Example 3, in which we were trying to find the shape of a curved mirror that reflects light from a source at the origin in a beam parallel to the x-axis. Recall that the geometric considerations in Example 3 led to the differential equation [see Equation (**5**)]

$$y' = \frac{-x + \sqrt{x^2 + y^2}}{y}. \tag{11}$$

Now, instead of proceeding as we did in Example 3, suppose we substitute $z = x^2 + y^2$. Then

$$\frac{dz}{dx} = 2x + 2y\frac{dy}{dx},$$

and replacing dy/dx by the right-hand side of Equation (**11**), we have

$$\begin{aligned}\frac{dz}{dx} &= 2x + 2y\left(\frac{-x + \sqrt{x^2 + y^2}}{y}\right)\\ &= 2x - 2x + 2\sqrt{x^2 + y^2},\end{aligned}$$

or

$$\frac{dz}{dx} = 2\sqrt{z}.$$

Separating variables gives us

$$\frac{dz}{2\sqrt{z}} = dx,$$

and integrating, we get

$$\sqrt{z} = x + c.$$

Replacing z by $x^2 + y^2$ and squaring both sides yields

$$x^2 + y^2 = x^2 + 2cx + c^2, \qquad \text{or} \qquad y^2 = 2cx + c^2.$$

Since c is arbitrary, we have the same family of parabolas as before. ◄

SELF-QUIZ

Answer True *or* False *in Problems I to IV.*

I. $\sqrt{x^2 + y^2}$ is homogenous of degree 0.

II. $x^n + y^n$ is homogeneous of degrees 0 if n is a positive integer.

III. $(x^n + y^n)/(x^n - y^n)$ is homogeneous of degree 0 if n is a positive integer.

IV. $(x^3 + y^3)/(y - x)$ is homogeneous of degree 0.

V. The substitution ________ will enable us to solve the differential equation $dy/dx = y/x + \sqrt{x^2 + y^2}$ by separating the variables.

(a) $x = \frac{1}{y}$ **(b)** $z = \frac{y}{x}$

(c) $z = \frac{x}{y}$ **(d)** $z = x^2 + y^2$

Answers to Self-Quiz

I. False ($\sqrt{t^2} = |t|$) **II.** False **III.** True **IV.** False **V.** (b), (c)

PROBLEMS 2.5

Find the general solution in Problems 1–10 by using the method given in Example 2. When an initial condition is given, find the particular solution that satisfies that condition and sketch its graph.

1. $xy' - y = \sqrt{xy}$

2. $\dfrac{dx}{dt} = \dfrac{x}{t} + \cosh\dfrac{x}{t}$

3. $x\dfrac{dy}{dx} = xe^{y/x} + y,\ y(1) = 0$

4. $2xyy' = x^2 + y^2,\ y(-1) = 1$

5. $y' = \dfrac{y^2 + xy}{x^2},\ y(1) = 1$

6. $(x + v)\dfrac{dx}{dv} = x$

7. $\dfrac{dy}{dx} = \dfrac{2xy - y^2}{2xy - x^2},\ y(1) = 2$

8. $3xyy' + x^2 + y^2 = 0$

9. $xy' - y = \sqrt{x^2 + y^2}$

10. $\dfrac{dy}{dx} = \dfrac{x^2y + 2xy^2 - y^3}{2y^3 - xy^2 + x^3}$

Use the method given in Example 4 to solve Problems 11–16.

11. $2y' = x^2 + 4xy + 4y^2 + 3$

12. $9y' + (x + y - 1)^2 = 0$

13. $(x + y)y' = 2x + 2y - 3$

14. $y' + 1 = \sqrt{x + y + 2},\ y(0) = 2$

15. $y' + \sin^2(x + y) = 0$

16. $y' = \dfrac{e^y}{e^x} - 1$

Find a substitution that provides a solution to each of Problems 17–22.

17. $xy' + y = (xy)^3$

18. $xy' = \sqrt{1 - x^2y^2} - y$

19. $xy' = e^{xy} - y,\ y(1) = 1$

20. $y' = y^2e^x - y$

21. $y' = (x + y)\ln(x + y) - 1$

22. $y' = y(\ln y + x^2 - 2x)$

23. Show, if $ad - bc \neq 0$, that there are constants h and k such that the substitutions $x = u + h$ and $y = v + k$ convert the quotient $\dfrac{ax + by + m}{cx + dy + n}$ into the quotient $\dfrac{au + bv}{cu + dv}$.

Use Problem 23 and the method given in Example 2 to find a solution for each of Problems 24–29.

24. $\dfrac{dy}{dx} = \dfrac{x - y - 5}{x + y - 1}$

25. $y' = \dfrac{x + 2y + 2}{y - 2x}$

26. $y' = \dfrac{y - x - 1}{y + x}$

27. $\dfrac{dx}{dt} = \dfrac{x + 1}{x + t}$

28. $\dfrac{dy}{dx} = \dfrac{2x - 3y + 4}{4x + y - 6}$

29. $\dfrac{dy}{dx} = \dfrac{3x - y - 3}{x + y - 5}$

30. Solve the equation

$$\frac{dy}{dx} = \frac{1 - xy^2}{2x^2y}$$

by making the substitution $v = y/x^n$ for an appropriate value of n.

31. Use the method of Problem 30 to solve

$$\frac{dx}{dt} = \frac{x - tx^2}{t + xt^2}$$

Another equation that can be solved directly is **Clairaut's* equation:**

$$y = xy' + f(y'). \qquad \textbf{(i)}$$

We can differentiate both sides with respect to x to obtain

$$y' = y' + xy'' + f'(y')y'',$$

where the last term is a result of the chain rule. Canceling like terms, we get

$$[x + f'(y')]y'' = 0.$$

Since one of the factors must vanish, two different solutions arise:

(a) If $y'' = 0$, then $y' = c$, and substituting this value into **(i)** produces the *general solution*

$$y = cx + f(c),$$

which is a collection of straight lines.

(b) If $x + f'(y') = 0$, then $x = -f'(y')$, and **(i)** may be rewritten as

$$y = f(y') - y'f'(y').$$

Here x and y are both expressed in terms of functions of y', so if we substitute $y' = t$, we obtain the parametrized curve

$$x = -f'(t), \qquad y = f(t) - tf'(t).$$

This curve is also a solution, called the **singular solution,** to Clairaut's equation. For example, the equation

$$y = xy' - e^{y'}$$

has the general solution $y = cx - e^c$ and the parametrized curve

$$x = e^t, \qquad y = e^t(1 - t),$$

as a singular solution. Generally we can proceed no further, but in this case we can eliminate the parameter t by letting $t = \ln x$, so that the singular solution is given by

$$y = x(1 - \ln x), \qquad x > 0.$$

In problems 32–36 find the general and singular solutions for each Clairaut equation.

32. $y = xy' + \ln y'$

33. $y = xy' + (y')^3$

34. $y = xy' - \sqrt{y'}$

35. $y = x\dfrac{dy}{dx} + \dfrac{1}{4}\left(\dfrac{dy}{dx}\right)^4$

36. $(y - xy')^2 - (y')^2 = 1$

2.6 COMPARTMENTAL ANALYSIS

A complicated physical or biological process can often be divided into several distinct stages. The entire process can then be described by the interactions between the individual stages. Each such stage is called a **compartment** or pool, and the contents of each compartment are assumed to be well mixed. Material from one

*Named for the French mathematician Alexis Claude Clairaut (1713–1765) (pronounced Claire-Ó). Clairaut, a mathematical prodigy, wrote a well-received paper on third-order curves at the age of eleven. This paper helped him win a seat in the French Academy of Sciences at the ineligible age of eighteen. In 1736 he traveled to Lapland with Pierre de Maupertius to measure the length of a degree of one of the earth's meridians. This measurement helped to confirm a belief of Newton and Huygens that the earth was flattened at the poles. Clairaut made many other contributions to mathematics and physics. Among them was his computation of the 1759 return of Halley's comet with an error of about a month.

compartment is transferred to another and is immediately incorporated into the latter. Because of the name we have given to the stages, the entire process is called a **compartmental system.*** An **open** system is one in which there are inputs to or outputs from the system through one or more compartments. A system that is not open is said to be **closed.**

In this section we investigate only the simplest such system: the one-compartment system. Additional work on more complicated systems will be found in later chapters.

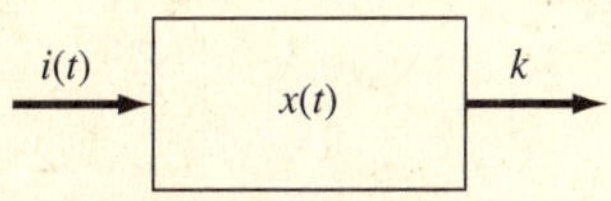

FIGURE 2.20
A one-compartment system

Figure 2.20 illustrates a one-compartment system consisting of a quantity $x(t)$ of material in the compartment, an input rate $i(t)$ at which material is being introduced to the system, and a **fractional transfer coefficient** k indicating the fraction of the material in the compartment that is being removed from the system per unit time. It is clear that the rate at which the quantity x is changing depends on the difference between the input and output at any time t, leading to the differential equation

$$\frac{dx}{dt} = i(t) - kx(t). \tag{1}$$

As we saw in Section 2.3 [Equation **(11),** p. 56], this linear equation has the solution

$$x(t) = e^{-kt}\left(\int i(t)e^{kt}\,dt + c\right). \tag{2}$$

This simple model applies to many different problems, as we illustrate below.

EXAMPLE 1 ▶ **Determining salt concentrations**

Consider a tank holding 100 gallons of water in which are dissolved 50 pounds of salt. Suppose that 2 gallons of brine, each containing 3 pounds of dissolved salt, run into the tank per minute, and the mixture, kept uniform by high-speed stirring, runs out of the tank at the rate of 2 gallons per minute. Find the amount of salt in the tank at any time t (see Figure 2.21).

Solution Let $x(t)$ be the number of pounds of salt at the end of t minutes. Since each gallon of brine that enters the compartment (tank) contains 3 pounds of salt, we know that $i(t) = 6$. On the other hand, $k = \frac{2}{100}$, since 2 of the 100 gallons in the tank are being removed each minute. Thus Equation **(1)** becomes

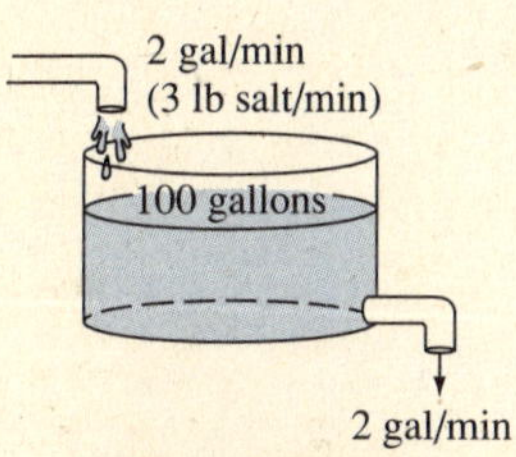

FIGURE 2.21
The tank always holds 100 gallons of brine

$$\overset{\text{input}\ \ \text{output}}{\frac{dx}{dt} = 6 - \frac{2}{100}x,}$$

or

$$\frac{dx}{dt} + \frac{1}{50}x = 6. \tag{3}$$

---

* This name is frequently used in mathematical biology. Engineers refer to such systems as **block diagrams.**

Multiplying both sides by the integrating factor $e^{t/50}$, we get

$$(e^{t/50}x)' = 6e^{t/50},$$

which has the solution

$$\begin{aligned} x(t) &= e^{-t/50}\left(6\int e^{t/50}\,dt + c\right), \\ &= 300 + ce^{-t/50}. \end{aligned}$$

At $t = 0$ we have

$$50 = x(0) = 300 + c,$$

so that

$$x(t) = 300 - 250e^{-t/50}.$$

Observe that x increases and the ratio of salt to water in the tank approaches the ratio of salt to water in the input stream (3 lb/gal) as time increases. ◀

The fractional transfer coefficient k may be a function of time, as we see in the following example.

EXAMPLE 2 ▶ **Determining salt concentrations when volume is not constant**
Suppose that, in Example 1, 3 gallons of brine, each containing 1 pound of salt, run into the tank each minute, and that all other facts are the same. Now $i(t) = 3$, but since the quantity of brine in the tank increases with time, the fraction that is being transferred is $k = 2/(100 + t)$. The numerator of k is the number of gallons being removed, and $100 + t$ is the number of gallons in the tank at time t. The equation describing the system is

$$\frac{dx}{dt} = 3 - \frac{2x}{100 + t}, \tag{4}$$

or

$$\frac{dx}{dt} + \frac{2x}{100 + t} = 3.$$

The integrating factor in this case is

$$e^{\int 2\,dt/(100+t)} = e^{\ln(100+t)^2} = (100 + t)^2,$$

so we have, successively,

$$\begin{aligned} [(100 + t)^2x]' &= 3(100 + t)^2, \\ (100 + t)^2x &= (100 + t)^3 + c, \\ x(t) &= (100 + t) + c(100 + t)^{-2}. \end{aligned}$$

Setting $t = 0$, we find that $c = -50(100)^2$, so

$$x(t) = 100 + t - 50\left(1 + \frac{t}{100}\right)^{-2}.$$

After 100 minutes, for example, we have

$$x(100) = 200 - \frac{50}{4} = 187.5 \text{ pounds}$$

of salt in the tank. ◀

The input function $i(t)$ may depend not only on time but also on the quantity present.

EXAMPLE 3 ▶ **Modelling a biological process**

Systems with periodic inputs and fractional transfer coefficients often occur in biological processes because of diurnal periods of activity. For example, ACTH (adrenocorticotropic hormone) secretion by the anterior pituitary follows a 24-hour cycle that drives the secretion of adrenal steroids in such a way that the levels of these steroids in the blood plasma peak near 8:00 A.M. and are at a minimum near 8:00 P.M. Let $k(t) = A + B \sin \omega t$, with $A > B$, in Equation **(1)**, which leads to

$$\frac{dx}{dt} = i(t) - (A + B \sin \omega t)x. \tag{5}$$

where $x(t)$ denotes that amount (in mg) of ACTH secreted at time t. Since

$$\int (A + B \sin \omega t)\, dt = At - \frac{B}{\omega} \cos \omega t + c,$$

we may use the integrating factor $e^{At+(B/\omega)(1-\cos \omega t)}$ on both sides of Equation **(5)**:

$$\begin{aligned}\frac{d}{dt}[e^{At+(B/\omega)(1-\cos wt)}x(t)] &= e^{At+(B/\omega)(1-\cos \omega t)}[x' + (A + B \sin \omega t)x] \\ &= e^{At+(B/w)(1-\cos \omega t)}i(t).\end{aligned} \tag{6}$$

Integrating both sides of Equation **(6)** from 0 to t, we have

$$e^{At+(B/\omega)(1-\cos \omega t)}x(t)\Big|_0^t = \int_0^t i(t)e^{At+(B/\omega)(1-\cos \omega t)}\, dt,$$

or

$$x(t) = e^{-At-(B/\omega)(1-\cos \omega t)}\left(x(0) + \int_0^t i(t)e^{At+(B/\omega)(1-\cos \omega t)}\, dt\right). \tag{7}$$

Since $1 - \cos \omega t = 2 \sin^2 \frac{\omega t}{2}$, we can write Equation **(7)** as

$$x(t) = e^{-At-2B\sin^2(\omega t/2)/\omega}\left(x(0) + \int_0^t i(t)e^{At+2B\sin^2(\omega t/2)/\omega}\, dt\right). \tag{8}$$

If $i(t) \equiv 0$, then $x(t)$ behaves as shown in Figure 2.22 where $x(0)e^{-At}$ is an upper bound, and the factor $e^{-2B\sin^2(\omega t/2)/\omega}$ oscillates between $e^{-2B/\omega}$ and 1. ◀

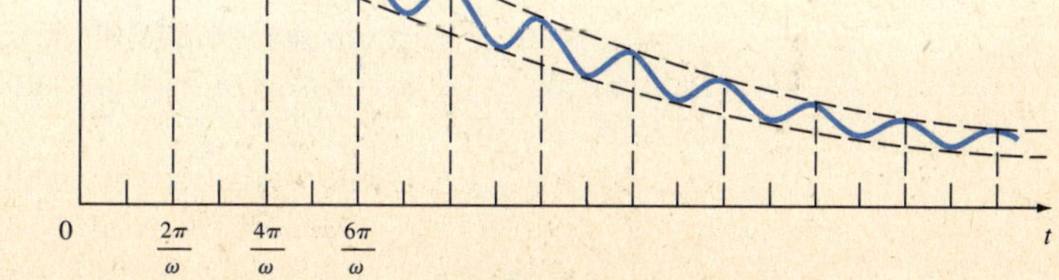

FIGURE 2.22
Graph of $x(t) = x(0)e^{-At-2B\sin^2(\omega t/2)/\omega}$

PROBLEMS 2.6

1. Carbon 14 (^{14}C) has a half-life of 5730 years and is uniformly distributed in the atmosphere in the form of carbon dioxide. Living plants absorb carbon dioxide and maintain a fixed ratio of ^{14}C to the stable element ^{12}C. At death, the disintegration of ^{14}C changes this ratio. Compare the concentrations of ^{14}C in two identical pieces of wood, one of them freshly cut, the other 2000 years old.

2. Radioactive iodine ^{131}I is often used as a tracer in medicine. Suppose that a given dose Q_0 is injected into the bloodstream at time $t = 0$ and is evenly distributed in the entire bloodstream before any loss occurs. If the daily removal rate of the iodine is k_1 percent by the kidney and k_2 percent by the thyroid gland, what percentage of the initial amount remains in the blood after one day?

3. Suppose that an infected individual is introduced into a population of size N, all of whom are susceptible to the disease. If we assume that the rate of infection is proportional to the product of the numbers of infectives and susceptibles present, what is the number of infectives at any time t? Call k the **specific infection rate.**

4. A tank initially contains 100 liters of fresh water. Brine containing 20 grams per liter of salt flows into the tank at the rate of 4 liters per minute, and the mixture, kept uniform by stirring, runs out at the same rate. How long does it take for the quantity of salt in the tank to become 1 kilogram?

5. Given the same data as in Problem 4, determine how long it takes for the quantity of salt in the tank to increase from 1 to 1.5 kilograms.

6. A tank contains 100 gallons of fresh water. Brine containing 2 pounds per gallon of salt runs into the tank at the rate of 4 gallons per minute, and the mixture, kept uniform by stirring, runs out at the rate of 2 gallons per minute. Find (a) the amount of salt present when the tank has 120 gallons of brine; (b) the concentration of salt in the tank at the end of 20 minutes.

7. A tank contains 50 liters of water. Brine containing x grams per liter of salt enters the tank at the rate of 1.5 liters per minute. The mixture, thoroughly stirred, leaves the tank at the rate of 1 liter per minute. If the concentration is to be 20 grams per liter at the end of 20 minutes, what is the value of x?

8. A tank holds 500 gallons of brine. Brine containing 2 pounds per gallon of salt flows into the tank at the rate of 5 gallons per minute, and the mixture, kept uniform, flows out at the rate of 10 gallons per minute. If the maximum amount of salt is found in the tank at the end of 20 minutes, what was the initial salt content of the tank?

9. Phosphate excretion in human metabolism is at a minimum at 3:00 A.M. and rises to a peak at 6:00 P.M. If the rate of excretion is

$$\frac{1}{3} - \frac{1}{6}\cos\frac{\pi}{12}(t - 6)$$

grams per hour at time t hours $(0 \le t \le 24)$, the body contains 400 grams of phosphate, and the patient is only allowed to drink water, what is the amount of phosphate in the patient's body at all times t for $0 \le t \le 24$?

10. Suppose that in Problem 9 the patient is allowed three meals during the day in such a way that the body takes in phosphate at a rate given by

$$i(t) = \begin{cases} \frac{1}{3}\text{ g/hr}, & 8 \le t \le 16, \\ 0\text{ g/hr}, & \text{otherwise.} \end{cases}$$

Obtain a formula for the amount of phosphate in the patient's body at all times t. When is it at a maximum?

11. Given a one-compartment system with k constant and $i(t) = A + B\sin\omega t$, $A > B$, find a solution of the system. How does it differ from that of the system in which the input is constant and the fractional transfer coefficient is periodic [see Equation **(8)**]?

2.7 SIMPLE ELECTRIC CIRCUITS

In this section we consider simple electric circuits containing a resistor and an inductor or capacitor in series with a source of electromotive force (emf). Such circuits are shown in Figure 2.23. Their action can be understood very easily without any special knowledge of electricity.

- An electromotive force E (volts), usually a battery or generator, drives an electric charge Q (coulombs) and produces a current I (amperes). The current is defined as the rate of flow of the charge, and we can write

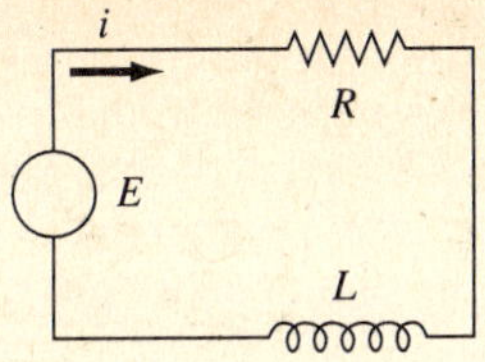

(a) An RL-circuit

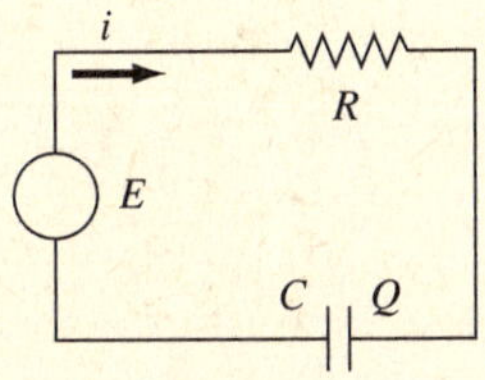

(b) An RC-circuit

FIGURE 2.23
Two electric circuits

$$I = \frac{dQ}{dt}. \tag{1}$$

- A resistor of resistance R (ohms) is a component of the circuit that opposes the current, dissipating the energy in the form of heat. It produces a drop in voltage given by **Ohm's law:**

$$E_R = RI. \tag{2}$$

- An inductor of inductance L (henrys) opposes any change in current by producing a voltage drop of

$$E_L = L\frac{dI}{dt}. \tag{3}$$

- A capacitor of capacitance C (farads) stores charge. In so doing, it resists the flow of further charge, causing a drop in the voltage of

$$E_C = \frac{Q}{C}. \tag{4}$$

Note that Equations **(2)**, **(3)** and **(4)**, and the "law" given below are mathematical models.

The quantities R, L, and C are usually constants associated with the particular component in the circuit; E may be a constant or a function of time. The fundamental principle guiding such circuits is given by

Kirchhoff's voltage law

The algebraic sum of all voltage drops around a closed circuit is zero.

In the circuit of Figure 2.23(a), the resistor and the inductor cause voltage drops of E_R and E_L, respectively. The emf however, *provides* a voltage of E (that is, a voltage drop of $-E$). Thus Kirchhoff's voltage law yields

$$E_R + E_L - E = 0.$$

Transposing E to the other side of the equation and using Equations **(2)** and **(3)** to replace E_R and E_L, we have

$$L\frac{dI}{dt} + RI = E. \tag{5}$$

The following two examples illustrate the use of Equation **(5)** in analyzing the circuit shown in Figure 2.23(a).

EXAMPLE 1* ▶ **Solving an RL-circuit with a constant voltage**

An inductance of 2 henrys and a resistance of 10 ohms are connected in series with an emf of 100 volts. If the current is zero when $t = 0$, what is the current at the end of 0.1 second?

Solution Since $L = 2$, $R = 10$, and $E = 100$, Equation **(5)** and the initial current yield the initial-value problem:

$$2\frac{dI}{dt} + 10I = 100, \qquad I(0) = 0. \tag{6}$$

* This example is typical in electrical engineering. It is, for example, very similar to Exercise 12a in C. A. Desoer and E. S. Kuh, *Basic Circuit Theory* (New York: McGraw Hill, 1969), p. 169.

Dividing both sides of Equation **(6)** by 2, we note that the resulting linear first-order equation has e^{5t} as an integrating factor; that is,

$$\frac{d}{dt}(e^{5t}I) = e^{5t}\left(\frac{dI}{dt} + 5I\right) = 50e^{5t}. \tag{7}$$

Integrating both ends of Equation **(7),** we get

$$e^{5t}I(t) = 10e^{5t} + c,$$

or

$$I(t) = 10 + ce^{-5t}. \tag{8}$$

Setting $t = 0$ in Equation **(8)** and using the initial condition $I(0) = 0$, we have

$$0 = I(0) = 10 + c,$$

which implies that $c = -10$. Substituting this value into Equation **(8)**, we obtain an equation for the current at all times t:

$$I(t) = 10(1 - e^{-5t}).$$

Thus, when $t = 0.1$, we have

$$I(0.1) = 10(1 - e^{-0.5}) \approx 3.93 \text{ amperes.}$$ ◀

EXAMPLE 2 ▶ **Solving an RL-circuit with a periodic voltage**

Suppose that the emf $E = 100 \sin 60t$ volts but all other values remain the same as those given in Example 1. In this case Equation **(5)** yields

$$2\frac{dI}{dt} + 10I = 100 \sin 60t, \qquad I(0) = 0. \tag{9}$$

Again dividing by 2 and multiplying both sides by the integrating factor e^{5t}, we have

$$\frac{d}{dt}(e^{5t}I) = e^{5t}\left(\frac{dI}{dt} + 5I\right) = 50e^{5t} \sin 60t. \tag{10}$$

Integrating both ends of Equation **(10)** and using Formula 168 of the integral table in Appendix 1, we obtain

$$\begin{aligned} I(t) &= e^{-5t}\left[50 \int (\sin 60t)\, e^{5t}\, dt + c\right] \\ &= e^{-5t}\left[50e^{5t}\left(\frac{5 \sin 60t - 60 \cos 60t}{3625}\right) + c\right] \\ &= \frac{2 \sin 60t - 24 \cos 60t}{29} + ce^{-5t}. \end{aligned}$$

Setting $t = 0$, we find that $c = \frac{24}{29}$ and

$$I(t) = \frac{2 \sin 60t - 24 \cos 60t}{29} + \frac{24}{29}e^{-5t}.$$

Thus

$$I(0.1) = \frac{2 \sin 6 - 24 \cos 6}{29} + \frac{24}{29}e^{-0.5} \approx -0.31 \text{ amperes.}$$ ◀

transient and steady-state current

In the previous example the term $24e^{-5t}/29$ is called the **transient current** because it approaches zero as t increases without bound. The other part of the current, $(2 \sin 60t - 24 \cos 60t)/29$, is called the **steady-state current**.

For the circuit in Figure 2.23(b) we have $E_R + E_C - E = 0$, or

$$RI + \frac{Q}{C} = E.$$

Using the fact that $I = dQ/dt$, we obtain the linear first-order equation

$$R\frac{dQ}{dt} + \frac{Q}{C} = E. \tag{11}$$

The next example illustrates how to use Equation **(11)**.

EXAMPLE 3 ▶ **Solving an RC-circuit with a constant voltage**

If a resistance of 2000 ohms and a capacitance of 5×10^{-6} farad are connected in series with an emf of 100 volts, what is the current at $t = 0.1$ second if $I(0) = 0.01$ ampere?

Solution Setting $R = 2000$, $C = 5 \times 10^{-6}$, and $E = 100$ in Equation **(11)**, we have

$$2000\left(\frac{dQ}{dt} + 100Q\right) = 100,$$

or

$$\frac{dQ}{dt} + 100Q = \frac{1}{20}, \tag{12}$$

from which we can determine $Q(0)$ since

$$\frac{1}{20} = Q'(0) + 100Q(0) = I(0) + 100Q(0).$$

Thus

$$\begin{aligned} Q(0) &= \frac{1}{100}\left[\frac{1}{20} - I(0)\right] = \frac{1}{100}\left(\frac{1}{20} - \frac{1}{100}\right) \\ &= \frac{1}{100}\left(\frac{4}{100}\right) = 4 \times 10^{-4} \text{ coulombs.} \end{aligned} \tag{13}$$

Multiplying both sides of Equation **(12)** by the integrating factor e^{100t}, we get

$$\frac{d}{dt}(e^{100t}Q) = \frac{e^{100t}}{20},$$

and integrating this equation yields

$$e^{100t}Q = \frac{e^{100t}}{2000} + c.$$

Dividing both sides by e^{100t} gives us

$$Q(t) = \frac{1}{2000} + ce^{-100t},$$

and setting $t = 0$, we find that

$$c = Q(0) - \frac{1}{2000} = (4 \times 10^{-4}) - (5 \times 10^{-4}) = -10^{-4}.$$

Thus the charge at all times t is

$$Q(t) = \frac{5 - e^{-100t}}{10^4},$$

and the current is

$$I(t) = Q'(t) = \frac{1}{100}e^{-100t}.$$

Thus $I(0.1) = 10^{-2}e^{-10} \approx 4.54 \times 10^{-7}$ amperes. ◀

Induction

Induction has been avoided in long-distance lines because it slows transmission. The delays are caused when the magnetic field induced by a current stores and then discharges the energy in a signal. However, adding inductance to a circuit allows it to transmit the signal without distortion, a very useful consideration when we realize that each line may carry dozens of messages simultaneously. Distortion affects the signal by permitting higher-frequency components to outpace slower ones; inductance allows the slower ones to catch up, as shown in Figure 2.24.

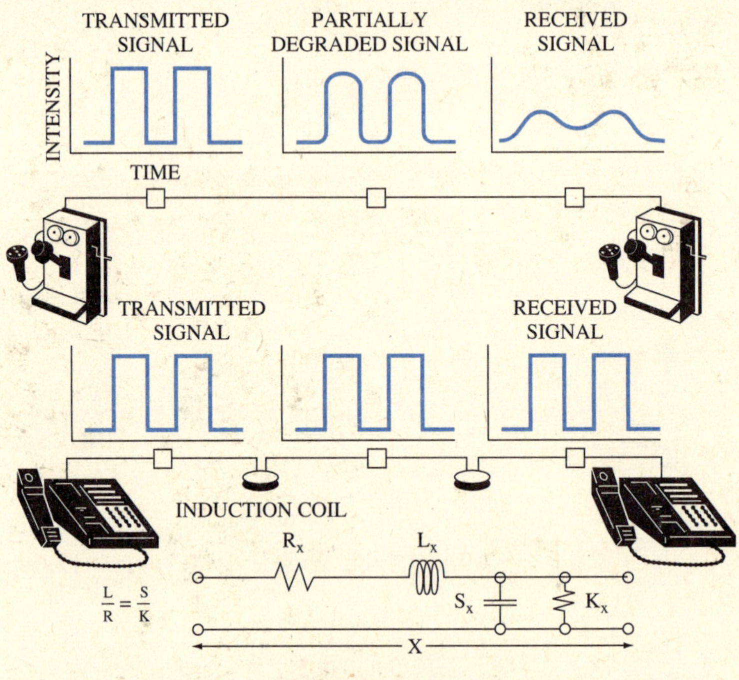

FIGURE 2.24
Distortion destroys a signal by causing its higher-frequency components to outpace its lower-frequency ones, turning a sharp pulse into a blur (*top*). Distortionless transmission (*middle*) incorporates induction loading to balance the equation (*bottom, with circuit diagram*) of inductance *L*, linear resistance *R*, capacitance *S*, and leakage resistance *K*, caused by leakage between a circuit's forward and return legs.
Source: Scientific American, June 1990; p. 125.

PROBLEMS 2.7

In Problems 1–5 assume that the RL circuit shown in Figure 2.23(a) has the given resistance (ohms, Ω), inductance (henrys, H), emf (volts, V), and initial current (amperes, amp). Find an expression for the current at all times t and calculate the current after 1 second.

1. $R = 10\ \Omega$, $L = 1\ H$, $E = 12$ V, $I(0) = 0$ amp
2. $R = 8\ \Omega$, $L = 1$ H, $E = 6$ V, $I(0) = 1$ amp
3. $R = 50\ \Omega$, $L = 2$ H, $E = 100$ V, $I(0) = 0$ amp
4. $R = 10\ \Omega$, $L = 5$ H, $E = 10 \sin t$ V, $I(0) = 1$ amp
5. $R = 10\ \Omega$, $L = 10$ H, $E = e^t$ V, $I(0) = 0$ amp

In Problems 6–10 use the given resistance, capacitance (farads, f), emf, and initial charge (coulombs) in the RC *circuit shown in Figure 2.23(b). Find an expression for the charge at all time* t.

6. $R = 1\ \Omega$, $C = 1$ f, $E = 12$ V, $Q(0) = 0$ coulomb

7. $R = 10\ \Omega$, $C = 0.001$ f, $E = 10 \cos 60t$ V, $Q(0) = 0$ coulomb

8. $R = 1\ \Omega$, $C = 0.01$ f, $E = \sin 60t$ V, $Q(0) = 0$ coulomb

9. $R = 100\ \Omega$, $C = 10^{-4}$ f, $E = 100$ V, $Q(0) = 1$ coulomb

10. $R = 200\ \Omega$, $C = 5 \times 10^{-5}$ f, $E = 1000$ V, $Q(0) = 1$ coulomb

11. The capacitor C in the circuit illustrated below is charged to 10 volts when the switch is closed. Obtain a differential equation for the capacitor voltage and find the voltage for all times t given that $R = 1000$ ohms and $C = 10^{-6}$ farad.*

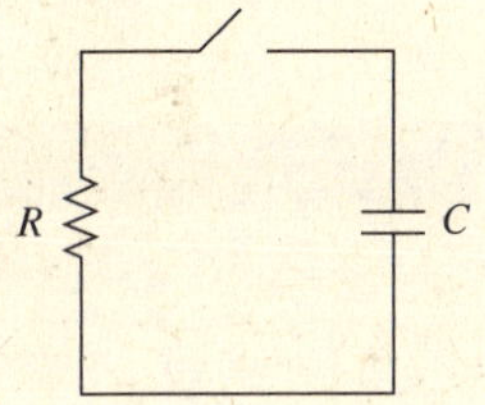

12. An inductance of 1 henry and a resistance of 2 ohms are connected in series with a battery of $6e^{-0.0001t}$ volt. No current is lowing initially. When does the current measure 0.5 ampere?

13. A variable resistance $R = 1/(5 + t)$ ohms and a capacitance of 5×10^{-6} farad are connected in series with an emf of 100 volts. If $Q(0) = 0$, what is the charge on the capacitor after 1 minute?

14. In the *RC* circuit [Figure 2.23(b)] with constant voltage E, how long will it take the current to decrease to one-half its original value?

15. Suppose that the voltage in an *RC* circuit is $E(t) = E_0 \cos wt$, where $2\pi/\omega$ is the period of the cycle. Assuming that the initial charge is zero, what are the charge and current as functions of R, C, w, and t?

16. Show that the current in Problem 15 consists of two parts: a steady-state term that has a period of $\frac{2\pi}{\omega}$ and a transient term that tends to zero as t increases.

17. Show that if R in Problem 16 is small, then the transient term can be quite large for small values of t. [This is why fuses can blow when a switch is flipped.]

18. Find the steady-state current, given that a resistance of 2000 ohms and a capacitance of 3×10^{-6} farad are connected in series with an alternating emf of $120 \cos 2t$ volts.

19. Find an expression for the current of a series *RL* circuit, where $R = 100$ ohms, $L = 2$ henrys, $I(0) = 0$ amp, and the emf voltage satisfies

$$E = \begin{cases} 6, & \text{for } 0 \le t \le 10, \\ 7 - e^{10-t}, & \text{for } \quad t \ge 10. \end{cases}$$

20. Repeat Problem 19 with $R = 100/(1 + t)$, all other values remaining the same.

2.8 SUCCESSIVE APPROXIMATIONS (OPTIONAL)

In Sections 2.1–2.5 we discussed a number of techniques that can be used to solve first-order differential equations or initial-value problems. In this section we describe an iterative technique, due to Picard, that provides an alternate approach to solving initial-value problems. The main use of this method, however, is theoretical: it forms the basis for the proof of Picard's theorem in Section 1.2; the existence-uniqueness theorem. The technique that we develop here will be used in Appendix 3 to prove Theorem 1.2.1.

Consider the initial-value problem

$$y' = f(x, y), \qquad y(x_0) = y_0. \tag{1}$$

If we integrate both sides of the differential equation in **(1)** from x_0 to x with respect to x, we have

$$\int_{x_0}^{x} y'(x)\,dx = \int_{x_0}^{x} f(x, y(x))\,dx,$$

* This example is Exercise 5.28 in Shearer et al., *System Dynamics* (Reading, Mass.: Addison-Wesley, 1971), p. 141. Reprinted with permission of Addison-Wesley Publishing Co.

or

$$y(x) - y(x_0) = \int_{x_0}^{x} f(t, y(t))\, dt. \tag{2}$$

(We have changed the variable of integration on the right-hand side to t to avoid confusion.)

Rewriting Equation **(2)**, we get (since $y(x_0) = y_0$)

$$y(x) = y_0 + \int_{x_0}^{x} f(t, y(t))\, dt. \tag{3}$$

Equation **(3)** is an alternative way of writing initial-value problem (1): note that if we set $x = x_0$ in Equation **(3)**, then $y(x_0) = y_0$, since the integral is zero. Furthermore, if we differentiate both sides of Equation **(3)**, we obtain the differential equation

$$y'(x) = f(x, y(x)). \tag{4}$$

Thus Equations **(1)** and **(3)** are equivalent ways of writing the same initial-value problem.

We now define a sequence of functions $\{y_n(x)\}$, called **Picard* iterations**, by successive formulas:

Definition Picard iterations

$$\begin{aligned} y_0(x) &= y_0 \\ y_1(x) &= y_0 + \int_{x_0}^{x} f[t, y_0(t)]\, dt \\ y_2(x) &= y_0 + \int_{x_0}^{x} f[t, y_1(t)]\, dt \\ &\vdots \\ y_n(x) &= y_0 + \int_{x_0}^{x} f[t, y_{n-1}(t)]\, dt \end{aligned} \tag{5}$$

We will show in Appendix 3 that under the conditions of the existence-uniqueness Theorem 1.2.1, the Picard iterations defined by Equation **(5)** converge uniformly to a solution of Equation **(3).** We illustrate the process of this iteration by a simple example.

EXAMPLE 1 ▶ **Applying Picard iterations to an initial-value problem**

Consider the initial-value problem

$$y'(x) = y(x), \qquad y(0) = 1. \tag{6}$$

As we know, Equation **(6)** has the unique solution $y(x) = e^x$. In this case, the function $f(x, y)$ in Equation **(1)** is given by $f(x, y(x)) = y(x)$, so the Picard iterations defined by Equation **(5)** yield, successively,

*Charles Emile Picard (1856–1941), one of the most eminent French mathematicians of the past century, made several outstanding contributions to mathematical analysis. Picard published the results we discuss in this section in 1890 and 1893: *Journal de Mathematique* (4) 6 (1890): 145–210; 9(1893): 217–271.

$$\begin{aligned}
y_0(x) &= y_0 = 1,\\
y_1(x) &= 1 + \int_0^x (1)\,dt = 1 + x,\\
y_2(x) &= 1 + \int_0^x (1 + t)\,dt = 1 + x + \frac{x^2}{2},\\
y_3(x) &= 1 + \int_0^x \left(1 + t + \frac{t^2}{2}\right) dt = 1 + x + \frac{x^2}{2!} + \frac{x^3}{3!},
\end{aligned}$$

and clearly,

$$y_n(x) = 1 + x + \frac{x^2}{2!} + \dots + \frac{x^n}{n!} = \sum_{k=0}^{n} \frac{x^k}{k!}.$$

Hence

$$\lim_{n\to\infty} y_n(x) = \sum_{k=0}^{\infty} \frac{x^k}{k!} = e^x,$$

since the series is the Maclaurin series of e^x. ◀

You should not be fooled by the relative ease with which we obtained the solution in Example 1. In general, the Picard iterations quickly give rise to formidable integrals that are often difficult or impossible to integrate.

EXAMPLE 2 ▶ **The Picard iterations may be impossible to obtain explicitly**

If we apply Picard iterations to the initial-value problem

$$y' = e^{x^2} y, \qquad y(0) = 1,$$

we get

$$\begin{aligned}
y_0(x) &= y_0 = 1,\\
y_1(x) &= 1 + \int_0^x e^{t^2}\,dt,
\end{aligned}$$

and this integral cannot be written in terms of elementary functions. ◀

We will use Picard iterations in Appendix 3 to prove Theorem 1.2.1. However, we can give a simple existence-uniqueness proof for linear first-order equations without using this powerful tool.

THEOREM 1 ◆ **Uniqueness theorem**

Suppose that $a(x)$ and $f(x)$ are continuous functions. Then the linear initial-value problem

$$\frac{dy}{dx} + a(x)y = f(x), \qquad y(x_0) = y_0 \tag{7}$$

has at most one solution. ◆

Remark

This theorem does not claim that a solution exists. However, a global existence result for this equation has already been proved. (One solution to Equation **(7)** is given by Equation **(2.3.11)** for an appropriate constant C. All integrals in the equation exist because $a(x)$ and $f(x)$ are assumed continuous.)

Proof Suppose that both $y_1(x)$ and $y_2(x)$ satisfy Equation **(7)**. Let $y_3(x) = y_1(x) - y_2(x)$. We must show that $y_3(x) \equiv 0$. But, since y_1 and y_2 satisfy Equation **(7)**,

$$\begin{aligned} y_3' + a(x)y_3 &= (y_1 - y_2)' + a(x)(y_1 - y_2) \\ &= [y_1' + a(x)y_1] - [y_2' + a(x)y_2] = f(x) - f(x) = 0. \end{aligned}$$

Also, $y_3(x_0) = y_1(x_0) - y_2(x_0) = y_0 - y_0 = 0$. Thus y_3 satisfies the initial-value problem

$$\frac{dy}{dx} + a(x)y = 0, \qquad y(x_0) = 0. \tag{8}$$

Multiplying both sides of the differential equation in Equation **(8)** by the integrating factor

$$e^{\int_{x_0}^{x} a(t)\,dt}$$

yields

$$\left[e^{\int_{x_0}^{x} a(t)\,dt} y_3(x)\right]' = e^{\int_{x_0}^{x} a(t)\,dt}[y_3' + a(x)\,y_3] = 0,$$

so

$$e^{\int_{x_0}^{x} a(t)\,dt} y_3(x) = C \text{ (a constant)}.$$

Thus

$$y_3(x) = Ce^{-\int_{x_0}^{x} a(t)\,dt}. \tag{9}$$

But by the initial condition in Equation **(8)**,

$$0 = y_3(x_0) = Ce^{-\int_{x_0}^{x_0} a(t)\,dt} = C, \tag{10}$$

so that $C = 0$. Substituting this value of C in Equation **(9)** proves that $y_3(x) \equiv 0$ is the only solution to Equation **(8)**. This completes the proof. ◆

PROBLEMS 2.8

Use Picard iterations to solve the following initial-value problems (in closed form).

1. $y' = -y,\ y(0) = 1$ **2.** $y' = 2y,\ y(0) = 5$ **3.** $y' = x + y,\ y(0) = 0$ **4.** $y' = 2xy,\ y(0) = 1$

2.9 FIRST-ORDER DIFFERENCE EQUATIONS (OPTIONAL)

Differential equations arise in physics and biology when we consider the instantaneous rate of change of one variable with respect to another. However, in many situations it may be more meaningful to study a process in a sequence of well-defined stages or steps. For example, the available data in some experiment may be organized in fixed increments, such as the length to the nearest inch, age in weeks, or populations measured every year. Problems of this nature are often posed as **difference equations**, that is, as equations involving the differences between values of the variable at different stages.

EXAMPLE 1 ▶ **Determining nitrogen concentration**

A patient in a hospital is suddenly administered oxygen. Let V be the volume of gas contained in the lungs after inspiration and V_D the amount present at the end of expiration (commonly called the *dead space*). Assuming uniform and complete mixing of the gases in the lungs, what is the concentration of nitrogen in the lungs at the end of the nth inspiration?

Solution The amount of nitrogen in the lungs at the end of the nth inspiration must equal the amount in the dead space at the end of the $(n - 1)$st expiration. If x_n is the concentration of nitrogen in the lungs at the end of the nth inspiration, then

$$Vx_n = V_D x_{n-1}. \tag{1}$$

Subtracting Vx_{n-1} from both sides of Equation **(1)**, we have

$$V(x_n - x_{n-1}) = (V_D - V)x_{n-1}. \tag{2}$$

The difference $x_n - x_{n-1}$ is the discrete analogue of a derivative, since it measures the change in concentration of nitrogen in the lungs. Thus Equation **(2)** is a discrete version of a first-order differential equation.

If we can find an expression (in n) for the concentrations x_n that satisfies **(1)** for all values of $n = 1, 2, 3, \ldots$, then we say we have a **solution** for difference Equation **(1).** It is easy to solve Equation **(1),** since

$$x_n = \frac{V_D}{V}x_{n-1} = \left(\frac{V_D}{V}\right)^2 x_{n-2} = \ldots = \left(\frac{V_D}{V}\right)^n x_0,$$

where x_0 is the concentration of nitrogen in the air. ◀

If N_1 and N_2 are, respectively, the largest and smallest values of n that occur in the equation, then the **order** of the difference equation is $N_1 - N_2$.

A difference equation is said to be **linear** if it can be written in the form

$$y_{n+k} + a_n y_{n+k-1} + \ldots + b_n y_{n+1} + c_n y_n = f_n,$$

consisting only of sums of the unknown terms $y_{n+k}, y_{n+k-1}, \ldots, y_n$. Here we consider only the first-order linear difference equation

$$y_{n+1} + a_n y_n = f_n.$$

EXAMPLE 2 ▶ **Compound interest**

We can use difference equations to derive the well-known compound interest formula. Let $P_0 = P$ be the *principal,* that is, the initial amount of the investment. If r is the annual rate of interest and is paid annually, then the amount in the account at the end of the first year is $P_1 = P_0 + rP_0 = (1 + r)P_0$. More generally, the amount at year $t + 1$ is based on the amount at year t:

$$P_{t+1} = P_t + rP_t = (1 + r)P_t. \tag{3}$$

Then, using Equation **(3)** repeatedly for different values of t, we have

$$P_t = (1 + r)P_{t-1} = (1 + r)^2 P_{t-2} = \ldots = (1 + r)^t P_0.$$

The formula $P_t = P_0(1 + r)^t$ is the **compound interest formula**. ◀

EXAMPLE 3 ▶ **Linear population growth**

In Section 1.1 we modeled Malthusian population growth by the differential equation $dN/dt = \alpha N$, where N is the population size at time t and α is the relative population growth rate. We assumed in this model that the population was changing *continuously*. However, births and deaths are distinct events, so that continuous change may not be a good assumption if the population being measured is small. Also, we generally do not measure populations continuously: the U.S. Census is taken every ten years. A biologist might measure a bacteria population every hour, but she would not measure it continuously.

Let P_n be the population at the end of the nth time period. We assume the time periods are equal and that the growth is at a constant rate of β. Then the change in population from period n to period $n + 1$ is

$$P_{n+1} - P_n = \beta P_n \quad \text{or} \quad P_{n+1} = (1 + \beta)P_n.$$

Thus,

$$P_n = (1 + \beta)^n P_0.$$

For example, if $\beta = 0.05$ and $P_0 = 1000$, then P_n grows as shown in Table 1, and the population would eventually overwhelm the earth. ◀

TABLE 1

n	$P_n = 1000(1.05)^n$
0	1000
10	1629
20	2653
50	11,467
100	131,501
200	17,292,581
500	3.9323×10^{13}
1000	1.5463×10^{24}

EXAMPLE 4 ▶ **The logistic difference equation and chaos**

To avoid the lack of reality in Example 3 caused by exponential growth, biologists use a discrete version of the logistic equation. They assume there is a **carrying capacity**, K, beyond which there are insufficient resources in the environment for the population N_t to grow. If β is the growth rate of population, the discrete logistic equation is given by:

$$N_{t+1} = rN_t\left(1 - \frac{N_t}{K}\right), \tag{4}$$

where $r = 1 + \beta$. Observe that N_t must never exceed K, because if it did N_{t+1} would be negative, and this would have no biological meaning.

To study this nonlinear first-order difference equation, it is useful to rescale and eliminate one of the parameters. Let $x_t = N_t/K$, that is, let x_t denote the percentage of the carrying capacity present at the end of period t. Making this substitution for all t, Equation **(4)** becomes

$$Kx_{t+1} = rKx_t(1 - x_t)$$

or

$$x_{t+1} = rx_t(1 - x_t). \tag{5}$$

Although Equation **(5)** resembles the continuous logistic equation (see Example 3 in Section 2.2) we shall soon see that it has a very different behavior.

Although it is not possible to solve Equation **(5)**, there is an easy geometric method to find the values $x_1, x_2, \ldots$, once we are given x_0 and r. This procedure

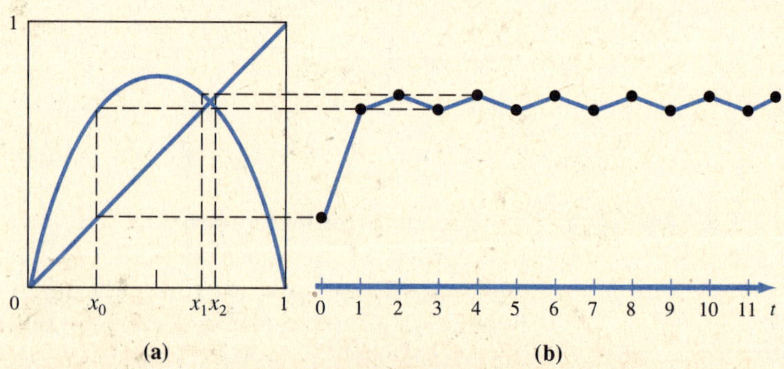

FIGURE 2.25
(a) Cobweb diagram **(b)** Time-sequence

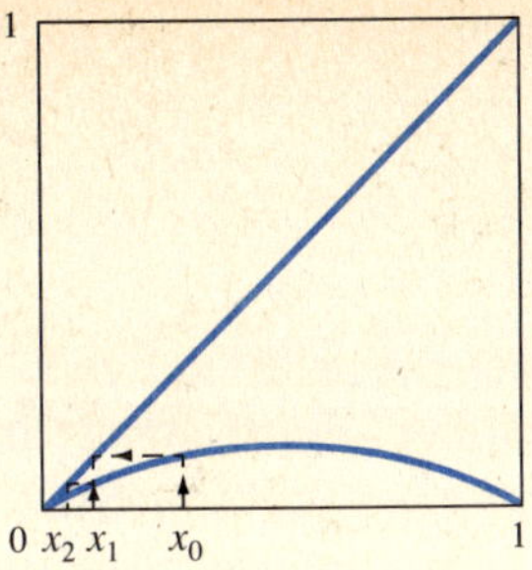

(a)

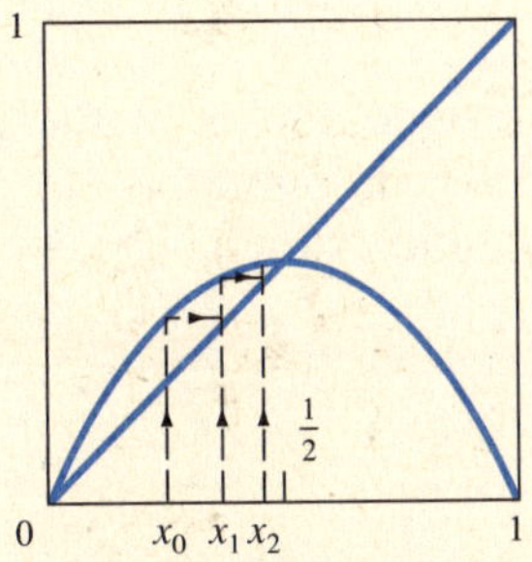

(b)

FIGURE 2.26
Cobweb diagrams for **(a)** $r = 0.5$ **(b)** $r = 2$

is illustrated in Figure 2.25 which consists of two parts: (a) the graph of $f(x) = rx(1 - x)$ and (b) a horizontal time-sequence axis showing the values of x_t, for $t = 0, 1, 2, \ldots$.

To interpret Figure 2.25, start at height x_0 at $t = 0$ on the horizontal time-sequence in part (b). Move *left* horizontally until you reach the 45° line in part (a), and then move down to the abscissa: this is the point x_0. Move vertically upward in part (a) until you reach the curve $f(x)$: This is the value of x_1. Move *right* horizontally until you are directly above $t = 1$ in part (b). Plot this point, and draw the line between $(0, x_0)$ and $(1, x_1)$. Repeat the above process to find $x_2, x_3, \ldots$.

Part (b) of Figure 2.25 is unnecessary, if you are willing to keep all of the information in part (a) of the graph. That is, the values $x_0, x_1, x_2, \ldots$ are all marked on the abscissa, and provide a visual summary of the changes in the population.

Figure 2.26(a) shows the cobweb diagram that is obtained when $r = 0.5$: Note that the values $x_0, x_1, x_2, \ldots$ are converging to 0, so the population is going extinct. Figure 2.26(b) plots $f(x)$ for $r = 2$: Here the population is converging to $\frac{1}{2}$. The graph in Figure 2.25 plots $f(x)$ for $r = 3.4$, and does not converge to any one value; instead it appears to oscillate between two values. An explanation of this behavior would take us far beyond the scope of this text, but leads to a phenomenon known as **chaos.**

To illustrate this phenomenon numerically, we return to Equation **(5)** with $r = \alpha$:

$$x_{t+1} = \alpha x_t(1 - x_t).$$

Using calculus, it is easy to show that $x(1 - x) = x - x^2$ takes its maximum values of $\frac{1}{4}$ when $x = \frac{1}{2}$. Thus, as long as $0 < \alpha < 4$ and $0 < x_0 < 1$, x_n will always be a number in the interval $(0, 1)$.

There is no way to find a formula for x_t in terms of x_0 and α. The best we can do is to iterate to see what happens. We will experiment with different values of the parameter α. In each case, we will assume that the population starts with 30% of its maximum; that is; $x_0 = 0.3$.

TABLE 2
$x_{n+1} = 2x_n(1 - x_n)$

n	x_n
0	0.3
1	0.42
2	0.4872
3	0.49967232
4	0.4999997853
5	0.5
6	0.5
10	0.5

Case 1 $\alpha = 2$

Using a programmable calculator, we obtain the values in Table 2. All numbers are given to ten decimal places—as displayed on the calculator. Since $2(0.5)(1 - 0.5) = 0.5$, we see that $x_5 = 0.5$ and $x_n = 0.5$ for $n > 5$. In fact, when $\alpha = 2$, the population fraction x_n will converge to 0.5 for any x_0 in the interval (0, 1). That is, the population quickly stabilizes at 50% of the carrying capacity.

Let us try another value, say $\alpha = 2.6$. The results are given in Table 3. Now it appears that $x_n \to 0.6153846154$ as $n \to \infty$. Let us try to explain these results. If x_n doesn't change, then

$$x_{t+1} = \alpha x_t(1 - x_t) = x_t.$$

Consider the equation

$$\alpha x(1 - x) = x. \qquad (6)$$

If Equation **(6)** holds for some number $\bar{x}$, then $\bar{x}$ is called a **fixed point** of the function $\alpha x(1 - x)$. We can solve Equation **(6)** as follows:

$$\alpha x(1 - x) = x$$
$$\alpha x - \alpha x^2 = x$$
$$\alpha x^2 = \alpha x - x = x(\alpha - 1).$$

TABLE 3
$x_{n+1} = 2.6x_n(1 - x_n)$

n	x_n
0	0.3
1	0.546
2	0.6444984
3	0.5957125522
4	0.6261816791
5	0.6086032780
10	0.6159010126
15	0.6153444000
20	0.6153877422
30	0.6153846343
40	0.6153846155
50	0.6153846154
100	0.6153846154

TABLE 4
$x_{t+1} = 3.1x_t(1 - x_t)$

t	x_t
0	0.3
1	0.651
2	0.7043169
3	0.6455892736
4	0.7092916666
5	0.6392106947
10	0.7277992157
11	0.6141333037
20	0.7563402704
21	0.5712979639
60	0.7465662424
61	0.5581030517
100	0.7645665200
101	0.5580141252
200	0.7645665200
201	0.5580141252
202	0.7645665200
203	0.5580141252

One solution is $x = 0$, which means extinction. If $x \neq 0$, we can divide by it to obtain

$$\alpha x = \alpha - 1 \quad \text{or} \quad x = \frac{\alpha - 1}{\alpha}.$$

This *seems* to imply that if $1 < \alpha < 4$ and $0 < x_0 < 1$, then the iterates given by Equation **(6)** will converge to the fixed point $\bar{x} = (\alpha - 1)/\alpha$. For example, if $\alpha = 2$, then $(\alpha - 1)/\alpha = 0.5$. If $\alpha = 2.6$, then $(\alpha - 1)/\alpha = \frac{1.6}{2.6} \approx 0.6153846154$, as in Table 3.

If you guessed that the problem was solved, you'd be wrong. Let us try another value, say $\alpha = 3.1$. Then we expect the iterates to converge to $\frac{2.1}{3.1} \approx 0.6774193548$. Starting again at $x_0 = 0.3$, we obtain the results in Table 4.

Now, the iterates are not converging to $(\alpha - 1)/\alpha$ or any one number. Rather, the even-numbered iterates converge to 0.76456652 while the odd-numbered iterates converge to 0.5580141252. This phenomenon is called **bifurcation.** The two numbers are called **limit points** of the sequence $\{x_n\}$.

But this is not the worst of it. If $\alpha = 3.5$, then, again starting at $x_0 = 0.3$, we hit *four* values infinitely often. You should verify on your calculator that for $t \geq 10$,

$$\begin{aligned} x_{4t} &= 0.382819683 \\ x_{4t+1} &= 0.8269407066 \\ x_{4t+2} &= 0.5008842103 \\ x_{4t+3} &= 0.8749972636. \end{aligned}$$

Now the population is returning to the same period in every fourth time period and there are four limit points.

The strange and unexpected behavior exhibited by the simple looking logistic difference equation is one of the earliest studied examples of a phenomenon known as **chaos.** One of the first scientists to discover chaos in the logistic difference equation in the early 1970s was the Australian biologist Robert May. He found that as he increased the parameter α, the bifurcations would continue at an ever increasing rate so that there would be 2, then 4, then 8, 16, and 32, . . . limit points. A further increase in x leads to other even periods that are not powers of 2. Increasing x even more leads to an odd number of limit points in descending order: . . . , 11, 9, 7, 5, 3. A further increase leads to chaos—the iterates x_n bounce around in the interval $(0, 1)$, never being the same value again. It is apparent that even this harmless looking but nonlinear difference equation has a very complicated structure indeed. ◀

Any mathematical explanation of chaos—even in this apparently simple case—would take us far beyond the scope of this text. May described his findings in a paper with George Oster: "Bifurcations and Dynamic Complexity in Simple Ecological Models," *The American Naturalist* 110 (1976), p. 573. The mathematics behind this phenomenon is explained in a paper by James York and Tien-Yien Li: "Period Three Implies Chaos," *American Mathematical Monthly* 82 (1975), pp. 985-92. For a very readable and entertaining history of the study of this and related phenomena, read *Chaos* by James Gleick, Viking Penguin, Inc., New York, 1987.

SELF-QUIZ

I. Which of the following is *not* a linear difference equation?
(a) $y_{n+1} = 2ny_n$ **(b)** $y_{n+1} = y_{n+1} - y_{n-1}$
(c) $y_{n+1} = y_n y_{n+1}$ **(d)** $\dfrac{y_{n+1}}{y_n} = 4, \quad y_0 \neq 0$

II. What is the general solution to $y_{n+1} = 3y_n$?
(a) $y_n = c + 3^n$ **(b)** $y_n = c \cdot 3^n$
(c) $y_n = c + 3n$ **(d)** $y_n = c^n 3^n$

III. The difference equation $y_{n+3} - y_{n+2} = y_{n+1} - y_n$ is of order ______.
(a) 1 **(b)** 2 **(c)** 3 **(d)** 4

Answers to Self-Quiz

I. (c) II. (b) III. (c)

PROBLEMS 2.9

In Problems 1–9 find the general solution of each difference equation and a particular solution when an initial condition is specified.

1. $y_{n+1} - y_n = 2^{-n}$
2. $y_{n+1} = \dfrac{n+5}{n+3}y_n$
3. $y_{n+1} - 3y_n = 3y_{n+1} - y_n$
4. $2y_{n+1} = y_n, y_0 = 1$
5. $(n + 1)y_{n+1} = (n + 2)y_n, y_0 = 1$
6. $y_{n+1} = ny_n$
7. $y_{n+1} - 5y_n = 2, y_0 = 2$
8. $y_{n+1} - ny_n = n!, y_0 = 5$
9. $y_{n+1} - e^{-2n}y_n = e^{-n^2}$

10. Radium transmutes at a rate of 1% every 25 years. Consider a sample of r_0 grams of radium. If r_n is the amount of radium remaining in the sample after $25n$ years, obtain a difference equation for r_n and find its solution. How much radium is left after 100 years?

11. A fair coin is marked "1" on one side and "2" on the other side. The coin is tossed repeatedly and a cumulative score of the outcomes is recorded. Define P_n to be the probability that at some time the cumulative score takes on the value n. Prove that $P_n = 1 - (\frac{1}{2})P_{n-1}$. Assuming that $P_0 = 1$, derive the formula for P_n.

12. In constructing a mathematical model of a population, it is assumed that the probability P_n that a couple produces exactly n offspring satisfies the equation $P_n = 0.7P_{n-1}$. Find P_n in terms of P_0 and determine P_0 from the fact that

$$P_0 + P_1 + P_2 + \ldots = 1.$$

13. An alternative model of Problem 12 is given by $P_n = (\frac{1}{n})P_{n-1}$. For this model, find P_n in terms of P_0 and prove that $P_0 = \frac{1}{e}$.

14. Let x_k denote the number of permutations of n objects taken k at a time. For every permutation of k objects, we can get a total of $n - k$ permutations of $k + 1$ objects by taking one of the remaining $n - k$ objects and placing it at the end. Thus $x_{k+1} = (n - k)x_k$. Prove that the number of permutations of n objects taken k at a time is $n!/(n - k)!$

15. If we let x_n in Problem 14 denote the number of combinations of the n objects (order does not count), then every permutation of $k + 1$ objects occurs (in different orders) $k + 1$ times. Thus

$$x_{k+1} = \frac{n-k}{k+1}x_k.$$

Prove that the number of combinations of n objects taken k at a time is

$$\frac{n!}{(n-k)!k!} \equiv \binom{n}{k}.$$

The expression on the right-hand side is called the **binomial coefficient**.

2.10 SOLVING FIRST-ORDER EQUATIONS USING SYMBOLIC MANIPULATORS

We continue showing how symbolic manipulating programs, such as MAPLE, MATHEMATICA, DERIVE, MATLAB, or MATHCAD, can be used in solving differential equations. In this section we will show their use in finding solutions of first-order differential equations. We will find the algebraic expression that is the solution to the problem, rather than a numerical approximation, as was done using Euler's method in Chapter 1.

Properties of symbolic manipulators

Programs such as MATHEMATICA, MAPLE, DERIVE, MATLAB, and MATHCAD are computer programs that can do mathematics symbolically. That is, they can integrate or differentiate a function and perform algebraic operations, using the symbols we use to define the function. You may have noticed while reading this chapter that many of the methods we developed for solving first-order differential equations involved integrations, differentiations, and algebraic manipulations of expressions. In what follows, we shall discuss what steps any of the above symbolic manipulating programs would have to take to solve successfully a particular differential equation.

Since all of these programs for symbolic manipulation are different, the exact commands to do any of these tasks will change from program to program. Consequently, we shall develop a set of *generic commands,* which we shall use to explain each process. Once we have described the generic algorithm for solving a particular type of differential equation, each symbolic manipulating program requires you to define a "script" or command to execute sequentially the steps in the algorithm.

We shall use the letter x to denote the independent variable in this section, but any other letter could have been used instead. The following basic generic commands will be used to construct each "script":

`INT(f(x), x)` = {integrate the function $f(x)$ with respect to x. The result will be antiderivative with constant equal to zero}.

`SOLVE(f(x,y)=c, y)` = {solve the equation $f(x,y) = c$ for the variable y in terms of the variable x}.

`DIF(f(x), x)` = {differentiate the function $f(x)$ with respect to x}.

`LIM(f(x), x, x0)` = {limit of $f(x)$ as $x \to x0$}.

`SUBST(f(x,y)[=c], x=x0, y=y0)` = {substitute $x = x0$ and $y = y0$ in the function $f(x,y)$ [or in the equation $f(x,y) = c$]}.

We shall also use the following notation: $\exp(x) = e^x$.

Separation of variables

The method of separation of variables was developed in Sections 2.1 and 2.2. Recall that this method can be used when the equation has the form

$$\frac{dy}{dx} = p(x)q(y), \tag{1}$$

that is, when the right side of the equation can be factored into the product of a function of x and a function of y. The technique involves treating the derivative as if it were a fraction, and using algebra to separate the dependent and independent variables, obtaining

$$\frac{dy}{q(y)} = p(x)\,dx.$$

Integrating this expression on both sides, and adding an arbitrary constant, yields the general solution to the differential equation:

$$\int \frac{dy}{q(y)} = \int p(x)\,dx + c.$$

If possible, we then solve the resulting equation for y in terms of x. This is exactly what we have to "script" for the symbolic manipulating program.

We define a function called SEPARABLE $(x, y, p(x), q(y), c)$ which we will use to solve the separable equation $y' = p(x)q(y)$. It will be useful to suppress the dependence of p on x and of q on y. Then, the set of instructions we need to define the solution to Equation **(1)** are as follows:

```
SEPARABLE(x,y,p,q,c):= SOLVE(INT(1/q,y)=INT(p,x)+c,y).
```

This script indicates that we are to perform the integrations obtaining the equation which is the first term of the SOLVE command, and then solve this function for y in terms of the x and c. The symbol := is used to indicate that we are defining the function to the left side of the symbol. We illustrate this process with an example.

EXAMPLE 1 ▶

Solve the initial-value problem

$$y' = 3x^2e^{-y}, \qquad y(0) = 2.$$

Solution This is Example 2.1.2. Notice that the equation is separable with $p(x) = 3x^2$ and $q(y) = e^{-y}$. Thus, the solution is given by the function

```
SEPARABLE(x,y,3*x^2,exp(-y),c)
```

and, if the script has been properly entered, the program should do the following steps:

```
SOLVE(INT(1/exp(-y),y)=INT(3*x^2,x)+c,y)=
SOLVE(exp(y)=x^3+c,y)=
```

yielding

```
y=ln(x^3+c).
```
◀

This is the general solution of the problem. However, since Example 1 is an initial-value problem, we need to determine the value of the constant c. This will be done with the following script.

Finding a particular solution

Solving an initial-value problem is an even easier script than that used in solving separable equations. To see why this is so, we shall continue using Example 1 as an illustration.

We know that the general solution to Example 1 is (GS): $y = \ln(x^3 + c)$, and that the initial condition is $y(0) = 2$. Thus, we need to substitute $x = 0$ and $y = 2$ into the general solution and solve for c.

```
IV(x,y,x0,y0,GS,c)  := SOLVE(SUBST(GS,x=x0,y=y0),c).
```

What this script does is as follows:

```
SOLVE(SUBST(y=ln(x³+c),x=0,y=2),c)=
SOLVE(2=ln(0³+c),c)=
SOLVE(2=ln(c),c)=
```

or

$$c = e^2.$$

Thus, the particular solution for the initial-value problem is $y = \ln(x^3 + e^2)$.

The script above will find the value of c even when we do not have a general solution for y as a function of x. In the cases where we do have a solution for y as a function of x, there is an alternate method for finding c. Observe that substituting $x = 0$ and $y = 2$ in the general solution is equivalent to letting $y \to 2$ as $x \to 0$, that is, the substitution amounts to taking a limit:

$$\lim_{x\to 0} \ln(x^3 + c) = 2,$$

so that

$$\ln(c) = 2 \quad \text{or} \quad c = e^2.$$

We script this alternate method by defining the function `IVALT(x,y,x0,y0,c)` where we use the general solution of the term y in the function:

```
IVALT(x,y,x0,y0,c):= SOLVE(LIM(y,x,x0)=y0,c).
```

This script would work as follows:

```
SOLVE(LIM(ln(x³+c),x,0)=2,c)=
SOLVE(ln(c)=2,c)=
c=e².
```

Linear Equations

Linear first-order differential equations are discussed in Section 2.3. Recall that the first-order linear equation

$$y' + p(x)y = q(x) \tag{2}$$

can be solved by multiplying both sides by the integrating factor $e^{\int p(x)\,dx}$, yielding

$$(e^{\int p(x)\,dx}\,y)' = q(x)\,e^{\int p(x)\,dx}.$$

Integrating both sides, we get

$$e^{\int p(x)\,dx}\,y = \int q(x)\,e^{\int p(x)\,dx}\,dx + c.$$

Thus, the algorithm we need to use to solve Equation **(2)** can be given as follows:

```
LINEAR(x, y, p, q, c):=
        SOLVE(y*exp(INT(p,x)=INT(q*exp(INT(p,x)),x)+c, y).
```

EXAMPLE 2 ▶

Solve the initial-value problem:
$y' + 2\,x\,y = x^3$, $y(1) = 1$.

Solution Compare this problem to Example 2.3.4. Since the problem is linear, with `p(x)=2*x` and `q(x)=x³`, we find its general solution by using `LINEAR(x,y, 2*x,x^3,c)`, which then does the following:

```
SOLVE(y*exp(INT(2*x,x)=INT(x³*exp(INT(2*x,x)+c,y)=
SOLVE(y*exp(x²)=INT(x³*exp(x²),x)+c,y)=
SOLVE(y*exp(x²)=((x² - 1)/2)*exp(x²)+c,y)=
```

or

$$y = \left(\frac{x^2 - 1}{2}\right) + c\,e^{-x^2}.$$

Once we have the general solution we can use either of the initial-value scripts to determine c: for example, use

```
IVALT(x,y,1,1,c)
```

where the general solution is used instead of y. This yields $c = e$, so that the particular solution is

$$y = ee^{-x^2} + \frac{1}{2}(x^2 - 1) = e^{1-x^2} + \frac{1}{2}(x^2 - 1).$$ ◀

Bernoulli equations

A Bernoulli equation has the form

$$\frac{dy}{dx} + p(x)y = q(x)y^k. \tag{3}$$

It can be rewritten as a linear equation by making the substitution $z = y^{1-k}$. Multiplying both sides of Equation **(3)** by $(1 - k)y^{-k}$, and using the fact that $z' = (1 - k)y^{1-k}y'$, we get the linear equation in z

$$z' + (1 - k)p(x)z = (1 - k)q(x).$$

Thus, solving the Bernoulli Equation **(3)** is equivalent to solving this linear equation for $z = y^{1-k}$. Hence, the algorithm can be given by

```
BERNOULLI(x,y,p,q,k,c):=
        LINEAR(x,y^(1-k),(1-k)*p,(1-k)*q,c)
```

Homogeneous equations

An equation of the form

$$y' = r(x, y), \tag{4}$$

where the function r is homogeneous of degree zero, that is, $r(tx, ty) = r(x, y)$, can be solved by substituting $y = x\,z$. Observe that $y' = x\,z' + z$, and letting $t = \frac{1}{x}$, $r(x, y) = r(1, z)$, so that Equation **(4)** becomes

$$xz' + z = r(1, z) \qquad \text{or} \qquad x\frac{dz}{dx} = r(1, z) - z,$$

which is the separable equation

$$\frac{dz}{dx} = \frac{r(1, z) - z}{x}. \tag{5}$$

Thus, the script for such equations involves using the algorithm for separable equations:

```
HOMOGENEOUS(x,y,r,c):=
    SUBST(SEPARABLE(x,z,1/x, SUBST(r-y,x=1,y=z),c),x=x,z=y/x)
```

Alternatively, we could also give the script as follows:

```
HOMOGENEOUS(x,y,r,c):=
    LIM(SEPARABLE(x,y,1/x,LIM((r,x,1)-y,c),y,y/x)
```

which solves Equation **(5)** with y instead of z and then substitutes $\frac{y}{x}$ for y.

It is also possible to build in a test for showing that the function is homogeneous. Since $r(tx, ty) = r(x, y)$, it follows that $r(x, xy) = r(1, y)$ an expression free of x. Thus, r is homogeneous of degree zero whenever

$$\frac{d}{dx} r(x, xy) = 0 \qquad \text{or} \qquad \texttt{DIF(LIM(r, y, xy), x)= 0.}$$

Exact equations

An equation $p(x, y)\, dx + q(x, y)\, dy = 0$ is called exact if

$$\frac{\partial p}{\partial y} = \frac{\partial q}{\partial x} \qquad \text{or} \qquad \texttt{DIF(p,y)=DIF(q, x),}$$

so it is easy to test for exactness. Once we know an equation is exact, we can assume the solution has the form $g(x, y) = c$, so that

$$dg = \frac{\partial g}{\partial x}\, dx + \frac{\partial g}{\partial y}\, dy = p\, dx + q\, dy = 0.$$

Thus $p = \partial g/\partial x$, implying that

$$g = \int p\, dx + h(y).$$

To determine the function $h(y)$, which vanishes when we differentiate with respect to x, note that

$$q = \frac{\partial g}{\partial y} = \frac{d}{dy}\left(\int p\, dx + h(y)\right) = \frac{d}{dy}\left(\int p\, dx\right) + h'(y).$$

Hence

$$h(y) = \int \left(q - \frac{d}{dy}\left(\int p\, dx\right)\right) dy.$$

Thus we can describe the script for this process as follows:

```
EXACT(x,y,p,q,c):=
        SOLVE(INT(p,x)+INT(q-DIF(INT(p,x),y),y)=c,y).
```

If the differential equation is not exact, it is possible to find the integrating factor to make it exact if the function

$$k = \frac{\dfrac{\partial p}{\partial y} - \dfrac{\partial q}{\partial x}}{q}$$

depends only on x, or if

$$K = \frac{\dfrac{\partial q}{\partial x} - \dfrac{\partial p}{\partial y}}{p}$$

depends only on y. These can be easily tested:

```
DIF((DIF(p,y)-DIF(q,x))/q,y)=0
DIF((DIF(q,x)-DIF(p,y))/p,x)=0
```

whenever k depends only on x, and similarly for K. The integrating factors are

$\exp(\int k\,dx)$ and $\exp(\int K\,dy)$, so their scripts involve the expressions

```
exp(INT((DIF(p, y)-DIF(q, x))/q, x)
```

and

```
exp(INT((DIF(q, x)-DIF(p, y))/p, x).
```

EXAMPLE 3 ▶

Solve the differential equation $(3x^2 - y^2)\,dy - 2xy\,dx = 0$.

Solution This is Example 2.4.4. First, let's see what happens when we test to see if this is an exact equation, with $p(x, y) = -2xy$ and $q(x, y) = (3x^2 - y^2)$:

```
DIF(p,y)=DIF(q,x)
DIF(-2*x*y,y)=DIF(3*x^2-y^2, x)
-2*x=6*x
```

confirming that it isn't exact. So we need to find an integrating factor:

```
DIF((DIF(q,x)-DIF(p,y))/p,x)=
DIF((6*x-(-2)*x)/(-2*x*y),x)=
DIF(-4/y,x)=0
```

so the integrating factor is $\exp\int K\,dy = \exp(-4\int dy/y) = y^{-4}$. Multiplying p and q by this term, we get the exact equation:

$$\left(\frac{3x^2}{y^4} - \frac{1}{y^2}\right) dy - \frac{2x}{y^3}\,dx = 0.$$

Hence, `EXACT(x,y,-2*x/y^3,(3*x^2 - y^2/y^4, c))` yields successively

```
SOLVE(INT(p,x)+INT(q-DIF(INT(p,x),y),y)=c,y)=
SOLVE(-x²/y³+∫(3x²/y⁴-1/y²+3x²/y⁴)dy=c,y)=
SOLVE(-x²/y³+1/y=c,y)=
```

or

$$cy^3 = y^2 - x^2. \quad ◀$$

PROBLEMS 2.10

Solve each of the following problems using a symbolic manipulating program such as MATHEMATICA, MAPLE, DERIVE, MATLAB, or MATCAD:

1. $x\,y' = y^2,\ y(1) = 1$

2. $y' = y\sqrt{1 - x}$

3. $y' - xy = 0,\ y(1) = 2$

4. $xy' - y = x,\ y(1) = 1$

5. $y' - (\cos x)y = x^2$

6. $xy' + (1 - x)y = xe^x,\ y(1) = e$

7. $y' = \dfrac{x - y}{x + 2y},\ y(0) = 1$

8. $xy' = 2y + \sqrt{y^2 + x^2}$

9. $y' = \dfrac{3y^4 + 3x^2y^3 - x^4y}{xy^3 - 2x^5}$

(*Hint:* Divide first by powers of x or y.)

10. $(2x^2y^3 - y^2)\,dx + (x^3y^2 - x)\,dy = 0$

SUMMARY OUTLINE OF CHAPTER 2

Separation of Variables p. 36
The differential equation $dy/dx = f(x, y)$ can be solved by the method of separation of variables if $f(x, y) = g(x)h(y)$. In this case, we can write

$$\int \frac{dy}{h(y)} = \int g(x)\,dx + C. \tag{1}$$

Logistic Equation pp. 49–50
The equation $dP/dt = P(\beta - \delta P)$ is called the **logistic equation.** The growth shown by this equation is called **logistic growth.** The solution of the logistic equation is

$$P(t) = \frac{\beta}{\delta + \left(\dfrac{\beta}{P(0)} - \delta\right)^{e^{-\beta t}}}.$$

Linear Differential Equation p. 53
An nth-order differential equation is **linear** if it can be written in the form

$$\frac{d^n y}{dx^n} + a_{n-1}(x)\frac{d^{n-1}y}{dx^{n-1}} + \ldots + a_1(x)\frac{dy}{dx} + a_0(x)y = f(x).$$

Hence a first-order linear equation has the form

$$\frac{dy}{dx} + a(x)y = f(x) \tag{2}$$

and a second–order linear equation can be written as

$$\frac{d^2y}{dx^2} + a(x)\frac{dy}{dx} + b(x)y = f(x).$$

Homogeneous and Nonhomogeneous Linear Differential Equations p. 54
If the function $f(x)$ is the zero function, the linear differential equation is said to be **homogeneous.** Otherwise, we say that the linear differential equation is **nonhomogeneous.**

Nonlinear Differential Equation p. 54
Any differential equation that cannot be written as a linear equation is said to be **nonlinear.**

Integrating Factor and Solution to a First-Order Linear Differential Equation pp. 55–56
The first-order Equation **(2)** can be solved by multiplying both sides of the equation by the **integrating factor** $e^{\int a(x)\,dx}$. The solutions to Equation **(2)** are

$$y = \left[\int f(x)e^{\int a(x)\,dx}\,dx + C\right]e^{-\int a(x)\,dx}.$$

Bernoulli's Equation p. 58
The equation

$$\frac{dy}{dx} + a(x)y = f(x)y^n$$

is known as **Bernoulli's equation.** It can be transformed into a linear equation and then solved by making the substitution $z = y^{1-n}$.

The **total differential** dg of a function of two variables $g(x, y)$ is defined by p. 64

$$dg = \frac{\partial g}{\partial x}dx + \frac{\partial g}{\partial y}dy.$$

Exact Differential Equation p. 64
The differential equation $M(x, y)\,dx + N(x, y)\,dy = 0$ is **exact** if we can find a differentiable function $g(x, y)$ such that

$$\frac{\partial g}{\partial x} = M \quad \text{and} \quad \frac{\partial g}{\partial y} = N$$

Test for Exactness p. 65
The equation $M(x, y)\,dx + N(x, y)\,dy = 0$ is exact if and only if

$$\frac{\partial M}{\partial y} = \frac{\partial N}{\partial x}$$

If $f(tx, ty) = f(x, y)$, then the function f is said to be **homogeneous of degree zero.** The equation

$$\frac{dy}{dx} = F\left(\frac{y}{x}\right)$$

can be solved by making the substitution $z = \frac{y}{x}$. pp. 68–69

Some Facts about Simple Electric Circuits pp. 80–81
An electromotive force E (volts), usually a battery or generator, drives an electric charge Q (coulombs) and produces a current I (amperes). The current is defined as the rate of flow of the charge, and we can write

$$I = \frac{dQ}{dt}.$$

A resistor of resistance R (ohms) is a component of the circuit that opposes the current, dissipating the energy in the form of heat. It produces a drop in voltage given by **Ohm's law:**

$$E_R = RI.$$

An inductor of inductance L (henrys) opposes any change in current by producing a voltage drop of

$$E_L = L\frac{dI}{dt}.$$

A capacitor of capacitance C (farads) stores charge. In so doing, it resists the flow of further charge, causing a drop in the voltage of

$$E_C = \frac{Q}{C}.$$

The quantities R, L, and C are usually constants associated with the particular component in the circuit; E may be a constant or a function of time. The fundamental principle guiding such circuits is given by **Kirchhoff's voltage law:**

The algebraic sum of all voltage drops around a closed circuit is zero.

REVIEW EXERCISES FOR CHAPTER 2

Find the general solution to each of Exercises 1–30. When an initial condition is given, find the particular solution that satisfies the condition.

1. $x\dfrac{dy}{dx} = y^2,\ y(1) = 1$

2. $\dfrac{dy}{dx} = y\sqrt{1 - x}$

3. $\dfrac{dy}{dx} = \dfrac{\sqrt{1 - y^2}}{x}$

4. $x\dfrac{dy}{dx} = \tan y,\ y(1) = \dfrac{\pi}{2}$

5. $\dfrac{dy}{dx} + x = x(y^2 + 1)$

6. $xy' = y(1 - 2y),\ y(1) = 2$

7. $yy' = \cos x,\ y(\pi) = 0$

8. $y' = xy(2 - 3y)$

9. $y = xy' + \dfrac{1}{y'}$
10. $y - xy' = \sqrt{1 - (y')^2}$
11. $y' - xy = 0,\ y(1) = 1$
12. $xy' - y = x,\ y(1) = 1$
13. $y' - (\sin x)y = \sin x$
14. $y' - \dfrac{1}{x}y = e^x$
15. $(1 + x^2)y' + xy = \sqrt{1 + x^2}$
16. $y' - (\cos x)y = x^2$
17. $xy' - 2y = x^2,\ y(1) = 1$
18. $xy' + (1 - x)y = xe^x,\ y(1) = e$
19. $y' - xy = \begin{cases} 1, & x \le 0 \\ 0, & x > 0 \end{cases}$
20. $y' + xy = \begin{cases} x, & x \le 1 \\ 1, & x > 1 \end{cases}$
21. $y' = \dfrac{x - y}{x + 2y},\ y(0) = 1$
22. $xy' = 2y + \sqrt{y^2 + x^2}$
23. $xy' = -y + \sqrt{xy + 1}$
24. $xy' = \sqrt{x^2y^2 - 1} - y,\ y(1) = 2$
25. $y' = y + xy^2$
26. $y' + xy = e^x y^3$
27. $(y - e^y \sec^2 x)\,dx + (x - e^y \tan x)\,dy = 0$
28. $(2x^2y^3 - y^2)\,dx + (x^3y^2 - x)\,dy = 0$
29. $\dfrac{dy}{dx} = \dfrac{3y^4 + 3x^2y^3 - x^4y}{xy^3 - 2x^5}$
30. $\dfrac{dy}{dx} + \dfrac{x + y\sqrt{x^2 + y^2}}{y + x\sqrt{x^2 + y^2}} = 0$
31. A paper mill is located where a river enters a lake of volume 10^9 cubic meters. The river, which has a constant flow of 1000 cubic meters per second, is the lake's only inlet. Assume that at time $t = 0$, the paper mill begins pumping pollutants into the river at the rate of 1 m^3/sec, and that the inflow and outflow of the lake are constant. How high a concentration of pollutant is there in the lake after 10 hours? After 100 hours? After one year?
32. Assume that the paper mill in Exercise 31 stops polluting the river at the end of 1 hour. Find an expression for the concentration of pollutant in the lake at all time t.
33. Assume the paper mill in Exercise 31 pollutes the river for 1 hour each day. Find an expression for the concentration of pollutant in the lake at all time t. What is the maximum concentration that the pollution reaches in the lake?
34. A $20' \times 12' \times 8'$ room contains five chain-smokers who are playing poker. An exhaust fan is removing 10 ft^3/min of smoky air, which is replaced by pure air seeping in under the door. Each chain-smoker is contributing 0.1 ft^3/min of smoke to the room. Find an expression for the concentration of smoke in the room at time t, assuming the room contains no smoke at time $t = 0$.
35. Newton's law of cooling can be used to estimate the time of death in a homicide.* Assume that at the time of death the body temperature is 98.6°F, when the corpse is discovered its temperature is 75°F, and 1 hour after discovery it is 70°F. If the constant ambient temperature is 68°F, how long before the corpse was discovered did the homicide occur?
36. **Stefan's law** of radiation states that the rate of change of temperature from a body at absolute temperature T is $T' = k(T^4 - T_0^4)$, where T_0 is the absolute temperature of the surrounding medium.
    (a) Solve this differential equation.
    (b) Show that if $T - T_0$ is small compared to T_0, then Newton's law of cooling is a close approximation to Stefan's law.
37. Another equation that has been proposed to model population growth is **Gompertz**† equation,

$$\frac{dP}{dt} = P(a - b \ln P).$$

Find a solution to this equation and determine its behavior as $t \to \infty$.

38. A. G. W. Cameron,‡ in an article concerning the processes in the primitive solar nebula, obtained the first-order differential equation

$$\frac{dx}{dt} = \frac{ax^{5/6}}{(b - Bt)^{3/2}},$$

where a, b, and B are constants. Solve this equation.

*D. A. Smith, "The Homicide Problem Revisited," *The Two Year College Mathematics Journal* 9 (1978): 141–145.

†Named after Benjamin Gompertz (1779–1865), an English mathematician.

‡"Accumulation Processes in the Primitive Solar Nebula," *Icarus* 18 (1973): 407–450.

39. A boat weighing 4000 pounds drifts away from a dock at 3 feet per second. What is the minimum distance it drifts if a 180-pound crewman exerts a force equal to his weight on a rope tied to the bow of the boat?

40. L. L. Thurstone* used the separable differential equation

$$y' = \frac{2k}{\sqrt{m}}[y(1-y)]^{3/2}$$

41. A 6-foot chain weighing 10 pounds per foot is placed on a frictionless table so that 1 foot of chain hangs over the edge of the table. Find an equation describing the amount of chain still on the table for all time $t \geq 0$. [*Hint:* The weight of the chain that is falling changes with time.]

42. A power cable, hanging from fixed towers, has a weight of w pounds per foot.

(a) Let $y(x)$ be the position of the cable x feet horizontally away from a tower and T_H be the horizontal component of tension in the cable. Show that

$$y'(x + \Delta x) - y'(x) = \frac{w}{T_H}\Delta s,$$

where Δs is the length of the cable over the horizontal interval of length Δx.

(b) Use part (a), the Pythagorean theorem, and limits to deduce the differential equation

$$y'' = \frac{w}{T_H}\sqrt{1 + (y')^2}.$$

(c) Solve the differential equation in part (b) to obtain an expression for $y'(x)$, assuming $y'(0) = 0$.

(d) If $y(0) = 0$, find $y(x)$.

43. Suppose a constant capacitor is connected in series to an emf whose voltage is a sine wave. Show that the current is 90° out of phase with the voltage.

44. Repeat Exercise 43 with the capacitor replaced by an inductor. What can you say in this case?

In Exercises 45–47 use Picard iterations to solve the given initial-value problem.

45. $y' = y + 1,\ y(0) = 0$

46. $y' = 3x^2y,\ y(0) = 1$

47. $y' = \dfrac{y}{x},\ y(1) = 1$

[*Hint:* $x_0 = 1$ in equation (2.8.5).]

* "The Learning Function," *Journal of General Psychology* 3 (1930): 469–493.

3 Second- and Higher-Order Linear Differential Equations

In Chapter 2 we solved a variety of first-order differential equations. In this chapter we discuss those second- and higher-order differential equations for which closed form solutions can readily be obtained.

3.1 THEORY OF LINEAR DIFFERENTIAL EQUATIONS

Definition Homogeneous and nonhomogeneous linear differential equations

A second-order differential equation is **linear** if it can be written in the form

$$y''(x) + a(x)y'(x) + b(x)y(x) = f(x), \tag{1}$$

where a, b, and f are functions of the independent variable x only.

The most general third-order linear equation can be written

$$y'''(x) + a(x)y''(x) + b(x)y'(x) + c(x)y(x) = f(x), \tag{2}$$

where $a(x)$, $b(x)$, $c(x)$, and $f(x)$ are functions of the independent variable x only. Equations **(1)** and **(2)** are special cases of the **general linear nth-order equation:**

$$y^{(n)}(x) + a_{n-1}(x)y^{(n-1)}(x) + \ldots + a_1(x)y'(x) + a_0(x)y(x) = f(x). \tag{3}$$

If the function $f(x)$ is identically zero, we say that Equations **(1), (2),** and **(3)** are **homogeneous.** Otherwise, they are **nonhomogeneous.**

EXAMPLE 1 ▶ **Homogeneous and nonhomogeneous linear differential equations**

(a) The equation $y'' + 2xy' + 3y = 0$ is homogeneous.

(b) The equation $y'' + 2xy' + 3y = e^x$ is nonhomogeneous. ◀

If the coefficient functions $a(x)$ and $b(x)$ are constants, $a(x) = a$ and $b(x) = b$, the equation is said to have **constant coefficients.** (As we see below, linear differential equations with constant coefficients are the easiest to solve.) If either $a(x)$ or $b(x)$ is not constant, the equation is said to have **variable coefficients.**

EXAMPLE 2 ▶ **Equations with constant and variable coefficients**

(a) The equation $y'' + 3y' - 10y = 0$ has constant coefficients.

(b) The equation $y'' + 3xy' - 10x^2y = 0$ has variable coefficients. ◀

In this section we will answer the questions: When does a second-order linear differential equation have solutions and, if it does, how do we know we have found all of them? Because there is a lot of material to cover, we break the section into four parts.

Existence and uniqueness of solutions

In Section 2.3 we discussed the first-order linear differential equation

$$y' + a(x)y = f(x).$$

If $a(x)$ and $f(x)$ are continuous functions in an interval $[x_1, x_2]$, $x_1 < x_2$, this differential equation has a general solution involving an arbitrary constant (see Equation **2.3.11**). [The solutions are given by $y = [\int f(x)e^{\int a(x)\,dx}\,dx + C]e^{-\int a(x)\,dx}$.] The arbitrary constant can be determined if an initial condition $y(x_0) = y_0$ is given at some value x_0 in $[x_1, x_2]$. In this case the initial-value problem has a unique solution. We can restate these basic facts as follows.

THEOREM ♦ **Existence-uniqueness theorem for first-order linear initial-value problems**
Suppose that a and f are continuous functions in the interval $x_1 \le x \le x_2$, the number x_0 is in $[x_1, x_2]$, and that the initial condition $y(x_0) = y_0$ is given. Then in $[x_1, x_2]$ there exists exactly one solution $y(x)$ to the differential equation

$$y'(x) + a(x)y(x) = f(x)$$

that satisfies the given initial condition. ♦

This is a very nice result, because it tells us that every linear first-order equation with a given initial condition has a unique solution (and it even tells us what that solution is!). We need only set about finding it. It turns out that this special property holds for linear initial-value problems of any order. The only difference is that in order to have a unique solution to a second-order equation, we must specify two initial conditions, for a third-order equation three conditions, and so on. One case of the following central theorem is proved in Appendix 3 (Theorem 8).

THEOREM 1 ♦ **Existence-uniqueness theorem for linear initial-value problems**
Suppose that $a_1, a_2, \ldots, a_n$, and f are continuous functions for $x_1 \le x \le x_2$. Let $c_0, c_1, c_2, \ldots, c_{n-1}$ be n constants and suppose that x_0 is in $[x_1, x_2]$. Suppose further that n initial conditions are given:

$$y(x_0) = c_0, \qquad y'(x_0) = c_1, \qquad y''(x_0) = c_2 \ldots, \qquad y^{(n-1)}(x_0) = c_{n-1}. \quad \textbf{(4)}$$

Then on $[x_1, x_2]$ there exists exactly one solution $y(x)$ to the differential equation

$$y^{(n)}(x) + a_{n-1}(x)y^{(n-1)}(x) + a_{n-2}(x)y^{(n-2)}(x) + \ldots + a_0(x)y(x) = f(x)$$

that satisfies the n given initial conditions. ♦

Note

The conditions given in Equation **(4)** all involve evaluations of the unknown function y and its derivatives at the *same* point x_0. This is a *crucial* requirement for the existence and uniqueness of a solution.

Boundary value problems

There is another type of problem involving the differential equation **(3)** in which conditions at more than one point are given. For example, we might specify $y(x_1) = k_1$, and $y'(x_2) = k_2, \ldots, y^{(n-1)}(x_{n-1}) = k_{n-1}$. Conditions of this sort are called **boundary conditions,** and a differential equation together with a set of boundary conditions is called a **boundary value problem.**

It is important to note that the existence and uniqueness of a solution, guaranteed by Theorem 1 for initial-value problems, does not hold for boundary value problems.

EXAMPLE 3 ▶ Boundary value problems with infinitely many solutions, no solution, and a unique solution

Observe that the boundary value problem

$$y'' + y = 0, \qquad y(0) = y(\pi) = 0,$$

has infinitely many solutions:

$$y = c \sin x, \qquad \text{for any constant } c.$$

To check, note that $\sin 0 = \sin \pi = 0$, so the boundary conditions are satisfied, and

$$(c \sin x)'' + c \sin x = -c \sin x + c \sin x = 0.$$

We will be able to show in Section 3.4 (see Problem 3.4.31) that the boundary value problem

$$y'' + y = 0, \qquad y(0) = 0, \qquad y(\pi) = 1$$

has *no* solution. On the other hand, the initial-value problem

$$y'' + y = 0, \qquad y(0) = 0, \qquad y'(0) = 1$$

has the unique solution $y = \sin x$; the initial conditions are satisfied since $\sin 0 = 0$, $(\sin x)' = \cos x$, and $\cos 0 = 1$, and

$$(\sin x)'' = \sin x = -\sin x + \sin x = 0. \quad ◀$$

If we apply Theorem 1 to the second-order Equation **(1),** we have the following result.

THEOREM ◆ Existence-uniqueness theorem for second-order, linear initial-value problems

Suppose that a, b, and f are continuous functions for $x_1 \le x \le x_2$, that x_0 is in $[x_1, x_2]$ and that the two initial conditions $y(x_0) = y_0$ and $y'(x_0) = y_1$ are given, where y_0 and y_1 are real numbers. Then in $[x_1, x_2]$ there exists exactly one solution $y(x)$ to the differential equation

$$y''(x) + a(x)y'(x) + b(x)y(x) = f(x) \tag{5}$$

that satisfies the two given initial conditions. ◆

For simplicity we limit most of our discussion in this chapter to second-order linear equations. We emphasize, however, that *every* result we prove can be extended to higher-order linear equations (see Section 3.8).

The requirement that the functions $a(x)$, $b(x)$, and $f(x)$ be continuous is also an essential requirement for the existence of a unique solution, as the following example demonstrates.

EXAMPLE 4 ▶ **An initial-value problem with infinitely many solutions**

Verify that the function

$$y = cx^3 + x$$

is a solution (for any constant c) of the initial-value problem

$$x^2y'' - 3xy' + 3y = 0, \qquad y(0) = 0, \qquad y'(0) = 1;$$

that is, the problem does not have a unique solution—it has infinitely many solutions.

Solution Since $y' = 3cx^2 + 1$, the initial conditions are satisfied. Then $y'' = 6cx$, and

$$\begin{aligned} x^2(6cx) - 3x(3cx^2 + 1) + 3(cx^3 + x) &= 6cx^3 - 9cx^3 - 3x + 3cx^3 + 3x \\ &= 0. \end{aligned}$$

Theorem 1 does not apply here, because in order to obtain a differential equation of the form **(5)**, we must divide $x^2y'' - 3xy' + 3y = 0$ by x^2:

$$y'' - \frac{3}{x}y' + \frac{3}{x^2}y = 0.$$

Then $a(x) = -3/x$ and $b(x) = 3/x^2$. Theorem 1 requires that $a(x)$ and $b(x)$ be continuous in an interval containing x_0 (here $x_0 = 0$). However, there is *no* interval containing 0 such that $-3/x$ and $3/x^2$ are continuous (neither function is even defined at zero.) ◀

There are special techniques for handling some problems of this sort, which we will discuss in Chapter 5.

Linear combinations and linear independence

We now know that the differential equation **(5)** has solutions if a, b, and f are continuous. What do these solutions look like? How many are there?

Before solving a second-order differential equation, it helps to know what we are seeking. A clue is provided by examining a first-order equation. Consider the equation

$$y' + 2y = 0. \tag{6}$$

In Section 1.1 we saw that one solution to this equation is $y = e^{-2x}$. In fact, $y = ce^{-2x}$ is a solution for any constant c, and every solution to Equation **(6)** has the form ce^{-2x}. We can summarize this result by noting that once we have found one nonzero solution to Equation **(6)**, we have found all of the solutions, since every other solution is a constant multiple of this one solution.

It turns out that similar results hold for all homogeneous second-order equations:

$$y'' + a(x)y' + b(x)y = 0. \tag{7}$$

The major difference is that now we have to find *two* solutions to Equation **(7)** where neither solution is a multiple of the other. We now make these ideas more precise.

Linear combination, linear independence and linear dependence

Let y_1 and y_2 be any two functions defined on an interval $[x_1, x_2]$. By a **linear combination** of y_1 and y_2 we mean a function $y(x)$ that can be written in the form

$$y(x) = c_1 y_1(x) + c_2 y_2(x),$$

for some constants c_1 and c_2. Two functions are **linearly independent** on an interval $[x_1, x_2]$, $x_1 < x_2$, whenever the relation

$$c_1 y_1(x) + c_2 y_2(x) = 0, \tag{8}$$

for all x in $[x_1, x_2]$, implies that $c_1 = c_2 = 0$. Otherwise they are **linearly dependent.**

EXAMPLE 5 ▶ **Two linearly independent functions**

Verify that the functions $y_1 = 1$ and $y_2 = x$ are linearly independent on the interval $[0, 1]$.

Solution To determine linear independence or dependence we must consider Equation **(8)**:

$$c_1y_1 + c_2y_2 = c_1 \cdot 1 + c_2 \cdot x = 0. \tag{9}$$

This equation must hold for all x in $[0, 1]$. If $x = 0$, we have

$$c_1 \cdot 1 + c_2 \cdot 0 = c_1 = 0.$$

But then at $x = 1$ we get $c_2 \cdot 1 = 0$. Hence Equation **(9)** holds for all x in $[0, 1]$ if and only if $c_1 = c_2 = 0$. This proves that $y_1 = 1$ and $y_2 = x$ are linearly independent. ◀

The notions of linear combination, linear independence and linear dependence extend easily to a collection of n functions $y_1(x), y_2(x), \ldots, y_n(x)$, with $n > 2$. We assume that all these functions are defined in $[x_1, x_2]$. A **linear combination** of these functions is any function of the form

$$y(x) = c_1y_1(x) + c_2y_2(x) + \ldots + c_ny_n(x),$$

where $c_1, c_2, \ldots, c_n$ are constants. We will say that the collection of functions $y_1(x), y_2(x), \ldots, y_n(x)$ are **linearly independent** on an interval $[x_1, x_2]$, $x_1 < x_2$, if the equation

$$c_1y_1(x) + c_2y_2(x) + \ldots + c_ny_n(x) = 0$$

holds for all x in $[x_1, x_2]$ only when $c_1 = c_2 = \ldots = c_n = 0$. Otherwise, we say that the collection of functions is **linearly dependent** on $[x_1, x_2]$. We will say more about this in Section 3.8.

There is an easy way to see that two functions y_1 and y_2 are linearly dependent. If $c_1 y_1(x) + c_2 y_2(x) = 0$ (where not both c_1 and c_2 are zero), we may suppose that $c_1 \neq 0$. Dividing the above expression by c_1, we obtain

$$y_1(x) + \frac{c_2}{c_1}y_2(x) = 0,$$

or

$$y_1(x) = -\frac{c_2}{c_1}y_2(x) = cy_2(x). \tag{10}$$

This leads to a useful fact.

Determining whether two functions are linearly dependent or independent

Two functions are linearly dependent on the interval $[x_0, x_1]$ if and only if one of the functions is a constant multiple of the other.

EXAMPLE 5 ▶ **Two linearly independent functions**

It is easy to see that the functions $y_1 = 1$ and $y_2 = x$ are linearly independent on $[0, 1]$, since x is not a *constant* multiple of 1. ◀

The general solution to a linear, homogeneous, second-order differential equation

The notions of linear combination and linear independence are central to the theory of linear homogeneous equations, as is illustrated by the results that follow.

THEOREM 2 ♦

Every homogeneous linear second-order differential equation

$$y'' + a(x)y' + b(x)y = 0 \tag{11}$$

has two linearly independent solutions.

Proof The existence part of Theorem 1 guarantees that we can find a solution $y_1(x)$ to Equation **(11)** satisfying

$$y_1(x_0) = 1 \quad \text{and} \quad y_1'(x_0) = 0.$$

Similarly, we can also find a solution $y_2(x)$ to Equation **(11)** satisfying

$$y_2(x_0) = 0 \quad \text{and} \quad y_2'(x_0) = 1.$$

Now consider the equation

$$c_1 y_1(x) + c_2 y_2(x) = 0 \tag{12}$$

and its derivative

$$c_1 y_1'(x) + c_2 y_2'(x) = 0. \tag{13}$$

Setting $x = x_0$ in Equation **(12)** yields $c_1 \cdot 1 + c_2 \cdot 0 = 0$, or $c_1 = 0$, whereas substituting $x = x_0$ in Equation **(13)** gives $c_1 \cdot 0 + c_2 \cdot 1 = c_2 = 0$. Thus Equation **(12)** holds only when $c_1 = c_2 = 0$, implying that solutions $y_1(x)$ and $y_2(x)$ are linearly independent. ♦

EXAMPLE 6 ▶ **Two linearly independent solutions to a second-order equation**

Verify that $\sin x$ and $\cos x$ are linearly independent solutions to $y'' + y = 0$.

Solution To check that $\sin x$ and $\cos x$ are solutions, we write

$$(\sin x)'' + \sin x = -\sin x + \sin x = 0,$$
$$(\cos x)'' + \cos x = -\cos x + \cos x = 0.$$

Now consider the equation

$$c_1 \sin x + c_2 \cos x = 0,$$

for all values x, and its derivative

$$c_1 \cos x - c_2 \sin x = 0.$$

If $x = 0$, the first equation yields $c_2 = 0$ whereas the second gives $c_1 = 0$. Thus $\sin x$ and $\cos x$ are linearly independent. Alternatively, they are independent according to Equation **(10)**, because

$$\frac{\sin x}{\cos x} = \tan x \neq \text{constant.} \quad ◀$$

Every linear, homogeneous, second-order differential equation has two linearly independent solutions. The next two theorems show us that this is all the information we need; that is, once we have two linearly independent solutions, we can construct them all.

THEOREM 3 ♦ **Principle of superposition for homogeneous equations**
Let $y_1(x)$ and $y_2(x)$ be any two solutions of the homogeneous equation

$$y'' + a(x)y' + b(x)y = 0. \tag{14}$$

Then any linear combination of them is also a solution of Equation **(14)**.

Proof Let $y(x) = c_1 y_1(x) + c_2 y_2(x)$. Then

$$\begin{aligned} y'' + ay' + by &= c_1 y_1'' + c_2 y_2'' + c_1 a y_1' + c_2 a y_2' + c_1 b y_1 + c_2 b y_2 \\ &= c_1(y_1'' + a y_1' + b y_1) + c_2(y_2'' + a y_2' + b y_2) \\ &= c_1 \cdot 0 + c_2 \cdot 0 = 0, \end{aligned}$$

since y_1 and y_2 are solutions of the homogeneous Equation **(14)**. ♦

THEOREM 4 ♦ **A second-order linear, homogeneous equation can have at most two linearly independent solutions**
Let $y_1(x)$ and $y_2(x)$ be linearly independent solutions to Equation **(14)** and let $y_3(x)$ be another solution. Then there exist unique constants c_1 and c_2 such that

$$y_3(x) = c_1 y_1(x) + c_2 y_2(x).$$

In other words, any solution of Equation **(14)** can be written as a linear combination of two given linearly independent solutions of Equation **(14)**. ♦

We stress the importance of this theorem. It indicates that once we have found two linearly independent solutions y_1 and y_2 of Equation **(14)**, we have essentially found *all* the solutions of Equation **(14)**. (We delay the proof of this theorem until the end of the section.)

Theorem 4 can be extended to nth-order equations. The following result will be useful in Section 3.8.

An important result

An nth-order linear, homogeneous differential equation with continuous coefficients can have at most n linearly independent solutions on an interval $[x_1, x_2]$.

General solution to a linear, second-order equation

The general solution of Equation **(14)** is given by the linear combination

$$y(x) = c_1 y_1(x) + c_2 y_2(x)$$

where c_1 and c_2 are arbitrary constants and $y_1(x)$ and $y_2(x)$ are linearly independent solutions of Equation **(14)**.

EXAMPLE 6 (revisited) ▶ **The general solution to a linear equation**

The general solution to $y'' + y = 0$ is

$$y = c_1 \cos x + c_2 \sin x. \quad ◀$$

If we look at the proof of Theorem 2 we see that Equations **(12)** and **(13)** yield a system of equations

$$\begin{aligned} c_1 y_1(x) + c_2 y_2(x) &= 0, \\ c_1 y_1'(x) + c_2 y_2'(x) &= 0, \end{aligned} \tag{15}$$

that can be used to determine whether the solutions $y_1(x)$ and $y_2(x)$ are linearly independent. The procedure involves substituting a value $x = x_0$ in Equation **(15),** which determines $y_1(x_0)$, $y_2(x_0)$, $y_1'(x_0)$, and $y_2'(x_0)$. We are then left with a linear system of two homogeneous equations in the two unknowns c_1 and c_2. This system* has only the trivial solution $c_1 = c_2 = 0$ if and only if the determinant

$$\begin{vmatrix} y_1(x_0) & y_2(x_0) \\ y_1'(x_0) & y_2'(x_0) \end{vmatrix} = y_1(x_0)y_2'(x_0) - y_1'(x_0)y_2(x_0)$$

does not equal zero. This fact leads to the following very useful definition.

Wronskian

Let $y_1(x)$ and $y_2(x)$ be any two solutions to the differential equation

$$y'' + a(x)y' + b(x)y = 0. \tag{16}$$

The **Wronskian**† of y_1 and y_2 is defined as

$$W(y_1, y_2)(x) = \begin{vmatrix} y_1(x) & y_2(x) \\ y_1'(x) & y_2'(x) \end{vmatrix} = y_1(x)y_2'(x) - y_1'(x)y_2(x).$$

HISTORICAL NOTE

The Wronskian is named after the Polish philosopher Jozef Maria Hoëne-Wronski (1778–1853). Although he was born in Poland, Wronski was educated in Germany and lived most of his life in France. Wronski's only significant contribution to mathematics was the determinant that bears his name. In fact, his principal interest was in the philosophy of mathematics.

Wronski was a somewhat bizarre figure in the history of mathematics. He founded a philosophical movement called *messianism* which greatly resembled a religious sect. It was based on a revelation of the universal and rational "Absolute." It is not clear whether the Absolute represented the absolute truth that could be found only in mathematics or was simply

---

* A homogeneous system

$$\begin{aligned} ax + by &= 0 \\ cx + dy &= 0 \end{aligned}$$

has nontrivial solutions for the unknowns x and y if and only if the determinant

$$\begin{vmatrix} a & b \\ c & d \end{vmatrix} = ad - bc$$

equals zero (see Theorem 3 in Appendix 4).

† Pronounced Rón-ski-en.

an ideal manifested in Wronski himself. Wronski's messiansim was most effective when it was used to question the validity of Lagrange's use of what we now call Taylor series. His *Réfutation* caused quite a stir in the early nineteenth century because it was very convincing. However, the "solution" Wronski proposed was extremely vague and was quickly consigned to the rubbish heap of history. The "philosophical" reputation of Wronski's theories soon followed. When he died in 1853, Wronski had long since gone insane.

The Wronskian is defined at all points x at which $y_1(x)$ and $y_2(x)$ are differentiable.

Since $y_1(x)$ and $y_2(x)$ are solutions to Equation **(16),** their second derivatives exist. Hence we can differentiate $W(y_1, y_2)(x)$ with respect to x:

$$\begin{aligned} W'(y_1, y_2)(x) &= [y_1(x)y_2'(x) - y_1'(x)y_2(x)]' \\ &= y_1'(x)y_2'(x) + y_1(x)y_2''(x) - y_1''(x)y_2(x) - y_1'(x)y_2'(x) \\ &= y_1(x)y_2''(x) - y_1''(x)y_2(x). \end{aligned}$$

Since y_1 and y_2 are solutions to Equation **(16),**

$$y_1'' + ay_1' + by_1 = 0 \quad \text{and} \quad y_2'' + ay_2' + by_2 = 0.$$

Multiplying the first of these equations by y_2 and the second by y_1 and subtracting, we obtain

$$y_1y_2'' - y_2y_1'' + a(y_1y_2' - y_2y_1') = 0,$$

which is just

$$W' + aW = 0. \tag{17}$$

But Equation **(17)** is a linear first-order equation similar to Equation **(2.3.2)** (p. 54) with solution [see Equation **(2.3.3)**]

An equation for the Wronskian

$$W(y_1, y_2)(x) = ce^{-\int a(x)\,dx} \tag{18}$$

for some arbitrary constant c. Equation **(18)** is known as **Abel's formula.**

Since an exponential is never zero, we see that $W(y_1, y_2)(x)$ is either always zero (when $c = 0$) or never zero (when $c \neq 0$). The importance of this fact is given by the following theorem.

THEOREM 5 ♦ Independence of two solutions

Two solutions $y_1(x)$ and $y_2(x)$ of Equation **(16)** are linearly independent on $[x_0, x_1]$, $x_0 < x_1$, if and only if $W(y_1, y_2)(x) \neq 0$ at any x in $[x_0, x_1]$. ♦

Theorem 5 is useful in at least three ways. First, it provides us with an easy way to determine whether or not two solutions are linearly independent. Second, it greatly simplifies the proof of Theorem 4—as we see later in this section. Third, the Wronskian can be easily extended to third- and higher-order equations with similar results. It is not easy to verify directly that three solutions are linearly independent, but the task is made easy by use of the Wronskian.

EXAMPLE 6 ▶ Using the Wronskian to verify linear independence

We have seen that $y_1(x) = \sin x$ and $y_2(x) = \cos x$ are linearly independent solutions of $y'' + y = 0$. The linear independence is easily verified using the

Wronskian:

$$W(y_1, y_2)(x) = \begin{vmatrix} \sin x & \cos x \\ \cos x & -\sin x \end{vmatrix} = -\sin^2 x - \cos^2 x = -1 \neq 0. \quad ◀$$

EXAMPLE 7 ▶ **Using the Wronskian to verify linearly independence**

(a) Verify that $y_1(x) = e^{-5x}$ and $y_2(x) = e^{2x}$ are linearly independent solutions of the differential equation

$$y'' + 3y' - 10y = 0. \tag{19}$$

(b) Use the general solution obtained from y_1 and y_2 to solve the initial-value problem

$$y'' + 3y' - 10y = 0, \qquad y(0) = 3, \qquad y'(0) = -1. \tag{20}$$

Solution **(a)** That $y_1(x)$ and $y_2(x)$ are solutions to the differential equation follows from

$$(e^{-5x})'' + 3(e^{-5x})' - 10(e^{-5x}) = 25e^{-5x} - 15e^{-5x} - 10e^{-5x} = 0,$$
$$(e^{2x})'' + 3(e^{2x})' - 10(e^{2x}) = 4e^{2x} + 6e^{2x} - 10e^{2x} = 0.$$

Also,

$$W(y_1, y_2)(x) = \begin{vmatrix} e^{-5x} & e^{2x} \\ -5e^{-5x} & 2e^{2x} \end{vmatrix} = 2e^{-3x} + 5e^{-3x} = 7e^{-3x} \neq 0,$$

so the solutions are linearly independent. (This is also true since $e^{-5x}/e^{2x} = e^{-7x} \neq$ constant.) Thus the general solution of Equation **(19)** is

$$y(x) = c_1 y_1(x) + c_2 y_2(x) = c_1 e^{-5x} + c_2 e^{2x}, \tag{21}$$

where c_1 and c_2 are arbitrary constants.

(b) To find the particular solution of initial-value problem **(20)**, we must determine the constants c_1 and c_2 so that **(21)** satisfies the initial conditions in **(20)**. Setting $x = 0$ in **(21)**, we have

$$c_1 + c_2 = y(0) = 3.$$

Differentiating **(21)**, we get

$$y'(x) = -5c_1 e^{-5x} + 2c_2 e^{2x},$$

and setting $x = 0$ gives

$$-5c_1 + 2c_2 = y'(0) = -1.$$

Hence

$$c_1 + c_2 = 3$$
$$-5c_1 + 2c_2 = -1.$$

The first equation implies that $c_1 = 3 - c_2$, so

$$-5c_1 + 2c_2 = -5(3 - c_2) + 2c_2 = -1,$$

or

$$-15 + 5c_2 + 2c_2 = -1,$$
$$7c_2 = 14,$$
$$c_2 = 2.$$

Hence $c_1 = 1$, and the particular solution of initial-value problem **(20)** is

$$y = e^{-5x} + 2e^{2x}. \quad ◀$$

The general solution to a linear, nonhomogeneous second-order differential equation

We now turn briefly to the nonhomogeneous equation

$$y'' + a(x)y' + b(x)y = f(x). \tag{22}$$

Let y_p be any particular solution to Equation **(22)**. If we know the general solution to the associated homogeneous equation

$$y'' + a(x)y' + b(x)y = 0, \tag{23}$$

we can find all solutions to Equation **(22)**.

THEOREM 6 ♦ The difference of two solutions to a nonhomogeneous equation is a solution to the associated homogeneous equation

Let y_p be a particular solution of Equation **(22)** and let y^* be any other solution. Then $y* - y_p$ is a solution of Equation **(23)**; that is,

$$y^*(x) = c_1 y_1(x) + c_2 y_2(x) + y_p(x),$$

for some constants c_1 and c_2 where y_1, y_2 are two linearly independent solutions of Equation **(23)**.

Thus in order to find all solutions to the nonhomogeneous equation, we need find only one solution to the nonhomogeneous equation and the general solution of the homogeneous equation.

Proof We have

$$\begin{aligned}(y^* - y_p)'' &+ a(y^* - y_p)' + b(y^* - y_p)\\ &= (y^{*\prime\prime} + ay^{*\prime} + by^*) - (y_p'' + ay_p' + by_p)\\ &= f - f = 0. \quad ♦\end{aligned}$$

> **General solution to a linear, nonhomogeneous second-order equation**
>
> Let $y_p(x)$ be one solution to the nonhomogeneous Equation **(22)** and let $y_1(x)$ and $y_2(x)$ be two linearly independent solutions to the homogeneous Equation **(23)**. Then the general solution to Equation **(22)** is given by
>
> $$y(x) = c_1y_1(x) + c_2y_2(x) + y_p(x),$$
>
> where c_1 and c_2 are arbitrary constants.

EXAMPLE 8 ▶ A particular solution to a nonhomogeneous equation

It is not difficult to verify that $\frac{1}{2}xe^x$ is a particular solution of $y'' - y = e^x$. Two linearly independent solutions of $y'' - y = 0$ are given by $y_1 = e^x$ and $y_2 = e^{-x}$. The general solution is therefore $y(x) = c_1y_1 + c_2y_2 + \frac{1}{2}xe^x = c_1e^x + c_2e^{-x} + \frac{1}{2}xe^x$. Note that y_1 and y_2 are linearly independent for all real numbers, since

$$W(y_1, y_2)(x) = e^x(-e^{-x}) - e^{-x}(e^x) = -2 \neq 0. \quad ◀$$

> **Principle of superposition for nonhomogeneous equation**
>
> Suppose that y_f is a solution to
>
> $$y'' + a(x)y' + b(x)y = f(x) \tag{24}$$
>
> and y_g is a solution to
>
> $$y'' + a(x)y' + b(x)y = g(x) \tag{25}$$
>
> Then $y_f + y_g$ is a solution to
>
> $$y'' + a(x)y' + b(x)y = f(x) + g(x) \tag{26}$$

Proof Let $y(x) = y_f(x) + y_g(x)$. Then

$$y'' + a(x)y' + b(x)y = (y_f + y_g)'' + a(x)(y_f + y_g)' + b(x)(y_f + y_g)$$

from Equations **(24)** and **(25)**

$$= (y_f'' + a(x)y_f' + b(x)y_f) + (y_g'' + a(x)y_g' + b(x)y_g) = f(x) + g(x) \quad \blacklozenge$$

EXAMPLE 9 ▶ **Using the principle of superposition**

It is not difficult to verify that one solution to $y'' - y = x^2$ is $-2 - x^2$ and one solution to $y'' - y = 2e^x$ is xe^x. It then follows from the principle of superposition that one solution to $y'' - y = x^2 + 2e^x$ is $-2 - x^2 + xe^x$. To verify this we set

$$y = -2 - x^2 + xe^x.$$

Then

$$y' = -2x + (x + 1)e^x$$
$$y'' = -2 + (x + 2)e^x.$$

So

$$y'' - y = [-2 + (x + 2)e^x] - [-2 - x^2 + xe^x]$$
$$= -2 + xe^x + 2e^x + 2 + x^2 - xe^x = x^2 + 2e^x. \quad ◀$$

In the next three sections we present methods for finding the general solution of homogeneous equations. In Sections 3.5 and 3.6 techniques for obtaining a solution y_p of a nonhomogeneous equation are developed.

General solutions

We have used the term **general solution** to describe a two-parameter (c_1 and c_2) family of solutions of either the homogeneous Equation **(23)** or the nonhomogeneous Equation **(22).** The general solution of an nth-order linear differential equation is a family of solutions involving n parameters (see Section 3.8). Any solution that is free of arbitrary parameters is called a **particular solution.**

There is another school of thought about the concept of a general solution. This viewpoint holds that a general solution must contain *all* solutions to the differential equation in some interval. No difficulties arise from this requirement when we are considering linear differential equations, since Theorems 4 and 6 guarantee that every solution is obtained by the linear combinations

$$c_1y_1(x) + c_2y_2(x) \quad \text{or} \quad c_1y_1(x) + c_2y_2(x) + y_p(x),$$

respectively. However, similar results are *not* known for nonlinear equations. Even if we have a two-parameter family of solutions to a nonlinear second-order equation (that is, a solution that involves two arbitrary constants), it is certainly not easy to determine if this family includes all solutions of the equation. The prespective on page 51 provides an example of infinitely many singular solutions that are not included in a one-parameter family of solutions. Higher-order nonlinear equations often possess singular solutions.

Thus, on a practical level, the requirement that a "general solution" include all solutions presents an extra complication. Observe, however, that both definitions yield the same outcome for linear equations.

Proofs of Theorems 4 and 5

THEOREM 5 ♦

Two solutions $y_1(x)$ and $y_2(x)$ of

$$y'' + a(x)y' + b(x)y = 0$$

are linearly independent on $[x_0, x_1]$, $x_0 < x_1$, if and only if $W(y_1, y_2)(x) \neq 0$ for all x in $[x_0, x_1]$.

Proof We first show that if $W(y_1, y_2)(x) = 0$ for some x_2 in $[x_0, x_1]$, then y_1 and y_2 are linearly dependent. Consider the system of equations

$$\begin{aligned} c_1 y_1(x_2) + c_2 y_2(x_2) &= 0, \\ c_1 y_1'(x_2) + c_2 y_2'(x_2) &= 0. \end{aligned} \tag{27}$$

The determinant of this system is

$$y_1(x_2)y_2'(x_2) - y_2(x_2)y_1'(x_2) = W(y_1, y_2)(x_2) = 0.$$

Thus, according to the theory of determinants (see Theorem 3 in Appendix 4), there exists a solution (c_1, c_2) for the system in equations **(27)** where c_1 and c_2 are not both equal to zero. Define $y(x) = c_1y_1(x) + c_2y_2(x)$. By Theorem 3, $y(x)$ is a solution of

$$y'' + a(x)y' + b(x)y = 0.$$

But since c_1 and c_2 satisfy system **(27)**,

$$y(x_2) = c_1 y_1(x_2) + c_2 y_2(x_2) = 0$$

and

$$y'(x_2) = c_1 y_1'(x_2) + c_2 y_2'(x_2) = 0.$$

Thus $y(x)$ solves the initial-value problem

$$y'' + a(x)y' + b(x)y = 0, \qquad y(x_2) = y'(x_2) = 0.$$

But this initial-value problem also has the solution $y_3 \equiv 0$ for all values of x in $x_0 \leq x \leq x_1$. By Theorem 1, the solution of this initial-value problem is unique, so necessarily $y(x) = y_3(x) \equiv 0$. Thus

$$y(x) = c_1 y_1(x) + c_2 y_2(x) = 0,$$

for all values of x in $x_0 \leq x \leq x_1$, which proves that y_1 and y_2 are linearly dependent.

We now assume that $W(y_1, y_2)(x) \neq 0$ for all x in $[x_0, x_1]$ and prove that y_1 and y_2 are linearly independent. Consider the equation

$$c_1 y_1(x) + c_2 y_2(x) = 0,$$

for all x in $[x_0, x_1]$. Differentiating with respect to x, we have

$$c_1 y_1'(x) + c_2 y_2'(x) = 0,$$

for all x in $[x_0, x_1]$. Select any x_2 in $[x_0, x_1]$ and consider the homogeneous system

$$\begin{aligned} c_1 y_1(x_2) + c_2 y_2(x_2) &= 0, \\ c_1 y_1'(x_2) + c_2 y_2'(x_2) &= 0. \end{aligned}$$

Its determinant,

$$\begin{vmatrix} y_1(x_2) & y_2(x_2) \\ y_1'(x_2) & y_2'(x_2) \end{vmatrix} = W(y_1, y_2)(x_2),$$

does not equal zero; thus $c_1 = c_2 = 0$. Hence $y_1(x)$ and $y_2(x)$ are linearly independent on $[x_0, x_1]$. ♦

THEOREM 4 ♦

Let $y_1(x)$ and $y_2(x)$ be linearly independent solutions on $[x_0, x_1]$, $x_0 < x_1$, to

$$y'' + a(x)y' + b(x)y = 0$$

and let $y_3(x)$ be another solution on this interval. Then there exist unique constants c_1 and c_2 such that

$$y_3(x) = c_1y_1(x) + c_2y_2(x).$$

Proof Let $y_3(x_0) = \alpha$ and $y_3'(x_0) = \beta$. Consider the linear system of equations in two unknowns c_1 and c_2:

$$\begin{aligned} y_1(x_0)c_1 + y_2(x_0)c_2 &= \alpha, \\ y_1'(x_0)c_1 + y_2'(x_0)c_2 &= \beta. \end{aligned} \tag{28}$$

As we saw earlier, the determinant of this system is $W(y_1, y_2)(x_0)$, which is nonzero since the solutions are linearly independent. Thus (by Theorem 5) there is a unique solution (c_1, c_2) to systems of equations **(28)** and a solution $y*(x) = c_1y_1(x) + c_2y_2(x)$ that satisfies the conditions $y*(x_0) = \alpha$ and $y*'(x_0) = \beta$. Since every initial-value problem has a unique solution (by Theorem 1), it must follow that $y_3(x) = y*(x)$ on the interval $x_0 \le x \le x_1$, and so the proof is complete. ♦

An aside: well-posed problems

In Section 1.2 (on page 18) we stated an existence-uniqueness result for first-order differential equations. On page 105 we gave an existence-uniqueness theorem of first-order linear equations and, in Theorem 1 in Section 3.1.1, we extended that result to nth-order linear differential equations.

There are, however, other questions that can be asked. Consider the initial-value problem of exponential growth:

$$y' = ay, \qquad y(0) = y_0. \tag{29}$$

As we have seen several times, the solution to this initial-value problem is

$$y(x) = y_0e^{ax}.$$

Suppose that we change the initial condition to $y(0) = y_1$. Then the solution is

$$\overline{y}(x) = y_1e^{ax}.$$

We then find that

$$|y(x) - \overline{y}(x)| = |y_0 - y_1|e^{ax}.$$

Clearly, if y_0 is close to y_1 and x is small, then $|y(x) - \overline{y}(x)|$ is also small. In this case we say that the solutions to the problem **(29) depend continuously on the**

initial condition. More generally, the solutions to an initial-value problem **depend continuously on initial conditions** if, whenever we vary the initial conditions by a small amount, the solutions change only by a small amount. We could make this notion more mathematically precise but that would lead us too far astray.

When an nth-order initial-value problem has a unique solution for a given set of n initial conditions and when the solutions depend continuously on the initial conditions, the problem is said to be **well-posed.** The linear initial-value problem (4), (5) is well-posed whenever the functions, $a_0, a_1, \ldots, a_{n-1}$ are continuous.

However, many nonlinear initial-value problems are not well-posed. In Example 1.2.6 on page 17 we saw an initial-value problem that was not well-posed because it had more than one solution. For certain nonlinear problems it happens that even a small change in the initial condition can lead to very great changes in the solutions—even for small values of the independent variable. This is one of the criteria for **chaos.**

In Section 2.9, we gave one example of chaos occurring in a nonlinear difference equation (see page 90). Chaos is a fascinating subject and many recent books and articles discuss it. We shall not, however, discuss continuous dependence on initial conditions, well-posedness, or chaos any further in this book. After all, you should have something to look forward to in your next differential equations course.

SELF-QUIZ

I. Which of the following problems is guaranteed by the existence-uniqueness theorem to have a unique solution?

(a) $y'' + \frac{1}{x}y' + 3y = x^2;\ y(0) = 0,\ y'(0) = 1$

(b) $y'' + \frac{1}{x}y' + 3y = x^2;\ y(1) = 1,\ y'(1) = 3$

(c) $y'' + \frac{1}{x}y' + 3y = x^2;\ y(1) = 1,\ y(2) = 3$

(d) $xy'' + y' + 4xy = 0;\ y(0) = 1,\ y'(0) = 2$

II. e^x and e^{2x} are two solutions to $y'' - 3y' + 2y = 0$. Which of the following are also solutions to $y'' - 3y' + 2y = 0$?
(a) $e^x + e^{2x}$ **(b)** $e^x - e^{2x}$ **(c)** $-8e^{2x}$
(d) $17e^x$ **(e)** $4e^x - 3.8e^{2x}$

III. Which of the following pairs of functions are linearly independent?

(a) e^x, e^{2x} **(b)** $\frac{-3}{x}, \frac{4}{x}$ **(c)** $\ln x, \ln x^3$

(d) $\sin x, \tan x$ **(e)** $3, x$

IV. Which of the following is the general solution to $y'' + y = 0$?
(a) $c_1e^x + c_2e^{-x}$ **(b)** $c_1 \sin x + c_2 \sin 2x$
(c) $c_1 \cos x + c_2 \cos(-x)$ **(d)** $c_1 \sin x + c_2 \cos x$
(e) $c_1 \cos x - c_2 \sin x$

V. The Wronskian for the functions $y_1(x) = x^2$ and $y_2(x) = x^5$ is
(a) $5x^6 - 2x^5$ **(b)** x^7 **(c)** $2x^3 - 5x^9$
(d) $2x - 5x^4$ **(e)** $3x^6$

VI. The general solution to $y'' + 4y = 0$ is $c_1 \cos 2x + c_2 \sin 2x$. One solution to $y'' + 4y = x$ is $y = \frac{1}{4}x$. One solution to $y'' + 4y = e^x$ is $y = e^x/5$. Which of the following are solutions to $y'' + 4y = x + e^x$?

(a) $\frac{1}{4}x + e^{x/5}$ **(b)** $x + 4e^{x/5}$ **(c)** $\frac{1}{4}x - e^{x/5}$

(d) $\frac{1}{4}x + e^{x/5} + 3\cos 2x - 5\sin 2x$

(e) $\frac{1}{4}x + \sin 2x$ **(f)** $e^{x/5} - 3\cos 2x$

(g) $\frac{1}{4}x + e^{x/5} + \sin 2x$ **(h)** $\frac{1}{4}x + e^{x/5} - \cos 2x$

Answers to Self-Quiz

I. b **II.** all of them **III.** a, d, e **IV.** d or e **V.** e **VI.** a, d, g, h

PROBLEMS 3.1

In Problems 1–10 determine whether the given equation is linear or nonlinear. If it is linear, state whether it is homogeneous or nonhomogeneous with constant or variable coefficients.

1. $y'' + 2x^3y' + y = 0$
2. $y'' + 2y' + y^2 = x$
3. $y'' + 3y' + yy' = 0$
4. $y'' + 3y' + 4y = 0$
5. $y'' + 3y' + 4y = \sin x$
6. $y'' + y(2 + 3y) = e^x$
7. $y'' + 4xy' + 2x^3y = e^{2x}$
8. $y'' + \sin(xe^x)y' + 4xy = 0$
9. $3y'' + 16y' + 2y = 0$
10. $yy'y'' = 1$

In Problems 11–16, determine whether the given pair of functions is linearly dependent or independent on R.

11. $y_1 = x, y_2 = x^2$
12. $y_1 = x^3, y_2 = -3x^3$
13. $y_1 = 2e^{3x}, y_2 = 3e^{2x}$
14. $y_1 = \ln(1 + x^2), y_2 = \ln(1 + x^2)^5$
15. $y_1 = x^2 - 1, y_2 = x^2 + 1$
16. $y = \sin x, y_2 = \cos x$

In Problems 17–24, use the Wronskian to show that the given functions are linearly independent on the given interval.

17. $\frac{1}{x}, \frac{1}{x^2}; [0.1, \infty)$
18. $e^x, \ln x; (1, \infty)$
19. $\sin x, \tan x; \left[0, \frac{\pi}{3}\right]$
20. $1, x^2, x^7; \mathbb{R}$
21. $x^n, x^m, x^p; \mathbb{R};$ no two of n, m, and p are equal integers.
22. $\sin x, \sin 2x; \mathbb{R}$
23. $\cos 3x, \cos 5x, \sin x; \mathbb{R}$
24. $\sin nx, \sin mx, \cos nx, \cos mx; \mathbb{R}; n \neq m$

In Problems 25–28 test each of the functions $1, x, x^2$, *and* x^3 *to see which functions satisfy the given differential equation. Then construct the general solution to the equation by writing a linear combination of the linearly independent solutions you have found.*

25. $y'' = 0$
26. $y''' = 0$
27. $xy'' - y' = 0$
28. $x^2y'' - 2xy' + 2y = 0$

In Problems 29–36, verify that the two given functions are linearly independent solutions of the given homogeneous equation and then find the general solution.

29. $y'' + 9y = 0; \sin 3x, \cos 3x$
30. $y'' - y' - 2y = 0; e^{2x}, e^{-x}$
31. $y'' + 2y' - 15y = 0; e^{3x}, e^{-5x}$
32. $y'' + 9y' + 14y = 0; e^{-2x}, e^{-7x}$
33. $y'' - 6y' + 9y = 0; e^{3x}, xe^{3x}$
34. $y'' + 4y' + 4y = 0; e^{-2x}, xe^{-2x}$
35. $y'' + 2y' + 2y = 0; e^{-x}\cos x, e^{-x}\sin x$
36. $y'' - 6y' + 13y = 0; e^{3x}\cos 2x, e^{3x}\sin 2x$

In Problems 37–40, a particular solution to a nonhomogeneous equation is given. Using the information found in Problems 29, 30, 32, and 35, find the general solution.

37. $y'' + 9y = 4x; y_p = \frac{4}{9}x$

38. $y'' - y' - 2y = e^x; y_p = -\frac{1}{2}e^x$

39. $y'' + 2y' - 15y = 15x^2; y_p = -x^2 - \frac{4}{15}x - \frac{8}{225}$

40. $y'' + 2y' + 2y = \cos x; y_p = \frac{1}{5}\cos x + \frac{2}{5}\sin x$

41. Let $y_1(x)$ be a solution of the homogeneous equation

$$y'' + a(x)y' + b(x)y = 0$$

on the interval $\alpha \le x \le \beta$. Suppose that the curve y_1 is tangent to the x-axis at some point of this interval. Prove that y_1 must be identically zero.

42. Let $y_1(x)$ and $y_2(x)$ be two nontrivial solutions of the homogeneous equation

$$y'' + a(x)y' + b(x)y = 0$$

on the interval $\alpha \le x \le \beta$. Suppose $y_1(x_0) = y_2(x_0) = 0$ for some point $\alpha \le x_0 \le \beta$. Show that y_2 is a constant multiple of y_1.

43. **(a)** Show that x and x^3 are linearly independent on $|x| < 1$ even though the Wronskian $W(x, x^3) = 0$ at $x = 0$.
(b) Show that x and x^3 are solutions to

$$x^2y'' - 3xy' + 3y = 0.$$

Does this contradict Theorem 5?

44. **(a)** Show that $y_1(x) = \sin x^2$ and $y_2(x) = \cos x^2$ are linearly independent solutions of

$$xy'' - y' + 4x^3y = 0.$$

(b) Calculate $W(y_1, y_2)(x)$ and show that it is zero when $x = 0$. Does this result contradict Theorem 5? [*Hint:* In Theorem 5, as elsewhere in this section, it is assumed that $a(x)$ and $b(x)$ are continuous on some interval.]

45. Show that $y_1(x) = \sin x$ and $y_2(x) = 4 \sin x - 2 \cos x$ are linearly independent solutions of $y'' + y = 0$. Write the solution $y_3(x) = \cos x$ as a linear combination of y_1 and y_2.

46. Prove that $e^x \sin x$ and $e^x \cos x$ are linearly independent solutions of the equation

$$y'' - 2y' + 2y = 0.$$

(a) Find a solution that satisfies the conditions $y(0) = 1$, $y'(0) = 4$.
(b) Find another pair of linearly independent solutions.

47. Assume that some nonzero solution of

$$y'' + a(x)y' + b(x)y = 0, \qquad y(0) = 0$$

vanishes at some point x_1, where $x_1 > 0$. Prove that any other solution vanishes at $x = x_1$. Assume that a and b are continuous on $[0, x_1]$.

48. Define the function $s(x)$ to be the unique solution of the initial-value problem

$$y'' + y = 0; \qquad y(0) = 0, \qquad y'(0) = 1,$$

and the function $c(x)$ as the solution of

$$y'' + y = 0; \qquad y(0) = 1, \qquad y'(0) = 0.$$

Without using trigonometry, prove that
(a) $\dfrac{ds}{dx} = c(x)$; **(b)** $\dfrac{dc}{dx} = -s(x)$;
(c) $s^2 + c^2 = 1$.

49. **(a)** Show that $y_1 = \sin \ln x^2$ and $y_2 = \cos \ln x^2$ are linearly independent solutions of

$$y'' + \frac{1}{x}y'' + \frac{4}{x^2}y = 0 \quad (x > 0).$$

(b) Calculate $W(y_1, y_2)(x)$.

3.2 USING ONE SOLUTION TO FIND ANOTHER: REDUCTION OF ORDER

As we saw in Theorem 3.1.4, it is easy to write down the general solution of the homogeneous equation

$$y'' + a(x)y' + b(x)y = 0, \tag{1}$$

provided we know two linearly independent solutions y_1 and y_2 of Equation **(1)**. The general solution is then given by

$$y = c_1 y_1 + c_2 y_2,$$

where c_1 and c_2 are arbitrary constants. Unfortunately, there is no general procedure for determining y_1 and y_2 unless a and b are constant functions or a and b take some special form (as in Section 3.7). However, a standard procedure does exist for finding a y_2 when a y_1 is known. This method is of considerable importance, since it is sometimes possible to find one solution by inspecting the equation or by trial and error.

We assume that y_1 is a nonzero solution of Equation **(1)** on some interval and seek another solution y_2 such that y_1 and y_2 are linearly independent. Suppose that y_2 can be found. Then, since y_1 and y_2 are linearly independent, $y_2 \neq ky_1$ for any constant k. So

$$\frac{y_2}{y_1} = v(x)$$

must be a nonconstant function of x, and $y_2 = vy_1$ must satisfy Equation **(1)**. Thus

$$(vy_1)'' + a(vy_1)' + b(vy_1) = 0. \tag{2}$$

But

$$(vy_1)' = vy_1' + v'y_1 \tag{3}$$

and

$$\begin{aligned}(vy_1)'' &= (vy_1' + v'y_1)' = vy_1'' + v'y_1' + v'y_1' + v''y_1 \\ &= vy_1'' + 2v'y_1' + v''y_1. \end{aligned} \tag{4}$$

Using Equations **(3)** and **(4)** in Equation **(2),** we have

$$(vy_1'' + 2v'y_1' + v''y_1) + a(vy_1' + v'y_1) + bvy_1 = 0,$$

or

$$v(y_1'' + ay_1' + by_1) + v'(2y_1' + ay_1) + v''y_1 = 0. \tag{5}$$

The first term in parentheses in Equation **(5)** is zero since y_1 is a solution of Equation **(1),** so we obtain

$$v''y_1 + v'(2y_1' + ay_1) = 0.$$

Set $z = v'$, so that $z' = v''$, and divide by y_1 to obtain

$$z' + \left(\frac{2y_1'}{y_1} + a(x)\right)z = 0, \tag{6}$$

a separable linear first-order equation. Thus we have *reduced the order* of our equation. Multiply by the integrating factor for Equation **(6)**

$$e^{\int (2y_1'/y_1 + a(x))\,dx} = e^{2\ln y_1 + \int a(x)\,dx} = y_1^2 e^{\int a(x)\,dx},$$

so that Equation **(6)** becomes

$$(y_1^2 e^{\int a(x)\,dx} z)' = 0. \tag{7}$$

A solution to Equation **(7)** is (if the constant of integration is 1)

$$z = v' = \frac{1}{y_1^2} e^{-\int a(x)\,dx}.$$

Since the exponential is never zero, v is nonconstant. To find v, we perform another integration and obtain

Finding a second solution, y_2, when one solution, y_1 is given

$$y_2(x) = y_1(x)v(x) = y_1(x) \int \frac{e^{-\int a(x)\,dx}}{y_1^2(x)}\,dx. \tag{8}$$

Remark

It is not advisable to memorize formula **(8).** It is only necessary to remember the substitution $y_2 = vy_1$ and substitute this into the original differential equation. The following examples illustrate this procedure.

EXAMPLE 1 ▶ **Finding a second solution when one is given**

Note that $y_1 = x$ is a solution of

$$x^2y'' - xy' + y = 0, \qquad x > 0. \tag{9}$$

Setting $y_2 = y_1v = xv(x)$, it follows that $y_2' = xv' + v$, $y_2'' = xv'' + 2v'$, and Equation **(9)** becomes

$$x^2(xv'' + 2v') - x(xv' + v) + (xv) = 0,$$

or

$$x^3v'' + x^2v' = 0.$$

Setting $z = v'$ and separating variables, we have

$$\frac{dz}{z} = \frac{-dx}{x} \quad \text{or} \quad \ln|z| = -\ln x, \qquad \text{since } x > 0$$

from which we obtain by exponentiation

$$v' = z = \frac{1}{x}.$$

Thus $v(x) = \ln x$, so $y_2 = y_1v = x \ln x$ and the general solution of Equation **(9)** is

$$y = c_1x + c_2x \ln x, \quad x > 0. \blacktriangleleft$$

EXAMPLE 2 ▶ **Finding a second solution to a Legendre equation**

Consider the **Legendre equation of order one:**

$$(1 - x^2)y'' - 2xy' + 2y = 0, \qquad -1 < x < 1. \tag{10}$$

Again, it is not hard to verify that $y_1(x) = x$ is a solution. Setting $y_2 = xv$, we have

$$\begin{aligned}(1 - x^2)(xv'' + 2v') - 2x(xv' + v) + 2xv &= x(1 - x^2)v'' + 2(1 - 2x^2)v' \\ &= 0;\end{aligned}$$

when we substitute $z = v'$ and divide by $x(1 - x^2)$, we can use partial fractions to obtain

$$\frac{dz}{dx} + \frac{2(1 - 2x^2)z}{x(1 - x^2)} = \frac{dz}{dx} + \left(\frac{2}{x} - \frac{2x}{1 - x^2}\right)z = 0.$$

Separating variables and integrating, we obtain

$$\ln|z| = \int \frac{dz}{z} = \int \left(\frac{2x}{1 - x^2} - \frac{2}{x}\right) dx = -\ln(1 - x^2) - \ln(x^2).$$

Thus, exponentiating and using partial fractions, we get

$$v' = \frac{1}{x^2(1 - x^2)} = \frac{1}{x^2} + \frac{1}{2}\left(\frac{1}{1 + x} + \frac{1}{1 - x}\right).$$

Hence

$$\begin{aligned}y_2 = y_1v &= x \int \left[\frac{1}{x^2} + \frac{1}{2}\left(\frac{1}{1 + x} + \frac{1}{1 - x}\right)\right] dx \\ &= x\left[\frac{-1}{x} + \frac{1}{2}\ln\left(\frac{1 + x}{1 - x}\right)\right],\end{aligned}$$

or

$$y_2 = \frac{x}{2}\ln\left(\frac{1 + x}{1 - x}\right) - 1, \qquad |x| < 1. \blacktriangleleft$$

SELF-QUIZ

I. Suppose that $y_1(x)$ is a solution to the differential equation $y'' + \frac{1}{x}y' + (\sin x)y = 0$. Then a second solution is given by ______.

(a) $y_2(x) = y_1(x) \int \ln \frac{x}{y_1^2(x)}\, dx$

(b) $y_2(x) = y_1(x) \int \frac{x}{y_1^2(x)}\, dx$

(c) $y_2(x) = y_1(x) \int \frac{1}{xy_1^2(x)}\, dx$

(d) $y_2(x) = y_1(x) \int \frac{e^x}{y_1^2(x)}\, dx$

(e) $y_2(x) = y_1(x) \int \frac{e^{1/x}}{y_1^2(x)}\, dx$

II. $y_1 = e^{cx}$ is a solution to $y'' - 2cy' + c^2y = 0$ where c is a real number. A second linearly independent solution on $\mathbb{R}$ is ______.

(a) $e^{cx} \int dx$

(b) $e^{cx} \int \frac{e^{-2cx}}{e^{2cx}}\, dx$

(c) $e^{cx} \int ce^{-2cx}\, dx$

(d) $e^{cx} \int \frac{e^{-c^2x}}{e^{2cx}}\, dx$

(e) $e^{cx} \int \frac{e^{-x^2/2}}{e^{2cx}}\, dx$

Answers to Self-Quiz

I. c **II.** a

PROBLEMS 3.2

In each of Problems 1–25 a second-order differential equation and one solution $y_1(x)$ are given. Verify that $y_1(x)$ is indeed a solution and find a second linearly independent solution.

1. $y'' - 2y' + y = 0,\ y_1(x) = e^x$

2. $y'' + 4y = 0,\ y_1(x) = \sin 2x$

3. $y'' - 4y = 0,\ y_1(x) = e^{2x}$

4. $y'' - 4y' + 4y = 0,\ y_1(x) = xe^{2x}$

5. $y'' + 5y' + 6y = 0,\ y_1(x) = e^{-2x}$

6. $y'' + 2y' + 2y = 0,\ y_1(x) = e^{-x} \cos x$

7. $y'' + y' + 7y = 0,\ y_1(x) = e^{-x/2} \cos \frac{3\sqrt{3}}{2}x$

8. $y'' + 10y' + 25y = 0,\ y_1(x) = xe^{-5x}$

9. $y'' - 30y' + 200y = 0,\ y_1(x) = e^{20x}$

10. $y'' + \left(\frac{3}{x}\right)y' = 0,\ y_1(x) = 1$

11. $y'' + y' + \frac{1}{4}y = 0;\ y_1(x) = e^{-x/2}$

12. $y'' + 0.2y' - 0.03y = 0;\ y_1(x) = e^{x/10}$

13. $x^2y'' + xy' - 4y = 0,\ y_1(x) = x^2,\ x > 0$

14. $y'' - 2xy' + 2y = 0,\ y_1(x) = x$

15. $x^2y'' - 2xy' + (x^2 + 2)y = 0,$ $(x > 0),\ y_1 = x \sin x$

16. $xy'' + (2x - 1)y' - 2y = 0,\ (x > 0),\ y_1 = e^{-2x}$

17. $xy'' + (x - 1)y' + (3 - 12x)y = 0,\ (x > 0),$ $y_1 = e^{3x}$

18. $xy'' - y' + 4x^3y = 0,\ (x > 0),\ y_1 = \sin(x^2)$

19. $xy'' - (x + n)y' + ny = 0$ (integer $n > 0$), $y_1(x) = e^x,\ x > 0$

20. $x^{1/3}y'' + y' + \left(\frac{1}{4}x^{-1/3} - \frac{1}{6x} - 6x^{-5/3}\right)y = 0,$ $y_1 = x^3e^{-3x^{2/3}/4},\ (x > 0)$

21. $(3 + x^2)y'' + 2xy' - 2y = 0;\ y_1(x) = x$

22. $y'' + \tan x\, y' = 0;\ y_1(x) = 1,\ 0 \le x \le \frac{\pi}{4}$

23. $y'' - 2 \cot x\, y' = 0;\ y_1(x) = 1,\ \frac{\pi}{4} \le x \le \frac{\pi}{2}$

24. $y'' + 2 \sec x\, y' = 0;\ y_1(x) = 1,\ 0 \le x \le \frac{\pi}{4}$

25. $y'' + 4 \csc x\, y' = 0;\ y_1(x) = 1,\ \frac{\pi}{4} \le x \le \frac{\pi}{2}$

In Problems 26–28, an equation and one solution are given. Find an expression involving an integral that gives a second, linearly independent solution. Do not try to evaluate the integral.

26. $2y'' + xy' - 2y = 0;\ y_1(x) = x^2 - 2$

27. $y'' - 2xy' - 2y = 0;\ y_1(x) = e^{x^2}$

28. $y'' + \sin x\, y' - \frac{\sin x}{x}y = 0;\ y_1(x) = x$

29. The **Bessel differential equation** is given by

$$x^2y'' + xy' + (x^2 - p^2)y = 0.$$

For $p = \frac{1}{2}$, verify that $y_1(x) = (\sin x)/\sqrt{x}$ is a solution for $x > 0$. Find a second linearly independent solution.

30. Letting $p = 0$ in the equation of Problem 29, we obtain the **Bessel differential equation of index zero,** which we will study in Chapter 5. One solution is the **Bessel function of order zero** denoted by $J_0(x)$. In terms of $J_0(x)$, find a second linearly independent solution.

3.3 HOMOGENEOUS EQUATIONS WITH CONSTANT COEFFICIENTS: REAL ROOTS

In this section we present a simple procedure for finding the general solution to the linear homogeneous equation with constant coefficients

$$y'' + ay' + by = 0. \tag{1}$$

Recall that for the comparable first-order equation $y' + ay = 0$ the general solution is $y(x) = ce^{-ax}$. It is then not implausible to "guess" that there may be a solution to Equation **(1)** of the form $y(x) = e^{\lambda x}$ for some number λ (real or complex). Setting $y(x) = e^{\lambda x}$, we obtain $y' = \lambda e^{\lambda x}$ and $y'' = \lambda^2 e^{\lambda x}$, so that Equation **(1)** yields

$$\lambda^2 e^{\lambda x} + a\lambda e^{\lambda x} + be^{\lambda x} = 0.$$

Since $e^{\lambda x} \neq 0$, we can divide this equation by $e^{\lambda x}$ to obtain

The characteristic equation

$$\lambda^2 + a\lambda + b = 0 \tag{2}$$

where a and b are real numbers. Equation **(2)** is called the **characteristic equation** of the differential equation **(1)**. It is clear that if λ satisfies Equation **(2)**, then $y(x) = e^{\lambda x}$ is a solution to Equation **(1)**. As we saw in Section 3.1, we need only obtain two linearly independent solutions. Equation **(2)** has the roots

$$\lambda_1 = \frac{-a + \sqrt{a^2 - 4b}}{2} \quad \text{and} \quad \lambda_2 = \frac{-a - \sqrt{a^2 - 4b}}{2}. \tag{3}$$

There are three possibilities: $a^2 - 4b > 0$, $a^2 - 4b = 0$, $a^2 - 4b < 0$.

Case 1: Roots real and unequal

If $a^2 - 4b > 0$, then λ_1 and λ_2 are distinct real numbers, given by **(3)**, and $y_1(x) = e^{\lambda_1 x}$ and $y_2 = e^{\lambda_2 x}$ are distinct solutions.

These two solutions are linearly independent because

$$\frac{y_1}{y_2} = e^{(\lambda_1 - \lambda_2)x},$$

which is clearly not a constant when $\lambda_1 \neq \lambda_2$. Thus we have proved the following theorem.

THEOREM 1 ♦ The general solution to a homogeneous equation in the case of real, unequal roots

If $a^2 - 4b > 0$, then the roots of the characteristic equation are real and unequal and the general solution to Equation **(1)** is given by

$$y(x) = c_1 e^{\lambda_1 x} + c_2 e^{\lambda_2 x} \tag{4}$$

where c_1 and c_2 are arbitrary constants and λ_1 and λ_2 are the real roots of Equation **(2)**. ♦

EXAMPLE 1 ▶ A homogeneous equation with real, distinct roots

Consider the equation

$$y'' + 3y' - 10y = 0.$$

The characteristic equation is $\lambda^2 + 3\lambda - 10 = 0$, $a^2 - 4b = 49$, and the roots are $\lambda_1 = 2$ and $\lambda_2 = -5$ (the order in which the roots are taken is irrelevant). The general solution is

$$y(x) = c_1 e^{2x} + c_2 e^{-5x}.$$

If we specify the initial conditions $y(0) = 1$ and $y'(0) = 3$, for example, then differentiating and substituting $x = 0$, we obtain the simultaneous equations

$$c_1 + c_2 = 1,$$
$$2c_1 - 5c_2 = 3,$$

which have the unique solution $c_1 = \frac{8}{7}$ and $c_2 = -\frac{1}{7}$. The unique solution to the initial-value problem is therefore

$$y(x) = \frac{1}{7}(8e^{2x} - e^{-5x}).$$

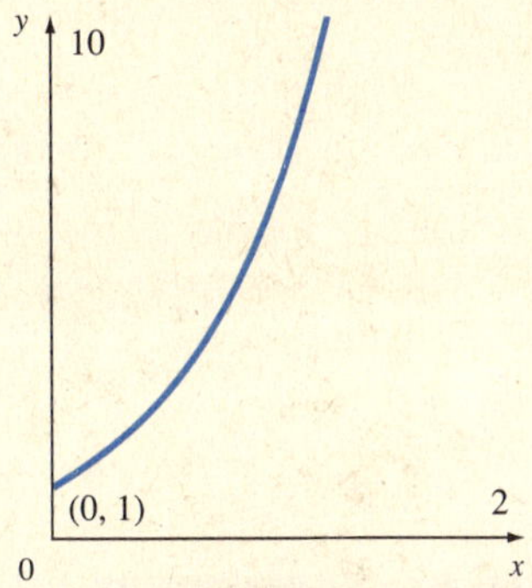

FIGURE 3.1
Graph of
$y = \frac{1}{7}(8e^{2x} - e^{-5x}),\quad x \geq 0$

Note that since $e^{2x} \to \infty$ and $e^{-5x} \to 0$, the solution $y(x) \to \infty$ as $x \to \infty$. It is sketched in Figure 3.1. ◀

Case 2: Roots real and equal

Suppose $a^2 - 4b = 0$. In this case Equation **(2)** has the double root $\lambda_1 = \lambda_2 = -\frac{a}{2}$. Thus $y_1(x) = e^{-ax/2}$ is a solution of Equation **(1)**. To find the second solution y_2, we make use of Equation **(3.2.8)**, p. 121, since one solution is known:

$$y_2(x) = y_1(x)\int \frac{e^{-ax}}{y_1^2(x)}\,dx = e^{-ax/2}\int \frac{e^{-ax}}{(e^{-ax/2})^2}\,dx$$
$$= e^{-ax/2}\int \frac{e^{-ax}}{e^{-ax}}\,dx = e^{-ax/2}\int dx = xe^{-ax/2}.$$

Since $y_2/y_1 = x$, it follows that y_1 and y_2 are linearly independent. Hence we have the following result.

THEOREM 2 ♦ **The general solution to a homogeneous equation in the case of one real root**

If $a^2 - 4b = 0$, then the roots of the characteristic equation are equal and the general solution to Equation **(1)** is given by

$$y(x) = c_1 e^{-ax/2} + c_2 xe^{-ax/2} = (c_1 + c_2 x)e^{-ax/2} \tag{5}$$

where c_1 and c_2 are arbitrary constants. ♦

EXAMPLE 2 ▶ **A homogeneous equation with one real root**

Consider the equation

$$y'' + 6y' + 9 = 0.$$

The characteristic equation is $\lambda^2 + 6\lambda + 9 = 0$, and $a^2 - 4b = 0$, yielding the unique double root $\lambda_1 = -\frac{a}{2} = -3$. The general solution is

$$y(x) = c_1 e^{-3x} + c_2 x\, e^{-3x}.$$

If we use the initial conditions $y(0) = 1$, $y'(0) = 7$, we obtain the simultaneous equations

$$c_1 = 1$$
$$-3c_1 + c_2 = 7,$$

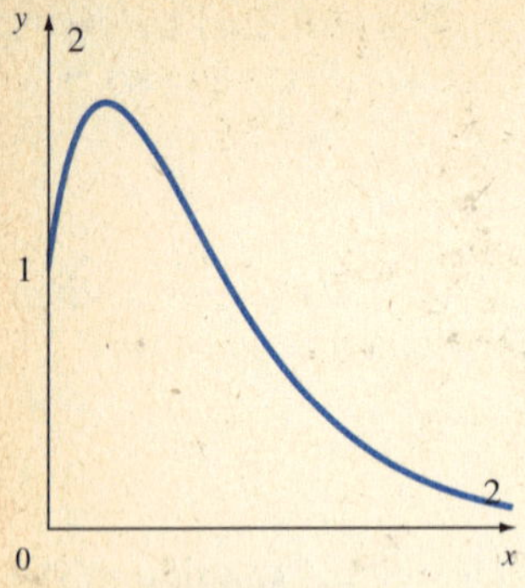

FIGURE 3.2
Graph of $y(x) = e^{-3x}(1 + 10x)$

which yield the unique solution (since $c_2 = 7 + 3c_1 = 10$).

$$y(x) = e^{-3x} + 10x\,e^{-3x} = e^{-3x}(1 + 10x).$$

Since $\lim_{x\to\infty} e^{-3x} = \lim_{x\to\infty} xe^{-3x} = 0$, the solution $y(x) \to 0$ as $x \to \infty$. It is sketched in Figure 3.2. ◀

We will deal with the more complicated situation of complex roots $(a_2 - 4b < 0)$ in Section 3.4.

SELF-QUIZ

I. Two linearly independent solutions to $y'' + y' - 6y = 0$ are ______.
(a) e^{2x}, e^{3x} **(b)** e^{-2x}, e^{3x} **(c)** e^{2x}, e^{-3x}
(d) e^{-2x}, e^{-3x} **(e)** e^{x}, e^{-6x}

II. Two linearly independent solutions to $y'' + 8y' + 16y = 0$ are ______.
(a) e^{8x}, e^{16x} **(b)** $e^{4x}, 2e^{4x}$ **(c)** $e^{-4x}, -4e^{-4x}$
(d) $e^{4x}, x\,e^{4x}$ **(e)** e^{-4x}, xe^{-4x}

III. Which condition ensures that the characteristic equation for $y'' + 6y' + ky = 0$ has two real roots?
(a) $k \geq 9$ **(b)** $k > 9$ **(c)** $k \leq 9$ **(d)** $k < 9$
(e) k is a real number.

IV. Which condition ensures that the characteristic equation for $y'' + ky' - 5y = 0$ has two real roots?
(a) $k^2 < 20$ **(b)** $k^2 > 20$ **(c)** $k > \sqrt{5}$
(d) $0 < k < \sqrt{5}$ **(e)** k is a real number.

Answers to Self-Quiz

I. c **II.** e **III.** d **IV.** e

PROBLEMS 3.3

In Problems 1–24 find the general solution of each equation. When initial conditions are specified, give the particular solution that satisfies them and sketch a graph using a graphing calculator or a computer algebra system.

1. $y'' - 4y = 0$
2. $x'' + x' - 6x = 0,\ x(0) = 0,\ x'(0) = 5$
3. $y'' - 3y' + 2y = 0$
4. $y'' + 5y' + 6y = 0,\ y(0) = 1,\ y'(0) = 2$
5. $4x'' + 20x' + 25x = 0,\ x(0) = 1,\ x'(0) = 2$
6. $y'' + 6y' + 9y = 0$
7. $x'' - x' - 6x = 0,\ x(0) = -1,\ x'(0) = 1$
8. $y'' - 8y' + 16y = 0,\ y(0) = 2,\ y'(0) = -1$
9. $y'' - 5y' = 0$
10. $y'' + 17y' = 0,\ y(0) = 1,\ y'(0) = 0$
11. $y'' + 2\pi y' + \pi^2 y = 0$
12. $y'' - 13y' + 42y = 0$
13. $z'' + 2z' - 15z = 0$
14. $w'' + 8w' + 12w = 0$
15. $y'' - 8y' + 16y = 0,\ y(0) = 1,\ y'(0) = 6$
16. $y'' + 2y' + y = 0,\ y(1) = \dfrac{2}{e},\ y'(1) = -\dfrac{3}{e}$
17. $y'' - 2y = 0$
18. $y'' + 6y' + 5y = 0$
19. $y'' - 5y = 0,\ y(0) = 3,\ y'(0) = -\sqrt{5}$
20. $y'' - 2y' - 2y = 0,\ y(0) = 1,\ y'(0) = 1 + 3\sqrt{3}$
21. $y'' + 3y' = 0,\ y(0) = 1,\ y'(0) = 1$
22. $y'' - 8y' + 16y = 0$
23. $y'' + 10y' + 16y = 0$
24. $y'' - 6y' - 16y = 0,\ y(1) = y'(1) = 1$

In Problems 25–30 match each initial-value problem with the graph of its solution.

25. $y'' - 7y' + 12y = 0,\ y(0) = -1,\ y'(0) = -2$
26. $y'' + y' - 12y = 0,\ y(0) = -1,\ y'(0) = -10$
27. $y'' - y' - 12y = 0,\ y(0) = -1,\ y'(0) = 10$
28. $y'' + 7y' + 12y = 0,\ y(0) = -1,\ y'(0) = 2$
29. $y'' - 7y' + 12y = 0,\ y(0) = -1,\ y'(0) = 2$
30. $y'' + 7y' + 12y = 0,\ y(0) = -1,\ y'(0) = -2$

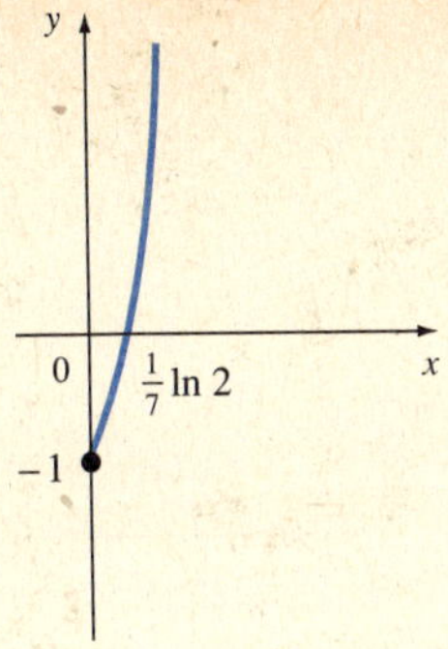

(a)

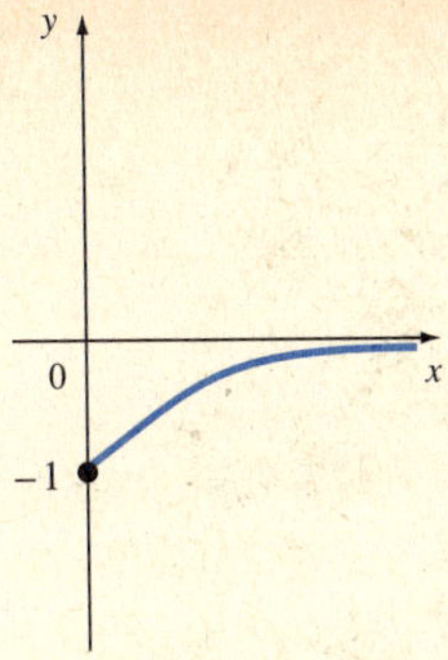

(b)

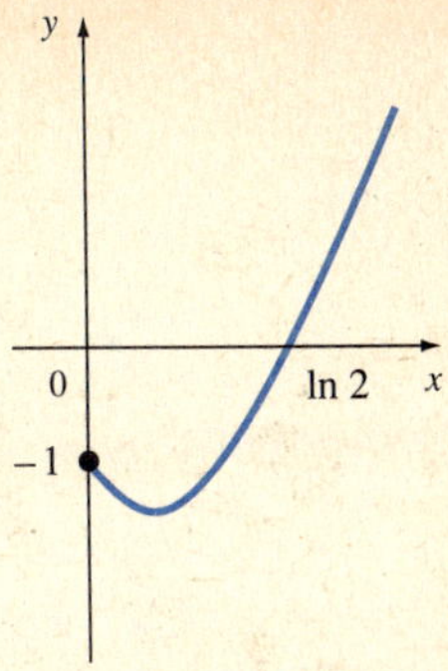

(c)

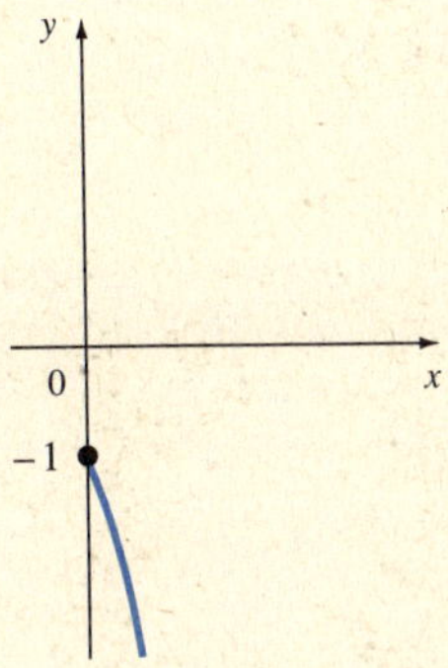

(d)

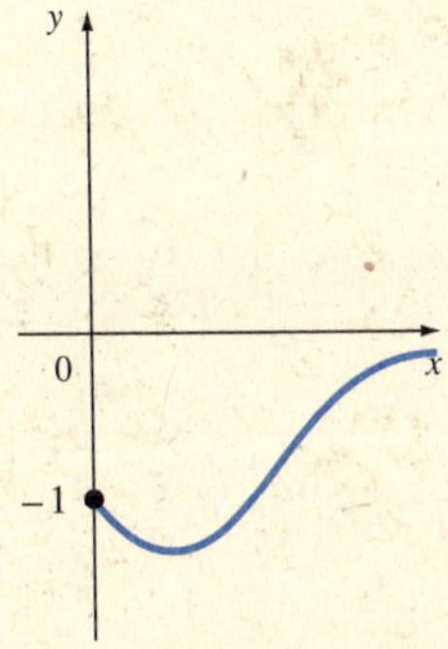

(e)

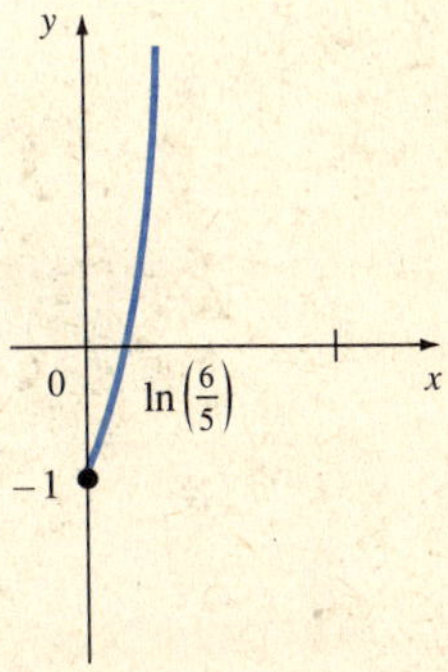

(f)

The Riccati Equation* *Linear second-order differential equations may also be used in finding the solution to the* ***Riccati equation:***

$$y' + y^2 + a(x)y + b(x) = 0. \qquad \textbf{(i)}$$

This nonlinear first-order equation frequently occurs in physical applications. To change it into a linear second-order equation, let $y = z'/z$. *Then* $y' = (z''/z) - (z'/z)^2$, *so Equation* **(i)** *becomes*

$$\frac{z''}{z} - \left(\frac{z'}{z}\right)^2 + \left(\frac{z'}{z}\right)^2 + a(x)\left(\frac{z'}{z}\right) + b(x) = 0.$$

Multiplying by z, we obtain the linear second-order equation

$$z'' + a(x)z' + b(x)z = 0. \qquad \textbf{(ii)}$$

If the general solution to Equation **(ii)** *can be found, the quotient* $y = z'/z$ *is the general solution to Equation* **(i)**.

31. Suppose $z = c_1z_1 + c_2z_2$ is the general solution to Equation **(ii)**. Explain why the quotient z'/z involves only one arbitrary constant. [*Hint:* Divide numerator and denominator by c_1.]

32. For arbitrary constants a, b, and c, find the substitution that changes the nonlinear equation

$$y' + ay^2 + by + c = 0$$

into a linear second-order equation with constant coefficients. What second-order equation is obtained?

Use the method above to find the general solution to the Riccati equations in Problems 33–38. If an initial condition is specified, give the particular solution that satisfies that condition.

33. $y' + y^2 - 1 = 0,\ y(0) = -\dfrac{1}{3}$

34. $\dfrac{dx}{dt} + x^2 + 1 = 0$

35. $y' + y^2 - 2y + 1 = 0$

36. $y' + y^2 + 3y + 2 = 0,\ y(0) = 1$

37. $y' + y^2 - y - 2 = 0$

38. $y' + y^2 + 2y + 1 = 0,\ y(1) = 0$

*Jacopo Francesco Riccati (1676–1754), an Italian mathematician, physicist, and philosopher, was responsible for bringing much of Newton's work on calculus to the attention of Italian mathematicians.

39. Suppose that the two roots λ_1 and λ_2 of the characteristic equation **(2)** are real and satisfy $\lambda_2 = \lambda_1 + h$.

(a) Verify that

$$\phi_h(x) = \frac{e^{\lambda_2 x} - e^{\lambda_1 x}}{\lambda_2 - \lambda_1}$$

is a solution of Equation **(1)**.

(b) Hold λ_1 fixed and evaluate $\phi_h(x)$, as $h \to 0$, with L'Hôpital's rule.

(c) Verify that parts (a) and (b) yield Theorem 2's conclusion.

3.4 HOMOGENEOUS EQUATIONS WITH CONSTANT COEFFICIENTS: COMPLEX ROOTS

The material in this section requires some familiarity with the basic properties of complex numbers. A review of these properties is given in Appendix 5.

One of the facts we use repeatedly in this section is the **Euler formula*** (see Appendix 5):

Euler's formula

$$e^{i\beta x} = \cos \beta x + i \sin \beta x.$$

We return to our examination of the homogeneous equation with constant coefficients

$$y'' + ay' + by = 0, \tag{1}$$

where $a^2 - 4b < 0$.

Case 3: complex conjugate roots

Suppose $a^2 - 4b < 0$. The roots of the characteristic equation to Equation **(1)** are

$$\lambda_1 = \alpha + i\beta, \qquad \lambda_2 = \alpha - i\beta, \tag{2}$$

where $\alpha = -\frac{a}{2}$ and $\beta = \sqrt{4b - a^2}/2$. Thus $y_1 = e^{\lambda_1 x}$ and $y_2 = e^{\lambda_2 x}$ are solutions to Equation **(1)**. However, in this case it is useful to recall that any linear combination of solutions is also a solution (see Theorem 3.1.3, p. 110) and to consider instead the solutions

$$y_1^* = \frac{e^{\lambda_1 x} + e^{\lambda_2 x}}{2} \quad \text{and} \quad y_2^* = \frac{e^{\lambda_1 x} - e^{\lambda_2 x}}{2i}.$$

Since $\cos(-\theta) = \cos\theta$ and $\sin(-\theta) = -\sin\theta$, we can rewrite y_1^* as

$$y_1^* = \frac{e^{(\alpha+i\beta)x} + e^{(\alpha-i\beta)x}}{2} = \frac{e^{\alpha x}}{2}(e^{i\beta x} + e^{-i\beta x})$$

Euler formula ↓

$$= \frac{e^{\alpha x}}{2}[\cos\beta x + i\sin\beta x + \cos(-\beta x) + i\sin(-\beta x)]$$

$$= e^{\alpha x}\cos\beta x.$$

* See the biographical sketch of Euler on page 29.

Similarly, $y_2^* = e^{\alpha x} \sin \beta x$, and the linear independence of y_1^* and y_2^* follows easily since

$$\frac{y_1^*}{y_2^*} = \cot \beta x, \quad \beta \neq 0,$$

which is not a constant. Alternatively, we can compute $W(y_1^*, y_2^*)(x)$:

$$\begin{aligned} W(y_1^*, y_2^*)(x) &= \begin{vmatrix} e^{\alpha x} \cos \beta x & e^{\alpha x} \sin \beta x \\ e^{\alpha x}(\alpha \cos \beta x - \beta \sin \beta x) & e^{\alpha x}(\alpha \sin \beta x + \beta \cos \beta x) \end{vmatrix} \\ &= e^{2\alpha x}(\alpha \cos \beta x \sin \beta x + \beta \cos^2 \beta x - \alpha \cos \beta x \sin \beta x \\ &\quad + \beta \sin^2 \beta x) \end{aligned}$$

$$\beta(\cos^2 \beta x + \sin^2 \beta x) = \beta$$

$$\downarrow$$

$$= \beta e^{2\alpha x} \neq 0.$$

Thus we have proved the following theorem.

THEOREM 1 ♦ **The general solution to a homogeneous equation in the case of complex conjugate roots**

If $a^2 - 4b < 0$, then the characteristic equation has complex conjugate roots, and the general solution to

$$y'' + ay' + by = 0 \tag{3}$$

is given by

$$y(x) = e^{\alpha x}(c_1 \cos \beta x + c_2 \sin \beta x) \tag{4}$$

where c_1 and c_2 are arbitrary constants and

$$\alpha = -\frac{a}{2}, \qquad \beta = \frac{\sqrt{4b - a^2}}{2}.$$ ♦

EXAMPLE 1 ▶ **Harmonic motion**

Solve the differential equation $y'' + y = 0$.

Solution The characteristic equation is $\lambda^2 + 1 = 0$, and it has the roots $\lambda = \pm i$. By Equation **(3)** we have $a = 0$ and $b = 1$, so that $\alpha = 0$ and $\beta = 1$. The general solution is

$$y(x) = c_1 \cos x + c_2 \sin x.$$

Observe that

$$y'(x) = -c_1 \sin x + c_2 \cos x,$$

so that $y(0) = c_1$ and $y'(0) = c_2$. Clearly, if either c_1 or c_2 is zero, the solution is sinusoidal. We shall see below that this is still the case even if neither c_1 nor c_2 is zero. ◀

EXAMPLE 2 ▶ **An initial-value problem with complex conjugate roots**

Consider the problem

$$y'' + y' + y = 0, \qquad y(0) = 1, \quad y'(0) = 3.$$

Solution We have $\lambda^2 + \lambda + 1 = 0$ with roots $\lambda_1 = (-1 + i\sqrt{3})/2$ and $\lambda_2 = (-1 - i\sqrt{3})/2$. Then $\alpha = -\frac{1}{2}$ and $\beta = \sqrt{3}/2$, so the general solution is

$$y(x) = e^{-x/2}\left(c_1 \cos \frac{\sqrt{3}}{2}x + c_2 \sin \frac{\sqrt{3}}{2}x\right).$$

To solve the initial-value problem, we differentiate, set $x = 0$, and solve the simultaneous equations

$$c_1 = 1,$$
$$\frac{\sqrt{3}}{2}c_2 - \frac{1}{2}c_1 = 3.$$

Thus $c_1 = 1$, $c_2 = 7/\sqrt{3}$, and

$$y(x) = e^{-x/2}\left(\cos \frac{\sqrt{3}}{2}x + \frac{7}{\sqrt{3}} \sin \frac{\sqrt{3}}{2}x\right). \quad ◀ \tag{5}$$

We will often obtain answers in the form **(5).** It makes things clearer if we can write the answer in terms of a single sine or cosine function. The following formulas will be very useful.

Harmonic identities

$$a \cos \omega t + b \sin \omega t = A \cos(\omega t - \delta) \tag{6}$$

$$a \cos \omega t + b \sin \omega t = A \sin(\omega t + \phi) \tag{7}$$

Here $A = \sqrt{a^2 + b^2}$, δ is the unique number in $[0, 2\pi)$ such that $\cos \delta = a/\sqrt{a^2 + b^2} = \frac{a}{A}$ and $\sin \delta = \frac{b}{A}$, so $\tan \delta = \frac{b}{a}$. ϕ is the unique number in $[0, 2\pi)$ such that $\cos \phi = \frac{b}{A}$ and $\sin \phi = \frac{a}{A}$, so $\tan \phi = \frac{a}{b}$.

Remark

amplitude
frequency
phase shift

In Equation **(6),** A is called the **amplitude** of the function $A \cos(\omega t - \delta)$, $\omega/2\pi$ is called the **frequency** of the function, and $|\delta|/\omega$ is called the **phase shift.**

We prove Equation **(6)** and leave Equation **(7)** as an exercise.

Proof To prove Equation (6), we first note that

$$\left(\frac{a}{\sqrt{a^2 + b^2}}\right)^2 + \left(\frac{b}{\sqrt{a^2 + b^2}}\right)^2 = \frac{a^2}{a^2 + b^2} + \frac{b^2}{a^2 + b^2} = 1$$

so

$$\left(\frac{a}{\sqrt{a^2 + b^2}}, \frac{b}{\sqrt{a^2 + b^2}}\right)$$

is a point on the unit circle. If δ is chosen as in Figure 3.3, then

$$\cos \delta = \frac{a}{\sqrt{a^2 + b^2}} = \frac{a}{A} \quad \text{and} \quad \sin \delta = \frac{b}{\sqrt{a^2 + b^2}} = \frac{b}{A}$$

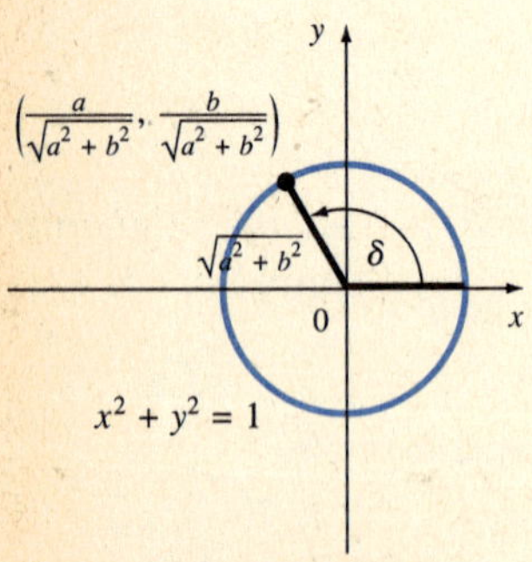

FIGURE 3.3
In this figure $\cos \delta = a/\sqrt{a^2 + b^2}$ and $\sin \delta = b/\sqrt{a^2 + b^2}$

Then

$$\cos(x - y) = \cos x \cos y + \sin x \sin y$$

$$\begin{aligned} A \cos(\omega t - \delta) &= A[\cos \omega t \cos \delta + \sin \omega t \sin \delta] \\ &= A \cos \delta \cos \omega t + A \sin \delta \sin \omega t \\ &= A\left(\frac{a}{A}\right) \cos \omega t + A\left(\frac{b}{A}\right) \sin \omega t \\ &= a \cos \omega t + b \sin \omega t. \quad ♦ \end{aligned}$$

EXAMPLE 3 ▶ **Rewriting the solution to Example 2 using the harmonic identities**

In Example 2 we obtained the solution

$$y(x) = e^{-x/2}\left(\cos\frac{\sqrt{3}}{2}x + \frac{7}{\sqrt{3}}\sin\frac{\sqrt{3}}{2}x\right).$$

In Equation **(6)**, we set $a = 1$ and $b = 7/\sqrt{3}$ to obtain

$$A = \sqrt{1 + \frac{49}{3}} = \sqrt{\frac{52}{3}} \approx 4.163 \text{ and } \delta = \cos^{-1}\frac{1}{\sqrt{\frac{52}{3}}} \approx 1.328.$$

(δ is in the first quadrant because $\cos\delta > 0$ and $\sin\delta > 0$).

Thus,

$$y(x) = e^{-x/2}\left(\cos\frac{\sqrt{3}}{2}x + \frac{7}{\sqrt{3}}\sin\frac{\sqrt{3}}{2}x\right)$$
$$\approx 4.163e^{-x/2}\cos\left(\frac{\sqrt{3}}{2}x - 1.328\right).$$

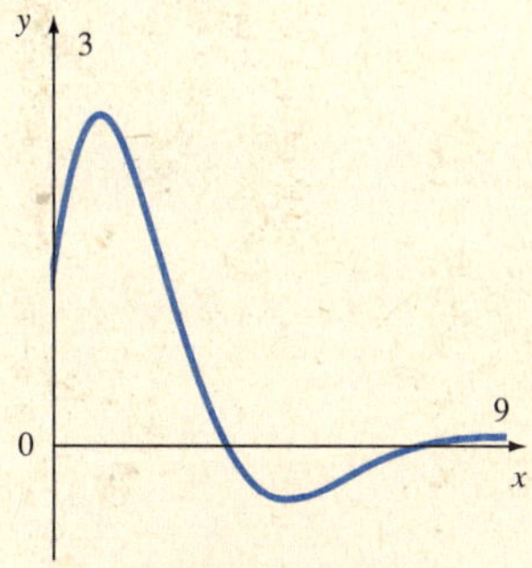

FIGURE 3.4
Graph of $y = 4.163e^{-x/2} \cdot \cos((\sqrt{3}/2)x - 1.328) = 4.163e^{-x/2} \cdot \sin((\sqrt{3}/2)x + 0.2426)$, $x \geq 0$

Using Equation **(7)**, we have $\cos\phi = (7/\sqrt{3})/(\sqrt{52/3}) = 7/\sqrt{52}$ so $\phi \approx 0.2426$ and

$$e^{-x/2}\left(\cos\frac{\sqrt{3}}{2}x + \frac{7}{\sqrt{3}}\sin\frac{\sqrt{3}}{2}x\right) \approx 4.163e^{-x/2}\sin\left(\frac{\sqrt{3}}{2}x + 0.2426\right).$$

A graph of this solution is given in Figure 3.4. ◀

EXAMPLE 4 ▶ **Simplifying the solution of an initial-value problem**

Solve the problem

$$y'' - 6y' + 13y = 0, \qquad y(0) = 3, \qquad y'(0) = -5. \tag{8}$$

Solution The characteristic equation is $\lambda^2 - 6\lambda + 13 = 0$, which has the complex conjugate roots $\lambda = \frac{1}{2}[6 \pm \sqrt{36 - 52}] = 3 \pm 2i$. Thus, the general solution is

$$y(x) = e^{3x}[c_1\cos 2x + c_2\sin 2x]. \tag{9}$$

Differentiating with respect to x we have

$$y'(x) = e^{3x}[(3c_1 + 2c_2)\cos 2x + (3c_2 - 2c_1)\sin 2x]. \tag{10}$$

Set $x = 0$ in Equations **(9)** and **(10)** and use the initial conditions to obtain

$$3 = y(0) = c_1$$
$$-5 = y'(0) = 3c_1 + 2c_2.$$

Hence $c_2 = -7$, so the particular solution to problem **(8)** is

$$y(x) = e^{3x}(3\cos 2x - 7\sin 2x). \tag{11}$$

If we wish to simplify the term in parenthesis in Equation **(11)** by using the

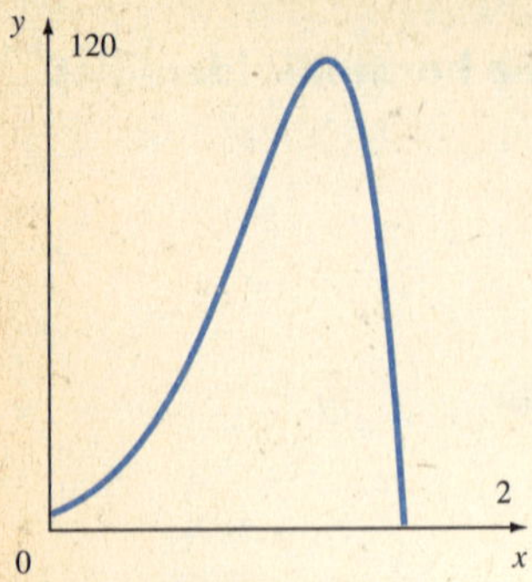

FIGURE 3.5 Graph of $y = \sqrt{58}e^{3x}\cos(2x - 1.166)$

harmonic identity **(6)**, we let $a = 3$ and $b = -7$, so that $A = \sqrt{3^2 + (-7)^2} = \sqrt{58}$ and

$$(\cos\delta, \sin\delta) = \left(\frac{3}{\sqrt{58}}, \frac{-7}{\sqrt{58}}\right).$$

Clearly δ is in the fourth quadrant (since x is positive and y is negative), so

$$\delta = \sin^{-1}\left(\frac{-7}{\sqrt{58}}\right) \approx -1.166 \text{ (radians)},$$

and we can rewrite Equation **(11)** as

$$y(x) = \sqrt{58}\, e^{3x}\cos(2x - 1.166).$$

A graph of this function is shown in Figure 3.5. ◀

SELF-QUIZ

I. What can be said about the solution $y(x)$ to $y'' + 4y = 0$; $y(0) = 1$, $y'(0) = 2$?
(a) $y(x) \to 0$ as $x \to \infty$ **(b)** $y(x) \to \infty$ as $x \to \infty$
(c) $y(x) \to$ a (nonzero constant) as $x \to \infty$
(d) none of the above

II. What can be said about the solution to $y'' + 2y' + 4y = 0$; $y(0) = 1$, $y'(0) = 2$?
(a) $y(x) \to 0$ as $x \to \infty$ **(b)** $y(x) \to \infty$ as $x \to \infty$
(c) $y(x) \to$ a (nonzero constant) as $x \to \infty$
(d) none of the ab

III. What can be said about the solution to $y'' - 2y' + 4y = 0$; $y(0) = 1$, $y'(0) = 2$?
(a) $y(x) \to 0$ as $x \to \infty$ **(b)** $y(x) \to \infty$ as $x \to \infty$
(c) $y(x) \to$ a (nonzero constant) as $x \to \infty$
(d) none of the above

IV. If $3\cos t + 4\sin t = 5\cos(t - \delta)$, then $\delta \approx$ ______.
(a) 0.6435 **(b)** 0.9273 **(c)** 0.7227 **(d)** 2.4981

V. If $-3\cos t + 4\sin t = 5\cos(t - \delta)$, then $\delta \approx$ ______.
(a) 0.6435 **(b)** 0.9273 **(c)** 2.4981 **(d)** 2.2143

VI. If $-3\cos t - 4\sin t = 5\cos(t - \delta)$, then $\delta \approx$ ______.
(a) −0.6435 **(b)** −0.9723 **(c)** 4.0689 **(d)** 3.7851

VII. If $3\cos t - 4\sin t = 5\cos(t - \delta)$, then $\delta \approx$ ______.
(a) 5.3559 **(b)** 5.6397 **(c)** 4.0689 **(d)** 3.7851

Answers to Self-Quiz

I. d **II.** a **III.** d **IV.** b **V.** d **VI.** c **VII.** a

PROBLEMS 3.4

In Problems 1–12 find the general solution of each equation. When initial conditions are specified, give the particular solution that satisfies them, and use the harmonic identities to rewrite it using Equation **(6)**. *Sketch the resulting function.*

1. $y'' + 2y' + 2y = 0$

2. $8y'' + 4y' + y = 0$, $y(0) = 0$, $y'(0) = 1$

3. $x'' + x' + 7x = 0$

4. $y'' + y' + 2y = 0$

5. $\dfrac{d^2x}{d\theta^2} + 4x = 0$, $x\left(\dfrac{\pi}{4}\right) = 1$, $x'\left(\dfrac{\pi}{4}\right) = 3$

6. $y'' + y = 0$, $y(\pi) = 2$, $y'(\pi) = -1$

7. $y'' + \dfrac{1}{4}y = 0$, $y(\pi) = 1$, $y'(\pi) = -1$

8. $y'' + 6y' + 12y = 0$

9. $y'' + 2y' + 5y = 0$

10. $y'' + 2y' + 5y = 0$, $y(0) = 1$, $y'(0) = -3$

11. $y'' + 2y' + 2y = 0$, $y(\pi) = e^{-\pi}$, $y'(\pi) = -2e^{-\pi}$

12. $y'' + 2y' + 5y = 0$, $y(\pi) = e^{-\pi}$, $y'(\pi) = 3e^{-\pi}$

In Problems 13–24, write each sum as a single cosine function and a single sine function. Use a value of δ or ϕ in the interval $[0, 2\pi$.)

13. $2 \cos t + 3 \sin t$

14. $-2 \cos t + 3 \sin t$

15. $-2 \cos t - 3 \sin t$

16. $2 \cos t - 3 \sin t$

17. $4 \cos 2\theta + 3 \sin 2\theta$

18. $-3 \cos 2\theta - 4 \sin 2\theta$

19. $-5 \cos \frac{\alpha}{2} + 12 \sin \frac{\alpha}{2}$

20. $12 \cos \frac{\alpha}{2} - 5 \sin \frac{\alpha}{2}$

21. $\frac{1}{2} \cos 4\beta - \frac{1}{4} \sin 4\beta$

22. $3 \cos \pi x + 2 \sin \pi x$

23. $-7 \cos \frac{\pi}{2}x - 3 \sin \frac{\pi}{2}x$

24. $x_0 \cos \omega_0 t + \frac{v_0}{\omega_0} \sin \omega_0 t$

In Problems 25–30, match the given initial-value problem with the graph of its solution.

25. $y'' - 2y' + 2y = 0,\ y(0) = 1,\ y'(0) = -2$

26. $y'' + 2y' + 2y = 0,\ y(0) = 1,\ y'(0) = -2$

27. $y'' + y = 0,\ y(0) = 1,\ y'(0) = -2$

28. $y'' + 4y = 0,\ y(0) = 1,\ y'(0) = -2$

29. $y'' - 2y' + 5y = 0,\ y(0) = 1,\ y'(0) = -2$

30. $y'' + 2y' + 5y = 0,\ y(0) = 1,\ y'(0) = -2$

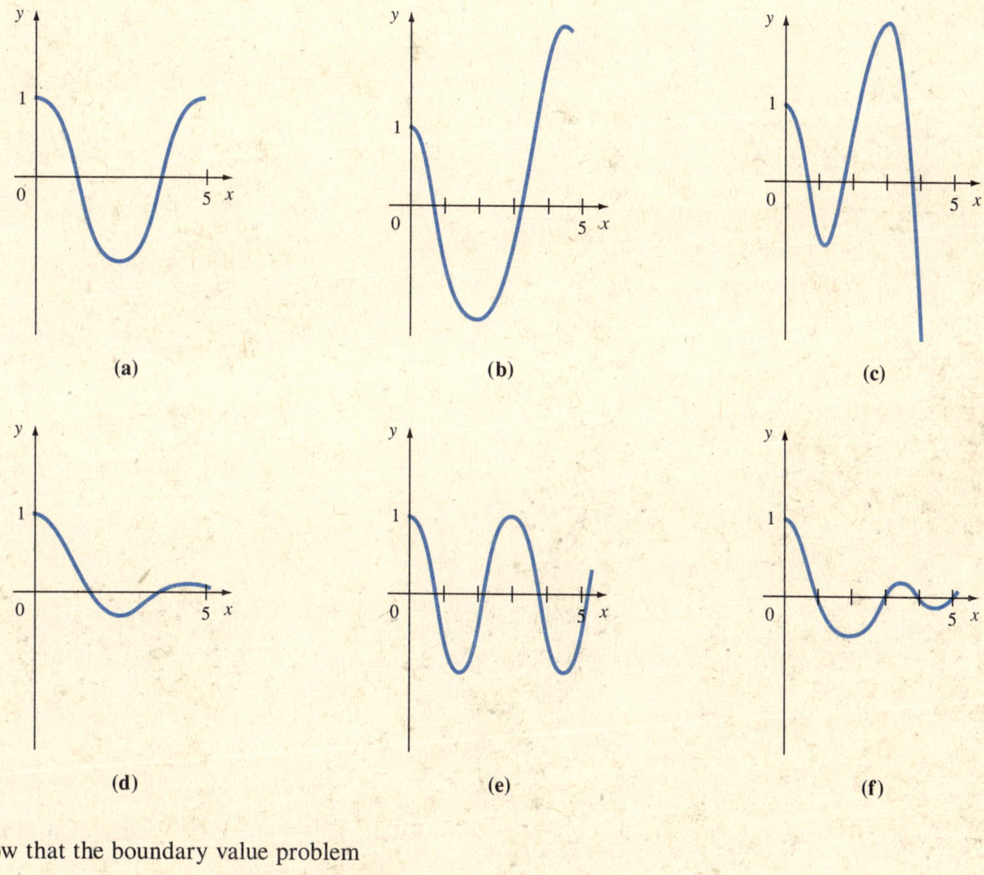

31. Show that the boundary value problem

$$y'' + y = 0, \qquad y(0) = 0, \qquad y(\pi) = 1$$

has no solutions.

3.5 NONHOMOGENEOUS EQUATIONS I: VARIATION OF PARAMETERS*

In this section we consider a procedure, due to J. L. Lagrange (1736–1813), for finding a particular solution of any nonhomogeneous linear equation

$$y'' + a(x)y' + b(x)y = f(x), \tag{1}$$

where the functions $a(x)$, $b(x)$, and $f(x)$ are continuous. To use this method it is necessary to know the general solution $c_1y_1(x) + c_2y_2(x)$ of the homogeneous equation

$$y'' + a(x)y' + b(x)y = 0. \tag{2}$$

If $a(x)$ and $b(x)$ are constants, the general solution to Equation **(2)** can always be obtained by the methods of Sections 3.3 and 3.4. If $a(x)$ and $b(x)$ are not both constants, it may be difficult or even impossible to find this general solution; however, if one solution y_1 of Equation **(2)** can be found, then the method of reduction of order (see Section 3.2) yields the general solution to Equation **(2)**.

Lagrange† noticed that any particular solution y_p of Equation **(1)** must have the property that y_p/y_1 and y_p/y_2 are not constants, suggesting that we look for a particular solution of Equation **(1)** of the form

$$y(x) = c_1(x)y_1(x) + c_2(x)y_2(x). \tag{3}$$

This replacement of constants or parameters by variables gives the method its name. Differentiating **(3)**, we obtain

$$y'(x) = c_1(x)y_1'(x) + c_2(x)y_2'(x) + c_1'(x)y_1(x) + c_2'(x)y_2(x).$$

To simplify this expression, it is convenient (but not necessary—see Problem 29) to set

$$c_1'(x)y_1(x) + c_2'(x)y_2(x) = 0. \tag{4}$$

Then

$$y'(x) = c_1(x)y_1'(x) + c_2(x)y_2'(x).$$

Differentiating once again, we obtain

$$y''(x) = c_1(x)y_1''(x) + c_2(x)y_2''(x) + c_1'(x)y_1'(x) + c_2'(x)y_2'(x).$$

Substitution of the expressions for $y(x)$, $y'(x)$, and $y''(x)$ into Equation **(1)** yields

$$\begin{aligned} y'' + a(x)y' + b(x)y &= c_1(x)(y_1'' + ay_1' + by_1) + c_2(x)(y_2'' + ay_2' + by_2) \\ &\quad + c_1'y_1' + c_2'y_2' \\ &= f(x). \end{aligned}$$

But y_1 and y_2 are solutions to the homogeneous equation, so the equation above reduces to

$$c_1'y_1' + c_2'y_2' = f(x). \tag{5}$$

---

* This procedure is also called the method of **variation of constants** or **Lagrange's method.**

† The reason he made this observation is evident from Problem 2.3.43 (p. 63). It also follows from the method of reduction of order used in Section 3.2.

This gives a second equation relating $c_1'(x)$ and $c_2'(x)$, and we have the simultaneous equations

Basic equations in the variation of parameters method

$$\begin{aligned} y_1c_1' + y_2c_2' &= 0 \\ y_1'c_1' + y_2'c_2' &= f(x). \end{aligned} \qquad (6)$$

The determinant of system **(6)** is the Wronskian

$$\begin{vmatrix} y_1 & y_2 \\ y_1' & y_2' \end{vmatrix} = W(y_1, y_2)(x) \neq 0. \qquad \text{since } y_1 \text{ and } y_2 \text{ are linearly independent}$$

Thus, for each value of x, $c_1'(x)$ and $c_2'(x)$ are uniquely determined and the problem has essentially been solved. We obtain, from **(6)**,

$$\begin{aligned} y_1y_2'c_1' + y_2y_2'c_2' &= 0, && \text{first equation multiplied by } y_2' \\ y_1'y_2c_1' + y_2y_2'c_2' &= y_2f(x). && \text{second equation multiplied by } y_2 \end{aligned}$$

So

$$(y_1y_2' - y_1'y_2)c_1' = -y_2f(x),$$

or

$$c_1' = \frac{-y_2f(x)}{W(y_1, y_2)(x)}.$$

A similar calculation yields an expression for c_2'. Thus we obtain

Solution of the basic equations

$$c_1'(x) = \frac{-f(x)y_2(x)}{y_1(x)y_2'(x) - y_1'(x)y_2(x)} = \frac{-f(x)y_2(x)}{W(y_1, y_2)(x)}, \qquad (7)$$

$$c_2'(x) = \frac{f(x)y_1(x)}{y_1(x)y_2'(x) - y_1'(x)y_2(x)} = \frac{f(x)y_1(x)}{W(y_1, y_2)(x)}. \qquad (8)$$

Finally, if we can integrate c_1' and c_2', we can substitute c_1 and c_2 into **(3)** to obtain a particular solution to the nonhomogeneous equation; that is,

$$y_p(x) = c_1(x)y_1(x) + c_2(x)y_2(x), \qquad (9)$$

where

$$c_1(x) = \int c_1'(x)\,dx, \qquad c_2(x) = \int c_2'(x)\,dx,$$

with $c_1'(x)$ and $c_2'(x)$ given by Equations **(7)** and **(8)**.

Remark

It is not advisable to try to memorize Equations **(7)** and **(8)**, particularly as it is almost always easier to solve system **(6)** directly. The key concept to this method is system **(6)**. We illustrate with three examples.

EXAMPLE 1 ▶ **Solving a differential equation by the variation of parameters method**

Solve $y'' - y = e^{2x}$ by the method of variation of parameters.

Solution The solutions to the homogeneous equation are $y_1 = e^{-x}$ and $y_2 = e^x$. Using these solutions, system **(6)** becomes

$$e^{-x}c_1' + e^x c_2' = 0,$$
$$-e^{-x}c_1' + e^x c_2' = e^{2x}.$$

Adding these two equations and then subtracting them, we get

$$2e^x c_2' = e^{2x} \quad \text{and} \quad 2e^{-x}c_1' = -e^{2x},$$

or

$$c_1' = \frac{-e^{3x}}{2} \quad \text{and} \quad c_2' = \frac{e^x}{2}.$$

Integrating these functions, we obtain $c_1(x) = -e^{3x}/6$ and $c_2(x) = e^x/2$. A particular solution is therefore

$$y_p = c_1(x)y_1(x) + c_2(x)y_2(x) = \frac{-e^{2x}}{6} + \frac{e^{2x}}{2} = \frac{e^{2x}}{3},$$

and the general solution is

$$y(x) = c_1e^x + c_2e^{-x} + \frac{e^{2x}}{3}. \quad ◀$$

EXAMPLE 2 ▶ **Solving an initial-value problem by the variation of parameters method**

Determine the solution of

$$y'' + y = 4 \sin x$$

that satisfies $y(0) = 3$ and $y'(0) = -1$.

Solution Here $y_1(x) = \cos x$, $y_2(x) = \sin x$, and **(6)** becomes

$$\cos x \cdot c_1' + \sin x \cdot c_2' = 0,$$
$$-\sin x \cdot c_1' + \cos x \cdot c_2' = 4 \sin x.$$

Solving these equations simultaneously for c_1' and c_2', we have

$$c_1' = -4 \sin^2 x \quad \text{and} \quad c_2' = 4 \sin x \cos x.$$

Since

$$\int \sin^2 x\, dx = \frac{1}{2}(x - \sin x \cos x),$$

we see that

$$c_1 = 2(\sin x \cos x - x), \qquad c_2 = 2 \sin^2 x,$$

and a particular solution is

$$\begin{aligned} y_p(x) &= c_1(x)y_1(x) + c_2(x)y_2(x) \\ &= 2(\sin x \cos x - x) \cos x + 2 \sin^2 x \sin x \\ &= 2 \sin x(\cos^2 x + \sin^2 x) - 2x \cos x = 2 \sin x - 2x \cos x. \end{aligned}$$

Thus the general solution is

$$\begin{aligned} y &= c_1 \cos x + c_2 \sin x + 2 \sin x - 2x \cos x \\ &= c_1 \cos x + c_2^* \sin x - 2x \cos x, \end{aligned}$$

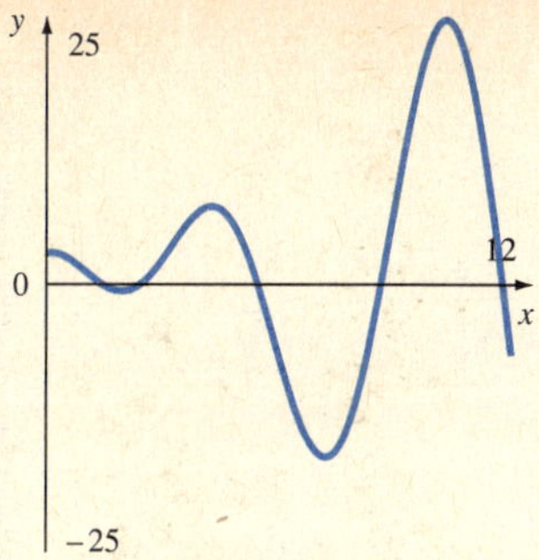

FIGURE 3.6
Graph of $y =$ $(3 - 2x)\cos x + \sin x$

where $c_2^* = c_2 + 2$. To solve the initial-value problem, we differentiate:

$$y' = -c_1 \sin x + c_2^* \cos x + 2x \sin x - 2 \cos x.$$

But

$$3 = y(0) = c_1 \quad \text{and} \quad -1 = y'(0) = c_2^* - 2,$$

so $c_2^* = 1$ and the unique solution is

$$y = 3 \cos x + \sin x - 2x \cos x = (3 - 2x) \cos x + \sin x.$$

A graph of this function is illustrated in Figure 3.6. ◀

EXAMPLE 3 ▶ **Using variation of parameters to solve an initial-value problem**

Solve the problem

$$y'' + y = \tan x, \qquad y(0) = 1, \qquad y'(0) = 1.$$

Solution The solutions to the homogeneous equation are $y_1 = \cos x$ and $y_2 = \sin x$. System **(6)** becomes

$$\cos x \cdot c_1' + \sin x \cdot c_2' = 0,$$
$$-\sin x \cdot c_1' + \cos x \cdot c_2' = \tan x,$$

for which we obtain by elimination

$$c_1'(x) = -\tan x \sin x = -\frac{\sin^2 x}{\cos x} = \frac{\cos^2 x - 1}{\cos x} = \cos x - \sec x,$$
$$c_2'(x) = \tan x \cos x = \sin x.$$

Hence

$$c_1(x) = \sin x - \ln|\sec x + \tan x|$$

and

$$c_2(x) = -\cos x.$$

Thus the particular solution is

$$y_p(x) = c_1(x) y_1(x) + c_2(x) y_2(x)$$
$$= \cos x \sin x - \cos x \ln|\sec x + \tan x| - \sin x \cos x$$
$$= -\cos x \ln|\sec x + \tan x|,$$

and the general solution is

$$y(x) = c_1 \cos x + c_2 \sin x - \cos x \ln|\sec x + \tan x|.$$

Differentiating the general solution with respect to x, we get

$$y'(x) = -c_1 \sin x + c_2 \cos x + \sin x \ln|\sec x + \tan x| - 1,$$

and setting $x = 0$ in these last two equations, we have by the initial conditions

$$1 = y(0) = c_1 - \ln|1| = c_1,$$
$$1 = y'(0) = c_2 - 1,$$

so that $c_2 = 2$, and the solution to the initial-value problem, shown in Figure 3.7, is

$$y(x) = \cos x + 2 \sin x - \cos x \ln|\sec x + \tan x|. \quad ◀$$

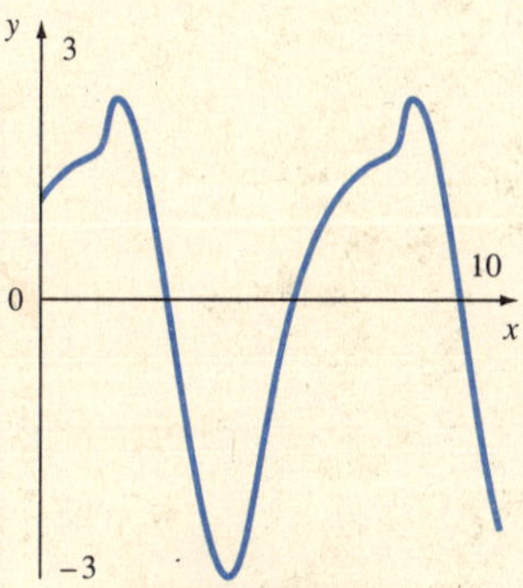

FIGURE 3.7
Graph of $y(x) =$ $\cos x + 2 \sin x -$ $\cos x \ln|\sec x + \tan x|$
Do you think Figure 3.7 is the solution on the interval $0 \le x \le 10$? Explain.

Perspective Green's functions

We may rewrite Equation **(9)** in integral form as

$$\begin{aligned} y_p(x) &= c_1(x)y_1(x) + c_2(x)y_2(x) \\ &= -y_1(x)\int^x \frac{y_2(t)f(t)}{W(y_1, y_2)(t)}\,dt + y_2(x)\int^x \frac{y_1(t)f(t)}{W(y_1, y_2)(t)}\,dt, \end{aligned}$$

or

$$y_p(x) = \int^x \left[\frac{y_1(t)y_2(x) - y_1(x)y_2(t)}{W(y_1, y_2)(t)}\right] f(t)\,dt. \tag{10}$$

If we set the lower limit of integration of Equation **(10)** to, for example, x_0, then

$$y_p(x) = \int_{x_0}^x \left[\frac{y_1(t)y_2(x) - y_1(x)y_2(t)}{W(y_1, y_2)(t)}\right] f(t)\,dt \tag{11}$$

is a particular solution to the initial-value problem

$$y'' + a(x)y' + b(x)y = f(x), \qquad y(x_0) = 0, \qquad y'(x_0) = 0.$$

This is so because it is a solution to the nonhomogeneous linear equation by the development earlier in this section, $y_p(x_0) = 0$ trivially from Equation **(11),** and

$$\begin{aligned} y_p'(x) &= \overbrace{\left[\frac{y_1(x)y_2(x) - y_1(x)y_2(x)}{W(y_1, y_2)(x)}\right]}^{=0} f(x) \\ &\quad + \int_{x_0}^x \left[\frac{y_1(t)y_2'(x) - y_1'(x)y_2(t)}{W(y_1, y_2)(t)}\right] f(t)\,dt, \end{aligned}$$

so that $y_p'(x_0) = 0$.

If we wish to solve the initial-value problem

$$y'' + a(x)y' + b(x)y = f(x), \qquad y(x_0) = y_0, \qquad y'(x_0) = y_0', \tag{12}$$

we need only select linearly independent solutions y_1 and y_2 of the homogeneous equation that satisfy the conditions

$$y_1(x_0) = 1, \qquad y_1'(x_0) = 0,$$

and

$$y_2(x_0) = 0, \qquad y_2'(x_0) = 1.$$

Then

$$\begin{aligned} y(x) &= y_0 y_1(x) + y_0' y_2(x) \\ &\quad + \int_{x_0}^x \left[\frac{y_1(t)y_2(x) - y_1(x)y_2(t)}{W(y_1, y_2)(t)}\right] f(t)\,dt \end{aligned} \tag{13}$$

is the unique solution to initial-value problem **(12).** The function in brackets in Equation **(13)** is called a **Green's function** for the differential equation in **(12).**

SELF-QUIZ

I. In using the variation of parameters method to the equation $y'' + y = \sec x$, which system of equations must be solved?

(a) $c_1' \sin x + c_2' \cos x = 0$
$c_1' \cos x - c_2' \sin x = \sec x$

(b) $c_1' \sin x + c_2' \cos x = 0$
$c_1' \cos x + c_2' \sin x = \sec x$

(c) $c_1' \sin x + c_2' \cos x = \sec x$
$c_1' \cos x - c_2' \sin x = 0$

(d) $c_1' \sin x + c_2' \cos x = \sec x$
$c_1' \cos x + c_2' \sin x = 0$

II. In using the variation of parameters method applied to the equation $y'' - 4y' + 4y = e^{x^2}$, which system of equations must be solved?

(a) $c_1' e^{2x} + c_2' x e^{2x} = e^{x^2}$
$2c_1' e^{2x} + (1 + 2x)c_2' e^{2x} = 0$

(b) $c_1' x e^{2x} + c_2' e^{2x} = 0$
$(1 + 2x)c_1' x e^{2x} + 2c_2' e^{2x} = e^{x^2}$

(c) $c_1' e^{2x} + c_2' x e^{2x} = 0$
$c_1' x e^{2x} + 2c_2' e^{2x} = e^{x^2}$

(d) $c_1' e^{2x} + c_2' x e^{2x} = 0$
$2c_1' e^{2x} + c_2' x e^{2x} = e^{x^2}$

Answers to Self-Quiz

(I) a **(II)** b

PROBLEMS 3.5

In Problems 1–24 find the general solution to each equation by the method of variation of parameters. If initial conditions are given, find the solution and sketch its graph, using a graphing calculator or a computer algebra system.

1. $y'' - 3y' + 2y = 10$

2. $y'' - y' - 6y = 20e^{-2x}$

3. $y'' + 4y = 3 \sin x,\ y(0) = 1,\ y'(0) = 2$

4. $y'' + y' = 3x^2,\ y(1) = 0,\ y'(1) = 1$

5. $y'' - 3y' + 2y = 6e^{3x}$

6. $y'' - 4y' + 4y = 6xe^{2x}$

7. $y'' - 2y' + y = -4e^x$

8. $y'' + y = 1 + x + x^2,\ y(0) = 1,\ y'(0) = -1$

9. $y'' - 7y' + 10y = 100x$

10. $y'' + 4y = 16x \sin 2x$

11. $y'' - y' = \sec^2 x - \tan x,\ y(0) = 3,\ y'(0) = 2$

12. $y'' + y = \cot x$

13. $y'' + 4y = \sec 2x$

14. $y'' + 4y = \sec x \tan x,\ y(0) = 1,\ y'(0) = 4$

15. $y'' - 2y' + y = \dfrac{e^x}{(1 - x)^2}$

16. $y'' - y = \sin^2 x$

17. $y'' - y = \dfrac{(2x - 1)e^x}{x^2},\ y(1) = 9,\ y'(1) = 5$

18. $y'' - 3y' - 4y = \dfrac{e^{4x}(5x - 2)}{x^3}$

19. $y'' - 4y' + 4y = \dfrac{e^{2x}}{(1 + x)},\ y(1) = \dfrac{2}{3},\ y'(1) = \dfrac{1}{2}$

20. $y'' + 2y' + y = e^{-x} \ln|x|$

21. $y'' - 9y = \operatorname{sech} 3x$

22. $y'' + 6y' + 9y = x^{-2}e^{-3x},\ y(1) = \dfrac{2}{e^3},\ y'(1) = \dfrac{-5}{e^3}$

23. $y'' - 4y' + 3y = \dfrac{e^x}{e^x + 1}$

24. $y'' - 4y' + 3y = e^x \sec x$

25. Find a particular solution of

$$y'' + \frac{1}{x}y' - \frac{y}{x^2} = \frac{1}{x^2 + x^3},\ (x > 0),$$

given that two solutions of the associated homogeneous equations are $y_1 = x$ and $y_2 = \frac{1}{x}$.

26. Find a particular solution of

$$y'' - \frac{2}{x}y' + \frac{2}{x^2}y = \frac{\ln|x|}{x}, \quad (x > 0)$$

given the two homogeneous solutions $y_1 = x$ and $y_2 = x^2$.

27. Verify that

$$y = \frac{1}{\omega}\int_0^x f(t) \sin \omega(x - t)\, dt$$

is a particular solution of $y'' + \omega^2 y = f(x)$. [*Hint:* Use Equation **(10)**.]

28. Find a particular solution to the initial-value problem

$$y'' - \omega^2 y = f(x), \qquad y(0) = y'(0) = 0.$$

29. This problem shows why there is no loss in generality in Equation **(4)** by setting

$$c_1' y_1 + c_2' y_2 = 0.$$

Suppose that we instead let $c_1' y_1 + c_2' y_2 = z(x)$, with $z(x)$ an undetermined function of x.

(a) Show that we then obtain the system

$$c_1'y_1 + c_2'y_2 = z,$$
$$c_1'y_1' + c_2'y_2' = f - z' - az.$$

(b) Show that the system in part (a) has the solution

$$c_1' = \frac{-y_2 f}{W(y_1, y_2)} + \frac{(e^{\int a(x)\,dx} z y_2)'}{e^{\int a(x)\,dx} W(y_1, y_2)},$$

$$c_2' = \frac{y_1 f}{W(y_1, y_2)} - \frac{(e^{\int a(x)\,dx} z y_1)'}{e^{\int a(x)\,dx} W(y_1, y_2)}.$$

(c) Integrate by parts to show that

$$\int \frac{(e^{\int a(x)\,dx} z y_i)'}{e^{\int a(x)\,dx} W(y_1, y_2)}\,dx = \frac{z y_i}{W(y_1, y_2)}, \quad i = 1, 2.$$

(d) Conclude that the particular solution obtained by letting $c_1'y_1 + c_2'y_2 = z$ is identical to that obtained by assuming Equation **(4)**.

30. Suppose one solution $y_1(x)$ of the homogeneous counterpart of the linear differential equation

$$y'' + a(x)y' + b(x)y = f(x) \qquad \textbf{(i)}$$

is known. Use the substitution $y_2 = vy_1$ of Section 3.2 to find the general solution of Equation **(i)**:

$$y = c_1 y_1(x) + c_2 y_1(x) \int^x \frac{h(t)}{y_1^2(t)}\,dt$$

$$+ y_1(x) \int^x \frac{h(t)}{y_1^2(t)} \left[\int^t \frac{y_1(s) f(s)}{h(s)}\,ds \right] dt,$$

where $h(t) = \exp(-\int^t a(u)\,du)$.

3.6 NONHOMOGENEOUS EQUATIONS II: UNDETERMINED COEFFICIENTS

In this section we present an alternate method for solving certain linear nonhomogeneous differential equations with constant coefficients. The method is not as general as the method of variation of parameters that we considered in Section 3.5. However, it is sometimes easier to solve specific problems by using it.

Consider the equation

$$y'' + ay' + by = f(x), \qquad \textbf{(1)}$$

where a and b are constants, and every term of the function $f(x)$ is a *product* of one or more of the following:

1. a polynomial in x (a constant k is a polynomial of degree zero)
2. an exponential function $e^{\alpha x}$
3. $\cos \beta x$, $\sin \beta x$.

The method we shall present applies only when $f(x)$ is a function of this type; if any term in $f(x)$ is not of this type, we cannot use the method of this section. For example, we cannot use the method if $f(x) = \tan x$.

Note that a multitude of cases are of this type:

1. $f(x) = 17$; (f is constant)
2. $f(x) = x\,e^{2x}$; (f is a polynomial times an exponential)
3. $f(x) = e^{4x} \cos x$; (an exponential times a cosine)
4. $f(x) = x \cos 2x + 3 \sin 2x$; (polynomial times a cosine plus a constant times a sine)
5. $f(x) = \frac{1}{2}e^{-3x} \cos 3x - \sqrt{5}\,(x^2 + 3)e^{2x} \sin x$; (each term is a product of all three types)

Observe that if we differentiate a product of the three types above, we get an expression whose terms are again of this type:

$$(P_n(x)e^{\alpha x} \cos \beta x)' = P_n'(x)e^{\alpha x} \cos \beta x + \alpha\, P_n(x)e^{\alpha x} \cos \beta x - \beta\, P_n(x)e^{\alpha x} \sin \beta x,$$

for any polynomial $P_n(x)$ in x. This suggests that we may be able to determine a particular solution to Equation **(1)** by substituting a function composed of terms of the same "form" as those in $f(x)$ for the dependent variable y.

The **method of undetermined coefficients** assumes that each term in the particular solution to Equation **(1)** is of the same "form" as each term in $f(x)$, but has polynomial terms with undetermined coefficients, thus giving the method its name. Thus, for example, if f contains the term x^2e^x, which is a quadratic polynomial times e^x, then we assume that a solution has a term having the form $(ax^2 + bx + c)e^x$—a quadratic polynomial times e^x. Here the constants a and b are to be determined. If we substitute the right combination of such terms, the method asserts that when we compare both sides of the resulting equation, it will be possible to "determine" the unknown coefficients. Before attempting to make this procedure precise, we illustrate it with several examples.

EXAMPLE 1 ▶ **Solving when $f(x)$ is a polynomial**

Solve the following equation:

$$y'' - y = x^2 \tag{2}$$

Solution Since $f(x) = x^2$ is a polynomial of degree two, we "guess" that Equation **(2)** has a solution $y_p(x)$ that is a polynomial of degree two. We try the *most general* polynomial of degree two:

$$y_p(x) = a + bx + cx^2.$$

Then $y_p' = b + 2cx$ and $y_p'' = 2c$, so if we substitute y_p'' and y_p into Equation **(2)** we obtain

$$2c - (a + bx + cx^2) = x^2. \tag{3}$$

Equating coefficients, in **(3)** we have

Coefficient of constant term	Coefficient of x	Coefficient of x^2
↓	↓	↓
$2c - a = 0,$	$-b = 0,$	$-c = 1,$

which immediately yields $a = -2$, $b = 0$, $c = -1$, and the particular solution

$$y_p(x) = -2 - x^2. \tag{4}$$

This particular solution is easily verified by substitution into Equation **(2)**. Finally, since the general solution of the homogeneous equation $y'' - y = 0$ is given by

$$y = c_1e^x + c_2e^{-x},$$

the general solution of Equation **(2)** is

$$y = c_1e^x + c_2e^{-x} - 2 - x^2. \quad ◀ \tag{5}$$

EXAMPLE 2 ▶ **Solving a nonhomogeneous equation when $f(x) = e^{\alpha x}\sin\beta x$**

Solve

$$y'' - 3y' + 2y = e^x \sin x. \tag{6}$$

Solution Here $f(x)$ is a product of all three types (with polynomial equal to 1). In collecting the terms for y we must choose not only terms of the form $e^x \sin x$, but also terms of the form $e^x \cos x$, since any derivative of the former will produce a term like the latter. For this reason, we guess that the particular solution has the form

$$y_p(x) = ae^x\sin x + be^x \cos x.$$

Then

$$y_p'(x) = (a - b)e^x\sin x + (a + b)e^x \cos x$$

and

$$y_p''(x) = 2ae^x \cos x - 2be^x \sin x.$$

Substituting these expressions into Equation **(6)** we have

$$e^x(2a \cos x - 2b \sin x) - 3e^x[(a - b)\sin x + (a + b)\cos x] + 2e^x(a \sin x + b \cos x) = e^x \sin x.$$

Dividing both sides by e^x and equating the coefficients of $\sin x$ and $\cos x$, we have

$$2a - 3(a + b) + 2b = 0,$$
$$-2b - 3(a - b) + 2a = 1,$$

which yield $a = -\frac{1}{2}$ and $b = \frac{1}{2}$ so that

$$y_p = \frac{e^x}{2}(\cos x - \sin x).$$

Again this result is easily verified by substitution. Finally, the general solution of Equation **(6)** is

$$y = c_1 e^{2x} + c_2 e^x + \frac{e^x}{2}(\cos x - \sin x).$$ ◀

If $f(x)$ had been the function $5\, e^x \cos x$, we would have used exactly the same guess for the particular solution.

EXAMPLE 3 ▶ **Solving a nonhomogeneous equation when $f(x)$ is a polynomial times an exponential**

Solve

$$y'' + y = x\, e^{2x}.$$

Solution Here $f(x)$ is the product of a first-degree polynomial and an exponential. The *most general* expression of this form is

$$y_p(x) = e^{2x}(a + bx).$$

Then

$$y_p'(x) = e^{2x}(2a + b + 2bx), \qquad y_p''(x) = e^{2x}(4a + 4b + 4bx),$$

and substitution yields

$$e^{2x}(4a + 4b + 4bx) + e^{2x}(a + bx) = xe^{2x}.$$

Dividing both sides by e^{2x} and equating like powers of x, we obtain the equations

$$5a + 4b = 0, \qquad 5b = 1.$$

Thus $a = -\frac{4}{25}$, $b = \frac{1}{5}$, and a particular solution is

$$y_p(x) = \frac{e^{2x}}{25}(5x - 4).$$

Therefore the general solution of this example is (since $y'' + y = 0$ is the equation of the harmonic oscillator)

$$y(x) = c_1 \sin x + c_2 \cos x + \frac{e^{2x}}{25}(5x - 4).$$ ◀

Difficulties arise in connection with problems of this type whenever any term of the guessed solution is a solution of the homogeneous equation

$$y'' + ay' + by = 0. \tag{7}$$

For example, in the equation

$$y'' + y = (1 + x + x^2) \sin x, \tag{8}$$

the function $f(x)$ is the sum of three functions, one of which ($\sin x$) is a solution to the homogeneous equation $y'' + y = 0$. As another example, in

$$y'' + y = (x + x^2) \sin x \tag{9}$$

the guessed solution is $y_p = (a_0 + a_1x + a_2x^2)\sin x + (b_0 + b_1x + b_1x^2) \cos x$ and, $a_0 \sin x + b_0 \cos x$ is a solution to the homogeneous equation $y'' + y = 0$. When this situation occurs, the method of undetermined coefficients must be modified. To see why, consider the following example.

EXAMPLE 4 ▶ **Modifying the method**

Find the solution to the equation

$$y'' - y = 2e^x. \tag{10}$$

Solution The general solution of $y'' - y = 0$ is

$$y(x) = c_1e^x + c_2e^{-x}.$$

Here $f(x) = 2e^x$ is a solution to the homogeneous equation. If we try to find a solution of the form Ae^x we get nowhere, since Ae^x is a solution to the homogeneous equation for every constant A and, therefore, it cannot possibly be a solution to the nonhomogeneous equation.

What do we do ? Recall that if λ is a double root of the characteristic equation for a homogeneous differential equation, then two solutions are $e^{\lambda x}$ and $xe^{\lambda x}$. This suggests that we try Axe^x instead of Ae^x as a possible solution to Equation **(10).** Thus we consider a particular solution of the form

$$y_p = Axe^x.$$

Then $y_p' = Ae^x(x + 1)$, $y_p''(x) = Ae^x(x + 2)$, and

$$y_p'' - y_p = Ae^x(x + 2) - Axe^x = 2Ae^x = 2e^x.$$

Hence $A = 1$ and $y_p = xe^x$. Thus the general solution is

$$y(x) = c_1e^x + c_2e^{-x} + xe^x. \quad ◀$$

The preceding example suggests the following rule.

Modifications of the method

If any term of the guessed solution $y_p(x)$ is a solution of the homogeneous Equation **(7),** multiply $y_p(x)$ by x repeatedly until no term of the product $x^ky_p(x)$ is a solution of Equation **(7).** Then use the product $x^ky_p(x)$ to solve Equation **(1).**

EXAMPLE 5 ▶ **Solution by modified method**

Find the solution to

$$y'' + y = \cos x \tag{11}$$

that satisfies $y(0) = 2$ and $y'(0) = -3$.

Solution The general solution to $y'' + y = 0$ is $y = c_1\cos x + c_2\sin x$. Since $f(x) = \cos x$ is a solution, we must use the modification of the method to find a particular solution to Equation **(11)**. Ordinarily we would guess a solution of the form $y_p = A\cos x + B\sin x$. Instead we multiply by x and try a solution of the form

$$y_p = Ax\cos x + Bx\sin x.$$

Note that no term of y_p is a solution to $y'' + y = 0$. Then

$$y_p' = A\cos x - Ax\sin x + B\sin x + Bx\cos x$$

and

From Equation **(11)**

$$\begin{aligned}\cos x = y_p'' + y_p &= (-2A\sin x - Ax\cos x + 2B\cos x - Bx\sin x)\\ &\quad + (Ax\cos x + Bx\sin x)\\ &= -2A\sin x + 2B\cos x.\end{aligned}$$

Therefore

$$-2A = 0, \qquad 2B = 1, \qquad B = \frac{1}{2},$$

and

$$y_p = \frac{1}{2}x\sin x.$$

Thus the general solution to Equation **(11)** is

$$y = c_1\cos x + c_2\sin x + \frac{1}{2}x\sin x.$$

We are not finished yet, as initial conditions were given. We have

$$y' = -c_1\sin x + c_2\cos x + \frac{1}{2}x\cos x + \frac{1}{2}\sin x.$$

Then

$$y(0) = c_1 = 2 \qquad \text{and} \qquad y'(0) = c_2 = -3,$$

which yields the unique solution

$$y(x) = 2\cos x - 3\sin x + \frac{1}{2}x\sin x. \quad ◀$$

Warning

As in this example, you should apply the initial conditions only to the general solution.

EXAMPLE 6 ▶ Modifying twice

Find the general solution of

$$y'' - 4y' + 4y = e^{2x}.$$

Solution The homogeneous equation $y'' - 4y' + 4y = 0$ has the independent solutions e^{2x} and xe^{2x}. Thus, multiplying $f(x) = e^{2x}$ by x *twice*, we look for a particular solution of the form $y_p = ax^2e^{2x}$. Then

$$y_p' = ae^{2x}(2x^2 + 2x)$$

and

$$y_p'' = ae^{2x}(4x^2 + 8x + 2),$$

so

$$\begin{aligned} y_p'' - 4y_p' + 4y_p &= ae^{2x}(4x^2 + 8x + 2 - 8x^2 - 8x + 4x^2) \\ &= 2ae^{2x} = e^{2x}, \end{aligned}$$

or $2a = 1$ and $a = \frac{1}{2}$. Thus $y_p = \frac{1}{2}x^2e^{2x}$, and the general solution is

$$y(x) = c_1e^{2x} + c_2xe^{2x} + \frac{1}{2}x^2e^{2x} = e^{2x}\left(c_1 + c_2x + \frac{1}{2}x^2\right). \blacktriangleleft$$

EXAMPLE 7 ▶ **Using superposition**

Consider the equation

$$y'' - y = x^2 + 2e^x.$$

Using the results of Examples 1 and 4 and the principle of superposition (see p. 114), we find immediately that a particular solution is given by

$$y_p(x) = -2 - x^2 + xe^x. \blacktriangleleft$$

EXAMPLE 8 ▶ **Solving when $f(x)$ is a product of a polynomial and sine**

Find the general solution to

$$y'' + y = x \sin x.$$

Solution The guessed solution is $y_p = (Ax + B)\cos x + (Cx + D)\sin x$. Since $B \cos x + D \sin x$ solves $y'' + y = 0$, the modification is required. We therefore multiply by x and try a solution of the form

$$y_p = (Ax^2 + Bx)\cos x + (Cx^2 + Dx)\sin x.$$

Then

$$\begin{aligned} y_p' &= [Cx^2 + (2A + D)x + B] \cos x + [-Ax^2 + (2C - B)x + D] \sin x, \\ y_p'' &= [-Ax^2 + (4C - B)x + 2A + 2D] \cos x \\ &\quad + [-Cx^2 - (4A + D)x + 2C - 2B] \sin x, \end{aligned}$$

and

$$y_p'' + y_p = [4Cx + 2A + 2D] \cos x + [-4Ax + 2C - 2B] \sin x \overset{\text{given}}{\downarrow}{=} x \sin x.$$

This yields $A = -\frac{1}{4}$, $B = 0$, $C = 0$, $D = \frac{1}{4}$, and the particular solution

$$y_p = -\frac{1}{4}x^2 \cos x + \frac{1}{4}x \sin x$$

Thus the general solution is

$$y = \left(c_1 - \frac{1}{4}x^2\right)\cos x + \left(c_2 + \frac{1}{4}x\right)\sin x. \blacktriangleleft$$

SELF-QUIZ

I. In solving $y'' + ay' + by = f(x)$, for which functions f below does the method of undetermined coefficients *not* apply. (There is more than one answer.)

(a) $f(x) = x^3 e^{4x}$
(b) $f(x) = x^{-1}$
(c) $f(x) = \sqrt{2}\, e^{3x} \cos 4x$
(d) $f(x) = \sqrt{x}\, e^x$
(e) $f(x) = \dfrac{\cos x}{\sin x}$
(f) $f(x) = x^3 e^x - e^{-2x} \cos 4x$

II. To solve $y'' + 2y' + y = e^{-x}$, we must multiply the guessed particular solution $y_p(x) = A\, e^{-x}$ by ________.

(a) nothing **(b)** x **(c)** x^2 **(d)** x^3

Answers to Self-Quiz

I. b, d, e **II.** c

PROBLEMS 3.6

In Problems 1–33 find the general solution of each differential equation. If initial conditions are given, find the particular solution that satisfies them, and sketch its solution using a graphing calculator or a computer algebra system.

1. $y'' + 4y = 3 \sin x$
2. $y'' - y' - 6y = 20e^{-2x}$, $y(0) = 0$, $y'(0) = 6$
3. $y'' - 3y' + 2y = 6e^{3x}$
4. $y'' + y' = 3x^2$, $y(0) = 4$, $y'(0) = 0$
5. $y'' - 2y' + y = -4e^x$
6. $y'' - 4y' + 4y = 6xe^{2x}$, $y(0) = 0$, $y'(0) = 3$
7. $y'' - 7y' + 10y = 100x$, $y(0) = 0$, $y'(0) = 5$
8. $y'' + y = 1 + x + x^2$
9. $y'' + y' = x^3 - x^2$
10. $y'' + 4y = 16x \sin 2x$
11. $y'' - 4y' + 5y = 2e^{2x} \cos x$
12. $y'' - y' - 2y = x^2 + \cos x$
13. $y'' + 6y' + 9y = 10e^{-3x}$
14. $y'' + 8y' + 16y = 3e^{4x}$
15. $y'' + 8y' + 16y = 2xe^{4x}$
16. $y'' + 8y' + 16y = -x^2 e^{4x}$
17. $y'' + 8y' + 16y = 5x^3 e^{4x}$; $y(0) = 1$, $y'(0) = -2$
18. $y'' + 6y' + 13y = \cos 2x$; $y\left(\dfrac{\pi}{2}\right) = 1$, $y'\left(\dfrac{\pi}{2}\right) = 0$
19. $y'' + 6y' + 13y = e^{-3x} \sin 2x$
20. $y'' + 6y' + 13y = xe^{-3x} \sin 2x$
21. $y'' + 6y' + 13y = 2xe^{-3x} \cos 2x + x^2 e^{-3x} \sin 2x$
22. $y'' - 5y' + 6y = x^2 e^x$
23. $y'' - 3y' + 2y = \cos x - \sin x$
24. $y'' + 9y = x^3 + x$
25. $y'' + 9y = x^4 - x^2 + 3$; $y(0) = 1$, $y'(0) = 0$
26. $y'' + 9y = x \cos 3x$; $y(0) = 1$, $y'(0) = 0$
27. $y'' + 9y = -3x \sin 3x$; $y(0) = 1$, $y'(0) = 0$
28. $y'' - 4y = e^{2x}(3 - 5x)$; $y(0) = 2$, $y'(0) = 1$
29. $y'' - 4y = e^{2x}(1 + x + x^2)$; $y(0) = 2$, $y'(0) = 1$
30. $y'' + 4y' + 3y = x^2 \sin 2x$
31. $y'' + 4y' + 3y = x^2 e^{-x} \cos 2x$
32. $y'' - y' - 2y = e^x - e^{-2x}$
33. $y'' - y' - 2y = 3 + e^x$

Use the principle of superposition to find the general solution of each of the equations in Problems 34–37.

34. $y'' + y = 1 + 2 \sin x$
35. $y'' - 2y' - 3y = x - x^2 + e^x$
36. $y'' + 4y = 3 \cos 2x - 7x^2$
37. $y'' + 4y' + 4y = xe^x + \sin x$

38. Show by the methods of this section that a particular solution of

$$y'' + 2ay' + b^2 y = A \sin \omega x \qquad (a, \omega > 0)$$

is given by

$$y = \frac{A \sin(\omega x - \alpha)}{\sqrt{(b^2 - \omega^2)^2 + 4\omega^2 a^2}},$$

where

$$\alpha = \tan^{-1} \frac{2a\omega}{(b^2 - \omega^2)}, \qquad (0 < \alpha < \pi).$$

39. Let $f(x)$ be a polynomial of degree n. Show that, if $b \neq 0$, there is always a solution that is a polynomial of degree n for the equation $y'' + ay' + by = f(x)$

40. Use the method indicated in Problem 39 to find a particular solution of

$$y'' + 3y' + 2y = 9 + 2x - 2x^2.$$

In Problems 41–44 find particular solutions to the given differential equation.

41. $y'' + y = (x + x^2)\sin x$

42. $y'' - y' = x^2$

43. $y'' - 2y' + y = x^2e^x$

44. $y'' - 4y' + 3y = x^3e^{3x}$

45. In many physical problems (see, for example, Section 6.3), the nonhomogeneous term $f(x)$ is specified by different formulas in different intervals of x. Find

(a) a general solution of the equation

$$y'' + y = \begin{cases} x, & 0 \le x \le 1, \\ 1, & x \ge 1; \end{cases}$$

[*Note:* This "solution" is not differentiable at $x = 1$.]

(b) a particular solution of part (a) that satisfies the initial conditions

$$y(0) = 0, \qquad y'(0) = 1.$$

3.7 EULER EQUATIONS

For most linear second-order equations with variable coefficients it is impossible to write solutions in terms of elementary functions. In most cases it is necessary to use techniques such as the power series method (Chapter 5) to obtain information about solutions. However, there is one class of such equations that do arise in applications for which solutions in terms of elementary functions can be obtained.

Euler equation

An equation of the form

$$x^2y'' + axy' + by = f(x), \qquad x \neq 0 \tag{1}$$

is called an **Euler equation.**

Note

Equation **(1)** can be written

$$y'' + \frac{a}{x}y' + \frac{b}{x^2}y = \frac{f(x)}{x^2},$$

which is not defined for $x = 0$. This is why we make the restriction that $x \neq 0$. We begin by solving the homogeneous Euler equation

$$x^2y'' + axy' + by = 0, \qquad x \neq 0. \tag{2}$$

If we can find two linearly independent solutions to Equation **(2)**, we can solve Equation **(1)** by the method of variation of parameters. There are two ways to solve Equation **(2)**; each one involves a certain trick. We give one method here and leave the other for the problem set (see Problem 18).

The first method involves guessing an appropriate solution to Equation **(2)**. We note that if $y = x^\lambda$ for some number λ, then $y' = \lambda x^{\lambda-1}$ and $y'' = \lambda(\lambda - 1)x^{\lambda-2}$. This is interesting because x^2y'', xy', and y all can be written as constant multiples of x^λ. Therefore we guess that there is a solution having the form $y = x^\lambda$. Substituting this into Equation **(2)**, we obtain

$$\lambda(\lambda - 1)x^\lambda + a\lambda x^\lambda + bx^\lambda = x^\lambda[\lambda(\lambda - 1) + a\lambda + b] = 0.$$

If $x \neq 0$, we can divide by x^λ to obtain the

Characteristic equation* for the Euler equation:

$$\lambda(\lambda - 1) + a\lambda + b = 0, \tag{3}$$

or

$$\lambda^2 + (a - 1)\lambda + b = 0. \tag{4}$$

As with constant-coefficient equations, there are three cases to consider.

Case 1

Characteristic Equation **(4)** has two real, distinct roots.

EXAMPLE 1 ▶ **An Euler equation when the roots are real and distinct**

Find the general solution to

$$x^2y'' + 2xy' - 12y = 0, \qquad x \neq 0.$$

Solution The characteristic equation is

$$\lambda(\lambda - 1) + 2\lambda - 12 = \lambda^2 + \lambda - 12 = 0 = (\lambda + 4)(\lambda - 3),$$

with roots $\lambda_1 = -4$ and $\lambda_2 = 3$. Thus two solutions (that are linearly independent) are

$$y_1 = x^{-4} = \frac{1}{x^4} \qquad \text{and} \qquad y_2 = x^3,$$

and the general solution is

$$y(x) = \frac{c_1}{x^4} + c_2x^3.$$

In Figure 3.8, we provide a graph of the solution for $c_1 = c_2 = 1$. ◀

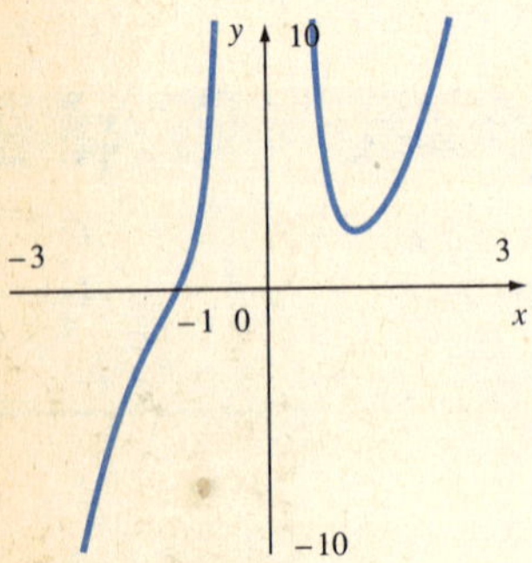

FIGURE 3.8
Graph of
$y = 1/x^4 + x^3, \quad x \neq 0$

We summarize the results of Example 1 in the following theorem.

THEOREM 1 ♦ **Solving an Euler equation when the roots are real and unequal**

If λ_1 and λ_2 are real and distinct, then the general solution to Equation **(2)** is

$$y(x) = c_1x^{\lambda_1} + c_2x^{\lambda_2}, \qquad x \neq 0. \quad ♦ \tag{5}$$

Case 2

The roots are real and equal ($\lambda_1 = \lambda_2$).

EXAMPLE 2 ▶ **An Euler equation when the roots are real and equal**

Find the general solution to

$$x^2y'' - 3xy' + 4y = 0, \qquad x > 0. \tag{6}$$

---

* The term "characteristic equation" is generally reserved for linear equations with *constant* coefficients. The only reason we can use this term in this case is that the substitution in Problem 18 on page 000 converts Equation **(1)** into a second-order constant-coefficient equation with characteristic Equation **(4)**. *The method of characteristic equations is generally not applicable to equations with variable coefficients.*

Solution The characteristic equation is

$$\lambda^2 - 4\lambda + 4 = (\lambda - 2)^2 = 0,$$

with the single root $\lambda = 2$. Thus one solution is $y_1(x) = x^2$. To find a second solution we use the method of reduction of order (see Section 3.2). Let $y_2 = vy_1 = x^2v$. Then $y_2' = x^2v' + 2xv$ and $y_2'' = x^2v'' + 4xv' + 2v$, so Equation **(6)** becomes

$$x^2(x^2v'' + 4xv' + 2v) - 3x(x^2v' + 2xv) + 4(x^2v) = 0,$$

or, with $z = v'$,

$$x^4v'' + x^3v' = x^4z' + x^3z = 0.$$

Separating variables, we have

$$\ln|z| = \int \frac{dz}{z} = -\int \frac{dx}{x} = -\ln x,$$

so that

$$v' = z = x^{-1}.$$

Then

$$y_2(x) = y_1v = x^2 \int \frac{dx}{x} = x^2 \ln x.$$

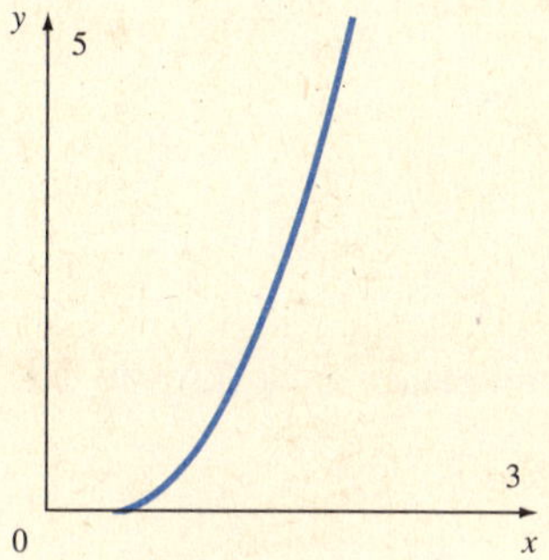

FIGURE 3.9
Graph of $y = x^2(1 + \ln x)$, $x > 0$

Thus the general solution to Equation **(6)** is

$$y(x) = c_1x^2 + c_2x^2\ln x = x^2(c_1 + c_2\ln x).$$

In Figure 3.9, we provide a graph of the solution for $c_1 = c_2 = 1$. ◀

We summarize the results of Example 2 in Theorem 2.

THEOREM 2 ◆ **Solving an Euler equation when the roots are real and equal**
If λ is the only root of characteristic Equation **(4)**, then the general solution to Equation **(2)** is

$$y(x) = x^\lambda(c_1 + c_2\ln|x|), \qquad x \neq 0. \quad ◆$$

Case 3

The roots are complex conjugates ($\lambda_1 = \alpha + i\beta$, $\lambda_2 = \alpha - i\beta$).

EXAMPLE 3 ▶ **An Euler equation with complex conjugate roots**
Find the general solution of

$$x^2y'' + 5xy' + 13y = 0, \qquad x > 0. \tag{7}$$

Solution The characteristic equation is

$$\lambda^2 + 4\lambda + 13 = 0,$$

and

$$\lambda = \frac{-4 \pm \sqrt{16 - 4(13)}}{2} = \frac{-4 \pm \sqrt{-36}}{2} = -2 \pm 3i.$$

Thus two linearly independent solutions are

$$y_1(x) = x^{-2+3i} \qquad \text{and} \qquad y_2(x) = x^{-2-3i}.$$

Using the material in Appendix 5, we can eliminate the imaginary exponents. First we note that

$$x^a = e^{\ln x^a} = e^{a \ln x}.$$

By the Euler formula ($e^{i\beta x} = \cos \beta x + i \sin \beta x$),

$$\begin{aligned} y_1(x) &= (x^{-2})(x^{3i}) = x^{-2}e^{3i\ln x} \\ &= x^{-2}[\cos(3 \ln x) + i \sin(3 \ln x)] \end{aligned}$$

and

$$\begin{aligned} y_2(x) &= (x^{-2})(x^{-3i}) = x^{-2}e^{3i\ln x} \\ &= x^{-2}[\cos(3 \ln x) - i \sin(3 \ln x)]. \end{aligned}$$

We now form two new solutions:

$$y_3(x) = \frac{1}{2}[y_1(x) + y_2(x)] = x^{-2} \cos(3 \ln x)$$

and

$$y_4(x) = \frac{1}{2i}[y_1(x) - y_2(x)] = x^{-2} \sin(3 \ln x).$$

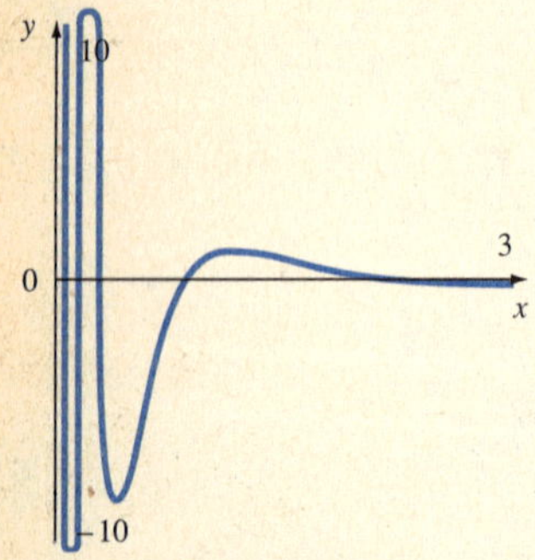

FIGURE 3.10
Graph of $y = (1/x^2)[\cos(3 \ln x) + \sin(3 \ln x)]$, $x > 0$

These new solutions contain no complex numbers and are easier to work with. The general solution to Equation **(7)** is

$$y(x) = x^{-2}[c_1 \cos(3 \ln x) + c_2 \sin(3 \ln x)].$$

A sketch of the solution is given in Figure 3.10 in the case $c_1 = c_2 = 1$. ◀

We summarize the result of Example 3 in the following theorem.

THEOREM 3 ♦ **Solving an Euler equation when the roots are complex conjugate**

If $\lambda_1 = \alpha + i\beta$ and $\lambda_2 = \alpha - i\beta$ are complex conjugate roots of the characteristic Equation **(4)**, then the general solution to Equation **(2)** is

$$y(x) = x^{\alpha}[c_1\cos(\beta \ln|x|) + c_2\sin(\beta \ln|x|)], \qquad x \neq 0. \quad ♦ \tag{8}$$

EXAMPLE 4 ▶ **Solving a nonhomogeneous Euler equation**

Find the general solution to

$$x^2y'' + 2xy' - 12y = \sqrt{x}, \qquad x > 0. \tag{9}$$

Solution In Example 1 we found the homogeneous solutions

$$y_1 = x^{-4} \qquad \text{and} \qquad y_2 = x^3.$$

Dividing both sides of Equation **(9)** by x^2, we obtain the standard form

$$y'' + \frac{2}{x}y' - \frac{12}{x^2}y = x^{-3/2}.$$

We now apply the method of variation of parameters, obtaining the system

$$\begin{aligned} x^{-4}c_1' + x^3c_2' &= 0, \\ -4x^{-5}c_1' + 3x^2c_2' &= x^{-3/2}. \end{aligned}$$

Solving these equations simultaneously, we get

$$c_1' = \frac{-x^{7/2}}{7} \quad \text{and} \quad c_2' = \frac{x^{-7/2}}{7}.$$

Hence

$$c_1(x) = -\frac{1}{7}\cdot\frac{2}{9}x^{9/2}, \qquad c_2(x) = -\frac{1}{7}\cdot\frac{2}{5}x^{-5/2},$$

so

$$y_p(x) = c_1(x)y_1(x) + c_2(x)y_2(x) = -\frac{1}{7}\left[\frac{2}{9}x^{9/2}\cdot x^{-4} + \frac{2}{5}x^{-5/2}\cdot x^3\right]$$

$$= \left(\frac{2}{9}+\frac{2}{5}\right)\frac{-x^{1/2}}{7} = -\frac{4}{45}x^{1/2}.$$

Thus the general solution is given by

$$y(x) = c_1x^{-4} + c_2x^3 - \frac{4}{45}x^{1/2}. \quad ◀$$

SELF-QUIZ

I. Two linearly independent solutions to $x^2y'' + 3xy' + 5y = 0$ are ______.
(a) x^{-1}, x^{-5}
(b) $\frac{\cos(2\ln|x|)}{x}, \frac{\sin(2\ln|x|)}{x}$
(c) $x^{-3}, x^{-3}\ln|x|$
(d) $x^{-3/2}\cos\left(\frac{\sqrt{21}\ln|x|}{2}\right), x^{-3/2}\sin\left(\frac{\sqrt{21}\ln|x|}{2}\right)$

II. Two linearly independent solutions to $x^2y'' + 5xy' + 3y = 0$ are ______.
(a) x^{-1}, x^{-3} **(b)** $x^{(-5+\sqrt{13})/2}, x^{(-5-\sqrt{13})/2}$
(c) $x^{-5/2}\cos\left(\frac{\sqrt{13}}{2}\ln|x|\right)$, $x^{-5/2}\sin\left(\frac{\sqrt{13}}{2}\ln|x|\right)$
(d) $x^{-5/2}, x^{-5/2}\ln|x|$

III. Two linearly independent solutions to $x^2y'' + 7y' + 9y = 0$ are ______.
(a) $x^{-7-\sqrt{13}/2}, x^{-7+\sqrt{13}/2}$ **(b)** x^3, x^{-3}
(c) $x^{-3}\sin(3\ln|x|), x^{-3}\cos(3\ln|x|)$
(d) $x^{-3}, x^{-3}\ln|x|$

Answers to Self-Quiz

I. b **II.** a **III.** d

PROBLEMS 3.7

In Problems 1–17 find the general solution to the given Euler equation for $x > 0$. Find the unique solution when initial conditions are given, and sketch its graph using a graphing calculator or a computer algebra system.

1. $x^2y'' + xy' - y = 0$
2. $x^2y'' - 5xy' + 9y = 0$
3. $x^2y'' - xy' + 2y = 0$
4. $x^2y'' - 2y = 0$, $y(1) = 3$, $y'(1) = 1$
5. $4x^2y'' - 4xy' + 3y = 0$, $y(1) = 0$, $y'(1) = 1$
6. $x^2y'' + 3xy' + 2y = 0$
7. $x^2y'' - 3xy' + 3y = 0$
8. $x^2y'' + 5xy' + 4y = 0$, $y(1) = 1$, $y'(1) = 3$
9. $x^2y'' + 5xy' + 5y = 0$
10. $4x^2y'' - 8xy' + 8y = 0$
11. $x^2y'' + 2xy' - 12y = 0$
12. $x^2y'' + xy' + y = 0$
13. $x^2y'' + 3xy' - 15y = \frac{1}{x}$
14. $x^2y'' + 3xy' + y = 3x^6$
15. $x^2y'' - 5xy' + 9y = x^3$
16. $x^2y'' + 3xy' - 15y = x^2e^x$
17. $x^2y'' + xy' + y = 10$

18. Show that the homogeneous Euler equation **(2)** can be transformed into the constant-coefficient equation $y'' + (a - 1)y' + by = 0$ by making the substitution $x = e^t$ $(t = \ln x)$. [*Hint:* By the chain rule,

$$\frac{dy}{dt} = \frac{dy}{dx}\frac{dx}{dt} = x\frac{dy}{dx}$$

and

$$\frac{d^2y}{dt^2} = \frac{d}{dx}\left(x\frac{dy}{dx}\right)\frac{dx}{dt} = x^2\frac{d^2y}{dx^2} + x\frac{dy}{dx}.\Big]$$

Use the method of Problem 18 to solve Problems 19–22.

19. $x^2y'' + 7xy' + 5y = x$

20. $x^2y'' + 3xy' - 3y = 5x^2$

21. $x^2y'' - 2y = \ln x$, $(x > 0)$

22. $4x^2y'' - 4xy' + 3y = \sin \ln(-x)$ $(x < 0)$

23. The equation $xy'' + 4y' = 0$, $x > 0$ arises in astronomy.* Obtain its general solution.

3.8 HIGHER-ORDER LINEAR DIFFERENTIAL EQUATIONS

In this section we extend the results of the chapter to linear differential equations of order higher than two. There is little theoretical difference between second- and higher-order systems, so we can be relatively brief. We state all theorems without proof.

The *general nonhomogeneous linear nth-order equation* is

$$y^{(n)}(x) + a_{n-1}(x)y^{(n-1)}(x) + \ldots + a_1(x)y'(x) + a_0(x)y(x) = f(x). \quad \textbf{(1)}$$

The associated homogeneous equation is

$$y^{(n)}(x) + a_{n-1}(x)y^{(n-1)}(x) + \ldots + a_1(x)y'(x) + a_0(x)y(x) = 0. \quad \textbf{(2)}$$

In Theorem 3.1.1 (p. 105) we stated that Equation **(1)** has a unique solution provided that all the functions in the equation are continuous and n initial conditions are specified. Now we concern ourselves with finding the general solutions to Equations **(1)** and **(2)**. To do so we follow the procedures we have developed for solving second-order equations.

linear independence

We say that the functions $y_1, y_2, \ldots, y_n$ are **linearly independent** in $[x_0, x_1]$ if the following condition holds:

$$c_1y_1(x) + c_2y_2(x) + \ldots + c_ny_n(x) = 0 \quad \text{for all } x \in [x_0, x_1]$$
$$\text{implies that } c_1 = c_2 = \ldots = c_n = 0.$$

linear combination
Wronskian

Otherwise the functions are **linearly dependent.** The expression $c_1y_1 + c_2y_2 + \ldots + c_ny_n$ is called a **linear combination** of the functions $y_1, y_2, \ldots, y_n$.

The Wronskian of $y_1, y_2, \ldots, y_n$ is defined by

$$W(y_1, y_2, \ldots, y_n)(x) = \begin{vmatrix} y_1 & y_2 & \cdots & y_n \\ y_1' & y_2' & \cdots & y_n' \\ y_1'' & y_2'' & \cdots & y_n'' \\ \vdots & \vdots & & \vdots \\ y_1^{(n-1)} & y_2^{(n-1)} & \cdots & y_n^{(n-1)} \end{vmatrix} \quad \textbf{(3)}$$

THEOREM 1 ♦ **Properties of the Wronskian**

Let $a_0, a_1, \ldots, a_{n-1}$ be continuous in $[x_0, x_1]$ and let $y_1, y_2, \ldots, y_n$ be n solutions of Equation **(2)**. Then

1. $W(y_1, y_2, \ldots, y_n)(x)$ is zero either for all $x \in [x_0, x_1]$ or for no $x \in [x_0, x_1]$.
2. $y_1, y_2, \ldots, y_n$ are linearly independent if and only if $W(y_1, y_2, \ldots, y_n)(x) \neq 0$. ♦

* Z. Kopal, "Stress History of the Moon and of Terrestrial Planets," *Icarus* 2 (1963): 381.

EXAMPLE 1 ▶ **Three linearly independent solutions**

The functions 1, x, and x^2 are solutions to the equation $y'''(x) = 0$. Determine whether they are linearly independent or dependent for all x.

Solution The easiest way to test for linear independence is to use the Wronskian:

$$W(y_1, y_2, y_3)(x) = \begin{vmatrix} 1 & x & x^2 \\ 0 & 1 & 2x \\ 0 & 0 & 2 \end{vmatrix} = 2 \neq 0,$$

so the functions are linearly independent. Alternatively, consider

$$c_1 \cdot 1 + c_2 x + c_3 x^2 = 0.$$

Setting $x = 0$, it follows that $c_1 = 0$. Setting $x = \pm 1$ yields the system

$$\begin{aligned} c_2 + c_3 &= 0, \\ -c_2 + c_3 &= 0, \end{aligned}$$

so $c_2 = c_3 = 0$, implying that the functions are linearly independent. ◀

The procedure for solving a linear nth-order equation is as follows:

Step 1 Find n linearly independent solutions, $y_1, y_2, \ldots, y_n$, to the homogeneous equation

$$y^{(n)} + a_{n-1}(x)y^{(n-1)} + \ldots + a_1(x)y' + a_0(x)y = 0.$$

general solution

The **general solution** to this equation is then

$$y(x) = c_1 y_1(x) + c_2 y_2(x) + \ldots + c_n y_n(x). \tag{4}$$

Step 2 Find one solution, $y_p(x)$, to the nonhomogeneous equation

$$y^{(n)} + a_{n-1}(x)y^{(n-1)} + \ldots + a_1(x)y' + a_0(x)y = f(x).$$

The **general solution** to this equation is then given by

$$y(x) = c_1 y_1(x) + c_2 y_2(x) + \ldots + c_n y_n(x) + y_p, \tag{5}$$

where $y_1, y_2, \ldots, y_n$ are the n linearly independent solutions of Step 1.

As in the case of second-order equations, we can generally obtain these solutions only when the coefficients $a_k(x)$ are all constants. In this case, Equations **(1)** and **(2)** are said to have **constant coefficients.** We only deal with the case where these constants are real.

The general nth-order linear, homogeneous constant-coefficient equation is

$$y^{(n)}(x) + a_{n-1}y^{(n-1)}(x) + \ldots + a_1 y'(x) + a_0 y(x) = 0. \tag{6}$$

Note that

$$\frac{d^n}{dx^n}e^{\lambda x} = \lambda^n e^{\lambda x}.$$

If we substitute $y = e^{\lambda x}$ into Equation **(6)** and then divide by $e^{\lambda x}$, we obtain the **characteristic equation**

$$\lambda^n + a_{n-1}\lambda^{n-1} + \ldots + a_1\lambda + a_0 = 0. \tag{7}$$

Equation **(7)** has n roots $\lambda_1, \lambda_2, \ldots, \lambda_n$. Some of these roots may be real and distinct, real and equal, distinct complex conjugate pairs, or equal complex conjugate pairs. If a root λ_k (real or complex) occurs m times, we say that it has

multiplicity m. The following rules tell us how to find the general solution to Equation **(6)**.

Procedure for solving linear homogeneous equations with constant coefficients

1. Obtain characteristic Equation **(7)**.
2. Find the roots $\lambda_1, \lambda_2, \ldots, \lambda_n$ of **(7)**. (This is usually the most difficult step.)
3. For each real root λ_k of multiplicity 1 (*single root*), one solution to Equation **(6)** is $y_k = e^{\lambda_k x}$.
4. For each real root λ_k of multiplicity $m > 1$, m solutions to Equation **(6)** are

$$y_1 = e^{\lambda_k x},\ y_2 = xe^{\lambda_k x},\ \ldots,\ y_m = x^{m-1}e^{\lambda_k x}.$$

5. If $\alpha + i\beta$ and $\alpha - i\beta$ are simple roots, then two solutions to Equation **(6)** are

$$y_1 = e^{\alpha x}\cos \beta x \qquad \text{and} \qquad y_2 = e^{\alpha x}\sin \beta x.$$

6. If $\alpha + i\beta$ and $\alpha - i\beta$ are roots of multiplicity $m > 1$, then $2m$ solutions to Equation **(6)** are

$$y_1 = e^{\alpha x}\cos \beta x,\ y_2 = xe^{\alpha x}\cos \beta x,\ \ldots,\ y_m = x^{m-1}e^{\alpha x}\cos \beta x,$$
$$y_{m+1} = e^{\alpha x}\sin \beta x,\ y_{m+2} = xe^{\alpha x}\sin \beta x,\ \ldots,\ y_{2m} = x^{m-1}e^{\alpha x}\sin \beta x.$$

7. If $y_1, y_2, \ldots, y_n$ are the n solutions obtained in Steps 3–6, then $y_1, y_2, \ldots, y_n$ are linearly independent and the general solution to Equation **(6)** is given by

$$y(x) = c_1y_1(x) + c_2y_2(x) + \ldots + c_ny_n(x).$$

EXAMPLE 2 ▸ **Solving a third-order equation**

Find the general solution of

$$y''' - 3y'' - 10y' + 24y = 0.$$

Solution The characteristic equation is

$$\lambda^3 - 3\lambda^2 - 10\lambda + 24 = (\lambda - 2)(\lambda + 3)(\lambda - 4) = 0,$$

with roots $\lambda_1 = 2$, $\lambda_2 = -3$, and $\lambda_3 = 4$. Since these roots are real and distinct, three linearly independent solutions are

$$y_1 = e^{2x}, \qquad y_2 = e^{-3x}, \qquad y_3 = e^{4x},$$

and the general solution is

$$y(x) = c_1e^{2x} + c_2e^{-3x} + c_3e^{4x}. \quad ◂$$

EXAMPLE 3 ▸ **Solving a fourth-order equation**

Find the general solution of

$$y^{(4)} - 4y''' + 6y'' - 4y' + y = 0.$$

Solution The characteristic equation is

$$\lambda^4 - 4\lambda^3 + 6\lambda^2 - 4\lambda + 1 = (\lambda - 1)^4 = 0,$$

with the single root $\lambda = 1$ of multiplicity 4. Thus four linearly independent solutions are

$$y_1 = e^x, \qquad y_2 = xe^x, \qquad y_3 = x^2e^x, \qquad y_4 = x^3e^x,$$

and the general solution is

$$y(x) = e^x(c_1 + c_2x + c_3x^2 + c_4x^3). \blacktriangleleft$$

EXAMPLE 4 ▶ **Solving a fifth-order equation**

Find the general solution of

$$y^{(5)} - 2y^{(4)} + 8y'' - 12y' + 8y = 0.$$

Solution The characteristic equation is

$$\lambda^5 - 2\lambda^4 + 8\lambda^2 - 12\lambda + 8 = 0,$$

which can be factored into

$$(\lambda + 2)(\lambda^2 - 2\lambda + 2)^2 = 0.$$

The solutions to $\lambda^2 - 2\lambda + 2 = 0$ are $\lambda = 1 \pm i$. Thus the roots are

$$\lambda_1 = -2 \text{ (simple)}, \qquad \lambda_2 = 1 + i, \qquad \lambda_3 = 1 - i,$$

with the complex roots λ_2 and λ_3 having multiplicity 2. Thus five linearly independent solutions are

$$y_1 = e^{-2x}, \qquad y_2 = e^x\cos x, \qquad y_3 = xe^x\cos x,$$
$$y_4 = e^x\sin x, \qquad y_5 = xe^x\sin x,$$

and the general solution is

$$y(x) = c_1e^{-2x} + (c_2 + c_3x)e^x\cos x + (c_4 + c_5x)e^x\sin x. \blacktriangleleft$$

Remark

In solving the last three characteristic equations we made the factoring look easy. Finding roots of a polynomial of degree greater than two is, in general, very difficult.

How do we find a particular solution to the nonhomogeneous Equation **(1)**? As with second-order equations, there are two methods: undetermined coefficients and variation of parameters. The method of undetermined coefficients is identical to the technique we used for second-order equations. The method of variation of parameters is discussed in Problems 29 and 30.

Finally, certain equations with variable coefficients can be solved. The higher-order Euler equation is discussed in Problems 31–34.

SELF-QUIZ

I. Which of the following are solutions to $y''' - 3y'' + 3y' - y = 0$?

(a) e^x **(b)** e^{-x}
(c) xe^x **(d)** xe^{-x}
(e) x^2e^x **(f)** x^2e^{-x}
(g) x^3e^x **(h)** x^3e^{-x}

II. Which of the following are solutions to $y^{(4)} - 4y'' + 4y = 0$?

(a) $e^{\sqrt{2}x}$ **(b)** $e^{-\sqrt{2}x}$
(c) $xe^{\sqrt{2}x}$ **(d)** $xe^{-\sqrt{2}x}$
(e) $\cos\sqrt{2}\,x$ **(f)** $\sin\sqrt{2}\,x$

III. Which of the following are solutions to a homogeneous differential equation whose characteristic equation is $(\lambda + 1)^3(\lambda^2 - 4)(\lambda^2 + 2\lambda + 2)$?

(a) e^{-x} **(b)** xe^{-x}
(c) x^2e^{-x} **(d)** e^{2x}
(e) e^{-2x} **(f)** $e^{-x}\cos x$
(g) $e^{-x}\sin x$

Answers to Self-Quiz

I. a, c, e **II.** a, b, c, d **III.** all of them

PROBLEMS 3.8

In Problems 1–16 find the general solution to the given equation. If initial conditions are given, find the particular solutions that satisfy them and sketch their graphs using a graphing calculator or a computer algebra system.

1. $y^{(4)} + 2y'' + y = 0$

2. $y''' - y'' - y' + y = 0$

3. $y''' - 3y'' + 3y' - y = 0$, $y(0) = 1$, $y'(0) = 2$, $y''(0) = 3$

4. $x''' + 5x'' - x' - 5x = 0$

5. $y''' - 9y' = 0$, $y(0) = 3$, $y'(0) = 0$, $y''(0) = 18$

6. $y''' - 6y'' + 3y' + 10y = 0$

7. $y^{(4)} = 0$

8. $y^{(4)} - 9y'' = 0$

9. $y^{(4)} - 5y'' + 4y = 0$

10. $y^{(5)} - 2y''' + y' = 0$

11. $y^{(4)} - 4y'' = 0$, $y(0) = 1$, $y'(0) = 3$, $y''(0) = 0$, $y'''(0) = 16$

12. $y^{(4)} - 4y''' - 7y'' + 22y' + 24y = 0$

13. $y''' - y'' + y' - y = 0$

14. $y''' - 3y'' + 4y' - 2y = 0$, $y(0) = 1$, $y'(0) = 2$, $y''(0) = 3$

15. $y''' - 27y = 0$

16. $y^{(5)} + 2y''' + y' = 0$, $y\left(\frac{\pi}{2}\right) = 0$, $y'\left(\frac{\pi}{2}\right) = 1$, $y''\left(\frac{\pi}{2}\right) = 0$, $y'''\left(\frac{\pi}{2}\right) = -3$, $y^{(4)}\left(\frac{\pi}{2}\right) = 0$

17. Show that the solutions y_1, y_2, and y_3 of the linear third-order differential equation

$$y''' + a_1(x)y'' + a_2(x)y' + a_3(x)y = 0$$

that satisfy the conditions

$$\begin{aligned} y_1(x_0) &= 1, & y_1'(x_0) &= 0, & y_1''(x_0) &= 0, \\ y_2(x_0) &= 0, & y_2'(x_0) &= 1, & y_2''(x_0) &= 0, \\ y_3(x_0) &= 0, & y_3'(x_0) &= 0, & y_3''(x_0) &= 1, \end{aligned}$$

respectively, are linearly independent.

18. Show that *any* solution of

$$y''' + a_1(x)y'' + a_2(x)y' + a_3(x)y = 0$$

can be expressed as a linear combination of the solutions y_1, y_2, y_3 given in Problem 17. [*Hint:* If $y(x_0) = c_1$, $y'(x_0) = c_2$, and $y''(x_0) = c_3$, consider the linear combination $c_1y_1 + c_2y_2 + c_3y_3$.]

19. Consider the third-order equation

$$y''' + a(x)y'' + b(x)y' + c(x)y = 0$$

and let $y_1(x)$ and $y_2(x)$ be two linearly independent solutions. Define $y_3(x) = v(x)y_1(x)$ and assume that $y_3(x)$ is a solution to the equation.

(a) Find a second-order differential equation that is satisfied by v'.

(b) Show that $(y_2/y_1)'$ is a solution of this equation.

(c) Use the result of part (b) to find a second linearly independent solution of the equation derived in part (a).

20. Consider the equation

$$y''' - \left(\frac{3}{x^2}\right)y' + \left(\frac{3}{x^3}\right)y = 0, \quad (x > 0).$$

(a) Show that $y_1(x) = x$ and $y_2(x) = x^3$ are two linearly independent solutions.

(b) Use the results of Problem 19 to get a third linearly independent solution.

21. Consider the third-order equation

$$y''' + a(x)y'' + b(x)y' + c(x)y = 0,$$

where a, b, and c are continuous functions of x in some interval I. Prove that if $y_1(x)$, $y_2(x)$, and $y_3(x)$ are solutions to the equation, then so is any linear combination of them.

22. In Problem 21, let

$$W(y_1, y_2, y_3)(x) = \begin{vmatrix} y_1 & y_2 & y_3 \\ y_1' & y_2' & y_3' \\ y_1'' & y_2'' & y_3'' \end{vmatrix}.$$

(a) Show that W satisfies the differential equation

$$W'(x) = -a(x)W.$$

(b) Prove that $W(y_1, y_2, y_3)(x)$ is either always zero or never zero.

23. **(a)** Prove that the solutions $y_1(x)$, $y_2(x)$, $y_3(x)$ of the equation in Problem 21 are linearly independent on $[x_0, x_1]$ if, and only if, $W(y_1, y_2, y_3) \neq 0$.
(b) Show that $\sin t$, $\cos t$, and e^t are linearly independent solutions of

$$y''' - y'' + y' - y = 0$$

on any interval (a, b) where $-\infty < a < b < \infty$.

24. Assume that $y_1(x)$ and $y_2(x)$ are two solutions to

$$y''' + a(x)y'' + b(x)y' + c(x)y = f(x).$$

Prove that $y_3(x) = y_1(x) - y_2(x)$ is a solution of the associated homogeneous equation.

In Problems 25–28 use the method of undetermined coefficients to find the general solution of the given equation.

25. $y''' - y'' - y' + y = e^x$

26. $y''' - y'' - y' + y = e^{-x}$

27. $y''' - 3y'' - 10y' + 24y = x + 3$

28. $y^{(4)} + 2y'' + y = 3 \cos x$

29. Consider the third-order equation

$$y''' + ay'' + by' + cy = f(x).$$

Let $y_1(x)$, $y_2(x)$, and $y_3(x)$ be three linearly independent solutions to the associated homogeneous equation. Assume that there is a solution of this equation of the form $y(x) = c_1(x)y_1(x) + c_2(x)y_2(x) + c_3(x)y_3(x)$.
(a) Following the steps used in deriving the method of variation of parameters for second-order equations, derive a method for solving third-order equations.
(b) Find a particular solution of the equation

$$y''' - 2y' - 4y = e^{-x} \tan x.$$

30. Use the method derived in Problem 29 to find a particular solution of

$$y''' + 5y'' + 9y' + 5y = 2e^{-2x}\sec x.$$

In Problems 31–33 guess that there is a solution of the form $y = x^\lambda$ to solve the given Euler equation.

31. $x^3y''' + 2x^2y'' - xy' + y = 0$

32. $x^3y''' - 12xy' + 24y = 0$

33. $x^3y''' + 4x^2y'' + 3xy' + y = 0$

34. Show that the substitution $x = e^t$ can be used to solve the third-order Euler equation

$$x^3y''' + x^2y'' - 2xy' + 2y = 0.$$

3.9 SECOND-ORDER DIFFERENCE EQUATIONS (OPTIONAL)

In Section 2.9 we presented an excursion on linear first-order difference equations. Here we derive a method for the solution of linear second-order difference equations. The technique we give is easily extended to higher-order linear difference equations.

First consider the linear homogeneous second-order difference equation with constant coefficients

$$y_{n+2} + ay_{n+1} + by_n = 0, \qquad b \neq 0. \tag{1}$$

We saw in Section 2.9 that the first-order equation $y_n = ay_{n-1}$ has the general solution $y_n = a^n y_0$. It is, then, not implausible to "guess" that there are solutions to Equation **(1)** of the form $y_n = \lambda^n$ for some λ (real or complex). Substituting $y_n = \lambda^n$ into Equation **(1),** we obtain

$$\lambda^{n+2} + a\lambda^{n+1} + b\lambda^n = 0.$$

Since this equation is true for all $n \geq 0$, it holds for $n = 0$, so we have

The characteristic equation for a second-order difference equation

$$\lambda^2 + a\lambda + b = 0. \tag{2}$$

This is the **characteristic equation** for the difference Equation **(1)** and is identical to the characteristic equation for the second-order differential equation derived in

Section 3.3. As before, the roots are

$$\lambda_1 = \frac{-a + \sqrt{a^2 - 4b}}{2} \quad \text{and} \quad \lambda_2 = \frac{-a - \sqrt{a^2 - 4b}}{2}. \tag{3}$$

Again there are three cases.

Case 1

λ_1 and λ_2 are distinct real roots of Equation **(2)**. Then the general solution of Equation **(1)** is given by

$$y_n = c_1\lambda_1^n + c_2\lambda_2^n \tag{4}$$

EXAMPLE 1 ▶ **A difference equation with two real roots**

Consider the equation

$$x_{n+2} - 5x_{n+1} - 6x_n = 0.$$

The characteristic equation is $\lambda^2 - 5\lambda - 6 = 0$, with the roots $\lambda_1 = 6$ and $\lambda_2 = -1$. The general solution is given by $y_n = c_1(6)^n + c_2(-1)^n$. If we specify the initial conditions $y_0 = 3$, $y_1 = 11$, for example, we obtain the system

$$c_1 + c_2 = 3,$$
$$6c_1 - c_2 = 11,$$

which has the unique solution $c_1 = 2$, $c_2 = 1$, and the specific solution

$$y_n = 2 \cdot 6^n + (-1)^n. \quad ◀$$

Case 2

If $a^2 - 4b = 0$, the two roots of Equation **(2)** are equal, and we have the solution $x_n = \lambda^n$ where $\lambda = -\frac{a}{2}$. A second linearly independent solution is $n\lambda^n$.

Check

If $y_n = n\lambda^n$, then

$$y_{n+2} + ay_{n+1} + by_n = (n + 2)\lambda^{n+2} + a(n + 1)\lambda^{n+1} + bn\lambda^n$$
$$= \lambda^n\left[n(\lambda^2 + a\lambda + b) + 2\lambda\left(\lambda + \frac{a}{2}\right)\right] = 0,$$

because $\lambda^2 + a\lambda + b = 0$ and $\lambda = -\frac{a}{2}$. Thus the general solution to Equation **(1)** is given by

$$y_n = (c_1 + c_2 n)\lambda^n. \tag{5}$$

EXAMPLE 2 ▶ **A difference equation with one real root**

Consider the equation

$$y_{n+2} - 6y_{n+1} + 9y_n = 0, \tag{6}$$

with the initial conditions $y_0 = 5$, $y_1 = 12$. The characteristic equation is $\lambda^2 - 6\lambda + 9 = 0$ with the double root $\lambda = 3$. The general solution is therefore

$$y_n = c_1 3^n + c_2 n 3^n = 3^n(c_1 + nc_2).$$

Using the initial conditions, we obtain

$$c_1 = 5,$$
$$3c_1 + 3c_2 = 12.$$

The unique solution is $c_1 = 5$, $c_2 = -1$, and we have the specific solution to Equation **(6)**: $y_n = 5 \cdot 3^n - n \cdot 3^n = 3^n(5 - n)$. ◀

Case 3

If $a^2 - 4b < 0$, the roots of $\lambda^2 + a\lambda + b = 0$ are the complex conjugates

$$\lambda_1 = \alpha + i\beta \qquad \text{and} \qquad \lambda_2 = \alpha - i\beta.$$

We write these numbers in polar form (see Appendix 5):

$$\lambda_1 = re^{i\theta} \qquad \text{and} \qquad \lambda_2 = re^{-i\theta},$$

where

$$r = \sqrt{\alpha^2 + \beta^2} \qquad \text{and} \qquad \tan\theta = \frac{\beta}{\alpha}$$

Equation **(1)** has two solutions,

$$x_n = r^n e^{in\theta} = r^n(\cos n\theta + i \sin n\theta)$$

and

$$y_n = r^n e^{-in\theta} = r^n(\cos n\theta - i \sin n\theta).$$

Two other solutions are

$$x_n^* = \frac{1}{2}(x_n + y_n) = r^n \cos n\theta$$

and

$$y_n^* = \frac{1}{2i}(x_n - y_n) = r^n \sin n\theta.$$

The general solution to Equation **(1)** is then

$$y_n = r^n(c_1\cos n\theta + c_2\sin n\theta). \tag{7}$$

EXAMPLE 3 ▶ **A difference equation with complex conjugate roots**

Let $y_{n+2} + y_n = 0$. The characteristic equation is $\lambda^2 + 1 = 0$, with the roots $\pm i$. Here $\alpha = 0$ and $\beta = 1$, so $r = 1$ and $\theta = \frac{\pi}{2}$. The general solution is given by

$$y_n = c_1\cos\frac{n\pi}{2} + c_2\sin\frac{n\pi}{2}. \tag{8}$$

This is the equation for **discrete harmonic motion,** which corresponds to the equation of ordinary harmonic motion as discussed in Example 3.4.1. If we specify $y_0 = 0$ and $y_1 = 1000$, we obtain

$$c_1 = 0, \qquad c_1\cos\frac{\pi}{2} + c_2\sin\frac{\pi}{2} = 1000,$$

with the solution $c_1 = 0$, $c_2 = 1000$, and the specific solution

$$y_n = 1000 \sin\frac{n\pi}{2}.$$ ◀

The theory of linear difference equations is very similar to the theory of linear differential equations. The definitions of linear combination and of linear independence of solutions are the same as the definitions in Section 3.1. The general solution of a homogeneous second-order difference equation is given by $c_1x_n + c_2y_n$, where x_n and y_n are linearly independent solutions to Equation **(1)**; that is, if we know two solutions, we know them all. The general solution to Equation **(1)** therefore always takes one of the forms **(4)**, **(5)**, or **(7)** depending on the nature of the roots of the characteristic equation.

We can also study the nonhomogeneous equation

$$y_{n+2} + ay_{n+1} + by_n = f_n. \tag{9}$$

It is not difficult to show that if y_n^* and y_n^{**} are two solutions to Equation **(9)**, then their difference $y_n^* - y_n^{**}$ is a solution to Equation **(1)**. Thus the general solution to the nonhomogeneous Equation **(9)** is given by

$$y_n = c_1x_n + c_2x_n^* + y_n^*,$$

where x_n and x_n^* are linearly independent solutions to Equation **(1)**.

Finally, there is a method of variation of parameters that can be used to find one solution to Equation **(9)**. Rather than say more about nonhomogeneous equations in this brief excursion, we provide an interesting application of homogeneous difference equations.

An application to games and quality control

Suppose that two people, A and B, play a certain unspecified game several times in succession. One of the players must win on each play of the game (ties are ruled out). The probability that A wins is $p\,(\neq 0)$, and the corresponding probability for B is q. Of course, $p + q = 1$. One dollar is bet by each competitor on each play of the game. Suppose that player A starts with k dollars and player B with j dollars. We let P_n denote the probability that A will bankrupt B (that is, win all B's money) when A has n dollars. Clearly $P_0 = 0$ (this is the probability that player A will wipe out player B when A has no money) and $P_{k+j} = 1$ (A has already eliminated B from further competition). To calculate P_n for other values of n between 0 and $k + j$, we have the following difference equation:

$$P_n = qP_{n-1} + pP_{n+1}. \tag{10}$$

To understand this equation, we note that if A has n dollars at one turn, his probability of having $n - 1$ dollars at the next turn is q, and similarly for the other term in Equation **(10)**.

Equation **(10)** can be written in our usual format as

$$pP_{n+1} - P_n + qP_{n-1} = 0,$$

or

$$P_{n+2} - \frac{1}{p}P_{n+1} + \frac{q}{p}P_n = 0.$$

The characteristic equation is

$$\lambda^2 - \frac{1}{p}\lambda + \frac{q}{p} = 0,$$

which has the roots

$$\frac{1}{2p}(1 \pm \sqrt{1 - 4pq}).$$

Since

$$\sqrt{1 - 4pq} = \sqrt{1 - 4p(1 - p)} = \sqrt{4p^2 - 4p + 1} = |1 - 2p|,$$

we discover that the roots of the characteristic equation are

$$\lambda_1 = \frac{1 - p}{p} = \frac{q}{p}, \qquad \lambda_2 = 1.$$

The general solution to Equation **(10)** is therefore

$$P_n = c_1\left(\frac{q}{p}\right)^n + c_2$$

when $p \neq q$, and

$$P_n = c_1 + c_2 n$$

when $p = q = \frac{1}{2}$ (by the rules for repeated roots). Keeping in mind the fact that $0 \leq P_n \leq 1$ for all n and using the boundary conditions $P_0 = 0$ and $P_{k+j} = 1$, we can finally obtain

$$P_n = \frac{1 - \left(\frac{q}{p}\right)^n}{1 - \left(\frac{q}{p}\right)^{k+j}}, \qquad \text{if } p \neq q, \tag{11}$$

and

$$P_n = \frac{n}{k + j}, \qquad \text{if } p = q = \frac{1}{2}, \tag{12}$$

for all n such that $0 \leq n \leq k + j$.

EXAMPLE 4 ▶ **Modeling a game of roulette**

In a game of roulette a gambler bets against a casino. Suppose that the gambler bets one dollar on red each time. On a typical roulette wheel there are eighteen reds, eighteen blacks, a zero, and a double zero. If the wheel is honest, the probability that the casino will win is $p = \frac{20}{38} \approx 0.5263$. Then $q \approx 0.4737$ and $\frac{q}{p} = 0.9$. Suppose that both the gambler and the casino start with k dollars. Then the probability that the gambler will be cleaned out is

$$P_k = \frac{1 - \left(\frac{q}{p}\right)^k}{1 - \left(\frac{q}{p}\right)^{2k}} = \frac{1}{1 + \left(\frac{q}{p}\right)^k} = \frac{1}{1 + (0.9)^k}.$$

TABLE 3.1

k	P_k	P_k^*
1	0.5263	0.5263
5	0.6287	0.8740
10	0.7415	0.9492
25	0.9330	0.9923
50	0.9949	0.9995
100	0.9999	0.9999

The second column in Table 3.1 shows the probability that the gambler will lose all his money for different initial amounts k. Note that the more money both start with, the more likely it is that the gambler will lose all his money. It is evident that as the gambler continues to play, the probability that he will be wiped out approaches certainty. This is why even a very small advantage enables the casino, in almost every case, to break the gambler if he continues to bet.

The third column in Table 3.1 gives the probability that the gambler will lose all his money if he has one dollar and the casino has k dollars:

$$P_k^* = \frac{1 - \left(\frac{q}{p}\right)^k}{1 - \left(\frac{q}{p}\right)^{k+1}}.$$

Thus, if the gambler has much less money than the casino, he is even more likely to lose it all. The combination of a small advantage, table limits, and huge financial resources makes running a casino a very profitable business indeed. ◀

EXAMPLE 5 ▶ **A model of species competition**

The previous example can be used as a simple model of competition for resources between two species in a single ecological niche. Suppose that two species, A and B, control a combined territory of N resource units. The resource units may be such things as acres of grassland, number of trees, and so on. If species A controls k units, then species B controls the remaining $N - k$ units. Suppose that during each unit of time there is competition for one unit of resource between the two species. Let species A have the probability p of being successful. If we let P_n denote the probability that species B will lose all its resources when species A has n units of resource, then P_n is determined by Equation **(10)**, with initial conditions $P_0 = 0$ and $P_N = 1$. By Equations **(2)** and **(3)**, the solution to this equation is

$$P_n = \begin{cases} \dfrac{1 - [(1 - p)/p]^n}{1 - [(1 - p)/p]^N}, & \text{if } p \neq \dfrac{1}{2}, \\ \dfrac{n}{N}, & \text{if } p = \dfrac{1}{2}. \end{cases}$$

As in the previous example, even a small competitive advantage virtually ensures the extinction of the weaker species. For example, if $p = 0.55$ and species A has n of a total of 100 units, then we have the probabilities in Table 3.2.

TABLE 3.2

n	P_n
1	0.1818
2	0.3306
3	0.4523
4	0.5519
5	0.6334
10	0.8656
25	0.9934

Note that even if the species with the competitive advantage starts with as few as 4 of the 100 units of resource, it has a better than even chance to supplant the weaker species. With one quarter of the resources, it is virtually certain to do so. ◀

A process similar to the one discussed here has been used with success to determine whether, in a manufacturing process, a batch of articles is satisfactory. Let us briefly describe this process here. More extensive details can be found in the paper by G. A. Barnard.*

To test whether a batch of articles is satisfactory, we introduce a scoring system. The score is initially set at N. If a randomly sampled item is found to be defective, we subtract k. If it is acceptable, we add 1. The procedure stops when the score reaches either $2N$ or becomes nonpositive. If $2N$, the batch is accepted; if nonpositive, it is rejected. Suppose that the probability of selecting an acceptable item is p, and $q = 1 - p$. Let P_n denote the probability that the batch will be rejected when the score is at n. Then after the next choice, the score will be either increased by 1 with probability p or decreased by k with probability q. Thus

$$P_n = pP_{n+1} + qP_{n-k},$$

which can be written as the $(k + 1)$st-order difference equation

$$P_{n+k+1} - \frac{1}{p}P_{n+k} + \frac{q}{p}P_n = 0,$$

with boundary conditions

$$P_{1-k} = \ldots = P_0 = 1, \qquad P_{2N} = 0.$$

* G. A. Barnard, "Sequential Tests in Industrial Statistics," Supplement to the *Journal of the Royal Statistical Society* 8, no. 1 (1946).

The case of $k = 1$ reduces to Example 5, with the result that a batch is almost certain to be accepted or rejected depending on whether p is greater or less than $\frac{1}{2}$.

SELF-QUIZ

I. Which of the following are solutions to $y_{n+2} + 4y_{n+1} + 4y_n = 0$?
(a) $y_n = 2^n$ **(b)** $y_n = (-2)^n$
(c) $y_n = n2^n$ **(d)** $y_n = n(-2)^n$

II. Which of the following are solutions to $y_{n+2} + 4y_{n+1} + 5y_n = 0$?
(a) $y_n = 4^n$ **(b)** $y_n = 5^n$
(c) $y_n = (-4)^n$
(d) $y_n = (-2)^n \cos\left[n \tan^{-1}\left(-\frac{1}{2}\right)\right]$
(e) $y_n = (-2)^n \sin\left[n \tan^{-1}\left(-\frac{1}{2}\right)\right]$
(f) $y_n(-2)^n \cos\left[n\left(\pi - \tan^{-1}\frac{1}{2}\right)\right]$
(g) $y_n = (-2)^n \sin\left[n\left(\pi - \tan^{-1}\frac{1}{2}\right)\right]$

III. Which of the following are solutions to $y_{n+2} + 4y_{n+1} - 5y_n = 0$?
(a) $y_n = 5^n$ **(b)** $y_n = (-5)^n$
(c) $y_n = 1$ **(d)** $y_n = (-1)^n$
(e) $y_n = (-2)^n \cos\left(n \tan^{-1}\frac{1}{2}\right)$
(f) $y_n = (-2)^n \sin\left(n \tan^{-1}\frac{1}{2}\right)$

Answers to Self-Quiz

I. b, d **II.** f, g **III.** b, c

PROBLEMS 3.9

In Problems 1–9 find the general solution to each given difference equation. When initial conditions are specified, find the unique solution that satisfies them.

1. $y_{n+2} - 3y_{n+1} - 4y_n = 0$

2. $y_{n+2} + 7y_{n+1} + 6y_n = 0, y_0 = 0, y_1 = 1$

3. $6y_{n+2} + 5y_{n+1} + y_n = 0, y_0 = 1, y_1 = 0$

4. $y_{n+2} + y_{n+1} - 6y_n = 0, y_0 = 1, y_1 = 2$

5. $y_{n+2} + 2y_{n+1} + y_n = 0$

6. $y_{n+2} - \sqrt{2}\, y_{n+1} + y_n = 0$

7. $y_{n+2} - 2y_{n+1} + 2y_n = 0, y_0 = 1, y_1 = 2$

8. $y_{n+2} + 8y_n = 0$

9. $y_{n+2} - 2y_{n+1} + 4y_n = 0, y_0 = 0, y_1 = 1$

10. The **Fibonacci numbers** are a sequence of numbers such that each one is the sum of its two predecessors. The first few Fibonacci numbers are 1, 1, 2, 3, 5, 8, 13, ...
(a) Formulate an initial-value difference equation that generates the Fibonacci numbers.
(b) Find the solution to this equation.
(c) Show that the ratio of successive Fibonacci numbers tends to $(1 + \sqrt{5})/2$ as $n \to \infty$. This ratio, known as the **golden ratio,** was often used in ancient Greek architecture whenever rectangular structures were constructed. It was believed that when the ratio of the sides of a rectangle was this number, the resulting structure was most pleasing to the eye.

11. In a study of infectious diseases, a record is kept of outbreaks of measles in a particular school. It is estimated that the probability of at least one new infection occurring in the nth week after an outbreak is $P_n = P_{n-1} - \frac{1}{5}P_{n-2}$. If $P_0 = 0$ and $P_1 = 1$, what is P_n? After how many weeks is the probability of the occurrence of a new case of measles less than 10%?

12. Two competing species of drosophila (fruit flies) are growing under favorable conditions. In each generation, species A increases its population by 60% and species B increases by 40%. If initially there are 1000 flies of each species, what is the total population after n generations?

13. Snow geese mate in pairs in late spring of each year. Each female lays an average of five eggs, approximately 20% of which are claimed by predators and foul weather. The goslings, 60% of which are female, mature rapidly and are fully developed by the time the annual migration begins. Hunters and disease claim about 300,000 geese and 200,000 ganders annually. Are snow geese in danger of extinction?

14. In the discussion that introduces Equation **(10),** assume that player A has a 10% competitive advantage. If player A starts with three dollars, how much money must player B start with to have a better than even chance to win all of A's money? To have an 80% chance?

15. Answer the questions in Problem 14 given that player A has
(a) a 2% advantage;
(b) a 20% advantage.

16. In the discussion that introduces Equation **(10)**, suppose that on each play of the game player A has the probability p to win one dollar, q to win two dollars, and r to lose one dollar, where $p + q + r = 1$. Let P_n be as before. Write a difference equation that determines P_n, assuming that each player starts with N dollars.

3.10 USING SYMBOLIC MANIPULATORS

This section will illustrate how programs such as MAPLE, MATHEMATICA, DERIVE, MATLAB, or MATHCAD can be used to find solutions of linear constant-coefficient second-order differential equations. As in Chapter 2, we shall use these programs to find the actual algebraic expression that is the solution to the problem, rather than a numerical approximation as was done in Chapter 1.

We shall extend one of the generic commands we described in Chapters 1 and 2, by allowing the SOLVE command to apply to two variables and equations. Thus, given equations EQ1(c_1, c_2) = 0 and EQ2(c_1, c_2) = 0,

`SOLVE([EQ1(c₁,c₂)=0, EQ2(c₁,c₂)=0], [c₁,c₂])` = {solves the system of two equations for the unknown terms c_1, c_2}.

We shall also need a command to branch on alternatives depending on whether a variable is positive, zero, or negative.

`IF(x, f₊, f₀, f₋)` = {if $x > 0$ return f_+, if $x = 0$ return f_0, else return f_-}.

Linear constant coefficient second-order equations

Our ultimate goal will be to provide a script for finding the general solution of any differential equation of the form

$$y'' + p\,y' + q\,y = r(x) \tag{1}$$

where p and q are constants, and $r(x)$ is any function of the independent variable x. Since the homogeneous case is easy, we begin there.

Homogeneous equations

If $r(x) \equiv 0$ in Equation **(1)**, we call the equation **homogeneous.** The general solution is easy to obtain once we know the eigenvalues, which are the roots of the quadratic equation $m^2 + p\,m + q = 0$, or

$$m = \frac{-p \pm \sqrt{p^2 - 4q}}{2}.$$

Denote the discriminant by $d = p^2 - 4q$. Then, as in Sections 3.3 and 3.4, we have three cases:

$d > 0$

In this case the general solution is given by

$$y = c_1 \exp\left(\frac{-p + \sqrt{d}}{2}x\right) + c_2 \exp\left(\frac{-p - \sqrt{d}}{2}x\right).$$

$d = 0$

In this case the general solution is given by

$$y = (c_1 + c_2\,x)\exp\left(\frac{-p\,x}{2}\right).$$

$d < 0$

Here the general solution is given by

$$y = \exp\left(\frac{-px}{2}\right)\left(c_1\cos\left(\frac{\sqrt{-d}\,x}{2}\right) + c_2\sin\left(\frac{\sqrt{-d}\,x}{2}\right)\right).$$

Thus, an algorithm for solving the homogeneous equation

$$y'' + p\,y' + q\,y = 0,$$

can be described as follows:

```
LIN2GEN(x,y,p,q):=
IF(p*p−4*q, y=c₁*exp((−p+sqrt(p*p−4*q))*x/2)+c₂*
exp((−p−sqrt(p*p−4*q))*x/2), y=(c₁+c₂*x)*exp(−p*x/2),
y=exp(−p*x/2)*(c₁*cos(sqrt(4*q−p*p)*x/2)+
c₂*sin(sqrt(4*q−p*p)*x/2)))
```

EXAMPLE 1 ▶

Solve $y'' + 3y' - 10y = 0$.

Solution This homogeneous equation is Example 3.3.1. Here $p = 3$, $q = -10$, and $r(x) = 0$, so `LIN2GEN(x, y, 3, −10)` yields successively:

```
IF(3²+40,y=c₁*exp((−3+√(9+40))*x/2)+c₂*exp((−3−√(9+40))*x/2),....)

IF(49,y=c₁*exp(4*x/2)+c₂*exp(−10*x/2),...)
```

so the general solution is

$$y = c_1\,e^{2x} + c_2\,e^{-5x}.$$ ◀

We can modify the algorithm `LIN2GEN` by recalling that in Section 3.2 we developed a technique for finding a second linearly independent solution once one solution was known. The second solution y_2, is given by the expression (see Equation **3.2.8**),

$$y_2(x) = y_1(x)\int\frac{e^{-\int p(x)\,dx}}{(y_1(x))^2}\,dx = y_1\int\frac{e^{-px}}{y_1^2}\,dx. \tag{2}$$

This suggests defining a new function which will return y_2, when given y_1 and p:

```
LIN2G1(x,y₁,p):= (y₁*INT(exp(−p*x)/y₁^2,x))
```

In terms of this function, we have the following alternate algorithm for solving linear second-order homogeneous equations:

```
LIN2ALT(x,y,p,q):=
        IF(p*p−4*q,y=c₁*exp((−p+sqrt(p*p−4*q))*x/2)
        +c₂*LIN2G1(x,exp((−p+sqrt(p*p−4*q))*x/2),p),
        y=c₁*exp(−p*x/2)+c₂*LIN2G1(x, exp(−p*x/2),p),
        y=c₁*exp(−p*x/2)*(cos(sqrt(4*q−p*p)*x/2)+c₂*
        LIN2G1(x,exp(−p*x/2)*cos(sqrt(4*q−p*p)*x/2),p))
```

We now turn to the task of solving nonhomogeneous equations.

Nonhomogeneous equations

We shall use the method of variation of parameters described in Section 3.5. If we examine Equations **(3.5.7–9)** we see that the general solution to the nonhomogeneous equation

$$y'' + p\,y' + q\,y = r(x) \tag{3}$$

is given, in terms of the two linearly independent solutions y_1 and y_2 of the homogeneous equation, by the expression

$$y = c_1 y_1 + c_2 y_2 - y_1 \int \frac{y_2 r}{W(y_1, y_2)}\,dx + y_2 \int \frac{y_1 r}{W(y_1, y_2)}\,dx, \tag{4}$$

where $W(y_1, y_2)$ is the Wronskian: $W(y_1, y_2) = y_1 y_2' - y_2 y_1'$.

Observe that if $r \equiv 0$, the two integrals will be zero, so we will get the solution of the homogeneous equation. Knowing the form of this general solution suggests the following strategy:

1. Find one solution of the homogeneous problem.
2. Use LIN2G1 to find the second linearly independent solution.
3. Place all the information into Equation **(4)** to find the general solution of either a homogeneous or a nonhomogeneous equation.

We now describe an algorithm for doing each of these tasks. The algorithm consists of five functions, each using the preceding functions. The user would enter only the last function.

```
1. LIN2(x,p,q):=
   IF(p*p-4*q, exp((-p+sqrt(p*p-4*q))*x/2), exp(-p*x/2),
   exp(-p*x/2)*cos(sqrt(4*q-p*p)*x/2))
2. LIN2G1(x,u,p):= (u*INT(exp(-INT(p,x)/u^2,x))
3. WRNSK(x,u,v):= u*DIF(v, x)-v*DIF(u,x)
   LIN2G2(x,u,v,r):=c1*u+c2*v-u*INT(v*r/WRNSK(x,u,v),x)
   +v*INT(u*r/WRNSK(x,u,v),x)
   LINEAR2(x,y,p,q,r):=
       (y=LIN2G2(x,LIN2(x,p,q),LIN2G1(x,LIN2(x,p,q),p),r)
```

We illustrate these steps with an example.

EXAMPLE 2 ▶

Solve the equation $x'' + x = t\,e^{2t}$.

Solution Here t is the independent variable and x is the dependent variable, so we would enter:

```
LINEAR2(t,x,0,1,t*exp(2*t))
```

because $p = 0$, $q = 1$, and $r(t) = t\,e^{2t}$. The algorithm would go through the following steps in finding the solution:

```
(x=LIN2G2(t,LIN2(t,0,1), LIN2G1(t,LIN2(t,0,1),0),t*exp(2*t))
LIN2(t,0,1):=IF(0²-4, . . . . , . . . . , exp(-0*t/2)*cos(sqrt(4-0²)*t/2))=cos(t)
LIN2G1(t,cos (t),0):= (cos(t)*INT(exp(-0*t)/cos(t)^2, t))
        =cos(t)*INT(sec²(t),t)=cos(t)*tan(t)=sin(t)
WRNSK(t,cos(t),sin(t)):=cos(t)*DIF(sin(t),t)-sin(t)*DIF(cos(t),t)
        =cos²(t)+sin²(t)=1
```

```
LIN2G2(t,cos(t),sin(t),t*exp(2*t)):=c₁*cos(t)+c₂*sin(t)
        -cos(t)*INT(sin(t)*t*exp(2*t)/WRNSK(t,cos(t),sin(t)),t)
        +sin(t)*INT(cos(t)*t*exp(2*t)/WRNSK(t,cos(t),sin(t)),t)
        =c₁*cos(t)+c₂*sin(t)-cos(t)*INT(sin(t)*t*exp(2*t), t)
        +sin(t)*INT(cos(t)*t*exp(2*t),t)
        =c₁*cos(t)+c₂*sin(t)+(5*t-4)/25
```

so that the answer is

$$x = c_1 \cos(t) + c_2 \sin(t) + \frac{5t - 4}{25} e^{2t}. \blacktriangleleft$$

Initial conditions

If you want to specify initial condition, first find the general solution, $y = GS$, of Equation **(3)** and then apply the following algorithm:

```
IV2(x,y,GS,x0,dy0,c1,c2):=
          SOLVE([SUBST(GS=y,x=x0,y=y0),SUBST(DIF(GS,x)=dy,
          x=x0,dy=dy0)],[c1,c2])
```

This script determines the values of c1 and c2 in the general solution that satisfy the initial conditions

$$y(x0) = y0, \qquad \text{and} \qquad y'(x0) = dy0.$$

We now give an example of its use.

EXAMPLE 3 ▶

Solve the initial-value problem

$$y'' + y = \cos(x), \qquad y(0) = 2, \qquad y'(0) = -3.$$

Solution This is Example 3.6.5. Since there is no y' term, $p = 0$, $q = 1$, and $r(x) = \cos(x)$. Then LINEAR2(x, y, 0, 1, cos(x)) yields the general solution

$$y = \text{c1} \cos(x) + \text{c2} \sin(x) + \frac{x}{2} \sin(x).$$

Since this is an initial-value problem, we wish to determine the values of c1 and c2 in the solution. So we write

```
IV2(x,y,c1*cos(x)+c2*sin(x)+x*sin(x)/2,0,2,-3,c1,c2)
```

The following takes place once this line is executed by the symbolic manipulator:

```
SOLVE([SUBST(y=c1*cos(x)+c2*sin(x)+x*sin(x)/2,x=x0,y=y0),
SUBST(DIF(c1*cos(x)+c2*sin(x)+x*sin(x)/2,x)=dy,x=x0,dy=dy0)],[c1,c2])
```

or

```
SOLVE([c1*cos(x0)+c2*sin(x0)+x0*sin(x0)/2=y0,
-c1*sin(x0)+c2*cos(x0)+sin(x0)/2+x0*cos(x0)/2=dy0],[c1,c2])
```

from which it follows that

```
[c1,c2]=[2,-3].
```

Thus, the particular solution of the initial-value problem is

$$y = 2\cos(x) - 3\sin(x) + \frac{x}{2} \sin(x). \blacktriangleleft$$

PROBLEMS 3.10

Solve each of the following problems using MATHEMATICA , MAPLE, DERIVE, MATLAB, or MATHCAD.

1. $y'' - 10y' + 21y = 0$

2. $y'' - 9y' - 21y = 0$

3. $y'' - 9y' + 21y = 0$

4. $y'' - 3y' + 4y = 0,\ y(0) = 0,\ y'(0) = 1$

5. $y'' - 9y' + 20y = 0,\ y(0) = 3,\ y'(0) = 2$

6. $y'' + 4y' + 4y = e^{-2x}/x^2$

7. $y'' + 4y = \tan(2x)$

8. $y'' + y = \sec(x)$

9. $y'' + 4y' + 3y = x^2e^{3x},\ y(0) = 1,\ y'(0) = 0$

10. $y'' + 4y' + 3y = x^2e^{3x},\ y(0) = 0,\ y'(0) = 1$

SUMMARY OUTLINE OF CHAPTER 3

Homogeneous and Nonhomogeneous Linear Differential Equations p. 104
The most general second-order linear equation can be written

$$y''(x) + a(x)y'(x) + b(x)y(x) = f(x) \tag{1}$$

whereas the most general third-order linear equation can be written

$$y'''(x) + a(x)y''(x) + b(x)y'(x) + c(x)y(x) = f(x), \tag{2}$$

where $a(x)$, $b(x)$, $c(x)$, and $f(x)$ are functions of the independent variable x only. Equations **(1)** and **(2)** are special cases of the **general linear nth-order equation:**

$$y^{(n)}(x) + a_{n-1}(x)y^{(n-1)}(x) + \cdots + a_1(x)y'(x) + a_0(x)y(x) = f(x). \tag{3}$$

If the function $f(x)$ is identically zero, we say that Equations **(1), (2),** and **(3)** are **homogeneous.** Otherwise, they are **nonhomogeneous.**

Constant and Variable Coefficients p. 104
If the coefficient functions $a_i(x)$ in Equation **(3)** are constants, the equation is said to have **constant coefficients.** Otherwise, it has **variable coefficients.**

Existence-uniqueness theorem for linear initial-value problems p. 105
Suppose that $a_1, a_2, \ldots, a_n$, and f are continuous functions for $x_1 \le x \le x_2$. Let $c_0, c_1, c_2, \ldots, c_{n-1}$ be n constants and suppose that x_0 is in $[x_1, x_2]$. Suppose further than n initial conditions are given:

$$y(x_0) = c_0, \quad y'(x_0) = c_1, \quad y''(x_0) = c_2, \ldots, \quad y^{(n-1)}(x_0) = c_{n-1}.$$

Then there exists exactly one solution $y(x)$ to the differential equation

$$y^{(n)}(x) + a_{n-1}(x)y^{(n-1)}(x) + a_{n-2}(x)y^{(n-2)}(x) + \cdots + a_0(x)y(x) = f(x)$$

that satisfies the n given initial conditions.

Existence-Uniquences Theorem for Second-Order, Linear Initial-Value Problems p. 106
If $a(x)$, $b(x)$, and $f(x)$ are continuous functions, then the equation

$$y''(x) + a(x)y'(x) + b(x)y(x) = f(x)$$

has a unique solution that satisfies the conditions

$$y(x_0) = y_0, \quad y'(x_0) = y_1$$

for any real numbers x_0, y_0, and y_1.

Linear Combination, Linear Independence, and Linear Dependence p. 108
Let y_1 and y_2 be any two functions. By a **linear combination** of y_1 and y_2 we mean a function $y(x)$ that can be written in the form

$$y(x) = c_1y_1(x) + c_2y_2(x),$$

for some constants c_1 and c_2. Two functions are **linearly independent** on an interval $[x_1, x_2]$ whenever the relation

$$c_1y_1(x) + c_2y_2(x) = 0,$$

for all x in $[x_1, x_2]$, implies that $c_1 = c_2 = 0$. Otherwise they are **linearly dependent.**

Two functions are linearly dependent on the interval $[x_0, x_1]$ if, and only if, one of the functions is a constant multiple of the other. p. 109

A **linear combination** of n functions is any function of the form p. 108

$$y(x) = c_1y_1(x) + c_2y_2(x) + \cdots + c_ny_n(x),$$

where $c_1, c_2, \ldots, c_n$ are constants. We will say that the collection of functions $y_1(x), y_2(x), \ldots, y_n(x)$ are **linearly independent** on an interval $[x_1, x_2]$ if the equation

$$c_1y_1(x) + c_2y_2(x) + \cdots + c_ny_n(x) = 0$$

is only true for all x in $[x_1, x_2]$ when $c_1 = c_2 = \cdots = c_n = 0$. Otherwise, we say that the collection of functions is **linearly dependent** on $[x_1, x_2]$.

Every homogeneous linear second-order differential equation

$$y'' + a(x)y' + b(x)y = 0$$

has two linearly independent solutions. p. 109

Principle of Superposition for Homogeneous Equations p. 110
Let $y_1(x)$ and $y_2(x)$ be any two solutions of the homogeneous equation

$$y'' + a(x)y' + b(x)y = 0. \tag{4}$$

Then any linear combination of them is also a solution.

An nth-order linear, homogeneous differential equation with continuous coefficients can have at most n linearly independent solutions. p. 110

General Solution to a Linear, Second-Order Equation p. 110
The general solution of Equation **(4)** is given by the linear combination

$$y(x) = c_1y_1(x) + c_2y_2(x)$$

where c_1 and c_2 are arbitrary constants and $y_1(x)$ and $y_2(x)$ are linearly independent solutions of Equation **(4).**

Wronskian
Let $y_1(x)$ and $y_2(x)$ be any two solutions to the differential equation p. 111

$$y'' + a(x)y' + b(x)y = 0.$$

The **Wronskian** of y_1 and y_2 is defined as

$$W(y_1, y_2)(x) = \begin{vmatrix} y_1(x) & y_2(x) \\ y_1'(x) & y_2'(x) \end{vmatrix} = y_1(x)y_2'(x) - y_1'(x)y_2(x).$$

$W(y_1, y_2)(x) = ce^{-\int a(x)\,dx}$ p. 112

Two solutions $y_1(x)$ and $y_2(x)$ of Equation **(4)** are linearly independent on $[x_0, x_1]$ if and only if $W(y_1, y_2)(x) \neq 0$. p. 112
Let y_p and y_q be two solutions to the nonhomogeneous equation p. 114

$$y'' + a(x)y' + b(x)y = f(x). \tag{5}$$

Then their difference $y_p - y_q$ is a solution to the associated homogeneous equation

$$y'' + a(x)y' + b(x)y = 0. \tag{6}$$

General Solution to a Linear, Nonhomogeneous Second-Order Equation p. 114
Let y_p be a solution to Equation **(5)** and let y_1 and y_2 be two linearly independent solutions to Equation **(6).** Then the general solution to Equation **(5)** is given by

$$y(x) = c_1y_1(x) + c_2y_2(x) + y_p(x)$$

where c_1 and c_2 are arbitrary constants.

Principle of Superposition for Nonhomogeneous Equations p. 114
Suppose that y_f is a solution to

$$y'' + a(x)y' + b(x)y = f(x)$$

and y_g is a solution to

$$y'' + a(x)y' + b(x)y = g(x).$$

Then $y_f + y_g$ is a solution to

$$y'' + a(x)y' + b(x)y = f(x) + g(x).$$

Finding a Second Solution to a Homogeneous Equation When One is Given p. 121
If y_1 is a solution to Equation **(6)**, then a second, linearly independent selection is given by

$$y_2(x) = y_1(x)\int \frac{e^{-\int a(x)\,dx}}{y_1^2(x)}\,dx$$

The **characteristic equation** of the equation p. 124

$$y'' + ay' + by = 0 \tag{7}$$

is $\lambda^2 + a\lambda + b = 0$. Here a and b are constants.

The general solution to a second-order, linear differential equation with constant coefficients

1. Let λ_1 and λ_2 be unequal real roots of Equation **(7)**; then the general solution to Equation **(7)** is

$$y = c_1 e^{\lambda_1 x} + c_2 e^{\lambda_2 x} \quad \text{p. 124}$$

where c_1 and c_2 are arbitrary constants.
2. If $\lambda_1 = \lambda_2$, then the general solution to Equation **(7)** is

$$y = e^{\lambda x}(c_1 + c_2 x) \quad \text{p. 125}$$

3. If $\lambda_1, \lambda_2 = \alpha \pm i\beta$, then the general solution is

$$y = e^{\alpha x}(c_1 \cos \beta x + c_2 \sin \beta x) \quad \text{p. 129}$$

Harmonic Identities p. 130

$$a \cos \omega t + b \sin \omega t = A \cos(\omega t - \delta)$$
$$a \cos \omega t + b \sin \omega t = A \sin(\omega t + \phi) \tag{8}$$

Here $A = \sqrt{a^2 + b^2}$, δ is the unique number such that $\cos \delta = a/\sqrt{a^2 + b^2} = \frac{a}{A}$ and $\sin \delta = \frac{b}{A}$, so $\tan \delta = \frac{b}{a}$. ϕ is the unique number such that $\cos \phi = \frac{b}{A}$ and $\sin \phi = \frac{a}{A}$, so $\tan \phi = \frac{a}{b}$.

Variation of Parameters Method p. 134
We seek a particular solution to the nonhomogeneous equation

$$y'' + ay' + by = f(x) \tag{9}$$

with the associated homogeneous equation

$$y'' + ay' + by = 0. \tag{10}$$

To find the solution to Equation **(9)**: pp. 134–135

Step 1
Write $y(x) = c_1(x)y_1(x) + c_2(x)y_2(x)$ where y_1 and y_2 are linearly independent solutions to Equation **(10)**.

Step 2
Obtain the basic equations:

$$y_1 c_1' + y_2 c_2' = 0$$
$$y_1' c_1' + y_2' c_2' = f(x).$$

Step 3

Solve the basic equations:

$$c_1'(x) = \frac{-f(x)y_2(x)}{y_1(x)y_2'(x) - y_1'(x)y_2(x)} = \frac{-f(x)y_2(x)}{W(y_1, y_2)(x)},$$

$$c_2'(x) = \frac{f(x)y_1(x)}{y_1(x)y_2'(x) - y_1'(x)y_2(x)} = \frac{f(x)y_1(x)}{W(y_1, y_2)(x)}.$$

Step 4

Integrate c_1' and c_2', if possible, to obtain $y = c_1y_1 + c_2y_2$.

Method of Undetermined Coefficients p. 140

Let

$$y'' + ay' + by = f(x), \tag{11}$$

where a and b are constants, and every term of the function $f(x)$ is a product of one or more of the following:

1. a polynomial in x, (a constant k is a polynomial of degree zero)
2. an exponential function $e^{\alpha x}$,
3. $\cos \beta x$, $\sin \beta x$.

The method of undetermined coefficients assumes that each term in the particular solution to Equation **(11)** is of the same "form" as each term in $f(x)$, but has polynomial terms with undetermined coefficients. Substitute the "guessed" particular solution y_p into Equation **(11)** to "determine" the unknown coefficients.

Modifications of the Method of Undetermined Coefficients p. 143

If any term of the guessed solution $y_p(x)$ is a solution of the homogeneous equation, multiply $y_p(x)$ by x repeatedly until no term of the product $x^k y_p(x)$ is a solution of the homogeneous equation. Then use the product $x^k y_p(x)$ to solve Equation **(11)**.

Euler Equation p. 147

An equation of the form

$$x^2y'' + axy' + by = f(x), \qquad x \neq 0 \tag{12}$$

is called an **Euler equation.**

Characteristic Equation for an Euler Equation p. 148

$$\lambda^2 + (a - 1)\lambda + b = 0 \tag{13}$$

General Solution of an Euler Equation

Let λ_1 and λ_2 be the roots of Equation **(13)**.

Case 1

If $\lambda_1 \neq \lambda_2$ are real, the general solution to Equation **(12)** is

$$y(x) = c_1x^{\lambda_1} + c_2x^{\lambda_2}, \qquad x \neq 0. \quad \text{p. 148}$$

Case 2

If $\lambda_1 = \lambda_2$, the general solution to Equation **(12)** is

$$y(x) = x^{\lambda}(c_1 + c_2 \ln|x|). \quad \text{p. 149}$$

Case 3

If $\lambda_1, \lambda_2 = \alpha \pm i\beta$, then the general solution to Equation **(12)** is

$$y(x) = x^{\alpha}[c_1 \cos(\beta \ln|x|) + c_2\sin(\beta \ln|x|)]. \quad \text{p. 150}$$

REVIEW EXERCISES FOR CHAPTER 3

In Exercises 1–5 a second-order differential equation and one solution $y_1(x)$ are given. Verify that $y_1(x)$ is indeed a solution and find a second linearly independent solution.

1. $y'' + 4y = 0;\ y_1(x) = \sin 2x$

2. $y'' - 6y' + 9y = 0;\ y_1(x) = e^{3x}$

3. $x^2y'' + xy' - 4y = 0;\ y_1(x) = x^2$

4. $y'' + \frac{1}{x}y' + \left(1 - \frac{1}{4x^2}\right)y = 0;\ y_1(x) = x^{-1/2}\sin x.$

5. $(1 - x^2)y'' - 2xy' + 2y = 0;\ y_1(x) = x$

In Exercises 6–34 find the general solution to the given equation. If initial conditions are given, find the particular solutions that satisfies them and sketch their graphs.

6. $y'' - 9y' + 20y = 0$

7. $y'' - 9y' + 20y = 0;\ y(0) = 3,\ y'(0) = 2$

8. $y'' - 3y' + 4y = 0$

9. $y'' - 3y' + 4y = 0;\ y(0) = 0,\ y'(0) = 1$

10. $y'' = 0$

11. $4y'' + 4y' + y = 0$

12. $y'' - 11y = 0$

13. $y'' - 2y' + 7y = 0$

14. $y'' - y' - 2y = \sin 2x$

15. $y''' - 6y'' + 11y' - 6y = 0$

16. $y'' - 2y' + y = xe^x$

17. $y'' - 2y' + y = x^2 - 1;\ y(0) = 2,\ y'(0) = 1$

18. $y'' + y = \sec x,\ 0 < x < \frac{\pi}{2}$

19. $y'' - 2y' + y = \frac{2e^x}{x^3}$

20. $y'' + 4y' + 4y = \frac{e^{-2x}}{x^2};\ x > 0$

21. $y'' + 6y' + 9y = e^{-3x}$

22. $y'' + 6y' + 9y = xe^{-3x}$

23. $y'' + 6y' + 9y = x^2e^{-3x}$

24. $y'' + 6y' + 9y = e^{-3x}\cos x$

25. $y'' + 2y' + 2y = x^2$

26. $y'' + 2y' + 2y = \cos x$

27. $y'' + 2y' + 2y = \sin 2x$

28. $y'' + 2y' + 2y = x\sin x$

29. $y'' - 5y' + 6y = e^{3x}$

30. $y'' - 5y' + 6y = xe^{2x}$

31. $x^2y'' + 5xy' + 4y = 0;\ x > 0$

32. $x^2y'' - 2xy' + 3y = 0;\ x > 0$

33. $y''' + y'' - 8y' - 12y = 0$

34. $y^{(4)} + 8y'' + 4y = 0$

35. Prove that if y_1 and y_2 are linearly independent solutions of

$$y'' + a(x)y' + b(x)y = 0,$$

then they cannot have a common point of inflection unless $a(x)$ and $b(x)$ vanish simultaneously at that point.

36. By setting $y = u(x)v(x)$, show that it is possible to select $v(x)$ so that

$$y'' + a(x)y' + b(x)y = 0$$

takes the form

$$u'' + c(x)u = 0.$$

[This is known as the **normal form** of a second-order linear differential equation.]

37. If y_1 and y_2 are linearly independent solutions of

$$y'' + a(x)y' + b(x)y = 0,$$

show that between consecutive zeros of y_1 there is exactly one zero of y_2.

4 SOME PHYSICAL APPLICATIONS OF LINEAR DIFFERENTIAL EQUATIONS

In Chapter 3 we developed several techniques for solving initial-value problems of the form

$$y'' + ay' + by = f(t), \qquad y(t_0) = y_0, \qquad y'(t_0) = y_0', \tag{1}$$

where a and b are constants and t is time. In this chapter we see that many physical problems, such as the motion of an object in a mechanical system or the flow of electric current in a simple series circuit, lead to a problem of form **(1)** when expressed mathematically. Thus it is only necessary to make proper interpretations of the terms in **(1)** to obtain solutions to physical problems in different disciplines.

4.1 VIBRATIONAL MOTION: SIMPLE HARMONIC MOTION

Differential equations were first studied in attempts to describe the motion of particles with mass subject to various forces. As a simple example, consider a spring, whose upper end is securely fastened, of natural (unstretched) length l_0 [see Figure 4.1(a)], to which we attach an object of mass m [see Figure 4.1(b)].* The addition of the mass m stretches the spring to length l. This static elongation of the spring is the result of two forces:

1. The force of gravity $F_g = mg$, acting downward.
2. The spring force F_s, acting upward.

In Figure 4.1(b), the object is in its equilibrium position, that is, at the point where the object remains at rest. For this to happen the two forces must be numerically equal. According to **Hooke's law**† the spring force exerted on the object is proportional to the difference between the length l of the spring and its natural or

* The most common systems of units are given in the table below.

Systems of Units	*Force*	*Length*	*Mass*	*Time*	g
International (SI)	newton (N)	meter (m)	kilogram (kg)	second (s)	9.81 m/sec^2
English	pound (lb)	foot (ft)	slug	second (sec)	32.2 ft/sec^2

1 N = 1 kg-m/s^2 = 0.22481 lb 1 kg = 0.06852 slug
1 m = 3.28084 ft 1 lb = 1 slug-ft/s^2 = 4.4482 N

† Robert Hooke (1638–1703), a British mathematician and physicist, was one of the first scientists to state the **inverse square law:** The force of gravitational attraction between two bodies is inversely proportional to the square of the distance between them.

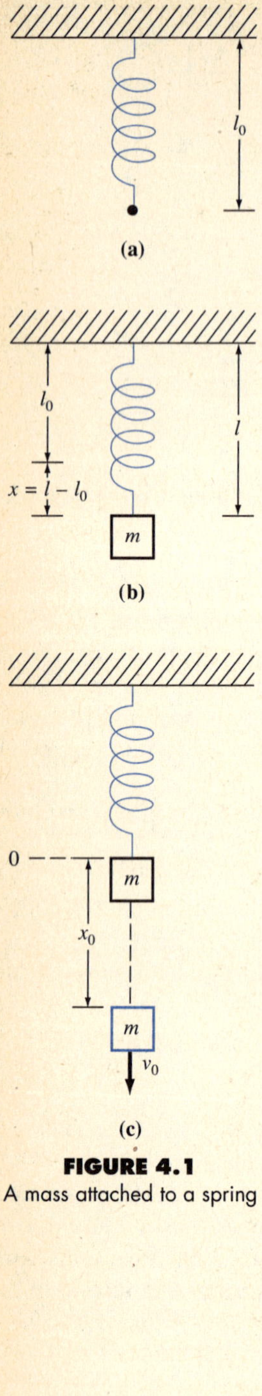

FIGURE 4.1
A mass attached to a spring

equilibrium length l_0. The positive constant of proportionality k is called the **spring constant.** Thus

$$F_s = k(l - l_0).$$

If the mass is at rest, then it is the force of gravity that pulls the spring beyond its natural length; this means that

$$mg = k(l - l_0)$$

as well. It is the case that $F_g = F_s$ only as long as the mass is left at rest.

In Figure 4.1(c) we denote the equilibrium position of the object on the spring, that is, the position where the spring has length l, by $x = 0$. Suppose that the object is given an initial displacement x_0 and an initial velocity v_0. Can we describe the movement of the object?

In answering the question we have just posed, we make three simplifying assumptions, the last two of which we remove in the next two sections.

Three basic assumptions

1. All motion is along a vertical line through the center of gravity of the object, which is treated as a point mass.
2. There is no damping force F_d due to the medium in which the mass is moving (such as air resistance).
3. No other forces (beyond the ones already mentioned) are applied to the mass.

Now suppose we have put the mass in Figure 4.1(c) in motion. What forces act on the mass? Clearly the gravitational force is constant, $F_g = mg$, but the spring force depends on the displacement x of the mass:

$$F_s = k(x + l - l_0)$$

As we increase x (moving downward in the same direction as the gravitational force), the spring force acting upward increases, so the total force on the mass is

$$F = F_g - F_s = mg - k(x + l - l_0)$$
$$= -kx,$$

since $mg = k(l - l_0)$.

Newton's second law of motion states that the force F acting on this mass moving with varying velocity v is equal to the time rate of change of the momentum mv; since the mass is constant,

$$F = \frac{d(mv)}{dt} = m\frac{d^2x}{dt^2}. \qquad v = \frac{dx}{dt}$$

Equating the two forces, we have

$$m\frac{d^2x}{dt^2} = -kx. \tag{1}$$

Two conditions were initially imposed on the mechanical system: At time $t = 0$ the mass has an initial displacement x_0 and an initial velocity v_0. Hence we have the initial-value problem

$$\frac{d^2x}{dt^2} + \frac{k}{m}x = 0, \qquad x(0) = x_0, \qquad x'(0) = v_0. \tag{2}$$

To find the solution of **(2)**, we note that the characteristic equation has the complex root $\pm i\omega_0$, where $\omega_0 = \sqrt{k/m}$, leading to the general solution

$$x(t) = c_1 \cos \omega_0 t + c_2 \sin \omega_0 t.$$

Using the initial conditions, we find that $c_1 = x_0$ and $c_2 = v_0/\omega_0$, so the solution **(2)** is given by

$$x(t) = x_0 \cos \omega_0 t + \left(\frac{v_0}{\omega_0}\right) \sin \omega_0 t. \tag{3}$$

Equation **(3)** describes the **free vibrations** or **free motion** of the mechanical system, since it is free of external influencing forces other than those imposed by gravity and the spring itself.

We would like to write $x(t)$ in the form

$$x(t) = A \sin(\omega_0 t + \phi),$$

so that we can graph (and understand) the superposition of the sine and cosine functions in Equation **(3).** To do so we use the harmonic identity established in Section 3.4 (page 130):

$$a \cos \omega t + b \sin \omega t = A \sin(\omega t + \phi),$$

where $A = \sqrt{a^2 + b^2}$, $\cos \phi = \frac{b}{A}$, $\sin \phi = \frac{a}{A}$. Here, $a = x_0$ and $b = v_0/\omega_0$, so we may rewrite Equation **(3)** as

$$x(t) = A \sin(\omega_0 t + \phi), \tag{4}$$

with A and ϕ determined by

$$A = \sqrt{x_0^2 + \left(\frac{v_0}{\omega_0}\right)^2}, \qquad \cos \phi = \frac{v_0}{A\omega_0}, \qquad \text{and} \qquad \sin \phi = \frac{x_0}{A}.$$

It is tempting to write

$$\tan \phi = \frac{\sin \phi}{\cos \phi} = \frac{\dfrac{x_0}{A}}{\dfrac{v_0}{A\omega_0}} = \frac{x_0 \omega_0}{v_0},$$

so that $\phi = \tan^{-1}(x_0 \omega_0 / v_0)$, but this is correct only if the angle ϕ is in the first or fourth quadrant, because that is the range of the arctangent.

The motion of the mass in Equation **(4)** is called **simple harmonic motion,** since it is sinusoidal.

From this equation it is clear that the mass oscillates between the extreme positions $\pm A$; A is called the **amplitude** of the motion. Since the sine term has period $2\pi/\omega_0$, this is the time required for each complete oscillation. The **natural frequency** f of the motion is the number of complete oscillations per unit time:*

$$f = \frac{\omega_0}{2\pi}. \tag{5}$$

Note that although the amplitude depends on the initial conditions, the frequency does not.

---

* Cycles/sec = hertz (Hz).

EXAMPLE 1 ▶ **Determining A and ϕ**

Suppose that $x_0 = 0.5$ meters, $k = 0.4\ N/m$, $m = 10$ kilograms, and $v_0 = 0.25$ meters per second. Then $\omega_0 = \sqrt{k/m} = \sqrt{0.04} = 0.2$ and Equation **(3)** becomes

$$x(t) = 0.5 \cos 0.2t + \frac{0.25}{0.2} \sin 0.2t = 0.5 \cos 0.2t + 1.25 \sin 0.2t.$$

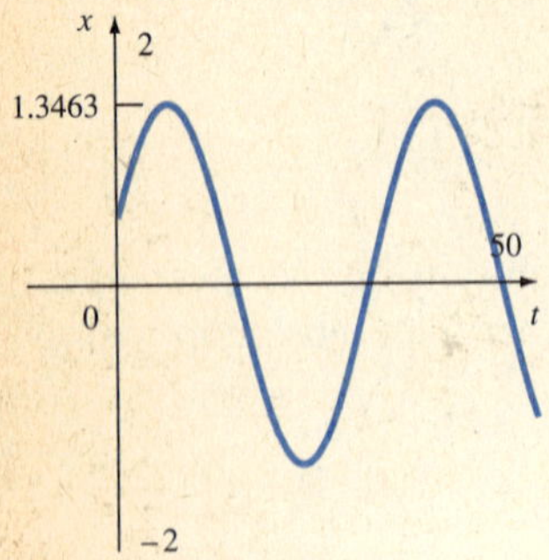

FIGURE 4.2
Graph of $x(t) = 1.3463 \sin(0.2t + 0.3805)$

Now

$$A = \sqrt{0.5^2 + 1.25^2} = \sqrt{1.8125} \approx 1.3463$$

and

$$\phi = \tan^{-1} \frac{(0.5)(0.2)}{0.25} = \tan^{-1} 0.4 \approx 0.3805 \text{ radians},$$

so we may write

$$x(t) \approx 1.3463 \sin(0.2t + 0.3805) \text{ meters.}$$

This function is sketched in Figure 4.2. ◀

EXAMPLE 2 ▶ **Finding an equation of motion**

Consider a spring fixed at its upper end and supporting a weight of 10 pounds at its lower end. Suppose the 10-pound weight stretches the spring by 6 inches. Find the equation of motion of the weight if it is drawn to a position 4 inches below its equilibrium position and released.

Solution By Hooke's law, since a force of 10 pounds stretches the spring by $\frac{1}{2}$ foot, $10 = k(\frac{1}{2})$ or $k = 20$ (lb/ft). We are given the initial values $x_0 = \frac{1}{3}$(ft) and $v_0 = 0$, so by Equation **(3)** and the identity* $\frac{k}{m} = \frac{gk}{w} \approx 64/\text{sec}^2$, we obtain

$$x(t) = \frac{1}{3} \cos 8t \text{ feet.}$$

Thus the amplitude is $\frac{1}{3}$ foot (= 4 in.), and the frequency is $f \approx \frac{4}{\pi}$ hertz. ◀

Foucault's Pendulum

The Foucault pendulum can be used to prove that the earth rotates. Once set in motion, the pendulum swings in a plane that is fixed in relation to the universe. Because of this, the pendulum can detect the earth's rotation. A Foucault pendulum can be observed in the rotunda of the Smithsonian Museum in Washington, D.C. It swings above a circular dial on which the hours of the day are marked. Small blocks are set at the hour marks, and the pendulum knocks them over at the precise time.

It is easy to construct a pendulum, but the trick is to keep it moving without coming into contact with it. Energy to continue the motion of the pendulum, without interfering with it, can be provided electromagnetically. Plans for building an inexpensive, easy to build, electronically controlled Foucault pendulum can be found in the article, "Apparatus for Teaching Physics: A Very Short, Portable Foucault Pendulum," by H. Kruglak in *Physics-Teacher,* vol. 21 (1983) pp. 477–79. A Foucault pendulum is shown in Figure 4.3.

*The identity $w = mg$ may be used to convert weight to mass. Keep in mind that pounds and newtons are units of weight (force) whereas slugs and kilograms are units of mass. The gravitational constant $g = 9.81$ m/s^2 = 32.2 ft/sec^2 (approximately).

FIGURE 4.3
The Foucault pendulum (Cameramann International, Ltd.)

PROBLEMS 4.1

In Problems 1–6 determine the equation of motion of a point mass m *attached to a coiled spring with spring constant k, initially displaced a distance x_0 from equilibrium, and released with velocity v_0 subject to no damping or other external forces.*

1. $m = 10$ kg, $k = 1000$ N/m, $x_0 = 1$ m, $v_0 = 0$
2. $m = 10$ kg, $k = 10$ N/m, $x_0 = 0$, $v_0 = 1$ m/s
3. $m = 10$ kg, $k = 10$ N/m, $x_0 = 3$ m, $v_0 = 4$ m/s
4. $m = 1$ kg, $k = 16$ N/m, $x_0 = 4$ m, $v_0 = 0$
5. $m = 1$ kg, $k = 25$ N/m, $x_0 = 0$, $v_0 = 3$ m/s
6. $m = 9$ kg, $k = 1$ N/m, $x_0 = 4$ m, $v_0 = 1$ m/s
7. One end of a rubber band is fixed at a point *A*. An object of 1-kilogram mass, attached to the other end, stretches the rubber band vertically to point *B* in such a way that the length *AB* is 16 centimeters greater than the natural length of the band. If the mass is further drawn to a position 8 centimeters below *B* and released, what is its velocity (if we neglect resistance) as it passes point *B*?
8. If the mass in Problem 7 is released at a position 8 centimeters above *B*, what is its velocity as it passes 1 centimeter above *B*?
9. A cylindrical block of wood of radius and height 1 foot and weighing 124.8 pounds floats with its axis vertical in water (62.4 lb/ft^3). If it is depressed so that the surface of the water is tangent to the block and is then released, what is its period of vibration and equation of motion? Neglect resistance. [*Hint:* The upward force on the block is equal to the weight of the water displaced by the block.]
10. A cubical block of wood, 1 foot on a side, is depressed so that its upper face lies along the surface of the water and is then released. The period

of vibration is found to be 1 second. Neglecting resistance, what is the weight of the block of wood?

11. An ideal pendulum consists of a weightless rod of length l attached at one end to a frictionless hinge and supporting a body of mass m at the other end. Suppose the pendulum is displaced an angle θ_0 and released, as illustrated below. The tangential acceleration of the ideal pendulum is $l\theta''$ and must be proportional, by Newton's second law of motion, to the tangential component of gravitational force.

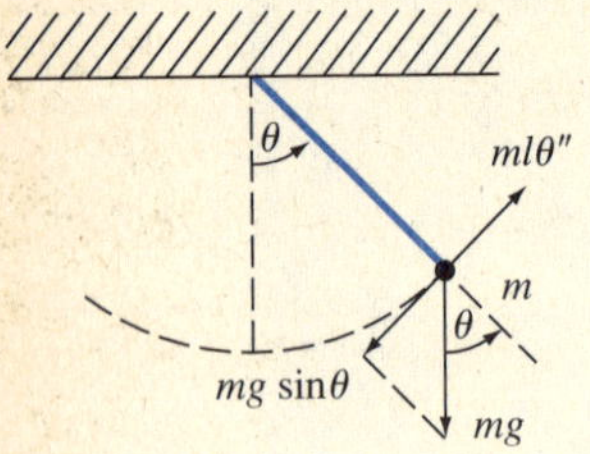

(a) Neglecting air resistance, show that the ideal pendulum satisfies the nonlinear initial-value problem

$$l\frac{d^2\theta}{dt^2} = -g\sin\theta, \quad \theta(0) = \theta_0, \quad \theta'(0) = 0. \quad \textbf{(i)}$$

(b) Assuming θ_0 is small, explain why Equation **(i)** may be approximated by the linear initial-value problem

$$\frac{d^2\theta}{dt^2} + \frac{g}{l}\theta = 0, \quad \theta(0) = \theta_0, \quad \theta'(0) = 0. \quad \textbf{(ii)}$$

(c) Solve Equation **(ii)** assuming that the rod is 6 inches long and that the initial displacement θ_0 is 0.5 radian. What is the frequency of the pendulum?

12. A grandfather clock has a pendulum that is 1 meter long. The clock ticks each time the pendulum reaches the rightmost extent of its swing. Neglecting friction and air resistance, and assuming that the motion is small, determine how many times the clock ticks in 1 minute.

13. A turbine of unknown weight is placed on a spring-supported mounting platform of unknown spring constant. What is the natural frequency of the system if the turbine lowers the platform by $\frac{1}{8}$ inch?

14. A cubical block of wood 1 foot on a side and weighing 41.6 pounds floats in water (62.4 lb/ft^3). Find its equation of motion, neglecting resistance, if it is depressed so that its top just touches the surface of the water and is then released.

15. Assume that a straight tube has been bored through the center of the earth and that a particle of weight w pounds is dropped into the tube. If the radius of the earth is 3960 miles and the gravitational attraction is proportional to the distance from its center, how long does it take (neglecting resistance) for the particle to drop halfway to the center? Does it pass through the tube to the other side?

16. A weight w is suspended by two springs, having spring constants k_1 and k_2, connected in series (see illustration). What is the **effective** spring constant k for such a system? [*Hint:* No derivatives are required.]

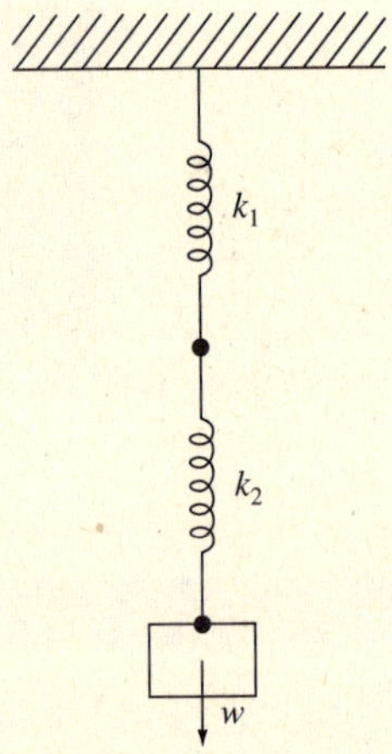

17. A small ball of weight w is placed in the middle of a tightly stretched, vertical string of length $2L$ and tension T_0. Show that for small horizontal displacements the ball undergoes simple harmonic motion. What is its period? [*Note:* Neglect gravity.]

4.2 VIBRATIONAL MOTION: DAMPED VIBRATIONS

Throughout the discussion in Section 4.1 we made the assumption that the only forces involved were gravity and the spring force. This assumption, however, is not very realistic. To take care of such things as friction in the spring and air resistance, we now assume that there is a damping force (that tends to slow things down), which can be thought of as the resultant of all other external forces acting on the object, which we still treat as a point mass. It is reasonable to assume that the magnitude of the damping force is proportional to the velocity of the particle (for example, the slower the movement, the smaller the air resistance) and directed opposite to the direction of velocity. Therefore we add the term $c(\frac{dx}{dt})$, where c is the damping constant that depends on all external factors, to Equation **(4.1.2)**. The equation of motion then becomes

$$\frac{d^2x}{dt^2} = -\frac{k}{m}x - \frac{c}{m}\frac{dx}{dt}, \qquad x(0) = x_0, \qquad x'(0) = v_0, \tag{1}$$

or

$$\frac{d^2x}{dt^2} + \frac{c}{m}\frac{dx}{dt} + \frac{k}{m}x = 0, \qquad x(0) = x_0, \qquad x'(0) = v_0. \tag{2}$$

(Of course, since c depends on external factors, it may very well not be a constant at all but may vary with time and position. In that case c is really $c(t, x)$, and the equation becomes much harder to analyze than the constant-coefficient case.)

To study Equation **(2)**, we first find the roots of the characteristic equation:

$$\frac{-c \pm \sqrt{c^2 - 4mk}}{2m}. \tag{3}$$

The nature of the general solution depends on the discriminant $\sqrt{c^2 - 4mk}$. If $c^2 > 4mk(> 0)$, then both roots are negative, since $\sqrt{c^2 - 4mk} < c$. In this case

$$x(t) = k_1 \exp\left(\frac{-c + \sqrt{c^2 - 4mk}}{2m}\right)t + k_2 \exp\left(\frac{-c - \sqrt{c^2 - 4mk}}{2m}\right)t \tag{4}$$

becomes small as t becomes large whatever the initial conditions may be. In the event that the discriminant vanishes,

$$x(t) = (k_1 + k_2 t)e^{(-c/2m)t}, \tag{5}$$

and the solution has a similar behavior as t increases. Equations **(4)** and **(5)** give the **(free) damped vibrations** of the mechanical system.

EXAMPLE 1 ► **Finding an equation of motion with damping**

A spring fixed at its upper end is stretched 6 inches by a 10-pound weight attached at its lower end. The spring-mass system is suspended in a viscous system (such as oil or water) so that the system is subjected to a damping force (pounds) of

$$5\frac{dx}{dt}.$$

Describe the motion of the system if the weight is drawn down an additional 4 inches and released.

Solution The differential equation is given by

$$\frac{d^2x}{dt^2} + 16\frac{dx}{dt} + 64x = 0. \tag{6}$$

since $\frac{k}{m} = \frac{gk}{w} \approx 32[10/(\frac{1}{2})]/10 = 64$ per second per second and $\frac{c}{m} = \frac{gc}{w} \approx 32(5)/10 = 16$ per second. The initial conditions are $x(0) = \frac{1}{3}$ and $x'(0) = 0$. Hence the roots of the characteristic equation are

$$\frac{-16 \pm \sqrt{(16)^2 - 4(64)}}{2} = -8,$$

and Equation **(6)** has the general solution

$$x(t) = e^{-8t}(k_1 + k_2 t).$$

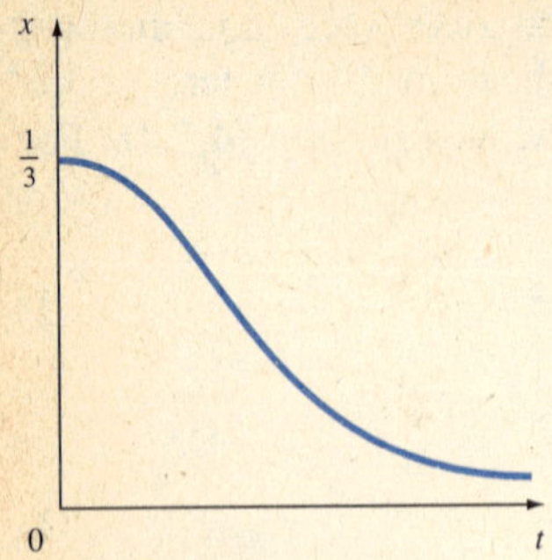

FIGURE 4.4
Damped motion: Graph of $x(t) = \frac{1}{3}e^{-8t}(1 + 8t)$

Applying the initial conditions yields

$$x(t) = \frac{1}{3}e^{-8t}(1 + 8t) \text{ feet,}$$

which has the graph shown in Figure 4.4. ◀

We observe that the solution does not oscillate. This system is called **critically damped,** because the discriminant $c^2 - 4mk = 0$. If there is any less damping (so $b^2 - 4ac < 0$), then oscillations will begin.

Damped vibrations

Damped vibrations occur whenever $c > 0$. If

1. $c^2 > 4mk$, the system experiences **overdamping.**
2. $c^2 = 4mk$, the system experiences **critical damping.**
3. $c^2 < 4mk$, the system experiences **underdamping.**

EXAMPLE 2 ▶ **Damping that leads to oscillations**

If $c^2 < 4mk$, the general solution of Equation **(2)** is

$$x(t) = e^{(-c/2m)t}\left(c_1 \cos \frac{\sqrt{4mk - c^2}}{2m}t + c_2 \sin \frac{\sqrt{4mk - c^2}}{2m}t\right), \tag{7}$$

which shows an oscillation with frequency

$$f = \frac{\sqrt{4mk - c^2}}{4\pi m}.$$

The factor $e^{(-c/2m)t}$ is called the **damping factor.** Letting $c = 4$ pounds per foot per second in Example 1 leads to the general solution

$$x(t) = e^{-32t/5}\left(c_1\cos \frac{24}{5}t + c_2\sin \frac{24}{5}t\right) \text{ feet.}$$

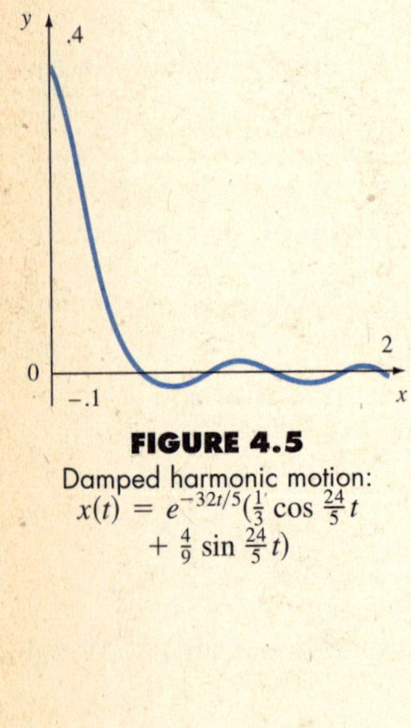

FIGURE 4.5
Damped harmonic motion: $x(t) = e^{-32t/5}(\frac{1}{3} \cos \frac{24}{5}t + \frac{4}{9} \sin \frac{24}{5}t)$

Note that $e^{-32t/5} \to 0$ as $t \to \infty$, so the damped motion decays to zero as time increases.

Using the initial values, we find that $c_1 = \frac{1}{3}$ and $c_2 = \frac{4}{9}$. The motion is illustrated in Figure 4.5. ◀

Damping forces also occur as the result of mechanical devices. A typical example is provided by automobile shock absorbers in which a piston is driven into a dashpot or damper (see Figure 4.6). Mechanical systems involving torsional motion, such as brakes on automobiles, also experience damped vibrations (see Figure 4.7). In this case the equation of motion is given by

$$I\frac{d^2\theta}{dt^2} + c\frac{d\theta}{dt} + k\theta = 0, \qquad \theta(0) = \theta_0, \qquad \theta'(0) = \theta_0', \tag{8}$$

where I is the moment of inertia, θ is the angular displacement, $c\, d\theta/dt$ is the damping torque, and $k\theta$ is the elastic torque due to twisting of the shaft.

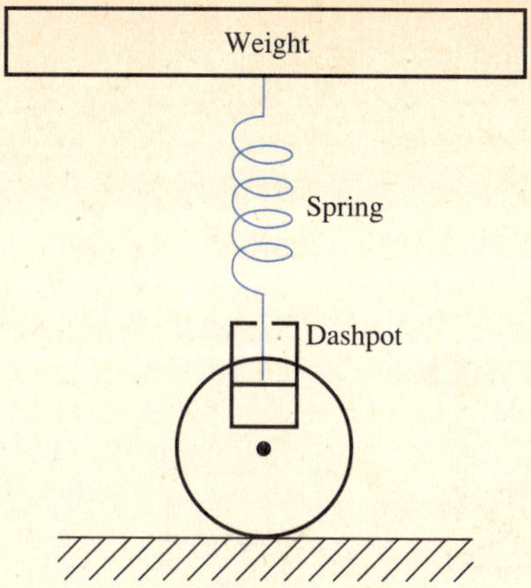

FIGURE 4.6
A piston is driven into a dashpot

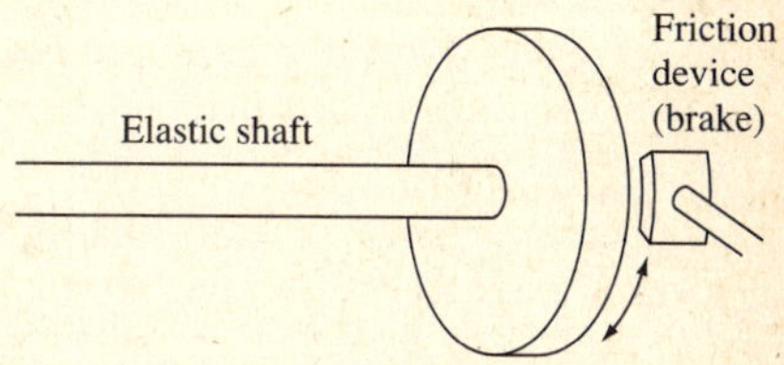

FIGURE 4.7
A brake applies a frictional force causing damping

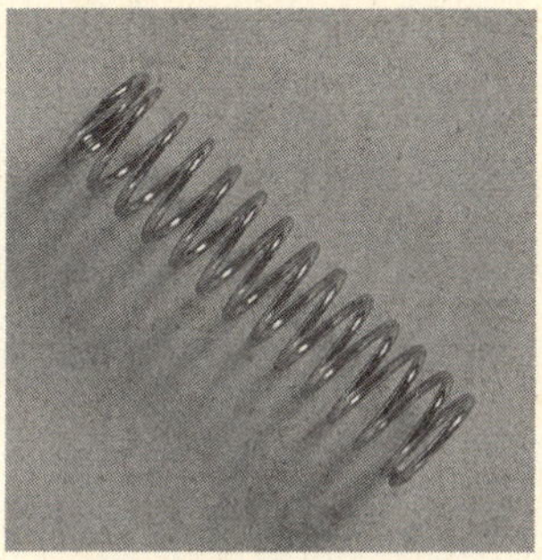

FIGURE 4.8
A spring (Scott, Foresman and Company)

Springs

Every day we come into contact with springs in the most unexpected places. Springs, such as that shown in Figure 4.8, are ubiquitous: they are found in mattresses, sofas, cars, bicycle brakes, airplanes, trains, snowmobiles, ski bindings, pogo sticks, and jeweled-movement watches. They add immeasurably to our comfort, by damping shocks and softening impacts. It may seem to you that everything that could be known about springs has long since been invented. However, that is definitely not the case; improvements are made all the time. Recently interest has been focused on plastic coil springs (see the article in *Plastics World,* vol. 49, (1991), p. 22, by Bernie Miller).

PROBLEMS 4.2

In Problems 1–6 determine the equation of motion of a point mass m attached to coiled spring with spring constant k, initially displaced a distance x_0 from equilibrium, and released with initial velocity v_0, subject to a damping constant c.

1. $m = 10$ kg, $k = 1000$ N/m, $x_0 = 1$ m, $v_0 = 0$, $c = 200$ kg/s $= 200$ N/(m/s)

2. $m = 10$ kg, $k = 10$ N/m, $x_0 = 0$, $v_0 = 1$ m/s, $c = 20$ kg/s

3. $m = 10$ kg, $k = 10$ N/m, $x_0 = 3$ m, $v_0 = 4$ m/s, $c = 10\sqrt{5}$ kg/s

4. $m = 1$ kg, $k = 16$ N/m, $x_0 = 4$ m, $v_0 = 0$, $c = 10$ kg/s

5. $m = 1$ kg, $k = 25$ N/m, $x_0 = 0$ m, $v_0 = 3$ m/s, $c = 8$ kg/s

6. $m = 9$ kg, $k = 1$ N/m, $x_0 = 4$ m, $v_0 = 1$ m/s, $c = 10$ kg/s

7. An object of 10-gram mass is suspended from a vibrating spring, the resistance in kg being numerically equal to half the velocity (in m/s) at any instant. If the period of the motion is 8 seconds, what is the spring constant (in kg/s^2)?

8. A weight w (lb) is suspended from a spring whose constant is 10 pounds per foot. The motion of the weight is subject to a resistance (lb) numerically equal to half the velocity (ft/sec). If the motion is to have a 1-second period, what are the possible values of w?

9. A weight of 48 pounds hangs from a spring with spring constant 50 pounds per foot. The damping in the system is 30% of critical. Determine the motion of the weight if it is pulled down 2 inches from its equilibrium position and released with an upward velocity of 1 foot per second.

10. Answer Problem 9 for damping that is 90% of critical.

11. Answer Problem 9 for critical damping.

12. Answer Problem 9 for damping that is 150% of critical.

13. Answer Problem 9 if 48 pounds of additional weight is suddenly applied to the system when it is at rest at equilibrium.

14. A weight, hung on an ideal spring in a vacuum (assume no air resistance) and set vibrating, has a period of 1 second. When the spring-weight system is placed in a viscous medium such that the resistance is proportional to the velocity, its (damped) period is found to be 2 seconds. Determine the differential equation corresponding to the damped vibrations.

15. A container weighing 2 pounds is half filled with 4 pounds of mercury. When hung on the end of a spring, it stretches it by 2 inches. The period of oscillation is found to be 0.3 seconds. If 4 pounds more mercury is added, the period becomes 0.4 seconds. Can the resistance be proportional to the velocity in this situation?

16. The disk in Figure 4.6 has a radius of 12 inches and weighs 100 pounds. The observed frequency of torsional vibration is 2π radians per second. When a different body is attached to the same shaft, the observed frequency of torsional vibration is 2.4π radians per second. Find the moment of inertia of the second body with respect to the axis of the shaft. [See Equation **(8)**.]

4.3 VIBRATIONAL MOTION: FORCED VIBRATIONS

The motion of the point mass considered in Sections 4.1 and 4.2 is determined by the inherent forces of the spring-weight system and the natural forces acting on the system. Accordingly, the vibrations are called **free** or **natural vibrations.** We now assume that the point mass is also subject to an external periodic force $F_0 \sin \omega t$, which is due to the motion of the object to which the upper end of the spring is attached (see Figure 4.6). In this case the mass undergoes **forced vibrations.**

We have seen in the previous sections that a spring-weight system, having a damping force proportional to the velocity, satisfies the initial-value problem

$$m\frac{d^2x}{dt^2} = -kx - c\frac{dx}{dt}, \qquad x(0) = x_0, \qquad x'(0) = v_0.$$

If we subject such a system to an additional periodic external force $F_0 \sin \omega t$, we obtain the nonhomogeneous second-order differential equation

$$m\frac{d^2x}{dt^2} = -kx - c\frac{dx}{dt} + F_0 \sin \omega t,$$

which we write in the form

$$\frac{d^2x}{dt^2} + \frac{c}{m}\frac{dx}{dt} + \frac{k}{m}x = \frac{F_0}{m}\sin \omega t. \tag{1}$$

We can solve Equation **(1)** by the method of undetermined coefficients. We already have solved the associated homogeneous equation

$$\frac{d^2x}{dt^2} + \frac{c}{m}\frac{dx}{dt} + \frac{k}{m}x = 0 \tag{2}$$

in Section 4.2—in three different cases (see Equations **(4.2.4)** and **(4.2.5)** on page 179, and Equation **(4.2.7)** on page 180). We assume that $\sin \omega t$ is not a solution of Equation **(2)**. Then, a particular solution to Equation **(2)** has the form

$$x_p(t) = b_1 \cos \omega t + b_2 \sin \omega t.$$

Substituting this function into Equation **(1)** yields the simultaneous equations

$$(\omega_0^2 - \omega^2)b_1 + \frac{c\omega}{m}b_2 = 0,$$

$$-\frac{c\omega}{m}b_1 + (\omega_0^2 - \omega^2)b_2 = \frac{F_0}{m}, \tag{3}$$

where $\omega_0 = \sqrt{k/m}$, from which we obtain

$$b_1 = \frac{-F_0 c\omega}{m^2(\omega_0^2 - \omega^2)^2 + (c\omega)^2} \qquad b_2 = \frac{F_0 m(\omega_0^2 - \omega^2)}{m^2(\omega_0^2 - \omega^2)^2 + (c\omega)^2}.$$

Using the same method we used to obtain Equation **(4.1.4)** on page 175, we have

$$x_P(t) = A\sin(\omega t + \phi), \tag{4}$$

where

$$A = \frac{\dfrac{F_0}{k}}{\sqrt{\left[1 - \left(\dfrac{\omega}{\omega_0}\right)^2\right]^2 + \left(2\dfrac{c}{c_0}\dfrac{\omega}{\omega_0}\right)^2}}, \quad \text{and} \quad \tan\phi = \frac{2\dfrac{c}{c_0}\dfrac{\omega}{\omega_0}}{\left(\dfrac{\omega}{\omega_0}\right)^2 - 1},$$

with $c_0 = 2m\omega_0$. Here A is the amplitude of the motion, ϕ is the **phase angle,** c/c_0 is the **damping ratio,** and ω/ω_0 is the **frequency ratio** of the motion.

The general solution is found by superimposing the periodic function **(4)** on the general solution [Equations **(4.2.4), (4.2.5),** or **(4.2.7)**] of the homogeneous equation **(2).** Since the solution of the homogeneous equation damps out as t increases, the general solution is very close to **(4)** for large values of t. Figure 4.9 illustrates two typical situations.

It is interesting to see what occurs if the damping constant c vanishes. In this case, Equation **(2)** becomes

$$\frac{d^2x}{dt^2} + \frac{k}{m}x = 0$$

with general solution $x(t) = c_1 \cos\omega_0 t + c_2 \sin\omega_0 t$, $\omega_0 = \sqrt{k/m}$. There are two cases.

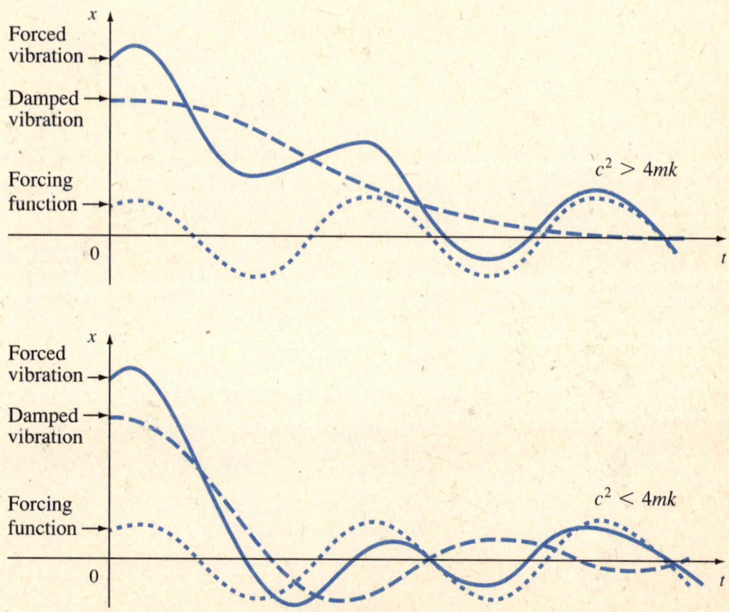

FIGURE 4.9
Damped and forced vibrations

Case 1

If $\omega^2 \neq \omega_0^2$, we superimpose the periodic function **(4)** on the general solution of the homogeneous equation $x'' + \omega_0^2 x = 0$, obtaining

$$x(t) = c_1 \cos \omega_0 t + c_2 \sin \omega_0 t + \frac{\dfrac{F_0}{k}}{1 - \left(\dfrac{\omega}{\omega_0}\right)^2} \sin \omega t. \tag{5}$$

Using the initial conditions, we find that

$$c_1 = x_0 \qquad \text{and} \qquad c_2 = \frac{v_0}{\omega_0} - \frac{\left(\dfrac{F_0}{k}\right)\left(\dfrac{\omega}{\omega_0}\right)}{1 - \left(\dfrac{\omega}{\omega_0}\right)^2},$$

so

$$x(t) = A \sin(\omega_0 t + \phi) + \frac{\dfrac{F_0}{k}}{1 - \left(\dfrac{\omega}{\omega_0}\right)^2} \sin \omega t,$$

where

$$A = \sqrt{c_1^2 + c_2^2} \qquad \text{and} \qquad \tan \phi = \frac{c_1}{c_2}.$$

Hence the motion in this case is simply the sum of two sinusoidal (sine) curves, as illustrated in Figure 4.10.

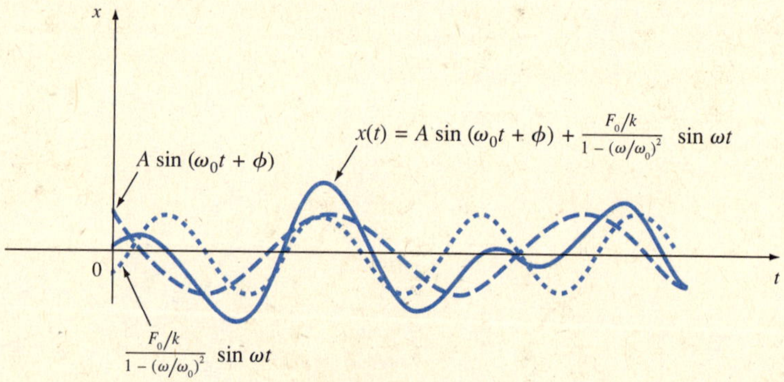

FIGURE 4.10
The solid blue curve is the sum of two sinusoidal terms

Case 2

If $\omega^2 = \omega_0^2$ then Equation **(1)** is

$$\frac{d^2x}{dt^2} + \omega^2 x = \frac{F_0}{m} \sin \omega t. \tag{6}$$

The general solution to the associated homogeneous equation

$$\frac{d^2x}{dt^2} + \omega^2 x = 0 \tag{7}$$

is

$$x(t) = c_1 \cos \omega t + c_2 \sin \omega t.$$

Since the nonhomogeneous term $(F_0/m)\sin \omega t$ is a solution to **(7)**, we seek a particular solution to Equation **(6)** of the form

$$x_p(t) = b_1 t\cos \omega t + b_2 t \sin \omega t. \tag{8}$$

Substituting Equation **(8)** into

$$\frac{d^2x}{dt^2} + \frac{k}{m}x = \frac{F_0}{m}\sin \omega t$$

we get

$$b_1 = \frac{-F_0}{2m\omega} \quad \text{and} \quad b_2 = 0,$$

so the general solution has the form

$$x(t) = c_1\cos \omega t + c_2\sin \omega t - \frac{F_0}{2m\omega}t \cos \omega t. \tag{9}$$

Note that as t increases, the vibrations described by the last term in Equation **(7)** increase without bound. The external force is said to be in **resonance** with the vibrating mass. It is evident that the displacement here becomes so large that the elastic limit of the spring is exceeded, leading to fracture or to a permanent distortion in the spring.

Suppose that c is positive but very close to zero, while $\omega^2 = \omega_0^2$. Note that Equation **(3)** yields $b_1 = -F_0/c\omega$ and $b_2 = 0$ when Equation **(2)** is substituted into Equation **(1)**. Superimposing

$$x_p(t) = \frac{-F_0}{c\omega}\cos \omega t$$

on the solution of the homogeneous equation [see Equation **(4.2.7)**, since c^2 is very small] and letting $c_0 = 2m\omega_0$, we obtain

$$x(t) = e^{(-c/c_0)\omega_0 t}\left[c_1 \cos \omega_0\sqrt{1-\left(\frac{c}{c_0}\right)^2}\,t + c_2 \sin \omega_0\sqrt{1-\left(\frac{c}{c_0}\right)^2}\,t\right] - \frac{F_0}{c\omega}\cos \omega t. \tag{10}$$

Since c/c_0 is very small, for small values of t we see that Equation **(10)** can be approximated by

$$x(t) \approx c_1 \cos \omega t + c_2 \sin \omega t - \frac{F_0}{2m\omega}\left(\frac{2m}{c}\right)\cos \omega t,$$

which bears a marked resemblance to Equation **(9)** *when* **(9)** *is evaluated at large values of t* (since $\frac{2m}{c}$ is large). Thus the *damped* spring problem approaches resonance. This phenomenon is extremely important in engineering, since resonance may produce undesirable effects such as metal fatigue and structural fracture, as well as desirable objectives such as sound and light amplification.

An example of resonance occurred when a column of soldiers marched across Broughton bridge, near Manchester, England, in 1831. Their rhythmic marching produced a periodic force of large amplitude closely approximating the natural frequency of the bridge, causing it to collapse. For that reason soldiers today are required to march out of step when crossing a bridge.

A more recent disaster was the collapse of the Tacoma (Washington) Narrows Bridge at Puget Sound on November 7, 1940 (see Figure 4.11). When it was

FIGURE 4.11
The collapse of the Tacoma Narrows Bridge on November 7, 1940. (UPI/Bettman)

completed, the bridge was the third largest suspension bridge in the world. From its opening on July 1, 1940, the bridge enjoyed enormous popularity because of its strange weaving motions and soon earned the nickname "Galloping Gertie." Many automobile passengers complained of motion sickness when crossing it.

On November 7, winds reached 42 miles per hour in the Tacoma area. At around 7:00 A.M. the bridge began to resonate violently. By 10 A.M. the oscillations were so great that one edge of the roadway alternated between being 28 feet higher and 28 feet lower than the other edge. Shortly after 11 A.M. the entire bridge crashed 190 feet into Puget Sound. The only vehicles on the bridge at the time of its collapse were a logging truck and the automobile of a reporter, Mr. Leonard Coatsworth. Both drivers abandoned their vehicles and crawled to safety. Mr. Coatsworth struggled over sections of the bridge, pitching at 45° angles, and made his way 500 yards to the support towers, 425 feet high. The only loss of life was Mr. Coatsworth's pet cocker spaniel, which went down with the car.

The bridge was built at a cost $6.4 million and, unsurprisingly, many explanations and excuses were given following its collapse. Some construction engineers opined that the introduction of stiffening plate or web girders in place of the older lattice or open girders was responsible. Clark W. Eldridge, chief engineer of the bridge, said that Washington State highway engineers had protested against the design, but that it was built anyway in the interest of economy. He blamed the ill-fated design on eastern engineers. "The employment of eastern engineers was because of a requirement by the money-lending agencies that engineers of a national reputation be employed." Perhaps the clearest explanation of what happened was provided in a *New York Times* editorial on November 9, 1940.

> Like all suspension bridges, that at Tacoma both heaved and swayed with a high wind. It takes only a tap to start a pendulum swinging. Time successive taps correctly and soon the pendulum swings with its maximum amplitude. So with

> this bridge. What physicists call resonance was established, with the result that the swaying and heaving exceeded the limits of safety. Steel buckled like cardboard. The concrete roadway cracked like ice. The whole structure broke, with the exception of the towers.

None of this collapse was anticipated, probably because few of the individuals who helped design the bridge understood resonance. Even afterward, there was eagerness to repeat the same mistake. Shortly after the collapse of the bridge, Governor Clarence D. Martin of the state of Washington declared, "We are going to build the same exact [sic] bridge, exactly as before." A well-known engineer who had had nothing to do with the original construction and who did understand what had happened responded, "If you build the exact same bridge exactly as before, it will fall into the exact same river exactly as before."

Linearity versus nonlinearity (again): thinking about a mathematical model

In the last three sections, we discussed one of the basic models in classical mechanics. In every example, we were led to a differential equation having the form

$$\frac{d^2x}{dt^2} + \frac{c}{m}\frac{dx}{dt} + \frac{k}{m}x \approx \frac{F_0}{m}\sin \omega t \qquad \textbf{(11)}$$

which can be written in the form

$$x'' + ax' + bx = f(t). \qquad \textbf{(12)}$$

Equations **(11)** and **(12)** are linear, and it may have seemed that the assumptions that led to these equations were reasonable. But are they? Let us consider our first equation (Equation **(4.1.2)**):

$$\frac{d^2x}{dt^2} + \frac{k}{m}x = 0. \qquad \textbf{(13)}$$

This equation followed from the three basic assumptions on page 174. The most reasonable of these assumptions is that all motion is along a vertical line. However, if we allow the object to move in two dimensions, then Equation **(13)** will almost certainly not be realistic. For example, in Problem 4.1.11 on page 178, we asked you to show that, even under idealized circumstances, the motion of a freely moving pendulum will be governed by the differential equation

$$\frac{d^2\theta}{dt^2} + \frac{g}{l}\sin\theta = 0. \qquad \textbf{(14)}$$

Equation **(14)** is nonlinear. We will show, in Chapter 10, that, in a limited sense, solutions to Equation **(14)** can be approximated by solutions to the linear equation

$$\frac{d^2\theta}{dt^2} + \frac{g}{l}\theta - 0;$$

but this does not alter the fact that, even in this idealized situation, the underlying differential equation is nonlinear.

In Section 4.2, we added friction to our model and obtained the equation

$$\frac{d^2x}{dt^2} + \frac{c}{m}\frac{dx}{dt} + \frac{k}{m}x = 0. \qquad \textbf{(15)}$$

Here, we assumed that the frictional force or damping force was proportional to the velocity $v = \frac{dx}{dt}$. But why should it be proportional to v? Kinetic energy, for example, is given by $\frac{1}{2}mv^2$. The faster an object moves, the greater its kinetic energy, and it might be argued that the frictional force should be proportional to kinetic energy. Then Equation **(15)** would become

$$\frac{d^2x}{dt^2} + c\left(\frac{dx}{dt}\right)^2 + \frac{k}{m}x = 0 \tag{16}$$

which is nonlinear. In fact, in hydrodynamics the damping force is indeed proportional to v^2. This leads to another consideration. If the velocity is small, then the damping term will be small, and approximating v^2 by v (which is larger if $v < 1$) might not lead to a significant error in the solution. However, if v is large, then replacing v^2 with v in the model might make the model almost useless—even if the resulting linear equation is easy to solve.

You might ask, is it reasonable that the constant c in Equations **(15)** or **(16)** be a constant? If for a pendulum, for example, the friction is greater at some parts of the cycle, then friction might be a nonlinear function of ω; or friction may depend on the temperature or the wind velocity which, in turn, might be functions of time.

Evidently, there are many different kinds of forces that act on moving objects in the real world. A 10-gram pebble falls faster than a 10-gram feather, although, in a vacuum, they would fall at the same speed. Each time extra forces are taken into account, our simplified linear model becomes less realistic.

So, again: Beware of linear models. Always ask whether the equations are linear because they truly mirror reality, or whether they are linear because linear equations are the easiest ones to solve. In the latter case, you should seek a more complicated, but also more realistic, nonlinear equation. Even though we cannot find solutions of such equations in closed form, we can approximate solutions numerically. And a numerical approximation to a reasonable equation is far better than an exact solution to an equation that was given only because it can easily be solved.

PROBLEMS 4.3

In Problems 1–6 determine the equation of motion of a point mass m attached to a coiled spring with spring constant k, initially displaced a distance x_0 from equilibrium, and released with velocity v_0 subject to a damping constant c and an external force $F_0 \sin \omega t$ newtons.

1. $m = 10$ kg, $k = 1000$ N/m, $x_0 = 1$ m, $v_0 = 0$, $c = 200$ kg/s, $F_0 = 1$ N, $\omega = 10$ rad/s.
2. $m = 10$ kg, $k = 10$ N/m, $x_0 = 0$, $v_0 = 1$ m/s, $c = 20$ kg/s, $F_0 = 1$ N, $\omega = 1$ rad/s.
3. $m = 10$ kg, $k = 10$ N/m, $x_0 = 3$ m, $v_0 = 4$ m/s, $c = 10\sqrt{5}$ kg/s, $F_0 = 1$ N, $\omega = 1$ rad/s.
4. $m = 1$ kg, $k = 16$ N/m, $x_0 = 4$ m, $v_0 = 0$, $c = 10$ kg/s, $F_0 = 4$ N, $\omega = 4$ rad/s.
5. $m = 1$ kg, $k = 25$ N/m, $x_0 = 0$, $v_0 = 3$ m/s, $c = 8$ kg/s, $F_0 = 1$ N, $\omega = 3$ rad/s.
6. $m = 9$ kg, $k = 1$ N/m, $x_0 = 4$ m, $v_0 = 1$ m/s, $c = 10$ kg/s, $F_0 = 2$ N, $\omega = \frac{1}{3}$ rad/s.
7. An object of 1-gram mass is hanging at rest on a spring that is stretched 25 centimeters by the weight. The upper end of the spring is given the periodic force 0.01 sin $2t$ newtons and air resistance has a magnitude (kg/s) 0.02162 times the velocity in meters per second. Find the equation of motion of the object.
8. A 100-pound weight is suspended from a spring with spring constant 20 pounds per inch. When the system is vibrating freely, we observe that in consecutive cycles the amplitude decreases by 40%. If a force of 20 cos ωt newtons acts on the system, find the amplitude and phase shift of the resulting steady-state motion if $\omega = 9$ radians per second.
9. Repeat Problem 8 for $\omega = 12$ radians per second.
10. A particle of weight w moves along the x-axis under the influence of a force $F = -kx$. Friction on the particle is proportional to the force between the particle and the surface on which it moves. Find the differential equation governing the motion of this particle.

11. Consider the forced vibrations of an undamped mechanical spring-weight system, where the external force is $F_0 \sin \omega t$ newtons.

(a) Show that if $\omega \neq \omega_0 (= \sqrt{k/m})$, then the solution is given by

$$x(t) = c_1 \cos \omega_0 t + c_2 \sin \omega_0 t + \frac{F_0}{k - m\omega^2} \sin \omega t.$$

(b) Discuss what happens if ω is close, but not equal, to ω_0. [*Hint:* Use the procedure for finding Equation **(4.1.4)**. The phenomenon that occurs is called **beats** as illustrated here. It occurs whenever an impressed frequency is close to a natural frequency of a mechanical system. The dotted curve is called the *envelope* of the graph of $x(t)$.

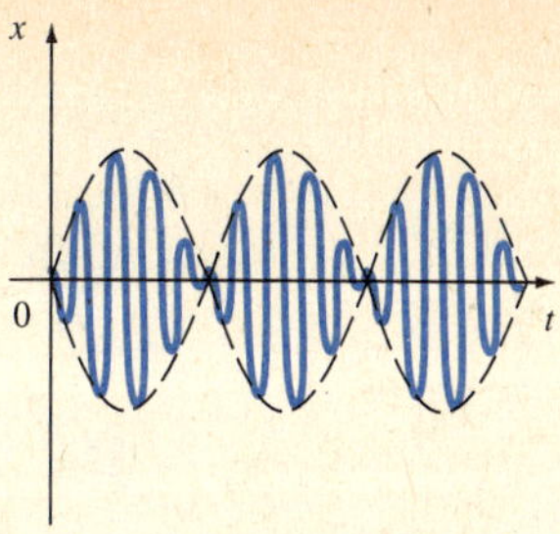

12. A 48-pound weight is suspended from a spring with spring constant 50 pounds per inch. In 10 cycles we see that the maximum displacement decreases by 50%. Suppose an external force of $F_0(\sin 15t + \sin 16t)$ newtons is applied to the spring-weight system. Will beats occur? Why?

4.4 ELECTRIC CIRCUITS

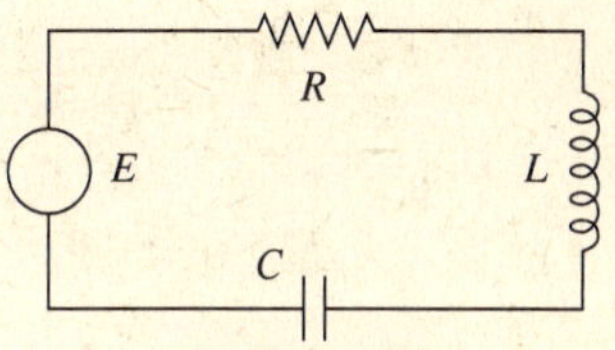

FIGURE 4.12
An RLC-circuit

We make use of the concepts of electric circuitry developed in Section 2.7 and the methods of Chapter 3 to study a simple electric circuit containing a resistor, an inductor, and a capacitor in series with an electromotive force (Figure 4.12). Suppose that R, L, C, and E are constants. Applying Kirchhoff's law, we obtain

$$L\frac{dI}{dt} + RI + \frac{Q}{C} = E. \tag{1}$$

Since $dQ/dt = I$, we may differentiate Equation **(1)** to get the second-order homogeneous differential equation

$$L\frac{d^2I}{dt^2} + R\frac{dI}{dt} + \frac{I}{C} = 0. \tag{2}$$

To solve this equation, we note that the characteristic equation

$$\lambda^2 + \frac{R}{L}\lambda + \frac{1}{CL} = 0$$

has the following roots:

$$\lambda_1 = \frac{-R + \sqrt{R^2 - \dfrac{4L}{C}}}{2L}, \qquad \lambda_2 = \frac{-R - \sqrt{R^2 - \dfrac{4L}{C}}}{2L},$$

or, rewriting the radical in dimensionless units,* we have

$$\lambda_1 = \frac{R}{2L}\left(-1 + \sqrt{1 - \frac{4L}{CR^2}}\right), \qquad \lambda_2 = \frac{R}{2L}\left(-1 - \sqrt{1 - \frac{4L}{CR^2}}\right). \tag{3}$$

Equation **(2)** may now be solved using the methods of Sections 3.3 and 3.4.

* 1 henry = 1 volt-sec/amp; 1 farad = 1 coulomb/volt; 1 ohm = 1 volt/amp; 1 coulomb = 1 amp-sec.

Remark

Equation **(1)** can be turned into a second-order differential equation even if E is a nondifferentiable function (like a square wave). Setting $I = dQ/dt$ and $dI/dt = d^2Q/dt^2$ in Equation **(1)**, we obtain

$$L\frac{d^2Q}{dt^2} + R\frac{dQ}{dt} + \frac{Q}{C} = E. \tag{1'}$$

EXAMPLE 1 ▶ **Solving an RLC-circuit with a constant voltage**

Let $L = 1$ henry (H), $R = 100$ ohms (Ω), $C = 10^{-4}$ farad (f), and $E = 1000$ volts (V) in the circuit shown in Figure 4.12. Suppose that no charge is present and no current is flowing at time $t = 0$ when E is applied. By Equation **(3)** we see that the characteristic equation has the roots $\lambda_1 = -50 + 50\sqrt{3}\,i$ and $\lambda_2 = -50 - 50\sqrt{3}\,i$, since

$$R^2 - \frac{4L}{C} = 10{,}000 - 4 \times 10^4 = -30{,}000$$

and $\frac{R}{2L} = 50$. Thus

$$I(t) = e^{-50t}(c_1\cos 50\sqrt{3}\,t + c_2\sin 50\sqrt{3}\,t).$$

Applying the initial condition $I(0) = 0$, we have $c_1 = 0$. Hence

$$I(t) = c_2e^{-50t}\sin 50\sqrt{3}\,t,$$

and

$$I'(t) = 50c_2e^{-50t}(\sqrt{3}\cos 50\sqrt{3}\,t - \sin 50\sqrt{3}\,t).$$

To establish the value of c_2, we must make use of Equation **(1)** and the initial condition $Q(0) = 0$. Since

$$\begin{aligned} Q(t) &= C\left(E - L\frac{dI}{dt} - RI\right) \\ &= 10^{-4}[1000 - 50c_2e^{-50t}(\sqrt{3}\cos 50\sqrt{3}\,t - \sin 50\sqrt{3}\,t + 2\sin 50\sqrt{3}\,t)] \\ &= \frac{1}{10} - \frac{c_2}{200}e^{-50t}(\sin 50\sqrt{3}\,t + \sqrt{3}\cos 50\sqrt{3}\,t), \end{aligned}$$

it follows that

$$Q(0) = \frac{1}{10} - \frac{c_2\sqrt{3}}{200} = 0 \qquad \text{or} \qquad c_2 = \frac{20}{\sqrt{3}}.$$

Hence

$$Q(t) = \frac{1}{10} - \frac{1}{10\sqrt{3}}e^{-50t}(\sin 50\sqrt{3}\,t + \sqrt{3}\cos 50\sqrt{3}\,t)$$

and

$$I(t) = \frac{20}{\sqrt{3}}e^{-50t}\sin 50\sqrt{3}\,t.$$

From these equations we observe that the current rapidly damps out and that the charge rapidly approaches its **steady-state value** of 0.1 coulomb. Here $I(t)$ is called the **transient current** (because its contribution to the total current decreases toward zero with increasing time). ◄

EXAMPLE 2 ► **Solving an RLC-circuit with a periodic voltage**

Let the inductance, resistance, and capacitance in Example 1 remain the same, but suppose that $E = 962 \sin 60t$ volts. By Equation **(1)** we have

$$\frac{dI}{dt} + 100I + 10^4Q = 962 \sin 60t, \tag{4}$$

and converting Equation **(4)** so that all expressions are in terms of $Q(t)$, we obtain

$$\frac{d^2Q}{dt^2} + 100\frac{dQ}{dt} + 10^4Q = 962 \sin 60t. \tag{5}$$

It is evident that Equation **(5)** has a particular solution of the form

$$Q_p(t) = A_1 \sin 60t + A_2 \cos 60t. \tag{6}$$

To determine the values A_1 and A_2, we substitute Equation **(6)** into Equation **(5)**, obtaining the simultaneous equations

$$6400A_1 - 6000A_2 = 962,$$
$$6000A_1 + 6400A_2 = 0.$$

Thus $A_1 = \frac{2}{25}$, $A_2 = -\frac{3}{40}$, and since the general solution of the homogeneous equation is the same as that of Equation **(2)**, the general solution of Equation **(5)** is

$$\begin{aligned} Q(t) &= e^{-50t}(c_1 \cos 50\sqrt{3}\, t + c_2 \sin 50\sqrt{3}\, t) \\ &\quad + \frac{2}{25}\sin 60t - \frac{3}{40}\cos 60t. \end{aligned} \tag{7}$$

Differentiating **(7)**, we obtain

$$\begin{aligned} I(t) &= 50e^{-50t}[(\sqrt{3}\, c_2 - c_1)\cos 50\sqrt{3}\, t - (c_2 + \sqrt{3}\, c_1)\sin 50\sqrt{3}\, t] \\ &\quad + \frac{24}{5}\cos 60t + \frac{9}{2}\sin 60t. \end{aligned}$$

Setting $t = 0$ and using the initial conditions, we obtain

$$Q(0) = c_1 - \frac{3}{40} = 0,$$

$$I(0) = 50(\sqrt{3}\, c_2 - c_1) + \frac{24}{5} = 0,$$

so $c_1 = \frac{3}{40}$ and $c_2 = -21/1000\sqrt{3}$. Therefore

$$\begin{aligned} Q(t) &= \frac{e^{-50t}}{1000}(75 \cos 50\sqrt{3}\, t - 7\sqrt{3} \sin 50\sqrt{3}\, t) \\ &\quad + \frac{80 \sin 60t - 75 \cos 60t}{1000}, \end{aligned}$$

$$\begin{aligned} I(t) &= -\frac{e^{-50t}}{5}(24 \cos 50\sqrt{3}\, t + 17\sqrt{3} \sin 50\sqrt{3}\, t) \\ &\quad + \frac{48 \cos 60t + 45 \sin 60t}{10}. \end{aligned}$$ ◄

PROBLEMS 4.4

1. Let $L = 10$ H, $R = 250\ \Omega$, $C = 10^{-3}$ f, and $E = 900$ V in Example 1. With the same assumptions, calculate the current and charge for all values of $t \geq 0$.

2. Suppose instead that $E = 50 \cos 30t$ in Problem 1. Find $Q(t)$ for $t \geq 0$.

In Problems 3–6 find the steady-state current in the RLC circuit of Figure 4.12 for the given values.

3. $L = 5$ H, $R = 10\ \Omega$, $C = 0.1$ f, $E = 25 \sin t$ V

4. $L = 10$ H, $R = 40\ \Omega$, $C = 0.025$ f, $E = 100 \cos 5t$ V

5. $L = 1$ H, $R = 7\ \Omega$, $C = 0.1$ f, $E = 100 \sin 10t$ V

6. $L = 2.5$ H, $R = 10\ \Omega$, $C = 0.08$ f, $E = 100 \cos 5t$ V

Find the transient current in the RLC circuit of Figure 4.12 for Problems 7–12.

7. Values from Problem 3.

8. Values from Problem 4.

9. Values from Problem 5.

10. Values from Problem 6.

11. $L = 20$ H, $R = 40\ \Omega$, $C = 10^{-3}$ f, $E = 500 \sin t$ V

12. $L = 24$ H, $R = 48\ \Omega$, $C = 0.375$ f, $E = 900 \cos t$ V

13. Given that $L = 1$ H, $R = 1200\ \Omega$, $C = 10^{-6}$ f, $I(0) = Q(0) = 0$, and $E = 100 \sin 600t$ V, determine the transient current and the steady-state current.

14. Find the ratio of the current flowing in the circuit of Problem 13 to that which would be flowing if there were no resistance, at $t = 0.001$ sec.

15. Consider the system governed by Equation **(1)** for the case where the resistance is zero and $E = E_0 \sin \omega t$. Show that the solution consists of two parts, a general solution with frequency $1/\sqrt{LC}$ and a particular solution with frequency ω. The frequency $1/\sqrt{LC}$ is called the **natural frequency** of the circuit. Note that if $\omega = 1/\sqrt{LC}$, the particular solution disappears.

16. To allow for different variations of the voltage, let us assume in Equation **(1)** that $E = E_0 e^{it}$ $(= E_0 \cos t + iE_0 \sin t)$. Assume also, as in Problem 15, that $R = 0$. Finally, for simplicity assume that $E_0 = L = C = 1$. Then $1 = \omega = 1/\sqrt{LC}$.
    (a) Show that Equation **(2)** becomes
    $$\frac{d^2I}{dt^2} + I = e^{it}.$$
    (b) Determine λ such that $I(t) = \lambda t e^{it}$ is a solution.
    (c) Calculate the general solution and show that the magnitude of the current increases without bound as t increases. This phenomenon is caused by resonance.

17. Let an inductance of L henrys, a resistance of R ohms, and a capacitance of C farads be connected in series with an emf of $E_0 \sin \omega t$ volts. Suppose $Q(0) = I(0) = 0$, and $4L > R^2C$.
    (a) Find the expressions for $Q(t)$ and $I(t)$.
    (b) What value of ω produces resonance?

18. Solve Problem 17 for $4L = R^2C$.

19. Solve Problem 17 for $4L < R^2C$.

4.5 SOLVING SPRING-MASS PROBLEMS USING A SYMBOLIC MANIPULATOR

This section will show you how a generic symbol manipulator can be used to solve the spring-mass problems of Sections 4.1 to 4.3. The symbolic manipulators can also be used to draw a graph of the solutions of the linear constant-coefficient second-order differential equations that arise in vibratory motion. We shall use the scripts that we constructed in Chapter 3:

```
LINEAR2(x,y,p,q,r)
```

—which finds the general solution to the linear constant coefficient differential equation

$y'' + p\,y' + q\,y = r(x)$,

where p and q are constants and $r(x)$ is a function of x.

When initial conditions are given, we use the general solution $y = GS$ of the differential equation,

```
IV2(x,y,GS,x0,y0,dy0,c1,c2)
```
—to obtain the values of c1 and c2 in the general solution so that y(x0)=y0, y′(x0)=dy0.

EXAMPLE 1 ▶

A spring attached by its upper end is stretched 6 inches by a 10-pound weight attached at its lower end. The spring-mass system is suspended in water and is subject to a damping force of five times its velocity. Describe the motion of the spring-mass system and find a particular solution if the weight is drawn down an additional 4 inches and released. Finally, draw a graph of the solution.

Solution This is Example 4.2.1. Here $m = \text{weight}/g = \frac{10}{32}$, $c = 5$, and $k = 10/(\frac{1}{2}) = 20$, so the equation of motion is

$$x'' + \left(\frac{c}{m}\right)x' + \left(\frac{k}{m}\right)x = x'' + 16x' + 64x = 0 \qquad x(0) = \frac{1}{3} \qquad x'(0) = 0.$$

To get the general solution we give the command: LINEAR2(t, x, 16, 64, 0) and obtain the general solution

$$x = e^{-8t}(\text{c1} + \text{c2}\ t).$$

Since this is an initial-value problem, we wish to determine the values of c1 and c2 in the solution. So we write the command:

```
IV2(t,x,(c1+c2*t)*exp(-8*t),0,1/3,0,c1,c2).
```

This yields the values [c1, c2] = $[\frac{1}{3}, \frac{8}{3}]$, so the particular solution is

$$x(t) = e^{-8t}\left(\frac{1}{3} + \frac{8}{3}t\right).$$

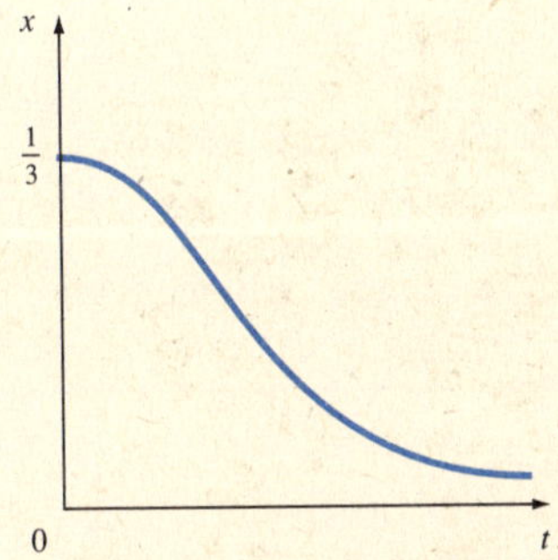

Each symbolic manipulator has its own commands for plotting. The figure at left resembles the output obtained by using DERIVE. ◀

PROBLEMS 4.5

In Problems 1–5, use MATHEMATICA, MAPLE, DERIVE, MATLAB, *or* MATHCAD *to determine the equation of motion of a point mass m attached to a coiled spring with spring constant k, initially displaced a distance x_0 from equilibrium, and released with velocity v_0 subject to a damping constant c and an external force F_0 sin ωt newtons. Graph the solution using your symbolic manipulator.*

1. $m = 10$ kg, $k = 1000$ N/m, $x_0 = 1$ m, $v_0 = 0$, $c = 200$ kg/s, $F_0 = 1$ N, $\omega = 10$ rad/s
2. $m = 10$ kg, $k = 10$ N/m, $x_0 = 0$, $v_0 = 1$ m/s, $c = 20$ kg/s, $F_0 = 1$ N, $\omega = 1$ rad/s
3. $m = 10$ kg, $k = 10$ N/m, $x_0 = 3$ m, $v_0 = 4$ m/s, $c = 10\sqrt{5}$ kg/s, $F_0 = 1$ N, $\omega = 1$ rad/s
4. $m = 1$ kg, $k = 16$ N/m, $x_0 = 4$ m, $v_0 = 0$, $c = 10$ kg/s, $F_0 = 4$ N, $\omega = 4$ rad/s
5. $m = 1$ kg, $k = 25$ N/m, $x_0 = 0$, $v_0 = 3$ m/s, $c = 8$ kg/s, $F_0 = 1$ N, $\omega = 3$ rad/s

REVIEW EXERCISES FOR CHAPTER 4

In Exercises 1–6 determine the equation of motion of a mass m attached to a coiled spring with spring constant k, initially displaced a distance x_0 from equilibrium, and released with velocity v_0 subject to
(a) *no damping or external forces,*
(b) *a damping constant c but no external force,*
(c) *an external force $F_0 \sin \omega t$ but no damping,*
(d) *both a damping constant c and external force $F_0 \sin \omega t$.*

1. $m = 20$ kg, $k = 1000$ N/m, $x_0 = 1$ m, $v_0 = 0$, $c = 200$ kg/s, $F_0 = 1$ N, $\omega = 10$ rad/s

2. $m = 25$ kg, $k = 40$ N/m, $x_0 = 0$, $v_0 = 1$ m/s, $c = 20$ kg/s, $F_0 = 1$ N, $\omega = 1$ rad/s

3. $m = 25$ kg, $k = 40$ N/m, $x_0 = 3$ m, $v_0 = 4$ m/s, $c = 10\sqrt{5}$ kg/s, $F_0 = 1$ N, $\omega = 1$ rad/s

4. $m = 1$ kg, $k = 36$ N/m, $x_0 = 4$ m, $v_0 = 0$, $c = 10$ kg/s, $F_0 = 4$ N, $\omega = 4$ rad/s

5. $m = 4$ kg, $k = 25$ N/m, $x_0 = 0$, $v_0 = 3$ m/s, $c = 8$ kg/s, $F_0 = 1$ N, $\omega = 3$ rad/s

6. $m = 9$ kg, $k = 81$ N/m, $x_0 = 4$ m, $v_0 = 1$ m/s, $c = 10$ kg/s, $F_0 = 2$ N, $\omega = \frac{1}{3}$ rad/s

7. Let an inductance of $L = 2$ henrys, a resistance of $R = 50$ ohms and a capacitance $C = 10^{-4}$ farad be connected in series with an emf of $E = 1000$ volts (see Figure 4.12). Suppose no charge is present and no current is flowing at time $t = 0$ when E is applied. Find the transient and steady-state solutions for the charge Q at all times t.

In Exercises 8–11 find the steady-state current in the RLC circuit of Figure 4.12 for the given values.

8. $L = 5$ H, $R = 20\ \Omega$, $C = 0.1$ f, $E = 25 \sin t$ V

9. $L = 10$ H, $R = 240\ \Omega$, $C = 0.025$ f, $E = 100 \cos t$ V

10. $L = 1$ H, $R = 9\ \Omega$, $C = 0.1$ f, $E = 100 \sin 10t$ V

11. $L = 2.5$ H, $R = 20\ \Omega$, $C = 0.08$ f, $E = 100 \cos 5t$ V

12. A 27-pound weight hangs from a spring of spring constant 18 pounds per inch. We find that during free motion the amplitude decreases to one tenth in six complete cycles of the motion. Find the equation describing the motion of the spring-weight system.

13. A spring fixed at its upper end supports a 10-pound weight that stretches the spring by $\frac{1}{2}$ foot. An external periodic force of $2 \cos 8t$ pounds is applied to the spring-weight system. Describe the motion, neglecting resistance.

14. Repeat Exercise 13 with a damping constant of $c = 0.01$.

15. Repeat Exercise 14 with an external periodic force of $2 \sin \frac{65}{8} t$ pounds.

5 Power Series Solutions of Differential Equations

In Chapter 3 we studied several methods for solving second- and higher-order differential equations. With the exception of the Euler equation and a few equations in which one solution was easily guessed, the techniques applied only to linear differential equations with *constant coefficients.* The case of linear differential equations with *variable coefficients* is much more complicated. Many of the most important differential equations in applied mathematics—for example, the Bessel equation and the Legendre equation—are of this type. In this chapter we consider a method that is often successful for obtaining solutions to such equations. Since the solutions so obtained are in the form of power series, the procedure used is known as the **power series method.** For the method to apply, the variable coefficients must be expressed as power series. The last two sections of the chapter are an introduction to Bessel functions and Legendre polynomials. In those sections we consider some of the properties of these special functions and a few of the standard procedures used in working with them.

5.1 THE POWER SERIES METHOD

In this section we review some of the basic properties of power series before discussing the power series method. We take it for granted that most readers have received some background in power series in an earlier course in calculus. A **power series** in $(x - a)$ is an infinite series of the form

$$\sum_{n=0}^{\infty} c_n(x-a)^n = c_0 + c_1(x-a) + c_2(x-a)^2 + \ldots, \tag{1}$$

where $c_0, c_1, \ldots$ are constants, called the **coefficients** of the series, a is a constant called the **center** of the series, and x is an independent variable. In particular, a power series centered at zero ($a = 0$) has the form

$$\sum_{n=0}^{\infty} c_n x^n = c_0 + c_1 x + c_2 x^2 + c_3 x^3 + \ldots. \tag{2}$$

Note that polynomials are also power series, since they have this form.

A series of form **(1)** can always be reduced to form **(2)** by the substitution $X = x - a$. This substitution is merely a translation of the coordinate system. It is easy to see that the behavior of Equation **(2)** near zero is exactly the same as the behavior of Equation **(1)** near a. For this reason we need only study the properties of series of form **(2).**

Every power series **(2)** has an **interval of convergence** consisting of all values x for which the series converges. The interval of convergence includes the interval $|x| < R$, where R is the **radius of convergence** of the power series **(2),** and *may* include one or both of the endpoints $x = \pm R$. The radius of convergence is often obtained* from the **ratio test:**

$$R = \lim_{n\to\infty} \left| \frac{c_n}{c_{n+1}} \right|,$$

when this limit exists. Suppose that $0 < R < \infty$. If $|x| < R$, the power series **(2)** **converges absolutely,** that is, $\Sigma_{n=0}^{\infty} |c_n||x|^n$ converges; if $|x| > R$, the series **(2)** **diverges.**

When $R = 0$, the interval of convergence consists only of $x = 0$; when $R = \infty$, the power series **(2)** converges (absolutely) for all x.

EXAMPLE 1 ▶ **A power series for which $R = 3$**

Determine the radius and interval of convergence for the series

$$\sum_{n=0}^{\infty} \frac{x^n}{(n + 1)3^n}.$$

Solution Here $c_n = 1/(n + 1)3^n$ so

$$R = \lim_{n\to\infty} \left| \frac{c_n}{c_{n+1}} \right| = \lim_{n\to\infty} \frac{\dfrac{1}{(n+1)3^n}}{\dfrac{1}{(n+2)3^{n+1}}} = \lim_{n\to\infty} \frac{(n+2)3^{n+1}}{(n+1)3^n}$$

$$= 3\lim_{n\to\infty} \frac{n+2}{n+1} = 3 \cdot 1 = 3.$$

Thus the radius of convergence is 3 and the series converges in $(-3, 3)$. If $x = 3$, the series is $\Sigma_{n=0}^{\infty} 1/(n + 1) = 1 + \frac{1}{2} + \frac{1}{3} + \ldots$ which is the divergent harmonic series. If $x = -3$, the series $\Sigma_{n=0}^{\infty} (-1)^n/(n + 1) = 1 - \frac{1}{2} + \frac{1}{3} - \frac{1}{4} + \ldots$ which converges by the alternating series test (it converges to ln 2). Thus the series converges for $-3 \leq x < 3$ and the interval of convergence is $[-3, 3)$. ◀

EXAMPLE 2 ▶ **A power series for which $R = 0$**

Determine the radius of convergence of $\Sigma_{n=0}^{\infty} n!x^n$.

Solution

$$R = \lim_{n\to\infty} \left| \frac{c_n}{c_n + 1} \right| = \lim_{n\to\infty} \frac{n!}{(n+1)!} = \lim_{n\to\infty} \frac{1}{n+1} = 0.$$

The series converges only for $x = 0$. ◀

EXAMPLE 3 ▶ **A power series for which $R = \infty$**

Determine the radius of convergence of $\Sigma_{n=0}^{\infty} x^n/n!$.

Solution

$$R = \lim_{n\to\infty} \left| \frac{c_n}{c_n + 1} \right| = \lim_{n\to\infty} \frac{\dfrac{1}{n!}}{\dfrac{1}{(n+1)!}} = \lim_{n\to\infty} \frac{(n+1)!}{n!} = \lim_{n\to\infty} (n + 1) = \infty.$$

* When the ratio test fails, the radius of convergence can often be obtained using the **root test.** See S. I. Grossman, *Calculus,* 5th ed. (Philadelphia: Harcourt, Brace, 1993), 679.

The series converges for all real x. As you have seen in your calculus course, the series converges to e^x. ◀

EXAMPLE 4 ▶ **A series which converges in (−1, 1)**

Determine the radius and interval of convergence for the series

$$\sum_{n=0}^{\infty} x^{2n} = 1 + x^2 + x^4 + \dots .$$

Solution Here $c_n = 1$ if n is even and $c_n = 0$ if n is odd. Thus $c_n/c_{n+1} = 0$ if n is odd or is undefined if n is even. The simplest thing to do in this case is to apply the ratio test directly. The ratio of successive terms is $x^{2n+2}/x^{2n} = x^2$ which is < 1 if $|x| < 1$ and is > 1 if $|x| > 1$. Thus the series converges in $(-1, 1)$ and $R = 1$. If $x = \pm 1$, then $x^2 = 1$ and

$$\sum_{n=0}^{\infty} x^{2n} = \sum_{n=0}^{\infty} 1 = 1 + 1 + 1 + \dots$$

which evidently diverges. Thus the interval of convergence is $(-1, 1)$. ◀

Every power series **(2)** represents a differentiable function

$$f(x) = \sum_{n=0}^{\infty} c_n x^n \tag{3}$$

in the interval $|x| < R$. Moreover, *the derivative $f'(x)$ is obtained by term-by-term differentiation*:

$$f'(x) = \left(\sum_{n=0}^{\infty} c_n x^n\right)' = \sum_{n=1}^{\infty} n c_n x^{n-1}, \tag{4}$$

and the termwise derivative $f'(x)$ has the same radius of convergence as the original series **(3).** The series in **(4)** may again be differentiated term by term to obtain the second derivative $f''(x)$, and this process may be repeated infinitely many times. Observe that $f(0) = c_0$, $f'(0) = c_1$, $f''(0) = 2!c_2$, and, in general, $f^{(n)}(0) = n!c_n$. Thus $c_n = f^{(n)}(0)/n!$, so **(3)** becomes the **Maclaurin series*** of $f(x)$:

$$f(x) = \sum_{n=0}^{\infty} \frac{f^{(n)}(0)}{n!} x^n. \tag{5}$$

Any function that has a Maclaurin series **(5),** or a **Taylor series,**†

$$f(x) = \sum_{n=0}^{\infty} \frac{f^{(n)}(a)}{n!} (x - a)^n, \tag{6}$$

which converges to $f(x)$ in the interval $|x| < R$ or $|x - a| < R$, is said to be **analytic** in the interval $|x| < R$, or $|x - a| < R$, respectively. The following Maclaurin series, analytic for all x, should be familiar:

$$e^x = \sum_{n=0}^{\infty} \frac{x^n}{n!} = 1 + x + \frac{x^2}{2!} + \frac{x^3}{3!} + \dots,$$

$$\sin x = \sum_{n=0}^{\infty} \frac{(-1)^n x^{2n+1}}{(2n+1)!} = x - \frac{x^3}{3!} + \frac{x^5}{5!} - \frac{x^7}{7!} + \dots,$$

* Named after the Scottish mathematician Colin Maclaurin (1698–1746).

† Named after the British mathematician Brook Taylor (1685–1731).

$$\cos x = \sum_{n=0}^{\infty} \frac{(-1)^n x^{2n}}{(2n)!} = 1 - \frac{x^2}{2!} + \frac{x^4}{4!} - \frac{x^6}{6!} + \dots,$$

$$\sinh x = \sum_{n=0}^{\infty} \frac{x^{2n+1}}{(2n+1)!} = x + \frac{x^3}{3!} + \frac{x^5}{5!} + \frac{x^7}{7!} + \dots,$$

$$\cosh x = \sum_{n=0}^{\infty} \frac{x^{2n}}{(2n)!} = 1 + \frac{x^2}{2!} + \frac{x^4}{4!} + \frac{x^6}{6!} + \dots.$$

There are functions $f(x)$ that have derivatives of all orders at a given point a and yet are not analytic. In these cases, the remainder term $R_N(x - a)$ of **Taylor's formula**

$$f(x) = \left[\sum_{n=0}^{N} \frac{f^{(n)}(a)}{n!}(x - a)^n\right] + R_N(x - a)$$

does not tend to zero, for all x in $|x - a| < R$, as N tends to infinity. An example of such a function is given in Problem 15.

The fundamental assumption made in solving a linear differential equation $f(x, y, y', y'', \dots) = 0$ by the power series method is that *the solution of the differential equation can be expressed in the form of a power series,* say,

$$y = \sum_{n=0}^{\infty} c_n x^n = c_0 + c_1 x + c_2 x^2 + \dots. \tag{7}$$

Power series expansions for $y', y'', \dots$ are then obtained by differentiating Equation **(7)** term by term:

$$y' = \sum_{n=1}^{\infty} n c_n x^{n-1} = c_1 + 2c_2 x + 3c_3 x^2 + \dots \tag{8}$$

$$y'' = \sum_{n=2}^{\infty} n(n-1) c_n x^{n-2} = 2c_2 + 3 \cdot 2c_3 x + 4 \cdot 3c_4 x^2 + \dots \tag{9}$$

$$\vdots$$

and these series are substituted into the given differential equation. Collecting the terms involving like powers of x, we then obtain an expression of the form

$$k_0 + k_1 x + k_2 x^2 + \dots = \sum_{n=0}^{\infty} k_n x^n = 0, \tag{10}$$

where the coefficients $k_0, k_1, k_2, \dots$ are expressions involving the unknown coefficients $c_0, c_1, c_2, \dots$. Since Equation **(10)** must hold for all values of x in the interval of convergence, all the coefficients $k_0, k_1, k_2, \dots$ must be zero. From the equations

$$k_0 = 0, \qquad k_1 = 0, \qquad k_2 = 0, \dots$$

we may be able to determine successively the coefficients $c_0, c_1, c_2, \dots$. To illustrate that the power series method does provide the required solution, in some cases, we solve three problems, two of which can be solved more easily by other methods.

EXAMPLE 5 ▶ **Solving an initial-value problem using power series**

Solve the initial-value problem

$$y' = y + x^2, \qquad y(0) = 1. \tag{11}$$

Solution Inserting Equations **(7)** and **(8)** into the differential equation, we have

$$c_1 + 2c_2x + 3c_3x^2 + 4c_4x^3 + \ldots = (c_0 + c_1x + c_2x^2 + c_3x^3 + \ldots) + x^2.$$

Collecting like powers of x yields

$$(c_1 - c_0) + (2c_2 - c_1)x + (3c_3 - c_2 - 1)x^2 + (4c_4 - c_3)x^3 + \ldots = 0.$$

Equating each of the coefficients to zero, we obtain the identities

$$c_1 - c_0 = 0, \qquad 2c_2 - c_1 = 0, \qquad 3c_3 - c_2 - 1 = 0,$$
$$4c_4 - c_3 = 0, \ldots,$$

from which we find that

$$c_1 = c_0, \qquad c_2 = \frac{c_1}{2} = \frac{c_0}{2!}, \qquad c_3 = \frac{c_2 + 1}{3} = \frac{c_0 + 2}{3!},$$
$$c_4 = \frac{c_3}{4} = \frac{c_0 + 2}{4!}, \ldots.$$

Substituting these values into Equation **(7)**, we obtain *the power series solution* to the differential equation in **(11)**:

$$y = c_0 + c_0x + \frac{c_0}{2!}x^2 + \frac{c_0 + 2}{3!}x^3 + \frac{c_0 + 2}{4!}x^4 + \frac{c_0 + 2}{5!}x^5 + \ldots.$$

Adding and subtracting $2[1 + x + x^2/2!]$, we can rewrite y as

$$y = (c_0 + 2)\left[1 + x + \frac{x^2}{2!} + \frac{x^3}{3!} + \frac{x^4}{4!} + \ldots\right] - 2\left[1 + x + \frac{x^2}{2!}\right].$$

To solve the initial-value problem, we set $x = 0$, obtaining

$$1 = y(0) = c_0 + 2 - 2 = c_0.$$

Thus the solution of the initial-value problem **(11)** is

$$y = 3\left[1 + x + \frac{x^2}{2!} + \frac{x^3}{3!} + \frac{x^4}{4!} + \ldots\right] - 2\left[1 + x + \frac{x_2}{2!}\right].$$

This example exhibits an unusual occurrence: looking carefully at the series in square brackets, we recognize the expansion for e^x, so we have the solution

$$y = 3e^x - x^2 - 2x - 2. \quad ◀$$

It is often difficult or impossible to recognize the power series solution of a differential equation as a known function. However, it is nevertheless a perfectly good solution which cannot be avoided. As we have noted before, many differential equations have solutions that cannot be written in terms of familiar functions.

Note

The equation $y' = y + x^2$ is a first-order, nonhomogeneous, linear differential equation with constant coefficients. It can be solved by the methods of Section 2.3. We solved it here to illustrate the power series method. The same comment applies to the equation in Example 6. Have patience. We will soon (starting with Example 7) solve differential equations for which the power series method provides the only way to get a closed form solution.

EXAMPLE 6 ▶ **Using power series to solve the equation of the harmonic oscillator**

Solve the differential equation

$$y'' + y = 0. \tag{12}$$

Solution Using Equations **(7)** and **(9),** we have

$$(2c_2 + 3 \cdot 2c_3x + 4 \cdot 3c_4x^2 + \ldots) + (c_0 + c_1x + c_2x^2 + \ldots) = 0.$$

Gathering like powers of x yields

$$(2c_2 + c_0) + (3 \cdot 2c_3 + c_1)x + (4 \cdot 3c_4 + c_2)x^2 + \ldots = 0.$$

Setting each of the coefficients to zero, we obtain

$$2c_2 + c_0 = 0, \quad 3 \cdot 2c_3 + c_1 = 0, \quad 4 \cdot 3c_4 + c_2 = 0,$$
$$5 \cdot 4c_5 + c_3 = 0, \ldots,$$

and

$$c_2 = -\frac{c_0}{2!}, \quad c_3 = -\frac{c_1}{3!}, \quad c_4 = -\frac{c_2}{4 \cdot 3} = \frac{c_0}{4!},$$
$$c_5 = -\frac{c_3}{5 \cdot 4} = \frac{c_1}{5!}, \ldots.$$

Substituting these values into the power series **(7)** for y yields

$$y = c_0 + c_1x - \frac{c_0}{2!}x^2 - \frac{c_1}{3!}x^3 + \frac{c_0}{4!}x^4 + \frac{c_1}{5!}x^5 + \ldots.$$

Splitting this series into two parts, we have the general power series solution to Equation **(12),**

$$y = c_0\left(1 - \frac{x^2}{2!} + \frac{x^4}{4!} - \ldots\right) + c_1\left(x - \frac{x^3}{3!} + \frac{x^5}{5!} - \ldots\right).$$

Here, again, we can recognize the power series: using the Maclaurin series for $\sin x$ and $\cos x$, we obtain the familiar general solution

$$y = c_0 \cos x + c_1 \sin x$$

to Equation **(12).** We observe that in this case the power series method produces two arbitrary constants c_0 and c_1. ◀

Remark

In the series solution of the first-order equation $y' = y + x^2$ (Example 5) our solution has one arbitrary constant (c_0). For the second-order equation $y'' + y = 0$ (Example 6) our series solution has two arbitrary constants (c_0 and c_1). This should not be surprising. From the basic existence-uniqueness theorem on page 105, we know that one initial condition is needed to obtain a unique solution for a first-order equation while two initial conditions are needed for uniqueness in a second-order equation. However, it *is* surprising that the use of power series has provided the general solution to $x'' + x = 0$ while the solutions obtained by use of the characteristic equations provides two particular solutions from which the general solution must be constructed.

So far we have considered only linear equations with constant coefficients. We turn now to linear equations with variable coefficients.

EXAMPLE 7 ▶ **Power series solutions of a differential equation with variable coefficients**

Solve the differential equation

$$y'' + xy' + y = 0. \tag{13}$$

Solution Using the power series method, we obtain the equation

$$\sum_{n=2}^{\infty} n(n-1)c_n x^{n-2} + x\sum_{n=1}^{\infty} nc_n x^{n-1} + \sum_{n=0}^{\infty} c_n x^n = 0. \tag{14}$$

We use the summation notation in this example in order to develop the skill in manipulating power series that will be required later on. *In order to gather all three power series into a single one, we need to rewrite each of the sums in Equation **(14)** so that the general term contains the same power of x.* Consider the first series:

$$T_1 = \sum_{n=2}^{\infty} n(n-1)c_n x^{n-2}.$$

To obtain the exponent k in place of $n - 2$, we make the substitution $k = n - 2$. Then $n = k + 2$, so every place we see an n, we replace it with $k + 2$. Since n ranges from 2 to ∞ and $k = n - 2$, k ranges from $2 - 2 = 0$ to ∞. Thus we have

$$T_1 = \sum_{k=0}^{\infty} (k+2)(k+1)c_{k+2}x^k.$$

(Annotations: $k = 0$ comes from $n = 2$; $(k+2)$ replaces n; $(k+1)$ replaces $n - 1 = (k + 2) - 1$; c_{k+2} replaces c_n; x^k replaces x^{n-2}.)

Next, we need to rewrite the second series so that the general term is x^k. Here

$$T_2 = x\sum_{n=1}^{\infty} nc_n x^{n-1} = \sum_{n=1}^{\infty} nc_n x \cdot x^{n-1} = \sum_{n=1}^{\infty} nc_n x^n = \sum_{k=1}^{\infty} kc_k x^k$$

(with $k = n$) if we set $k = n$. Similarly, setting $k = n$ in the third series, we have

$$T_3 = \sum_{n=0}^{\infty} c_n x^n = \sum_{k=0}^{\infty} c_k x^k.$$

(with $k = n$)

Thus Equation **(14)** becomes

$$\sum_{k=0}^{\infty} (k+2)(k+1)c_{k+2}x^k + \sum_{k=1}^{\infty} kc_k x^k + \sum_{k=0}^{\infty} c_k x^k = 0.$$

Note that the second sum can also be allowed to range from zero to infinity, since $kc_k x^k = 0$ when $k = 0$. (Had this not happened, we would have to treat the terms corresponding to the index $k = 0$ separately. See, for example, Example 4 in Section 5.2 in which we have to treat the case $k = 0$ separately.)

Gathering like terms in x produces the equation

$$\sum_{k=0}^{\infty} [(k+2)(k+1)c_{k+2} + (k+1)c_k]x^k = 0.$$

Setting the terms in brackets equal to zero, we obtain the general recursion formula

$$(k+2)(k+1)c_{k+2} + (k+1)c_k = 0.$$

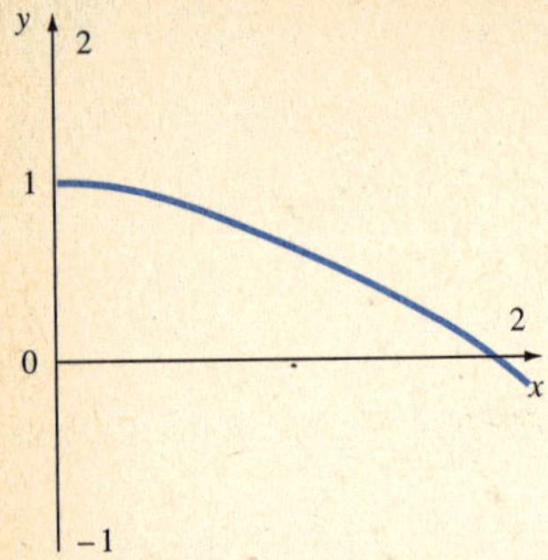

$y = 1 - \frac{x^2}{2} + \frac{x^4}{2\cdot 4} - \frac{x^4}{2\cdot 4\cdot 6}$

(a)

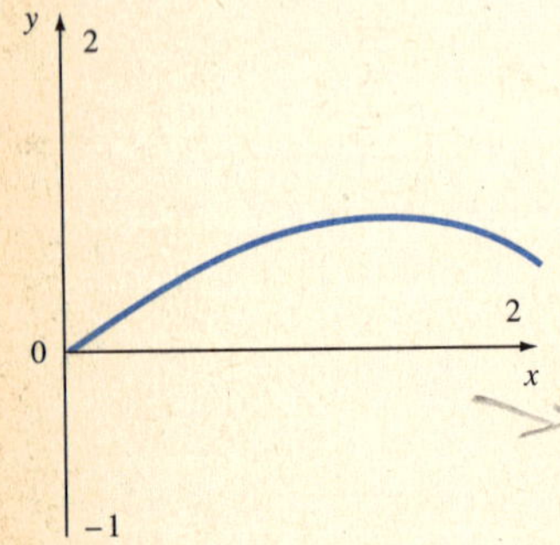

$y = x - \frac{x^3}{3} + \frac{x^5}{3\cdot 5} - \frac{x^7}{3\cdot 5\cdot 7}$

(b)

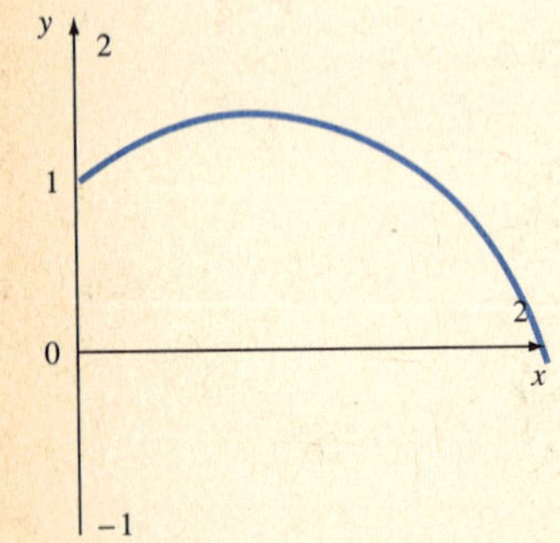

$y = \left(1 - \frac{x^2}{2} + \frac{x^4}{2\cdot 4} - \frac{x^6}{2\cdot 4\cdot 6}\right) + \left(x - \frac{x^3}{3} + \frac{x^5}{3\cdot 5} - \frac{x^7}{3\cdot 5\cdot 7}\right)$

(c)

FIGURE 5.1
Three approximations to solutions of Example 7

Therefore

$$(k+2)c_{k+2} = -c_k \quad \text{or} \quad c_{k+2} = -\frac{c_k}{k+2}$$

and

$$c_2 = -\frac{c_0}{2}, \qquad c_3 = -\frac{c_1}{3}, \qquad c_4 = -\frac{c_2}{4} = \frac{c_0}{2\cdot 4},$$

$$c_5 = -\frac{c_3}{5} = \frac{c_1}{3\cdot 5}, \qquad c_6 = -\frac{c_4}{6} = -\frac{c_0}{2\cdot 4\cdot 6}, \ldots.$$

Hence the power series for y can be written in the form

$$\begin{aligned} y &= c_0 + c_1 x - \frac{c_0}{2}x^2 - \frac{c_1}{3}x^3 + \frac{c_0}{2\cdot 4}x^4 + \frac{c_1}{3\cdot 5}x^5 - \ldots \\ &= c_0\left(1 - \frac{x^2}{2} + \frac{x^4}{2\cdot 4} - \frac{x^6}{2\cdot 4\cdot 6} + \ldots\right) \\ &\quad + c_1\left(x - \frac{x^3}{3} + \frac{x^5}{3\cdot 5} - \frac{x^7}{3\cdot 5\cdot 7} + \ldots\right) \end{aligned}$$

by separating the terms that involve c_0 and c_1.

In this example we cannot obtain closed form solutions to Equation **(13)**. However, we can use *truncations* of the power series above to obtain graphs which approximate the solutions. Figure 5.1(a) shows the graph of

$$y(x) = 1 - \frac{x^2}{2} + \frac{x^4}{2\cdot 4} - \frac{x^6}{2\cdot 4\cdot 6},$$

which approximates the solution to the initial-value problem

$$y'' + x\,y' + y = 0, \qquad y(0) = 1, \qquad y'(0) = 0. \qquad \textbf{Choose } c_0 = 1 \textbf{ and } c_1 = 0$$

Similarly, Figure 5.1(b) illustrates the graph of

$$y(x) = x - \frac{x^3}{3} + \frac{x^5}{3\cdot 5} - \frac{x^7}{3\cdot 5\cdot 7},$$

which approximates the solution to the initial-value problem

$$y'' + x\,y' + y = 0, \qquad y(0) = 0, \qquad y'(0) = 1, \qquad \textbf{Choose } c_0 = 0 \textbf{ and } c_1 = 1$$

while Figure 5.1(c) provides the graph of the function

$$y(x) = \left(1 - \frac{x^2}{2} + \frac{x^4}{2\cdot 4} - \frac{x^6}{2\cdot 4\cdot 6}\right) + \left(x - \frac{x^3}{3} + \frac{x^5}{3\cdot 5} - \frac{x^7}{3\cdot 5\cdot 7}\right),$$

which approximates the solution to the initial-value problem

$$y'' + x\,y' + y = 0, \qquad y(0) = 1, y'(0) = 1. \qquad \textbf{Choose } c_0 = c_1 = 1$$

The accuracy of these approximations depend on the domain of the graph. Suppose we are interested in graphing the solution over the interval $[0, c]$. If $c \le 0.1$, then we need not include very many of the terms in the power series, because the terms become very small. For example, the x^6 term is approximately 2×10^{-8} while the x^7 terms $\approx 10^{-9}$. If $c > 1$, then we may need to use many more terms in the power series to obtain an accurate graph: for example, the x^8 term exceeds 2.6×10^{-3}. ◀

Perspective The Taylor series method

The Taylor series and the derivatives of a differential equation can also be used to construct its power series solution.

Example 5 (revisited)
Consider the initial-value problem

$$y' = y + x^2, \qquad y(0) = 1.$$

Differentiating both sides of the differential equation repeatedly and successively evaluating each derivative at the initial value of $x = 0$, we have

$$\begin{aligned} y'(0) &= y + x^2|_{x=0} = y(0) + (0)^2 = 1, \\ y''(0) &= y' + 2x|_{x=0} = y'(0) + 2(0) = 1, \\ y'''(0) &= y'' + 2|_{x=0} = y''(0) + 2 = 3, \\ y^{(4)}(0) &= y'''|_{x=0} = y'''(0) = 3, \ldots . \end{aligned}$$

Substituting these derivatives in the Taylor series

$$y(x) = \sum_{n=0}^{\infty} \frac{y^{(n)}(a)}{n!}(x - a)^n$$

with $a = 0$, we have

$$\begin{aligned} y(x) &= 1 + x + \frac{x^2}{2!} + \frac{3x^3}{3!} + \frac{3x^4}{4!} + \ldots \\ &= (1 + 2) - 2 + (x + 2x) - 2x \\ &\quad + \left(\frac{x^2}{2!} + \frac{2x^2}{2!}\right) - x^2 + \frac{3x^3}{3!} + \frac{3x^4}{4!} + \ldots \\ &= 3\left(1 + x + \frac{x^2}{2!} + \frac{x^3}{3!} + \frac{x^4}{4!} + \ldots\right) - 2 - 2x - x^2 \\ &= 3e^x - 2 - 2x - x^2, \end{aligned}$$

which is the result that we obtained before. ◀

Example 7 (revisited)
From Theorem 3.1.2, we know that the homogeneous differential equation

$$y'' + xy' + y = 0 \tag{15}$$

has two linearly independent solutions y_1 and y_2 satisfying

$$y_1(0) = 1, \qquad y_1'(0) = 0, \quad \text{and} \quad y_2(0) = 0, \qquad y_2'(0) = 1.$$

Thus, differentiating repeatedly, we have

$$\begin{aligned} y'' &= -xy' - y, \\ y''' &= -xy'' - 2y', \\ y^{(4)} &= -xy''' - 3y'', \\ &\vdots \\ y^{(n+1)} &= -xy^{(n)} - ny^{(n-1)}. \end{aligned}$$

Using the initial conditions for the solution y_1, we can successively evaluate $y_1''(0), y_1'''(0), \ldots$:

$$\begin{aligned}
y_1''(0) &= -xy_1' - y_1|_{x=0} = -1, \\
y_1'''(0) &= -xy_1'' - 2y_1'|_{x=0} = 0, \\
y_1^{(4)}(0) &= -xy_1''' - 3y_1''|_{x=0} = 3, \\
y_1^{(5)}(0) &= -xy_1^{(4)} - 4y_1'''|_{x=0} = 0, \\
y_1^{(6)}(0) &= xy_1^{(5)} - 5y_1^{(4)}|_{x=0} = -15, \ldots,
\end{aligned}$$

so the Taylor series **(6)** (with $a = 0$) becomes

$$y_1(x) = 1 - \frac{x^2}{2!} + \frac{3x^4}{4!} - \frac{15x^5}{6!} + \ldots = 1 - \frac{x^2}{2} + \frac{x^4}{2 \cdot 4} - \frac{x^6}{2 \cdot 4 \cdot 6} + \ldots.$$

Similarly, using the initial conditions for y_2, we have successively

$$\begin{aligned}
y_2''(0) &= -xy_2' - y_2|_{x=0} = 0, \\
y_2'''(0) &= -xy_2'' - 2y_2'|_{x=0} = -2, \\
y_2^{(4)}(0) &= -xy_2^{(3)} - 3y_2''|_{x=0} = 0, \\
y_2^{(5)}(0) &= -xy_2^{(4)} - 4y_2^{(3)}|_{x=0} = 8, \ldots
\end{aligned}$$

yielding

$$y_2(x) = x - \frac{2x^3}{3!} + \frac{8x^5}{5!} - \ldots = x - \frac{x^3}{3} + \frac{x^5}{3 \cdot 5} - \ldots. \blacktriangleleft$$

These are the same two linearly independent power series solutions that were obtained in Example 7 using the power series method.

SELF-QUIZ

I. $1 + 2x + \frac{4x^2}{2!} + \frac{8x^3}{3!} + \frac{16x^4}{4!} + \ldots$ is the Maclaurin series for ______.

(a) e^x **(b)** e^{2x} **(c)** e^{x^2}
(d) $\cos x$ **(e)** $\cos 2x$

II. $1 - x + \frac{x^3}{3!} - \frac{x^5}{5!} + \ldots$ is the Maclaurin series for ______.

(a) $\sin x$ **(b)** $\cos x$ **(c)** $\sin(x - 1)$
(d) $\cos(x - 1)$ **(e)** $1 - \sin x$ **(f)** $1 - \cos x$

III. $1 - x^2 + \frac{x^4}{2!} - \frac{x^6}{3!} + \ldots$ is the Maclaurin series for ______.

(a) $\sin x$ **(b)** $\cos x$ **(c)** $\sin x^2$
(d) $\cos x^2$ **(e)** e^{x^2} **(f)** e^{-x^2}

IV. If $y = \sum_{n=0}^{\infty} c_n x^n$ is the solution to $y'' + xy' + y = 0$, $y(0) = 1$, $y'(0) = 1$, then $c_0 = 1$, $c_1 = 0$, and $c_2 =$ ______.

(a) 1 **(b)** -1 **(c)** $\frac{1}{2}$
(d) $-\frac{1}{2}$ **(e)** $\frac{2}{3}$ **(f)** $-\frac{2}{3}$

Answer True *or* False.

V. A function f is said to be analytic in $|x - a| < R$ if it has a convergent Taylor series in that interval.

Answers to Self-Quiz

I. b **II.** e **III.** f **IV.** d **V.** False (the series must converge to f)

PROBLEMS 5.1

In Problems 1–8 find the Taylor series centered at a and its corresponding interval of convergence for the given function.

1. $f(x) = e^x$, $a = 1$

2. $f(x) = \sin^2 x$, $a = 0$

3. $f(x) = \cos x$, $a = \dfrac{\pi}{4}$

4. $f(x) = \sinh x$, $a = \ln 2$

5. $f(x) = e^{bx}$, $a = -1$

6. $f(x) = xe^x$, $a = 1$

7. $f(x) = (x - 1)\ln x$, $a = 1$

8. $f(x) = \begin{cases} \dfrac{\sin x}{x}, & x \neq 0, \quad a = 0 \\ 1, & x = 0 \end{cases}$

In Problems 9–12 derive the given Taylor series.

9. $\ln(1 + x) = x - \dfrac{x^2}{2} + \dfrac{x^3}{3} - \dfrac{x^4}{4} + \dots, \quad |x| < 1$

10. $\sin^{-1}x = x + \dfrac{1}{2}\cdot\dfrac{x^3}{3} + \dfrac{1}{2}\cdot\dfrac{3}{4}\cdot\dfrac{x^5}{5} + \dfrac{1}{2}\cdot\dfrac{3}{4}\cdot\dfrac{5}{6}\cdot\dfrac{x^7}{7} + \dots, \quad |x| < 1$

11. $\ln x = (x - 1) - \dfrac{(x-1)^2}{2} + \dfrac{(x-1)^3}{3} - \dfrac{(x-1)^4}{4} + \dots, \quad 0 < x < 2$

12. $\dfrac{1}{2 - x} = 1 + (x - 1) + (x - 1)^2 + (x - 1)^3 + \dots, \quad 0 < x < 2$

13. Show that

$$\frac{1}{1 + x} = 1 - x + x^2 - x^3 + \dots, \quad |x| < 1.$$

Then prove that

(a) $\ln(1 + x) = x - \dfrac{x^2}{2} + \dfrac{x^3}{3} - \dfrac{x^4}{4} + \dots, \quad |x| < 1,$

(b) $\tan^{-1}x = x - \dfrac{x^3}{3} + \dfrac{x^5}{5} - \dfrac{x^7}{7} + \dots, \quad |x| < 1.$

(c) $\dfrac{1}{(1 + x)^2} = 1 - 2x + 3x^2 - 4x^3 + \dots, \quad |x| < 1.$

14. Show that the series

$$\sum_{n=1}^{\infty} \frac{x^n}{n} = x + \frac{x^2}{2} + \frac{x^3}{3} + \frac{x^4}{4} + \dots$$

diverges at $x = 1$ by proving that the partial sums satisfy the inequality

$$S_{2^k}(1) = \sum_{n=1}^{2^k} \frac{1}{n} \geq 1 + \frac{k}{2}.$$

15. Consider the function

$$f(x) = \begin{cases} e^{-1/x^2}, & x \neq 0, \\ 0, & x = 0. \end{cases}$$

(a) Show that f has derivatives of all orders at $x = 0$ and that

$$f'(0) = f''(0) = \dots = 0.$$

[*Hint*: Use the limit $f'(0) = \lim_{x\to 0} \frac{f(x) - f(0)}{x}$.]

(b) Conclude that $f(x)$ does not have a Taylor series expansion at $x = 0$, even though it is infinitely differentiable there. Thus f is not analytic at $x = 0$ (why?).

16. Using Taylor's formula, prove the **binomial formula**

$$(1 + x)^p = 1 + px + \frac{p(p-1)}{1\cdot 2}x^2 + \frac{p(p-1)(p-2)}{1\cdot 2\cdot 3}x^3 + \dots.$$

Where does it converge?

In Problems 17–24 solve the initial-value problem **(a)** *by the power series method and* **(b)** *by one of the techniques found in Chapters 2 or 3. Show that the two answers you obtain are the same.*

17. $y' = y + 1$, $y(0) = 1$

18. $y' = y + x$, $y(0) = 2$

19. $y' - 2y = x^2$, $y(1) = 1$

20. $y' - \dfrac{3}{x}y = x^3$, $y(1) = 4$

21. $y'' + 4y = 2x$, $y(0) = 0$, $y'(0) = 2$

22. $y'' + y' - 6y = 0$, $y(0) = 0$, $y'(0) = 5$

23. $y'' + y' = 3x^2$, $y(0) = 4$, $y''(0) = 0$

24. $y'' + y = 1 + x + x^2$, $y(0) = 1$, $y'(0) = -1$

In Problems 25–36 find the general power series solution of each equation by the power series method. When initial conditions are specified, give the solution that satisfies them. Then try to recognize a solution in terms of elementary functions (this will not *be possible in some problems).*

25. $(1 + x^2)y'' + 2xy' - 2y = 0$

26. $xy'' - xy' + y = e^x$, $y(0) = 1$, $y'(0) = 2$

27. $xy'' - x^2y' + (x^2 - 2)y = 0$, $y(0) = 0$, $y'(0) = 1$

28. $(1 - x)y'' - y' + xy = 0$, $y(0) = y'(0) = 1$

29. $y'' - 2xy' + 4y = 0$, $y(0) = 1$, $y'(0) = 0$

30. $(1 - x^2)y'' - xy' + y = 0$, $y(0) = 0$, $y'(0) = 1$

31. $y'' - xy' + y = -x\cos x$, $y(0) = 0$, $y'(0) = 2$

32. $y'' - xy' + xy = 0,\ y(0) = 2,\ y'(0) = 1$

33. $(1 - x)^2y'' - (1 - x)y' - y = 0,$
$y(0) = y'(0) = 1$

34. $y'' - 2xy' + 2y = 0$

35. $y'' - 2xy' - 2y = x,\ y(0) = 1,\ y'(0) = -\frac{1}{4}$

36. $y'' - x^2y = 0$

37. The **Airy equation,***

$$y'' - xy = 0,$$

has applications in the theory of diffraction. Find the general solution of this equation.

38. The **Hermite equation,**†

$$y'' - 2xy' + 2py = 0,$$

where p is constant, arises in quantum mechanics in connection with the Schrödinger‡ equation for a harmonic oscillator. Show that if p is a positive integer, one of the two linearly independent solutions of the Hermite equation is a polynomial, called the **Hermite polynomial** $H_p(x)$.

Use the Taylor series method to find the general solution of the differential equation in Problems 39–44.

39. $y'' - 2xy' + 2y = 0$

40. $x^2y'' - xy' + y = 0$, at $a = 1$

41. $x^2y'' + xy' - 4y = 0$, at $a = 1$

42. $x^2y'' - 2xy' + (x^2 + 2)y = 0$, at $a = 1$

43. $(1 - x^2)y'' - 2xy' + 6y = 0$

44. $(1 - x^2)y'' - 2xy' + 2y = 0$

5.2 ORDINARY AND SINGULAR POINTS

The power series method sometimes fails to yield a solution for one equation while working very well for an apparently similar equation.

EXAMPLE 1 ▶ Applying the power series method to three Euler equations

Solve the equations

$$x^2y'' + axy' + by = 0,^{\S} \tag{1}$$

for the following three sets of values for the coefficients a and b:

1. $a = -2,\ b = 2;$
2. $a = -1,\ b = 1;$
3. $a = 1,\ b = 1.$

Solution Set

$$y = \sum_{n=0}^{\infty} c_nx^n, \qquad y' = \sum_{n=1}^{\infty} nc_nx^{n-1}, \qquad y'' = \sum_{n=2}^{\infty} n(n-1)c_nx^{n-2}.$$

Then, since $n(n - 1) = 0$ at $n = 0$ and $n = 1$,

$$\begin{aligned} x^2y'' + axy' + by &= x^2\sum_{n=0}^{\infty} n(n-1)c_nx^{n-2} + ax\sum_{n=0}^{\infty} nc_nx^{n-1} + b\sum_{n=0}^{\infty} c_nx^n \\ &= \sum_{n=0}^{\infty} n(n-1)c_nx^n + \sum_{n=0}^{\infty} anc_nx^n + \sum_{n=0}^{\infty} bc_nx^n \\ &= \sum_{n=0}^{\infty} [n(n-1) + an + b]c_nx^n = 0. \end{aligned} \tag{2}$$

* Sir George Biddell Airy (1801–1892) was Lucasian Professor of Mathematics, director of the observatory, and Plumian Professor of Astronomy at Cambridge University in England until 1835. Then he was appointed director of the Greenwich Observatory (Astronomer royal). He remained there until his retirement in 1881. He did much work in lunar and solar photography, planetary motion, optics and other areas.

† Charles Hermite (1882–1901) was a French mathematician known for his contributions in algebra and number theory.

‡ Erwin Schrödinger (1887–1961) was an Austrian physicist. He was awarded the Nobel prize in 1933 (jointly with P. A. M. Dirac) for his work in quantum mechanics.

§ This is Euler's equation. We discussed equations of this type in Section 3.7.

1. Substitute $a = -2$ and $b = 2$ into Equation **(2)** to obtain

$$\sum_{n=0}^{\infty} (n^2 - 3n + 2)c_n x^n = 0.$$

Equating each of the coefficients to zero, we have

$$(n - 2)(n - 1)c_n = 0,$$

implying that $c_n = 0$ for all $n \neq 1$ or 2. Hence

$$y = c_1 x + c_2 x^2$$

is the general solution to Equation **(1)** with $a = -2$ and $b = 2$.

2. Equation **(2)** yields, with $a = -1$ and $b = 1$,

$$\sum_{n=0}^{\infty} (n^2 - 2n + 1)c_n x^n = 0,$$

so $(n - 1)^2 c_n = 0$. Thus $c_n = 0$ for all $n \neq 1$, yielding the solution

$$y = c_1 x.$$

Since the general solution of a second-order linear differential equation involves *two* linearly independent solutions, the power series method has given us only half of the general solution. We can use the method in Section 3.2 to find the other solution. Substitute $y = xv$ into Equation **(1)** (with $a = -1$ and $b = 1$) to obtain

$$x^2(xv)'' - x(xv)' + (xv) = 0,$$

or

$$x^3 v'' + x^2 v' = 0.$$

Setting $z = v'$, we obtain the first-order separable differential equation

$$x^3 \frac{dz}{dx} + x^2 z = 0.$$

Thus

$$\frac{dz}{z} = -\frac{dx}{x},$$

so $z = \frac{c}{x}$ and $v = c \ln|x|$. Hence the general solution to **(2)** is

$$y = Ax + Bx \ln|x|.$$

Observe that $\ln|x|$ is not defined at $x = 0$, so it is not possible to find a Maclaurin series for this solution.

3. Equation **(2)** gives us the series

$$\sum_{n=0}^{\infty} (n^2 + 1)c_n = 0.$$

Equating the coefficients to zero, we have $(n^2 + 1)c_n = 0$, so $c_n = 0$ for every n. Hence the power series method fails completely in helping us find the general solution

$$y = A\cos(\ln|x|) + B\sin(\ln|x|)$$

of the Euler Equation **(1)** (check!). ◀

To begin to understand the reason for this anomaly we use the following result. (The rest of the explanation will be given in Section 5.3.)

THEOREM 1* ♦ **The power series method works when the coefficient functions are analytic**

There is a unique Maclaurin series $y(x)$ satisfying the initial-value problem

$$y'' + a(x)y' + b(x)y = 0, \qquad y(0) = \alpha, \qquad y'(0) = \beta,$$

provided $a(x)$ and $b(x)$ can both be represented by Maclaurin series converging in an interval $|x| < R$. The power series $y(x)$ also converges in $|x| < R$. ♦

Note that this theorem guarantees the success of the power series method whenever $a(x)$ and $b(x)$ are analytic† at $x = 0$.

Definition Ordinary point and singular point

We call $x = 0$ an **ordinary point** of the differential equation

$$y'' + a(x)y' + b(x)y = 0 \tag{3}$$

when both $a(x)$ and $b(x)$ are analytic at $x = 0$. If $x = 0$ is not an ordinary point, it is called a **singular point** of the differential equation.

If we write each of the second-order homogeneous equations in Example 1 in the form of Equation **(3),** we obtain

$$y'' + \frac{a}{x}y' + \frac{b}{x^2}y = 0.$$

Neither of the terms $a(x) = a/x$ or $b(x) = b/x^2$ is defined at $x = 0$, and we have no power series representation for $a(x)$ or $b(x)$ that converges in an open interval containing $x = 0$.

Remark

As we just pointed out, Theorem 1 gives sufficient but not necessary conditions for the power series method to work. That is, if the conditions of Theorem 1 hold, then the differential equation has a solution $y(x)$ for $|x| < R$ that can be found using the power series method. However, the theorem does *not* state that if the conditions fail to hold, then the power solution method will not work.

Thus, Theorem 1 does not apply to Example 1 because $x = 0$ is a singular point of the differential Equation **(1).** This means that there is no guarantee that a series solution of Equation **(1)** exists; it may, or it may not.

---

*A proof may be found in E. A. Coddington, *An Introduction to Ordinary Differential Equations* (Englewood Cliffs, N.J.: Prentice-Hall, 1961), § 3.9.

†Recall that a real valued function f is analytic at x_0 if it can be expressed as a Taylor series centered at x_0 which converges to $f(x)$ for every x in an interval centered at x_0.

EXAMPLE 2 ▶ **A differential equation for which $x = 0$ is an ordinary point**

Determine whether Theorem 1 applies and obtain, if possible, a Maclaurin series solution to the differential equation

$$(1 - x^2)y'' - 2xy' + 2y = 0. \tag{4}$$

Solution Rewriting Equation **(4)** in the form

$$y'' - \frac{2x}{1-x^2}y' + \frac{2}{1-x^2}y = 0,$$

we observe that

$$a(x) = \frac{-2x}{1-x^2} = -2x(1 + x^2 + x^4 + \ldots)$$

and

$$b(x) = \frac{2}{1-x^2} = 2(1 + x^2 + x^4 + \ldots)$$

are Maclaurin series that converge in the interval $|x| < 1$. Hence $x = 0$ is an ordinary point of Equation **(4)** and the power series method will yield the general solution.

Note, however, that $x = \pm 1$ are singular points for Equation **(4)**, so that a Taylor series solution of Equation **(4)** centered at $x = \pm 1$ may not exist.

We now obtain the Maclaurin series solution of Equation **(4)** guaranteed by Theorem 1. Setting

$$y = \sum_{n=0}^{\infty} c_n x^n$$

and multiplying the power series representations of y, y', and y'' by the appropriate coefficients of Equation **(4)**, we get

$$(1 - x^2)\sum_{n=2}^{\infty} n(n-1)c_n x^{n-2} - 2x\sum_{n=1}^{\infty} nc_n x^{n-1} + 2\sum_{n=0}^{\infty} c_n x^n = 0.$$

But

$$(1 - x^2)\sum_{n=2}^{\infty} n(n-1)c_n x^{n-2}$$

$$= \sum_{n=2}^{\infty} n(n-1)c_n x^{n-2} - \sum_{n=2}^{\infty} n(n-1)c_n x^n$$

$k = n - 2$ ↓

$$= \sum_{k=0}^{\infty} (k+2)(k+1)c_{k+2}x^k - \sum_{n=2}^{\infty} n(n-1)c_n x^n$$

$n(n-1) = 0$ at $n = 0$ and $n = 1$ ↓

$$= \sum_{k=0}^{\infty} (k+2)(k+1)c_{k+2}x^k - \sum_{n=0}^{\infty} n(n-1)c_n x^n$$

$k = n$ ↓

$$= \sum_{n=0}^{\infty} (n+2)(n+1)c_{n+2}x^n - \sum_{n=0}^{\infty} n(n-1)c_n x^n.$$

Thus we obtain

$$\sum_{n=0}^{\infty} [(n+2)(n+1)c_{n+2} - n(n-1)c_n]x^n - \sum_{n=0}^{\infty} 2nc_nx^n + \sum_{n=0}^{\infty} 2c_nx^n = 0,$$

or

$$\begin{aligned}\sum_{n=0}^{\infty} &[(n+2)(n+1)c_{n+2} + (-n(n-1) - 2n + 2)c_n]x^n \\ &= \sum_{n=0}^{\infty} [(n+2)(n+1)c_{n+2} + (-n^2 - n + 2)c_n]x^n \\ &= \sum_{n=0}^{\infty} [(n+2)(n+1)c_{n+2} - (n+2)(n-1)c_n]x^n \\ &= \sum_{n=0}^{\infty} (n+2)[(n+1)c_{n+2} - (n-1)c_n]x^n = 0. \qquad (5)\end{aligned}$$

Every example involving power series involves algebraic manipulations like the ones above. Here we have provided every detail. In subsequent examples we leave some of the details to you.

We note that Equation **(5)** holds only if, for each n,

$$(n+1)c_{n+2} = (n-1)c_n,$$

or

$$c_{n+2} = \frac{n-1}{n+1}c_n.$$

Setting $n = 1$, we immediately see that $c_3 = 0$. Hence $c_3 = c_5 = c_7 = \cdots = 0$. If even values of n are chosen, we have

$$c_2 = \frac{0-1}{0+1}c_0 = -c_0, \qquad c_4 = \frac{2-1}{2+1}c_2 = \frac{1}{3}c_2 = -\frac{1}{3}c_0,$$

$$c_6 = \frac{3}{5}c_4 = -\frac{3}{5}\cdot\frac{1}{3}c_0 = -\frac{1}{5}c_0, \qquad c_8 = \frac{5}{7}c_6 = -\frac{1}{7}c_0, \ldots,$$

and, choosing c_1 arbitrarily, we find the general solution

$$y = c_1x + c_0\left(1 - x^2 - \frac{x^4}{3} - \frac{x^6}{5} - \cdots\right).$$

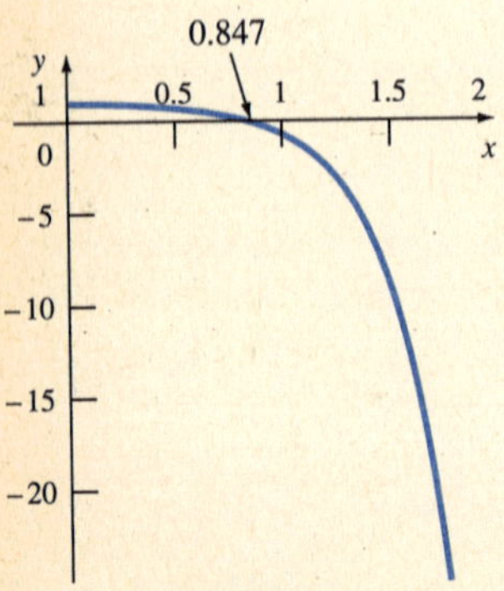

FIGURE 5.2
Graph of $f(x) = 1 - x^2 - x^4/3 - x^6/5 - x^8/7$

A graph of the function

$$y = 1 - x^2 - \frac{x^4}{3} - \frac{x^6}{5} - \frac{x^8}{7}$$

is given in Figure 5.2. This graph approximates the solution to the initial-value problem

$$(1 - x^2)y'' - 2xy' + 2y = 0, \qquad y(0) = 1,\ y'(0) = 0,$$

to three decimal places in the interval $[0, 0.5]$, but is off by more than 0.111 at $x = 1$ (why?). Note that since we start at $x = 0$, the graph is only given for $x > 0$. [However, because f is an even function, its graph for $x < 0$ is the reflection about the y-axis of the graph for $x > 0$.] ◄

EXAMPLE 3 ▶ **A differential equation for which $x = 0$ is a singular point**

Determine whether Theorem 1 applies and obtain, if possible, a Maclaurin series solution to the differential equation

$$y'' + \frac{2}{x}y' + y = 0. \tag{6}$$

Solution Here $a(x) = \frac{2}{x}$ which is not defined at $x = 0$. Thus $x = 0$ is a singular point of Equation **(6),** and we should anticipate the possibility of trouble in using the power series method. We have $xy'' + 2y' + xy = 0$ and

$$x\sum_{n=2}^{\infty} n(n-1)c_n x^{n-2} + 2\sum_{n=1}^{\infty} nc_n x^{n-1} + x\sum_{n=0}^{\infty} c_n x^n = 0,$$

or

$$\sum_{n=2}^{\infty} n(n-1)c_n x^{n-1} + \sum_{n=1}^{\infty} 2nc_n x^{n-1} + \sum_{n=0}^{\infty} c_n x^{n+1} = 0. \tag{7}$$

Setting $k = n - 1$, the first series is

$$\sum_{k=1}^{\infty} (k+1)kc_{k+1}x^k,$$

while the second is

taking out the first term

$$\sum_{k=0}^{\infty} 2(k+1)c_{k+1}x^k = 2c_1 + \sum_{k=1}^{\infty} 2(k+1)c_{k+1}x^k.$$

Finally, setting $k = n + 1$, the third series is

$$\sum_{k=1}^{\infty} c_{k-1}x^k$$

and Equation **(7)** becomes

$$2c_1 + \sum_{k=1}^{\infty} \{[(k+1)k + 2(k+1)]c_{k+1} + c_{k-1}\}x^k$$

$$= 2c_1 + \sum_{k=1}^{\infty} [(k+1)(k+2)c_{k+1} + c_{k-1}]x^k = 0.$$

Therefore $c_1 = 0$ and

$$c_{k+1} = -\frac{c_{k-1}}{(k+1)(k+2)}.$$

Thus $c_3 = c_5 = c_7 = \cdots = 0$. If odd values of k are chosen (so that $k + 1$ is even), we have

$$c_2 = -\frac{c_0}{3 \cdot 2} = -\frac{c_0}{3!}, \qquad c_4 = -\frac{c_2}{5 \cdot 4} = \frac{c_0}{5 \cdot 4 \cdot 3!} = \frac{c_0}{5!}, \ldots,$$

and we obtain one solution to Equation **(6):**

$$y = c_0\left(1 - \frac{x^2}{3!} + \frac{x^4}{5!} - \frac{x^6}{7!} + \cdots\right) = \frac{c_0}{x}\left(x - \frac{x^3}{3!} + \frac{x^5}{5!} - \cdots\right)$$

$$= c_0\frac{\sin x}{x}.$$

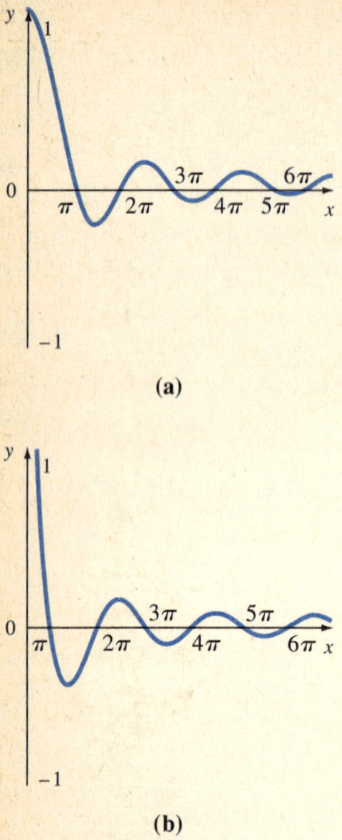

FIGURE 5.3
Graphs of **(a)** $y = x^{-1}\sin x$, **(b)** $y = x^{-1}\cos x$

However, the power series method does not yield the general solution. We can find that solution by the method in Example **(12)**. Setting $y = (v \sin x)/x$, we have, after some algebra,

$$(\sin x)v'' + 2(\cos x)v' = 0,$$

or, letting $z = v'$,

$$\sin x \frac{dz}{dx} = -2z \cos x.$$

Hence

$$\frac{dz}{z} = -\frac{2 \cos x \, dx}{\sin x},$$

or

$$\ln|z| = -2 \ln|\sin x| + c,$$

or

$$v' = \frac{k}{\sin^2 x} = k \csc^2 x.$$

Thus $v = -k \cot x$, so that another solution of Equation **(6)** is

$$y = \left(\frac{\sin x}{x}\right) \cot x = \left(\frac{\sin x}{x}\right) \frac{\cos x}{\sin x} = \frac{\cos x}{x}$$

and the general solution is

$$y = A\frac{\sin x}{x} + B\frac{\cos x}{x}.$$

Graphs of $y = x^{-1} \sin x$ and $y = x^{-1} \cos x$ are given in Figure 5.3. ◀

Note that 0 is a singular point in (b) since $x^{-1} \cos x$ is not defined at 0. However, 0 is not a singular point in (a) because $\lim_{x \to 0} (\sin x)/x = 1$ so that $x^{-1} \sin x$ can be extended to a new function that takes the value 1 at $x = 0$ and is equal to $(\sin x)/x$ for $x \neq 0$.

EXAMPLE 4 ▶ **A nonhomogeneous differential equation for which 0 is an ordinary point**

Find two linearly independent solutions to the equation

$$-y'' + xy' + y = x\sin x. \tag{8}$$

Solution Clearly $x = 0$ is an ordinary point of Equation **(8)**. We set

$$y = \sum_{n=0}^{\infty} c_n x^n.$$

Then

$$y' = \sum_{n=1}^{\infty} nc_n x^{n-1} \text{ so } xy' = \sum_{n=0}^{\infty} nc_n x^n$$

$$y'' = \sum_{n=2}^{\infty} n(n-1)c_n x^{n-2}$$

$$\overset{k = n-2}{\downarrow} \qquad\qquad \overset{n = k}{\downarrow}$$

$$= \sum_{k=0}^{\infty} (k+2)(k+1)c_{k+2} x^k = \sum_{n=0}^{\infty} (n+2)(n+1)c_{n+2} x^n.$$

Also, from page 197,

$$\sin x = \sum_{n=0}^{\infty} \frac{(-1)^n x^{2n+1}}{(2n+1)!}$$

so

$$x\sin x = \sum_{n=0}^{\infty} \frac{(-1)^n x^{2n+2}}{(2n+1)!} \overset{k=2n}{=} \sum_{\substack{k=0\\ k \text{ even}}}^{\infty} \frac{(-1)^{k/2} x^{k+2}}{(k+1)!} \overset{n=k+2}{=} \sum_{\substack{n=2\\ n \text{ even}}}^{\infty} \frac{(-1)^{(n-2)/2} x^n}{(n-1)!}.$$

We write the series $x\sin x$ in this somewhat unusual way so we will have like powers of x when we equate the series for the two sides of Equation **(8)**. That is, the "n's" on both sides of Equation **(8)** will represent the same numbers.

Inserting the series for y, xy', y'', and $x\sin x$ into Equation **(8)**, we obtain

$$-y'' + xy' + y = \sum_{n=0}^{\infty} c_n(n+1)x^n - \sum_{n=0}^{\infty} (n+2)(n+1)c_{n+2}x^n = \sum_{\substack{n=2\\ n \text{ even}}}^{\infty} \frac{(-1)^{(n-2)/2} x^n}{(n-1)!}. \quad \textbf{(9)}$$

There are three cases to consider:

Case 1: $n = 0$ Here Equation **(9)** yields

$$c_0 - 2c_2 = 0 \quad \text{or} \quad c_2 = \frac{c_0}{2}. \quad \textbf{(10)}$$

Case 2: n odd Here

$$c_n(n+1) - (n+2)(n+1)c_{n+2} = 0,$$

or

$$c_{n+2} = \frac{c_n}{n+2}. \quad \textbf{(11)}$$

Case 3: n even ($n \geq 2$) This is more complicated. We have

$$c_n(n+1) - (n+2)(n+1)c_{n+2} = \frac{(-1)^{(n-2)/2}}{(n-1)!}$$

or

$$c_{n+2} = \frac{-\dfrac{(-1)^{(n-2)/2}}{(n-1)!} + c_n(n+1)}{(n+2)(n+1)} = \frac{\dfrac{(-1)^{n/2}}{(n-1)!} + c_n(n+1)}{(n+2)(n+1)}. \quad \textbf{(12)}$$

From Equations **(11)** and **(12)** we see that c_0 and c_1 can be chosen arbitrarily. To find two linearly independent solutions we will first choose $c_0 = 1$ and $c_1 = 0$ and then choose $c_0 = 0$ and $c_1 = 1$. Other choices can be made but these are the easiest to work with.

First solution Since $c_1 = 0$, we see from Equation **(11)** that all odd-numbered terms are 0. We then compute

from Equation (10)

$$c_2 = \frac{c_0}{2} = \frac{1}{2}$$

$n = 2$

$$c_4 = \frac{\dfrac{(-1)^1}{1!} + c_2(3)}{(4)(3)} = \frac{-1 + \dfrac{3}{2}}{12} = \frac{1}{24}$$

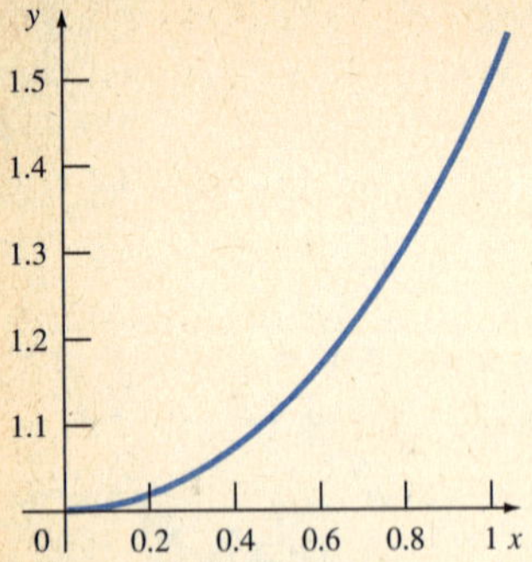

FIGURE 5.4 Graph of $y = 1 + x^2/2 + x^4/24 + x^6/80 + 19x^8/13{,}440$

$$c_6 \overset{n=4}{=} \frac{\dfrac{(-1)^2}{3!} + 5c_4}{(6)(5)} = \frac{\dfrac{1}{6} + \dfrac{5}{24}}{30} = \frac{1}{80}$$

$$c_8 \overset{n=6}{=} \frac{\dfrac{(-1)^3}{5!} + 7c_6}{(8)(7)} = \frac{\dfrac{-1}{120} + \dfrac{7}{80}}{56} = \frac{19}{13{,}440}$$

It would be difficult to obtain a nice, compact formula for the coefficients and so we stop here. The solution is

$$y_1(x) = 1 + \frac{x^2}{2} + \frac{x^4}{24} + \frac{x^6}{80} + \frac{19x^8}{13{,}440} + \cdots.$$

The graph of the first five terms of this solution is in Figure 5.4.

Second Solution $(c_0 = 0, c_1 = 1)$

odd terms:

$$c_3 \overset{n=1}{=} \frac{c_1}{1+2} = \frac{c_1}{3} = \frac{1}{3}$$

$$c_5 = \frac{c_3}{5} = \frac{1}{3 \cdot 5} = \frac{1}{15}$$

$$c_7 = \frac{c_5}{7} = \frac{1}{3 \cdot 5 \cdot 7} = \frac{1}{105}$$

$$\vdots$$

even terms:

$$c_0 = 0$$

$$c_2 = \frac{c_0}{2} = 0$$

$$c_4 = \frac{-1 + 3c_2}{12} = -\frac{1}{12}$$

$$c_6 = \frac{\dfrac{1}{6} + 5c_4}{30} = \frac{\dfrac{1}{6} - \dfrac{5}{12}}{30} = -\frac{1}{120}$$

$$c_8 = \frac{-\dfrac{1}{120} + 7c_6}{56} = \frac{-\dfrac{1}{120} - \dfrac{7}{120}}{56} = -\frac{1}{840}$$

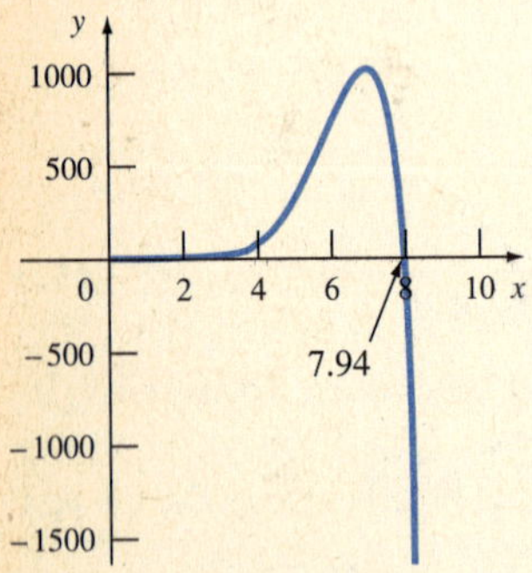

FIGURE 5.5 Graph of $y = x + x^3/3 - x^4/12 + x^5/15 - x^6/120 + x^7/105 - x^8/840$

We stop here to obtain

$$y_2(x) = x + \frac{x^3}{3} - \frac{x^4}{12} + \frac{x^5}{15} - \frac{x^6}{120} + \frac{x^7}{105} - \frac{x^8}{840} + \cdots.$$

A graph of the first seven terms of this series is given in Figure 5.5. ◀

At this point it is worth a reminder that solutions by power series, like those obtained in Example 4, are very useful—even if they are not very pretty. If you have access to a computer then you can easily program it to obtain any number of coefficients—using the recursion formulas **(11)** and **(12).** All you have to tell the computer are the values of $c_0 = y(0)$ and $c_1 = y'(0)$. Then the computer can draw graphs and give you values of $y(x)$ at any desired value of x to as many decimal places of accuracy as you need. Writing the solution in a closed form—as you did in Chapter 3—may be more aesthetically pleasing but, with a computer, you can get the same information using the power series method.

Remark

In all the examples in the first two sections of this chapter, we sought a power series of the form $\Sigma_{n=0}^{\infty} c_n x^n$. These are power series around $x_0 = 0$. However, it is sometimes useful (and necessary) to look for power series of the form $\Sigma_{n=0}^{\infty} c_n(x - x_0)^n$, where $x_0 \neq 0$. For example, we saw in Example 1 **(3)** that the power series method failed to yield a solution to the differential equation $y'' + (1/x)y' + (1/x^2)y = 0$. Here both $1/x$ and $1/x^2$ can be written as power series of the form $\Sigma_{n=0}^{\infty} c_n(x - 1)^n$ with radius of convergence 1. [$1/x = \Sigma_{n=0}^{\infty} (-1)^n(x - 1)^n$ which converges for $0 < x < 2$.] Then the power series method *will* yield a solution of the form $y(x) = \Sigma_{n=0}^{\infty} d_n(x - 1)^n$ which converges for $|x - 1| < 1$ $(0 < x < 2)$.

SELF-QUIZ

Answer True *or* False.

I. $x = 0$ is an ordinary point of $xy'' + y' + y = 0$.

II. $x = 0$ is an ordinary point of $y'' + xy' + y = 0$.

III. $x = 0$ is an ordinary point of $2y'' + 5y' + \ln(1 - x)y = 0$.

IV. $x = 0$ is an ordinary point of $\ln(1 - x)y'' + 3y' - 6y = 0$.

V. The power series method is guaranteed to yield the solution to the equation

$$(\sin x)y'' + (\cos x)y' - 3y = 0, \quad y(0) = 1, \quad y'(0) = 0.$$

VI. The power series method is guaranteed to yield the solution to

$$(\cos x)y'' + (\sin x)y' - 3y = 0, \quad y(0) = 1, \quad y'(0) = 0.$$

VII. If $x = 0$ is an ordinary point for the equation $y'' + a(x)y' + b(x)y = 0$, then the power series method is guaranteed to produce two linearly independent series solutions.

Answers to Self-Quiz

I. False **II.** True **III.** True **IV.** False **V.** False **VI.** True **VII.** True

PROBLEMS 5.2

In Problems 1–16 find two linearly independent power series about the ordinary point $x = 0$ that are solutions to the given differential equation.

1. $y'' - xy' + y = 0$

2. $y'' - xy' + 2y = 0$

3. $y'' - 3xy = 0$

4. $y'' + 4xy' - y = 0$

5. $y'' - xy' + xy = 0$

6. $y'' - 2x^2y = 0$

7. $y'' + x^2y' + 2xy = 0$

8. $y'' + x^2y' - xy = 0$

9. $(1 + x^2)y'' + 2xy' - 2y = 0$

10. $(2 + x)y'' - y' + xy = 0$

11. $(1 - x)^2y'' - (1 - x)y' - y = 0$

12. $(x^2 + 1)y'' - 6y = 0$

13. $(2x^2 + 1)y'' + 2xy' - 18y = 0$

14. $(x^2 + 2)y'' + 3xy' + y = 0$

15. $y'' - xy' + y = -x \cos x$

16. $y'' - 2xy' - 2y = x$

17. Find a solution to the initial-value problem

$$y''' - xy = 0, \qquad y(0) = 1, \qquad y'(0) = 0, \qquad y''(0) = 0.$$

18. Solve the **Airy equation**

$$y'' - xy = 0, \qquad y(1) = 1, \qquad y'(1) = 0.$$

19. Solve the initial-value problem

$$y'' - xy' - y = 0, \qquad y(0) = 1, \qquad y'(0) = 0.$$

In Problems 20–23 find the first four nonzero terms in the power series solution to the given initial-value problem.

20. $y'' + (\sin x)y = 0,\ y(0) = 1,\ y'(0) = 0$

21. $y'' - e^x y = 0,\ y(0) = y'(0) = 1$

22. $y'' + (\cos x)y = 0,\ y(0) = 1,\ y'(0) = 0$

23. $y'' + (\cos x)y = 0,\ y(0) = 0,\ y'(0) = 1$

In Problems 24–32 use the power series method to obtain at least one solution about the singular point $x = 0$. Then use the method of reduction of order (Section 3.2) to find the general solution.

24. $x^2y'' + 2xy' - 2y = 0$

25. $xy'' + (1 - 2x)y' - (1 - x)y = 0$

26. $xy'' + 2y' - xy = 0$

27. $x^2y'' + x(x - 1)y' - (x - 1)y = 0$

28. $xy'' + (1 - x)y' - y = 0$

29. $xy'' - (1 - x)y' - 2y = 0$

30. $x(x - 1)y'' - (1 - 3x)y' + y = 0$

31. $x(x - 1)y'' + 3y' - 2y = 0$

32. $x^2y'' - x(1 - x)y' + y = 0$

33. Does the power series method (at $x = 0$) yield a solution to the equation
(a) $x^2y' = y$?
(b) $x^3y' = y$?

34. Show that the power series method fails at $x = 0$ for

$$x^2y'' + x^2y' + y = 0$$

35. Show that the power series method fails at $x = 0$ for

$$x^3y'' + xy' + y = 0$$

36. Show that the power series method fails at $x = 0$ for

$$x^4y'' + 2x^3y' - y = 0$$

5.3 THE METHOD OF FROBENIUS: CASE 1

If we look at the solution of Example 5.2.3 (p. 211), we see that the solution $(\cos x)/x$ was not obtained by the power series method. In fact, $(\cos x)/x$ cannot be written as a power series in x. However, it *can* be written as a power series in x times a power of x:

$$\frac{\cos x}{x} = x^{-1}\left(1 - \frac{x^2}{2!} + \frac{x^4}{4!} - \cdots\right).$$

This suggests that we should try to find solutions of the form

$$y = x^r(c_0 + c_1x + c_2x^2 + c_3x^3 + \cdots), \tag{1}$$

where r is some real or complex number, whenever $x = 0$ is a singular point of the differential equation. For one class of singular points, this modification of the power series method does yield solutions.*

Definition Regular singular point

We call $x = 0$ a **regular singular point** of the differential equation

$$y'' + a(x)y' + b(x)y = 0 \tag{2}$$

if both the functions $xa(x)$ and $x^2b(x)$ have convergent Maclaurin series in an open interval containing $x = 0$.

* But again, we point out that it will usually be impossible to associate a known function with the series you have generated.

Observe that $x = 0$ is a regular singular point for the differential equation in Example 5.2.3 because $xa(x) = 2$ and $x^2b(x) = x^2$, both of which are convergent Maclaurin series (for all x) with only one nonzero coefficient.

Definition Irregular point

A singular point that is not regular is called **irregular**.

EXAMPLE 1 ▶ **Two differential equations with irregular singular points**

The point $x = 0$ is an irregular singular point of the two equations

$$y'' + \frac{1}{x^2}y' + y = 0$$

and

$$y'' + \frac{1}{x}y' + \frac{1}{x^3}y = 0$$

because in the first case $xa(x) = x^{-1}$, while in the second case $x^2b(x) = x^{-1}$. The function of x^{-1} is not defined at $x = 0$, so it cannot have a convergent Maclaurin series. ◀

The method of Frobenius

To simplify the explanation of the modified power series method, called the **method of Frobenius,*** we assume that $x = 0$ is a regular singular point of the equation

$$y'' + a(x)y' + b(x)y = 0 \tag{3}$$

and that Equation **(3)** has a solution of the form

$$y = x^r(c_0 + c_1x + c_2x^2 + \ldots) = \sum_{n=0}^{\infty} c_nx^{r+n}, \qquad x > 0, \tag{4}$$

where r is some real or complex number. We can assume that $c_0 = 1$, since any constant multiple of a solution is again a solution of the homogeneous differential equation. In addition, the choice $c_0 = 1$ simplifies much of the following discussion. The restriction $x > 0$ is necessary to prevent difficulties for certain values of r, such as $r = \frac{1}{2}$ and $-\frac{1}{4}$, since we are not interested in imaginary solutions. [If we need to find a solution valid for $x < 0$, we can change variables by substituting $X = -x$ into Equation **(3)** and solve the resulting equation for $X > 0$.]

Since

$$y' = \sum_{n=0}^{\infty} c_n(r + n)x^{r+n-1}$$

and

$$y'' = \sum_{n=0}^{\infty} c_n(r + n)(r + n - 1)x^{r+n-2},$$

* See the Historical Note on page 223.

Equation **(3)** can be rewritten as

$$\sum_{n=0}^{\infty} c_n(r+n)(r+n-1)x^{r+n-2} + a(x)\sum_{n=0}^{\infty} c_n(r+n)x^{r+n-1}$$
$$+ b(x)\sum_{n=0}^{\infty} c_n x^{r+n} = 0,$$

or, factoring an x and an x^2 in the second and third series, respectively, we obtain

$$\sum_{n=0}^{\infty} c_n[(r+n)(r+n-1) + (r+n)xa(x) + x^2b(x)]x^{r+n-2} = 0. \qquad \textbf{(5)}$$

As $x = 0$ is a regular singular point, both $xa(x)$ and $x^2b(x)$ can be expressed as convergent power series in x:

$$xa(x) = a_0 + a_1x + a_2x^2 + \ldots,$$
$$x^2b(x) = b_0 + b_1x + b_2x^2 + \ldots.$$

But $n \geq 0$, so x^{r-2} is the smallest power of x in Equation **(5)**. Since the coefficients of a power series whose sum is zero must vanish, we have, for $n = 0$,

$$c_0[r(r-1) + a_0r + b_0] = 0.$$

Definition Indicial equation

By hypothesis $c_0 = 1$, so we obtain the **indicial equation**

$$r(r-1) + a_0r + b_0 = 0, \qquad \textbf{(6)}$$

whose roots, r_1 and r_2, are called the **exponents** of the differential equation **(3)**.

In what follows we see that one of the solutions of Equation **(3)** is always of form **(4)** and that there are three possible forms for the second linearly independent solution corresponding to the following cases:

Case 1: r_1 and r_2 differ but not by an integer.
Case 2: $r_1 = r_2$.
Case 3: r_1 and r_2 differ by a nonzero integer.

We shall treat Case 1 in the remainder of this section and defer Cases 2 and 3 to Section 5.4.

THEOREM 1 ♦ **Form of solution for case 1**

Let $x = 0$ be a regular singular point of the differential equation

$$y'' + a(x)\,y' + b(x)\,y = 0, \qquad x \neq 0,$$

and suppose the roots of the indicial equation r_1 and r_2 differ, but not by an integer. Then, Frobenius's method will yield two solutions, for $x > 0$, of the forms

$$y_1(x) = x^{r_1}(c_0 + c_1x + c_2x^2 + \ldots), \qquad c_0 = 1,$$
$$y_2(x) = x^{r_2}(c_0^* + c_1^*x + c_2^*x^2 + \ldots), \qquad c_0^* = 1.$$

That y_1 and y_2 are linearly independent follows easily from the fact that y_1/y_2 cannot be constant, since if it were, the roots r_1 and r_2 would coincide. The coefficients c_1, c_2, . . . are obtained by setting the coefficients of each power of x equal to zero in Equation **(5)**. We find c_1^*, c_2^*, ... in a similar manner. The procedure is demonstrated in the following two examples. ♦

Remark

This theorem should clarify the terms *regular* and *irregular* singular points. For a regular singular point, Frobenius's method works. For an irregular singular point, Frobenius's method does not work in general (although it does work in some special cases).

EXAMPLE 2 ▶ **A regular singular point where the indicial exponents differ by $\frac{1}{2}$**

Solve the equation

$$x\,y'' + \frac{1}{2}y' - y = 0, \qquad x > 0. \tag{7}$$

Solution We substitute

$$y = x^r(c_0 + c_1x + c_2x^2 + \ldots) = x^r \sum_{n=0}^{\infty} c_n x^n = \sum_{n=0}^{\infty} c_n x^{r+n}$$

into Equation **(7)** to obtain

$$\sum_{n=0}^{\infty}(r+n)(r+n-1)c_nx^{r+n-1} + \frac{1}{2}\sum_{n=0}^{\infty}(r+n)c_nx^{r+n-1}$$
$$-\sum_{n=0}^{\infty}c_nx^{r+n} = \sum_{n=0}^{\infty}(r+n)(r+n-1)\,c_nx^{r+n-1} + \frac{1}{2}\sum_{n=0}^{\infty}(r+n)c_nx^{r+n-1}$$

set $k = n + 1$ and then set $n = k$

$$-\sum_{n=0}^{\infty}c_{n-1}x^{r+n-1} = c_0\underbrace{\left[r(r-1) + \frac{1}{2}r\right]}_{\text{indicial equation}}x^{r-1}$$

$$+\sum_{n=1}^{\infty}\left\{(r+n)\left[(r+n-1)+\frac{1}{2}\right]c_n - c_{n-1}\right\}x^{r+n-1} = 0. \tag{8}$$

The indicial equation is $r^2 - \frac{1}{2}r = r(r - \frac{1}{2}) = 0$, with roots $r = 0$ and $r = 1/2$ that do not differ by an integer. First let $r = 0$, assuming a solution of the form

$$y = \sum_{n=0}^{\infty} c_n x^n;$$

then Equation **(8)** becomes (since $r = 0$)

$$\sum_{n=1}^{\infty}\left\{n\left[(n-1)+\tfrac{1}{2}\right]c_n - c_{n-1}\right\}x^{n-1} = 0.$$

This leads to the recurrence equation $n(n - \frac{1}{2})c_n = c_{n-1}$, or

$$c_n = \frac{c_{n-1}}{n(n-\frac{1}{2})}, \qquad \text{for } n \geq 1.$$

Then

$$c_1 = \frac{2c_0}{1!}, \qquad c_2 = \frac{c_1}{2 \cdot \frac{3}{2}} = \frac{2c_1}{2 \cdot 3} = \frac{2^2 c_0}{2!\,(1 \cdot 3)},$$

$$c_3 = \frac{2c_2}{3 \cdot 5} = \frac{2^3 c_0}{3!\,(1 \cdot 3 \cdot 5)}, \qquad c_4 = \frac{2c_3}{4 \cdot 7} = \frac{2^4 c_0}{4!\,(1 \cdot 3 \cdot 5 \cdot 7)}, \ldots$$

so that

$$\begin{aligned} y_1(x) &= c_0 + \frac{2c_0}{1!}x + \frac{2^2 c_0}{2!\,(1 \cdot 3)}x^2 + \frac{2^3 c_0}{3!\,(1 \cdot 3 \cdot 5)}x^3 \\ &\quad + \frac{2^4 c_0}{4!\,(1 \cdot 3 \cdot 5 \cdot 7)}x^4 + \ldots \\ &= c_0\left(\sum_{n=0}^{\infty} \frac{(2x)^n}{n!\,(1 \cdot 3 \cdot \ldots \cdot (2n-1))}\right) \\ &= c_0\left(\sum_{n=0}^{\infty} \frac{2^n(2x)^n}{(2 \cdot 4 \cdot \ldots \cdot 2n)(1 \cdot 3 \cdot \ldots \cdot (2n-1))}\right) \\ &= c_0\left(\sum_{n=0}^{\infty} \frac{(4x)^n}{(2n)!}\right) = c_0\left(\sum_{n=0}^{\infty} \frac{(2\sqrt{x})^{2n}}{(2n)!}\right) = c_0 \cosh(2\sqrt{x}). \end{aligned}$$

The last equation follows from the Maclaurin series for $\cosh x$ on page 198. Next, we set $r = \frac{1}{2}$ in Equation **(8)** obtaining

$$\sum_{n=1}^{\infty} \left\{ n\left(n + \frac{1}{2}\right)c_n - c_{n-1} \right\} x^{n-1/2} = 0.$$

Thus

$$c_n = \frac{c_{n-1}}{n\left(n + \frac{1}{2}\right)},$$

so that

$$c_1 = \frac{2c_0}{1 \cdot 3}, \qquad c_2 = \frac{2c_1}{2 \cdot 5} = \frac{2^2 c_0}{2!\,(1 \cdot 3 \cdot 5)},$$

$$c_3 = \frac{2c_2}{3 \cdot 7} = \frac{2^3 c_0}{3!\,(1 \cdot 3 \cdot 5 \cdot 7)}, \qquad c_4 = \frac{2c_3}{4 \cdot 9} = \frac{2^4 c_0}{4!\,(1 \cdot 3 \cdot 5 \cdot 7 \cdot 9)}, \ldots$$

and

$$\begin{aligned} y_2(x) &= \sqrt{x}\left(c_0 + \frac{2c_0}{1 \cdot 3}x + \frac{2^2 c_0}{2!\,(1 \cdot 3 \cdot 5)}x^2 + \frac{2^3 c_0}{3!\,(1 \cdot 3 \cdot 5 \cdot 7)}x^3 \right. \\ &\quad \left. + \frac{2^4 c_0}{4!\,(1 \cdot 3 \cdot 5 \cdot 7 \cdot 9)}x^4 + \ldots\right) \\ &= c_0\sqrt{x}\left(\sum_{n=0}^{\infty} \frac{(2x)^n}{n!\,(1 \cdot 3 \cdot \ldots \cdot (2n+1))}\right) \\ &= c_0\sqrt{x}\left(\sum_{n=0}^{\infty} \frac{2^n(2x)^n}{(2 \cdot 4 \cdot \ldots \cdot 2n)(1 \cdot 3 \cdot \ldots \cdot (2n+1))}\right) \\ &= \frac{c_0}{2}\left(\sum_{n=0}^{\infty} \frac{(2\sqrt{x})^{2n+1}}{(2n+1)!}\right) = \frac{c_0}{2}\sinh(2\sqrt{x}). \end{aligned}$$

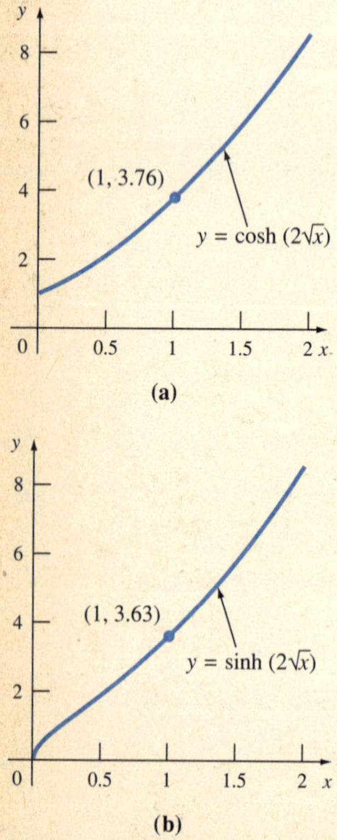

FIGURE 5.6
Graphs of $\cosh(2\sqrt{x})$ and $\sinh(2\sqrt{x})$

The last equation follows from the Maclaurin series for $\sinh x$ on page 198. Hence, the general solution of Equation **(7)** is

$$y = A\cosh(2\sqrt{x}) + B\sinh(2\sqrt{x}).$$

In Figure 5.6 we provide graphs of $\cosh(2\sqrt{x})$ and $\sinh(2\sqrt{x})$ for $x > 0$. ◀

The roots of the indicial equation may also be complex, as the following example illustrates.

EXAMPLE 3 ▶ **A regular singular point with complex exponents**

Find the general solution of the equation

$$y'' + \frac{1}{x}y' + \frac{1}{x^2}y = 0, \qquad x > 0. \tag{9}$$

Solution Substitute $y = \Sigma_{n=0}^{\infty} c_n x^{r+n}$ into Equation **(9)** to obtain

$$\sum_{n=0}^{\infty} c_n[(r+n)(r+n-1) + (r+n) + 1]x^{r+n-2} = 0. \tag{10}$$

The indicial equation (obtained by setting $n = 0$ in the expression in brackets) is

$$r(r-1) + r + 1 = r^2 + 1 = 0,$$

with the roots $r_1 = i$, $r_2 = -i$,[†] which do not differ by an integer. Setting $r = i$ in Equation **(10)**, we have

$$\sum_{n=0}^{\infty} c_n[(i+n)(i+n-1) + (i+n) + 1]\,x^{i+n-2} = 0.$$

Equating all coefficients of this series to zero, we have, after combining terms, and using the fact that $(i+n)^2 = i^2 + 2in + n^2 = n^2 + 2in - 1$,

$$0 = c_n[(i+n)^2 + 1] = c_n(n^2 + 2in) = c_n n(n + 2i),$$

which holds only if $c_n = 0$ for $n > 0$. Note that

$$e^{\ln u} = u \qquad \text{and that} \qquad e^{iu} = \cos u + i\sin u,$$

for any $u > 0$. This suggests that

$$x^i = e^{\ln x^i} = e^{i\ln x} = \cos\ln x + i\sin\ln x.$$

Thus, setting $c_0 = 1$ (any constant gives us a solution),

$$y_1(x) = x^i = e^{i\,(\ln x)} = [\cos(\ln x) + i\sin(\ln x)].$$

Similarly, substituting $r = -i$ into Equation **(10)** yields the series

$$\sum_{n=0}^{\infty} c_n^*[(-i+n)(-i+n-1) + (-i+n) + 1]x^{-i+n-2} = 0,$$

whose coefficients satisfy the condition

$$c_n^*\,[n^2 - 2in] = 0.$$

Thus $c_n^* = 0$ for $n > 0$. Hence, setting $c_0^* = 1$,

$$y_2(x) = x^{-i} = [\cos(\ln x) - i\sin(\ln x)].$$

[†] We remind you that a review of properties of complex numbers is given in Appendix 5.

Finally, since linear combinations of solutions are solutions, the real and imaginary parts of y_1 and y_2,

$$y_1^*(x) = \frac{1}{2}(y_1 + y_2) = \cos(\ln x),$$

$$y_2^*(x) = \frac{1}{2i}(y_1 - y_2) = \sin(\ln x)$$

are solutions of Equation **(9)**. That y_1^* and y_2^* are linearly independent follows since $y_2^*/y_1^* = \tan(\ln x)$, which is nonconstant. Hence Equation **(9)** has the general solution (see Example 5.2.1 **(3)**, p. 208)

$$y = A\cos(\ln x) + B\sin(\ln x), \qquad x > 0.$$

Graphs of $y_1 = \sin(\ln x)$, $x > 0$ are given in Figure 5.7. The graph of $\cos(\ln x)$ is similar. ◀

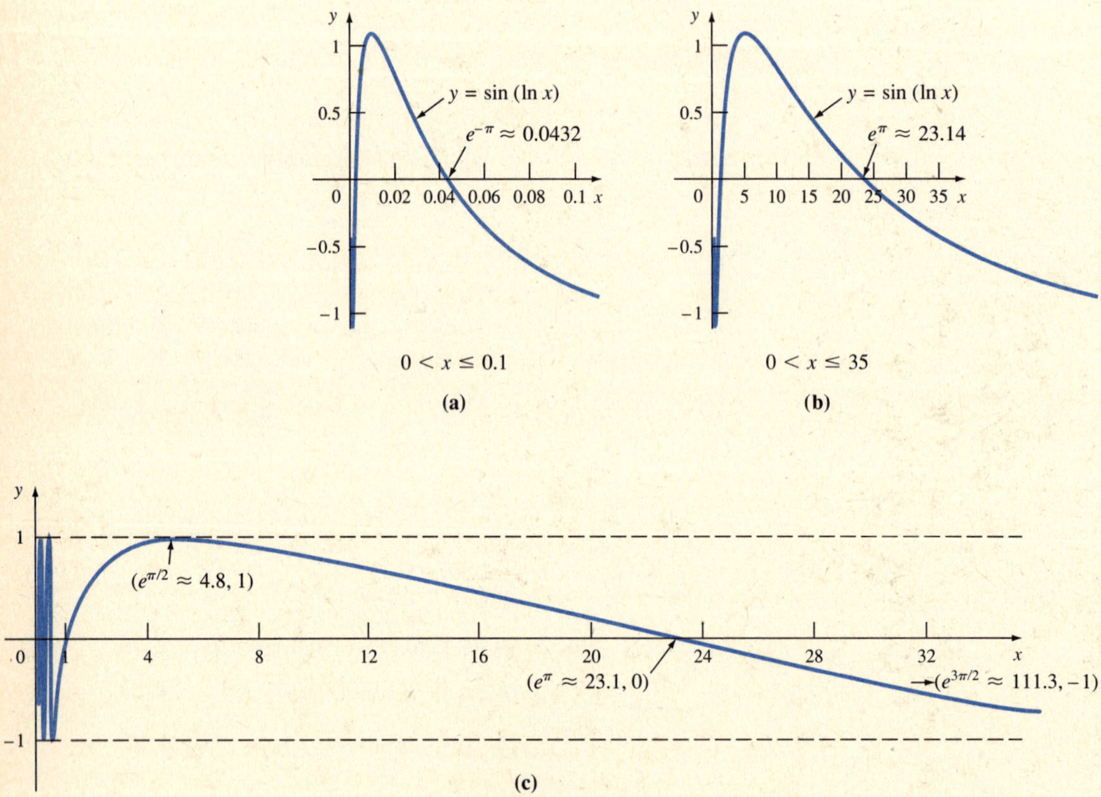

FIGURE 5.7
Graphs of $y = \sin(\ln x)$, $x > 0$

HISTORICAL NOTE

The method of Frobenius is named after the German mathematician Georg Ferdinand Frobenius (1848–1917). Frobenius did some important work on the convergence of infinite series and worked, in particular, on the type of series that bears his name. In addition, he published a method for defining a certain kind of sum for a divergent series. However, Frobenius is best known for his work in abstract algebra. In 1878 he gave the first complete proof of the famous Cayley-Hamilton theorem of linear algebra—twenty years after it was first stated. Throughout his life he did pioneering work in linear algebra and group theory.

SELF-QUIZ

Answer True *or* False *in Problems I to IV*

I. $x = 0$ is a regular singular point of $xy'' + 2y' + 3x^2y = 0$.

II. $x = 0$ is a regular singular point of $x^2y'' + 2y' + 3xy = 0$.

III. $x = 0$ is a regular singular point of $y'' + 2\ln|x|\,y' - 5y = 0$.

IV. The equation $x^2y'' + y' + y = 0$ has a regular singular point at the origin.

V. The exponents of the indicial equation of $x^2y'' + 3xy' - 3y = 0$ are ______.
(a) $-3, 1$ **(b)** $3, -1$ **(c)** $-1, -3$
(d) $1, 3$ **(e)** $\left(\dfrac{-3 \pm \sqrt{21}}{2}\right)$

VI. The exponents of the indicial equation of $y'' + \dfrac{\sin x}{x}y' - \dfrac{6\cos x}{x^2}y = 0$ are ______.
(a) $\pm\sqrt{6}$ **(b)** $\pm i\sqrt{6}$ **(c)** $3, 2$
(d) $-3, -2$ **(e)** $-3, 2$ **(f)** $3, -2$

VII. The exponents of the indicial equation of $y'' + \ln|x|\,y' + \sqrt{2}\,y = 0$ are ______.
(a) $0, 1$ **(b)** $0, \sqrt{2}$ **(c)** $0, -\sqrt{2}$
(d) $0, -1$ **(e)** $x = 0$ is irregular

Answers to Self-Quiz

I. True **II.** False **III.** False **IV.** False **V.** a **VI.** f **VII.** e

PROBLEMS 5.3

In Problems 1–20 find the general solution to the given differential equation by the method of Frobenius at $x = 0$.

1. $y'' + \dfrac{1}{2x}y' + \dfrac{1}{4x}y = 0$

2. $4xy'' + 2y' + y = 0$

3. $2xy'' + y' + xy = 0$

4. $4xy'' + 3y' - 3y = 0$

5. $3xy'' + y' - y = 0$

6. $2xy'' + y' + y = 0$

7. $2xy'' + y' + 2y = 0$

8. $2xy'' - y' + y = 0$

9. $9x^2y'' + x^2y' + 2y = 0$

10. $2x^2y'' + 3xy' + (x - 1)y = 0$

11. $2xy'' + (1 + x)y' - 2y = 0$

12. $2xy'' - (x - 3)y' - y = 0$

13. $y'' + \dfrac{x+1}{2x}y' + \dfrac{3}{2x}y = 0$

14. $y'' + \dfrac{1}{2x}y' - \dfrac{x+1}{2x^2}y = 0$

15. $2x^2y'' + 3xy' - (1 + x)y = 0$

16. $(x - 1)y'' - \dfrac{4x^2 - 3x + 1}{2x}y' + \dfrac{2x^2 - x + 2}{2x}y = 0$

17. $2xy'' - y' + xy = 0$

18. $2x^2y'' - xy' - (x^2 - 1)y = 0$

19. $x^2y'' + \left(x + \dfrac{2}{9}\right)y = 0$

20. $x^2y'' + xy' + (1 + x)y = 0$

21. Find the first four terms of the two linearly independent solutions y_1 and y_2 for the equation

$$(\sin x)y'' + y = 0.$$

[*Hint:* Expand $\sin x$ as a Maclaurin series.]

22. The differential equation

$$x^2y'' + (4x - 1)y' + 2y = 0$$

has $x = 0$ as an irregular singular point.

(a) Suppose that Equation **(4)** is inserted into this equation. Show that $r = 0$ and that the corresponding "solution" by the method of Frobenius is

$$y = \sum_{n=0}^{\infty} (n + 1)!x^n.$$

(b) Prove that the series in (a) has radius of convergence $R = 0$. Hence, even though a Frobenius series may formally satisfy a differential equation, it may not be a valid solution at an irregular singular point.

23. Consider the differential equation

$$y'' - \frac{1}{x^2}y' + \frac{1}{x^3}y = 0.$$

(a) Show that $x = 0$ is an irregular singular point of this equation.

(b) Use the fact that $y_1 = x$ is a solution to find a second independent solution.

(c) Show that the solution y_2 cannot be expressed as a series of form **(4)**. Thus this solution cannot be found by the method of Frobenius.

5.4 FROBENIUS'S METHOD: CASES 2 AND 3

In this section we continue our study of the method of Frobenius.

Case 2: $r_1 = r_2$

Here we set $r = r_1$ and determine the coefficients $c_1, c_2, \ldots$ as in Case 1. We can then use the method of reduction of order to find the second linearly independent solution, since one solution is known. Consider the following example.

EXAMPLE 1 ▶ **A regular singular point where the exponents are equal**

Solve the equation

$$y'' + y' + \frac{1}{4x^2}y = 0, \qquad x > 0. \tag{1}$$

Solution If we substitute $y = \sum_{n=0}^{\infty} c_n x^{r+n}$ into Equation **(1)** we obtain

$$\sum_{n=0}^{\infty} c_n(r + n)(r + n - 1)x^{r+n-2} + \sum_{n=0}^{\infty} c_n(r + n)x^{r+n-1} + \sum_{n=0}^{\infty} \frac{1}{4}c_n x^{r+n-2} = 0.$$

We need to express all sums in terms of the same power of x. If $k = n + 1$, then $n = k - 1$ and $x^{r+n-1} = x^{r+k-2}$. Thus, with $k = n + 1$, the second sum above can be written

$$\sum_{k=1}^{\infty} c_{k-1}(r + k - 1)x^{r+k-2} \overset{\text{setting } n = k}{=} \sum_{n=1}^{\infty} c_{n-1}(r + n - 1)x^{r+n-2},$$

and we have (taking out the $n = 0$ terms from the first and third sums)

$$c_0\left[r(r - 1) + \frac{1}{4}\right]x^{r-2} + \sum_{n=1}^{\infty} \left\{c_n\left[(r + n)(r + n - 1) + \frac{1}{4}\right] + c_{n-1}(r + n - 1)\right\}x^{r+n-2} = 0. \tag{2}$$

The indicial equation is

$$r(r-1)+\frac{1}{4}=r^2-r+\frac{1}{4}=\left(r-\frac{1}{2}\right)^2=0,$$

which has the double root $r=\frac{1}{2}$. Substitute $r=\frac{1}{2}$ into Equation **(2)** to obtain

$$\sum_{n=1}^{\infty}\left\{c_n\left[\left(n+\frac{1}{2}\right)\left(n-\frac{1}{2}\right)+\frac{1}{4}\right]+c_{n-1}\left(n-\frac{1}{2}\right)\right\}x^{n-3/2}$$
$$=\sum_{n=1}^{\infty}\left[n^2c_n+\left(n-\frac{1}{2}\right)c_{n-1}\right]x^{n-3/2}=0.$$

This leads to the recurrence equation

$$c_n=-\frac{\left(n-\frac{1}{2}\right)c_{n-1}}{n^2},\quad \textbf{for } n\geq 1.$$

Hence

$$c_1=-\frac{c_0}{2},\ c_2=-\frac{c_1\left(\frac{3}{2}\right)}{2^2}=\frac{3c_0}{2^2\cdot 2^2},\ c_3=-\frac{5c_2}{2\cdot 3^2}=-\frac{3\cdot 5c_0}{2^3\cdot 2^2\cdot 3^2},$$
$$c_4=-\frac{7c_3}{2\cdot 4^2}=\frac{3\cdot 5\cdot 7c_0}{2^4\cdot 2^2\cdot 3^2\cdot 4^2},\ldots,$$

so that

$$y_1(x)=x^{1/2}\left(c_0-\frac{c_0}{2}x+\frac{3c_0}{2^2\cdot 2^2}x^2-\frac{3\cdot 5c_0}{2^3\cdot 2^2\cdot 3^2}x^3+\frac{3\cdot 5\cdot 7c_0}{2^4\cdot 2^2\cdot 3^2\cdot 4^2}x^4-\ldots\right)$$
$$=c_0x^{1/2}\left[1-\left(\frac{x}{2}\right)+\frac{3}{2^2}\left(\frac{x}{2}\right)^2-\frac{3\cdot 5}{2^2\cdot 3^2}\left(\frac{x}{2}\right)^3+\frac{3\cdot 5\cdot 7}{2^2\cdot 3^2\cdot 4^2}\left(\frac{x}{2}\right)^4-\ldots\right]$$
$$\underset{\downarrow}{\mathbf{c_0=1}}$$
$$=x^{1/2}\sum_{n=0}^{\infty}\frac{(2n)!}{(n!)^3}\left(\frac{-x}{4}\right)^n,\qquad x>0.$$

If a_n denotes the absolute value of the nth term of the series above, then, since $x>0$,

$$a_n=\frac{(2n)!}{(n!)^3}\left(\frac{x}{4}\right)^n\quad\text{and}\quad\lim_{n\to\infty}\left|\frac{a_n}{a_{n+1}}\right|=\lim_{n\to\infty}\frac{(n+1)^3}{(2n+2)(2n+1)}\left(\frac{4}{x}\right)=\infty.$$

Thus the series converges (absolutely) for every $x>0$.

We can use the method of reduction of order to produce the second linearly independent solution y_2. Recall that to find y_2 we set $y_2=vy_1$, where y_1 is the solution above. Hence

$$y_2''+y_2'+\frac{1}{4x^2}y_2=v''y_1+v'(2y_1'+y_1)+v\left(y_1''+y_1'+\frac{1}{4x^2}y_1\right)$$
$$=v''y_1+v'(2y_1'+y_1)=0$$

because y_1 satisfies the differential equation. Thus

$$\frac{v''}{v'}=-2\frac{y_1'}{y_1}-1=\frac{-2\left(\frac{1}{2\sqrt{x}}\right)\left[1-3\left(\frac{x}{2}\right)+\frac{3\cdot 5}{2^2}\left(\frac{x}{2}\right)^2-\ldots\right]}{\sqrt{x}\left[1-\left(\frac{x}{2}\right)+\frac{3}{2^2}\left(\frac{x}{2}\right)^2+\ldots\right]}-1$$

After finding the first few terms by long division (carried out exactly as in the division of one polynomial by another), we have

$$\frac{v''}{v'} = \frac{-1}{x}\left(1 - x + \frac{x^2}{4} - \ldots\right) - 1 = \frac{-1}{x} - \frac{x}{4} + \ldots . \quad \textbf{(3)}$$

Integrating both sides of Equation **(3)**, we obtain

$$\ln v' = -\ln x - \frac{x^2}{8} + \ldots,$$

or

$$v' = \frac{1}{x}\exp\left(-\frac{x^2}{8} + \ldots\right)$$
$$= \frac{1}{x}\left[1 + \left(\frac{-x^2}{8} + \ldots\right) + \frac{1}{2!}\left(\frac{-x^2}{8} + \ldots\right)^2 + \ldots\right]$$

by expanding the exponential as a power series. Integrating once more we get

$$v = \ln x - \frac{x^2}{16} + \ldots,$$

so that

$$y_2 = y_1 \ln x + \sqrt{x}\left(\sum_{n=1}^{\infty} c_n^* x^n\right). \quad ◀ \quad \textbf{(4)}$$

The result of Example 1 suggests the following.

THEOREM 1 ♦ **Form of solution when indicial equation has a double root**
If $x = 0$ is a regular singular point of the differential equation

$$y'' + a(x)y' + b(x)\,y = 0, \qquad x > 0, \quad \textbf{(5)}$$

and the two roots of the indicial equation are equal to r, then the two linearly independent solutions of Equation **(5)** have the form

$$y_1 = x^r \sum_{n=0}^{\infty} c_n x^n \quad \text{and} \quad y_2 = y_1 \ln x + x^r \sum_{n=1}^{\infty} c_n^* x^n. \ ♦ \quad \textbf{(6)}$$

This theorem provides an alternate technique for obtaining the solution y_2: instead of using the method of reduction of order, simply substitute the expression for y_2 in Equation **(6)** directly into the differential equation. We illustrate this approach by completing our solution of Example 1.

EXAMPLE 1 (continued) ▶ We obtain the second linearly independent solution by setting $r = 1/2$ and

$$y_2 = y_1 \ln x + \sum_{n=1}^{\infty} c_n^* x^{n+1/2}.$$

Then

$$y_2' = \frac{y_1}{x} + y_1' \ln x + \sum_{n=1}^{\infty}\left(n + \frac{1}{2}\right)c_n^* x^{n-1/2},$$

and

$$y_2'' = -\frac{y_1}{x^2} + \frac{2y_1'}{x} + y_1'' \ln x + \sum_{n=1}^{\infty}\left(n^2 - \frac{1}{4}\right)c_n^* x^{n-3/2}.$$

Substituting these expressions into the differential Equation **(1)**, we have

$$y_2'' + y_2' + \frac{1}{4x^2}y_2 = \frac{2y_1'}{x} - \frac{y_1}{x^2} + \frac{y_1}{x} + \left(y_1'' + y_1' + \frac{1}{4x^2}y_1\right)\ln x + \sum_{n=1}^{\infty}\left(n^2 - \frac{1}{4}\right)c_n^* x^{n-3/2} + \sum_{n=1}^{\infty}\left(n + \frac{1}{2}\right)c_n^* x^{n-1/2} + \frac{1}{4x^2}\sum_{n=1}^{\infty} c_n^* x^{n+1/2} = 0. \tag{7}$$

Since y_1 is a solution to Equation **(1)**, $y_1'' + y_1' + (1/4x^2)y_1 = 0$, so that Equation **(7)** becomes

$$\frac{2y_1'}{x} - \frac{y_1}{x^2} + \frac{y_1}{x} + \sum_{n=1}^{\infty}\left(n^2 - \frac{1}{4}\right)c_n^* x^{n-3/2} + \sum_{n=1}^{\infty}\left(n + \frac{1}{2}\right)c_n^* x^{n-1/2} + \frac{1}{4x^2}\sum_{n=1}^{\infty} c_n^* x^{n+1/2} = 0.$$

Taking the terms involving y_1 to the right-hand side and substituting $n = N - 1$ in the second sum, we get

$$\sum_{n=1}^{\infty}\left(n^2 - \frac{1}{4}\right)c_n^* x^{n-3/2} + \sum_{N=2}^{\infty}\left(N - \frac{1}{2}\right)c_{N-1}^* x^{N-3/2} + \frac{1}{4}\sum_{n=1}^{\infty} c_n^* x^{n-3/2} = -\frac{2y_1'}{x} + \frac{y_1}{x^2} - \frac{y_1}{x},$$

or, setting $n = N$,

$$c_1^* x^{-1/2} + \sum_{n=2}^{\infty}\left(n^2 c_n^* + \left(n - \frac{1}{2}\right)c_{n-1}^*\right)x^{n-3/2} = -\frac{2y_1'}{x} + \frac{y_1}{x^2} - \frac{y_1}{x}. \tag{8}$$

We have already computed the series for y_1 with $c_0 = 1$, so we can compute as many terms as we want on the right-hand side of Equation **(8)**:

$$y_1 = x^{1/2}\sum_{n=0}^{\infty}\frac{(2n)!}{(n!)^3}\left(\frac{-x}{4}\right)^n = \sum_{n=0}^{\infty}\frac{(-1)^n(2n)!}{4^n(n!)^3}x^{n+1/2},$$

$$y_1' = \sum_{n=0}^{\infty}\frac{(-1)^n(2n+1)!}{2\cdot 4^n(n!)^3}x^{n-1/2},$$

so that

$$\begin{aligned}
-\frac{2y_1'}{x} + \frac{y_1}{x^2} - \frac{y_1}{x}
&= -\frac{2}{x}\sum_{n=0}^{\infty}\frac{(-1)^n(2n+1)!}{2\cdot 4^n(n!)^3}x^{n-1/2} + \left(\frac{1}{x^2} - \frac{1}{x}\right)\sum_{n=0}^{\infty}\frac{(-1)^n(2n)!}{4^n(n!)^3}x^{n+1/2} \\
&= -\sum_{n=0}^{\infty}\frac{(-1)^n(2n+1)!}{4^n(n!)^3}x^{n-3/2} + \sum_{n=0}^{\infty}\frac{(-1)^n(2n)!}{4^n(n!)^3}x^{n-3/2} - \sum_{n=0}^{\infty}\frac{(-1)^n(2n)!}{4^n(n!)^3}x^{n-1/2} \\
&= (-1+1)x^{-3/2} + \left(-\frac{(-1)3!}{4} + \frac{(-1)2!}{4} - 1\right)x^{-1/2} \\
&\quad + \sum_{n=2}^{\infty}\left(-\frac{(-1)^n(2n+1)!}{4^n(n!)^3} + \frac{(-1)^n(2n)!}{4^n(n!)^3} - \frac{(-1)^{n-1}(2n-2)!}{4^{n-1}((n-1)!)^3}\right)x^{n-3/2}
\end{aligned}$$

or

$$-\frac{2y_1'}{x}+\frac{y_1}{x^2}-\frac{y_1}{x}=\sum_{n=2}^{\infty}\frac{(-1)^{n-1}(2n-2)!}{4^{n-1}n!\,(n-1)!\,(n-2)!}x^{n-3/2}.$$

Thus, Equation **(8)** becomes

$$c_1^* x^{-1/2}+\sum_{n=2}^{\infty}\left[n^2c_n^*+\left(n-\frac{1}{2}\right)c_{n-1}^*\right]x^{n-3/2}$$
$$=-\frac{x^{1/2}}{4}+\frac{x^{3/2}}{8}-\frac{5x^{5/2}}{128}+\cdots.$$

Since there is no $x^{-1/2}$ term on the right-hand side, it follows that $c_1^*=0$. Then

$$4c_2^*+\frac{3}{2}c_1^*=-\frac{1}{4},\quad 9c_3^*+\frac{5}{2}c_2^*=\frac{1}{8},\quad 16c_4^*+\frac{7}{2}c_3^*=-\frac{5}{128},\ldots,$$

so that $c_2^*=-1/16$, $c_3^*=1/32$, $c_4^*=-19/2^{11}$, ... Hence,

$$y_2=y_1\ln x+\sqrt{x}\left(-\frac{x^2}{16}+\frac{x^3}{32}-\frac{19x^4}{2^{11}}+\ldots\right). \blacktriangleleft$$

Note

The computations in Example 1 are very complicated. Unfortunately, without using a computer program that manipulates symbols, there is no other way to reduce the work.

Case 3: r_1 and r_2 differ by a nonzero integer

Suppose that $r_1>r_2$. Then one solution of Equation **(3)** has the form

$$y_1=x^{r_1}(c_0+c_1x+c_2x^2+\ldots),\qquad c_0=1,\quad x>0,$$

as in Case 1. In some instances it is not possible to determine y_2 as was done in Case 1, because the procedure regenerates the same series expansion we obtained for y_1 (in this case, the first r_1-r_2 coefficients c_n^* vanish). When this occurs, we proceed as in Case 2. These two possibilities are illustrated in the following two examples.

EXAMPLE 2 ▶ **A regular singular point where the exponents differ by 5**

Solve

$$xy''+(x-4)y'-2y=0. \tag{9}$$

Solution Setting $y=\sum_{n=0}^{\infty}c_nx^{r+n}$ in Equation **(9)** yields $xy''+(x-4)y'-2y=$

$$\sum_{n=0}^{\infty}(r+n)(r+n-1)c_nx^{r+n-1}+\sum_{n=0}^{\infty}(r+n)c_nx^{r+n}-\sum_{n=0}^{\infty}4(r+n)c_nx^{r+n-1}$$
$$-\sum_{n=0}^{\infty}2c_nx^{r+n}=c_0[r(r-1)-4r]x^{r-1}$$
$$+\sum_{n=0}^{\infty}[(r+n-2)c_n+(r+n+1)(r+n-4)c_{n+1}]x^{r+n}=0. \tag{10}$$

Thus the indicial equation is $r^2-5r=r(r-5)=0$, so that the roots are 0 and 5. Selecting the smaller root $r=0$, Equation **(10)** becomes

$$\sum_{n=0}^{\infty}[(n-2)c_n+(n+1)(n-4)c_{n+1}]x^n=0.$$

Since the term in brackets is zero, we have

$$\begin{aligned}
-2\,c_0 - 4\,c_1 &= 0,\\
-c_1 - 6\,c_2 &= 0,\\
0\,c_2 - 6\,c_3 &= 0, \text{ implying that } c_3 = 0\\
c_3 - 4\,c_4 &= 0,\\
2\,c_4 + 0\,c_5 &= 0, \text{ implying that } c_4 = 0\\
3\,c_5 + 6\,c_6 &= 0, \ldots
\end{aligned}$$

Thus, $c_1 = -c_0/2$, $c_2 = -c_1/6 = c_0/12$, and using the recursion equation

$$c_{n+1} = -\frac{(n-2)c_n}{(n+1)(n-4)}, \quad n > 4,$$

we get

$$c_6 = -\frac{3c_5}{6 \cdot 1} = \frac{5!\,3!\,c_5}{2!\,6!\,1!}, \quad c_7 = -\frac{4c_6}{7 \cdot 2} = \frac{5!\,4!\,c_5}{2!\,7!\,2!}, \ldots,$$

which gives us the general solution

$$y = c_0\left(1 - \frac{x}{2} + \frac{x^2}{12}\right) + \frac{5!}{2!}c_5 x^5\left(\sum_{n=0}^{\infty} \frac{(-1)^n (n+2)!\,x^n}{(n+5)!\,n!}\right). \quad ◀$$

If we had substituted the larger root $r = 5$, we would only have obtained the second term in the solution (check!). Thus, using the smaller root of the indicial equation sometimes yields the general solution; using the larger root always yields a solution, but never the general solution because the power series begins with terms of the form x^{r_1} with $r_1 > r_2$. We warn, however, that using r_2 may sometimes not yield a solution of the form

$$x^{r_2}\left(\sum_{n=0}^{\infty} c_n x^n\right),$$

as the next example illustrates.

EXAMPLE 3 ▶ A regular singular point where the exponents differ by 2

Solve

$$x^2 y'' + xy' - (x^2 + 1)y = 0. \tag{11}$$

Solution Substituting $y = \sum_{n=0}^{\infty} c_n x^{r+n}$ into Equation **(11)** and changing the index of summation as in Example 2, we obtain

$$\begin{aligned}
&c_0[r(r-1) + r - 1]x^r + c_1[r(r+1) + (r+1) - 1]x^{r+1}\\
&\quad + \sum_{n=2}^{\infty} \{c_n[(r+n)(r+n-1) + (r+n) - 1] - c_{n-2}\}x^{r+n} = 0.
\end{aligned} \tag{12}$$

The indicial equation is

$$r(r-1) + r - 1 = r^2 - 1 = 0,$$

with roots $r_1 = 1$ and $r_2 = -1$, which differ by the integer 2. Setting $r = -1$ in Equation **(12)**, we obtain

$$-c_1 + \sum_{n=2}^{\infty} [c_n(n^2 - 2n) - c_{n-2}]x^{n-1} = 0,$$

and so $c_1 = 0$ and $n(n-2)c_n = c_{n-2}$. Setting $n = 2$, we see that $c_0 = 0$, contradicting the assumption that $c_0 = 1$ (see p. 217).

Setting $r = 1$ in Equation **(12)** leads to

$$3c_1x^2 + \sum_{n=2}^{\infty} [c_n(n^2 + 2n) - c_{n-2}]x^{n+1} = 0,$$

so $c_1 = 0$ and

$$c_n = \frac{c_{n-2}}{n(n+2)}, \qquad \textbf{for } n \geq 2.$$

Thus all the coefficients with odd-numbered subscripts are zero, and

$$c_2 = \frac{c_0}{2 \cdot 4}, \qquad c_4 = \frac{c_2}{4 \cdot 6} = \frac{c_0}{2 \cdot 4^2 \cdot 6}, \qquad c_6 = \frac{c_4}{6 \cdot 8} = \frac{c_0}{2 \cdot 4^2 \cdot 6^2 \cdot 8}, \ldots.$$

Hence, if $c_0 = 1$,

$$\begin{aligned} y_1(x) &= x\left(c_0 + \frac{c_0}{2 \cdot 4}x^2 + \frac{c_0}{2 \cdot 4^2 \cdot 6}x^4 + \frac{c_0}{2 \cdot 4^2 \cdot 6^2 \cdot 8}x^6 + \ldots\right) \\ &= x\left(1 + \frac{1}{1!\,2!}\left(\frac{x}{2}\right)^2 + \frac{1}{2!\,3!}\left(\frac{x}{2}\right)^4 + \frac{1}{3!\,4!}\left(\frac{x}{2}\right)^6 + \ldots\right). \end{aligned}$$

Since we have *one* solution, the other solution can be obtained by the method of reduction of order:

$$\frac{v''}{v'} = -\frac{2y_1'}{y_1} - \frac{1}{x} = -\frac{3}{x} - \frac{x}{2} + \ldots.$$

Integrating both sides of this equation, we have

$$\ln v' = -3 \ln x - \frac{x^2}{4} + \ldots,$$

or

$$v' = x^{-3} \exp\left(-\frac{x^2}{4} + \ldots\right) = x^{-3} - \frac{1}{4}x^{-1} + \ldots.$$

Integrating once more, we get

$$v = -\frac{1}{2}x^{-2} - \frac{1}{4}\ln x + \ldots,$$

so

$$y_2 = vy_1 = -\frac{1}{4}y_1 \ln x - \frac{1}{2}x^{-1} + \ldots, \qquad x > 0. ◀$$

In general, if the method of Case 1 fails for r_2, the procedure above yields the solution

$$y_2 = c_{-1}^*(\ln x)y_1 + x^{r_2}(c_0^* + c_1^*x + \cdots), \qquad x > 0,$$

where c_{-1}^* may equal zero.

We gather all the facts we have proved in this and the previous section.

Solving a differential equation with a regular singular point at $x = 0$: the method of Frobenius

Let $x = 0$ be a regular singular point of the differential equation

$$y'' + a(x)y' + b(x)y = 0, \qquad x \neq 0, \tag{13}$$

and let r_1 and r_2 be the roots of the indicial equation

$$r(r-1) + a_0 r + b_0 = 0,$$

where a_0 and b_0 are given by the power series expansions

$$xa(x) = a_0 + a_1 x + a_2 x^2 + \dots,$$
$$x^2 b(x) = b_0 + b_1 x + b_2 x^2 + \dots.$$

Then Equation **(13)** has two linearly independent solutions y_1 and y_2 whose form depends on r_1 and r_2 as follows:

Case 1 If r_1 and r_2 differ but not by an integer, then

$$y_1(x) = |x|^{r_1}\left(\sum_{n=0}^{\infty} c_n x^n\right), \qquad c_0 = 1,$$
$$y_2(x) = |x|^{r_2}\left(\sum_{n=0}^{\infty} c_n^* x^n\right), \qquad c_0^* = 1.$$

(The absolute-value signs are needed to avoid the assumption that $x > 0$.)

Case 2 If $r_1 = r_2 = r$, then

$$y_1(x) = |x|^{r}\left(\sum_{n=0}^{\infty} c_n x^n\right), \qquad c_0 = 1,$$
$$y_2(x) = |x|^{r}\left(\sum_{n=0}^{\infty} c_n^* x^n\right) + y_1(x)\ln|x|.$$

Case 3 If $r_1 - r_2$ is a positive integer, then

$$y_1(x) = |x|^{r_1}\left(\sum_{n=0}^{\infty} c_n x^n\right), \qquad c_0 = 1,$$
$$y_2(x) = |x|^{r_2}\left(\sum_{n=0}^{\infty} c_n^* x^n\right) + c_{-1}^* y_1(x)\ln|x|, \qquad c_0^* = 1,$$

and c_{-1}^* may equal zero.

Furthermore, if the power series expansions for $xa(x)$ and $x^2b(x)$ are valid for $|x| < R$, then the solutions y_1 and y_2 are valid for $0 < |x| < R$. The proof of this fact is left as an exercise (see Problems 28 and 29).

SELF-QUIZ

1. $y(x) = |x|^2\left(\sum_{n=0}^{\infty} c_n x^n\right) + |x|^2\left(\sum_{n=0}^{\infty} c_n^* x^n\right)\ln|x|$ is the form of a solution to ________.

(a) $y'' - \frac{2}{x}y' + \frac{3}{4x^2}y = 0$

(b) $y'' - \frac{3}{x}y' + \frac{4}{x^2}y = 0$

(c) $y'' - \frac{4}{x}y' + \frac{4}{x^2}y = 0$

(d) none of these

II. $y_1(x) = |x|^{1/2}\left(\sum_{n=0}^{\infty} c_n x^n\right)$ and $y_2(x) = |x|^{3/2}\left(\sum_{n=0}^{\infty} c_n^* x^n\right)$ are the forms of two linearly independent solutions to ______.

(a) $y'' - \frac{2}{x}y' + \frac{3}{4x^2}y = 0$

(b) $y'' - \frac{3}{x}y' + \frac{4}{x^2}y = 0$

(c) $y'' - \frac{4}{x}y' + \frac{4}{x^2}y = 0$

(d) none of these

III. $y_1(x) = |x|^4\left(\sum_{n=0}^{\infty} c_n x^n\right)$ and $y_2(x) = |x|\left(\sum_{n=0}^{\infty} c_n^* x^n\right) + c_{-1}^* y_1(x)\ln|x|$ are the forms of two linearly independent solutions to ______.

(a) $y'' - \frac{2}{x}y' + \frac{3}{4x^2}y = 0$

(b) $y'' - \frac{3}{x}y' + \frac{4}{x^2}y = 0$

(c) $y'' - \frac{4}{x}y' + \frac{4}{x^2}y = 0$

(d) none of these

Answer True *or* False.

IV. The incidial equation for $x(x-1)y'' + (x-2)y' + (x-3)y = 0$ is $r^2 - 1 = 0$.

V. If the indicial equation for the regular singular point $x = 0$ has a double root, then one of the series solutions must involve a $\ln x$ term.

Answers to Self-Quiz

I. b **II.** d **III.** c **IV.** False **V.** True

PROBLEMS 5.4

In Problems 1–24 find the general solution to the given differential equation by the method of Frobenius at $x = 0$.

1. $y'' + \frac{2}{x}y' - \frac{2}{x^2}y = 0$

2. $xy'' - y' + 4x^3y = 0$

3. $y'' + \frac{3}{x}y' + 4x^2y = 0$

4. $x^2y'' + Axy' + By = 0$, A and B constants

5. $xy'' + 2y' + xy = 0$

6. $x^2y'' + x(x+1)y' - y = 0$

7. $x^2y'' + x(x-1)y' - y = 0$

8. $x^2y'' - x(x+1)y' + y = 0$

9. $xy'' + 2y' + y = 0$

10. $x^2y'' + \frac{4}{3}xy' + \left(x - \frac{2}{9}\right)y = 0$

11. $xy'' + (1-2x)y' - (1-x)y = 0$

12. $x(x+1)^2y'' + (1-x^2)y' - (1-x)y = 0$

13. $y'' - 2y' + \left(1 + \frac{1}{4x^2}\right)y = 0$

14. $x^2y'' - xy' - \left(x^2 + \frac{5}{4}\right)y = 0$

15. $x(x-1)y'' - (1-3x)y' + y = 0$

16. $x^2(x^2-1)y'' - x(x^2+1)y' + (x^2+1)y = 0$

17. $y'' + \frac{y'}{x} - y = 0$

18. $y'' + \frac{2(1-2x)}{x(1-x)}y' - \frac{2}{x(1-x)}y = 0$

19. $y'' + \frac{6}{x}y' + \left(\frac{6}{x^2} - 1\right)y = 0$

20. $y'' + \frac{4}{x}y' + \left(1 + \frac{2}{x^2}\right)y = 0$

21. $x^2y'' + x(x-1)y' - (x-1)y = 0$

22. $xy'' - (3+x)y' + 2y = 0$

23. $y'' + \frac{1}{4x^2}y = 0$

24. $x(x-1)y'' + y' - 2y = 0$

25. If $r_1 = r_2$, verify that the technique in Example 1 always yields

$$v(x) = \ln x - (a_1 + 2c_1)x + \dots,$$

where

$$xa(x) = a_0 + a_1x + a_2x^2 + a_3x^3 + \dots$$

and $y_2 = vy_1$ with

$$y_1 = x^r(c_0 + c_1x + c_2x^2 + \dots), \qquad c_0 = 1.$$

26. Show that the method of Frobenius fails for the equation

$$x^4y'' + 2x^3y' - y = 0, \qquad x > 0,$$

which has $x = 0$ as an irregular singular point. Find a solution by assuming that

$$y = \sum_{n=0}^{\infty} c_n x^{-n}.$$

27. Prove that the second linearly independent solution of the equation

$$y'' + \frac{1}{x}y' + y = 0$$

has the form

$$y_2 = y_1 \ln x + \sum_{n=1}^{\infty} \frac{(-1)^{n+1}}{2^{2n}(n!)^2}\left(1 + \frac{1}{2} + \cdots + \frac{1}{n}\right)x^{2n}, \quad x > 0.$$

28. Let $x_0 = 0$ and suppose that the Maclaurin series

$$a(x) = \sum_{n=0}^{\infty} a_n x^n, \qquad b(x) = \sum_{n=0}^{\infty} b_n x^n$$

are valid for $|x| < R$.

(a) Show that if Equation **(13)** has a power series solution

$$y(x) = \sum_{n=0}^{\infty} c_n x^n, \qquad c_0 = 1,$$

then the coefficients c_n satisfy the recursion formula

$$(n+1)(n+2)c_{n+2} = -\sum_{k=0}^{n} [(k+1)a_{n-k}c_{k+1} + b_{n-k}c_k].$$

(b) Use the root test formula for finding the radius of convergence to show that for any r such that $0 < r < R$ there is a constant $M > 0$ such that

$$(n+1)(n+2)|c_{n+2}| \leq \frac{M}{r^n}\sum_{k=0}^{\infty} [(k+1)|c_{k+1}| + |c_k|]r^k + M|c_{n+1}|r.$$

(c) Use the ratio test to prove that the series

$$\sum_{n=0}^{\infty} |c_n| x^n$$

converges for $|x| < r$. Then by the comparison test for series, it follows that the series $y(x)$ converges for $|x| < r$. Since r is arbitrary, it follows that the series representation of the solution is valid for $|x| < R$. Assume that $\lim_{n\to\infty} |C_{n+1}/C_n|$ exists.

29. Modify the argument in Problem 28 to justify the last statement of the method of Frobenius on p. 231.

5.5 SPECIAL FUNCTIONS

Bessel functions

The differential equation

$$x^2y'' + xy' + (x^2 - p^2)y = 0, \tag{1}$$

which is known as the **Bessel equation of order** p (≥ 0), is one of the most important differential equations in applied mathematics. The equation was first investigated in 1703 by Jakob Bernoulli (see p. 60) in connection with the oscillatory behavior of a hanging chain, and later by the German mathematician Friedrich Wilhelm Bessel (1784–1846) in his studies of planetary motion. Since then, the Bessel functions have been used in the studies of elasticity, fluid motion, potential theory, diffusion, and the propagation of waves.

To find a solution to Bessel's equation we use the method of Frobenius. We substitute

$$y(x) = \sum_{n=0}^{\infty} c_n x^{r+n}, \qquad x \neq 0, \qquad c_0 = 1, \tag{2}$$

into Equation **(1)** obtaining

$$\sum_{n=0}^{\infty} c_n(r+n)(r+n-1)x^{r+n} + \sum_{n=0}^{\infty} c_n(r+n)x^{r+n} - \sum_{n=0}^{\infty} p^2c_nx^{r+n} + \sum_{n=2}^{\infty} c_{n-2}x^{r+n} = 0,$$

or

$$c_0(r^2 - p^2)x^r + c_1[(r+1)^2 - p^2]x^{r+1} + \sum_{n=2}^{\infty} \{c_n[(n+r)^2 - p^2] + c_{n-2}\}x^{r+n} = 0. \tag{3}$$

The indicial equation is $r^2 - p^2 = 0$, with roots $r_1 = p$ (≥ 0) and $r_2 = -p$. Setting $r = p$ in Equation **(3)** yields

$$(1 + 2p)c_1x^{p+1} + \sum_{n=2}^{\infty} [n(n + 2p)c_n + c_{n-2}]x^{n+p} = 0,$$

indicating that $c_1 = 0$ and that

$$c_n = -\frac{c_{n-2}}{n(n + 2p)}, \qquad \text{for } n \geq 2. \tag{4}$$

Hence all the coefficients with odd-numbered subscripts c_{2j+1} are zero, since by Equation **(4)** they can all be expressed as a multiple of c_1. Letting $n = 2j + 2$, we see that the coefficients with even-numbered subscripts satisfy the equation

$$c_{2(j+1)} = -\frac{c_{2j}}{2^2(j + 1)(p + j + 1)}, \qquad \text{for } j \geq 0,$$

which yields

$$c_2 = -\frac{c_0}{2^2(p + 1)},$$

$$c_4 = -\frac{c_2}{2^2 \cdot 2(p + 2)} = \frac{c_0}{2^4 2!\,(p + 1)(p + 2)},$$

$$c_6 = -\frac{c_4}{2^2 \cdot 3(p + 3)} = -\frac{c_0}{2^6 3!\,(p + 1)(p + 2)(p + 3)}, \ldots$$

Hence series **(2)** becomes

$$y_1(x) = |x|^p\left[c_0 - \frac{c_0}{2^2(p + 1)}x^2 + \frac{c_0}{2^4 2!\,(p + 1)(p + 2)}x^4 - \ldots\right], \tag{5}$$

or

$$y_1 = c_0|x|^p \sum_{n=0}^{\infty} (-1)^n \frac{x^{2n}}{2^{2n}n!\,(p + 1)(p + 2)\cdots(p + n)}. \tag{6}$$

If $c_n = 1/2^{2n}n!\,(p + 1)(p + 2)\cdots(p + n)$, then

$$\lim_{n\to\infty}\left|\frac{c_n}{c_{n+1}}\right| = \lim_{n\to\infty}\frac{2^{2(n+1)}(n + 1)!\,(p + 1)(p + 2)\ldots(p + n)(p + n + 1)}{2^{2n}(n!)(p + 1)(p + 2)\ldots(p + n)}$$
$$= 4\lim_{n\to\infty}(n + 1)(p + n + 1) = \infty$$

so the series **(6)** converges for every real x.

We can write y_1 in another way, and in so doing, can introduce a function which is of great importance in applied mathematics.

The gamma function

If $p > -1$, we define the **gamma function,** denoted by $\Gamma(p)$, by

$$\Gamma(p + 1) = \int_0^{\infty} e^{-t}t^p\,dt.$$

Integrating $\Gamma(p + 1)$ by parts, we have

$$\Gamma(p + 1) = \int_0^{\infty} e^{-t}t^p\,dt = -e^{-t}t^p\Big|_0^{\infty} + p\int_0^{\infty} e^{-t}t^{p-1}\,dt.$$

The first expression on the right is zero, and the integral on the right-hand side is $\Gamma(p)$. We thus have the basic property of gamma functions:

$$\Gamma(p + 1) = p\Gamma(p).$$

Since

$$\Gamma(1) = \int_0^\infty e^{-t}\,dt = -e^{-t}\Big|_0^\infty = 1,$$

it follows that $\Gamma(2) = \Gamma(1) = 1!$, $\Gamma(3) = 2\Gamma(2) = 2!, \ldots$, and in general, $\Gamma(n + 1) = n!$. Thus the gamma function is the extension to real numbers $p > -1$ of the factorial function.

It is customary in Equation **(6)** to let $c_0 = [2^p\Gamma(p + 1)]^{-1}$. Then Equation **(6)** becomes

$$J_p(x) = \left|\frac{x}{2}\right|^p \sum_{n=0}^{\infty} (-1)^n \frac{\left(\frac{x}{2}\right)^{2n}}{n!\,\Gamma(p + n + 1)}, \qquad x \neq 0, \tag{7}$$

which is called the **Bessel function of the first kind of order *p*.** Thus $J_p(x)$ is the first solution of Equation **(1).**

To find the second solution, we must consider the difference $r_1 - r_2 = 2p$. By Case 1 of Section 5.3, if p is not a multiple of $1/2$, we can again apply the method of Frobenius with $r = -p$ to find the second solution.

From the results in Section 5.4, the second solution may include $\ln x$ terms when p is a multiple of $1/2$. We shall investigate all of these possibilities. Let $r = -p$ in Equation **(3)** and assume first that p is not an integer. Then we obtain

$$(1 - 2p)c_1x^{-p+1} + \sum_{n=2}^{\infty} [n(n - 2p)c_n + c_{n-2}]x^{n-p} = 0, \tag{8}$$

indicating that $c_1 = 0$ if $p \neq \frac{1}{2}$ and that

$$c_n = -\frac{c_{n-2}}{n(n - 2p)}. \tag{9}$$

Note that when p is not a multiple of $1/2$, all the coefficients c_n with odd-numbered subscripts are zero. Furthermore, even if p is a multiple of $1/2$, say $p = (2m + 1)/2$, then c_{2m+1} is arbitrary and the recurrence relation **(9)** yields the coefficients

$$c_{2m+3} = -\frac{c_{2m+1}}{2^2(p + 1)}, \qquad c_{2m+5} = \frac{c_{2m+1}}{2^4 2!\,(p + 1)(p + 2)}, \ldots$$

Hence the odd-numbered coefficients generate the series

$$\begin{aligned} &|x|^{-p}\left[c_{2m+1}x^{2m+1} - \frac{c_{2m+1}}{2^2(p + 1)}x^{2m+3} + \frac{c_{2m+1}x^{2m+5}}{2^4 2!\,(p + 1)(p + 2)} - \cdots\right] \\ &= c_{2m+1}|x|^p\left[1 - \frac{x^2}{2^2(p + 1)} + \frac{x^4}{2^4 2!\,(p + 1)(p + 2)} - \cdots\right], \end{aligned}$$

which is a multiple of $J_p(x)$. Thus we can ignore the odd-numbered coefficients and concentrate on using Equation **(9)** to calculate the coefficients with even-numbered subscripts. Then

$$c_2 = -\frac{c_0}{2^2(1 - p)}, \qquad c_4 = -\frac{c_2}{2^2 \cdot 2(2 - p)} = \frac{c_0}{2^4 2!\,(1 - p)(2 - p)}, \ldots,$$

and the second solution, when p is not a multiple of $1/2$, is the convergent series

$$J_{-p}(x) = \left|\frac{x}{2}\right|^{-p} \sum_{n=0}^{\infty} (-1)^n \frac{(x)^{2n}}{2^{2n} n!\, \Gamma(n - p + 1)}. \tag{10}$$

THEOREM 1 ♦ Solving the Bessel equation of order *p*

If p is not a multiple of $1/2$, then

$$y(x) = AJ_p(x) + BJ_{-p}(x)$$

is the general solution of the Bessel equation for all values $x \neq 0$. ◀

If p is an integer, then the term $(n - 2p)$ in the recurrence relation **(9)** is zero for the integer $n = 2p$. Hence c_{2p-2} is zero, and iterating Equation **(9)** repeatedly, we see that $c_{2p-2} = c_{2p-4} = \cdots = c_2 = c_0 = 0$. But this contradicts the assumed form **(2)** of the solution. Thus the method of Frobenius cannot be used when p is a positive integer, and the second linearly independent solution of Equation **(1)** must be calculated by the method in Example 5.4.3 (p. 229). After an extremely long (but straightforward) calculation, we obtain

$$y_2(x) = J_p(x) \ln|x| - \frac{1}{2}\left[\sum_{k=0}^{p-1} \frac{(p-k-1)!}{k!}\left(\frac{x}{2}\right)^{2k-p} + \frac{h_p}{p!}\left(\frac{x}{2}\right)^p + \sum_{k=0}^{\infty} \frac{(-1)^k [h_k + h_{p+k}]}{k!\,(p+k)!}\left(\frac{x}{2}\right)^{2k+p}\right], \tag{11}$$

where

$$h_p = 1 + \frac{1}{2} + \frac{1}{3} + \cdots + \frac{1}{p} \tag{12}$$

and p is a positive integer.

It is customary to replace Equation **(11)** by the linear combination of solutions

$$Y_p(x) = \frac{2}{\pi}[y_2(x) + (\gamma - \ln 2) J_p(x)], \qquad p = 0, 1, 2, \ldots,$$

where

$$\gamma = \lim_{p\to\infty} (h_p - \ln p) = 0.5772156649\ldots$$

is the **Euler constant.** This particular solution is obviously independent of $J_p(x)$ and is called the **Bessel function of the second kind of order *p*,** or **Neumann's function of order *p*.*** It is defined by the formula

$$Y_p(x) = \frac{2}{\pi} J_p(x)\left(\ln\frac{x}{2} + \gamma\right) - \frac{1}{\pi}\left[\sum_{k=0}^{p-1} \frac{(p-k-1)!}{k!}\left(\frac{x}{2}\right)^{2k-p} + \frac{h_p}{p!}\left(\frac{x}{2}\right)^p + \sum_{k=0}^{\infty} (-1)^k \frac{[h_k + h_{p+k}]}{k!\,(p+k)!}\left(\frac{x}{2}\right)^{2k+p}\right],$$

for all integers $p = 0, 1, 2, \ldots$.

The function Y_p may be extended to all real numbers $p \geq 0$ (see Problem 24) by letting

$$Y_p(x) = \frac{1}{\sin p\pi}[J_p(x) \cos p\pi - J_{-p}(x)], \qquad p \neq 0, 1, 2, \ldots.$$

Using this definition of Y_p, we have the following result valid for all p:

* Carl Gottfried Neumann (1832–1925) was a German mathematician.

THEOREM 2 ◆ The general solution of the Bessel equation of order p is

$$y(x) = AJ_p(x) + BY_p(x), \qquad x \neq 0. \quad ◆$$

Graphs of the functions J_0, J_1, J_2, Y_0, Y_1, and Y_2 are given in Figures 5.8 and 5.9.

In Figure 5.10 we provide the graphs of $J_{-1}(x)$ and $J_{-2}(x)$ to show the behavior of these functions for x near 0.

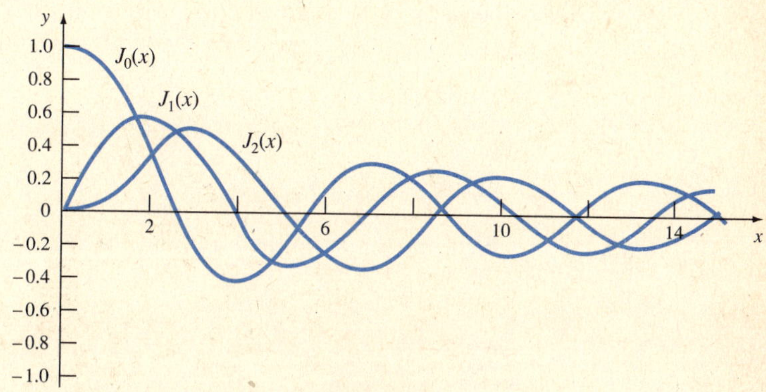

FIGURE 5.8
Graphs of J_0, J_1, and J_2

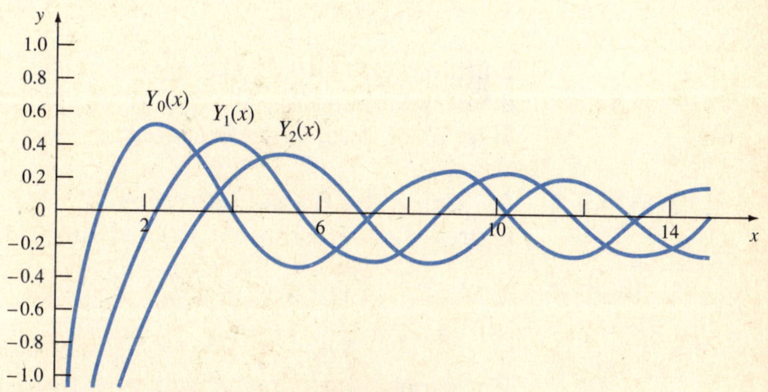

FIGURE 5.9
Graphs of Y_0, Y_1, and Y_2.

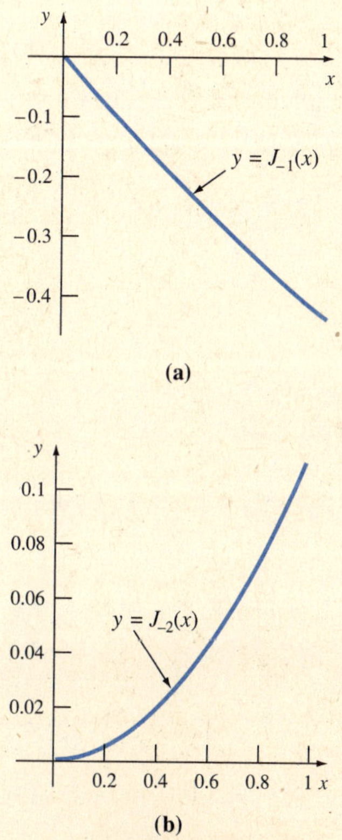

FIGURE 5.10
Graphs of $J_{-1}(x)$ and $J_{-2}(x)$ for $0 \leq x \leq 1$

Properties of Bessel functions

Now that we have the expansions for $J_p(x)$ and $Y_p(x)$, we can derive a number of important expressions involving Bessel functions and their derivatives. For simplicity, we assume $x > 0$. The first two identities are immediate consequences of Equation **(7)**:

$$\frac{d}{dx}[x^p J_p(x)] = x^p J_{p-1}(x), \tag{13}$$

$$\frac{d}{dx}[x^{-p} J_p(x)] = -x^{-p} J_{p+1}(x). \tag{14}$$

To prove Equation **(13)**, we differentiate the product x^pJ_p term by term:

$$\frac{d}{dx}\sum_{n=0}^{\infty}(-1)^n\frac{2^p\left(\frac{x}{2}\right)^{2n+2p}}{n!\,\Gamma(p+n+1)}=\sum_{n=0}^{\infty}(-1)^n\frac{2^{p-1}\left(\frac{x}{2}\right)^{2n+2p-1}2(n+p)}{n!\,\Gamma(p+n+1)}$$

$$=x^p\sum_{n=0}^{\infty}(-1)^n\frac{\left(\frac{x}{2}\right)^{2n+p-1}}{n!\,\Gamma(p+n)}=x^pJ_{p-1}(x),$$

since $\Gamma(p+n+1)=(p+n)\Gamma(p+n)$. The proof of Equation **(14)** is similar (see Problem 4). Expanding the left-hand sides of Equations **(13)** and **(14)**, we have

$$x^pJ_p'+px^{p-1}J_p=x^pJ_{p-1}$$

and

$$x^{-p}J_p'-px^{-p-1}J_p=-x^{-p}J_{p+1},$$

which may be simplified to yield the identities

$$xJ_p'=xJ_{p-1}-pJ_p, \tag{15}$$

$$xJ_p'=pJ_p-xJ_{p+1}. \tag{16}$$

Subtracting Equation **(16)** from Equation **(15)**, we obtain the recursion relation

$$xJ_{p+1}-2pJ_p+xJ_{p-1}=0. \tag{17}$$

Adding the two together yields

$$2J_p'=J_{p-1}-J_{p+1}. \tag{18}$$

Equations **(13)** through **(18)** are extremely important in solving problems involving the Bessel functions, since they allow us to express Bessel functions of higher order in terms of lower-order functions.

EXAMPLE 1 ▶ **Using the Bessel identities**

Express $J_3(x)$ in terms of $J_0(x)$ and $J_1(x)$.

Solution We let $p=2$ in Equation **(17)**.

$$xJ_3=4J_2-xJ_1.$$

If we apply Equation **(17)** with $p=1$ to J_2, we obtain

$$xJ_2=2J_1-xJ_0.$$

Thus

$$J_3(x)=\frac{4}{x}J_2-J_1=\frac{4}{x^2}(2J_1-xJ_0)-J_1=\left(\frac{8}{x^2}-1\right)J_1(x)-\frac{4}{x}J_0(x).$$ ◀

EXAMPLE 2 ▶ **Integrals involving Bessel functions**

Evaluate the integral

$$\int x^4J_1(x)\,dx. \tag{19}$$

Solution Integrating **(19)** by parts, we have, by Equation **(13)**

$$\int x^2[x^2J_1(x)]\,dx \overset{(x^2J_2)'=x^2J_1}{=} x^2(x^2J_2)-\int x^2J_2\cdot 2x\,dx=x^4J_2(x)-2\int x^3J_2\,dx.$$

Applying Equation **(13)** to the last integral, we obtain

$$\int x^4 J_1(x)\,dx = x^4 J_2(x) - 2x^3 J_3(x) + C. \quad ◀$$

Many differential equations with variable coefficients can be reduced to Bessel equations.

EXAMPLE 3 ▶ **Solving differential equations by converting them to Bessel equations**

Solve the equation

$$y'' + k^2xy = 0 \tag{20}$$

Solution The following substitution reduces Equation **(20)** to a Bessel equation: Let $u = y/\sqrt{x}$ and $z = 2kx^{3/2}/3$. Then

$$\frac{du}{dz} = \frac{\dfrac{du}{dx}}{\dfrac{dz}{dx}} = \frac{y'}{kx} - \frac{y}{2kx^2}$$

and

$$\frac{d^2u}{dz^2} = \frac{\dfrac{d}{dx}\left(\dfrac{du}{dz}\right)}{\dfrac{dz}{dx}} = \frac{y''}{k^2x^{3/2}} - \frac{3y'}{2k^2x^{5/2}} + \frac{y}{k^2x^{7/2}}.$$

Hence

$$z^2\frac{d^2u}{dz^2} + z\frac{du}{dz} = \frac{4}{9}x^{3/2}y'' + \frac{1}{9}\frac{y}{\sqrt{x}},$$

and using Equation **(20)** for y'', we obtain

$$z^2\frac{d^2u}{dz^2} + z\frac{du}{dz} = -\frac{4}{9}k^2x^3\left(\frac{y}{\sqrt{x}}\right) + \frac{1}{9}\frac{y}{\sqrt{x}} = -\left(z^2 - \frac{1}{9}\right)u,$$

or

$$z^2\frac{d^2u}{dz^2} + z\frac{du}{dz} + \left(z^2 - \frac{1}{9}\right)u = 0, \tag{21}$$

the Bessel equation of order one-third. Since Equation **(21)** has the general solution

$$u(z) = AJ_{1/3}(z) + BJ_{-1/3}(z),$$

Equation **(20)** has the solution

$$y(x) = \sqrt{x}\left[AJ_{1/3}\left(\frac{2}{3}kx^{3/2}\right) + BJ_{-1/3}\left(\frac{2}{3}kx^{3/2}\right)\right]. \quad ◀$$

Legendre polynomials

Another very important differential equation that arises in many applications is the **Legendre differential equation**

$$(1 - x^2)y'' - 2xy' + p(p + 1)y = 0, \tag{22}$$

where p is a given real number. Any solution of Equation **(22)** is called a **Legendre function.***

Dividing Equation **(22)** by $(1 - x^2)$, we obtain the equation

$$y'' - \frac{2x}{1 - x^2}y' + \frac{p(p + 1)}{1 - x^2}y = 0,$$

and we observe, using the geometric series

$$\frac{1}{1 - x^2} = 1 + x^2 + x^4 + x^6 + \ldots,$$

that the coefficient functions

$$a(x) = -\frac{2x}{1 - x^2} = -2x(1 + x^2 + x^4 + \ldots)$$

$$b(x) = \frac{p(p + 1)}{1 - x^2} = p(p + 1)(1 + x^2 + x^4 + \ldots)$$

have convergent power series representations in the interval $|x| < 1$. By Theorem 5.2.1, it follows that Equation **(22)** must have a power series representation valid in the interval $|x| < 1$. Substituting $y = \Sigma_{n=0}^{\infty} c_n x^n$ and its derivatives into Equation **(22)**, we have

$$(1 - x^2)\sum_{n=2}^{\infty} c_n n(n - 1)x^{n-2} - 2x\sum_{n=1}^{\infty} c_n n x^{n-1} + p(p + 1)\sum_{n=0}^{\infty} c_n x^n = 0,$$

or

$$\sum_{n=0}^{\infty}\{(n + 2)(n + 1)c_{n+2} - c_n[n(n + 1) - p(p + 1)]\}x^n = 0. \tag{23}$$

Setting the coefficients of the sum **(23)** to zero, we obtain the recurrence relation

$$(n + 2)(n + 1)c_{n+2} = c_n(n^2 + n - p^2 - p) = c_n(n - p)(n + p + 1).$$

Thus we have

$$c_{n+2} = -\frac{(p - n)(p + n + 1)}{(n + 2)(n + 1)}c_n. \tag{24}$$

Therefore

$$c_2 = -\frac{p(p + 1)}{2!}c_0, \qquad c_3 = -\frac{(p - 1)(p + 2)}{3!}c_1,$$

$$c_4 = -\frac{(p - 2)(p + 3)}{4 \cdot 3}c_2 = \frac{(p - 2)p(p + 1)(p + 3)}{4!}c_0, \ldots.$$

Inserting these values for the coefficients into the power series expansion for $y(x)$ yields

$$y(x) = c_0 y_1(x) + c_1 y_2(x), \tag{25}$$

where

$$y_1(x) = 1 - p(p + 1)\frac{x^2}{2!} + (p - 2)p(p + 1)(p + 3)\frac{x^4}{4!} - \ldots, \tag{26}$$

* Named after a famous French mathematician Adrien Marie Legendre (1752–1833). See the biographical sketch on page 243.

$$y_2(x) = x - (p-1)(p+2)\frac{x^3}{3!} + (p-3)(p+1)(p+2)(p+4)\frac{x^5}{5!} - \dots . \tag{27}$$

In many applications, the parameter p in the Legendre equation is a nonnegative integer. When this occurs, the right-hand side of Equation **(24)** is zero for $n = p$, implying that $c_{p+2} = c_{p+4} = c_{p+6} = \dots = 0$. Thus one of the Equations **(26)** or **(27)** reduces to a polynomial of degree p (for even p, it is y_1; for odd p, it is y_2). These polynomials, multiplied by an appropriate constant, are called the **Legendre polynomials.** It is customary to set

$$c_p = \frac{(2p)!}{2^p(p!)^2}, \qquad p = 0, 1, 2, \dots, \tag{28}$$

so, by Equation **(24),**

$$c_{p-2} = -\frac{p(p-1)}{2(2p-1)}c_p = -\frac{(2p-2)!}{2^p(p-1)!\,(p-2)!},$$

$$c_{p-4} = -\frac{(p-2)(p-3)}{4(2p-3)}c_{p-2} = \frac{(2p-4)!}{2^p 2!\,(p-2)!\,(p-4)!}, \dots,$$

and in general

$$c_{p-2k} = \frac{(-1)^k(2p-2k)!}{2^p k!\,(p-k)!\,(p-2k)!}.$$

Then the **Legendre polynomials of degree p** are given by

$$P_p(x) = \sum_{k=0}^{M} \frac{(-1)^k(2p-2k)!}{2^p k!\,(p-k)!\,(p-2k)!}x^{p-2k}, \qquad p = 0, 1, 2, \dots, \tag{29}$$

where M is the largest integer not greater than $p/2$. In particular, we have

$$P_0(x) = 1, \qquad P_1(x) = x, \qquad P_2(x) = \frac{1}{2}(3x^2 - 1),$$

$$P_3(x) = \frac{1}{2}(5x^3 - 3x), \qquad P_4(x) = \frac{1}{8}(35x^4 - 30x^2 + 3), \dots .$$

As these particular results illustrate, as a consequence of choice **(28)** of the value of c_p, we have $P_p(1) = 1$ and $P_p(-1) = (-1)^p$ for all integers $p \geq 0$.

Properties of Legendre polynomials

The following equation is obtained by repeated differentiation of x^{2p-2k} and use of the binomial theorem:

Rodrigues's formula

$$P_p(x) = \frac{1}{2^p p!}\frac{d^p}{dx^p}(x^2 - 1)^p, \qquad p = 0, 1, 2, \dots . \tag{30}$$

This formula, called the **Rodrigues formula,*** provides an easy way of computing successive Legendre polynomials.

* Named after the French mathematician and banker Olinde Rodrigues (1794–1851).

EXAMPLE 4 ▶ **Using Rodrigues's formula to find a Legendre polynomial**
Show that $P_2(x) = (3x^2 - 1)/2$.

Solution By the Rodrigues formula,

$$P_2(x) = \frac{1}{2^2 2!}\frac{d^2}{dx^2}(x^4 - 2x^2 + 1) = \frac{1}{8}(12x^2 - 4) = \frac{1}{2}(3x^2 - 1). \blacktriangleleft$$

We can use the Rodrigues formula to obtain several useful recurrence relations:

$$(p+1)P_p = P'_{p+1} - xP'_p, \qquad (2p+1)P_p = P'_{p+1} - P'_{p-1}. \tag{31}$$

Subtract the first identity in **(31)** from the second one, yielding

$$pP_p = xP'_p - P'_{p-1}, \qquad p = 1, 2, \ldots. \tag{32}$$

Finally, we note that from **(31)** and **(32)** we can get

$$\begin{aligned}(p+1)P_{p+1} &- (2p+1)xP_p + pP_{p-1} \\ &= (xP'_{p+1} - P'_p) - x(P'_{p+1} - P'_{p-1}) + (P'_p - xP'_{p-1}) = 0,\end{aligned}$$

so we can eliminate all derivatives and obtain the relation

Recursion relationship for Legendre polynomials

$$(p+1)P_{p+1} + pP_{p-1} = (2p+1)xP_p, \qquad p = 1, 2, \ldots. \tag{33}$$

Equation **(33)** can be used to generate all the Legendre polynomials. We illustrate this iterative technique in the next example.

EXAMPLE 5 ▶ **Computing Legendre polynomials**
Starting with $P_0 = 1$ and $P_1 = x$, calculate the polynomials P_2, P_3, and P_4.

Solution By Equation **(33)**

$$P_{p+1} = \frac{(2p+1)xP_p - pP_{p-1}}{p+1},$$

so

$$P_2 = \frac{3xP_1 - P_0}{2} = \frac{3x^2 - 1}{2},$$

$$P_3 = \frac{5xP_2 - 2P_1}{3} = \frac{15x^3 - 5x - 4x}{6} = \frac{5x^3 - 3x}{2},$$

$$P_4 = \frac{7xP_3 - 3P_2}{4} = \frac{35x^4 - 21x^2 - 9x^2 + 3}{8} = \frac{35x^4 - 30x^2 + 3}{8}. \blacktriangleleft$$

Remark

We point out without proof that the polynomial solutions to Legendre's equation are the only solutions that are bounded on $[-1, 1]$. This fact is often useful in situations involving the solution of certain types of boundary value problems.

ADRIEN MARIE LEGENDRE (1752–1833)

(Bettman Archive)

Adrien Marie Legendre is known in the history of elementary mathematics principally for his very popular *Eléments de géométrie,* in which he attempted a pedagogical improvement of Euclid's *Elements* by considerably rearranging and simplifying many of the propositions. This work was very favorably received in America and became the prototype of the geometry textbooks in this country. In fact, the first English translation of Legendre's geometry was made in 1819 by John Farrar of Harvard University. Three years later another English translation was made, by the famous Scottish litterateur Thomas Carlyle, who early in life was a teacher of mathematics. Carlyle's translation, as later revised by Charles Davies, and later still by J. H. Van Amringe, ran through thirty-three American editions. In later editions of his geometry, Legendre attempted to prove the parallel postulate (see his Section 13-6). Legendre's chief work in higher mathematics centered around number theory, elliptic functions, the method of least squares, and integrals. He was also an assiduous computer of mathematical tables. Legendre's name is today connected with the second-order differential equation

$$(1 - x^2)y'' - 2xy' + p(p + 1)\, y = 0,$$

which is of considerable importance in applied mathematics. Functions satisfying this differential equation are called **Legendre functions** (of order p). When p is a nonnegative integer, the equation has polynomial solutions of special interest called **Lengendre polynomials.** Legendre's name is also associated with the symbol $(c\,|\,p)$ of number theory. The **Legendre symbol** $(c\,|\,p)$ is equal to ± 1 according to whether the integer c, which is prime to p, is or is not a quadratic residue of the odd prime p. [For example, $(6\,|\,19) = 1$ since the congruence $x^2 \equiv 6 \pmod{19}$ has a solution, and $(39\,|\,47) = -1$ since the congruence $x^2 \equiv 39 \pmod{47}$ has no solution.]

In addition to his *Eléments de géométrie,* which appeared in 1794, Legendre published a two-volume 859-page work, *Essai sur la théorie des nombres* (1797–1798), which was the first treatise devoted exclusively to number theory. He later wrote a three-volume treatise, *Exercises du calcul intégral* (1811–1819), that, for comprehensiveness and authoritativeness, rivaled the similar work of Euler. Legendre later expanded parts of this work into another three-volume treatise, *Traité des fonctions elliptiques et des intégrals eulériennes* (1825–1832). In geodesy, Legendre achieved considerable fame for his triangulation of France.

SELF-QUIZ

I. Which of the following is *not* a Bessel Equation?
(a) $x^2 y'' + x y' + x^2 y = 0$
(b) $x^2 y'' + x y' + (x^2 + 1)y = 0$
(c) $x^2 y'' + x y' + (x^2 - 1)y = 0$
(d) $x^2 y'' + x y' + (x^2 - 0.073)y = 0$

II. $\dfrac{d}{dx} x^3 J_3(x) =$ ______.
(a) $3x^2 J_3(x)$ **(b)** $x^4 J_3(x)$
(c) $x^3 J_2(x)$ **(d)** $3x^2 J_2(x)$

III. Which of the following expressions can be used to find the Legendre polynomial $P_3(x)$?
(a) $\dfrac{1}{48}\dfrac{d^3}{dx^3}(x^2 - 1)^3$ **(b)** $\dfrac{1}{48}(x^2 - 1)^3$
(c) $\dfrac{1}{48}3(x^2 - 1)^2(2x)$ **(d)** $\dfrac{1}{48}\dfrac{d^2}{dx^2}(x^2 - 1)^2$

IV. The second-degree Legendre polynomial $P_2(x) =$ ______.

(a) x^2 **(b)** $\frac{1}{2}x^2$

(c) $1 + x + \frac{1}{2}x^2$ **(d)** $\frac{1}{2} + \frac{3}{2}x^2$

(e) $-\frac{1}{2} + \frac{3}{2}x^2$

Answers to Self-Quiz

I. b **II.** c **III.** a **IV.** e

PROBLEMS 5.5

Bessel functions

1. Prove that $J_p(x)$ and $J_{-p}(x)$ are linearly independent.
2. Express $J_4(x)$ in terms of $J_0(x)$ and $J_1(x)$.
3. Show that
   (a) $4J_p''(x) = J_{p+2}(x) - 2J_p(x) + J_{p-2}(x)$;
   (b) $-8J_p'''(x) = J_{p+3}(x) - 3J_{p+1}(x) + 3J_{p-1}(x) - J_{p-3}(x)$.
4. Prove that $[x^{-p}J_p(x)]' = -x^{-p}J_{p+1}(x)$.
5. Show that $[x^pY_p(x)]' = x^pY_{p-1}(x)$.
6. Prove that $[x^{-p}Y_p(x)]' = -x^{-p}Y_{p+1}(x)$.
7. Using the equations of Problems 5 and 6, show that
   (a) $x(Y_{p+1} + Y_{p-1}) = 2pY_p$;
   (b) $2Y_p' = Y_{p-1} - Y_{p+1}$.
8. Prove the following identities:
   (a) $\int J_1(x)\,dx = -J_0(x) + C$
   (b) $\int x^2J_1(x)\,dx = 2xJ_1(x) - x^2J_0(x) + C$
   (c) $\int xJ_0(x)\,dx = xJ_1(x) + C$
   (d) $\int x^3J_0(x)\,dx = (x^3 - 4x)J_1(x) + 2x^2J_0(x) + C$
   (e) $\int J_0(x)\cos x\,dx = xJ_0(x)\cos x + xJ_1(x)\sin x + C$
   (f) $\int J_0(x)\sin x\,dx = xJ_0(x)\sin x - xJ_1(x)\cos x + C$
   (g) $\int J_1(x)\cos x\,dx = xJ_1(x)\cos x - (x\sin x + \cos x)J_0(x) + C$
   (h) $\int J_1(x)\sin x\,dx = xJ_1(x)\sin x + (x\cos x - \sin x)J_0(x) + C$
9. Verify the following identities:
   (a) $\int x^2J_0(x)\,dx = x^2J_1(x) + xJ_0(x) - \int J_0(x)\,dx + C$
   (b) $\int x^{-1}J_1(x)\,dx = -J_1(x) + \int J_0(x)\,dx + C$
   (c) $\int xJ_1(x)\,dx = -xJ_0(x) + \int J_0(x)\,dx + C$
   (d) $\int x^3J_1(x)\,dx = 3x^2J_1(x) - (x^3 - 3x)J_0(x) - 3\int J_0(x)\,dx + C$
10. Show that $\int J_0(\sqrt{x})\,dx = 2\sqrt{x}\,J_1(\sqrt{x}) + C$.
11. Show that $\int xJ_0(x)\sin x\,dx = \frac{1}{3}\{x^2[J_0(x)\sin x - J_1(x)\cos x] + xJ_1(x)\sin x\} + C$.
12. Show that $\int xJ_1(x)\cos x\,dx = \frac{1}{3}\{x^2[J_1(x)\cos x - J_0(x)\sin x] + 2xJ_1(x)\sin x\} + C$.
13. Show that $\int J_0(x)\,dx = 2[J_1(x) + J_3(x) + J_5(x) + \ldots] + C$.

In Problems 14–21 reduce each given equation to a Bessel equation and solve it.

14. $x^2y'' + xy' + (a^2x^2 - p^2)y = 0$
15. $4x^2y'' + 4xy' + (x^2 - p^2)y = 0$
16. $x^2y'' + xy' + 4(x^4 - p^2)y = 0$
17. $xy'' - y' + xy = 0$
18. $x^2y'' + \left(x^2 + \frac{1}{4}\right)y = 0$
19. $xy'' + (1 + 2k)y' + xy = 0$
20. $y'' + k^2x^2y = 0$
21. $y'' + k^2x^4y = 0$

22. Prove that the second linearly independent solution of the equation

$$y'' + \frac{1}{x}y' + y = 0$$

has the form

$$y_2(x) = J_0(x)\ln x + \sum_{n=1}^{\infty} \frac{(-1)^{n+1}}{2^{2n}(n!)^2}\left(1 + \frac{1}{2} + \cdots + \frac{1}{n}\right)x^{2n}, \quad x > 0.$$

23. Obtain Equation **(11)** for $p = 1$ by the methods of this section.

24. Prove that if $p \geq 0$ is an integer,

$$\lim_{q \to p} \frac{J_q(x)\cos q\pi - J_{-q}(x)}{\sin q\pi} = Y_p(x).$$

This is the extension of Y_p to all real values $p \geq 0$.

25. **(a)** Expand $e^{(x/2)[t-(1/t)]}$ as a power series in t by multiplying the series for $e^{xt/2}$ and $e^{-x/2t}$.

(b) Show that the coefficient of t^n in the expansion obtained in part (a) is $J_n(x)$.

(c) Conclude that

$$e^{(x/2)[t-(1/t)]} = J_0(x) + \sum_{n=1}^{\infty} J_n(x)[t^n + (-t)^{-n}].$$

This function is called the **generating function** of the Bessel functions.

(d) Set $t = e^{i\theta}$ in the expression in part (c) and obtain the identities

$$\cos(x\sin\theta) = J_0(x) + 2\sum_{n=1}^{\infty} J_{2n}(x)\cos 2n\theta$$

and

$$\sin(x\sin\theta) = 2\sum_{n=1}^{\infty} J_{2n-1}(x)\sin(2n-1)\theta.$$

26. Show that $J_0(x) = \left(\frac{1}{\pi}\right)\int_0^{\pi} \cos(x\cos t)\,dt$.

27. Show that $J_n(x) = \left(\frac{1}{\pi}\right)\int_0^{\pi} \cos(nt - x\sin t)\,dt$.

28. **(a)** Prove for all integers $n > 0$ that

$$\int_0^{\infty} J_{n+1}(x)\,dx = \int_0^{\infty} J_{n-1}(x)\,dx$$

by means of Equation **(18)**, given the approximation

$$J_n(x) \approx \sqrt{\frac{2}{\pi x}}\cos\left(x - \frac{\pi}{4} - \frac{n\pi}{2}\right) \quad \text{for } x \text{ large.}$$

(b) Prove a similar fact for Y_n given the approximation

$$Y_n(x) \approx \sqrt{\frac{2}{\pi x}}\sin\left(x - \frac{\pi}{4} - \frac{n\pi}{2}\right).$$

(c) Show that $\int_0^{\infty} J_n(x)\,dx = 1$.

(d) Prove that $\int_0^{\infty} \left[\frac{J_n(x)}{x}\right] dx = \frac{1}{n}$.

29. In Section 4.1 we considered the vibrations of mass-spring systems whose springs had a constant elastic force per unit elongation (the spring constant k). In practice, the elastic force per unit elongation $k(t)$ of a spring decays with time. Suppose $k(t) = k_0 e^{-at} + k_1$, with $a > 0$, so that Equation **(4.1.2)** becomes

$$\frac{d^2x}{dt^2} + \frac{k_0 e^{-at} + k_1}{m}x = 0, \; x(0) = x_0, x'(0) = v_0.$$

(a) Use the substitution $u = ce^{-at/2}$, $c = \frac{2}{a}\sqrt{k_0/m}$, to convert this differential equation into a Bessel equation.

(b) Solve the Bessel equation, and obtain a solution of the given initial-value problem.

Legendre polynomials

30. Prove Rodrigues's formula starting with Equation **(29)**.

31. Calculate P_5, P_6, P_7, and P_8 by means of Equation **(33)**.

32. Prove that series **(26)** has a radius of convergence $R = 1$.

33. Prove that series **(27)** has a radius of convergence $R = 1$.

34. Calculate P_4 by means of the Rodrigues's formula.

35. Prove that $P_{2p+1}(0) = 0$, for all integers $p \geq 0$.

36. Prove that

$$P_{2p}(0) = \frac{(-1)^p(2p)!}{2^{2p}(p!)^2},$$

for all $p \geq 0$.

37. Prove that, for all integers $p \geq 0$,

(a) $P'_{2p}(0) = 0$;

(b) $P'_{2p+1}(0) = \dfrac{(-1)^p(2p+1)!}{2^{2p}(p!)^2}$.

38. Show that, for all integers $p > 0$,

(a) $\int_0^1 P_p(x)\,dx = \dfrac{1}{p+1}P_{p-1}(0)$;

(b) $\int_0^1 P_{2p}(x)\,dx = 0$;

(c) $\int_0^1 P_{2p+1}(x)\,dx = (-1)^p\dfrac{(2p)!}{2^{2p+1}p!\,(p+1)!}$.

(d) Compute these integrals for $p = 0$.

39. Show that the functions in Equations **(26)** and **(27)** are linearly independent.

40. Prove Equations **(31)** and **(32)** using Rodrigues' formula.

41. Use the binomial theorem (see Problem 5.1.16) to prove that

$$\frac{1}{\sqrt{1-2xz+z^2}} = P_0(x) + P_1(x)z + P_2(x)z^2 + \cdots + P_n(x)z^n + \ldots,$$

where $P_n(x)$ is the nth Legendre polynomial. This identity is called the **generating function** for Legendre polynomials.

Other special functions.

42. Consider the **Laguerre equation***

$$xy'' + (1-x)y' + py = 0. \qquad \textbf{(i)}$$

(a) Show that if p is a nonnegative integer, there is a polynomial solution to Equation **(i)** of the form

$$L_p(x) = \sum_{n=0}^{p} \frac{(-1)^n p!x^n}{(p-n)!(n!)^2}.$$

The functions $L_p(x)$ are known as the **Laguerre polynomials.**

(b) Calculate $L_0(x)$, $L_1(x)$, $L_2(x)$, $L_3(x)$, and $L_4(x)$.

43. Consider the **Hermite equation**

$$y'' - 2xy' + 2py = 0. \qquad \textbf{(ii)}$$

(a) Use the method of Frobenius to show that all solutions of Equation **(ii)** are of the form

$$c_0\left[1 + \sum_{n=1}^{\infty} \frac{2^n(-p)(2-p)\ldots(2n-2-p)x^{2n}}{(2n)!}\right] + c_1\left[x + \sum_{n=1}^{\infty} \frac{2^n(1-p)(3-p)\ldots(2n-1-p)x^{2n+1}}{(2n+1)!}\right],$$

where c_0 and c_1 are arbitrary constants.

(b) Show that Equation **(ii)** has a polynomial solution of degree p for a nonnegative integer p. These polynomials, denoted by $H_p(x)$, are called the **Hermite polynomials of degree *p*.**

(c) Show that

$$H_p(x) = \sum_{n=0}^{M} \frac{(-1)^n p!(2x)^{p-2n}}{n!(p-2n)!},$$

where M is the greatest integer $\le p/2$.

(d) Calculate H_0, H_1, H_2, H_3, and H_4.

5.6 AN EXCURSION: OSCILLATORY SOLUTIONS OF LINEAR SECOND-ORDER DIFFERENTIAL EQUATIONS WITH VARIABLE COEFFICIENTS

In this chapter we showed how some linear differential equations with variable coefficients could be solved by power series methods. In this excursion we show how interesting results about such equations can be found without solving the equations.

All solutions of the linear differential equation $y'' + \omega^2 y = 0$ are of the form $y = A\sin(\omega x + \delta)$ (see p. 130). These functions oscillate with period $2\pi/\omega$. In addition, each solution has an infinite number of zeros

$$x = -\frac{\delta}{\omega}, \frac{\pm\pi - \delta}{\omega}, \frac{\pm 2\pi - \delta}{\omega}, \ldots.$$

If a linear second-order differential equation has constant coefficients, then all its solutions can be found. The solutions oscillate if and only if the roots of the characteristic equation are complex conjugates. But what if we have variable coefficients? Can we determine when they have oscillatory solutions? It is this question that we explore here.

Consider the differential equation

$$y'' + p(x)y = 0. \qquad \textbf{(1)}$$

oscillation We say that $y(x)$ **oscillates** on the interval $[c, \infty)$ if y is nonconstant and has infinitely many zeros on $[c, \infty)$.

*Edmond Laguerre (1834–1886) was a French mathematician whose research was in geometry and infinite series.

EXAMPLE 1 ▶ **An oscillatory function**

$y = A\sin(\omega x + \delta)$ oscillates on $[c, \infty)$ for any real number c. ◀

THEOREM 1* ♦ **Sturm separation theorem**

Let y_1 and y_2 be linearly independent solutions of Equation **(1)** and let x_1 and x_2 be consecutive zeros of y_1. Then y_2 has exactly one zero in the interval (x_1, x_2).

Proof Consider the Wronskian of y_1 and y_2 (see p. 111):

$$W(y_1, y_2)(x) = y_1(x)y_2'(x) - y_2(x)y_1'(x) \neq 0,$$

since y_1 and y_2 are independent. Now

$$W(y_1, y_2)(x_1) = y_1(x_1)y_2'(x_1) - y_2(x_1)y_1'(x_1) \overset{y_1(x_1)=0}{=} -y_2(x_1)y_1'(x_1) \tag{2}$$

and

$$W(y_1, y_2)(x_2) = -y_2(x_2)y_1'(x_2). \tag{3}$$

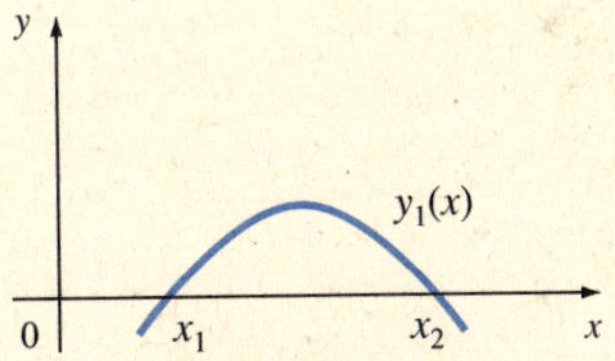

FIGURE 5.11
$y_1(x)$ is positive in (x_1, x_2).

Since $y_1(x) \neq 0$ in the interval (x_1, x_2) and y_1 is continuous, we conclude that y_1 is either always positive or always negative on that interval. Assume that y_1 is positive. The proof is similar in the other case. The situation is illustrated in Figure 5.11.

From **(2)** and **(3)**, we know that $y_2(x_1) \neq 0$ and $y_2(x_2) \neq 0$. Assume that $y_2(x_1) > 0$. Figure 5.11 shows that $y_1'(x_1) \geq 0$. But $y_1'(x_1) = 0$ is impossible, because then y_1 would satisfy the initial-value problem

$$y'' + p(x)y = 0, \qquad y_1(x_1) = y_1'(x_1) = 0, \tag{4}$$

which (by the existence-uniqueness theorem, p. 105) would imply $y_1 \equiv 0$. Hence $y_1'(x_1) > 0$, so from **(2)**

$$W(y_1, y_2)(x_1) < 0. \tag{5}$$

Since W is continuous and nonzero (because y_1 and y_2 are linearly independent), we must have

$$W(y_1, y_2)(x_2) = -y_2(x_2)y_1'(x_2) < 0.$$

[Otherwise, if $W(y_1, y_2)(x_2) > 0$, then W would have a zero in (x_1, x_2), contradicting linear independence.] But, from Figure 5.11, $y_1'(x_2) < 0$, implying that $y_2(x_2) < 0$.

Since $y_2(x_1) > 0$ and $y_2(x_2) < 0$, we conclude from the intermediate-value theorem that y_2 has at least one zero in (x_1, x_2). However, y_2 cannot have two or more zeros in (x_1, x_2). If $x_3 < x_4$ were two such zeros, then reversing the role of y_1 and y_2 in the proof above shows that y_1 has a zero in (x_3, x_4); call it x_5. But then $x_1 < x_5 < x_2$ are three zeros of y_1, contradicting the hypothesis that x_1 and x_2 are consecutive zeros of y_1. Thus y_2 has exactly one zero in (x_1, x_2). ♦

Theorem 1 shows that if one solution of Equation **(1)** oscillates, then every other nontrivial solution of Equation **(3)** oscillates. Conversely, if one nontrivial solution of Equation **(1)** does not oscillate, no solution oscillates. Thus we can talk about Equation **(1)** being **oscillatory** or **nonoscillatory.**

Recall that all solutions of $y'' + \omega^2 y = 0$ are of the form $y = A\sin(\omega x + \delta)$ and oscillate with period $\frac{2\pi}{\omega}$. If we increase the size of ω, the oscillations become

---

*This theorem was first proved by the Swiss mathematician Jacques Charles François Sturm (1803–1855).

more frequent. Is this also true for second-order linear equations with variable coefficients? The following result, called the **Sturm comparison theorem,** compares the rates of oscillation of solutions of two equations

$$y'' + p(x)y = 0, \tag{6}$$

$$z'' + q(x)z = 0. \tag{7}$$

THEOREM 2 ♦ Sturm comparison theorem

Let y be a solution of Equation **(6)** with consecutive zeros at x_1 and x_2. Suppose $q(x) \geq p(x)$, with p and q continuous and strict inequality holding for at least one point of the interval $[x_1, x_2]$. Then any solution z of Equation **(7)** that has a zero at x_1 has another zero in the interval (x_1, x_2). Roughly speaking, the larger q is, the more rapidly solutions oscillate.

Proof We can assume that $y(x) > 0$ on the interval (x_1, x_2); the proof is similar in the other case. Then, by the comments we made about the initial-value problem **(4),** it follows that $y'(x_1) > 0$ and $y'(x_2) < 0$ (see Figure 5.12).

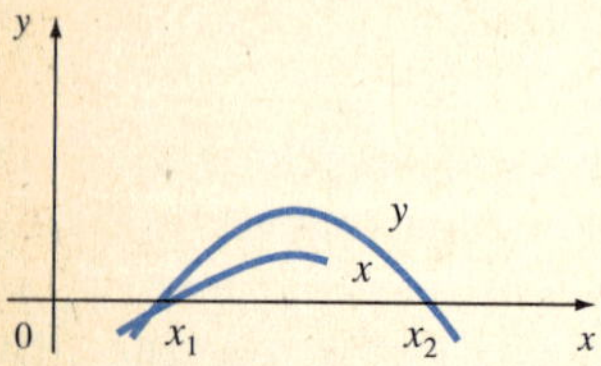

FIGURE 5.12
$y'(x_1) > 0$ and $y'(x_2) < 0$.

Suppose $z'(x_1) > 0$; a similar proof holds in the other case. Multiply Equation **(6)** by $-z$ and Equation **(7)** by y to obtain

$$-zy'' - pyz = 0,$$
$$yz'' + qyz = 0.$$

Adding these two equations, we have

$$yz'' - zy'' + (q - p)yz = 0. \tag{8}$$

Observe that

$$(yz' - zy')' = yz'' + y'z' - y'z' - zy'' = yz'' - zy''.$$

Hence, if we integrate **(8)** from x_1 to x_2, we get

$$[y(x)z'(x) - z(x)y'(x)]\Big|_{x_1}^{x_2} + \int_{x_1}^{x_2} [q(x) - p(x)]y(x)z(x)\,dx = 0,$$

or, since $y(x_1) = y(x_2) = z(x_1) = 0$,

$$-z(x_2)y'(x_2) + \int_{x_1}^{x_2} [q(x) - p(x)]y(x)z(x)\,dx = 0.$$

Thus

$$\int_{x_1}^{x_2} [q(x) - p(x)]y(x)z(x)\,dx = z(x_2)y'(x_2). \tag{9}$$

Now assume $z(x) > 0$ on (x_1, x_2). Then the integral on the left side of **(9)** is positive while the right side is not [since $y'(x_2) < 0$]. From this contradiction we can infer the truth of the theorem. ♦

COROLLARY ♦

If $q(x) \geq \omega^2 > 0$ on $[c, \infty)$, then

$$y'' + q(x)y = 0$$

oscillates.

Proof In Example 1 we showed that $y'' + \omega^2 y = 0$ had the oscillatory solution $y = A\sin(\omega x + \delta)$. By Theorem 2, it follows that the solutions to $y'' + qy = 0$ oscillate as well. ♦

EXAMPLE 2 ▶ **Oscillatory solutions of a Bessel equation**

Consider the Bessel equation of order zero:

$$x^2y'' + xy' + x^2y = 0. \tag{10}$$

Dividing both sides by x^2, we get

$$y'' + \frac{1}{x}y' + y = 0. \tag{11}$$

Substituting

$$y = ze^{-(1/2)\int dx/x} = \frac{z}{\sqrt{x}}$$

we have

$$y' = \frac{z'}{\sqrt{x}} - \frac{z}{2x^{3/2}} = \frac{1}{\sqrt{x}}\left(z' - \frac{z}{2x}\right),$$

$$y'' = \frac{z''}{\sqrt{x}} - \frac{z'}{x^{3/2}} + \frac{3z}{4x^{5/2}} = \frac{1}{\sqrt{x}}\left(z'' - \frac{z'}{x} + \frac{3z}{4x^2}\right),$$

and Equation **(11)** becomes

$$0 = y'' + \frac{1}{x}y' + y = \frac{1}{\sqrt{x}}\left[\left(z'' - \frac{z'}{x} + \frac{3z}{4x^2}\right) + \frac{1}{x}\left(z' - \frac{z}{2x}\right) + z\right],$$

or

$$z'' + \left(1 + \frac{1}{4x^2}\right)z = 0. \tag{12}$$

But $1 + 1/4x^2 > 1$ on $[c, \infty)$ for $c > 0$, so all nontrivial solutions z to Equation **(12)** oscillate by the corollary. Since $\sqrt{x} > 0$, $y = z/\sqrt{x}$ must also oscillate. Thus, without solving the Bessel equation of order zero, we can show that each of its nontrivial solutions oscillates. This result is not at all obvious even after the equation has been solved. ◀

THEOREM 3 ♦

If $p(x)$ is continuous on $[a, \infty)$ and

$$\int_a^\infty p(x)\,dx = +\infty, \tag{13}$$

then every nontrivial solution of

$$y'' + p(x)y = 0 \tag{14}$$

oscillates on $[a, \infty)$.

Proof* Suppose that some solution y of Equation **(14)** has at most a finite number of zeros on $[a, \infty)$. Then there is a number $b \geq a$ such that $y(x) \neq 0$ on $[b, \infty)$. Let

$$z(x) = \frac{y'(x)}{y(x)}.$$

* This is a special case of the proof by W. J. Coles, "A Simple Proof of a Well-Known Oscillation Theorem," *Proceedings of the American Mathematical Society* 19 (1968): 507.

Then

$$z' = \frac{yy'' - (y')^2}{y^2} = \frac{y''}{y} - z^2$$

and we obtain the Riccati equation

$$z'(x) + z^2(x) = \frac{y''(x)}{y(x)} = -p(x). \tag{15}$$

If we integrate both sides of Equation **(15)** from b to c ($> b$), we get

$$z(x)\Big|_b^c + \int_b^c z^2(x)\,dx = -\int_b^c p(x)\,dx,$$

or

$$z(c) + \int_b^c z^2(x)\,dx = z(b) - \int_b^c p(x)\,dx. \tag{16}$$

Since $z(b)$ is finite and the integral on the right side of Equation **(16)** tends to infinity as c approaches infinity, the left side of Equation **(16)** is negative for all c no smaller than some number x_0. Hence, replacing c on the left side of Equation **(16)** by $x \geq x_0$ and changing variables of integration to avoid confusion,

$$0 \leq \int_b^x z^2(t)\,dt \leq -z(x). \tag{17}$$

Set $R(x) = \int_b^x z^2(t)\,dt$. Then, squaring both sides of Equation **(17)**, we get

$$R^2(x) = \left[\int_b^x z^2(t)\,dt\right]^2 \leq z^2(x). \tag{18}$$

But $R'(x) = z^2(x)$, so inequality Equation **(18)** can be rewritten as

$$R^2(x) \leq R'(x).$$

Separating variables, we have

$$\int_{x_0}^x dx \leq \int_{R(x_0)}^{R(x)} \frac{dR}{R^2},$$

or

$$x - x_0 \leq \frac{1}{R(x_0)} - \frac{1}{R(x)}.$$

Therefore

$$x \leq x_0 + \frac{1}{R(x_0)}, \tag{19}$$

since R is positive. But the left side of **(19)** can be made arbitrarily large, while the right side is bounded, leading to a contradiction. Hence the assumption that y has only finitely many zeros is false. ♦

EXAMPLE 3 ▶ **An equation with oscillatory solutions**

The nontrival solutions of

$$y'' + \frac{1}{x}y = 0, \qquad x \geq 1,$$

are oscillatory since

$$\int_1^{\infty} \frac{dx}{x} = +\infty. \blacktriangleleft$$

This excursion presents only a brief glimpse of a well-developed theory on the oscillation or nonoscillation of linear second-order differential equations.

5.7 USING SYMBOLIC MANIPULATORS

This section will show you how MATHEMATICA, MAPLE, DERIVE, MATLAB, or MATHCAD can be used to find series solutions of linear, nonconstant coefficient, second-order differential equations. We will see that two previously defined commands are essential for this task. Recall that:

`ITERATE(f(v), v, v₀, n)` = {do n iterations of the function $v_{k+1} = f(v_k)$ beginning at v_0 and save the values $v_0, v_1, \ldots, v_n$ in a vector. The symbol v can be either a variable or a vector of variables.}.

`SOLVE(EQ(c)=0, c)` = {solves the given equation $EQ(c) = 0$ for the unknown term c}.

Taylor series

First we will present two scripts for solving differential equations using the Taylor series method, given in the perspective of Section 5.1. The first script describes the method for the first-order initial-value problem

$$y' = r(x, y), \qquad y(x_0) = y_0. \tag{1}$$

Recall from Section 5.1 that the method consists in calculating the higher derivatives of y, evaluating them at x_0, and substituting the resulting values into the series

$$y(x) = \sum_{n=0}^{\infty} \frac{y^{(n)}(x_0)}{n!}(x - x_0)^n. \tag{2}$$

The first derivative is immediate from Equation **(1)** since $y'(x_0) = r(x_0, y_0)$. To find the second and higher derivatives, apply the chain rule:

$$y''(x) = \frac{d}{dx} r(x, y) = \frac{\partial r}{\partial x} + \frac{\partial r}{\partial y}\frac{dy}{dx} = [1, r] \cdot \left[\frac{\partial r}{\partial x}, \frac{\partial r}{\partial y}\right],$$

where the notation $\mathbf{u} = [u_1, u_2, \ldots, u_n]$ is used to indicate a vector in n-dimensional space, and if $\mathbf{v} = [v_1, v_2, \ldots, v_n]$ then the **dot product** of **u** and **v** is given by

$$\mathbf{u} \cdot \mathbf{v} = u_1 v_1 + u_2 v_2 + \cdots + u_n v_n.$$

Following this process, if $y'' = R(x, y)$, then $y''' = [1, r] \cdot [\partial R/\partial x, \partial R/\partial y]$, and so forth. Observe that in each case we are taking the dot product of the fixed vector $[1, r]$ and the **gradient** of the previous derivative

$$\text{gradient of } r(x, y) = \text{grad } r = \nabla r = \left[\frac{\partial r}{\partial x}, \frac{\partial r}{\partial y}\right].$$

Finally, we evaluate these derivatives at x_0 by substituting $x = x_0$ and $y = y_0$.

If we analyze the steps in this process, we realize there are three items we must keep track of:

1. The order of the derivative k, so we can divide by $k!$.
2. The function $y^{(k-1)}$ since we plan to take its gradient.
3. The cummulative sum of the terms to that point.

This suggests that we use a vector in the ITERATE process, so that each entry of the vector keeps track of each of these items. Hence, consider the vector

$$\mathbf{v} = [v_1, v_2, v_3] = \left[k, y^{(k-1)}, \sum_{n=0}^{k} \frac{f^{(n)}(x_0)}{n!}(x - x_0)^n\right].$$

A script for finding the first n terms of the Taylor series solution of any first-order initial-value problem **(1)** is given by the following algorithm:

```
TAYLOR1(x,y,r(x,y),x₀,y₀,n):=
        ITERATE([v₁+1,[1, r]•GRAD(v₂), v₃+SUBST(v₂,x=x₀,y=y₀)*
        (x−x₀)^v₁/v₁!], v,[1,r,y₀],n).
```

We could have used the command `LIM(v₂, [x, y], [x₀, y₀])` instead of the `SUBST` command above to achieve the same purpose.

It is just as easy to find the Taylor series solution of a second- or higher-order differential equation. Consider the variable coefficient second-order linear initial-value problem

$$y'' + p(x)y' + q(x)\,y = 0, \qquad y(x_0) = y_0, \qquad y'(x_0) = z_0. \tag{3}$$

The idea is to rewrite this second-order equation as a system of first-order equations:

$$\begin{aligned} y' &= z, & y(x_0) &= y_0, \\ z' &= -q(x)y - p(x)z, & z(x_0) &= z_0. \end{aligned} \tag{4}$$

We know the values $y(x_0) = y_0$ and $y'(x_0) = z(x_0) = z_0$. To find the second and third-derivatives $y''(x_0)$ and $y'''(x_0)$, observe that

$$\begin{aligned} y'' &= z' = -qy - pz = [1, z, -qy - pz] \cdot [0, 0, 1] \\ &= [1, z, -qy - pz] \cdot \nabla(z), \\ y''' &= z'' = \frac{d}{dx}[-qy - pz] = -q'y - qy' - p'z - pz' \\ &= -q'y - p'z - qz + p(qy + pz) \\ &= [1, z, -qy - pz] \cdot [-q'y - p'z, -q, -p] \\ &= [1, z, -qy - pz] \cdot \nabla(-qy - pz). \end{aligned}$$

In effect, we are again taking the dot product of the gradient of the previous derivative by a fixed vector. Thus, the algorithm is now

```
TAYLOR2(x, y, z, p(x), q(x), x₀, y₀, z₀, n):=
      ITERATE([v₁+1, [1,z,−q*y−p*z]•GRAD(v₂),v₃+
         SUBST(v₂,x=x₀,y=y₀,z=z₀)*(x−x₀)^v₁/v₁!],v,[1,z,y₀],n).
```

We illustrate this algorithm with an example.

EXAMPLE 1 ▶

Find the first three terms of the Taylor series solution of the initial-value problem

$$y'' + xy' + y = 0, \qquad y(0) = 1, \qquad y'(0) = 0.$$

Solution This is Example 5.1.7. Here, $p(x) = x$, $q(x) = 1$, $x_0 = 0$, $y_0 = 1$, $z_0 = 0$, so we give the command

```
TAYLOR2(x, y, z, x, 1, 0, 1, 0, 3):=
```

which sequentially produces the vectors:

```
{[1, z, y0]=[1, z, 1],
[1+1, [1, z,-y-x*z]•GRAD(z),1+SUBST(z, x=0,y=1,z=0)*x/1!}=
          =[2,-y-x*z,1],
 [2+1, [1,z,-y-x*z]•GRAD(-y-x*z), 1+SUBST(-y-x*z, x=0,y=1,z=0)
*x^2/2!]= [3, [1,z,-y-x*z]•[-z,-1,-x], 1-x^2/2!]=
         =[3, -2*z+x*y+x^2*z, 1 - x^2/2!]
```

and the v_3 term carries the first three terms (one of which is zero) of the Taylor series solution. If we find the first ten terms, the algorithm would yield

$$v_3 = 1 - \frac{x\hat{\ }2}{2} + \frac{x\hat{\ }4}{8} - \frac{x\hat{\ }6}{48} + \frac{x\hat{\ }8}{384} - \frac{x\hat{\ }10}{3840}.$$

This answer can be rewritten in the form

$$y_1(x) = 1 - \frac{x^2}{2} + \frac{x^4}{2\cdot 4} - \frac{x^6}{2\cdot 4\cdot 6} + \frac{x^8}{2\cdot 4\cdot 6\cdot 8} - \frac{x^{10}}{2\cdot 4\cdot 6\cdot 8\cdot 10} + \cdots$$

which is the answer arrived at in Example 5.1.7.

Similarly, if the initial conditions are changed to $y(0) = 0$, $y'(0) = 1$, then

```
TAYLOR2(x, y, z, x, 1, 0, 0, 1, 10)
```

yields

$$y_2(x) = x - \frac{x^3}{3} + \frac{x^5}{15} - \frac{x^7}{105} + \frac{x^9}{945} - \cdots$$

which is the $y_2(x)$ power series in that example.

Finally, since *any* solution to this differential equation can be expressed in terms of these two power series solutions (see Section 3.1), the general solution to the differential equation is $y(x) = c_1y_1(x) + c_2y_2(x)$. ◄

The Taylor series method succeeds in finding a power series only in fairly limited situations. Generally, you should not apply this method unless $p(x)$ and $q(x)$ are polynomials.

Frobenius's method

Recall from Section 5.3, that the value of r in the substitution

$$y(x) = \sum_{n=0}^{\infty} c_n x^{n+r} \tag{5}$$

is obtained by finding the roots of the indicial equation for the differential equation

$$m(x)y'' + p(x)y' + q(x)y = 0. \tag{6}$$

In what follows we will assume that the singular point is at $x_0 = 0$, and that $m(x)$, $xp(x)$, and $x^2q(x)$ can all be expanded as Maclaurin series. We have written Equation **(6)** in this form, instead of in the way it was discussed in Section 5.3, because symbol manipulators are more able to deal with polynomials than with rational expressions.

Rewrite Equation **(5)** in the form $y(x) = x^r Z(x)$, where $Z(x)$ is the power series, and substitute this expression into Equation **(6)**. Since $y = x^r Z$,

$$qy = x^r qZ = (x^{r-2}Z)(x^2 q),$$
$$py' = p(rx^{r-1}Z + x^r Z') = (x^{r-2}Z)(rxp) + x^r pZ',$$

and

$$my'' = mr(r-1)x^{r-2}Z + 2mrx^{r-1}Z' + mx^r Z'',$$

so that Equation **(6)** becomes

$$x^{r-2}\{x^2 mZ'' + x(2rm + xp)Z' + (r(r-1)m + rxp + x^2 q)Z\} = 0. \qquad \textbf{(7)}$$

Thus, if $m(x) = m_0 + m_1 x + m_2 x^2 + \ldots$, $xp(x) = p_0 + p_1 x + p_2 x^2 + \ldots$, and $x^2 q(x) = q_0 + q_1 x + q_2 x^2 + \ldots$, the term with the lowest power in x has the coefficient

$$m_0 r(r-1) + p_0 r + q_0 = 0,$$

which, if $m_0 = 1$, is the usual indicial equation (see Equation **(5.3.6)**). Therefore, a script for finding the indicial equation is

```
INDICIAL(x, m(x), p(x), q(x)):=
    SOLVE(r*(r−1)+r*LIM(x*p/m, x, 0)+LIM(x^2*q/m,x,0)=0,r).
```

EXAMPLE 2 ▶

Find the roots of the indicial equation (about zero) for the differential equation

$$xy'' + \left(x - \frac{1}{4}\right)y' + \frac{1}{4x}y = 0.$$

Solution We give the command

```
INDICIAL(x,x,x−1/4,1/(4*x)),
```

and get sequentially

```
SOLVE(r*(r−1)+r*LIM(x−1/4,x,0)+LIM(1/4,x,0)=0,r)
```

$$\texttt{SOLVE(r^2-(5/4)*r+(1/4)=0,r)} = \left[r=\frac{1}{4}, r=1 \right]. \quad ◀$$

Once the roots of the indicial equation are known, we can proceed to use Frobenius's method. It is generally best to use the smallest root first, as on occasion it will yield the complete general solution to the problem.

The method of proof consists of substituting a partial sum for Z,

$$Z(x) = \sum_{n=0}^{N} c_n x^n, \qquad \textbf{(8)}$$

in the term in braces of Equation **(7)** and solving for the unknown coefficients c_2, $c_3, \ldots, c_N$ in terms of c_0 and c_1. We can give a script for this procedure:

```
Z(x,N):=SUM(c(n)*x^n,n,0,N)={the partial sum Σ(n=0..N) c(n)*x^n}
FROBENIUS(x, r, m, p, q, N):=
   SOLVE(x*x*m*DIF(DIF(Z(x,N), x), x)+(x^2*p+2*r*x*m)*
     DIF(Z(x, N), x)+(m*r^2 − m*r+x*p*r+x^2*q)*Z(x, N)= 0, c),
```

where $c = [c(1), c(2), \ldots, c(N)]$. We illustrate with an example.

EXAMPLE 3 ▶

Find the general solution of the Bessel equation

$$x^2y'' + xy' + \left(x^2 - \frac{1}{4}\right)y = 0.$$

Solution First we find the roots (about 0) of the indicial equation by giving the command

```
INDICIAL(x,x^2,x,x^2-1/4)
```

which yields the solution $[r = 1/2, r = -1/2]$. Next, we select $r = -\frac{1}{2}$ and give the command

```
FROBENIUS(x,-1/2,x^2,x,x^2-1/4,10)
```

which will give, for different powers of x, the vector:

$$[0, 0, 0, 0, 0, 2\,c(2) + c(0), 6\,c(3) + c(1), 12\,c(4) + c(2), \ldots].$$

If the symbolic manipulator can solve for $c(k)$, $k = 2, \ldots, N$ in terms of $c(0)$ and $c(1)$, (some cannot), we obtain

$$C(2) = -\frac{C(0)}{2!},\ C(3) = -\frac{C(1)}{3!},\ C(4) = -\frac{C(2)}{12} = \frac{C(0)}{4!},\ C(5) = -\frac{C(3)}{20}$$
$$= \frac{C(1)}{5!},\ C(6) = -\frac{C(4)}{30} = -\frac{C(0)}{6!},\ C(7) = -\frac{C(1)}{7!}$$

from which it follows that the general solution is

$$y(x) = \frac{1}{\sqrt{x}}\left\{C(0)\left[1 - \frac{x^2}{2!} + \frac{x^4}{4!} - \frac{x^6}{6!} + \cdots\right]\right.$$
$$\left. + C(1)\left[x - \frac{x^3}{3!} + \frac{x^5}{5!} - \frac{x^7}{7!} + \ldots\right]\right\}$$

or

$$y(x) = C(0)\,\frac{\cos(x)}{\sqrt{x}} + C(1)\,\frac{\sin(x)}{\sqrt{x}}.$$ ◀

PROBLEMS 5.7

Use the Taylor series method and a symbolic manipulating program to find the first ten terms of the Taylor series solution of problems 1–5.

1. $y'' - xy' + y = 0,\ y(0) = 0,\ y'(0) = 2$

2. $y'' - 2x\,y' + 2y = 0,\ y(0) = 0,\ y'(0) = 1$

3. $y'' + 2x\,y' + 2y + 0,\ y(0) = 0,\ y'(0) = 1$

4. $y'' - x\,y' + x\,y = 0,\ y(0) = 2,\ y'(0) = 1$

5. $y'' - x^2\,y = 0,\ y(0) = 1,\ y'(0) = 0$

Use Frobenius's method to find the first eight terms of each solution of problem 6–9.

6. $x\,y'' + 2y' + x\,y = 0$

7. $x\,y'' + y = 0$

8. $4x\,y'' + 2y' - y = 0$

9. $x^2\,y'' + (x - 2)y = 0$

Find one solution to the following problem using Frobenius's method.

10. $x^2\,y'' - x\,y' + (x^2 + 1)y = 0$

SUMMARY OUTLINE OF CHAPTER 5

Power series: $\sum_{n=0}^{\infty} c_n(x-a)^n$ p. 195

coefficients, c_n
center of the series, a
interval of convergence, radius of convergence, R
ratio test: $R = \lim_{n\to\infty} \left|\frac{c_n}{c_{n+1}}\right|$
absolute convergence: when $\sum_0^{\infty} |c_n|\,|x|^n$ converges;
converges when $|x| < R$;
diverges when $|x| > R$

Maclaurin series: $\sum_{n=0}^{\infty} \frac{f^{(n)}(0)}{n!}x^n$. Taylor series: $\sum_{n=0}^{\infty} \frac{f^{(n)}(a)}{n!}(x-a)n$.

Analytic function
Taylor's Formula

Taylor Series Method p. 203

Binomial Formula: $(1+x)^p = 1 + px + \frac{p(p-1)}{1\cdot 2}x^2 + \frac{p(p-1)(p-2)}{1\cdot 2\cdot 3}x^3 + \ldots$ p. 205

Airy's Equation: $y'' - xy = 0$ p. 206

Power Series Method: Let $y = \sum_0^{\infty} c_n x^n$. p. 208

When is it guaranteed to work for $y'' + a(x)y' + b(x)y = 0$? When $a(x)$, $b(x)$ are analytic functions.
ordinary points: $x = 0$ is ordinary if $a(x)$ and $b(x)$ are analytic at $x = 0$.
singular points: points that are not ordinary points

Method of Frobenius Let $y = \sum_0^{\infty} c_n x^{n+r}$. p. 217

regular singular point: when $xa(x)$ and $x^2 b(x)$ are analytic functions at $x = 0$.
irregular point: a singular point that is not a regular singular point
indicial equation: $r(r-1) + a_0 r + b_0 = 0$
exponents of the indicial equation: the roots of the indicial equation
differing, but not by an integer
equal exponents
differing by an integer

Bessel's Equation of Order p: $x^2y'' + xy' + (x^2 - p^2)y = 0$ p. 233
of the first kind: $J_p(x)$
of the second kind: $Y_p(x)$
gamma function $\Gamma(p)$
Euler constant: $\gamma = 0.5772156649\ldots$

Legendre's Equation: $(1-x^2)y'' - 2x\,y' + p(p+1)y = 0$ p. 239
Legendre function: any solution of Legendre's equation
Legendre polynomial of degree p: $P_p(x)$; a polynomial solution of Legendre's equation
Rodrigues's Formula:

$$P_p(x) = \frac{1}{2^p p!}\frac{d^p}{dx^p}(x^2-1)^p$$

Hermite's Equation: $y'' - 2xy' + 2py = 0$ p. 246

Laguerre's Equation: $xy'' + (1-x)y' + py = 0$ p. 246

Oscillations p. 246
oscillatory equation: has infinitely many zeros on $[c, \infty)$
nonoscillatory equation

Sturm Separation Theorem p. 247

Sturm Comparison Theorem p. 248

REVIEW EXERCISES FOR CHAPTER 5

1. Show that when $-1 \le x < 1$,
$$\ln(1-x) = -\left(x + \frac{x^2}{2} + \frac{x^3}{3} + \frac{x^4}{4} + \dots\right).$$

2. Use Taylor's theorem to show that
$$\sin x = \sin a + \cos a(x-a) - \frac{\sin a}{2!}(x-a)^2 - \frac{\cos a}{3!}(x-a)^3 + \dots.$$

3. Verify that
$$\ln\frac{1+x}{1-x} = 2\left(x + \frac{x^3}{3} + \frac{x^5}{5} + \dots + \frac{x^{2n-1}}{2n-1} + \dots\right), \quad |x| < 1.$$

4. Show that
$$\ln\frac{x^2}{x^2-1} = \frac{1}{x^2} + \frac{1}{2x^4} + \frac{1}{3x^6} + \dots,$$
for $|x| > 1$. [*Hint:* Consider the series $\ln(1+x) = x - x^2/2 + x^3/3 - x^4/4 + \dots$, $|x| < 1$.]

5. Use Taylor's theorem to show that $e^{\tan x} = 1 + x + x^2/2! + 3x^3/3! + 9x^4/4! + 37x^5/5! + \dots$, for $|x| < \frac{\pi}{2}$.

In Exercises 6–11 use power series to find the solution of the given initial-value problem.

6. $xy'' + xy' - y = -e^{-x}$, $y(0) = 1$, $y'(0) = 0$
7. $xy'' - y' + (1-x)y = x^2(1-x)$, $y(0) = y'(0) = 1$
8. $(1+x)y'' + (2 - x - x^2)y' + y = 0$, $y(0) = 0$, $y'(0) = 1$
9. $xy'' - 2y' + xy = x^2 - 2 - 2\sin x$, $y(0) = 0$, $y'(0) = 1$
10. $y'' - 3x^2y' - 6xy = 0$, $y(0) = 1$, $y'(0) = 0$
11. $y'' + 2xy' + 2y = 0$, $y(0) = 1$, $y'(0) = 0$

In Exercises 12–20 use the method of Frobenius to solve the given differential equations.

12. $4xy'' + 2y' - y = 0$
13. $2x(1-2x)y'' + (1+4x^2)y' - (1+2x)y = 0$
14. $2x(1-2x)y'' + (1+x)y' - y = 0$
15. $x(1-x)y'' + 2(1-2x)y' - 2y = 0$
16. $2x(1-x)y'' + (1-x)y' + 3y = 0$
17. $2x^2y'' + x(1-x)y' - y = 0$
18. $xy'' + y = 0$
19. $x^2y'' + (x^2-2)y = 0$
20. $x^2y'' - xy' + (x^2+1)y = 0$
21. Prove the following identities if $J_0(a) = J_0(b) = 0$:
    (a) $\int_0^1 xJ_0(ax)J_0(bx)\,dx = 0,\ a \ne b$
    (b) $\int_0^1 xJ_0^2(ax)\,dx = \frac{1}{2}J_1^2(ax)$
    (c) $x^2J_0'(x) = x[J_1(x) - J_0(x)] - (x^2-1)J_1(x)$
    (d) $[xJ_0(x)J_1(x)]' = x[J_0^2(x) - J_1^2(x)]$
22. Find solutions in terms of Bessel functions for the following differential equations:
    (a) $x^2y'' + (x^2-2)y = 0$
    (b) $x^2y'' - xy' + (1 + x^2 - k^2)y = 0$
23. **Modified Bessel functions:** The function $I_p(x) = i^{-p}J_p(ix)$, $i^2 = -1$ is called the **modified Bessel function of the first kind of order p.** Show that $I_p(x)$ is a solution of the differential equation
$$x^2y'' + xy' - (x^2+p^2)y = 0,$$
and obtain its series representation by the method of Frobenius.
24. Show that another solution of the differential equation in Problem 23 is the **modified Bessel function of the second kind of order p:** (p not an integer)
$$K_p(x) = \frac{\pi}{2\sin p\pi}[I_{-p}(x) - I_p(x)].$$
25. Prove (see Problem 5.5.25) that
$$e^{(x/2)[t+(1/t)]} = I_0(x) + \sum_{n=1}^{\infty} I_n(x)[t^n + t^{-n}].$$
26. Prove the following identities:
    (a) $\frac{d}{dx}[x^pI_p(x)] = x^pI_{p-1}(x)$
    (b) $\frac{d}{dx}[x^{-p}I_p(x)] = x^{-p}I_{p+1}(x)$
    (c) $x(I_{p-1} - I_{p+1}) = 2pI_p$

27. Gauss's* hypergeometric equation is given by

$$x(1-x)y'' + [c - (a+b+1)x]y' - aby = 0,$$

where a, b, and c are constants. Show that it has a solution of the form

$$y_1(x) = 1 + \frac{ab}{c}x + \frac{a(a+1)b(b+1)}{c(c+1)}\frac{x^2}{2!} + \frac{a(a+1)(a+2)b(b+1)(b+2)}{c(c+1)(c+2)}\frac{x^3}{3!} + \cdots.$$

This series is called the **hypergeometric series** and is denoted by the symbol $F(a, b, c; x)$.

28. Prove, using the result of Exercise 27, that
(a) $F(-a, b, b; -x) = (1+x)^a$;
(b) $xF(1, 1, 2; -x) = \ln(1+x)$.

29. Show that the **Chebyshëv equation**†

$$(1-x^2)y'' - xy' + p^2y = 0$$

has a polynomial solution when p is an integer.

30. Prove the following identities for integral values of n:

(a) $\displaystyle\int_{-1}^{1} xP_n(x)P_{n-1}(x)\,dx = \frac{2x}{4n^2-1}$

(b) $\displaystyle\int_{-1}^{1} (1-x^2)[P_n'(x)]^2\,dx = \frac{2n(n+1)}{2n+1}$

(c) $\displaystyle(m+n+1)\int_0^1 x^m P_n(x)\,dx = m\int_0^1 x^{m-1}P_{n-1}(x)\,dx$

31. Find a solution for the differential equation

$$y'' + 2(\cot x)y' + n(n+1)y = 0.$$

[*Hint:* Make the trigonometric substitution $t = \cos x$.]

32. Show that the substitution

$$y(x) = z(x)\exp\left(-\frac{1}{2}\int a(x)\,dx\right)$$

transforms the linear equation

$$y'' + a(x)y' + b(x)y = 0$$

into an equation of the form

$$z'' + p(x)z = 0.$$

---

* Carl Friedrich Gauss (1777–1855) is sometimes referred to as the greatest mathematician. He spent his entire prolific life in Germany. An infant prodigy, he made brilliant discoveries in geometry, algebra, number theory, statistics, and astronomy.

† Pafnuti Lvovich Chebyshëv (1821–1894) was the leading Russian mathematician of his time.

6 LAPLACE TRANSFORMS

6.1 INTRODUCTION: DEFINITION AND BASIC PROPERTIES OF THE LAPLACE TRANSFORM

One of the most efficient methods of solving certain ordinary and partial differential equations is to use Laplace* transforms. The effectiveness of the Laplace transform is due to its ability to convert a differential equation into an algebraic equation whose solution yields the solution of the differential equation when the transformation is inverted.

Definition Laplace transform

Let $f(t)$ be a real-valued function that is defined for $t \geq 0$. Suppose that $f(t)$ is multiplied by e^{-st} and the result is integrated with respect to t from zero to infinity. If the integral

$$\mathcal{L}\{f(t)\} = \int_0^\infty e^{-st} f(t)\, dt = \lim_{A \to \infty} \int_0^A e^{-st} f(t)\, dt \tag{1}$$

converges, the resulting function of s† is called the **Laplace transform** of the function f. We stress that the original function f is a function of the variable t, whereas its Laplace transform F is a function of the variable s, and write

$$F(s) = \mathcal{L}\{f(t)\}.$$

EXAMPLE 1 ► **The Laplace transform of a constant**

Find the Laplace transform of the function $f(t) \equiv k$ (a constant).

Solution

$$\mathcal{L}\{k\} = \int_0^\infty e^{-st} k\, dt = k \lim_{A \to \infty} \int_0^A e^{-st}\, dt = k \lim_{A \to \infty} \left(\frac{e^{-st}}{-s} \bigg|_0^A \right)$$

$$= k \lim_{A \to \infty} \frac{1 - e^{-sA}}{s} = \frac{k}{s}, \text{ provided that } s > 0. \blacktriangleleft$$

---

* Named after the French mathematician Pierre Simon de Laplace (1749–1827). See the biographical sketch on page 267.

† In general, the variable s is complex, but in this book we will only consider real values of s.

In this chapter we will use Laplace transforms to solve differential equations. In order to do so, we must be able to recover a function from its transform. The following theorem enables us to do so when f is a continuous function. The proof of this theorem involves complex variable theory and so is omitted.

THEOREM 1 ♦ **Uniqueness of Laplace transforms**

Suppose that f is continuous on $[0, \infty)$ and has a Laplace transform $F(s) = \mathcal{L}\{f(t)\}$. Then there is no other continuous function having the same Laplace transform; that is, if g is continuous and $F(s) = \mathcal{L}\{g(t)\}$, then $f(t) = g(t)$ for $t \geq 0$. This means that there is a one-to-one correspondence between continuous functions (which have Laplace transforms) and their Laplace transforms. ♦

We now define the inverse Laplace transform.

Definition Inverse transform

It should now be apparent why the word *transform* is associated with this operation. The operation transforms the original function $f(t)$ into a new function $F(s) = \mathcal{L}\{f(t)\}$. Since we are interested in reversing the procedure, we call the original function f the **inverse transform** of $F = \mathcal{L}\{f\}$ and denote it by $f = \mathcal{L}^{-1}\{F\}$. Note that the last statement translates into the mathematical symbols $\mathcal{L}^{-1}\{F\} = \mathcal{L}^{-1}\{\mathcal{L}\{f\}\} = f$; that is

$$\text{if} \quad \mathcal{L}\{f(t)\} = F(s), \quad \text{then } \mathcal{L}^{-1}\{F(s)\} = f(t). \tag{2}$$

Note

This definition makes sense because of Theorem 1; that is, if $F(s)$ is given, then there is only one continuous function f such that $\mathcal{L}\{f(t)\} = F(s)$.

EXAMPLE 2 ▶ **The Laplace transform of e^{at}**

Let $f(t) = e^{at}$, where a is constant. Then

$$\mathcal{L}\{e^{at}\} = \int_0^\infty e^{-st}e^{at}\,dt = \int_0^\infty e^{-(s-a)t}\,dt = \lim_{A\to\infty} \left.\frac{e^{-(s-a)t}}{a-s}\right|_0^A,$$

and the integral converges, if $s - a > 0$, to

$$\mathcal{L}\{e^{at}\} = \frac{1}{s-a}.$$

Thus the Laplace transform of the function e^{at} is the function $F(s) = \frac{1}{s-a}$ for $s > a$. Note that we can also conclude, from **(2)**, that

$$\mathcal{L}^{-1}\{F(s)\} = \mathcal{L}^{-1}\left\{\frac{1}{s-a}\right\} = e^{at}. \quad ◀$$

The last example suggests that $\mathcal{L}\{f(t)\}$ may not be defined for all values of s; but if it is defined, then it will exist for suitably large values of s. Indeed, $\mathcal{L}\{e^{at}\}$ is defined only for values $s > a$.

EXAMPLE 3 ▶ **A function with no Laplace transform**

Let $f(t) = \frac{1}{t}$. Then

$$\int_0^\infty \frac{e^{-st}}{t}\,dt = \int_0^1 \frac{e^{-st}}{t}\,dt + \int_1^\infty \frac{e^{-st}}{t}\,dt. \tag{3}$$

But for t in the interval $0 \le t \le 1$, $e^{-st} \ge e^{-s}$ if $s > 0$. Thus,

$$\int_0^\infty \frac{e^{-st}}{t}\,dt \ge e^{-s}\int_0^1 \frac{dt}{t} + \int_1^\infty \frac{e^{-st}}{t}\,dt.$$

However,

$$\int_0^1 t^{-1}\,dt = \lim_{A\to 0}\int_A^1 t^{-1}\,dt = \lim_{A\to 0} \ln t\Big|_A^1$$

$$= \lim_{A\to 0}(\ln 1 - \ln A) = \lim_{A\to 0}(-\ln A) = +\infty.$$

Therefore the integral **(3)** diverges, so $f(t) = \frac{1}{t}$ has no Laplace transform. ◀

This example and the preceding discussion pinpoint the need for answers to the following questions:

1. Which functions f have Laplace transforms?
2. Can two discontinuous functions f and g have the same Laplace transform?

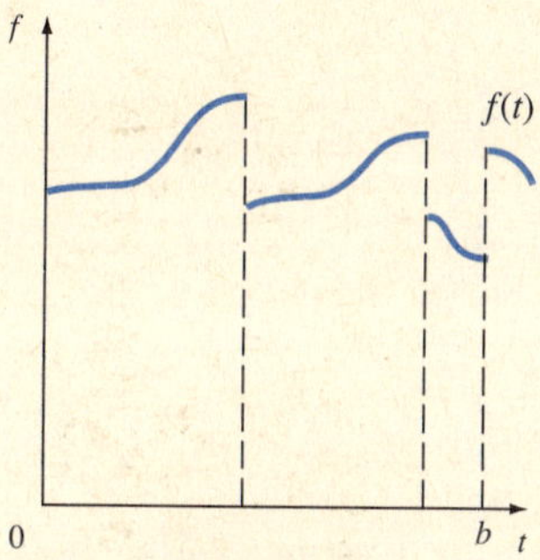

FIGURE 6.1
A piecewise continuous function

In answering these questions, we will be satisfied if a class of functions, large enough to contain virtually all functions that arise in practice, can be found for which the Laplace transforms exist and whose inverses are unique. We will see in Section 6.2 that certain differential equations can be solved by first taking the Laplace transform of both sides of the equation, then finding the Laplace transform of the solution, and finally arriving at the answer by taking the unique inverse transform of the transform of the solution. The uniqueness of the Laplace transform is guaranteed by Theorem 1 in the case f is continuous. The need for a unique inverse arises from the fact that we need to reverse the transformation to find the solution of a given problem, and this step is not possible if the inverses within the given class of functions are not unique.

In order to give simple conditions that guarantee the existence of a Laplace transform, we require the following definitions:

jump discontinuity

A function f has a **jump discontinuity** at a point t^* if the function has different (finite) limits as it approaches t^* from the left and from the right or if the two limits are equal but different from $f(t^*)$. Note that $f(t^*)$ may or may not be equal to either

$$\lim_{t\to t^{*+}} f(t) \quad \text{or} \quad \lim_{t\to t^{*-}} f(t).$$

piecewise continuous function

In fact, $f(t^*)$ may not even be defined. A function f defined on $(0, \infty)$ is **piecewise continuous** if it is continuous on every finite interval $0 \le t \le b$, except possibly at finitely many points where it has jump discontinuities. Such a function is illustrated in Figure 6.1. The class of piecewise continuous functions includes every continuous function, as well as many important discontinuous functions such as the unit step function, square waves, and the staircase function, which we will encounter later in this chapter.

THEOREM 2 ♦ **Existence theorem**

Let f be piecewise continuous on $t \ge 0$ and satisfy the condition

$$|f(t)| \le Me^{at}, \tag{4}$$

for $t \ge T$, where a, M, and T are fixed nonnegative constants. Then $\mathcal{L}\{f(t)\}$ exists for all $s > a$.

Proof Since f is piecewise continuous, $e^{-st}f(t)$ has a finite integral over any finite interval on $t \geq 0$, and

$$|\mathscr{L}\{f(t)\}| = \left|\int_0^\infty e^{-st}f(t)\,dt\right| \leq \left|\int_0^T e^{-st}f(t)\,dt\right| + \int_T^\infty e^{-st}|f(t)|\,dt. \quad \textbf{(5)}$$

The first integral on the right-hand side of Equation **(5)** exists and, by Equation **(4)**,

$$\int_T^\infty e^{-st}|f(t)|\,dt \leq M\int_T^\infty e^{-(s-a)t}\,dt = \left.\frac{Me^{-(s-a)t}}{a-s}\right|_T^\infty,$$

which converges to $Me^{-(s-a)t}/(s-a)$ as t approaches infinity if $s > a$. Thus it follows by the comparison theorem of integrals that $\mathscr{L}\{f(t)\}$ exists. ♦

The conditions in Theorem 2 are easy to test.

EXAMPLE 4 ▶ **A function that has a Laplace transform and one that does not**

The function $f(t) = t^n$ has a Laplace transform because

$$t^n \leq n!\,e^t, \qquad \text{for any } t > 0 \text{ and integer } n \geq 0.$$

To see this, observe that

$$e^t = 1 + t + \frac{t^2}{2!} + \cdots + \frac{t^n}{n!} + \cdots,$$

so $t^n/n! \leq e^t$, or $t^n \leq n!\,e^t$, for $t > 0$.

On the other hand,

$$e^{t^2} > Me^{at}, \quad \text{since} \quad t^2 > \ln M + at$$

for sufficiently large t, regardless of the choice of M and a. Exponentiating both sides, we get

$$e^{t^2} > e^{\ln M + at} = e^{\ln M}e^{at} = Me^{at}.$$

Thus e^{t^2} does not satisfy Theorem 2. It is not hard to show that e^{t^2} has no Laplace transform.

This follows from the fact that $\int_0^A e^{-st}e^{t^2}\,dt = \int_0^A e^{t^2-st}\,dt$. If $s < 0$, then

$$\int_0^A e^{t^2-st}\,dt \geq \int_0^A e^{-st}\,dt = \tfrac{1}{s}(1 - e^{-As}) \to -\infty \text{ as } A \to \infty,$$

(since $-As > 0$). If $s > 0$, then for $t > 2s$, $t^2 - st > 2st - st = st$. So, if $A > 2s$,

$$\int_0^A e^{t^2-st}\,dt > \int_{2s}^A e^{t^2-st}\,dt > \int_{2s}^A e^{st}\,dt = \frac{1}{s}(e^{sA} - e^{2s^2}) \to \infty \text{ as } A \to \infty.$$

This shows that the limit in **(1)** diverges for every nonzero s (you should prove it for $s = 0$) so that e^{t^2} does not have a Laplace transform. ◀

Note

e^{t^2} does not have a Laplace transform because we just proved it. You *cannot* infer that e^{t^2} does not have a Laplace transform because the conditions of Theorem 2 are not satisfied. There are functions that have Laplace transforms but for which the conditions of Theorem 2 fail to hold.

It should also be noted that there are functions having Laplace transforms that do not satisfy the hypotheses of Theorem 1. For example, $f(t) = 1/\sqrt{t}$ is infinite at $t = 0$; but setting $x^2 = st$, we have

$$\mathscr{L}\{t^{-1/2}\} = \int_0^\infty e^{-st}t^{-1/2}\,dt = \frac{2}{\sqrt{s}}\int_0^\infty e^{-x^2}\,dx = \sqrt{\frac{\pi}{s}},$$

according to Formula 216 in Appendix 1.

The proof of uniqueness would take us too far afield at this point. However, it can be shown that two functions having the same Laplace transform cannot differ at every point in an interval of positive length,* although they may differ at several isolated points. Such differences are generally of no importance in applications. Hence the Laplace transform has an essentially unique inverse. In particular, different continuous functions have different Laplace transforms.

One of the most important properties of the Laplace transform is stated in the following theorem.

THEOREM 3 ♦ **Linearity property of the Laplace transform**

If $\mathscr{L}\{f(t)\}$ and $\mathscr{L}\{g(t)\}$ exist, then $\mathscr{L}\{af(t) + bg(t)\}$ exists and

$$\mathscr{L}\{af(t) + bg(t)\} = a\mathscr{L}\{f(t)\} + b\mathscr{L}\{g(t)\}.$$

Proof By definition

$$\begin{aligned}\mathscr{L}\{af(t) + bg(t)\} &= \int_0^\infty e^{-st}\,[af(t) + bg(t)]\,dt \\ &= a\int_0^\infty e^{-st}f(t)\,dt + b\int_0^\infty e^{-st}g(t)\,dt \\ &= a\mathscr{L}\{f(t)\} + b\mathscr{L}\{g(t)\} = aF(s) + bG(s). \quad ♦\end{aligned}$$

In other words, the Laplace transform of a linear combination of functions is the linear combination of their Laplace transforms. Conversely, we have the following.

COROLLARY ♦ **Linearity property of the inverse Laplace transform**

If $F(s) = \mathscr{L}\{f(t)\}$ and $G(s) = \mathscr{L}\{g(t)\}$, then

$$\begin{aligned}\mathscr{L}^{-1}\{aF(s) + bG(s)\} &= a\mathscr{L}^{-1}\{F(s)\} + b\mathscr{L}^{-1}\{G(s)\} \\ &= af(t) + bg(t). \quad ♦\end{aligned}$$

The proof of the corollary is left as an exercise (Problem 79).

The linearity property allows us to deal with a linear equation term by term in order to obtain its Laplace transform.

It is essential that we begin to recognize the Laplace transforms of many different functions as quickly as possible. Table 6.1 gives the transforms of seven basic functions. A much more extensive list is given in Section 6.2. Note that $0! = 1$.

---

*See, for example, I. S. Sokolnikoff and R. M. Redheffer, *Mathematics of Physics and Modern Engineering* (New York: McGraw-Hill, 1966), p. 217.

TABLE 6.1 Laplace Transforms of Some Basic Functions

$f(t)$	$\mathcal{L}\{f(t)\}$	*Domain of Definition*	
e^{at}	$\dfrac{1}{s-a}$	$s > a$ (real)	**(6)**
k (a constant)	$\dfrac{k}{s}$	$s > 0$	**(7)**
t^n	$\dfrac{n!}{s^{n+1}}$	$s > 0$	**(8)**
$\sin bt$	$\dfrac{b}{s^2+b^2}$	$s > 0$	**(9)**
$\cos bt$	$\dfrac{s}{s^2+b^2}$	$s > 0$	**(10)**
$\sinh bt$	$\dfrac{b}{s^2-b^2}$	$s > \lvert b\rvert$	**(11)**
$\cosh bt$	$\dfrac{s}{s^2-b^2}$	$s > \lvert b\rvert$	**(12)**

EXAMPLE 5 ▶ **The Laplace transform of t^n**

Let n be a positive integer. Then Equation **(8)** is obtained by integrating

$$\mathcal{L}\{t^n\} = \int_0^\infty e^{-st}t^n\,dt$$

by parts with $u = t^n$ and $dv = e^{-st}\,dt$. We have $du = nt^{n-1}\,dt$ and $v = -(\frac{1}{s})e^{-st}$:

$$\mathcal{L}\{t^n\} = \frac{-e^{-st}t^n}{s}\bigg|_0^\infty + \frac{n}{s}\int_0^\infty e^{-st}t^{n-1}\,dt$$

$$= \frac{n}{s}\mathcal{L}\{t^{n-1}\}, \qquad \text{for } s > 0,$$

because $\lim_{t\to\infty} e^{-st}t^n/s = 0$, if $s > 0$, by n applications of L'Hôpital's rule. Thus

$$\mathcal{L}\{t^n\} = \frac{n}{s}\mathcal{L}\{t^{n-1}\} = \frac{n(n-1)}{s^2}\mathcal{L}\{t^{n-2}\} = \cdots = \frac{n!}{s^n}\mathcal{L}\{t^0\}.$$

But $t^0 = 1$, so by Equation **(7)** we have $\mathcal{L}\{1\} = \frac{1}{s}$ and

$$\mathcal{L}\{t^n\} = \frac{n!}{s^{n+1}}. \quad ◀$$

EXAMPLE 6 ▶ **The Laplace transform of sin *at***

The Laplace transform of sin at can be obtained by integrating by parts. Let $u =$ sin at and $dv = e^{-st}\,dt$; then $du = a\cos at\,dt$ and $v = (-1/s)e^{-st}$, so

$$\mathcal{L}\{\sin at\} = \int_0^\infty e^{-st}\sin at\,dt = \frac{-e^{-st}\sin at}{s}\bigg|_0^\infty + \frac{a}{s}\int_0^\infty e^{-st}\cos at\,dt,$$

or

$$\mathcal{L}\{\sin at\} = \frac{a}{s}\mathcal{L}\{\cos at\}. \tag{13}$$

We must integrate this last integral again by parts with $u = \cos at$ and $dv = e^{-st}\,dt$; we have $du = -a \sin at\,dt$ and $v = -(1/s)e^{-st}$, yielding

$$\mathcal{L}\{\sin at\} = \frac{a}{s}\int_0^\infty e^{-st}\cos at\,dt = \frac{a}{s}\left[-\frac{e^{-st}\cos at}{s}\bigg|_0^\infty - \frac{a}{s}\int_0^\infty e^{-st}\sin at\,dt\right],$$

or

$$\mathcal{L}\{\sin at\} = \frac{a}{s^2} - \frac{a^2}{s^2}\mathcal{L}\{\sin at\}. \tag{14}$$

Moving the last term on the right-hand side of Equation **(14)** to the left-hand side, we get

$$\left(1 + \frac{a^2}{s^2}\right)\mathcal{L}\{\sin at\} = \frac{a}{s^2},$$

or

$$\frac{s^2 + a^2}{s^2}\mathcal{L}\,\{\sin at\} = \frac{a}{s^2}.$$

Multiplying both sides by $s^2/(s^2 + a^2)$, we obtain

$$\mathcal{L}\{\sin at\} = \frac{a}{s^2 + a^2}.$$

Equation **(10)** follows at once from Equations **(9)** and **(13),** since

$$\mathcal{L}\{\cos at\} = \frac{s}{a}\mathcal{L}\,\{\sin at\} = \frac{s}{s^2 + a^2}.$$

(Explain why these computations are valid only if $s > 0$.) ◀

Problem 41 illustrates another method for finding the Laplace transform of cos *at* and sin *at* by using the Euler formula (see Appendix 5)

$$e^{iat} = \cos(at) + i\sin(at).$$

EXAMPLE 7 ▶ **The Laplace transform of sinh *at***

Since $\sinh at = (e^{at} - e^{-at})/2$, we can use Theorem 2 and Equation **(6)** to obtain

$$\mathcal{L}\{\sinh at\} = \mathcal{L}\left\{\frac{e^{at} - e^{-at}}{2}\right\} = \frac{1}{2}(\mathcal{L}\{e^{at}\} - \mathcal{L}\{e^{-at}\})$$

$$= \frac{1}{2}\left(\frac{1}{s-a} - \frac{1}{s+a}\right) = \frac{a}{s^2 - a^2}.$$

Similarly,

$$\mathcal{L}\{\cosh at\} = \mathcal{L}\left(\frac{e^{ut} + e^{-at}}{2}\right\} = \frac{1}{2}(\mathcal{L}\{e^{at}) + \mathcal{L}\{e^{-at}\})$$

$$= \frac{1}{2}\left(\frac{1}{s-a} + \frac{1}{s+a}\right) = \frac{s}{s^2 - a^2}.$$ ◀

EXAMPLE 8 ▶ **A Laplace transform**

Compute $\mathcal{L}\{3t^5 - t^8 + 4 - 5e^{2t} + 6\cos 3t\}$.

Solution Using the linearity property and the results in Table 6.1, we have

$$\begin{aligned}
&\mathcal{L}\{3t^5 - t^8 + 4 - 5e^{2t} + 6\cos 3t\} \\
&\quad = 3\mathcal{L}\{t^5\} - \mathcal{L}\{t^8\} + \mathcal{L}\{4\} - 5\mathcal{L}\{e^{2t}\} + 6\mathcal{L}\{\cos 3t\} \\
&\quad = 3\frac{5!}{s^6} - \frac{8!}{s^9} + \frac{4}{s} - \frac{5}{s-2} + \frac{6s}{s^2+9} \\
&\quad = \frac{360}{s^6} - \frac{40{,}320}{s^9} + \frac{4}{s} - \frac{5}{s-2} + \frac{6s}{s^2+9}.
\end{aligned}$$ ◀

There are many other facts that can be used to facilitate the computation of Laplace transforms. One of these is given below. Other facts will be given in the exercises and in the next three sections.

The next theorem presents a quick way of computing the Laplace transform $\mathcal{L}\{e^{at}f(t)\}$ when $\mathcal{L}\{f(t)\}$ is known.

THEOREM 4 ◆ **First shifting property of the Laplace transform**

Suppose that $F(s) = \mathcal{L}\{f(t)\}$ exists for $s > b$. If a is a real number, then

$$\mathcal{L}\{e^{at}f(t)\} = F(s-a), \qquad \text{for } s > a + b. \tag{15}$$

Rewriting **(15)** in terms of the inverse Laplace transform, we have

$$\mathcal{L}^{-1}\{F(s-a)\} = e^{at}f(t). \tag{16}$$

Proof By definition,

$$\begin{aligned}
\mathcal{L}\{e^{at}f(t)\} &= \int_0^\infty e^{-st}e^{at}f(t)\,dt \int_0^\infty e^{-(s-a)t}f(t)\,dt \\
&= \mathcal{L}\{f(t)\}|_{s-a} = F(s-a),
\end{aligned}$$

if $s - a > b$ or $s > a + b$. As the formula suggests, *to find* $\mathcal{L}\{e^{at}f(t)\}$ *we simply replace each* s *in* $\mathcal{L}\{f(t)\}$ *by* $s - a$. ◆

EXAMPLE 9 ▶ **Using the first shifting theorem**

Compute $\mathcal{L}\{e^{2t}\cos 3t\}$.

Solution Since

$$F(s) = \mathcal{L}\{\cos 3t\} = \frac{s}{s^2+9}$$

and $a = 2$, we have

$$\mathcal{L}\{e^{2t}\cos 3t\} = F(s-2) = \frac{s-2}{(s-2)^2+9}.$$ ◀

If we apply the first shifting property to the Laplace transforms in Table 6.1, we obtain the results shown in Table 6.2.

In Section 6.2 we will have frequent need to evaluate inverse Laplace transforms. The next example demonstrates the use of completing the square in finding inverse Laplace transforms.

EXAMPLE 10 ▶ **An inverse transform**

Compute

$$\mathcal{L}^{-1}\left\{\frac{s+9}{s^2+6s+13}\right\}.$$

TABLE 6.2 More Laplace Transforms

$f(t)$	$\mathscr{L}\{f(t)\}$	*Domain of Definition*
$e^{at}t^n$	$\dfrac{n!}{(s-a)^{n+1}}$	$s > a$
$e^{at}\sin bt$	$\dfrac{b}{(s-a)^2+b^2}$	$s > a$
$e^{at}\cos bt$	$\dfrac{s-a}{(s-a)^2+b^2}$	$s > a$
$e^{at}\sinh bt$	$\dfrac{b}{(s-a)^2-b^2}$	$s > a + \lvert b\rvert$
$e^{at}\cosh bt$	$\dfrac{s-a}{(s-a)^2-b^2}$	$s > a + \lvert b\rvert$

Solution Example 9 provides a hint. We begin by completing the square in the denominator:

$$\frac{s+9}{s^2+6s+13} = \frac{s+9}{(s+3)^2+4}.$$

But $4 = 2^2$ and $s + 9 = (s + 3) + 6$, so we can write

$$\frac{s+9}{s^2+6s+13} = \frac{(s+3)+6}{(s+3)^2+2^2}.$$

By the linearity property of the inverse Laplace transform (see the corollary to Theorem 2), the inverse transform of a sum equals the sum of the inverse transforms. Hence

$$\mathscr{L}^{-1}\left\{\frac{(s+3)+6}{(s+3)^2+2^2}\right\} = \mathscr{L}^{-1}\left\{\frac{(s+3)}{(s+3)^2+2^2}\right\} + 3\mathscr{L}^{-1}\left\{\frac{2}{(s+3)^2+2^2}\right\},$$

and using Table 6.2 (or the first shifting property and Table 6.1), we have

$$\mathscr{L}^{-1}\left\{\frac{(s+3)+6}{(s+3)^2+2^2}\right\} = e^{-3t}\cos 2t + 3e^{-3t}\sin 2t. \quad ◀$$

PIERRE-SIMON DE LAPLACE (1749–1827)

Pierre-Simon de Laplace (pronounced La pláhss) was born of poor parents in 1749. His early mathematical ability won him good teaching posts, and as a political opportunist he ingratiated himself with whichever party happened to be in power during the uncertain days of the French Revolution. His most outstanding work was done in the fields of celestial mechanics, probability, differential equations, and geodesy. He published two monumental works, *Traité de mécanique céleste* (five volumes, 1799–1825) and *Théorie analytique des probabilités* (1812), each of which was preceded by an extensive nontechnical exposition. The five-volume *Traité de mécanique céleste,* which earned him the title of "the Newton of France," embraced all previous discoveries in celestial mechanics along with Laplace's own contributions and marked the author as the unrivaled master in the subject. It may be of interest to repeat a couple of anecdotes often told in connection with this work. When Napoleon teasingly remarked that God was not mentioned in his treatise, Laplace replied, "Sire, I did not need that hypothesis." And when he translated Laplace's treatise into English, the American astronomer Nathaniel Bowditch remarked, "I never come across one of Laplace's 'thus it plainly appears' without feeling sure that I have hours of hard work before me to fill up the chasm and find out and show how it plainly appears." Laplace's name is connected with the *nebular hypothesis* of cosmogony and with the so-called *Laplace equation* of potential theory (though neither of these contributions originated with Laplace), with the so-called *Laplace transform* that later became the key to the operational calculus of Heaviside, and with the *Laplace expansion* of a determinant. Laplace died in 1827, exactly one hundred years after the death of Isaac Newton. According to one report, his last words were "What we know is slight; what we don't know is immense."

The following story about Laplace offers a valuable suggestion to one applying for a position. When Laplace arrived as a young man in Paris seeking a professorship of mathematics, he submitted his recommendations by prominent people to d'Alembert but was not received. Returning to his lodgings, Laplace wrote d'Alembert a brilliant letter on the general principles of mechanics. This opened the door, and d'Alembert replied, "Sir, you notice that I paid little attention to your recommendations. You don't need any; you have introduced yourself better." A few days later Laplace was appointed professor of mathematics at the Military School of Paris.

Laplace was very generous to beginners in mathematical research. He called these beginners his stepchildren, and there are several instances in which he withheld publication of a discovery to allow a beginner the opportunity to publish first.

We close our brief account of Laplace with two quotations due to him: "All the effects of nature are only mathematical consequences of a small number of immutable laws." "In the final analysis, the theory of probability is only common sense expressed in numbers."

SELF-QUIZ

Answer Problems I and II true or false.

I. If a function is not piecewise continuous for $t \geq 0$ then it cannot have a Laplace transform.

II. Every polynomial has a Laplace transform defined for all $s > 0$.

III. $\mathcal{L}\{\sin \frac{t}{2}\} =$ ______.

(a) $\dfrac{2}{s^2+4}$ **(b)** $\dfrac{s}{s^2+4}$

(c) $\dfrac{2}{4s^2+1}$ **(d)** $\dfrac{s}{4s^2+1}$

IV. $\mathcal{L}\{t^2 - 3t + 5\} =$ ______.

(a) $\dfrac{2}{s^3} - \dfrac{3}{s^2} + \dfrac{5}{s}$ **(b)** $\dfrac{3}{s^3} - \dfrac{3}{s^2} + \dfrac{5}{s}$

(c) $\dfrac{6}{s^3} - \dfrac{6}{s^2} + \dfrac{5}{s}$ **(d)** $\dfrac{1!}{s^3} - \dfrac{3!}{s^2} + \dfrac{5!}{s}$

V. If $\mathcal{L}\{f(t)\} = \dfrac{1}{s^2+4s+5}$, then $\mathcal{L}\{e^{2t}f(t)\} =$ ______.

(a) $\dfrac{e^{2s}}{s^2+4s+5}$

(b) $\dfrac{1}{(s-2)(s^2+4s+5)}$

(c) $\dfrac{1}{(s^2+4s+5)-2}$

(d) $\dfrac{1}{(s-2)^2+4(s-2)+5}$

VI. $\mathcal{L}^{-1}\left\{\dfrac{s-1}{s^2-2s+10}\right\} =$ ______.

(a) $e^t \cos \sqrt{10}\, t$ **(b)** $e^t \cos 3t$

(c) $e^{-t} \cos \sqrt{10}\, t$ **(d)** $e^{-t} \cos 3t$

Answers to Self-Quiz

I. False **II.** True **III.** c **IV.** a **V.** d **VI.** b

PROBLEMS 6.1

Find the Laplace transforms of the functions in Problems 1–40, where a, b, and c are real constants. For what values of s are the transforms defined?

1. $5t + 2$
2. $4t - 5$
3. $9t^2 - 7$
4. $10t^2 - 5t$
5. $t^2 + 8t - 16$
6. $23t^3 + 8t - 7$
7. $\frac{t^3}{8} + \frac{t^2}{4} + \frac{t}{2} + 1$
8. $\frac{t^5}{71} - \frac{t^3}{4} + 3$
9. $at + b$
10. $at^2 + bt + c$
11. e^{5t+2}
12. e^{4t-3}
13. $e^{t/2}$
14. $e^{-t/5}$
15. $e^{-t-1/2}$
16. e^{at+b}
17. $\sin(3t)$
18. $\cos\left(-\frac{t}{4}\right)$
19. $\cos(7t)$
20. $\sin\left(\frac{t}{3}\right)$
21. $\sin(5t + 2)$
22. $\cos(5t - 9)$
23. $\cos(at + b)$
24. $\sin(at + b)$
25. $\cosh(t/2)$
26. $\sinh\left(-\frac{t}{4}\right)$
27. $\cosh(5t - 2)$
28. $\sinh(3t + 5)$
29. $\sinh(at + b)$
30. $\cosh(at + b)$
31. te^t
32. t^2e^{3t}
33. $(t^3 - 1)e^{-t}$
34. $e^{5t}(2t^2 - 3t)$
35. $e^t \sin t$
36. $e^{-t} \cos 3t$
37. $e^{4t} \cos 2t$
38. $e^{-t} \sinh 3t$
39. $e^{-t}(\sin t + \cos t)$
40. $e^{3t}(2t + \cosh t)$

---

41. Recall that $e^{iat} = \cos at + i \sin at$ (see Equation **(14)** in Appendix 5).
   (a) Show that $\mathcal{L}\{e^{iat}\} = \frac{1}{s - ia}$; $s > 0$.
   (b) Show that $\frac{1}{s - ia} = \frac{s + ia}{s^2 + a^2}$.
   (c) Use parts (a) and (b) to derive (without integration by parts) the formulas for $\mathcal{L}\{\sin at\}$ and $\mathcal{L}\{\cos at\}$. [*Hint:* Equate real and imaginary parts.]

In Problems 42–55 find f(t) where F(s) = $\mathcal{L}\{f(t)\}$ is given. If necessary, complete the square in the denominator.

42. $\frac{7}{s^2}$
43. $\frac{18}{s^3} + \frac{7}{s}$
44. $\frac{a_1}{s} + \frac{a_2}{s^2} + \frac{a_3}{s^3} + \cdots + \frac{a_{n+1}}{s^{n+1}}$
45. $\frac{s + 1}{s^2 + 1}$
46. $\frac{7}{s - 3}$
47. $\frac{s - 2}{s^2 - 2}$
48. $\frac{s - 2}{s^2 + 3}$
49. $\frac{1}{(s - 1)^2}$
50. $\frac{3}{s^2 + 2s + 2}$
51. $\frac{3}{s^2 + 4s + 9}$
52. $\frac{s + 12}{s^2 + 10s + 35}$
53. $\frac{2s - 1}{s^2 + 2s + 8}$
54. $\frac{7s - 8}{s^2 + 9s + 25}$
55. $\frac{cs + d}{s^2 + 2as + b}$ $b > a^2 > 0$; a, b, c, d are real

In Problems 56–61 express each given hyperbolic function in terms of exponentials and apply the first shifting theorem to prove the given equality.

56. $\mathcal{L}\{\cosh^2 at\} = \frac{s^2 - 2a^2}{s(s^2 - 4a^2)}$
57. $\mathcal{L}\{\sinh^2 at\} = \frac{2a^2}{s(s^2 - 4a^2)}$
58. $\mathcal{L}\{\cosh at \sin at\} = \frac{a(s^2 + 2a^2)}{s^4 + 4a^4}$
59. $\mathcal{L}\{\cosh at \cos at\} = \frac{s^3}{s^4 + 4a^4}$
60. $\mathcal{L}\{\sinh at \sin at\} = \frac{2a^2s}{s^4 + 4a^4}$
61. $\mathcal{L}\{\sinh at \cos at\} = \frac{a(s^2 - 2a^2)}{s^4 + 4a^4}$

Using the method above, find the Laplace transforms in Problems 62–67.

62. $\mathscr{L}\{\cosh at \cosh bt\}$

63. $\mathscr{L}\{\sinh at \sinh bt\}$

64. $\mathscr{L}\{\cosh at \sin bt\}$

65. $\mathscr{L}\{\cosh at \cos bt\}$

66. $\mathscr{L}\{\sinh at \sin bt\}$

67. $\mathscr{L}\{\sinh at \cos bt\}$

68. Suppose that $F(s) = \mathscr{L}\{f(t)\}$ exists for $s > a$. Show that

$$\mathscr{L}\{tf(t)\} = -F'(s), \qquad \text{for } s > a. \qquad \textbf{(i)}$$

[*Hint:* Assume that you can interchange the derivative and integral on the right-hand side of Equation **(i)**.]

69. Use Equation **(i)** to show that

$$\mathscr{L}\{t^n f(t)\} = (-1)^n \frac{d^n}{ds^n} F(s), \qquad \text{for } s > a. \qquad \textbf{(ii)}$$

*Use Equations **(i)** and **(ii)** to compute the Laplace transform of the functions given in Problems 70–78. Assume that* a *and* b *are real.*

70. te^t

71. $t^3 e^{-t}$

72. $t \sin t$

73. $t^2 \cos 3t$

74. $te^t \sin t$

75. $te^{at} \cos bt$

76. $te^{at} \sin bt$

77. $3te^{-t} \cosh t$

78. $te^{-t} \sinh 2t$

79. Suppose that $f(t) = \mathscr{L}^{-1}\{F(s)\}$ and that $g(t) = \mathscr{L}^{-1}\{G(s)\}$. Prove the linearity property of the inverse Laplace transform:

$$af(t) + bg(t) = \mathscr{L}^{-1}\{aF(s) + bG(s)\},$$

where a and b are any real constants.

80. The **gamma function** is defined by (see also page 234)

$$\Gamma(x) = \int_0^\infty e^{-u} u^{x-1}\, du, \qquad x > 0.$$

(a) Show that $\Gamma(x + 1) = \int_0^\infty e^{-u} u^x \, du$.

(b) By integrating by parts, show that $\Gamma(x + 1) = x\Gamma(x)$.

(c) Show that $\Gamma(1) = 1$.

(d) Using the results of parts (b) and (c), show that if n is a positive integer, then $\Gamma(n + 1) = n!$.

(e) By making the substitution $u = st$ in part (a), show that

$$\mathscr{L}\{t^x\} = \frac{\Gamma(x + 1)}{s^{x+1}}, \qquad s > 0, x > -1.$$

81. It can be shown that $\Gamma(\frac{1}{2}) = \sqrt{\pi}$. Use this fact and the results of Problem 80 to compute the following:

(a) $\mathscr{L}\left\{\frac{1}{\sqrt{t}}\right\}$ **(b)** $\mathscr{L}\{\sqrt{t}\}$ **(c)** $\mathscr{L}\{t^{5/2}\}$

6.2 SOLVING INITIAL-VALUE PROBLEMS BY LAPLACE TRANSFORM METHODS

In this section we show how the theory developed in Section 6.1 can be applied to solve linear initial-value problems. We see that the Laplace transform converts linear initial-value problems with constant coefficients into algebraic equations whose solutions are the Laplace transforms of the solutions to the initial-value problems.

The most important property of Laplace transforms for solving differential equations concerns the transform of the derivative of a function f. We prove below that differentiation of f roughly corresponds to multiplication of the transform by s.

THEOREM 1 ♦ Differentiation property

Suppose that there exist nonnegative constants a, M, and T such that $f(t)$ satisfies the condition

$$|f(t)| \leq Me^{at} \qquad \textbf{(1)}$$

for $t \geq T$, and suppose that $f'(t)$ is piecewise continuous for $t \geq 0$. Then the Laplace transform of $f'(t)$ exists for all $s > a$, and

$$\mathscr{L}\{f'(t)\} = s\mathscr{L}\{f(t)\} - f(0). \qquad \textbf{(2)}$$

Proof Since f is differentiable, it is also continuous. Hence it satisfies the conditions of the existence theorem (Theorem 6.1.1) and has a Laplace transform. Suppose, first, that $f'(t)$ is continuous on $t \geq 0$. Then integrating $\mathcal{L}\{f'(t)\}$ by parts, we set $u = e^{-st}$ and $dv = f'(t)\,dt$; then $du = -se^{-st}\,dt$, $v = f(t)$, and

$$\mathcal{L}\{f'(t)\} = \int_0^\infty e^{-st}f'(t)\,dt = e^{-st}f(t)\Big|_0^\infty + s\int_0^\infty e^{-st}f(t)\,dt. \tag{3}$$

Since $f(t)$ satisfies Equation **(1)**, the first term on the right-hand side in Equation **(3)** vanishes at the upper limit when $s > a$, and by definition we obtain $\mathcal{L}\{f'(t)\} = s\mathcal{L}\{f(t)\} - f(0)$. When $f'(t)$ is piecewise continuous, the proof is similar. We simply break up the range of integration into parts on each of which $f'(t)$ is continuous and integrate by parts as in Equation **(3)**. All first terms cancel out or vanish except $-f(0)$, and the second terms combine to yield $s\mathcal{L}\{f(t)\}$. ♦

Theorem 1 may be extended to apply to piecewise continuous functions $f(t)$ (see Problem 50).

Equation **(2)** may be applied repeatedly to obtain the Laplace transform of higher-order derivatives:

$$\mathcal{L}\{f''(t)\} = s\mathcal{L}\{f'(t)\} - f'(0) = s[s\mathcal{L}\{f(t)\} - f(0)] - f'(0),$$

or

$$\mathcal{L}\{f''(t)\} = s^2\mathcal{L}\{f(t)\} - sf(0) - f'(0). \tag{4}$$

Similarly,

$$\mathcal{L}\{f'''(t)\} = s^3\mathcal{L}\{f(t)\} - s^2f(0) - sf'(0) - f''(0),$$

leading by induction to the following extension of Theorem 1.

THEOREM 2 ♦ **The Laplace transform of the kth derivative**
Let $f^{(k)}(t)$ satisfy inequality (1) for $k = 0, 1, 2, \ldots, n-1$, and suppose that $f^{(n)}(t)$ is piecewise continuous on $t \geq 0$. Then $\mathcal{L}\{f^{(n)}(t)\}$ exists and is given by

$$\mathcal{L}\{f^{(n)}(t)\} = s^n\mathcal{L}\{f(t)\} - s^{n-1}f(0) - s^{n-2}f'(0) - \ldots - f^{(n-1)}(0),$$

or

$$\mathcal{L}\{f^{(n)}(t)\} = s^n\mathcal{L}\{f(t)\} - \sum_{j=0}^{n-1} s^{n-j-1}f^{(j)}(0). \quad ♦ \tag{5}$$

Theorems 1 and 2 are important, since they are used to reduce the Laplace transform of a differential equation into an equation involving only the transform of the solution. Several such applications are considered in this section. In addition, these theorems are also useful in determining the transforms of certain functions.

EXAMPLE 1 ▸ **Using the differentiation property to find a Laplace transform**
Compute $\mathcal{L}\{\sin^2 at\}$.

Solution Let $f(t) = \sin^2 at$. Then

$$f'(t) = 2a\sin at\cos at = a\sin 2at,$$

so

$$\frac{2a^2}{s^2 + 4a^2} = \mathcal{L}\{f'\} = s\mathcal{L}\{f\} - f(0).$$

Since $f(0) = 0$, it follows that

$$\mathcal{L}\{\sin^2 at\} = \frac{2a^2}{s(s^2 + 4a^2)}, \qquad s > 0.$$

Alternatively, since $\sin^2 \theta = (1 - \cos 2\theta)/2$, we have

$$\sin^2 at = \frac{1 - \cos 2at}{2}$$

and

$$\begin{aligned}\mathcal{L}\{\sin^2 at\} &= \mathcal{L}\left\{\frac{1}{2}\right\} - \frac{1}{2}\mathcal{L}\{\cos 2at\} = \frac{1}{2s} - \frac{s}{2(s^2 + 4a^2)} \\ &= \frac{s^2 + 4a^2}{2s(s^2 + 4a^2)} - \frac{s^2}{2s(s^2 + 4a^2)} = \frac{4a^2}{2s(s^2 + 4a^2)} = \frac{2a^2}{2(s^2 + 4a^2)}.\end{aligned}$$

◀

EXAMPLE 2 ▸ Using the differentiation property of second derivatives

Compute $\mathcal{L}\{t \sin at\}$.

Solution Suppose that $f(t) = t \sin at$. Then

$$f'(t) = \sin at + at \cos at,$$
$$f''(t) = 2a \cos at - a^2 t \sin at.$$

Thus, since $f(0) = f'(0) = 0$,

$$2a\mathcal{L}\{\cos at\} - a^2\mathcal{L}\{f(t)\} = \mathcal{L}\{f''(t)\} = s^2\mathcal{L}\{f(t)\},$$

so

$$(s^2 + a^2)\mathcal{L}\{f(t)\} = 2a\mathcal{L}\{\cos at\} = \frac{2as}{s^2 + a^2},$$

or

$$\mathcal{L}\{f(t)\} = \frac{2as}{(s^2 + a^2)^2}, \qquad s > 0. \quad ◀$$

An alternate technique for computing this answer is provided later in this section (p. 278).

We now apply Theorems 1 and 2 to solve initial-value problems. In what follows, we take Laplace transforms without worrying about their existence. Any solution so obtained must be checked by substitution into the original equation.

EXAMPLE 3 ▸ Solving an initial-value problem

Find the solution of the initial-value problem

$$y'' - 4y = 0, \qquad y(0) = 1, \qquad y'(0) = 2. \tag{6}$$

Solution Taking the Laplace transform of both sides of the differential equation in **(6)** and using the differentiation property, we transform Equation **(6)** into the algebraic equation

$$[s^2\mathcal{L}\{y\} - sy(0) - y'(0)] - 4\mathcal{L}\{y\} = [s^2\mathcal{L}\{y\} - s - 2] - 4\mathcal{L}\{y\} = 0;$$

thus $(s^2 - 4)\mathcal{L}\{y\} = s + 2$ and

$$\mathcal{L}\{y\} = \frac{s + 2}{s^2 - 4} = \frac{1}{s - 2}.$$

By Table 6.1 (p. 264) we have

$$y(t) = e^{2t},$$

which satisfies all the conditions in **(6)**. ◀

EXAMPLE 4 ▶ **Using Laplace transforms to solve an initial-value problem**

Solve the initial-value problem

$$y'' + 4y = 0, \qquad y(0) = 1, \qquad y'(0) = 2. \tag{7}$$

Solution Using the differentiation property, we obtain

$$[s^2\mathcal{L}\{y\} - sy(0) - y'(0)] + 4\mathcal{L}\{y\} = s^2\mathcal{L}\{y\} - s - 2 + 4\mathcal{L}\{y\} = 0.$$

Solving for $\mathcal{L}\{y\}$, we have

$$\mathcal{L}\{y\} = \frac{s+2}{s^2+4} = \left(\frac{s}{s^2+4}\right) + \left(\frac{2}{s^2+4}\right).$$

By referring to Table 6.1 we find that

$$y(t) = \cos 2t + \sin 2t,$$

which can readily be verified to be the solution of **(7)**. In calculating the inverse transform, we used the linearity of $\mathcal{L}^{-1}$. ◀

The preceding example indicates the necessity of writing $\mathcal{L}\{y\}$ as a linear combination of terms for which the inverse Laplace transforms are known.

EXAMPLE 5 ▶ **Solving nonhomogeneous initial-value problems**

Find the solution of the initial-value problem

$$y'' - 3y' + 2y = 4t - 6, \qquad y(0) = 1, \qquad y'(0) = 3. \tag{8}$$

Solution Taking the Laplace transform of both sides and using the differentiation property, we have, from Table 6.1,

$$[s^2\mathcal{L}\{y\} - s - 3] - 3[s\mathcal{L}\{y\} - 1] + 2\mathcal{L}\{y\} = \frac{4}{s^2} - \frac{6}{s},$$

so

$$(s^2 - 3s + 2)\mathcal{L}\{y\} = s + \frac{4}{s^2} - \frac{6}{s} = \frac{s^3 - 6s + 4}{s^2}.$$

Hence, factoring the numerator and denominator, we obtain

$$\begin{aligned}\mathcal{L}\{y\} &= \frac{s^3 - 6s + 4}{s^2(s^2 - 3s + 2)} \\ &= \frac{(s-2)(s^2+2s-2)}{s^2(s-2)(s-1)} = \frac{s^2+2s-2}{s^2(s-1)}.\end{aligned}$$

But

$$\frac{s^2+2s-2}{s^2(s-1)} = \frac{s^2}{s^2(s-1)} + \frac{2s-2}{s^2(s-1)} = \frac{1}{s-1} + \frac{2}{s^2},$$

so

$$\mathcal{L}\{y\} = \frac{1}{s-1} + \frac{2}{s^2}.$$

Using Table 6.1, we obtain the solution

$$y = e^t + 2t$$

to the initial-value problem in Equation **(8)**. ◀

We now discuss the most general second-order linear initial-value problem with constant coefficients. Suppose that we wish to solve the nonhomogeneous differential equation with constant coefficients

$$y'' + ay' + by = f(t), \qquad y(0) = y_0, \qquad y'(0) = y_1. \tag{9}$$

The general existence-uniqueness theorem (Theorem 3.1.1 on p. 105) states that the initial-value problem **(9)** has a unique solution if $f(t)$ is continuous. Assuming that this is the case, and taking the Laplace transforms of both sides, we obtain

$$\mathcal{L}\{y''\} + a\mathcal{L}\{y'\} + b\mathcal{L}\{y\} = \mathcal{L}\{f\}.$$

Now by Theorems 1 and 2 (differentiation properties) we have

$$[s^2\mathcal{L}\{y\} - sy(0) - y'(0)] + a[s\mathcal{L}\{y\} - y(0)] + b\mathcal{L}\{y\} = \mathcal{L}\{f\}.$$

Then

$$[s^2 + as + b]\mathcal{L}\{y\} - [sy(0) + ay(0) + y'(0)] = \mathcal{L}\{f\},$$

so

$$\mathcal{L}\{y\} = \frac{(s + a)y(0) + y'(0) + \mathcal{L}\{f\}}{s^2 + as + b}. \tag{10}$$

Three facts are evident from Equation **(10)**:

1. Initial conditions must be given so that the first two terms in the numerator of Equation **(10)** are determined.
2. The function f must have a Laplace transform so that the last term in the numerator is determined.
3. We must be able to find $\mathcal{L}^{-1}$ of the right-hand side of Equation **(10)** in order to determine the solution y of the initial-value problem **(9)**.

Thus Laplace transform methods are primarily used for the solution of linear initial-value problems with constant coefficients.

It should be clear that the major difficulty in solving problem **(9)** lies in finding the inverse transform of the right-hand side of Equation **(10)**. There is a general formula that provides the solution as an integral, but some knowledge of complex variable theory is required to take full advantage of this formula. Fortunately, many of the transforms you will encounter in solving initial-value problems can be inverted using techniques from calculus. We illustrate with some examples.

EXAMPLE 6 ▶ **Using partial functions to solve an initial-value problem**

Solve the initial-value problem

$$y'' - 5y' + 4y = e^{2t}, \qquad y(0) = 1, \qquad y'(0) = 0.$$

Solution Making use of the differentiation property and Table 6.1, we have

$$[s^2\mathcal{L}\{y\} - sy(0) - y'(0)] - 5[s\mathcal{L}\{y\} - y(0)] + 4\mathcal{L}\{y\} = \mathcal{L}\{e^{2t}\},$$

or

$$[s^2\mathcal{L}\{y\} - s] - 5[s\mathcal{L}\{y\} - 1] + 4\mathcal{L}\{y\} = \frac{1}{s - 2},$$

so

$$(s^2 - 5s + 4)\mathcal{L}\{y\} = s - 5 + \frac{1}{s-2} = \frac{s^2 - 7s + 11}{s-2}.$$

Then

$$\mathcal{L}\{y\} = \frac{s^2 - 7s + 11}{(s-2)(s^2 - 5s + 4)} = \frac{s^2 - 7s + 11}{(s-2)(s-1)(s-4)}. \tag{11}$$

Review of partial fractions

At this point we pause. Remember that when you studied techniques of integration in calculus, you integrated functions like the right-hand side of Equation **(11)** by using the method of **partial fractions.** This method is useful here. We seek constants A, B, and C such that

$$\frac{A}{s-2} + \frac{B}{s-1} + \frac{C}{s-4} = \frac{s^2 - 7s + 11}{(s-2)(s-1)(s-4)}. \tag{12}$$

Why? Because we know that $\mathcal{L}^{-1}\{1/(s-2)\} = e^{2t}$, so $\mathcal{L}^{-1}\{A/(s-2)\} = Ae^{2t}$, and so on.

There is an easy method for finding these constants:

$$A = \left.\frac{s^2 - 7s + 11}{(s-1)(s-4)}\right|_{s=2} = -\frac{1}{2};$$

$$B = \left.\frac{s^2 - 7s + 11}{(s-2)(s-4)}\right|_{s=1} = \frac{5}{3};$$

$$C = \left.\frac{s^2 - 7s + 11}{(s-2)(s-1)}\right|_{s=4} = -\frac{1}{6}.$$

Observe that we eliminate the denominator $(s - a)$ of each term on the left-hand side of Equation **(12)** from the right-hand side of Equation **(12)** and evaluate the resulting equation at $s = a$ to obtain the desired constant.

To understand why this procedure works, let us multiply both sides of Equation **(12)** by $(s - 2)$. Then we have

$$A + (s-2)\left(\frac{B}{s-1} + \frac{C}{s-4}\right) = \frac{s^2 - 7s + 11}{(s-1)(s-4)}. \tag{13}$$

Setting $s = 2$ on both sides eliminates all but the constant A on the left-hand side of Equation **(13)**, and therefore

$$A = \left.\frac{s^2 - 7s + 11}{(s-1)(s-4)}\right|_{s=2}.$$

Returning to our problem, we see that

$$\mathcal{L}\{y\} = \frac{\left(-\frac{1}{2}\right)}{s-2} + \frac{\left(\frac{5}{3}\right)}{s-1} + \frac{\left(-\frac{1}{6}\right)}{s-4},$$

which implies, according to Table 6.1, that

$$y(t) = -\frac{e^{2t}}{2} + \frac{5e^{t}}{3} - \frac{e^{4t}}{6}. \blacktriangleleft$$

EXAMPLE 7 ▶ **Using partial fractions to solve an equation**

Solve

$$y'' + 2y' + 2y = t, \qquad y(0) = y'(0) = 1.$$

Solution Using the differentiation property, we obtain

$$[s^2\mathcal{L}\{y\} - s - 1] + 2[s\mathcal{L}\{y\} - 1] + 2\mathcal{L}\{y\} = \frac{1}{s^2},$$

or

$$(s^2 + 2s + 2)\mathcal{L}\{y\} = \frac{1}{s^2} + s + 3 = \frac{s^3 + 3s^2 + 1}{s^2};$$

thus

$$\mathcal{L}\{y\} = \frac{s^3 + 3s^2 + 1}{s^2(s^2 + 2s + 2)}. \tag{14}$$

The expression $s^2 + 2s + 2$ does not have real roots, since $2^2 - 4(1)(2) = -4 < 0$, so we write the right-hand side of Equation **(14)** as

$$\frac{s^3 + 3s^2 + 1}{s^2(s^2 + 2s + 2)} = \frac{As + B}{s^2 + 2s + 2} + \frac{C}{s} + \frac{D}{s^2}. \tag{15}$$

Why do we do this? Because

$$\frac{s + 1}{s^2 + 2s + 2} = \frac{s + 1}{(s + 1)^2 + 1},$$

and so

$$\mathcal{L}^{-1}\left\{\frac{s + 1}{s^2 + 2s + 1}\right\} = \mathcal{L}^{-1}\left\{\frac{s + 1}{(s + 1)^2 + 1}\right\} = e^{-t}\cos t$$

as a consequence of the first shifting theorem (see Table 6.2). Similarly,

$$\mathcal{L}^{-1}\left\{\frac{1}{(s + 1)^2 + 1}\right\} = e^{-t}\sin t.$$

Also, by Table 6.1,

$$\mathcal{L}^{-1}\left\{\frac{1}{s}\right\} = 1 \quad \text{and} \quad \mathcal{L}^{-1}\left\{\frac{1}{s^2}\right\} = t.$$

Further review of partial fractions

There are procedures for finding the constants A, B, C, and D in Equation **(15),** but these are more complicated than the method we used in Example 6. We can find these constants directly by combining terms:

$$\frac{(As + B)s^2 + C(s^2 + 2s + 2)s + D(s^2 + 2s + 2)}{s^2(s^2 + 2s + 2)} = \frac{s^3 + 3s^2 + 1}{s^2(s^2 + 2s + 2)}.$$

Equating coefficients of like powers of s, we obtain

$$\begin{aligned} A \quad + C \qquad &= 1 \quad &&\text{these are the coefficients of } s^3 \\ B + 2C + \ D &= 3 &&\text{these are the coefficients of } s^2 \\ 2C + 2D &= 0 &&\text{these are the coefficients of } s \\ 2D &= 1 &&\text{these are the constant terms} \end{aligned}$$

From the last equation we see that $D = \frac{1}{2}$, so working backward, we obtain $C = -\frac{1}{2}$, $B = \frac{7}{2}$, and $A = \frac{3}{2}$. Then

$$\begin{aligned} y &= \mathcal{L}^{-1}\left\{\frac{s^3 + 3s^2 + 1}{s^2(s^2 + 2s + 2)}\right\} \\ &= \mathcal{L}^{-1}\left\{\frac{\frac{3}{2}s + \frac{7}{2}}{s^2 + 2s + 2} + \frac{\left(-\frac{1}{2}\right)}{s} + \frac{\left(\frac{1}{2}\right)}{s^2}\right\} \\ &= \mathcal{L}^{-1}\left\{\frac{\left(\frac{3}{2}\right)(s+1)}{(s+1)^2 + 1} + \frac{2}{(s+1)^2 + 1} + \frac{\left(-\frac{1}{2}\right)}{s} + \frac{\left(\frac{1}{2}\right)}{s^2}\right\}, \end{aligned}$$

or

$$y = \frac{3}{2}e^{-t}\cos t + 2e^{-t}\sin t - \frac{1}{2} + \frac{1}{2}t,$$

which is the solution to our differential equation. ◀

The methods used in the last two examples apply to the problem of inverting a Laplace transform obtained in trying to solve an initial-value problem with constant coefficients. In some special cases, we can use these techniques to solve linear problems with variable coefficients. First, however, we need to prove the identities that were stated in Problems 6.1.68 and 6.1.69.

Consider the derivative

$$\frac{d}{ds}\mathcal{L}\{f(t)\} = \frac{d}{ds}\int_0^\infty e^{-st}f(t)\,dt. \tag{16}$$

If we reverse the order in which the operations of differentiation and integration are performed on the right-hand side of Equation **(16)**, we obtain

$$\frac{d}{ds}\mathcal{L}\{f(t)\} = \int_0^\infty \frac{d}{ds}e^{-st}f(t)\,dt = \int_0^\infty -te^{-st}f(t)\,dt = -\mathcal{L}\{tf(t)\}.$$

We have proved the following.

THEOREM 3 ♦

If $\mathcal{L}\{f(t)\}$ exists for $s > a$, then $\mathcal{L}\{tf(t)\}$ exists for $s > a$ and

$$\mathcal{L}\{tf(t)\} = -\frac{d}{ds}\mathcal{L}\{f(t)\}. \quad ♦ \tag{17}$$

Using Equation **(17)** repeatedly, we obtain

$$\begin{aligned} \mathcal{L}\{t^n f(t)\} &= -\frac{d}{ds}\mathcal{L}\{t^{n-1}f(t)\} = (-1)^2\frac{d^2}{ds^2}\mathcal{L}\{t^{n-2}f(t)\} \\ &= \cdots = (-1)^n\frac{d^n}{ds^n}\mathcal{L}\{f(t)\}. \end{aligned} \tag{18}$$

Remark

In Equation **(16)** we reversed the order of differentiation and integration. For most functions this procedure can be mathematically justified. To do so, however, may be difficult. On the other hand, if the method succeeds in providing a verifiable solution to our problem, then the method has worked for us and we need not worry whether what we did was technically correct. It is easier to check that the final solution is a solution to the differential equation than to verify that the "reversal" is legitimate.

EXAMPLE 2 (revisited) ▶ Compute $\mathscr{L}\{t \sin at\}$.

Solution By Equation **(17)**,

$$\mathscr{L}\{t \sin at\} = -\frac{d}{ds}\mathscr{L}\{\sin at\} = -\frac{d}{ds}\left(\frac{a}{s^2 + a^2}\right) = \frac{2as}{(s^2 + a^2)^2}. \quad ◀$$

The following example illustrates how Equations **(17)** and **(18)** can be used in conjunction with the differentiation property to solve some initial-value problems with variable coefficients.

EXAMPLE 8 ▶ **Using Theorem 3 to solve a variable coefficient problem**

Solve the equation with variable coefficients

$$ty'' - ty' - y = 0, \qquad y(0) = 0, \qquad y'(0) = 3.$$

Solution If we let $Y(s) = \mathscr{L}\{y(t)\}$, then, by the differentiation property and Equation **(17),**

$$\mathscr{L}\{ty''\} = -\frac{d}{ds}\mathscr{L}\{y''\} = -\frac{d}{ds}\{s^2Y(s) - sy(0) - y'(0)\}$$
$$= -s^2Y' - 2sY + y(0) = -s^2Y' - 2sY$$

and

$$\mathscr{L}\{ty'\} = -\frac{d}{ds}\mathscr{L}\{y'\} = -\frac{d}{ds}\{sY - y(0)\} = -sY' - Y.$$

Substituting these expressions into the Laplace transform of the original equation yields

$$-s^2Y' - 2sY + sY' + Y - Y = 0.$$

Rearranging and canceling terms, we have

$$(s^2 - s)Y' + 2sY = 0.$$

We now divide both sides by $s^2 - s = s(s - 1)$ to obtain

$$Y' + \frac{2}{s - 1}Y = 0.$$

Separating variables, we have

$$\frac{dY}{Y} = -\frac{2}{s - 1}\,ds,$$

and an integration yields

$$\ln|Y| = -2\ln|s - 1| + C,$$

or

$$Y(s) = \frac{c}{(s - 1)^2}.$$

Thus, by Table 6.2,

$$y(t) = cte^t.$$

Note that $y(0) = 0$. To find c, we differentiate and use the second initial condition to obtain

$$3 = y'(0) = c(t + 1)e^t\big|_{t=0} = c.$$

Thus the unique solution to the initial-value problem is given by

$$y(t) = 3te^t. \blacktriangleleft$$

We caution the reader not to expect to be able to solve all variable coefficient equations by this method. It works only when

1. the coefficients $a_i(t)$ are polynomials in t;
2. the differential equation involving $Y(s)$ can be solved; and
3. the inverse transform of $Y(s)$ can be found.

It is rare that all these conditions can be met (see Problem 49).

The following brief list of Laplace transforms will be useful in doing the exercises.

TABLE 6.3 Short table of Laplace transforms

$f(t)$	$\mathcal{L}\{f(t)\}$	$f(t)$	$\mathcal{L}\{f(t)\}$
c	$\frac{c}{s}$	e^{at}	$\frac{1}{s-a}$
t^n	$\frac{n!}{s^{n+1}}$	$e^{at}t^n$	$\frac{n!}{(s-a)^{n+1}}$
$\sin bt$	$\frac{b}{s^2+b^2}$	$e^{at}\sin bt$	$\frac{b}{(s-a)^2+b^2}$
$\cos bt$	$\frac{s}{s^2+b^2}$	$e^{at}\cos bt$	$\frac{s-a}{(s-a)^2+b^2}$
$\sinh bt$	$\frac{b}{s^2-b^2}$	$e^{at}\sinh bt$	$\frac{b}{(s-a)^2-b^2}$
$\cosh bt$	$\frac{s}{s^2-b^2}$	$e^{at}\cosh bt$	$\frac{s-a}{(s-a)^2-b^2}$

$\mathcal{L}\{f'(t)\} = s\mathcal{L}\{f(t)\} - f(0)$, $\quad \mathcal{L}\{f''(t)\} = s^2\mathcal{L}\{f(t)\} - sf(0) - f'(0)$

$\mathcal{L}\{tf(t)\} = -\frac{d}{ds}\mathcal{L}\{f(t)\}$, $\quad \mathcal{L}\{t^nf(t)\} = (-1)^n\frac{d^n}{ds^n}\mathcal{L}\{f(t)\}$.

SELF-QUIZ

I. If $\mathcal{L}\{f(t)\} = \dfrac{1}{s^2+5}$ and $f(0) = 0$, then $\mathcal{L}\{f'(t)\} =$ ________.

(a) $\dfrac{s}{s^2+5}$ **(b)** $\dfrac{\sqrt{5}}{s^2+5}$

(c) $\dfrac{s-\sqrt{5}}{s^2+5}$ **(d)** $\dfrac{1}{s(s^2+5)}$

II. If $\mathcal{L}\{f(t)\} = g(s)$ and $f(0) = 1$, $f'(0) = -3$, then $\mathcal{L}\{f''(t)\} =$ ________.

(a) $s^2g(s) - 1$ **(b)** $s^2g(s) - s + 3$

(c) $s^2g(s) - s - 3$ **(d)** $s^2g(s) + 3s - 1$

III. If y satisfies $y'' + 3y = 0$, $y(0) = 2$, $y'(0) = -1$, then $\mathcal{L}\{y\} =$ ________.

(a) $\dfrac{2s-1}{s^2+3}$ **(b)** $\dfrac{2s-1}{s^2-3}$

(c) $\dfrac{1-2s}{s^2+3}$ **(d)** $\dfrac{1-2s}{s^2-3}$

IV. Find the form of the partial fraction decomposition of $\dfrac{2s+3}{s^2-1}$.

(a) $\dfrac{As+B}{s^2-1}$ **(b)** $\dfrac{A}{s-1}+\dfrac{B}{s+1}$

(c) $\dfrac{A}{s-1}+\dfrac{Bs+C}{s^2-1}$

(d) $\dfrac{A}{s-1}+\dfrac{B}{s+1}+\dfrac{C}{s^2-1}$

V. Find the form of the partial fraction decomposition of $\dfrac{3s^2-2s+7}{s^2(s-5)}$.

(a) $\dfrac{As+B}{s^2}+\dfrac{c}{s-5}$ **(b)** $\dfrac{A}{s^2}+\dfrac{B}{s-5}$

(c) $\dfrac{A}{s}+\dfrac{B}{s^2}+\dfrac{C}{s-5}$ **(d)** $\dfrac{A}{s}+\dfrac{Bs+C}{s^2}+\dfrac{D}{s-5}$

VI. Find the form of the partial decomposition of $\dfrac{2s^3+4s+1}{(s+2)(s^2-4s+10)}$.

(a) $\dfrac{A}{s+2}+\dfrac{Bs+C}{s^2-4s+10}$

(b) $2+\dfrac{A}{s+2}+\dfrac{Bs+C}{s^2-4s+10}$

(c) $\dfrac{A}{s+2}+\dfrac{B}{s^2-4s+10}$

(d) $2+\dfrac{A}{s+2}+\dfrac{B}{s^2-4s+10}+\dfrac{C}{(s+2)(s^2-4s+10)}$

Answers to Self-Quiz

I. a **II.** b **III.** a **IV.** b **V.** a, c (although (a) will leave us where we started) **VI.** b

PROBLEMS 6.2

In Problems 1–20 solve the given initial-value problems.

1. $y''+y=0,\ y(0)=1,\ y'(0)=0$
2. $y''+2y'=0,\ y(0)=0,\ y'(0)=1$
3. $y''-a^2y=0,\ y(0)=A,\ y'(0)=B$
4. $y''-ay'=0,\ y(0)=1,\ y'(0)=a$
5. $y''+2y'+5y=0,\ y(0)=y'(0)=1$
6. $y''-2y'+4y=0,\ y(0)=y'(0)=1$
7. $y''-4y'+3y=1,\ y(0)=1,\ y'(0)=4$
8. $y''-3y'-4y=8,\ y(0)=0,\ y'(0)=-2$
9. $y''-9y=t,\ y(0)=1,\ y'(0)=2$
10. $y''-y'-6y=2t^2,\ y(0)=4,\ y'(0)=1$
11. $y'''+y=0,\ y(0)=y''(0)=1,\ y'(0)=-1$
12. $y^{(4)}-16y=0,\ y(0)=y''(0)=0,$ $y'(0)=y'''(0)=1$
13. $y^{(4)}-y=0,\ y(0)=y''(0)=0,$ $y'(0)=y'''(0)=1$
14. $y'''-4y'-2y=e^{4t},\ y(0)=0,\ y'(0)=0,$ $y''(0)=\dfrac{1}{2}$
15. $y''+k^2y=\cos kt,\ y(0)=0,\ y'(0)=k$ [*Hint:* See Example 2.]
16. $y''+9y=\cos t,\ y(0)=y'(0)=0$
17. $y''+a^2y=\sin at,\ y(0)=a,\ y'(0)=a^2$
18. $y''-4y=te^{-t},\ y(0)=y'(0)=2$
19. $y^{(4)}-y=\cos t,\ y(0)=y''(0)=1,$ $y'(0)=y'''(0)=0$
20. $y^{(4)}-y=\sinh t,\ y(0)=y''(0)=0,$ $y'(0)=y'''(0)=1$

In Problems 21–29 find the Laplace transform of each function by using the differentiation property or Equation **(18)**.

21. $\cos^2 at$
22. $t\cos a$
23. $t^2\sin at$
24. $t^2\cos at$
25. $t\sin^2 t$
26. $t\cos^2 t$
27. $t\sin^2 at$
28. $t^2\cos^2 3t$
29. $t^2\sin^2 2t$
30. By reversing the order of integration, show that, if $F(s)=\mathcal{L}\{f(t)\}$, then

$$\int_s^\infty F(s)\,ds=\mathcal{L}\left\{\frac{f(t)}{t}\right\}. \tag{i}$$

In particular, it follows that

$$\mathcal{L}^{-1}\left\{\int_s^\infty F(s)\,ds\right\}=\frac{f(t)}{t}.$$

31. Let $g(t) = \int_0^t f(u)\,du$. Using calculus and the differentiation property (Theorem 1), show that
(a) $g'(t) = f(t)$ at all points of continuity of $f(t)$;
(b) $\mathscr{L}\{f(t)\} = s\mathscr{L}\{g(t)\} - g(0)$;
(c) $g(0) = 0$.
Finally, using parts (b) and (c), conclude that

$$\mathscr{L}\left\{\int_0^t f(u)\,du\right\} = \frac{1}{s}\mathscr{L}\{f(t)\}. \qquad \textbf{(ii)}$$

In Problems 32–44 use Equations **(i)** *and* **(ii)** *to compute the Laplace transform of the given function.*

32. $\dfrac{\cos t - 1}{t}$

33. $\dfrac{\sin t}{t}$

34. $\dfrac{\sinh t}{t}$

35. $\dfrac{\sin 3t}{t}$

36. $\dfrac{\sinh kt}{t}$

37. $\dfrac{\sin kt}{t}$

38. $\dfrac{1 - \cos at}{t}$

39. $\dfrac{1 - \cosh at}{t}$

40. $\displaystyle\int_0^t \frac{\sin ku}{u}\,du$

41. $\displaystyle\int_0^t \frac{1 - \cosh au}{u}\,du$

42. $\displaystyle\int_0^t \frac{1 - \cos au}{u}\,du$

43. $\operatorname{erf}(t) = \dfrac{2}{\sqrt{\pi}}\displaystyle\int_0^t e^{-u^2}\,du$

44. $\dfrac{e^{-k^2/4t}}{\sqrt{\pi t}}$

Find the inverse Laplace transform of the functions in Problems 45–48. Use derivatives and integrals.

45. $\ln\left(1 + \dfrac{a^2}{s^2}\right)$

46. $\ln\dfrac{s - a}{s - b}$

47. $\arctan\dfrac{1}{s}$

48. $\dfrac{1}{s}\arctan\dfrac{1}{s}$

49. Consider the equation

$$y'' + ty = 0, \qquad y(0) = 0, \qquad y'(0) = 1.$$

(a) Obtain a differential equation for $Y(s) = \mathscr{L}\{y(t)\}$.
(b) Solve the differential equation and find $Y(s)$. [Note that it is not possible to invert this transform by the methods we have discussed.]

50. Let $f(t)$ be continuous, except for a jump discontinuity at $t = a$ (>0), and let it satisfy all other conditions of Theorem 1. Prove that

$$\mathscr{L}\{f'(t)\} = s\mathscr{L}\{f(t)\} - f(0) - e^{-as}[f(a+0) - f(a-0)],$$

where

$$f(a+0) = \lim_{h\to 0^+} f(a+h)$$

and

$$f(a-0) = \lim_{h\to 0^-} f(a+h).$$

[Note that $f(a+0) - f(a-0) = 0$ if f is continuous at a.]

6.3 STEP FUNCTIONS, PERIODIC FUNCTIONS, CONVOLUTIONS, AND INTEGRAL EQUATIONS

In Section 6.2 we saw how to solve linear differential equations in which the forcing function $f(t)$ was continuous. In a great number of applications, however, the forcing function is discontinuous. In this section and in Section 6.4 we will give some examples of differential equations with discontinuous forcing functions.

Unit step function

The following function, which is extremely important for practical applications, is known as the **unit step function** or **Heaviside function*** [see Figure 6.2(a)].

*Named after the British physicist Oliver Heaviside (1850–1925). Heaviside had difficulty getting his research published because he made use of unusual methods in solving problems.

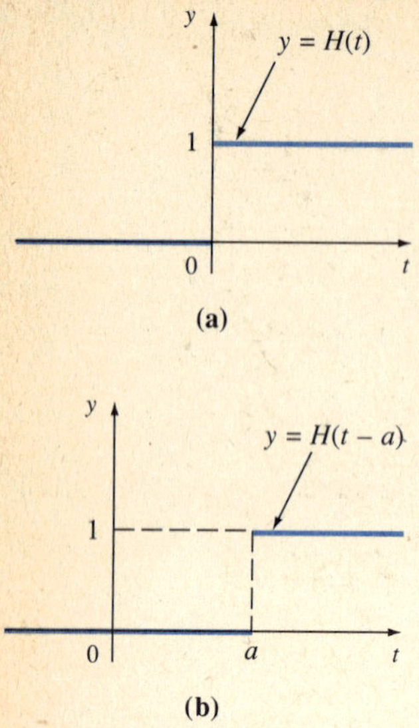

FIGURE 6.2
The Heaviside functions **(a)** $H(t)$, **(b)** $H(t - a)$

Heaviside function

$$H(t) = \begin{cases} 0, & t < 0, \\ 1, & t > 0. \end{cases} \tag{1}$$

In particular, if a is any fixed constant, we can shift the Heaviside function by a units [see Figure 6.2(b)]. by defining

The shifted Heaviside function

$$H(t - a) = \begin{cases} 0, & t < a, \\ 1, & t > a. \end{cases} \tag{2}$$

Then $H(t - a)$ has a jump discontinuity at $t = a$. Note that $H(t - a)$ is not defined at $t = a$.

A practical use of such a function might be to model a light switch that is switched on at time $t = a$.

For $a \geq 0$ and $s > 0$, we obtain

The Laplace transform of a Heaviside function

$$\mathscr{L}\{H(t - a)\} = \int_0^\infty e^{-st} H(t - a)\, dt = \int_a^\infty e^{-st}\, dt = \frac{e^{-as}}{s}. \tag{3}$$

The following theorem shows that multiplying a function by a unit step function has the effect of multiplying its transform by an exponential function.

THEOREM 1 ♦ **The second shifting property of Laplace transforms**
Let $a > 0$. Then, if $f(t)$ has a Laplace transform, so does $f(t - a)H(t - a)$ and

$$\mathscr{L}\{f(t - a)H(t - a)\} = e^{-as}\mathscr{L}\{f(t)\}, \tag{4}$$

or, in terms of inverse transforms with $F(s) = \mathscr{L}\{f(t)\}$,

$$\mathscr{L}^{-1}\{e^{-as}F(s)\} = f(t - a)H(t - a). \tag{5}$$

Proof A graph illustrating a function $f(t - a)H(t - a)$ is given in Figure 6.3.
Using the definition and the substitution $x = t - a$, we find that

$$\mathscr{L}\{f(t - a)H(t - a)\} = \int_0^\infty e^{-st} f(t - a)H(t - a)\, dt = \int_a^\infty e^{-st} f(t - a)\, dt$$

$$\underset{\text{let } x = t - a}{=} \int_0^\infty e^{-s(x+a)} f(x)\, dx = e^{-as}\mathscr{L}\{f(t)\}. \quad ♦$$

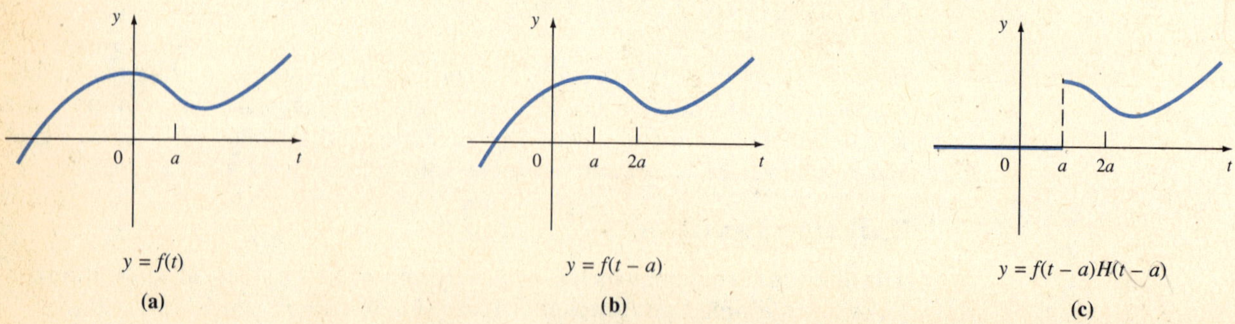

FIGURE 6.3
An arbitrary function f is graphed in **(a)**. The graph of $y = f(t - a)$ in **(b)** is simply a shift to the right by a units of the original graph. For the graph in **(c)**, truncate the graph in **(b)** at $t = a$, replacing it by $y = 0$ for $t < a$.

The next four examples illustrate the use of the second shifting property.

EXAMPLE 1 ▶ **Using the second shifting theorem**

$\mathcal{L}\{\sin a(t-b)H(t-b)\} = e^{-bs}\mathcal{L}\{\sin at\} = ae^{-bs}/(s^2+a^2)$. ◀

EXAMPLE 2 ▶ **Applying the second shifting theorem to a function with a jump discontinuity**

Compute $\mathcal{L}\{f(t)\}$ when

$$f(t) = \begin{cases} e^t, & 0 \le t < 2\pi, \\ e^t + \cos t, & t > 2\pi. \end{cases}$$

Solution The function $f(t)$ has a jump discontinuity at $t = 2\pi$. We may write

$$f(t) = e^t + H(t-2\pi)\cos(t-2\pi),$$

since

$$H(t-2\pi)\cos(t-2\pi) = \begin{cases} 0 & t < 2\pi, \\ \cos(t-2\pi) = \cos t, & t > 2\pi. \end{cases}$$

Thus, by the second shifting property (Theorem 1),

$$\mathcal{L}\{f(t)\} = \mathcal{L}\{e^t\} + e^{-2\pi s}\mathcal{L}\{\cos t\} = \frac{1}{s-1} + \frac{se^{-2\pi s}}{s^2+1}. \quad ◀$$

EXAMPLE 3 ▶ **The Laplace transform of a truncated quadratic**

Compute $\mathcal{L}\{t^2H(t-3)\}$.

Solution In order to use the second shifting property, we must write t^2 as a polynomial in $(t-3)$. Since $(t-3)^2 = t^2 - 6t + 9$, we have

$$t^2 = (t-3)^2 + 6t - 9 = (t-3)^2 + 6(t-3) + 9$$

so

$$\mathcal{L}\{t^2H(t-3)\} = \mathcal{L}\{(t-3)^2H(t-3)\} + 6\mathcal{L}\{(t-3)H(t-3)\} + 9\mathcal{L}\{H(t-3)\}$$

from (4) and (3) ↓

$$= e^{-3s}\mathcal{L}\{t^2\} + 6e^{-3s}\mathcal{L}\{t\} + e^{-3s}\mathcal{L}\{9\}$$

$$= e^{-3s}\left\{\frac{2}{s^3} + \frac{6}{s^2} + \frac{9}{s}\right\}. \quad ◀$$

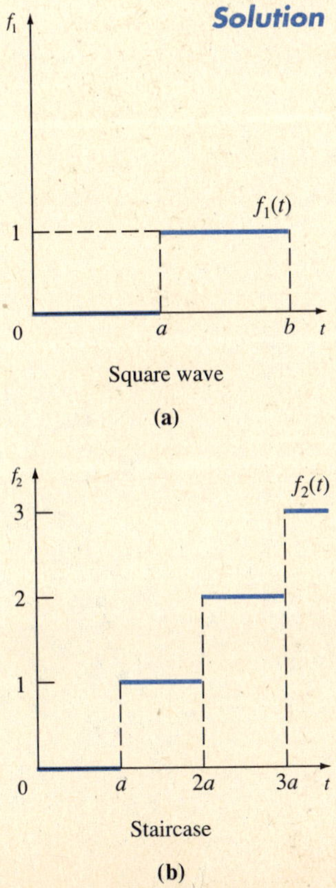

FIGURE 6.4
The graphs of two piecewise continuous functions

The unit step function can be used as a building block in the construction of other functions; for example,

$$f_1(t) = H(t-a) - H(t-b), \qquad a < b,$$

is a square wave between a and b [see Figure 6.4(a)], whereas

$$f_2(t) = H(t-a) + H(t-2a) + H(t-3a), \qquad a > 0,$$

yields a three-step staircase [see Figure 6.4(b)]. By the linearity property of Laplace transforms (Theorem 6.1.2) we obtain

$$\mathcal{L}\{f_1(t)\} = \frac{1}{s}(e^{-as} - e^{-bs}),$$

and

$$\mathscr{L}\{f_2(t)\} = \frac{1}{s}(e^{-as} + e^{-2as} + e^{-3as}).$$

EXAMPLE 4 ▶ **The Laplace transform of an infinite staircase**

Compute the Laplace transform of the infinite staircase [obtained by continuing the staircase in Figure 6.4(b) forever]:

$$f(t) = H(t) + H(t - a) + H(t - 2a) + H(t - 3a) + \dots, \quad a > 0. \quad \textbf{(6)}$$

Solution Since $e^{-as} < 1$, if $as > 0$ we can use the formula for the sum of a geometric series:

$$\sum_{n=0}^{\infty} x^n = 1 + x + x^2 + \dots = \frac{1}{1 - x}, \qquad |x| < 1. \quad \textbf{(7)}$$

Then, for $s > 0$, by Equation **(3),**

$$\mathscr{L}\{f(t)\} = \frac{1}{s}(1 + e^{-as} + e^{-2as} + \dots) = \frac{1}{s(1 - e^{-as})}. \quad ◀ \quad \textbf{(8)}$$

EXAMPLE 5 ▶ **An initial-value problem with discontinuous forcing function f(t)**

Solve the initial-value problem

$$y'' + y = f(t), \qquad y(0) = y'(0) = 0,$$

where

$$f(t) = \begin{cases} 1, & 0 \le t \le 1, \\ 0, & t > 1. \end{cases}$$

Solution We can rewrite $f(t)$ using Heaviside functions as $f(t) = H(t) - H(t - 1)$. Taking Laplace transforms, we get

$$s^2\mathscr{L}\{y\} + \mathscr{L}\{y\} = \frac{1}{s} - \frac{e^{-s}}{s}, \quad \text{or} \quad (s^2 + 1)\mathscr{L}\{y\} = \frac{1 - e^{-s}}{s}.$$

Hence

$$\mathscr{L}\{y\} = \frac{1 - e^{-s}}{s(s^2 + 1)},$$

and since

$$\frac{1}{s(s^2 + 1)} = \frac{1}{s} - \frac{s}{s^2 + 1},$$

we have

$$\mathscr{L}\{y\} = \frac{1}{s} - \frac{s}{s^2 + 1} - \frac{e^{-s}}{s} + \frac{se^{-s}}{s^2 + 1}.$$

Using the second shifting property (Theorem 1), we obtain

$$\begin{aligned} y(t) &= 1 - \cos t - H(t - 1) + H(t - 1)\cos(t - 1) \\ &= 1 - \cos t - H(t - 1)[1 - \cos(t - 1)]. \quad ◀ \end{aligned}$$

Laplace transforms of periodic functions

THEOREM 2 ♦ **Periodicity property of the Laplace transform**

Let $f(t)$ be piecewise continuous in $[0, \omega]$ and periodic with period $\omega (\omega > 0)$; that is, $f(t + \omega) = f(t)$, for each $t \ge 0$. Then $f(t)$ has the Laplace transform

$$F(s) = \mathcal{L}\{f(t)\} = \frac{\int_0^\omega e^{-st}f(t)\,dt}{1 - e^{-\omega s}}, \tag{9}$$

valid for every $s > 0$.

Proof By definition,

$$\begin{aligned}
F(s) &= \int_0^\infty e^{-st}f(t)\,dt = \int_0^\omega e^{-st}f(t)\,dt + \int_\omega^\infty e^{-st}f(t)\,dt \\
&= \int_0^\omega e^{-st}f(t)\,dt + \int_0^\infty e^{-st}f(t)H(t-\omega)\,dt \\
&= \int_0^\omega e^{-st}f(t)\,dt + \int_0^\infty e^{-st}\underbrace{f(t-\omega)}_{f \text{ is periodic}}H(t-\omega)\,dt \\
&= \int_0^\omega e^{-st}f(t)\,dt + \mathcal{L}\{f(t-\omega)H(t-\omega)\} \\
&= \int_0^\omega e^{-st}f(t)\,dt + e^{-\omega s}F(s)
\end{aligned}$$

so

$$F(s) - e^{-\omega s}F(s) = \int_0^\omega e^{-st}f(t)\,dt$$

from which the theorem follows. ♦

EXAMPLE 6 ▶ **The Laplace transform of |sin at|**

Find the Laplace transform of the function

$$f(t) = |\sin at|, \qquad a > 0.$$

Solution Note that $f(t)$ has period $\omega = \frac{\pi}{a}$. By Theorem 2 we have

$$\mathcal{L}\{|\sin at|\} = \frac{\int_0^{\pi/a} e^{-st}\sin at\,dt}{1 - e^{-\pi s/a}},$$

since $|\sin at| = \sin at$ in $[0, \frac{\pi}{a}]$. Now using Formula 168 in Appendix 1, we have

$$\begin{aligned}
\int_0^{\pi/a} e^{-st}\sin at\,dt &= \frac{e^{-st}}{s^2+a^2}(-s\sin at - a\cos at)\Big|_0^{\pi/a} \\
&= \frac{a(e^{-\pi s/a}+1)}{s^2+a^2},
\end{aligned}$$

so

$$\mathcal{L}\{|\sin at|\} = \frac{a}{s^2+a^2}\,\frac{1+e^{-\pi s/a}}{1-e^{-\pi s/a}},$$

which can be simplified by using hyperbolic functions to

$$\mathcal{L}\{|\sin at|\} = \frac{a}{s^2+a^2}\coth\left(\frac{\pi s}{2a}\right). \quad ◀$$

EXAMPLE 7 ▶ **The Laplace transform of a periodic square wave**

Compute the Laplace transform of the periodic square wave

$$f(t) = H(t) - 2H(t-a) + 2H(t-2a) - 2H(t-3a) + \dots, \qquad a > 0.$$

Solution Note that f is periodic of period $2a$ (see Figure 6.5). We have, from Equation **(9)**,

$$F(s) = \frac{\int_0^{2a} e^{-st}f(t)\,dt}{1 - e^{-2as}}.$$

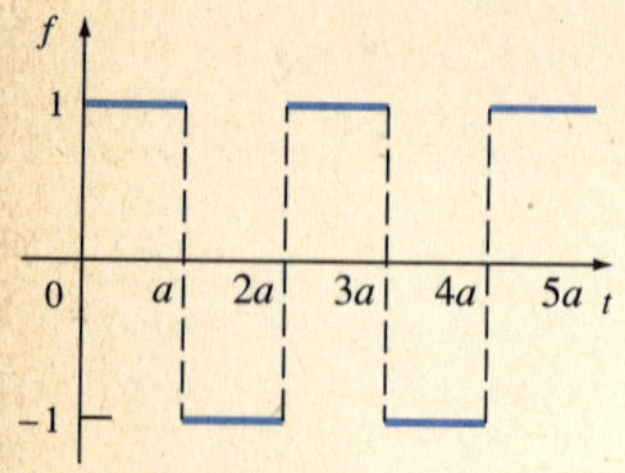

FIGURE 6.5
A periodic square wave

But $f(t) = 1$ for $0 < t < a$ and $f(t) = -1$ for $a < t < 2a$, so

$$\int_0^{2a} e^{-st}f(t) = \int_0^a e^{-st}\,dt - \int_a^{2a} e^{-st}\,dt = -\frac{1}{s}e^{-st}\Big|_0^a + \frac{1}{s}e^{-st}\Big|_a^{2a}$$
$$= \frac{1}{s}(-e^{-as} + 1 + e^{-2as} - e^{-as}) = \frac{1}{s}(1 - e^{-as})^2.$$

Since $1 - e^{-2as} = (1 - e^{-as})(1 + e^{-as})$, we obtain

$$F(s) = \frac{1}{s}\frac{(1 - e^{-as})(1 - e^{-as})}{(1 - e^{-as})(1 + e^{-as})} = \frac{1 - e^{-as}}{s(1 + e^{-as})}. \quad \textbf{(10)}$$

We can write this in another way as

multiply top and bottom by $e^{as/2}$

$$\frac{1}{s}\frac{(1 - e^{-as})}{(1 + e^{-as})} = \frac{1}{s}\left(\frac{e^{as/2} - e^{-as/2}}{e^{as/2} + e^{-as/2}}\right) = \frac{1}{s}\tanh\frac{as}{2}. \quad ◀$$

EXAMPLE 8 ▶ **Solving an initial-value problem whose nonhomogeneous term is a periodic square wave**

Solve the initial-value problem

$$y'' + 2y' + 5y = f(t), \qquad y(0) = y'(0) = 0,$$

where $f(t)$ is the periodic square wave in Figure 6.6.

Solution Taking Laplace transforms and using Equation **(10)** in Example 7 with $a = \pi$, we have

$$(s^2 + 2s + 5)\mathcal{L}\{y\} = \frac{1 - e^{-\pi s}}{s(1 + e^{-\pi s})},$$

or

$$\mathcal{L}\{y\} = \frac{1}{s(s^2 + 2s + 5)}\frac{1 - e^{-\pi s}}{1 + e^{-\pi s}}.$$

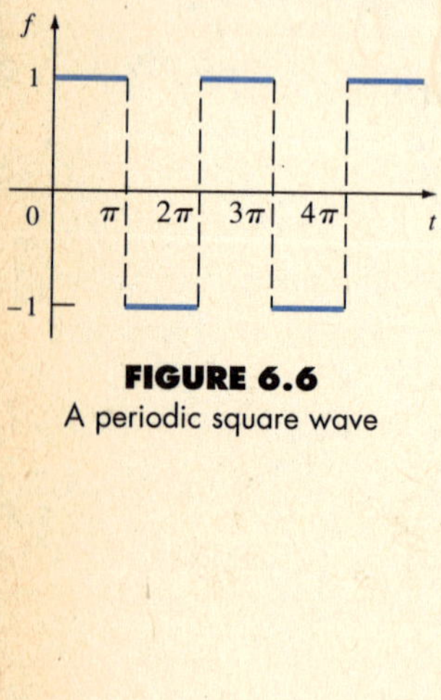

FIGURE 6.6
A periodic square wave

But

$$\frac{1}{s(s^2 + 2s + 5)} = \frac{1}{5}\left[\frac{1}{s} - \frac{s + 2}{s^2 + 2s + 5}\right] = \frac{1}{5}\left[\frac{1}{s} - \frac{s + 2}{(s + 1)^2 + 2^2}\right]$$

and, by the geometric series **(7)** in Example 4,

$$\frac{1 - e^{-\pi s}}{1 + e^{-\pi s}} = (1 - e^{-\pi s})(1 - e^{-\pi s} + e^{-2\pi s} - e^{-3\pi s} + \ldots)$$
$$= 1 - 2e^{-\pi s} + 2e^{-2\pi s} - 2e^{-3\pi s} + \ldots.$$

Thus

$$\mathcal{L}\{y\} = \frac{1}{5}\left[\frac{1}{s} - \frac{s + 2}{(s + 1)^2 + 2^2}\right](1 - 2e^{-\pi s} + 2e^{-2\pi s} - 2e^{-3\pi s} + \ldots).$$

By the first and second shifting theorems we have

$$\mathcal{L}^{-1}\left\{\frac{1}{5}\left[\frac{1}{s} - \frac{(s+1)+1}{(s+1)^2+2^2}\right]\right\} = \frac{1}{5}\left[1 - \overbrace{e^{-t}\left(\cos 2t + \frac{1}{2}\sin 2t\right)}^{\equiv g(t)}\right]$$

$$= \frac{1}{5}[1 - g(t)]$$

and

$$\mathcal{L}^{-1}\left\{\frac{2}{5}\left[\frac{1}{s} - \frac{(s+1)+1}{(s+1)^2+2^2}\right]e^{-k\pi s}\right\} = \frac{2}{5}[1 - g(t-k\pi)]H(t-k\pi).$$

But

$$g(t-k\pi) = e^{-(t-k\pi)}[\cos 2(t-k\pi) + \frac{1}{2}\sin 2(t-k\pi)]$$

$$= e^{k\pi}g(t),$$

so

$$y(t) = \frac{1}{5}[1 - g(t)] - \frac{2}{5}[1 - e^{\pi}g(t)]H(t-\pi) + \frac{2}{5}[1 - e^{2\pi}g(t)]H(t-2\pi)$$

$$-\frac{2}{5}[1 - e^{3\pi}g(t)]H(t-3\pi) + \ldots$$

$$= \frac{1}{5}[1 - 2H(t-\pi) + 2H(t-2\pi) - 2H(t-3\pi) + \ldots]$$

$$-\frac{g(t)}{5}[1 - 2e^{\pi}H(t-\pi) + 2e^{2\pi}H(t-2\pi) - 2e^{3\pi}H(t-3\pi) + \ldots].$$

Then

$$y(t) = \frac{1}{5}(f(t) - g(t)[1 - 2e^{\pi}H(t-\pi) + 2e^{2\pi}H(t-2\pi) - \ldots]).$$

Hence, if $n\pi < t < (n+1)\pi$,

$$y(t) = \frac{1}{5}[(-1)^n - g(t)(1 - 2e^{\pi} + \cdots + (-1)^n 2e^{n\pi})]$$

$$= \frac{1}{5}\left((-1)^n - g(t)\left[2\left(\frac{1 + (-1)^n e^{(n+1)\pi}}{1 + e^{\pi}}\right) - 1\right]\right)$$

$$= \frac{1}{5}\left((-1)^n + g(t) - 2g(t)\left(\frac{1 + (-1)^n e^{(n+1)\pi}}{1 + e^{\pi}}\right)\right).$$ ◄

EXAMPLE 9 ► **Modeling the effect of public health campaigns on the spread of AIDS***

Public health officials in many countries have, for several years, mounted advertising campaigns which have the goal of changing people's behavior in order to limit the spread of AIDS, to reduce the percentage of people who smoke, to cut alcohol and drug abuse, and so on.

In this example we discuss a simple model for determining the effectiveness of such a campaign. In the model we have a **susceptible** population. This is the group

*This example is based on the paper of the same title by David Tudor in *SIAM Review* 34, no. 2 (June 1992): 300–303.

of people who can get the disease, are potential smokers, and so on. The percentage of susceptibles who are affected (the percentage who become infected with the disease, who begin to smoke, etc.) is denoted by $y(t)$.

In 1959 H. Muench* discussed a simple differential equation model for the spread of a disease:

$$\frac{dy}{dt} = r(1 - y), \qquad 0 \le y \le 1$$

where r is a positive constant. Note that as y increases, the *rate* of increase of y decreases. The solution to this linear equation is

$$y(t) = 1 - [1 - y(0)]e^{-rt}.$$

Since $e^{-rt} \to 0$ as $t \to \infty$, we see that $y(t) \to 1$; that is, eventually all the susceptibles become infected.

To avoid this dire outcome, public health officials launch a series of advertising campaigns which are designed to increase public awareness and to suggest ways to protect against the disease (or to persuade people not to smoke, etc.). If the campaign is mounted at time t_1 and is effective from then on, then the rate of increase of $y(t)$ is reduced by some amount α from time t_1 onwards, and the differential equation modeling this increase becomes

$$\frac{dy}{dt} = r(1 - y) - \alpha H(t - t_1)$$

where α is a constant and $0 < \alpha < r$. To solve this equation, we take Laplace transforms:

$$s\mathcal{L}\{y\} - y(0) = \frac{r}{s} - r\mathcal{L}\{y\} - \alpha\frac{e^{-t_1 s}}{s}$$

$$(s + r)\mathcal{L}\{y\} = \frac{r}{s} + y(0) - \frac{\alpha}{s}e^{-t_1 s}$$

$$\mathcal{L}\{y\} = \frac{r}{s(s + r)} + \frac{y(0)}{s + r} - \frac{\alpha e^{-t_1 s}}{s(s + r)}.$$

Using partial fractions, we see that

$$\frac{1}{s(s + r)} = \frac{1}{r}\left[\frac{1}{s} - \frac{1}{s + r}\right]$$

so

$$\mathcal{L}\{y\} = \frac{1}{s} - \frac{1}{s + r} + \frac{y(0)}{s + r} - \frac{\alpha}{r}\frac{e^{-t_1 s}}{s} + \frac{\alpha}{r}\frac{e^{-t_1 s}}{(s + r)}.$$

Now $\mathcal{L}^{-1}\{e^{-t_1 s}/s\} = H(t - t_1)$ by Equation **(3)**. To compute $\mathcal{L}^{-1}\{e^{-t_1 s}/(s + r)\}$ we observe that, by Equation **(3)**,

$$\mathcal{L}\{H(t - t_1)e^{-r(t-t_1)}\} = e^{-t_1 s}\mathcal{L}\{e^{-rt}\} = \frac{e^{-t_1 s}}{s + r}.$$

---

* H. Muench, *Catalytic Models in Epidemiology,* Harvard University Press, Cambridge, MA, 1959.

Thus $\mathcal{L}^{-1}\{e^{-t_1 s}/(s + r)\} = H(t - t_1)e^{-r(t-t_1)}$ and

$$y(t) = 1 - e^{-rt} + y(0)e^{-rt} - \frac{\alpha}{r}H(t - t_1) + \frac{\alpha}{r}H(t - t_1)e^{-r(t-t_1)}$$

$$= 1 - [1 - y(0)]e^{-rt} - \frac{\alpha}{r}H(t - t_1)[1 - e^{-r(t-t_1)}].$$

We see that as $t \to \infty$, $y(t) \to 1 - \frac{\alpha}{r}$. This means that, assuming the advertising campaign retains its effectiveness, the percentage of susceptibles who fall prey to the disease (or who start smoking) reaches a limit of less than 100%. For example, if $r = 0.5$ and $\alpha = 0.3$, then $\frac{\alpha}{r} = 0.6$ and $1 - \frac{\alpha}{r} = 0.4$ which means that "only" 40% of the susceptible population become infected.

We discuss an extension of this model in Problem 50. ◀

Convolutions

It often occurs that in the process of solving a linear differential equation by transforms, we end up with a transform that is the product of two other transforms. Although we proved in Problem 6.1.79 on page 270 that

$$\mathcal{L}^{-1}\{F + G\} = \mathcal{L}^{-1}\{F\} + \mathcal{L}^{-1}\{G\},$$

it is not true that $\mathcal{L}^{-1}\{FG\} = \mathcal{L}^{-1}\{F\}\mathcal{L}^{-1}\{G\}$; for example, if $F(s) = 1/s$ and $G(s) = 1/s^2$, then $F(s)G(s) = 1/s^3$, but

$$\mathcal{L}^{-1}\{FG\} = \mathcal{L}^{-1}\left\{\frac{1}{s^3}\right\} = \frac{t^2}{2}, \qquad \mathcal{L}^{-1}\{F\} = 1, \qquad \mathcal{L}^{-1}\{G\} = t$$

and, clearly, $\mathcal{L}^{-1}\{FG\} \neq \mathcal{L}^{-1}\{F\}\mathcal{L}^{-1}\{G\}$.

There is, however, a mathematical expression that is equal to the product of two Laplace transforms. To show this, we need the following definition.

Definition Convolution

If f and g are piecewise continuous functions, then the **convolution** of f and g, written $(f * g)$, is defined by

$$(f * g)(t) = \int_0^t f(t - u)g(u)\, du.$$

The notation $(f * g)(t)$ indicates that the convolution $f * g$ is a function of the independent variable t.

Using the change of variables $v = t - u$, we see that

$$(f * g)(t) = -\int_t^0 f(v)g(t - v)\, dv = \int_0^t g(t - v)f(v)\, dv = (g * f)(t).$$

Hence $(f * g)(t) = (g * f)(t)$, and we can take the convolution in either order without altering the result. We may now state an important result.

THEOREM 3 ♦ **Convolution theorem for Laplace transforms**

If $F(s) = \mathcal{L}\{f(t)\}$ and $G(s) = \mathcal{L}\{g(t)\}$ exist, then $\mathcal{L}\{(f * g)(t)\}$ exists and

$$\mathcal{L}\{(f * g)(t)\} = F(s)G(s).$$

Proof By definition,

$$F(s)G(s) = \left[\int_0^\infty e^{-su}f(u)\,du\right]\left[\int_0^\infty e^{-sv}g(v)\,dv\right]$$
$$= \int_0^\infty \int_0^\infty e^{-s(u+v)}f(u)g(v)\,dv\,du. \tag{11}$$

If we make the change of variables $t = u + v$, with u fixed, then $dt = dv$ and the integral **(11)** is equal to

$$F(s)G(s) = \int_0^\infty \int_0^\infty e^{-st}f(u)g(t-u)\,dt\,du. \tag{12}$$

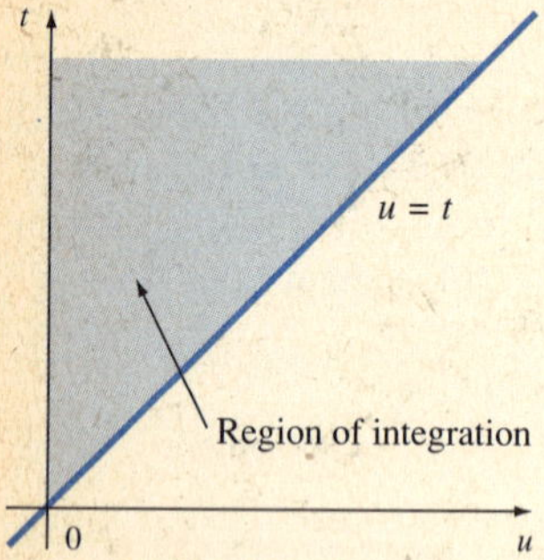

FIGURE 6.7
Region of integration

Changing the order of integration* and noting that

$$\int_0^\infty \int_u^\infty dt\,du = \int_0^\infty \int_0^t du\,dt$$

(see Figure 6.7), we have the integral in **(12)** equal to

$$F(s)G(s) = \int_0^\infty \int_0^t e^{-st}f(u)g(t-u)\,du\,dt$$
$$= \int_0^\infty e^{-st}\left[\int_0^t g(t-u)f(u)\,du\right]dt$$
$$= \int_0^\infty e^{-st}(g*f)(t)\,dt = \int_0^\infty e^{-st}(f*g)(t)\,dt$$
$$= \mathscr{L}\{f*g\}. \quad \blacklozenge$$

COROLLARY ♦

If $F(s) = \mathscr{L}\{f(t)\}$ and $G(s) = \mathscr{L}\{g(t)\}$, then

$$\mathscr{L}^{-1}\{F(s)G(s)\} = (f*g)(t). \quad \blacklozenge \tag{13}$$

Our first applications of this theorem and corollary are in the computation of inverse transforms.

EXAMPLE 10 ► **Using convolutions to invert a Laplace transform**

Compute $\mathscr{L}^{-1}\{s/(s^2+1)^2\}$.

Solution Since

$$\mathscr{L}\{\cos t\} = \frac{s}{s^2+1} \quad \text{and} \quad \mathscr{L}\{\sin t\} = \frac{1}{s^2+1},$$

we have, by Equation **(13)**,

$$\mathscr{L}^{-1}\left\{\frac{s}{(s^2+1)^2}\right\} = \mathscr{L}^{-1}\left\{\frac{s}{s^2+1}\cdot\frac{1}{s^2+1}\right\} = \cos t * \sin t.$$

By Theorem 3, this is the convolution of $\cos t$ and $\sin t$. But

$$\sin t * \cos t = \int_0^t \sin(t-u)\cos u\,du$$

* This may not always be possible. Conditions under which reversing the order of integration is permissible are found in most advanced calculus texts.

$$= \int_0^t (\sin t \cos u - \cos t \sin u) \cos u \, du$$

$$= \sin t \int_0^t \cos^2 u \, du - \cos t \int_0^t \sin u \cos u \, du,$$

$$= \left[\sin t \left(\frac{\sin u \cos u + u}{2} \right) - \cos t \frac{\sin^2 u}{2} \right] \Bigg|_{u=0}^{u=t}$$

$$= \frac{t \sin t}{2}. \quad ◄$$

EXAMPLE 11 ► **Using convolutions to solve an initial-value problem**

Solve the initial-value problem

$$y'' + y = f(t), \qquad y(0) = 0, \qquad y'(0) = 1, \tag{14}$$

where

$$f(t) = \begin{cases} 1, & 0 < t < 1, \\ 0, & t > 1. \end{cases}$$

Solution Since $f(t) = H(t) - H(t-1)$, taking Laplace transforms of the differential equation we have

$$s^2 \mathscr{L}\{y\} - 1 + \mathscr{L}\{y\} = \frac{1 - e^{-s}}{s},$$

or

$$\mathscr{L}\{y\} = \frac{1 + s - e^{-s}}{s(s^2+1)} = \frac{1}{s} - \frac{s-1}{s^2+1} - \frac{e^{-s}}{s} \cdot \frac{1}{s^2+1}.$$

Using the convolution theorem for Laplace transforms (Theorem 3), we get

$$y(t) = 1 - \cos t + \sin t - \sin t * H(t-1).$$

But by the definition of the Heaviside function,

$$\begin{aligned} \sin t * H(t-1) &= \int_0^t \sin(t-u) H(u-1) \, du \\ &= H(t-1) \int_1^t \sin(t-u) \, du \\ &= H(t-1) \cos(t-u) \big|_1^t \\ &= H(t-1)[1 - \cos(t-1)], \end{aligned}$$

so the solution to the initial-value problem **(14)** is

$$y(t) = 1 - \cos t + \sin t - H(t-1)[1 - \cos(t-1)]. \quad ◄$$

Volterra integral equations

Although the convolution theorem is obviously very useful in calculating inverse transforms, it also has important applications in a very different area. In 1931 the Italian mathematician Vito Volterra* published a book that contained a fairly sophisticated model of population growth. It would be beyond the scope of this book

---

* V. Volterra, *Leçons sur la théorie mathématique de la lutte pour la vie* (Paris: Gauthier-Villars, 1931).

to go into a derivation of Valterra's model. However, a central equation in this model is of the form

$$x(t) = f(t) + \int_0^t a(t - u)x(u)\, du. \tag{15}$$

An equation of this type, where $f(t)$ and $a(t)$ can be assumed to be continuous, is called a **Volterra integral equation.** Since the publication of Volterra's papers, many diverse phenomena in thermodynamics, electric systems theory, nuclear reactor theory, and chemotherapy have been modeled with Volterra integral equations.

It is quite easy to see how Laplace transforms can be used to solve an equation in the form of Equation **(15).** Taking transforms on both sides of Equation **(15),** using the convolution theorem, and denoting transforms by the appropriate capital letters, we obtain

$$X(s) = F(s) + A(s)X(s),$$

or

$$X(s)[1 - A(s)] = X(s) - A(s)X(s) = F(s),$$

so

$$X(s) = \frac{F(s)}{1 - A(s)}. \tag{16}$$

Looking at Equation **(16),** we immediately see that if $F(s)$ and $A(s)$ are defined for $s \geq s_0$, then $X(s)$ is similarly defined as long as $A(s) \neq 1$. Once $X(s)$ is known, we may (if possible) calculate the solution $x(t) = \mathscr{L}^{-1}\{X(s)\}$.

EXAMPLE 12 ▶ Solving a Volterra integral equation

Consider the integral equation

$$x(t) = t^2 + \int_0^t \sin(t - u)x(u)\, du. \tag{17}$$

Taking transforms, we have

$$X(s) = \frac{2}{s^3} + \frac{1}{s^2 + 1}X(s),$$

or

$$X(s) = \frac{\frac{2}{s^3}}{1 - \frac{1}{s^2 + 1}} = \frac{2(s^2 + 1)}{s^5} = \frac{2}{s^3} + \frac{2}{s^5}.$$

Hence the solution to Equation **(17)** is given by

$$x(t) = t^2 + \frac{1}{12}t^4. \quad ◀$$

There are other applications of the very useful convolution theorem given in the excercises. The student, however, should always keep in mind that the greatest difficulty in using any of these methods is that it is frequently difficult to calculate inverse transforms. Unfortunately, most problems that arise lead to inverting transforms that do not fit into familiar patterns. For this reason, methods have been devised for estimating such inverses. The interested reader should consult a more advanced book on Laplace transforms, such as the excellent book by Widder.*

SELF-QUIZ

I. Which of the following has the graph below?
(a) $H(t-3)$ **(b)** $2H(t-3)$
(c) $3H(t-2)$ **(d)** $3H(t+2)$
(e) $2H(t+3)$ **(f)** none of these

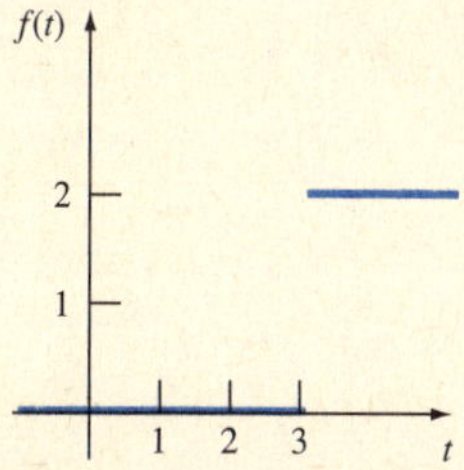

II. Which of the following is the Laplace transform of the function graphed above?
(a) $\dfrac{2e^{-3s}}{s}$ **(b)** $\dfrac{3e^{-2s}}{s}$ **(c)** $\dfrac{2e^{3s}}{s}$
(d) $\dfrac{3e^{2s}}{s}$ **(e)** $\dfrac{e^{-3s}}{s}$ **(f)** none of these

III. Which of the following functions has the Laplace transform $2e^s/(s^2+4)$?
(a) $2\,H(t-1)\sin t$ **(b)** $2\,H(t+1)\sin t$
(c) $H(t-1)\sin 2t$ **(d)** $H(t+1)\sin 2t$
(e) $H(t-1)\sin 2(t-1)$
(f) $H(t+1)\sin 2(t+1)$

IV. Let $f(t)=\begin{cases}1, & 0\le t<1\\ 0, & 1\le t<2\end{cases}$ and suppose $f(t+2)=f(t)$. Then $\mathcal{L}\{f(t)\}=$ ______.
(a) $\dfrac{1}{s(1-e^{-s})}$ **(b)** $\dfrac{1}{s(1+e^{-2s})}$
(c) $\dfrac{1}{s(1-e^{-2s})}$ **(d)** $\dfrac{1}{s(1+e^{-s})}$
(e) $\dfrac{1-e^{-s}}{s(1-e^{-2s})}$

V. Let $F(s)=\mathcal{L}\{f(t)\}$ and $G(s)=\mathcal{L}\{g(t)\}$. Then $\mathcal{L}\{\int_0^t f(t-u)\,g(u)\,du\}=$ ______.
(a) $\displaystyle\int_0^s F(s-u)\,G(u)\,du$ **(b)** $\displaystyle\int_0^s G(s-u)\,F(u)\,du$
(c) $F(s)\,G(s)$ **(d)** $F(s)+G(s)$

Answers to Self-Quiz

I. b **II.** a **III.** f **IV.** d = e **V.** c

PROBLEMS 6.3

In Problems 1–8 use the second shifting theorem to show the given equation.

1. $\mathcal{L}\{tH(t-1)\}=e^{-s}\left(\dfrac{1}{s^2}+\dfrac{1}{s}\right)$

2. $\mathcal{L}\{t^2H(t-1)\}=e^{-s}\left(\dfrac{2}{s^3}+\dfrac{2}{s^2}+\dfrac{1}{s}\right)$

3. $\mathcal{L}\{e^tH(t-1)\}=\dfrac{e^{-(s-1)}}{s-1}$

4. $\mathcal{L}\{e^{at}H(t-b)\}=\dfrac{e^{-b(s-a)}}{s-a}$

5. $\mathcal{L}\left\{\sin t\cdot H\left(t-\dfrac{\pi}{2}\right)\right\}=\dfrac{se^{-\pi s/2}}{s^2+1}$

6. $\mathcal{L}\{\cos a(t-b)\cdot H(t-b)\}=\dfrac{e^{-bs}s}{s^2+a^2}$

7. $\mathcal{L}\{\sinh a(t-b)\cdot H(t-b)\}=\dfrac{e^{-bs}a}{s^2-a^2}$

8. $\mathcal{L}\{\cosh a(t-b)\cdot H(t-b)\}=\dfrac{e^{-bs}s}{s^2-a^2}$

In Problems 9–11 represent the graphed functions in terms of unit step functions and find their respective Laplace transforms.

9.

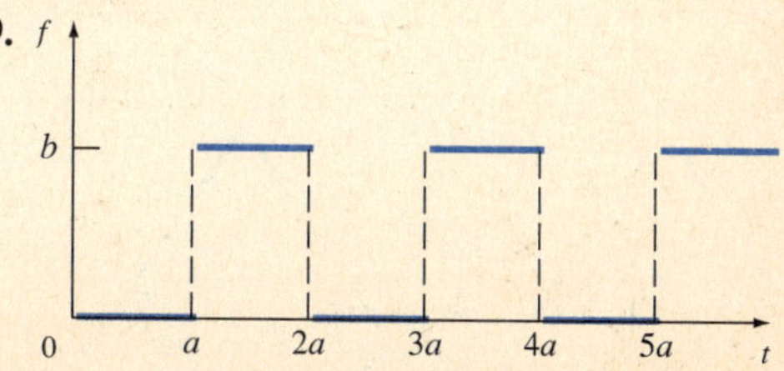

* D.V. Widder, *The Laplace Transformation* (Princeton, N.J.: Princeton University Press, 1941).

10.

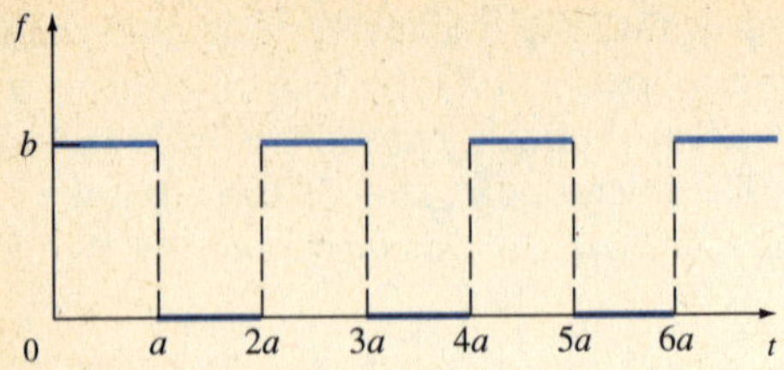

11.

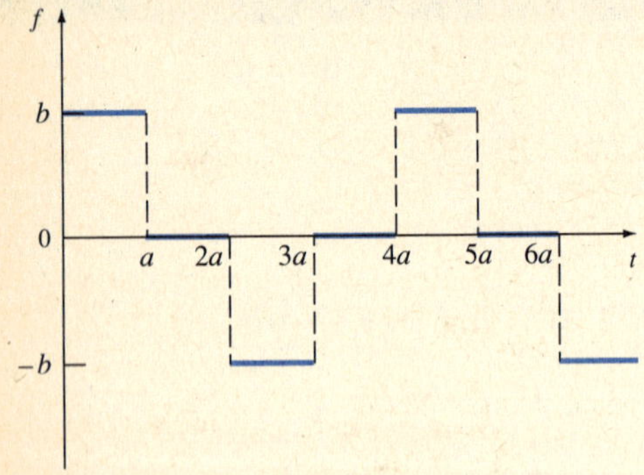

12. Let $g(t)$ be the function shown below.

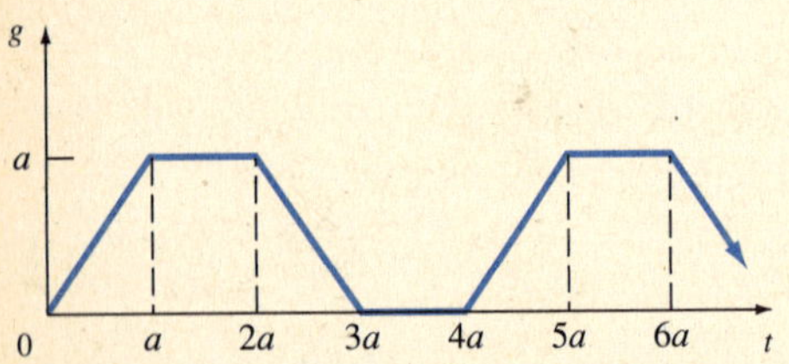

(a) Show that $g'(t)$ is piecewise continuous and that $g'(t) = f(t)$, where f is the step function of Problem 11 with $b = 1$.

(b) Use the differentiation property to compute $\mathscr{L}\{g\}$.

In Problems 13 and 14 use the method of Problem 12 to find the Laplace transform of the given functions.

13.

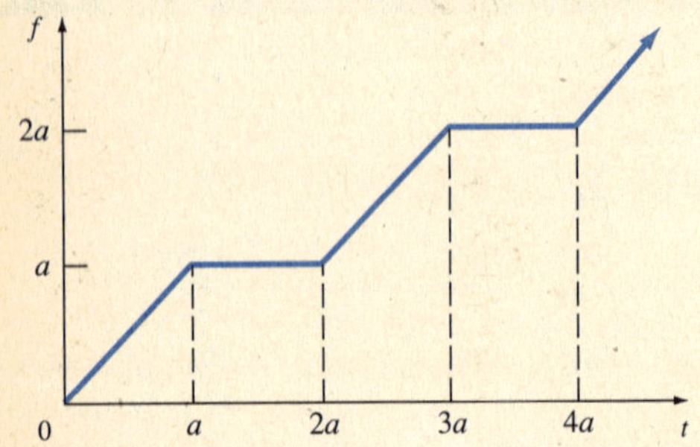

14.

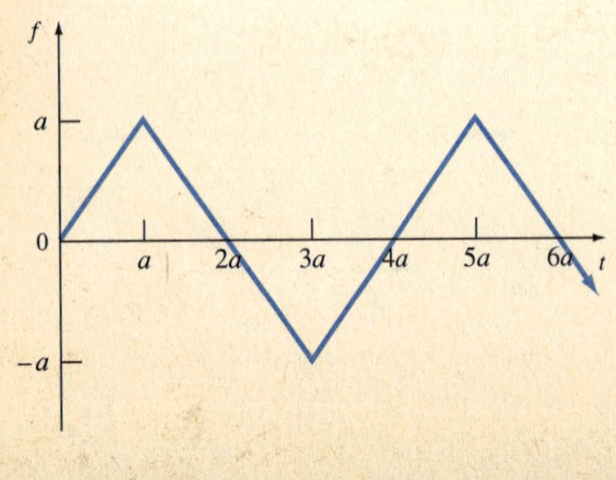

15. Compute the Laplace transform of the **half-wave rectifier** whose graph is given below.

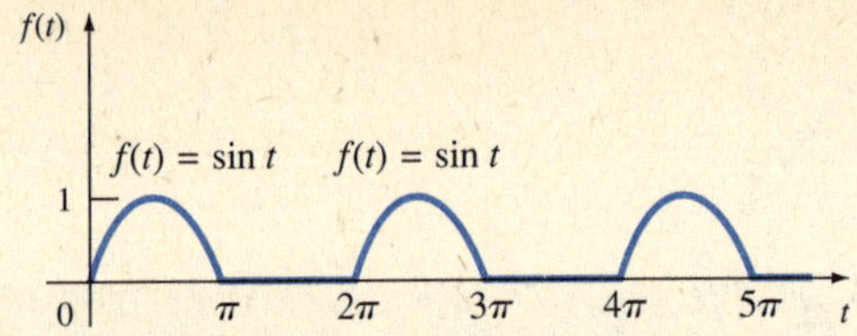

16. Compute the Laplace transform of the **sawtooth wave** given below.

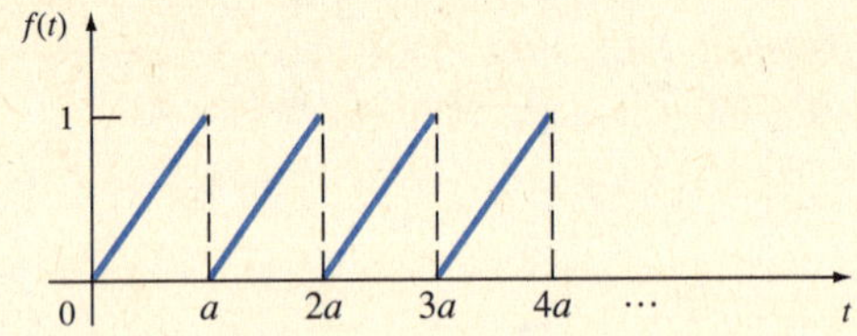

17. Let $f(t) = \begin{cases} \sin t, & t < 4\pi, \\ \sin t + \cos t, & t > 4\pi. \end{cases}$

Compute $\mathscr{L}\{f(t)\}$.

18. Let $f(t) = \begin{cases} \cos t, & t < \dfrac{3\pi}{2}, \\ \cos t + \sin t, & t > \dfrac{3\pi}{2}. \end{cases}$

Compute $\mathscr{L}\{f(t)\}$.

In Problems 19–24 compute the inverse of the given Laplace transform.

19. $\dfrac{e^{-\pi s}}{1 + s^2}$

20. $\dfrac{se^{-3\pi s}}{1 + s^2}$

21. $\dfrac{s - se^{-\pi s}}{1 + s^2}$

22. $\dfrac{1 - e^{-2s}}{s^2}$

23. $\dfrac{1 + e^{-4s}}{s^5}$

24. $\dfrac{e^{-2s}}{s^2 - 1}$ [*Hint:* Use partial fractions.]

25. Let

$$f(t) = \begin{cases} 1, & t < 1, \\ 3, & 1 < t < 7, \\ 5, & t > 7. \end{cases}$$

Write $f(t)$ as a step function and compute $\mathscr{L}\{f(t)\}$.

26. Let

$$f(t) = \begin{cases} -2, & 0 < t < 1, \\ 0, & 1 < t < 10, \\ 2, & t > 10. \end{cases}$$

Write $f(t)$ as a step function and compute $\mathscr{L}\{f(t)\}$.

In Problems 27–32 use the convolution theorem to calculate the inverse Laplace transforms of the given functions.

27. $F(s) = \dfrac{3}{s^4(s^2 + 1)}$ **28.** $F(s) = \dfrac{1}{(s^2 + 1)^2}$

29. $F(s) = \dfrac{1}{s(s^2 + a^2)}$ **30.** $F(s) = \dfrac{1}{(s^2 + 1)^3}$

31. $F(s) = \dfrac{e^{-3s}}{s^3}$ **32.** $F(s) = \dfrac{e^{-10s}}{s^5}$

In Problems 33–36 find the Laplace transform of each given convolution integral.

33. $f(t) = \displaystyle\int_0^t (t - u)^3 \sin u \, du$

34. $f(t) = \displaystyle\int_0^t e^{-(t-u)} \cos 2u \, du$

35. $f(t) = \displaystyle\int_0^t (t - u)^3 u^5 \, du$

36. $f(t) = \displaystyle\int_0^t \sinh 4(t - u) \cosh 5u \, du$

In Problems 37–43 solve the given initial-value problem with discontinuous forcing function $f(t)$. Draw a graph of the solution.

37. $y'' + y = f(t)$, $y(0) = y'(0) = 0$, where

$$f(t) = \begin{cases} t, & 0 < t < \pi \\ 0, & t > \pi. \end{cases}$$

38. $y'' - y = f(t)$, $y(0) = 1$, $y'(0) = 0$, where

$$f(t) = \begin{cases} 1, & 0 < t < 1 \\ 0, & t > 1. \end{cases}$$

39. $y'' + 2y' + 10y = f(t)$, $y(0) = y'(0) = 0$, where $f(t)$ is the function in Problem 38.

40. $y'' + 2y' + 5y = f(t)$, $y(0) = 0$, $y'(0) = 1$, where $f(t)$ is defined in Problem 37.

41. $y'' + 2y' + 10y = f(t)$, $y(0) = 1$, $y'(0) = 0$, where $f(t)$ is defined in Problem 38.

42. $y'' - 2y' + 2y = f(t)$, $y(0) = y'(0) = 0$, where

$$f(t) = \begin{cases} 0, & t < \dfrac{\pi}{2}, \\ 1, & \dfrac{\pi}{2} < t < \dfrac{3\pi}{2}, \\ 2, & t > \dfrac{3\pi}{2}. \end{cases}$$

43. $y'' + 4y' + 13y = f(t)$, $y(0) = y'(0) = 0$, where

$$f(t) = \begin{cases} 1, & t < \pi, \\ 0, & t > \pi. \end{cases}$$

44. Solve the Volterra integral equation

$$x(t) = t + \frac{1}{6}\int_0^t (t - u)^3 x(u) \, du.$$

45. Solve the Volterra integral equation

$$x(t) = e^{-t} - 2\int_0^t \cos(t - u) x(u) \, du.$$

46. A model rocket of 1-kilogram mass blasts off from a playground field with a propulsion force of 10 newtons. Assume that the motor runs out of propellant after 10 seconds, that the propellant has 500 grams mass, and that the propellant is consumed at a constant rate.

(a) Write the equations of motion for this situation assuming a damping force (due to air resistance) proportional to the velocity of the rocket.

(b) Solve the equation in part (a) using the Laplace transform methods of this section.

47. Use the convolution theorem to show that

$$\mathcal{L}\left\{\int_0^t f(u) \, du\right\} = \frac{F(s)}{s}, \text{ where } F(s) = \mathcal{L}\{f(t)\}.$$

[*Hint:* $\int_0^t f(u) \, du = (1 * f)(t)$.]

48. Compute $\mathcal{L}\{\int_0^{t-a} f(u) \, du\}$. [*Hint:* $\int_0^{t-a} f(u) \, du = \int_0^t f(u)H(t - a - u) \, du$.]

49. If $f(0) = g(0) = 0$, show that

(a) $f' * g = f * g'$;

(b) $(f * g)' = \frac{1}{2}[f' * g + f * g']$.

50. In Example 9, suppose that several advertising campaigns are mounted at times $t_1, t_2, \ldots, t_n$. Then, assuming that the campaigns retain their effectiveness, a model for the spread of the infection is given by

$$\frac{dy}{dt} = r(1 - y) - \sum_{k=1}^{n} \alpha_k H(t - t_k)$$

where $\alpha_k > 0$ and $\Sigma_{k=1}^{n} \alpha_k < r$.

(a) Solve this differential equation.

(b) Determine $\lim_{t \to \infty} y(t)$.

6.4 THE DIRAC DELTA FUNCTION

The **Dirac* delta function** (also called the **unit impulse function**), $\delta(t-a)$, is loosely described as a "function" that is zero everywhere except at $t = a$ and has the property that

$$\int_{-\infty}^{\infty} \delta(t-a)\,dt = 1. \tag{1}$$

As an illustration, we describe the Dirac delta function for $a = 0$. For any $\epsilon > 0$, consider the approximate delta functions (see Figure 6.8)

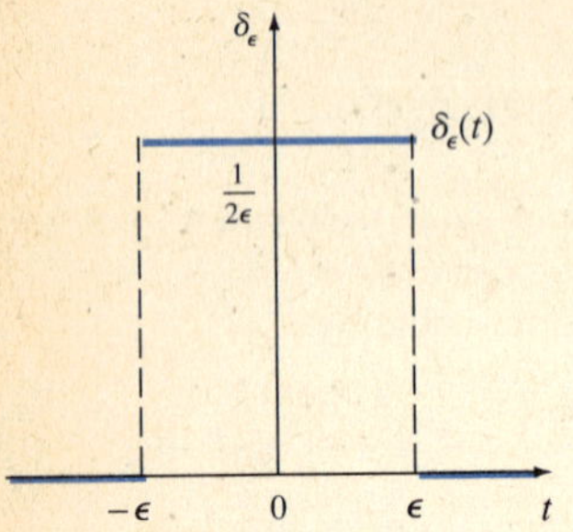

FIGURE 6.8
An approximate delta function

$$\delta_\epsilon(t) = \begin{cases} \dfrac{1}{2\epsilon}, & -\epsilon \le t \le \epsilon, \\ 0, & |t| > \epsilon. \end{cases}$$

Clearly, $\delta_\epsilon(t)$ is piecewise continuous and

$$\int_{-\infty}^{\infty} \delta_\epsilon(t)\,dt = \int_{-\epsilon}^{\epsilon} \frac{1}{2\epsilon}\,dt = 1.$$

Then $\delta(t)$ may be defined by

$$\delta(t) = \lim_{\epsilon\to 0} \delta_\epsilon(t).$$

Of course, $\delta(t)$ is not a function. However, it is the limit of piecewise continuous functions. There are several theorems of advanced calculus that, in some situations, allow us to treat a limit of piecewise continuous functions as if it were a legitimate function itself. We will not prove this fact here but we will, nevertheless, make use of it in some of our discussions of the delta function.

We need not have defined the Dirac delta function as a limit of step functions. We could have used instead the continuous functions

$$\delta_\epsilon(t) = \begin{cases} \dfrac{1}{\epsilon}\left[1 - \dfrac{|t|}{\epsilon}\right], & \text{for } |t| \le \epsilon, \\ 0, & \text{elsewhere,} \end{cases}$$

shown in Figure 6.9(a), or the smooth functions

$$\delta_\epsilon(t) = \begin{cases} \dfrac{15}{16\epsilon^5}(t^2 - \epsilon^2)^2, & \text{for } |t| \le \epsilon, \\ 0, & \text{elsewhere,} \end{cases}$$

illustrated in Figure 6.9(b), or any similar collection of functions having an integral of 1 and being positive in $|t| < \epsilon$. However, if we had done so, the calculations that follow would have been much more cumbersome.

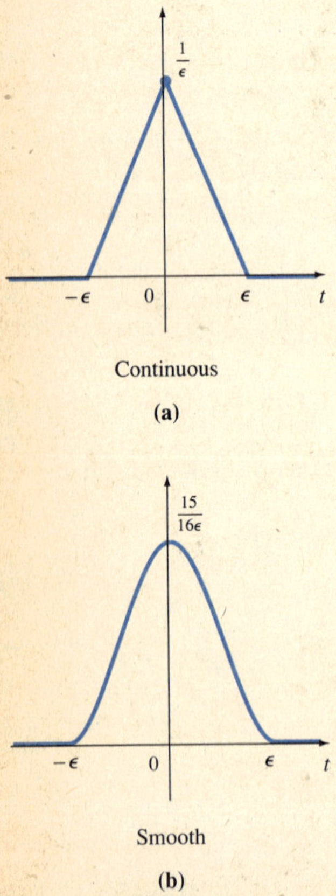

FIGURE 6.9
Other possible choices for $\delta_\epsilon(t)$

We now prove an important property of the Dirac delta function assuming that we can make the following interchange of limit and integral. Let $f(t)$ be any continuous function. Then

$$\int_{-\infty}^{\infty} \delta(t)f(t)\,dt = \int_{-\infty}^{\infty} \lim_{\epsilon\to 0} \delta_\epsilon(t)f(t)\,dt = \lim_{\epsilon\to 0} \int_{-\infty}^{\infty} \delta_\epsilon(t)f(t)\,dt. \tag{2}$$

*This function was first discussed by the British physicist Paul A. M. Dirac (pronounced Di-ráck with the *i* pronounced as in bridge), 1902–1984, in 1932. In 1933 he received the Nobel Prize in Physics (jointly with E. Schrödinger) for his work in quantum theory.

But $\delta_\epsilon(t) = 0$ for $|t| > \epsilon$, so that **(2)** becomes

$$\int_{-\infty}^{\infty} \delta(t)f(t)\,dt = \lim_{\epsilon\to 0}\int_{-\epsilon}^{\epsilon} \delta_\epsilon(t)f(t)\,dt. \tag{3}$$

We now remind you of the mean value theorem for integrals.

Mean value theorem for integrals

Suppose that f and g are continuous in the interval $[a, b]$, and that $g(t) \neq 0$ on $[a, b]$. Then there exists a number $\bar{t}$ in (a, b) such that

$$\int_a^b f(t)g(t)\,dt = f(\bar{t})\int_a^b g(t)\,dt.$$

Applying the mean value theorem for integrals to the integral on the right side of **(3)**, with $g(t) = \delta_\epsilon(t) = \frac{1}{2\epsilon}$ on the interval $[a, b] = [-\epsilon, \epsilon]$, we obtain

$$\int_{-\epsilon}^{\epsilon} \delta_\epsilon(t)f(t)\,dt = f(\bar{t})\int_{-\epsilon}^{\epsilon} \frac{1}{2\epsilon}\,dt = f(\bar{t}),$$

where $|\bar{t}| < \epsilon$. Hence, **(3)** yields

$$\int_{-\infty}^{\infty} \delta(t)f(t) = \lim_{\epsilon\to 0}\int_{-\epsilon}^{\epsilon} \delta_\epsilon(t)f(t)\,dt = \lim_{\epsilon\to 0} f(\bar{t}) = f(0),$$

because $|\bar{t}| < \epsilon$ and $\epsilon \to 0$. Thus, because the delta function $\delta(t) = 0$ for $t \neq 0$, the integral *evaluates the function f at t* $= 0$. More generally, we have

$$\int_{-\infty}^{\infty} \delta(t-a)f(t)\,dt = f(a). \tag{4}$$

Using **(4)** we can calculate the Laplace transform of the Dirac delta function:

$$\mathcal{L}\{\delta(t-a)\} = \int_0^{\infty} e^{-st}\delta(t-a)\,dt = e^{-as}, \qquad \text{for } a > 0. \tag{5}$$

There is an alternate way of arriving at this result. Since $\delta(t-a) = 0$ for $t \neq a$, it follows that

$$H(t-a) = \int_{-\infty}^{t} \delta(u-a)\,du, \tag{6}$$

since the integral in **(6)** is zero if $t < a$ and equals 1 when $t > a$. By the integration property (see Problem 6.2.31 on page 281), if $a \geq 0$,

$$\frac{1}{s}\mathcal{L}\{\delta(t-a)\} = \mathcal{L}\left\{\int_0^t \delta(u-a)\,du\right\} = \mathcal{L}\{H(t-a)\} = \frac{e^{-as}}{s}.$$

Thus **(5)** holds for $a \geq 0$ and $s > 0$.

One of the many useful properties of the Dirac delta function arises in connection with convolutions:

$$\delta(t-a) * f(t) = \int_0^t \delta((t-a)-u)f(u)\,du = \begin{cases} 0, & \text{if } t < a, \\ f(t-a), & \text{if } t \geq a, \end{cases}$$

because $t - a$ is negative when $t < a$. Hence, using the Heaviside function, we have

$$\delta(t-a) * f(t) = f(t-a)H(t-a). \tag{7}$$

We use formula **(7)** in the next two examples.

EXAMPLE 1 ▶ **Solving an initial-value problem with a Heaviside forcing function**

Solve the initial-value problem

$$y'' + 4y = 1 - H(t-1), \qquad y(0) = 1, \qquad y'(0) = 0.$$

Solution Using the differentiation property (Equation **(6.2.4)** on page 271) and Equation **(6.3.3)** on page 282, we obtain the Laplace transform of the initial-value problem:

$$s^2\mathcal{L}\{y\} - s + 4\mathcal{L}\{y\} = \frac{1}{s} - \frac{e^{-s}}{s},$$

$$(s^2+4)\mathcal{L}\{y\} = s + \frac{1}{s} - \frac{e^{-s}}{s},$$

or

$$\mathcal{L}\{y\} = \frac{s}{s^2+4} + \frac{1}{s(s^2+4)} - \frac{e^{-s}}{s(s^2+4)}.$$

By partial fractions, we have

$$\frac{1}{s(s^2+4)} = \frac{1}{4}\left(\frac{1}{s} - \frac{s}{s^2+4}\right),$$

so that

$$\mathcal{L}\{y\} = \frac{3}{4}\left(\frac{s}{s^2+4}\right) + \frac{1}{4s} - e^{-s}\frac{1}{4}\left(\frac{1}{s} - \frac{s}{s^2+4}\right).$$

But $\mathcal{L}\{\delta(t-1)\} = e^{-s}$ and $\mathcal{L}\{1-\cos 2t\} = 1/s - s/(s^2+4)$, so, by the convolution theorem on page 289,

$$y(t) = \frac{3}{4}\cos 2t + \frac{1}{4} - \delta(t-1) * \frac{1}{4}[1-\cos 2t].$$

But, from **(7)**,

$$\delta(t-1) * \frac{1}{4}[1-\cos 2t] = \frac{1}{4}[1-\cos 2(t-1)]H(t-1),$$

so that

$$y(t) = \frac{3}{4}\cos 2t + \frac{1}{4} - \frac{1}{4}[1-\cos 2(t-1)]H(t-1).$$ ◀

EXAMPLE 2 ▶ **Solving an initial-value problem with a delta forcing function**

Solve

$$y'' + 2y' + y = \delta(t-1), \qquad y(0) = 2, \qquad y'(0) = 3, \tag{8}$$

Solution The Laplace transform of Equation **(8)** is

$$[s^2\mathcal{L}\{y\} - 2s - 3] + 2[s\mathcal{L}\{y\} - 2] + \mathcal{L}\{y\} = e^{-s},$$

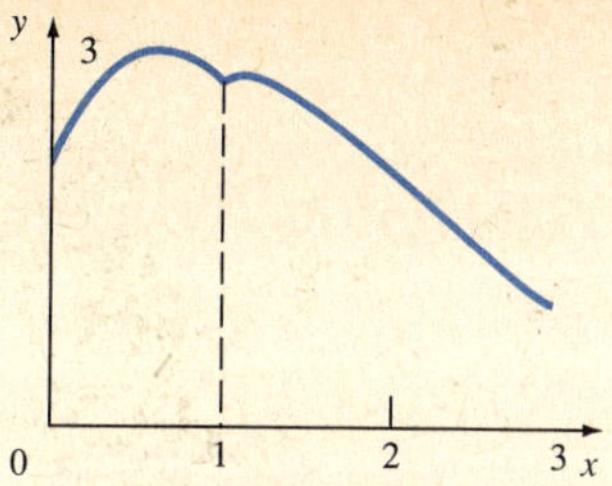

FIGURE 6.10
$y(t) = 2e^{-t} + 5te^{-t} + (t-1)e^{-(t-1)} H(t-1)$

or

$$(s^2 + 2s + 1)\mathcal{L}\{y\} = 2s + 7 + e^{-s}.$$

Hence we get

$$\mathcal{L}\{y\} = \frac{2s + 7 + e^{-s}}{s^2 + 2s + 1} = \frac{2(s+1)}{(s+1)^2} + \frac{5}{(s+1)^2} + \frac{e^{-s}}{(s+1)^2}$$
$$= \frac{2}{s+1} + \frac{5}{(s+1)^2} + \frac{e^{-s}}{(s+1)^2}.$$

Since $\mathcal{L}\{te^{-t}\} = (s+1)^{-2}$ (use Equation **(6.2.17)** on page 277), it follows from **(7)** and the convolution theorem (6.3.3) that

$$y(t) = 2e^{-t} + 5te^{-t} + \delta(t-1) * [te^{-t}]$$
$$= 2e^{-t} + 5te^{-t} + (t-1)e^{-(t-1)} H(t-1). \quad \textbf{(9)}$$

A graph of this solution is given in Figure 6.10. ◀

SELF-QUIZ

I. $\int_{-\infty}^{\infty} \delta(t+2)\, dt =$ ______.
(a) 2 **(b)** -2 **(c)** 1 **(d)** -1
(e) none of these

II. $\int_{-\infty}^{\infty} \delta(t+2)e^t\, dt =$ ______.
(a) e^{2t} **(b)** e^{-2t} **(c)** e^2 **(d)** e^{-2}
(e) none of these

III. $\int_0^t \sin(t-u)\delta\left(u - \frac{\pi}{6}\right) du = \begin{cases} 0, & 0 \le t < \frac{\pi}{6} \\ f(t), & t \ge \frac{\pi}{6} \end{cases}$
where $f(t) =$ ______.

(a) $\sin\left(t - \frac{\pi}{6}\right)$ **(b)** $\sin\left(t + \frac{\pi}{2}\right)$ **(c)** $\sin t$
(d) $\cos\left(t - \frac{\pi}{6}\right)$ **(e)** $\cos\left(t + \frac{\pi}{6}\right)$

IV. If $\mathcal{L}\{y\} = \frac{se^{-\pi s/2}}{s^2+1}$, then $y = \delta\left(t - \frac{\pi}{2}\right) * \cos t =$ ______.

(a) $\delta\left(t - \frac{\pi}{2}\right) \cos t$ **(b)** $\delta\left(t - \frac{\pi}{2}\right) \sin t$
(c) $\delta(t) \cos\left(t - \frac{\pi}{2}\right)$ **(d)** $\delta(t) \sin\left(t - \frac{\pi}{2}\right)$
(e) $\cos\left(t - \frac{\pi}{2}\right) H\left(t - \frac{\pi}{2}\right)$
(f) $\sin\left(t - \frac{\pi}{2}\right) H\left(t - \frac{\pi}{2}\right)$

Answers to Self-Quiz
I. c **II.** d **III.** a **IV.** e

PROBLEMS 6.4

In Problems 1–20 use the convolution theorem and Equation **(5)** *to solve the given initial-value problem with discontinuous forcing function.*

1. $y' + y = 2H(t-3)$, $y(0) = 0$
2. $y' - 3y = 3H(t-3)$, $y(0) = 3$
3. $y' + 2y = tH(t-4)$, $y(0) = 2$
4. $y' + 2y = e^{-2t}H(t-1)$, $y(0) = 2$
5. $y' + 2y = e^{-2t}[H(t-1) - H(t-2)]$, $y(0) = 3$
6. $y' - y = e^t[1 - H(t-3)]$, $y(0) = 1$
7. $y'' + y = H(t-1)$, $y(0) = y'(0) = 0$
8. $y'' + y = tH(t-2)$, $y(0) = 1$, $y'(0) = 0$
9. $y'' + 2y' + y = H(t-1) + H(t-2) + H(t-3)$, $y(0) = 0$, $y'(0) = 1$
10. $y'' + y = \sin t \cdot H(t-2)$, $y(0) = y'(0) = 0$
11. $y'' + 9y = \sin 3t \cdot H(t-3)$, $y(0) = 1$, $y'(0) = -1$
12. $y'' - y' + 6y = \delta(t-2)$, $y(0) = 1$, $y'(0) = -2$
13. $y'' - 4y' + 13y = \delta(t-1)$, $y(0) = 0$, $y'(0) = 3$
14. $y'' + 2y' + y = \delta(t-2) + 1$, $y(0) = 1$, $y'(0) = 0$

15. $y'' + y = \delta(t - 2\pi)$, $y(0) = 1$, $y'(0) = 0$
16. $y'' - 4y' + 8y = \delta(t - \pi) + \delta(t - 2\pi)$, $y(0) = 0$, $y'(0) = 1$
17. $y'' + 6y' + 10y = e^{-3t} H(t - 1) + t\,\delta(t - 2)$, $y(0) = 1$, $y'(0) = 0$
18. $y'' - 2y' - 3y = e^{-t} H(t - 1) + e^{3t}\delta(t - 3)$, $y(0) = -1$, $y'(0) = 0$
19. $y'' + 2y' - 3y = e^{t}\,\delta(t - 1) - e^{-3t} H(t - 4)$, $y(0) = 2$, $y'(0) = -1$
20. $y'' + y = \sin t \cdot H(t - \pi) + 2\cos t \cdot \delta(t - 2\pi)$, $y(0) = 1$, $y'(0) = 0$
21. Solve the initial-value problem

$$y'' + y = \sum_{n=0}^{\infty} \delta(t - 2\pi n), \qquad y(0) = y'(0) = 0.$$

Show that the solution becomes unbounded as t increases.

22. In Example 9 in Section 6.3 (page 287), suppose that the advertising campaign is effective only at the moment, t_1, when it is mounted; then the spread of infection can be modeled by

$$\frac{dy}{dt} = r(1 - y) - \alpha\delta(t - t_1).$$

(a) Solve this equation.
(b) Show that $\lim_{t\to\infty} y(t) = 1$, so that the effects of the campaign eventually die out.

23. In Problem 22 suppose that a series of campaigns are carried out at times $t_1, t_2, \ldots, t_n$. The equation then becomes

$$\frac{dy}{dt} = r(1 - y) - \sum_{k=1}^{n} \alpha_k\delta(t - t_k).$$

Answer the questions in Problem 22 using this equation.

24. Solve the initial-value problem

$$y'' + y = \sum_{n=0}^{\infty} \delta(t - \pi n), \qquad y(0) = y'(0) = 0.$$

Show that the solution remains bounded as t increases.

6.5 APPLICATIONS TO ELECTRIC CIRCUITS (OPTIONAL)

There are many physical models that give rise to differential equations with discontinuous forcing functions. In this section we discuss several models involving electric circuits. We begin with the *RLC* circuit discussed in Section 4.4.

Consider the differential equation

$$L\frac{dI}{dt} + RI + \frac{Q}{C} = E. \tag{1}$$

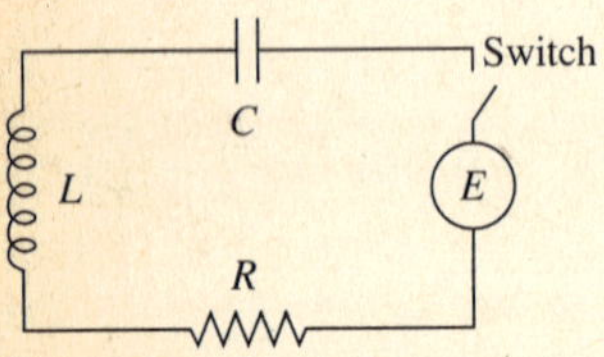

FIGURE 6.11
An RLC circuit

[See Equation **(4.4.1)** p. 189.] This is the equation that describes the circuit shown (again) in Figure 6.11.

We assume that the voltage source, which may be a battery, is controlled by a switch that, initially, is turned off, and that $Q(0) = I(0) = 0$. At some later time t_1, the switch is turned on and the voltage is then equal to some constant value E_0. In this situation it is easy to see that

$$E(t) = E_0 H(t - t_1). \tag{2}$$

But

$$\frac{dQ}{dt} = I, \qquad Q(0) = 0,$$

so that $s\mathcal{L}\{Q(t)\} = \mathcal{L}\{I(t)\}$. For simplicity we set $\hat{Q} = \mathcal{L}\{Q(t)\}$, $\hat{I} = \mathcal{L}\{I(t)\}$, and $\hat{E} = \mathcal{L}\{E(t)\}$, so that Equation **(1)** becomes

$$Ls\hat{I} + R\hat{I} + \frac{\hat{I}}{sC} = \hat{E},$$

or

$$\hat{I}(s) = T(s)\hat{E}(s), \tag{3}$$

where the function $T(s)$ is given by

$$T(s) = \frac{\frac{s}{L}}{s^2 + \frac{R}{L}s + \frac{1}{LC}}. \tag{4}$$

and

$$\hat{E}(s) = \mathscr{L}\{E_0 H(t - t_1)\} = \frac{E_0}{s}e^{-t_1 s}.$$

We can use this to solve for $\hat{I}(s)$ in Equation **(3).**

Remark

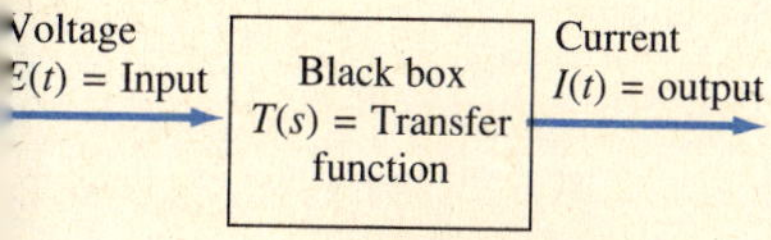

FIGURE 6.12
A black box

$E(t)$ is not defined at $t = t_1$ and, evidently, is not differentiable at that point. Therefore the use of Equation **(3)** is not strictly valid. However, we can get around this difficulty by observing that, according to Equation **(6.4.6)** on page 297, $H'(t - t_1) = \delta(t - t_1)$. Of course, $\delta(t - t_1)$ is not a function in the traditional sense but, since $\delta(t - t_1)$ has a Laplace transform [see Equation **(6.4.5)**], we will not worry about this difficulty.

black boxes

Engineers find this compartment or "black box" concept to be very useful in their work. In a **black box** something goes in (the input) and is transformed into something that comes out (the output). In this example, we have the black box setup shown in Figure 6.12. In Equation **(3),** we have a relationship between the transforms of the input and the output. The function $T(s)$ is called a **transfer function** and describes, precisely, the inner workings of the black box subject to the driving function $E(t)$. Equation **(3)** tells us exactly what we get out in terms of what we put in and gives us a simple equation relating input to output.

transfer functions

EXAMPLE 1 ▶ **Describing an RLC circuit after a switch is turned on**

An RLC circuit with $R = 10$ ohms, $L = 1$ henry, and $C = 0.01$ farad is hooked up to a battery that delivers a steady voltage of 20 volts when switched on. If the switch, initially off, is turned on after 10 seconds, find the current for all future values of t. Assume that Q and I are zero when the switch is turned on.

Solution Using Equations **(3)** and **(4)** and the values given above, we have

$$\begin{aligned}\hat{I}(s) &= \frac{s}{s^2 + 10s + 100}\left(\frac{E_0}{s}e^{-10s}\right)\\ &= \frac{E_0 e^{-10s}}{s^2 + 10s + 100} = \frac{20e^{-10s}}{(s+5)^2 + 75}.\end{aligned}$$

But

$$\begin{aligned}f(t) &= \mathscr{L}^{-1}\left\{\frac{20}{(s+5)^2 + 75}\right\}\\ &= \frac{20}{\sqrt{75}}\mathscr{L}^{-1}\left\{\frac{\sqrt{75}}{(s+5)^2 + 75}\right\} = \frac{20}{\sqrt{75}}e^{-5t}\sin\sqrt{75}\,t.\end{aligned}$$

so that

$$I(t) = \delta(t - 10) * f(t),$$

which by Equation **(6.4.7)** yields

$$I(t) = \frac{20}{\sqrt{75}} e^{-5(t-10)} \sin\sqrt{75}\,(t - 10)\,H(t - 10). \blacktriangleleft$$

Note that there is no current if $t < 10$.

EXAMPLE 2 ▶ **An *LC* circuit with a piecewise smooth *EMF***

Consider the LC circuit given in Figure 6.13 with $I(0) = I'(0) = 0$. The voltage is given by

$$E(t) = \begin{cases} 25t, & 0 \le t \le 4, \\ 100, & t > 4. \end{cases}$$

Find the current for all values of $t \ge 0$.

Solution Using Equation **(1)** we have (since $\frac{1}{0.04} = 25$)

$$\frac{dI}{dt} + 25Q = E$$

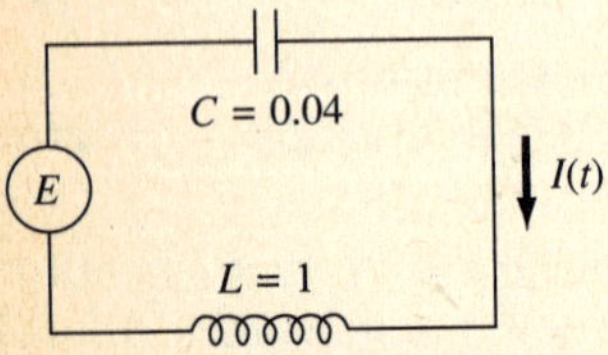

FIGURE 6.13
An LC circuit

and, differentiating,

$$\frac{d^2I}{dt^2} + 25I = E'(t) = \begin{cases} 25, & 0 \le t \le 4, \\ 0, & t > 4, \end{cases}$$

or

$$\frac{d^2I}{dt^2} + 25I = 25 - 25H(t - 4), \qquad I(0) = I'(0) = 0.$$

Taking Laplace transforms and using $I(0) = I'(0) = 0$, we have

$$s^2\hat{I}(s) + 25\hat{I}(s) = \frac{25}{s} - \frac{25e^{-4s}}{s} = -\frac{25}{s}(e^{-4s} - 1),$$

so

$$\hat{I}(s) = -\frac{25(e^{-4s} - 1)}{s(s^2 + 25)}.$$

Note that

$$\frac{25}{s(s^2 + 25)} = \frac{1}{s} - \frac{s}{s^2 + 25}$$

and

$$\hat{I}(s) = (e^{-4s} - 1)\left(\frac{-1}{s} + \frac{s}{s^2 + 25}\right)$$

$$= -\frac{e^{-4s}}{s} + e^{-4s}\frac{s}{s^2 + 25} + \frac{1}{s} - \frac{s}{s^2 + 25}.$$

Thus

$$I(t) = 1 - \cos 5t + H(t - 4)[\cos 5(t - 4) - 1]. \blacktriangleleft$$

EXAMPLE 3 ▶ **Analyzing a parallel circuit**

Consider the parallel electric shown in Figure 6.14, where the arrows denote the direction of current flow over each component of the circuit. We assume that

$I(t) = CE_0\delta(t - t_1)$. Clearly there is no current except at $t = t_1$. Show that if $E(0) = 0$, then $I(t)$ yields enough current to charge the capacitor to the voltage E_0 immediately.

Solution We refer to Kirchhoff's laws given in Section 2.7. We have the system

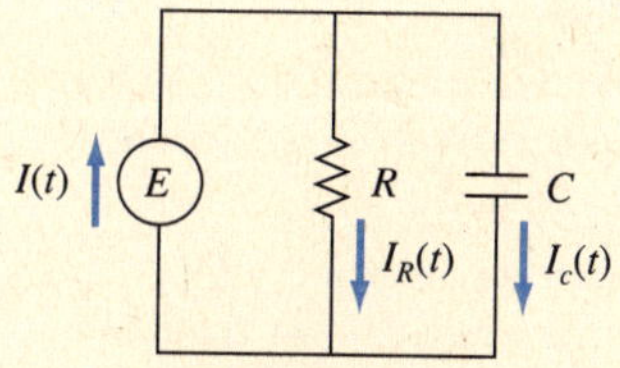

FIGURE 6.14
A parallel circuit

$$E = RI_R,$$
$$E = \frac{1}{C}Q_C, \tag{5}$$

or, differentiating,

$$E' = \frac{1}{C}\frac{dQ_C}{dt} = \frac{1}{C}I_C(t). \tag{6}$$

The total current is given by

$$I = I_R + I_C. \tag{7}$$

Thus

$$I(t) = I_R(t) + I_C(t) = \frac{E}{R} + CE',$$

or

$$E'(t) + \frac{1}{RC}E(t) = \frac{1}{C}I(t) = E_0\delta(t - t_1).$$

Taking transforms and using the fact that $E(0) = 0$, we have

$$s\hat{E}(s) + \frac{1}{RC}\hat{E}(s) = E_0e^{-st_1},$$

or

$$\hat{E}(s) = \frac{E_0e^{-st_1}}{s + \frac{1}{RC}}.$$

Thus

$$E(t) = \delta(t - t_1) * E_0e^{-(1/RC)t},$$

or

$$E(t) = E_0e^{-(1/RC)(t-t_1)}H(t - t_1). \quad ◀$$

Note that the voltage on the capacitor is zero before time $t = t_1$. At that time it jumps to the value E_0.

EXAMPLE 4 ▶

In Example 3, what happens to the current after time t_1?

Solution For $t > t_1$, $I(t) = 0$. Thus for $t > t_1$ the only current is the current through the resistor and capacitor. From Equation **(5)**, we have, for $t > t_1$,

$$I_R(t) = \frac{1}{R}E(t) = \frac{1}{R}E_0e^{-(1/RC)(t-t_1)}H(t - t_1)$$
$$= \frac{E_0}{R}e^{-(1/RC)(t-t_1)}.$$

Thus the current through the resistor decreases exponentially after the time $t = t_1$; that is, the current "bleeds off" through the resistor after the instant at which the capacitor is charged. ◀

PROBLEMS 6.5

1. Find the current for all t in the *RLC* circuit of Example 1, if $R = 20$ ohms, $L = \frac{1}{2}$ henry, and $C = 0.002$ farad, and a steady voltage 50-volt battery, which is initially off, is turned on 30 seconds later.

2. Answer the question in Problem 1 if $R = 15$ ohms, $L = 2$ henrys, $C = 0.04$ farad, and E is a steady-voltage battery of 25 volts that is turned on 1 minute later.

In Problems 3–5 find the current for all values of t in the LC circuit of Example 2 using the given data. Graph the solution.

3. $L = 1$ henry, $C = \frac{1}{16}$ farad,

$$E(t) = \begin{cases} 16t, & 0 \le t \le 5, \\ 80, & t > 5. \end{cases} \text{ volts}$$

4. $L = 1$ henry, $C = 0.04$ farad,

$$E(t) = \begin{cases} 0, & 0 \le t < 2, \\ 20t, & 2 \le t \le 4, \\ 80, & t > 4. \end{cases} \text{ volts}$$

5. $L = 1$ henry, $C = 0.1$ farad,

$$E(t) = \begin{cases} 10t, & 0 \le t \le 2, \\ 20, & 2 \le t \le 4, \\ 20t, & t > 4. \end{cases} \text{ volts}$$

6. In Example 3 find the voltage across the resistor at $t = 20$ seconds if $E_0 = 10$ volts, $t_1 = 10$ seconds, $R = 10$ ohms, and $C = 0.1$ farad.

7. An undamped spring supports an object of 1-kilogram mass. The spring constant of the spring is 4 newtons per meter. Suppose that a force $f(t)$ is applied to the object, where

$$f(t) = \begin{cases} 2t, & 0 \le t < \dfrac{\pi}{2} \\ 0, & t \ge \dfrac{\pi}{2} \end{cases} \text{ newtons.}$$

   (a) Find the equation of motion of the mass.
   (b) Assuming that, initially, the mass is displaced downward 1 meter before $f(t)$ is applied, find the position of the object for all $t > 0$.

8. Answer the questions in Problem 7 if

$$f(t) = \begin{cases} 0, & t < \dfrac{\pi}{2} \\ 4, & t > \dfrac{\pi}{2} \end{cases} \text{ newtons.}$$

6.6 FINDING LAPLACE TRANSFORMS WITH A SYMBOLIC MANIPULATOR

We now demonstrate how MATHEMATICA, MAPLE, DERIVE, MATLAB, or MATHCAD may be used to find Laplace transforms.

The most obvious application of a symbolic manipulator to Laplace transforms is in calculating the Laplace transform of a function. By definition **(6.1.1)**

$$\mathcal{L}\{f(t)\} = \int_0^\infty e^{-st} f(t)\, dt,$$

so that

```
LAPLACE(t,f(t)):=
LIM(INT(f*exp(−s*t),t),t,∞)−LIM(INT(f*exp(−s*t),t),t,0).
```

In principle, this script should work without difficulty, but some of the symbolic manipulators may balk at the first limit, since they will not know s is positive. You may need to resort to some trickery to get the program you are working with to give you the desired answer. One way to do this is to drop the first limit entirely.

```
LAPLACE(t,f(t)):= −LIM(INT(f*exp(−s*t),t),t,0).
```

EXAMPLE 1 ▶

Find the Laplace transform of the function $f(t) = e^{at}\cos(bt)$.

Solution Give the command LAPLACE(t, exp(a*t)*cos(b*t)) obtaining sequentially:

```
-LIM(INT(exp((a-s)*t)*cos(b*t),t),t,0)=
-LIM(exp((a-s)*t)*((a-s)*cos(b*t)+b*sin(b*t)/((a-s)^2+b^2),t,0)=
-((a-s)/((a-s)^2+b^2))=(s-a)/((a-s)^2+b^2).
```
◀

PROBLEMS 6.6

Find the Laplace transforms of the following functions using a symbolic manipulator such as MATHEMATICA, MAPLE, DERIVE, MATLAB, *or* MATHCAD.

1. $f(t) = e^{3t}\sin(5t)$

2. $f(t) = t^3 e^{4t}$

3. $f(t) = t^3\cosh(5t)$

4. $f(t) = t^7 e^{-3t}\sin(4t)$

5. $f(t) = t^3\cosh(5t^2)$

6. $f(t) = H(t-3)e^{5t-2}\cos(3t+1)$

SUMMARY OUTLINE OF CHAPTER 6

Definition of Laplace Transform: $\mathcal{L}\{f(t)\} = F(s) = \int_0^\infty e^{-st}f(t)\,dt$ p. 259
- Inverse Laplace transform: $\mathcal{L}^{-1}\{F(s)\} = f(t)$ p. 260
- Existence theorem for Laplace transforms: $|f(t)| \le Me^{at}$
- piecewise continuity, jump discontinuity
- Linearity: $\mathcal{L}\{af(t) + bg(t)\} = a\mathcal{L}\{f(t)\} + b\mathcal{L}\{g(t)\}$ p. 263
- First shifting theorem: $\mathcal{L}\{e^{at}f(t)\} = F(s-a)$ p. 266
- Differentiation property: $\mathcal{L}\{f'(t)\} = s\mathcal{L}\{f(t)\} - f(0)$ p. 270
- Higher derivatives, solution of initial-value problems p. 271

Partial Fractions p. 275

Derivative of Laplace Transform: $\mathcal{L}\{tf(t)\} = -\frac{d}{ds}\mathcal{L}\{f(t)\}$ p. 277

Unit Step Function = Heaviside Function: $H(t-a)$ p. 282
- Laplace transform of Heaviside function: $\mathcal{L}\{H(t-a)\} = e^{-as}/s$ p. 282
- Second shifting property: $\mathcal{L}\{f(t-a)H(t-a)\} = e^{-as}\mathcal{L}\{f(t)\}$ p. 282
- Using the Heaviside function to build other functions p. 283

Periodic Functions: Their Laplace transform p. 285
- If $f(t+\omega) = f(t)$, then $\mathcal{L}\{f(t)\} = \int_0^\omega e^{-st}f(t)\,dt/(1-e^{-\omega s})$ p. 285
- square waves p. 286

Convolutions: $f * g = \int_0^t f(t-u)g(u)\,du$ p. 289
- convolution property of Laplace transforms: $\mathcal{L}\{f*g\} = \mathcal{L}\{f\}\mathcal{L}\{g\}$ p. 289

Volterra Integral Equations: $x(t) = f(t) + \int_0^t a(t-u)x(t)\,dt$ p. 292

Dirac's Delta Function: $\delta(t-a)$ p. 296
- Point evaluation: $\int_{-\infty}^{\infty} \delta(t-a)f(t)\,dt = f(a)$ p. 297
- Laplace transform: $\mathcal{L}\{\delta(t-a)\} = e^{-as}$ p. 297
- convolution formula: $\delta(t-a) * f(t) = f(t-a)H(t-a)$ p. 298

Black Boxes, Transfer Functions p. 301

REVIEW EXERCISES FOR CHAPTER 6

In Exercises 1–17 find the Laplace transform of the given function.

1. $3t - 2$

2. $t^3 + 4t^2 - 2t + 1$

3. e^{2t-1}

4. $\cos(2t+1)$

5. te^{-t} **6.** $e^{-t}\cos 2t$ **7.** $\sinh(3t - 4)$ **8.** t^3e^{2t}

9. $te^t \cos t$ **10.** $\cosh^2 2t$ **11.** $\cos^2 t$ **12.** $\int_0^t u \cos u \, du$

13. $t^2 \cos 2t$ **14.** $\cos t \cdot H(t - 2\pi)$ **15.** $\delta(t - 3)$

16. The function shown in the figure below:

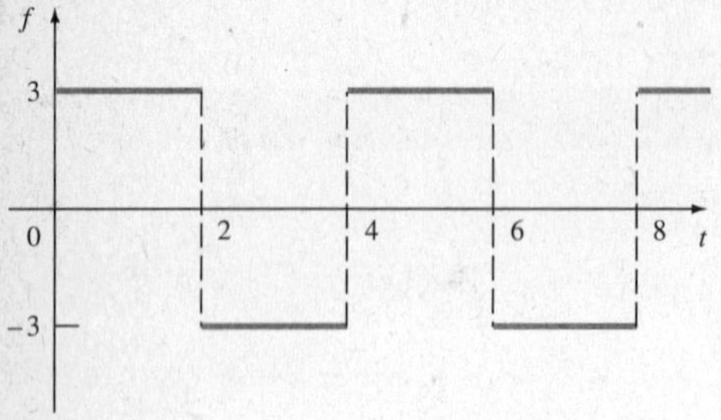

17. $f(t) = \begin{cases} \sin t, & t < 2\pi \\ \sin t + \cos t, & t > 2\pi \end{cases}$

In Exercises 18–32 find the inverse of the given Laplace transform.

18. $\dfrac{3}{s^2}$ **19.** $\dfrac{-14}{s}$

20. $\dfrac{s + 2}{s^2 + 4}$ **21.** $\dfrac{1}{(s - 2)^2}$

22. $\dfrac{s - 3}{s^2 - 3}$ **23.** $\dfrac{1}{s^2 + 4s + 5}$

24. $\dfrac{s + 3}{s^2 + 8s + 17}$ **25.** $\dfrac{s}{s^2 - 3s + 2}$

26. $\dfrac{s + 2}{s(s^2 + 4)}$ **27.** $\dfrac{s^2 + 8s - 3}{(s^2 + 2s + 1)(s^2 + 1)}$

28. $\ln \dfrac{s - 3}{s - 2}$ **29.** $\ln\left(1 + \dfrac{4}{s^2}\right)$

30. $\dfrac{se^{-(\pi/2)s}}{1 + s^2}$ **31.** $\dfrac{s - se^{-(\pi/2)s}}{1 + s^2}$

32. $\dfrac{e^{-3s}}{s^2 - 1}$

In Exercises 33–41 solve the given initial-value problem by Laplace transform methods.

33. $y'' + y = 0, y(0) = 1, y'(0) = 3$

34. $y'' - 4y = 0, y(0) = 2, y'(0) = -5$

35. $y'' - 5y' + 6y = 0, y(0) = 2, y'(0) = 1$

36. $y'' + 4y' + 4y = 0, y(0) = -1, y'(0) = 2$

37. $y'' - y = te^{2t}, y(0) = 0, y'(0) = 1$

38. $y'' + 9y = \sin t, y(0) = y'(0) = 0$

39. $y'' + 2y' + 2y = H(t - 3), y(0) = y'(0) = 0$

40. $y' - 6y = H(t - 2), y(0) = 0$

41. $y'' - 4y' + 4y = \delta(t - 1), y(0) = 0, y'(0) = 1$

In Exercises 42–45 find the Laplace transform of the given convolution integral.

42. $\int_0^t (t - u)^2 \sin u \, du$

43. $\int_0^t (t - u)^4 u^7 \, du$

44. $\int_0^t e^{15(t-u)} u^{25} \, du$

45. $f(t) = \int_0^t e^{17(t-u)} u^{19} \, du$

46. Solve the initial-value problem

$$y'' + 6y' + 18y = f(t), \qquad y(0) = y'(0) = 0,$$

where

$$f(t) = \begin{cases} 2, & t < \pi, \\ 0, & t > \pi. \end{cases}$$

47. Use the convolution theorem to compute the inverse Laplace transform of

(a) $F(s) = \dfrac{2}{s^3(s^2 + 1)}$

(b) $F(s) = \dfrac{1}{s(s^2 + 4)}$

48. A 50-kilogram mass is suspended from a spring with spring constant 20 newtons per meter. When the system is vibrating freely, the maximum displacement of each consecutive cycle decreases by 20%. Assume that a force equal to $10 \cos \omega t$ newtons acts on the system. Find the amplitude of the resultant steady-state motion if
(a) $\omega = 8$ radians per second;
(b) $\omega = 10$ radians per second;
(c) $\omega = 12$ radians per second;
(d) $\omega = 14$ radians per second;
(e) $\omega = 18$ radians per second.

49. In Figure 6.11, find $I(t)$ if $E(t) = \sin t$, $Q(0) = I(0) = 0$, $L = 2$ henrys, $R = 20$ ohms, and $C = 0.02$ farad.

50. Let $L = 1$ henry, $R = 100$ ohms, $C = 10^{-4}$ farad, and $E = 1000 \sin t$ volts in the circuit in Figure 6.11. Suppose that no charge and no current are initially present. Find the current and charge at all times t.

51. Let L, R, and C be as in Problem 49, but let $E(t) = 10 \sin 10t$. If $Q(0) = I(0) = 0$, find $Q(t)$.

7 INTRODUCTION TO SYSTEMS OF DIFFERENTIAL EQUATIONS AND APPLICATIONS

This chapter presents two different approaches to the solution of systems of linear differential equations.

1. The **method of elimination** and its allied **method of determinants** (Section 7.3) provide an efficient and elementary procedure for solving systems of two first-order linear differential equations in two unknown functions. As the name suggests, the technique involves eliminating one of the unknown functions by converting the system of two first-order equations into a single linear second-order equation in the remaining dependent variable. This second-order equation may then be solved using the methods in Chapter 3. Although the method is easily extendable to systems of n linear first-order equations in n unknown variables, the task of eliminating $(n - 1)$ dependent variables becomes unwieldy for $n > 2$.
2. The **method of Laplace transforms** (Section 7.4) converts a system of n linear first-order differential equations into a system of n linear equations, which may then be solved by the techniques of linear algebra. Once the transform of each unknown function has been determined, we may use the procedures in Chaper 6 to obtain a solution to the system. This last part may be the most difficult step in this procedure for large n, as the partial-fraction decomposition may be extremely tedious to obtain.

In Sections 7.5 through 7.8 we provide a number of interesting applications of second-order systems. The last application, a model of epidemics, contains material that is used in much current research in the field of mathematical biology. The other applications are of central importance to students of the physical and biological sciences and engineering.

7.1 SYSTEMS OF FIRST-ORDER EQUATIONS

In the differential equations we have discussed so far in this book, there was exactly one unknown function. A typical problem was to find the solution of, say, $x'' + a(t)x' + b(t)x = f(t)$, where a, b, and f are given. Here the solution consists of the single function $x(t)$.

In many applications, however, there is more than one unknown function. In these cases we are often led to a number of interrelated differential equations.

These equations together form what we call a **system** of differential equations. One typical system is given below:

$$\begin{aligned} x_1' &= f_1(t, x_1, x_2, \ldots, x_n), \\ x_2' &= f_2(t, x_1, x_2, \ldots, x_n), \\ &\vdots \\ x_n' &= f_n(t, x_1, x_2, \ldots, x_n), \end{aligned} \tag{1}$$

where each f_i, $i = 1, 2, \ldots, n$, is a function of $n + 1$ variables. The problem is to find n differentiable functions $x_1(t), x_2(t), \ldots, x_n(t)$ that satisfy **(1)** for all t in some interval $\alpha < t < \beta$. Any collection of n differential functions $x_1(t), x_2(t), \ldots, x_n(t)$ that satisfy the system **(1)** for all values t in some interval $\alpha < t < \beta$ is called a **solution** of the system **(1)** on that interval. System **(1)** is called a **first-order** system because it involves only first derivatives of the functions $x_1, x_2, \ldots, x_n$.

solution of a system of differential equations

first-order system

We will see in this section that certain higher-order differential equations or systems of higher-order differential equations can be written as first-order systems of form **(1)**. Therefore we limit our considerations to such first-order systems, that is, to systems that can be written in form **(1)**.

In much the same way that differential equations are classified as linear or nonlinear, systems are either linear or nonlinear.

Linear system

A first-order system is said to be **linear** if it can be written in the form

$$\begin{aligned} \frac{dx_1}{dt} &= a_{11}(t)x_1 + a_{12}(t)x_2 + \cdots + a_{1n}(t)x_n + f_1(t), \\ \frac{dx_2}{dt} &= a_{21}(t)x_1 + a_{22}(t)x_2 + \cdots + a_{2n}(t)x_n + f_2(t), \\ &\vdots \\ \frac{dx_n}{dt} &= a_{n1}(t)x_1 + a_{n2}(t)x_2 + \cdots + a_{nn}(t)x_n + f_n(t), \end{aligned} \tag{2}$$

where a_{ij} and f_i are functions that are continuous on some common interval $\alpha < t < \beta$. Any system for which this cannot be done is said to be **nonlinear.** The functions a_{ij} are called the **coefficients** of the linear system **(2)**. If every a_{ij} is constant, system **(2)** is said to have **constant coefficients.** If every function f_i satisfies $f_i(t) \equiv 0$ for $\alpha < t < \beta$, $i = 1, 2, \ldots, n$, then system **(2)** is said to be **homogeneous;** otherwise it is called a **nonhomogeneous** system.

EXAMPLE 1 ▶ **A linear system arising in a chemical flow problem**

Suppose that a chemical solution flows from one container, at a rate proportional to its volume, into a second container. It flows out from the second container at a constant rate. Let $x_1(t)$ and $x_2(t)$ denote the volumes of solution in the first and second containers, respectively, at time t. (The containers may be, for example, cells, in which case we are describing a diffusion process across a cell wall.) To establish the necessary equations, we note that the change in volume equals the difference between input and output in each container. The change in volume is the derivative of volume with respect to time. Since no chemical is flowing into the first container, the change in its volume equals the output:

$$\frac{dx_1}{dt} = -c_1x_1,$$

where c_1 is a positive constant of proportionality. The amount of solution c_1x_1 flowing out of the first container is the input of the second container. Let c_2 be the constant output of the second container. Then the change in volume in the second container equals the difference between its input and output:

$$\frac{dx_2}{dt} = c_1x_1 - c_2.$$

Thus we can describe the flow of solution by means of two differential equations. Since more than one differential equation is involved, we have obtained the linear nonhomogeneous system with constant coefficients

$$\begin{aligned} \frac{dx_1}{dt} &= -c_1x_1, \\ \frac{dx_2}{dt} &= c_1x_1 - c_2, \end{aligned} \tag{3}$$

(c_1 and c_2 are positive constants). By a solution of system **(3)** we mean a pair of functions $(x_1(t), x_2(t))$ that simultaneously satisfy the two equations in **(3)**. ◀

Initial-value problem

In addition to the given system of differential equations **(1)** or **(2)**, initial conditions of the form

$$x_1(t_0) = x_{10}, x_2(t_0) = x_{20}, \ldots, x_n(t_0) = x_{n0} \tag{4}$$

may also be given for some t_0 satisfying $\alpha < t_0 < \beta$. We call system **(1)** or **(2)** together with the initial conditions **(4)** an **initial-value problem.**

EXAMPLE 1 (continued) ▶ **Solving the chemical flow problem when initial conditions are given**

Recall system **(3)** of Example 1:

$$\begin{aligned} \frac{dx_1}{dt} &= -c_1x_1, \\ \frac{dx_2}{dt} &= c_1x_1 - c_2. \end{aligned}$$

It is easy to solve this system by solving the two equations successively (this is not usually possible). If we denote the initial volumes in the two containers by $x_1(0)$ and $x_2(0)$, respectively (these are the initial conditions), we see that the first equation has the solution

$$x_1(t) = x_1(0)e^{-c_1t}. \tag{5}$$

Substituting Equation **(5)** into the second equation of **(3)**, we obtain the equation

$$\frac{dx_2}{dt} = c_1x_1(0)e^{-c_1t} - c_2,$$

which, upon integration from 0 to t, yields the solution

$$x_2(t) = x_2(0) + x_1(0)(1 - e^{-c_1 t}) - c_2 t. \tag{6}$$

Equations **(5)** and **(6)** together constitute the unique solution of system **(3)** that satisfies the given initial conditions. ◀

As the definitions above indicate, many of the terms we have used in studying a single differential equation are also used to describe systems of differential equations. This is deliberate, because the system reduces to a single differential equation if we set $n = 1$ in **(1)** or **(2)**.

As in the case with initial-value problems involving a single differential equation (see Theorem 1.2.1), it is necessary to impose conditions on the functions f_1, $f_2, \ldots, f_n$ in order to guarantee that the initial-value problem **(1)** and **(4)** has a unique solution. The following result, which is proved in Appendix 3 (Theorem 7), provides sufficient conditions for a local solution to the initial-value problem **(1)** and **(4)**.

THEOREM 1 ♦ **Local existence-uniqueness theorem**

Let $f_1, f_2, \ldots, f_n$ and each of the first partials $\partial f_1/\partial x_1, \ldots, \partial f_1/\partial x_n, \partial f_2/\partial x_1, \ldots, \partial f_n/\partial x_n$ be continuous in an $(n + 1)$-dimensional rectangular region R given by $\alpha < t < \beta, \quad c_1 < x_1 < d_1, \ldots, \quad c_n < x_n < d_n$ containing the point $(t_0, x_{10}, \ldots, x_{n0})$. Then, in some interval $|t - t_0| < h$ contained in (α, β), there is a unique solution of system **(1)** that also satisfies the initial conditions **(4)**. ♦

For linear systems the existence-uniqueness theorem is simpler and provides a global solution. We prove the following result in Appendix 3 (Theorem 8).

THEOREM 2 ♦ **Global existence-uniqueness theorem**

If $a_{11}, a_{12}, \ldots, a_{nn}, f_1, \ldots, f_n$ are continuous functions on the interval $\alpha < t < \beta$ containing the point t_0, then system **(2)** has a unique solution that satisfies the initial conditions **(4)**. Furthermore, this solution is valid throughout the interval $\alpha < t < \beta$. ♦

Observe that in Theorem 1 we were only guaranteed a solution in a subinterval of $\alpha < t < \beta$, whereas in Theorem 2 the solution is valid throughout the interval in which the hypotheses hold. In particular, if $a_{11}, a_{12}, \ldots, a_{nn}, f_1, \ldots, f_n$ are continuous functions on the real numbers $\mathbb{R}$, the unique solution is valid on all of $\mathbb{R}$.

Warning

Without stating it, we will always assume that a_{ij} and f_i are continuous functions of t on some common open interval containing t_0. If they are all constants, the solution is unique for all real numbers $\mathbb{R}$. For example, the solution we obtained in Example 1 is valid for all t in the real numbers $\mathbb{R}$, since the functions $a_{11} = -c_1$, $a_{12} = 0, f_1 = 0, a_{21} = c_1, a_{22} = 0, f_2 = -c_2$ are constant and therefore continuous on $\mathbb{R}$.

Writing a higher-order differential equation as a first-order system

We mentioned earlier that certain higher-order differential equations, such as

$$y^{(n)} = F(t, y, y', y'', \ldots, y^{(n-1)}), \tag{7}$$

could be written as an equivalent system of first-order equations of form **(1)**. The procedure is as follows: We define n new variables $x_1, x_2, \ldots, x_n$ by setting

$$x_1 = y, x_2 = y', x_3 = y'', \ldots, x_n = y^{(n-1)}. \tag{8}$$

Then, since $y' = x_1' = x_2, y'' = x_2' = x_3, \ldots, y^{(n-1)} = x_{n-1}' = x_n$ and $y^{(n)} = x_n'$, we have the following first-order system

$$\begin{aligned} x_1' &= x_2, \\ x_2' &= x_3, \\ &\vdots \\ x_{n-1}' &= x_n, \\ x_n' &= F(t, x_1, x_2, \ldots, x_n), \end{aligned} \tag{9}$$

which is a special case of system **(1)**. The technique we have developed above can easily be applied to certain higher-order systems, as the following example illustrates.

EXAMPLE 2 ▶ **A mass-spring system leads to a linear, first-order system**

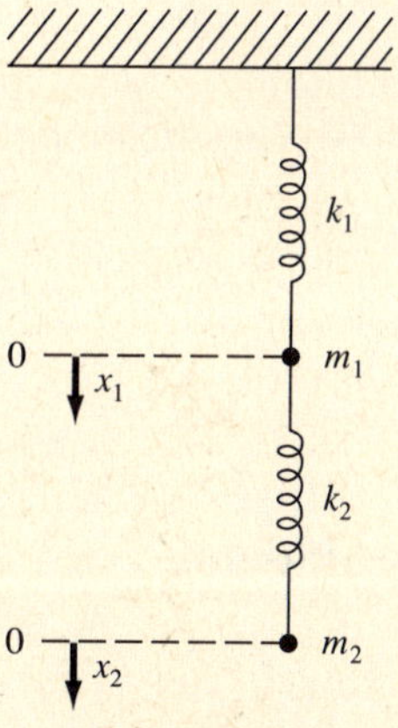

FIGURE 7.1
Two masses are suspended in series on two springs

Consider the mass-spring system of Figure 7.1, which is a direct generalization of the system described in Section 4.1 (see p. 173). In this example we have two objects of mass m_1 and m_2 suspended by springs in series with spring constants k_1 and k_2. If the vertical displacements from equilibrium of the two point masses are denoted by $x_1(t)$ and $x_2(t)$, respectively, then using assumptions (a) and (b) (Hooke's law, p. 174), we find that the net forces acting on the two masses are given by

$$\begin{aligned} F_1 &= -k_1x_1 + k_2(x_2 - x_1), \\ F_2 &= -k_2(x_2 - x_1). \end{aligned}$$

Here the positive direction is downward. Note that the first spring is compressed when $x_1 < 0$ and the second spring is compressed when $x_1 > x_2$. The equations of motion are a system of two linear second-order differential equations:

$$\begin{aligned} m_1\frac{d^2x_1}{dt^2} &= -k_1x_1 + k_2(x_2 - x_1) = -(k_1 + k_2)x_1 + k_2x_2, \\ m_1\frac{d^2x_2}{dt^2} &= -k_2(x_2 - x_1) = k_2x_1 - k_2x_2. \end{aligned} \tag{10}$$

To rewrite system **(10)** as a first-order system, we define the new variables $x_3 = x_1'$ and $x_4 = x_2'$. Then $x_3' = x_1''$, $x_4' = x_2''$ and **(10)** can be expressed as the system of four first-order equations

$$\begin{aligned} x_1' &= x_3, \\ x_2' &= x_4, \\ x_1'' = x_3' &= \left(\frac{k_1 + k_2}{m_1}\right)x_1 + \left(\frac{k_2}{m_1}\right)x_2, \\ x_2'' = x_4' &= \left(\frac{k_2}{m_2}\right)x_1 - \left(\frac{k_2}{m_2}\right)x_2. \end{aligned} \tag{11}$$

This is a linear homogeneous system with constant coefficients. ◀

We now show that linear differential equations (and systems) of *any* order can be converted, by the introduction of new variables, into a linear system of first-order differential equations. This concept is very important, since it means that the study of first-order linear systems provides a unified theory for all linear differential equations and systems. From a practical point of view, it means, for example, that once we know how to solve first-order linear systems with constant coefficients, we will be able to solve any constant-coefficient linear differential equation or system.

THEOREM 3 ♦ **Rewriting an nth-order equation as a first-order system**

The linear nth-order differential equation

$$x^{(n)} + a_1(t)x^{(n-1)} + a_2(t)x^{(n-2)} + \cdots + a_{n-1}(t)x' + a_n(t)x = f(t) \tag{12}$$

can be rewritten as a system of n first-order linear equations.

Proof Define $x_1 = x, x_2 = x', x_3 = x'', \ldots, x_n = x^{(n-1)}$. Then we have the linear nonhomogeneous system

$$\begin{aligned} x_1' &= x_2, \\ x_2' &= x_3, \\ &\vdots \\ x_{n-1}' &= x_n, \\ x_n' &= -a_nx_1 - a_{n-1}x_2 - \cdots - a_1x_n + f. \end{aligned} \tag{13}$$ ♦

Suppose that n initial conditions are specified for the nth-order Equation **(12)**:

$$x(t_0) = c_1, x'(t_0) = c_2, \ldots, x^{(n-1)}(t_0) = c_n.$$

These initial conditions can be immediately transformed into an initial condition for system **(13)**:

$$x_1(t_0) = c_1, x_2(t_0) = c_2, \ldots, x_n(t_0) = c_n. \tag{14}$$

EXAMPLE 3 ▶ **Writing a third-order equation with variable coefficients as a first-order system**

Consider the initial-value problem

$$t^3x''' + 4t^2x'' - 8tx' + 8x = 0, \qquad x(2) = 3, x'(2) = -6, x''(2) = 14.$$

Defining $x_1 = x, x_2 = x', x_3 = x''$, we obtain the system (since $x_3' = x'''$)

$$\begin{aligned} x_1' &= x_2, \\ x_2' &= x_3, \\ x_3' &= -\frac{8}{t^3}x_1 + \frac{8}{t^2}x_2 - \frac{4}{t}x_3, \end{aligned}$$

with the initial condition $x_1(2) = 3, x_2(2) = -6, x_3(2) = 14$. ◀

We will see, in the sections that follow, that almost every concept that we are already familiar with for differential equations extends to systems. As in the case of differential equations, nonlinear systems are extremely difficult to solve, so virtually all our discussion will concern linear first-order systems.

SELF-QUIZ

I. Which of the following systems are linear?
(a) $x' = 2x,\ y' = xy$ **(b)** $x' = 2x + 3y + \sin t,\ y' = -3x + 5y - e^t$
(c) $x' = (\cos t)x + y,\ y' = 3x - ye^t$ **(d)** $x' = e^x + y,\ y' = y - x$

II. If $x'' + 2x' - x = 0$ is transformed into a first-order system, then which system results?
(a) $x' = y,\ y' = -x + 2y$ **(b)** $x' = y,\ y' = 2x - y$
(c) $x' = y,\ y' = x - 2y$ **(d)** $x' = y,\ y' = -x + 2y$

III. If $x'' - 2tx' + e^t x = \cos t$ is transformed into a first-order system, which system results?

(a) $x' = 2ty,$ $y' = e^t x + \cos t$

(b) $x' = 2ty - e^t x,$ $y' = x$

(c) $x' = y,$ $y' = e^t x - 2ty + \cos t$

(d) $x' = y,$ $y' = -e^t x + 2ty + \cos t$

Answers to Self-Quiz

I. b, c **II.** c **III.** d

PROBLEMS 7.1

In Problems 1–7 transform each given equation into a system of first-order equations.

1. $x'' + 2x' + 3x = 0$

2. $x'' - 6tx' + 3t^3 x = \cos t$

3. $x''' - x'' + (x')^2 - x^3 = t$

4. $x^{(4)} - \cos x(t) = t$

5. $x''' + xx'' - x'x^4 = \sin t$

6. $xx'x''x''' = t^5$

7. $x''' - 3x'' + 4x' - x = 0$

8. A mass m moves in xyz-space according to the following equations of motion:

$$mx'' = f(t, x, y, z),$$
$$my'' = g(t, x, y, z),$$
$$mz'' = h(t, x, y, z).$$

Transform these equations into a system of six first-order equations.

9. Consider the uncoupled system

$$x_1' = x_1, \quad x_2' = x_2.$$

(a) What is the general solution of this system?

(b) Show that there is no second-order equation equivalent to this system. [*Hint:* Show that any second-order equation has solutions that are not solutions of this system.] This shows that first-order systems are more general than higher-order equations in the sense that any of the latter can be written as a first-order system, but not vice versa.

10. Let $x = x_1(t)$, $y = y_1(t)$ and $x = x_2(t)$, $y = y_2(t)$ be two solutions of the linear homogeneous system

$$x' = a_{11}(t)x + a_{12}(t)y,$$
$$y' = a_{21}(t)x + a_{22}(t)y. \qquad \textbf{(i)}$$

Show that

$$x = c_1 x_1(t) + c_2 x_2(t), \quad y = c_1 y_1(t) + c_2 y_2(t),$$

is also a solution of **(i)**, for any constants c_1 and c_2. This is the **principle of superposition for systems.**

11. Consider the nonhomogeneous system

$$x' = a_{11}(t)x + a_{12}(t)y + f_1(t),$$
$$y' = a_{21}(t)x + a_{22}(t)y + f_2(t). \qquad \textbf{(ii)}$$

Show that if $x = x_1(t)$, $y = y_1(t)$ and $x = x_2(t)$, $y = y_2(t)$ are two solutions of the nonhomogeneous system **(ii)**, then $x = x_1(t) - x_2(t)$, $y = y_1(t) - y_2(t)$ is a solution of the homogeneous system **(i)**.

7.2 LINEAR SYSTEMS: THEORY

In this section we consider the linear system of two first-order equations

$$x' = a_{11}(t)x + a_{12}(t)y + f_1(t),$$
$$y' = a_{21}(t)x + a_{22}(t)y + f_2(t), \qquad (1)$$

and the associated homogeneous system (i.e., $f_1 = f_2 = 0$)

$$x' = a_{11}(t)x + a_{12}(t)y,$$
$$y' = a_{21}(t)x + a_{22}(t)y. \qquad (2)$$

The point of view here emphasizes the similarities between such systems and the linear second-order equations discussed in Section 3.1. That there is a parallel between the two theories should not be surprising, since we have already shown in Section 7.1 that any linear second-order equation can always be transformed into a system of form **(1).**

Throughout our discussion we assume that the functions a_{ij} and f_i are all continuous on a common interval I. As we stated in Section 7.1, this guarantees the following result, provided t_0 is a point in that interval (see Appendix 3):

THEOREM 1 ♦ **Existence-uniqueness theorem**

If the functions $a_{11}(t)$, $a_{12}(t)$, $a_{21}(t)$, $a_{22}(t)$, $f_1(t)$, and $f_2(t)$ are continuous on the open interval I, then given any numbers t_0, x_0, and y_0, with t_0 in I, there exists exactly one solution $(x(t), y(t))$ of system **(1)** that satisfies $x(t_0) = x_0$ and $y(t_0) = y_0$. ♦

Solution to a system

By a **solution** of system **(1)** [or **(2)**] we mean a *pair* of functions $(x(t), y(t))$ that possess first derivatives and that satisfy the given equation on an interval I.

Warning

Whenever we compare two or more solutions or pairs of functions, we assume that all the functions are continuous on a common interval I.

Linear combination

The pair of functions $(x_3(t), y_3(t))$ is a **linear combination** of the pairs $(x_1(t), y_1(t))$ and $(x_2(t), y_2(t))$ if there exist constants c_1 and c_2 such that the following two equations hold:

$$\begin{aligned} x_3(t) &= c_1x_1(t) + c_2x_2(t), \\ y_3(t) &= c_1y_1(t) + c_2y_2(t). \end{aligned} \tag{3}$$

The next theorem is the systems analogue of Theorem 3.1.3 (p. 110). Its proof is left as an exercise.

THEOREM 2 ♦ **A linear combination of solutions to a homogeneous system is a solution**

If the pairs $(x_1(t), y_1(t))$ and $(x_2(t), y_2(t))$ are solutions of the homogeneous system **(2),** then any linear combination of them is also a solution of system **(2).** ♦

EXAMPLE 1 ▶ **A linear combination of solutions to a homogeneous system is a solution**

Consider the system

$$\begin{aligned} x' &= -x + 6y, \\ y' &= x - 2y. \end{aligned} \tag{4}$$

It is easy to verify that the pairs $(-2e^{-4t}, e^{-4t})$ and $(3e^{t}, e^{t})$ are solutions of system **(4).** Hence, by Theorem 2, the pair $(-2c_1e^{-4t} + 3c_2e^{t}, c_1e^{-4t} + c_2e^{t})$ is a solution of **(4)** for any constants c_1 and c_2. ◀

Linear independence

We define two pairs of functions $(x_1(t), y_1(t))$ and $(x_2(t), y_2(t))$ to be **linearly independent** on an interval I if, whenever the equations

$$\begin{aligned} c_1x_1(t) + c_2x_2(t) &= 0, \\ c_1y_1(t) + c_2y_2(t) &= 0 \end{aligned} \tag{5}$$

hold *for all values of t in I*, then $c_1 = c_2 = 0$.

In Example 1, the two given pairs of solutions are linearly independent since $c_1e^{-4t} + c_2e^t$ vanishes for all t only when $c_1 = c_2 = 0$.

Wronskian

Given two solutions $(x_1(t), y_1(t))$ and $(x_2(t), y_2(t))$ of system **(2),** we define the **Wronskian** of the two solutions by the following determinant:

$$W(t) = \begin{vmatrix} x_1(t) & y_1(t) \\ x_2(t) & y_2(t) \end{vmatrix} = x_1(t)y_2(t) - x_2(t)y_1(t). \tag{6}$$

We can then prove the next theorem.

THEOREM 3 ♦ **The general solution to a homogeneous system**

Let $(x_1(t), y_1(t))$ and $(x_2(t), y_2(t))$ be solutions of the homogeneous system **(2)** satisfying $W(t) \neq 0$ for every t in some interval I. If $(x^*(t), y^*(t))$ is any other solution of system **(2),** there exist constants c_1 and c_2 such that

$$\begin{aligned} x^* &= c_1x_1 + c_2x_2, \\ y^* &= c_1y_1 + c_2y_2. \end{aligned} \tag{7}$$

Thus $(c_1x_1(t) + c_2x_2(t), c_1y_1(t) + c_2y_2(t))$ is the **general solution** of system **(2).**

Proof Let t_0 in I be given and consider the linear system of two equations in the unknown quantities c_1 and c_2:

$$\begin{aligned} c_1x_1(t_0) + c_2x_2(t_0) &= x^*(t_0), \\ c_1y_1(t_0) + c_2y_2(t_0) &= y^*(t_0). \end{aligned} \tag{8}$$

The determinant of this system is $W(t_0)$, which is nonzero by assumption. Thus there is a unique pair of constants (c_1, c_2) satisfying **(8).** By Theorem 2,

$$(c_1x_1(t) + c_2x_2(t), c_1y_1(t) + c_2y_2(t))$$

is a solution of **(2).** But, by **(8),** this solution satisfies the same initial conditions at t_0 as the solution $(x^*(t), y^*(t))$. By the uniqueness part of Theorem 1, these solutions must be identical for all t in I. ♦

EXAMPLE 2 ▶ **Writing the general solution of a homogeneous system**

In Example 1 the Wronskian $W(t)$ is

$$W(t) = \begin{vmatrix} -2e^{-4t} & e^{-4t} \\ 3e^t & e^t \end{vmatrix} = -2e^{-3t} - 3e^{-3t} = -5e^{-3t} \neq 0.$$

Hence we need look no further for the general solution of system **(4).** The general solution is given by

$$(x(t), y(t)) = (-2c_1e^{-4t} + 3c_2e^t, c_1e^{-4t} + c_2e^t).$$

A particular solution can be obtained by assigning values to c_1 and c_2. For example, set $c_1 = 2$ and $c_2 = -1$ to obtain the particular solution

$$(-4e^{-4t} - 3e^t, 2e^{-4t} - e^t).$$

This should be checked by differentiating and inserting the results into system **(4).** ◀

In view of the condition required in Theorem 3 that the Wronskian $W(t)$ never vanish on I, we must consider the properties of the Wronskian more carefully. Let (x_1, y_1) and (x_2, y_2) be two solutions of the homogeneous system **(2).** Since $W(t) = x_1y_2 - x_2y_1$, we have

$$\begin{aligned} W'(t) &= x_1y_2' + x_1'y_2 - x_2y_1' - x_2'y_1 \\ &= x_1(a_{21}x_2 + a_{22}y_2) + y_2(a_{11}x_1 + a_{12}y_1) - x_2(a_{21}x_1 + a_{22}y_1) \\ &\quad - y_1(a_{11}x_2 + a_{12}y_2). \end{aligned}$$

Multiplying these expressions through and canceling like terms, we obtain

$$\begin{aligned} W' &= a_{11}x_1y_2 + a_{22}x_1y_2 - a_{11}x_2y_1 - a_{22}x_2y_1 \\ &= (a_{11} + a_{22})(x_1y_2 - x_2y_1) = (a_{11} + a_{22})W. \end{aligned}$$

Thus, separating variables and integrating,

An expression for the Wronskian

$$W(t) = W(t_0)\exp\left(\int_{t_0}^{t} [a_{11}(u) + a_{22}(u)]\, du\right). \quad \textbf{(9)}$$

We have shown the following theorem to be true.

THEOREM 4 ♦ The Wronskian of a homogeneous system is either always zero or never zero

Let (x_1, y_1) and (x_2, y_2) be two solutions of the homogeneous system **(2).** Then the Wronskian $W(t)$ is either always zero or never zero in the interval I (since $e^x \neq 0$ for any x). ♦

We are now ready to state the theorem that links linear independence with a nonvanishing Wronskian (see Theorem 3.1.5).

THEOREM 5 ♦ Two solutions of a homogeneous system are linearly independent if and only if their Wronskian is nonzero

Two solutions $(x_1(t), y_1(t))$ and $(x_2(t), y_2(t))$ of the homogeneous system **(2)** are linearly independent on an interval I if and only if $W(t) \neq 0$ in I.

Proof Let the solutions be linearly independent and suppose that $W(t) = 0$ in I. Then $x_1y_2 = x_2y_1$ or $x_1/x_2 = y_1/y_2 = f(t)$, for some function $f(t)$. Now

$$\begin{aligned} f' &= \left(\frac{x_1}{x_2}\right)' = \frac{x_2x_1' - x_1x_2'}{x_2^2} = \frac{x_2(a_{11}x_1 + a_{12}y_1) - x_1(a_{11}x_2 + a_{12}y_2)}{x_2^2} \\ &= \frac{a_{12}(x_2y_1 - x_1y_2)}{x_2^2} = -\frac{a_{12}W}{x_2^2} = 0, \end{aligned}$$

since $W = 0$. Thus $f'(t) = 0$ so $f(t) = c$, a constant. Then $x_1 = cx_2$ and $y_1 = cy_2$, so the solutions are dependent, which is a contradiction. Hence $W(t) \neq 0$. Conversely, let $W(t) \neq 0$ in I. If the solutions were dependent, then there would exist constants c_1 and c_2, not both zero, such that

$$c_1x_1 + c_2x_2 = 0,$$
$$c_1y_1 + c_2y_2 = 0.$$

Assuming that $c_1 \neq 0$, we then have $x_1 = cx_2$, $y_1 = cy_2$, where $c = -c_2/c_1$. But then

$$W(t) = x_1y_2 - x_2y_1 = cx_2y_2 - cx_2y_2 = 0,$$

which is again a contradiction. Therefore the solutions are linearly independent. ♦

General solution

We may redefine what we mean by a **general solution** in terms of these concepts: Let (x_1, y_1) and (x_2, y_2) be solutions of the homogeneous linear system

$$\begin{aligned} x' &= a_{11}x + a_{12}y, \\ y' &= a_{21}x + a_{22}y. \end{aligned} \qquad \textbf{(10)}$$

Then $(c_1x_1 + c_2x_2, c_1y_1 + c_2y_2)$ is the general solution of system **(10)** provided that $W(t) \neq 0$; that is, provided that the solutions (x_1, y_1) and (x_2, y_2) are linearly independent.

Finally, let us consider the nonhomogeneous system **(1).** The following theorem is the direct analogue of Theorem 3.1.6. Its proof is left as an exercise.

THEOREM 6 ♦ **The general solution of a nonhomogeneous system**

Let (x^*, y^*) be the general solution of system **(1),** and let (x_p, y_p) be any solution of **(1).** Then $(x^* - x_p, y^* - y_p)$ is the general solution of the homogeneous equation **(10).** In other words, the general solution of system **(1)** can be written as the sum of the general solution of the homogeneous system **(10)** and any particular solution of the nonhomogeneous system **(1).** ♦

EXAMPLE 3 ▶ **The general solution of a nonhomogeneous system**

Consider the system

$$\begin{aligned} x' &= 3x + 3y + t, \\ y' &= -x - y + 1. \end{aligned} \qquad \textbf{(11)}$$

Note first that $(1, -1)$ and $(-3e^{2t}, e^{2t})$ are solutions to the homogeneous system

$$\begin{aligned} x' &= 3x + 3y, \\ y' &= -x - y. \end{aligned}$$

A particular solution to system **(11)** is $(-\frac{1}{4}(t^2 + 9t + 3), \frac{1}{4}(t^2 + 7t))$. The general solution to **(11)** is, therefore,

$$(x(t), y(t)) = \left(c_1 - 3c_2e^{2t} - \frac{1}{4}(t^2 + 9t + 3), -c_1 + c_2e^{2t} + \frac{1}{4}(t^2 + 7t)\right).$$

◀

We close this section by noting that the theorems in this section can easily be generalized to apply to systems of three or more equations.

SELF-QUIZ

I. The Wronskian of $(e^t, 2e^t)$ and $(e^{2t}, -3e^{2t})$ is ______.

(a) $5e^{3t}$ **(b)** $-5e^{3t}$ **(c)** $5e^{-3t}$
(d) $-5e^{-3t}$ **(e)** $-e^{3t}$

II. If $(e^{-t}\cos 2t, -e^{-t}\sin 2t)$ and $(4e^{-t}\cos 2t, 3e^{-t}\sin 2t)$ are solutions to a homogeneous system, then which of the following are also solutions?
(a) $(5e^{-t}\cos 2t, 2e^{-t}\sin 2t)$
(b) $(e^{-t}\cos 2t, 3e^{-t}\sin 2t)$
(c) $(4e^{-t}\cos 2t, -e^{-t}\sin 2t)$
(d) $(23e^{-t}\cos 2t, 18e^{-t}\sin 2t)$
(e) $(-6e^{-t}\cos 2t, 5e^{-t}\sin 2t)$

III. Which of the following pairs of solutions are *not* linearly independent?
(a) $(e^{-t}, 2e^{-t}), (te^{-t}, -2te^{-t})$
(b) $(e^{-t}, 4e^{-t}), (-2te^{-t}, -8te^{-t})$
(c) $(-e^{-t}, 3e^{-t}), (2te^{-t}, 3te^{-t})$
(d) $(-4e^{-t}, 6e^{-t}), (2te^{-t}, -3te^{-t})$

IV. Suppose that $(e^t, 2e^t)$ and $(2e^{2t}, -3e^{2t})$ are two linearly independent solutions of the homogeneous system **(2)** while $(\sin 3t, \cos 3t)$ is a solution to the nonhomogeneous system **(1)**; which of the following is also a solution to the nonhomogeneous system?
(a) $(e^t + \sin 3t, 2e^t + \cos 3t)$
(b) $(e^t + 2\sin 3t, 2e^t + 2\cos 3t)$
(c) $(2e^{2t} + 5\sin 3t, -3e^{2t} + 5\sin 3t)$
(d) $(8e^t + \sin 3t, 16e^t + \cos 3t)$

Answers to Self-Quiz

I. b **II.** all of them **III.** b, d **IV.** a, d

PROBLEMS 7.2

1. (a) Show that $(e^{-3t}, -e^{-3t})$ and $((1-t)e^{-3t}, te^{-3t})$ are solutions to

$$x' = -4x - y,$$
$$y' = x - 2y.$$

(b) Calculate the Wronskian of these solutions and verify that the solutions are linearly independent on $\mathbb{R}$.
(c) Write the general solution to the system.

2. (a) Show that $(e^{2t}\cos 2t, -2e^{2t}\sin 2t)$ and $(e^{2t}\sin 2t, 2e^{2t}\cos 2t)$ are solutions of the system

$$x' = 2x + y,$$
$$y' = -4x + 2y.$$

(b) Calculate the Wronskian of these solutions and show that they are linearly independent on $\mathbb{R}$.
(c) Show that $(\frac{1}{4}te^{2t}, -\frac{11}{4}e^{2t})$ is a solution of the nonhomogeneous system

$$x' = 2x + y + 3e^{2t},$$
$$y' = -4x + 2y + te^{2t}.$$

(d) Combining (a) and (c), write the general solution of the nonhomogeneous equation in (c).

3. (a) Show that $(\sin t^2, 2t\cos t^2)$ and $(\cos t^2, -2t\sin t^2)$ are solutions of the system

$$x' = y,$$
$$y' = -4t^2x + \frac{1}{t}y.$$

(b) Show where the solutions are linearly independent.
(c) Show that $W(0) = 0$.
(d) Explain the apparent contradiction of Theorem 5.

4. **(a)** Show that $(\sin \ln t^2, (\frac{2}{t})\cos \ln t^2)$ and $(\cos \ln t^2, -(\frac{2}{t}) \sin \ln t^2)$ are linearly independent solutions of the system

$$x' = y,$$

$$y' = -\frac{4}{t^2}x - \frac{1}{t}y.$$

(b) Calculate the Wronskian $W(t)$.

5. Prove Theorem 2.

6. Prove Theorem 6. [*Hint:* See the proof of Theorem 3.1.6 on page 114.]

7.3 THE METHOD OF ELIMINATION AND THE METHOD OF DETERMINANTS FOR LINEAR SYSTEMS WITH CONSTANT COEFFICIENTS

In this section we discuss an elementary method for solving a system of simultaneous first-order linear differential equations by converting the system into a single higher-order linear differential equation that may then be solved by the methods we have seen in previous chapters. It is easiest to motivate this method with an example.

EXAMPLE 1 ▶ **Solving a feedback system by the method of elimination**

Let tank X contain 100 gallons of brine in which 100 pounds of salt is dissolved and tank Y contain 100 gallons of water. Suppose that water flows into tank X at the rate of 2 gallons per minute, and that the mixture flows from tank X into tank Y at 3 gallons per minute. Each minute, 1 gallon is pumped from Y back to X (establishing **feedback**) while 2 gallons are flushed away. We wish to find the amount of salt in both tanks at all time t (see Figure 7.2).

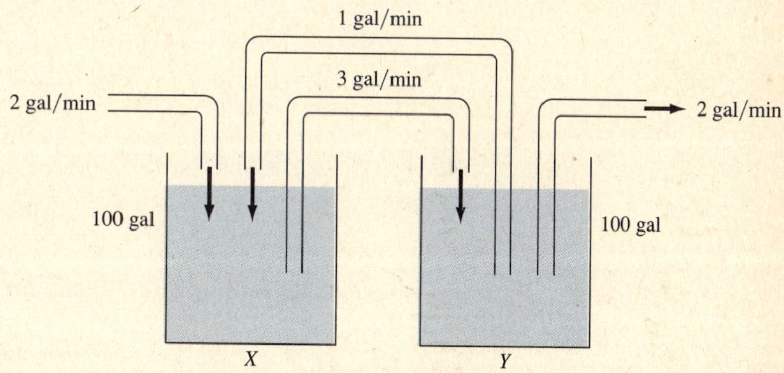

FIGURE 7.2
Two gallons enter and leave the system each minute

Solution If we let $x(t)$ and $y(t)$ represent the number of pounds of salt in tanks X and Y at time t and note that the change in weight equals the difference between input and output, we can again derive a system of linear first-order equations. Tanks X and Y initially contain $x(0) = 100$ and $y(0) = 0$ pounds of salt, respectively, at time $t = 0$. The quantities $\frac{x}{100}$ and $\frac{y}{100}$ are, respectively, the amounts of salt contained in each gallon of water taken from tanks X and Y at time t. Three gallons are being removed from tank X and added to tank Y, while only 1 of the 3 gallons removed from tank Y is put in tank X. Thus we have the system

$$\begin{aligned} \frac{dx}{dt} &= -3\frac{x}{100} + \frac{y}{100}, \qquad x(0) = 100. \\ \frac{dy}{dt} &= 3\frac{x}{100} - 3\frac{y}{100}, \qquad y(0) = 0. \end{aligned} \tag{1}$$

Since both equations in system **(1)** involve *both* dependent variables, we cannot immediately solve for one of the variables, as we did in Example 7.1.1. Instead, we use the operation of differentiation to eliminate one of the dependent variables. Suppose we begin by solving the second equation for $x(t)$ in terms of the dependent variable y and its derivative:

$$\frac{3}{100}x = \frac{3}{100}y + \frac{dy}{dt},$$

or

$$x = y + \frac{100}{3}\frac{dy}{dt}. \tag{2}$$

We now differentiate Equation **(2)** and equate the result to the first equation in **(1)**:

$$\frac{dx}{dt} = \frac{dy}{dt} + \frac{100}{3}\frac{d^2y}{dt^2} = \frac{-3}{100}x + \frac{y}{100}. \tag{3}$$

But, from the second equation of **(1)**,

$$\frac{3}{100}x = \frac{dy}{dt} + \frac{3}{100}y,$$

so, inserting this into the last expression in **(3)**, we obtain

$$\frac{100}{3}\frac{d^2y}{dt^2} + \frac{dy}{dt} = \overbrace{-\frac{dy}{dt} - \frac{3}{100}y}^{-\frac{3}{100}x} + \frac{1}{100}y.$$

This yields the second-order equation

$$\frac{100}{3}\frac{d^2y}{dt^2} + 2\frac{dy}{dt} + \frac{2y}{100} = 0. \tag{4}$$

The initial conditions for Equation **(4)** are obtained directly from system **(1)**, since $y(0) = 0$ and

$$y'(0) = 3\frac{x(0)}{100} - 3\frac{y(0)}{100} = 3.$$

Multiplying both sides of Equation **(4)** by $\frac{3}{100}$, we have the initial-value problem

$$y'' + \frac{6}{100}y' + \frac{6}{(100)^2}y = 0, \qquad y(0) = 0,\ y'(0) = 3. \tag{5}$$

The characteristic equation for **(5)** is

$$\lambda^2 + \frac{6}{100}\lambda + \frac{6}{(100)^2} = 0, \tag{6}$$

which has the roots

$$\left.\begin{matrix}\lambda_1\\ \lambda_2\end{matrix}\right\} = \frac{\dfrac{-6}{100} \pm \sqrt{\dfrac{36}{(100)^2} - \dfrac{24}{(100)^2}}}{2} = \frac{-6 \pm \sqrt{12}}{200},$$

or

$$\lambda_1 = \frac{-3 + \sqrt{3}}{100}, \qquad \lambda_2 = \frac{-3 - \sqrt{3}}{100}.$$

Thus the general solution is

$$y(t) = c_1 e^{[(-3+\sqrt{3})t]/100} + c_2 e^{[(-3-\sqrt{3})t]/100}.$$

Using the initial conditions, we obtain the simultaneous equations

$$\begin{aligned} c_1 + \qquad\qquad c_2 &= 0,\\ \frac{-3 + \sqrt{3}}{100} c_1 - \frac{3 + \sqrt{3}}{100} c_2 &= 3. \end{aligned}$$

These have the unique solution $c_1 = -c_2 = 50\sqrt{3}$. Hence

$$y(t) = 50\sqrt{3}\,[e^{[(-3+\sqrt{3})t]/100} - e^{[(-3-\sqrt{3})t]/100}],$$

and substituting this function into the right-hand side of Equation **(2)**, we obtain, after some algebra,

$$x(t) = 50[e^{[(-3+\sqrt{3})t]/100} + e^{[(-3-\sqrt{3})t]/100}].$$

Note that, as is evident from the problem, the amounts of salt in the two tanks approach zero as time tends to infinity. ◀

The technique we have used in solving Example 1 is called the **method of elimination,** since all but one dependent variable is eliminated by repeated differentiation. The method is quite elementary but requires many calculations. We will introduce a direct way of obtaining the characteristic equation without doing the elimination procedure. Nevertheless, because of its simplicity, elimination can be a very useful tool.

EXAMPLE 2 ▶ **Solving an initial-value system by elimination**

Consider the system

$$\begin{aligned} x' &= x + y, \qquad & x(0) &= 1,\\ y' &= -3x - y, \qquad & y(0) &= 0. \end{aligned} \tag{7}$$

Differentiating the first equation and substituting the second equation for y', we have

$$x'' = x' + \overbrace{(-3x - y)}^{y'} = x' - 3x - \overbrace{(x' - x)}^{y \text{ from the first equation of (7)}},$$

or

$$x'' + 2x = 0.$$

Thus

$$x(t) = c_1 \cos \sqrt{2}\, t + c_2 \sin \sqrt{2}\, t.$$

Then, according to the first equation of system **(7)**,

$$\begin{aligned} y = x' - x &= -\sqrt{2}\, c_1 \sin \sqrt{2}\, t + \sqrt{2}\, c_2 \cos \sqrt{2}\, t \\ &\quad - c_1 \cos \sqrt{2}\, t - c_2 \sin \sqrt{2}\, t \\ &= (c_2\sqrt{2} - c_1) \cos \sqrt{2}\, t - (\sqrt{2}\, c_1 + c_2) \sin \sqrt{2}\, t. \end{aligned}$$

Using the initial conditions, we find that

$$x(0) = c_1 = 1, \qquad y(0) = c_2\sqrt{2} - c_1 = 0, \quad \text{or} \quad c_2 = \frac{1}{\sqrt{2}}.$$

Therefore the unique solution of system **(7)** is given by the pair of functions

$$\begin{aligned} x(t) &= \cos \sqrt{2}\, t + \frac{1}{\sqrt{2}} \sin \sqrt{2}\, t, \\ y(t) &= -\left(\sqrt{2} + \frac{1}{\sqrt{2}}\right) \sin \sqrt{2}\, t = -\frac{3}{\sqrt{2}} \sin \sqrt{2}\, t. \end{aligned}$$ ◀

EXAMPLE 3 ▶ **Solving a nonhomogeneous system by elimination**

Consider the system

$$\begin{aligned} x' &= 2x + y + t, \\ y' &= x + 2y + t^2. \end{aligned}$$

Proceeding as before, we obtain

$$\begin{aligned} x'' &= 2x' + y' + 1 = 2x' + (x + 2y + t^2) + 1 \\ &= 2x' + x + (2x' - 4x - 2t) + t^2 + 1, \end{aligned}$$

or

$$x'' - 4x' + 3x = t^2 - 2t + 1 = (t - 1)^2. \tag{8}$$

The solution to the homogeneous part of Equation **(8)** is $x(t) = c_1e^t + c_2e^{3t}$. A particular solution of Equation **(8)** is easily found to be $\frac{1}{3}t^2 + \frac{2}{9}t + \frac{11}{27}$, so the general solution Equation **(8)** is

$$x(t) = c_1e^t + c_2e^{3t} + \frac{1}{3}t^2 + \frac{2}{9}t + \frac{11}{27}.$$

As before, since $y = x' - 2x - t$, we obtain

$$y(t) = c_1e^t + 3c_2e^{3t} + \frac{2}{3}t + \frac{2}{9} - 2c_1e^t - 2c_2e^{3t} - \frac{2}{3}t^2 - \frac{4}{9}t - \frac{22}{27} - t,$$

or

$$y(t) = -c_1e^t + c_2e^{3t} - \frac{2}{3}t^2 - \frac{7}{9}t - \frac{16}{27}.$$ ◀

The method illustrated in the last three examples can easily be generalized to apply to linear systems with three or more equations. A linear system of n first-order equations usually reduces to an nth-order linear differential equation, because it generally requires one differentiation to eliminate each variable $x_2, \ldots, x_n$ from the system. However, while the method is easy in theory, it is extremely cumbersome to implement in practice. So, instead of giving tedious examples, we discuss a much more efficient method based on the theory of determinants. If your knowledge of determinants is a bit rusty, read Appendix 4 before continuing.

The method of determinants

We now develop a more efficient way of obtaining the characteristic equation of the differential equation that is obtained by eliminating all but one dependent variable of a homogeneous system (nonhomogeneous systems are discussed in Problems 18–23 and 32–42).

Consider the homogeneous system

$$\begin{aligned} x' &= a_{11}x + a_{12}y, \\ y' &= a_{21}x + a_{22}y, \end{aligned} \qquad \textbf{(9)}$$

where the a_{ij} are constants. Our main tool for solving second-order linear homogeneous equations with constant cofficients involved obtaining a characteristic equation by "guessing" that the solution had the form $y = e^{\lambda x}$.

Parallel to the method of Section 3.3, we guess that there is a solution to system **(9)** of the form $(\alpha e^{\lambda t}, \beta e^{\lambda t})$, where α, β, and λ are constants yet to be determined. Substituting $x(t) = \alpha e^{\lambda t}$ and $y(t) = \beta e^{\lambda t}$ into system **(9),** we obtain

$$\begin{aligned} x' &= \alpha\lambda e^{\lambda t} = a_{11}\alpha e^{\lambda t} + a_{12}\beta e^{\lambda t}, \\ y' &= \beta\lambda e^{\lambda t} = a_{21}\alpha e^{\lambda t} + a_{22}\beta e^{\lambda t}. \end{aligned}$$

After dividing by $e^{\lambda t}$, we obtain the linear system

$$\begin{aligned} (a_{11} - \lambda)\alpha + a_{12}\beta &= 0, \\ a_{21}\alpha + (a_{22} - \lambda)\beta &= 0. \end{aligned} \qquad \textbf{(10)}$$

We would like to find values for λ such that the system of Equation **(10)** has a solution (α, β) where α and β are not both zero. According to the theory of determinants (see Appendix 4), such a solution occurs whenever the determinant of the system

$$\text{Det} = \begin{vmatrix} a_{11} - \lambda & a_{12} \\ a_{21} & a_{22} - \lambda \end{vmatrix} = (a_{11} - \lambda)(a_{22} - \lambda) - a_{21}a_{12} \qquad \textbf{(11)}$$

equals zero. Solving the equation Det $= 0$, we obtain the quadratic equation

Characteristic equation

$$\lambda^2 - (a_{11} + a_{22})\lambda + (a_{11}a_{22} - a_{21}a_{12}) = 0. \qquad \textbf{(12)}$$

We define this to be the **characteristic equation** of system **(9).** That we are using the same term again is no accident, as we now demonstrate. Suppose we differentiate the first equation in system **(9)** and eliminate the function $y(t)$:

$$x'' = a_{11}x' + a_{12}\overbrace{(a_{21}x + a_{22}y)}^{y'}.$$

Then

$$x'' - a_{11}x' - a_{12}a_{21}x = a_{22}a_{12}y = \overbrace{a_{22}(x' - a_{11}x)}^{a_{12}y = x' - a_{11}x},$$

and gathering like terms, we obtain the homogeneous equation

$$x'' - (a_{11} + a_{22})x' + (a_{11}a_{22} - a_{12}a_{21})x = 0. \qquad \textbf{(13)}$$

The characteristic equation for **(13)** is exactly the same as **(12).** Hence the algebraic steps needed to obtain **(13)** can be avoided by setting the determinant Det $= 0$.

As in Sections 3.3 and 3.4, there are three cases to consider, depending on whether the two roots λ_1 and λ_2 of the characteristic equation are real and distinct, real and equal, or complex conjugates. We deal with the three cases separately. Rather than discuss the theory, we demonstrate each case with an example.

Case 1: distinct real roots

EXAMPLE 4 ▶ A system with real, distinct roots

Consider the system

$$\begin{aligned} x' &= -x + 6y, \\ y' &= x - 2y. \end{aligned} \tag{14}$$

Here $a_{11} = -1$, $a_{12} = 6$, $a_{21} = 1$, $a_{22} = -2$, and Equation **(12)** becomes

$$\text{Det} = \begin{vmatrix} -1-\lambda & 6 \\ 1 & -2-\lambda \end{vmatrix} = (\lambda + 2)(\lambda + 1) - 6 = \lambda^2 + 3\lambda - 4 = 0,$$

which has the roots $\lambda_1 = -4$, $\lambda_2 = 1$.

At this point we can write down the general solution of *either* dependent variable:

$$x(t) = c_1e^{-4t} + c_2e^t, \tag{15}$$

or

$$y(t) = k_1e^{-4t} + k_2e^t. \tag{16}$$

It does not matter which expression, **(15)** or **(16)**, we choose. However, once we pick one, we must determine the other dependent variable in terms of the first. For example, if we choose **(15)**, then using the first equation in **(14)**, we have

$$\begin{aligned} y &= \frac{1}{6}(x' + x) = \frac{1}{6}(-4c_1e^{-4t} + c_2e^t + c_1e^{-4t} + c_2e^t) \\ &= \frac{1}{6}(-3c_1e^{-4t} + 2c_2e^t) = -\frac{1}{2}c_1e^{-4t} + \frac{1}{3}c_2e^t. \end{aligned}$$

Thus the general solution is

$$\begin{aligned} x(t) &= c_1e^{-4t} + c_2e^t, \\ y(t) &= -\frac{1}{2}c_1e^{-4t} + \frac{1}{3}c_2e^t, \end{aligned}$$

which can be written as

$$(x(t), y(t)) = \left(c_1e^{-4t} + c_2e^t, -\frac{1}{2}c_1e^{-4t} + \frac{1}{3}c_2e^t\right).$$

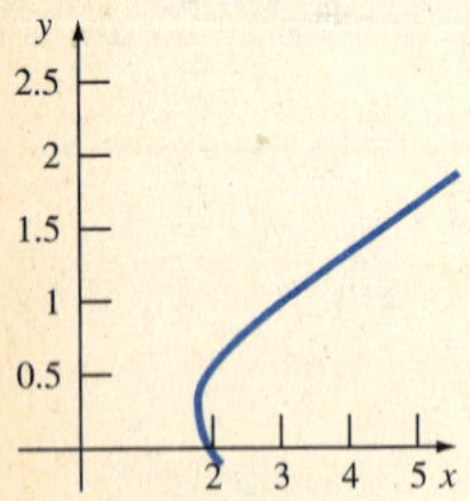

FIGURE 7.3
Graph of $(x(t), y(t)) = (e^{-4t} + e^t, -\frac{1}{2}e^{-4t} + \frac{1}{3}e^t)$, $0 \le t \le 2$

In Figure 7.3 we provide a parametric, computer-drawn sketch of the solution in the case $c_1 = c_2 = 1$ (using MATHEMATICA). ◀

Case 2: one real root

EXAMPLE 5 ▶ A system with one real root

Consider the system

$$\begin{aligned} x' &= -4x - y, \\ y' &= x - 2y. \end{aligned} \tag{17}$$

Equation **(12)** is

$$\text{Det} = \begin{vmatrix} -4-\lambda & -1 \\ 1 & -2-\lambda \end{vmatrix} = (\lambda+4)(\lambda+2)+1 = \lambda^2+6\lambda+9 = 0,$$

which has the double root $\lambda_1 = \lambda_2 = -3$. Hence the general solution to **(17)** for x is given by

$$x(t) = c_1e^{-3t} + c_2te^{-3t} = (c_1 + c_2t)e^{-3t}.$$

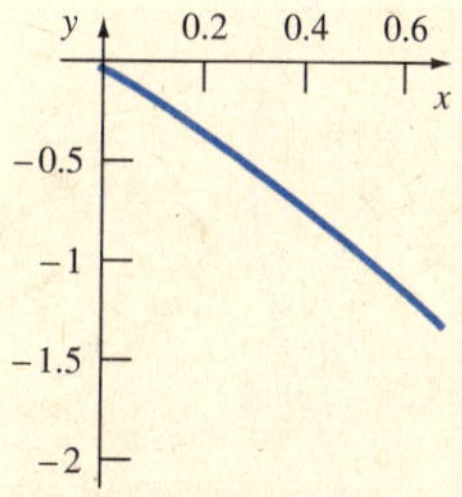

FIGURE 7.4
Graph of $(x(t), y(t)) = (e^{-3t}(1+t), e^{-3t}(-2-t))$

From the first equation in **(17),**

$$\begin{aligned} y = -x' - 4x &= -e^{-3t}(-3c_1 - 3c_2t + c_2) - e^{-3t}(4c_1 + 4c_2t) \\ &= e^{-3t}(-c_1 - c_2 - c_2t). \end{aligned}$$

Thus the general solution to **(17)** is

$$(x(t), y(t)) = (e^{-3t}(c_1 + c_2t), e^{-3t}(-c_1 - c_2 - c_2t)).$$

A computer-drawn sketch of the solution in the case $c_1 = c_2 = 1$ is given in Figure 7.4. ◀

Case 3: complex conjugate roots

EXAMPLE 6 ▶ **A system with complex conjugate roots**

Consider the system

$$\begin{aligned} x' &= 4x + y, \\ y' &= -8x + 8y, \end{aligned} \tag{18}$$

with

$$\text{Det} = \begin{vmatrix} 4-\lambda & 1 \\ -8 & 8-\lambda \end{vmatrix} = \lambda^2 - 12\lambda + 40 = 0.$$

The roots of the characteristic equation are $\lambda_1 = 6 + 2i$ and $\lambda_2 = 6 - 2i$. Then the general solution to **(18)** for y is

$$y(t) = e^{6t}(c_1 \cos 2t + c_2 \sin 2t).$$

Using the second equation of **(18),** we have

$$8x = 8y - y' \quad \text{or} \quad x = y - \frac{1}{8}y',$$

so

$$\begin{aligned} x(t) &= e^{6t}(c_1 \cos 2t + c_2 \sin 2t) \\ &\quad - \frac{1}{8}\underbrace{[6e^{6t}(c_1 \cos 2t + c_2 \sin 2t) + e^{6t}(-2c_1 \sin 2t + 2c_2 \cos 2t)]}_{y'} \\ &= e^{6t}\left[\left(c_1 - \frac{6}{8}c_1 - \frac{2}{8}c_2\right)\cos 2t + \left(c_2 - \frac{6}{8}c_2 + \frac{2}{8}c_1\right)\sin 2t\right] \\ &= e^{6t}\left[\frac{1}{4}(c_1 - c_2)\cos 2t + \frac{1}{4}(c_1 + c_2)\sin 2t\right]. \end{aligned}$$

Thus the general solution to **(18)** is

$$\begin{aligned} &(x(t), y(t)) \\ &\quad = \left(\frac{e^{6t}}{4}[(c_1 - c_2)\cos 2t + (c_1 + c_2)\sin 2t],\ e^{6t}(c_1\cos 2t + c_2\sin 2t)\right). \end{aligned}$$

Computer-drawn sketches of the solution are given in Figure 7.5 for $c_1 = 2$ and $c_2 = 1$. Because of the oscillatory behavior of the solutions, the appearance of the graphs changes as the range of t-values changes. In Figure 7.5 we provide four different graphs corresponding to four different ranges.

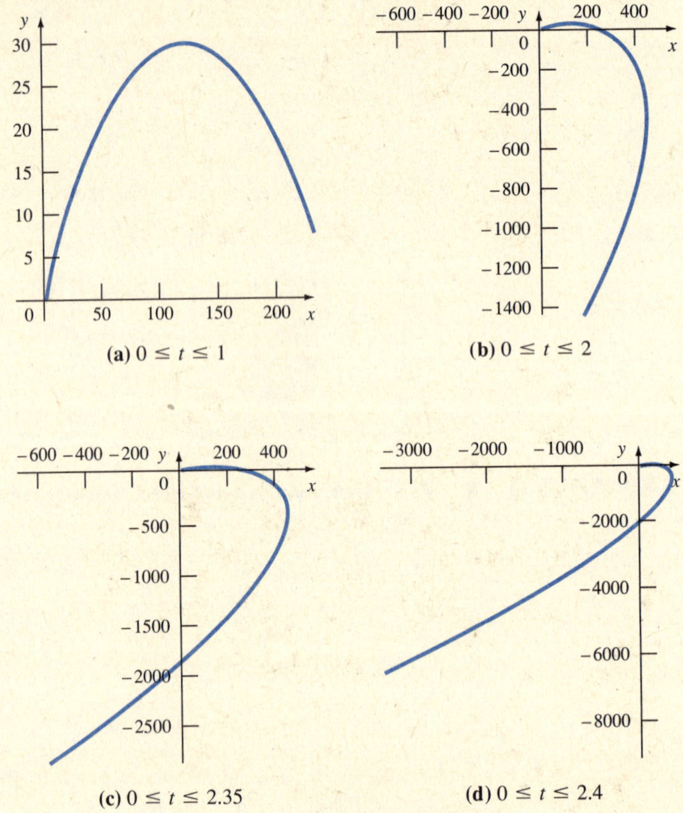

FIGURE 7.5
Graphs of $(x(t), y(t)) = ((e^{-6t}/4)\,[\cos 2t + 3\sin 2t],\ e^{6t}[2\cos 2t + \sin 2t])$

The method of elimination, particularly when we use determinants to obtain the characteristic equation of the system, provides a quick way to find the general solution to a homogeneous system of two equations in two unknown functions. The procedure can be extended to systems of more than two equations, but the algebra involved is often extremely tedious. We summarize the results of Examples 4, 5, and 6.

The method of determinants: solving a linear, homogeneous system with constant coefficients

To find the general solution of the system

$$x' = a_{11}x + a_{12}y,$$
$$y' = a_{21}x + a_{22}y,$$

in the case $a_{12} \neq 0$, first find two numbers λ_1 and λ_2 that satisfy the characteristic equation

$$\begin{vmatrix} a_{11} - \lambda & a_{12} \\ a_{21} & a_{22} - \lambda \end{vmatrix} = 0.$$

Case 1 If $\lambda_1 \neq \lambda_2$ are real, then

$$x(t) = c_1 e^{\lambda_1 t} + c_2 e^{\lambda_2 t}, \tag{19}$$

and $y(t)$ can be obtained from the equation

$$y = \frac{1}{a_{12}}(x' - a_{11}x). \tag{20}$$

Case 2 If $\lambda_1 = \lambda_2$, then

$$x(t) = e^{\lambda_1 t}(c_1 + c_2 t), \tag{21}$$

and $y(t)$ can be obtained from Equation **(20).**

Case 3 If $\lambda_1 = a + ib$ and $\lambda_2 = a - ib$, then

$$x(t) = e^{at}(c_1 \cos bt + c_2 \sin bt), \tag{22}$$

and again $y(t)$ can be obtained from Equation **(20).**

Remark 1

If $a_{12} = 0$ and $a_{21} \neq 0$, then $y(t)$ can be written in form **(19), (21),** or **(22)** and $x(t)$ can then be obtained from the equation

$$x = \frac{1}{a_{21}}(y' - a_{22}y). \tag{23}$$

Remark 2

If $a_{12} = a_{21} = 0$, then the system is **uncoupled** and $x(t)$ and $y(t)$ can be obtained immediately:

$$x(t) = c_1 e^{a_{11}t} \quad \text{and} \quad y(t) = c_2 e^{a_{22}t}.$$

SELF-QUIZ

I. Which are roots of the characteristic equation of the following system?

$$x' = -2x - 2y$$
$$y' = -5x + y$$

(a) $-2, 1$ **(b)** $4, -3$ **(c)** $-4, 3$
(d) $4, 3$ **(e)** $-4, -3$

II. Which are the roots of the characteristic equation of the following system?

$$x' = 2x - y$$
$$y' = 5x - 2y$$

(a) $1, -1$ **(b)** 1 **(c)** -1
(d) $i, -i$ **(e)** $2, -2$

III. Which are the roots of the characteristic equation of the following system?

$$x' = -3x + 2y$$
$$y' = \qquad - 3y$$

(a) -3 **(b)** $-3, 2$ **(c)** $3, -2$ **(d)** $3i, -3i$
(e) $1, -5$

Answer True or False.

IV. If the roots of the characteristic equation are negative real numbers, then all solutions to the system approach 0 as $t \to \infty$.

V. If one of the roots of the characteristic equations is a positive real number, then *all* solutions to the system approach infinity as $t \to \infty$.

Answers to Self-Quiz

I. c **II.** d **III.** a **IV.** True **V.** False

PROBLEMS 7.3

In Problems 1–8 find the general solution of each system of equations. When initial conditions are given, find the unique solution.

1. $x' = x + 2y,$
$y' = 3x + 2y$

2. $x' = x + 2y + t - 1, x(0) = 0,$
$y' = 3x + 2y - 5t - 2, y(0) = 4$

3. $x' = -4x - y,$
$y' = x - 2y$

4. $x' = x + y, x(0) = 1,$
$y' = y, y(0) = 0$

5. $x' = 8x - y,$
$y' = 4x + 12y$

6. $x' = 2x + y + 3e^{2t},$
$y' = -4x + 2y + te^{2t}$

7. $x' = 3x + 3y + t,$
$y' = -x - y + 1$

8. $x' = 4x + y, x\left(\frac{\pi}{4}\right) = 0,$
$y' = -8x + 8y, y\left(\frac{\pi}{4}\right) = 1$

9. By elimination, find a solution to the following nonlinear system:
$x' = x + \sin x \cos x + 2y,$
$y' = (x + \sin x \cos x + 2y) \sin^2 x + x.$

Use the method of elimination to solve the systems in Problems 10 and 11.

10. $x_1' = x_1,$
$x_2' = 2x_1 + x_2 - 2x_3,$
$x_3' = 3x_1 + 2x_2 + x_3$

11. $x_1' = x_1 + x_2 + x_3,$
$x_2' = 2x_1 + x_2 - x_3,$
$x_3' = -8x_1 - 5x_2 - 3x_3$

In Problems 12–17 use the method of determinants to find two linearly independent solutions for each given system.

12. $x' = 4x - 3y,$
$y' = 5x - 4y$

13. $x' = 7x + 6y,$
$y' = 2x + 6y$

14. $x' = -x + y,$
$y' = -5x + 3y$

15. $x' = x + y,$
$y' = -x + 3y$

16. $x' = -4x - y,$
$y' = x - 2y$

17. $x' = 4x - 2y,$
$y' = 5x + 2y$

18. Consider the nonhomogeneous system

$$\begin{aligned} x' &= a_{11}x + a_{12}y + f_1, \\ y' &= a_{21}x + a_{22}y + f_2. \end{aligned} \qquad \text{(i)}$$

Let (x_1, y_1) and (x_2, y_2) be two linearly independent solution pairs of the homogeneous system **(9)**. Show that

$$\begin{aligned} x_p(t) &= v_1(t)x_1(t) + v_2(t)x_2(t), \\ y_p(t) &= v_1(t)y_1(t) + v_2(t)y_2(t), \end{aligned}$$

is a particular solution of system **(i)** if v_1 and v_2 satisfy the equations

$$\begin{aligned} v_1'x_1 + v_2'x_2 &= f_1, \\ v_1'y_1 + v_2'y_2 &= f_2. \end{aligned}$$

This process for finding a particular solution of the nonhomogeneous system **(i)** is called the **method of variation of parameters for systems.** Note the close parallel between this method and the method given in Section 3.5.

In Problems 19–23 use the method of variation of parameters to find a particular solution for each given nonhomogeneous system.

19. $x' = 2x + y + 3e^{2t},$
$y' = -4x + 2y + te^{2t}$

20. $x' = 3x + 3y + t,$
$y' = -x - y + 1$

21. $x' = -2x + y,$
$y' = -3x + 2y + 2 \sin t$

22. $x' = -x + y + \cos t,$
$y' = -5x + 3y$

23. $x' = 3x - 2y + t$
$y' = 2x - 2y + 3e^t$

24. In Example 1, when does tank Y contain a maximum amount of salt? How much salt is in tank Y at that time?

25. Suppose that in Example 1 the rate of flow from tank Y to tank X is 2 gallons per minute (instead of 1) and all other facts are unchanged. Find the differential equations for the amount of salt in each tank at all times t.

26. Tank X contains 500 gallons of brine in which 500 pounds of salt are dissolved. Tank Y contains 500 gallons of water. Water flows into tank X at the rate of 30 gallons per minute, and the mixture flows into Y at the rate of 40 gallons per minute. From Y the solution is pumped back into X at the rate of 10 gallons per minute and into a third tank at the rate of 30 gallons per minute. Find the maximum amount of salt in Y. When does this concentration occur?

27. Suppose that in Problem 26 tank X contains 1000 gallons of brine. Solve the problem, given that all other conditions are unchanged.

28. Consider the mass-spring system illustrated below. Here three objects of mass m_1, m_2, and m_3 are suspended in series by three springs with spring constants k_1, k_2, and k_3, respectively. Formulate a system of second-order differential equations that describes this system.

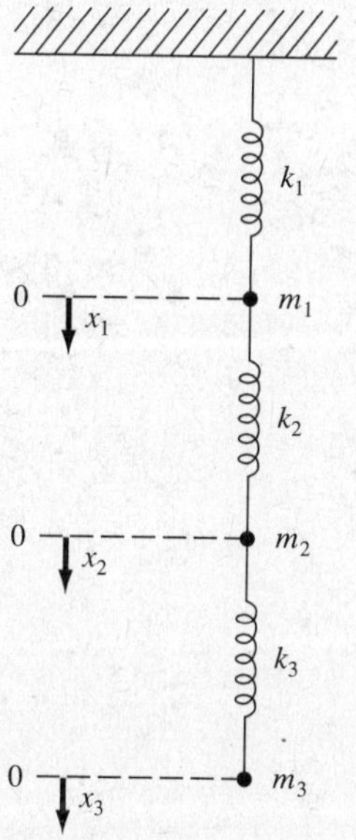

29. Find a single fourth-order linear differential equation in terms of the dependent variable x_1 for system **(7.1.11)** on p. 311. Find a solution to the system if $m_1 = 1$ kg, $m_2 = 2$ kg, $k_1 = 5$ N/m, $k_2 = 4$ N/m.

30. In a study concerning the distribution of radioactive potassium ^{42}K between red blood cells and the plasma of the human blood, C. W. Sheppard and W. R. Martin added ^{42}K to freshly drawn blood.* They discovered that although the total amount of potassium (stable and radioactive) in the red cells and in the plasma remained practically constant during the experiment, the radioactivity was gradually transmitted from the plasma to the red cells. Thus the behavior of the radioactivity is that of a linear closed two-compartment system. If the fractional transfer coefficient from the plasma to the cells is $k_{12} = 30.1\%$ per hour, while $k_{21} = 1.7\%$ per hour, and the initial radioactivity was 800 counts per minute in the plasma and 25 counts per minute in the red cells, what is the number of counts per minute in the red cells after 300 minutes?

31. Temperature inversions and low wind speeds often trap air pollutants in mountain valleys for an extended period of time. Gaseous sulfur compounds are often a significant air pollution problem, and their study is complicated by their rapid oxidation. Hydrogen sulfide (H_2S) oxidizes into sulfur dioxide (SO_2), which in turn oxidizes into a sulphate. The following model has been proposed for determining the concentrations $x(t)$ and $y(t)$ of H_2S and SO_2, respectively, in a fixed airshed.† Let

$$\frac{dx}{dt} = -\alpha x + \gamma, \qquad \frac{dy}{dt} = \alpha x - \beta y + \delta,$$

where the constants α and β are the conversion rates of H_2S into SO_2 and SO_2 into sulphate, respectively, and γ and δ are the production rates of H_2S and SO_2, respectively. Solve the equations sequentially and estimate the concentration levels that could be reached under a prolonged air pollution episode.

The Method of Undetermined Coefficients for Systems

32. Consider the nonhomogeneous system **(i)** on page 328. Define the differential operators.

$$L_1 = D - a_{11} \qquad L_2 = -a_{12}$$
$$L_3 = -a_{21} \qquad L_4 = D - a_{22},$$

where, for example, $Df = df/dt$. Show that system **(i)** can be written in the form

$$\begin{aligned} L_1x + L_2y &= f_1, \\ L_3x + L_4y &= f_2. \end{aligned} \qquad \textbf{(ii)}$$

**Journal of General Physiology* 33 (1950): 703–722.

† R. L. Bohac, "A Mathematical Model for the Conversion of Sulphur Compounds in the Missoula Valley Airshed," *Proceedings of the Montana Academy of Science* (1974).

33. By multiplying the first equation of **(ii)** by L_4 and the second equation by $-L_2$, show that the following equation is obtained:

$$(L_1L_4 - L_2L_3)x = L_4f_1 - L_2f_3. \qquad \textbf{(iii)}$$

34. Rewrite Equation **(iii)** as a linear, nonhomogeneous second-order differential equation in x only.

35. Show that system **(ii)** can also be written as

$$(L_1L_4 - L_2L_3)y = L_1f_2 - L_3f_1 \qquad \textbf{(iv)}$$

36. Rewrite **(iv)** as a second-order differential equation in y only.

37. The expression $L_1L_4 - L_2L_3 = \begin{vmatrix} L_1 & L_2 \\ L_3 & L_4 \end{vmatrix}$ is called the **operator determinant** of the system **(ii)**. Show that the equation $L_1L_4 - L_2L_4 = 0$, treated as a polynomial in D, has the same zeros as the characteristic Equation **(12)**.

38. Consider the system

$$x' = 3x + 3y + t,$$
$$y' = -x - y + 1 \qquad \textbf{(v)}$$

(a) Use the technique of Problems 32–37 and the method of undetermined coefficients to obtain the general solution to **(v)**.

(b) Each general solution involves two arbitrary constants for a total of four arbitrary constants. By inserting $x(t)$ and $y(t)$ into the first (or second) equation of **(v)**, find equations relating these four constants and show that there are really only two arbitrary constants.

In Problems 39–42 use the technique of Problem 33 to find the general solution to each system.

39. $x' = 2x + y + 3e^{2t}$
$y' = -4x + 2y + te^{2t}$

40. $x' = -2x + y,$
$y' = -3x + 2y + 2\sin t$

41. $x' = -x + y + \cos t,$
$y' = -5x + 3y$

42. $x' = 3x - 2y + t,$
$y' = 2x - 2y + 3e^t$

7.4 LAPLACE TRANSFORM METHODS FOR SYSTEMS

Laplace transform techniques are very useful for solving systems of differential equations with given initial conditions. Consider the system

$$\frac{dx}{dt} = a_{11}x + a_{12}y + f(t),$$
$$\frac{dy}{dt} = a_{21}x + a_{22}y + g(t), \qquad \textbf{(1)}$$

with initial conditions $x(0) = x_0$, $y(0) = y_0$. Taking the Laplace transform of both equations in system **(1)** and letting the corresponding capital letters represent the Laplace transforms of the functions x, y, f, and g (that is, we let $X = L\{x\}$, $Y = L\{y\}$, and so on), we obtain

$$sX - x_0 = a_{11}X + a_{12}Y + F,$$
$$sY - y_0 = a_{21}X + a_{22}Y + G. \qquad \textbf{(2)}$$

Gathering all the terms involving X and Y on the left-hand side, we obtain from **(2)** *the system of simultaneous equations*

$$(s - a_{11})X - a_{12}Y = F + x_0,$$
$$-a_{21}X + (s - a_{22})Y = G + y_0. \qquad \textbf{(3)}$$

If $F + x_0$ and $G + y_0$ are not both identically zero, we may solve Equations **(3)** simultaneously, obtaining

$$X = \frac{(s - a_{22})(F + x_0) + a_{12}(G + y_0)}{s^2 - (a_{11} + a_{22})s + (a_{11}a_{22} - a_{12}a_{21})},$$
$$Y = \frac{(s - a_{11})(G + y_0) + a_{21}(F + x_0)}{s^2 - (a_{11} + a_{22})s + (a_{11}a_{22} - a_{12}a_{21})}. \qquad \textbf{(4)}$$

Note that the denominators of the fractions in Equation **(4)** are identical. The denominator is called the **characteristic polynomial** for the system (see Sections 3.3 and 7.3). We may rewrite Equations **(4)** in the form

$$X = \frac{(s - a_{22})(F + x_0) + a_{12}(G + y_0)}{(s - \lambda_1)(s - \lambda_2)},$$
$$Y = \frac{(s - a_{11})(G + y_0) + a_{21}(F + x_0)}{(s - \lambda_1)(s - \lambda_2)}, \qquad \textbf{(5)}$$

where λ_1 and λ_2 are the solutions of the characteristic equation

$$\lambda^2 - (a_{11} + a_{22})\lambda + (a_{11}a_{22} - a_{12}a_{21}) = 0.$$

We can now find the solution (x, y) of system **(1)** with given initial conditions by inverting (if possible) Equations **(5).**

It should be apparent from the discussion above that the Laplace transform allows us to convert a system of differential equations with given initial conditions into a system of simultaneous algebraic equations. This method clearly can be generalized to apply to systems of n linear first-order differential equations with constant coefficients, thus yielding a corresponding system of n simultaneous linear algebraic equations. As in the case of a single differential equation (see Section 6.2), Laplace transform methods are primarily intended for the solution of systems of linear differential equations with constant coefficients and given initial conditions.

EXAMPLE 1 ▶ **Solving an initial-value system using Laplace transforms**

Solve the system

$$\frac{dx}{dt} = 3x - 4y, \qquad x(0) = 1,$$
$$\frac{dy}{dt} = 2x + 3y, \qquad y(0) = 0. \qquad \textbf{(6)}$$

Solution Let $X = \mathscr{L}\{x(t)\}$, $Y = \mathscr{L}\{y(t)\}$, then **(6)** is transformed into

$$sX - 1 = 3X - 4Y,$$
$$sY = 2X + 3Y,$$

which may be rewritten as

$$(s - 3)X + 4Y = 1,$$
$$-2X + (s - 3)Y = 0.$$

Solving these equations simultaneously for X and Y we get

$$X = \frac{s - 3}{(s - 3)^2 + 8}, \qquad Y = \frac{2}{(s - 3)^2 + 8},$$

so that $x(t) = e^{3t} \cos \sqrt{8}\, t$ and $y(t) = (2/\sqrt{8})\, e^{3t} \sin \sqrt{8}t$. ◀

EXAMPLE 2 ▶ Solving nonhomogeneous systems using Laplace transforms

Consider the initial-value problem

$$\begin{aligned} x' &= x - y - e^{-t}, \qquad & x(0) &= 1, \\ y' &= 2x + 3y + e^{-t}, \qquad & y(0) &= 0. \end{aligned} \tag{7}$$

Using the differentiation property of Laplace transforms, we have

$$\begin{aligned} sX - 1 &= X - Y - \frac{1}{s+1}, \\ sY &= 2X + 3Y + \frac{1}{s+1}. \end{aligned} \tag{8}$$

System **(8)** may be rewritten as

$$\begin{aligned} (s-1)X + Y &= \frac{s}{(s+1)}, \\ -2X + (s-3)Y &= \frac{1}{(s+1)}, \end{aligned}$$

from which we find that

$$\begin{aligned} X &= \frac{s^2 - 3s - 1}{(s+1)[(s-2)^2 + 1]} = \frac{A_1}{s+1} + \frac{A_2(s-2) + A_3}{(s-2)^2 + 1}, \\ Y &= \frac{3s - 1}{(s+1)[(s-2)^2 + 1]} = \frac{B_1}{s+1} + \frac{B_2(s-2) + B_3}{(s-2)^2 + 1}. \end{aligned}$$

Using partial fractions decomposition, we obtain

$$A_1 = \frac{3}{10}, \qquad A_2 = \frac{7}{10}, \qquad A_3 = -\frac{11}{10},$$

$$B_1 = -\frac{2}{5} \qquad B_2 = \frac{2}{5}, \quad \text{and} \quad B_3 = \frac{9}{5}.$$

Therefore we have the solution

$$(x, y) = \left(\frac{3}{10}e^{-t} + \frac{e^{2t}}{10}(7\cos t - 11\sin t), -\frac{2}{5}e^{-t} + \frac{e^{2t}}{5}(2\cos t + 9\sin t)\right).$$

◀

SELF-QUIZ

I. Which algebraic system results from transforming the following system?

$$\begin{aligned} x' &= x + 2y + t - 1, \qquad & x(0) &= 0 \\ y' &= 3x + 2y - 5t - 2, \qquad & y(0) &= 4 \end{aligned}$$

(a) $sX = X + 2Y$
$sY = 3X + 2Y$

(b) $sX = X + 2Y + \dfrac{1}{s^2} - \dfrac{1}{s}$
$sY = 3X + 2Y - \dfrac{5}{s^2} - \dfrac{2}{s}$

(c) $sX = X + 2Y + \dfrac{1}{s^2} - \dfrac{1}{s}$
$sY = 3X + 2Y - \dfrac{5}{s^2} - \dfrac{2}{s} + 4$

(d) $sX = X + 2Y - \dfrac{1}{s^2} + \dfrac{1}{s}$
$sY - 4 = 3X + 2Y + \dfrac{5}{s^2} - \dfrac{2}{s}$

II. Which algebraic system results from transforming the following system?

$$x' = 2x + y + 3e^{2t}, \qquad x(0) = 6$$
$$y' = -4x + 2y + te^{2t}, \qquad y(0) = -3$$

(a) $sX = 2X + Y + \dfrac{3}{s-2} + 6$

$sY = -4X + 2Y + \dfrac{1}{(s-2)^2} - 3$

(b) $sX = 2X + Y + \dfrac{3}{s-2} - 6$

$sY = -4X + 2Y + \dfrac{1}{(s-2)^2} + 3$

(c) $sX = 2X + Y + 6$

$sY = -4X + 2Y - 3$

(d) $sX = 2X + Y + \dfrac{3}{s-2}$

$sY = -4X + 2Y + \dfrac{1}{(s-2)^2}$

Answers to Self-Quiz

I. c **II.** a

PROBLEMS 7.4

In Problems 1–12 solve the initial-value problems involving systems of differential equations by the Laplace transform method.

1. $x' = y,\ x(0) = 1$
$y' = x,\ y(0) = 0$

2. $x' = -3x + 4y,\ x(0) = 3$
$y' = -2x + 3y,\ y(0) = 2$

3. $x' = 4x - 2y,\ x(0) = 2$
$y' = 5x + 2y,\ y(0) = -2$

4. $x' = x - 2y,\ x(0) = 1$
$y' = 4x + 5y,\ y(0) = -2$

5. $x' = -3x + 4y + \cos t,\ x(0) = 0$
$y' = -2x + 3y + t,\ y(0) = 1$

6. $x' = 4x - 2y + e^t,\ x(0) = 1$
$y' = 5x + 2y - t,\ y(0) = 0$

7. $x' = x - 2y + t^2,\ x(0) = 1$
$y' = 4x + 5y - e^t,\ y(0) = -1$

8. $x'' + x + y = 0,\ x(0) = x'(0) = 0$
$x' + y' = 0,\ y(0) = 1$

9. $x'' = y + \sin t,\ x(0) = 1,\ x'(0) = 0$
$y'' = -x' + \cos t,\ y(0) = -1,\ y'(0) = -1$

10. $x'' = 2y + 2,\ x(0) = 2,\ x'(0) = 2$
$y' = -x + 5e^{2t} + 1,\ y(0) = 1$

11. $x = z',\ x(0) = y(0) = z(0) = 1$
$x' + y + z = 1$
$-x + y' + z = 2\sin t$

12. $x'' = y - z,\ x(0) = x'(0) = 0$
$y'' = x' + z',\ y(0) = -1,\ y'(0) = 1$
$z'' = -(1 + x + y),\ z(0) = 0,\ z'(0) = 1$

13. Solve B. M. Planck's equations of the motion of a particle projected from the earth at latitude β, where ω is the angular velocity of the earth's rotation and g is the acceleration due to gravity:

$$\frac{d^2x}{dt^2} = 2\omega \sin\beta \frac{dy}{dt},$$
$$\frac{d^2y}{dt^2} = -2\omega\left(\sin\beta \frac{dx}{dt} + \cos\beta \frac{dz}{dt}\right),$$
$$\frac{d^2z}{dt^2} = 2\omega \cos\beta \frac{dy}{dt} - g.$$

7.5 ELECTRIC CIRCUITS WITH SEVERAL LOOPS

In this section we make use of the concepts developed in Sections 2.5 and 4.4 to study models of electric networks with two or more coupled closed circuits. The two fundamental principles governing such models are the two laws of Kirchhoff:

(i) The algebraic sum of all voltage drops around any closed circuit is zero.
(ii) The algebraic sum of the currents flowing into any junction in the network is zero.

EXAMPLE 1 ▶ **An electric circuit with two loops**

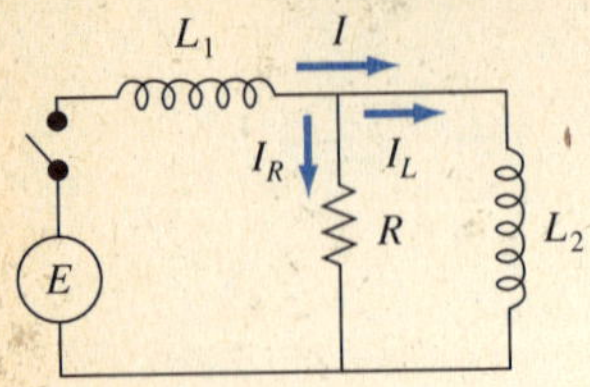

FIGURE 7.6
An electric circuit with two inductors

Consider the two-loop electric circuit in Figure 7.6. Suppose we wish to find the current in each loop as a function of time, given that all currents are zero when the switch is closed at $t = 0$. Let I be the current flowing through the inductor L_1 and let I_R and I_L denote the current flowing through the resistor and the inductor L_2, respectively. By Kirchhoff's current law (ii), $I = I_R + I_L$, so when we apply Kirchhoff's voltage law (i) to each loop, we obtain the system

$$L_1 \frac{dI}{dt} + RI_R = E, \tag{1}$$

$$L_2 \frac{dI_L}{dt} - RI_R = 0. \tag{2}$$

Replacing I by $I_R + I_L$ in **(1)**, we have the system of simultaneous linear differential equations

$$L_1 \frac{dI_R}{dt} + L_1 \frac{dI_L}{dt} + RI_R = E, \tag{3}$$

$$L_2 \frac{dI_L}{dt} - RI_R = 0. \tag{4}$$

If we multiply **(4)** by L_1/L_2, we obtain

$$L_1 \frac{dI_L}{dt} = \frac{L_1 R}{L_2} I_R,$$

which we can substitute into **(3)** to eliminate the I_L variable, yielding

$$L_1 \frac{dI_R}{dt} + \left(\frac{L_1}{L_2} + 1\right) RI_R = E. \tag{5}$$

If L_1, L_2, and R are constant, we can solve the linear first-order differential equation by the method in Section 2.3. Assume that $L_1 = 1$ henry, $L_2 = \frac{1}{2}$ henry, $R = 20$ ohms, and $E = 50$ volts. Then

$$\frac{dI_R}{dt} + 60I_R = 50.$$

Multiplying both sides by the integrating factor e^{60t}, we have

$$\frac{d}{dt}(e^{60t} I_R) = 50e^{60t},$$

and an integration yields

$$e^{60t} I_R = \frac{5}{6} e^{60t} + k_1.$$

Since $I_R(0) = 0$, it follows that $k_1 = -\frac{5}{6}$ and

$$I_R(t) = \frac{5}{6}(1 - e^{-60t}). \tag{6}$$

We can find I_L by substituting **(6)** into **(4)** to get

$$\frac{1}{2} \frac{dI_L}{dt} = \frac{100}{6}(1 - e^{-60t}),$$

from which it follows that

$$I_L(t) = \frac{100}{3}\left(t + \frac{e^{-60t}}{60}\right) + k_2.$$

Setting $t = 0$, so that $I_L(0) = 0$, we obtain $k_2 = -\frac{5}{9}$ and

$$I_L(t) = \frac{100}{3}t + \frac{5}{9}(e^{-60t} - 1). \quad ◀ \tag{7}$$

EXAMPLE 2 ▶ **An RLC-circuit with two loops**

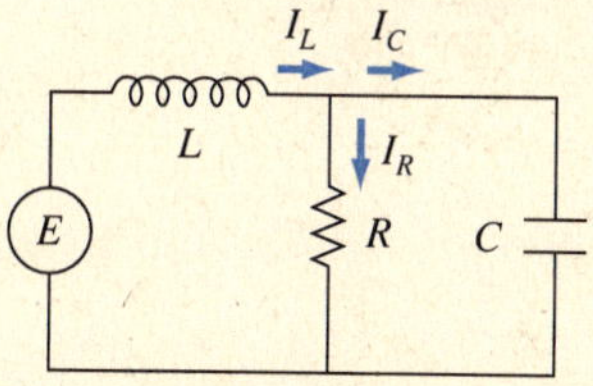

FIGURE 7.7
An RLC-circuit with two loops

Consider the circuit in Figure 7.7. There are two loops. By Kirchhoff's voltage law, we obtain

$$L\frac{dI_L}{dt} + RI_R = E, \tag{8}$$

$$\frac{Q_C}{C} - RI_R = 0. \tag{9}$$

Since $I = \frac{dQ}{dt}$, the second equation may be rewritten as

$$\frac{I_C}{C} - R\frac{dI_R}{dt} = 0. \tag{10}$$

By Kirchhoff's current law, we have

$$I_L = I_C + I_R,$$

which, if substituted into **(10)**, yields, together with **(8)**, the nonhomogeneous system of linear first-order differential equations

$$\begin{aligned}\frac{dI_L}{dt} &= -\frac{R}{L}I_R + \frac{E}{L},\\ \frac{dI_R}{dt} &= \frac{I_L}{RC} - \frac{I_R}{RC}.\end{aligned} \tag{11}$$

The characteristic equation of this system is

$$\begin{aligned}\text{Det} = \begin{vmatrix} -\lambda & -\frac{R}{L} \\ \frac{1}{RC} & -\lambda - \frac{1}{RC}\end{vmatrix} &= \lambda\left(\lambda + \frac{1}{RC}\right) + \frac{1}{LC}\\ &= \lambda^2 + \frac{\lambda}{RC} + \frac{1}{LC} = 0.\end{aligned} \tag{12}$$

The roots of **(12)** are $(-L \pm \sqrt{L^2 - 4R^2LC})/2RLC$.

The value of the discriminant

$$L^2 - 4R^2LC = L(L - 4R^2C)$$

is now important. For simplicity, assume that $R = 100$ ohms, $C = 1.5 \times 10^{-4}$ farad, $E = 100$ volts, and $L = 8$ henrys. Then

$$L - 4R^2C = 8 - 4(100)^2 \cdot 1.5(10^{-4}) = 2,$$

so the roots of **(12)** are $\lambda_1 = -50$ and $\lambda_2 = -\frac{50}{3}$.

Consider the homogeneous system

$$\begin{aligned}\frac{dI_L}{dt} &= -\frac{R}{L}I_R,\\ \frac{dI_R}{dt} &= \frac{I_L}{RC} - \frac{I_R}{RC}.\end{aligned} \tag{13}$$

From Equations **(7.3.19)** and **(7.3.20)** on page 327, the general solution to **(13)** is given by

$$I_L(t) = c_1e^{-50t} + c_2e^{-(50/3)t}$$

and

$$\begin{aligned}I_R(t) &= -\frac{L}{R}\frac{dI_L}{dt} = -\frac{8}{100}\left[-50c_1e^{-50t} - \frac{50}{3}c_2e^{-(50/3)t}\right]\\ &= 4c_1e^{-50t} + \frac{4}{3}c_2e^{-(50/3)t}.\end{aligned} \tag{14}$$

Since $E = 100$ volts is constant, when we use the method of undetermined coefficients we assume a particular solution of the form

$$(I_L, I_R)_p = (A, B), \tag{15}$$

where A and B are constants we must determine. Substituting **(15)** into the nonhomogeneous system **(11)**, we get

$$0 = -\frac{100}{8}B + \frac{100}{8},$$

$$0 = \frac{A - B}{(100)1.5 \times 10^{-4}},$$

so $B = 1 = A$. Finally, we find the constants c_1 and c_2 by using the initial conditions $I_L(0) = 0 = I_R(0)$ in the equations

$$I_L = c_1e^{-50t} + c_2e^{-(50/3)t} + 1,$$

$$I_R = 4c_1e^{-50t} + \frac{4}{3}c_2e^{-(50/3)t} + 1.$$

We obtain

$$I_L(0) = c_1 + c_2 + 1 = 0,$$

$$I_R(0) = 4c_1 + \frac{4}{3}c_2 + 1 = 0,$$

with solution $c_1 = \frac{1}{8}$ and $c_2 = -\frac{9}{8}$. Thus the unique solution to the initial-value problem is

$$(I_L, I_R) = \left(\frac{1}{8}e^{-50t} - \frac{9}{8}e^{(-50/3)t} + 1, \frac{1}{2}e^{-50t} - \frac{3}{2}e^{(-50/3)t} + 1\right). \blacktriangleleft$$

EXAMPLE 3 ▶ **A circuit whose characteristic equation has a double root**

If $L = 6$ henrys in Example 2 and all other facts remain the same, then $L - 4R^2C = 0$ and **(12)** has the double root $\lambda_1 = \lambda_2 = -\frac{1}{2RC} = -\frac{100}{3}$. Hence, from equation **(7.3.21)** on page 327, the general solution to the homogeneous system **(13)** is

$$I_L(t) = (c_1 + c_2t)e^{-(100/3)t}$$

and

$$I_R(t) = -\frac{L}{R}\frac{dI_L}{dt} = -\frac{6}{100}e^{(-100/3)t}\left[\frac{-100}{3}(c_1 + c_2 t) + c_2\right]$$
$$= \left[\left(2c_1 - \frac{3}{50}c_2\right) + 2c_2 t\right]e^{-(100/3)t}.$$

As in Example 2, we find the particular solutions to **(11)** to be

$$(I_L, I_R) = (1, 1).$$

Hence the general solution to **(11)** is

$$I_L(t) = (c_1 + c_2 t)e^{(-100/3)t} + 1,$$
$$I_R(t) = \left[\left(2c_1 - \frac{3}{50}c_2\right) + 2c_2 t\right]e^{(-100/3)t} + 1.$$

Using the initial conditions, we have

$$I_L(0) = c_1 + 1 = 0,$$
$$I_R(0) = 2c_1 - \frac{3}{50}c_2 + 1 = 0,$$

or

$$c_1 = -1 \quad \text{and} \quad c_2 = -\frac{50}{3}.$$

Thus the unique solution to the initial-value problem is

$$(I_L(t), I_R(t)) = \left(\left(-1 - \frac{50}{3}t\right)e^{(-100/3)t} + 1, \left(-1 - \frac{100}{3}t\right)e^{(-100/3)t} + 1\right).$$

◀

EXAMPLE 4 ▶ **A circuit with oscillatory transient currents**

Suppose, in Example 2, that $L = 3$ henrys and all other facts remain unchanged. Then $L - 4R^2C = -3$, so the characteristic equation **(12)** has the roots $\frac{100(-1 \pm i)}{3}$. Then, from equation **(7.3.22)** on page 327 solutions to the homogeneous system **(13)** are

$$I_L(t) = e^{-(100/3)t}\left(c_1 \cos\frac{100}{3}t + c_2 \sin\frac{100}{3}t\right)$$

and

$$I_R(t) = \frac{L}{R}\left(-\frac{dI_L}{dt}\right) = \frac{3}{100}\left(-\frac{dI_L}{dt}\right)$$
$$= \frac{3}{100}e^{(-100/3)t}\left[\frac{100}{3}\left(c_1 \cos\frac{100}{3}t + c_2 \sin\frac{100}{3}t + c_1 \sin\frac{100}{3}t - c_2 \cos\frac{100}{3}t\right)\right]$$
$$= e^{(-100/3)t}\left[(c_1 - c_2)\cos\frac{100}{3}t + (c_1 + c_2)\sin\frac{100}{3}t\right].$$

The nonhomogeneous term in system **(11)** is $\frac{E}{L}$, a constant. Thus it is reasonable to find a particular solution to **(11)** of the form $I_L = A$ and $I_R = B$, where A and B are constants. Inserting these values into **(11)**, we obtain, as in Example 3,

$$I_L = I_R = 1.$$

The general solution to **(11)** is, therefore,

$$I_L(t) = e^{(-100/3)t}\left(c_1 \cos \frac{100}{3}t + c_2 \sin \frac{100}{3}t\right) + 1$$

$$I_R(t) = e^{(-100/3)t}\left[(c_1 - c_2) \cos \frac{100}{3}t + (c_1 + c_2) \sin \frac{100}{3}t\right] + 1.$$

Finally, setting $I_R(0) = I_L(0) = 0$, we obtain

$$c_1 + 1 = 0,$$
$$c_1 - c_2 + 1 = 0,$$

with solution $c_1 = -1$, $c_2 = 0$.

Thus the unique solution to our initial-value problem is

$$I_L(t) = 1 - e^{(-100/3)t} \cos \frac{100}{3}t$$

$$I_R(t) = 1 - e^{(-100/3)t}\left(\cos \frac{100}{3}t + \sin \frac{100}{3}t\right). \blacktriangleleft$$

PROBLEMS 7.5

1. Let $R = 100$ ohms, $L = 4$ henrys, $C = 10^{-4}$ farad, and $E = 100$ volts in the network in Figure 7.7. Assume that the currents I_R and I_L are both zero at time $t = 0$. Find the currents when $t = 0.001$ second.
2. Let $L = 1$ henry in Problem 1 and assume that all the other facts are unchanged. Find the currents when $t = 0.001$ second and 0.1 second.
3. Let $L = 8$ henrys in Problem 1 and assume that all the other facts are unchanged. Find the currents when $t = 0.001$ second and 0.1 second.
4. Assume that $E = 100e^{-1000t}$ volts and that all the other values are unchanged in Problem 1. Do
   (a) Problem 1.
   (b) Problem 2.
   (c) Problem 3.
5. Repeat Problem 4 for $E = 100 \sin 60\pi t$ volts.
6. Find the current at time t in each loop of the network shown below, given that $E = 100$ volts, $R = 10$ ohms, and $L = 10$ henrys.

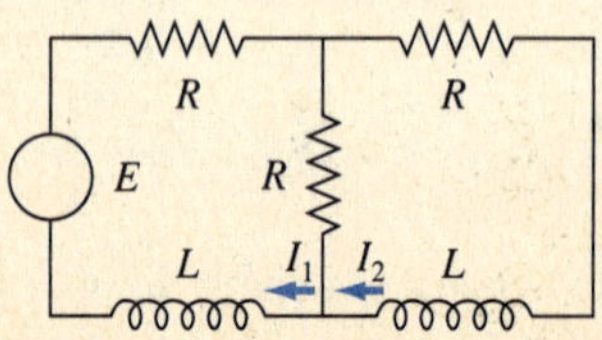

7. Repeat Problem 6 for $E = 10 \sin t$ volts.
8. Consider the air-core transformer network shown below with $E = 10 \cos t$ volts, $R = 1$ ohm, $L = 2$ henrys, and mutual inductance $L_* = -1$ (which depends on the relative modes of winding of the two coils involved). Treating the mutual inductance as an inductance for each circuit, find the two circuit currents at all times t assuming they are zero at $t = 0$.

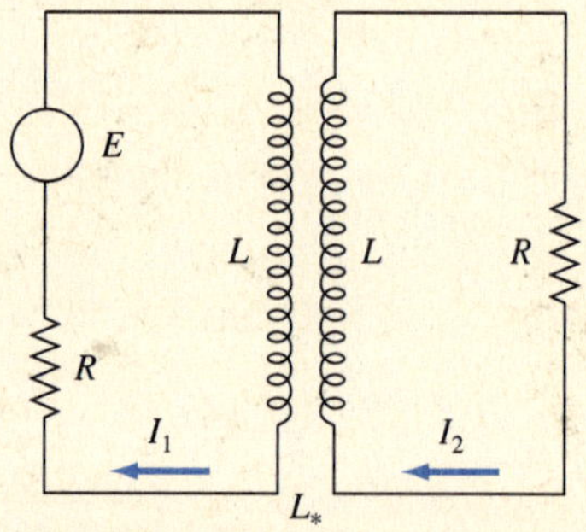

9. Consider the following circuit. Find the current in the top loop if all the resistances are $R = 100$ ohms and the inductances are $L = 1$ henry. Assume that $E = 100$ volts and that there is no current flowing when the switch is closed at time $t = 0$.

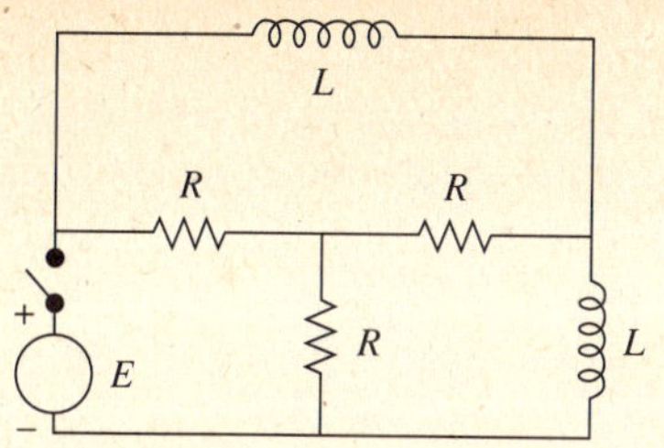

10. Repeat Problem 9 for the circuit shown below, where $C = 10^{-4}$ farad.

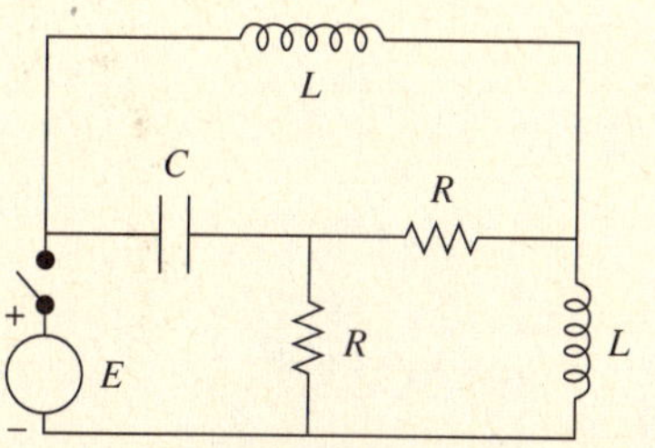

11. Consider the **string insulator** illustrated here. Assume that the voltage at $E_0 = A \cos \omega t$. Find the voltage at the kth junction E_k for each k. [*Hint:* Express E_k as a function of the voltages at previous junctions.]

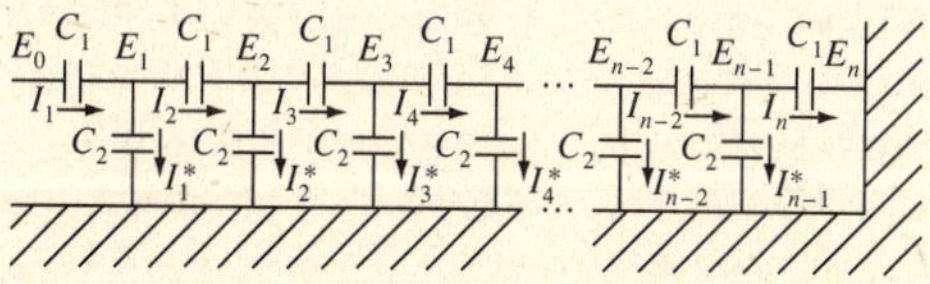

12. Consider the following **low-pass filter,** where we *assume* that the voltage in the line is alternating and that the voltage at each junction is given by $E_k = A_k \cos \omega t$ for fixed ω. Show that the only nontrivial solutions satisfy

$$\omega_N = \frac{2}{\sqrt{CL}} \sin \frac{N\pi}{2(n+1)}, \quad N = 1, 2, \ldots, n.$$

[Thus waves of frequency higher than $\omega_n < 2/\sqrt{CL}$ are damped out as t increases.]

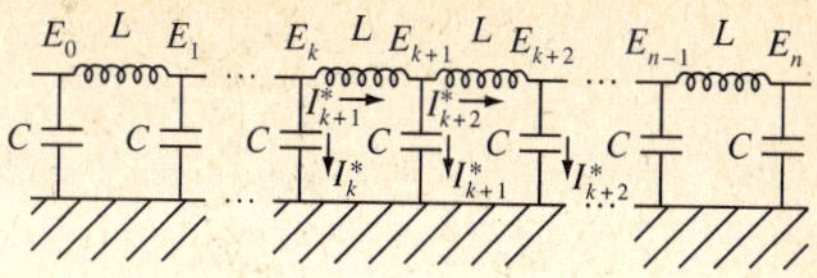

13. Consider the network shown below, called a **high-pass filter.** Show that it damps out all waves below a certain **cut-off** frequency.

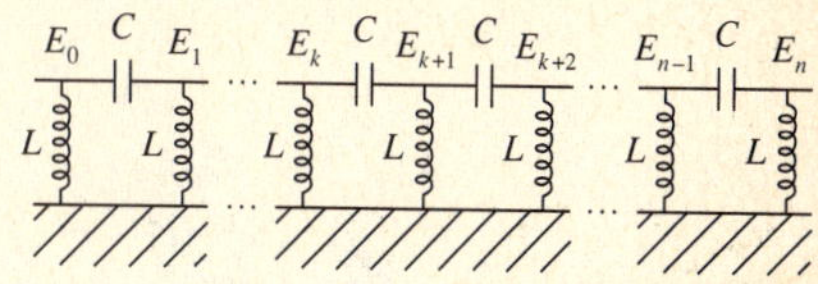

14. The illustrated network is called a **band-pass filter,** since frequencies outside a certain band are damped out. Find the two cut-off frequencies.

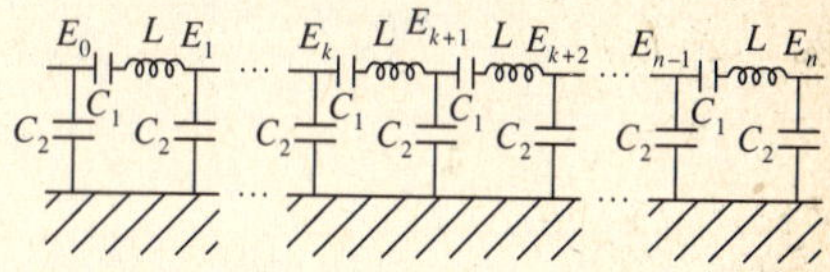

15. R. FitzHugh* proposed the electric circuit shown below as a model for the transmission of current in a myelinated axon. Find the voltage at all junctions x_k in the network. Assume that $x_0 = A_0 \cos \omega t$.

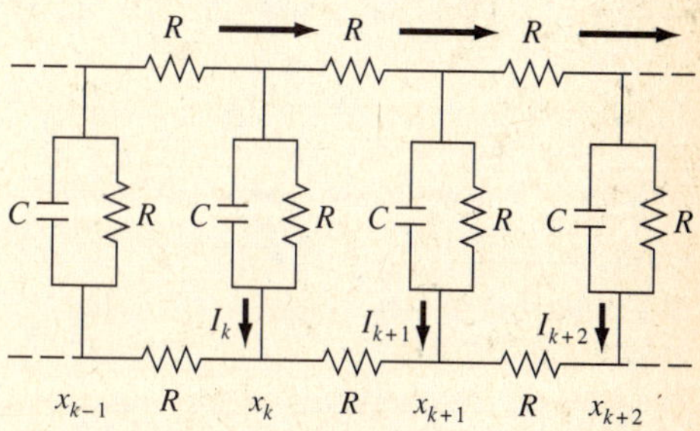

$k = 0, \pm 1, \pm 2 \ldots$

7.6 CHEMICAL MIXTURE AND POPULATION BIOLOGY PROBLEMS

In Section 7.3 we presented an example of a chemical mixture problem involving feedback. The problem set of that section included similar situations, as well as applications of systems to medicine and air pollution. In this section we present further examples related to these topics and to population biology. We solve the first example by the method of determinants.

* "Computation of Impulse Initiation and Saltatory Conduction in Myelinated Fibres: Theoretical Basis of the Velocity-Diameter Relation," *Biophysics Journal 2* (1962):11–21.

EXAMPLE 1 ▶ **A system with feedback**

Let tank X contain 10 gallons of brine in which 10 pounds of salt is dissolved and tank Y contain 20 gallons of water. Suppose water flows into tank X at the rate of 2 gallons per minute, and that the mixture flows from tank X into tank Y at 4 gallons per minute. From Y, 2 gallons are pumped back to X each minute (establishing **feedback**) while 2 gallons are flushed away. (a) Find the amount of salt in both tanks at all time t (see Figure 7.8). (b) At what time is the amount of salt in tank Y a maximum?

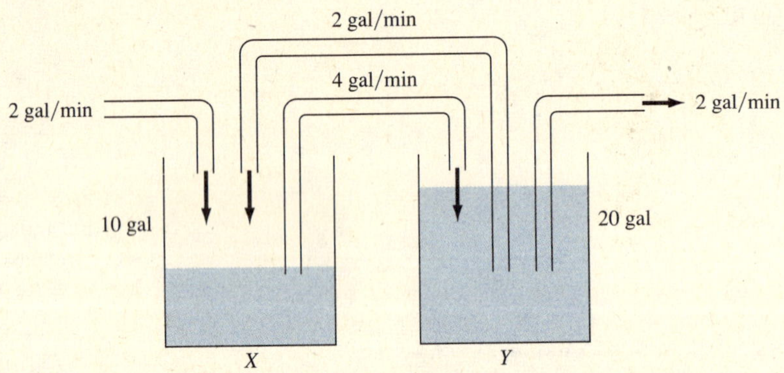

FIGURE 7.8
Two gallons enter and leave the system each minute. With feedback, the salt concentrations in the two tanks is constantly changing

Solution **(a)** If we let $x(t)$ and $y(t)$ represent the number of pounds of salt in tanks X and Y at time t and note that the change in weight equals the difference between input and output, we can again derive a system of linear first-order equations. Tanks X and Y initially contain $x(0) = 10$ and $y(0) = 0$ pounds of salt, respectively, at time $t = 0$. The quantities

$$\frac{x}{10} \quad \text{and} \quad \frac{y}{20}$$

are, respectively, the amounts of salt contained in each gallon of water taken from tanks X and Y at time t. Thus we have the system

$$\begin{aligned} \frac{dx}{dt} &= -4\frac{x}{10} + 2\frac{y}{20}, \qquad x(0) = 10, \\ \frac{dy}{dt} &= 4\frac{x}{10} - 4\frac{y}{20}, \qquad y(0) = 0. \end{aligned} \tag{1}$$

From **(1)** we obtain the determinant

$$\text{Det} = \begin{vmatrix} -\frac{2}{5} - \lambda & \frac{1}{10} \\ \frac{2}{5} & -\frac{1}{5} - \lambda \end{vmatrix} = \left(\lambda + \frac{1}{5}\right)\left(\lambda + \frac{2}{5}\right) - \frac{1}{25} = 0,$$

or

$$\lambda^2 + \frac{3}{5}\lambda + \frac{1}{25} = 0. \tag{2}$$

The characteristic equation **(2)** has the roots

$$\lambda_1 = \frac{-3 + \sqrt{5}}{10} \quad \text{and} \quad \lambda_2 = \frac{-3 - \sqrt{5}}{10}.$$

From equation (7.3.19), on page 327,

$$x(t) = c_1 e^{(-3+\sqrt{5})t/10} + c_2 e^{(-3-\sqrt{5})t/10}.$$

From the first equation of **(1)**,

$$y(t) = 10\frac{dx}{dt} + 4x$$

after some algebra

$$\downarrow$$

$$= (1 + \sqrt{5})c_1 e^{(-3+\sqrt{5})t/10} + (1 - \sqrt{5})c_2 e^{(-3-\sqrt{5})t/10}.$$

Using the initial conditions, we obtain

$$x(0) = c_1 + c_2 = 10,$$
$$y(0) = (1 + \sqrt{5})c_1 + (1 - \sqrt{5})c_2 = 0.$$

Then $c_1 = 5 - \sqrt{5}$, $c_2 = 5 + \sqrt{5}$, and the unique solution to the initial-value problem is

$$(x(t), y(t)) = ((5 - \sqrt{5})e^{(-3+\sqrt{5})t/10} + (5 + \sqrt{5})e^{(-3-\sqrt{5})t/10},$$
$$4\sqrt{5}\, e^{(-3+\sqrt{5})t/10} - 4\sqrt{5}\, e^{(-3-\sqrt{5})t/10}).$$

(b) $y(t)$ is a maximum when $y'(t) = 0$. But

$$y'(t) = 4\sqrt{5}\left[\frac{-3 + \sqrt{5}}{10} e^{(-3+\sqrt{5})t/10} + \frac{3 + \sqrt{5}}{10} e^{(-3-\sqrt{5})t/10}\right] = 0.$$

Multiplying through by $(10/4\sqrt{5})e^{3t/10}$ and simplifying, we obtain

$$(-3 + \sqrt{5})e^{\sqrt{5}t/10} = (-3 - \sqrt{5})e^{-\sqrt{5}t/10},$$

so

$$e^{\sqrt{5}t/5} = \frac{-3 - \sqrt{5}}{-3 + \sqrt{5}} = \frac{7 + 3\sqrt{5}}{2}.$$

Hence

$$t = \sqrt{5}\ln\left(\frac{7 + 3\sqrt{5}}{2}\right) \approx 4.3 \text{ minutes.} \blacktriangleleft$$

EXAMPLE 2 ▶ **A system of species competition**

Consider two species that inhabit the same ecosystem. Denote one of the species by x and the other by y and assume that $x(t)$ and $y(t)$ are the sizes of the respective populations at time t.

If the two populations were isolated from each other, and resources were plentiful, it would be reasonable to assume that the growth rate of their populations would be proportional to their number;

$$\frac{dx}{dt} = ax \quad \text{and} \quad \frac{dy}{dt} = by,$$

where a and b are positive constants (the average growth rate per individual of each population).

However, if the two populations *compete* for the same resource (say they both consume the same forage), then the growth rate of each population is influenced by the size of the other population. In such a case we might arrive at the linear **competition system**

$$\frac{dx}{dt} = ax - Ay, \qquad \frac{dy}{dt} = by - Bx, \tag{3}$$

where A and B are positive constants (the average rate of competition per individual of the other population). The characteristic equation for system **(3)** is

$$\lambda^2 - (a + b)\lambda + (ab - AB) = 0. \tag{4}$$

If **(4)** has the roots $\lambda_1 \neq \lambda_2$, then

$$x(t) = c_1 e^{\lambda_1 t} + c_2 e^{\lambda_2 t}$$

and

$$y(t) = \frac{1}{A}\left(ax - \frac{dx}{dt}\right) = \frac{1}{A}[(a - \lambda_1)c_1 e^{\lambda_1 t} + (a - \lambda_2)c_2 e^{\lambda_2 t}].$$

The constants c_1 and c_2 can be determined from the initial populations $x(0)$ and $y(0)$. ◀

EXAMPLE 3 ▶ **A feedback system involving the production of glucose and insulin**
Many biological systems are controlled by the production of enzymes or hormones that stimulate or inhibit the secretion of some compound. For example, the pancreatic hormone glucagon stimulates the release of glucose from the liver to the plasma. A rise in blood glucose inhibits the secretion of glucagon but causes an increase in the production of the hormone insulin. Insulin, in turn, aids in the removal of glucose from the blood and in its conversion to glycogen in the muscle tissue. Let G and I be the deviations of plasma glucose and plasma insulin from the normal (fasting) level, respectively. We then have the system

$$\frac{dG}{dt} = -k_{11}G - k_{12}I, \qquad \frac{dI}{dt} = k_{21}G - k_{22}I, \tag{5}$$

where the positive constants k_{ij} are model parameters, some of which may be determined experimentally. This system is known to exhibit a strongly damped oscillatory behavior; direct injection of glucose into the blood produces a fall of blood glucose to a level below fasting in about one and a half hours followed by a rise slightly above the fasting level in about three hours. Hence the characteristic equation of system **(5)**,

$$\text{Det} = \begin{vmatrix} -k_{11} - \lambda & -k_{12} \\ k_{21} & -k_{22} - \lambda \end{vmatrix} = (k_{11} + \lambda)(k_{22} + \lambda) + k_{12}k_{21}$$

$$= \lambda^2 + (k_{11} + k_{22})\lambda + (k_{11}k_{22} + k_{12}k_{21}) = 0,$$

must have complex conjugate roots $-a \pm ib$, with $a = (k_{11} + k_{22})/2$ and $b = \sqrt{k_{12}k_{21} - (k_{11} - k_{22})^2/4}$, since only complex roots can lead to oscillatory behavior. Thus our solution has the form

$$(G(t), I(t)) = (e^{-at}(c_1 \cos bt + c_2 \sin bt), e^{-at}(d_1 \cos bt + d_2 \sin bt)),$$

where c_1, c_2, d_1, and d_2 are determined by the initial conditions and one of the equations in **(5)**; and $b = \frac{2\pi}{3}$ if we measure time in hours.*

Assume now that a glucose injection is administered at a time when plasma insulin and glucose are at fasting levels and that the glucose is diffused completely in the blood before the insulin level begins to increase ($t = 0$). Then $G(0) = G_0$ equals the ratio of the volume of glucose administered to blood volume, and $I(0) = 0$. Since $G(t)$ is at a maximum when $t = 0$, it follows that $c_1 = G_0$ and $c_2 = 0$. Hence

$$G(t) = G_0 e^{-at} \cos bt. \tag{6}$$

But

$$d_1 = I(0) = 0, \tag{7}$$

and from the first equation in **(5)**

$$\begin{aligned} \frac{dG}{dt} &= G_0 e^{-at}(-a \cos bt - b \sin bt) \\ &= -k_{11} G_0 e^{-at} \cos bt - k_{12} e^{-at}(d_2 \sin bt), \end{aligned}$$

or, equating sine and cosine terms,

$$aG_0 = k_{11} G_0 \quad \text{and} \quad bG_0 = k_{12} d_2. \tag{8}$$

Thus

$$I(t) = \frac{bG_0}{k_{12}} e^{-at} \sin bt$$

and $k_{11} = a = (k_{11} + k_{22})/2$, so $k_{11} = k_{22}$.

If the minimum level $G(\frac{3}{2})(< 0)$ is known, then by Equation **(6)**,

$$e^{-3a/2} = \frac{\left|G\left(\frac{3}{2}\right)\right|}{G_0},$$

so

$$k_{11} = a = -\frac{2}{3} \ln \frac{\left|G\left(\frac{3}{2}\right)\right|}{G_0}.$$

If we determine the plasma insulin at any given time $t_0 > 0$, we can then evaluate the parameters k_{12} and k_{21}. ◀

* The function $A \cos bt$ has period $\frac{2\pi}{b}$. If the period is three hours, then $\frac{2\pi}{b} = 3$ and $b = \frac{2\pi}{3}$.

PROBLEMS 7.6

1. Suppose that in Example 1 only $1\frac{1}{2}$ gallons per minute are pumped back from tank Y to tank X, while $2\frac{1}{2}$ gallons per minute are flushed away. If $2\frac{1}{2}$ gallons per minute of water flows into X, when does Y contain the maximum amount of salt? How much salt does it contain?

2. Tank X contains 500 gallons of brine in which 500 pounds of salt are dissolved. Tank Y contains 1500 gallons of water. Water flows into tank X at the rate of 30 gallons per minute, and the mixture flows into Y at the rate of 40 gallons per minute. From Y, the solution is pumped back into X at the rate of 10 gallons per minute and into a third tank at the rate of 30 gallons per minute. Find the maximum amount of salt in Y. When does this concentration occur?

3. Suppose that in Problem 2 tank X contains 2000 gallons of brine. Solve the problem, given that all other conditions are unchanged.

4. In some instances, populations in an ecosystem interact in such a way that each species, by its presence, benefits the other. For example, a pollinating insect may feed on the pollen of a plant species. Such an interaction is often called **mutualism.**
   (a) Obtain a system of equations similar to those in Example 2 for a linear mutualism system, with all constants positive.
   (b) Analyze the system in (a) and find its solution assuming given initial conditions.

5. In some biological populations, one species preys exclusively on another. If $x(t)$ is the population of the prey at time t and $y(t)$ is the population of the predator, the system

$$\frac{dx}{dt} = ax - Ay, \qquad \frac{dy}{dt} = Bx - by, \tag{i}$$

describes a linear **predator-prey** model. Here a is the growth rate per individual of the prey, A is the predation rate per predator, b is the death rate per individual of the predators, and Bx provides a measurement of the food supply that regulates the growth rate of the predators. Solve **(i)** for arbitrary initial values of x and y.

6. Linear nonhomogeneous systems of differential equations

$$x' = a_{11}x + a_{12}y + b_1, \qquad y' = a_{21}x + a_{22}y + b_2, \tag{ii}$$

have been used to describe periodic fluctuations of the concentration of hormones in the blood.* If a_{11}, a_{12}, a_{21}, a_{22}, b_1, and b_2 are constants with $b_2 = 0$, find the general solution to **(ii)**.

7. Repeat Problem 6 with $b_1, b_2 \neq 0$.†

8. Lewis F. Richardson describes a model of an arms race between two nations X and Y.‡ Only one weapon (missile or bomb) is involved; $x(t)$ denotes the number of weapons nation X has at time t, and $y(t)$ denotes those of Y at that time. Then

$$\frac{dx}{dt} = -a_1x + a_2y + a_3, \qquad \frac{dy}{dt} = b_1x - b_2y + b_3, \tag{iii}$$

where a_1 and b_2 are "expense and fatigue" coefficients, b_1 and a_2 are "defense" coefficients, and a_3 and b_3 involve the effects of all other factors on the production of weapons. If all the coefficients are positive constants, obtain a general solution to **(iii)**.

9. Repeat Problem 8 assuming that $b_3 = 0$ and $a_3 = a(1 - \cos(\pi t/4))$.

10. In an experiment of cholesterol turnover in humans, radioactive cholesterol $-4 - {}^{14}C$ was injected intravenously and the total plasma cholesterol and radioactivity were measured. It was discovered that the turnover of cholesterol behaves like a two-compartment system.§ The compartment consisting of the organs and blood has a rapid turnover, while the turnover in the other compartment is much slower. Assume that the body intakes and excretes all cholesterol through the first

---

* L. Danziger and G. L. Elmergreen, *Bulletin of Mathematical Biophysics* 19 (1957): 9–18.

† Such a model has been used by E. Ackerman, L. C. Gatewood, J. W. Rosevear, and G. D. Molnar, "Blood Glucose Regulation and Diabetes," in *Concepts and Models of Biomathematics,* ed. F. Heinmets (New York: Marcel Dekker, 1969).

‡ *Generalized Foreign Politics.* British Journal of Psychology Monograph Supplement 23 (1939).

§ D. S. Goodman and R. P. Noble, "Turnover of Plasma Cholesterol in Man," *Journal of Clinical Investigations* 47 (1968):231–241.

compartment. Let $x(t)$ and $y(t)$ denote the deviations from normal cholesterol levels in each compartment. Suppose that the daily fractional transfer coefficient from compartment x is 0.134, of which 0.036 is the input to compartment y, and that the transfer coefficient from compartment y is 0.02.

(a) Describe the problem discussed above as a system of homogeneous linear differential equations.

(b) Obtain the general solution of the system.

7.7 MECHANICAL SYSTEMS

In Section 7.1 we discussed a mass-spring system consisting of two masses and two springs, with one mass and spring pair suspended from another and the latter suspended from a fixed object. In this section we present some additional examples and problems of this type.

EXAMPLE 1 ▶ A mass-spring system

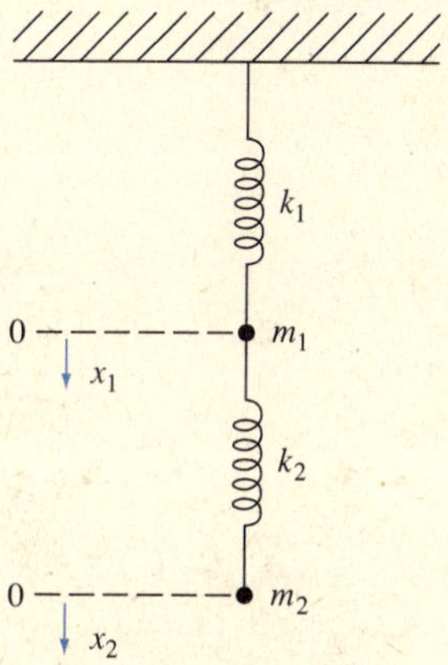

FIGURE 7.9 Two masses are suspended in series on two springs

Consider the mass-spring system in Figure 7.9, where the point masses m_1 and m_2 are connected in series by springs with spring constants k_1 and k_2. If we denote the vertical displacements from equilibrium of the two masses by $x_1(t)$ and $x_2(t)$, respectively, then using the assumptions of Hooke's law (p. 174), we find that the net forces acting on the two masses are given by

$$F_1 = -k_1x_1 + k_2(x_2 - x_1),$$
$$F_2 = -k_2(x_2 - x_1).$$

Here the positive direction is downward. Note that the first spring is compressed when $x_1 < 0$ and the second spring is compressed when $x_1 > x_2$. The equations of motion are

$$\begin{aligned} m_1\frac{d^2x_1}{dt^2} &= -k_1x_1 + k_2(x_2 - x_1) = -(k_1 + k_2)x_1 + k_2x_2, \\ m_2\frac{d^2x_2}{dt^2} &= -k_2(x_2 - x_1) = k_2x_1 - k_2x_2, \end{aligned} \tag{1}$$

a system of two second-order linear differential equations with constant coefficients.

To rewrite **(1)** as a system of first-order linear equations we define the variables $x_3 = x_1'$ and $x_4 = x_2'$. Then $x_3' = x_1''$, $x_4' = x_2''$, and **(1)** can be expressed as the system of four first-order equations

$$\begin{aligned} x_1' &= x_3, \\ x_2' &= x_4, \\ x_1'' = x_3' &= -\left(\frac{k_1 + k_2}{m_1}\right)x_1 + \left(\frac{k_2}{m_1}\right)x_2, \\ x_2'' = x_4' &= \left(\frac{k_2}{m_2}\right)x_1 - \left(\frac{k_2}{m_2}\right)x_2. \end{aligned} \tag{2}$$

The characteristic equation for this system is given by

$$\text{Det} = \begin{vmatrix} -\lambda & 0 & 1 & 0 \\ 0 & -\lambda & 0 & 1 \\ -\left(\dfrac{k_1 + k_2}{m_1}\right) & \dfrac{k_2}{m_1} & -\lambda & 0 \\ \dfrac{k_2}{m_2} & -\dfrac{k_2}{m_2} & 0 & -\lambda \end{vmatrix} = 0. \tag{3}$$

For simplicity, we assume that $k_1 = 9m_1/2$, $k_2 = 2m_2$, and $7m_1 = 4m_2$. Then **(3)** becomes

See Appendix 4

$$0 = \begin{vmatrix} -\lambda & 0 & 1 & 0 \\ 0 & -\lambda & 0 & 1 \\ -8 & \dfrac{7}{2} & -\lambda & 0 \\ 2 & -2 & 0 & -\lambda \end{vmatrix} = \lambda^4 + 10\lambda^2 + 9 = (\lambda^2 + 9)(\lambda^2 + 1). \tag{4}$$

The four roots are $\lambda_1 = 3i$, $\lambda_2 = -3i$, $\lambda_3 = i$, $\lambda_4 = -i$. Hence, from the theory in Section 7.3, we know that there are constants $a_1, a_2, a_3, a_4, b_1, b_2, b_3$, and b_4 such that

$$\begin{aligned} x_1 &= a_1 \cos 3t + a_2 \sin 3t + a_3 \cos t + a_4 \sin t, \\ x_2 &= b_1 \cos 3t + b_2 \sin 3t + b_3 \cos t + b_4 \sin t. \end{aligned} \tag{5}$$

Substituting **(5)** into **(1)**, we get

$$a_1 = -\frac{7}{2}b_1, \qquad a_2 = -\frac{7}{2}b_2, \qquad a_3 = \frac{1}{2}b_3, \qquad a_4 = \frac{1}{2}b_4. \tag{6}$$

If initial conditions are given

$$\begin{aligned} x_1(0) &= x_{10}, \qquad x_1'(0) = x_3(0) = x_{30}, \\ x_2(0) &= x_{20}, \qquad x_2'(0) = x_4(0) = x_{40}, \end{aligned}$$

we obtain the system of equations

$$\begin{aligned} a_1 \quad + a_3 \quad &= x_{10}, \\ 3a_2 \quad + a_4 &= x_{30}, \\ b_1 \quad + b_3 \quad &= x_{20}, \\ 3b_2 \quad + b_4 &= x_{40}. \end{aligned}$$

Together with **(6)**, this system determines all the coefficients in **(5)**:

$$b_1 = \frac{x_{20} - 2x_{10}}{8}, \qquad b_2 = \frac{x_{40} - 2x_{30}}{24},$$

$$b_3 = \frac{7x_{20} + 2x_{10}}{8}, \qquad b_4 = \frac{7x_{40} + 2x_{30}}{8}. \quad ◀$$

PROBLEMS 7.7

1. Show that the oscillations exhibited by the masses m_1 and m_2 in Example 1 can be represented by the superposition of *two* cosines.
2. Solve Example 1 with $k_1 = 5m_1/2$, $k_2 = 2m_2$, and $3m_1 = 4m_2$.
3. Solve Example 1 with $k_1 = 3m_1/2$, $k_2 = 2m_2$, and $m_1 = 4m_2$.
4. When a water wave travels past any point, the water rises as the wave approaches the point and recedes as it moves past the point. If a ship is in the trough of a wave, its angular acceleration from the vertical is given by

$$\frac{d^2\psi}{dt^2} = -b^2\psi + A \sin \omega t. \tag{i}$$

(a) Rewrite **(i)** as a system of first-order differential equations.
(b) Solve the resulting system by the method of determinants.
(c) Solve the system in part (a) by finding its principal solution.

5. A rotating, straight slender shaft may be dynamically unstable at high speeds. When the period ω of rotation is nearly equal to one of its periods of lateral vibration, the shaft is then said to **whirl.** A whirling shaft satisfies the differential equation

$$EI\frac{d^4y}{dx^4} = m\omega^2 y, \qquad \textbf{(ii)}$$

where $y = y(x)$ is the distance of the shaft from its geometric axis, E denotes Young's modulus of elasticity, I is the moment of inertia of the shaft, and m is the mass. Assume that the shaft is hinged at both ends, has length L, and satisfies

$$y(0) = y(L) = y''(0) = y''(L) = 0.$$

(a) Express **(ii)** as a system of first-order differential equations.
(b) Show that nontrivial solutions exist if and only if $\omega = n^2\pi^2\sqrt{EI/m}/L^2$.
(c) Find the rotational speed necessary to produce whirling of a steel shaft $1\frac{1}{2}$ inches in diameter and 8 feet long. [*Hint*: $E = 4.32 \times 10^9$, $I = \pi r^4/4$, $m = \frac{6}{32.2}$.]

6. Consider the coupled mechanical system shown below, where two masses m_1 and m_2 rest on a frictionless plane and are attached to fixed walls by two springs with spring constants k_1 and k_3. The masses m_1 and m_2 are connected by a spring with spring constant k_2.

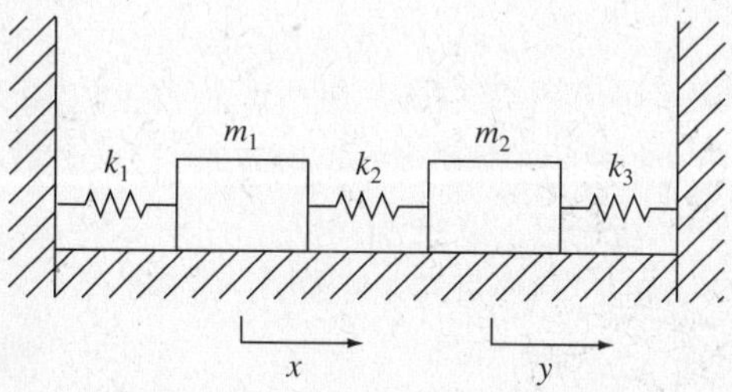

(a) Obtain the system of differential equations describing the motion of this mechanical system.
(b) Solve the system in part (a), and show that the motion of each mass is a superposition of two simple harmonic motions.

7. Determine the equations of motion of a double pendulum consisting of two simple pendulums of masses m_1 and m_2 and lengths l_1 and l_2, respectively, shown here. You may assume that the angular displacements are so small that $\sin\theta \approx \theta$ and $\sin\phi \approx \phi$.

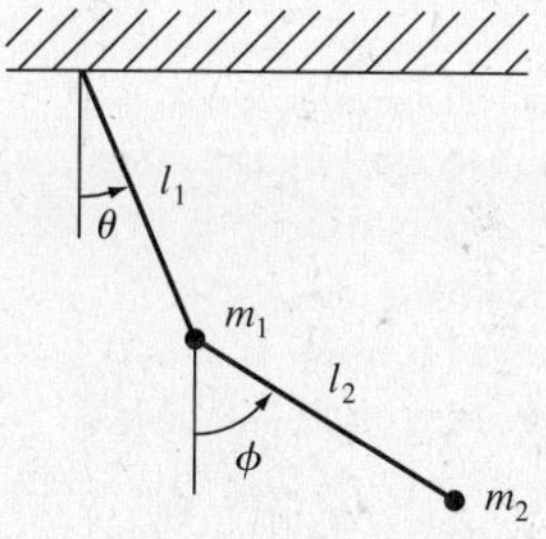

8. Using the system in Problem 7, neglecting the terms containing $(\theta')^2$ and $(\phi')^2$, and replacing $\cos(\theta - \phi)$ by 1, show that one obtains the system

$$\begin{aligned} l_1(m_1 + m_2)\theta'' + m_2 l_2\phi'' + g(m_1 + m_2)\theta &= 0, \\ l_1\theta'' \qquad\qquad + l_2\phi'' + g\phi &= 0. \end{aligned}$$

9. Find the general solution of the system in Problem 8.

7.8 A MODEL FOR EPIDEMICS (OPTIONAL)

In recent years many mathematicians and biologists have attempted to find reasonable mathematical models to describe the growth of an epidemic in a population. One such model has been used by the Center for Disease Control in Atlanta, Georgia, to help form a public testing policy to limit the spread of gonorrhea.* In this section we provide a simple model to describe what may happen in an epidemic.

An **epidemic** is the spread of an infectious disease through a community that affects a significant proportion of the population. The epidemic may begin when a certain number of infected individuals enter the community. This could result, for example, from new people moving to the community or residents returning from a trip. We make the following assumptions:

1. Everyone in the community is initially susceptible to the disease; that is, no one is immune.
2. The disease is spread only by direct contact between a susceptible person (hereafter called a **susceptible**) and an infected person (called an **infective**).
3. Everyone who has had the disease and has **recovered** is immune. The people who die from the disease are also considered to be in the recovered class in this model.
4. After the disease has been introduced into the community, the total population N of the community remains fixed.
5. The infectives are introduced into the community (that is, the epidemic "starts") at time $t = 0$.

In order to model the spread of the disease, we define three variables:

$x(t)$ is the number of susceptibles at time t;

$y(t)$ is the number of infectives at time t;

$z(t)$ is the number of recovered persons at time t.

Then, by assumption (4), we have

$$x(t) + y(t) + z(t) = N. \tag{1}$$

Also, we have

$$y(0) = \text{number of initial infectives}, \tag{2}$$

$$x(0) = N - y(0), \tag{3}$$

$$z(0) = 0. \tag{4}$$

Equation **(4)** states the obvious fact that no one has yet recovered or died from the disease at the time the epidemic begins.

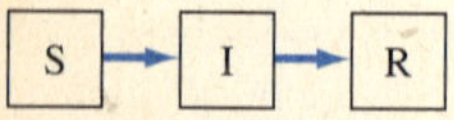

FIGURE 7.10
An SIR model

To get a better idea of what is going on, look at Figure 7.10. Here we see that a susceptible can become infective and an infective can recover. These are the only possibilities. For reasons that are obvious, the model we are describing is, in the literature, usually referred to as an **SIR model.**

What mechanism regulates the rate at which the disease spreads? Most likely the disease spreads more rapidly (that is, susceptibles become infectives) if the number of infectives or the number of susceptibles increases, because chances of

*J. A. Yorke, H. W. Hethcote, and A. Nold, "Dynamics and Control of the Transmission of Gonorrhea," *Sexually Transmitted Diseases* 5(2) (1978):51–56.

contact between infectives and susceptibles increase. Thus it is reasonable to assume that the rate of change of the number of susceptibles is proportional both to the number of susceptibles and to the number of infectives. In mathematical terms, we have

$$x'(t) = -\alpha x(t)y(t), \tag{5}$$

where α is a constant of proportionality and the minus sign indicates that the number of susceptibles is decreasing. Equation **(5)** is called the **law of mass action.**

On the other hand, it is reasonable to assume that the rate at which people recover or die from the disease is proportional to the number of infectives. This gives us the equation

$$z'(t) = \beta y(t). \tag{6}$$

Equations **(5)** and **(6)** constitute the system of equations defining our epidemic model. The constant α is often called the **infection rate** and β is called the **removal rate.**

We begin our analysis by dividing Equation **(5)** by Equation **(6):**

$$\frac{x'(t)}{z'(t)} = -\frac{\alpha x(t)y(t)}{\beta y(t)} = -\frac{\alpha}{\beta}x(t). \tag{7}$$

Rearranging the terms in **(7)** yields

$$\frac{x'(t)}{x(t)} = -\frac{\alpha}{\beta}z'(t),$$

and after integrating both sides we have

$$\int \frac{x'(t)}{x(t)}\,dt = -\frac{\alpha}{\beta}\int z'(t)\,dt + C,$$

or

$$\ln x(t) = -\frac{\alpha}{\beta}z(t) + C. \tag{8}$$

Setting $t = 0$ in Equation **(8)** and nothing that $z(0) = 0$, we have

$$\ln x(0) = 0 + C \qquad \text{or} \qquad C = \ln x(0).$$

Thus

$$\ln x(t) = -\frac{\alpha}{\beta}z(t) + \ln x(0),$$

so

$$\ln\frac{x(t)}{x(0)} = -\frac{\alpha}{\beta}z(t) \quad \text{or} \quad \frac{x(t)}{x(0)} = e^{(-\alpha/\beta)z(t)},$$

and, finally,

$$x(t) = x(0)e^{(-\alpha/\beta)z(t)}. \tag{9}$$

A major question about any epidemic is, can it be controlled? It is clear that the quantity $y'(t)$ is a measure of how bad the epidemic is. The bigger $y'(t)$ is, the greater the number of people becoming infected. We can say that the epidemic has been **controlled** if, beyond some point, $y'(t) \le 0$. Certainly, at some point $y'(t)$ must become negative. This follows from the fact that the population size N is fixed. Then, if everyone becomes infected, $y'(t)$ will be negative as infectives die or recover, and there cannot possibly be any new infectives (there is no one left to become infected). Thus the epidemic is controlled if $y'(t) \le 0$ for every $t > t_0$.

To analyze this situation, from Equation **(1)** we have

$$y(t) = N - x(t) - z(t),$$

so

$$y'(t) = -x'(t) - z'(t) = ax(t)y(t) - \beta y(t), \tag{10}$$

or

$$y'(t) = \alpha y(t)\left[x(t) - \frac{\beta}{\alpha}\right]. \tag{11}$$

Thus $y'(t) \le 0$ whenever $x(t) \le \frac{\beta}{\alpha}$. The term $\frac{\beta}{\alpha}$ is called the **relative removal rate.** We can now answer our question. Since $x' \le 0$ by Equation **(5),** if $x(0) \le \frac{\beta}{\alpha}$, then $x(t) \le \frac{\beta}{\alpha}$ for all t; that is

the epidemic will be controlled at the start if the condition $x(0) \le \frac{\beta}{\alpha}$ holds.

Biologically, this means that

> an epidemic will not ensue if the initial number of susceptibles (N minus the number of initial infectives) does not exceed the relative removal rate.

EXAMPLE 1 ▶ **Conditions leading to an epidemic**

Let $N = 1000$, $y(0) = 50$, $\alpha = 0.0001$, and $\beta = 0.01$. Then $x(0) = 950$ and $\frac{\beta}{\alpha} = 100$, so $x(0) > \frac{\beta}{\alpha}$ and an epidemic will ensue. ◀

EXAMPLE 2 ▶ **Conditions that do not lead to an epidemic**

Let $N = 1000$, $y(0) = 50$, $\alpha = 0.0001$, and $\beta = 0.1$. Then $x(0) = 950$, $\frac{\beta}{\alpha} = 1000$, and an epidemic will not ensue. ◀

We now ask another question. Suppose that $x(0) > \frac{\beta}{\alpha}$ and an epidemic does ensue. How many people eventually become infected? Equivalently, after the epidemic has run its course, how many susceptibles remain? Since $z(t) \le N$, Equation **(9)** provides a *positive* lower bound for the number of susceptibles at every time t:

$$x(t) = x(0)e^{(-\alpha/\beta)z(t)} \ge x(0)e^{-(\alpha/\beta)N} > 0. \tag{12}$$

Denote the number of susceptibles that never get infected by x^*. The fact that x^* is positive [and greater than or equal to the right side of Equation **(12)**] has some interesting consequences. By Equation **(5)** note that $x'(t) < 0$ whenever $x(t)$ and $y(t)$ are positive. Since $x(t)$ does not decrease any lower than x^*, this implies that $x' \to 0$ as $t \to \infty$. But $x \ge x^*$, hence $y \to 0$ as $t \to \infty$. Call $W = N - x^*$ the **extent** of the epidemic: it is the total number of individuals infected. We can obtain a formula for W as follows: Divide Equation **(10)** by Equation **(5)** to obtain

$$\frac{y'(t)}{x'(t)} = \frac{\alpha x(t)y(t) - \beta y(t)}{-\alpha x(t)y(t)} = \frac{\beta}{\alpha x(t)} - 1.$$

Rearranging and integrating both sides, we have

$$\int y'(t)\,dt = \frac{\beta}{\alpha}\int \frac{x'(t)}{x(t)}\,dx - \int x'(t)\,dt + C$$

or

$$y(t) = \frac{\beta}{\alpha}\ln x(t) - x(t) + C,$$

so that

$$x(t) + y(t) - \frac{\beta}{\alpha} \ln x(t) = C.$$

Setting $t = 0$ and recalling that $N = x(0) + y(0)$, we get

$$C = N - \frac{\beta}{\alpha} \ln x(0),$$

or

$$x(t) + y(t) - N = \frac{\beta}{\alpha} \ln \frac{x(t)}{x(0)}. \tag{13}$$

Letting $t \to \infty$, we have $x(t) \to x^*$ and $y(t) \to 0$, so Equation **(13)** becomes

$$x^* - N = \frac{\beta}{\alpha} \ln \frac{x^*}{x(0)},$$

or

$$-W = \frac{\beta}{\alpha} \ln \frac{N - W}{x(0)}. \tag{14}$$

Multiplying both sides of Equation **(14)** by $\frac{\alpha}{\beta}$ and exponentiating, we have

$$e^{(-\alpha/\beta)W} = \frac{N - W}{x(0)},$$

from which we obtain

Determining the extent, *W*, of an epidemic

$$N - x(0)e^{(-\alpha/\beta)W} - W = 0. \tag{15}$$

If α and β are > 0, it is impossible to find an explicit solution for W in terms of α, β, and N. The best we can do is solve it numerically. This is a reasonable thing to do using Newton's method from calculus, as we show in the next example.

EXAMPLE 3 ▶ **Using Newton's method to compute the extent of the epidemic**
Let $N = 1000$, $y(0) = 50$, $\alpha = 0.0001$, and $\beta = 0.09$. Compute the extent of the epidemic.

Solution We first note that $x(0) = 950$ and $\frac{\beta}{\alpha} = 900$, so $x(0) > \frac{\beta}{\alpha}$ and an epidemic will ensue. Since $\frac{\alpha}{\beta} = \frac{1}{900}$, we must find a root of the equation

$$f(W) = 1000 - 950e^{-W/900} - W = 0. \tag{16}$$

Then

$$f'(W) = \frac{950}{900}e^{-W/900} - 1,$$

and by Newton's method we obtain the iterates

$$W_{n+1} = W_n - \frac{f(W_n)}{f'(W_n)},$$

or

$$W_{n+1} = W_n - \frac{1000 - 950e^{-W_n/900} - W_n}{\left(\frac{950}{900}\right)e^{-W_n/900} - 1}. \tag{17}$$

TABLE 7.1

n	W_n	$e^{-W_n/900}$	(a) $1000 - 950e^{-W_n/900} - W_n$	(b) $\frac{950}{900}e^{-W_n/900} - 1$	$\frac{(a)}{(b)}$	$W_{n+1} = W_n - \frac{(a)}{(b)}$
0	100.0000000	0.8948393168	−49.9026490400	−0.0554473878	−900.0000000000	1000.0000000
1	1000.0000000	0.3291929878	−312.7333384000	−0.6525185129	479.2712119000	520.7287882
2	520.7287882	0.5606897580	−53.3840582900	−0.4081608110	130.7917293000	389.9370589
3	389.9370589	0.6483896843	−5.9072589340	−0.3155886666	18.7182226700	371.2188362
4	371.2188362	0.6620161196	−0.1341497754	−0.3012052071	0.4453766808	370.7734595
5	370.7734595	0.6623438079	−0.0000770206	−0.3008593139	0.0002560020	370.7732035
6	370.7732035	0.6623439963	−0.0000000004	−0.3008591150	−0.0000000013	370.7732035

We have no idea, initially, what W is (although $W > y(0) = 50$), so we start with a guess: $W_0 = 100$. We then obtain the iterates given in Table 7.1. After six iterations we obtain the value $W_6 = 370.7732035$, which is correct to ten significant figures, as can be verified by substituting it into Equation **(16).** Thus $W \approx 371$, which means that by the time the epidemic has run its course, 371 individuals have been infected. ◀

Before leaving this section, we mention some of the limitations of this model. The constants α and β often vary with time. Also, the recovered individuals may, after a time, lose their immunity (or never acquire it) and reenter the susceptible state. This could give rise to periodic epidemics in which individuals get the disease many times. This lack of immunity holds for many diseases, notably gonorrhea and certain types of influenza. Finally, there may be many factors other than relative population sizes that control the sizes of the three classes. Such factors might include weather, the available food supply, living conditions, and the presence of other diseases in the community. However, even a simple model like this one can give us the kind of insight needed to study more complicated situations. In a few cases this has been done with great success.*

PROBLEMS 7.8

In Problems 1–5 values for N, y(0), α, and β are given. Determine whether an epidemic will occur and, if so, find the extent of the epidemic.

1. $N = 1000$, $y(0) = 100$, $\alpha = 0.001$, $\beta = 0.4$

2. $N = 1000$, $y(0) = 100$, $\alpha = 0.0001$, $\beta = 0.1$

3. $N = 10{,}000$, $y(0) = 1500$, $\alpha = 10^{-5}$, $\beta = 0.2$

4. $N = 10{,}000$, $y(0) = 1500$, $\alpha = 10^{-5}$, $\beta = 0.02$

5. $N = 25{,}000$, $y(0) = 5000$, $\alpha = 10^{-5}$, $\beta = 0.1$

* If you are interested in learning more about epidemic models, consult the following: Paul Waltman, *Deterministic Threshold Models in the Theory of Epidemics. Lecture Notes in Biomathematics*, vol. 1. (New York: Springer-Verlag, 1974); Klaus Dietz, "Epidemics and Rumours: A Survey," *Journal of the Royal Statistical Society* Series A., 130 (1967):505–528.

6. Let $f(W)$ be given by Equation **(15)**. Show that there is exactly one positive value of W for which $f(W) = 0$. [*Hint:* Show that $f(0) > 0$ and that $f''(W) < 0$ for all W and that $f(W) < 0$ if W is sufficiently large.]

7. Prove that the maximum number of infectives, $y_{\max}$, is given by

$$y_{\max} = N + \frac{\beta}{\alpha}\left[\ln\left(\frac{\beta}{\alpha x(0)}\right) - 1\right]$$
$$\geq N - \frac{\beta}{\alpha}\left[1 + \ln\left(\frac{\alpha N}{\beta}\right)\right] > 0,$$

whenever $x(0) > \frac{\beta}{\alpha}$.

7.9 SOLVING TWO-DIMENSIONAL LINEAR SYSTEMS

This section will show you how a generic symbol manipulator can be used to solve the chemical mixture problems of Sections 7.3 and 7.6. The symbolic manipulators can also be used to solve nonhomogeneous two-dimensional systems, as described in Section 7.2, or two-loop electric circuits, as in Section 7.5. We shall use two scripts that we described and constructed in Chapter 3. Recall first that

`LINEAR2(t, x, p, q, r)`—finds the general solution $x(t)$ to the linear constant coefficient differential equation

$$x'' + px' + qx = r(t),$$

where p and q are constants and $r(t)$ is a function of the independent variable t.

Consider the linear two-dimensional system

$$x' = ax + by + f(t),$$
$$y' = cx + dy + g(t),$$

where a, b, c, and d are constants, and $f(t)$, $g(t)$, $x(t)$, and $y(t)$ are functions of the independent variable t.

If $b = 0$ and $c = 0$, then the system is **uncoupled** (see Remark 2 in Section 7.3) and each equation can be solved independently as a first-order linear differential equation. If either b or c is nonzero, we can use elimination (see Section 7.3) to rewrite the system as a second-order linear differential equation. For example, differentiate the first equation and substitute the second for y':

$$x'' = ax' + by' + f' = ax' + b(cx + dy + g) + f'$$
$$= ax' + bcx + d(by) + bg + f'.$$

Now solve the first equation of the system for the term (by) and substitute it above:

$$x'' = ax' + bcx + d(x' - ax - f) + bg + f'$$
$$= (a + d)x' + (bc - ad)x + (bg - df + f')$$

or

$$x'' - (a + d)x' + (ad - bc)x = bg - df + f'. \tag{1}$$

A similar computation to eliminate x instead of y yields the second-order equation:

$$y'' - (a + d)y' + (ad - bc)y = cf - ag + g'. \tag{2}$$

We can now apply LINEAR2(t, x, p, q, r) to Equation **(1)** to obtain the general solution for $x(t)$: set $p = -(a + d)$, $q = ad - bc$, and $r(t) = bg(t) - df(t) + f'(t)$. But what about the associated general solution for $y(t)$? When b is nonzero, this quickly follows from the first equation in the system:

$$y(t) = \frac{x'(t) - ax(t) - f(t)}{b}. \tag{3}$$

If $b = 0$ but $c \neq 0$, we cannot use Equation **(3)**; instead apply LINEAR2(t, y, p, q, r) to Equation **(2)** with p and q as before and $r(t) = cf(t) - ag(t) + g'(t)$, to get the general solution for $y(t)$, and use the second equation of the system

$$x(t) = \frac{y'(t) - dy(t) - g(t)}{c}, \qquad \textbf{(4)}$$

to obtain the associated general solution in x. Consequently, any script that we write for the solution of such a system *must check whether* $b = 0$ *or* $c = 0$. We now describe a script that does this, and returns the statement "uncoupled: use first order methods" when both b and c are zero.

`LINSYS2(t, x, y, a, b, c, d, f, g)`—solves the constant coefficient system

$$x' = ax + by + f(t),$$
$$y' = cx + dy + g(t),$$

where a, b, c, d are constants and $f(t)$, $g(t)$ are any functions of t.

The algorithm uses some of the subfunctions of LINEAR2(t, x, p, q, r) defined in Section 3.10. It consists of the following building blocks which should be entered in the order shown so that each function may use preceding functions:

```
SYS2G1(t,a,b,c,d,f,g):= LIN2G2(t, LIN2(t,-a-d,a*d-b*c),
    LIN2G1(t,LIN2(t,-a-d, a*d-b*c),-a-d), b*g-d*f+DIF(f,t))
SYS2G2(t,a,b,c,d,f,g):= (DIF(SYS2G1(t,a,b,c,d,f,g), t)
    -a*SYS2G1(t,a,b,c,d,f,g)-f)/b
SYS2G(t,x,y,a,b,c,d,f,g):= ([x=SYS2G1(t,a,b,c,d,f,g),
    y=SYS2G2(t,a,b,c,d,f,g)])
LINSYS2(t,x,y,a,b,c,d,f,g):= IF(b, SYS2G(t,x,y,a,b,c,d,f,g),
    IF(c, SYS2G(t,y,x,d,c,b,a,g,f), "uncoupled:use first order methods",
    SYS2G(t,y,x,d,c,b,a,g,f)), SYS2G(t,x,y,a,b,c,d,f,g))
```

The algorithm first tests b. If $b \neq 0$, it gives the vector SYS2G(t, x, y, a, b, c, d, f, g) which uses the right side of the LINEAR2(t, x, p, q, r) with p = −a − d, q = a*d − b*c, and r = b*g − d*f + f′(t) to compute the general solution for x as the first entry, and the equation y = (x′ − a*x − f)/b as the second entry. Thus, the solution consists of the solutions of Equations **(1)** and **(3)**. If $b = 0$, it tests for c. When $c = 0$ it returns "uncoupled: use first order methods". Otherwise, it reverses the role of x and y, and solves Equations **(2)** and **(4)**. The reader should check this last assertion.

EXAMPLE 1 ▶

Solve the system of equations from Example 7.3.1 involving flow of brine between two tanks:

$$x' = -3\frac{x}{100} + \frac{y}{100}, \qquad x(0) = 100,$$

$$y' = 3\frac{x}{100} - 3\frac{y}{100}, \qquad y(0) = 0.$$

Solution We begin by obtaining the general solution. Since the system is homogeneous, $f = g = 0$. Thus, we give the command

```
LINSYS2(t,x,y,-3/100,1/100,3/100,-3/100,0,0)
```

Since $b = \frac{1}{100}$, the algorithm eliminates the y variable and obtains the second-order differential equation $x'' + 6x'/100 + 6x/100^2 = 0$. It solves this equation using LINEAR2 obtaining the general solution exp(−3∗t/100)∗[c_1∗exp(√3∗t/100) + c_2∗exp(−√3∗t/100)] for x. It then uses the first equation of the system to calculate $y(t)$:

```
y=100x'+3x=√3*exp(-3*t/100)*[c₁*exp(√3*t/100)-c₂*exp(-√3*t/100)].
```

We now have to describe an algorithm for determining the value of c_1 and c_2 for the given initial conditions. This is almost identical with the algorithm IV2: we already have the general solutions $x = GS_1$ and $y = GS_2$ as a vector of equations, so we use SOLVE again

```
IVSYS2(t,[x=GS₁,y=GS₂],t0,x0,y0,c₁,c₂):=
    SOLVE(SUBST([x=GS₁,y=GS₂],t=t0,x=x0,y=y0),[c₁,c₂])
```

Essentially, this algorithm substitutes t, x, and y in each equation by $t0$, $x0$, $y0$ respectively, and solves the resulting system of two equations in the unknowns c_1, c_2.

Hence, to complete this example, we issue the command

```
IVSYS2(t,[x=GS₁,y=GS₂],0,100,0,c₁,c₂)
```

where GS_1 and GS_2 are the general solutions given above for x and y respectively. In this case it follows that $[c_1, c_2] = [50, 50]$, so the final answer is

```
x(t)=50*exp(-3*t/100)*[exp(√3*t/100)+exp(-√3*t/100)]
y(t)=50√3*exp(-3*t/100]*[exp(√3*t/100)- exp(-√3*t/100)].
```

◀

EXAMPLE 2 ▶

Solve the system in Example 7.2.3:

$$x' = 3x + 3y + t, \qquad y' = -x - y + 1.$$

Solution Here we give the command LINSYS2(t, x, y, 3, 3, −1, −1, t, 1) and obtain:

```
x" - 2*x'=4+t
x=c₁+c₂*exp(2*t) - t*(9+t)/4
y=(x' - 3*x - t)/3=-c₁ - (1/3)*c₂*exp(2*t)+(t^2+7*t - 3)/4
```

as the general solution. ◀

PROBLEMS 7.9

Solve each of the following systems in Problems 1–10 using MATHEMATICA, MAPLE, DERIVE, MATLAB, or MATHCAD. When initial values are given obtain the particular solution.

1. $x' = -4x - y$
$y' = x - 2y$

2. $x' = 2x + y$
$y' = -4x + 2y$

3. $x' = x + y,\ x(0) = 1$
$y' = -3x - y,\ y(0) = 0$

4. $x' = 3x + t,\ x(0) = 3$
$y' = -x + 7y,\ y(0) = -2$

5. $u' = u + 3v + 2s$
$v' = 4u - v + e^s$

6. $x' = 2x + y + t,\ x(1) = 5$
$y' = x + 2y + t^2,\ y(1) = 2$

7. $x' = 2x + y + 3e^{2t}$
$y' = -4x + 2y + te^{2t}$

8. $x' = 7x + te^t$
$y' = -3y - \sin t$

9. $u' = -u + v + \cos s$
$v' = -5u + 3v$

10. $x' = 3x - 2y + t,\ x(2) = 0$
$y' = 2x - 2y + 3e^t,\ y(2) = -1$

11. Prove that SYS2G(t, y, x, d, c, b, a, g, f) solves the Equations (2) and (4).

12. Explain what you would mean by a boundary value problem for a system of equations. What difficulties do you anticipate in designing an algorithm BVSYS2 to obtain its solutions?

SUMMARY OUTLINE OF CHAPTER 7

A Typical System of Differential Equations p. 308

$$\begin{aligned} x_1' &= f_1(t, x_1, x_2, \ldots, x_n), \\ x_2' &= f_2(t, x_1, x_2, \ldots, x_n), \\ &\vdots \\ x_n' &= f_n(t, x_1, x_2, \ldots, x_n), \end{aligned} \tag{1}$$

where each f_i, $i = 1, 2, \ldots, n$, is a function of $n + 1$ variables.

Solution to a System p. 308

Any collection of n differentiable functions $x_1(t), x_2(t), \ldots, x_n(t)$ that satisfy the system **(1)** for all values t in some interval $\alpha < t < \beta$ is called a **solution** of the system **(1)** on that interval.

First-Order System p. 308

System **(1)** is called a **first-order** system because it involves only first derivatives of the functions $x_1, x_2, \ldots, x_n$.

Linear, First-Order System p. 308

A first-order system is said to be **linear** if it can be written in the form

$$\begin{aligned} \frac{dx_1}{dt} &= a_{11}(t)x_1 + a_{12}(t)x_2 + \cdots + a_{1n}(t)x_n + f_1(t), \\ \frac{dx_2}{dt} &= a_{21}(t)x_1 + a_{22}(t)x_2 + \cdots + a_{2n}(t)x_n + f_2(t), \\ &\vdots \\ \frac{dx_n}{dt} &= a_{n1}(t)x_1 + a_{n2}(t)x_2 + \cdots + a_{nn}(t)x_n + f_n(t), \end{aligned} \tag{2}$$

where a_{ij} and f_i are functions that are continuous on some common interval $\alpha < t < \beta$. Any system for which this cannot be done is said to be **nonlinear.** The functions a_{ij} are called the **coefficients** of the linear system **(2).** If every a_{ij} is constant, system (2) is said to have **constant coefficients.** If every function f_i satisfies $f_i(t) \equiv 0$ for $\alpha < t < \beta$, $i = 1, 2, \ldots, n$, then system **(2)** is said to be **homogeneous;** otherwise it is called a **nonhomogeneous** system.

Initial-Value Problem p. 309

In addition to the given system of differential equations **(1)** or **(2),** initial conditions of the form

$$x_1(t_0) = x_{10},\ x_2(t_0) = x_{20}, \ldots, x_n(t_0) = x_{n0} \tag{3}$$

may also be given for some t_0 satisfying $\alpha < t_0 < \beta$. We call system **(1)** or **(2)** together with the initial condition **(3)** an **initial-value problem.**

Local Existence-Uniqueness Theorem p. 310

Let $f_1, f_2, \ldots, f_n$ and each of the first partials $\partial f_1/\partial x_1, \ldots, \partial f_1/\partial x_n, \partial f_2/\partial x_1, \ldots, \partial f_n/\partial x_n$ be continuous in an $(n + 1)$-dimensional rectangular region R given by $\alpha < t < \beta$, $c_1 < x_1 < d_1, \ldots, c_n < x_n < d_n$ containing the point $(t_0, x_{10}, \ldots, x_{n0})$. Then, in some interval $|t - t_0| < h$ contained in (α, β), there is a unique solution of system **(1)** that also satisfies the initial conditions **(3).**

Global Existence-Uniqueness Theorem p. 310

If $a_{11}, a_{12}, \ldots, a_{nn}, f_1, \ldots, f_n$ are continuous functions on the interval $\alpha < t < \beta$ containing the point t_0, then system **(2)** has a unique solution that satisfies the initial conditions **(3).** Furthermore, this solution is valid throughout the interval $\alpha < t < \beta$.

Writing a Higher-Order Equation as a First-Order System pp. 310–311

The equation

$$y^{(n)} = F(t, y, y', y'', \ldots, y^{(n-1)}),$$

can be written as a system of first-order equations of form **(1).** The procedure is as follows: we define n new variables $x_1, x_2, \ldots, x_n$ by setting

$$x_1 = y, x_2 = y', x_3 = y'', \ldots, x_n = y^{(n-1)}.$$

Then, since $y' = x_1' = x_2$, $y'' = x_2' = x_3, \ldots, y^{(n-1)} = x_{n-1}' = x_n$ and $y^{(n)} = x_n'$, we have the following first-order system

$$\begin{aligned} x_1' &= x_2, \\ x_2' &= x_3, \\ &\vdots \\ x_{n-1}' &= x_n, \\ x_n' &= F(t, x_1, x_2, \ldots, x_n), \end{aligned}$$

which is a special case of system **(1).**

Linear, Homogeneous Systems in Two Unknown Functions p. 313

These are systems that can be written in the form

$$\begin{aligned} x' &= a_{11}(t)x + a_{12}(t)y \\ y' &= a_{21}(t)x + a_{22}(t)y \end{aligned} \qquad \textbf{(4)}$$

Linear, Nonhomogeneous Systems in Two Unknown Functions p. 313

$$\begin{aligned} x' &= a_{11}(t)x + a_{12}(t)y + f_1(t) \\ y' &= a_{21}(t)x + a_{22}(t)y + f_2(t) \end{aligned} \qquad \textbf{(5)}$$

Existence-Uniqueness Theorem p. 314

If the functions $a_{11}(t)$, $a_{12}(t)$, $a_{21}(t)$, $a_{22}(t)$, $f_1(t)$, and $f_2(t)$ are continuous on I, then given any numbers t_0, x_0, and y_0, with t_0 in I, there exists exactly one solution $(x(t), y(t))$ of system **(1)** that satisfies $x(t_0) = x_0$ and $y(t_0) = y_0$.

Solution to the System (4) or (5) p. 314

By a **solution** of system **(4)** [or **(5)**] we mean a *pair* of functions $(x(t), y(t))$ that possess first derivatives and that satisfy the given equations.

Linear Combination p. 314

The pair of functions $(x_3(t), y_3(t))$ is a **linear combination** of the pairs $(x_1(t), y_1(t))$ and $(x_2(t), y_2(t))$ if there exist constants c_1 and c_2 such that the following two equations hold:

$$\begin{aligned} x_3(t) &= c_1x_1(t) + c_2x_2(t) \\ y_3(t) &= c_1y_1(t) + c_2y_2(t) \end{aligned}$$

If the pairs $(x_1(t), y_1(t))$ and $(x_2(t), y_2(t))$ are solutions of the homogeneous system **(5),** then any linear combination of them is also a solution of system **(5).**

Linear Independence p. 315

We define two pairs of functions $(x_1(t), y_1(t))$ and $(x_2(t), y_2(t))$ to be **linearly independent** on an interval I if, whenever the equations

$$\begin{aligned} c_1x_1(t) + c_2x_2(t) &= 0, \\ c_1y_1(t) + c_2y_2(t) &= 0 \end{aligned}$$

hold for all values of t in I, then $c_1 = c_2 = 0$.

Wronskian pp. 315–316

Given two solutions $(x_1(t), y_1(t))$ and $(x_2(t), y_2(t))$ of system **(4),** we define the **Wronskian** of the two solutions by the following determinant:

$$W(t) = \begin{vmatrix} x_1(t) & y_1(t) \\ x_2(t) & y_2(t) \end{vmatrix} = x_1(t)y_2(t) - x_2(t)y_1(t).$$

Moreover,

$$W(t) = W(t_0)\exp\left(\int_{t_0}^{t} [a_{11}(u) + a_{22}(u)]\, du\right).$$

Two solutions $(x_1(t), y_1(t))$ and $(x_2(t), y_2(t))$ of the homogeneous system **(4)** are linearly independent on an interval I if and only if $W(t) \neq 0$ in I.

General Solution to a Homogeneous System p. 317

Let (x_1, y_1) and (x_2, y_2) be solutions of the homogeneous linear system **(4).** Then $(c_1x_1 + c_2x_2, c_1y_1 + c_2y_2)$ is the **general solution** of system **(4)** provided that $W(t) \neq 0$; that is, provided that the solutions (x_1, y_1) and (x_2, y_2) are linearly independent.

General Solution to a Nonhomogeneous System p. 317

Let (x^*, y^*) be the general solution of system **(1),** and let (x_p, y_p) be any solution of **(5).** Then $(x^* - x_p, y^* - y_p)$ is the general solution of the homogeneous equation **(4).** In other words, the general solution of system **(5)** can be written as the sum of the general solution of the homogeneous system **(4)** and any particular solution of the nonhomogeneous system **(5).**

Characteristic Equation p. 323

Consider the homogeneous system with constant coefficients

$$\begin{aligned} x' &= a_{11}x + a_{12}y \\ y' &= a_{21}x + a_{22}y \end{aligned} \tag{6}$$

The characteristic equation of **(6)** is given by

$$\begin{vmatrix} a_{11} - \lambda & a_{12} \\ a_{21} & a_{22} - \lambda \end{vmatrix} = 0$$

or

$$\lambda^2 - (a_{11} + a_{22})\lambda + (a_{11}a_{22} - a_{21}a_{12}) = 0$$

Method of Determinants for Solving a Linear, Homogeneous System with Constant Coefficients pp. 326–327

To find the general solution of the system

$$\begin{aligned} x' &= a_{11}x + a_{12}y, \\ y' &= a_{21}x + a_{22}y, \end{aligned}$$

in the case $a_{12} \neq 0$, first find two numbers λ_1 and λ_2 that satisfy the characteristic equation

$$\begin{vmatrix} a_{11} - \lambda & a_{12} \\ a_{21} & a_{22} - \lambda \end{vmatrix} = 0.$$

Case 1 If $\lambda_1 \neq \lambda_2$ are real, then pp. 324, 327

$$x(t) = c_1e^{\lambda_1 t} + c_2e^{\lambda_2 t},$$

and $y(t)$ can be obtained from the equation

$$y = \frac{1}{a_{12}}(x' - a_{11}x). \tag{7}$$

Case 2 if $\lambda_1 = \lambda_2$, then pp. 324, 327

$$x(t) = e^{\lambda_1 t}(c_1 + c_2t),$$

and $y(t)$ can be obtained from Equation **(7).**

Case 3 If $\lambda_1 = a + ib$ and $\lambda_2 = a - ib$, then pp. 324, 327

$$x(t) = e^{at}(c_1 \cos bt + c_2 \sin bt),$$

and again $y(t)$ can be obtained from Equation **(7).**

Laplace Transform Method for Systems pp. 330–331

Consider the system

$$\frac{dx}{dt} = a_{11}x + a_{12}y + f(t),$$
$$\frac{dy}{dt} = a_{21}x + a_{22}y + g(t), \tag{8}$$

with initial conditions $x(0) = x_0$, $y(0) = y_0$. Taking the Laplace transform of both equations in system **(8)** and letting the corresponding capital letters represent the Laplace transforms of the functions x, y, f, and g (that is, we let $X = L\{x\}$, $Y = L\{y\}$, and so on), we obtain

$$sX - x_0 = a_{11}X + a_{12}Y + F,$$
$$sY - y_0 = a_{21}X + a_{22}Y + G. \tag{9}$$

System **(9)** can be solved as follows:

$$X = \frac{(s - a_{22})(F + x_0) + a_{12}(G + y_0)}{(s - \lambda_1)(x - \lambda_2)},$$

$$Y = \frac{(s - a_{11})(G + y_0) + a_{21}(F + x_0)}{(s - \lambda_1)(s - \lambda_2)},$$

where λ_1 and λ_2 are the solutions of the characteristic equation

$$\lambda^2 - (a_{11} + a_{22})\lambda + (a_{11}a_{22} - a_{12}a_{21}) = 0.$$

Then $x(t)$ and $y(t)$ can be found by inverting (if possible) the Laplace transforms $X(s)$ and $Y(s)$.

REVIEW EXERCISES FOR CHAPTER 7

In Exercises 1–3 transform the equation into a first-order system.

1. $x''' - 6x'' + 2x' - 5x = 0$

2. $x'' - 3x' + 4t^2x = \sin t$

3. $xx'' + x'x''' = \ln t$

Find the general solution (or particular solution when initial conditions are given) by the method of elimination or the method of determinants.

4. $x' = x + y,$
$y' = 9x + y$

5. $x' = 4x - y,$
$y' = x + 2y$

6. $x' = x - 4y$
$y' = x + y$

7. $x' = x + 2y,$
$y' = 4x + 3y$

8. $x' = 3x + 2y,$
$y' = -5x + y$

9. $x' = -x - 3e^{-2t},$
$y' = -2x - y - 6e^{-2t}$

10. $x' = -4x - 6y + 9e^{-3t}, x(0) = -9,$
$y' = x + y - 5e^{-3t}, y(0) = 4$

In Exercises 11–14 solve the given system using Laplace transforms.

11. $x' = x + 2y, x(0) = 1$
$y' = 2x - 3y, y(0) = 2$

12. $x' = 2x + 3y + t, x(0) = 2$
$y' = 4x - 5y + 1, y(0) = 0$

13. $x' = 2x - 5y + e^t, x(0) = 1$
$y' = 3x - 7y + t^2, y(0) = -1$

14. $x' = 2x - y, x(0) = 1$
$y' = 5x - 2y, y(0) = 3$

15. A direct-current transmission line of length L connecting a power source to a distant receiver is subject to (a) drops in voltage due to the resistance of the line, and (b) leakage of current along the line due to imperfect insulation. If $E(x)$ and $I(x)$ are the voltage and current at a distance x from the power source, R is the resistance (ohms) per unit length of line, and G is the leakance (conductance) (mhos) per unit length, find

(a) a system of differential equations describing the voltage and current in the transmission line;
(b) the solution to part (a);
(c) the current and voltage at the end of the line.

16. J. P. Brady and C. Marmasse obtained the initial-value problem

$$ay'' + y' + by = \beta, \qquad y(0) = \alpha, \qquad y'(0) = \beta,$$

to describe avoidance learning in rats.* Here $y(t)$ is the value of the learning curve of the rat at time t, α and β are the initial values, and a and b are constants.
(a) Obtain a system of first-order equations equivalent to the given initial-value problem.
(b) Solve the system in part (a).

17. Tank X contains 150 gallons of pure water and tank Y 150 gallons of brine in which 60 pounds of salt has been dissolved. Liquid circulates from each tank to the other at the rate of 9 gallons per minute, with the mixture kept uniform in each tank by stirring.
(a) How much salt is in each tank at any time t?
(b) How much salt is in each tank as $t \to \infty$?
(c) When does tank X contain the maximum amount of salt?

18. Find the current I_2 in the indicated loop for the illustrated circuit, where $R = 10$ ohms and $L = 1$ henry, assuming $I_1 = I_2 = 0$ when the switch is closed at time $t = 0$. Assume $E = 10$ volts.

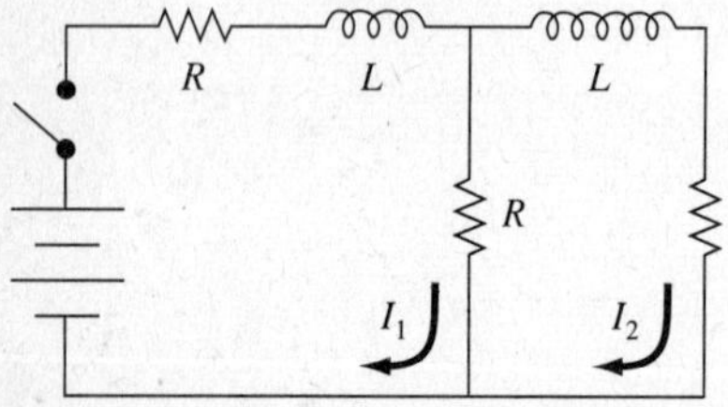

19. S. Grossberg obtained the initial-value problem

$$\frac{dx}{dt} = a(b - x) - cy, \qquad x(0) = b,$$

$$\frac{dy}{dt} = d(x - y) - (a + c)y, \qquad y(0) = \frac{bd}{a + d},$$

where a, b, c, d are positive constants.† Show that y decays monotonically to a positive minimum.

20. A transformer having a spark gap in the primary circuit (see illustration) is called an **oscillation transformer.** Suppose that the capacitor C_1 is initially charged to e_0 volts, and that this is the only source of energy. If C_1 is allowed to discharge through the primary circuit, the current jumps the spark gap G and continues around the circuit through the coil L_1. Find the currents in the circuits at all subsequent times. Let M denote the mutual inductance.

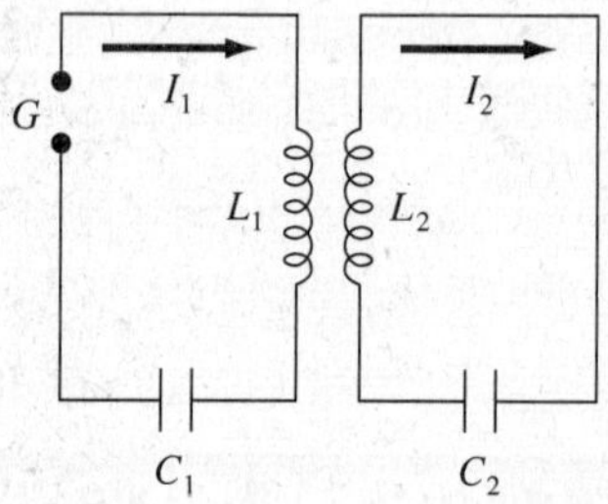

* *Psychological Record* 12 (1962):361–368.
† *Journal of Theoretical Biology* (1969):325–364.

8 NUMERICAL METHODS

8.1 ERROR ANALYSIS

In every chapter of this book we have performed numerical computations: solving differential equations, applying the methods of power series and Frobenius, and using Laplace transforms. With few exceptions, we have limited our examples to problems involving a small number of variables—not because most applications have only two or three variables but because the computations would have been too tedious otherwise.

With the recent and widespread use of calculators and computers, the situation has been altered. The remarkable strides made in the last few years in the theory of numerical methods for solving certain computational problems have made it possible to perform, quickly and accurately, the calculations mentioned in the first paragraph.

The use of the computer presents new difficulties, however. Computers do not store numbers such as $\frac{2}{3}$, $7\frac{3}{8}$, $\sqrt{2}$, and π. Rather, every computer uses what is called **floating-point arithmetic.** In this system every number is represented in the form

$$x = \pm 0.d_1d_2 \cdots d_k \times 10^n, \qquad d_1 \neq 0, \tag{1}$$

where $d_1, d_2, \ldots, d_k$ are single-digit integers and n is an integer. Any number written in this form is called a **floating-point number.** In Equation **(1),** the number $\pm 0.d_1 d_2 \cdots d_k$ is called the **mantissa** and the number n is called the **exponent.** The number k is called the **number of significant digits** in the expression.

EXAMPLE 1 ▶ **Four floating-point numbers**

The following numbers are expressed in floating-point form:

(a) $\frac{1}{4} = 0.25$
(b) $2378 = 0.2378 \times 10^4$
(c) $-0.000816 = -0.816 \times 10^{-3}$
(d) $83.27 = 0.8327 \times 10^2$ ◀

If the number of significant digits were unlimited, we would have no problem. Almost every time numbers are introduced into a computer, however, errors begin to accumulate. This can happen in one of two ways:

1. **Truncation:** All significant digits after k digits are simply "cut off." For example, if truncation is used, $\frac{2}{3} = 0.666666...$ is stored (with $k = 8$) as $\frac{2}{3} = 0.66666666 \times 10^0$.
2. **Rounding:** If $d_{k+1} \geq 5$, then 1 is added to d_k and the resulting number is truncated. Otherwise, the number is simply truncated. For example, with rounding (and $k = 8$), $\frac{2}{3}$ is stored as $\frac{2}{3} = 0.66666667 \times 10^0$.

We can illustrate how some numbers are stored with truncation and rounding by using eight significant digits:

Number	*Truncated Number*	*Rounded Number*
$\frac{8}{3}$	0.26666666×10^1	0.26666667×10^1
π	0.31415926×10^1	0.31415927×10^1
$-\frac{1}{57}$	$-0.17543859 \times 10^{-1}$	$-0.17543860 \times 10^{-1}$

Individual round-off or truncation errors do not seem very significant. When thousands of computational steps are involved, however, the *accumulated* round-off error can be devastating. Thus, in discussing any numerical scheme, it is necessary to know not only whether you will get the right answer, theoretically, but also how badly the round-off errors will accumulate. To keep track of things, we define two types of error. If x is the actual value of a number and x^* is the number that appears in the computer, then the **absolute error** ϵ_a is defined by

Absolute error

$$\epsilon_a = |x^* - x|. \tag{2}$$

More interesting in most situations (as we see in Example 2) is the **relative error** ϵ_r, defined by

Relative error

$$\epsilon_r = \left|\frac{x^* - x}{x}\right|. \tag{3}$$

EXAMPLE 2 ▶ **Finding absolute and relative errors**

Let $x = 2$ and $x^* = 2.1$. Then $\epsilon_a = 0.1$ and $\epsilon_r = \frac{0.1}{2} = 0.05$. If $x_1 = 2000$ and $x_1^* = 2000.1$, then again, $\epsilon_a = 0.1$. But now $\epsilon_r = \frac{0.1}{2000} = 0.00005$. Most people would agree that the 0.1 error in the first case is more significant than the 0.1 error in the second. ◀

Much of numerical analysis is concerned with questions of **convergence** and **stability.** If x is the answer to a problem and our computational method gives us approximating values x_n, then the method converges if, theoretically, x_n approaches x as n gets large. If, moreover, it can be shown that the round-off errors do not accumulate in such a way as to make the answer unreliable, then the method is stable.

It is easy to give an example of a procedure in which round-off error can be quite large. Suppose we wish to compute $y = \frac{1}{x - 0.66666665}$. For $x = \frac{2}{3}$, if the computer truncates, then $x = 0.66666666$ and $y = \frac{1}{0.00000001} = 10^8 = 10 \times 10^7$. If the computer rounds, then $x = 0.66666667$ and $y = \frac{1}{0.00000002} = 5 \times 10^7$. The difference here is enormous. The correct answer is $1/(\frac{2}{3} - \frac{66666665}{100000000}) = 60{,}000{,}000 = 6 \times 10^7$.

In this chapter we examine several numerical procedures for approximating the solution of a differential equation. In certain instances we also discuss convergence and stability. This is but a very superficial view of numerical methods, however; entire books and courses are devoted to the subject.

SELF-QUIZ

I. If π is written as a floating-point number, then its exponent is ________.

(a) 0 **(b)** 1 **(c)** 2 **(d)** -1

II. Your calculator shows that $\sqrt{122} = 11.045361$. If this answer is written as a floating-point number, we will write ________.

(a) 11.045361×10^0 **(b)** 1.1045361×10^1
(c) 0.11045361×10^2 **(d)** 0.011045361×10^3

III. How would you write $-\sqrt{130} = -11.401754$ as a truncated floating-point number with six significant digits ________?

(a) -11.40175×10^0
(b) -1.14017×10^1
(c) -0.114017×10^2
(d) -0.1140175×10^2
(e) -11.401×10^0
(f) -1.1401×10^1

IV. Suppose you write π as a floating-point number with five significant digits. Then the absolute error is ________.

(a) 0.73×10^{-5} **(b)** -0.73×10^{-5}
(c) -0.27×10^{-5} **(d)** 0.27×10^{-5}

V. What is the relative error in Question IV?

(a) 0.73×10^{-5} **(b)** 0.27×10^{-5}
(c) 0.23×10^{-5} **(d)** -0.27×10^{-5}

Answers to Self-Quiz

I. b **II.** c **III.** c **IV.** a **V.** c

PROBLEMS 8.1

In Problems 1–13 convert the number to a floating-point number with eight decimal places of accuracy. Either truncate (T) *or round off* (R) *as indicated.*

1. $\frac{1}{3}$(T) **2.** $\frac{7}{8}$ **3.** -0.000035 **4.** $\frac{7}{9}$(R)

5. $\frac{7}{9}$(T) **6.** $\frac{33}{7}$(T) **7.** $\frac{85}{11}$(R) **8.** $-18\frac{5}{6}$(T)

9. $-18\frac{5}{6}$(R) **10.** 237,059,628(T) **11.** 237,059,628(R) **12.** -23.7×10^{15}

13. 8374.2×10^{-24}

In Problems 14–21 the number x and an approximation x^* are given. Find the absolute and relative errors ϵ_a and ϵ_r.

14. $x = 5$, $x^* = 0.49 \times 10^1$

15. $x = 500$, $x^* = 0.4999 \times 10^3$

16. $x = 3720$, $x^* = 0.3704 \times 10^4$

17. $x = \frac{1}{8}$, $x^* = 0.12 \times 10^0$

18. $x = \frac{1}{800}$, $x^* = 0.12 \times 10^{-2}$

19. $x = -5\frac{5}{6}$, $x^* = -0.583 \times 10^1$

20. $x = 0.70465$, $x^* = 0.70466 \times 10^0$

21. $x = 70465$, $x^* = 0.70466 \times 10^5$

8.2 EULER METHODS FOR FIRST-ORDER DIFFERENTIAL EQUATIONS

In Section 1.4 we described the Euler method of approximating the solution of the first-order initial-value problem

$$\frac{dy}{dx} = f(x, y), \qquad y(x_0) = y_0, \tag{1}$$

at the points

$$x_0, x_1 = x_0 + h, x_2 = x_0 + 2h, \ldots, x_n = x_0 + nh$$

Euler method

for h, some nonzero real number. The method, defined by the iterative formula

$$y_{n+1} = y_n + hf(x_n, y_n), \tag{2}$$

with $y_0 = y(x_0)$, approximates the solution by following the tangent to the solution curve passing through (x_n, y_n) for a small horizontal distance. Note that if we replace $f(x_n, y_n)$ by the derivative $y'_n = y'(x_n)$, then Equation **(2)** resembles the first two terms of the Taylor series for $y(x)$ at the value x_n. The following example illustrates how the work can be organized in tabular form.

EXAMPLE 1 ▶ **Solving an initial-value problem using Euler's method**

Find an approximate value for the solution of the initial-value problem

$$y' = x + y^2, \qquad y(1) = 0,$$

at $x = 1, 1.1, 1.2, 1.3, 1.4, 1.5$.

Solution We arrange our work in columns: The first column contains the values of x at $1, 1.1, \ldots, 1.5$; the second contains the values of y beginning with the initial condition; the third, the computed value of $f(x_n, y_n) = x_n + y_n^2$ for the given x and y on that row; and the last the computed value of y_{n+1} according to Equation **(2)**. The value in the last column is then transferred to the y entry in the next row to compute the entries in the third and fourth columns (see Table 8.1).

TABLE 8.1 Euler method with $h = 0.1$, for $y' = x + y^2$, $y(1) = 0$

x_n	y_n	$f(x_n, y_n)$	$y_n + hf(x_n, y_n)$
1.00	0.00	1.00	0.10
1.10	0.10	1.11	0.21
1.20	0.21	1.24	0.34
1.30	0.34	1.41	0.48
1.40	0.48	1.63	0.64
1.50	0.64		

◀

As we saw in Section 1.4 (see Example 1.4.1 on p. 26), we can usually reduce the discretization error by reducing the step size. For example, if instead of a step size of $h = 0.1$ we use $h = 0.05$, our solution would be that shown in Table 8.2, while $h = 0.025$ would yield Table 8.3. Observe that in each case the approximate value of $y(1.5)$ increases. This is the behavior that one would anticipate if the solution curves in the direction field are all increasing in the interval $1 \leq x \leq 1.5$. Figure 8.1 illustrates this assertion: the approximation with step size $\frac{h}{2}$ lies above the one of step size h, because the slope of the solution curves, near the true solution

TABLE 8.2 Euler method with $h = 0.5$, for $y' = x + y^2$, $y(1) = 0$

x_n	y_n	$f(x_n, y_n)$	$y_n + hf(x_n, y_n)$
1.00	0.00000	1.0000	0.05000
1.05	0.05000	1.0525	0.10263
1.10	0.10263	1.1105	0.15815
1.15	0.15815	1.1750	0.21690
1.20	0.21690	1.2470	0.27925
1.25	0.27925	1.3280	0.34565
1.30	0.34565	1.4195	0.41663
1.35	0.41663	1.5236	0.49281
1.40	0.49281	1.6429	0.57495
1.45	0.57495	1.7806	0.66398
1.50	0.66398		

TABLE 8.3 Euler method with $h = 0.025$, for $y' = x + y^2$, $y(1) = 0$

x_n	y_n	$f(x_n, y_n)$	$y_n + hf(x_n, y_n)$
1.000	0.00000	1.0000	0.02500
1.025	0.02500	1.0256	0.05064
1.050	0.05064	1.0526	0.07695
1.075	0.07695	1.0809	0.10398
1.100	0.10398	1.1108	0.13175
1.125	0.13175	1.1424	0.16031
1.150	0.16031	1.1757	0.18970
1.175	0.18970	1.2110	0.21997
1.200	0.21997	1.2484	0.25118
1.225	0.25118	1.2881	0.28339
1.250	0.28339	1.3303	0.31664
1.275	0.31664	1.3753	0.35103
1.300	0.35103	1.4232	0.38661
1.325	0.38661	1.4745	0.42347
1.350	0.42347	1.5293	0.46170
1.375	0.46170	1.5882	0.50140
1.400	0.50140	1.6514	0.54269
1.425	0.54269	1.7195	0.58568
1.450	0.58568	1.7930	0.63050
1.475	0.63050	1.8725	0.67732
1.500	0.67732		

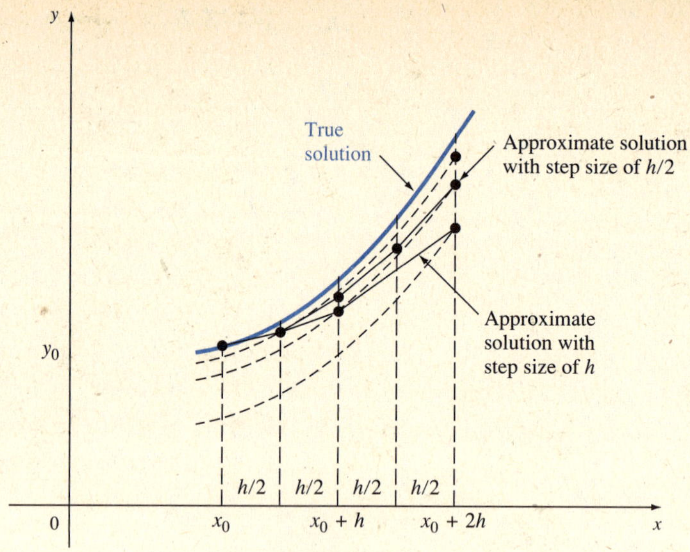

FIGURE 8.1
Reducing discretization error by reducing step size

at $x_0 + \frac{h}{2}$, exceeds the slope at x_0. Thus two steps of size $\frac{h}{2}$ incur less error, in this situation, than one step of size h.

Improved Euler method

The improved Euler method was developed to reduce the discretization error that occurs in the Euler method. The procedure *averages* the slopes at the left and right endpoints of each step and uses this average slope to compute the next y-value. To see why this is a reasonable procedure, integrate both sides of Equation **(1)** between x_0 and x_1:

$$y(x_1) - y(x_0) = \int_{x_0}^{x_1} \frac{dy}{dx}\, dx = \int_{x_0}^{x_1} f(x, y(x))\, dx,$$

or

$$y(x_1) = y(x_0) + \int_{x_0}^{x_1} f(x, y(x))\, dx. \tag{3}$$

If we approximate the integral in Equation **(3)** by the **trapezoidal rule** of calculus, we obtain

$$y(x_1) \approx y(x_0) + \frac{h}{2}[f(x_0, y(x_0)) + f(x_1, y(x_1))]. \tag{4}$$

Since $y(x_1)$ is not known, we replace it by the value found by the Euler method, which we call z_1; then Equation **(4)** can be replaced by the system of equations

$$z_1 = y_0 + hf(x_0, y_0),$$

$$y_1 = y_0 + \frac{h}{2}[f(x_0, y_0) + f(x_1, z_1)].$$

Figure 8.2 illustrates the effect of averaging the slopes at the endpoint of the interval and provides a geometric interpretation of each step of the general procedure.

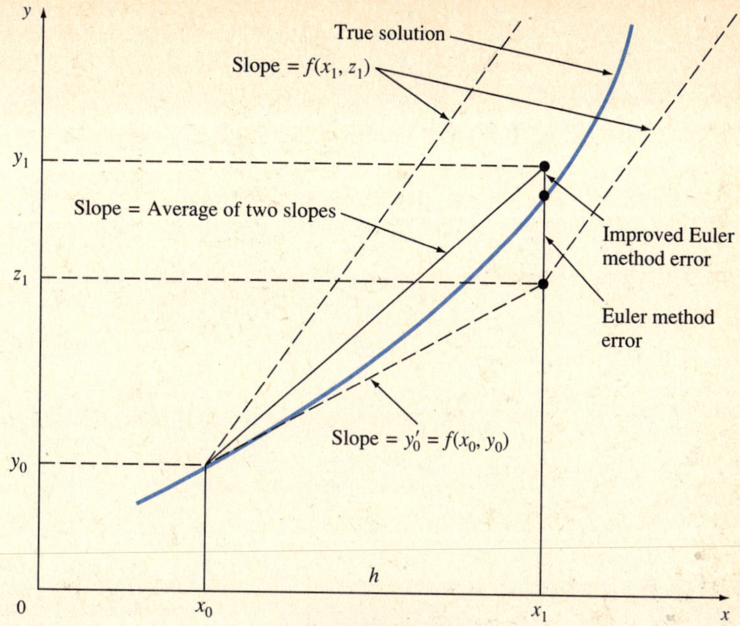

FIGURE 8.2
Improved Euler method

Improved Euler method

$$z_{n+1} = y_n + hf(x_n, y_n),$$

$$y_{n+1} = y_n + \frac{h}{2}[f(x_n, y_n) + f(x_{n+1}, z_{n+1})]. \quad \textbf{(5)}$$

EXAMPLE 2 ▶ **Solving an initial-value problem with the improved Euler method**
Consider the initial-value problem that we solved using the Euler method in Example 1.4.2 (p. 27):

$$\frac{dy}{dx} = y + x^2, \qquad y(0) = 1. \quad \textbf{(6)}$$

We wish to determine $y(1)$. We saw in Section 1.4 that with $h = 0.2$, the Euler method yields the approximation $y(1) \approx 2.77$, which is off by about 12% from the exact solution $y(1) = 3e - 5 \approx 3.1548$. Using $x_0 = 0$, $y_0 = 1$, and $h = 0.2$, Equation **(5)** yields

$$z_1 = y_0 + hf(x_0, y_0) = 1 + 0.2(1 + 0^2) = 1.2,$$

$$y_1 = y_0 + \frac{h}{2}[f(x_0, y_0) + f(x_1, z_1)] = 1 + 0.1[(1 + 0^2) + 1.2 + 0.2^2]$$

$$= 1 + 0.1(2.24) = 1.224 \approx 1.22,$$

$$z_2 = y_1 + hf(x_1, y_1) = 1.22 + 0.2[1.22 + (0.2)^2] = 1.472 \approx 1.47,$$

$$y_2 = y_1 + \frac{h}{2}[f(x_1, y_1) + f(x_2, z_2)]$$

$$= 1.22 + 0.1[1.22 + (0.2)^2 + 1.47 + (0.4)^2] = 1.509 \approx 1.51,$$

and so on. Table 8.4 shows the approximate values of the solution of Equation **(6)**; the absolute error this time is less than 1%.

TABLE 8.4 Improved Euler method with $h = 0.2$, for $y' = y + x^2$, $y(0) = 1$

x_n	y_n	$f(x_n, y_n) = y_n + x_n^2$	z_{n+1}	$f(x_{n+1}, z_{n+1}) = z_{n+1} + x_{n+1}^2$	y_{n+1}
0.0	1.00	1.00	1.20	1.24	1.22
0.2	1.22	1.26	1.47	1.63	1.51
0.4	1.51	1.67	1.84	2.20	1.90
0.6	1.90	2.26	2.35	2.99	2.43
0.8	2.43	3.07	3.04	4.04	3.14
1.0	3.14				

If we reduce the step size to $h = 0.1$, we obtain the results in Table 8.5. The actual value obtained using the improved Euler method to five significant figures is $y(1) \approx 3.1504$, an absolute error of 0.14%. A comparison of both methods and step sizes is shown in Table 8.6

TABLE 8.5 Improved Euler method with $h = 0.1$, for $y' = y + x^2$, $y(0) = 1$

x_n	y_n	$f(x_n, y_n)$	z_{n+1}	$f(x_{n+1}, z_{n+1})$	y_{n+1}
0.00	1.00	1.00	1.10	1.11	1.11
0.10	1.11	1.12	1.22	1.26	1.22
0.20	1.22	1.26	1.35	1.44	1.36
0.30	1.36	1.45	1.50	1.66	1.52
0.40	1.52	1.68	1.68	1.93	1.70
0.50	1.70	1.95	1.89	2.25	1.91
0.60	1.91	2.27	2.13	2.62	2.15
0.70	2.15	2.64	2.41	3.05	2.43
0.80	2.43	3.07	2.74	3.55	2.77
0.90	2.77	3.58	3.12	4.12	3.15
1.00	3.15				

TABLE 8.6 A comparison of results using Euler and improved Euler methods with $h = 0.2$, and $h = 0.1$ for the initial-value problem $y' = y + x^2$, $y(0) = 1$

x	*Euler*	*Improved Euler*	*Exact*
$h = 0.2$			
0.0	1.000000	1.00000000	1.00000000
0.2	1.200000	1.22400000	1.22420828
0.4	1.448000	1.51408000	1.51547409
0.6	1.769600	1.90237760	1.90635640
0.8	2.195520	2.42810067	2.43662279
1.0	2.762624	3.13908282	3.15484548
$h = 0.1$			
0.0	1.00000000	1.00000000	1.00000000
0.1	1.10000000	1.10550000	1.10551275
0.2	1.21100000	1.22412750	1.22420828
0.3	1.33610000	1.35936089	1.35957642
0.4	1.47871000	1.51504378	1.51547409
0.5	1.64258100	1.69542338	1.69616381
0.6	1.83183910	1.90519283	1.90635640
0.7	2.05102301	2.14953808	2.15125812
0.8	2.30512531	2.43418958	2.43662279
0.9	2.59963784	2.76547948	2.76880934
1.0	2.94060163	3.15040483	3.15484548

EXAMPLE 3 ▶ **Applying the improved Euler method in Example 1**

If we apply the improved Euler method to the initial-value problem in Example 1,

$$y' = x + y^2, \qquad y(1) = 0,$$

with $h = 0.1$, we obtain the values in Table 8.7. Here $y(1) \approx 0.6918$ to four decimal places. If we change the step size to $h = 0.05$, we obtain $y(1) \approx 0.6916$, suggesting that the calculation in Table 8.7 probably has an absolute error of no more than 0.05%. A comparison of the improved Euler and Euler methods for both step sizes is given in Table 8.8.

An exact solution of this initial-value problem can be obtained: Use the substitution $y = -z'/z$ to transform the given Riccati equation into the second-order

TABLE 8.7 Improved Euler method with $h = 0.1$, for $y' = x + y^2$, $y(1) = 0$

x_n	y_n	$f(x_n, y_n)$	z_{n+1}	$f(x_{n+1}, z_{n+1})$	y_{n+1}
1.00	0.00	1.00	0.10	1.11	0.11
1.10	0.11	1.11	0.22	1.25	0.22
1.20	0.22	1.25	0.35	1.42	0.36
1.30	0.36	1.43	0.50	1.65	0.51
1.40	0.51	1.66	0.68	1.96	0.69
1.50	0.69				

TABLE 8.8 A comparison of results using Euler and improved Euler methods with $h = 0.1$, and $h = 0.05$, for the initial-value problem $y' = x + y^2$, $y(1) = 0$

x	*Euler*	*Improved Euler*
$h = 0.1$		
1.0	0.000000000	0.000000000
1.1	0.100000000	0.105500000
1.2	0.211000000	0.223402573
1.3	0.335452100	0.356966908
1.4	0.476704911	0.510823653
1.5	0.639429668	0.691781574
$h = 0.05$		
1.00	0.000000000	0.000000000
1.05	0.050000000	0.051312500
1.10	0.102625000	0.105398434
1.15	0.158151595	0.162573808
1.20	0.216902191	0.223209962
1.25	0.279254519	0.287746157
1.30	0.345653673	0.356705840
1.35	0.416627496	0.430717867
1.40	0.492806420	0.510544562
1.45	0.574949328	0.597119307
1.50	0.663977664	0.691597674

equation $z'' + xz = 0$, which can then be solved in terms of Bessel functions by the methods of Section 5.4. However, the resulting quotient of Bessel functions is difficult to evaluate. Numerical methods provide information when either no exact solution is known or the solution is difficult to evaluate. Comparisons between various methods, for varying step sizes, may be used to justify the selection of a given value. ◀

In Section 8.5, we will discuss the discretization error of the Euler method. Problems 26 and 27 provide the basic technique for finding the step-by-step discretization error for the Euler and improved Euler methods.

SELF-QUIZ

I. Suppose we wish to solve the initial-value problem $y' = x^2 + y^2$, $y(0) = -1$, by Euler's method with $h = 0.1$. Then $y_0 =$ ______.
(a) 1 **(b)** 0 **(c)** −1 **(d)** 2

II. What would y_1 be for Question I?
(a) −0.9 **(b)** −1 **(c)** 0 **(d)** 1.1

III. If we wanted to solve the initial-value problem in Question I using the improved Euler method, then $y_1 =$ ______.

(a) $1 + \frac{0.1}{2}\{(0^2 + 1^2) + (0.1^2 + (-0.9)^2)\}$

(b) $-1 + \frac{0.1}{2}\{(0^2 + 1^2) + (0.1^2 + (-0.9)^2)\}$

(c) $-0.9 + \frac{0.1}{2}\{(0^2 + 1^2) + (0.1^2 + (-0.9)^2)\}$

(d) $-1 + \frac{0.1}{2}\{(0.1^2 + (-1)^2) + (0.2^2 + (-0.9)^2)\}$

Answers to Self-Quiz

I. c **II.** a **III.** b

PROBLEMS 8.2

In Problems 1 through 10 solve each problem exactly using the methods of Chapter 2. Then
(a) *Use the Euler method with the indicated value of* h;
(b) *Use the improved Euler method with the given value of* h.

Compare the accuracy of the two methods with the exact answer. What is each relative error?

1. $\frac{dy}{dx} = x + y$, $y(0) = 1$. Find $y(1)$ with $h = 0.2$.

2. $\frac{dy}{dx} = x - y$, $y(1) = 2$. Find $y(3)$ with $h = 0.4$.

3. $\frac{dy}{dx} = \frac{x - y}{x + y}$, $y(2) = 1$. Find $y(1)$ with $h = -0.2$.

4. $\frac{dy}{dx} = \frac{y}{x} + \left(\frac{y}{x}\right)^2$, $y(1) = 1$. Find $y(2)$ with $h = 0.2$.

5. $\frac{dy}{dx} = x\sqrt{1 + y^2}$, $y(1) = 0$. Find $y(3)$ with $h = 0.4$.

6. $\frac{dy}{dx} = x\sqrt{1 - y^2}$, $y(1) = 0$. Find $y(2)$ with $h = 0.125$.

7. $\frac{dy}{dx} = \frac{y}{x} - \frac{5}{2}x^2y^3$, $y(1) = \frac{1}{\sqrt{2}}$. Find $y(2)$ with $h = 0.125$.

8. $\frac{dy}{dx} = \frac{-y}{x} + x^2y^2$, $y(1) = \frac{2}{9}$. Find $y(3)$ with $h = \frac{1}{3}$.
[*Hint:* This is a Bernoulli equation.]

9. $\frac{dy}{dx} = ye^x$, $y(0) = 2$. Find $y(2)$ with $h = 0.2$.

10. $\frac{dy}{dx} = xe^y$, $y(0) = 0$. Find $y(1)$ with $h = 0.1$.

In Problems 11 through 20 use (a) the Euler method, or (b) the improved Euler method to graph approximately the solution of the given initial-value problem by plotting the points (x_k, y_k) over the indicated range, where $x_k = x_0 + kh$.

11. $y' = xy^2 + y^3$, $y(0) = 1$, $h = 0.02$, $0 \le x \le 0.1$

12. $y' = x + \sin(\pi y)$, $y(1) = 0$, $h = 0.2$, $1 \le x \le 2$

13. $y' = x + \cos(\pi y)$, $y(0) = 0$, $h = 0.4$, $0 \le x \le 2$

14. $y' = \cos(xy)$, $y(0) = 0$, $h = \dfrac{\pi}{4}$, $0 \le x \le \pi$

15. $y' = \sin(xy)$, $y(0) = 1$, $h = \dfrac{\pi}{4}$, $0 \le x \le 2\pi$

16. $y' = \sqrt{x^2 + y^2}$, $y(0) = 1$, $h = 0.5$, $0 \le x \le 5$

17. $y' = \sqrt{y^2 - x^2}$, $y(0) = 1$, $h = 0.1$, $0 \le x \le 1$

18. $y' = \sqrt{x + y^2}$, $y(0) = 1$, $h = 0.2$, $0 \le x \le 1$

19. $y' = \sqrt{x + y^2}$, $y(1) = 2$, $h = -0.2$, $0 \le x \le 1$

20. $y' = \sqrt{x^2 + y^2}$, $y(1) = 5$, $h = -0.2$, $0 \le x \le 1$

21. Let

$$\frac{dy}{dx} = \frac{x - y}{x + y}, \qquad y(2) = 0.$$

Use the improved Euler method to approximate $y(1)$ with $h = -0.2$. Compare your answer with the exact value and explain why the method failed.

22. Use the improved Euler method to approximate $y(3)$ with $h = 0.4$ for the initial-value problem

$$\frac{dy}{dx} + \frac{3}{x}y = x^2y^2, \qquad y(1) = 2.$$

Compare your answer to the exact answer and explain why the numerical technique failed.

23. Repeat Problem 22 for the initial-value problem

$$\frac{dy}{dx} + \frac{y}{x} = x^3y^3, \qquad y(1) = 1.$$

24. Let $y' = e^{xy}$, $y(0) = 1$, and obtain a value for $y(4)$ with $h = 0.5$ and $h = 0.1$ using the improved Euler method. What difficulties are encountered? How much confidence do you have in your answer?

25. The **three-term Taylor series method** is given by

$$y_{n+1} = y_n + hy'_n + \frac{h^2}{2}y''_n,$$

where $y'_n = f(x_n, y_n)$, $x_n = x_0 + nh$, and

$$y''_n = \frac{d}{dx}[f(x, y)]\bigg|_{(x_n, y_n)} = \frac{\partial f}{\partial x}(x_n, y_n) + \frac{\partial f}{\partial y}(x_n, y_n)\, y'_n.$$

Use the three-term Taylor series method to find $y(1)$ for the initial-value problem

$$y' = y + x^2, \qquad y(0) = 1,$$

with $h = 0.2$ and $h = 0.1$. Compare your results with those in Table 8.6.

26. The discretization error ϵ_E in the Euler method is given by $h\epsilon_E = y - y_1$, where $y = y(x_0 + h)$ is the exact solution and $y_1 = y_0 + hy'_0$, where $y_0 = y(x_0)$. Show that $|\epsilon_E| \le kh$. [*Hint:* Use Taylor's theorem to rewrite $y(x_0 + h) = y(x_0) + hy'(x_0) + (h^2/2)\, y''(\xi)$.]

27. Show that the discretization error ϵ_{IE} for the improved Euler method satisfies $|\epsilon_{IE}| \le kh^2$. [*Hint:* Expand $y(x_0 + h)$ and $f(x_0 + h, y_0 + hy'_0)$ using Taylor's theorem.]

28. What is the discretization error for the three-term Taylor series method in Problem 25?

---

*The following TRUE BASIC and FORTRAN programs were used to compute the tables for the improved Euler method in Example 2. Modify these programs to obtain a numerical solution for Problems 29–32 using the improved Euler method.**

TRUE BASIC

```
LET x=0                         ! initial x value
LET y=1                         ! initial y value
LET xlast=1                     ! last x value
LET h=.2                        ! step size
PRINT " x "," y "
DO
```

* There is now a great deal of software available to solve differential equations numerically. Here and in two other sections you are encouraged to write your own programs. In Section 8.8 we discuss how to use symbolic manipulators to solve them.

```
        LET dydx=y+x*x                          ! dy/dx
        LET z=y+h*dydx                          ! Euler step
        LET x=x+h                               ! update x
        LET dzdx=z+x*x                          ! derivative at (x+h,z)
        LET y=y+h*(dydx+dzdx)/2                 ! improved Euler algorithm
        PRINT x,y
LOOP until x>=xlast
END
```

FORTRAN

```
      X=0.0
      Y=1.0
      XLAST=1.0
      H=0.2
      N=(XLAST-X)/H
      WRITE (*, '(/7X,A,12X,A)') 'X','Y'
      DO 11 I=0, N
         DYDX=Y+X*X
         Z=Y+H*DYDX
         X=X+H
         DZDX=Z+X*X
         Y=Y+H*(DYDX+DZDX)/2.0
         WRITE (*,'(1X,F10.4,1X,F14.6)')X,Y
   11 CONTINUE
      END
```

29. $y' = x + y^2$, $y(1) = -2$, $h = 0.01$. Find $y(2)$.

30. $y' = \sqrt{x^2 + y^2}$, $y(0) = 1$, $h = 0.1$. Find $y(5)$.

31. $y' = \sqrt{x + y^2}$, $y(1) = 2$, $h = 0.02$. Find $y(2)$.

32. $y' = y\,e^x$, $y(0) = 2$, $h = 0.02$. Find $y(2)$.

Observe that in the two programs above, we calculate the derivative twice, once as dydx and once as dzdx. It it often convenient to let both of these calculations be done by a single subroutine. The program below illustrates this usage.

```
LET x=0                                 ! initial x value
LET y=1                                 ! initial y value
LET xlast=1                             ! last x value
LET h=.1                                ! step size
PRINT " x   "," y "
DO
    CALL difeq (x,y,dydx)               ! derivative at (x,y)
    LET z=y+h*dydx                      ! Euler step
    LET x=x+h                           ! update x
    CALL difeq (x,z,dzdx)               ! derivative at (x+h,z)
    LET y=y+h*(dydx+dzdx)/2             ! improved Euler algorithm
    PRINT x,y
LOOP until x>=xlast
END
SUB difeq(x,y,dydx)
LET dydx=y+x*x
END SUB
```

33. Write and run a FORTRAN program that mimics the TRUE BASIC program given above.

8.3 THE RUNGE-KUTTA METHOD

This powerful method gives accurate results without a large number of steps (that is, without the need to make the step size too small). The efficiency is obtained by using a version of Simpson's rule (a frequently used method of numerical integration) in evaluating the following integral:

$$y(x_1) = y(x_0) + \int_{x_0}^{x_1} f(x, y(x))\,dx, \tag{1}$$

which yields the solution at x_1 of the initial-value problem

$$y' = f(x, y), \qquad y(x_0) = y_0. \tag{2}$$

If $f(x, y(x))$ is a function of x alone, then **Simpson's rule** of calculus states that

$$\begin{aligned}\int_{x_0}^{x_0+h} f(x)\,dx &\approx \frac{\frac{h}{2}}{3}\left[f(x_0) + 4f\left(x_0 + \frac{h}{2}\right) + f(x_0 + h)\right] \\ &\approx \frac{1}{6}\left[hf(x_0) + 4hf\left(x_0 + \frac{h}{2}\right) + hf(x_0 + h)\right].\end{aligned} \tag{3}$$

We modify this rule (a justification is provided later in this section) to adjust for the fact that f also depends on y with the following approximation:

$$\int_{x_0}^{x_0+h} f(x, y(x))\,dx \approx \frac{1}{6}(m_1 + 2m_2 + 2m_3 + m_4), \tag{4}$$

where

$$\begin{aligned} m_1 &= hf(x_0, y_0), \\ m_2 &= hf\left(x_0 + \frac{h}{2}, y_0 + \frac{m_1}{2}\right), \\ m_3 &= hf\left(x_0 + \frac{h}{2}, y_0 + \frac{m_2}{2}\right), \\ m_4 &= hf(x_0 + h, y_0 + m_3). \end{aligned} \tag{5}$$

Note that if f depends only on x, then $m_2 = m_3$ and Equation **(4)** becomes Simpson's rule. Setting $x_1 = x_0 + h, x_2 = x_0 + 2h, \ldots, x_n = x_0 + nh$, we obtain the following recursive equation, called the **Runge-Kutta method.**

Runge-Kutta method

$$y_{n+1} = y_n + \frac{1}{6}(m_1 + 2m_2 + 2m_3 + m_4), \tag{6}$$

where

$$\begin{aligned} m_1 &= hf(x_n, y_n), \\ m_2 &= hf\left(x_n + \frac{h}{2}, y_n + \frac{m_1}{2}\right), \\ m_3 &= hf\left(x_n + \frac{h}{2}, y_n + \frac{m_2}{2}\right), \\ m_4 &= hf(x_n + h, y_n + m_3). \end{aligned} \tag{7}$$

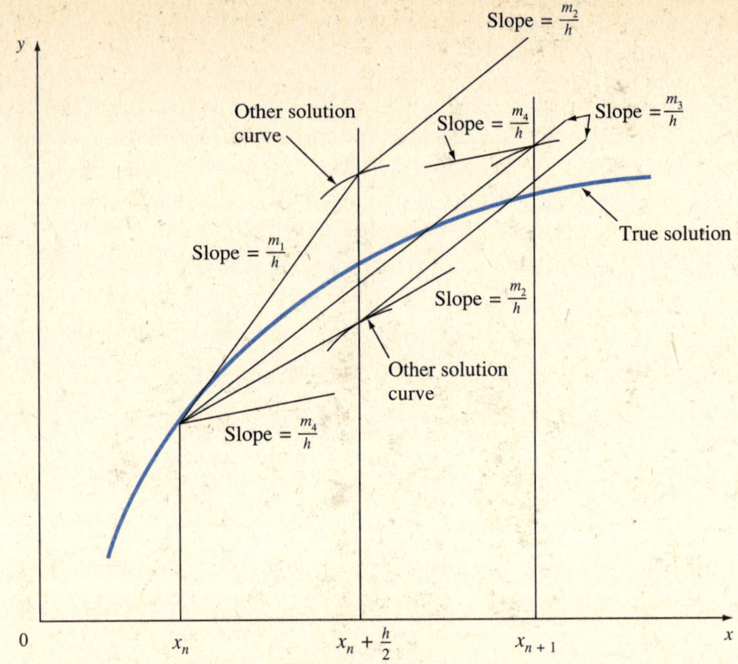

FIGURE 8.3
Runge-Kutta method: weighted average of slopes m_1, m_2, m_3, m_4

Note that the values m_1/h, m_2/h, m_3/h, and m_4/h are four slopes between $x_n \leq x \leq x_{n+1}$, so that Equation **(6)** is a weighted average of these slopes—a procedure similar to the one we used in the improved Euler method. (See Figure 8.3.)

The following two examples illustrate the use of Equations **(6)** and **(7)**.

EXAMPLE 1 ▶ **Solving an initial-value problem using the Runge-Kutta method**
Consider the initial-value problem in Example 8.2.2 (and Example 1.4.2) in which we wish to evaluate $y(1)$ given

$$\frac{dy}{dx} = y + x^2, \qquad y(0) = 1.$$

If we apply the Runge-Kutta method with $h = 1$, then $y(1) = y_1$, which is obtained by first calculating the expressions in Equation **(7)**:

$$m_1 = f(0, 1) = 1,$$
$$m_2 = f\left(\frac{1}{2}, \frac{3}{2}\right) = \frac{7}{4},$$
$$m_3 = f\left(\frac{1}{2}, \frac{15}{8}\right) = \frac{17}{8},$$
$$m_4 = f\left(1, \frac{25}{8}\right) = \frac{33}{8}.$$

Thus, using Equation **(6)**,

$$y_1 = 1 + \frac{1}{6}\left(1 + \frac{7}{2} + \frac{17}{4} + \frac{33}{8}\right) = \frac{151}{48} \approx 3.146.$$

In one step this method got us even closer to the correct value than the improved Euler method did in five steps.

If we use five steps ($h = 0.2$) to calculate $y(1)$, we can arrange our work in tabular form as shown in Table 8.9. A comparison of the Runge-Kutta method for step sizes of $h = 0.2$ and $h = 0.1$ with the Euler and improved Euler methods is given in Table 8.10. The $y(1)$ entry for both step sizes, when displayed to four significant figures, is $y(1) \approx 3.1548$, which is the exact answer to that many significant figures (see Table 8.10).

TABLE 8.9 Runge-Kutta method with $h = 0.2$, for $y' = y + x^2$, $y(0) = 1$

x_n	y_n	m_1	m_2	m_3	m_4
0.00	1.00	0.20	0.22	0.22	0.25
0.20	1.22	0.25	0.29	0.29	0.34
0.40	1.52	0.34	0.39	0.39	0.45
0.60	1.91	0.45	0.52	0.53	0.62
0.80	2.44	0.62	0.71	0.72	0.83
1.00	3.15				

Note: These values were obtained with much higher accuracy on a computer and then rounded to two decimal places.

TABLE 8.10 A comparison of results using the Euler, improved Euler, and Runge-Kutta methods with $h = 0.2$, and $h = 0.1$ for the initial-value problem $y' = y + x^2$, $y(0) = 1$

x	*Euler*	*Improved Euler*	*Runge-Kutta*	*Exact*
$h = 0.2$				
0.0	1.000000	1.00000000	1.00000000	1.00000000
0.2	1.200000	1.22400000	1.22420667	1.22420828
0.4	1.448000	1.51408000	1.51546869	1.51547409
0.6	1.769600	1.90237760	1.90634413	1.90635640
0.8	2.195520	2.42810067	2.43659938	2.43662279
1.0	2.762624	3.13908282	3.15480515	3.15484548
$h = 0.1$				
0.0	1.00000000	1.00000000	1.00000000	1.00000000
0.1	1.10000000	1.10550000	1.10551271	1.10551275
0.2	1.21100000	1.22412750	1.22420815	1.22420828
0.3	1.33610000	1.35936089	1.35957618	1.35957642
0.4	1.47871000	1.51504378	1.51547370	1.51547409
0.5	1.64258100	1.69542338	1.69616320	1.69616381
0.6	1.83183910	1.90519283	1.90635551	1.90635640
0.7	2.05102301	2.14953808	2.15125689	2.15125812
0.8	2.30512531	2.43418958	2.43662112	2.43662279
0.9	2.59963784	2.76547948	2.76880714	2.76880934
1.0	2.94060163	3.15040483	3.15484264	3.15484548

EXAMPLE 2 ▶ **Using the Runge-Kutta method on a computer**
Consider the initial-value problem

$$y' = x + y^2, \qquad y(1) = 0$$

(see Examples 8.2.1 and 8.2.3). We wish to find approximate values of the solution at $x = 1, 1.1, 1.2, 1.3, 1.4, 1.5$. Setting $h = 0.1$ and applying the Runge-Kutta method, Equations **(6)** and **(7),** we obtain the values in Table 8.11.

A comparison of the Runge-Kutta method with the Euler and improved Euler methods is made in Table 8.12. A computer program for the Runge-Kutta method is given at the end of this section.

TABLE 8.11 Runge-Kutta method with $h = 0.1$, for $y' = x + y^2$, $y(1) = 0$

x_n	y_n	m_1	m_2	m_3	m_4
1.00	0.00	0.10	0.11	0.11	0.11
1.10	0.11	0.11	0.12	0.12	0.12
1.20	0.22	0.12	0.13	0.13	0.14
1.30	0.36	0.14	0.15	0.15	0.17
1.40	0.51	0.17	0.18	0.18	0.20
1.50	0.69				

TABLE 8.12 A comparison of results using Euler, improved Euler, and Runge-Kutta methods with $h = 0.1$, and $h = 0.05$, for the initial-value problem $y' = x + y^2$, $y(1) = 0$

x	*Euler*	*Improved Euler*	*Runge-Kutta*
$h = 0.1$			
1.0	0.000000000	0.000000000	0.000000000
1.1	0.100000000	0.105500000	0.105360367
1.2	0.211000000	0.223402573	0.223135961
1.3	0.335452100	0.356966908	0.356601567
1.4	0.476704911	0.510823653	0.510424419
1.5	0.639429668	0.691781574	0.691496736
$h = 0.05$			
1.00	0.000000000	0.000000000	0.000000000
1.05	0.050000000	0.051312500	0.051293290
1.10	0.102625000	0.105398434	0.105360321
1.15	0.158151595	0.162573808	0.162517274
1.20	0.216902191	0.223209962	0.223135802
1.25	0.279254519	0.287746157	0.287655681
1.30	0.345653673	0.356705840	0.356601169
1.35	0.416627496	0.430717867	0.430602377
1.40	0.492806420	0.510544562	0.510423556
1.45	0.574949328	0.597119307	0.597001050
1.50	0.663977664	0.691597674	0.691494999

◀

In the Euler method we used one value of the derivative $y' = f(x, y)$ for each iteration. In the improved Euler method we used two values. In the Runge-Kutta formulas, we made use of four values of the derivative for each iteration. We now show how to find the four "best" values in a sense to be made precise later.

To begin, we need to recall the Taylor series expansions of a function of one variable:

$$y(x_0 + h) = y(x_0) + hy'(x_0) + \frac{h^2}{2!}y''(x_0) + \frac{h^3}{3!}y'''(x_0) + \ldots, \tag{8}$$

and of two variables:

$$\begin{aligned} f(x_0 + mh, y_0 + nh) = f(x_0, y_0) &+ h(mf_x + nf_y) \\ &+ \frac{h^2}{2!}(m^2 f_{xx} + 2mnf_{xy} + n^2 f_{yy}) \\ &+ \frac{h^3}{3!}(m^3 f_{xxx} + 3m^2 nf_{xxy} + 3mn^2 f_{xyy} + n^3 f_{yyy}) \\ &+ \ldots, \end{aligned} \tag{9}$$

where the partials are all evaluated at the point (x_0, y_0). Since

$$y' = f(x, y), \tag{10}$$

we find that

$$\begin{aligned} y'' &= f_x + (f_y)y' = f_x + ff_y, \\ y''' &= f_{xx} + 2ff_{xy} + f^2 f_{yy} + f_y(f_x + ff_y), \end{aligned}$$

and so on. (Note that $f_{xy} = f_{yx}$ when these derivatives are continuous.) Thus **(8)** can be written in the form

$$\begin{aligned} y_1 - y_0 = hf + \frac{h^2}{2}(f_x + ff_y) + \frac{h^3}{6}[f_{xx} + 2ff_{xy} + f^2 f_{yy} \\ + f_y(f_x + ff_y)] + \ldots. \end{aligned} \tag{11}$$

We need to select several points (x, y) so that the Taylor series expansions **(9)** of the corresponding $f(x, y)$ terms coincide with the terms on the right-hand side of **(11)**. Suppose we let

$$\begin{aligned} m_1 &= hf(x_0, y_0), \\ m_2 &= hf(x_0 + nh, y_0 + nm_1), \\ m_3 &= hf(x_0 + ph, y_0 + pm_2), \\ m_4 &= hf(x_0 + qh, y_0 + qm_3). \end{aligned}$$

(The reason for doing this is made clear shortly.) Using **(9)**, we may write these values as

$$\begin{aligned} m_1 &= hf, \\ m_2 &= h\left[f + nh(f_x + ff_y) + \frac{(nh)^2}{2}(f_{xx} + 2ff_{xy} + f^2 f_{yy}) + \ldots\right], \\ m_3 &= h\Big\{f + ph(f_x + ff_y) \\ &\quad + \frac{h^2}{2}[p^2(f_{xx} + 2ff_{xy} + f^2 f_{yy}) + 2npf_y(f_x + ff_y)] + \ldots\Big\}, \end{aligned}$$

$$m_4 = h\Big\{f + qh(f_x + ff_y) + \frac{h^2}{2}[q^2(f_{xx} + 2ff_{xy} + f^2f_{yy}) + 2\,pqf_y(f_x + ff_y)] + \ldots\Big\},$$

where all functions are evaluated at the point (x_0, y_0).

We now consider an expression of the form

$$am_1 + bm_2 + cm_3 + dm_4$$

and try to equate it to the right-hand side of **(11).** This has the effect of giving us a numerical scheme that agrees with the solution to **(10)** up to and including third-order terms. Then the error is no greater than terms like kh^4, and so on. Matching like expressions, we find that

$$\begin{aligned}
&\textbf{coefficient of } hf && a + b + c + d = 1,\\
&\textbf{coefficient of } h^2(f_x + ff_y) && bn + cp + dq = \frac{1}{2},\\
&\textbf{coefficient of } \tfrac{h^3}{2}(f_{xx} + 2ff_{xy} + f^2f_{yy}) && bn^2 + cp^2 + dq^2 = \frac{1}{3},\\
&\textbf{coefficient of } h^3[f_y(f_x + ff_y)] && cn + dpq = \frac{1}{6}.
\end{aligned} \tag{12}$$

Any solution of these equations produces a method in which there is no error up to the third-order terms. Suppose we take $n = p = \frac{1}{2}$ and $q = 1$. Then **(12)** reduces to the system of equations

$$\begin{aligned}
a + b + c + d &= 1,\\
b + c + 2d &= 1,\\
3b + 3c + 12d &= 4,\\
3c + 6d &= 2,
\end{aligned}$$

which has the solution $a = d = \frac{1}{6}$, $b = c = \frac{1}{3}$. Thus the Runge-Kutta formula [Equations **(6)** and **(7)**] agrees with $y_1 - y_0$ for all terms up to and including the terms in h^3. Actually, with quite a bit more work, one can show that they agree in the h^4 terms, too. Thus the error (if any) involves only terms in h^5 and higher. Hence, for small h, we should expect to get very good results.

Other formulas are also readily derivable. Suppose we choose $n = \frac{1}{3}$, $p = \frac{2}{3}$, $q = 1$. *Then* **(12)** yields

$$\begin{aligned}
a + b + c + d &= 1,\\
2b + 4c + 6d &= 3,\\
b + 4c + 9d &= 3,\\
4c + 12d &= 3,
\end{aligned}$$

which has a solution $a = d = \frac{1}{8}$, $b = c = \frac{3}{8}$. The formula, known as the **Kutta-Simpson $\frac{3}{8}$-rule,** may be written

$$y_1 = y_0 + \frac{1}{8}(m_1 + 3m_2 + 3m_3 + m_4),$$

where

$$
\begin{aligned}
m_1 &= hf(x_0, y_0),\\
m_2 &= hf\left(x_0 + \frac{h}{3}, y_0 + \frac{m_1}{3}\right),\\
m_3 &= hf\left(x_0 + \frac{2h}{3}, y_0 + \frac{2m_2}{3}\right),\\
m_4 &= hf(x_0 + h, y_0 + m_3).
\end{aligned}
\tag{13}
$$

Similarly, the choice $n = \frac{1}{3}$, $p = \frac{2}{3}$, $q = 1$ also yields a solution $a = c = 0$, $b = \frac{3}{4}$, $d = \frac{1}{4}$; hence

$$y_1 = y_0 + \frac{1}{4}(3m_2 + m_4),$$

where m_2 and m_4 are defined as in **(13).** Since the number of possible choices of n, p, and q is infinite, the reader may be amused by solving **(12)** for a, b, c, and d with whatever values of n, p, and q that he or she selects.

*TRUE BASIC**

```
LET x=0
LET y=1
LET xlast=1
LET h=.2
PRINT " x "," y "
DO
   CALL runkt(x,y,h)
   PRINT x,y
LOOP until x>=xlast
END

SUB runkt (x,y,h)
    LET yy=y
    LET sy=y
    LET xx=x
    FOR k=1 to 4
        CALL difeq(x,y,dydx)
        LET sy=sy+h*dydx/6
        IF (k=2) or (k=3) then
           LET sy=sy+h*dydx/6
        END IF
        LET x=xx+h/2
        LET y=yy+h*dydx/2
        IF k>=3 then
           LET x=x+h/2
           LET y=y+h*dydx/2
        END IF
    NEXT k
    LET y=sy
END SUB

SUB difeq(x,y,dydx)
    LET dydx=y+x*x
END SUB
```

*FORTRAN**

```
      X=0.0
      Y=1.0
      XLAST=1.0
      H=0.2
      N=(XLAST-X)/H
      WRITE (*,'(/7X,A,12X,A)')'X','Y'
      DO 11 I=0,N
         CALL RUNKT (X,Y,H)
         WRITE (*,'(1X,F10.4,1X,F14.6)') X,Y
11    CONTINUE
      END
C
      SUBROUTINE RUNKT (X,Y,H)
      YY=Y
      SY=Y
      XX=X
      DO 100 K=1,4
         CALL DIFEQ(X,Y,DYDX)
         SY=SY+H*DYDX/6.0
         IF ((K.EQ.2).OR.(K.EQ.3)) THEN
            SY=SY+H*DYDX/6.0
         ENDIF
         X=XX+H/2.0
         Y=YY+H*DYDX/2.0
         IF (K.GE.3) THEN
            X=X+H/2.0
            Y=Y+H*DYDX/2.0
         ENDIF
100   CONTINUE
      Y=SY
      END
C
      SUBROUTINE DIFEQ (X,Y,DYDX)
      DYDX=Y+X*X
      END
```

*See the footnote on page 371.

SELF-QUIZ

Answer true or false.

I. All Runge-Kutta methods approximate the next value of y by averaging several slopes.

II. The Kutta-Simpson $\frac{3}{8}$-rule agrees with $y_1 - y_0$ for all terms up to to h^3.

III. The Runge-Kutta formula $y_1 = y_0 + \frac{1}{20}\{m_1 + 12m_2 + 3m_3 + 4m_4\}$ satisfies Equation **(12)** with $n = p = \frac{1}{2}$ and $q = 1$.

Answers to Self-Quiz

I. True **II.** True **III.** False ($n = \frac{1}{3}$, and $p = \frac{2}{3}$)

PROBLEMS 8.3

In Problems 1 through 10 solve each problem exactly using the methods of Chapter 2. Then use the Runge-Kutta method with the given h. Compare the accuracy of this method with the exact answer.

1. $\frac{dy}{dx} = x + y$, $y(0) = 1$. Find $y(1)$ with $h = 0.2$

2. $\frac{dy}{dx} = x - y$, $y(1) = 2$. Find $y(3)$ with $h = 0.4$.

3. $\frac{dy}{dx} = \frac{x - y}{x + y}$, $y(2) = 1$. Find $y(1)$ with $h = -0.2$.

4. $\frac{dy}{dx} = \frac{y}{x} + \left(\frac{y}{x}\right)^2$, $y(1) = 1$. Find $y(2)$ with $h = 0.2$.

5. $\frac{dy}{dx} = x\sqrt{1 + y^2}$, $y(1) = 0$. Find $y(3)$ with $h = 0.4$.

6. $\frac{dy}{dx} = x\sqrt{1 - y^2}$, $y(1) = 0$. Find $y(2)$ with $h = 0.125$.

7. $\frac{dy}{dx} = \frac{y}{x} - \frac{5}{2}x^2y^3$, $y(1) = 1/\sqrt{2}$. Find $y(2)$ with $h = 0.125$.

8. $\frac{dy}{dx} = \frac{-y}{x} + x^2y^2$, $y(1) = \frac{2}{9}$. Find $y(3)$ with $h = \frac{1}{3}$.

9. $\frac{dy}{dx} = ye^x$, $y(0) = 2$. Find $y(2)$ with $h = 0.2$.

10. $\frac{dy}{dx} = xe^y$, $y(0) = 0$. Find $y(1)$ with $h = 0.1$.

In Problems 11 through 20 use the Runge-Kutta method to graph approximately the solution of the given initial-value problem by plotting the points (x_k, y_k) over the indicated range, where $x_k = x_0 + kh$.

11. $y' = xy^2 + y^3$, $y(0) = 1$, $h = 0.02$, $0 \le x \le 0.1$

12. $y' = x + \sin(\pi y)$, $y(1) = 0$, $h = 0.2$, $1 \le x \le 2$

13. $y' = x + \cos(\pi y)$, $y(0) = 0$, $h = 0.4$, $0 \le x \le 2$

14. $y' = \cos(xy)$, $y(0) = 0$, $h = \frac{\pi}{4}$, $0 \le x \le \pi$

15. $y' = \sin(xy)$, $y(0) = 1$, $h = \frac{\pi}{4}$, $0 \le x \le 2\pi$

16. $y' = \sqrt{x^2 + y^2}$, $y(0) = 1$, $h = 0.5$, $0 \le x \le 5$

17. $y' = \sqrt{y^2 - x^2}$, $y(0) = 1$, $h = 0.1$, $0 \le x \le 1$

18. $y' = \sqrt{x + y^2}$, $y(0) = 1$, $h = 0.2$, $0 \le x \le 1$

19. $y' = \sqrt{x + y^2}$; $y(1) = 2$, $h = -0.2$, $0 \le x \le 1$

20. $y' = \sqrt{x^2 + y^2}$, $y(1) = 5$, $h = -0.2$, $0 \le x \le 1$

21. Let

$$\frac{dy}{dx} = \frac{x - y}{x + y}, \qquad y(2) = 0.$$

Use the Runge-Kutta method to approximate $y(1)$ with $h = -0.2$. Compare your answer with the exact value and explain why the method fails.

22. Use the Runge-Kutta method to approximate $y(3)$ with $h = 0.4$ for the initial-value problem

$$\frac{dy}{dx} + \frac{3}{x} = x^2y^2, \qquad y(1) = 2.$$

Compare your answer to the exact answer and explain why the numerical technique fails.

23. Repeat Problem 22 for the initial-value problem

$$\frac{dy}{dx} + \frac{y}{x} = x^3y^3, \qquad y(1) = 1$$

24. Let $y' = e^{xy}$, $y(0) = 1$, and obtain a value for $y(4)$ with $h = 0.5$ using the Runge-Kutta method. What difficulties are encountered? How much confidence do you have in your answer?

25. Set $d = 0$ in Equation **(12)** and let $n = \frac{1}{3}, p = \frac{2}{3}$. Then find the coefficients a, b, and c for these choices. The resulting formula is called **Heun's formula.**

26. Set $a = d = 0, b = \frac{3}{4}, c = \frac{1}{4}$ in Equation **(12)** and find n and p.

27. Set $a = \frac{2}{8}, b = c = \frac{3}{8}, d = 0$ in Equation **(12)** and find n and p.

28. Prove that the Runge-Kutta formula agrees with equation **(11)** up to and including the h^4 terms.

29. Explain why the choice $n = \frac{1}{3}, p = \frac{2}{3}, q = 1$ in Equation **(12)** yielded two sets of solutions a, b, c, and d. Are other solutions possible? Does the choice $n = p = \frac{1}{2}, q = 1$ allow for multiple solutions?

8.4 PREDICTOR-CORRECTOR FORMULAS

The simplest type of predictor-corrector formula is one that we have already used, namely, the improved Euler method. Recall that with this method we first obtain a value for y_n by the Euler method (the *predicting* part of the process) and then improve on the accuracy of the method by applying a trapezoidal rule (the *correcting* phase of the process). In this section we indicate how such methods are derived and provide several procedures that perform their tasks with a high degree of accuracy.

We begin by developing a procedure for obtaining **quadrature formulas.** These formulas are designed to obtain the approximate value of a definite integral by using equally spaced values of the integrand. Two well-known quadrature formulas are the trapezoidal rule and Simpson's rule,* which can be written in the forms

$$\textbf{Trapezodial rule: } y_1 - y_0 = \frac{h}{2}(y_0' + y_1'), \tag{1}$$

$$\textbf{Simpson's rule: } y_2 - y_0 = \frac{h}{3}(y_0' + 4y_1' + y_2'), \tag{2}$$

respectively, where h is the step size, $x_n = x_0 + nh$, $y_n = y(x_n)$, and $y_n' = f(x_n, y_n)$. Note that, unlike the previously discussed methods, here we have slightly shifted the point of view by representing the right-hand side as a sum of derivatives instead of values of the function f.

In Equation **(1)** we are using only the points y_0' and y_1', whereas in Equation **(2)** we also use y_2'. For this reason, Equation **(1)** is called a **2-point quadrature formula,** whereas Equation **(2)** is a **3-point quadrature formula.** In general, the more points involved, the higher the accuracy. In the present case, Equation **(1)** provides the exact integral only for straight lines, whereas Equation **(2)** is also exact for quadratic polynomials (see Problems 15 and 16).

We now develop a method for obtaining 3-point quadrature formulas. The method is easily adapted to n-point quadrature formulas.

3-point quadrature formulas

Assume that the function $y(x)$ is a quadratic polynomial

$$y(x) = a_0 + a_1x + a_2x^2. \tag{3}$$

Let h and x_0 be fixed real numbers and define $x_k = x_0 + kh$, for all integers k. Let $y_k = y(x_k)$. We wish to find coefficients A_0, A_1, A_2 such that

$$y_j - y_i = h(A_0y_0' + A_1y_1' + A_2y_2'), \tag{4}$$

* These rules can be found in most calculus books.

where $i, j = 0, 1, 2$ and $i \neq j$. This leads to an integration scheme that is exact for quadratic polynomials. For example, if $i = 0$, then the left-hand side of Equation **(4)** becomes

$$\begin{aligned} y_j - y_0 &= a_1(x_j - x_0) + a_2(x_j^2 - x_0^2) \\ &= a_1 jh + a_2[2x_0 jh + (jh)^2] = j(a_1 h + 2a_2 x_0 h) + j^2(a_2 h^2). \end{aligned} \quad \textbf{(5)}$$

On the other hand, using the facts that $y' = a_1 + 2a_2 x$ and $x_k = x_0 + kh$, we find that the right-hand side of Equation **(4)** becomes

$$\begin{aligned} &h[A_0 y_0' + A_1 y_1' + A_2 y_2'] \\ &\quad = h[A_0(a_1 + 2a_2 x_0) + A_1(a_1 + 2a_2 x_1) + A_2(a_1 + 2a_2 x_2)] \\ &\quad = (A_0 + A_1 + A_2)(a_1 h + 2a_2 x_0 h) + (2A_1 + 4A_2)(a_2 h^2). \end{aligned} \quad \textbf{(6)}$$

Equating Equations **(5)** and **(6)**, we obtain the simultaneous equations

$$\begin{aligned} A_0 + A_1 + A_2 &= j, \\ 2A_1 + 4A_2 &= j^2. \end{aligned} \quad \textbf{(7)}$$

Since this system is underdetermined, it has an infinite number of solutions A_0, A_1, A_2, each leading to a different quadrature formula. For example, if $j = 1$, we can pick $A_0 = \frac{5}{6}$, $A_1 = -\frac{1}{6}$, $A_2 = \frac{1}{3}$ and arrive at the formula

$$y_1 - y_0 = \frac{h}{6}(5y_0' - y_1' + 2y_2'). \quad \textbf{(8)}$$

If $j = 2$, then selecting $A_0 = \frac{1}{3}$, $A_1 = \frac{4}{3}$, $A_2 = \frac{1}{3}$ yields Simpson's rule **(2)**. In addition, quadrature formulas may be added or subtracted to yield new formulas. For example, subtracting **(8)** from **(2)** yields the **Adams-Bashforth formula,**

$$y_2 - y_1 = \frac{h}{2}(-y_0' + 3y_1'). \quad \textbf{(9)}$$

Predictor-corrector methods

Let us now illustrate how predictor-corrector formulas are used. Suppose we are given the initial-value problem

$$\frac{dy}{dx} = f(x, y), \qquad y(x_0) = y_0. \quad \textbf{(10)}$$

The value y_0 is given, y_0' can be obtained by evaluating Equation **(10)** at $x = x_0$, $y = y_0$. We next perform an improved Euler method computation to obtain y_1 [and y_1' by means of Equation **(10)**]. At this point it is often desirable to apply repeatedly the trapezoidal rule to the process until the value of y_1 stabilizes (that is, remains unchanged to a given number of decimal places; see Example 1). Then since y_0, y_0', y_1, and y_1' are all known, we may use **(9)** to *predict* the value of y_2. This value is then used in Equation **(10)** to obtain y_2', and the trapezoidal rule, Equation **(1)**, is used to *correct* the y_2 value previously determined. The process is now repeated to predict and correct y_3 and y_3' in terms of the known values y_1, y_1', y_2, y_2'. We use the Adams-Bashforth formula

$$y_{n+1} = y_n + \frac{h}{2}(-y_{n-1}' + 3y_n') \quad \textbf{(11)}$$

to *predict* the value of y_{n+1}, Equation **(10)** to obtain y'_{n+1}, and the trapezoidal rule

$$y_{n+1} = y_n + \frac{h}{2}(y'_{n+1} + y'_n) \tag{12}$$

to *correct* the value of y_{n+1}, and Equation **(10)** again to obtain y'_{n+1}.

EXAMPLE 1 ▶ Using a predictor-corrector formula to solve an initial-value problem numerically

Let $dy/dx = y + x^2$, $y(0) = 1$, and suppose we wish to find $y(1)$, with $h = 0.2$, by using the predictor-corrector formulas **(11)** and **(12)**.

The predictor formula **(11)** requires that we know the values of y'_0, y_1, and y'_1 before it can be used to generate y_2. At this point, the only one of these three values that we know is

$$y'_0 = f(x_0, y_0) = y_0 + x_0^2 = 1,$$

since $x_0 = 0$, $y_0 = y(x_0) = y(0) = 1$. Therefore, before we can begin using predictor formula **(11)** we must determine y_1. [Once we know y_1 we can compute $y'_1 = f(x_1, y_1)$.] The usual method for obtaining y_1 is to use the improved Euler method *repeatedly*. We illustrate this procedure below:

x_0	y_0	y'_0	$\bar{y}_1$	$\bar{y}'_1$	y_1
0.0	1.0	1.0	1.20	1.24	1.22
			1.22	1.26	1.23
			1.23	1.27	1.23

The first entry in the $\bar{y}_1$ column is obtained by the Euler formula: $y_1 = y_0 + hy'_0$. The entries in the $\bar{y}'_1$ column are obtained from the differential equation $\bar{y}'_1 = f(x_1, \bar{y}_1) = \bar{y}_1 + x_1^2$. The last column is determined by the improved Euler formula

$$y_1 = y_0 + \frac{h}{2}(y'_0 + \bar{y}'_1),$$

and the resulting value for y_1 is transferred to the next row in the $\bar{y}_1$ column. We then repeat the steps in the $\bar{y}'_1$ and y_1 columns. We see that the process stabilizes with $y_1 = 1.23$ and $y'_1 = 1.27$ (provided that we are using two-decimal-place accuracy).

We now apply Equations **(11)** and **(12)** where we have arranged the calculations in Table 8.13. The result is better than the answer obtained by the improved Euler method (with slightly more work). A comparison with the other methods we have discussed in this chapter for this initial-value problem is given in Table 8.14. A similar comparison for Example 8.3.2 is given in Table 8.15. ◀

TABLE 8.13 Predictor-corrector method with $h = 0.2$, for $y' = y + x^2$, $y(0) = 1$

			Predictor		*Corrector*
x_n	y_n	$y'_n = f(x_n, y_n)$	$y_{n+1} = y_n + \frac{h}{2}(-y'_{n-1} + 3y'_n)$	y'_{n+1}	$y_{n+1} = y_n + \frac{h}{2}(y'_{n+1} + y'_n)$
0.0	1.00	1.00			
0.2	1.23	1.27	$1.23 + (0.1)(-1.00 + 3.81) = 1.51$	1.67	$1.23 + (0.1)(1.67 + 1.27) = 1.52$
0.4	1.52	1.68	$1.52 + (0.1)(-1.27 + 5.04) = 1.90$	2.26	$1.52 + (0.1)(2.26 + 1.68) = 1.91$
0.6	1.91	2.27	$1.91 + (0.1)(-1.68 + 6.81) = 2.42$	3.06	$1.91 + (0.1)(3.06 + 2.27) = 2.44$
0.8	2.44	3.08	$2.44 + (0.1)(-2.27 + 9.24) = 3.14$	4.14	$2.44 + (0.1)(4.14 + 3.08) = 3.16$
1.0	3.16				

TABLE 8.14 A comparison of results using the Euler, improved Euler, Runge-Kutta, and predictor-corrector methods with $h = 0.2$ and $h = 0.1$, for the initial-value problem $y' = y + x^2$, $y(0) = 1$

x	*Euler*	*Improved Euler*	*Runge-Kutta*	*Predictor-Corrector*	*Exact*
$h = 0.2$					
0.0	1.000000	1.00000000	1.00000000	1.00000000	1.00000000
0.2	1.200000	1.22400000	1.22420667	1.22666667	1.22420828
0.4	1.448000	1.51408000	1.51546869	1.52000000	1.51547409
0.6	1.769600	1.90237760	1.90634413	1.91373333	1.90635640
0.8	2.195520	2.42810067	2.43659938	2.44789200	2.43662279
1.0	2.762624	3.13908282	3.15480515	3.17136982	3.15484548
$h = 0.1$					
0.0	1.00000000	1.00000000	1.00000000	1.00000000	1.00000000
0.1	1.10000000	1.10550000	1.10551271	1.10578950	1.10551275
0.2	1.21100000	1.22412750	1.22420815	1.22473687	1.22420828
0.3	1.33610000	1.35936089	1.35957618	1.36040661	1.35957642
0.4	1.47871000	1.51504378	1.51547370	1.51666348	1.51547409
0.5	1.64258100	1.69542338	1.69616320	1.69777879	1.69616381
0.6	1.83183910	1.90519283	1.90635551	1.90847335	1.90635640
0.7	2.05102301	2.14953808	2.15125689	2.15396479	2.15125812
0.8	2.30512531	2.43418958	2.43662112	2.44001982	2.43662279
0.9	2.59963784	2.76547948	2.76880714	2.77301204	2.76880934
1.0	2.94060163	3.15040483	3.15484264	3.15998578	3.15484548

TABLE 8.15 A comparison of results using the Euler, improved Euler, Runge-Kutta, and predictor-corrector methods with $h = 0.1$ and $h = 0.05$, for the initial-value problem $y' = x + y^2$, $y(1) = 0$

x	*Euler*	*Improved Euler*	*Runge-Kutta*	*Predictor-Corrector*
$h = 0.1$				
1.0	0.000000000	0.000000000	0.000000000	0.000000000
1.1	1.100000000	0.105500000	0.105360367	0.100505000
1.2	0.211000000	0.223402573	0.223135961	0.218364951
1.3	0.335452100	0.356966908	0.356601567	0.351874547
1.4	0.476704911	0.510823653	0.510424419	0.505718937
1.5	0.639429668	0.691781574	0.691496736	0.686823519
$h = 0.05$				
1.00	0.000000000	0.000000000	0.000000000	0.000000000
1.05	0.050000000	0.051312500	0.051293290	0.050062600
1.10	0.102625000	0.105398434	0.105360321	0.104145660
1.15	1.158151595	0.162573808	0.162517274	0.161316022
1.20	0.216902191	0.223209962	0.223135802	0.221945882
1.25	0.279254519	0.287746157	0.287655681	0.286475712
1.30	0.345653673	0.356705840	0.356601169	0.355430697
1.35	0.416627496	0.430717867	0.430602377	0.429442224
1.40	0.492806420	0.510544562	0.510423556	0.509276341
1.45	0.574949328	0.597119307	0.597001050	0.595871978
1.50	0.663977664	0.691597674	0.691494999	0.690393049

A very accurate predictor-corrector method due to Milne uses a 4-point quadrature formula as a predictor and Simpson's rule as a corrector:

$$\text{Predictor: } y_{n+4} - y_n = \frac{4h}{3}(2y'_{n+1} - y'_{n+2} + 2y'_{n+3}),$$

$$\text{Corrector: } y_{n+4} - y_{n+2} = \frac{h}{3}(y'_{n+2} + 4y'_{n+3} + y'_{n+4}). \tag{13}$$

To apply this method, it is necessary to have good values for y_0, y'_0, y_1, y'_1, y_2, y'_2, y_3, and y'_3.

Predictor-corrector methods have many advantages from the point of view of accuracy and the amount of work involved. The extra step involved in "correcting" dramatically improves the accuracy, without requiring an inordinate amount of extra work.

SELF-QUIZ

Answer true or false.

I. A 4-point quadrature formula would be based on fitting a cubic polynomial through four points.

II. The trapezoidal rule in Equation **(1)**, $y_1 - y_0 = \frac{h}{2}(y'_0 + y'_1)$, is just another name for the improved Euler method.

Answers to Self-Quiz

I. True **II.** False

PROBLEMS 8.4

In Problems 1 through 10 solve each problem exactly using the methods of Chapter 2. Then

(a) *use the predictor-corrector method of Example 1 [formulas* **(11)** *and* **(12)***] and the indicated value of h to find an approximate solution to the given value of y;*

(b) *use the predictor-corrector method in Equation* **(13)** *with the given h to find the indicated value of y. Use the values generated in part (a) to initialize Milne's method.*

Compare the accuracy of these methods with the exact answer.

1. $y' = x + y$, $y(0) = 1$. Find $y(1)$ with $h = 0.2$.

2. $y' = x - y$, $y(1) = 2$. Find $y(3)$ with $h = 0.4$.

3. $y' = \dfrac{x - y}{x + y}$, $y(2) = 1$. Find $y(1)$ with $h = -0.2$.

4. $y' = \left(\dfrac{y}{x}\right) + \left(\dfrac{y}{x}\right)^2$, $y(1) = 1$. Find $y(2)$ with $h = 0.2$.

5. $y' = x\sqrt{1 + y^2}$, $y(1) = 0$. Find $y(3)$ with $h = 0.4$.

6. $y' = x\sqrt{1 - y^2}$, $y(1) = 0$. Find $y(2)$ with $h = \dfrac{1}{8}$.

7. $y' = \left(\dfrac{y}{x}\right) - \left(\dfrac{5x^2y^3}{2}\right)$, $y(1) = 1/\sqrt{2}$. Find $y(2)$ with $h = \dfrac{1}{8}$.

8. $y' = \left(-\dfrac{y}{x}\right) + x^2y^2$, $y(1) = \dfrac{2}{9}$. Find $y(3)$ with $h = \dfrac{1}{3}$.

9. $y' = ye^x$, $y(0) = 2$. Find $y(2)$ with $h = 0.2$.

10. $y' = xe^y$, $y(0) = 0$. Find $y(1)$ with $h = 0.1$.

11. Obtain the trapezoidal rule **(1)** by using the 3-point quadrature formulas **(4)** and **(7)**.

12. Obtain these 3-point quadrature formulas:

(a) $y_1 - y_0 = \dfrac{h}{12}(5y'_0 + 8y'_1 - y'_2)$

(b) $y_2 - y_0 = \dfrac{h}{8}(y'_0 + 14y'_1 + y'_2)$

(c) $y_2 - y_0 = \dfrac{h}{4}(y'_0 + 6y'_1 + y'_2)$

13. Obtain the 4-point quadrature formula

$$y_3 - y_0 = \frac{h}{2}(-y_0' + 5y_1' + 2y_2').$$

14. Use the formula obtained in Problem 13 and Simpson's rule as a predictor-corrector to solve the intial-value problem

$$y' = x + y, y(0) = 1,$$

for $y(1)$ with $h = 0.2$. Use the improved Euler method to find y_1 and y_2.

15. Let $y(x) = ax + b$. Show that the trapezoidal rule **(1)** provides the exact value for $\int_0^1 y(x)\,dx$ for step sizes $h = \frac{1}{2}$ and $\frac{1}{5}$.

16. Let $y(x) = ax^2 + bx + c$. Show that Simpson's rule **(2)** provides the exact value for $\int_0^1 y(x)\,dx$ for step sizes $h = \frac{1}{4}$ and $\frac{1}{8}$.

17. Show that the equations for 4-point quadrature formulas analogous to **(7)** are

$$A_0 + A_1 + A_2 + A_3 = j,$$
$$2A_1 + 4A_2 + 6A_3 = j^2,$$
$$3A_1 + 12A_2 + 27A_3 = j^3.$$

Use these equations to derive Milne's Equation **(13)**.

18. Obtain the underdetermined system of equations for 5-point quadrature formulas analogous to those in Problem 17.

19. Use the three-term Taylor series method (see Problem 8.2.25) as a predictor and Simpson's rule as a corrector to find an approximate solution of $y(1)$ for the initial-value problem

$$y' = x + y, \quad y(0) = 1,$$

with $h = 0.2$. [Compare this result with Problems 1 and 14.]

8.5 ERROR ANALYSIS FOR THE EULER METHOD (OPTIONAL)

In this section we discuss only the discretization errors encountered in the use of the Euler method. Round-off errors depend not only on the method and the number of steps in the calculation but also on the type of instrument (hand-calculator, computer, pencil and paper) used for computing the answer. Round-off error is not discussed in this section (it will be discussed in Section 8.6), although it should never be ignored.

Let us again consider the first-order initial-value problem

$$y' = f(x, y), \qquad y(x_0) = y_0, \tag{1}$$

and use the iteration scheme

$$y_{n+1} = y_n + hf(x_n, y_n), \tag{2}$$

where h is a fixed step size.

We assume for the remainder of this section that $f(x, y)$ possesses continuous first partial derivatives. Then, on any finite interval, $\frac{\partial f(x,y)}{\partial y}$ is bounded by some constant that we denote by L (a continuous function is always bounded on a closed, bounded interval). Since $y'(x) = f(x, y)$, we obtain, by the chain rule,

$$y''(x) = \frac{\partial f}{\partial x}(x, y) + \frac{\partial f}{\partial y}(x, y)y'(x),$$

which must be continuous since it is the sum of continuous functions. Hence $y''(x)$ must be bounded on the interval $x_0 \le x \le a$. So we assume that $|y''(x)| < M$ for some positive constant M.

We now wish to estimate the error e_n at the nth step of the iteration defined by Equation **(2)**. Since $y(x_n)$ is the exact value of the solution $y(x)$ at the point $x_n = x_0 + nh$, and y_n is the approximate value at that point, the error at the nth step is given by

$$e_n = y_n - y(x_n). \tag{3}$$

Note that $y_0 = y(x_0)$, so $e_0 = 0$.

Now $y(x_{n+1}) = y(x_n + h)$ and $y''(x)$ is continuous. So we may use Taylor's theorem with remainder to obtain

$$y(x_{n+1}) = y(x_n + h) = y(x_n) + hy'(x_n) + \frac{h^2}{2}y''(\xi_n), \tag{4}$$

where $x_n \le \xi_n \le x_{n+1}$. We may now state the main result of this section.

THEOREM 1 ♦ **Convergence of the discretization error for Euler's method**

Let $f(x, y)$ have continuous first partial derivatives and let y_n be the approximate solution of Equation **(1)** generated by the Euler method **(2).** Suppose that $y(x)$ is defined and the inequalities

$$\left|\frac{\partial f}{\partial y}(x, y)\right| < L \quad \text{and} \quad |y''(x)| < M$$

hold on the bounded interval $x_0 \le x \le a$. Then the error $e_n = y_n - y(x_n)$ satisfies the inequality

$$|e_n| \le \frac{hM}{2L}(e^{(x_n - x_0)L} - 1) = \frac{hM}{2L}(e^{nhL} - 1). \tag{5}$$

In particular, since $x_n - x_0 \le a - x_0$ (which is finite), $|e_n|$ tends to zero as h tends to zero.

Proof A subtraction of Equation **(4)** from Equation **(2)** yields

$$y_{n+1} - y(x_{n+1}) = y_n - y(x_n) + h[f(x_n, y_n) - y'(x_n)] - \frac{h^2}{2}y''(\xi_n),$$

or

$$e_{n+1} = e_n + h[f(x_n, y_n) - f(x_n, y(x_n))] - \frac{h^2}{2}y''(\xi_n). \tag{6}$$

By the mean value theorem of differential calculus,

$$\begin{aligned} f(x_n, y_n) - f(x_n, y(x_n)) &= \frac{\partial f}{\partial y}(x_n, \hat{y}_n)[y_n - y(x_n)] \\ &= \frac{\partial f}{\partial y}(x_n, \hat{y}_n)e_n, \end{aligned} \tag{7}$$

where $\hat{y}_n$ is between y_n and $y(x_n)$. We substitute **(7)** into **(6)** to obtain

$$e_{n+1} = e_n + h\frac{\partial f}{\partial y}(x_n, \hat{y}_n)e_n - \frac{h^2}{2}y''(\xi_n). \tag{8}$$

But $|\frac{\partial f}{\partial y}| \le L$ and $|y''| \le M$, so taking the absolute value of both sides of Equation **(8)** and using the triangle inequality, we obtain

$$|e_{n+1}| \le |e_n| + hL|e_n| + \frac{h^2}{2}M = (1 + hL)|e_n| + \frac{h^2}{2}M. \tag{9}$$

We now consider the difference equation (see Section 2.9)

$$r_{n+1} = (1 + hL)r_n + \frac{h^2}{2}M, \qquad r_0 = 0, \tag{10}$$

and claim that if r_n is the solution to Equation **(10),** then $|e_n| \le r_n$. We show this by induction. It is true for $n = 0$, since $e_0 = r_0 = 0$. We assume it is true for $n = k$

and prove it for $n = k + 1$; that is, we assume that $|e_m| \le r_m$, for $m = 0, 1, \ldots, k$. Then

$$r_{k+1} = (1 + hL)r_k + \frac{h^2}{2}M \ge (1 + hL)|e_k| + \frac{h^2}{2}M \ge |e_{k+1}|,$$

and the claim is proved [the last step follows from the inequality **(9)**]. We can solve Equation **(10)** by proceeding inductively: $r_1 = h^2M/2 = [(1 + hL) - 1]\frac{hM}{2L}$, $r_2 = [(1 + hL) + 1]h^2\frac{M}{2} = [(1 + hL)^2 - 1]\frac{hM}{2L}$. We find that

$$r_n = \frac{hM}{2L}(1 + hL)^n - \frac{hM}{2L}.$$

Now $e^{hL} = 1 + hL + h^2L^2/2! + \ldots$, so

$$1 + hL \le e^{hL} \qquad \text{and} \qquad (1 + hL)^n \le (e^{hL})^n = e^{nhL}.$$

Thus

$$|e_n| \le r_n \le \frac{hM}{2L}e^{nhL} - \frac{hM}{2L} = \frac{hM}{2L}(e^{nhL} - 1). \tag{11}$$

But $x_n = x_0 + nh$, so $x_n - x_0 = nh$ and **(11)** becomes

$$|e_n| \le \frac{hM}{2L}(e^{(x_n - x_0)L} - 1).$$

Thus the theorem is proved. ♦

Theorem 1 not only shows that the errors get small as h tends to zero; it also tells us *how fast* the errors decrease. If we define the constant k by

$$k = \frac{M}{2L}|e^{(a - x_0)L} - 1|, \tag{12}$$

we have

$$|e_n| \le kh. \tag{13}$$

Thus the error is bounded by a *linear* function of h. (Note that $|e_n|$ is bounded by a term that depends only on h, not on n.) Roughly speaking, this implies that the error decreases at a rate proportional to the decrease in the step size. If, for example, we halve the step size, then we can expect at least to halve the error. Actually, since the estimates used in arriving at **(13)** were very crude, we can often do better, as in Example 1.4.2, where we halved the step size and decreased the error by a factor of four. Nevertheless, it is useful to have an upper bound for the error. It should be noted, however, that this bound may be difficult to obtain, since it is frequently difficult to find a bound for $y''(x)$.

EXAMPLE 1 ▶ **Finding the discretization error**

Consider the equation $y' = y$, $y(0) = 1$. We have

$$f(x, y) = y \quad \text{and} \quad \left|\frac{\partial f}{\partial y}(x, y)\right| = 1 = L.$$

Since the solution of the problem is $y(x) = e^x$, we have $|y''| \le e^1 = M$ on the interval $0 \le x \le 1$. Then Equation **(12)** becomes

$$k = \frac{e}{2}|e - 1| = \frac{e^2 - e}{2} \approx 2.34,$$

so that the error is proportional to the step size:

$$|e_n| \le 2.34h.$$

Therefore, using a step size of, say, $h = 0.1$, we can expect to have an error at each step of less than 0.234 (see Table 8.16). We note that the greatest actual error is about half of the maximum possible error according to Equation **(13)**.

TABLE 8.16 Discretization error using Euler's method for the initial-value problem $y' = y$, $y(0) = 1$

x_n	$y'_n = f(x_n, y_n)$	$y_{n+1} = y_n + hy'_n$	$y(x_n) = e^{x_n}$	$e_n = y_n - y(x_n)$
0.0	1.00	1.10	1.00	0.00
0.1	1.10	1.21	1.11	−0.01
0.2	1.21	1.33	1.22	−0.01
0.3	1.33	1.46	1.35	−0.02
0.4	1.46	1.61	1.49	−0.03
0.5	1.61	1.77	1.65	−0.04
0.6	1.77	1.95	1.82	−0.05
0.7	1.95	2.15	2.01	−0.06
0.8	2.15	2.37	2.23	−0.08
0.9	2.37	2.61	2.46	−0.09
1.0	2.61		2.72	−0.11

It turns out that it is possible to derive error estimates like **(11)** or **(13)** for every method we discuss in this chapter for solving differential equations numerically. Actually, to derive these estimates would take us beyond the scope of this book,* but we should mention that for the Runge-Kutta method, which was discussed in Section 8.3, the discretization error e_n is of the form

$$|e_n| \le kh^4,$$

for some appropriate constant k. Thus, halving the step size, for example, has the effect of decreasing the bound on the error by a factor of $2^4 = 16$. However, the price for this greater accuracy is to have to calculate $f(x, y)$ at four points [see Equations **(8.3.7)**] for each step in the iteration.

PROBLEMS 8.5

1. Consider the differential equation $y' = -y$, $y(0) = 1$. We wish to find $y(1)$.
   (a) Calculate an upper bound on the error of the Euler method as a function of h.
   (b) Calculate this bound for $h = 0.1$ and $h = 0.2$.
   (c) Perform the iterations for $h = 0.2$ and $h = 0.1$ and compare the actual error with the maximum error.

2. Consider the equation of Problem 1. If we ignore round-off error, how many iterations have to be performed in order to guarantee that the calculation of $y(1)$ obtained by the Euler method is correct to (a) five decimal places? (b) six decimal places?

3. Answer the questions in Problem 2 for the equation
$$y' = 3y - x^2, \quad y(1) = 2$$
if we wish to find $y(1.5)$.

* For a more detailed analysis, see for example, C. W. Gear, *Numerical Initial Value Problems in Ordinary Differential Equations* (Englewood Cliffs, N.J.: Prentice-Hall, 1971).

8.6 A PERSPECTIVE: NUMERICAL INSTABILITY CAUSED BY PROPAGATION OF ROUND-OFF ERROR

In this section we show how a theoretically very accurate method can produce results that are useless. A *multistep method* is one that involves information about the solution at more than one point. Consider the multistep method given by the equation

$$y_{n+1} = y_{n-1} + 2hf(x_n, y_n). \tag{1}$$

Here it is necessary to use both the nth and the $(n - 1)$st iterate to obtain the $(n + 1)$st iterate. It can be shown* that this method has the following error estimate:

$$|e_n| = |y_n - y(x_n)| \leq kh^2.$$

Since the error for the Euler method is $|e_n| \leq kh$, we would theoretically expect more accuracy in solving our initial-value problem by using Equation **(1)** than by using the Euler method. However, this does not always turn out to be the case.

EXAMPLE 1 ▶ **Solving an initial-value problem by two methods, one unstable**

Consider the initial-value problem

$$y' = -y + 2, \qquad y(0) = 1.$$

The solution to this equation is easily obtained: $y(x) = 2 - e^{-x}$. Let us obtain $y(5)$ by the Euler method and the method of Equation **(1).** To use the latter, we need two intitial values y_0 and y_1. Since we know the solution, we use the exact value $y_1 = y(x_1) = 2 - e^{-x_1}$. Table 8.17 illustrates the computation with a step size $h = 0.25$. The second column is the correct value of $y(x_n)$ to four decimal places. Column three gives the Euler iterates, and column four gives the iterates obtained by the two-step method **(1).** Column five is the Euler error, $e_n^{(\text{Euler})} = y_n^{(\text{Euler})} - y(x_n)$, and column six is the error of the two-step method, $e_n^{(2s)} = y_n^{(2s)} - y(x_n)$.

It is evident that the two-step method **(1)** produces a smaller error for small values of x_n than the Euler method. However, as x_n increases, the error in the Euler method decreases, whereas the error in the two-step method not only increases but does so with oscillating sign. This phenomenon is called **numerical instability.** As we see below, it is due to a propagation of round-off errors.

Let us now explain what leads to this instability. In the example, $f(x_n, y_n) = -y_n + 2$, so Equation **(1)** is

$$y_{n+1} = y_{n-1} + 2h(2 - y_n),$$

or

$$y_{n+1} + 2hy_n - y_{n-1} = 4h, \qquad y_0 = 1. \tag{2}$$

This is a linear second-order **nonhomogeneous difference equation** (see Section 3.10). Its solution is obtained in a manner analogous to that used in solving linear nonhomogeneous differential equations. First we find the general solution of the **homogeneous** difference equation

$$y_{n+1} + 2hy_n - y_{n-1} = 0. \tag{3}$$

*See S. D. Conte, *Elements of Numerical Analysis* (New York: McGraw-Hill, 1965), section 6.6.

TABLE 8.17 A comparison of the error using the Euler method and that of Equation (1)

x_n	$y(x_n) = 2 - e^{-x_n}$ (exact)	$y_n^{(E)} = y_{n-1}^{(E)} + h(2 - y_{n-1}^{(E)})$	$y_n^{(2s)} = y_{n-2}^{(2s)} + 2h(2 - y_{n-1}^{(2s)})$	$e_n^{(E)}$	$e_n^{(2s)}$
0.00	1.0000	1.0000	1.0000	0.0000	0.0000
0.25	1.2212	1.2500	1.2212	0.0288	0.0000
0.50	1.3935	1.4375	1.3894	0.0440	−0.0041
0.75	1.5276	1.5781	1.5265	0.0505	−0.0011
1.00	1.6321	1.6836	1.6262	0.0515	−0.0059
1.25	1.7135	1.7627	1.7134	0.0492	−0.0001
1.50	1.7769	1.8220	1.7695	0.0453	−0.0074
1.75	1.8262	1.8665	1.8287	0.0403	+0.0025
2.00	1.8647	1.8999	1.8552	0.0352	−0.0095
2.25	1.8946	1.9249	1.9011	0.0303	+0.0065
2.50	1.9179	1.9437	1.9047	0.0258	−0.0132
2.75	1.9361	1.9578	1.9488	0.0217	+0.0127
3.00	1.9502	1.9684	1.9303	0.0182	−0.0199
3.25	1.9612	1.9763	1.9837	0.0151	+0.0225
3.50	1.9698	1.9822	1.9385	0.0124	−0.0313
3.75	1.9765	1.9867	2.0145	0.0102	+0.0380
4.00	1.9817	1.9900	1.9313	0.0083	−0.0504
4.25	1.9857	1.9925	2.0489	0.0068	+0.0632
4.50	1.9889	1.9944	1.9069	0.0055	−0.0820
4.75	1.9913	1.9958	2.0955	0.0045	+0.1042
5.00	1.9933	1.9969	1.8952	0.0036	−0.0981

To do this, we substitute $y_n = \lambda^n$ in Equation **(3),** obtaining

$$\lambda^{n-1}(\lambda^2 + 2h\lambda - 1) = 0.$$

Since we are not interested in trivial solutions, assume $\lambda \neq 0$. Thus the two roots of the **characteristic equation**

$$\lambda^2 + 2h\lambda - 1 = 0 \tag{4}$$

given by

$$\lambda_1 = \frac{-2h + \sqrt{4h^2 + 4}}{2} = -h + \sqrt{1 + h^2} \quad \text{and} \quad \lambda_2 = -h - \sqrt{1 + h^2}$$

yield the solutions $y_{n,1} = (\lambda_1)^n$ and $y_{n,2} = (\lambda_2)^n$ to Equation **(3).** Hence Equation **(3)** has the general solution

$$y_n = c_1\lambda_1^n + c_2\lambda_2^n.$$

Finally, we need to find *any* particular solution to the nonhomogeneous Equation **(2).** Since $4h$ is constant, using the method of undetermined coefficients, we set $y_{n,p} = A$. Then Equation **(2)** becomes

$$A + 2hA - A = 4h,$$

so that $A = 2$. Hence the general solution of Equation **(2)** is given by

$$y_n = c_1\lambda_1^n + c_2\lambda_2^n + 2. \tag{5}$$

By the binomial theorem (see Problem 5.1.18),

$$(1 + h^2)^{1/2} = 1 + \frac{1}{2}h^2 - \frac{1}{8}h^4 + \frac{1}{16}h^6 - \dots,$$

where the omitted terms are higher powers of h. Hence the roots of the characteristic Equation **(4)** can be written as

$$\lambda_1 = 1 - h + \alpha(h) \quad \text{and} \quad \lambda_2 = -1 - h - \alpha(h), \tag{6}$$

where

$$\alpha(h) = \frac{h^2}{2} - \frac{h^4}{8} + \frac{h^6}{16} - \dots .$$

Substituting Equation **(6)** into Equation **(5)** yields

$$y_n = c_1[1 - h + \alpha(h)]^n + c_2(-1)^n[1 + h + \alpha(h)]^n + 2. \tag{7}$$

From calculus we know that

$$\lim_{k\to\infty}\left(1 + \frac{1}{k}\right)^k = \lim_{h\to 0}(1 + h)^{1/h} = e.$$

Therefore, since $x_n = 0 + nh = nh$, we have

$$\lim_{h\to 0}(1 - h)^n = \lim_{h\to 0}(1 - h)^{x_n/h} = e^{-x_n} \quad \text{and} \quad \lim_{h\to 0}(1 + h)^n = e^{x_n}.$$

Hence as $h \to 0$ we may ignore the higher-order terms $\alpha(h)$ in **(7)** to obtain

$$y_n = c_1 e^{-x_n} + 2 + c_2(-1)^n e^{x_n}. \tag{8}$$

Here lies the problem. The exact solution of the problem requires that $c_1 = -1$ and $c_2 = 0$. However, even a small round-off error may cause c_2 to be nonzero, and this error grows exponentially while the real solution is approaching the constant two. This is the phenomenon we observed in Table 8.17. Note that the $(-1)^n$ in **(8)** causes the errors to oscillate (as we also observed). ◀

The problem arose because we approximated a *first*-order differential equation by a *second*-order difference equation. Such approximations do not always lead to this kind of instability, but it is a possibility that cannot be ignored. In general, to analyze the effectiveness of a given method, we must not only estimate the discretization error but also show that the method is not numerically unstable (that is, *that it is numerically stable*), for the given problem.

8.7 NUMERICAL SOLUTION OF SYSTEMS AND BOUNDARY VALUE PROBLEMS

The methods developed in Sections 8.2, 8.3, and 8.4 can be extended very easily to apply to higher-order equations and systems of equations. Consider the system

$$\frac{dx}{dt} = f(t, x, y), \qquad \frac{dy}{dt} = g(t, x, y), \tag{1}$$

where t is the independent variable and x and y are dependent variables.

Euler methods

For the Euler and improved Euler methods, it is necessary only to reinterpret the formulas.

$$y_{n+1} = y_n + hf(x_n, y_n)$$

and

$$y_{n+1} = y_n + \frac{h}{2}[f(x_n, y_n) + f(x_{n+1}, z_{n+1})], \quad \text{with} \quad z_{n+1} = y_n + hf(x_n, y_n).$$

Here y_n, z_{n+1}, and f should be interpreted as vectors. For the system given in **(1)** that means that Euler's method is reinterpreted as:

$$\begin{aligned} x_{n+1} &= x_n + hf(t_n, x_n, y_n) = x_n + hx'_n, \\ y_{n+1} &= y_n + hg(t_n, x_n, y_n) = y_n + hy'_n, \end{aligned} \tag{2}$$

while the improved Euler method is given by first solving **(2)**, and calling the results $\tilde{x}_{n+1}$ and $\tilde{y}_{n+1}$, and then calculating

$$\begin{aligned} x_{n+1} &= x_n + \frac{h}{2}\left\{f(t_n, x_n, y_n) + f(t_n + h, \tilde{x}_{n+1}, \tilde{y}_{n+1})\right\} = x_n + \frac{h}{2}\left\{x'_n + \tilde{x}'_{n+1}\right\}, \\ y_{n+1} &= y_n + \frac{h}{2}\left\{g(t_n, x_n, y_n) + g(t_n + h, \tilde{x}_{n+1}, \tilde{y}_{n+1})\right\} = y_n + \frac{h}{2}\left\{y'_n + \tilde{y}'_{n+1}\right\}. \end{aligned} \tag{3}$$

We illustrate the first system with an example.

EXAMPLE 1 ▶ Solving a system using Euler's method

Consider the initial-value problem

$$\begin{aligned} \frac{dx}{dt} &= -3x + 4y, \qquad x(0) = 1, \\ \frac{dy}{dt} &= -2x + 3y, \qquad y(0) = 2. \end{aligned} \tag{4}$$

Suppose we are seeking the values $x(1)$ and $y(1)$. In this problem, t is the independent variable, and x and y are the dependent variables. If we wish to use the Euler method, formula **(1)** translates into the equations

$$\begin{aligned} x_{n+1} &= x_n + hx'_n = x_n + h(-3x_n + 4y_n), \\ y_{n+1} &= y_n + hy'_n = y_n + h(-2x_n + 3y_n). \end{aligned}$$

The initial values are $x_0 = 1, y_0 = 2$, and, for $h = 0.2$, the procedure is essentially the same as before (see Table 8.18).

The solution of **(4)** is given by

$$x(t) - 3e^t - 2e^{-t}, \qquad y(t) = 3e^t - e^{-t},$$

so $x(1) = 3e - 2e^{-1} \approx 7.419$ and $y(1) = 3e - e^{-1} \approx 7.787$, implying that our method has an error of about 10%. The accuracy may be improved by selecting smaller values of h. ◀

TABLE 8.18

t_n	x_n	y_n	x'_n	y'_n	$x_{n+1} = x_n + hx'_n$	$y_{n+1} = y_n + hy'_n$
0.0	1.00	2.00	5.00	4.00	2.00	2.80
0.2	2.00	2.80	5.20	4.40	3.04	3.68
0.4	3.04	3.68	5.60	4.96	4.16	4.67
0.6	4.16	4.67	6.20	5.69	5.40	5.81
0.8	5.40	5.81	7.04	6.63	6.81	7.14
1.0	6.81	7.14				

No additional difficulty is caused by having a nonhomogeneous or nonlinear system of equations.

Predictor-corrector methods

EXAMPLE 2 ▶ **Adapting the Adams-Bashforth method to a system**

Adapt predictor-corrector formulas **(11)** and **(12)** in Section 8.4 to the system in Example 1.

We assume from the calculations in Table 8.18 that $x_0 = 1$, $x'_0 = 5$, $x_1 = 2$, $x'_1 = 5.2$, $y_0 = 2$, $y'_0 = 4$, $y_1 = 2.8$, $y'_1 = 4.4$, and $h = 0.2$. With these values we first predict

$$\begin{aligned} x_{n+2} &= x_{n+1} + \frac{h}{2}(-x'_n + 3x'_{n+1}), \\ y_{n+2} &= y_{n+1} + \frac{h}{2}(-y'_n + 3y'_{n+1}), \end{aligned} \tag{5}$$

and then use these values to compute x'_{n+2}, y'_{n+2}, using the system of differential equations in Equations **(4)**. Then, to correct these values, we use the trapezoidal rules:

$$\begin{aligned} x_{n+2} &= x_{n+1} + \frac{h}{2}(x'_{n+2} + x'_{n+1}), \\ y_{n+2} &= y_{n+1} + \frac{h}{2}(y'_{n+2} + y'_{n+1}), \end{aligned} \tag{6}$$

and recalculate x'_{n+2}, y'_{n+2} with **(4)**.

It is clear that such methods are laborious, but they are easily carried out on a computer. The calculations shown in Table 8.19 agree far better with the exact values, $x(1) \approx 7.419$ and $y(1) \approx 7.787$, than those of Example 1.

TABLE 8.19

					Predictor				*Corrector*	
t_{n+1}	x_{n+1}	y_{n+1}	x'_{n+1}	y'_{n+1}	x_{n+2}	y_{n+2}	x'_{n+2}	y'_{n+2}	x_{n+2}	y_{n+2}
0.0	1.00	2.00	5.00	4.00						
0.2	2.00	2.80	5.20	4.40	3.06	3.72	5.70	5.04	3.09	3.74
0.4	3.09	3.74	5.69	5.04	4.28	4.81	6.40	5.87	4.30	4.83
0.6	4.30	4.83	6.42	5.89	5.66	6.09	7.38	6.95	5.68	6.11
0.8	5.68	6.11	7.40	6.97	7.26	7.61	8.66	8.31	7.29	7.64
1.0	7.29	7.64								

Runge-Kutta method

The Runge-Kutta formula for a system of differential equations is a direct generalization of Equations **(8.3.6)** and **(8.3.7.)**:

$$x_1 = x_0 + \frac{1}{6}(m_1 + 2m_2 + 2m_3 + m_4),$$
$$y_1 = y_0 + \frac{1}{6}(n_1 + 2n_2 + 2n_3 + n_4), \tag{7}$$

where

$$m_1 = hf(t_0, x_0, y_0), \qquad n_1 = hg(t_0, x_0, y_0),$$
$$m_2 = hf\left(t_0 + \frac{h}{2}, x_0 + \frac{m_1}{2}, y_0 + \frac{n_1}{2}\right), \quad n_2 = hg\left(t_0 + \frac{h}{2}, x_0 + \frac{m_1}{2}, y_0 + \frac{n_1}{2}\right),$$
$$m_3 = hf\left(t_0 + \frac{h}{2}, x_0 + \frac{m_2}{2}, y_0 + \frac{n_2}{2}\right), \quad n_3 = hg\left(t_0 + \frac{h}{2}, x_0 + \frac{m_2}{2}, y_0 + \frac{n_2}{2}\right), \tag{8}$$
$$m_4 = hf(t_0 + h, x_0 + m_3, y_0 + n_3), \qquad n_4 = hg(t_0 + h, x_0 + m_3, y_0 + n_3).$$

It should now be apparent how this procedure is generalized for systems involving more dependent variables. We apply the formulas above to system **(4)** in Example 1 with $h = 1$:

$$m_1 = 5, \qquad n_1 = 4,$$
$$m_2 = \frac{11}{2}, \qquad n_2 = 5,$$
$$m_3 = \frac{27}{4}, \qquad n_3 = 6,$$
$$m_4 = \frac{35}{4}, \qquad n_4 = \frac{17}{2}.$$

We obtain

$$x_1 \approx 7.375, \qquad y_1 \approx 7.750.$$

Even though this process involves more complicated computations at each step, it involves less work than predictor-corrector methods. Since it is also quite accurate, it is thus the preferred method for hand calculations. A computer program that does the Runge-Kutta method for arbitrary systems is given at the end of this section.

All of these methods can be applied to higher-order differential equations by converting each higher-order equation into a system of first-order equations using the procedure outlined in Section 7.1.

Boundary value problems

Consider the differential equation

$$y'' = f(x, y, y'), \tag{9}$$

with boundary conditions $y(a) = y_a$ and $y(b) = y_b$. We now describe a procedure, sometimes called the *shooting method,* that is often used to solve such problems. The idea is to convert Equation **(9)** into an initial-value problem with initial conditions $y(a) = y_a$ and $y'(a) = M_0$, where the number M_0 is arbitrarily selected. Using one of the previously described techniques (Euler method, Runge-Kutta, or predictor-corrector method), we now calculate the value $y(b) = N_0$ for the initial-value problem. This number will undoubtedly be different from the required value y_b, so we again solve Equation **(9)** as an initial-value problem with $y(a) = y_a$ and

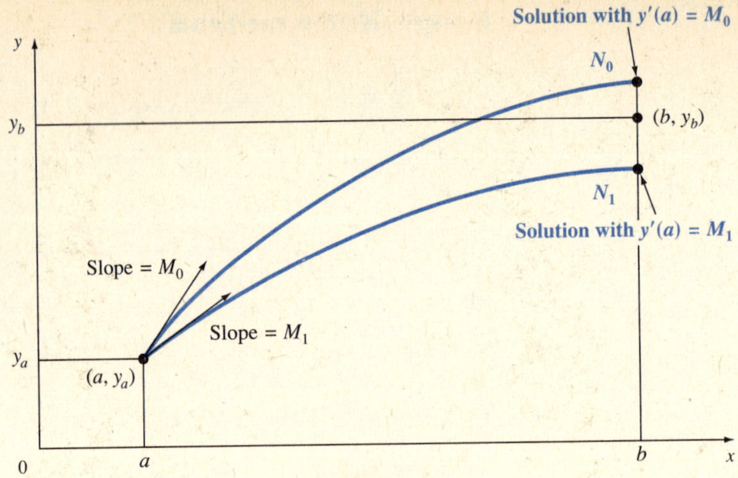

FIGURE 8.4
Shooting method: interpolate slopes and points

$y'(a) = M_1$, obtaining $y(b) = N_1$ (see Figure 8.4). The assumption is now made that $y(b)$ varies linearly with the values $y'(a)$, so the next value we choose for $y'(a)(= M_2)$ is selected by solving the equation

$$\frac{M_2 - M_1}{M_0 - M_1} = \frac{y_b - N_1}{N_0 - N_1}, \tag{10}$$

or

$$M_2 = M_1 + \frac{y_b - N_1}{N_0 - N_1}(M_0 - M_1). \tag{11}$$

Equation **(11)** yields very accurate information if the user is prepared to repeat the process several times. Again, it is clear that since this process is very laborious, it is best performed by using a computer.

Finally, it should be noted that although shooting methods are useful for finding approximate solutions for many boundary value problems, they do not work, for a variety of reasons, for all such problems. Other methods, some very complex, are available to deal with these situations.

Perspective Stiff equations

As soon as we begin to deal with a system of two or more first-order equations (or a higher-order equation), we encounter the phenomenon of **stiffness.** Stiffness arises in any problem in which the dependent variable can be expressed in terms of two or more very different scales in the independent variable. For example, consider the differential equation

$$y'' - 15y' - 100y = 0. \tag{12}$$

This second-order equation is equivalent to the system

$$y' = z,$$
$$z' = 100y + 15z.$$

The general solution has the form

$$y = c_1e^{20x} + c_2e^{-5x}, \tag{13}$$

since the characteristic equation is $\lambda^2 - 15\lambda - 100 = (\lambda - 20)(\lambda + 5) = 0$. Observe that the general solution depends (for increasing x) on a rapidly growing term and a rapidly decaying term.

Suppose we are interested in solving the initial-value problem consisting of Equation **(12)** and the initial conditions

$$y(0) = 1, \qquad y'(0) = -5.$$

In this case the exact solution is $y = e^{-5x}$. However, as Table 8.20 shows, each method begins with a numerical solution that decays as e^{-5x} but then rapidly deviates from the exact solution.

TABLE 8.20 A comparison of the Runge-Kutta method and the exact solution $y = e^{-5x}$, with $h = 0.02$ and $h = 0.2$, for the initial-value problem $y'' - 15y' - 100y = 0$, $y(0) = 1$, $y'(0) = -5$

x	y, *Runge-Kutta Method*	y, *Exact*
$h = 0.02$		
0.00	1.000000000	1.000000000
0.02	0.862070833	0.904837418
0.04	0.635804065	0.818730753
0.06	0.300813510	0.740818221
0.08	−0.165364430	0.670320046
0.10	−0.786953158	0.606530660
0.12	−1.589425340	0.548811636
0.14	−2.598755640	0.496585304
0.16	−3.840388700	0.449328964
0.18	−5.337870700	0.406569660
0.20	−7.111090350	0.367879441
$h = 0.2$		
0.0	1.00000000	1.000000000
0.2	−7.29166667	0.367879441
0.4	−42.5121528	0.1353335283
0.6	50.8496093	0.0497870684
0.8	1105.45961	0.0183156389
1.0	1836.41827	6.737947×10^{-3}
1.2	−20190.9270	$2.47875217 \times 10^{-3}$
1.4	−96487.7822	$9.11881964 \times 10^{-4}$
1.6	198302.943	$3.35462627 \times 10^{-4}$
1.8	2733507.62	$1.23409804 \times 10^{-4}$
2.0	2889103.65	$4.53999298 \times 10^{-5}$

The reason for this difficulty is simple: Each step in every numerical procedure introduces a small error ϵ, so the *numerical solution* is now a linear combination of both e^{-5x} and e^{20x}:

$$y_{\text{numerical}} \approx e^{-5x} + \epsilon e^{20x}.$$

No matter how small ϵ is, the ϵe^{20x} term quickly dominates the other term (compare this situation to that in Section 8.6). Stiff equations can be handled if precautions are taken to deal with them*; modern computer codes include such precautions.

* For further information about this topic, see C. W. Gear, *Numerical Initial Value Problems in Ordinary Differential Equations* (Englewood Cliffs, N.J.: Prentice-Hall, 1971).

PROBLEMS 8.7

In Problems 1–8 find $x(1)$ and $y(1)$ with $h = 0.2$ using (a) the Euler method, (b) the improved Euler method, and (c) the Runge-Kutta method for each initial-value system. Check your accuracy by calculating the exact value.

1. $x' = 4x - 2y$, $x(0) = 1$
   $y' = 5x + 2y$, $y(0) = 2$
2. $x' = x + y$, $x(0) = 1$
   $y' = x - y$, $y(0) = 0$
3. $x' = x + 2y$, $x(0) = 0$
   $y' = 3x + 2y$, $y(0) = 1$
4. $x' = -4x - y$, $x(0) = 0$
   $y' = x - 2y$, $y(0) = 1$
5. $x' = 2x + y + t$, $x(0) = 1$
   $y' = x + 2y + t^2$, $y(0) = 0$
6. $x' = x + 2y + t - 1$, $x(0) = 0$
   $y' = 3x + 2y - 5t - 2$, $y(0) = 4$
7. $x' = 3x + 3y + t$, $x(0) = 0$
   $y' = -x - y + 1$, $y(0) = 2$
8. $x' = 4x - 3y + t$, $x(0) = 1$
   $y' = 5x - 4y - 1$, $y(0) = -1$
9. Solve the initial-value problem

$$y'' = y' + xy^2, \qquad y(0) = 1, \qquad y'(0) = 0,$$

   for $y(1)$ with $h = 0.2$ by the Euler method.

10. Use the "shooting method" to determine the value of $y'(0)$ for the boundary value problem

$$y'' = -y^2, \qquad y(0) = y(1) = 0,$$

   in such a way that y is positive over the interval $0 < x < 1$. Use the Runge-Kutta method [Equations **(7)** and **(8)**] with $h = 0.1$.

11. Find the maximum value of y over the interval $0 \le x \le 1$ for the boundary value problem

$$y'' + yy' + 1 = 0, \qquad y(0) = y(1) = 0.$$

12. Develop a BASIC (or FORTRAN) program to perform the "shooting method" using the improved Euler method for an equation of the form

$$y'' = f(x, y, y'), \qquad y(a) = y_a,\ y(b) = y_b.$$

   Use it to find the maximum value of y over the interval $0 \le x \le 1$ for the boundary value problem

$$y'' + \sin y = 0, \qquad y(0) = y(1) = 0.$$

*TRUE BASIC Runge-Kutta system solver**

```
DIM x(10)                     ! increase dimension if n > 10
LET n=2                       ! number of dependent variables
PRINT " t ";                  ! column headings
FOR i=1 to n
    READ x(i)
    PRINT " x(";i;") ";
NEXT i
PRINT
DATA 1, 2                     ! initial dependent variable values
LET t=0                       ! initial t value
LET tlast=1                   ! last t value
LET h=.2                      ! step size
DO
```

*See the footnote on page 371.

```
     CALL runktn(n,t,x(),h)              ! Runge-Kutta subroutine
     PRINT t;
     FOR i=1 to n
        PRINT x(i);
     NEXT i
     PRINT
 LOOP while t<tlast
 END

 SUB runktn (n,t,x(),h)
     DIM xx(10), sx(10), dx(10)          ! increase dimension if n > 10
     FOR i=1 to n
        LET xx(i)=x(i)
        LET sx(i)=x(i)
     NEXT i
     LET tt=t
     LET h2=h/2
     LET h6=h/6
     LET k=1
     DO
       CALL difeqn (t,n,x(),dx())
       FOR i=1 to n
          LET sx(i)+h6*dx(i)
       NEXT i
       IF (k=2) or (k=3) then
          FOR i=1 to n
             LET sx(i)=sx(i)+h6*dx(i)
          NEXT i
       END IF
       LET k=k+1
       IF k=5 then
          FOR i=1 to n
             LET x(i)=sx(i)
          NEXT i
          EXIT DO
       END IF
       LET t=tt+h2
       FOR i=1 to n
          LET x(i)=xx(i)+h2*dx(i)
       NEXT i
       IF k=4 then
          LET t=t+h2
          FOR i=1 to n
             LET x(i)=x(i)+h2*dx(i)
          NEXT i
          END IF
       LOOP
END SUB

SUB difeqn (t,n,x(),dx())
    LET dx(1)= -3*x(1)+4*x(2)
    LET dx(2)= -2*x(1)+3*x(2)
END SUB
```

8.8 NUMERICAL METHODS USING SYMBOLIC MANIPULATORS

This section will show you how MATHEMATICA, MAPLE, DERIVE, MATLAB, or MATHCAD can be used to find numerical solutions to first-order differential equations and systems by the improved Euler and Runge-Kutta algorithms. A previously defined command is essential for this task. Recall that:

```
ITERATE(f(v), v, v0, n)
```

= {do n iterations of the function $v_{k+1} = f(v_k)$
beginning at v_0 and save the values $v_0, v_1, \ldots, v_n$ in a vector.
The symbol v can be either a variable or a vector of variables.}.

The improved Euler and Runge-Kutta methods are useful techniques for finding the numerical solution of a single first-order differential equation

$$y' = f(x, y) \tag{1}$$

or system of first-order differential equations

$$\begin{aligned} y_1' &= f_1(x, y_1, y_2, \ldots, y_n) \\ y_2' &= f_2(x, y_1, y_2, \ldots, y_n) \\ &\cdots \\ y_n' &= f_n(x, y_1, y_2, \ldots, y_n). \end{aligned} \tag{2}$$

Here x is the independent variable and y or $\mathbf{y} = (y_1, y_2, \ldots, y_n)$ is the dependent variable. Thus, system **(2)** can be written in vector form as

$$\mathbf{y}' = \mathbf{f}(x, \mathbf{y}), \tag{3}$$

where $\mathbf{f} = (f_1, f_2, \ldots, f_n)$. The algorithms we will describe work both for the single equation as well as for the system. Care must be given in entering the data so that your symbolic manipulator is given the correct description of the problem.

Improved Euler method

The improved Euler method is given by the system of difference equations:

$$\begin{aligned} x_{n+1} &= x_n + h \\ z_{n+1} &= y_n + hf(x_n, y_n) \\ y_{n+1} &= y_n + h\left[\frac{f(x_n, y_n) + f(x_{n+1}, z_{n+1})}{2}\right], \end{aligned} \tag{4}$$

where h is the step size. Essentially, the z_{n+1} term is an Euler step, or predictor, while the y_{n+1} term corrects the Euler step by averaging two slopes. Since we want our algorithm to work equally well for systems as it does for single equations, let $\mathbf{v} = (x, \mathbf{y})$, where $\mathbf{y}$ is an n-vector (n may be equal to 1), and let $\mathbf{f}$ be the vector of equations in **(3)**. Observe, using these notations, that **(4)** is equivalent to the two commands

```
IE(f, v, u, h):=u+h*[1, LIM(f, v, u)] {the first two equations in (4)}
IEULER(v, f, v0, h, n):=ITERATE((u+IE(f, v, u, h)
           +h*[1, LIM(f, v, IE(f, v, u, h)])/2, u, v0, n)
```

Here, IEULER uses the IE command to do the Euler step. Note that the first term in the ITERATE command is the vector

```
([xn, yn]+([xn, yn]+[h, h*f(xn, yn)])+[h, h*f(xn+1, zn+1)])/2=yn+1
```

.

Runge-Kutta method

The algorithm for the Runge-Kutta method, whether for a single first-order equation **(1)** or a system **(3)** is given by

$$\mathbf{y}_{n+1} = \mathbf{y}_n + \frac{(\mathbf{m}_0 + 2\mathbf{m}_1 + 2\mathbf{m}_2 + \mathbf{m}_3)}{6} \tag{5}$$

where

$$\mathbf{m}_0 = h\mathbf{f}(x_n, \mathbf{y}_n),$$

$$\mathbf{m}_1 = h\mathbf{f}\left(x_n + \frac{1}{2}h, \mathbf{y}_n + \frac{1}{2}\mathbf{m}_0\right),$$

$$\mathbf{m}_2 = h\mathbf{f}\left(x_n + \frac{1}{2}h, \mathbf{y}_n + \frac{1}{2}\mathbf{m}_1\right),$$

$$\mathbf{m}_3 = h\mathbf{f}(x_n + h, \mathbf{y}_n + \mathbf{m}_2).$$

The script for this procedure consists of five functions:

```
RK3(p, v, u, c1, c2, c3):=(c1+LIM(p, v, u+c3)+2*(c2+c3))/6
RK2(p, v, u, c1, c2):=RK3(p, v, u, c1, c2, LIM(p, v, u+c2/2))
RK1(p, v, u, c1):=RK2(p, v, u, c1, LIM(p, v, u+c1/2))
RK0(p, v, u):=RK1(p, v, u, LIM(p, v, u))
RK(v, f, v₀, h, n):=ITERATE(u+RK0(h*[1, f], v, u), u, v₀, n)
```

The command RK calls RK0 initially with $\mathbf{u} = \mathbf{v}_0$ and $\mathbf{p} = \mathbf{v}_0 + h*[1, \mathbf{f}(\mathbf{v}_0)] = \mathbf{c1} = \mathbf{m}_0$, which then calls RK1 for which $\mathbf{p} = \mathbf{c2} = \mathbf{m}_1$, and so forth, until the algorithm in **(5)** is complete. Once that is done, the ITERATE command replaces $\mathbf{v}_0$ by $\mathbf{v}$ and the process continues.

To illustrate the use of these commands, we solve Example 8.3.2.

EXAMPLE 1 ▶ **Solving an earlier problem using symbolic manipulation**

Solve Example 8.3.2,

$$y' = x + y^2, \qquad y(0) = 1, h = 0.1, n = 5,$$

using both IEULER and RK.

Solution Since this is a single first-order differential equation, the two commands needed to perform the iterations are

```
IEULER([x, y], x+y^2, [0, 1], 0.1, 5)
```

and

```
RK([x, y], x+y^2, [0, 1], 0.1, 5).
```

The two scripts would return to four decimal places, respectively.

x-value	*y (Improved Euler)*	*y (Runge-Kutta)*
0	1	1
0.1	1.1155	1.1165
0.2	1.2708	1.2736
0.3	1.4820	1.4880
0.4	1.7768	1.7893
0.5	2.2070	2.2345

◀

EXAMPLE 2 ▶ Using the IEULER and RK scripts

Find $x(1)$ and $y(1)$ with $h = 0.25$ for the system below using the improved Euler and Runge-Kutta algorithms:

$$x' = x - y^2 + t, \qquad x(0) = 1,$$
$$y' = x^2 + y - t, \qquad y(0) = 2.$$

Solution Here $\mathbf{f} = [x - y^2 + t, x^2 + y - t]$, so we would give the commands

```
IEULER([t,x,y], [x-y*y+t, x*x+y-t], [0,1,2], 0.25, 4)
```

and

```
RK([t,x,y], [x - y*y+t, x*x+y-t], [0,1,2], 0.25, 4).
```

These commands would produce the following results:

t-value	$[x, y]$ *(improved Euler)*	$[x, y]$ *(Runge-Kutta)*
0	[1, 2]	[1, 2]
0.25	[−0.2578,2.6953]	[−0.2206,2.6410]
0.5	[−2.7444,3.9009]	[−2.8890,3.9329]
0.75	[−11.2232,12.3213]	[−26.8733,28.7988]
1.0	[−310.517,370.571]	$[-3.56 \times 10^{10}, 7.07 \times 10^{10}]$

Clearly, with such a difference in the answers arising from the two methods, further analysis is necessary. One might begin by choosing a smaller step size. ◀

PROBLEMS 8.8

Find the required answer for each of the following problems using the improved Euler and/or Runge-Kutta method and MATHEMATICA, MAPLE, DERIVE, MATLAB, or MATHCAD:

1. $y' = x + y$, $y(0) = 1$, $h = 0.2$, $n = 5$

2. $y' = x - y$, $y(1) = 2$, $h = 0.4$, $n = 5$

3. $\dfrac{dy}{dx} = xy^2 + y^3$, $y(0) = 1$, $h = 0.02$, $n = 20$

4. $\dfrac{ds}{dt} = s + \cos(\pi t)$, $y(1) = 0$, $h = 0.05$, $n = 20$

5. $y' = \sqrt{x^2 + y^2}$, $y(0) = 1$, $h = 0.05$, $n = 20$

6. $y' = \sqrt{x + y^2}$, $y(1) = 2$, $h = 0.05$, $n = 20$

7. $y' = \dfrac{x - y}{x + y}$, $y(2) = 1$, $h = -0.05$, $n = 20$

8. $y' = \dfrac{x + y^2}{x + 2y^2}$, $y(1) = 2$, $h = 0.1$, $n = 20$

9. $x' = 4x - 2y$, $x(0) = 1$, $h = 0.1$, $n = 10$
$y' = 5x + 2y$, $y(0) = 2$

10. $x' = 5x + 2y - t$, $x(0) = 1$, $h = 0.1$, $n = 20$
$y' = 2x - 2y + 1$, $y(0) = 0$

11. Let $\frac{dy}{dx} = \frac{x - y}{x + y}$, $y(2) = 0$ and use the improved Euler or Runge-Kutta method to find $y(1)$ with $h = -0.05$. Then solve this problem using the HOMOGENEOUS command of Section 2.10, and compare the two answers. If they are different, explain why.

12. Find $y(1)$ and $y'(1)$ for the initial-value problem $y'' = y' + xy^2$, $y(0) = 1$, $y'(0) = 0$, with $h = 0.05$ by either the improved Euler or Runge-Kutta method.

SUMMARY OUTLINE OF CHAPTER 8

Floating-Point Arithmetic p. 361
- floating-point number
- mantissa
- exponent
- number of significant digits

Truncation and rounding errors p. 362

Absolute error: ϵ_a p. 362

Relative error: ϵ_r p. 362

Stability p. 362

Convergence p. 362

Euler method: $y_{n+1} = y_n + hf(x_n, y_n)$ p. 364

Improved Euler method

$z_{n+1} = y_n + hf(x_n, y_n)$ p. 367

$$y_{n+1} = y_n + \frac{h}{2}\{f(x_n, y_n) + f(x_{n+1}, z_{n+1})\}, \; x_{n+1} = x_n + h$$

Three-term Taylor series method p. 371

Runge-Kutta method

$$y_{n+1} = y_n + \frac{1}{6}(m_1 + 2m_2 + 2m_3 + m_4) \quad \text{p. 373}$$

where $m_1 = hf(x_n, y_n)$, $m_2 = hf\left(x_n + \frac{h}{2}, y_n + \frac{m_1}{2}\right)$,

$m_3 = hf\left(x_n + \frac{h}{2}, y_n + \frac{m_2}{2}\right)$, $m_4 = hf(x_n + h, y_n + m_3)$.

Kutta-Simpson $\frac{3}{8}$-rule p. 378

Heun's formula p. 381

Quadrature formulas p. 381

Adams-Bashforth formula: $y_2 - y_1 = \frac{h}{2}(-y_0' + 3y_1')$ p. 382

Predictor-Corrector methods p. 382

Milne's method, p. 385

Convergence of discretization error theorem p. 387
If $f(x, y)$ has continuous first partials, $|f_y| < L$, $|y''| < M$ hold on $[x_0, a]$ and y_n is the approximate solution generated by Euler's method, then the discretization error $e_n = y_n - y(x_n)$ satisfies $|e_n| \le \frac{hM}{2L}(e^{nhL} - 1)$. In particular, $e_n \to 0$ as $h \to 0$

Numerical instability p. 390

Numerical solution of systems: pp. 392–393
Euler's method
Improved Euler method
Runge-Kutta method

Shooting methods for boundary value problems p. 396

Stiff equations p. 396

REVIEW EXERCISES FOR CHAPTER 8

In Exercises 1–6 solve the given initial-value problem using the methods of Chapter 2. Then use the (a) improved Euler method or (b) the Runge-Kutta method and the given value of h to obtain an approximate solution at the indicated value of x. Compare the numerical answer with the exact answer.

1. $\frac{dy}{dx} = \frac{e^x}{y}$, $y(0) = 2$. Find $y(3)$ with $h = \frac{1}{2}$.

2. $\frac{dy}{dx} = \frac{e^y}{x}$, $y(1) = 0$. Find $y\left(\frac{1}{2}\right)$ with $h = -0.1$.

3. $\dfrac{dy}{dx} = \dfrac{y}{\sqrt{1+x^2}}$, $y(0) = 1$. Find $y(3)$ with $h = \dfrac{1}{2}$.

4. $xy\dfrac{dy}{dx} = y^2 - x^2$, $y(1) = 2$. Find $y(3)$ with $h = \dfrac{1}{2}$.

5. $\dfrac{dy}{dx} = y - xy^3$, $y(0) = 1$. Find $y(3)$ with $h = \dfrac{1}{2}$.

6. $\dfrac{dy}{dx} = \dfrac{2xy}{3x^2 - y^2}$, $y\left(-\dfrac{3}{8}\right) = -\dfrac{3}{4}$. Find $y(6)$ with $h = \dfrac{3}{8}$.

7. Consider the differential equation in Exercise 6 with the initial condition $y(0) = -1$. Use the Runge-Kutta method or the improved Euler method to calculate $y(6)$ with $h = 1$. Why does the numerical solution differ from the exact answer?

8. Suppose the initial condition in Exercise 4 is $y(1) = 1$. Can any of the methods of Section 8.2 or 8.4 provide the correct answer for $y(3)$?

9. Consider the initial-value problem

$$y' = 1 + y^2, \qquad y(0) = 0.$$

Can any of the methods in Section 8.2 be used to obtain $y(2)$?

In Exercises 10–13 find the value of the dependent function(s) at 2 with $h = 0.2$ using (a) the Euler method, (b) the Runge-Kutta method, and (c) the improved Euler method as predictor, and the trapezoidal rule as corrector to solve numerically the following initial-value problems. Compare the numerical answer to the exact solution.

10. $x^2y'' - xy' + y = 0$,
$y(1) = 1$, $y'(1) = 0$

11. $x^2y'' + xy' + y = 0$,
$y(1) = 1$, $y'(1) = 0$

12. $x' = x - 2y$, $x(0) = 1$
$y' = 2x + 5y$, $y(0) = 0$

13. $x' = x + 2y$, $x(0) = 1$
$y' = 2x + 5y$, $y(0) = 0$

9 MATRIX METHODS FOR SYSTEMS OF DIFFERENTIAL EQUATIONS

In this chapter, we use matrices to provide a more cohesive treatment of systems of differential equations. This material is more difficult and more comprehensive than the introductory material in Chapter 7.

Although a brief review of matrices is given in Section 9.1, additional familiarity with linear algebra is recommended. One important topic in linear algebra—eigenvalues and eigenvectors—is discussed in detail in Section 9.4.

9.1 REVIEW OF MATRICES

matrix An $m \times n$ **matrix*** A is a rectangular array of mn numbers arranged in m rows and n columns:

$$A = \begin{pmatrix} a_{11} & a_{12} & \cdots & a_{1j} & \cdots & a_{1n} \\ a_{21} & a_{22} & \cdots & a_{2j} & \cdots & a_{2n} \\ \vdots & \vdots & & \vdots & & \vdots \\ a_{i1} & a_{i2} & \cdots & a_{ij} & \cdots & a_{in} \\ \vdots & \vdots & & \vdots & & \vdots \\ a_{m1} & a_{m2} & \cdots & a_{mj} & \cdots & a_{mn} \end{pmatrix}. \tag{1}$$

The ijth component of A, denoted by a_{ij}, is the number appearing in the ith row and jth column of A. We will sometimes write the matrix A as $A = (a_{ij})$. Usually matrices will be denoted by capital letters.

square matrix If A is an $m \times n$ matrix with $m = n$, then A is called a **square matrix.** An $m \times n$ matrix with all components equal to zero is called the $m \times n$ **zero matrix.**

An $m \times n$ matrix is said to have the **size** $m \times n$. Two matrices $A = (a_{ij})$ and $B = (b_{ij})$ are **equal** if (a) they have the same size and (b) corresponding components are equal.

* The term "matrix" was first used in 1850 by the British mathematician James Joseph Sylvester (1814–1897) to distinguish matrices from determinants. In fact, the term "matrix" was intended to mean "mother of determinants."

EXAMPLE 1 ▶ **Five matrices**

Five matrices of different sizes are given below:

$$\begin{pmatrix} 1 & 3 \\ 4 & 2 \end{pmatrix},\ 2 \times 2 \text{ (square)} \qquad \begin{pmatrix} -1 & 3 \\ 4 & 0 \\ 1 & -2 \end{pmatrix},\ 3 \times 2$$

$$\begin{pmatrix} -1 & 4 & 1 \\ 3 & 0 & 2 \end{pmatrix},\ 2 \times 3 \qquad \begin{pmatrix} 1 & 6 & -2 \\ 3 & 1 & 4 \\ 2 & -6 & 5 \end{pmatrix},\ 3 \times 3 \text{ (square)}$$

$$\begin{pmatrix} 0 & 0 & 0 & 0 \\ 0 & 0 & 0 & 0 \end{pmatrix},\ 2 \times 4 \text{ zero matrix}$$ ◀

vectors A **row vector** is a matrix with one row. A **column vector** is a matrix with one column. Vectors are usually denoted by lowercase boldface letters.

EXAMPLE 2 ▶ **Five vectors**

(a) $\mathbf{v} = (1, 2)$ row vector

(b) $\mathbf{v} = \begin{pmatrix} 1 \\ 3 \\ -2 \end{pmatrix}$ column vector

(c) $\mathbf{0} = \begin{pmatrix} 0 \\ 0 \\ 0 \end{pmatrix}$ zero vector

(d) $\mathbf{v} = (2, -1, 5, 7, 4)$ row vector

(e) $\mathbf{v} = \begin{pmatrix} 1 \\ -1 \\ 2 \\ 3 \\ 5 \end{pmatrix}$ column vector ◀

Note

Unless stated otherwise, we will always assume that the numbers in a matrix or vector are real.

Matrices and vectors can be added together and multiplied by scalars (real numbers).

Addition of matrices*

Let $A = (a_{ij})$ and $B = (b_{ij})$ be two $m \times n$ matrices. Then the sum of A and B is the $m \times n$ matrix $A + B$ given by

$$A + B = (a_{ij} + b_{ij}) = \begin{pmatrix} a_{11} + b_{11} & a_{12} + b_{12} & \cdots & a_{1n} + b_{1n} \\ a_{21} + b_{21} & a_{22} + b_{22} & \cdots & a_{2n} + b_{2n} \\ \vdots & \vdots & & \vdots \\ a_{m1} + b_{m1} & a_{m2} + b_{m2} & \cdots & a_{mn} + b_{mn} \end{pmatrix}. \quad \textbf{(2)}$$

---

* The algebra of matrices, that is, the rules by which matrices can be added and multiplied, was developed by the English mathematician Arthur Cayley (1821–1895) in 1857. Matrices arose with Cayley in connection with linear transformations of the type

$$x^* = ax + by, \qquad y^* = cx + dy,$$

where a, b, c, d are real numbers, and which may be thought of as mapping the point (x, y) into the point (x^*, y^*). Clearly, the transformation above is completely determined by the four coefficients a, b, c, d, and so the transformation can be symbolized by the square array

$$\begin{pmatrix} a & b \\ c & d \end{pmatrix},$$

which we have called a **(square) matrix.**

In other words, $A + B$ is the $m \times n$ matrix obtained by adding the corresponding components of A and B.

Warning

The sum of two matrices is defined only when both matrices have the same size. Thus, for example, it is not possible to add together the matrices

$$\begin{pmatrix} 1 & 2 & 3 \\ 4 & 5 & 6 \end{pmatrix} \quad \text{and} \quad \begin{pmatrix} -1 & 0 \\ 2 & -5 \\ 4 & 7 \end{pmatrix}.$$

EXAMPLE 3 ▶ **The sum of two matrices**

$$\begin{pmatrix} 2 & 4 & -6 & 7 \\ 1 & 3 & 2 & 1 \\ -4 & 3 & -5 & 5 \end{pmatrix} + \begin{pmatrix} 0 & 1 & 6 & -2 \\ 2 & 3 & 4 & 3 \\ -2 & 1 & 4 & 4 \end{pmatrix} = \begin{pmatrix} 2 & 5 & 0 & 5 \\ 3 & 6 & 6 & 4 \\ -6 & 4 & -1 & 9 \end{pmatrix}. \blacktriangleleft$$

Multiplication of a matrix by a scalar

If $A = (a_{ij})$ is an $m \times n$ matrix and if α is a scalar, then the $m \times n$ matrix αA is given by

$$\alpha A = (\alpha a_{ij}) = \begin{pmatrix} \alpha a_{11} & \alpha a_{12} & \cdots & \alpha a_{1n} \\ \alpha a_{21} & \alpha a_{22} & \cdots & \alpha a_{2n} \\ \vdots & \vdots & & \vdots \\ \alpha a_{m1} & \alpha a_{m2} & \cdots & \alpha a_{mn} \end{pmatrix}. \tag{3}$$

In other words, $\alpha A = (\alpha a_{ij})$ is the matrix obtained by multiplying each component of A by α.

EXAMPLE 4 ▶ **Multiplying a matrix by various scalars**

Let

$$A = \begin{pmatrix} 1 & -3 & 4 & 2 \\ 3 & 1 & 4 & 6 \\ -2 & 3 & 5 & 7 \end{pmatrix}.$$

Then

$$2A - \begin{pmatrix} 2 & -6 & 8 & 4 \\ 6 & 2 & 8 & 12 \\ -4 & 6 & 10 & 14 \end{pmatrix},$$

$$-3A = \begin{pmatrix} -3 & 9 & -12 & -6 \\ -9 & -3 & -12 & -18 \\ 6 & -9 & -15 & -21 \end{pmatrix}, \quad \text{and} \quad 0A = \begin{pmatrix} 0 & 0 & 0 & 0 \\ 0 & 0 & 0 & 0 \\ 0 & 0 & 0 & 0 \end{pmatrix}. \blacktriangleleft$$

We now define the multiplication of two matrices. Quite obviously, we could define the product of two $m \times n$ matrices $A = (a_{ij})$ and $B = (b_{ij})$ to be the $m \times n$ matrix whose ijth component is $a_{ij}b_{ij}$. However, for just about all the important applications involving matrices, another kind of product is needed.

Product of two matrices

Let $A = (a_{ij})$ be an $m \times n$ matrix whose ith row is denoted by $\mathbf{a}_i$. Let $B = (b_{ij})$ be an $n \times p$ matrix whose jth column is denoted by $\mathbf{b}_j$. Then the product of A and B is an $m \times p$ matrix $C = (c_{ij})$, where

$$c_{ij} = a_{i1}b_{1j} + a_{i2}b_{2j} + \cdots + a_{in}b_{nj}. \tag{4}$$

The product in **(4)** is called the **dot product** of the vectors $\mathbf{a}_i$ and $\mathbf{b}_j$ and is denoted by $\mathbf{a}_i \cdot \mathbf{b}_j$. Thus **(4)** can be written as

$$c_{ij} = \mathbf{a}_i \cdot \mathbf{b}_j. \tag{5}$$

Warning

Two matices can be multiplied together only if the number of columns of the first is equal to the number of rows of the second. Otherwise, the vectors $\mathbf{a}_i$ and $\mathbf{b}_j$ have different numbers of components and the dot product in **(4)** and **(5)** is not defined.

EXAMPLE 5 ▶ Multiplying two matrices

If

$$A = \begin{pmatrix} 1 & 3 \\ -2 & 4 \end{pmatrix} \quad \text{and} \quad B = \begin{pmatrix} 3 & -2 \\ 5 & 6 \end{pmatrix},$$

calculate AB and BA.

Solution Let $C = (c_{ij}) = AB$. Then

$$c_{11} = \mathbf{a}_1 \cdot \mathbf{b}_1 = (1 \quad 3) \cdot \begin{pmatrix} 3 \\ 5 \end{pmatrix} = 3 + 15 = 18;$$

$$c_{12} = \mathbf{a}_1 \cdot \mathbf{b}_2 = (1 \quad 3) \cdot \begin{pmatrix} -2 \\ 6 \end{pmatrix} = -2 + 18 = 16;$$

$$c_{21} = (-2 \quad 4) \cdot \begin{pmatrix} 3 \\ 5 \end{pmatrix} = -6 + 20 = 14;$$

$$c_{22} = (-2 \quad 4) \cdot \begin{pmatrix} -2 \\ 6 \end{pmatrix} = 4 + 24 = 28.$$

Thus,

$$C = AB = \begin{pmatrix} 18 & 16 \\ 14 & 28 \end{pmatrix}.$$

Similarly, leaving out the intermediate steps, we see that

$$C' = BA = \begin{pmatrix} 3 & -2 \\ 5 & 6 \end{pmatrix}\begin{pmatrix} 1 & 3 \\ -2 & 4 \end{pmatrix} = \begin{pmatrix} 3+4 & 9-8 \\ 5-12 & 15+24 \end{pmatrix} = \begin{pmatrix} 7 & 1 \\ -7 & 39 \end{pmatrix}.$$

Note that $AB \neq BA$; in fact none of the corresponding entries are equal. Thus matrix multiplication is **not commutative.** ◀

Remark

Observe that if A is an $m \times n$ matrix and B is an $n \times m$ matrix, then the product AB is an $m \times m$ matrix and BA is an $n \times n$ matrix. Thus, not only are the entries different, but the sizes of the two products are unequal if $m \neq n$.

Although matrix multiplication is not commutative, it is associative.

THEOREM 1 ♦ **Associative law of matrix multiplication**

Let A be an $m \times n$ matrix, B an $n \times p$ matrix, and C a $p \times q$ matrix. Then $(AB)C$ and $A(BC)$ are $m \times q$ matrices and

$$(AB)C = A(BC) \quad ♦$$

The proof of this theorem is left as an exercise (see Problem 61).

Definition 1: Identity matrix

The $n \times n$ **identity matrix** is the $n \times n$ matrix with 1s down the **main diagonal*** and 0s everywhere else; that is,

$$I_n = (\delta_{ij}), \qquad \text{where } \delta_{ij} = \begin{cases} 1, & \text{if } i = j, \\ 0, & \text{if } i \neq j. \end{cases} \tag{6}$$

EXAMPLE 6 ► **The 3 × 3 and 5 × 5 identity matrices**

$$I_3 = \begin{pmatrix} 1 & 0 & 0 \\ 0 & 1 & 0 \\ 0 & 0 & 1 \end{pmatrix} \quad \text{and} \quad I_5 = \begin{pmatrix} 1 & 0 & 0 & 0 & 0 \\ 0 & 1 & 0 & 0 & 0 \\ 0 & 0 & 1 & 0 & 0 \\ 0 & 0 & 0 & 1 & 0 \\ 0 & 0 & 0 & 0 & 1 \end{pmatrix} \quad ◄$$

THEOREM 2 ♦ **Multiplying by the identity matrix**

Let A be a square $n \times n$ matrix. Then

$$AI_n = I_nA = A.$$

In other words, I_n commutes with every $n \times n$ matrix and leaves it unchanged after multiplication on the left or right.

Note that I_n behaves for $n \times n$ matrices in the way the number 1 behaves for real numbers (since $1 \cdot a = a \cdot 1 = a$ for every real number a).

Proof Let c_{ij} be the ijth element of AI_n. Then

$$c_{ij} = a_{i1}\delta_{1i} + a_{i2}\delta_{2j} + \cdots + a_{ij}\delta_{jj} + \cdots + a_{in}\delta_{nj}.$$

But, from **(6)**, this sum is equal to a_{ij}. Thus $AI_n = A$. In a similar fashion we can show that $I_nA = A$, and this proves the theorem. ♦

Notation

From now on we will write the identity matrix simply as I, since if A is $n \times n$, the products IA and AI are defined only if I is also $n \times n$.

---

* The main diagonal of $A = (a_{ij})$ consists of the components a_{11}, a_{22}, a_{33}, and so on. Unless otherwise stated, we refer to the main diagonal simply as the **diagonal.**

Definition 2: The inverse of a matrix

Let A and B be $n \times n$ matrices. Suppose that

$$AB = BA = I.$$

Then B is called the **inverse** of A and is written as A^{-1}. We then have

$$AA^{-1} = A^{-1}A = I.$$

If A has an inverse, then A is said to be **invertible.**

Remark

From this definition it immediately follows that $(A^{-1})^{-1} = A$ if A is invertible.

Remark

This definition does *not* state that every squre matrix has an inverse. In fact, there are many square matrices that have no inverse.

In Definition 2 we defined *the* inverse of a matrix. This statement suggests that inverses are unique. This is indeed the case, as the following theorem shows.

THEOREM 3 ♦ Uniqueness of the inverse

If a square matrix A is invertible, then its inverse is unique.

Proof Suppose that B and C are two inverses for A. We can show that $B = C$. By definition, we have $AB = BA = I$ and $AC = CA = I$. Then $B(AC) = BI = B$ and $(BA)C = IC = C$. But $B(AC) = (BA)C$, since matrix multiplication is associative, by the associative law of matrix multiplication. Hence $B = C$ and the theorem is proved. ♦

Another important fact about inverses is given below.

THEOREM 4 ♦ The product of invertible matrices is invertible

Let A and B be invertible $n \times n$ matrices. Then AB is invertible and

$$(AB)^{-1} = B^{-1}A^{-1}.$$

Proof To prove this result, we refer to Definition 2; that is, $B^{-1}A^{-1} = (AB)^{-1}$ if and only if $B^{-1}A^{-1}(AB) = (AB)(B^{-1}A^{-1}) = I$. But this follows, since

$$(B^{-1}A^{-1})(AB) = B^{-1}(A^{-1}A)B = B^{-1}IB = B^{-1}B = I$$

and

$$(AB)(B^{-1}A^{-1}) = A(BB^{-1})A^{-1} = AIA^{-1} = AA^{-1} = I. \quad ♦$$

There are two basic questions that come to mind once we have defined the inverse of a matrix:

1. What matrices do have inverses?
2. If a matrix has an inverse, how can we compute it?

There are two essentially different ways to compute the inverse of a square matrix.* The method we describe below involves less work when dealing with large matrices ($n > 3$).

* The other method involves the computation of det A and n^2 $(n - 1) \times (n - 1)$ determinants.

Computation of inverses by Gauss-Jordan elimination

Consider the system

$$\begin{aligned} 2x_1 + 4x_2 &= 6 \\ 3x_1 + 5x_2 &= 2. \end{aligned} \tag{7}$$

We solve it as follows:

Step 1 Divide the first row by 2:

$$\begin{aligned} x_1 + 2x_2 &= 3 \\ 3x_1 + 5x_2 &= 2. \end{aligned}$$

Step 2 Multiply the first row by -3, and add it to the second row:

$$\begin{aligned} x_1 + 2x_2 &= 3 \\ -x_2 &= -7. \end{aligned}$$

Step 3 Multiply the second row by -1:

$$\begin{aligned} x_1 + 2x_2 &= 3 \\ x_2 &= 7. \end{aligned}$$

Step 4 Multiply the second row by -2, and add it to the first row:

$$\begin{aligned} x_1 &= -11 \\ x_2 &= 7. \end{aligned}$$

The solution is now displayed.

Check

$$\begin{aligned} 2(-11) + 4(7) &= 6 \\ 3(-11) + 5(7) &= 2 \end{aligned}$$

We see that performing certain **elementary row operations** on the original systems leads to an equivalent system in which the solution is displayed.

Elementary row operations

Operation	Symbol
(i) Multiply the ith row by a nonzero number c	$M_i(c)$
(ii) Multiply the ith row by c and add it to the jth row	$A_{ij}(c)$
(iii) Permute (interchange) the ith and jth rows	P_{ij}

We can write system **(7)** as an **augmented matrix:**

$$\left(\begin{array}{cc|c} 2 & 4 & 6 \\ 3 & 5 & 2 \end{array}\right)$$

The first line reads $2x_1 + 4x_2 = 6$, and so on. We repeat the solution using augmented matrices and our elementary row operations notation:

$$\left(\begin{array}{cc|c} 2 & 4 & 6 \\ 3 & 5 & 2 \end{array}\right) \xrightarrow{M_1(\frac{1}{2})} \left(\begin{array}{cc|c} 1 & 2 & 3 \\ 3 & 5 & 2 \end{array}\right) \xrightarrow{A_{1,2}(-3)} \left(\begin{array}{cc|c} 1 & 2 & 3 \\ 0 & -1 & -7 \end{array}\right)$$

$$\xrightarrow{M_2(-1)} \left(\begin{array}{cc|c} 1 & 2 & 3 \\ 0 & 1 & 7 \end{array}\right) \xrightarrow{A_{2,1}(-2)} \left(\begin{array}{cc|c} 1 & 0 & -11 \\ 0 & 1 & 7 \end{array}\right)$$

Again, we "see" the solution $x_1 = -11$, $x_2 = 7$. The process we have carried out is called **row reduction.**

System **(7)** can also be written in the matrix form

$$A\mathbf{x} = \mathbf{b} \tag{8}$$

where

$$A = \begin{pmatrix} 2 & 4 \\ 3 & 5 \end{pmatrix}, \qquad \mathbf{x} = \begin{pmatrix} x_1 \\ x_2 \end{pmatrix}, \qquad \text{and} \qquad \mathbf{b} = \begin{pmatrix} 6 \\ 2 \end{pmatrix}.$$

Suppose A is invertible. Then we can multiply both sides of **(8)** on the left by A^{-1} to obtain

$$\begin{aligned} A^{-1}A\mathbf{x} &= A^{-1}\mathbf{b} \\ I\mathbf{x} &= A^{-1}\mathbf{b} \\ \mathbf{x} &= A^{-1}\mathbf{b} \end{aligned}$$

That is, in the row reduction process

$$A\mathbf{x} = \mathbf{b} = I\mathbf{b}$$

has been transformed into

$$I\mathbf{x} = A^{-1}\mathbf{b}$$

and $A^{-1}\mathbf{b}$ is the unique solution to the system. This suggests the following procedure for finding the inverse of a matrix. More details and proofs can be found in any standard linear algebra text.* The following procedure is called **Gauss-Jordan elimination.**

Procedure for computing the inverse of a square matrix A

Step 1 Write the augmented matrix $(A|I)$.
Step 2 Use row reduction to reduce A to either
(a) the identity matrix or
(b) a matrix with a row of zeros.

* See, for example, S. I. Grossman, *Elementary Linear Algebra,* 5th ed. (Philadelphia, PA: Saunders, 1994), Chapter 1.

Step 3 Decide if A is invertible.

(a) If A can be reduced to the identity matrix I, then A^{-1} is the matrix to the right of the vertical bar.

(b) If the row reduction leads to a row of zeros to the left of the vertical bar, then A is not invertible.

EXAMPLE 7 ▶ **Computing the inverse of a 3 × 3 matrix**

Let

$$A = \begin{pmatrix} 2 & 4 & 6 \\ 4 & 5 & 6 \\ 3 & 1 & -2 \end{pmatrix}.$$

Calculate A^{-1} if it exists.

Solution *Step 1*

$$\left(\begin{array}{ccc|ccc} \multicolumn{3}{c|}{\overbrace{\qquad\qquad}^{A}} & \multicolumn{3}{c}{\overbrace{\qquad\quad}^{I}} \\ 2 & 4 & 6 & 1 & 0 & 0 \\ 4 & 5 & 6 & 0 & 1 & 0 \\ 3 & 1 & -2 & 0 & 0 & 1 \end{array}\right)$$

Step 2

$$\xrightarrow{M_1(\frac{1}{2})} \left(\begin{array}{ccc|ccc} 1 & 2 & 3 & \frac{1}{2} & 0 & 0 \\ 4 & 5 & 6 & 0 & 1 & 0 \\ 3 & 1 & -2 & 0 & 0 & 1 \end{array}\right)$$

$$\xrightarrow[A_{1,3}(-3)]{A_{1,2}(-4)} \left(\begin{array}{ccc|ccc} 1 & 2 & 3 & \frac{1}{2} & 0 & 0 \\ 0 & -3 & -6 & -2 & 1 & 0 \\ 0 & -5 & -11 & -\frac{3}{2} & 0 & 1 \end{array}\right)$$

$$\xrightarrow{M_2(-\frac{1}{3})} \left(\begin{array}{ccc|ccc} 1 & 2 & 3 & \frac{1}{2} & 0 & 0 \\ 0 & 1 & 2 & \frac{2}{3} & -\frac{1}{3} & 0 \\ 0 & -5 & -11 & -\frac{3}{2} & 0 & 1 \end{array}\right)$$

$$\xrightarrow[A_{2,3}(5)]{A_{2,1}(-2)} \left(\begin{array}{ccc|ccc} 1 & 0 & -1 & -\frac{5}{6} & \frac{2}{3} & 0 \\ 0 & 1 & 2 & \frac{2}{3} & -\frac{1}{3} & 0 \\ 0 & 0 & -1 & \frac{11}{6} & -\frac{5}{3} & -1 \end{array}\right)$$

$$\xrightarrow{M_3(-1)} \left(\begin{array}{ccc|ccc} 1 & 0 & -1 & -\frac{5}{6} & \frac{2}{3} & 0 \\ 0 & 1 & 2 & \frac{2}{3} & -\frac{1}{3} & 0 \\ 0 & 0 & 1 & -\frac{11}{6} & \frac{5}{3} & -1 \end{array}\right)$$

$$\xrightarrow[A_{3,2}(-2)]{A_{3,1}(1)} \left(\begin{array}{ccc|ccc} 1 & 0 & 0 & -\frac{8}{3} & \frac{7}{3} & -1 \\ 0 & 1 & 0 & \frac{13}{3} & -\frac{11}{3} & 2 \\ 0 & 0 & 1 & -\frac{11}{6} & \frac{5}{3} & -1 \end{array}\right).$$

(The left block is labeled I and the right block A^{-1}.)

Step 3 Since A has been reduced to I, we conclude that

$$A^{-1} = \begin{pmatrix} -\frac{8}{3} & \frac{7}{3} & -1 \\ \frac{13}{3} & -\frac{11}{3} & 2 \\ -\frac{11}{6} & \frac{5}{3} & -1 \end{pmatrix}.$$

After doing all these computations, it is always a good idea to check your answer.

Check

$$A^{-1}A = \begin{pmatrix} -\frac{8}{3} & \frac{7}{3} & -1 \\ \frac{13}{3} & -\frac{11}{3} & 2 \\ -\frac{11}{6} & \frac{5}{3} & -1 \end{pmatrix}\begin{pmatrix} 2 & 4 & 6 \\ 4 & 5 & 6 \\ 3 & 1 & -2 \end{pmatrix} = \begin{pmatrix} 1 & 0 & 0 \\ 0 & 1 & 0 \\ 0 & 0 & 1 \end{pmatrix}$$

Similarly, $AA^{-1} = I$. ◀

EXAMPLE 8 ▶ **A 3 × 3 matrix that is not invertible**

Let

$$A = \begin{pmatrix} 1 & -3 & 4 \\ 2 & -5 & 7 \\ 0 & -1 & 1 \end{pmatrix}.$$

Calculate A^{-1} if it exists.

Solution We proceed as before:

$$\left(\begin{array}{ccc|ccc} 1 & -3 & 4 & 1 & 0 & 0 \\ 2 & -5 & 7 & 0 & 1 & 0 \\ 0 & -1 & 1 & 0 & 0 & 1 \end{array}\right) \xrightarrow{A_{1,2}(-2)} \left(\begin{array}{ccc|ccc} 1 & -3 & 4 & 1 & 0 & 0 \\ 0 & 1 & -1 & -2 & 1 & 0 \\ 0 & -1 & 1 & 0 & 0 & 1 \end{array}\right)$$

$$\xrightarrow[A_{2,3}(1)]{A_{2,1}(3)} \left(\begin{array}{ccc|ccc} 1 & 0 & 1 & -5 & 3 & 0 \\ 0 & 1 & -1 & -2 & 1 & 0 \\ 0 & 0 & 0 & -2 & 1 & 1 \end{array}\right).$$

There is now a row of zeros to the left of the vertical bar. We conclude that A is not invertible. ◀

It is possible to determine whether a matrix is invertible by computing its determinant.

THEOREM 5 ♦ **Determinants and invertibility**

A square matrix A is invertible if and only if $\det A \neq 0$. ♦

EXAMPLE 9 ► **A matrix is invertible if and only if its determinant is nonzero**

In Examples 7 and 8 we have

$$\det\begin{pmatrix} 2 & 4 & 6 \\ 4 & 5 & 6 \\ 3 & 1 & -2 \end{pmatrix} = 6 \neq 0 \quad \text{and} \quad \det\begin{pmatrix} 1 & -3 & 4 \\ 2 & -5 & 7 \\ 0 & -1 & 1 \end{pmatrix} = 0.$$

Thus the first matrix is invertible while the second one is not. ◄

It is especially easy to compute the inverse of a 2×2 matrix. The following theorem can be proved by matrix multiplication.

THEOREM 6 ♦ **The inverse of a 2 × 2 matrix**

Let

$$A = \begin{pmatrix} a_{11} & a_{12} \\ a_{21} & a_{22} \end{pmatrix}$$

be a 2×2 matrix. Then A is invertible if and only if $\det A = a_{11}a_{22} - a_{12}a_{21} \neq 0$. If $\det A \neq 0$, then

$$A^{-1} = \frac{1}{\det A}\begin{pmatrix} a_{22} & -a_{12} \\ -a_{21} & a_{11} \end{pmatrix} = \begin{pmatrix} \dfrac{a_{22}}{a_{11}a_{22} - a_{12}a_{21}} & \dfrac{-a_{12}}{a_{11}a_{22} - a_{12}a_{21}} \\ \dfrac{-a_{21}}{a_{11}a_{22} - a_{12}a_{21}} & \dfrac{a_{11}}{a_{11}a_{22} - a_{12}a_{21}} \end{pmatrix} \quad (9)$$ ♦

Remark

Theorem 6 shows us that finding the inverse of an invertible 2×2 matrix is very simple. We just reverse the diagonal components, multiply each off-diagonal component by -1, and divide the result by $\det A$.

EXAMPLE 10 ► **Computing the inverse of a 2 × 2 matrix**

Compute the inverse (if it exists) of

$$A = \begin{pmatrix} 2 & 5 \\ -7 & 4 \end{pmatrix}.$$

Solution Since $\det A = 43$, A is invertible. Then, from Equation **(9)**,

$$A^{-1} = \frac{1}{43}\begin{pmatrix} 4 & -5 \\ 7 & 2 \end{pmatrix} = \begin{pmatrix} \dfrac{4}{43} & -\dfrac{5}{43} \\ \dfrac{7}{43} & \dfrac{2}{43} \end{pmatrix}.$$

Check

$$A^{-1}A = \frac{1}{43}\begin{pmatrix} 4 & -5 \\ 7 & 2 \end{pmatrix}\begin{pmatrix} 2 & 5 \\ -7 & 4 \end{pmatrix} = \frac{1}{43}\begin{pmatrix} 43 & 0 \\ 0 & 43 \end{pmatrix} = \begin{pmatrix} 1 & 0 \\ 0 & 1 \end{pmatrix} = I.$$

Similarly, $AA^{-1} = I$; this shows that our answer is correct. ◄

Matrix functions, derivatives, and integrals

We have discussed matrix theory here in order to apply it to the solution of systems of differential equations. In such a context, the matrices we encounter often have components that are functions rather than real numbers. For that reason, we provide some elementary definitions.

vector function

An n-component **vector function**

$$\mathbf{v}(t) = \begin{pmatrix} v_1(t) \\ v_2(t) \\ \vdots \\ v_n(t) \end{pmatrix} \tag{10}$$

is an n-vector each of whose components is a function (usually assumed to be continuous).

matrix function

An $n \times n$ **matrix function** $A(t)$ is an $n \times n$ matrix

$$A(t) = \begin{pmatrix} a_{11}(t) & a_{12}(t) & \cdots & a_{1n}(t) \\ a_{21}(t) & a_{22}(t) & \cdots & a_{2n}(t) \\ \vdots & \vdots & & \vdots \\ a_{n1}(t) & a_{n2}(t) & \cdots & a_{nn}(t) \end{pmatrix} \tag{11}$$

each of whose n^2 components is a function. We may add and multiply vector and matrix functions in the same way that we add and multiply constant vectors and matrices. Thus if, for example,

$$A(t) = \begin{pmatrix} a_{11}(t) & a_{12}(t) \\ a_{21}(t) & a_{22}(t) \end{pmatrix} \quad \text{and} \quad B(t) = \begin{pmatrix} b_{11}(t) & b_{12}(t) \\ b_{21}(t) & b_{22}(t) \end{pmatrix},$$

then

$$A(t)B(t) = \begin{pmatrix} a_{11}(t)b_{11}(t) + a_{12}(t)b_{21}(t) & a_{11}(t)b_{12}(t) + a_{12}(t)b_{22}(t) \\ a_{21}(t)b_{11}(t) + a_{22}(t)b_{21}(t) & a_{21}(t)b_{12}(t) + a_{22}(t)b_{22}(t) \end{pmatrix}.$$

derivatives and integrals

We may also differentiate and integrate vector and matrix functions, componentwise. Thus if $\mathbf{v}(t)$ is given by **(10),** then

$$\mathbf{v}'(t) = \begin{pmatrix} v_1'(t) \\ v_2'(t) \\ \vdots \\ v_n'(t) \end{pmatrix} \quad \text{and} \quad \int_{t_0}^{t} \mathbf{v}(s)\,ds = \begin{pmatrix} \int_{t_0}^{t} v_1(s)\,ds \\ \int_{t_0}^{t} v_2(s)\,ds \\ \vdots \\ \int_{t_0}^{t} v_n(s)\,ds \end{pmatrix}$$

Similarly, if $A(t)$ is given by **(11),** then

$$A'(t) = \begin{pmatrix} a_{11}'(t) & a_{12}'(t) & \cdots & a_{1n}'(t) \\ a_{21}'(t) & a_{22}'(t) & \cdots & a_{2n}'(t) \\ \vdots & \vdots & & \vdots \\ a_{n1}'(t) & a_{n2}'(t) & \cdots & a_{nn}'(t) \end{pmatrix}$$

and

$$\int_{t_0}^{t} A(s)\, ds = \begin{pmatrix} \int_{t_0}^{t} a_{11}(s)\, ds & \int_{t_0}^{t} a_{12}(s)\, ds & \cdots & \int_{t_0}^{t} a_{1n}(s)\, ds \\ \int_{t_0}^{t} a_{21}(s)\, ds & \int_{t_0}^{t} a_{22}(s)\, ds & \cdots & \int_{t_0}^{t} a_{2n}(s)\, ds \\ \vdots & \vdots & & \vdots \\ \int_{t_0}^{t} a_{n1}(s)\, ds & \int_{t_0}^{t} a_{n2}(s)\, ds & \cdots & \int_{t_0}^{t} a_{nn}(s)\, ds \end{pmatrix}.$$

EXAMPLE 11 ► The derivative and integral of a vector function

Let

$$\mathbf{v}(t) = \begin{pmatrix} t \\ t^2 \\ \sin t \\ e^t \end{pmatrix}.$$

Then

$$\mathbf{v}'(t) = \begin{pmatrix} 1 \\ 2t \\ \cos t \\ e^t \end{pmatrix} \quad \text{and} \quad \int_0^t \mathbf{v}(s)\, ds = \begin{pmatrix} \frac{t^2}{2} \\ \frac{t^3}{3} \\ 1 - \cos t \\ e^t - 1 \end{pmatrix}. \blacktriangleleft$$

SELF-QUIZ

I. Which of the following is $2A - 4B$ if $A = (2 \quad 0 \quad 0)$ and $B = (3 \quad 1)$?
(a) $(-8 \quad -4)$ **(b)** $(5 \quad 0 \quad 1)$
(c) $(16 \quad -4 \quad 0)$
(d) This operation cannot be performed.

II. Which of the following must be the second row of matrix B if $3A - B = 2C$ for

$$A = \begin{pmatrix} 1 & -1 & 1 \\ 0 & 0 & 3 \\ 4 & 2 & 0 \end{pmatrix} \quad \text{and} \quad C = \begin{pmatrix} 1 & 0 & 0 \\ 0 & 1 & 0 \\ 0 & 0 & 1 \end{pmatrix}?$$

(a) $-3 \quad 2 \quad 6$ **(b)** $0 \quad -2 \quad 9$
(c) $3 \quad -2 \quad 6$ **(d)** $0 \quad 2 \quad -9$

III. Which of the following would be the size of the product matrix AB when a 2×4 matrix A is multiplied by a 4×3 matrix B?
(a) 2×3 **(b)** 3×2 **(c)** 4×4
(d) This product cannot be found.

IV. Which of the following is true of matrices A and B if AB is a column vector?
(a) B is a column vector.
(b) A is a row vector.
(c) A and B are square vectors.
(d) The numbers of rows in A must equal the number of columns in B.

V. Which of the following is true about a product AB if A is a 4×5 matrix?
(a) B must have 4 rows and the result will have 5 columns.
(b) B must have 5 columns and the result will be a square matrix.
(c) B must have 4 columns and the result will have 5 rows.
(d) B must have 5 rows and the result will have 4 rows.

VI. Which of the following is true?
(a) Every square matrix has an inverse.
(b) A square matrix has an inverse if its row reduction leads to a row of zeros.
(c) A square matrix is invertible if it has an inverse.
(d) A square matrix B is the inverse of A if $AI = B$.

VII. Which of the following is invertible?

(a) $\begin{pmatrix} 1 & 3 \\ -3 & -9 \end{pmatrix}$ **(b)** $\begin{pmatrix} 6 & -1 \\ 1 & -\frac{1}{6} \end{pmatrix}$

(c) $\begin{pmatrix} 2 & -3 \\ 1 & -1 \end{pmatrix}$ **(d)** $\begin{pmatrix} 1 & 0 \\ 2 & 0 \end{pmatrix}$

VIII. Which of the following is true of an invertible matrix A?
(a) The product of A and I is A^{-1}.
(b) A is a 2×3 matrix.
(c) $A = A^{-1}$.
(d) A is a square matrix.

Answers to Self-Quiz

I. d **II.** b **III.** a **IV.** a **V.** d **VI.** c **VII.** c **VIII.** d

PROBLEMS 9.1

In Problems 1–12 perform the indicated computation with

$$A = \begin{pmatrix} 1 & 3 \\ 2 & 5 \\ -1 & 2 \end{pmatrix}, \quad B = \begin{pmatrix} -2 & 0 \\ 1 & 4 \\ -7 & 5 \end{pmatrix}, \quad \text{and} \quad C = \begin{pmatrix} -1 & 1 \\ 4 & 6 \\ -7 & 3 \end{pmatrix}.$$

1. $3A$

2. $A + B$

3. $A - C$

4. $2C - 5A$

5. $0B$ (0 is the scalar zero)

6. $-7A + 3B$

7. $A + B + C$

8. $C - A - B$

9. $2A - 3B + 4C$

10. $7C - B + 2A$

11. Find a matrix D such that $2A + B - D$ is the 3×2 zero matrix.

12. Find a matrix E such that $A + 2B - 3C + E$ is the 3×2 zero matrix.

In Problems 13–20 perform the indicated computation with

$$A = \begin{pmatrix} 1 & -1 & 2 \\ 3 & 4 & 5 \\ 0 & 1 & -1 \end{pmatrix}, \quad B = \begin{pmatrix} 0 & 2 & 1 \\ 3 & 0 & 5 \\ 7 & -6 & 0 \end{pmatrix}, \quad \text{and} \quad C = \begin{pmatrix} 0 & 0 & 2 \\ 3 & 1 & 0 \\ 0 & -2 & 4 \end{pmatrix}.$$

13. $A - 2B$

14. $3A - C$

15. $A + B + C$

16. $2A - B + 2C$

17. $C - A - B$

18. $4C - 2B + 3A$

19. Find a matrix D such that $A + B + C + D$ is the 3×3 zero matrix.

20. Find a matrix E such that $3C - 2B + 8A - 4E$ is the 3×3 zero matrix.

In Problems 21–34 perform the indicated computation.

21. $\begin{pmatrix} 2 & 3 \\ -1 & 2 \end{pmatrix}\begin{pmatrix} 4 & 1 \\ 0 & 6 \end{pmatrix}$

22. $\begin{pmatrix} 3 & -2 \\ 1 & 4 \end{pmatrix}\begin{pmatrix} -5 & 6 \\ 1 & 3 \end{pmatrix}$

23. $\begin{pmatrix} 1 & -1 \\ 1 & 1 \end{pmatrix}\begin{pmatrix} -1 & 0 \\ 2 & 3 \end{pmatrix}$

24. $\begin{pmatrix} -5 & 6 \\ 1 & 3 \end{pmatrix}\begin{pmatrix} 3 & -2 \\ 1 & 4 \end{pmatrix}$

25. $\begin{pmatrix} -4 & 5 & 1 \\ 0 & 4 & 2 \end{pmatrix}\begin{pmatrix} 3 & -1 & 1 \\ 5 & 6 & 4 \\ 0 & 1 & 2 \end{pmatrix}$

26. $\begin{pmatrix} 7 & 1 & 4 \\ 2 & -3 & 5 \end{pmatrix}\begin{pmatrix} 1 & 6 \\ 0 & 4 \\ -2 & 3 \end{pmatrix}$

27. $\begin{pmatrix} 1 & 6 \\ 0 & 4 \\ -2 & 3 \end{pmatrix}\begin{pmatrix} 7 & 1 & 4 \\ 2 & -3 & 5 \end{pmatrix}$

28. $\begin{pmatrix} 1 & 4 & -2 \\ 3 & 0 & 4 \end{pmatrix}\begin{pmatrix} 0 & 1 \\ 2 & 3 \end{pmatrix}$

29. $\begin{pmatrix} 1 & 4 & 6 \\ -2 & 3 & 5 \\ 1 & 0 & 4 \end{pmatrix}\begin{pmatrix} 2 & -3 & 5 \\ 1 & 0 & 6 \\ 2 & 3 & 1 \end{pmatrix}$

30. $\begin{pmatrix} 2 & -3 & 5 \\ 1 & 0 & 6 \\ 2 & 3 & 1 \end{pmatrix}\begin{pmatrix} 1 & 4 & 6 \\ -2 & 3 & 5 \\ 1 & 0 & 4 \end{pmatrix}$

31. $(1 \quad 4 \quad 0 \quad 2)\begin{pmatrix} 3 & -6 \\ 2 & 4 \\ 1 & 0 \\ -2 & 3 \end{pmatrix}$

32. $\begin{pmatrix} 3 & 2 & 1 & -2 \\ -6 & 4 & 0 & 3 \end{pmatrix}\begin{pmatrix} 1 \\ 4 \\ 0 \\ 2 \end{pmatrix}$

33. $\begin{pmatrix} 3 & -2 & 1 \\ 4 & 0 & 6 \\ 5 & 1 & 9 \end{pmatrix}\begin{pmatrix} 1 & 0 & 0 \\ 0 & 1 & 0 \\ 0 & 0 & 1 \end{pmatrix}$

34. $\begin{pmatrix} 1 & 0 & 0 \\ 0 & 1 & 0 \\ 0 & 0 & 1 \end{pmatrix}\begin{pmatrix} 3 & -2 & 1 \\ 4 & 0 & 6 \\ 5 & 1 & 9 \end{pmatrix}$

35. Let A be a square matrix. Then A^2 is defined simply as AA. Calculate

$$\begin{pmatrix} 2 & -1 \\ 4 & 6 \end{pmatrix}^2.$$

36. Calculate A^2, where

$$A = \begin{pmatrix} 1 & -2 & 4 \\ 2 & 0 & 3 \\ 1 & 1 & 5 \end{pmatrix}.$$

37. Calculate A^3, where

$$A = \begin{pmatrix} -1 & 2 \\ 3 & 4 \end{pmatrix}.$$

38. Calculate A^2, A^3, A^4, and A^5, where

$$A = \begin{pmatrix} 0 & 1 & 0 & 0 \\ 0 & 0 & 1 & 0 \\ 0 & 0 & 0 & 1 \\ 0 & 0 & 0 & 0 \end{pmatrix}.$$

In Problems 39–48 use the methods of this section to determine whether the given matrix is invertible. If it is, compute the inverse.

39. $\begin{pmatrix} 3 & 2 \\ 1 & 2 \end{pmatrix}$

40. $\begin{pmatrix} 3 & 6 \\ -4 & -8 \end{pmatrix}$

41. $\begin{pmatrix} 0 & 1 \\ 1 & 0 \end{pmatrix}$

42. $\begin{pmatrix} 1 & 1 & 1 \\ 0 & 2 & 3 \\ 5 & 5 & 1 \end{pmatrix}$

43. $\begin{pmatrix} 3 & 2 & 1 \\ 0 & 2 & 2 \\ 0 & 1 & -1 \end{pmatrix}$

44. $\begin{pmatrix} 1 & 1 & 1 \\ 0 & 1 & 1 \\ 0 & 0 & 1 \end{pmatrix}$

45. $\begin{pmatrix} 1 & 2 & 3 \\ 1 & 1 & 2 \\ 0 & 1 & 2 \end{pmatrix}$

46. $\begin{pmatrix} 3 & 1 & 0 \\ 1 & -1 & 2 \\ 1 & 1 & 1 \end{pmatrix}$

47. $\begin{pmatrix} 2 & -1 & 4 \\ -1 & 0 & 5 \\ 19 & -7 & 3 \end{pmatrix}$

48. $\begin{pmatrix} 1 & 6 & 2 \\ -2 & 3 & 5 \\ 7 & 12 & -4 \end{pmatrix}$

49. Use the methods of this section to find the inverses of the following matrices with complex entries:

(a) $\begin{pmatrix} i & 2 \\ 1 & -i \end{pmatrix}$ **(b)** $\begin{pmatrix} 1-i & 0 \\ 0 & 1+i \end{pmatrix}$ **(c)** $\begin{pmatrix} 1 & i & 0 \\ -i & 0 & 1 \\ 0 & 1+i & 1-i \end{pmatrix}$

50. Show that for every real number θ the matrix

$$\begin{pmatrix} \sin\theta & \cos\theta & 0 \\ \cos\theta & -\sin\theta & 0 \\ 0 & 0 & 1 \end{pmatrix}$$

is invertible and find its inverse.

51. Calculate the inverse of

$$A = \begin{pmatrix} 2 & 0 & 0 \\ 0 & 3 & 0 \\ 0 & 0 & 4 \end{pmatrix}.$$

52. A square matrix $A = (a_{ij})$ is called **diagonal** if all its elements off the main diagonal are zero; that is, $a_{ij} = 0$ if $i \neq j$. [The matrix of Problem 51 is diagonal.] Show that a diagonal matrix is invertible if and only if each of its diagonal components is nonzero.

53. Let

$$A = \begin{pmatrix} a_{11} & 0 & \cdots & 0 \\ 0 & a_{22} & \cdots & 0 \\ & & \ddots & \\ 0 & 0 & \cdots & a_{nn} \end{pmatrix}$$

be a diagonal matrix such that each of its diagonal components is nonzero. Calculate A^{-1}.

54. Calculate the inverse of

$$A = \begin{pmatrix} 2 & 1 & -1 \\ 0 & 3 & 4 \\ 0 & 0 & 5 \end{pmatrix}.$$

55. Show that the matrix

$$A = \begin{pmatrix} 1 & 0 & 0 \\ -2 & 0 & 0 \\ 4 & 6 & 1 \end{pmatrix}$$

is not invertible.

56. A square matrix is called **upper (lower) triangular** if all its elements below (above) the main diagonal are zero. [The matrix of Problem 54 is upper triangular and the matrix of Problem 55 is lower triangular.] Show that an upper or lower triangular matrix is invertible if and only if each of its diagonal elements is nonzero.

In Problems 57–60 find the derivative and integral of each vector or matrix function.

57. $x(t) = (t, \sin t)$

58. $A(t) = \begin{pmatrix} \sqrt{t} & t^2 \\ e^{2t} & \sin 2t \end{pmatrix}$

59. $y(t) = \begin{pmatrix} e^t \\ \cos t \\ \tan t \end{pmatrix}$

60. $B(t) = \begin{pmatrix} \ln t & e^t \sin t & e^t \cos t \\ t^{5/2} & -\cos t & -\sin t \\ \dfrac{1}{t} & te^{t^2} & t^2 e^{t^3} \end{pmatrix}$

---

61. Prove Theorem 1. [*Hint:* Use a double sum to show that the ijth component of $(AB)C$ is equal to the ijth component of $A(BC)$.]

9.2 MATRICES AND LINEAR SYSTEMS OF DIFFERENTIAL EQUATIONS

Consider the following linear system of n first-order equations

$$\begin{aligned} x_1' &= a_{11}(t)x_1 + a_{12}(t)x_2 + \cdots + a_{1n}(t)x_n + f_1(t), \\ x_2' &= a_{21}(t)x_1 + a_{22}(t)x_2 + \cdots + a_{2n}(t)x_n + f_2(t), \\ &\vdots \\ x_n' &= a_{n1}(t)x_1 + a_{n2}(t)x_2 + \cdots + a_{nn}(t)x_n + f_n(t), \end{aligned} \tag{1}$$

which is **nonhomogeneous** if at least one of the functions $f_i(t)$, $i = 1, 2, \ldots, n$ is not the zero function, and the **associated homogeneous system**

$$\begin{aligned} x_1' &= a_{11}(t)x_1 + a_{12}(t)x_2 + \cdots + a_{1n}(t)x_n, \\ x_2' &= a_{21}(t)x_1 + a_{22}(t)x_2 + \cdots + a_{2n}(t)x_n, \\ &\vdots \\ x_n' &= a_{n1}(t)x_1 + a_{n2}(t)x_2 + \cdots + a_{nn}(t)x_n. \end{aligned} \tag{2}$$

We assume that all the coefficient functions a_{ij} and the functions f_i are continuous on some common interval containing the point t_0.

We now define the vector function $\mathbf{x}(t)$, the matrix function $A(t)$, and the vector function $\mathbf{f}(t)$ as follows:

$$\mathbf{x}(t) = \begin{pmatrix} x_1(t) \\ x_2(t) \\ \vdots \\ x_n(t) \end{pmatrix}, \quad A(t) = \begin{pmatrix} a_{11}(t) & a_{12}(t) & \ldots & a_{1n}(t) \\ a_{21}(t) & a_{22}(t) & \ldots & a_{2n}(t) \\ \vdots & \vdots & & \vdots \\ a_{n1}(t) & a_{n2}(t) & \ldots & a_{nn}(t) \end{pmatrix},$$

$$\mathbf{f}(t) = \begin{pmatrix} f_1(t) \\ f_2(t) \\ \vdots \\ f_n(t) \end{pmatrix}. \tag{3}$$

Then, using **(3)**, we can rewrite system **(1)** in a vector form.

Vector differential equation

$$\mathbf{x}'(t) = A(t)\mathbf{x}(t) + \mathbf{f}(t) \tag{4}$$

System **(2)** becomes

$$\mathbf{x}'(t) = A(t)\mathbf{x}(t). \tag{5}$$

It is clear from what has already been said that any linear differential equation or system can be written in form **(5)** if it is homogeneous, and in form **(4)** if it is nonhomogeneous. The reason for writing a system in these forms is that, besides the obvious advantage of compactness of notation, equations **(4)** and **(5)** behave very much like first-order linear differential equations, as we will see. It will be very easy to work with systems of equations in this way once we get used to the notation.

EXAMPLE 1 ▶ **Writing a system as a vector differential equation**

Consider the system

$$\begin{aligned} x_1' &= (2t)x_1 + (\sin t)x_2 - e^t x_3 - e^t, \\ x_2' &= -t^3 x_1 + e^{\sin t} x_2 - (\ln t)x_3 + \cos t, \\ x_3' &= 2x_1 - 5tx_2 + 2tx_3 + \tan t. \end{aligned}$$

It can be rewritten as

$$\begin{pmatrix} x_1 \\ x_2 \\ x_3 \end{pmatrix}' = \begin{pmatrix} 2t & \sin t & -e^t \\ -t^3 & e^{\sin t} & -\ln t \\ 2 & -5t & 2t \end{pmatrix} \begin{pmatrix} x_1 \\ x_2 \\ x_3 \end{pmatrix} + \begin{pmatrix} -e^t \\ \cos t \\ \tan t \end{pmatrix}. \quad ◀$$

EXAMPLE 2 ▶ **Writing a second-order equation in vector form**

Consider the general second-order linear differential equation

$$x'' + a(t)x' + b(t)x = f(t).$$

Defining $x_1 = x$ and $x_2 = x'$, we obtain the equivalent system

$$\begin{aligned} x_1' &= x_2, \\ x_2' &= -b(t)x_1 - a(t)x_2 + f(t), \end{aligned}$$

which can be rewritten as

$$\mathbf{x}' = A(t)\mathbf{x} + \mathbf{f}(t),$$

where

$$\mathbf{x} = \begin{pmatrix} x_1 \\ x_2 \end{pmatrix}, \quad A(t) = \begin{pmatrix} 0 & 1 \\ -b(t) & -a(t) \end{pmatrix}, \quad \mathbf{f}(t) = \begin{pmatrix} 0 \\ f(t) \end{pmatrix}. \quad ◀$$

EXAMPLE 3 ▶ **Writing a third-order equation in vector form**

Consider the third-order equation with constant coefficients

$$x''' - 6x'' + 11x' - 6x = 0. \tag{6}$$

Defining $x_1 = x$, $x_2 = x'$, and $x_3 = x''$, we obtain the system

$$\begin{aligned} x_1' &= x_2, \\ x_2' &= x_3, \\ x_3' &= 6x_1 - 11x_2 + 6x_3, \end{aligned}$$

or

$$\mathbf{x}' = A\mathbf{x}, \tag{7}$$

where

$$\mathbf{x} = \begin{pmatrix} x_1 \\ x_2 \\ x_3 \end{pmatrix} \quad \text{and} \quad A = \begin{pmatrix} 0 & 1 & 0 \\ 0 & 0 & 1 \\ 6 & -11 & 6 \end{pmatrix}. \blacktriangleleft$$

Let us now consider the initial-value problem

$$\mathbf{x}'(t) = A(t)\mathbf{x}(t) + \mathbf{f}(t), \qquad \mathbf{x}(t_0) = \mathbf{x}_0, \tag{8}$$

where

$$\mathbf{x}(t) = \begin{pmatrix} x_1(t) \\ x_2(t) \\ \vdots \\ x_n(t) \end{pmatrix} \quad \text{and} \quad \mathbf{x}_0 = \begin{pmatrix} x_{10} \\ x_{20} \\ \vdots \\ x_{n0} \end{pmatrix}.$$

solution to a vector differential equation

We say that a vector function

$$\boldsymbol{\phi}(t) = \begin{pmatrix} \phi_1(t) \\ \vdots \\ \phi_n(t) \end{pmatrix}$$

is a **solution** to Equation **(8)** if $\boldsymbol{\phi}$ is differentiable and satisfies the differential equation and the given initial condition in some interval containing t_0. The following theorem is proved in Appendix 3.

THEOREM 1 ♦ **Existence-uniqueness theorem**

Let $A(t)$ and $\mathbf{f}(t)$ be continuous matrix and vector functions, respectively, on some interval $[a, b]$ containing t_0 (that is, the component functions of both $A(t)$ and $\mathbf{f}(t)$ are continuous). Then there exists a unique vector function $\boldsymbol{\phi}(t)$ that is a solution to the initial-value problem **(8)** on the entire interval $[a, b]$. ♦

EXAMPLE 4 ▶ **The unique solution to an initial-value problem**

Referring to Example 3, we find by the methods of Chapter 3 that e^t, e^{2t}, and e^{3t} are solutions of Equation **(6).** Since a solution vector for this problem is

$$\boldsymbol{\phi}_1(t) = \begin{pmatrix} x(t) \\ x'(t) \\ x''(t) \end{pmatrix},$$

we see that three vector solutions of Equation **(7)** are

$$\boldsymbol{\phi}_1(t) = \begin{pmatrix} e^t \\ e^t \\ e^t \end{pmatrix} = e^t\begin{pmatrix} 1 \\ 1 \\ 1 \end{pmatrix}, \qquad \boldsymbol{\phi}_2(t)\begin{pmatrix} e^{2t} \\ 2e^{2t} \\ 4e^{2t} \end{pmatrix} = e^{2t}\begin{pmatrix} 1 \\ 2 \\ 4 \end{pmatrix},$$

and

$$\boldsymbol{\phi}_3(t) = \begin{pmatrix} e^{3t} \\ 3e^{3t} \\ 9e^{3t} \end{pmatrix} = e^{3t}\begin{pmatrix} 1 \\ 3 \\ 9 \end{pmatrix}. \qquad \textbf{(Check!)}$$

Thus the general solution is

$$\begin{aligned}\boldsymbol{\phi}(t) &= c_1\boldsymbol{\phi}_1(t) + c_2\boldsymbol{\phi}_2(t) + c_3\boldsymbol{\phi}_3(t) \\ &= \begin{pmatrix} c_1e^t + c_2e^{2t} + c_3e^{3t} \\ c_1e^t + 2c_2e^{2t} + 3c_3e^{3t} \\ c_1e^t + 4c_2e^{2t} + 9c_3e^{3t} \end{pmatrix}.\end{aligned}$$

If we specify the initial condition

$$\mathbf{x}(0) = \begin{pmatrix} 2 \\ -3 \\ 5 \end{pmatrix},$$

then it can be verified that $c_1 = 16$, $c_2 = -23$, and $c_3 = 9$, so that the unique solution vector is

$$\boldsymbol{\phi} = 16\boldsymbol{\phi}_1 - 23\boldsymbol{\phi}_2 + 9\boldsymbol{\phi}_3,$$

or

$$\begin{pmatrix} 16e^t - 23e^{2t} + 9e^{3t} \\ 16e^t - 46e^{2t} + 27e^{3t} \\ 16e^t - 92e^{2t} + 81e^{3t} \end{pmatrix}. \quad ◄$$

EXAMPLE 5 ► **The unique solution to an initial-value problem**

The system

$$\begin{aligned} x_1' &= -4x_1 - x_2, \qquad & x_1(0) = 1, \\ x_2' &= x_1 - 2x_2, \qquad & x_2(0) = 2, \end{aligned}$$

can be written as

$$\begin{pmatrix} x_1 \\ x_2 \end{pmatrix}' = \begin{pmatrix} -4 & -1 \\ 1 & -2 \end{pmatrix}\begin{pmatrix} x_1 \\ x_2 \end{pmatrix}, \qquad \begin{pmatrix} x_1(0) \\ x_2(0) \end{pmatrix} = \begin{pmatrix} 1 \\ 2 \end{pmatrix}.$$

It can easily be verified that

$$\boldsymbol{\phi}_1(t) = \begin{pmatrix} e^{-3t} \\ -e^{-3t} \end{pmatrix} \quad \text{and} \quad \boldsymbol{\phi}_2(t) = \begin{pmatrix} (1 - t)e^{-3t} \\ te^{-3t} \end{pmatrix}$$

are solution vectors. It can also be immediately verified that the unique solution vector that satisfies the given initial conditions is

$$\boldsymbol{\phi}(t) = \begin{pmatrix} (1 - 3t)e^{-3t} \\ (2 + 3t)e^{-3t} \end{pmatrix}. \quad ◄$$

The central problem of the remainder of this chapter is to derive properties of vector solutions and, where possible, to calculate them. In the next section we will show how all solutions to the homogeneous system $\mathbf{x}' = A\mathbf{x}$ can be represented in a convenient form, and in Section 9.6 we will show how information about the solutions to this homogeneous system can be used to find a particular solution to the nonhomogeneous system $\mathbf{x}' = A\mathbf{x} + \mathbf{f}$.

SELF-QUIZ

I. The equation $x'' - 2x' + 3x = e^t$ can be written in the form $\mathbf{x}' = A\mathbf{x} + \mathbf{f}(t)$ where $A =$ ______.

(a) $\begin{pmatrix} 1 & 0 \\ 3 & -2 \end{pmatrix}$ **(b)** $\begin{pmatrix} 1 & 0 \\ -3 & 2 \end{pmatrix}$ **(c)** $\begin{pmatrix} 0 & 1 \\ 3 & -2 \end{pmatrix}$ **(d)** $\begin{pmatrix} 0 & 1 \\ -3 & 2 \end{pmatrix}$

II. In Question I, $\mathbf{f}(t) =$ ______.

(a) $\begin{pmatrix} e^t \\ e^t \end{pmatrix}$ **(b)** $\begin{pmatrix} e^t \\ 0 \end{pmatrix}$ **(c)** $\begin{pmatrix} 0 \\ e^t \end{pmatrix}$ **(d)** $\begin{pmatrix} 0 \\ 0 \end{pmatrix}$

III. The equation $x''' - t^2x'' + tx' + 3x = 0$ can be written as $\mathbf{x}' = A(t)\mathbf{x}$ where $A(t) =$ ______.

(a) $\begin{pmatrix} 0 & 1 & 0 \\ 0 & 0 & 1 \\ -3 & -t & t^2 \end{pmatrix}$ **(b)** $\begin{pmatrix} 0 & 1 & t^2 \\ 0 & -t & 1 \\ -3 & 0 & 0 \end{pmatrix}$

(c) $\begin{pmatrix} 0 & 1 & 0 \\ 0 & 0 & 1 \\ 3 & t & -t^2 \end{pmatrix}$ **(d)** $\begin{pmatrix} -3 & -t & t^2 \\ 0 & 1 & 0 \\ 0 & 0 & 1 \end{pmatrix}$

Answers to Self-Quiz

I. d **II.** c **III.** a

PROBLEMS 9.2

In Problems 1–6 write each given equation or system in the matrix vector form **(4), (5),** *or* **(8).**

1. $x_1' = 2x_1 + 3x_2,$
$x_2' = 4x_1 - 6x_2$

2. $x_1' = (\cos t)x_1 - (\sin t)x_2 + e^{t^2},$
$x_2' = e^t x_1 + 2tx_2 - \ln t,$
$x_1(2) = 3,\ x_2(2) = 7$

3. $x''' - 2x'' + 4tx' - x = \sin t$

4. $x^{(4)} + 2x''' - 3x'' + 4x' - 7x = 0,$
$x(0) = 1,\ x'(0) = 2,\ x''(0) = 3,\ x'''(0) = 4$

5. $x_1' = 2tx_1 - 3t^2x_2 + (\sin t)x_3,$
$x_2' = 2x_1 - 4x_3 - \sin t,$
$x_3' = 17x_2 + 4tx_3 + e^t$

6. $x''' + a(t)x'' + b(t)x' + c(t)x = f(t),$
$x(t_0) = d_1,\ x'(t_0) = d_2,\ x''(t_0) = d_3$

In Problems 7–14 verify that each given vector function is a solution to the given system.

7. $\mathbf{x}' = \begin{pmatrix} 1 & 1 \\ -3 & -1 \end{pmatrix}\mathbf{x},$

$\boldsymbol{\phi}(t) = -\begin{pmatrix} \cos\sqrt{2}\,t \\ \sqrt{2}\sin\sqrt{2}\,t - \cos\sqrt{2}\,t \end{pmatrix}$

8. $\mathbf{x}' = \begin{pmatrix} 2 & 1 \\ 1 & 2 \end{pmatrix}\mathbf{x} + \begin{pmatrix} t \\ t^2 \end{pmatrix},$

$\boldsymbol{\phi}(t) = \begin{pmatrix} e^t + \frac{1}{3}t^2 + \frac{2}{9}t + \frac{11}{27} \\ -e^t - \frac{2}{3}t^2 - \frac{7}{9}t - \frac{16}{27} \end{pmatrix}$

9. $\mathbf{x}' = \begin{pmatrix} -1 & 6 \\ 1 & -2 \end{pmatrix}\mathbf{x},\quad \boldsymbol{\phi}(t) = \begin{pmatrix} 3e^t \\ e^t \end{pmatrix}$

10. $\mathbf{x}' = \begin{pmatrix} -4 & -1 \\ 1 & -2 \end{pmatrix}\mathbf{x},\quad \boldsymbol{\phi}(t) = \begin{pmatrix} (1+t)e^{-3t} \\ (-2-t)e^{-3t} \end{pmatrix}$

11. $\mathbf{x}' = \begin{pmatrix} 4 & 1 \\ -8 & 8 \end{pmatrix}\mathbf{x},$

$\boldsymbol{\phi}(t) = \begin{pmatrix} e^{6t}(\cos 2t - \frac{1}{2}\sin 2t) \\ e^{6t}(\cos 2t - 3\sin 2t) \end{pmatrix}$

12. $\mathbf{x}' = \begin{pmatrix} 1 & 1 & 1 \\ -1 & -1 & 0 \\ -1 & 0 & 1 \end{pmatrix}\mathbf{x},$

$\boldsymbol{\phi}(t) = \begin{pmatrix} \sin t - \cos t \\ \cos t \\ \cos t \end{pmatrix}$

13. $\mathbf{x}' = \begin{pmatrix} 1 & -1 & 1 & -1 \\ 0 & -1 & 2 & -2 \\ 0 & 0 & 2 & -3 \\ 0 & 0 & 0 & -2 \end{pmatrix}\mathbf{x},$

$\boldsymbol{\phi}(t) = \begin{pmatrix} e^{-2t} \\ 2e^{-2t} \\ 3e^{-2t} \\ 4e^{-2t} \end{pmatrix}$

14. $\mathbf{x}' = \begin{pmatrix} 3 & 2 & 1 \\ -1 & 0 & -1 \\ 1 & 1 & 2 \end{pmatrix}\mathbf{x},$

$\boldsymbol{\phi}(t) = \begin{pmatrix} e^{2t} + te^{2t} \\ -te^{2t} \\ te^{2t} \end{pmatrix}$

15. Let $\boldsymbol{\phi}_1(t)$ and $\boldsymbol{\phi}_2(t)$ be any two vector solutions of the homogeneous system $\mathbf{x}' = A(t)\mathbf{x}$. Show that $\boldsymbol{\phi}(t) = c_1\boldsymbol{\phi}_1(t) + c_2\boldsymbol{\phi}_2(t)$ is also a solution.

16. Let $\boldsymbol{\phi}_1(t)$ and $\boldsymbol{\phi}_2(t)$ be vector solutions of the nonhomogeneous system **(4).** Show that their difference,

$$\boldsymbol{\phi}(t) = \boldsymbol{\phi}_1(t) - \boldsymbol{\phi}_2(t),$$

is a solution to the homogeneous system (5).

9.3 FUNDAMENTAL AND PRINCIPAL MATRIX SOLUTIONS OF A HOMOGENEOUS SYSTEM OF DIFFERENTIAL EQUATIONS

In this section we discuss properties of the homogeneous system

$$\mathbf{x}' = A(t)\mathbf{x}, \tag{1}$$

where $\mathbf{x}(t)$ is an n-vector and $A(t)$ is an $n \times n$ matrix of functions. Throughout this section we assume that the components of $A(t)$ are functions continuous on some common interval containing the point t_0.

Linearly independent and fundamental sets of solutions

Let $\boldsymbol{\phi}_1(t), \boldsymbol{\phi}_2(t), \ldots, \boldsymbol{\phi}_m(t)$ be m vector solutions of system **(1).** We say that they are **linearly independent** on an interval if the equation (for all t in the common open interval)

$$c_1\boldsymbol{\phi}_1(t) + c_2\boldsymbol{\phi}_2(t) + \cdots + c_m\boldsymbol{\phi}_m(t) = \mathbf{0}$$

holds only for $c_1 = c_2 = \ldots = c_m = 0$. Since system **(1)** is equivalent to an nth-order equation, it is natural for us to seek n linearly independent solutions to the system. Any set of n linearly independent solutions of **(1)** is called a **fundamental set of solutions.**

In Section 7.3 we used the method of determinants to obtain the general solution for linear homogeneous systems with constant coefficients. If a general solution is known, it is an easy task to obtain a fundamental set of solutions: simply set *one* of the arbitrary coefficients c_j equal to 1 and set all the rest of them equal to zero for $j = 1, 2, \ldots, n$.

EXAMPLE 1 ▸ **A fundamental set of solutions where the characteristic equation has two real roots**

Let $\mathbf{x}' = A\mathbf{x}$, where

$$A = \begin{pmatrix} -1 & 6 \\ 1 & -2 \end{pmatrix}.$$

In Example 7.3.4 on page 324 we verified that

$$x_1(t) = c_1e^{-4t} + c_2e^{t},$$
$$x_2(t) = -\frac{1}{2}c_1e^{-4t} + \frac{1}{3}c_2e^{t},$$

is the general solution to this system. We can arrange this general solution in vector form as

$$\mathbf{x}(t) = c_1\begin{pmatrix} e^{-4t} \\ -\frac{1}{2}e^{-4t} \end{pmatrix} + c_2\begin{pmatrix} e^{t} \\ \frac{1}{3}e^{t} \end{pmatrix}.$$

Thus, setting $c_1 = 1$ and $c_2 = 0$, we obtain the solution vector

$$\boldsymbol{\phi}_1(t) = \begin{pmatrix} e^{-4t} \\ -\frac{1}{2}e^{-4t} \end{pmatrix},$$

while $c_1 = 0$ and $c_2 = 1$ yields

$$\boldsymbol{\phi}_2 = \begin{pmatrix} e^{t} \\ \frac{1}{3}e^{t} \end{pmatrix}.$$

The vectors $\boldsymbol{\phi}_1$ and $\boldsymbol{\phi}_2$ form a fundamental set of solutions since

$$W(t) = \begin{vmatrix} e^{-4t} & e^{t} \\ -\frac{1}{2}e^{-4t} & \frac{1}{3}e^{t} \end{vmatrix} = \left(\tfrac{1}{3} + \tfrac{1}{2}\right)e^{-3t} \neq 0. \quad ◀$$

EXAMPLE 2 ▶ A fundamental set of solutions where the characteristic equation has one real root

Let $\mathbf{x}' = A\mathbf{x}$, where

$$A = \begin{pmatrix} -4 & -1 \\ 1 & -2 \end{pmatrix}.$$

In Example 7.3.5 on page 324 we obtained the general solution (which we now write in vector form)

$$\mathbf{x}(t) = \begin{pmatrix} c_1e^{-3t} + c_2te^{-3t} \\ (-c_1 - c_2)e^{-3t} - c_2te^{-3t} \end{pmatrix},$$

or

$$\mathbf{x}(t) = c_1\begin{pmatrix} e^{-3t} \\ -e^{-3t} \end{pmatrix} + c_2\begin{pmatrix} te^{-3t} \\ (-1 - t)e^{-3t} \end{pmatrix} = c_1\boldsymbol{\phi}_1(t) + c_2\boldsymbol{\phi}_2(t).$$

The vectors $\boldsymbol{\phi}_1$ and $\boldsymbol{\phi}_2$ form a fundamental set of solutions since

$$W(t) = \begin{vmatrix} e^{-3t} & te^{-3t} \\ -e^{-3t} & (-1 - t)e^{-3t} \end{vmatrix} = -e^{-6t} \neq 0. \quad ◀$$

EXAMPLE 3 ▶ A fundamental set of solutions where the characteristic equation has no real roots

Consider the system $\mathbf{x}' = A\mathbf{x}$, where

$$A = \begin{pmatrix} 4 & 1 \\ -8 & 8 \end{pmatrix}.$$

In Example 7.3.6 on page 325 we obtained the general solution (written in vector form)

$$\mathbf{x}(t) = \begin{pmatrix} \frac{1}{4}e^{6t}[(c_1 - c_2)\cos 2t + (c_1 + c_2)\sin 2t] \\ e^{6t}(c_1 \cos 2t + c_2 \sin 2t) \end{pmatrix}$$

$$= c_1\begin{pmatrix} \frac{1}{4}e^{6t}(\cos 2t + \sin 2t) \\ e^{6t}\cos 2t \end{pmatrix} + c_2\begin{pmatrix} \frac{1}{4}e^{6t}(-\cos 2t + \sin 2t) \\ e^{6t}\sin 2t \end{pmatrix}.$$

Thus, by first setting $c_1 = 1$ and $c_2 = 0$ and then setting $c_1 = 0$ and $c_2 = 1$, we get the solutions

$$\boldsymbol{\phi}_1(t) = \begin{pmatrix} \frac{1}{4}e^{6t}(\cos 2t + \sin 2t) \\ e^{6t}\cos 2t \end{pmatrix}, \qquad \boldsymbol{\phi}_2(t) = \begin{pmatrix} \frac{1}{4}e^{6t}(-\cos 2t + \sin 2t) \\ e^{6t}\sin 2t \end{pmatrix},$$

and the Wronskian of these solutions is

$$W(t) = \begin{vmatrix} \frac{1}{4}e^{6t}(\cos 2t + \sin 2t) & \frac{1}{4}e^{6t}(-\cos 2t + \sin 2t) \\ e^{6t}\cos 2t & e^{6t}\sin 2t \end{vmatrix} = \tfrac{1}{4}e^{12t} \neq 0.$$

Thus $\boldsymbol{\phi}_1$ and $\boldsymbol{\phi}_2$ form a fundamental set of solutions. ◀

EXAMPLE 4 ▶ **A fundamental set of solutions for a system with three unknown functions**

Let $\mathbf{x}' = A\mathbf{x}$, where

$$A = \begin{pmatrix} 1 & 1 & -2 \\ -1 & 2 & 1 \\ 0 & 1 & -1 \end{pmatrix}.$$

Then it is not difficult to show that

$$\boldsymbol{\phi}_1(t) = \begin{pmatrix} e^{-t} \\ 0 \\ e^{-t} \end{pmatrix} = e^{-t}\begin{pmatrix} 1 \\ 0 \\ 1 \end{pmatrix}, \qquad \boldsymbol{\phi}_2(t) = \begin{pmatrix} 3e^{t} \\ 2e^{t} \\ e^{t} \end{pmatrix} = e^{t}\begin{pmatrix} 3 \\ 2 \\ 1 \end{pmatrix},$$

$$\boldsymbol{\phi}_3(t) = \begin{pmatrix} e^{2t} \\ 3e^{2t} \\ e^{2t} \end{pmatrix} = e^{2t}\begin{pmatrix} 1 \\ 3 \\ 1 \end{pmatrix}$$

are solutions of this system. To show that they are a fundamental set of solutions, we consider their Wronskian:

$$W(t) = \begin{vmatrix} e^{-t} & 3e^{t} & e^{2t} \\ 0 & 2e^{t} & 3e^{2t} \\ e^{-t} & e^{t} & e^{2t} \end{vmatrix}.$$

By Theorem 7.2.4 on page 316 (which extends to larger systems), the Wronskian is either always zero or never zero. Thus, we need only check the value of $W(t)$ at a single point, say $t = 0$:

$$W(0) = \begin{vmatrix} 1 & 3 & 1 \\ 0 & 2 & 3 \\ 1 & 1 & 1 \end{vmatrix} = 6 \neq 0.$$

Hence $\boldsymbol{\phi}_1$, $\boldsymbol{\phi}_2$, and $\boldsymbol{\phi}_3$ are linearly independent, so they form a fundamental set of solutions. ◀

Matrix solution and its Wronskian

Let $\boldsymbol{\phi}_1, \boldsymbol{\phi}_2, \ldots, \boldsymbol{\phi}_n$ be n-vector solutions of $\mathbf{x}' = A(t)\mathbf{x}$. Let $\boldsymbol{\Phi}(t)$ be the matrix whose columns are the vectors $\boldsymbol{\phi}_1, \boldsymbol{\phi}_2, \ldots, \boldsymbol{\phi}_n$; that is,

$$\Phi(t) = (\boldsymbol{\phi}_1(t), \ldots, \boldsymbol{\phi}_n(t)) = \begin{pmatrix} \phi_{11}(t) & \phi_{12}(t) & \cdots & \phi_{1n}(t) \\ \phi_{21}(t) & \phi_{22}(t) & \cdots & \phi_{2n}(t) \\ \vdots & \vdots & & \vdots \\ \phi_{n1}(t) & \phi_{n2}(t) & \cdots & \phi_{nn}(t) \end{pmatrix}. \quad (2)$$

Such a matrix is called **matrix solution** of the system $\mathbf{x}' = A\mathbf{x}$. Equivalently, *an $n \times n$ matrix function $\Phi(t)$ is a matrix solution of* $\mathbf{x}' = A\mathbf{x}$ *if and only if each of its columns is a solution vector of* $\mathbf{x}' = A\mathbf{x}$.

We define the **Wronskian** of a matrix solution $\Phi(t)$ of the system $\mathbf{x}' = A(t)\mathbf{x}$ by

$$W(t) = \det \Phi(t). \quad (3)$$

This is precisely the same definition of the Wronskian that we have used in Examples 1–4.

Fundamental matrix solution

If $W(t) \neq 0$, then the vectors $\boldsymbol{\phi}_1, \boldsymbol{\phi}_2, \ldots, \boldsymbol{\phi}_n$ form a fundamental set of solutions (that is, they are linearly independent), and $\Phi(t)$ is called a **fundamental matrix solution.**

In what follows, we show that fundamental matrix solutions play a central role in the theory of linear systems of differential equations.

EXAMPLE 5 ▸ Four fundamental matrix solutions

In the four previous examples of this section, the fundamental matrix solutions are, respectively,

1. $\begin{pmatrix} e^{-4t} & e^{t} \\ -\frac{1}{2}e^{-4t} & \frac{1}{3}e^{t} \end{pmatrix}$
2. $\begin{pmatrix} e^{-3t} & te^{-3t} \\ -e^{-3t} & (-1-t)e^{-3t} \end{pmatrix}$
3. $\begin{pmatrix} \frac{1}{4}e^{6t}(\cos 2t + \sin 2t) & \frac{1}{4}e^{6t}(-\cos 2t + \sin 2t) \\ e^{6t}\cos 2t & e^{6t}\sin 2t \end{pmatrix}$
4. $\begin{pmatrix} e^{-t} & 3e^{t} & e^{2t} \\ 0 & 2e^{t} & 3e^{2t} \\ e^{-t} & e^{t} & e^{2t} \end{pmatrix}$ ◂

Fundamental matrix solutions are not unique, due to the fact that a solution vector may be multiplied by any constant and still remain a solution. In addition, any linear combination of solutions is still a solution (see Problem 6). However, there is one particular fundamental matrix solution that is unique.

Principal matrix solution

The fundamental matrix solution $\mathbf{\Psi}(t)$ for the system

$$\mathbf{x}' = A(t)\mathbf{x}$$

that satisfies $\mathbf{\Psi}(t_0) = I$, where I is the identity matrix, is called the **principal matrix solution.***

Remark

In the definition above we are assuming that the point t_0 is contained within the interval of continuity that each of the coefficient functions $a_{ij}(t)$ have in common.

The next result proves that principal matrix solutions (and hence also fundamental matrix solutions) always exist for a linear homogeneous system.

THEOREM 1 ♦ **Existence–uniqueness theorem of the principal matrix solution**
Let $A(t)$ be continuous on some interval $[a, b]$. Then for any t_0, $a \le t_0 \le b$, there exists a unique fundamental matrix solution $\mathbf{\Psi}(t)$ of the system $\mathbf{x}' = A(t)\mathbf{x}$ satisfying the condition $\mathbf{\Psi}(t_0) = I$.

Proof Let $\boldsymbol{\delta}_i$, $i = 1, 2, \ldots, n$, denote the n-column vector that has a one in the ith position (row) and a zero everywhere else:

$$\boldsymbol{\delta}_1 = \begin{pmatrix} 1 \\ 0 \\ 0 \\ \vdots \\ 0 \end{pmatrix}, \quad \boldsymbol{\delta}_2 = \begin{pmatrix} 0 \\ 1 \\ 0 \\ \vdots \\ 0 \end{pmatrix}, \quad \ldots, \quad \boldsymbol{\delta}_n = \begin{pmatrix} 0 \\ 0 \\ \vdots \\ 0 \\ 1 \end{pmatrix}.$$

By the basic existence-uniqueness theorem (Theorem 9.2.1) with $\mathbf{f}(t) \equiv \mathbf{0}$, there exists a unique vector solution $\boldsymbol{\phi}_i(t)$ of $\mathbf{x}' = A\mathbf{x}$ that satisfies $\boldsymbol{\phi}_i(t_0) = \boldsymbol{\delta}_i$, $i = 1, 2, \ldots, n$. Define the matrix function

$$\mathbf{\Psi}(t) = (\boldsymbol{\phi}_1(t), \boldsymbol{\phi}_2(t), \ldots, \boldsymbol{\phi}_n(t)). \tag{4}$$

Then $\mathbf{\Psi}(t)$ is the matrix whose columns are the vector solutions $\boldsymbol{\phi}_i$, $i = 1, 2, \ldots, n$. Clearly $\mathbf{\Psi}(t)$ is a matrix solution and $\mathbf{\Psi}(t_0) = (\boldsymbol{\phi}_1(t_0), \ldots, \boldsymbol{\phi}_n(t_0)) = (\boldsymbol{\delta}_1, \boldsymbol{\delta}_2, \ldots, \boldsymbol{\delta}_n) = I$. It remains to be shown that $\mathbf{\Psi}(t)$ is a fundamental matrix solution. This is easy to do. We simply note that $\det \mathbf{\Psi}(t_0) = \det I = 1 \ne 0$. ♦

The calculation of a fundamental or principal matrix solution is generally impossible if $A(t)$ is nonconstant. If $A(t)$ is a constant matrix, then, as we will see in Section 9.5, the principal matrix solution can always, in principle, be obtained. [No pun intended.]

What does it mean to say that $\mathbf{\Phi}(t)$ is a matrix solution of $\mathbf{x}' = A(t)\mathbf{x}$? By definition, we know that each column $\boldsymbol{\phi}_i(t)$ of $\mathbf{\Phi}(t)$ is a solution to $\mathbf{x}' = A(t)\mathbf{x}$; that is,

$$\boldsymbol{\phi}_i' = A(t)\boldsymbol{\phi}_i.$$

Hence, by matrix multiplication it follows that

$$\mathbf{\Phi}' = A(t)\mathbf{\Phi} \tag{5}$$

* In some books this is called a **state transition matrix.**

EXAMPLE 6 ▶ **Illustration of the fact that $\Phi' = A(t)\Phi$ if Φ is a matrix solution**

We have seen in Example 1 that

$$\boldsymbol{\phi}_1(t) = \begin{pmatrix} e^{-4t} \\ -\frac{1}{2}e^{-4t} \end{pmatrix} \quad \text{and} \quad \boldsymbol{\phi}_2(t) = \begin{pmatrix} e^{t} \\ \frac{1}{3}e^{t} \end{pmatrix}$$

are solutions to $\mathbf{x}' = A\mathbf{x}$, where

$$A = \begin{pmatrix} -1 & 6 \\ 1 & -2 \end{pmatrix}.$$

Let

$$\Phi(t) = (\boldsymbol{\phi}_1(t), \boldsymbol{\phi}_2(t)) = \begin{pmatrix} e^{-4t} & e^{t} \\ -\frac{1}{2}e^{-4t} & \frac{1}{3}e^{t} \end{pmatrix}.$$

Then, since matrix differentiation involves differentiation of each component,

$$\Phi'(t) = \begin{pmatrix} -4e^{-4t} & e^{t} \\ 2e^{-4t} & \frac{1}{3}e^{t} \end{pmatrix},$$

and

$$A(t)\Phi(t) = \begin{pmatrix} -1 & 6 \\ 1 & -2 \end{pmatrix}\begin{pmatrix} e^{-4t} & e^{t} \\ -\frac{1}{2}e^{-4t} & \frac{1}{3}e^{t} \end{pmatrix} = \begin{pmatrix} -4e^{-4t} & e^{t} \\ 2e^{-4t} & \frac{1}{3}e^{t} \end{pmatrix}.$$

Hence $\Phi'(t) = A(t)\Phi(t)$. ◀

THEOREM 2 ♦ **Multiplying a matrix solution by a constant matrix yields another matrix solution**

Let Φ be a matrix solution of $\mathbf{x}' = A(t)\mathbf{x}$ and let C be any constant square matrix. Then $\Phi_1 = \Phi C$ is also a matrix solution of $\mathbf{x}' = A(t)\mathbf{x}$.

Proof Since a matrix solution of $\mathbf{x}' = A(t)\mathbf{x}$ is also a solution of Equation **(5)**, we must show that Φ_1 is also a solution of Equation **(5)**. But $\Phi_1' = (\Phi C)' = \Phi' C + \Phi C' = \Phi' C$, since $C' = 0$, C being constant. Finally, since Φ is a solution,

$$\Phi_1' = \Phi' C = A\Phi C = A\Phi_1. \ ♦$$

EXAMPLE 7 ▶ **Illustration of Theorem 2**

We have seen in Example 6 that

$$\Phi(t) = \begin{pmatrix} e^{-4t} & e^{t} \\ -\frac{1}{2}e^{-4t} & \frac{1}{3}e^{t} \end{pmatrix}$$

is a matrix solution of $\mathbf{x}' = A(t)\mathbf{x}$, with

$$A = \begin{pmatrix} -1 & 6 \\ 1 & -2 \end{pmatrix}.$$

Let

$$C_1 = \begin{pmatrix} 1 & 2 \\ 3 & 4 \end{pmatrix} \quad \text{and} \quad C_2 = \begin{pmatrix} 1 & 2 \\ 2 & 4 \end{pmatrix}.$$

Then

$$\Phi_1 = \Phi C_1 = \begin{pmatrix} e^{-4t} + 3e^t & 2e^{-4t} + 4e^t \\ -\frac{1}{2}e^{-4t} + e^t & -e^{-4t} + \frac{4}{3}e^t \end{pmatrix}$$

and

$$\Phi_2 = \Phi C_2 = \begin{pmatrix} e^{-4t} + 2e^t & 2e^{-4t} + 4e^t \\ -\frac{1}{2}e^{-4t} + \frac{2}{3}e^t & -e^{-4t} + \frac{4}{3}e^t \end{pmatrix}$$

are also matrix solutions (verify this!). Note that although Φ and Φ_1 are fundamental matrix solutions, Φ_2 is not since

$$\det \Phi_2(0) = \begin{vmatrix} 3 & 6 \\ \frac{1}{6} & \frac{1}{3} \end{vmatrix} = 0. \blacktriangleleft$$

Obtaining principal matrix solutions from fundamental matrix solutions

Theorem 2 gives us an easy way of finding the principal matrix solution when a fundamental matrix solution is known. To see this, let $\Phi(t)$ be a fundamental matrix solution. Since $\det \Phi(t_0) \neq 0$, $\Phi(t_0)$ is invertible, and we define $C = \Phi^{-1}(t_0)$ and $\Psi(t) = \Phi(t)C$. By Theorem 2, $\Psi(t)$ is a matrix solution and $\Psi(t_0) = \Phi(t_0)C = \Phi(t_0)\Phi^{-1}(t_0) = I$, so $\Psi(t)$ is the principal matrix solution.

Thus

Obtaining the principal matrix solution from a fundamental matrix solution

if $\Phi(t)$ is a fundamental matrix solution for $\mathbf{x}' = A(t)\mathbf{x}$, then

$$\Psi(t) = \Phi(t)\Phi^{-1}(t_0)$$

is the principal matrix solution (associated with the point t_0).

EXAMPLE 8 ▶ **Finding the principal matrix solution for a system**

In Example 7 we showed that

$$\Phi(t) = \begin{pmatrix} e^{-4t} & e^t \\ -\frac{1}{2}e^{-4t} & \frac{1}{3}e^t \end{pmatrix}$$

is a fundamental matrix solution for the system

$$\mathbf{x}' = \begin{pmatrix} -1 & 6 \\ 1 & -2 \end{pmatrix}\mathbf{x}.$$

Assume $t_0 = 0$, so that

$$\Phi(t_0) = \Phi(0) = \begin{pmatrix} 1 & 1 \\ -\frac{1}{2} & \frac{1}{3} \end{pmatrix} \quad \text{and} \quad \Phi^{-1}(0) = \begin{pmatrix} \frac{2}{5} & -\frac{6}{5} \\ \frac{3}{5} & \frac{6}{5} \end{pmatrix}.$$

Then

$$\Psi(t) = \Phi(t)\Phi^{-1}(0) = \frac{1}{5}\begin{pmatrix} 2e^{-4t} + 3e^{t} & -6e^{-4t} + 6e^{t} \\ -e^{-4t} + e^{t} & 3e^{-4t} + 2e^{t} \end{pmatrix}$$

is the principal matrix solution. Note that

$$\Psi(0) = \frac{1}{5}\begin{pmatrix} 5 & 0 \\ 0 & 5 \end{pmatrix} = I. \blacktriangleleft$$

THEOREM 3 ♦ Any matrix solution is a constant (matrix) multiple of a fundamental matrix solution

Let $\Phi(t)$ be a fundamental matrix solution and let $X(t)$ be any other matrix solution of the system $\mathbf{x}' = A(t)\mathbf{x}$. Then there exists a constant matrix C such that $X(t) = \Phi(t)C$; that is, *any solution vector of* $\mathbf{x}' = A(t)\mathbf{x}$ *can be written as a linear combination of vectors in a fundamental set* (compare this theorem with Theorem 7.2.3, p. 315).

Before giving the proof, we should warn the reader that it is important to state on which side we are multiplying the matrix Φ by C, since matrix multiplication is not, in general, commutative.

Proof Since $\Phi(t)$ is a fundamental matrix solution, det $\Phi(t) \neq 0$ and $\Phi^{-1}(t)$ exists for every t. We will show that

$$\frac{d}{dt}[\Phi^{-1}(t)X(t)] = 0.$$

This will imply that $\Phi^{-1}(t)X(t)$ is a constant matrix C, proving the theorem.

First, we calculate

$$\frac{d}{dt}[\Phi^{-1}(t)].$$

Using the product rule of differentiation (which can easily be shown to apply to matrix products), we have

$$0 = \frac{dI}{dt} = \frac{d}{dt}(\Phi\Phi^{-1}) = \frac{d\Phi}{dt}\Phi^{-1} + \Phi\frac{d\Phi^{-1}}{dt}, \tag{6}$$

or, after multiplying both sides of **(6)** on the left by Φ^{-1} and solving for $d\Phi^{-1}/dt$,

$$\frac{d\Phi^{-1}}{dt} = -\Phi^{-1}\frac{d\Phi}{dt}\Phi^{-1}. \tag{7}$$

Note the analogy between **(7)** and the identity

$$\frac{d}{dt}\left(\frac{1}{f(t)}\right) = -\frac{f'(t)}{[f(t)]^2}.$$

Now, by the product formula of derivatives,

$$\frac{d}{dt}(\Phi^{-1}X) = \left(\frac{d}{dt}\Phi^{-1}\right)X + \Phi^{-1}\frac{dX}{dt}, \tag{8}$$

and since both Φ and X are solutions of $\mathbf{x}' = A(t)\mathbf{x}$, Equation **(8)** becomes

$$\begin{aligned}\frac{d}{dt}(\Phi^{-1}X) &= \left(-\Phi^{-1}\frac{d\Phi}{dt}\Phi^{-1}\right)X + \Phi^{-1}(AX) \\ &= -\Phi^{-1}A\Phi\Phi^{-1}X + \Phi^{-1}AX \\ &= -\Phi^{-1}AX + \Phi^{-1}AX = 0. \quad \blacklozenge\end{aligned}$$

EXAMPLE 9 ► Illustrating Theorem 3

Consider the system $\mathbf{x}' = A\mathbf{x}$, where

$$A = \begin{pmatrix} 1 & -2 \\ 2 & -3 \end{pmatrix}.$$

It is easy to verify that

$$\Phi_1(t) = \begin{pmatrix} e^{-t} & (2t+2)e^{-t} \\ e^{-t} & (2t+1)e^{-t} \end{pmatrix} = e^{-t}\begin{pmatrix} 1 & 2t+2 \\ 1 & 2t+1 \end{pmatrix}$$

is a fundamental matrix solution. Another matrix solution is

$$\Phi_2(t) = e^{-t}\begin{pmatrix} 4t+7 & 8t+1 \\ 4t+5 & 8t-3 \end{pmatrix}.$$

There is a matrix C such that $\Phi_2 = \Phi_1 C$. But $\Phi_1^{-1}(t)\Phi_2(t) = C$ holds for every value of t, in particular for $t = 0$. Thus

$$C = \Phi_1^{-1}(0)\Phi_2(0) = \begin{pmatrix} -1 & 2 \\ 1 & -1 \end{pmatrix}\begin{pmatrix} 7 & 1 \\ 5 & -3 \end{pmatrix} = \begin{pmatrix} 3 & -7 \\ 2 & 4 \end{pmatrix}. \quad ◄$$

EXAMPLE 10 ► Another illustration of Theorem 3

Consider the system $\mathbf{x}' = A\mathbf{x}$, where

$$A = \begin{pmatrix} 3 & -1 & 1 \\ -1 & 5 & -1 \\ 1 & -1 & 3 \end{pmatrix}.$$

A fundamental matrix solution is

$$\Phi(t) = \begin{pmatrix} e^{2t} & e^{3t} & e^{6t} \\ 0 & e^{3t} & -2e^{6t} \\ -e^{2t} & e^{3t} & e^{6t} \end{pmatrix}.$$

Another matrix solution is

$$X(t) = \begin{pmatrix} e^{2t} + 2e^{3t} + 3e^{6t} & e^{2t} - 3e^{3t} - 2e^{6t} & 2e^{2t} + 5e^{3t} + 7e^{6t} \\ 2e^{3t} - 6e^{6t} & -3e^{3t} + 4e^{6t} & 5e^{3t} - 14e^{6t} \\ -e^{2t} + 2e^{2t} + 3e^{6t} & -e^{2t} - 3e^{3t} - 2e^{6t} & -2e^{2t} + 5e^{3t} + 7e^{6t} \end{pmatrix}.$$

As in the previous example, a matrix C such that $X(t) = \Phi(t)C$ is given by

$$C = \Phi^{-1}(0)X(0) = \frac{1}{6}\begin{pmatrix} 3 & 0 & -3 \\ 2 & 2 & 2 \\ 1 & -2 & 1 \end{pmatrix}\begin{pmatrix} 6 & -4 & 14 \\ -4 & 1 & -9 \\ 4 & -6 & 10 \end{pmatrix}$$

$$= \frac{1}{6}\begin{pmatrix} 6 & 6 & 12 \\ 12 & -18 & 30 \\ 18 & -12 & 42 \end{pmatrix} = \begin{pmatrix} 1 & 1 & 2 \\ 2 & -3 & 5 \\ 3 & -2 & 7 \end{pmatrix}. \quad ◄$$

THEOREM 4 ♦ Any vector solution is a constant (vector) multiple of a fundamental matrix solution

Let $\Phi(t)$ be a fundamental matrix solution and let $\mathbf{x}(t)$ be any solution of $\mathbf{x}' = A(t)\mathbf{x}$. Then there exists a constant vector $\mathbf{c}$ such that

$$\mathbf{x}(t) = \Phi(t)\mathbf{c}. \tag{9}$$

Proof This theorem is an immediate consequence of Theorem 3 if we form the matrix solution $X(t) = (\mathbf{x}(t), \mathbf{x}(t), \ldots, \mathbf{x}(t))$ whose n columns are each the vector solution $\mathbf{x}(t)$. Then a matrix C exists such that $X(t) = \Phi(t)C$. Since each column of X is identical, every column of C is the same vector $\mathbf{c}$. ♦

general solution Theorem 4 asserts that the **general solution** of the system $\mathbf{x}' = A(t)\mathbf{x}$ is obtained by multiplying any fundamental matrix solution $\Phi(t)$ by an arbitrary vector of constants $\mathbf{c}$; that is,

$$\Phi(t)c = (\boldsymbol{\phi}_1(t), \boldsymbol{\phi}_2(t), \ldots, \boldsymbol{\phi}_n(t))\begin{pmatrix} c_1 \\ c_2 \\ \vdots \\ c_n \end{pmatrix} = c_1\boldsymbol{\phi}_1(t) + c_2\boldsymbol{\phi}_2(t) + \ldots + c_n\boldsymbol{\phi}_n(t).$$

This completely justifies our previous use of the term "general solution."

In this section we defined both fundamental and principal matrix solutions (relative to a point t_0) to $\mathbf{x}'(t) = A(t)\mathbf{x}(t)$. Why go to the extra trouble of computing the principal matrix solution? The answer is that in Section 9.6, when we discuss nonhomogeneous systems, we will need to compute the inverse of a fundamental matrix solution. In general, this involves a great deal of algebra. However, the computation of the inverse of a principal matrix solution is sometimes very easy. The following theorem is proved in Section 9.7 (Theorem 9.7.3).

THEOREM 5 ♦ **The inverse of the principal matrix solution**

Let $\Psi(t)$ be the principal matrix solution for the system $\mathbf{x}'(t) = A\mathbf{x}(t)$, relative to the point t_0, where A is a constant matrix. Then

$$\Psi^{-1}(t) = \Psi(-t). \quad ♦ \tag{10}$$

EXAMPLE 11 ► **Computing the inverse of a principal matrix solution**

In Example 8 we found that

$$\Psi(t) = \frac{1}{5}\begin{pmatrix} 2e^{-4t} + 3e^{t} & -6e^{-4t} + 6e^{t} \\ -e^{-4t} + e^{t} & 3e^{-4t} + 2e^{t} \end{pmatrix} \tag{11}$$

is the principal matrix solution relative to $t_0 = 0$ for the system

$$\mathbf{x}' = \begin{pmatrix} -1 & 6 \\ 1 & -2 \end{pmatrix}\mathbf{x}.$$

Then, from Equation **(10),** we obtain

Replace t by $-t$ in (11)

$$\Psi^{-1}(t) = \frac{1}{5}\begin{pmatrix} 2e^{4t} + 3e^{-t} & -6e^{4t} + 6e^{-t} \\ -e^{4t} + e^{t} & 3e^{4t} + 2e^{-t} \end{pmatrix}. \tag{12}$$

It is not difficult to verify that **(12)** is the inverse matrix for **(11).** ◄

SELF-QUIZ

I. Which of the following could *not* be a fundamental matrix solution for a system of differential equations?

(a) $\begin{pmatrix} e^t & e^{2t} \\ 2e^t & e^{2t} \end{pmatrix}$ **(b)** $\begin{pmatrix} e^t & e^{2t} \\ 2e^t & 2e^{2t} \end{pmatrix}$

(c) $\begin{pmatrix} 4e^t & -2e^{2t} \\ -6e^t & 3e^{2t} \end{pmatrix}$ **(d)** $\begin{pmatrix} e^t & -e^{2t} \\ -e^t & -e^{2t} \end{pmatrix}$

II. For $t_0 = 0$, which of the following *could* be the principal matrix solution for a system of differential equations?

(a) $\begin{pmatrix} e^{3t}(\cos t + \sin t) & e^{3t}(-\cos t + \sin t) \\ e^{3t}\cos t & e^{3t}\sin t \end{pmatrix}$

(b) $\begin{pmatrix} e^{3t}(\cos t + \sin t) & -4e^{3t}\sin t \\ 12e^{3t}\sin t & -e^{3t}(\sin t - \cos t) \end{pmatrix}$

(c) $\begin{pmatrix} e^{3t}\sin t & e^{3t}\cos t \\ e^{3t}\cos t & e^{3t}\sin t \end{pmatrix}$

(d) $\begin{pmatrix} e^{3t}(\cos t + 5\sin t) & -e^{3t}\sin t \\ -e^{3t}\sin t & \frac{1}{2}e^{3t}(2\cos t - 17\sin t) \end{pmatrix}$

III. Suppose that $\Phi(t)$ is a fundamental matrix solution for a system of differential equations. Then $\Phi(t)C$ is another fundamental matrix solution for which of the following matrices?

(a) $C = \begin{pmatrix} 6 & 4 \\ 3 & -2 \end{pmatrix}$ **(b)** $C = \begin{pmatrix} 6 & 4 \\ -3 & 2 \end{pmatrix}$

(c) $C = \begin{pmatrix} 6 & 4 \\ 3 & 2 \end{pmatrix}$ **(d)** $C = \begin{pmatrix} -6 & -4 \\ -3 & 2 \end{pmatrix}$

IV. If $\Phi(t)$ is a fundamental matrix solution for a system, then the principal matrix solution at t_0 is given by

(a) $\Phi(t)\Phi^{-1}(t_0)$ **(b)** $\Phi^{-1}(t_0)\Phi(t)$

(c) $\Phi(t)\Phi^{-1}(t)$ **(d)** $\Phi^{-1}(t)\Phi(t)$

V. Suppose that

$$\Psi(t) = \begin{pmatrix} e^{2t}(\cos t + \sin t) & -4e^{2t}\sin t \\ 5e^{2t}\sin t & e^{2t}(\cos t + 2\sin t) \end{pmatrix}$$

is the principal matrix solution for a system of differential equations. Then $\Psi^{-1}(t) =$ ______.

(a) $\begin{pmatrix} -e^{-2t}(\cos t + \sin t) & 4e^{-2t}\sin t \\ -5e^{-2t}\sin t & -e^{-2t}(\cos t + 2\sin t) \end{pmatrix}$

(b) $\begin{pmatrix} e^{-2t}(\sin t - \cos t) & 4e^{-2t}\sin t \\ 5e^{2t}\sin t & e^{-2t}(2\sin t - \cos t) \end{pmatrix}$

(c) $\begin{pmatrix} e^{-2t}(\cos t - \sin t) & 4e^{-2t}\sin t \\ -5e^{-2t}\sin t & e^{-2t}(\cos t - 2\sin t) \end{pmatrix}$

Answers to Self-Quiz

I. b, c **II.** b, d **III.** a, b, d **IV.** a (both c. and d. = I) **V.** c

PROBLEMS 9.3

In Problems 1–5 decide whether each given set of solution vectors constitutes a fundamental set of the given system by (a) determining whether the vectors are linearly independent, and (b) using the method of Wronskians to determine whether or not $W(t)$ is zero.

1. $\mathbf{x}' = \begin{pmatrix} 2 & 5 \\ 0 & 2 \end{pmatrix}\mathbf{x}, \quad \boldsymbol{\phi}_1(t) = \begin{pmatrix} e^{2t}(1 + 10t) \\ 2e^{2t} \end{pmatrix}, \quad \boldsymbol{\phi}_2(t) = \begin{pmatrix} e^{2t}(-3 + 20t) \\ 4e^{2t} \end{pmatrix}$

2. $\mathbf{x}' = \begin{pmatrix} 4 & -13 \\ 2 & -6 \end{pmatrix}\mathbf{x}, \quad \boldsymbol{\phi}_1(t) = \begin{pmatrix} e^{-t}(13\cos t - 26\sin t) \\ e^{-t}(7\cos t - 9\sin t) \end{pmatrix}, \quad \boldsymbol{\phi}_2(t) = \begin{pmatrix} e^{-t}(26\cos t - 52\sin t) \\ e^{-t}(14\cos t - 18\sin t) \end{pmatrix}$

3. $\mathbf{x}' = \begin{pmatrix} 1 & 1 \\ 4 & 1 \end{pmatrix}\mathbf{x}, \quad \boldsymbol{\phi}_1(t) = \begin{pmatrix} e^{3t} - e^{-t} \\ 2e^{3t} + 2e^{-t} \end{pmatrix}, \quad \boldsymbol{\phi}_2(t) = \begin{pmatrix} 2e^{3t} \\ 4e^{3t} \end{pmatrix}$

4. $\mathbf{x}' = \begin{pmatrix} 1 & -1 & 4 \\ 3 & 2 & -1 \\ 2 & 1 & -1 \end{pmatrix}\mathbf{x}, \quad \boldsymbol{\phi}_1(t) = \begin{pmatrix} e^{t} + 2e^{-2t} + 3e^{3t} \\ -4e^{t} - 2e^{-2t} + 6e^{3t} \\ -e^{t} - 2e^{-2t} + 3e^{3t} \end{pmatrix}, \quad \boldsymbol{\phi}_2(t) = \begin{pmatrix} -2e^{t} + 2e^{-2t} \\ 8e^{t} - 2e^{-2t} \\ 2e^{t} - 2e^{-2t} \end{pmatrix},$

$\boldsymbol{\phi}_3(t) = \begin{pmatrix} 3e^{t} - 6e^{-2t} + 3e^{3t} \\ -12e^{t} + 6e^{-2t} + 6e^{3t} \\ -3e^{t} + 6e^{-2t} + 3e^{3t} \end{pmatrix}$

5. $\mathbf{x}' = \begin{pmatrix} 3 & 2 & 1 \\ -1 & 0 & -1 \\ 1 & 1 & 2 \end{pmatrix}\mathbf{x}, \quad \boldsymbol{\phi}_1(t) = \begin{pmatrix} -e^{t} + te^{2t} + e^{2t} \\ e^{t} - te^{2t} \\ te^{2t} \end{pmatrix}, \quad \boldsymbol{\phi}_2(t) = \begin{pmatrix} 2e^{2t} + te^{2t} \\ -e^{2t} - te^{2t} \\ e^{2t} + te^{2t} \end{pmatrix},$

$\boldsymbol{\phi}_3(t) = \begin{pmatrix} -e^{t} + 3e^{2t} + 2te^{2t} \\ e^{t} - e^{2t} - 2te^{2t} \\ e^{2t} + 2te^{2t} \end{pmatrix}$

6. Let $\boldsymbol{\phi}_1(t), \boldsymbol{\phi}_2(t), \ldots, \boldsymbol{\phi}_m(t)$ be m solutions of the homogeneous system **(1)**. Show that

$$\boldsymbol{\phi}(t) = c_1\boldsymbol{\phi}_1(t) + c_2\boldsymbol{\phi}_2(t) + \cdots + c_m\boldsymbol{\phi}_m(t)$$

is also a solution.

In each of Problems 7–9, two matrix functions Φ_1 and Φ_2 are given. Find a matrix C such that $\Phi_2(t) = \Phi_1(t)C$.

7. $\Phi_1(t) = e^{6t}\begin{pmatrix} \cos 2t - \frac{1}{2}\sin 2t & \sin 2t + \frac{1}{2}\cos 2t \\ \cos 2t - 3\sin 2t & \sin 2t + 3\cos 2t \end{pmatrix}$, $\Phi_2(t) = e^{6t}\begin{pmatrix} \frac{1}{2}\cos 2t - \frac{3}{2}\sin 2t & -\frac{3}{2}\cos 2t + 2\sin 2t \\ -2\cos 2t - 4\sin 2t & \cos 2t + 7\sin 2t \end{pmatrix}$

8. $\Phi_1(t) = \begin{pmatrix} \sin e^t & \cos e^t \\ e^t\cos e^t & -e^t\sin e^t \end{pmatrix}$,

$\Phi_2(t) = \begin{pmatrix} \cos e^t & 3\sin e^t + 2\cos e^t \\ -e^t\sin e^t & 3e^t\cos e^t - 2e^t\sin e^t \end{pmatrix}$

9. $\Phi_1(t) = \begin{pmatrix} e^{-t} & 3e^t & e^{2t} \\ 0 & 2e^t & 3e^{2t} \\ e^{-t} & e^t & e^{2t} \end{pmatrix}$,

$\Phi_2(t) = \begin{pmatrix} e^{-t} + e^{2t} & -e^{-t} + 3e^t & e^{-t} + 3e^t \\ 3e^{2t} & 2e^t & 2e^t \\ e^{-t} + e^{2t} & -e^{-t} + e^t & e^{-t} + e^t \end{pmatrix}$

10. Let $\Phi_1(t)$ be a fundamental matrix solution of $\mathbf{x}' = A\mathbf{x}$. Then $\Phi_2 = \Phi_1 C$ is a matrix solution for any constant matrix C. Show that Φ_2 is a fundamental matrix solution if and only if C is nonsingular.

11. Let $\Phi(t)$ be a matrix solution of the system $\mathbf{x}' = A(t)\mathbf{x}$ where $A(t)$ is a 3×3 matrix. Prove that

$$W(t) = W(t_0)\exp\left(\int_{t_0}^{t} [a_{11}(s) + a_{22}(s) + a_{33}(s)]\, ds\right).$$

12. Using the result of Problem 11, prove that

$$W(t) = W(t_0)\exp\left(\int_{t_0}^{t} \operatorname{tr} A(s)\, ds\right),$$

where the trace of A, written $\operatorname{tr} A(t)$, is the sum of the diagonal elements of the matrix $A(t)$:

$$\operatorname{tr} A(t) = a_{11}(t) + a_{22}(t) + \cdots + a_{nn}(t),$$

for the case of $A(t)$ being an $n \times n$ matrix.

In Problems 13–16 (a) find the principal matrix solution $\Psi(t)$ for each given fundamental matrix solution $\Phi(t)$, assuming that $t_0 = 0$; (b) compute $\Psi^{-1}(t)$.

13. $\Phi(t) = e^{6t}\begin{pmatrix} \cos 2t - \frac{1}{2}\sin 2t & \sin 2t + \frac{1}{2}\cos 2t \\ \cos 2t - 3\sin 2t & \sin 2t + 3\cos 2t \end{pmatrix}$

14. $\Phi(t) = \begin{pmatrix} \sin e^t & \cos e^t \\ e^t\cos e^t & -e^t\sin e^t \end{pmatrix}$

15. $\Phi(t) = \begin{pmatrix} e^{-t} & 3e^t & e^{2t} \\ 0 & 2e^t & 3e^{2t} \\ e^{-t} & e^t & e^{2t} \end{pmatrix}$

16. $\Phi(t) = \begin{pmatrix} 2e^t + te^t & te^t + e^t & e^t \\ 0 & e^t & e^t \\ -3e^t - te^t & -te^t - 2e^t & -e^t \end{pmatrix}$

17. Consider the system

$$\mathbf{x}' = \begin{pmatrix} 3 & -2 \\ 2 & -1 \end{pmatrix}\mathbf{x}.$$

It is easy to verify that

$$\Phi(t) = \begin{pmatrix} 2te^t + e^t & 2te^t \\ 2te^t & -e^t + 2te^t \end{pmatrix}$$

is a fundamental matrix solution. Find a solution that satisfies each of the following initial conditions:

(a) $\mathbf{x}(0) = \begin{pmatrix} 1 \\ 2 \end{pmatrix}$ **(b)** $\mathbf{x}(0) = \begin{pmatrix} -2 \\ 3 \end{pmatrix}$

(c) $\mathbf{x}(1) = \begin{pmatrix} 0 \\ 1 \end{pmatrix}$ **(d)** $\mathbf{x}(-1) = \begin{pmatrix} 2 \\ 1 \end{pmatrix}$

(e) $\mathbf{x}(3) = \begin{pmatrix} 3 \\ 3 \end{pmatrix}$ **(f)** $\mathbf{x}(a) = \begin{pmatrix} b \\ c \end{pmatrix}$

18. In Example 4 we saw that a fundamental matrix solution to the system

$$\mathbf{x}' = \begin{pmatrix} 1 & 1 & -2 \\ -1 & 2 & 1 \\ 0 & 1 & -1 \end{pmatrix}\mathbf{x}$$

is

$$\Phi(t) = \begin{pmatrix} e^{-t} & 3e^{t} & e^{2t} \\ 0 & 2e^{t} & 3e^{2t} \\ e^{-t} & e^{t} & e^{2t} \end{pmatrix}.$$

Find the particular solutions that satisfy the following conditions:

(a) $\mathbf{x}(0) = \begin{pmatrix} 1 \\ -1 \\ 2 \end{pmatrix}$ **(b)** $\mathbf{x}(0) = \begin{pmatrix} 3 \\ 1 \\ 2 \end{pmatrix}$

(c) $\mathbf{x}(1) = \begin{pmatrix} 1 \\ 0 \\ 1 \end{pmatrix}$ **(d)** $\mathbf{x}(-1) = \begin{pmatrix} 2 \\ -3 \\ 5 \end{pmatrix}$

19. Consider the second-order equation

$$x'' + a(t)x' + b(t)x = 0. \qquad \textbf{(i)}$$

(a) Write Equation **(i)** in the form $\mathbf{x}' = A(t)\mathbf{x}$.
(b) Given that

$$\Phi(t) = \begin{pmatrix} \phi_1 & \phi_2 \\ \phi_1' & \phi_2' \end{pmatrix}$$

is a fundamental matrix solution, show that

$$\det \Phi(t) = \det \Phi(t_0) \exp\left(-\int_{t_0}^{t} a(s)\, ds\right).$$

(c) Show that the formula in part (b) can be rearranged as

$$\phi_2' - \frac{\phi_1'}{\phi_1}\phi_2 = \frac{\det \Phi(t_0)}{\phi_1} \exp\left(-\int_{t_0}^{t} a(s)\, ds\right). \qquad \textbf{(ii)}$$

Therefore, if one solution $\phi_1(t)$ of Equation **(i)** is known, then another solution can be calculated by solving this equation with $\det \Phi(t_0) = 1$.

20. Given that $\phi_1(t) = \sin(\ln t)$ is a solution, find a second linearly independent solution of

$$x'' + \frac{1}{t}x' + \frac{1}{t^2}x = 0.$$

21. Given that $\phi_1(t) = e^{t^2}$ is a solution of

$$x'' - 2tx' - 2x = 0,$$

find a second linearly independent solution.

22. Given that $\phi_1(t) = \sin t^2$ is a solution of

$$tx'' - x' + 4t^3x = 0,$$

find a second linearly independent solution.

9.4 EIGENVALUES AND EIGENVECTORS

In the next section we will present a method for computing the principal matrix solution of the system

$$\mathbf{x}' = A\mathbf{x}, \qquad \textbf{(1)}$$

where the matrix A is constant. The method involves the use of the eigenvalues and corresponding eigenvectors of the matrix A.

Eigenvalues and eigenvectors

Let A be an $n \times n$ matrix with real* components. A number λ (real or complex) is called an **eigenvalue** of A if there is a *nonzero* vector $\mathbf{v}$ with real or complex entries such that

$$A\mathbf{v} = \lambda\mathbf{v}. \tag{2}$$

The vector $\mathbf{v} \neq \mathbf{0}$ is called an **eigenvector of A corresponding to the eigenvalue λ.**

Note

Eigen is the German word for "own" or "proper." Eigenvalues are also called **proper values** or **characteristic values** and eigenvectors are called **proper vectors** or **characteristic vectors.**

Remark

As we will see (for example, in Example 6), a matrix with real components can have complex eigenvalues and eigenvectors. That is why, in the definition, we have asserted that λ and the components of $\mathbf{v}$ may be complex.

EXAMPLE 1 ► **Eigenvalues and eigenvectors of a 2 × 2 matrix**

Let

$$A = \begin{pmatrix} 10 & -18 \\ 6 & -11 \end{pmatrix}.$$

Then

$$A\begin{pmatrix} 2 \\ 1 \end{pmatrix} = \begin{pmatrix} 10 & -18 \\ 6 & -11 \end{pmatrix}\begin{pmatrix} 2 \\ 1 \end{pmatrix} = \begin{pmatrix} 2 \\ 1 \end{pmatrix}.$$

Thus $\lambda_1 = 1$ is an eigenvalue of A with corresponding eigenvector $\mathbf{v}_1 = \binom{2}{1}$. Similarly,

$$A\begin{pmatrix} 3 \\ 2 \end{pmatrix} = \begin{pmatrix} 10 & -18 \\ 6 & -11 \end{pmatrix}\begin{pmatrix} 3 \\ 2 \end{pmatrix} = \begin{pmatrix} -6 \\ -4 \end{pmatrix} = -2\begin{pmatrix} 3 \\ 2 \end{pmatrix},$$

so $\lambda_2 = -2$ is an eigenvalue of A with corresponding eigenvector $\mathbf{v}_2 = \binom{3}{2}$. As we soon see, these are the only eigenvalues of A. ◄

EXAMPLE 2 ► **The eigenvalues and eigenvectors of *I***

Let $A = I$. Then for any $\mathbf{v}$, $A\mathbf{v} = I\mathbf{v} = \mathbf{v}$. Thus 1 is the only eigenvalue of A and every nonzero $\mathbf{v}$ is an eigenvector of I. ◄

We compute the eigenvalues and eigenvectors of many matrices in this section. But first we need to prove some facts that can simplify our computations.

---

* This definition is also valid if A has complex components, but since the matrices we deal with have real components, the definition is sufficient for our purposes.

Suppose that λ is an eigenvalue of A. Then there exists a nonzero vector

$$\mathbf{v} = \begin{pmatrix} x_1 \\ x_2 \\ \cdot \\ \cdot \\ \cdot \\ x_n \end{pmatrix} \neq \mathbf{0}$$

such that $A\mathbf{v} = \lambda\mathbf{v} = \lambda I\mathbf{v}$.

Rewriting this, we have

$$(\lambda I - A)\mathbf{v} = \mathbf{0}. \tag{3}$$

If A is an $n \times n$ matrix, Equation **(3)** is a homogeneous system of n equations in the unknowns $x_1, x_2, \ldots, x_n$. Since, by assumption, the system has nontrivial solutions, we concluded that $\det(\lambda I - A) = 0$ (see Theorem 2 in Appendix 4). Conversely, if $\det(\lambda I - A) = 0$, then Equation **(3)** has nontrivial solutions and λ is an eigenvalue of A. On the other hand, if $\det(\lambda I - A) \neq 0$, then Equation **(3)** has only the solution $\mathbf{v} = \mathbf{0}$, so λ is *not* an eigenvalue of A. Summing up these facts, we have the following.

THEOREM 1 ♦ **Eigenvalues are zeros of a polynomial**

Let A be an $n \times n$ matrix. Then λ is an eigenvalue of A if and only if

$$p(\lambda) = \det(\lambda I - A) = 0. \quad ♦ \tag{4}$$

Characteristic equation and polynomial

Equation **(4)** is called the **characteristic equation** of A and $p(\lambda)$ is called the **characteristic polynomial** of A. [Compare this with Equation **(7.3.12)** on p. 323.]

As becomes apparent in the examples, $p(\lambda)$ is a polynomial of degree n in λ; for example, if

$$A = \begin{pmatrix} a & b \\ c & d \end{pmatrix},$$

then

$$\lambda I - A = \begin{pmatrix} \lambda & 0 \\ 0 & \lambda \end{pmatrix} - \begin{pmatrix} a & b \\ c & d \end{pmatrix} = \begin{pmatrix} \lambda - a & -b \\ -c & \lambda - d \end{pmatrix}$$

and

$$\begin{aligned} p(\lambda) &= \det(\lambda I - A) = (\lambda - a)(\lambda - d) - bc \\ &= \lambda^2 - (a + d)\lambda + (ad - bc). \end{aligned}$$

Similarly, if

$$A = \begin{pmatrix} a_{11} & a_{12} & \cdots & a_{1n} \\ a_{21} & a_{22} & \cdots & a_{2n} \\ \vdots & \vdots & & \vdots \\ a_{n1} & a_{n2} & \cdots & a_{nn} \end{pmatrix}$$

then

$$p(\lambda) = \det(\lambda I - A) = \begin{vmatrix} \lambda - a_{11} & -a_{12} & \cdots & -a_{1n} \\ -a_{21} & \lambda - a_{22} & \cdots & -a_{2n} \\ \vdots & \vdots & & \vdots \\ -a_{n1} & -a_{n2} & \cdots & \lambda - a_{nn} \end{vmatrix},$$

and $p(\lambda)$ can be written in the form

$$p(\lambda) = \lambda^n + b_{n-1}\lambda^{n-1} + \cdots + b_1\lambda + b_0 = 0. \tag{5}$$

By the fundamental theorem of algebra, any polynomial of degree n with real or complex coefficients has exactly n roots (counting multiplicities). By this we mean, for example, that the polynomial $(\lambda - 1)^5$ has five roots, all equal to the number 1. Since any eigenvalue of A is a root of the characteristic equation of A, an $n \times n$ matrix has n eigenvalues, some of which may be repeated. If $\lambda_1, \lambda_2, \ldots, \lambda_m$ are the distinct roots of Equation **(5)** with multiplicities $r_1, r_2, \ldots, r_m$, respectively, then Equation **(5)** may be factored to obtain

Algebraic multiplicities

$$p(\lambda) = (\lambda - \lambda_1)^{r_1}(\lambda - \lambda_2)^{r_2} \ldots (\lambda - \lambda_m)^{r_m} = 0. \tag{6}$$

The numbers $r_1, r_2, \ldots, r_m$ are called the **algebraic multiplicities** of the eigenvalues $\lambda_1, \lambda_2, \ldots, \lambda_m$, respectively.

THEOREM 2 ♦ **A linear combination of eigenvectors is an eigenvector**

Let $\mathbf{v}_1$ and $\mathbf{v}_2$ be eigenvectors of A corresponding to the same eigenvalue λ. Then $\mathbf{v} = c_1\mathbf{v}_1 + c_1\mathbf{v}_2$ is an eigenvector of A corresponding to λ; that is,

any linear combination of eigenvectors corresponding to the same eigenvalue is an eigenvector corresponding to that eigenvalue.

Proof

$$A\mathbf{v} = A(c_1\mathbf{v}_1 + c_2\mathbf{v}_2) = c_1 A\mathbf{v}_1 + c_2 A\mathbf{v}_2 = c_1\lambda\mathbf{v}_1 + c_2\lambda\mathbf{v}_2$$
$$= \lambda(c_1\mathbf{v}_1 + c_2\mathbf{v}_2) = \lambda\mathbf{v}. \quad ♦$$

According to Theorem 2, corresponding to each eigenvalue of a square matrix A there are an infinite number of eigenvectors. Clearly, then, listing every eigenvector is infeasible. Rather, our aim is to find as many linearly independent eigenvectors as possible corresponding to each eigenvalue. The following result is very useful and is given without proof.

THEOREM 3 ♦

If λ is a simple eigenvalue of A (that is, the algebraic multiplicity of λ is 1), then there is only one linearly independent eigenvector corresponding to λ. ♦

We now prove another useful result.

THEOREM 4 ♦ **Eigenvectors corresponding to distinct eigenvalues are linearly independent**

Let A be an $n \times n$ matrix and let $\lambda_1, \lambda_2, \ldots, \lambda_m$ be distinct eigenvalues of A with corresponding eigenvectors $\mathbf{v}_1, \mathbf{v}_2, \ldots, \mathbf{v}_m$. Then $\mathbf{v}_1, \mathbf{v}_2, \ldots, \mathbf{v}_m$ are linearly independent; that is, *eigenvectors corresponding to distinct eigenvalues are linearly independent.**

Proof We prove this by mathematical induction. We start with $m = 2$. Suppose that

$$c_1\mathbf{v}_1 + c_2\mathbf{v}_2 = \mathbf{0}. \tag{7}$$

Then, multiplying both sides of Equation **(7)** by A, we have

$$\mathbf{0} = A(c_1\mathbf{v}_1 + c_2\mathbf{v}_2) = c_1A\mathbf{v}_1 + c_2A\mathbf{v}_2,$$

or

$$c_1\lambda_1\mathbf{v}_1 + c_2\lambda_2\mathbf{v}_2 = \mathbf{0}. \tag{8}$$

We then multiply Equation **(7)** by λ_1 and subtract it from Equation **(8)** to obtain

$$(c_1\lambda_1\mathbf{v}_1 + c_2\lambda_2\mathbf{v}_2) - (c_1\lambda_1\mathbf{v}_1 + c_2\lambda_1\mathbf{v}_2) = \mathbf{0},$$

or

$$c_2(\lambda_2 - \lambda_1)\mathbf{v}_2 = \mathbf{0}.$$

Since $\mathbf{v}_2 \neq \mathbf{0}$ (by the definition of an eigenvector) and since $\lambda_2 \neq \lambda_1$, we conclude that $c_2 = 0$. Then inserting $c_2 = 0$ in Equation **(7)**, we obtain $c_1 = 0$, which proves the theorem in the case $m = 2$. Now suppose that the theorem is true for $m = k$; that is, assume that any k eigenvectors corresponding to distinct eigenvalues are linearly independent. We prove the theorem for $m = k + 1$. So we assume that

$$c_1\mathbf{v}_1 + c_2\mathbf{v}_2 + \cdots + c_k\mathbf{v}_k + c_{k+1}\mathbf{v}_{k+1} = \mathbf{0}. \tag{9}$$

Then, multiplying both sides of Equation **(9)** by A and using the fact that $A\mathbf{v}_i = \lambda_i\mathbf{v}_i$, we obtain

$$c_1\lambda_1\mathbf{v}_1 + c_2\lambda_2\mathbf{v}_2 + \cdots + c_k\lambda_k\mathbf{v}_k + c_{k+1}\lambda_{k+1}\mathbf{v}_{k+1} = \mathbf{0}. \tag{10}$$

We multiply both sides of Equation **(9)** by λ_{k+1} and subtract it from Equation **(10)**:

$$c_1(\lambda_1 - \lambda_{k+1})\mathbf{v}_1 + c_2(\lambda_2 - \lambda_{k+1})\mathbf{v}_2 + \cdots + c_k(\lambda_k - \lambda_{k+1})\mathbf{v}_k = \mathbf{0}.$$

But, by the induction assumption, $\mathbf{v}_1, \mathbf{v}_2, \ldots, \mathbf{v}_k$ are linearly independent. Thus

$$c_1(\lambda_1 - \lambda_{k+1}) = c_2(\lambda_2 - \lambda_{k+1}) = \ldots = c_k(\lambda_k - \lambda_{k+1}) = 0,$$

and, since $\lambda_i \neq \lambda_{k+1}$ for $i = 1, 2, \ldots, k$, we conclude that $c_1 = c_2 = \ldots = c_k = 0$. But, from Equation **(9)**, this means that $c_{k+1} = 0$.Thus the theorem is true for $m = k + 1$ and the proof is complete. ♦

We now proceed to calculate eigenvalues and corresponding eigenvectors. We do this using a three-step procedure.

To compute eigenvalues and eigenvectors

1. Find $p(\lambda) = \det(\lambda I - A)$. **(11)**
2. Find the roots $\lambda_1, \lambda_2, \ldots, \lambda_m$ of $p(\lambda) = 0$. **(12)**
3. Corresponding to each eigenvalue λ_i, solve the homogeneous system $(\lambda_i I - A)\mathbf{v} = \mathbf{0}$ or $(A - \lambda_i I)\mathbf{v} = \mathbf{0}$. **(13)**

---

* The definition of linear independence of vectors is the same as the definition of linear independence of functions (see pp. 108 and 152).

Remark

Step (2) is often the hardest one to carry out.

EXAMPLE 3 ► **A 2 × 2 matrix with real, distinct eigenvalues**
Let

$$A = \begin{pmatrix} 4 & 2 \\ 3 & 3 \end{pmatrix}.$$

Then

$$\det(\lambda I - A) = \begin{vmatrix} \lambda - 4 & -2 \\ -3 & \lambda - 3 \end{vmatrix} = (\lambda - 4)(\lambda - 3) - 6$$
$$= \lambda^2 - 7\lambda + 6 = (\lambda - 1)(\lambda - 6) = 0.$$

Thus the eigenvalues of A are $\lambda_1 = 1$ and $\lambda_2 = 6$. For $\lambda_1 = 1$, we solve $(A - I)\mathbf{v} = \mathbf{0}$ or

$$\begin{pmatrix} 3 & 2 \\ 3 & 2 \end{pmatrix}\begin{pmatrix} x_1 \\ x_2 \end{pmatrix} = \begin{pmatrix} 0 \\ 0 \end{pmatrix}.$$

Clearly, any eigenvector corresponding to $\lambda_1 = 1$ satisfies $3x_1 + 2x_2 = 0$, or $x_1 = -2x_2/3$. One such eigenvector is

$$\mathbf{v}_1 = \begin{pmatrix} 2 \\ -3 \end{pmatrix}.$$

According to Theorem 3, we cannot find any other linearly independent eigenvector corresponding to $\lambda_1 = 1$.

Similarly, the equation $(A - 6I)\mathbf{v} = \mathbf{0}$ means that

$$\begin{pmatrix} -2 & 2 \\ 3 & -3 \end{pmatrix}\begin{pmatrix} x_1 \\ x_2 \end{pmatrix} = \begin{pmatrix} 0 \\ 0 \end{pmatrix},$$

or $x_1 = x_2$. Thus $\mathbf{v}_2 = \begin{pmatrix} 1 \\ 1 \end{pmatrix}$ is an eigenvector corresponding to $\lambda_2 = 6$. Note that $\mathbf{v}_1$ and $\mathbf{v}_2$ are linearly independent by Theorem 4. ◄

EXAMPLE 4 ► **A 3 × 3 matrix with real, distinct eigenvalues**
Let

$$A = \begin{pmatrix} 1 & -1 & 4 \\ 3 & 2 & -1 \\ 2 & 1 & -1 \end{pmatrix}.$$

Then

$$\det(\lambda I - A) = \begin{vmatrix} \lambda - 1 & 1 & -4 \\ -3 & \lambda - 2 & 1 \\ -2 & -1 & \lambda + 1 \end{vmatrix} = \lambda^3 - 2\lambda^2 - 5\lambda + 6$$
$$= (\lambda - 1)(\lambda + 2)(\lambda - 3) = 0.$$

Thus the eigenvalues of A are $\lambda_1 = 1$, $\lambda_2 = -2$, and $\lambda_3 = 3$. Since each eigenvalue is simple, we need only find one eigenvector corresponding to each eigenvalue. Corresponding to $\lambda = 1$, we have

$$(A - I)\mathbf{v} = \begin{pmatrix} 0 & -1 & 4 \\ 3 & 1 & -1 \\ 2 & 1 & -2 \end{pmatrix}\begin{pmatrix} x_1 \\ x_2 \\ x_3 \end{pmatrix} = \begin{pmatrix} 0 \\ 0 \\ 0 \end{pmatrix}. \qquad \textbf{(14)}$$

We write out the equations in **(14)**:

$$\begin{aligned} 0x_1 - x_2 + 4x_3 &= 0, \\ 3x_1 + x_2 - x_3 &= 0, \\ 2x_1 + x_2 - 2x_3 &= 0. \end{aligned} \qquad \textbf{(15)}$$

We solve this system by Gauss-Jordan elimination. The process is to write system **(15)** as an augmented matrix and then row reduce:

$$\left(\begin{array}{ccc|c} 0 & -1 & 4 & 0 \\ 3 & 1 & -1 & 0 \\ 2 & 1 & -2 & 0 \end{array}\right) \xrightarrow[A_{1,3}(1)]{A_{1,2}(1)} \left(\begin{array}{ccc|c} 0 & -1 & 4 & 0 \\ 3 & 0 & 3 & 0 \\ 2 & 0 & 2 & 0 \end{array}\right)$$

$$\xrightarrow{M_2(\frac{1}{3})} \left(\begin{array}{ccc|c} 0 & -1 & 4 & 0 \\ 1 & 0 & 1 & 0 \\ 2 & 0 & 2 & 0 \end{array}\right) \xrightarrow{A_{2,3}(-2)} \left(\begin{array}{ccc|c} 0 & -1 & 4 & 0 \\ 1 & 0 & 1 & 0 \\ 0 & 0 & 0 & 0 \end{array}\right).$$

This denotes the system of equations

$$\begin{aligned} -x_2 + 4x_3 &= 0, \\ x_1 \qquad + x_3 &= 0 \end{aligned}$$

Thus $x_1 = -x_3$, $x_2 = 4x_3$, and an eigenvector corresponding to $\lambda_1 = 1$ is

$$\mathbf{v}_1 = \begin{pmatrix} -1 \\ 4 \\ 1 \end{pmatrix}.$$

Because of Theorem 3, we know that any other eigenvector of A corresponding to $\lambda_1 = 1$ is a multiple of $\mathbf{v}_1$.

For $\lambda_2 = -2$, we have

$$[A - (-2I)]\mathbf{v} = (A + 2I)\mathbf{v} = \mathbf{0},$$

or

$$\begin{pmatrix} 3 & -1 & 4 \\ 3 & 4 & -1 \\ 2 & 1 & 1 \end{pmatrix} \begin{pmatrix} x_1 \\ x_2 \\ x_3 \end{pmatrix} = \begin{pmatrix} 0 \\ 0 \\ 0 \end{pmatrix}.$$

This leads to

$$\left(\begin{array}{ccc|c} 3 & -1 & 4 & 0 \\ 3 & 4 & -1 & 0 \\ 2 & 1 & 1 & 0 \end{array}\right) \xrightarrow[A_{1,3}(1)]{A_{1,2}(4)} \left(\begin{array}{ccc|c} 3 & -1 & 4 & 0 \\ 15 & 0 & 15 & 0 \\ 5 & 0 & 5 & 0 \end{array}\right)$$

$$\xrightarrow{M_2(\frac{1}{15})} \left(\begin{array}{ccc|c} 3 & -1 & 4 & 0 \\ 1 & 0 & 1 & 0 \\ 5 & 0 & 5 & 0 \end{array}\right) \xrightarrow[A_{2,3}(-5)]{A_{2,1}(-4)} \left(\begin{array}{ccc|c} -1 & -1 & 0 & 0 \\ 1 & 0 & 1 & 0 \\ 0 & 0 & 0 & 0 \end{array}\right).$$

Thus $x_2 = -x_1$, $x_3 = -x_1$, and an eigenvector corresponding to $\lambda_2 = -2$ is

$$\mathbf{v}_2 = \begin{pmatrix} 1 \\ -1 \\ -1 \end{pmatrix}.$$

Finally, for $\lambda_3 = 3$, we have

$$(A - 3I)\mathbf{v} = \begin{pmatrix} -2 & -1 & 4 \\ 3 & -1 & -1 \\ 2 & 1 & -4 \end{pmatrix}\begin{pmatrix} x_1 \\ x_2 \\ x_3 \end{pmatrix} = \begin{pmatrix} 0 \\ 0 \\ 0 \end{pmatrix}$$

and

$$\left(\begin{array}{ccc|c} -2 & -1 & 4 & 0 \\ 3 & -1 & -1 & 0 \\ 2 & 1 & -4 & 0 \end{array}\right) \xrightarrow[A_{3,2}(1)]{A_{3,1}(1)} \left(\begin{array}{ccc|c} 0 & 0 & 0 & 0 \\ 5 & 0 & -5 & 0 \\ 2 & 1 & -4 & 0 \end{array}\right)$$

$$\xrightarrow{M_2(\frac{1}{5})} \left(\begin{array}{ccc|c} 0 & 0 & 0 & 0 \\ 1 & 0 & -1 & 0 \\ 2 & 1 & -4 & 0 \end{array}\right) \xrightarrow{A_{2,3}(-4)} \left(\begin{array}{ccc|c} 0 & 0 & 0 & 0 \\ 1 & 0 & -1 & 0 \\ -2 & 1 & 0 & 0 \end{array}\right).$$

Thus $x_3 = x_1$, $x_2 = 2x_1$, and an eigenvector corresponding to $\lambda_3 = 3$ is

$$\mathbf{v}_3 = \begin{pmatrix} 1 \\ 2 \\ 1 \end{pmatrix}.$$

Remark

In this and every other example, there are always an infinite number of choices for each eigenvector. We arbitrarily choose a simple one by setting one or more of the x_i's equal to 1. Other normalizations are sometimes useful. ◀

In the last example we found three eigenvectors that, according to Theorem 4, must be linearly independent. There is an easy way to verify this. The following result is standard in the theory of determinants.

THEOREM 5 ♦

The $n \times 1$ vectors $\mathbf{v}_1, \mathbf{v}_2, \ldots, \mathbf{v}_n$ are linearly independent if and only if the determinant of the matrix whose columns are $\mathbf{v}_1, \mathbf{v}_2, \ldots, \mathbf{v}_n$ is nonzero. ♦

EXAMPLE 5 ▶ **Using Theorem 5**

In Example 4 we can verify that the vectors

$$\mathbf{v}_1 = \begin{pmatrix} -1 \\ 4 \\ 1 \end{pmatrix}, \quad \mathbf{v}_2 = \begin{pmatrix} 1 \\ -1 \\ -1 \end{pmatrix}, \quad \text{and} \quad \mathbf{v}_3 = \begin{pmatrix} 1 \\ 2 \\ 1 \end{pmatrix}$$

are linearly independent because

$$\begin{vmatrix} -1 & 1 & 1 \\ 4 & -1 & 2 \\ 1 & -1 & 1 \end{vmatrix} = -6 \neq 0. \quad ◀$$

EXAMPLE 6 ▶ **A matrix with complex conjugate eigenvalues**

Let

$$A = \begin{pmatrix} 3 & -5 \\ 1 & -1 \end{pmatrix}.$$

Then

$$\det(\lambda I - A) = \begin{vmatrix} \lambda - 3 & 5 \\ -1 & \lambda + 1 \end{vmatrix} = \lambda^2 - 2\lambda + 2 = 0,$$

has eigenvalues

$$\left.\begin{matrix}\lambda_1\\\lambda_2\end{matrix}\right\} = \frac{-(-2) \pm \sqrt{4 - 4(1)(2)}}{2} = \frac{2 \pm \sqrt{-4}}{2} = \frac{2 \pm 2i}{2} = 1 \pm i.$$

Thus $\lambda_1 = 1 + i$ and $\lambda_2 = 1 - i$. Then

$$[A - (1 + i)I]\mathbf{v} = \begin{pmatrix} 2 - i & -5 \\ 1 & -2 - i \end{pmatrix}\begin{pmatrix} x_1 \\ x_2 \end{pmatrix} = \begin{pmatrix} 0 \\ 0 \end{pmatrix}$$

and we obtain $(2 - i)x_1 - 5x_2 = 0$ and $x_1 + (-2 - i)x_2 = 0$.* Thus $x_1 = (2 + i)x_2$, which yields the eigenvector (corresponding to $\lambda_1 = 1 + i$)

$$\mathbf{v}_1 = \begin{pmatrix} 2 + i \\ 1 \end{pmatrix}.$$

Similarly,

$$[A - (1 - i)I]\mathbf{v} = \begin{pmatrix} 2 + i & -5 \\ 1 & -2 + i \end{pmatrix}\begin{pmatrix} x_1 \\ x_2 \end{pmatrix} = \begin{pmatrix} 0 \\ 0 \end{pmatrix},$$

or $x_1 + (-2 + i)x_2 = 0$, which yields $x_1 = (2 - i)x_2$, and we obtain the eigenvector (corresponding to $\lambda_2 = 1 - i$)

$$\mathbf{v}_2 = \begin{pmatrix} 2 - i \\ 1 \end{pmatrix}. \quad ◀$$

Remark

This example illustrates that a real matrix may have complex eigenvalues and eigenvectors. It should be pointed out that some texts define eigenvalues of real matrices to be the *real* roots of the characteristic equation. With this definition, the matrix of the last example has *no* eigenvalues. This might make the computations simpler, but it also significantly reduces the usefulness of the theory of eigenvalues and eigenvectors. We will see an important illustration of the use of complex eigenvalues in Section 9.5 in our computation of principal matrix solutions.

EXAMPLE 7 ▶ A 2 × 2 matrix with only one linearly independent eigenvector

Let

$$A = \begin{pmatrix} 4 & 1 \\ 0 & 4 \end{pmatrix}.$$

Then

$$\det(\lambda I - A) = \begin{vmatrix} \lambda - 4 & -1 \\ 0 & \lambda - 4 \end{vmatrix} = (\lambda - 4)^2 = 0,$$

so $\lambda = 4$ is an eigenvalue of algebraic multiplicity 2 and we have

$$(A - 4I)\mathbf{v} = \begin{pmatrix} 0 & 1 \\ 0 & 0 \end{pmatrix}\begin{pmatrix} x_1 \\ x_2 \end{pmatrix} = \begin{pmatrix} x_2 \\ 0 \end{pmatrix}.$$

Thus $x_2 = 0$ and x_1 is arbitrary. Therefore the only linearly independent eigenvector is

$$\mathbf{v} = \begin{pmatrix} 1 \\ 0 \end{pmatrix}.$$

Here is an instance where a 2×2 matrix has only one linearly independent eigenvector. ◀

---

* Note that

$$(-2 - i)\begin{pmatrix} 2 - i \\ 1 \end{pmatrix} = \begin{pmatrix} -5 \\ -2 - i \end{pmatrix}.$$

EXAMPLE 8 ▸ An eigenvalue of algebraic multiplicity two with two corresponding linearly independent eigenvectors

Let

$$A = \begin{pmatrix} 3 & 2 & 4 \\ 2 & 0 & 2 \\ 4 & 2 & 3 \end{pmatrix}.$$

Then

$$\det(\lambda I - A) = \begin{vmatrix} \lambda - 3 & -2 & -4 \\ -2 & \lambda & -2 \\ -4 & -2 & \lambda - 3 \end{vmatrix}$$
$$= \lambda^3 - 6\lambda^2 - 15\lambda - 8 = (\lambda + 1)^2(\lambda - 8) = 0,$$

so the eigenvalues are $\lambda_1 = 8$ and $\lambda_2 = -1$ (with algebraic multiplicity 2). For $\lambda_1 = 8$, we obtain

$$(A - 8I)\mathbf{v} = \begin{pmatrix} -5 & 2 & 4 \\ 2 & -8 & 2 \\ 4 & 2 & -5 \end{pmatrix}\begin{pmatrix} x_1 \\ x_2 \\ x_3 \end{pmatrix} = \begin{pmatrix} 0 \\ 0 \\ 0 \end{pmatrix},$$

or, row reducing,

$$\left(\begin{array}{rrr|r} -5 & 2 & 4 & 0 \\ 2 & -8 & 2 & 0 \\ 4 & 2 & -5 & 0 \end{array}\right) \xrightarrow[A_{1,3}(-1)]{A_{1,2}(4)} \left(\begin{array}{rrr|r} -5 & 2 & 4 & 0 \\ -18 & 0 & 18 & 0 \\ 9 & 0 & -9 & 0 \end{array}\right)$$

$$\xrightarrow{M_2(\frac{1}{18})} \left(\begin{array}{rrr|r} -5 & 2 & 4 & 0 \\ -1 & 0 & 1 & 0 \\ 9 & 0 & -9 & 0 \end{array}\right) \xrightarrow{A_{2,3}(9)} \left(\begin{array}{rrr|r} 0 & 2 & -1 & 0 \\ -1 & 0 & 1 & 0 \\ 0 & 0 & 0 & 0 \end{array}\right).$$

Hence $x_3 = 2x_2$, $x_1 = x_3$, and we obtain an eigenvector

$$\mathbf{v}_1 = \begin{pmatrix} 2 \\ 1 \\ 2 \end{pmatrix}.$$

For $\lambda_2 = -1$, we have

$$(A + I)\mathbf{v} = \begin{pmatrix} 4 & 2 & 4 \\ 2 & 1 & 2 \\ 4 & 2 & 4 \end{pmatrix}\begin{pmatrix} x_1 \\ x_2 \\ x_3 \end{pmatrix} = \begin{pmatrix} 0 \\ 0 \\ 0 \end{pmatrix},$$

which gives us the single equation $2x_1 + x_2 + 2x_3 = 0$ or $x_2 = -2x_1 - 2x_3$. If $x_1 = 1$ and $x_3 = 0$, we obtain

$$\mathbf{v}_2 = \begin{pmatrix} 1 \\ -2 \\ 0 \end{pmatrix}.$$

If $x_1 = 0$ and $x_3 = 1$, we obtain

$$\mathbf{v}_3 = \begin{pmatrix} 0 \\ -2 \\ 1 \end{pmatrix}.$$

Here there are two linearly independent eigenvectors corresponding to an eigenvalue of multiplicity 2. ◂

EXAMPLE 9 ▶ **An eigenvalue of algebraic multiplicity three but with only one corresponding linearly independent eigenvector**

Let

$$A = \begin{pmatrix} -5 & -5 & -9 \\ 8 & 9 & 18 \\ -2 & -3 & -7 \end{pmatrix}.$$

Then

$$\det(\lambda I - A) = \begin{vmatrix} \lambda + 5 & 5 & 9 \\ -8 & \lambda - 9 & -18 \\ 2 & 3 & \lambda + 7 \end{vmatrix} = \lambda^3 + 3\lambda^2 + 3\lambda + 1$$
$$= (\lambda + 1)^3 = 0.$$

Thus $\lambda = -1$ is an eigenvalue of algebraic multiplicity 3. To find eigenvectors, we set

$$(A + I)\mathbf{v} = \begin{pmatrix} -4 & -5 & -9 \\ 8 & 10 & 18 \\ -2 & -3 & -6 \end{pmatrix}\begin{pmatrix} x_1 \\ x_2 \\ x_3 \end{pmatrix} = \begin{pmatrix} 0 \\ 0 \\ 0 \end{pmatrix}$$

and row reduce to obtain, successively,

$$\left(\begin{array}{ccc|c} -4 & -5 & -9 & 0 \\ 8 & 10 & 18 & 0 \\ -2 & -3 & -6 & 0 \end{array}\right) \xrightarrow[A_{3,2}(4)]{A_{3,1}(-2)} \left(\begin{array}{ccc|c} 0 & 1 & 3 & 0 \\ 0 & -2 & -6 & 0 \\ -2 & -3 & -6 & 0 \end{array}\right)$$

$$\xrightarrow[A_{1,3}(3)]{A_{1,2}(2)} \left(\begin{array}{ccc|c} 0 & 1 & 3 & 0 \\ 0 & 0 & 0 & 0 \\ -2 & 0 & 3 & 0 \end{array}\right).$$

This yields $x_2 = -3x_3$ and $2x_1 = 3x_3$. Setting $x_3 = 2$, we obtain only one linearly independent eigenvector:

$$\mathbf{v}_1 = \begin{pmatrix} 3 \\ -6 \\ 2 \end{pmatrix}. \quad ◀$$

EXAMPLE 10 ▶ **An eigenvalue of algebraic multiplicity three with two corresponding linearly independent eigenvectors**

Let

$$A = \begin{pmatrix} -1 & -3 & -9 \\ 0 & 5 & 18 \\ 0 & -2 & -7 \end{pmatrix}.$$

Then

$$\det(\lambda I - A) = \begin{vmatrix} \lambda + 1 & 3 & 9 \\ 0 & \lambda - 5 & -18 \\ 0 & 2 & \lambda + 7 \end{vmatrix} = (\lambda + 1)^3 = 0.$$

Thus, as in Example 9, $\lambda = -1$ is an eigenvalue of algebraic multiplicity 3. To find eigenvectors, we compute

$$(A + I)\mathbf{v} = \begin{pmatrix} 0 & -3 & -9 \\ 0 & 6 & 18 \\ 0 & -2 & -6 \end{pmatrix} = \begin{pmatrix} x_1 \\ x_2 \\ x_3 \end{pmatrix} = \begin{pmatrix} 0 \\ 0 \\ 0 \end{pmatrix}.$$

Thus $-2x_2 - 6x_3 = 0$ or $x_2 = -3x_3$, and x_1 is arbitrary. Setting $x_1 = 0$ and $x_3 = 1$, we obtain

$$\mathbf{v}_1 = \begin{pmatrix} 0 \\ -3 \\ 1 \end{pmatrix}.$$

Setting $x_1 = 1$ and $x_3 = 0$ yields

$$\mathbf{v}_2 = \begin{pmatrix} 1 \\ 0 \\ 0 \end{pmatrix}. \blacktriangleleft$$

Here there are two linearly independent eigenvectors corresponding to an eigenvalue of multiplicity 3.

EXAMPLE 11 ▶ **An eigenvalue of algebraic multiplicity three with three corresponding linearly independent eigenvectors**

Let

$$A = \begin{pmatrix} -1 & 0 & 0 \\ 0 & -1 & 0 \\ 0 & 0 & -1 \end{pmatrix} = -I.$$

The characteristic equation is $(\lambda + 1)^3 = 0$, and $\lambda = -1$ is an eigenvalue of algebraic multiplicity 3. Clearly, for any 3×1 vector $\mathbf{v}$,

$$A\mathbf{v} = -I\mathbf{v} = -\mathbf{v} = (-1)\mathbf{v}$$

and $\mathbf{v}(\neq \mathbf{0})$ is an eigenvector. Three linearly independent eigenvectors are

$$\mathbf{v}_1 = \begin{pmatrix} 1 \\ 0 \\ 0 \end{pmatrix}, \quad \mathbf{v}_2 = \begin{pmatrix} 0 \\ 1 \\ 0 \end{pmatrix}, \quad \text{and} \quad \mathbf{v}_3 = \begin{pmatrix} 0 \\ 0 \\ 1 \end{pmatrix}. \blacktriangleleft$$

Remark

Examples 9, 10, and 11 illustrate the fact that if λ is an eigenvalue of algebraic multiplicity 3, then there can be one, two, or three linearly independent eigenvectors corresponding to λ. A similar result is true if the multiplicity is n, where $n > 1$.

SELF-QUIZ

Answer Questions I–III true *or* false.

I. Eigenvalues of a matrix are defined only if the matrix is square.

II. If a matrix has real coefficients, then its eigenvalues must be real.

III. If λ is an eigenvalue of algebraic multiplicity 2, then there must be two linear independent eigenvectors corresponding to λ.

IV. The eigenvalues of

$$\begin{pmatrix} 2 & -1 & 0 \\ 0 & -4 & 5 \\ 0 & 0 & 3 \end{pmatrix}$$

are

(a) 2, −1, 0 **(b)** 0, −4, 5 **(c)** 2, −4, 3
(d) 0, 5, 3

V. If the characteristic equation of A is $(\lambda - 2)(\lambda + 3)^3 = 0$, then the possible number of linearly independent eigenvectors of A corresponding to $\lambda = -3$ is ______.
(a) 0 **(b)** 1
(c) 2 **(d)** 3 **(e)** 4

Answers to Self-Quiz

I. True **II.** False **III.** False **IV.** c **V.** b, c, d

PROBLEMS 9.4

In Problems 1–20 calculate the eigenvalues and eigenvectors of the given matrix. If the algebraic multiplicity of an eigenvalue is greater than 1, determine the number of linearly independent eigenvectors that correspond to it.

1. $\begin{pmatrix} -2 & -2 \\ -5 & 1 \end{pmatrix}$ **2.** $\begin{pmatrix} -12 & 7 \\ -7 & 2 \end{pmatrix}$ **3.** $\begin{pmatrix} 2 & -1 \\ 5 & -2 \end{pmatrix}$

4. $\begin{pmatrix} -3 & 0 \\ 0 & -3 \end{pmatrix}$ **5.** $\begin{pmatrix} -3 & 2 \\ 0 & -3 \end{pmatrix}$ **6.** $\begin{pmatrix} 3 & 2 \\ -5 & 1 \end{pmatrix}$

7. $\begin{pmatrix} 1 & -1 & 0 \\ -1 & 2 & -1 \\ 0 & -1 & 1 \end{pmatrix}$ **8.** $\begin{pmatrix} 1 & 1 & -2 \\ -1 & 2 & 1 \\ 0 & 1 & -1 \end{pmatrix}$ **9.** $\begin{pmatrix} 5 & 4 & 2 \\ 4 & 5 & 2 \\ 2 & 2 & 2 \end{pmatrix}$

10. $\begin{pmatrix} 1 & 2 & 2 \\ 0 & 2 & 1 \\ -1 & 2 & 2 \end{pmatrix}$ **11.** $\begin{pmatrix} 0 & 1 & 0 \\ 0 & 0 & 1 \\ 1 & -3 & 3 \end{pmatrix}$ **12.** $\begin{pmatrix} -3 & -7 & -5 \\ 2 & 4 & 3 \\ 1 & 2 & 2 \end{pmatrix}$

13. $\begin{pmatrix} 1 & -1 & -1 \\ 1 & -1 & 0 \\ 1 & 0 & -1 \end{pmatrix}$ **14.** $\begin{pmatrix} 7 & -2 & -4 \\ 3 & 0 & -2 \\ 6 & -2 & -3 \end{pmatrix}$ **15.** $\begin{pmatrix} 4 & 6 & 6 \\ 1 & 3 & 2 \\ -1 & -5 & -2 \end{pmatrix}$

16. $\begin{pmatrix} 4 & 1 & 0 & 1 \\ 2 & 3 & 0 & 1 \\ -2 & 1 & 2 & -3 \\ 2 & -1 & 0 & 5 \end{pmatrix}$ **17.** $\begin{pmatrix} a & 0 & 0 & 0 \\ 0 & a & 0 & 0 \\ 0 & 0 & a & 0 \\ 0 & 0 & 0 & a \end{pmatrix}$ **18.** $\begin{pmatrix} a & b & 0 & 0 \\ 0 & a & 0 & 0 \\ 0 & 0 & a & 0 \\ 0 & 0 & 0 & a \end{pmatrix}$, $b \neq 0$

19. $\begin{pmatrix} a & b & 0 & 0 \\ 0 & a & c & 0 \\ 0 & 0 & a & 0 \\ 0 & 0 & 0 & a \end{pmatrix}$, $bc \neq 0$ **20.** $\begin{pmatrix} a & b & 0 & 0 \\ 0 & a & c & 0 \\ 0 & 0 & a & d \\ 0 & 0 & 0 & a \end{pmatrix}$, $bcd \neq 0$

21. Show that for any real numbers a and b, the matrix

$$A = \begin{pmatrix} a & b \\ -b & a \end{pmatrix}$$

has the eigenvectors

$$\begin{pmatrix} 1 \\ i \end{pmatrix} \text{ and } \begin{pmatrix} 1 \\ -i \end{pmatrix}.$$

In Problems 22–28 assume that the matrix A has the eigenvalues $\lambda_1, \lambda_2, \ldots, \lambda_k$.

22. Show that the eigenvalues of A^t are $\lambda_1, \lambda_2, \ldots, \lambda_k$.*

23. Show that the eigenvalues of αA are $\alpha\lambda_1, \alpha\lambda_2, \ldots, \alpha\lambda_k$.

24. Show that A^{-1} exists if and only if $\lambda_1, \lambda_2, \ldots, \lambda_k \neq 0$.

25. If A^{-1} exists, show that the eigenvalues of A^{-1} are $1/\lambda_1, 1/\lambda_2, \ldots, 1/\lambda_k$.

26. Show that the matrix $A - \alpha I$ has the eigenvalues $\lambda_1 - \alpha, \lambda_2 - \alpha, \ldots, \lambda_k - \alpha$.

27. Show that the eigenvalues of A^2 are $\lambda_1^2, \lambda_2^2, \ldots, \lambda_k^2$.

28. Show that the eigenvalues of A^m are $\lambda_1^m, \lambda_2^m, \ldots, \lambda_k^m$ for $m = 1, 2, 3, \ldots$.

*A^t = the **transpose** of A = the matrix obtained by interchanging the rows and columns of A.

29. Let λ be an eigenvalue of A with corresponding eigenvector $\mathbf{v}$. Let $p(\lambda) = a_0 + a_1\lambda + a_2\lambda^2 + \cdots + a_n\lambda^n$. Define the matrix $p(A)$ by $p(A) = a_0 I + a_1 A + a_2 A^2 + \cdots + a_n A^n$. Show that $p(A)\mathbf{v} = p(\lambda)\mathbf{v}$.

30. Using the result of Problem 29, show that if $\lambda_1, \lambda_2, \ldots, \lambda_k$ are eigenvalues of A, then $p(\lambda_1), p(\lambda_2), \ldots, p(\lambda_k)$ are eigenvalues of $p(A)$.

31. Show that if A is an upper triangular matrix, then the eigenvalues of A are the diagonal components of A.

32. Let

$$A_1 = \begin{pmatrix} 2 & 0 & 0 & 0 \\ 0 & 2 & 0 & 0 \\ 0 & 0 & 2 & 0 \\ 0 & 0 & 0 & 2 \end{pmatrix}, \quad A_2 = \begin{pmatrix} 2 & 1 & 0 & 0 \\ 0 & 2 & 0 & 0 \\ 0 & 0 & 2 & 0 \\ 0 & 0 & 0 & 2 \end{pmatrix},$$

$$A_3 = \begin{pmatrix} 2 & 1 & 0 & 0 \\ 0 & 2 & 1 & 0 \\ 0 & 0 & 2 & 0 \\ 0 & 0 & 0 & 2 \end{pmatrix}, \quad \text{and} \quad A_4 = \begin{pmatrix} 2 & 1 & 0 & 0 \\ 0 & 2 & 1 & 0 \\ 0 & 0 & 2 & 1 \\ 0 & 0 & 0 & 2 \end{pmatrix}.$$

Show that, for each matrix, $\lambda = 2$ is an eigenvalue of algebraic multiplicity 4. In each case determine the number of linearly independent eigenvectors that correspond to $\lambda = 2$.

33. Let A be a real $n \times n$ matrix. Show that if λ_1 is a complex eigenvalue of A with corresponding eigenvector $\mathbf{v}_1$, then $\overline{\lambda}_1$ is an eigenvalue of A with corresponding eigenvector $\overline{\mathbf{v}}_1$. [$\overline{\lambda}_1$ is the complex conjugate of λ_1.]

9.5 COMPUTATION OF THE PRINCIPAL MATRIX SOLUTION

In this section we provide a method for finding the principal matrix solution $\Psi(t)$ for the homogeneous initial-value problem

$$\mathbf{x}'(t) = A\mathbf{x}(t), \qquad \mathbf{x}(t_0) = \mathbf{x}_0, \tag{1}$$

where A is a constant matrix. There are many procedures for computing $\Psi(t)$; another method will be given in Section 9.7. Our procedure in this section is to compute $\Psi(t)$ in two steps:

Procedure for finding the principal matrix solution

1. Find a fundamental solution $\Phi(t)$.
2. Find the principal matrix solution $\Psi(t)$ by computing (see p. 431)

$$\Psi(t) = \Phi(t)\Phi^{-1}(t_0). \tag{2}$$

Usually step (1) is the most difficult. However, if the $n \times n$ matrix A has n linearly independent eigenvectors, then the task is no more difficult than computing the eigenvalues and eigenvectors of A, as the following theorem suggests.

THEOREM 1 ♦ Finding a fundamental matrix solution

Suppose that the constant $n \times n$ matrix A has n linearly independent eigenvectors $\mathbf{v}_1, \mathbf{v}_2, \ldots, \mathbf{v}_n$ corresponding to the eigenvalues $\lambda_1, \lambda_2, \ldots, \lambda_n$ (not necessarily distinct), respectively. Then there exists a fundamental set of solutions to system **(1)** of the form

$$\boldsymbol{\phi}_1(t) = \mathbf{v}_1 e^{\lambda_1 t}, \qquad \boldsymbol{\phi}_2(\mathrm{t}) = \mathbf{v}_2 e^{\lambda_2 t}, \ldots, \boldsymbol{\phi}_n = \mathbf{v}_n e^{\lambda_n t}, \tag{3}$$

so the matrix $\Phi(t)$ whose columns are $\boldsymbol{\phi}_1(t), \ldots, \boldsymbol{\phi}_n(t)$ is a fundamental matrix solution for Equation **(1)**.

Proof We first show that $\boldsymbol{\phi}_i(t)$ is a solution of Equation **(1)** for $i = 1, 2, \ldots, n$. We have

$$\begin{aligned} \boldsymbol{\phi}_i'(t) &= (\mathbf{v}_i e^{\lambda_i t})' = \lambda_i \mathbf{v}_i e^{\lambda_i t} \\ &= A\mathbf{v}_i e^{\lambda_i t} = A\boldsymbol{\phi}_i(t), \end{aligned}$$

since $\mathbf{v}_i$ is an eigenvector of A corresponding to the eigenvalue λ_i. To show that $\Phi(t)$

is a fundamental matrix solution, we simply note that

$$\det \Phi(0) = \det(\mathbf{v}_1, \mathbf{v}_2, \ldots, \mathbf{v}_n) \neq 0,$$

(by Theorem 9.4.5), since the eigenvectors $\mathbf{v}_i$ are linearly independent. ♦

Using Theorem 1 we can compute a fundamental matrix solution and, therefore, the principal matrix solution for any t_0 when A has n linearly independent eigenvectors. This is always the case when A has n distinct eigenvalues since, according to Theorem 9.4.4, the eigenvectors corresponding to these eigenvalues are linearly independent. The only problem occurs when A has eigenvalues of algebraic multiplicity greater than 1 *and* not as many linearly independent eigenvectors as the multiplicity. We deal with this problem later in this section.

We divide our procedure into three cases. *In all cases we choose* $t_0 = 0$.

Case 1

A has n linearly independent eigenvectors and the eigenvalues of A are all real. Here we simply use Theorem 1.

EXAMPLE 1 ▶ **Finding the principal matrix solution when the eigenvalues are real and distinct**

Find the principal matrix solution for the system

$$\begin{pmatrix} x_1 \\ x_2 \end{pmatrix}' = \begin{pmatrix} 4 & 2 \\ 3 & 3 \end{pmatrix} \begin{pmatrix} x_1 \\ x_2 \end{pmatrix}. \tag{4}$$

Solution In Example 9.4.3 on page 442 we found that the eigenvalues of this system are $\lambda_1 = 1$ and $\lambda_2 = 6$ with corresponding linearly independent eigenvectors

$$\mathbf{v}_1 = \begin{pmatrix} 2 \\ -3 \end{pmatrix} \quad \text{and} \quad \mathbf{v}_2 = \begin{pmatrix} 1 \\ 1 \end{pmatrix}.$$

Therefore a fundamental set of solutions of Equation **(4)** is

$$\boldsymbol{\phi}_1(t) = \begin{pmatrix} 2 \\ -3 \end{pmatrix} e^t = \begin{pmatrix} 2e^t \\ -3e^t \end{pmatrix} \quad \text{and} \quad \boldsymbol{\phi}_2(t) = \begin{pmatrix} 1 \\ 1 \end{pmatrix} e^{6t} = \begin{pmatrix} e^{6t} \\ e^{6t} \end{pmatrix},$$

so a fundamental matrix solution is

$$\Phi(t) = \begin{pmatrix} 2e^t & e^{6t} \\ -3e^t & e^{6t} \end{pmatrix}.$$

Here

$$\Phi(0) = \begin{pmatrix} 2 & 1 \\ -3 & 1 \end{pmatrix}, \qquad \Phi^{-1}(0) = \frac{1}{5}\begin{pmatrix} 1 & -1 \\ 3 & 2 \end{pmatrix},$$

and, from Equation **(2)**, the principal matrix solution is given by

$$\Psi(t) = \frac{1}{5}\begin{pmatrix} 2e^t & e^{6t} \\ -3e^t & e^{6t} \end{pmatrix}\begin{pmatrix} 1 & -1 \\ 3 & 2 \end{pmatrix} = \frac{1}{5}\begin{pmatrix} 2e^t + 3e^{6t} & -2e^t + 2e^{6t} \\ -3e^t + 3e^{6t} & 3e^t + 2e^{6t} \end{pmatrix}.$$ ◀

EXAMPLE 2 ▶ **Finding the principal matrix solution when the eigenvalues are real and distinct**

Find the principal matrix solution for the system

$$\begin{pmatrix} x_1 \\ x_2 \\ x_3 \end{pmatrix}' = \begin{pmatrix} 1 & -1 & 4 \\ 3 & 2 & -1 \\ 2 & 1 & -1 \end{pmatrix} \begin{pmatrix} x_1 \\ x_2 \\ x_3 \end{pmatrix}. \tag{5}$$

Solution As was shown in Example 9.4.4 on page 442, the eigenvalues of A are $\lambda_1 = 1$, $\lambda_2 = -2$, and $\lambda_3 = 3$ with corresponding eigenvectors

$$\mathbf{v}_1 = \begin{pmatrix} -1 \\ 4 \\ 1 \end{pmatrix}, \quad \mathbf{v}_2 = \begin{pmatrix} 1 \\ -1 \\ -1 \end{pmatrix}, \quad \mathbf{v}_3 = \begin{pmatrix} 1 \\ 2 \\ 1 \end{pmatrix}.$$

Therefore a fundamental set of solutions to system **(5)** is

$$\boldsymbol{\phi}_1(t) = \begin{pmatrix} -1 \\ 4 \\ 1 \end{pmatrix} e^t = \begin{pmatrix} -e^t \\ 4e^t \\ e^t \end{pmatrix}, \quad \boldsymbol{\phi}_2(t) = \begin{pmatrix} 1 \\ -1 \\ -1 \end{pmatrix} e^{-2t} = \begin{pmatrix} e^{-2t} \\ -e^{-2t} \\ -e^{-2t} \end{pmatrix},$$

and

$$\boldsymbol{\phi}_3(t) = \begin{pmatrix} 1 \\ 2 \\ 1 \end{pmatrix} e^{3t} = \begin{pmatrix} e^{3t} \\ 2e^{3t} \\ e^{3t} \end{pmatrix},$$

yielding the fundamental matrix solution

$$\Phi(t) = \begin{pmatrix} -e^t & e^{-2t} & e^{3t} \\ 4e^t & -e^{-2t} & 2e^{3t} \\ e^t & -e^{-2t} & e^{3t} \end{pmatrix}.$$

Here

$$\Phi(0) = \begin{pmatrix} -1 & 1 & 1 \\ 4 & -1 & 2 \\ 1 & -1 & 1 \end{pmatrix}, \qquad \Phi^{-1}(0) = -\frac{1}{6}\begin{pmatrix} 1 & -2 & 3 \\ -2 & -2 & 6 \\ -3 & 0 & -3 \end{pmatrix}, \qquad \text{and}$$

$$\begin{aligned} \Psi(t) &= -\frac{1}{6}\begin{pmatrix} -e^t & e^{-2t} & e^{3t} \\ 4e^t & -e^{-2t} & 2e^{3t} \\ e^t & -e^{-2t} & e^{3t} \end{pmatrix}\begin{pmatrix} 1 & -2 & 3 \\ -2 & -2 & 6 \\ -3 & 0 & -3 \end{pmatrix} \\ &= -\frac{1}{6}\begin{pmatrix} -e^t - 2e^{-2t} - 3e^{3t} & 2e^t - 2e^{-2t} & -3e^t + 6e^{-2t} - 3e^{3t} \\ 4e^t + 2e^{-2t} - 6e^{3t} & -8e^t + 2e^{-2t} & 12e^t - 6e^{-2t} - 6e^{3t} \\ e^t + 2e^{-2t} - 3e^{3t} & -2e^t + 2e^{-2t} & 3e^t - 6e^{-2t} - 3e^{3t} \end{pmatrix}. \end{aligned}$$

◀

EXAMPLE 3 ▶ **Finding the principle matrix solution when the 3 × 3 matrix *A* has three linearly independent eigenvectors and using it to solve an initial-value problem**

(a) Find the principal matrix solution for the system

$$\begin{pmatrix} x_1 \\ x_2 \\ x_3 \end{pmatrix}' = \begin{pmatrix} 3 & 2 & 4 \\ 2 & 0 & 2 \\ 4 & 2 & 3 \end{pmatrix}\begin{pmatrix} x_1 \\ x_2 \\ x_3 \end{pmatrix}. \tag{6}$$

(b) Find the particular solution that satisfies

$$\mathbf{x}(0) = \begin{pmatrix} 2 \\ -1 \\ 3 \end{pmatrix}.$$

Solution **(a)** In Example 9.4.8 we found the simple eigenvalue $\lambda_1 = 8$ with eigenvector

$$\mathbf{v}_1 = \begin{pmatrix} 2 \\ 1 \\ 2 \end{pmatrix}$$

and the eigenvalue $\lambda_2 = -1$ of multiplicity 2 with the eigenvectors

$$\mathbf{v}_2 = \begin{pmatrix} 1 \\ -2 \\ 0 \end{pmatrix} \quad \text{and} \quad \mathbf{v}_3 = \begin{pmatrix} 0 \\ -2 \\ 1 \end{pmatrix}.$$

Therefore a fundamental set of solutions to system **(6)** is

$$\boldsymbol{\phi}_1(t) = e^{8t}\begin{pmatrix} 2 \\ 1 \\ 2 \end{pmatrix} = \begin{pmatrix} 2e^{8t} \\ e^{8t} \\ 2e^{8t} \end{pmatrix}, \qquad \boldsymbol{\phi}_2(t) = e^{-t}\begin{pmatrix} 1 \\ -2 \\ 0 \end{pmatrix} = \begin{pmatrix} e^{-t} \\ -2e^{-t} \\ 0 \end{pmatrix}, \qquad \text{and}$$

$$\boldsymbol{\phi}_3(t) = e^{-t}\begin{pmatrix} 0 \\ -2 \\ 1 \end{pmatrix} = \begin{pmatrix} 0 \\ -2e^{-t} \\ e^{-t} \end{pmatrix}.$$

A fundamental matrix solution is

$$\Phi(t) = \begin{pmatrix} 2e^{8t} & e^{-t} & 0 \\ e^{8t} & -2e^{-t} & -2e^{-t} \\ 2e^{8t} & 0 & e^{-t} \end{pmatrix}.$$

Here

$$\Phi(0) = \begin{pmatrix} 2 & 1 & 0 \\ 1 & -2 & -2 \\ 2 & 0 & 1 \end{pmatrix}, \qquad \Phi^{-1}(0) = -\frac{1}{9}\begin{pmatrix} -2 & -1 & -2 \\ -5 & 2 & 4 \\ 4 & 2 & -5 \end{pmatrix}, \qquad \text{and}$$

$$\Psi(t) = -\frac{1}{9}\begin{pmatrix} 2e^{8t} & e^{-t} & 0 \\ e^{8t} & -2e^{-t} & -2e^{-t} \\ 2e^{8t} & 0 & e^{-t} \end{pmatrix}\begin{pmatrix} -2 & -1 & -2 \\ -5 & 2 & 4 \\ 4 & 2 & -5 \end{pmatrix}$$

$$= -\frac{1}{9}\begin{pmatrix} -4e^{8t} - 5e^{-t} & -2e^{8t} + 2e^{-t} & -4e^{8t} + 4e^{-t} \\ -2e^{8t} + 2e^{-t} & -e^{8t} - 8e^{-t} & -2e^{8t} + 2e^{-t} \\ -4e^{8t} + 4e^{-t} & -4e^{8t} + 4e^{-t} & -4e^{8t} - 5e^{-t} \end{pmatrix}.$$

Remark

This example illustrates the fact that the technique suggested in Theorem 1 works even when A has fewer than n distinct eigenvalues (here $n = 3$), provided that there are n linearly independent eigenvectors.

(b) Let $\boldsymbol{\phi}(t) = \Psi(t)\mathbf{x}_0$. Then $\boldsymbol{\phi}(0) = \Psi(0)\mathbf{x}_0 = \mathbf{I}\mathbf{x}_0 = \mathbf{x}_0$, since $\Psi(t)$ is the principal matrix solution. Thus the solution to the initial-value problem is

$$\boldsymbol{\phi}(t) = \Psi(t)\mathbf{x}_0$$

$$= -\frac{1}{9}\begin{pmatrix} -4e^{8t} - 5e^{-t} & -2e^{8t} + 2e^{-t} & -4e^{8t} + 4e^{-t} \\ -2e^{8t} + 2e^{-t} & -e^{8t} - 8e^{-t} & -2e^{8t} + 2e^{-t} \\ -4e^{8t} + 4e^{-t} & -4e^{8t} + 4e^{-t} & -4e^{8t} - 5e^{-t} \end{pmatrix}\begin{pmatrix} 2 \\ -1 \\ 3 \end{pmatrix},$$

or

$$\boldsymbol{\phi}(t) = -\frac{1}{9}\begin{pmatrix} -18e^{8t} \\ -9e^{8t} + 18e^{-t} \\ -16e^{8t} - 11e^{-t} \end{pmatrix} = \begin{pmatrix} 2e^{8t} \\ e^{8t} - 2e^{-t} \\ \frac{16}{9}e^{8t} + \frac{11}{9}e^{-t} \end{pmatrix}. \blacktriangleleft$$

Remark

We proved the following fact in the last example.

Solving an initial-value problem

The unique solution to the initial-value problem

$$\mathbf{x}' = A\mathbf{x}, \qquad \mathbf{x}(t_0) = \mathbf{x}_0$$

is given by **(7)**

$$\boldsymbol{\phi}(t) = \Psi(t)\mathbf{x}_0,$$

where $\Psi(t)$ is the principal matrix solution relative to the initial point t_0.

Case 2

A has n linearly independent eigenvectors and some of the eigenvalues are complex conjugates. This case is not very different from case 1, except that now we have to deal with complex numbers. Several facts simplify our answers. From the Euler formula (see Equation **(14)** in Appendix 5),

$$e^{(\alpha+i\beta)t} = e^{\alpha t}e^{i\beta t} = e^{\alpha t}(\cos \beta t + i \sin \beta t) \tag{8}$$

and

$$e^{(\alpha-i\beta)t} = e^{\alpha t}e^{-i\beta t} = e^{\alpha t}(\cos \beta t - i \sin \beta t). \tag{9}$$

Adding **(8)** and **(9)** and dividing by 2 yields

$$e^{\alpha t} \cos \beta t = \frac{1}{2}[e^{(\alpha+i\beta)t} + e^{(\alpha-i\beta)t}]. \tag{10}$$

Subtracting **(9)** from **(8)** and dividing by $2i$ results in

$$e^{\alpha t} \sin \beta t = \frac{1}{2i}[e^{(\alpha+i\beta)t} - e^{(\alpha-i\beta)t}]. \tag{11}$$

If A has real entries, then the eigenvalues of A occur in complex conjugate pairs. If $\alpha \pm i\beta$ is such a pair, then the corresponding columns of the fundamental matrix solution have entries of the form $ce^{(\alpha+i\beta)t}$ and $de^{(\alpha-i\beta)t}$. When we multiply the fundamental matrix solution on the right by $\Phi^{-1}(0)$ it would be nice if, in the resulting principal matrix solution, only terms involving $e^{\alpha t}\cos \beta t$ and $e^{\alpha t}\sin \beta t$ occurred. In fact, this always happens! The following remarkable fact will be proved in Section 9.7.

THEOREM 2 ◆ **The principal matrix solution when A has complex conjugate eigenvalues**

Let the real matrix A have the complex conjugate eigenvalues $\alpha \pm i\beta$. Then the principal matrix solution of $\mathbf{x}' = A\mathbf{x}$ contains no complex terms; that is, the terms $e^{(\alpha+i\beta)t}$ and $e^{(\alpha-i\beta)t}$ always appear in pairs in such a way that they can be expressed as a *real* linear combination of the terms $e^{\alpha t}\cos \beta t$ and $e^{\alpha t}\sin \beta t$. ◆

We illustrate this with an example.

EXAMPLE 4 ▶ **Finding the principal matrix solution when the eigenvalues are complex conjugates**

Find the principal matrix solution of the system

$$\begin{pmatrix} x_1 \\ x_2 \end{pmatrix}' = \begin{pmatrix} 3 & -5 \\ 1 & -1 \end{pmatrix}\begin{pmatrix} x_1 \\ x_2 \end{pmatrix}. \tag{12}$$

Solution As was shown in Example 9.4.6, the eigenvalues of this system are $\lambda_1 = 1 + i$ and $\lambda_2 = 1 - i$, with corresponding eigenvectors

$$\mathbf{v}_1 = \begin{pmatrix} 2+i \\ 1 \end{pmatrix} \quad \text{and} \quad \mathbf{v}_2 = \begin{pmatrix} 2-i \\ 1 \end{pmatrix}.$$

Note that $v_1 = \overline{v}_2$

A fundamental set of solutions to system **(12)** is

$$\boldsymbol{\phi}_1(t) = \begin{pmatrix} 2+i \\ 1 \end{pmatrix}e^{(1+i)t} = \begin{pmatrix} (2+i)e^{(1+i)t} \\ e^{(1+i)t} \end{pmatrix},$$

and

$$\boldsymbol{\phi}_2(t) = \begin{pmatrix} 2-i \\ 1 \end{pmatrix}e^{(1-i)t} = \begin{pmatrix} (2-i)e^{(1-i)t} \\ e^{(1-i)t} \end{pmatrix},$$

with fundamental matrix solution

$$\Phi(t) = \begin{pmatrix} (2+i)e^{(1+i)t} & (2-i)e^{(1-i)t} \\ e^{(1+i)t} & e^{(1-i)t} \end{pmatrix}.$$

Here

$$\Phi(0) = \begin{pmatrix} 2+i & 2-i \\ 1 & 1 \end{pmatrix} \quad \text{and} \quad \Phi^{-1}(0) = \frac{1}{2i}\begin{pmatrix} 1 & -2+i \\ -1 & 2+i \end{pmatrix},$$

so

$$\Psi(t) = \frac{1}{2i}\begin{pmatrix} (2+i)e^{(1+i)t} & (2-i)e^{(1-i)t} \\ e^{(1+i)t} & e^{(1-i)t} \end{pmatrix}\begin{pmatrix} 1 & -2+i \\ -1 & 2+i \end{pmatrix}.$$

Before continuing, we note that

$$(2+i)(-2+i) = -4 + i^2 = -5 \quad \text{and} \quad (2-i)(2+i) = 4 - i^2 = 5.$$

Then

$$\Psi(t) = \frac{1}{2i}\begin{pmatrix} (2+i)e^{(1+i)t} - (2-i)e^{(1-i)t} & -5e^{(1+i)t} + 5e^{(1-i)t} \\ e^{(1+i)t} - e^{(1-i)t} & (-2+i)e^{(1+i)t} + (2+i)e^{(1-i)t} \end{pmatrix}$$

$$= \begin{pmatrix} 2\left(\dfrac{e^{(1+i)t} - e^{(1-i)t}}{2i}\right) + \left(\dfrac{e^{(1+i)t} + e^{(1-i)t}}{2}\right) & -5\left(\dfrac{e^{(1+i)t} - e^{(1-i)t}}{2i}\right) \\ \left(\dfrac{e^{(1+i)t} - e^{(1-i)t}}{2i}\right) & -2\left(\dfrac{e^{(1+i)t} - e^{(1-i)t}}{2i}\right) + \left(\dfrac{e^{(1+i)t} + e^{(1-i)t}}{2}\right) \end{pmatrix}$$

from (10) and (11)

$$= \begin{pmatrix} 2e^t \sin t + e^t \cos t & -5e^t \sin t \\ e^t \sin t & -2e^t \sin t + e^t \cos t \end{pmatrix} = e^t\begin{pmatrix} 2\sin t + \cos t & -5\sin t \\ \sin t & \cos t - 2\sin t \end{pmatrix}.$$ ◀

Case 3: The matrix A has fewer than n linearly independent eigenvectors

This situation is more complicated than the preceding one. To motivate our procedure, we begin with an example.

EXAMPLE 5 ▶ **Finding the principal matrix solution when the 2×2 matrix A has only one linearly independent eigenvector**

Find the principal matrix solution for the system

$$\begin{pmatrix} x_1 \\ x_2 \end{pmatrix}' = \begin{pmatrix} 7 & -1 \\ 9 & 1 \end{pmatrix}\begin{pmatrix} x_1 \\ x_2 \end{pmatrix}. \qquad (13)$$

Solution

$$\det(\lambda I - A) = \begin{vmatrix} \lambda - 7 & 1 \\ -9 & \lambda - 1 \end{vmatrix} = (\lambda - 7)(\lambda - 1) + 9 = \lambda^2 - 8\lambda + 16$$
$$= (\lambda - 4)^2.$$

Thus $\lambda = 4$ is the only eigenvalue of A. It has algebraic multiplicity 2. Then

$$(A - \lambda I)\begin{pmatrix} x_1 \\ x_2 \end{pmatrix} = \begin{pmatrix} 3 & -1 \\ 9 & -3 \end{pmatrix}\begin{pmatrix} x_1 \\ x_2 \end{pmatrix} = \begin{pmatrix} 0 \\ 0 \end{pmatrix},$$

which implies that $x_2 = 3x_1$ and that $\mathbf{v}_1 = \binom{1}{3}$ is an eigenvector. There is no other linearly independent eigenvector. One solution to **(13)** is, therefore,

$$\boldsymbol{\phi}_1(t) = e^{4t}\begin{pmatrix} 1 \\ 3 \end{pmatrix} = \begin{pmatrix} e^{4t} \\ 3e^{4t} \end{pmatrix}.$$

Recall that when the characteristic equation of a second-order differential equation has a double root λ, then the general solution to the equation takes the form

$$y(t) = (c_1 + c_2 t)e^{\lambda t}.$$

Hence, in order to find a second linearly independent solution, it is reasonable to seek two *vectors*, $\mathbf{u}$ and $\mathbf{w}$, such that

$$\boldsymbol{\phi}_2 = e^{4t}\mathbf{u} + te^{4t}\mathbf{w} \qquad (14)$$

is a solution to **(13).** (It might be tempting to seek a solution having the form $te^{4t}\mathbf{w}$, but this does not work. Try it.) Inserting **(14)** into the equation $\mathbf{x}' = A\mathbf{x}$ yields

$$\boldsymbol{\phi}_2' = 4e^{4t}\mathbf{u} + (1 + 4t)e^{4t}\mathbf{w} = A\boldsymbol{\phi}_2 = e^{4t}A\mathbf{u} + te^{4t}A\mathbf{w}.$$

We equate coefficients of e^{4t} and te^{4t} to obtain

$$4\mathbf{u} + \mathbf{w} = A\mathbf{u} \qquad (15)$$

and

$$4\mathbf{w} = A\mathbf{w}. \qquad (16)$$

Rewriting **(15)** and **(16)** leads to the equations

$$(A - 4I)\mathbf{w} = \mathbf{0}, \qquad (17)$$
$$(A - 4I)\mathbf{u} = \mathbf{w}. \qquad (18)$$

From Equation **(17),** we see that $\mathbf{w}$ is an eigenvector of A. Any vector $\mathbf{u}$ that satisfies Equation **(18)** is called a **generalized eigenvector** of A. In this case, if

generalized eigenvector

$$\mathbf{u} = \begin{pmatrix} u_1 \\ u_2 \end{pmatrix} \quad \text{and} \quad \mathbf{w} = \begin{pmatrix} 1 \\ 3 \end{pmatrix},$$

we obtain, from Equation **(18)**,

$$\begin{pmatrix} 3 & -1 \\ 9 & -3 \end{pmatrix}\begin{pmatrix} u_1 \\ u_2 \end{pmatrix} = \begin{pmatrix} 1 \\ 3 \end{pmatrix}.$$

Thus

$$\begin{aligned} 3u_1 - u_2 &= 1 \\ 9u_1 - 3u_2 &= 3. \end{aligned}$$

Then $u_1 = (1 + u_2)/3$, and (setting $u_2 = 2$) a generalized eigenvector of A is $\mathbf{u} = \binom{1}{2}$. Thus a second linearly independent solution to **(13)** is given by

$$\boldsymbol{\phi}_2 = e^{4t}(\mathbf{u} + t\mathbf{w}) = e^{4t}\left[\begin{pmatrix} 1 \\ 2 \end{pmatrix} + t\begin{pmatrix} 1 \\ 3 \end{pmatrix}\right] = e^{4t}\begin{pmatrix} 1 + t \\ 2 + 3t \end{pmatrix}.$$

A fundamental matrix solution is given by

$$\Phi(t) = e^{4t}\begin{pmatrix} 1 & 1 + t \\ 3 & 2 + 3t \end{pmatrix}.$$

Finally,

$$\Phi(0) = \begin{pmatrix} 1 & 1 \\ 3 & 2 \end{pmatrix} \quad \text{and} \quad \Phi^{-1}(0) = \begin{pmatrix} -2 & 1 \\ 3 & -1 \end{pmatrix},$$

so

$$\Psi(t) = e^{4t}\begin{pmatrix} 1 & 1 + t \\ 3 & 2 + 3t \end{pmatrix}\begin{pmatrix} -2 & 1 \\ 3 & -1 \end{pmatrix} = e^{4t}\begin{pmatrix} 1 + 3t & -t \\ 9t & 1 - 3t \end{pmatrix}. \quad ◀$$

In the last example we sought a solution to the equation $(A - \lambda I)\mathbf{u} = \mathbf{w}$, where $\mathbf{w}$ was an eigenvector of A. The following theorem asserts that, under certain conditions, this equation always has a solution. The proof is difficult and is omitted.

THEOREM 3 ◆ The existence of a generalized eigenvector

Let λ be an eigenvalue of A of algebraic multiplicity 2. Suppose that $\mathbf{w}$ is an eigenvector of A corresponding to λ and that there is no other linearly independent eigenvector of A corresponding to λ. Then the equation $(A - \lambda I)\mathbf{u} = \mathbf{w}$ has a solution. ◆

The technique of Example 5 can be generalized. We indicate how in the next example.

EXAMPLE 6 ► Finding the principal matrix solution when the 3 × 3 matrix *A* has only one linearly independent eigenvector

Find the principal matrix solution for the system

$$\begin{pmatrix} x_1 \\ x_2 \\ x_3 \end{pmatrix}' = \begin{pmatrix} -5 & -5 & -9 \\ 8 & 9 & 18 \\ -2 & -3 & -7 \end{pmatrix}\begin{pmatrix} x_1 \\ x_2 \\ x_3 \end{pmatrix}. \tag{19}$$

Solution In Example 9.4.9 on page 447 we saw that A had the single eigenvalue $\lambda = -1$ of algebraic multiplicity 3 and the single (independent) eigenvector

$$\mathbf{v}_1 = \begin{pmatrix} 3 \\ -6 \\ 2 \end{pmatrix}.$$

Thus one solution is

$$\boldsymbol{\phi}_1 = e^{-t}\begin{pmatrix} 3 \\ -6 \\ 2 \end{pmatrix}.$$

Using the multiple root theory of Section 3.8 as a model, we seek two other solutions having the forms

$$\boldsymbol{\phi}_2 = e^{-t}(\mathbf{u}_1 + t\mathbf{u}_2) \tag{20}$$

and

$$\boldsymbol{\phi}_3 = e^{-t}\left(\mathbf{w}_1 + t\mathbf{w}_2 + \frac{t^2}{2}\mathbf{w}_3\right). \tag{21}$$

Substituting **(20)** into $\mathbf{x}' = A\mathbf{x}$, we obtain

$$(A + I)\mathbf{u}_2 = \mathbf{0},$$
$$(A + I)\mathbf{u}_1 = \mathbf{u}_2.$$

Then

$$\mathbf{u}_2 = \begin{pmatrix} 3 \\ -6 \\ 2 \end{pmatrix}$$

(because it is an eigenvector), and

$$\mathbf{u}_1 = \begin{pmatrix} x_1 \\ x_2 \\ x_3 \end{pmatrix}$$

satisfies

$$\begin{pmatrix} -4 & -5 & -9 \\ 8 & 10 & 18 \\ -2 & -3 & -6 \end{pmatrix}\begin{pmatrix} x_1 \\ x_2 \\ x_3 \end{pmatrix} = \begin{pmatrix} 3 \\ -6 \\ 2 \end{pmatrix}. \tag{22}$$

One solution to **(22)** is found by row reduction:

$$\left(\begin{array}{ccc|c} -4 & -5 & -9 & 3 \\ 8 & 10 & 18 & -6 \\ -2 & -3 & -6 & 2 \end{array}\right) \xrightarrow[A_{3,2}(4)]{A_{3,1}(-2)} \left(\begin{array}{ccc|c} 0 & 1 & 3 & -1 \\ 0 & -2 & -6 & 2 \\ -2 & -3 & -6 & 2 \end{array}\right)$$

$$\xrightarrow[A_{1,3}(3)]{A_{1,2}(2)} \left(\begin{array}{ccc|c} 0 & 1 & 3 & -1 \\ 0 & 0 & 0 & 0 \\ -2 & 0 & 3 & -1 \end{array}\right).$$

Hence $x_2 = -1 - 3x_3$ and $-2x_1 = -1 - 3x_3$. One solution (obtained by setting $x_3 = 1$) is

$$\mathbf{u}_1 = \begin{pmatrix} 2 \\ -4 \\ 1 \end{pmatrix}.$$

Then

$$\boldsymbol{\phi}_2 = e^{-t}\left[\begin{pmatrix} 2 \\ -4 \\ 1 \end{pmatrix} + t\begin{pmatrix} 3 \\ -6 \\ 2 \end{pmatrix}\right] = e^{-t}\begin{pmatrix} 2 + 3t \\ -4 - 6t \\ 1 + 2t \end{pmatrix}.$$

Inserting Equation **(21)** into $\mathbf{x}' = A\mathbf{x}$ yields, after simplification,

$$(A + I)\mathbf{w}_3 = \mathbf{0},$$
$$(A + I)\mathbf{w}_2 = \mathbf{w}_3,$$
$$(A + I)\mathbf{w}_1 = \mathbf{w}_2.$$

Leaving out the details, we obtain (since $\mathbf{w}_3$ is an eigenvector)

$$\mathbf{w}_3 = \begin{pmatrix} 3 \\ -6 \\ 2 \end{pmatrix}, \quad \mathbf{w}_2 = \begin{pmatrix} 2 \\ -4 \\ 1 \end{pmatrix}, \quad \text{and} \quad \mathbf{w}_1 = \begin{pmatrix} 1 \\ -3 \\ 1 \end{pmatrix},$$

so

$$\boldsymbol{\phi}_3(t) = e^{-t}\left(\mathbf{w}_1 + t\mathbf{w}_2 + \frac{t^2}{2}\mathbf{w}_3\right) = e^{-t}\begin{pmatrix} 1 + 2t + \frac{3}{2}t^2 \\ -3 - 4t - 3t^2 \\ 1 + t + t^2 \end{pmatrix}.$$

Thus a fundamental matrix solution is

$$\Phi(t) = e^{-t}\begin{pmatrix} 3 & 2 + 3t & 1 + 2t + \frac{3}{2}t^2 \\ -6 & -4 - 6t & -3 - 4t - 3t^2 \\ 2 & 1 + 2t & 1 + t + t^2 \end{pmatrix}.$$

Finally,

$$\Phi(0) = \begin{pmatrix} 3 & 2 & 1 \\ -6 & -4 & -3 \\ 2 & 1 & 1 \end{pmatrix} \quad \text{and} \quad \Phi^{-1}(0) = \begin{pmatrix} 1 & 1 & 2 \\ 0 & -1 & -3 \\ -2 & -1 & 0 \end{pmatrix},$$

so

$$\Psi(t) = e^{-t}\begin{pmatrix} 3 & 2 + 3t & 1 + 2t + \frac{3}{2}t^2 \\ -6 & -4 - 6t & -3 - 4t - 3t^2 \\ 2 & 1 + 2t & 1 + t + t^2 \end{pmatrix}\begin{pmatrix} 1 & 1 & 2 \\ 0 & -1 & -3 \\ -2 & -1 & 0 \end{pmatrix}$$

$$= e^{-t}\begin{pmatrix} 1 - 4t - 3t^2 & -5t - \frac{3}{2}t^2 & -9t \\ 8t + 6t^2 & 1 + 10t + 3t^2 & 18t \\ -2t - 2t^2 & -3t - t^2 & 1 - 6t \end{pmatrix}. \blacktriangleleft$$

Remark

The technique in the last two examples works (although it is tedious) whenever A has a multiple eigenvalue and one eigenvector. If it has a multiple root and more than one eigenvector, and the number of linearly independent eigenvectors is less than the algebraic multiplicity of the eigenvalue (as in Example 9.4.10), then some other technique must be used.

There are some very elegant techniques that work in all cases. Two of these involve the Jordan canonical form of a matrix and the Cayley-Hamilton theorem. We do not discuss Jordan canonical form here, primarily because it involves more advanced matrix theory and because our methods work in all but the most exceptional cases. We will discuss the Cayley-Hamilton theorem in Section 9.7.

SELF-QUIZ

I. Suppose that the 2 × 2 matrix A has eigenvalues 3, −2 with corresponding eigenvectors $\binom{1}{2}$ and $\binom{-2}{1}$, respectively. A fundamental matrix solution for the system $\mathbf{x}' = A\mathbf{x}$ is ______.

(a) $\begin{pmatrix} e^t & e^{-2t} \\ e^{2t} & e^t \end{pmatrix}$ **(b)** $\begin{pmatrix} e^{3t} & -2e^{-2t} \\ 2e^{3t} & e^{-2t} \end{pmatrix}$

(c) $\begin{pmatrix} -2e^{3t} & e^{-2t} \\ e^{3t} & 2e^{-2t} \end{pmatrix}$ **(d)** $\begin{pmatrix} 1 & -2 \\ 2 & 1 \end{pmatrix}$

II. Suppose that the 3 × 3 matrix A has the eigenvalue −1 with corresponding eigenvector $\begin{pmatrix} -2 \\ 0 \\ 1 \end{pmatrix}$ and the eigenvalue 5 with corresponding eigenvectors $\begin{pmatrix} 0 \\ 1 \\ 3 \end{pmatrix}$ and $\begin{pmatrix} 3 \\ -1 \\ 0 \end{pmatrix}$. Then a fundamental matrix solution for the system $\mathbf{x}' = A\mathbf{x}$ is ______.

(a) $\begin{pmatrix} -2 & 0 & 3 \\ 0 & 1 & -1 \\ 1 & 3 & 0 \end{pmatrix}$

(b) $\begin{pmatrix} -2e^{-t} & 0 & 3e^{-t} \\ 0 & e^{5t} & -e^{5t} \\ e^{5t} & 3e^{5t} & 0 \end{pmatrix}$

(c) $\begin{pmatrix} -2e^{-t} & 0 & 3e^{5t} \\ 0 & e^{5t} & -e^{5t} \\ e^{-t} & 3e^{5t} & 0 \end{pmatrix}$

III. Suppose that the 2 × 2 matrix A has the single eigenvalue −2 and one linearly independent eigenvector $\binom{-1}{2}$. Suppose further that $\binom{3}{4}$ is a generalized eigenvector corresponding to $\lambda = -2$. Then a fundamental matrix solution for the system $\mathbf{x}' = A\mathbf{x}$ is ______.

(a) $e^{-2t}\begin{pmatrix} -1 & 3-t \\ 2 & 4+2t \end{pmatrix}$

(b) $e^{-2t}\begin{pmatrix} -1 & 3 \\ 2 & 4 \end{pmatrix}$

(c) $e^{-2t}\begin{pmatrix} -1 & -1+3t \\ 2 & 2+4t \end{pmatrix}$

Answers to Self-Quiz

I. b **II.** c **III.** a

PROBLEMS 9.5

Use the methods of this section to calculate the principal matrix solution of the system $\mathbf{x}' = A\mathbf{x}$*, where* A *is the given constant matrix. If initial conditions are given, find the solution of the initial-value problem.*

1. $\begin{pmatrix} -2 & -2 \\ -5 & 1 \end{pmatrix}$

2. $\begin{pmatrix} -12 & 7 \\ -7 & 2 \end{pmatrix}$

3. $\begin{pmatrix} 2 & -1 \\ 5 & -2 \end{pmatrix}$

4. $\begin{pmatrix} 3 & 2 \\ -5 & 1 \end{pmatrix}$

5. $\begin{pmatrix} 3 & -2 \\ 8 & -5 \end{pmatrix}$

6. $\begin{pmatrix} 1 & 1 \\ 1 & -1 \end{pmatrix}$

7. $\begin{pmatrix} 3 & -2 \\ 8 & -5 \end{pmatrix}$, $\mathbf{x}(0) = \begin{pmatrix} \frac{3}{4} \\ 1 \end{pmatrix}$

8. $\begin{pmatrix} 3 & -2 \\ 8 & -5 \end{pmatrix}$, $\mathbf{x}(0) = \begin{pmatrix} 1 \\ 2 \end{pmatrix}$

9. $\begin{pmatrix} 3 & -2 \\ 8 & -5 \end{pmatrix}$, $\mathbf{x}(0) = \begin{pmatrix} 2 \\ 5 \end{pmatrix}$

10. $\begin{pmatrix} 4 & 1 \\ -8 & 8 \end{pmatrix}$, $\mathbf{x}(0) = \begin{pmatrix} 1 \\ 0 \end{pmatrix}$

11. $\begin{pmatrix} 1 & -1 & 0 \\ -1 & 2 & -1 \\ 0 & -1 & 1 \end{pmatrix}$, $\mathbf{x}(0) = \begin{pmatrix} 1 \\ -2 \\ 3 \end{pmatrix}$

12. $\begin{pmatrix} 1 & 1 & -2 \\ -1 & 2 & 1 \\ 0 & 1 & -1 \end{pmatrix}$, $\mathbf{x}(0) = \begin{pmatrix} 2 \\ 5 \\ 7 \end{pmatrix}$

13. $\begin{pmatrix} 4 & 6 & 6 \\ 1 & 3 & 2 \\ -1 & -5 & -2 \end{pmatrix}$

14. $\begin{pmatrix} 7 & -2 & -4 \\ 3 & 0 & -2 \\ 6 & -2 & -3 \end{pmatrix}$

15. $\begin{pmatrix} 5 & 4 & 2 \\ 4 & 5 & 2 \\ 2 & 2 & 2 \end{pmatrix}$

16. $\begin{pmatrix} -3 & 0 & 2 \\ 1 & -1 & 0 \\ -2 & -1 & 0 \end{pmatrix}$

17. $\begin{pmatrix} 4 & 1 & 0 & 1 \\ 2 & 3 & 0 & 1 \\ -2 & 1 & 2 & -3 \\ 2 & -1 & 0 & 5 \end{pmatrix}$

18. $\begin{pmatrix} 0 & -1 & -2 \\ 1 & 0 & 1 \\ 2 & -1 & 0 \end{pmatrix}$

19. $\begin{pmatrix} 1 & -1 & 0 \\ -1 & 2 & -1 \\ 0 & -1 & 1 \end{pmatrix}$, $\mathbf{x}(0) = \begin{pmatrix} 1 \\ 0 \\ 1 \end{pmatrix}$

20. $\begin{pmatrix} 5 & 4 & 2 \\ 4 & 5 & 2 \\ 2 & 2 & 2 \end{pmatrix}$, $\mathbf{x}(0) = \begin{pmatrix} 1 \\ 2 \\ 3 \end{pmatrix}$

21. $\begin{pmatrix} 4 & 6 & 6 \\ 1 & 3 & 2 \\ -1 & -5 & -2 \end{pmatrix}$, $\mathbf{x}(0) = \begin{pmatrix} -1 \\ 0 \\ 2 \end{pmatrix}$

22. $\begin{pmatrix} 4 & 1 & 0 & 1 \\ 2 & 3 & 0 & 1 \\ -2 & 1 & 2 & -3 \\ 2 & -1 & 0 & 5 \end{pmatrix}$, $\mathbf{x}(0) = \begin{pmatrix} 4 \\ 5 \\ 6 \\ 2 \end{pmatrix}$

9.6 NONHOMOGENEOUS SYSTEMS

We now present a method for solving the nonhomogeneous system

$$\mathbf{x}' = A(t)\mathbf{x} + \mathbf{f}(t), \tag{1}$$

given that a fundamental matrix solution $\Phi(t)$ for the homogeneous system

$$\mathbf{x}' = A(t)\mathbf{x} \tag{2}$$

is known. Such a solution can be found if $A(t)$ is a constant matrix (by the methods of Section 9.5).

Throughout this section we assume that the components of $A(t)$ and $\mathbf{f}(t)$ are functions continuous on some common interval containing the point t_0.

THEOREM 1 ♦

Let $\boldsymbol{\phi}_p(t)$ and $\boldsymbol{\phi}_q(t)$ be two solutions of system **(1)**. Then their difference,

$$\boldsymbol{\phi}(t) = \boldsymbol{\phi}_p(t) - \boldsymbol{\phi}_q(t),$$

is a solution of Equation **(2)**.

Proof $\boldsymbol{\phi}' = (\boldsymbol{\phi}_p - \boldsymbol{\phi}_q)' = (A\boldsymbol{\phi}_p + \mathbf{f}) - (A\boldsymbol{\phi}_q + \mathbf{f}) = A(\boldsymbol{\phi}_p - \boldsymbol{\phi}_q) = A\boldsymbol{\phi}$. Thus, as in the case of linear scalar equations (Chapter 3), it is necessary to find only one particular solution of Equation **(1)**.

If $\boldsymbol{\phi}_p(t)$ is such a solution, then *the general solution of the nonhomogeneous system* **(1)** *is of the form*

General solution to a nonhomogeneous system

$$\boldsymbol{\phi}(t) = \Phi(t)\mathbf{c} + \boldsymbol{\phi}_p(t), \tag{3}$$

where **c** *is a vector of arbitrary constants and* $\Phi(t)$ *is a fundamental matrix solution of the homogeneous equation* **(2)**. That Equation **(3)** is a solution can be verified as follows:

$$\begin{aligned} \boldsymbol{\phi}'(t) &= \Phi'(t)\mathbf{c} + \boldsymbol{\phi}_p'(t) \\ &= [A(t)\Phi(t)\mathbf{c}] + [A(t)\boldsymbol{\phi}_p(t) + \mathbf{f}(t)], \end{aligned}$$

since Φ is a solution of the associated matrix equation and $\boldsymbol{\phi}_p$ is a particular solution of Equation **(2).** Combining terms, we have

$$\boldsymbol{\phi}'(t) = A(t)[\Phi(t)\mathbf{c} + \boldsymbol{\phi}_p(t)] + \mathbf{f}(t) = A(t)\boldsymbol{\phi}(t) + \mathbf{f}(t),$$

and $\boldsymbol{\phi}$ is a solution of Equation **(2).** ◀

variation of parameters

We now derive a *variation-of-parameters* formula for the nonhomogeneous system

$$\mathbf{x}' = A(t)\mathbf{x} + \mathbf{f}(t). \tag{4}$$

All variation-of-parameters formulas begin by assuming that a solution to the homogeneous equation $\mathbf{x}' = A(t)\mathbf{x}$ is known. Assuming that Φ is a fundamental matrix solution of the homogeneous equation, we seek a particular solution to **(4)** of the form

$$\boldsymbol{\phi}_p(t) = \Phi(t)\mathbf{c}(t), \tag{5}$$

where $\mathbf{c}(t)$ is a vector function in t. Differentiating both sides of **(5)** with respect to t, we have

$$\boldsymbol{\phi}_p' = \Phi'\mathbf{c} + \Phi\mathbf{c}' = A\Phi\mathbf{c} + \Phi\mathbf{c}' = A\boldsymbol{\phi}_p + \Phi\mathbf{c}'.$$

Since $\boldsymbol{\phi}_p$ is a particular solution of **(4),** it follows that $\Phi\mathbf{c}' = \mathbf{f}$. But every fundamental matrix solution has an inverse, so we can integrate $\mathbf{c}' = \Phi^{-1}\mathbf{f}$, obtaining

$$\boldsymbol{\phi}_p(t) = \Phi(t)\mathbf{c}(t) = \Phi(t)\int \Phi^{-1}(t)\mathbf{f}(t)\,dt. \tag{6}$$

This is the **variation-of-parameters formula** for a particular solution to the nonhomogeneous system **(4).** Thus, the general solution to system **(4)** has the form

General solution to a nonhomogeneous system

$$\boldsymbol{\phi}(t) = \Phi(t)\mathbf{c} + \Phi(t)\int \Phi^{-1}(t)\mathbf{f}(t)\,dt, \tag{7}$$

where $\mathbf{c}$ is an arbitrary *constant* vector.

For the initial-value problem

$$\mathbf{x}' = A(t)\mathbf{x} + \mathbf{f}(t), \qquad \mathbf{x}(t_0) = \mathbf{x}_0, \tag{8}$$

it is convenient to choose a particular solution $\boldsymbol{\phi}_p(t)$ that vanishes at t_0. This can be done by selecting the limits of integration in **(7)** to be from t_0 to t; that is,

$$\boldsymbol{\phi}(t) = \Phi(t)\mathbf{c} + \Phi(t)\int_{t_0}^{t} \Phi^{-1}(s)\mathbf{f}(s)\,ds.$$

Substituting $t = t_0$ in this equation, we obtain

$$\mathbf{x}_0 = \boldsymbol{\phi}(t_0) = \Phi(t_0)\mathbf{c},$$

which implies that $\mathbf{c} = \Phi^{-1}(t_0)\mathbf{x}_0$. Hence the solution of the initial-value problem **(8)** is

$$\boldsymbol{\phi}(t) = \Phi(t)\Phi^{-1}(t_0)\mathbf{x}_0 + \Phi(t)\int_{t_0}^{t} \Phi^{-1}(s)\mathbf{f}(s)\,ds = \boldsymbol{\phi}_h(t) + \boldsymbol{\phi}_p(t), \tag{9}$$

where $\boldsymbol{\phi}_h$ and $\boldsymbol{\phi}_p$ are the homogeneous and particular solutions, respectively. Note that if $\Psi(t)$ is the principal matrix solution of $\mathbf{x}' = A\mathbf{x}$, then $\Psi(t_0) = \Psi^{-1}(t_0) = I$.

Thus **(9)** takes the simpler form (multiply inside the integral by $\mathbf{\Phi}(t_0)$ and outside by its reverse)

$$\boldsymbol{\phi}(t) = \Psi(t)\mathbf{x}_0 + \Psi(t)\int_{t_0}^{t} \Psi^{-1}(s)\mathbf{f}(s)\,ds \tag{10}$$

when the principal matrix solution $\Psi(t)$ of $\mathbf{x}' = A(t)\mathbf{x}$ is used.

If $A(t) = A$, a constant matrix, things become simpler. Computing the inverse of a matrix function is generally quite tedious. However, if $A(t)$ is constant, then, according to Theorem 9.5.5,

$$\Psi^{-1}(t) = \Psi(-t). \tag{11}$$

Inserting **(11)** into **(10)**, we obtain the form of the variation-of-parameters formula that we will use most frequently:

The solution to a nonhomogeneous initial-value problem

$$\boldsymbol{\phi}(t) = \Psi(t)\mathbf{x}_0 + \Psi(t)\int_{t_0}^{t} \Psi(-s)\mathbf{f}(s)\,ds. \tag{12}$$

THEOREM 2 ♦ Summary: Solving a nonhomogeneous initial-value problem

Let $\Psi(t)$ be the principal matrix solution of the homogeneous system

$$\mathbf{x}' = A(t)\mathbf{x}, \qquad \mathbf{x}(t_0) = \mathbf{x}_0. \tag{13}$$

Then the solution to the initial-value problem

$$\mathbf{x}' = A(t)\mathbf{x} + \mathbf{f}(t), \qquad \mathbf{x}(t_0) = \mathbf{x}_0, \tag{14}$$

is given by

$$\boldsymbol{\phi}(t) = \Psi(t)\mathbf{x}_0 + \Psi(t)\int_{t_0}^{t} \Psi^{-1}(s)\mathbf{f}(s)\,ds.$$

If $A(t)$ is constant, then

$$\boldsymbol{\phi}(t) = \Psi(t)\mathbf{x}_0 + \Psi(t)\int_{t_0}^{t} \Psi(-s)\mathbf{f}(s)\,ds. \quad \blacklozenge \tag{15}$$

EXAMPLE 1 ▶ Solving a nonhomogeneous initial-value problem

Find the unique solution to the system

$$\mathbf{x}' = \begin{pmatrix} x_1 \\ x_2 \end{pmatrix}' = \begin{pmatrix} 4 & 2 \\ 3 & 3 \end{pmatrix}\begin{pmatrix} x_1 \\ x_2 \end{pmatrix} + \begin{pmatrix} e^t \\ e^{2t} \end{pmatrix} = A\mathbf{x} + \mathbf{f}(t), \qquad \mathbf{x}(0) = \begin{pmatrix} 1 \\ 2 \end{pmatrix}.$$

Solution The principal matrix solution for the homogeneous system is (see Example 9.5.1)

$$\Psi(t) = \frac{1}{5}\begin{pmatrix} 2e^t + 3e^{6t} & -2e^t + 2e^{6t} \\ -3e^t + 3e^{6t} & 3e^t + 2e^{6t} \end{pmatrix}.$$

Then

$$\Psi^{-1}(s) = \Psi(-s) = \frac{1}{5}\begin{pmatrix} 2e^{-s} + 3e^{-6s} & -2e^{-s} + 2e^{-6s} \\ -3e^{-s} + 3e^{-6s} & 3e^{-s} + 2e^{-6s} \end{pmatrix}.$$

By Equation **(12)**, the solution is given by

$$\begin{aligned}
\boldsymbol{\phi}(t) &= \Psi(t)\mathbf{x}_0 + \Psi(t)\int_0^t \Psi(-s)\mathbf{f}(s)\,ds \\
&= \frac{1}{5}\begin{pmatrix} 2e^t + 3e^{6t} & -2e^t + 2e^{6t} \\ -3e^t + 3e^{6t} & 3e^t + 2e^{6t} \end{pmatrix}\begin{pmatrix} 1 \\ 2 \end{pmatrix} \\
&\quad + \frac{1}{5}\Psi(t)\int_0^t \begin{pmatrix} 2e^{-s} + 3e^{-6s} & -2e^{-s} + 2e^{-6s} \\ -3e^{-s} + 3e^{-6s} & 3e^{-s} + 2e^{-6s} \end{pmatrix}\begin{pmatrix} e^s \\ e^{2s} \end{pmatrix} ds \\
&= \frac{1}{5}\begin{pmatrix} -2e^t + 7e^{6t} \\ 3e^t + 7e^{6t} \end{pmatrix} + \frac{1}{5}\Psi(t)\int_0^t \begin{pmatrix} 2 + 3e^{-5s} - 2e^s + 2e^{-4s} \\ -3 + 3e^{-5s} + 3e^s + 2e^{-4s} \end{pmatrix} ds \\
&= \frac{1}{5}\begin{pmatrix} -2e^t + 7e^{6t} \\ 3e^t + 7e^{6t} \end{pmatrix} + \frac{1}{5}\Psi(t)\begin{pmatrix} (2s - \frac{3}{5}e^{-5s} - 2e^s - \frac{1}{2}e^{-4s})\big|_0^t \\ (-3s - \frac{3}{5}e^{-5s} + 3e^s - \frac{1}{2}e^{-4s})\big|_0^t \end{pmatrix} \\
&= \frac{1}{5}\begin{pmatrix} -2e^t + 7e^{6t} \\ 3e^t + 7e^{6t} \end{pmatrix} + \frac{1}{25}\begin{pmatrix} 2e^t + 3e^{6t} & -2e^t + 2e^{6t} \\ -3e^t + 3e^{6t} & 3e^t + 2e^{6t} \end{pmatrix} \\
&\quad \times \begin{pmatrix} 2t - \frac{3}{5}e^{-5t} - 2e^t - \frac{1}{2}e^{-4t} + \frac{31}{10} \\ -3t - \frac{3}{5}e^{-5t} + 3e^t - \frac{1}{2}e^{-4t} - \frac{19}{10} \end{pmatrix} \\
&= \begin{pmatrix} -\frac{2}{5}e^t + \frac{7}{5}e^{6t} \\ \frac{3}{5}e^t + \frac{7}{5}e^{6t} \end{pmatrix} + \begin{pmatrix} \frac{2}{5}te^t - \frac{1}{2}e^{2t} + \frac{7}{25}e^t + \frac{11}{50}e^{6t} \\ \frac{-3}{5}te^t + \frac{1}{2}e^{2t} - \frac{18}{25}e^t + \frac{11}{50}e^{6t} \end{pmatrix} \\
&= \begin{pmatrix} \frac{2}{5}te^t - \frac{1}{2}e^{2t} - \frac{3}{25}e^t + \frac{81}{50}e^{6t} \\ \frac{-3}{5}te^t + \frac{1}{2}e^{2t} - \frac{3}{25}e^t + \frac{81}{50}e^{6t} \end{pmatrix}.
\end{aligned}$$

Note, as a check, that $\boldsymbol{\phi}(0) = \binom{1}{2}$. ◀

EXAMPLE 2 ▶ Solving a nonhomogeneous equation by first writing it as a system

Consider the initial-value problem

$$x'' + x = 2\cos t, \qquad x(0) = 5, \qquad x'(0) = 2.$$

Using the substitution $x_1 = x$, $x_2 = x'$, we can write this in matrix form

$$\begin{pmatrix} x_1 \\ x_2 \end{pmatrix}' = \begin{pmatrix} 0 & 1 \\ -1 & 0 \end{pmatrix}\begin{pmatrix} x_1 \\ x_2 \end{pmatrix} + \begin{pmatrix} 0 \\ 2\cos t \end{pmatrix},$$

where the homogeneous system has the prinicipal matrix solution

$$\Psi(t) = \begin{pmatrix} \cos t & \sin t \\ -\sin t & \cos t \end{pmatrix}.$$

Then

$$\Psi^{-1}(t) = \Psi(-t) = \begin{pmatrix} \cos t & -\sin t \\ \sin t & \cos t \end{pmatrix}$$

$\cos(-t) = \cos t$ and $\sin(-t) = -\sin t$

and

$$\int_0^t \Psi^{-1}(s)\mathbf{f}(s)\,ds = \int_0^t \begin{pmatrix} \cos s & -\sin s \\ \sin s & \cos s \end{pmatrix}\begin{pmatrix} 0 \\ 2\cos s \end{pmatrix} ds$$

$$= \int_0^t \begin{pmatrix} -2\sin s\cos s \\ 2\cos^2 s \end{pmatrix} ds = \begin{pmatrix} \cos^2 t - 1 \\ t + \sin t\cos t \end{pmatrix}.$$

Thus, since the solution $\boldsymbol{\phi}(t)$ satisfies the initial conditions

$$\boldsymbol{\phi}(0) = \begin{pmatrix} 5 \\ 2 \end{pmatrix},$$

we obtain, from Equation **(12)**,

$$\boldsymbol{\phi}(t) = \begin{pmatrix} \cos t & \sin t \\ -\sin t & \cos t \end{pmatrix}\begin{pmatrix} 5 \\ 2 \end{pmatrix} + \begin{pmatrix} \cos t & \sin t \\ -\sin t & \cos t \end{pmatrix}\begin{pmatrix} \cos^2 t - 1 \\ t + \sin t\cos t \end{pmatrix}.$$

$$= \begin{pmatrix} 5\cos t + 2\sin t + t\sin t \\ -4\sin t + 2\cos t + t\cos t \end{pmatrix},$$

where we have used the fact that

$$\cos^3 t + \sin^2 t\cos t = \cos t(\cos^2 t + \sin^2 t) = \cos t. \blacktriangleleft$$

EXAMPLE 3 ▶ Solving an initial-value problem with variable coefficients

Consider the system

$$\begin{pmatrix} x_1 \\ x_2 \\ x_3 \end{pmatrix}' = \begin{pmatrix} 0 & 1 & 0 \\ 0 & 0 & 1 \\ -\frac{2}{t^3} & \frac{2}{t^2} & \frac{1}{t} \end{pmatrix}\begin{pmatrix} x_1 \\ x_2 \\ x_3 \end{pmatrix} + \begin{pmatrix} 2t^2 \\ -t^3 \\ t^5 \end{pmatrix}, \tag{16}$$

with initial conditions $x_1(1) = 2$, $x_2(1) = 0$, and $x_3(1) = -1$. A fundamental matrix solution (check!) is

$$\Phi(t) = \begin{pmatrix} t & \frac{1}{t} & t^2 \\ 1 & -\frac{1}{t^2} & 2t \\ 0 & \frac{2}{t^3} & 2 \end{pmatrix}.$$

Note that this solution is valid only for $t > 0$, since $\Phi(t)$ is not defined at $t = 0$. Then

$$\Phi^{-1}(t) = \frac{1}{6}\begin{pmatrix} \frac{6}{t} & 0 & -3t \\ 2t & -2t^2 & t^3 \\ -\frac{2}{t^2} & \frac{2}{t} & 2 \end{pmatrix}.$$

Hence, by **(9)**, the solution to the initial-value problem **(16)** is

$$\boldsymbol{\phi}(t) = \Phi(t)\Phi^{-1}(1)\begin{pmatrix} 2 \\ 0 \\ -1 \end{pmatrix} + \Phi(t)\int_1^t \Phi^{-1}(s)\mathbf{f}(s)\,ds = \Phi(t)\mathbf{c} + \boldsymbol{\phi}_{\mathrm{p}}(t).$$

Setting $t = 1$, we have

$$\Phi^{-1}(1) = \frac{1}{6}\begin{pmatrix} 6 & 0 & -3 \\ 2 & -2 & 1 \\ -2 & 2 & 2 \end{pmatrix},$$

so

$$\Phi(t)\Phi^{-1}(1)\begin{pmatrix} 2 \\ 0 \\ 1 \end{pmatrix} = \frac{1}{6}\begin{pmatrix} 15t + \dfrac{3}{t} - 6t^2 \\ 15 - \dfrac{3}{t^2} - 12t \\ \dfrac{6}{t^3} - 12 \end{pmatrix}.$$

After a great deal of arithmetic, we arrive at

$$\mathbf{\phi}(t) = \frac{1}{6}\begin{pmatrix} 15t + \dfrac{3}{t} - 6t^2 \\ 15 - \dfrac{3}{t^2} - 12t \\ \dfrac{6}{t^3} - 12 \end{pmatrix} + \frac{1}{6}\begin{pmatrix} t & \dfrac{1}{t} & t^2 \\ 1 & -\dfrac{1}{t^2} & 2t \\ 0 & \dfrac{2}{t^3} & 2 \end{pmatrix}\int_1^t \begin{pmatrix} 12s - 3s^6 \\ 4s^3 + 2s^5 + s^8 \\ -4 - 2s^2 + 2s^5 \end{pmatrix} ds$$

$$= \begin{pmatrix} \dfrac{2}{27}t^8 - \dfrac{1}{14}t^7 - \dfrac{1}{18}t^5 - \dfrac{1}{2}t^3 + \dfrac{13}{18}t^2 + \dfrac{5}{2}t - \dfrac{13}{14} + \dfrac{7}{27t} \\ -\dfrac{17}{189}t^7 + \dfrac{1}{9}t^6 - \dfrac{5}{18}t^4 - \dfrac{1}{2}t^2 - \dfrac{5}{9}t^2 + \dfrac{11}{7} - \dfrac{7}{27t^2} \\ \dfrac{4}{27}t^6 - \dfrac{1}{9}t^3 - t - \dfrac{5}{9} + \dfrac{14}{27t^3} \end{pmatrix}. \blacktriangleleft$$

An answer this complex is difficult to check. However, you should at least verify that the initial conditions are satisfied.

SELF-QUIZ

Answer true *or* false.

I. Let $\Psi(t)$ be the principal matrix solution for the system $\mathbf{x}' = A(t)\mathbf{x}$. Then the solution to the initial-value problem $\mathbf{x}' = A(t)\mathbf{x} + \mathbf{f}(t)$, $\mathbf{x}(t_0) = \mathbf{x}_0$ is

$$\mathbf{\phi}(t) = \Psi(t)\mathbf{x}_0 + \Psi(t)\int_{t_0}^t \Psi(-s)\mathbf{f}(s)\,ds$$

II. Let $\Phi(t)$ be a fundamental matrix solution for the system $\mathbf{x}' = A(t)\mathbf{x}$. Then the solution to the initial-value problem $\mathbf{x}' = A(t)\mathbf{x} + \mathbf{f}(t)$, $\mathbf{x}(t_0) = \mathbf{x}_0$ is

$$\mathbf{\phi}(t) = \Phi(t)\Phi^{-1}(t_0)\mathbf{x}_0 + \Phi(t)\int_{t_0}^t \Phi^{-1}(s)\mathbf{f}(s)\,ds$$

Answers to Self-Quiz

I. False (True only if $A(t)$ is a constant matrix.) **II.** True

PROBLEMS 9.6

In each of Problems 1–9 calculate the principal matrix solution for the associated homeogeneous system and then use the variation-of-parameters formula **(12)** *to obtain a particular solution to the given nonhomogeneous system. Where initial conditions are given, find the unique solution that satisfies them.*

1. $\mathbf{x}' = \begin{pmatrix} -2 & -2 \\ -5 & 1 \end{pmatrix}\mathbf{x} + \begin{pmatrix} e^t \\ e^{2t} \end{pmatrix}$

2. $\mathbf{x}' = \begin{pmatrix} -12 & 7 \\ -7 & 2 \end{pmatrix}\mathbf{x} + \begin{pmatrix} t \\ 2 \end{pmatrix}$, $x_1(0) = 1, x_2(0) = 0$

3. $\mathbf{x}' = \begin{pmatrix} 2 & -1 \\ 5 & -2 \end{pmatrix}\mathbf{x} + \begin{pmatrix} \sin t \\ \cos t \end{pmatrix}$, $x_1(0) = 0, x_2(0) = 1$

4. $\mathbf{x}' = \begin{pmatrix} 3 & 2 \\ -5 & 1 \end{pmatrix}\mathbf{x} + \begin{pmatrix} 2\sin 3t \\ \cos 3t \end{pmatrix}$

5. $\mathbf{x}' = \begin{pmatrix} 1 & 1 & -2 \\ -1 & 2 & 1 \\ 0 & 1 & -1 \end{pmatrix}\mathbf{x} + \begin{pmatrix} e^t \\ e^{2t} \\ e^{3t} \end{pmatrix}$, $x_1(0) = 0, x_2(0) = 1, x_3(0) = -1$

6. $\mathbf{x}' = \begin{pmatrix} 0 & 1 & 0 \\ 0 & 0 & 1 \\ 1 & -3 & 3 \end{pmatrix}\mathbf{x} + \begin{pmatrix} t^2 \\ 0 \\ 1 \end{pmatrix}$

7. $\mathbf{x}' = \begin{pmatrix} 4 & 6 & 6 \\ 1 & 3 & 2 \\ -1 & -5 & -2 \end{pmatrix}\mathbf{x} + \begin{pmatrix} 1 - 4t - 6t^2 - 6t^4 \\ t - 3t^2 - 2t^4 \\ t + 5t^2 + 4t^3 + 2t^4 \end{pmatrix}$

8. $\mathbf{x}' = \begin{pmatrix} 1 & -1 & -1 \\ 1 & -1 & 0 \\ 1 & 0 & -1 \end{pmatrix}\mathbf{x} + \begin{pmatrix} 2 \\ e^{-t} \\ e^{-t} \end{pmatrix}$, $x_1(0) = 1, x_2(0) = -1, x_3(0) = 0$

9. $\mathbf{x}' = \begin{pmatrix} 2 & -5 \\ 1 & -2 \end{pmatrix}\mathbf{x} + \begin{pmatrix} 0 \\ \cot t \end{pmatrix}$, $0 < t < \pi$

In each of Problems 10–12, one homogeneous solution to a given system is given. Use the method of Problem 9.3.19 to obtain a fundamental matrix solution. Then use this solution to find the general solution of the nonhomogeneous system.

10. $\mathbf{x}' = \begin{pmatrix} 0 & 1 \\ -\frac{1}{4t^2} & 0 \end{pmatrix}\mathbf{x} + \begin{pmatrix} \sqrt{t} \\ 2 \end{pmatrix}$, $\boldsymbol{\phi}_1(t) = \begin{pmatrix} \sqrt{t} \\ \frac{1}{2\sqrt{t}} \end{pmatrix}$, $t > 0$.

11. $\mathbf{x}' = \begin{pmatrix} 0 & 1 \\ -\frac{1}{t^2} & -\frac{3}{t} \end{pmatrix}\mathbf{x} + \begin{pmatrix} t \\ e^t \end{pmatrix}$, $\boldsymbol{\phi}_1(t) = \begin{pmatrix} \frac{1}{t} \\ -\frac{1}{t^2} \end{pmatrix}$, $t > 0$

12. $\mathbf{x}' = \begin{pmatrix} \frac{-t}{1-t^2} & \frac{1}{1-t^2} \\ \frac{1}{1-t^2} & \frac{-t}{1-t^2} \end{pmatrix}\mathbf{x} + \begin{pmatrix} t^2 \\ t^3 \end{pmatrix}$, $\boldsymbol{\phi}_1(t) = \begin{pmatrix} 1 \\ t \end{pmatrix}$, $-1 < t < 1$

13. Let $\boldsymbol{\phi}_1(t)$ be a solution to $\mathbf{x}'(t) = A\mathbf{x}(t) + \mathbf{b}_1(t)$, $\boldsymbol{\phi}_2(t)$ a solution to $\mathbf{x}'(t) = A\mathbf{x}(t) + \mathbf{b}_2(t), \ldots,$ and $\boldsymbol{\phi}_n(t)$ a solution to $\mathbf{x}'(t) = A\mathbf{x}(t) + \mathbf{b}_n(t)$. Prove that $\boldsymbol{\phi}_1(t) + \boldsymbol{\phi}_2(t) + \cdots + \boldsymbol{\phi}_n(t)$ is a solution to

$$\mathbf{x}'(t) = A\mathbf{x}(t) + \mathbf{b}_1(t) + \mathbf{b}_2(t) + \cdots + \mathbf{b}_n(t).$$

This again is called the **principle of superposition.**

9.7 A PERSPECTIVE: THE MATRIX EXPONENTIAL

In this section we show how the principal matrix solution to $\mathbf{x}' = A\mathbf{x}$, where A is a constant matrix, can be represented in an exponential form. This form allows us to prove two important results stated earlier.

Consider the system

$$\mathbf{x}'(t) = A\mathbf{x}(t), \tag{1}$$

where A is a constant $n \times n$ matrix, and the related scalar equation

$$x'(t) = ax(t). \tag{2}$$

Equation **(2)** is Equation **(1)** in the case $n = 1$. Equation **(2)** has a solution

$$\psi(t) = e^{at}. \tag{3}$$

The function $\psi(t)$ is the principal 1×1 matrix solution to Equation **(2)** since $\psi(0) = 1$. By analogy, we define the matrix function e^{At} in the case where A is an $n \times n$ matrix.

Recall that the exponential function can be defined as the power series

$$e^{at} = 1 + at + \frac{(at)^2}{2!} + \ldots + \frac{(at)^m}{m!} + \ldots. \tag{4}$$

We use this expansion to define the matrix function:

Definition of e^{At}

$$e^{At} = I + At + \frac{(At)^2}{2!} + \frac{(At)^3}{3!} + \ldots + \frac{(At)^m}{m!} + \ldots. \tag{5}$$

Note that since powers of the matrix A are $n \times n$ matrices, the right side of **(5)** is an $n \times n$ matrix if the series converges.

The question of convergence for the series in **(5)** for an arbitrary $n \times n$ matrix A is settled in the following theorem, which we give without proof.

THEOREM 1 ♦ **Convergence of e^{At}**

The series

$$e^{At} = I + At + \frac{(At)^2}{2!} + \frac{(At)^3}{3!} + \ldots \tag{6}$$

converges for all t, can be differentiated term by term, and is the principal matrix solution of the system $\mathbf{x}' = A\mathbf{x}$. ♦

Note

We are assuming in this section that $t_0 = 0$ so that the principal matrix solution $\Psi(t) = e^{At}$ satisfies $\Psi(0) = I$. If $t_0 \neq 0$, then the principal matrix solution is given by $\Psi(t) = e^{A(t-t_0)}$. There are many techniques for computing e^{At}. For example

Computing e^{At} when A is a diagonal matrix

If

$$A = \begin{pmatrix} \lambda_1 & 0 & \cdots & 0 \\ 0 & \lambda_2 & \cdots & 0 \\ \vdots & \vdots & & \vdots \\ 0 & 0 & \cdots & \lambda_n \end{pmatrix},$$

then

$$e^{At} = \begin{pmatrix} e^{\lambda_1 t} & 0 & \cdots & 0 \\ 0 & e^{\lambda_2 t} & \cdots & 0 \\ \vdots & \vdots & & \vdots \\ 0 & 0 & \cdots & e^{\lambda_n t} \end{pmatrix}. \tag{7}$$

EXAMPLE 1 ▶ Computing e^{At} when A is a diagonal matrix

Let

$$A = \begin{pmatrix} 1 & 0 & 0 \\ 0 & 2 & 0 \\ 0 & 0 & 3 \end{pmatrix}.$$

Then

$$A^2 = \begin{pmatrix} 1 & 0 & 0 \\ 0 & 2^2 & 0 \\ 0 & 0 & 3^2 \end{pmatrix}, \quad A^3 = \begin{pmatrix} 1 & 0 & 0 \\ 0 & 2^3 & 0 \\ 0 & 0 & 3^3 \end{pmatrix}, \ldots, \quad A^m = \begin{pmatrix} 1 & 0 & 0 \\ 0 & 2^m & 0 \\ 0 & 0 & 3^m \end{pmatrix},$$

so

$$\begin{aligned} e^{At} &= I + At + \frac{A^2t^2}{2} + \ldots + \frac{A^mt^m}{m} + \ldots \\ &= \begin{pmatrix} 1 & 0 & 0 \\ 0 & 1 & 0 \\ 0 & 0 & 1 \end{pmatrix} + \begin{pmatrix} t & 0 & 0 \\ 0 & 2t & 0 \\ 0 & 0 & 3t \end{pmatrix} \\ &\quad + \ldots + \begin{pmatrix} \frac{t^m}{m!} & 0 & 0 \\ 0 & \frac{(2t)^m}{m!} & 0 \\ 0 & 0 & \frac{(3t)^m}{m!} \end{pmatrix} + \ldots = \begin{pmatrix} e^t & 0 & 0 \\ 0 & e^{2t} & 0 \\ 0 & 0 & e^{3t} \end{pmatrix}. \end{aligned}$$ ◀

The "brute force" method of calculating e^{At} by using definition **(5)** is useful only for particularly simple matrices A, such as the diagonal matrix in Example 1. Indeed, the procedure we used in Example 1 can be easily extended to obtain our assertion in **(7)**.

When the $n \times n$ matrix A has n linearly independent eigenvectors $\boldsymbol{\phi}_1, \boldsymbol{\phi}_2, \ldots, \boldsymbol{\phi}_n$, we know that $A\boldsymbol{\phi}_i = \lambda_i \boldsymbol{\phi}_i$, where λ_i is the eigenvalue corresponding to the eigenvector $\boldsymbol{\phi}_i$. Hence, if we form the matrix $\Phi = (\boldsymbol{\phi}_1, \boldsymbol{\phi}_2, \ldots, \boldsymbol{\phi}_n)$ whose columns are the n eigenvectors of A, we have

$$\begin{aligned} A\Phi &= A(\boldsymbol{\phi}_1, \boldsymbol{\phi}_2, \ldots, \boldsymbol{\phi}_n) = (A\boldsymbol{\phi}_1, A\boldsymbol{\phi}_2, \ldots, A\boldsymbol{\phi}_n) \\ &= (\lambda_1\boldsymbol{\phi}_1, \lambda_2\boldsymbol{\phi}_2, \ldots, \lambda_n\boldsymbol{\phi}_n) \\ &= \Phi \begin{pmatrix} \lambda_1 & 0 & \cdots & 0 \\ 0 & \lambda_2 & \cdots & 0 \\ \vdots & \vdots & & \vdots \\ 0 & 0 & \cdots & \lambda_n \end{pmatrix} = \Phi D, \end{aligned} \tag{8}$$

where D is a diagonal matrix. Since the columns of Φ are linearly independent, Φ is invertible, so

$$A = \Phi D \Phi^{-1}. \tag{9}$$

Then, using **(5)**, we have

$$e^{At} = I + At + A^2\frac{t^2}{2} + A^3\frac{t^3}{3!} + \ldots$$

$$= \Phi\Phi^{-1} + \Phi D\Phi^{-1}t + \Phi D\underbrace{\Phi^{-1}\Phi}_{=I} D\Phi^{-1}\frac{t^2}{2!} + \ldots$$

$$= \Phi I\Phi^{-1} + \Phi D\Phi^{-1}t + \Phi D^2\Phi^{-1}\frac{t^2}{2!} + \Phi D^3\Phi^{-1}\frac{t^3}{3!} + \ldots$$

$$= \Phi\left(I + Dt + D^2\frac{t^2}{2!} + D^3\frac{t^3}{3!} + \ldots\right)\Phi^{-1} = \Phi e^{Dt}\Phi^{-1}.$$

Thus the matrix exponential for A is obtained by multiplying the matrix exponential for D (which is easily obtained from **(7)**) by Φ on the left and Φ^{-1} on the right:

$$e^{At} = \Phi e^{Dt}\Phi^{-1}. \quad \textbf{(10)}$$

EXAMPLE 2 ▶ **Computing e^{At} when the 2×2 matrix A has two linearly independent eigenvectors**

In Example 9.4.3 we found that for the system

$$\mathbf{x}' = \begin{pmatrix} 4 & 2 \\ 3 & 3 \end{pmatrix}\mathbf{x} = A\mathbf{x}, \quad \textbf{(11)}$$

the matrix A has the eigenvalues $\lambda_1 = 1$ and $\lambda_2 = 6$ with corresponding eigenvectors

$$\boldsymbol{\phi}_1 = \begin{pmatrix} 2 \\ -3 \end{pmatrix} \quad \text{and} \quad \boldsymbol{\phi}_2 = \begin{pmatrix} 1 \\ 1 \end{pmatrix}. \quad \textbf{(12)}$$

Setting $\Phi = (\boldsymbol{\phi}_1, \boldsymbol{\phi}_2)$, we see that

$$A\Phi = \begin{pmatrix} 4 & 2 \\ 3 & 3 \end{pmatrix}\begin{pmatrix} 2 & 1 \\ -3 & 1 \end{pmatrix} = \begin{pmatrix} 2 & 6 \\ -3 & 6 \end{pmatrix},$$

$$\Phi D = \begin{pmatrix} 2 & 1 \\ -3 & 1 \end{pmatrix}\begin{pmatrix} 1 & 0 \\ 0 & 6 \end{pmatrix} = \begin{pmatrix} 2 & 6 \\ -3 & 6 \end{pmatrix},$$

so $A\Phi = \Phi D$. Since

$$\Phi^{-1} = \frac{1}{5}\begin{pmatrix} 1 & -1 \\ 3 & 2 \end{pmatrix},$$

it follows by Equation **(10)** that

$$e^{At} = \Phi e^{Dt}\Phi^{-1} \qquad e^{Dt} = \begin{pmatrix} e^{\lambda_1 t} & 0 \\ 0 & e^{\lambda_2 t} \end{pmatrix}.$$

$$= \frac{1}{5}\begin{pmatrix} 2 & 1 \\ -3 & 1 \end{pmatrix}\begin{pmatrix} e^t & 0 \\ 0 & e^{6t} \end{pmatrix}\begin{pmatrix} 1 & -1 \\ 3 & 2 \end{pmatrix}$$

$$= \frac{1}{5}\begin{pmatrix} 2 & 1 \\ -3 & 1 \end{pmatrix}\begin{pmatrix} e^t & -e^t \\ 3e^{6t} & 2e^{6t} \end{pmatrix} = \frac{1}{5}\begin{pmatrix} 2e^t + 3e^{6t} & -2e^t + 2e^{6t} \\ -3e^t + 3e^{6t} & 3e^t + 2e^{6t} \end{pmatrix}. \quad \textbf{(13)}$$

This is exactly the principal matrix solution for system **(11)** that we determined in Example 9.5.1. ◀

When the $n \times n$ matrix A does not have n linearly independent eigenvectors, the procedure in **(10)** cannot be used. We now give a procedure, using the Cayley-Hamilton theorem, that can be used in this case.

THEOREM 2 ♦ **The Cayley-Hamilton theorem***

Every square matrix satisfies its own characteristic equation; that is, if $p(\lambda) = 0$ is the characteristic equation of A, then $p(A) = 0$. ♦

We are now ready to calculate e^{At} for any square matrix A. Although the procedure may appear to be complicated, it is not difficult to use.

Before explaining the procedure, we define the notation we plan to use. Let $p(\lambda) = \det(\lambda I - A)$ be the characteristic polynomial of the $n \times n$ matrix A, and suppose that

$$p(\lambda) = (\lambda - \lambda_1)^{r_1}(\lambda - \lambda_2)^{r_2} \ldots (\lambda - \lambda_k)^{r_k}, \tag{14}$$

where $\lambda_1, \lambda_2, \ldots, \lambda_k$ are the eigenvalues of A of multiplicities $r_1, r_2, \ldots, r_k$, respectively. Using partial fractions, we can write

$$\frac{1}{p(\lambda)} = \frac{a_1(\lambda)}{(\lambda - \lambda_1)^{r_1}} + \frac{a_2(\lambda)}{(\lambda - \lambda_2)^{r_2}} + \cdots + \frac{a_k(\lambda)}{(\lambda - \lambda_k)^{r_k}}, \tag{15}$$

where, for each polynomial $a_i(\lambda)$,

$$\deg a_i(\lambda) \le r_i - 1. \tag{16}$$

Multiplying both sides of **(15)** by $p(\lambda)$, we get

$$1 = a_1(\lambda)q_1(\lambda) + a_2(\lambda)q_2(\lambda) + \cdots + a_k(\lambda)q_k(\lambda), \tag{17}$$

where $q_i(\lambda)$ is the polynomial consisting of all but the $(\lambda - \lambda_i)^{r_i}$ term of $p(\lambda)$:

$$p(\lambda) = q_i(\lambda)(\lambda - \lambda_i)^{r_i}. \tag{18}$$

We are now ready to begin the procedure for calculating e^{At} for any $n \times n$ matrix A. We illustrate each step in the procedure by applying it to the matrix

$$A = \begin{pmatrix} 1 & 2 & 3 \\ 0 & 0 & 4 \\ 0 & 0 & 0 \end{pmatrix}.$$

Step 1 *Find the characteristic polynomial $p(\lambda)$ and use it to determine the polynomials $a_i(\lambda)$ and $q_i(\lambda)$ in Equations* ***(17)*** *and* ***(18)***.

In this case we have

$$p(\lambda) = \det(\lambda I - A) = \begin{vmatrix} \lambda - 1 & -2 & -3 \\ 0 & \lambda & -4 \\ 0 & 0 & \lambda \end{vmatrix} = \lambda^2(\lambda - 1),$$

so $\lambda_1 = 0$ and $\lambda_2 = 1$, with $r_1 = 2$ and $r_2 = 1$, respectively. By Equation **(18)** we easily obtain

$$q_1(\lambda) = \lambda - 1 \quad \text{and} \quad q_2(\lambda) = \lambda^2. \tag{19}$$

Using **(16)**, we note that $\deg a_1(\lambda) \le 1$ and $\deg a_2(\lambda) \le 0$, so the general forms of the polynomials a_1 and a_2 are

$$a_1(\lambda) = a\lambda + b \quad \text{and} \quad a_2(\lambda) = c.$$

---

* Named after William Rowan Hamilton (1805–1865) and Arthur Cayley (1821–1895). Cayley published the first discussion of this famous theorem in 1858. Independently, Hamilton discovered the result in his work on quaternions. A proof of the Cayley-Hamilton theorem can be found in S. I. Grossman, *Elementary Linear Algebra,* 5th ed. (Philadelphia, PA: Saunders, 1994), Section 6.8.

Substituting these polynomials into Equation **(17)** and equating like powers of λ, we obtain

$$a_1(\lambda) = -\lambda - 1 \qquad \text{and} \qquad a_2(\lambda) = 1. \tag{20}$$

At this point we need to develop some additional facts. Recall the Cayley-Hamilton theorem. Since A satisfies its characteristic equation, **(18)** becomes

$$0 = p(A) = q_i(A)(A - \lambda_i I)^{r_i}. \tag{21}$$

Equation **(17)** is also satisfied by the matrix A, since it was derived from the characteristic equation. Thus

$$I = a_1(A)q_1(A) + a_2(A)q_2(A) + \cdots + a_k(A)q_k(A). \tag{22}$$

We can easily check **(22)** for our particular example:

$$I = (-A - I)(A - I) + A^2.$$

Using (7), we observe that $e^{\lambda_i tI} = e^{\lambda_i t}I$, so we have

$$e^{At} = e^{\lambda_i tI}e^{(A-\lambda_i I)t} = e^{\lambda_i t}\sum_{j=0}^{\infty}\frac{(A - \lambda_i I)^j t^j}{j!}. \tag{23}$$

Multiplying both ends of Equation **(23)** on the left by $q_i(A)$, we obtain

$$q_i(A)e^{At} = e^{\lambda_i t}\sum_{j=0}^{\infty}\frac{q_i(A)(A - \lambda_i I)^j t^j}{j!}. \tag{24}$$

By Equation **(21),** all the terms in the series for $j \geq r_i$ are equal to the zero matrix, so Equation **(24)** reduces to

$$q_i(A)e^{At} = e^{\lambda_i t}\sum_{j=0}^{r_i-1}\frac{q_i(A)(A - \lambda_i I)^j t^j}{j!}. \tag{25}$$

Multiplying each side of Equation **(25)** on the left by $a_i(A)$ yields

$$a_i(A)q_i(A)e^{At} = e^{\lambda_i t}\sum_{j=0}^{r_i-1}\frac{a_i(A)q_i(A)(A - \lambda_i I)^j t^j}{j!}. \tag{26}$$

Finally, summing Equation **(26)** over all the indices i and using Equation **(22),** we conclude that

$$e^{At} = Ie^{At} = \sum_{i=1}^{k} a_i(A)q_i(A)e^{At},$$

or

$$e^{At} = \sum_{i=1}^{k}\left\{e^{\lambda_i t}a_i(A)q_i(A)\sum_{j=0}^{r_i-1}\frac{(A - \lambda_i I)^j t^j}{j!}\right\}.$$

Although this equation looks formidable, for most applications it is very easy to use.

Step 2 *Using the polynomials $a_i(\lambda)$, $q_i(\lambda)$, and the eigenvalues λ_i with multiplicities r_i that were obtained in step 1, compute*

$$e^{At} = \sum_{i=1}^{k}\left\{e^{\lambda_i t}a_i(A)q_i(A)\sum_{j=0}^{r_i-1}\frac{(A - \lambda_i I)^j t^j}{j!}\right\}. \tag{27}$$

For the particular matrix A we are considering, we have

$$\lambda_1 = 0, \qquad r_1 = 2, \qquad \lambda_2 = 1, \qquad r_2 = 1,$$

and

$$q_1(A) = A - I, \qquad q_2(A) = A^2, \qquad a_1(A) = -A - I, \qquad a_2(A) = I.$$

Substituting these values in Equation **(27)**, we get

$$\begin{aligned} e^{At} &= \left\{e^{0t}(-A - I)(A - I)\sum_{j=0}^{1}\frac{A^j t^j}{j!}\right\} + \left\{e^t I A^2 \sum_{j=0}^{0}\frac{(A-I)^j t^j}{j!}\right\} \\ &= (A + I)(I - A)(I + At) + e^t A^2 \\ &= (I - A^2)(I + At) + e^t A^2. \end{aligned}$$

Since

$$A^2 = \begin{pmatrix} 1 & 2 & 3 \\ 0 & 0 & 4 \\ 0 & 0 & 0 \end{pmatrix}\begin{pmatrix} 1 & 2 & 3 \\ 0 & 0 & 4 \\ 0 & 0 & 0 \end{pmatrix} = \begin{pmatrix} 1 & 2 & 11 \\ 0 & 0 & 0 \\ 0 & 0 & 0 \end{pmatrix},$$

we have

$$\begin{aligned} e^{At} &= \begin{pmatrix} 0 & -2 & -11 \\ 0 & 1 & 0 \\ 0 & 0 & 1 \end{pmatrix}\begin{pmatrix} 1+t & 2t & 3t \\ 0 & 1 & 4t \\ 0 & 0 & 1 \end{pmatrix} + e^t \begin{pmatrix} 1 & 2 & 11 \\ 0 & 0 & 0 \\ 0 & 0 & 0 \end{pmatrix} \\ &= \begin{pmatrix} 0 & -2 & -8t - 11 \\ 0 & 1 & 4t \\ 0 & 0 & 1 \end{pmatrix} + \begin{pmatrix} e^t & 2e^t & 11e^t \\ 0 & 0 & 0 \\ 0 & 0 & 0 \end{pmatrix} \\ &= \begin{pmatrix} e^t & 2(e^t - 1) & 11e^t - 8t - 11 \\ 0 & 1 & 4t \\ 0 & 0 & 1 \end{pmatrix}. \end{aligned}$$

Calculation of e^{At} for any $n \times n$ matrix A

1. Find the characteristic polynomial for the matrix A,

$$\begin{aligned} p(\lambda) &= \det(\lambda I - A) \\ &= (\lambda - \lambda_1)^{r_1}(\lambda - \lambda_2)^{r_2} \ldots (\lambda - \lambda_k)^{r_k}; \end{aligned}$$

use it first to determine the polynomials

$$q_i(\lambda) = \frac{p(\lambda)}{(\lambda - \lambda_i)^{r_i}}$$

and then the polynomials $a_i(\lambda)$ that satisfy

$$a_1(\lambda)q_1(\lambda) + a_2(\lambda)q_2(\lambda) + \cdots + a_k(\lambda)q_k(\lambda) = 1. \qquad \textbf{(28)}$$

2. Replace each λ^m by A^m in the expressions for $a_i(\lambda)$ and $q_i(\lambda)$ and compute

$$e^{At} = \sum_{i=1}^{k}\left\{e^{\lambda_i t}a_i(A)q_i(A)\sum_{j=0}^{r_i - 1}\frac{(A - \lambda_i I)^j t^j}{j!}\right\}.$$

Two final facts about the principal matrix solution for a system $\mathbf{x}' = A\mathbf{x}$, where A is a constant matrix, can easily be obtained using the notion of a matrix exponential.

THEOREM 3 ♦ **(Theorem 9.3.5)**

Let $\Psi(t)$ be the principal matrix solution for the system $\mathbf{x}'(t) = A\mathbf{x}(t)$, $\mathbf{x}(t_0) = \mathbf{x}_0$, where A is a constant matrix. Then

$$\Psi^{-1}(t) = \Psi(-t).$$

Proof If $t_0 = 0$, we need to show that

$$(e^{At})^{-1} = e^{-At}.$$

But

$$e^{At}e^{-At} = \left(I + At + \frac{A^2t^2}{2!} + \frac{A^3t^3}{3!} + \ldots\right)\left(I - At + \frac{A^2t^2}{2!} - \frac{A^3t^3}{3!} + \ldots\right)$$
$$= \left[I + (A - A)t + \left(\frac{A^2}{2!} + \frac{A^2}{2!} - A^2\right)t^2 + \ldots\right] = I,$$

since all terms except I cancel. Similarly, $e^{-At}e^{At} = I$, which shows that $e^{-At} = (e^{At})^{-1}$. See the note on p. 468 for the case $t_0 \neq 0$. ♦

THEOREM 4 ♦ **(Theorem 9.5.2)**

Let the real matrix A have the complex eigenvalues $\alpha \pm i\beta$. Then the principal matrix solution of $\mathbf{x}' = A\mathbf{x}$ contains no complex terms.

Proof This is easy. We have

$$\Psi(t) = e^{At} = I + At + \frac{A^2t^2}{2!} + \frac{A^3t^3}{3!} + \ldots.$$

Since A is real, A^k is real for every k and the sum of real matrices is a real matrix. ♦

PROBLEMS 9.7

In Problems 1–9 use the method in Example 1 or Example 2 to compute e^{At}.

1. $A = \begin{pmatrix} 1 & 0 \\ 0 & 3 \end{pmatrix}$

2. $A = \begin{pmatrix} 1 & -1 \\ 2 & 3 \end{pmatrix}$

3. $A = \begin{pmatrix} -2 & -2 \\ -5 & 1 \end{pmatrix}$

4. $A = \begin{pmatrix} -12 & 7 \\ -7 & 2 \end{pmatrix}$

5. $A = \begin{pmatrix} 2 & -1 \\ 5 & -2 \end{pmatrix}$

6. $A = \begin{pmatrix} 0 & 1 & 0 \\ 0 & 0 & 1 \\ 1 & -3 & 3 \end{pmatrix}$

7. $A = \begin{pmatrix} 1 & -1 & 0 \\ -1 & 2 & -1 \\ 0 & -1 & 1 \end{pmatrix}$

8. $A = \begin{pmatrix} 4 & 6 & 6 \\ 1 & 3 & 2 \\ -1 & -5 & -2 \end{pmatrix}$

9. $A = \begin{pmatrix} 5 & 4 & 2 \\ 4 & 5 & 2 \\ 2 & 2 & 2 \end{pmatrix}$

10. The Cayley-Hamilton theorem can be used to calculate the inverse of a matrix if A^{-1} exists. If

$$p(\lambda) = \lambda^n + a_{n-1}\lambda^{n-1} + \ldots + a_1\lambda + a_0,$$

then

$$p(A) = A^n + a_{n-1}A^{n-1} + \ldots + a_1A + a_0I = 0,$$

and

$$A^{-1}p(A) = A^{n-1} + a_{n-1}A^{n-2} + \ldots + a_2A + a_1I + a_0A^{-1} = 0.$$

Thus

$$A^{-1} = \frac{1}{a_0}(-A^{n-1} - a_{n-1}A^{n-2} - \ldots - a_2A - a_1I).$$

Use this procedure to find the inverse of

$$A = \begin{pmatrix} 1 & -1 & 4 \\ 3 & 2 & -1 \\ 2 & 1 & -1 \end{pmatrix}.$$

*In Problems 11–20 use the two-step procedure **(28)** to find e^{At} for each given matrix A.*

11. $A = \begin{pmatrix} 4 & 1 \\ 0 & 4 \end{pmatrix}$ **12.** $A = \begin{pmatrix} 2 & -1 \\ -4 & 2 \end{pmatrix}$ **13.** $A = \begin{pmatrix} 1 & 2 \\ 0 & 1 \end{pmatrix}$ **14.** $A = \begin{pmatrix} 3 & 2 \\ -5 & 1 \end{pmatrix}$

15. $A = \begin{pmatrix} 3 & -2 \\ 8 & -5 \end{pmatrix}$ **16.** $A = \begin{pmatrix} 1 & 1 \\ 1 & -1 \end{pmatrix}$ **17.** $A = \begin{pmatrix} 1 & -1 & 0 \\ -1 & 2 & -1 \\ 0 & -1 & 1 \end{pmatrix}$

18. $A = \begin{pmatrix} 1 & 1 & -2 \\ -1 & 2 & 1 \\ 0 & 1 & -1 \end{pmatrix}$. **19.** $A = \begin{pmatrix} 4 & 6 & 6 \\ 1 & 3 & 2 \\ -1 & -5 & -2 \end{pmatrix}$ **20.** $A = \begin{pmatrix} 7 & -2 & -4 \\ 3 & 0 & -2 \\ 6 & -2 & -3 \end{pmatrix}$

9.8 SOLVING MATRIX DIFFERENTIAL EQUATIONS WITH A SYMBOLIC MANIPULATOR

We now demonstrate how MATHEMATICA, MAPLE, DERIVE, MATLAB, or MATHCAD may be used to solve matrix differential equations. Each of these symbolic manipulators has commands for finding the inverse of a matrix and solving a system of equations. We shall not dwell on these commands, as they are not directly applicable to differential equations and systems. Instead we shall concentrate on the commands that find the eigenvalues and eigenvectors of an $n \times n$ matrix **A**:

`EIGENVALUES(A)` = {finds n eigenvalues for the matrix **A**}

`EIGENVECTOR(A, λ)` = {finds an eigenvector **v** corresponding to the eigenvalue λ satisfying $\mathbf{Av} = \lambda\mathbf{v}$ and $\|\mathbf{v}\| = 1$}

Consider the system

$$\mathbf{x}' = A\mathbf{x}, \tag{1}$$

where A is a matrix of coefficients. It is easy to show that the solution to **(1)** is given by

$$\mathbf{x} = e^{At}\mathbf{c}, \tag{2}$$

where **c** is a vector of arbitrary constants and e^{At} is the exponential matrix

$$e^{At} = I + At + A^2\frac{t^2}{2!} + A^3\frac{t^3}{3!} + \dots + A^n\frac{t^n}{n!} + \dots, \tag{3}$$

because

$$\begin{aligned} \mathbf{x}' = (e^{At})'\,\mathbf{c} &= \left(A + A^2t + A^3\frac{t^2}{2!} + \dots\right)\mathbf{c} \\ &= A\left(I + At + A^2\frac{t^2}{2!} + \dots\right)\mathbf{c} = A\mathbf{x}. \end{aligned}$$

For simplicity, we shall assume that A is symmetric, that is that A equals its transpose ($A = A^{\mathrm{T}}$), so that all of the eigenvalues $\lambda_1, \lambda_2, \dots, \lambda_n$ of A are real. Suppose, further, that all of the eigenvalues are different. Let $\mathbf{v_1}, \mathbf{v_2}, \dots, \mathbf{v_n}$ be normalized (column) eigenvectors corresponding to these eigenvalues. Then $A\mathbf{v_k} = \lambda_k\mathbf{v_k}$, for each $k = 1, 2, \dots, n$. It is not hard to prove that $\mathbf{v_j^T v_k} = 0$ for $j \neq k$:

$$\mathbf{v_j^T}\, A\mathbf{v_k} = \mathbf{v_j^T}\, \lambda_k\mathbf{v_k} = \lambda_k\, \mathbf{v_j^T v_k}$$

and

$$(\mathbf{v_j^T} A \mathbf{v_k})^{\mathbf{TT}} = (\mathbf{v_k^T} A \mathbf{v_j})^{\mathbf{T}} = (\lambda_j \, \mathbf{v_k^T v_j})^{\mathbf{T}} = \lambda_j \mathbf{v_j^T v_k}$$

and since both terms are identical, it follows that $\lambda_j = \lambda_k$ unless $\mathbf{v_j^T v_k} = 0$.

The identities $A\mathbf{v_k} = \lambda_k \mathbf{v_k}$ imply that

$$AV = V\Lambda, \tag{4}$$

where

$$V = [\mathbf{v_1 v_2 \ldots v_n}] \qquad \text{and} \qquad \Lambda = \text{diag}(\lambda_1, \lambda_2, \ldots, \lambda_n),$$

that is, V is the matrix whose columns are all of the normalized eigenvectors and Λ is a diagonal matrix whose entries are the eigenvalues of A. Since the eigenvectors are normalized, we see that $V^{\mathbf{T}}V = \mathbf{I}$, implying that $V^{\mathbf{T}}$ is the inverse of V.

The point to all of this matrix theory is that it provides us with an easy way to write the matrix exponential **(3)**: using **(4)** we have $A = V\Lambda V^{\mathbf{T}}$ so that

$$\begin{aligned}
A^2 &= V\Lambda V^T \, V\Lambda V^T = V\Lambda^2 V^T, \\
A^3 &= V\Lambda V^T \, V\Lambda V^T \, V\Lambda V^T = V\Lambda^3 V^T, \\
&\ldots \\
A^n &= V\Lambda V^T \, V\Lambda V^T \ldots V\Lambda V^T = V\Lambda^n V^T,
\end{aligned}$$

and

$$\begin{aligned}
e^{At} &= I + At + A^2\frac{t^2}{2!} + A^3\frac{t^3}{3!} + \ldots + A^n\frac{t^n}{n!} + \ldots \\
&= VIV^T + V\Lambda V^T t + V\Lambda^2 V^T\frac{t^2}{2!} + \ldots + V\Lambda^n V^T\frac{t^n}{n!} + \ldots \\
&= V\left(I + \Lambda t + \Lambda^2\frac{t^2}{2!} + \ldots + \Lambda^n\frac{t^n}{n!} + \ldots\right)V^T = Ve^{\Lambda t}V^T.
\end{aligned}$$

Thus

$$\mathbf{x} = Ve^{\Lambda t}V^{\mathbf{T}}\mathbf{c} = Ve^{\Lambda t}\mathbf{k}, \tag{5}$$

where $\mathbf{k}$ is an arbitrary vector of constants. This provides the general solution to **(1)**.

To see how this can be done using a symbolic manipulator, consider the following example.

EXAMPLE 1 ▶

Find the general solution of the matrix differential equation

$$\mathbf{x}' = \begin{bmatrix} 5 & 2 \\ 2 & 2 \end{bmatrix}\mathbf{x}. \tag{6}$$

Solution Suppose the matrix A is written as rows:

$$A = [[5, 2], [2, 2]].$$

First we find the eigenvalues by giving the command

```
EIGENVALUES ([[5, 2], [2, 2]])
```

The symbolic manipulator would return the roots of the characteristic polynomial

$$(5 - \lambda)(2 - \lambda) - 4 = \lambda^2 - 7\lambda + 6 = (\lambda - 6)(\lambda - 1) = 0,$$

so the eigenvalues are $\lambda_1 = 6$ and $\lambda_2 = 1$. Now we find corresponding eigenvectors by giving the two commands:

```
EIGENVECTOR ([[5, 2], [2, 2]], 6)
    and EIGENVECTOR([[5, 2], [2, 2]], 1).
```

These are the column vectors

$$\begin{bmatrix} \frac{2}{\sqrt{5}} \\ \frac{1}{\sqrt{5}} \end{bmatrix} \text{ and } \begin{bmatrix} \frac{1}{\sqrt{5}} \\ -\frac{2}{\sqrt{5}} \end{bmatrix}.$$

Hence, the matrices V and Λ are given by

$$V = \begin{pmatrix} \frac{2}{\sqrt{5}} & \frac{1}{\sqrt{5}} \\ \frac{1}{\sqrt{5}} & -\frac{2}{\sqrt{5}} \end{pmatrix} \text{ and } \Lambda = \begin{bmatrix} 6 & 0 \\ 0 & 1 \end{bmatrix}.$$

Thus, the solution $\mathbf{x}$ is given by **(5)** as

$$\mathbf{x} = Ve^{\Lambda t}\mathbf{k} = \begin{pmatrix} \frac{2}{\sqrt{5}} & \frac{1}{\sqrt{5}} \\ \frac{1}{\sqrt{5}} & -\frac{2}{\sqrt{5}} \end{pmatrix} \begin{bmatrix} e^{6t} & 0 \\ 0 & e^{t} \end{bmatrix} \begin{bmatrix} k_1 \\ k_2 \end{bmatrix}$$

$$= \begin{pmatrix} \frac{2}{\sqrt{5}} & \frac{1}{\sqrt{5}} \\ \frac{1}{\sqrt{5}} & -\frac{2}{\sqrt{5}} \end{pmatrix} \begin{bmatrix} k_1e^{6t} \\ k_2e^{t} \end{bmatrix} = \frac{1}{\sqrt{5}} \begin{bmatrix} 2k_1e^{6t} + k_2e^{t} \\ k_1e^{6t} - 2k_2e^{t} \end{bmatrix}.$$

Each of these matrix operations can be automated by a script in your symbolic manipulator. ◄

PROBLEMS 9.8

Solve each of the following problems using the EIGENVALUES and EIGENVECTOR commands in MATHEMATICA, MAPLE, DERIVE, MATLAB, or MATHCAD.

1. $\mathbf{x}' = \begin{pmatrix} 3 & -1 \\ -1 & 3 \end{pmatrix}\mathbf{x}$

2. $\mathbf{x}' = \begin{pmatrix} 4 & 4 \\ 4 & 10 \end{pmatrix}\mathbf{x}$

3. $\mathbf{x}' = \begin{pmatrix} 6 & \sqrt{10} \\ \sqrt{10} & 3 \end{pmatrix}\mathbf{x}$

4. $\mathbf{x}' = \begin{pmatrix} 5 & \sqrt{3} \\ \sqrt{3} & 3 \end{pmatrix}\mathbf{x}$

5. $\mathbf{x}' = \begin{pmatrix} 5 & \sqrt{6} \\ \sqrt{6} & 4 \end{pmatrix}\mathbf{x}$

6. $\mathbf{x}' = \begin{pmatrix} 5 & \sqrt{2} \\ \sqrt{2} & 4 \end{pmatrix}\mathbf{x}$

7. $\mathbf{x}' = \begin{pmatrix} 4 & 1 & 1 \\ 1 & 4 & 1 \\ 1 & 1 & 6 \end{pmatrix}\mathbf{x}$

8. $\mathbf{x}' = \begin{pmatrix} 4 & 1 & 1 \\ 1 & 4 & 1 \\ 1 & 1 & 5 \end{pmatrix}\mathbf{x}$

9. $\mathbf{x}' = \begin{pmatrix} 4 & 1 & 1 \\ 1 & 4 & 1 \\ 1 & 1 & 4 \end{pmatrix}\mathbf{x}$

10. $\mathbf{x}' = \begin{pmatrix} 2 & 1 & 1 \\ 1 & 2 & 1 \\ 1 & 1 & 2 \end{pmatrix}\mathbf{x}$

11. Write a script that does all of the work involved in solving these problems automatically with your symbolic manipulator.

SUMMARY OUTLINE OF CHAPTER 9

An **m × n matrix** is a rectangular array of mn numbers arranged in m rows and n columns. If $m = n$, the matrix is a **square matrix.** p. 405

A **row vector** is a matrix with one row. p. 406

A **column vector** is a matrix with one column. p. 406

Addition of Matrices If A and B are $m \times n$ matrices, then $A + B$ is the $m \times n$ matrix obtained by adding the corresponding components of A and B. p. 406

Dot Product If $\mathbf{a} = (a_1, a_2, \ldots, a_n)$ and $\mathbf{b} = \begin{pmatrix} b_1 \\ b_2 \\ \vdots \\ b_n \end{pmatrix}$, then $\mathbf{a} \cdot \mathbf{b} = a_1b_1 + a_2b_2 + \cdots + a_nb_n$. p. 408

Matrix Product If A is an $m \times n$ matrix and B is an $n \times p$ matrix, then AB is an $m \times p$ matrix whose ijth component is the dot product of the ith row of A and the jth column of B. p. 408

An **Identity Matrix** is an $n \times n$ matrix, denoted by I, with 1s down the main diagonal and 0s everywhere else. For every $n \times n$ matrix A, $AI = IA = A$. p. 409

Matrix Inverse If A is an $n \times n$ matrix, then the **inverse** of A, if it exists, is a matrix A^{-1}, where $AA^{-1} = A^{-1}A = I$, the identity matrix. p. 410

If $A = \begin{pmatrix} a & b \\ c & d \end{pmatrix}$ and $D = ad - bc \neq 0$, then A is invertible and $A^{-1} = \dfrac{1}{D}\begin{pmatrix} d & -b \\ -c & a \end{pmatrix}$. p. 415

Elementary Row Operations

Operation	**Symbol** p. 411
(i) Multiply the ith row by a nonzero number c	$M_i(c)$
(ii) Multiply the ith row by c and add it to the jth row	$A_{ij}(c)$
(iii) Permute (interchange) the ith and jth rows	P_{ij}

Procedure for Computing the Inverse of a Square Matrix A p. 412

Step 1 Write the augmented matrix $(A \mid I)$.
Step 2 Use row reduction to reduce A to either
- **(a)** the identity matrix or
- **(b)** a matrix with a row of zeros.

Step 3 Decide if A is invertible.
- **(a)** If A can be reduced to the identity matrix I, then A^{-1} is the matrix to the right of the vertical bar.
- **(b)** If the row reduction leads to a row of zeros to the left of the vertical bar, then A is not invertible.

Vector Function An n-component **vector function** p. 416

$$\mathbf{v}(t) = \begin{pmatrix} v_1(t) \\ v_2(t) \\ \vdots \\ v_n(t) \end{pmatrix} \tag{1}$$

is an n-vector each of whose components is a function (usually assumed to be continuous).

Matrix Function An $n \times n$ **matrix function** $A(t)$ is an $n \times n$ matrix p. 416

$$A(t) = \begin{pmatrix} a_{11}(t) & a_{12}(t) & \cdots & a_{1n}(t) \\ a_{21}(t) & a_{22}(t) & \cdots & a_{2n}(t) \\ \vdots & \vdots & & \vdots \\ a_{n1}(t) & a_{n2}(t) & \cdots & a_{nn}(t) \end{pmatrix} \tag{2}$$

Derivatives and Integrals If $\mathbf{v}(t)$ is given by **(1)**, then p. 416

$$\mathbf{v}'(t) = \begin{pmatrix} v_1'(t) \\ v_2'(t) \\ \vdots \\ v_n'(t) \end{pmatrix} \quad \text{and} \quad \int_{t_0}^{t} \mathbf{v}(s)\, ds = \begin{pmatrix} \int_{t_0}^{t} v_1(s)\, ds \\ \int_{t_0}^{t} v_2(s)\, ds \\ \vdots \\ \int_{t_0}^{t} v_n(s)\, ds \end{pmatrix}.$$

If $A(t)$ is given by **(2)**, then

$$A'(t) = \begin{pmatrix} a_{11}'(t) & a_{12}'(t) & \cdots & a_{1n}'(t) \\ a_{21}'(t) & a_{22}'(t) & \cdots & a_{2n}'(t) \\ \vdots & \vdots & & \vdots \\ a_{n1}'(t) & a_{n2}'(t) & \cdots & a_{nn}'(t) \end{pmatrix}$$

and

$$\int_{t_0}^{t} A(s)\, ds = \begin{pmatrix} \int_{t_0}^{t} a_{11}(s)\, ds & \int_{t_0}^{t} a_{12}(s)\, ds & \cdots & \int_{t_0}^{t} a_{1n}(s)\, ds \\ \int_{t_0}^{t} a_{21}(s)\, ds & \int_{t_0}^{t} a_{22}(s)\, ds & \cdots & \int_{t_0}^{t} a_{2n}(s)\, ds \\ \vdots & \vdots & & \vdots \\ \int_{t_0}^{t} a_{n1}(s)\, ds & \int_{t_0}^{t} a_{n2}(s)\, ds & \cdots & \int_{t_0}^{t} a_{nn}(s)\, ds \end{pmatrix}.$$

Vector Differential Equations p. 420
Consider the following linear systems of n first-order equations

$$\begin{aligned} x_1' &= a_{11}(t)x_1 + a_{12}(t)x_2 + \cdots + a_{1n}(t)x_n + f_1(t), \\ x_2' &= a_{21}(t)x_1 + a_{22}(t)x_2 + \cdots + a_{2n}(t)x_n + f_2(t), \\ &\ \ \vdots \\ x_n' &= a_{n1}(t)x_1 + a_{n2}(t)x_2 + \cdots + a_{nn}(t)x_n + f_n(t), \end{aligned} \tag{3}$$

which is **nonhomogeneous** if at least one of the functions $f_i(t)$, $i = 1, 2, \ldots, n$ is not the zero function, and the **associated homogeneous system**

$$\begin{aligned} x_1' &= a_{11}(t)x_1 + a_{12}(t)x_2 + \cdots + a_{1n}(t)x_n, \\ x_2' &= a_{21}(t)x_1 + a_{22}(t)x_2 + \cdots + a_{2n}(t)x_n, \\ &\ \ \vdots \\ x_n' &= a_{n1}(t)x_1 + a_{n2}(t)x_2 + \cdots + a_{nn}(t)x_n. \end{aligned} \tag{4}$$

Define the vector function $\mathbf{x}(t)$, the matrix function $A(t)$, and the vector function $\mathbf{f}(t)$ as follows:

$$\mathbf{x}(t) = \begin{pmatrix} x_1(t) \\ x_2(t) \\ \vdots \\ x_n(t) \end{pmatrix}, \quad \mathbf{A}(t) = \begin{pmatrix} a_{11}(t) & a_{12}(t) & \cdots & a_{1n}(t) \\ a_{21}(t) & a_{22}(t) & \cdots & a_{2n}(t) \\ \vdots & \vdots & & \vdots \\ a_{n1}(t) & a_{n2}(t) & \cdots & a_{nn}(t) \end{pmatrix}, \quad \mathbf{f}(t) = \begin{pmatrix} f_1(t) \\ f_2(t) \\ \vdots \\ f_n(t) \end{pmatrix}.$$

Then system **(3)** can be written

$$\mathbf{x}'(t) = A(t)\mathbf{x}(t) + \mathbf{f}(t) \tag{5}$$

System **(4)** becomes

$$\mathbf{x}'(t) = A(t)\mathbf{x}(t) \tag{6}$$

Vector Initial-Value Problem p. 422

$$\mathbf{x}'(t) = A(t)\mathbf{x}(t) + \mathbf{f}(t), \qquad \mathbf{x}'(t_0) = \mathbf{x_0} \tag{7}$$

Existence-Uniqueness Theorem p. 422
Let $A(t)$ and $\mathbf{f}(t)$ be continuous matrix and vector functions, respectively, on some interval $[a, b]$ containing t_0 (that is, the component functions of both $A(t)$ and $\mathbf{f}(t)$ are continuous). Then there exists a unique vector function $\boldsymbol{\phi}(t)$ that is a solution to the initial-value problem **(7)** on the entire interval $[a, b]$.

Linearly Independent and Fundamental Sets of Solutions p. 425
Let $\boldsymbol{\phi}_1(t), \boldsymbol{\phi}_2(t), \ldots, \boldsymbol{\phi}_m(t)$ be m vector solutions of system **(6).** We say that they are **linearly independent** on an interval if the equation (for all t in the common interval).

$$c_1\boldsymbol{\phi}_1(t) + c_2\boldsymbol{\phi}_2(t) + \cdots + c_m\boldsymbol{\phi}_m(t) = \mathbf{0}$$

holds only for $c_1 = c_2 = \ldots = c_m = 0$. Since system **(1)** is equivalent to an nth-order equation, it is natural for us to seek n linearly independent solutions to the system. Any set of n linearly independent solutions of **(6)** is called a **fundamental set of solutions.**

Matrix Solution and its Wronskian p. 428
Let $\boldsymbol{\phi}_1, \boldsymbol{\phi}_2, \ldots, \boldsymbol{\phi}_n$ be n-vector solutions of $\mathbf{x}' = A(t)\mathbf{x}$. Let $\Phi(t)$ be the matrix whose columns are the vectors $\boldsymbol{\phi}_1, \boldsymbol{\phi}_2, \ldots, \boldsymbol{\phi}_n$; that is,

$$\Phi(t) = (\boldsymbol{\phi}_1(t), \ldots, \boldsymbol{\phi}(t)) = \begin{pmatrix} \phi_{11}(t) & \phi_{12}(t) & \cdots & \phi_{1n}(t) \\ \phi_{21}(t) & \phi_{22}(t) & \cdots & \phi_{2n}(t) \\ \vdots & \vdots & & \vdots \\ \phi_{n1}(t) & \phi_{n2}(t) & \cdots & \phi_{nn}(t) \end{pmatrix}.$$

Such a matrix is called a **matrix solution** of the system $\mathbf{x}' = A\mathbf{x}$. Equivalently, *an $n \times n$ matrix function $\Phi(t)$ is a matrix solution of $\mathbf{x}' = A\mathbf{x}$ if and only if each of its columns is a solution vector of $\mathbf{x}' = A\mathbf{x}$.*

We define the **Wronskian** of a matrix solution $\Phi(t)$ of the system $\mathbf{x}' = A(t)\mathbf{x}$ by

$$W(t) = \det \Phi(t).$$

Fundamental Matrix Solution p. 428
If $W(t) \neq 0$, then the vectors $\boldsymbol{\phi}_1, \boldsymbol{\phi}_2, \ldots, \boldsymbol{\phi}_n$ form a fundamental set of solutions (that is, they are linearly independent), and $\Phi(t)$ is called a **fundamental matrix solution.**

Principal Matrix Solution p. 429
The fundamental matrix solution $\Psi(t)$ for the system

$$\mathbf{x}' = A(t)\mathbf{x}$$

that satisfies $\Psi(t_0) = I$, where I is the identity matrix, is called the **principal matrix solution.**

Let $A(t)$ be continuous on some interval $[a, b]$. Then for any t_0, $a \leq t_0 \leq b$, there exists a unique fundamental matrix solution $\Psi(t)$ of the system, $\mathbf{x}' = A(t)\mathbf{x}$ satisfying the condition $\Psi(t_0) = I$.

Obtaining the Principal Matrix Solution from a Fundamental Matrix Solution p. 431
If $\Phi(t)$ is a fundamental matrix solution for $\mathbf{x}' = A(t)\mathbf{x}$, then

$$\Psi(t) = \Phi(t)\Phi^{-1}(t_0)$$

is the principal matrix solution (associated with the point t_0).

Let $\Phi(t)$ be a fundamental matrix solution and let $\mathbf{x}(t)$ be any solution of $\mathbf{x}' = A(t)\mathbf{x}$. Then there exists a constant vector $\mathbf{c}$ such that

$$\mathbf{x}(t) = \Phi(t)\mathbf{c}.$$

pp. 432–433

General Solution p. 436
The **general solution** of the system $\mathbf{x}' = A(t)\mathbf{x}$ is obtained by multiplying any fundamental matrix solution $\Phi(t)$ by an arbitrary vector of constants $\mathbf{c}$; that is,

$$\Phi(t)\mathbf{c} = (\boldsymbol{\phi}_1(t), \boldsymbol{\phi}_2(t), \ldots, \boldsymbol{\phi}_n(t))\begin{pmatrix} c_1 \\ c_2 \\ \vdots \\ c_n \end{pmatrix} = c_1\boldsymbol{\phi}_1(t) + c_2\boldsymbol{\phi}_2(t) + \ldots + c_n\boldsymbol{\phi}_n(t).$$

Solving an Initial-Value Problem p. 454
The unique solution to the initial-value problem

$$\mathbf{x}' = A\mathbf{x}, \qquad \mathbf{x}(t_0) = \mathbf{x}_0$$

is given by

$$\boldsymbol{\phi}(t) = \Psi(t)\mathbf{x}_0,$$

where $\Psi(t)$ is the principal matrix solution relative to the initial point t_0.

General Solution to a Nonhomogeneous System p. 462
Let $\Phi(t)$ be a fundamental matrix solution of **(6).** Then the **general solution** to $\mathbf{x}' = A(t)\mathbf{x} + \mathbf{f}(t)$ is given by

$$\boldsymbol{\phi}(t) = \Phi(t)\mathbf{c} + \Phi(t)\int \Phi^{-1}(t)\mathbf{f}(t)\,dt$$

Solution to a Nonhomogeneous Initial-Value Problem p. 463
Let $\Psi(t)$ be the principal matrix solution of the homogeneous system

$$\mathbf{x}' = A(t)\mathbf{x}, \qquad \mathbf{x}(t_0) = \mathbf{x}_0.$$

Then the solution to the initial-value problem

$$\mathbf{x}' = A(t)\mathbf{x} + \mathbf{f}(t), \qquad \mathbf{x}(t_0) = \mathbf{x}_0,$$

is given by

$$\boldsymbol{\phi}(t) = \Psi(t)\mathbf{x}_0 + \Psi(t)\int_{t_0}^{t} \Psi^{-1}(s)\mathbf{f}(s)\,ds.$$

If $A(t)$ is constant, then

$$\boldsymbol{\phi}(t) = \Psi(t)\mathbf{x}_0 + \Psi(t)\int_{t_0}^{t} \Psi(-s)\mathbf{f}(s)\,ds.$$

REVIEW EXERCISES FOR CHAPTER 9

In Exercises 1–4 write the given system of equations in vector matrix form.

1. $x_1' = 3x_1 - 4x_2,$
$x_2' = -2x_1 + 7x_2,$

2. $x_1' = (\sin t)x_1 + e^t x_2,$
$x_2' = -x_1 + (\tan t)x_2$

3. $x_1' = x_1 + x_2 + e^t,$
$x_2' = -3x_1 + 2x_2 + e^{2t}$

4. $x_1' = -tx_1 + t^2x_2 + t^3,$
$x_2' = -\sqrt{t}x_1 + \sqrt[3]{t}x_2 + t^{3/5}$

5. Consider the system

$$\mathbf{x}' = \begin{pmatrix} 4 & 2 \\ 3 & 3 \end{pmatrix}\mathbf{x}.$$

A fundamental matrix solution is

$$\Phi(t) = \begin{pmatrix} 2e^t & e^{6t} \\ -3e^t & e^{6t} \end{pmatrix}.$$

Find a solution that satisfies each of the following initial conditions.

(a) $\mathbf{x}(0) = \begin{pmatrix} 2 \\ 3 \end{pmatrix}$ **(b)** $\mathbf{x}(0) = \begin{pmatrix} -1 \\ 0 \end{pmatrix}$ **(c)** $\mathbf{x}(0) = \begin{pmatrix} 0 \\ 0 \end{pmatrix}$ **(d)** $\mathbf{x}(0) = \begin{pmatrix} 7 \\ -2 \end{pmatrix}$ **(e)** $\mathbf{x}(0) = \begin{pmatrix} a \\ b \end{pmatrix}$

In Exercises 6–11 calculate the eigenvalues and eigenvectors of the given matrix.

6. $\begin{pmatrix} 5 & -1 \\ 8 & 1 \end{pmatrix}$

7. $\begin{pmatrix} 2 & 5 \\ 0 & 2 \end{pmatrix}$

8. $\begin{pmatrix} 1 & 0 & 0 \\ 3 & 7 & 0 \\ -2 & 4 & -5 \end{pmatrix}$

9. $\begin{pmatrix} 1 & -1 & 0 \\ 1 & 2 & 1 \\ -2 & 1 & -1 \end{pmatrix}$

10. $\begin{pmatrix} 5 & -2 & 0 & 0 \\ 4 & -1 & 0 & 0 \\ 0 & 0 & 3 & -1 \\ 0 & 0 & 2 & 3 \end{pmatrix}$

11. $\begin{pmatrix} -2 & 1 & 0 \\ 0 & -2 & 1 \\ 0 & 0 & -2 \end{pmatrix}$

In Exercises 12–16 find the principal matrix solution at $t = 0$ of the given system.

12. $\mathbf{x}' = \begin{pmatrix} -3 & 4 \\ -2 & 3 \end{pmatrix}\mathbf{x}$

13. $\mathbf{x}' = \begin{pmatrix} 3 & -1 \\ -2 & 4 \end{pmatrix}\mathbf{x}$

14. $\mathbf{x}' = \begin{pmatrix} -3 & -4 \\ -2 & 1 \end{pmatrix}\mathbf{x}$

15. $\mathbf{x}' = \begin{pmatrix} -1 & -18 & -7 \\ 1 & -13 & -4 \\ -1 & 25 & 8 \end{pmatrix}\mathbf{x}$

16. $\mathbf{x}' = \begin{pmatrix} 2 & 1 & 0 \\ -2 & -1 & 2 \\ 1 & 1 & 1 \end{pmatrix}\mathbf{x}$

17. Solve the system

$$\mathbf{x}' = \begin{pmatrix} 2 & 1 \\ -4 & 2 \end{pmatrix}\mathbf{x} + \begin{pmatrix} 3 \\ t \end{pmatrix}e^{2t}, \ \mathbf{x}(0) = \begin{pmatrix} 3 \\ 2 \end{pmatrix}$$

18. Solve the system

$$\mathbf{x}' = \begin{pmatrix} 2 & 1 & 0 \\ -2 & -1 & 2 \\ 1 & 1 & 1 \end{pmatrix}\mathbf{x} + \begin{pmatrix} 0 \\ 1 \\ e^t \end{pmatrix}, \ \mathbf{x}(0) = \begin{pmatrix} 1 \\ 2 \\ 3 \end{pmatrix}$$

10 NONLINEAR EQUATIONS AND STABILITY

10.1 INTRODUCTION

In the preceding chapters we have seen that there are large classes of differential equations and systems having solutions defined in some interval. However, if an equation is nonlinear, then there is usually not any way to find its solution. For this reason, it is necessary to seek methods for describing the nature of a solution without explicitly solving the equation.

First, it is necessary to ask, What kind of information about a solution is it useful to have? We indicate a partial answer by considering our old standby: second-order linear equations with constant coefficients.

EXAMPLE 1 ▶ **Asymptotic behavior of three second-order equations**

Consider the three equations

(a) $x'' + 3x' + 2x = 0,$
(b) $x'' - 3x' + 2x = 0,$
(c) $x'' + x = 0.$

The general solutions to these equations are

(a) $x(t) = c_1e^{-t} + c_2e^{-2t},$
(b) $x(t) = c_1e^{t} + c_2e^{2t},$
(c) $x(t) = c_1\cos t + c_2\sin t.$

It is clear that all solutions of (a) approach zero as t tends to infinity, all nonzero solutions of (b) approach infinity as t tends to infinity, and all nonzero solutions of (c) remain bounded but do not approach any constant as t tends to infinity. Furthermore, the solutions of (c) are periodic of period 2π. ◀

Solutions of nonlinear equations, too, may approach zero, become unbounded, or remain bounded as t becomes large. They also may be periodic. It is fair to say that a major portion of modern research in the theory of ordinary differential equations is concerned with finding conditions that ensure that the solution of a nonlinear equation has one of these properties. There is no general method for analyzing *all* nonlinear equations. In this chaper we discuss some of the oldest known and most elementary ways of obtaining this information.

One of these methods is to consider the nonlinear equation as a *perturbation* of some linear equation, that is, to attempt to approximate a nonlinear equation by a "related" linear equation. We illustrate this method with some examples.

EXAMPLE 2 ▶ A pendulum problem

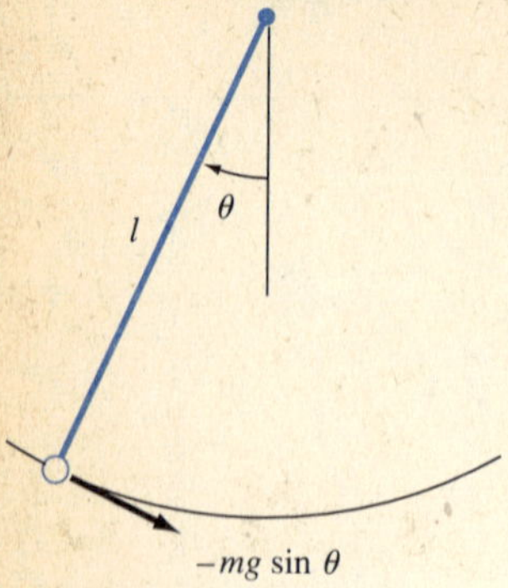

FIGURE 10.1
A swinging pendulum

Consider the freely swinging (frictionless) pendulum of length l shown in Figure 10.1. In Problem 4.1.11 we indicated that Newton's law of motion yields the nonlinear second-order equation (see p. 178)

$$\frac{d^2\theta}{dt^2} + \omega^2 \sin \theta = 0, \tag{1}$$

where $\omega^2 = g/l$. However, since

$$\lim_{\theta \to 0} \frac{\sin \theta}{\theta} = 1,$$

we may approximate Equation **(1)** for *small* values of θ by the linear equation

$$\frac{d^2\theta}{dt^2} + \omega^2 \theta = 0. \tag{2}$$

The general solution of Equation **(2)** is periodic:

$$\theta(t) = c_1 \cos \omega t + c_2 \sin \omega t.$$

How similar in behavior are the solutions of these two equations? This question will be answered in Section 10.3. ◀

EXAMPLE 3 ▶ The logistic equation

Consider the nonlinear first-order scalar equation

$$x' = -x + x^2. \tag{3}$$

This equation has two constant solutions,

$$x = 0 \qquad \text{and} \qquad x = 1,$$

easily verified by substitution into Equation **(3)**. For x close to zero, the nonlinear term x^2 is relatively small compared to the linear term $-x$, since

$$\lim_{x \to 0} \frac{x^2}{x} = 0.$$

Thus we wish to compare the solutions of Equation **(3)** with those of the linear equation

$$x' = -x, \tag{4}$$

whose general solution is

$$x(t) = x(0)e^{-t}.$$

The nonlinear equation **(3)** can be solved by a separation of variables:

$$\int \frac{dx}{-x + x^2} = \int dt = t + C.$$

By using partial fractions we have

$$\int \frac{dx}{-x + x^2} = \int \left(\frac{1}{x - 1} - \frac{1}{x} \right) dx = \ln|x - 1| - \ln|x| = \ln\left|\frac{x - 1}{x}\right|,$$

which implies that

$$\frac{x - 1}{x} = Ce^t \tag{5}$$

for some new constant C.

We may assume that $x(0) \neq 0$, because if $x(0) = 0$ we already have the unique solution $x(t) \equiv 0$. Then for $t = 0$, Equation **(5)** yields

$$\frac{x(0) - 1}{x(0)} = C.$$

Thus

$$\frac{x(t) - 1}{x(t)} = \frac{x(0) - 1}{x(0)} e^{t},$$

or, after some simple algebra,

$$x(t) = \frac{x(0)}{x(0)(1 - e^{t}) + e^{t}}. \tag{6}$$

This solution is defined as long as the denominator is not zero; that is, as long as

$$x(0)(1 - e^{t}) + e^{t} \neq 0,$$

or

$$e^{t} \neq \frac{x(0)}{x(0) - 1} = \frac{1}{C},$$

or

$$t \neq \ln \frac{x(0)}{x(0) - 1} = -\ln C.$$

Note that for small values of t, $1 - e^{t}$ is close to zero, so the solution $x(t)$, given by **(6)**, is close to $x(0)e^{-t}$, which is the solution to the linear equation **(4)**.

Now suppose that $0 < x(0) < 1$. Then by **(5)**, $C < 0$. Thus $Ce^{t} \neq 1$ for all $t \geq 0$, and the solution of Equation **(3)** [given by **(6)**] approaches zero as t tends to infinity. Hence for $0 < x(0) < 1$, Equation **(4)** is a "good" approximation of Equation **(3)** in the sense that the solutions of the two equations exhibit the same asymptotic behavior. When $x(0) \geq 1$, we are no longer "near" the zero solution and Equation **(4)** is not a good approximation. If $x(0) = 1$, we obtain the constant solution $x \equiv 1$. If $x(0) > 1$, then $0 < C < 1$ and $-\ln C > 0$, so solution **(5)** approaches ∞ as t tends to $-\ln C$. This situation is illustrated in Figure 10.2. In the

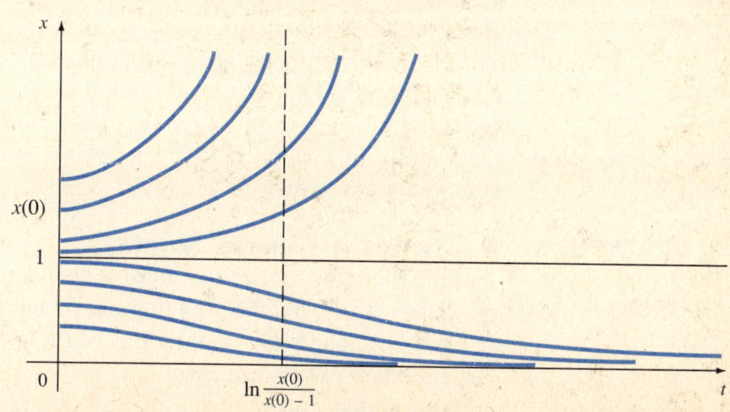

FIGURE 10.2
Solutions of logistic equation

terminology to be introduced in the next section, we may say that the solution $x \equiv 0$ of Equation **(3)** is **asymptotically stable,** since nearby solutions approach it, whereas the solution $x \equiv 1$ is **unstable,** since nearby solutions move away from it. ◀

In earlier chapters we introduced a terminology for classifying certain types of differential equations. Although nonlinear equations and systems can take many different forms, they can be roughly classified into two different categories. To illustrate this, we consider the system of two first-order equations

$$\begin{aligned} x' &= f(t, x, y), \\ y' &= g(t, x, y), \end{aligned} \tag{7}$$

where f and g are assumed to be continuously differentiable functions of t, x, and y over some region

$$D\colon a < t < b,\ c < x < d,\ e < y < h$$

in three-dimensional space. The existence and uniqueness theorems in Appendix 3 guarantee that there is a unique solution (defined over some interval in t) that passes through any initial point (t_0, x_0, y_0) in D.

autonomous system

System **(7)** is said to be **autonomous** (time independent) if the functions f and g do not depend on t. Otherwise, **(7)** is said to be **nonautonomous.** Hence **(7)** is autonomous if $f(t, x, y) = f(x, y)$ and $g(t, x, y) = g(x, y)$.

EXAMPLE 4 ▶ Autonomous and nonautonomous systems

The system

$$\begin{aligned} x' &= -x^2 + y, \\ y' &= -x + y^2 \end{aligned}$$

is autonomous, whereas the system

$$\begin{aligned} x' &= ty, \\ y' &= -x \end{aligned}$$

is nonautonomous. ◀

phase plane orbit

In previous chapters we wrote the solutions of a system of two equations as a pair $(x(t), y(t))$, where each of the solution functions depends on the independent variable t. A graph of such a solution would require three dimensions for t, x, and y. However, it is often of interest to treat t as a parameter and express the solution as a curve in the xy-plane. Then the xy-plane is called the **phase plane** of the system, and a curve that expresses the relation between x and y is called an **orbit** (or **phase portrait**) of the system. It is often possible to derive a great deal of information from an examination of the orbits of the system. It is much easier to do this for autonomous systems.

EXAMPLE 5 ▶ The Lotka-Volterra system

Assume that an ecosystem contains a predator species that feeds exclusively on a prey species and that the prey population has an ample food supply at all times. Let $y(t)$ and $x(t)$ denote the populations of the predator and prey species, respectively. Since food is readily available, the birth rate of the prey species is very likely to be a constant independent of time. The death rate, however, certainly depends on the number of predators.

On the other hand, the birth rate of the predator species is affected by the uncertain food supply, whereas its death rate may well be constant. Writing the growth rates per individual for the two species, we have

$$\frac{1}{x}\frac{dx}{dt} = \beta_1 + \delta_{12}y,$$
$$\frac{1}{y}\frac{dy}{dt} = \delta_{21}x + \beta_2,$$

or

$$\begin{aligned}\frac{dx}{dt} &= \beta_1 x + \delta_{12}xy,\\ \frac{dy}{dt} &= \beta_2 y + \delta_{21}xy,\end{aligned} \tag{8}$$

where β_1, $\delta_{21} > 0$ and δ_{12}, $\beta_2 < 0$ are constants of proportionality. System **(8)** is usually called the **Lotka-Volterra system** for the predator-prey model.

The model we have just described can be made more realistic by including terms for competition for resources. In this case the growth rate per individual in each species is also affected (by starvation) by the number of individuals of that species with whom it is competing for the resources (food, etc.). We obtain the system of equations (with δ_{11}, $\delta_{22} < 0$)

$$\begin{aligned}x' &= \beta_1 x + \delta_{11}x^2 + \delta_{12}xy,\\ y' &= \beta_2 y + \delta_{21}xy + \delta_{22}y^2.\end{aligned} \tag{9}$$

We will examine this system more closely in Section 10.3. At this point, we merely remark that an orbit of this system would give us very useful information. Increases or decreases in the population of one of the two species affect the population of the other species. An orbit graphically depicts this effect. Consider, for example, the orbit shown in Figure 10.3. We can derive a great deal of information from such an orbit. When $x = a$ units, an increase in x causes a decrease in y until $y = c$. Then both x and y increase until x reaches its maximum sustainable population of b units. The second population continues to rise until it reaches a population of d units, and so on. Moreover, both populations are periodic; that is, all population levels recur continually. Note that all this information can be obtained (if the orbits can be drawn) *without actually solving the system*! Note that this also works when the system is nonautonomous. ◀

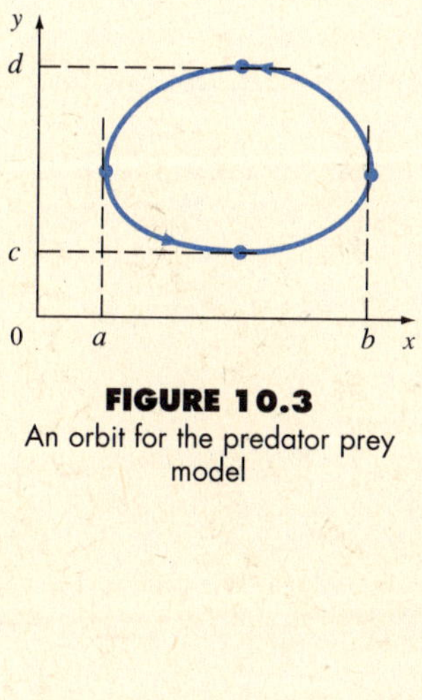

FIGURE 10.3
An orbit for the predator prey model

We illustrate the calculation of orbits in the next few examples of linear systems and return to the discussion of nonlinear systems (such as the Lokta-Volterra systems) in Section 10.3.

EXAMPLE 6 ▶ **The harmonic oscillator**

Consider the equation of the harmonic oscillator

$$x'' + x = 0,$$

with initial conditions $x(0) = 1$, $x'(0) = 0$. We may rewrite it as the autonomous system

$$\begin{aligned}x' &= y, & x(0) &= 1,\\ y' &= -x, & y(0) &= 0.\end{aligned}$$

There are two ways of finding the orbit. First, we observe that the unique solution of the initial-value problem is the solution $(\cos t, -\sin t)$. Since $\cos^2 t + \sin^2 t = 1$,

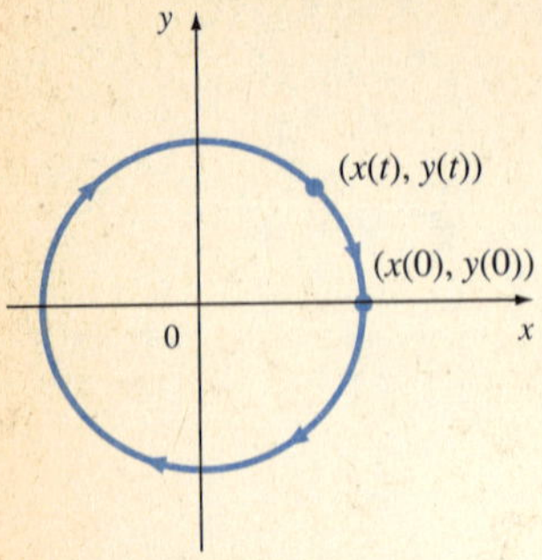

FIGURE 10.4
An orbit of the harmonic oscillator

the orbit satisfies the equation

$$x^2 + y^2 = 1,$$

which is the unit circle in the xy-plane (see Figure 10.4). The arrows in the figure indicate the direction in which the solutions move about the orbit as t increases. Note that as t increases, $\cos t$ (the x-coordinate) moves from 1 (when $t = 0$) to 0 (when $t = \frac{\pi}{2}$), to -1 (when $t = \pi$), to 0 (when $t = \frac{3\pi}{2}$), and back to 1 (at $t = 2\pi$). Similarly, $-\sin t$ (the y-coordinate) moves from 0 to -1, to 0, to $+1$, and back to 0. This phenomenon explains the direction indicated by the arrows in Figure 10.4. Therefore, starting at the point (1, 0) (corresponding to $t = 0$), x decreases as y decreases.

We can also find this orbit without solving the system. By the chain rule, we have

$$\frac{dy}{dx} = \frac{\frac{dy}{dt}}{\frac{dx}{dt}} = -\frac{x}{y}.$$

Separating the variables, we find that

$$y\,dy = -x\,dx,$$

or, after an integration,

$$x^2 + y^2 = C.$$

To evaluate the constant C, we note that at $t = 0$, $x = 1$ and $y = 0$, so that $C = x^2(0) + y^2(0) = 1$. (Thus the radius of the orbit depends on the initial conditions.)

Finally, to determine the direction on the orbit, notice that for $y > 0$ the function $x(t)$ increases, since $x' = y$, while $y(t)$ decreases when $x > 0$ because $y' = -x$. Thus the motion is clockwise around the orbit.

Now consider the harmonic oscillator with the new initial condition

$$x(t_0) = 1, \qquad y(t_0) = 0.$$

The unique solution pair is

$$(x(t), y(t)) = (\cos(t - t_0), -\sin(t - t_0)).$$

But then

$$x^2 + y^2 = \cos^2(t - t_0) + \sin^2(t - t_0) = 1,$$

which is the *same* orbit as above. In other words, *the orbit is independent of the initial value of t* [but not, of course, of the initial values $x(t_0)$ and $y(t_0)$]. This is a property shared by *all autonomous systems.* This fact is assumed for the remainder of this chapter. Its proof can be found in most advanced differential equations texts.* The property does not hold for nonautonomous systems, as is easily illustrated by the next example. ◀

*See, for example, H. K. Wilson, *Ordinary Differential Equations* (Reading, Mass.: Addison-Wesley, 1972).

EXAMPLE 7 ▶ **A nonautonomous system may have crossing orbits**

Consider the nonautonomous system

$$x' = \frac{1}{t}x, \qquad x(t_0) = 1,$$

$$y' = y, \qquad y(t_0) = 2.$$

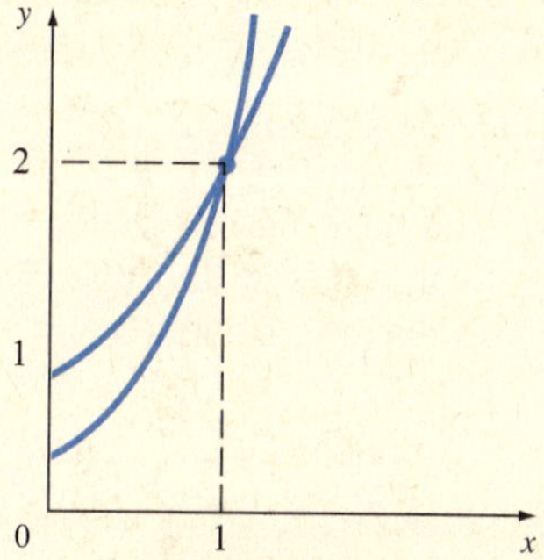

FIGURE 10.5
Orbits of a nonautonomous system

Since these equations are uncoupled, it is easy to calculate the solution pair

$$(x(t), y(t)) = \left(\frac{t}{t_0}, 2e^{t-t_0}\right).$$

Since $t = t_0x$, the orbit is

$$y = 2e^{t_0(x-1)}.$$

Thus different values of the initial value of t lead to different orbits. Figure 10.5 shows the orbits for $t_0 = 1$, $t_0 = 2$. Note that the two orbits intersect at the point $(1, 2)$. Such intersections are impossible for autonomous systems (see Problem 10). ◀

We deal with autonomous systems exclusively in the remainder of this chapter. As we saw in Example 2, autonomous systems can and do arise naturally. A discussion of the properties of nonautonomous systems is more complicated (one reason being the necessity to worry about the initial value of t). Such discussions can be found in many intermediate and advanced textbooks on differential equations.

SELF-QUIZ

I. Which of the following systems are autonomous?

(a) $x' = x + \sin^2 y$
$y' = 17\,e^x + y^2$

(b) $x' = t + y$
$y' = x - y$

(c) $x' = x + ty$
$y' = x - y$

(d) $x' = 17x + y^2$
$y' = x^3 - \pi y$

Answer True or False.

II. If $x' = f(x, y)$, $y' = g(x, y)$ is an autonomous system, then two of its orbits may intersect.

Answer to Self-Quiz

I. a, d, **II.** False

PROBLEMS 10.1

In each of Problems 1–6, (a) find a related linear equation (as in Examples 2 and 3), (b) find all constant solutions of the nonlinear equation, (c) solve the nonlinear equation, and (d) determine the behavior of the solutions of the nonlinear equation for values of $x(0)$ in the indicated range.

1. $x' = x - x^2$, $-\infty < x(0) < \infty$

2. $x' = -2x + 3x^2$, $x(0) \geq 0$

3. $x' = 2x + 3x^2$, $x(0) \geq 0$

4. $x' = 2x - 3x^2$, $-\infty < x(0) < \infty$

5. $x' = x(x - 1)(x - 2)$, $x(0) \geq 0$

6. $x' = -x(x - 1)(x - 2)$, $x(0) \geq 0$

7. **(a)** Draw the orbits for the initial-value problem

$$x' = y \qquad x(0) = a,$$

$$y' = -x, \qquad y(0) = b.$$

(b) Show that these orbits are identical to those for the same system with the initial conditions

$$x(t_0) = a, \qquad y(t_0) = b.$$

8. Show that the orbits for the equation

$$x'' + \omega^2 x = 0$$

are ellipses centered at the origin.

9. Find the orbits for the system

$$x' = tx, \qquad x(t_0) = 1,$$
$$y' = -y, \qquad y(t_0) = 1,$$

and graph these orbits for $t_0 = 0$, $t_0 = 1$ and $t_0 = 2$.

10. Suppose that the differential equations $x' = f(x, y)$ and $y' = g(x, y)$ have a unique solution whenever an initial condition is given for each variable. Show that no two orbits of the autonomous system

$$x' = f(x, y), \qquad y' = g(x, y) \qquad \textbf{(i)}$$

can ever intersect.

11. Use the result of Problem 10 to show that if (x_0, y_0) is a point having the property that $f(x_0, y_0) = g(x_0, y_0) = 0$ and if $(x(t), y(t))$ is a solution pair of the system **(i)** such that $f(x(t), y(t)) \neq 0$ for some value of t, then there is no value of t for which $(x(t), y(t)) = (x_0, y_0)$.

10.2 CRITICAL POINTS, STABILITY, AND PHASE PORTRAITS FOR LINEAR SYSTEMS

The general autonomous system of two first-order equations is given by

$$\begin{aligned} x' &= f(x, y), \\ y' &= g(x, y). \end{aligned} \qquad \textbf{(1)}$$

Since orbits are independent of t_0, we assume that $t_0 = 0$.

critical point

A point (x_0, y_0) is called a **critical point** of system **(1)** if

$$f(x_0, y_0) = g(x_0, y_0) = 0.$$

Any critical point (x_0, y_0) is a constant or **steady state** solution of **(1),** since the derivative of a constant is zero:

$$x_0' = 0 = f(x_0, y_0),$$
$$y_0' = 0 = g(x_0, y_0).$$

A critical point (x_0, y_0) of system **(1)** is a point of **equilibrium,** since once at this point we can never leave it,* the derivatives of both $x(t)$ and $y(t)$ being zero there. Physically, a critical point is often a point at which the potential energy is at a minimum. For instance, in Example 10.1.2, if we use the substitution $\mu = \frac{d\theta}{dt}$, we can write Equation **(10.1.1)** as

$$\theta' = \mu,$$
$$\mu' = -\omega^2 \sin\theta.$$

From Figure 10.1 it is clear that the potential energy is a minimum when $\theta = 0$. The point $(0, 0)$ is a point of equilibrium of the system. Of course there are other critical points (see the next example). This situation will be discussed in great detail in Example 10.3.4.

EXAMPLE 1 ▶ Critical points of a system

Consider the system

$$x' = y,$$
$$y' = -\omega^2 \sin x,$$

which was just obtained from Equation **(10.1.1).** This system has infinitely many critical points, since $(k\pi, 0)$ is critical for all integers k. ◀

---

*Actually, you never reach an equilibrium point unless you start there. You may, as the following examples illustrate, *approach* an equilibrium point as $t \to \infty$.

EXAMPLE 2 ▶ **Critical points**

Consider the system

$$x' = -x^2 + y,$$
$$y' = x - y^2.$$

The two critical points are (0, 0) and (1, 1). (Why?) ◀

The notion of stability is central to any discussion of the behavior of differential equations. Roughly, a solution $\phi(t)$ to a system of equations is **stable** if, whenever we start "close" to $\phi(t)$, we stay close to $\phi(t)$ for all future values of t. It is **asymptotically stable** if it is stable and the solutions that start close to $\phi(t)$ approach $\phi(t)$ as t tends to ∞. Finally, $\phi(t)$ is **unstable** if it is not stable.

We already saw an example of a system that is stable. In Example 10.1.6 the orbits (see Figure 10.4) were circles centered at the origin of radius $\sqrt{x^2(0) + y^2(0)}$. If the initial conditions are changed by a small amount, the radius of the circular orbit is changed by a small amount, and thus the new orbit stays close to the original one.

To define "closeness" more precisely, we make use of the Pythagorean distance between two points in the plane. If (x_1, y_1) and (x_2, y_2) are the two points, then the distance between them is

$$d = [(x_1 - x_2)^2 + (y_1 - y_2)^2]^{1/2}.$$

We denote by $(x(t, x^*, y^*), y(t, x^*, y^*))$ the unique solution pair to system **(1)** that satisfies the initial conditions

$$x(0) = x^*, \qquad y(0) = y^*.$$

Now we can give formal definitions of the above concepts.

Stability

The constant solution (or critical point) (x_0, y_0) is said to be

1. **stable** if for every number $\epsilon > 0$ there is a number $\delta > 0$ such that whenever

$$[(x_0 - x^*)^2 + (y_0 - y^*)^2]^{1/2} < \delta,$$

we have

$$\{[x_0 - x(t, x^*, y^*)]^2 + [y_0 - y(t, x^*, y^*)]^2\}^{1/2} < \epsilon,$$

for all $t \geq 0$ (that is, if you start close, you stay close);

2. **asymptotically stable** if it is stable and there exists a number $A > 0$ such that whenever

$$[(x_0 - x^*)^2 + (y_0 - y^*)^2]^{1/2} < A,$$

we have

$$\lim_{t \to \infty}\{[x_0 - x(t, x^*, y^*)]^2 + [y_0 - y(t, x^*, y^*)]^2\} = 0$$

(that is, if you start close enough, the solution approaches the critical point as t tends to ∞);

3. **unstable** if it is not stable (that is, no matter how close to the constant solution you start, there are solutions that move away from the constant solution).

The requirement that the solution be stable is part of the definition of asymptotic stability. There are examples, which we do not cite here, of nonlinear systems for which all solutions tend to zero but which first get very large, no matter how close to zero they start. The zero solutions for such systems are unstable.

The foregoing concepts are illustrated in Figure 10.6, where the critical point is taken to be the origin.

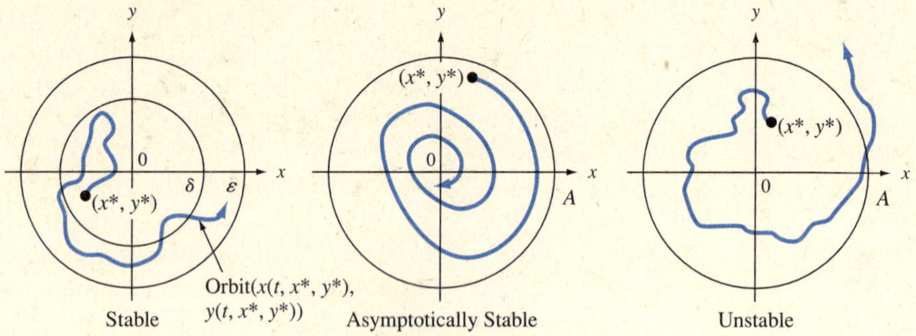

FIGURE 10.6
Illustration of three possibilities for stability

The situation for linear systems is easier: **For a linear systems, the critical point (x_0, y_0) is asymptotically stable whenever**

$$\lim_{t\to\infty} \{[x_0 - x(t, x^*, y^*)]^2 + [y_0 - y(t, x^*, y^*)]^2\} = 0;$$

that is, this limit suffices to prove both stability and asymptotic stability. We prove this fact in Theorem 1 (p. 498).

EXAMPLE 3 ▶ **A stable solution**

Consider again the harmonic oscillator

$$x' = y, \qquad y' = -x.$$

Clearly (0, 0) is a constant solution. The solution pair with the initial conditions $x(0) = x^*$, $y(0) = y^*$ is

$$(x(t, x^*, y^*), y(t, x^*, y^*)) = (x^*\cos t + y^*\sin t, -x^*\sin t + y^*\cos t),$$

so

$$[x(t, x^*, y^*) - 0]^2 + [y(t, x^*, y^*) - 0]^2$$
$$= (x^*\cos t + y^*\sin t)^2 + (-x^*\sin t + y^*\cos t)^2 = x^{*2} + y^{*2}.$$

Thus $\delta = \epsilon$ satisfies the definition of stability and the zero solution is stable. Note that the zero solution is not asymptotically stable. ◀

EXAMPLE 4 ▶ **An asymptotically stable solution**

Consider the system

$$x' = y,$$
$$y' = -2x - 3y.$$

Again (0, 0) is a critical point. The solution pair that satisfies $x(0) = x^*$, $y(0) = y^*$ is given by

$$(x(t, x^*, y^*), y(t, x^*, y^*))$$
$$= ((2x^* + y^*)e^{-t} - (x^* + y^*)e^{-2t}, -(2x^* + y^*)e^{-t} + 2(x^* + y^*)e^{-2t}),$$

which tends to zero as t tends to ∞ no matter how large x^* and y^* are. Hence the zero solution is asymptotically stable. As in this example, any situation in which the number A in the definition of asymptotic stability can be arbitrarily large yields the **global asymptotic stability** of the critical point. ◀

EXAMPLE 5 ▶ **An unstable solution**

The system

$$x' = y,$$
$$y' = -2x + 3y$$

has (0, 0) as a critical point. However, the zero solution is unstable, since for any $x^* \neq 0$ and $y^* \neq 0$ (no matter how small), the solution pair is

$$(x(t, x^*, y^*), y(t, x^*, y^*))$$
$$= ((2x^* - y^*)e^t + (y^* - x^*)e^{2t}, (2x^* - y^*)e^t + 2(y^* - x^*)e^{2t}),$$

which becomes arbitrarily large as t tends to ∞. ◀

The three examples considered above are all linear. The analysis of stability properties for nonlinear systems is much more difficult, since, in general, solutions to such systems cannot be found. We will consider a class of nonlinear systems in the next section, but in the remainder of this section we classify all possible linear systems* so that we may have some basis of comparison when we get to the nonlinear ones.

The linear systems we consider is

$$\begin{aligned} x' &= a_{11}x + a_{12}y, \\ & \qquad\qquad a_{11}a_{22} - a_{12}a_{21} \neq 0, \\ y' &= a_{21}x + a_{22}y, \end{aligned} \tag{2}$$

where the coefficients a_{ij} are real constants. As in Chapter 7, we derive the characteristic equation of the system

$$\lambda^2 - (a_{11} + a_{22})\lambda + (a_{11}a_{22} - a_{21}a_{12}) = 0, \tag{3}$$

with the roots λ_1 and λ_2.† The orbits of system **(2)** depend on the nature of these two roots. We therefore consider each case separately. We note that (0, 0) is the only constant solution (critical point) of the system (why?), so we restrict our attention to the nature of the orbits around the origin. Moreover, none of the roots of the characteristic equation **(3)** are zero.

In the following discussion we examine the different possibilities for the roots of **(3)** and draw representative orbits of the system for each case. Thus, merely knowing the roots of the characteristic equation is enough to determine the nature of the orbits.

---

* Except those in which at least one of the roots of the characteristic equation **(3)** is zero.

† For those of you who covered the material in Section 9.4, λ_1 and λ_2 are the eigenvalues of the matrix

$$A = \begin{pmatrix} a_{11} & a_{12} \\ a_{21} & a_{22} \end{pmatrix}.$$

Although we avoid the eigenvalue terminology to make this chapter accessible to students who have not covered all of Chapter 9, the material *should* be regarded in terms of eigenvalues. Note, too, that $\det A = a_{11}a_{22} - a_{12}a_{21} \neq 0$ so 0 is not an eigenvalue of A. This means that the system **(2)** does not have any constant solutions other than (0, 0).

Case 1: λ_1 and λ_2 are real, distinct, and of the same sign

We may assume, for simplicity, that $\lambda_1 > \lambda_2$. Then each solution pair has the form

$$(x(t), y(t)) = (c_1\alpha_1 e^{\lambda_1 t} + c_2\alpha_2 e^{\lambda_2 t}, c_1\beta_1 e^{\lambda_1 t} + c_2\beta_2 e^{\lambda_2 t}), \tag{4}$$

where c_1 and c_2 are arbitrary.*

Case 1(a) $\boldsymbol{\lambda_2 < \lambda_1 < 0}$ (both roots are negative). Clearly all solutions tend to $(0, 0)$ as t tends to ∞. First we assume that $c_1 = 0$ and $c_2 \neq 0$. Then $y = (\beta_2/\alpha_2)x$, which means that the orbit is a straight line with slope β_2/α_2. If $c_1 \neq 0$ and $c_2 = 0$, then the situation is similar and we obtain the line $y = (\beta_1/\alpha_1)x$. To obtain the other orbits, we assume that c_1 and c_2 are both nonzero. Then

$$\frac{y(t)}{x(t)} = \frac{c_1\beta_1 e^{\lambda_1 t} + c_2\beta_2 e^{\lambda_2 t}}{c_1\alpha_1 e^{\lambda_1 t} + c_2\alpha_2 e^{\lambda_2 t}},$$

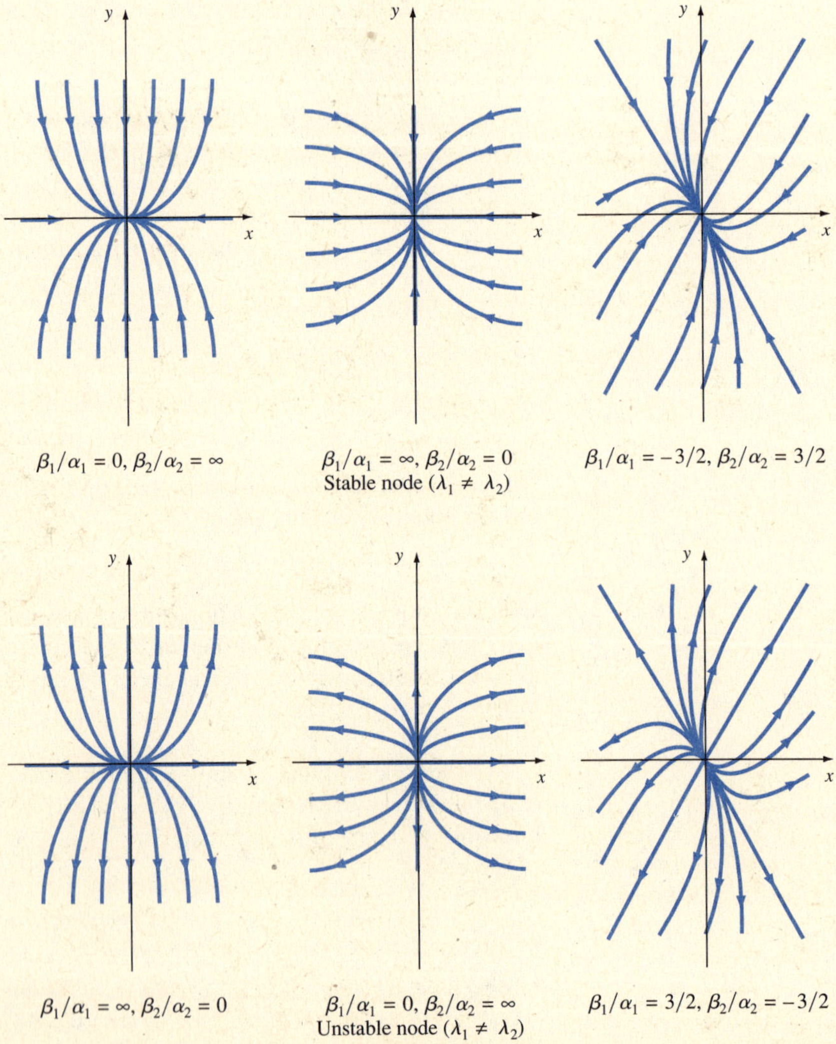

FIGURE 10.7
Behavior at six nodes

---

*The coefficients α_1, α_2, β_1, and β_2 are related: if $a_{12} \neq 0$, then

$$\frac{\beta_1}{\alpha_1} = \frac{\lambda_1 - a_{11}}{a_{12}} \quad \text{and} \quad \frac{\beta_2}{\alpha_2} = \frac{\lambda_2 - a_{11}}{a_{12}}.$$

and dividing the numerator and denominator by $e^{\lambda_1 t}$, we have

$$\frac{y(t)}{x(t)} = \frac{c_1\beta_1 + c_2\beta_2 e^{(\lambda_2-\lambda_1)t}}{c_1\alpha_1 + c_2\alpha_2 e^{(\lambda_2-\lambda_1)t}} \to \frac{c_1\beta_1}{c_1\alpha_1} = \frac{\beta_1}{\alpha_1}$$

as t tends to ∞. Thus these orbits approach the origin with the slope β_1/α_1. Similarly, as t tends to $-\infty$, all solutions but two become asymptotic to lines with slope β_2/α_2. Thus situation is illustrated in the top three graphs in Figure 10.7 for three different values of the slopes β_1/α_1 and β_2/α_2. It is clear that in this case the zero solution is asymptotically stable. Here the origin is called a **stable node.**

Case 1(b) $\boldsymbol{\lambda_1 > \lambda_2 > 0}$ (both roots are positive). Then all solutions (except the zero solution) approach ∞ as t tends to ∞. Hence the zero solution is unstable. The orbits are the same as in the previous case except that the direction of motion is reversed. The origin here is called an **unstable node.** As t tends to $-\infty$, all but two orbits approach zero with the slope β_2/α_2, and as t tends to ∞, all but two orbits become asymptotic to lines with slope β_1/α_1 (see the bottom graphs in Figure 10.7).

Case 2: λ_1 and λ_2 are real with opposite signs

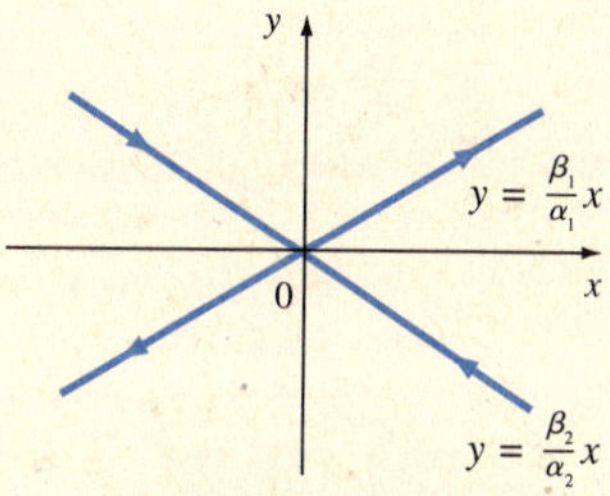

FIGURE 10.8
Four orbits into the critical point

We assume that $\lambda_1 > 0 > \lambda_2$. The situation here is very different from that of case 1. If $c_1 = 0$ and $c_2 \neq 0$, we obtain, as before,

$$\frac{y}{x} = \frac{\beta_2}{\alpha_2} \quad \text{or} \quad y = \frac{\beta_2}{\alpha_2}x.$$

As t tends to ∞, $x(t)$ and $y(t)$ approach zero. If $c_1 \neq 0$ and $c_2 = 0$, then $y = (\beta_1/\alpha_1)x$ and both x and y approach ∞ as t tends to ∞ and approach zero as t tends to $-\infty$. These four orbits are sketched in Figure 10.8. When both c_1 and c_2 are nonzero, the situation is more complicated. Again

$$\frac{y(t)}{x(t)} = \frac{c_1\beta_1 e^{\lambda_1 t} + c_2\beta_2 e^{\lambda_2 t}}{c_1\alpha_1 e^{\lambda_1 t} + c_2\alpha_2 e^{\lambda_2 t}} = \frac{c_1\beta_1 + c_2\beta_2 e^{(\lambda_2-\lambda_1)t}}{c_1\alpha_1 + c_2\alpha_2 e^{(\lambda_2-\lambda_1)t}},$$

which approaches β_1/α_1 as t tends to ∞. Hence all orbits are asymptotic to the line $y = (\beta_1/\alpha_1)x$ as t tends to ∞. Also

$$\frac{y(t)}{x(t)} = \frac{c_1\beta_1 e^{(\lambda_1-\lambda_2)t} + c_2\beta_2}{c_1\alpha_1 e^{(\lambda_1-\lambda_2)t} + c_2\alpha_2},$$

which approaches β_2/α_2 as t tends to $-\infty$. Hence all orbits are asymptotic to the line $y = (\beta_2/\alpha_2)x$ as t tends to $-\infty$. Finally, we observe that both $x(t)$ and $y(t)$ approach ∞ as t tends to $\pm\infty$, and, by uniqueness, no orbit can pass through the origin. The orbits are therefore as shown in Figure 10.9.

It is clear here that the origin is unstable. In this situation, the origin is called (for obvious reasons) a **saddle point.** We note that a saddle point has the property that exactly two orbits approach the origin and all others are "repelled" by it. The physical behavior corresponding to a saddle point is illustrated in Example 10.3.4.

Case 3: $\lambda_1 = \lambda_2 = \lambda$

Here either $\lambda < 0$ or $\lambda > 0$.

Case 3(a): $\boldsymbol{\lambda < 0}$ There are two ways in which the characteristic equation **(3)** can yield a double root. One possibility is

$$a_{11} = a_{22} \neq 0, \qquad a_{21} = a_{12} = 0. \tag{5}$$

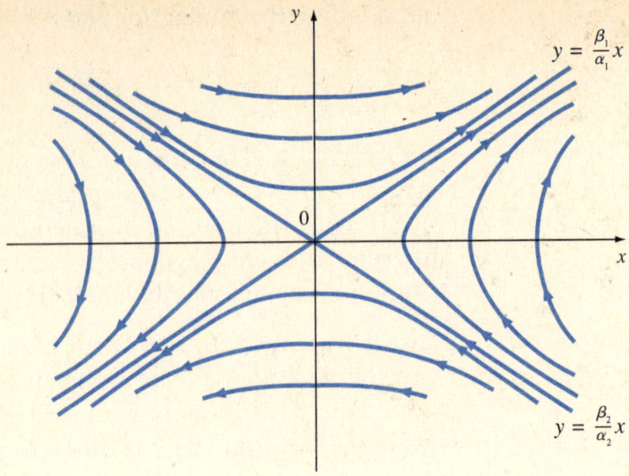

Saddle point

FIGURE 10.9

Then the characteristic equation is

$$\lambda^2 - 2a_{11}\lambda + a_{11}^2 = 0,$$

so $\lambda = a_{11}$ is the double root. Then systems **(2)** becomes

$$x' = \lambda x,$$
$$y' = \lambda y.$$

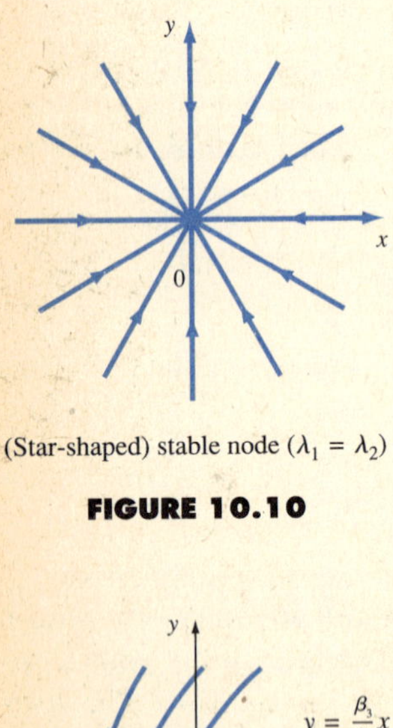

(Star-shaped) stable node ($\lambda_1 = \lambda_2$)

FIGURE 10.10

The solutions are obviously of the form

$$(x(t), y(t)) = (c_1 e^{\lambda t}, c_2 e^{\lambda t}),$$

so

$$\frac{y}{x} = \frac{c_2}{c_1} \quad \text{or} \quad y = \frac{c_2}{c_1}x.$$

Thus all orbits are straight lines with the slope c_2/c_1. Since $\lambda < 0$, all solutions approach zero as t tends to ∞, and the zero solution is asymptotically stable. The situation is graphed in Figure 10.10. The origin in this case is also called a node. Sometimes the nodes shown in Figure 10.7 are called **improper nodes,** whereas the node in Figure 10.10 is called a **proper node.** We will not use this terminology. Also, we should add that the node of Figure 10.10 is sometimes called a **star-shaped node.**

If $\lambda < 0$ is a double root but equalities **(5)** do not hold, then the equations are coupled and the general solution is, according to Equation **(7.3.21**, p. 327),

$$(x(t), y(t)) = ([c_1\alpha_1 + c_2(\alpha_2 + \alpha_3 t)]e^{\lambda t}, [c_1\beta_1 + c_2(\beta_2 + \beta_3 t)]e^{\lambda t}). \tag{6}$$

Stable node ($\lambda_1 = \lambda_2$)

FIGURE 10.11

Then

$$\frac{y}{x} = \frac{c_1\beta_1 + c_2\beta_2 + c_2\beta_3 t}{c_1\alpha_1 + c_2\alpha_2 + c_2\alpha_3 t} = \frac{\dfrac{c_1\beta_1}{t} + \dfrac{c_2\beta_2}{t} + c_2\beta_3}{\dfrac{c_1\alpha_1}{t} + \dfrac{c_2\alpha_2}{t} + c_2\alpha_3},$$

which approaches β_3/α_3 as t tends to $\pm\infty$. Since both $x(t)$ and $y(t)$ approach zero as t tends to ∞, the zero solution is asymptotically stable. Also, all orbits are asymptotic to the line $y = (\beta_3/\alpha_3)x$ as t tends to $\pm\infty$, as illustrated in Figure 10.11.

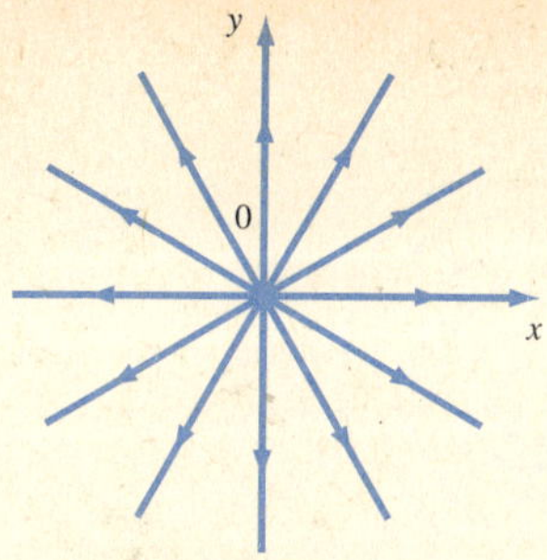

(Star-shaped) unstable node ($\lambda_1 = \lambda_2$)

FIGURE 10.12

Case 3(b): $\boldsymbol{\lambda > 0}$ The orbits are the same but with the arrows reversed, since now all solutions approach ∞ as t tends to ∞ (see Figures 10.12 and 10.13.) In these two cases the origin is unstable. Cases 3(a) and 3(b) provide another situation in which nodes arise.

Case 4: λ_1 and λ_2 are complex conjugates but not pure imaginary

Then $\lambda_1 = a + ib$ and $\lambda_2 = a - ib$, where neither a nor b is zero.

Case 4(a): $\boldsymbol{a < 0}$ According to Equation **(7.3.22),** p. 327, all solutions have the form

$$(x(t), y(t)) = (e^{at}[c_1(A_1\cos bt - A_2\sin bt) + c_2(A_1\sin bt + A_2\cos bt)], \\ e^{at}[c_1(B_1\cos bt - B_2\sin bt) + c_2(B_1\sin bt + B_2\cos bt)]). \quad \textbf{(7)}$$

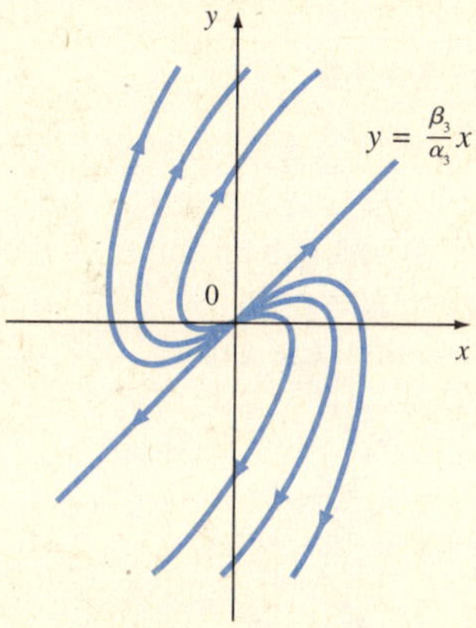

Unstable node ($\lambda_1 = \lambda_2$)

FIGURE 10.13

To simplify the notation, we define

$$k_1 = c_1A_1 + c_2A_2, \qquad k_2 = -c_1A_2 + c_2A_1,$$
$$k_3 = c_1B_1 + c_2B_2, \qquad k_4 = -c_1B_2 + c_2B_1.$$

Equation **(7)** becomes

$$(x(t), y(t)) = (e^{et}(k_1\cos bt + k_2\sin bt), e^{at}(k_3\cos bt + k_4\sin bt)). \quad \textbf{(8)}$$

We now define $A = \sqrt{k_1^2 + k_2^2}$ and $B = \sqrt{k_3^2 + k_4^2}$. Then we define α_1 and α_2 by

$$\cos\alpha_1 = \frac{k_1}{A}, \qquad \cos\alpha_2 = \frac{k_3}{B},$$
$$\sin\alpha_1 = -\frac{k_2}{A}, \qquad \sin\alpha_2 = -\frac{k_4}{B}, \quad \textbf{(9)}$$

so $k_1 = A\cos\alpha_1,\ k_2 = -A\sin\alpha_1,\ k_3 = B\cos\alpha_2,\ \ k_4 = -B\sin\alpha_4$, and

$$(x(t), y(t) = (Ae^{at}(\cos\alpha_1\cos bt - \sin\alpha_1\sin bt), \\ Be^{at}(\cos\alpha_2\cos bt - \sin\alpha_2\sin bt)) \quad \textbf{(10)} \\ = (Ae^{at}\cos(bt + \alpha_1), Be^{at}\cos(bt + \alpha_2)).$$

Then

$$\frac{y}{x} = \frac{B\cos(bt + \alpha_1)}{A\cos(bt + \alpha_2)},$$

which is defined whenever $\cos(bt + \alpha_2) \neq 0$. Since this expression is periodic, it is clear that as t tends to ∞, the ratio $\frac{y}{x}$ does not approach a limit but the orbits must circle around the origin. Since $a < 0$, $x(t)$ and $y(t)$ approach zero as t tends to ∞. Hence the orbits must spiral in toward the origin (see Figure 10.14). The zero solution is asymptotically stable, and the origin is called a **stable focus** (or **spiral point**).

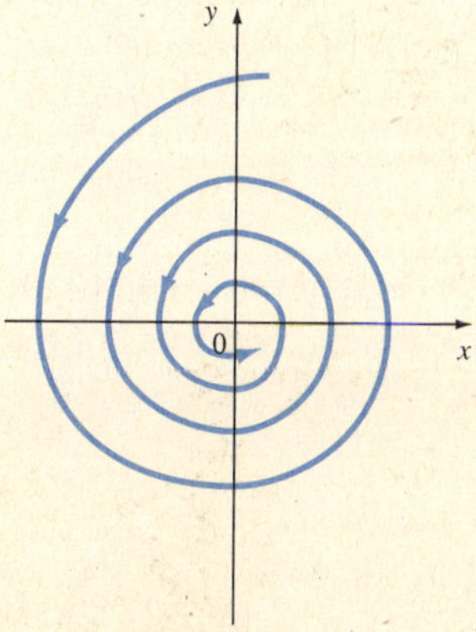

Stable focus (for $b > 0$)

FIGURE 10.14

Case 4(b): $\boldsymbol{a > 0}$ Here the analysis is as before except that all solutions approach ∞ as t tends to ∞ (see Figure 10.15). The origin is unstable.

Case 5: λ_1 and λ_2 are pure imaginary

Here $\lambda_1 = ib$ and $\lambda_2 = -ib$, and we can use the same analysis as above with $a = 0$: We have

$$(x(t), y(t)) = (A\cos(bt + \alpha_1), B\cos(bt + \alpha_2)). \quad \textbf{(11)}$$

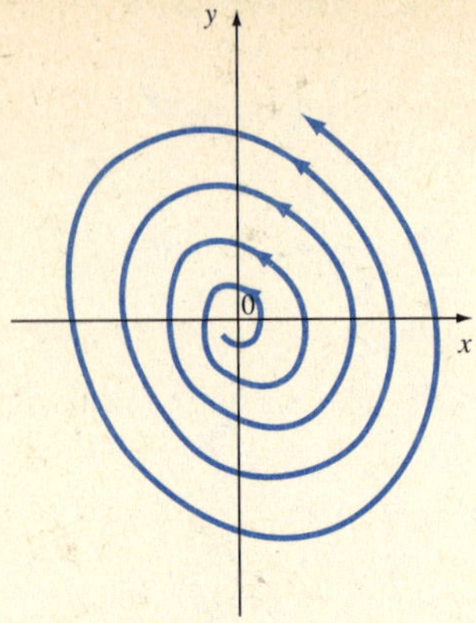

Unstable focus (for $b > 0$)

FIGURE 10.15

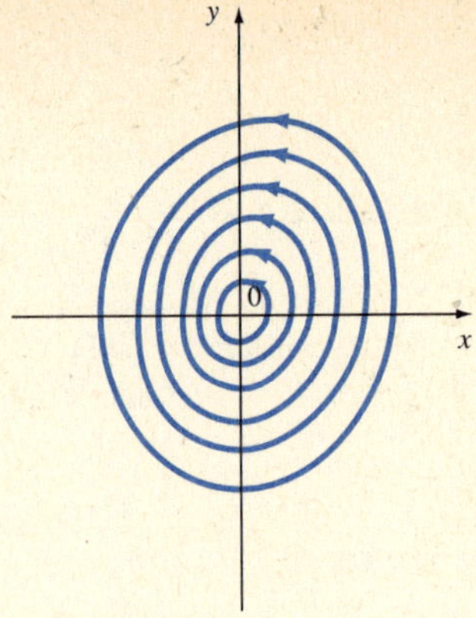

Center (for $b > 0$)

FIGURE 10.16

Clearly $x(t)$ and $y(t)$ are periodic with period $\frac{2\pi}{b}$, so every orbit beginning at the point (x^*, y^*) when $t = t^*$ returns to the *same point* when $t = t^* + \frac{2\pi}{b}$. Thus the orbits are closed curves. To get a feeling for the nature of these curves, we set $k_2 = k_3 = 0$ in Equation **(8),** so

$$x(t) = k_1 \cos bt \quad \text{and} \quad y(t) = k_4 \sin bt.$$

Then

$$\frac{x^2}{k_1^2} + \frac{y^2}{k_4^2} = \cos^2 bt + \sin^2 bt = 1,$$

which is the equation of an ellipse centered about (0, 0) with the x-axis as its axis of symmetry. If k_2 and k_3 are nonzero, we obtain ellipses that have a rotated axis of symmetry. In this situation the zero solution is stable but not asymptotically stable, since solutions do not approach zero. Then the origin is called a **center** (see Figure 10.16).

The entire preceding analysis is summarized in the theorem below.

THEOREM 1 ♦ Eigenvalues determine the type of critical point

Consider the system

$$x' = a_{11}x + a_{12}y,$$
$$y' = a_{21}x + a_{22}y,$$

where the a_{ij} are real constants and $a_{11}a_{22} - a_{12}a_{21} \neq 0$, so that the origin (0, 0) is the only critical point (see Problem 11). Let λ_1 and λ_2 be the two roots of the characteristic equation **(3).** Then

1. the origin is stable if λ_1 and λ_2 are pure imaginary;
2. the origin is asymptotically stable if Re $\lambda_1 < 0$ and Re $\lambda_2 < 0$;
3. the origin is unstable in all other cases. Moreover, the behavior of the orbits near the origin is as indicated in Table 10.1.

TABLE 10.1

λ_1, λ_2	*Type of Critical Point*
real, distinct, negative	stable node
real, distinct, positive	unstable node
real, distinct, opposite signs	saddle point (unstable)
real, equal, negative	stable node
real, equal, positive	unstable node
complex conjugate, not pure imaginary, negative real parts	stable focus
complex conjugate, not pure imaginary, positive real parts	unstable focus
pure imaginary	center (stable)

◆

EXAMPLE 6 ▶ **A center**

Consider the system

$$x' = x - 3y,$$
$$y' = x - y.$$

The characteristic equation is $\lambda^2 + 2 = 0$ with the roots $\pm\sqrt{2}\,i$. Therefore the zero solution is stable and the origin is a center. ◀

EXAMPLE 7 ▶ **A saddle point**

Consider the system

$$x' = 4x - y,$$
$$y' = 6x - 3y.$$

The characteristic equation is $\lambda^2 - \lambda - 6 = 0$ with the roots 3 and -2. Hence the origin is a saddle point and the zero solution is unstable. ◀

EXAMPLE 8 ▶ **A stable node**

Consider the system

$$x' = -4x - y,$$
$$y' = x - 2y.$$

The characteristic equation is $\lambda^2 + 6\lambda + 9 = 0$ with the double root $\lambda = -3$. Hence the origin is a stable node and the zero solution is asymptotically stable. ◀

EXAMPLE 9 ▶ **An RLC circuit**

Consider the *RLC* circuit shown in Figure 10.17. If $E = 0$, then from Section 4.4 we obtain the differential equation

$$\frac{d^2I}{dt^2} + \frac{R}{L}\frac{dI}{dt} + \frac{I}{CL} = 0 \tag{12}$$

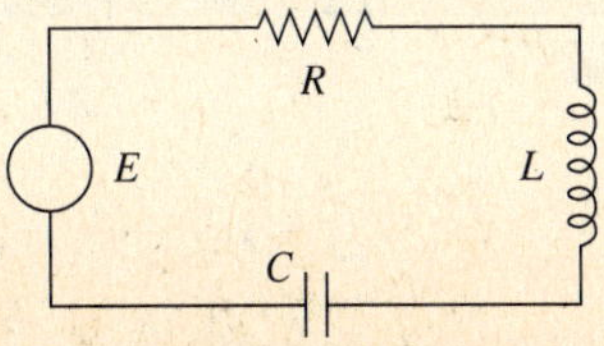

FIGURE 10.17
An RLC circuit

for the description of this circuit. Writing the equation as a system, we obtain

$$I' = y,$$
$$y' = -\frac{1}{CL}I - \frac{R}{L}y.$$

The characteristic equation is

$$\lambda^2 + \frac{R}{L}\lambda + \frac{1}{CL} = 0$$

with the roots

$$\lambda_1 = \frac{-R + \sqrt{R^2 - \dfrac{4L}{C}}}{2L} \quad \text{and} \quad \lambda_2 = \frac{-R - \sqrt{R^2 - \dfrac{4L}{C}}}{2L}. \tag{13}$$

There are three cases to consider, according to whether $R^2 - 4L/C$ is greater than zero, less than zero, or equal to zero.

1. If $R^2 - \frac{4L}{C} > 0$, then λ_1 and λ_2 are real, distinct, and negative, so the origin is a stable node.
2. If $R^2 - \frac{4L}{C} = 0$, then $\lambda_1 = \lambda_2 = -\frac{R}{2L}$, so the zero solution is again a stable node.
3. If $R^2 - \frac{4L}{C} < 0$, then λ_1 and λ_2 are complex conjugates with negative real parts, so the origin is a stable focus.

In all three cases the critical point (0, 0) is asymptotically stable, and the transient current tends to zero as t approaches infinity. Of course, the emf E is generally nonzero, but it affects only the steady-state current. By superposition, the current $I(t)$ is the sum of the steady-state and transient currents. ◀

SELF-QUIZ

I. The critical point of the system

$$x' = x - y + 1$$
$$y' = y - 2$$

is ________.

(a) $(-1, 2)$ **(b)** $(1, 2)$
(c) $(1, -2)$ **(d)** $(-1, -2)$

II. Which of the following are critical points of the system

$$x' = x(7 - x - y),$$
$$y' = y(8 + 2x - y)?$$

(a) $(0, 0)$ **(b)** $(7, 0)$
(c) $(0, 8)$ **(d)** $(-4, 0)$

III. Suppose a linear system has a critical point at (0, 0) with the characteristic equation $(\lambda - 3)(\lambda + 4) = 0$. Then the origin is a(n) ________.

(a) stable focus **(b)** unstable node
(c) center **(d)** saddle point

IV. What type of critical point does the system

$$x' = x - 5y$$
$$y' = x - y$$

have at the origin?

(a) stable focus **(b)** unstable focus
(c) center **(d)** saddle point

Answers to Self-Quiz

I. b **II.** a, b, c **III.** d **IV.** c

PROBLEMS 10.2

In Problems 1–8 describe the nature of the critical point (0, 0) *of each system and sketch the orbits.*

1. $x' = 4x - 3y,$ $y' = 5x - 4y$

2. $x' = -2x + 3y,$ $y' = x - 3y$

3. $x' = -x + y,$ $y' = -5x + 3y$

4. $x' = x + y$ $y' = -x - 3y$

5. $x' = -4x - 2y,$ $y' = 5x + 2y$

6. $x' = 4x - 3y,$ $y' = 8x + 5y$

7. $x' = 2x + y,$ $y' = -4x + 2y$

8. $x' = 2x$ $y' = 2y$

9. Consider the system

$$x' = a_{11}x + a_{12}y,$$
$$y' = a_{21}x + a_{22}y.$$

Let $T = a_{11} + a_{22}$, $D = a_{11}a_{22} - a_{12}a_{21}$, and $C = T^2 - 4D$. Show that the origin is
(a) a node if $D > 0$ and $C \geq 0$;
(b) a saddle point if $D < 0$;
(c) a focus if $T \neq 0$ and $C < 0$;
(d) a center if $T = 0$ and $D > 0$.

10. Show that the zero solution in Problem 9 is
(a) asymptotically stable if $D > 0$ and $T < 0$;
(b) stable if $T = 0$ and $D > 0$;
(c) unstable if $T > 0$ or $D < 0$.

---

11. Show that (0, 0) is the only critical point of system **(2)** if and only if $\lambda = 0$ is *not* a root of the characteristic equation **(3)**. [*Hint:* See Appendix 4.]

12. Let $L = 2$ henrys, $R = 50$ ohms, and $C = 0.0003$ farad. Locate graphically the position on the orbit of Equation **(12)** at $t = 1$ if $I(0) = 0$ and $I'(0)$ is one of the values 10, 100, and 1000.

13. Let $L = 0.44$ henry, $R = 1200$ ohms, and $C = 10^{-6}$ farad. Locate graphically the position of the solution on the orbit at $t = 1$ if $I(0) = 0$ and $I'(0)$ is one of the values 10, 100, and 1000.

14. Consider the mass-spring system that is damped such that Equation **(4.2.2)** holds; that is,

$$x'' + \frac{c}{m}x' + \frac{k}{m}x = 0.$$

Following the analysis of Example 9, describe the behavior of the orbits near the origin for different positive values of m, c, and k.

---

15. Consider the system

$$x' = 0,$$
$$y' = -x + y.$$

(a) Solve the system.
(b) Show that $\lambda_1 = 0$ and $\lambda_2 = 1$ are the roots of the characteristic equation **(3)**.
(c) Find all the critical points of the system.
(d) Show that all the orbits are straight lines.
(e) Show that the origin is an unstable equilibrium point.

16. Consider the system

$$x' = 0,$$
$$y' = -x - y.$$

(a) Solve the system.
(b) Show that the roots of **(3)** for this system are $\lambda_1 = 0$ and $\lambda_2 = -1$.
(c) Graph the orbits.
(d) Show that the zero solution is stable but *not* asymptotically stable.

10.3 STABILITY OF NONLINEAR SYSTEMS

In this section we discuss the stability properties of the autonomous nonlinear system

$$x' = f(x, y),$$
$$y' = g(x, y), \tag{1}$$

where $f(0, 0) = g(0, 0) = 0$ so that the origin is a critical point. We assume that

$f(x, y)$ and $g(x, y)$ can each be expanded in a Taylor series:

$$f(x, y) = f(0, 0) + \frac{\partial f}{\partial x}(0, 0)x + \frac{\partial f}{\partial y}(0, 0)y + \frac{\partial^2 f}{\partial x^2}(0, 0)\frac{x^2}{2} + \frac{\partial^2 f}{\partial x\,\partial y}(0, 0)xy + \frac{\partial^2 f}{\partial y^2}(0, 0)\frac{y^2}{2} + \dots,$$

$$g(x, y) = g(0, 0) + \frac{\partial g}{\partial x}(0, 0)x + \frac{\partial g}{\partial y}(0, 0)y + \frac{\partial^2 g}{\partial x^2}(0, 0)\frac{x^2}{2} + \frac{\partial^2 g}{\partial x\,\partial y}(0, 0)xy + \frac{\partial^2 g}{\partial y^2}(0, 0)\frac{y^2}{2} + \dots, \tag{2}$$

where the omitted terms all involve higher powers of x and y. Now we define

$$a_{11} = \frac{\partial f}{\partial x}(0, 0), \qquad a_{12} = \frac{\partial f}{\partial y}(0, 0), \qquad a_{21} = \frac{\partial g}{\partial x}(0, 0), \qquad a_{22} = \frac{\partial g}{\partial y}(0, 0)$$

and assume that

$$a_{11}a_{22} - a_{12}a_{21} \neq 0 \tag{3}$$

(see Problem 10.2.11). Using the fact that $f(0, 0) = g(0, 0) = 0$, we can write $f(x, y)$ and $g(x, y)$ as

Writing a system near a critical point as a linear system plus higher-order terms

$$f(x, y) = a_{11}x + a_{12}y + f_1(x, y),$$
$$g(x, y) = a_{21}x + a_{22}y + g_1(x, y),$$

where

$$\lim_{x, y \to 0} \frac{f_1(x, y)}{\sqrt{x^2 + y^2}} = \lim_{x, y \to 0} \frac{g_1(x, y)}{\sqrt{x^2 + y^2}} = 0. \tag{4}$$

The last property simply says that the point $(f_1(x, y), g_1(x, y))$ approaches the point $(0, 0)$ "faster" than the point (x, y) does. In one dimension, this is easily visualized by noting, for example, that the function $f_1(x) = x^2$ goes to zero faster than x, or

$$\lim_{x \to 0} \frac{x^2}{x} = 0.$$

We now show that condition **(4)** holds for the function x^2. Once this has been done, it will be clear that the condition holds for xy, y^2, and higher powers of x and y (such as x^3, x^2y, xy^2, and y^3, etc.). Now,

$$\lim_{x, y \to 0} \frac{x^2}{\sqrt{x^2 + y^2}} = 0,$$

because $\sqrt{x^2 + y^2}$ tends to zero as (x, y) tends to zero and

$$0 \le \frac{x^2}{\sqrt{x^2 + y^2}} \le \frac{x^2 + y^2}{\sqrt{x^2 + y^2}} = \sqrt{x^2 + y^2} \to 0.$$

Similarly,

$$\lim_{x, y \to 0} \frac{xy}{\sqrt{x^2 + y^2}} = 0,$$

because $x^2 + y^2 \geq 2|xy|$ since $(|x| - |y|)^2 \geq 0$; thus

$$0 \leq \frac{|xy|}{\sqrt{x^2 + y^2}} \leq \frac{\frac{1}{2}(x^2 + y^2)}{\sqrt{x^2 + y^2}} \to 0.$$

Note that condition **(3)** is satisfied if the determinant

$$\begin{vmatrix} \dfrac{\partial f}{\partial x} & \dfrac{\partial f}{\partial y} \\ \dfrac{\partial g}{\partial x} & \dfrac{\partial g}{\partial y} \end{vmatrix}$$

is nonzero at (0, 0).

EXAMPLE 1 ▶ **A typical nonlinear system**

The system

$$x' = 2x + 3y + x^3,$$
$$y' = x - 2y - y^{3/2}$$

satisfies the conditions **(3)** and **(4)** since (prove this)

$$\lim_{x, y \to 0} \frac{x^3}{\sqrt{x^2 + y^2}} = \lim_{x, y \to 0} \frac{y^{3/2}}{\sqrt{x^2 + y^2}} = 0. \blacktriangleleft$$

In the rest of this section we consider the system

A nonlinear system

$$x' = a_{11}x + a_{12}y + f_1(x, y),$$
$$y' = a_{21}x + a_{22}y + g_1(x, y), \qquad \textbf{(5)}$$

where **(3)** and **(4)** are satisfied, and the **associated linear system**

An associated linear system

$$x' = a_{11}x + a_{12}y,$$
$$y' = a_{21}x + a_{22}y. \qquad \textbf{(6)}$$

The following theorem enables us to determine the nature of the critical point (0, 0) of Equation **(5)** by indicating the behavior of the solutions of the associated linear system **(6)** near the origin. The proof of this theorem is difficult and beyond the scope of this text.*

THEOREM 1 ♦ **Perturbation theorem**

Let λ_1 and λ_2 be the roots of the characteristic equation of the associated linear system **(6)**.

(a) The nonlinear system **(5)** has the same type of critical point at the origin as the associated linear system **(6)** whenever

(i) $\lambda_1 \neq \lambda_2$ and (0, 0) is a node of system **(6)**,
(ii) $\lambda_1 = \lambda_2$ and (0, 0) is not a star-shaped node of system **(6)**,

* For a proof, see J. K. Hale, *Ordinary Differential Equations* (New York: Wiley, 1969).

(iii) (0, 0) is a saddle point of system **(6)**,
(iv) (0, 0) is a focus of system **(6)**.

(b) The origin is not necessarily the same type of critical point for the two systems in two cases:

(v) If $\lambda_1 = \lambda_2$ and (0, 0) is a star-shaped node of system **(6)**, then (0, 0) is either a node or a focus of system **(5)**.
(vi) If (0, 0) is a center of system **(6)**, then (0, 0) is either a center or a focus of system **(5)**. ♦

The next theorem relates the stability of the nonlinear system to that of the associated linear system. As before, the proof is omitted. The proof of part (a), however, is suggested in Problem 10.4.11.

THEOREM 2 ♦ Stability theorem

(a) If the zero solution of system **(6)** is asymptotically stable, then the zero solution of system **(5)** is asymptotically stable.
(b) If the zero solution of system **(6)** is unstable, then the zero solution of system **(5)** is unstable.
(c) If the zero solution of system **(6)** is stable but not asymptotically stable, then the zero solution of system **(5)** may be asymptotically stable, stable, or unstable: that is, anything can happen. ♦

Remark

Part (b) of this theorem holds even when **(3)** is not satisfied. We use this fact in Example 3.

Theorems 1 and 2 are often called **perturbation** theorems, because we may consider the nonlinear system **(5)** as a small perturbation of the associated linear system **(6)** *near the origin.*

EXAMPLE 2 ▶ The Lotka-Volterra system

In Example 10.1.5 we discussed the Lotka-Volterra system as a model for the interaction of two competing species:

$$\begin{aligned} x' &= \beta_1 x + \delta_{11}x^2 + \delta_{12}xy, \\ y' &= \beta_2 y + \delta_{21}xy + \delta_{22}y^2. \end{aligned} \tag{7}$$

We can think of this system as a perturbation of the associated uncoupled linear system

$$\begin{aligned} x' &= \beta_1 x, \\ y' &= \beta_2 y. \end{aligned} \tag{8}$$

System **(8)** has the obvious solution

$$(x(t), y(t)) = (c_1 e^{\beta_1 t}, c_2 e^{\beta_2 t}),$$

and

$$\lim_{x,y\to 0} \frac{x^2}{\sqrt{x^2+y^2}} = \lim_{x,y\to 0} \frac{xy}{\sqrt{x^2+y^2}} = \lim_{x,y\to 0} \frac{y^2}{\sqrt{x^2+y^2}} = 0.$$

Hence Theorems 1 and 2 apply in the following cases:

1. If β_1 and β_2 are negative, then the critical point $(0, 0)$ is asymptotically stable for both systems **(7)** and **(8)**. In both systems the origin is a stable node. This means that both populations become extinct, in the absence of other factors, if the initial populations are small; this is, if we start near $(0, 0)$. Thus small initial populations cannot sustain themselves.
2. If β_1 and β_2 have opposite signs, then the origin in both cases is a saddle point, and the zero solution is unstable. This situation implies that one of the populations becomes extinct while the other grows without bound (since the asymptotes of the saddle point are the x- and y-axes).
3. If β_1 and β_2 are unequal and positive, then the origin in both cases is an unstable node. Hence both populations increase if they start near $(0, 0)$. However, the orbits of **(7)** need not increase without bound, since as we move away from the origin the orbits may approach other critical points of **(7)** exhibiting a different type of behavior.

This is not a complete analysis of the system **(7)**, since we can expect different and perhaps more interesting types of behavior near the other critical points of the system. To find these other critical points, it is useful to rewrite Equation **(7)** in the form

$$x' = x(\beta_1 + \delta_{11}x + \delta_{12}y),$$
$$y' = y(\beta_2 + \delta_{21}x + \delta_{22}y).$$

isoclines Then, the **isoclines** (or **nullclines**) of the system are the curves where $x' = 0$ and $y' = 0$. Here, the $x' = 0$ isoclines are the lines $x = 0$ and $\delta_{11}x + \delta_{12}y = -\beta_1$, and all orbits cross these lines vertically, while the $y' = 0$ isoclines are the lines $y = 0$ and $\delta_{21}x + \delta_{22}y = -\beta_2$, which are crossed horizontally by the orbits. A critical point occurs whenever an $x' = 0$ isocline intersects a $y' = 0$ isocline. In Figure 10.18 we have drawn the isoclines in the first quadrant for the four situations of Equation **(7)** where $\beta_i < 0 < \delta_{ij}$ for $i, j = 1, 2$. It is easy to determine the direction field when x and y are very small positive numbers because the β_i terms dominate the expressions in parentheses. Hence, $x' < 0$ and $y' < 0$ near the origin. This determines the directions of motion on the isoclines, shown as arrows in Figure 10.18. In Figure 10.18(a) where $\beta_1/\beta_2 < \delta_{11}/\delta_{21}, \delta_{12}/\delta_{22}$, there are only three critical points in the first quadrant: the origin, $(-\beta_1/\delta_{11}, 0)$, and $(0, -\beta_2/\delta_{22})$. This is also true in Figure 10.18(b), where $\beta_1/\beta_2 > \delta_{11}/\delta_{21}, \delta_{12}/\delta_{22}$. Figures 10.18(c) and (d), where $\delta_{11}/\delta_{21} < \beta_1/\beta_2 < \delta_{12}/\delta_{22}$ and $\delta_{12}/\delta_{22} < \beta_1/\beta_2 < \delta_{11}/\delta_{21}$, respectively, include an interior equilibrium point at

$$\left(\frac{\beta_2\delta_{12} - \beta_1\delta_{22}}{\delta_{11}\delta_{22} - \delta_{21}\delta_{12}}, \frac{\beta_1\delta_{21} - \beta_2\delta_{11}}{\delta_{11}\delta_{22} - \delta_{21}\delta_{12}}\right) \tag{9}$$

in addition to the other three equilibrium points. We wish to find the stability at the remaining critical points.

Equation **(2)** provides the technique by which the associated linear system **(6)** is obtained at these other critical points: one expands $f(x, y)$ and $g(x, y)$ in their Taylor series centered this time not at the origin, but at the critical point (x_0, y_0) one is interested in. Thus, **(6)** now becomes

$$(x - x_0)' = a_{11}(x - x_0) + a_{12}(y - y_0)$$
$$(y - y_0)' = a_{21}(x - x_0) + a_{22}(y - y_0)$$

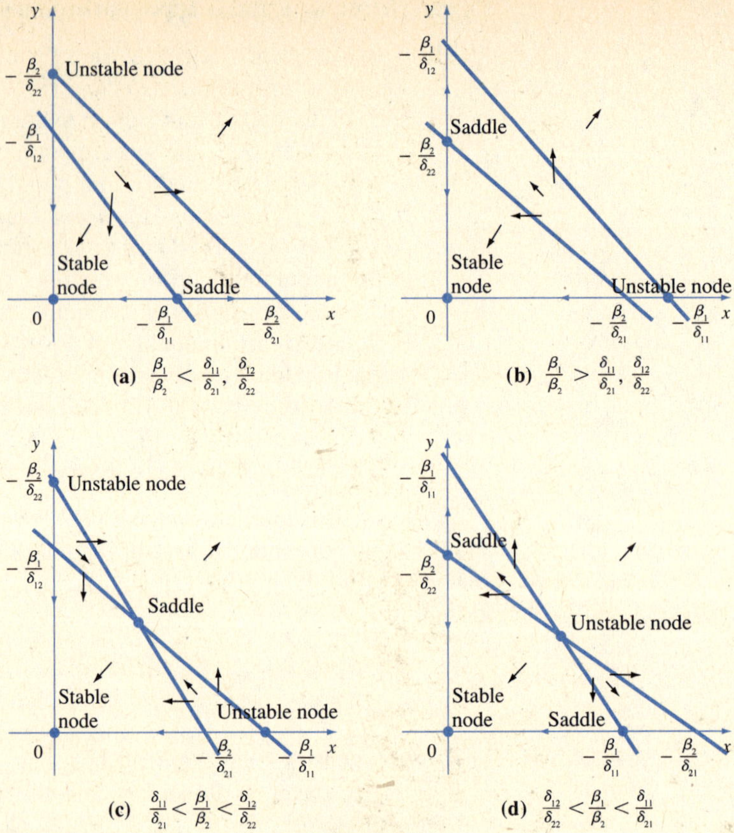

FIGURE 10.18
Dynamics of Lotka-Volterra system for $\beta_i < 0 < \delta_{ij}$

since $(x - x_0)' = x'$ and $(y - y_0)' = y'$, where

$$a_{11} = \frac{\partial f}{\partial x}(x_0, y_0), \quad a_{12} = \frac{\partial f}{\partial y}(x_0, y_0), \quad a_{21} = \frac{\partial g}{\partial x}(x_0, y_0), \quad a_{22} = \frac{\partial g}{\partial y}(x_0, y_0).$$

Writing the **Jacobian** for the system **(7)**

$$\begin{pmatrix} \frac{\partial f}{\partial x} & \frac{\partial f}{\partial y} \\ \frac{\partial g}{\partial x} & \frac{\partial g}{\partial y} \end{pmatrix} = \begin{pmatrix} \beta_1 + 2\delta_{11}x + \delta_{12}y & \delta_{12}x \\ \delta_{21}y & \beta_2 + \delta_{21}x + 2\delta_{22}y \end{pmatrix},$$

and evaluating the terms at the critical point $(-\beta_1/\delta_{11}, 0)$, we obtain the associated linear system, written in matrix form, as

$$\begin{pmatrix} x + \frac{\beta_1}{\delta_{11}} \\ y \end{pmatrix}' = \begin{pmatrix} -\beta_1 & -\frac{\beta_1\delta_{12}}{\delta_{11}} \\ 0 & \beta_2 - \frac{\beta_1\delta_{21}}{\delta_{11}} \end{pmatrix} \begin{pmatrix} x + \frac{\beta_1}{\delta_{11}} \\ y \end{pmatrix}$$

or

$$\left(x + \frac{\beta_1}{\delta_{11}}\right)' = -\beta_1\left(x + \frac{\beta_1}{\delta_{11}}\right) - \left(\frac{\beta_1\delta_{12}}{\delta_{11}}\right)y,$$

$$y' = \left(\beta_2 - \frac{\beta_1\delta_{21}}{\delta_{11}}\right)y.$$

Therefore, the eigenvalues of this system are $\lambda_1 = -\beta_1$ and $\lambda_2 = (\beta_2\delta_{11} - \beta_1\delta_{21})/\delta_{11}$. Since $\beta_i < 0 < \delta_{ij}$, it follows that $\lambda_1 > 0$, and the sign of λ_2 depends on the sign of the numerator. The numerator will be positive if and only if $\delta_{11}/\delta_{21} < \beta_1/\beta_2$, so that the critical point $(-\beta_1/\delta_{11}, 0)$ is an unstable node in Figures 10.18(b), (c) and a saddle point in Figures 10.18 (a), (d).

Similarly, you should check that the critical point $(0, -\beta_2/\delta_{22})$ is an unstable node in Figure 10.18 (a), (c) and a saddle point in Figure 10.18 (b), (d). Also check that the interior critical point is a saddle point in Figure 10.18 (c) and an unstable node in Figure 10.18 (d). ◄

EXAMPLE 3 ▶ **The dynamics of a nonlinear system**

Describe the dynamics of the system of equations

$$\begin{aligned} x' &= -2xy = f(x, y), \\ y' &= -x + y + xy - y^3 = g(x, y). \end{aligned} \tag{10}$$

Solution The $x' = 0$ isocline consists of the coordinate axes $x = 0$ and $y = 0$. Orbits cross these lines vertically. The $y' = 0$ isoclines are given by the equation

$$xy - y^3 - x + y = 0,$$

which can be rewritten as $x(y - 1) - y(y^2 - 1) = (y - 1)(x - y(y + 1)) = 0$. Thus, there are two isoclines: the line $y = 1$ and the parabola $x = y(y + 1)$. Orbits cross these curves horizontally. The isoclines are illustrated in Figure 10.19. Observe that the $y' = 0$ isoclines cross the $x' = 0$ isoclines in three places: (0, 0), (0, 1), and (0, −1). We need to determine the behavior at these three critical points.

Next we discuss the direction field in Figure 10.19: pick any point (x, y) and calculate x' and y' from system **(10)**. For example, at $(\frac{1}{2}, \frac{1}{2})$ we have $x' = -\frac{1}{2}$ and $y' = \frac{1}{8}$, so the motion is upward and to the left. It is useful to calculate the direction field on the isoclines: for example, on the $y' = 0$ isocline $y = 1$, the motion is

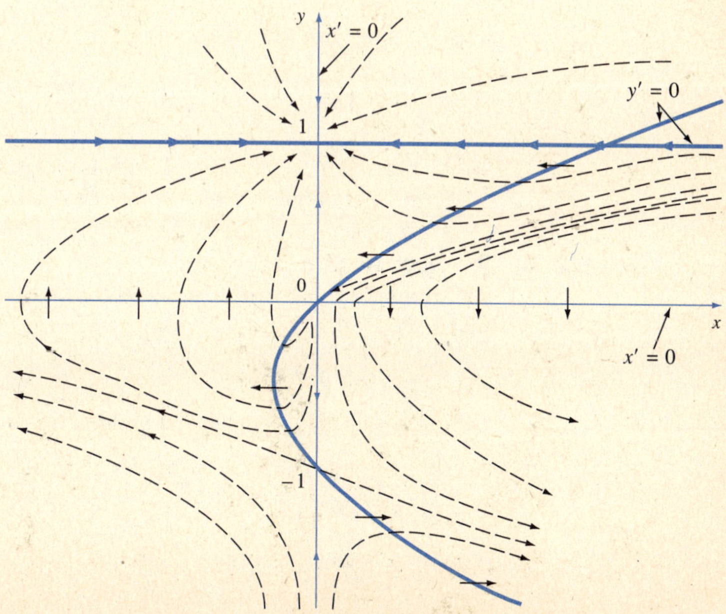

FIGURE 10.19
Phase plane for Example 3

downward in the first quadrant and upward in the second. Knowing the direction field can give us valuable clues concerning the critical points: observe that the orbits converge to the point (0, 1), indicating that it is probably a stable node. On the other hand, two orbits go to (0, -1) and two leave it, indicating that (0, -1) is a saddle point.

To confirm these statements and discover the nature of the critical point at the origin, we calculate the Jacobian:

$$\begin{pmatrix} \dfrac{\partial f}{\partial x} & \dfrac{\partial f}{\partial y} \\ \dfrac{\partial g}{\partial x} & \dfrac{\partial g}{\partial y} \end{pmatrix} = \begin{pmatrix} -2y & -2x \\ -1 + y & 1 + x - 3y^2 \end{pmatrix}. \tag{11}$$

Case 1: (0, 0)

Substituting $x = y = 0$ into the right side of **(11)**, we obtain the associated linear system

$$\begin{pmatrix} x \\ y \end{pmatrix}' = \begin{pmatrix} 0 & 0 \\ -1 & 1 \end{pmatrix}\begin{pmatrix} x \\ y \end{pmatrix},$$

with eigenvalues $\lambda_1 = 0$, $\lambda_2 = 1$. Hence the origin is unstable (see Problem 10.2.15), so the origin is unstable by the remark following Theorem 2.

Case 2: (0, 1)

Here the Jacobian yields the associated linear system

$$\begin{pmatrix} x \\ y - 1 \end{pmatrix}' = \begin{pmatrix} -2 & 0 \\ 0 & -2 \end{pmatrix}\begin{pmatrix} x \\ y - 1 \end{pmatrix},$$

with eigenvalues $\lambda_1 = -2$, $\lambda_2 = -2$. Thus (0, 1) is a stable star-shaped node.

Case 3: (0, -1)

The Jacobian in **(11)** yields

$$\begin{pmatrix} x \\ y + 1 \end{pmatrix}' = \begin{pmatrix} 2 & 0 \\ -2 & -2 \end{pmatrix}\begin{pmatrix} x \\ y + 1 \end{pmatrix},$$

so the eigenvalues are $\lambda_1 = 2$ and $\lambda_2 = -2$. Thus, (0, -1) is a saddle point, and is therefore unstable. ◀

EXAMPLE 4 ▶ **Motion of a pendulum**

In Example 10.1.2 we discussed the equation of motion of a frictionless pendulum given by

$$\theta'' + \omega^2 \sin\theta = 0. \tag{12}$$

Defining $\mu = \theta'$, we have the system

$$\begin{aligned} \theta' &= \mu, \\ \mu' &= -\omega^2 \sin\theta. \end{aligned} \tag{13}$$

This system has an infinite number of critical points of the form $(k\pi, 0)$, $k = 0$, ± 1, ± 1, ± 2, ± 3, ... The associated linear system near the critical point (0, 0) is

$$\begin{aligned} \theta' &= \mu, \\ \mu' &= -\omega^2 \theta, \end{aligned} \tag{14}$$

since the Taylor series of $\sin\theta$ is

$$\sin\theta = \theta - \frac{\theta^3}{3!} + \frac{\theta^5}{5!} - \dots .$$

This is the familiar system of the harmonic oscillator. The origin of **(14)** is a center, so we cannot conclude anything about the stability of the origin of **(13).** The results are more definitive for the critical point $(\pi, 0)$. Expanding $\sin\theta$ around $\theta = \pi$, we have

$$\begin{aligned}\sin\theta &= \sin\pi + (\sin\theta')\big|_{\theta=\pi}(\theta - \pi) + \dots\\ &= (\pi - \theta) - \frac{1}{3!}(\pi - \theta)^3 + \frac{1}{5!}(\pi - \theta)^5 - \dots .\end{aligned}$$

Thus the associated linear system is

$$\begin{aligned}\theta' &= \mu,\\ \mu' &= \omega^2(\theta - \pi).\end{aligned} \qquad \textbf{(15)}$$

Letting $z = \theta - \pi$, the roots of the characteristic equation $\lambda^2 - \omega^2 = 0$ are $\lambda_1 = \omega$, $\lambda_2 = -\omega$, so $(\pi, 0)$ is a saddle point. Hence we may conclude that the critical point $(\pi, 0)$ of system **(13)** is an unstable saddle point. Figure 10.1 illustrates intuitively what is happening. When $\theta = \pi$, the pendulum is pointing vertically upward, which is a position at which the potential energy is at a maximum and the kinetic energy is zero. Clearly such a position is unstable, and a small displacement causes large deviations from the initial position. On the other hand, it is just as clear that $\theta = 0$ (the vertically downward position) is a stable center. An initial displacement of θ_0 of a frictionless pendulum leads to periodic oscillations with a maximum displacement of θ_0. It is evident that all critical points of the form $(k\pi, 0)$ are of the same type as $(0, 0)$ if k is even and of the same type as $(\pi, 0)$ if k is odd. Before drawing the orbits, we can show that the zero solution of **(13)** really is a center. By Theorem 1 it must be a center or a focus. But the system is **conservative** (there is no loss or gain of energy), whereas a stable focus would imply that the energy was decreasing to zero and an unstable focus would imply an increase in the total energy. Hence a focus is ruled out, and the origin must be a center. This is all illustrated in Figure 10.20.

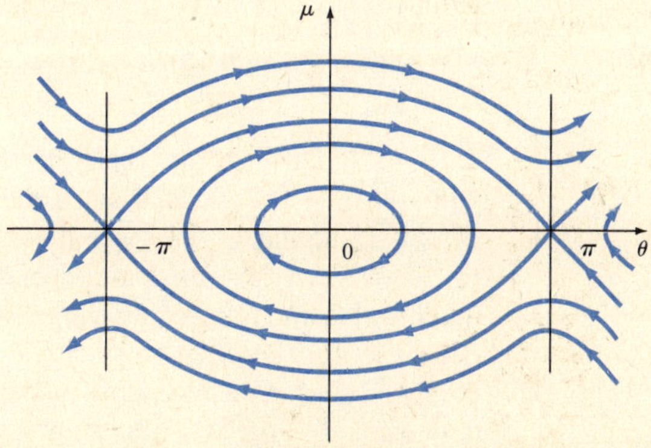

FIGURE 10.20
Phase plane for a frictionless pendulum

EXAMPLE 5 ▶ **A pendulum with friction**

Here we consider the pendulum of the previous example, this time including the effect of friction. We assume that the frictional force is proportional to the angular velocity θ'. Denoting this constant of proportionality by $\epsilon > 0$, we obtain the equation

$$\theta'' + \epsilon\theta' + \omega^2 \sin\theta - 0,$$

or, equivalently,

$$\begin{aligned} \theta' &= \mu, \\ \mu' &= -\omega^2 \sin\theta - \epsilon\mu. \end{aligned} \tag{16}$$

Again the origin is a critical point. However, in this case it is clear from physical considerations that the origin is asymptotically stable, since the friction causes initial deviations from the vertical to damp out. We can prove this easily by considering the zero solution of the associated linear system.

$$\begin{aligned} \theta' &= \mu, \\ \mu' &= -\omega^2\theta - \epsilon\mu. \end{aligned} \tag{17}$$

The characteristic equation is $\lambda^2 + \epsilon\lambda + \omega^2 = 0$, with the roots

$$\lambda_1 = \frac{-\epsilon + \sqrt{\epsilon^2 - 4\omega^2}}{2} \quad \text{and} \quad \lambda_2 = \frac{-\epsilon - \sqrt{\epsilon^2 - 4\omega^2}}{2}.$$

Hence the origin of **(17)** is a stable node (if $\epsilon > 2\omega$), a star-shaped stable node (if $\epsilon = 2\omega$), or a stable focus (if $\epsilon < 2\omega$). In all cases, by Theorem 2, the origin of system **(16)** is asymptotically stable. The last case is the most interesting. The "focal behavior" near the origin implies that the pendulum will continue to oscillate, but with constantly decreasing amplitudes. ◀

In this chapter we have concentrated on describing the dynamics of two-dimensional systems. The dynamics of autonomous systems in three or higher dimensions are much more complicated than those we have described so far. A particular phenomenon that does not occur for autonomous systems in two-dimensions, but does occur in three and higher dimensions, is **chaos.** Figure 10.21 illustrates a trajectory of the **Lorenz** equations

$$\begin{aligned} x' &= 10(y - x), \\ y' &= rx - y - xz, \\ z' &= xy - \frac{8}{3}z, \end{aligned} \tag{18}$$

for $r = 28$. Observe that there are three critical points: $(0, 0, 0)$, $(6\sqrt{2}, 6\sqrt{2}, 27)$, and $(-6\sqrt{2}, -6\sqrt{2}, 27)$, since the first equation yields $x = y$, the second gives $27x = xz$, so that $z = 27$ when $x \neq 0$, and the third yields $x^2 = 72$. If we calculate the Jacobian at these critical points, the characteristic equation for the origin yields one positive and two negative eigenvalues, so it is a saddle point. The characteristic equations for the other two critical points have one negative eigenvalue and two complex conjugate eigenvalues with positive real parts. Thus, there is a one-dimensional "inset" leading into each critical point, and a two-dimensional "outset" (corresponding to the two complex eigenvalues). Trajectories near such a critical point are drawn in along the inset, then spun out along the outset until they are drawn to the other critical point's inset. It turns out that the crossings from one's outset to the inset of the other occur unpredictably, so that two nearby

FIGURE 10.21
A trajectory for the Lorenz equations **(18)**

trajectories may behave quite differently. This sensitivity to initial conditions and exponential separation of nearby trajectories are both indications of chaos.

E. N. Lorenz obtained this system while studying the problem of weather formation. The chaotic behavior exhibited by this system implies that the associated fluid equations are turbulent, indicating that long-range weather forecasting may be impossible.

SELF-QUIZ

I. Which of the following are critical points of the Lotka-Volterra competition system

$$x' = x(1 - x - y)$$
$$y' = y(4 - 2x - 8y)?$$

(a) $(0, 0)$ **(b)** $(1, 0)$
(c) $\left(0, \frac{1}{2}\right)$ **(d)** $\left(\frac{2}{3}, \frac{1}{3}\right)$

II. The associated linear system for the origin for the system in Question I is ______.

(a) $x' = x$, $y' = y$ **(b)** $x' = x(1 - x)$, $y' = y(4 - 8y)$
(c) $x' = x$, $y' = 4y$ **(d)** $x' = -xy$, $y' = -2xy$

III. The associated linear system for the critical point $(1, 0)$ in Question I is ______.

(a) $x' = -x - y$, $y' = -2y$ **(b)** $x' = -(x - 1) - y$, $y' = -2y$
(c) $x' = -x - (y - 1)$, $y' = -2(y - 1)$ **(d)** $x' = x(1 - x)$, $y' = 0$

IV. The associated linear system at the origin for the system

$$x' = \sin x - \ln(1 + y)$$
$$y' = 1 - \cos x + \tan y$$

is ______.

(a) $x' = x - y$, $y' = y$ **(b)** $x' = x - y$, $y' = x + y$
(c) $x' = x + y$, $y' = -x + y$ **(d)** $x' = x - (1 + y)$, $y' = y$

Answers to Self-Quiz

I. all of them **II.** c **III.** b **IV.** a

PROBLEMS 10.3

In Problems 1–6 verify that in each case $(0, 0)$ *is a critical point and determine the asymptotic behavior of solutions near that point.*

1. $x' = 3\sin x + e^y - 1,$
$y' = xy - y$

2. $x' = \ln(1 + y) + \cosh x - 1,$
$y' = \tan y + 2x$

3. $x' = 1 - y - e^{-x},$
$y' = y - \sin x$

4. $x' = 2\cos y - \frac{1}{2}y + e^x - 3 - \sin 2x,$
$y' = 2\tan x - y$

5. $x' = (\sin x)(\cos x) + y^2,$
$y' = y^2 - x - x^3 + 3y$

6. $x' = -y + \epsilon x(1 - y^2),\ \epsilon > 0,$
$y' = x$

In Problems 7–13 determine the critical points of each nonlinear equation; find the associated linear system for each of these critical points; determine, if possible, the nature of each initial point and its stability properties; and sketch the orbits near each such point.

7. $x' = x + x^3,$
$y' = y + y^3$

8. $x' = -\sin x + x^2,$
$y' = \sin y$

9. $x' = x - x^2 + xy,$
$y' = 2y - xy - 6y^2$

10. $x' = -e^y + 1,$
$y' = e^x - 1$

11. $x' = -xy^2 + y^2 - 7xy - x^2 - 6x,$
$y' = x^2 + y$

12. $x' = 1 - xy,$
$y' = x - y^3$

13. $x' = 2y,$
$y' = -2x - y + y^4$

14. **(a)** Convert the equation

$$x'' + ax' + bx + x^2 = 0, \qquad a, b > 0,$$

into a system.

(b) Show that the origin is a stable focus or node and that the point $(-b, 0)$ is a saddle point.

(c) Conclude that the orbits have the form shown below. Sketch in the zero isoclines of your system.

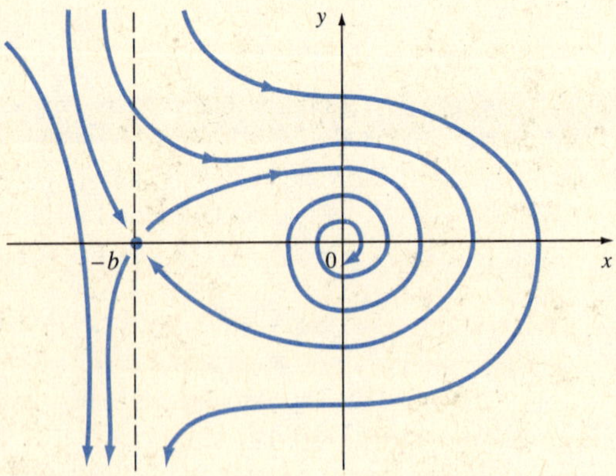

15. The **Van der Pol equation,**

$$x'' + \epsilon(x^2 - 1)x' + x = 0, \qquad \textbf{(i)}$$

arises in the study of a vacuum tube with three internal elements (triode). Draw its phase plane and show that if $\epsilon < 0$, then the origin is asymptotically stable. [It can also be shown that, if $\epsilon > 0$, then Equation **(i)** has a periodic solution that is approached by all other solutions as t tends to ∞. This phenomenon is called a **limit cycle.***]

* See, for example, H. K. Wilson, *Ordinary Differential Equations* (Reading, Mass.: Addison-Wesley, 1971).

16. Consider the following special case of the Lotka-Volterra system **(7)**:

$$\begin{aligned} x' &= -x - 2x^2 + xy, \\ y' &= -y + 7xy - 2y^2, \end{aligned} \qquad \textbf{(ii)}$$

where x and y are measured in hundreds of organisms.

(a) Show that system **(ii)** has four critical points but that only two of them have any biological meaning (negative populations are not permissible).

(b) Show that if both populations are small initially, both species become extinct (use the phase plane).

(c) Show that one biologically meaningful critical point is a saddle point, indicating that larger initial populations may lead to continued existence without the threat of extinction.

10.4 A PERSPECTIVE: LYAPUNOV'S METHOD

In the preceding section we introduced a method for deriving the stability of nonlinear systems by comparing them with linear ones. This procedure, while often useful, has two obvious drawbacks. First, in the case of a center, no information can be obtained. Second, and much more serious, the method fails to yield information concerning the behavior of nonlinear systems that cannot easily be treated as perturbations of linear systems. Landmark work to resolve these difficulties was done by the Russian mathematician A. A. Lyapunov.* Lyapunov reasoned intuitively as follows: Suppose we have a system of differential equations that arises from a description of a physical system; if a critical point corresponds to a point of minimum potential energy of the system, and if the energy in the system is constant or decreasing, then it is reasonable to "guess" that the critical point is stable; on the other hand, if the critical point corresponds to a maximum of potential energy, then the point is unstable.

EXAMPLE 1 ▶ **Vibrations of a spring**

Consider the vibrations of a mass attached to a coiled spring, the upper end of which is securely fastened (see Section 4.1). If we make the natural assumption that damping forces (friction, air resistance, etc.) are present, then the equation of motion is of the form [from Equation **(4.2.1)** on page 179]

$$x'' + \epsilon x' + \mu x = 0. \qquad \textbf{(1)}$$

For simplicity, we assume $\mu = 1$, so system **(1)** can be written

$$\begin{aligned} x' &= y, \\ y' &= -x - \epsilon y. \end{aligned} \qquad \textbf{(2)}$$

From the discussion in Section 10.3 it is obvious that the origin of this system is asymptotically stable. We use this simple problem to motivate the subsequent discussion.

The potential energy of the system is proportional to the square of x, the distance from the mass to the origin, whereas the kinetic energy is proportional to the square of the velocity x'. This is so because potential energy is usually given as an integral of position. Here the position is represented by x, so the potential

---

* Lyapunov's original paper on this subject appeared in 1892. His work was essentially unknown until a French translation appeared in 1907 under the title "Problème général de la stabilité du mouvement."

energy is proportional to $\int x\,dx = x^2/2$. Kinetic energy is given by the formula

$$KE = \frac{1}{2}mv^2 = \frac{1}{2}m\left(\frac{dx}{dt}\right)^2 = \frac{1}{2}my^2.$$

Thus the function $V(x, y)$, defined by

$$V(x, y) = \frac{1}{2}(x^2 + y^2), \tag{3}$$

where $y = x'$, is representative of the total energy of the system. Now, at the origin potential energy is a minimum, so the origin can be expected to be stable. The function $V(x, y)$ defined in **(3)** will be used shortly to prove that the origin is indeed stable. ◀

We can generalize these remarks. Let us consider the autonomous system

$$\begin{aligned} x' &= f(x, y), \\ y' &= g(x, y), \end{aligned} \tag{4}$$

where we assume that the origin is a critical point. Let $V(x, y)$ be a continuous real-valued function on the xy-plane with continuous first partial derivatives. Let D be a region containing the origin and suppose that $V(0, 0) = 0$ and $V(x, y) > 0$ for all other points (x, y) in D. Then $V(x, y)$ is said to be **positive definite** in D. If $V(0, 0) = 0$ and $V(x, y) < 0$ for all other points (x, y) in D, we say that $V(x, y)$ is **negative definite** in D. If $V(x, y) \geq 0$, or $V(x, y) \leq 0$, then the function is said to be **positive semidefinite,** or **negative semidefinite,** in D. If $V(x, y)$ satisfies none of these conditions, then V is said to be **indefinite** in D.

EXAMPLE 2 ▶ **Five functions illustrating definite, semidefinite, and indefinite properties**

(a) The function $V(x, y)$ given in **(3)** is positive definite for all real values of x and y, since $x^2 + y^2$ is the square of the distance from (x, y) to the origin.
(b) The function $-(x^2 + y^2)$ is negative definite.
(c) The function $V(x, y) = x^2$ is positive semidefinite, since $V(0, a) = 0$ for any real value of a.
(d) The function $V(x, y) = -y^2$ is negative semidefinite.
(e) The function $V(x, y) = xy$ is indefinite, since $V(a, a) = a^2 > 0$ and $V(a, -a) = -a^2 < 0$ for all numbers $a > 0$. ◀

Often the total energy of a system is a polynomial in x and y, in which case the following theorem is useful.

THEOREM 1 ♦ **Definiteness of quadratic functions**

The function

$$V(x, y) = x^2 + axy + by^2$$

is (a) positive definite if and only if $4b - a^2 > 0$; (b) positive semidefinite if and only if $4b - a^2 \geq 0$.

Proof If $4b - a^2 > 0$, then completing the square, we find that

$$x^2 + axy + by^2 = \left(x + \frac{a}{2}y\right)^2 + \left(b - \frac{a^2}{4}\right)y^2. \tag{5}$$

If $y \neq 0$, then $[b - (a^2/4)]y^2 > 0$ by assumption and $V(x, y)$ is positive definite. On the other hand, suppose that $[b - (a^2/4)] \leq 0$. Then choosing any nonzero y and $x = -\frac{a}{2}y$, we find expression **(5)** to be less than or equal to zero, which is a contradiction, since V is positive definite. Thus $4b - a^2 > 0$, and the first part of the theorem is proved. The second part follows in the same way. ♦

Note that $V(x, y)$ is negative definite if and only if $-V(x, y)$ is positive definite. Thus Theorem 1 is also a theorem about negative definiteness (and semidefiniteness).

EXAMPLE 3 ▶ **Several quadratic functions**

(a) The fuction $x^2 - xy + 2y^2$ is positive definite, since $4b - a^2 = 7 > 0$.
(b) The function $V(x, y) = -x^2 + 4xy - 4y^2$ is negative semidefinite, since $-V(x, y) = x^2 - 4xy + 4y^2$, $4b - a^2 = 0$, and $-V$ is positive semidefinite.
(c) The function $x^2 + 4xy - 4y^2$ is indefinite, since neither V nor $-V$ belongs to any of the other four categories. ◀

Now let $V(x, y)$ be a continuously differentiable, positive definite function. We define the **derivative of V along the orbits of system (4)** by

The derivative along orbits

$$V'(x, y) = \frac{\partial V}{\partial x}x' + \frac{\partial V}{\partial y}y' = \frac{\partial V}{\partial x}f(x, y) + \frac{\partial V}{\partial y}g(x, y). \qquad \textbf{(6)}$$

Lyapunov function

When this association between $V(x, y)$ and system **(4)** holds, $V(x, y)$ is called a **Lyapunov function** for the system. We emphasize that in order for it to be a Lyapunov function for system **(4),** $V(x, y)$ must be continuously differentiable and positive definite and must have its derivative along the orbits defined by **(6).** We should also emphasize that $V(x, y)$ is a Lyapunov function for **(4)** only when $V'(x, y)$ is defined with respect to **(4)** according to **(6).** There are always many choices for a Lyapunov function; often, however, most of these functions are not useful. The great importance of defining the derivative of V along the orbits of **(4)** is given in the next theorem.

THEOREM 2 ♦ **Lyapunov's theorem**

Let $V(x, y)$ be a Lyapunov function for the system **(4).** Then

(a) if $V'(x, y)$ is negative semidefinite, the origin is stable;
(b) if $V'(x, y)$ is negative definite, the origin is asymptotically stable;
(c) if $V'(x, y)$ is positive definite, the origin is unstable.

Proof*

(a) Let $\epsilon > 0$ be given. We must show that there is a $\delta > 0$ such that if (x_0, y_0) is in D and $\sqrt{x_0^2 + y_0^2} < \delta$, then the solution $(x(t, x_0, y_0), y(t, x_0, y_0))$ (see Section 10.2) satisfies the inequality

$$\sqrt{[x(t, x_0, y_0)]^2 + [y(t, x_0, y_0)]^2} < \epsilon$$

for all $t \geq 0$. We define

$$m = \min_{\sqrt{x^2+y^2}=\epsilon} V(x, y) > 0.$$

*This proof is difficult and may be omitted without loss of continuity.

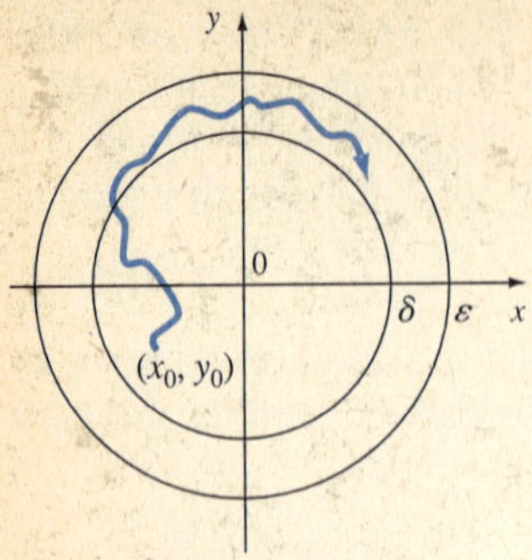

FIGURE 10.22
Continuity of V

This minimum exists since $V(x, y)$ is continuous on the circle $x^2 + y^2 = \epsilon^2$. Furthermore, by the continuity of V and the fact that $V(0, 0) = 0$, there exists a $\delta > 0$ such that $V(x, y) < m$ whenever $\sqrt{x_0^2 + y_0^2} < \delta$ (see Figure 10.22). Suppose that there is a number $t^* > 0$ for which $x(t^*, x_0, y_0)^2 + y(t^*, x_0, y_0)^2 = \epsilon^2$. Since $dV/dt = V'(x, y) \le 0$, we have $V(x(t, x_0, y_0), y(t, x_0, y_0)) \le V(x_0, y_0)$ for all $t \ge 0$. Then

$$V(x(t^*, x_0, y_0), y(t^*, x_0, y_0)) \le V(x_0, y_0)$$
$$< m \le V(x(t^*, x_0, y_0), y(t^*, x_0, y_0)),$$

since m is a minimum on the circle $x^2 + y^2 = \epsilon^2$. This contradiction establishes (a), since it shows that $x(t, x_0, y_0)^2 + y(t, x_0, y_0)^2 < \epsilon^2$ for all $t > 0$.

(b) By hypothesis, $V(x, y)$ decreases as t increases. Since V is positive for all x and y and decreasing along the orbits (with $x(t) = x(t, x_0, y_0)$ and $y(t) = y(t, x_0, y_0)$)

$$\lim_{t\to\infty} V(x(t), y(t)) = \lambda \ge 0$$

exists. We must show that $\lambda = 0$. Then, since $(0, 0)$ is the only point for which V is zero, we necessarily have

$$\lim_{t\to\infty} x(t) = \lim_{t\to\infty} y(t) = 0.$$

Suppose $\lambda > 0$; then, as before, there is a $\eta > 0$ such that $V(x_0, y_0) < \lambda$ whenever $\sqrt{x_0^2 + y_0^2} < \eta$, meaning that the orbit $(x(t), y(t))$ never enters the circular region described by $x^2 + y^2 < \eta^2$. Let ϵ be as in part (a) and let

$$m_1 = \min_{\eta \le \sqrt{x^2+y^2} \le \epsilon} -V'(x, y).$$

Since $\eta > 0$ and V' is negative definite, we have $m_1 > 0$. Then, since $m_1 \le -V'(x, y)$, we have $V'(x(t), y(t)) \le -m_1$ for all $t \ge 0$. Therefore, we obtain upon integration

$$\int_0^t V'(x(s), y(s))\, ds \le \int_0^t -m_1\, ds,$$

or

$$V(x(t), y(t)) - V(x_0, y_0) \le -m_1 t,$$

or

$$V(x(t), y(t)) \le V(x_0, y_0) - m_1 t.$$

But the right-hand side of the last inequality becomes negative when t approaches ∞, thereby contradicting the positive definiteness of V unless $m_1 = 0$. But $m_1 = 0$ if and only if $\eta = 0$, which holds if and only if $\lambda = 0$. Therefore, (b) is proved.

(c) The proof of part (c) is similar and therefore left as an exercise (see Problem 6). ♦

The rest of this section indicates the wide range of applicability of Theorem 2.

EXAMPLE 4 ▶ **A Lyapunov function**

In Example 1, the function $V(x, y) = \frac{1}{2}(x^2 + y^2)$ is a Lyapunov function. In addition, by Equation **(2)**,

$$V'(x, y) = \frac{\partial V}{\partial x}x' + \frac{\partial V}{\partial y}y' = xy + y(-x - \epsilon y) = -\epsilon y^2,$$

which is negative semidefinite. Hence Theorem 2 implies that the origin is stable.

Observe that Theorem 2 does not provide the best possible result, since we know the origin to be asymptotically stable. Stronger results are known and can be found in many books on advanced differential equations.* ◀

EXAMPLE 5 ▶ **The nonlinear spring**

In our example of the movement of a mass attached to a frictionless spring, we assumed that the restoring force was proportional to the distance of the mass to the origin. Generally, we may assume the restoring force to be a nonlinear function $-f(x)$ of the distance to the origin. The equation of motion is then

$$x'' + f(x) = 0. \tag{7}$$

To analyze this equation, we make the assumptions that

$$f(0) = 0 \quad \text{and} \quad xf(x) > 0 \quad \text{for } x \neq 0; \tag{8}$$

that is, $f(x)$ and x have the same sign. Equation **(7)**, together with condition **(8)**, is often called the equation of a **nonlinear spring.** We now write Equation **(7)** as

$$\begin{aligned} x' &= y, \\ y' &= -f(x). \end{aligned} \tag{9}$$

Clearly the origin is the only critical point of the system. The kinetic energy of this system is proportional to the square of the velocity $x'(=y)$, whereas the potential energy at any point x is given by

$$F(x) = \int_0^x f(x)\, dx.$$

Hence the total energy of the system is proportional to

$$V(x, y) = F(x) + \frac{y^2}{2}.$$

Since $f(x)$ and x have the same sign, $F \geq 0$ and $V(x, y)$ is positive definite. Moreover, $V'(x, y)$ is negative semidefinite since

$$V'(x, y) = F'(x)x' + yy' = f(x)y + y(-f(x)) = 0,$$

implying by Theorem 2 that the origin is stable. In this case, as in the linear case $x'' + x = 0$, it can be shown that the origin is a center. ◀

EXAMPLE 6 ▶ **Using Lyapunov's theorem to show asymptotic stability**

Consider the system

$$\begin{aligned} x' &= -x - \frac{x^3}{3} - x \sin y, \\ y' &= -y - \frac{y^3}{3}. \end{aligned} \tag{10}$$

Here the origin is the only critical point. We define $V(x, y) = \frac{1}{2}(x^2 + y^2)$. Then

$$\begin{aligned} V'(x, y) &= x\left(-x - \frac{x^3}{3} - x \sin y\right) + y\left(-y - \frac{y^3}{3}\right) \\ &= -x^2 - \frac{x^4}{3} - y^2 - \frac{y^4}{3} - x^2 \sin y. \end{aligned}$$

*See, for example, J. P. LaSalle and S. Lefshetz, *Stability by Lyapunov's Direct Method* (New York: Academic Press, 1961).

Now $|x^2 \sin y| \le |x^2|$, so $x^2 + x^2 \sin y \ge 0$. Hence

$$V'(x, y) = -\frac{x^4}{3} - y^2 - \frac{y^4}{3} - (x^2 + x^2 \sin y) \le -\frac{x^4}{3} - y^2 - \frac{y^4}{3}.$$

Therefore V' is negative definite and the origin is asymptotically stable. ◀

In Example 6 the Lyapunov function was pulled out of a hat. Actually, the problem of finding a Lyapunov function for a particular system is very difficult, especially if there is no way of estimating the total energy of the system. Although a Lyapunov function cannot be found in every case, it is often worthwhile in the case of a system of two first-order equations to try to find one of the form $V(x, y) = x^2 + axy + by^2$. By Theorem 1, such a Lyapunov function is positive definite if $4b > a^2$, as illustrated by the next example.

EXAMPLE 7 ▶ **Using the theorems to show instability**

Consider the system

$$\begin{aligned} x' &= xy^2 + x^2y + x^3, \\ y' &= y^3 - x^3. \end{aligned} \qquad \textbf{(11)}$$

Assume that $V(x, y) = x^2 + axy + by^2$ and $4b > a^2$. Then by Theorem 1, V is positive definite and

$$\begin{aligned} V' = {} & 2x(xy^2 + x^2y + x^3) + ax(y^3 - x^3) \\ & + ay(xy^2 + x^2y + x^3) + 2by(y^3 - x^3). \end{aligned}$$

If we choose $a = 0$ and $b = 1$, then

$$V' = 2x^2y^2 + 2x^3y + 2x^4 + 2y^4 - 2yx^3 = 2(x^2y^2 + x^4 + y^4),$$

which is positive definite. By Theorem 2, the origin is unstable. ◀

PROBLEMS 10.4

1. Show that the zero solution of the system

$$x' = xy^2 - \frac{x^3}{2},$$
$$y' = -\frac{y^3}{2} + \frac{yx^2}{5}$$

is asymptotically stable. [*Hint:* Try a Lyapunov function of the form $V(x, y) = ax^2 + by^2$.]

2. Show that the zero solution of the following system is stable:

$$x' = -6x^2y,$$
$$y' = -3y^3 + 6x^3$$

3. Consider the undamped pendulum of Example 10.1.2:

$$\theta'' + \omega^2 \sin\theta = 0, \qquad -\frac{\pi}{2} \le \theta \le \frac{\pi}{2}.$$

Prove by means of an appropriate Lyapunov function that the origin is stable. [*Hint:* Use the result of Example 5.]

4. Consider the system

$$x' = y - xf(x, y),$$
$$y' = -x - yf(x, y),$$

where $f(0, 0) = 0$ and $f(x, y)$ has a convergent power series expansion in a region D around the origin. Show that the zero solution is

(a) stable if $f(x, y) \ge 0$ in some region around $(0, 0)$;

(b) asymptotically stable if $f(x, y)$ is positive definite in some region around $(0, 0)$;

(c) unstable if in every region around $(0, 0)$ there are points (x, y) such that $f(x, y) < 0$.

5. Using the results of Problem 4, examine the stability properties of the zero solutions of the following systems:

(a) $x' = y - x(y^3 \sin^2 x)$,
$y' = -x - y(y^3 \sin^2 x)$;

(b) $x' = y - x(x^4 + y^6)$,
$y' = -x - y(x^4 + y^6)$;

(c) $x' = y - x(\sin^2 y)$,
$y' = -x - y(\sin^2 y)$.

6. Prove part (c) of Theorem 2.

7. Consider the equation

$$x'' + f(x, x') + g(x) = 0. \qquad \textbf{(i)}$$

Assume that f and g possess continuous first derivatives, that $f(0, 0) = g(0) = 0$, and that $yf(x, y) > 0$ when $y \neq 0$ and $xg(x) > 0$ for $x \neq 0$. After writing **(i)** as a system, prove that the origin is asymptotically stable.

8. Use the result of Problem 7 to prove the stability of the solutions of the equation

$$x'' + x'^3 + x^5 = 0.$$

[Note that the associated linear system does us no good here since the zero solution of this system is a center.]

9. Consider the linear system

$$\begin{aligned} x' &= a_{11}x + a_{12}y, \\ y' &= a_{21}x + a_{22}y. \end{aligned} \qquad \textbf{(ii)}$$

Assume the the zero solution is asymptotically stable (that is, the real parts of the roots of the characteristic equation are negative). Let $(x_1(t), y_1(t))$ and $(x_2(t), y_2(t))$ be the solutions of **(ii)** that satisfy the conditions

$$(x_1(0), y_1(0)) = (1, 0)$$

and

$$(x_2(0), y_2(0)) = (0, 1).$$

Define the function $V(x, y)$ by

$$\begin{aligned} V(x, y) = x^2 &\int_0^\infty [x_1^2(t) + y_1^2(t)]\,dt \\ &+ 2xy \int_0^\infty [x_1(t)x_2(t) + y_1(t)y_2(t)]\,dt \\ &+ y^2 \int_0^\infty [x_2^2(t) + y_2^2(t)]\,dt. \end{aligned} \qquad \textbf{(iii)}$$

(a) Show that $V(x, y)$ is positive definite.
(b) Show that $V'(x, y)$ is negative definite along the orbits of **(ii)**.

10. Use the result of Problem 9 to find a Lyapunov function that could be used to prove the asymptotic stability of the following systems:

(a) $x' = -4x - y$,
$y' = x - 2y$;

(b) $x' = -5x - 3y$,
$y' = 4x + 2y$.

11. Consider the system

$$\begin{aligned} x' &= a_{11}x + a_{12}y + f(x, y), \\ y' &= a_{21}x + a_{22}y + g(x, y), \end{aligned} \qquad \textbf{(iv)}$$

where f and g are "small" near the origin in the sense of Equation **(10.3.4)**. Assume that the zero solution of the associated linear system is asymptotically stable. Prove that the zero solution of **(iv)** is asymptotically stable. [*Hint:* Prove that there is a region D around $(0, 0)$ such that the function $V(x, y)$ defined in Problem 9 is a Lyapunov function for **(iv)** and that $V'(x, y) < 0$ *in D*.]

10.5 SOLVING NONLINEAR SYSTEMS WITH A SYMBOLIC MANIPULATOR

This section will show you how a generic symbol manipulator can be used to solve phase portraits for nonlinear two-dimensional systems. In designing the script for this purpose we shall need the following generalization of the SOLVE command:

`SOLVE(EQ(c)=0,c):` = {solves the system of equations $\mathbf{EQ(c)} = \mathbf{0}$ for the vector of unknowns $\mathbf{c}$}.

Consider the two-dimensional nonlinear autonomous system

$$\begin{aligned} x' &= f(x, y) \\ y' &= g(x, y). \end{aligned} \tag{1}$$

To analyze the phase portrait of **(1)** we need to do the following: (a) find the steady states or equilibrium points of system **(1),** (b) determine their stability and type, and (c) draw the phase portrait by finding the isoclines. All of these tasks can be done by a symbolic manipulator.

The equilibrium points can be determined by the command

`SOLVE ([f, g]=[0, 0], [x, y]),`

which solves the two equations $f(x, y) = 0$ and $g(x, y) = 0$ simultaneously for x and y. This is often the most difficult part of the problem, and a symbolic manipulator may have trouble doing this job, because the task is to determine where two curves intersect.

Suppose (x_0, y_0) is an equilibrium point. Using Taylor's theorem for a function of two variables, we can rewrite the functions $f(x, y)$ and $g(x, y)$ in the form

$$\begin{aligned} f(x, y) &= f(x_0, y_0) + \frac{\partial f}{\partial x}(x_0, y_0)(x - x_0) + \frac{\partial f}{\partial y}(x_0, y_0)(y - y_0) + \ldots \\ g(x, y) &= g(x_0, y_0) + \frac{\partial g}{\partial x}(x_0, y_0)(x - x_0) + \frac{\partial g}{\partial y}(x_0, y_0)(y - y_0) + \ldots \end{aligned} \tag{2}$$

where the rest of the terms consist of higher partial derivatives of f and g multiplied by powers of $(x - x_0)$ and $(y - y_0)$. Since (x_0, y_0) is an equilibrium point, it follows that $f(x_0, y_0) = g(x_0, y_0) = 0$, so, using **(2),** we can rewrite **(1)** in vector form as

$$\begin{pmatrix} (x - x_0)' \\ (y - y_0)' \end{pmatrix} = \begin{pmatrix} f(x, y) \\ g(x, y) \end{pmatrix} = \begin{pmatrix} \dfrac{\partial f}{\partial x} & \dfrac{\partial f}{\partial y} \\ \dfrac{\partial g}{\partial x} & \dfrac{\partial g}{\partial y} \end{pmatrix} \begin{pmatrix} x - x_0 \\ y - y_0 \end{pmatrix} + \ldots . \tag{3}$$

The matrix on the right side of **(3)** is the **Jacobian** and the partial derivatives are evaluated at the point (x_0, y_0). For sufficiently small $|x - x_0|$ and $|y - y_0|$ it follows that the higher order terms of the Taylor expansion are insignificant, so the solution near the equilibrium point (x_0, y_0) can be approximated by the solution to the linear system

$$\begin{pmatrix} (x - x_0)' \\ (y - y_0)' \end{pmatrix} = \begin{pmatrix} \dfrac{\partial f}{\partial x} & \dfrac{\partial f}{\partial y} \\ \dfrac{\partial g}{\partial x} & \dfrac{\partial g}{\partial y} \end{pmatrix} \begin{pmatrix} x - x_0 \\ y - y_0 \end{pmatrix}. \tag{4}$$

Thus, using the results of Section 10.2, the Jacobian tells us how fast the functions f and g are growing in the x and y directions at the equilibrium point. If both eigenvalues of the Jacobian are negative or have negative real parts, then the equilibrium point (x_0, y_0) is **stable.** If even one of the eigenvalues is positive or has a positive real part, the equilibrium point is **unstable.**

A script for finding the eigenvalues of the Jacobian might be as follows:

```
JACOBIAN([f,g],[x,y],[x₀,y₀]):=EIGENVALUES(SUBST([
    [DIF(f,x),DIF(f,y)],[DIF(g,x),DIF(g,y)]],x=x₀,y=y₀))
```

The type of the equilibrium point is determined as follows: If λ_1 and λ_2 are the eigenvalues of **(4)**:

$$\lambda_1, \lambda_2 \text{ real} \quad \begin{cases} \lambda_1 < 0, \lambda_2 < 0 & \text{then } (x_0, y_0) \text{ is a } \textbf{stable node.} \\ \lambda_1 < 0 < \lambda_2 & \text{then } (x_0, y_0) \text{ is a } \textbf{saddle point.} \\ 0 < \lambda_1, 0 < \lambda_2 & \text{then } (x_0, y_0) \text{ is an } \textbf{unstable node.} \end{cases}$$

$$\left.\begin{matrix}\lambda_1\\ \lambda_2\end{matrix}\right\} = a \pm bi \text{ complex} \quad \begin{cases} a < 0 & \text{then } (x_0, y_0) \text{ is a } \textbf{stable focus.} \\ a = 0 & \text{then } (x_0, y_0) \text{ is a } \textbf{center.} \\ a > 0 & \text{then } (x_0, y_0) \text{ is an } \textbf{unstable focus.} \end{cases}$$

Our final task is to draw the phase portrait. Using the graphics package of your symbolic manipulator graph the **isoclines:** $f(x, y) = 0$ and $g(x, y) = 0$. All motion is vertical on the isocline $f(x, y) = 0$ since $x' = 0$ on this curve. All motion on $g(x, y) = 0$ is horizontal since $y' = 0$ on this curve. Equilibrium points are located wherever the two curves intersect, as there is no motion at these points since $x' = y' = 0$.

It is not hard to determine whether the motion of the orbit or trajectory at other points on the isocline $f(x, y) = 0$ is upward or downward. Calculate $g(x, y)$ using the coordinates of that point:

$$\text{if } g(x, y) \text{ is } \begin{Bmatrix}\text{positive}\\ \text{negative}\end{Bmatrix} \text{ the motion is } \begin{Bmatrix}\text{upward}\\ \text{downward}\end{Bmatrix} \text{ since } y' \begin{Bmatrix} > 0 \\ < 0 \end{Bmatrix}.$$

Similarly, determine rightward or leftward motion at points on the $g(x, y) = 0$ isocline by calculating $f(x, y)$. Since there is very little uniformity in the way symbolic manipulators do graphics, we shall not illustrate the drawing of a phase portrait.

EXAMPLE 1 ▶

Find the equilibrium points, their stability, and their type for the Lotka-Volterra competition model

$$\begin{aligned} x' &= rx\left(1 - \frac{x}{k} - \frac{ay}{k}\right) \\ y' &= Ry\left(1 - \frac{y}{K} - \frac{Ax}{K}\right). \end{aligned} \tag{5}$$

Solution If we set the right sides of (**5**) equal to zero and then draw a graph, we obtain two straight lines for each equation. The first equation yields the lines

$$x = 0 \quad \text{and} \quad x + a\,y = k,$$

while the second equation gives us

$$y = 0 \quad \text{and} \quad y + Ax = K.$$

Each line of each pair will intersect the two lines of the other pair at a single point, so that there are exactly four equilibrium points. To find them we give the command

```
SOLVE([r*x*(1−x/k−a*y/k),R*y*(1−y/K−A*x/K)]
    =[0,0],[x,y]).
```

Three of the equilibrium points are on the axes: (0, 0) (0, K), and (k, 0). The other equilibrium point is at the intersection of the lines $x + a\,y = k$ and $y + A\,x = K$: $(x_0, y_0) = (\frac{k - aK}{1 - aA}, \frac{K - Ak}{1 - aA})$. The quadrant this point is located in depends on the coefficients a, k, A, K.

We now determine the stability and type of these equilibrium points by giving the four commands:

```
JACOBIAN([r*x*(1−x/k−a*y/k),R*y*(1−y/K−A*x/K)], [x,y],[0,0]
JACOBIAN([r*x*(1−x/k−a*y/k),R*y*(1−y/K−A*x/K], [x,y],[0,K])
JACOBIAN([r*x*(1−x/k−a*y/k),R*y*(1−y/K−A*x/K)], [x,y],[k,0])
JACOBIAN([r*x*(1−x/k−a*y/k),R*y*(1−y/K−A*x/K)], [x,y],[x₀,y₀]).
```

The symbolic manipulator calculates the Jacobian

$$\mathbf{J} = \begin{pmatrix} r\left(1 - \frac{y}{k} - \frac{ay}{k}\right) - \frac{rx}{k} & -\frac{rax}{k} \\ -\frac{RAy}{K} & R\left(1 - \frac{y}{K} - \frac{Ax}{K}\right) - \frac{Ry}{K} \end{pmatrix}$$

and evaluates it at the four equilibrium points getting:

(i) $\mathbf{J}(0, 0) = \begin{pmatrix} r & 0 \\ 0 & R \end{pmatrix}$ with eigenvalues $r > 0$ and $R > 0$, an unstable node,

(ii) $\mathbf{J}(0, \mathrm{K}) = \begin{pmatrix} r\left(1 - \frac{aK}{k}\right) & 0 \\ -RA & -R \end{pmatrix}$ with eigenvalues $r(\frac{k - aK}{k})$ and $-R$,

so $(0, K)$ is either a saddle (if $k > aK$) or a stable node (if $k < aK$).

(iii) $\mathbf{J}(k, 0) = \begin{pmatrix} -r & -ra \\ 0 & R\left(1 - \frac{Ak}{K}\right) \end{pmatrix}$ with eigenvalues $-r$ and $R(\frac{K - Ak}{K})$,

so $(k,0)$ is either a saddle (if $K > Ak$) or a stable node (if $K < Ak$).

(iv) $\mathbf{J}(x_0, y_0) = \begin{pmatrix} -\frac{rx_0}{k} & -\frac{rax_0}{k} \\ -\frac{RAy_0}{K} & -\frac{Ry_0}{K} \end{pmatrix}$ which has the characteristic polynomial

$$\lambda^2 + \left(\frac{rx_0}{k} + \frac{Ry_0}{K}\right)\lambda + \frac{rRx_0 y_0(1 - aA)}{kK} = 0.$$

This characteristic equation has a positive discriminant, so the eigenvalues are real. Hence (x_0, y_0) is either a stable node or a saddle, depending on whether or not $(rx_0/k + Ry_0/K)$ exceeds the discriminant, respectively. ◀

PROBLEMS 10.5

Use the methods of this section and MATHEMATICA, MAPLE, DERIVE, MATLAB, or MATHCAD to analyze the following nonlinear systems. Find the equilibrium points and analyze their type and stability for Problems 1–4.

1. $x' = x - y^2$
$y' = x^2 + y$

2. $x' = x - 2y$
$y' = x - y^2$

3. $x' = x - 3$
$y' = x^2 - y$

4. $x' = 3x - y + 4$
$y' = xy - 1$

5. A model on branching reduces to the nonlinear system

$$p' = a\,p - b\,n$$
$$n' = n(c\,p - d).$$

Find the equilibrium points and analyze their stability and type.

6. The Geirer-Meinhardt model for morphogenesis is given by the system

$$x' = a\,y + b\,x^2 - x\,y$$
$$y' = c\,y(b\,x^2 - y).$$

Find the equilibrium points and determine their stability and type for all choices of the parameters a, b, c.

7. The **Van der Pol equation** is given by the second-order nonlinear equation

$$x'' - a(x^2 - 1)x' + x = 0.$$

Substitute $y = x' - a(x^3/3 - x)$ and rewrite the Van der Pol equation as a system of two first-order equations. Then find its equilibrium points and determine their type for all values of the parameter a.

SUMMARY OUTLINE OF CHAPTER 10

Asymptotically Stable pp. 486, 491

Unstable pp. 486, 491

Stable p. 491

Autonomous System p. 486
nonautonomous

Phase Plane p. 486
orbit or phase portrait

Lotka-Volterra System pp. 486, 504, 521

Critical Point or Equilibrium Point pp. 490, 496–499
- stable and unstable node
  - proper and improper nodes
  - star-shaped nodes
- stable and unstable focus or spiral point
- center
- saddle point

Global Asymptotic Stability p. 493

Associated Linear System p. 503

Perturbation Theorems pp. 503, 504

Stability Theorem p. 504

Isoclines or Nullclines p. 505

Conservative System p. 509

Chaos p. 510

Lorenz Equations p. 510

Van der Pol Equation pp. 512, 513

Positive Definite, Negative Definite, Indefinite p. 514
positive semidefinite, negative semidefinite

Lyapunov Function p. 515

REVIEW EXERCISES FOR CHAPTER 10

In Exercises 1–3, (a) find a related linear equation, (b) find all constant solutions of the nonlinear equation, (c) solve the nonlinear equation, and (d) determine the behavior of the solutions of the nonlinear equation for values of $x(0)$ in the indicated range.

1. $x' = x + x^2,\ x(0) > 0$

2. $x' = 4x - x^2,\ -\infty < x(0) < \infty$

3. $x' = -x(x^2 - 1),\ x(0) > 0$

In Exercises 4–8 describe the nature of the critical point (0, 0) of each system and sketch the orbits.

4. $x' = -2x - 2y,$
$y' = -5x + y$

5. $x' = -12x + 7y,$
$y' = -7x + 2y$

6. $x' = 2x - y,$
$y' = 5x - 2y$

7. $x' = -2x,$
$y' = -2y$

8. $x' = 3x + 2y$
$y' = -5x + y$

In Exercises 9–10 verify that (0, 0) is a critical point and determine the asymptotic behavior of solutions near that point.

9. $x' = 4(\sin x)e^{2y},$
$y' = 2xy - 3y$

10. $x' = 3\cos y - 2y - 4e^x + 1 - \sin x,$
$y' = \tan x + y$

In Exercises 11–13, (a) determine the critical points of each nonlinear equation, (b) find the associated linear system for each of these critical points, (c) determine, if possible, the nature of each critical point, and (d) draw the phase plane.

11. $x' = -x + 2x^3,$
$y' = 2y + y^3$

12. $x' = -x - x\sin x,$
$y' = \sin y$

13. $x' = -2x - xy + 2x^2,$
$y' = 5y + xy - 2y^2$

11 FOURIER SERIES AND BOUNDARY VALUE PROBLEMS

11.1 INTRODUCTION TO TRIGONOMETRIC SERIES

Until this chapter we have considered differential equations whose solutions are determined by initial conditions. Here we discuss another equally important way to specify a particular solution of a differential equation; namely, specifying certain values of the function and its derivatives at two or more points. The differential equation together with the given conditions at two or more points is called a **boundary value problem** (this was defined earlier in Section 1.2).

In this section we present some examples of boundary value problems, together with a discussion of some of the powerful methods that are used to solve them. We see how the concepts of eigenvalues and eigenfunctions arise naturally in these situations and discuss the importance of Fourier series techniques in solving nonhomogeneous boundary value problems.

Since a boundary value problem requires that conditions be given at two or more points, we can assume the differential equation involved to be at least of second order, because a first-order equation is usually completely specified by a single condition (this follows from the existence-uniqueness result that was cited in Section 1.2).

EXAMPLE 1 ▶ **Harmonic motion**

Consider the simple harmonic motion of a point mass m attached to a coiled spring with spring constant k. Applying Newton's law of motion and Hooke's law to this situation, we obtained in Section 4.1 [see Equation **(4.1.2),** p. 174], the differential equation

$$x'' + p^2 x = 0, \qquad p^2 = \frac{k}{m}, \tag{1}$$

where $x = x(t)$ denotes the displacement at time t of the mass from its equilibrium position. Suppose the mass is initially ($t = 0$) at its equilibrium position when it is given an unknown initial velocity $v_0 > 0$. How can we guarantee that the mass will again be at its equilibrium position after precisely one second? Obviously, such a mechanism provides a simple, though crude, timepiece (see Figure 11.1).

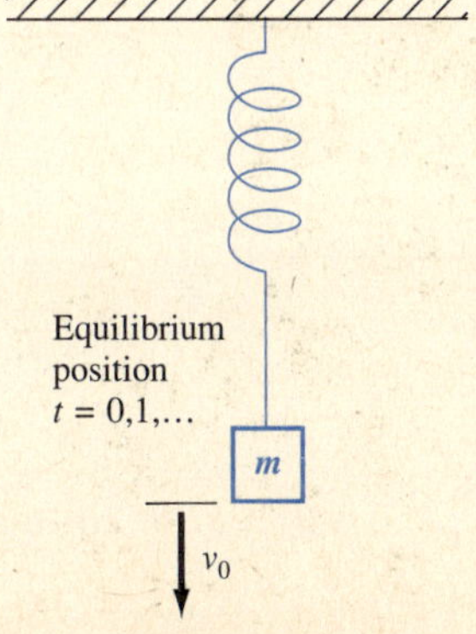

FIGURE 11.1
Harmonic motion

Our purpose now is to discover some of the properties of the unknown solution $x(t)$. If the time is measured in seconds, then we have the boundary conditions

$$x(0) = x(1) = 0. \tag{2}$$

The boundary conditions together with the differential equation **(1)** specify the boundary value problem we must solve.

Recall from Chapter 3 that to find the general solution of the differential equation **(1)** we need only solve the characteristic equation

$$\lambda^2 + p^2 = 0.$$

Since the roots of the characteristic equation are $\lambda = \pm ip$, the general solution of Equation **(1)** is given by

$$x(t) = A \cos pt + B \sin pt. \tag{3}$$

Setting $t = 0$ in **(3)** and using the boundary condition $x(0) = 0$, we observe that $A = 0$, so the solution must consist only of multiples of sin pt. For the boundary condition $x(1) = 0$ to hold, we must also have

$$B \sin p = x(1) = 0.$$

There are two cases to consider: either $B = 0$ or $\sin p = 0$. If $B = 0$, then solution **(3)** is constantly zero, which means that the mass remains at its equilibrium position at all times t. This is impossible, since the mass was given an initial velocity $v_0 > 0$. Therefore $\sin p = 0$, which is possible only if p is a nonzero multiple of π:

$$p = \pm\pi, \pm 2\pi, \pm 3\pi, \ldots .$$

[If $p = 0$, we see that **(3)** is again constantly zero.] Thus we have arrived at a surprising conclusion: There are infinitely many solutions of the form

$$x(t) = B \sin(n\pi)t, \tag{4}$$

where n is a nonzero integer, for the boundary value problem **(1), (2).** Furthermore, the constant B cannot be specified without additional information. Indeed, B depends on v_0 and n, since

$$v_0 = x'(0) = n\pi B,$$

or $B = v_0/n\pi$.

Actually, in doing the calculations above, we used one fact that was not part of the boundary value problem **(1), (2).** This was the assumption, given in the statement of the problem, that $v_0 > 0$. This assumption was used to disallow the *trivial solutions* $x(t) \equiv 0$ for all t. However, the results we have obtained illustrate one of the basic facts about boundary value problems: *Nontrivial solutions exist only for certain values of the parameter p.* All such nontrivial solutions **(4)** are called **eigenfunctions** of the problem, and the corresponding values of p^2 that yield these eigenfunctions are called **eigenvalues** of the problem. Rephrasing our work in this context, we see that

$$x(t) = \left(\frac{v_0}{n\pi}\right) \sin n\pi t \tag{5}$$

is an eigenfunction of this problem corresponding to the eigenvalue

$$\frac{k}{m} = p^2 = (n\pi)^2. \tag{6}$$

It is interesting to discover the physical significance of this solution. Suppose we are given a spring with (fixed) spring constant k. Solving **(6)** for m, we obtain the masses

$$m_n = \frac{k}{n^2\pi^2}, \qquad n = 1, 2, 3, \ldots, \tag{7}$$

each corresponding to a different value of n. Since k and π are fixed quantities, it is *only* for these masses that a solution of the boundary value problem exists. For any other choice, no matter what initial velocity is given, the mass will not be back at its equilibrium position after exactly one second. Thus the construction of our clock depends on the selection of one of the masses in **(7)**. Furthermore, **(5)** indicates that if mass m_n is chosen, it will be at its equilibrium position n times during the time period $0 < t \leq 1$. Since increasing n decreases both the weight m_n and the amplitude $(v_0/n\pi)$ of the oscillations, we arrive at the conclusion that *the lighter the mass, the more rapid the oscillation*. This fact is easy to observe in practice. ◀

EXAMPLE 2 ▶ **An epidemic model**

A population is subject to an influenza epidemic. Let $s(t)$ and $i(t)$ be, respectively, the numbers of susceptible and infected individuals at time t. Suppose that the rate of change of the number of susceptible individuals is proportional to the number of infected individuals, whereas the rate of change of the number of infected individuals is proportional to the number of susceptible individuals. (The difference between the change in the numbers of infected and susceptible individuals consists of those who have gained immunity or died.) The epidemic lasts a certain period of time, say, two months. [With this assumption, we can set $i(0) = i(2) = 0$.] We wish to discover the number of infected individuals at all time t (in months).

From the hypotheses above, we obtain the system of equations

$$\frac{ds}{dt} = -pi, \qquad \frac{di}{dt} = qs, \tag{8}$$

where p and q are unknown positive constants of proportionality. Differentiating the second of the equations in system **(8)** with respect to t, and substituting the first equation for $\frac{ds}{dt}$, we obtain the second-order equation

$$\frac{d^2 i}{dt^2} = q\frac{ds}{dt} = qpi,$$

or

$$i'' + k^2 i = 0, \qquad k^2 = pq, \tag{9}$$

with the boundary conditions

$$i(0) = i(2) = 0. \tag{10}$$

This boundary value problem is very similar to that of Example 1. We again find the general solution

$$i(t) = A\cos kt + B\sin kt.$$

Setting $t = 0$ and using the first boundary condition, we find that $0 = i(0) = A$. The second boundary condition $t = 2$ implies that

$$B\sin 2k = i(2) = 0.$$

Since we are not interested in the trivial solution $i(t) \equiv 0$, it follows that $B \neq 0$, so $2k$ must be a nonzero multiple of π. Hence the eigenvalues of this problem are

$$k^2 = \left(\frac{\pi}{2}\right)^2, (\pi)^2, \left(\frac{3\pi}{2}\right)^2, \ldots, \left(\frac{n\pi}{2}\right)^2, \ldots,$$

and the corresponding eigenfunctions $i_k(t)$ are given by

$$i_k(t) = B \sin \frac{n\pi t}{2}, \qquad k^2 = \left(\frac{n\pi}{2}\right)^2. \tag{11}$$

If the number of infected individuals is assumed to be positive during the entire epidemic $0 < t < 2$, then $n = 1$. This is a reasonable limitation since negative numbers of infected individuals do not have physical meaning. ◀

The boundary value problems in these two examples will also arise in several different problems in Chapter 12 (see Examples 12.8.1 and 12.9.1).

EXAMPLE 3 ▶ **A forced spring problem**

We return to Example 1, but we assume that the spring-mass system is also subject to a periodic external force $K \sin \omega t$, where $K > 0$ and $\omega > 0$. Using the theory of forced vibrations developed in Section 4.3, we obtain the nonhomogeneous second-order equation

$$m \frac{d^2x}{dt^2} = -kx + K \sin \omega t.$$

Using the same boundary conditions as in Example 1, we are led to the nonhomogeneous boundary value problem

$$x'' + \frac{k}{m}x = \frac{K}{m} \sin \omega t, \qquad x(0) = x(1) = 0. \tag{12}$$

There are two cases to consider: either $\omega = n\pi$ for some integer n or $\omega \neq n\pi$ for every integer n.

First, we use the method of undetermined coefficients (Section 3.6) to find a particular solution. We suppose that

$$x(t) = A \sin \omega t + B \cos \omega t. \tag{13}$$

Since $x(0) = 0$, it follows that $B = 0$. Differentiating Equation **(13)** twice and substituting into Equation **(12)**, we have

$$-A\omega^2 \sin \omega t + \frac{k}{m} A \sin \omega t = \frac{K}{m} \sin \omega t.$$

Canceling the sin ωt on both sides of this equation and solving for A, we obtain

$$A = \frac{\frac{K}{m}}{\left(\frac{k}{m}\right) - \omega^2} = \frac{K}{k - m\omega^2}.$$

Case 1

If $\omega = n_0\pi$, then

$$x_p(t) = \frac{K}{k - mn_0^2\pi^2} \sin n_0\pi t \tag{14}$$

is a particular solution of Equation **(12)** provided that

$$\omega^2 = (n_0\pi)^2 \neq \frac{k}{m}. \tag{15}$$

From Example 1 we know that the homogeneous boundary value problem

$$x'' + \frac{k}{m}x = 0, \qquad x(0) = x(1) = 0, \tag{16}$$

has the eigenfunction

$$x_n(t) = c \sin n\pi t$$

whenever

$$\frac{k}{m} = n^2\pi^2, \qquad n = 1, 2, 3, \ldots,$$

and the trivial solution for all other values. Hence

$$x(t) = \begin{cases} c \sin n\pi t + \dfrac{K}{m\pi^2(n^2 - n_0^2)} \sin n_0\pi t, & \dfrac{k}{m} = n^2\pi^2, \quad n \neq n_0, \\[2ex] \dfrac{K}{k - m\pi^2 n_0^2} \sin n_0\pi t, & \dfrac{k}{m} \neq n^2\pi^2. \end{cases}$$

If $\frac{k}{m} = n_0^2\pi^2 = \omega^2$, the method of undetermined coefficient requires that we multiply Equation **(13)** by t to find a particular solution. In this case $A = 0$ and $B = -\frac{K}{2\omega m}$, so

$$x_p(t) = -\frac{K}{2\omega m} t \cos \omega t = \frac{-Kt}{2\pi n_0 m} \cos n_0\pi t.$$

But then $x_p(1) = -K \cos n_0\pi/2\pi n_0 m \neq 0$, so the boundary conditions cannot be satisfied. Thus there is *no* solution if $\frac{k}{m} = n_0^2\pi^2$.

Case 2

If $\omega \neq n\pi$, we see that the function we have obtained by the method of undetermined coefficients,

$$x_p(t) = \frac{K}{k - m\omega^2} \sin \omega t, \tag{17}$$

does not satisfy the boundary condition $x_p(1) = 0$. Since only linear combinations of periodic functions of the form

$$x_n(t) = \sin n\pi t \tag{18}$$

satisfy the boundary conditions $x(0) = x(1) = 0$, we try to approximate **(17)** as closely as possible by a sum of such functions. We can show, using the methods of Section 11.4, that the trigonometric series

$$2\pi \sin \omega \sum_{n=1}^{\infty} \frac{(-1)^n n}{\omega^2 - (n\pi)^2} \sin n\pi t, \tag{19}$$

called a **Fourier series,** converges to the function $\sin \omega t$ at all points in $0 \leq t < 1$, and to zero at $t = 1$. Making use of this fact, we seek a solution of the form

$$x(t) = 2\pi \sin \omega \sum_{n=1}^{\infty} \frac{(-1)^n n c_n}{\omega^2 - (n\pi)^2} \sin n\pi t. \tag{20}$$

Substituting **(20)** into **(12)** and using the Fourier series, **(19),** we can easily show that $c_n = K/(k - mn^2\pi^2)$. Using these values in **(20),** we obtain the general solution of the boundary value problem **(12).** ◄

At this point one may wonder whether very many functions can be represented as a Fourier series. The question is significant, since forcing functions can be quite arbitrary. Actually, the class of functions that have Fourier series representations includes almost every practical forcing function. Although the Fourier series in **(19)** may appear complicated, it is really quite easy to work with once one becomes more familiar with it. We will see that the nth term in the series is independent of the other terms and behaves essentially like the nth coordinate of a certain vector. This phenomenon, called **orthogonality,** is one of the central themes of this chapter.

EXAMPLE 4 ▶ A pendulum problem

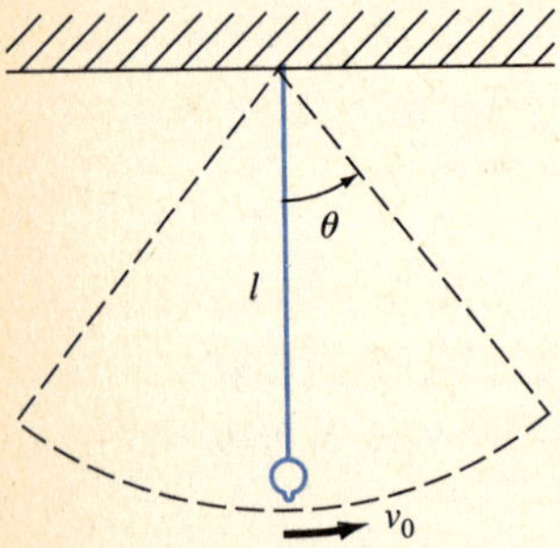

FIGURE 11.2
A pendulum

Consider the ideal pendulum consisting of a weightless rod of length l supported at one end and attached to a particle of positive weight at the other (Figure 11.2). In Problem 4.1.11 we found that the angular motion $\theta(t)$ of the pendulum is given by the nonlinear differential equation

$$\frac{d^2\theta}{dt^2} + \frac{g}{l}\sin\theta(t) = 0.$$

Suppose that initially $\theta(0) = 0$ (the rod is vertical) and that we give the pendulum an unknown initial velocity $\theta'(0) = v_0 > 0$. How can we guarantee that the mass will return to its initial position after precisely one second, that is, at $\theta(1) = 0$? The situation, which is essentially that of constructing a pendulum clock, reduces to the solution of the nonlinear boundary value problem

$$\theta'' + \frac{g}{l}\sin\theta = 0, \qquad \theta(0) = \theta(1) = 0.$$

In practice, the situation is much more complicated, since it involves damping effects due to friction and a forcing term (such as tension on a spring) to prevent the pendulum from stopping. We do not investigate nonlinear boundary value problems in this book; instead, we refer the interested reader to more advanced books on the subject.* ◀

Finally, it should be noted that boundary value problems can involve differential equations of order greater than two. For example, in studying the deflection of a beam, one arrives at the fourth-order boundary value problem†

$$\frac{d^4y}{dx^4} = k^2y,$$

with specified boundary conditions $y(0)$, $y'(0)$, $y(L)$, and $y'(L)$, where L is the length of the beam.

*See, for example, P. B. Bailey, L. F. Shampine, and P. E. Waltman, *Nonlinear Two-Point Boundary Value Problems* (New York: Academic Press, 1968).

†This equation is derived in C. R. Wylie, *Advanced Engineering Mathematics* (New York: McGraw-Hill, 1966), p. 323.

SELF-QUIZ

I. Which of the following are boundary value problems?

(a) $y' = e^y$, $y(0) = 1$

(b) $y'' = y$, $y(0) = y(1) = 2$

(c) $y'' = e^y$, $y(0) = y(1) = 2$

(d) $y'' = e^y$, $y(0) = y'(0) = 2$

II. Which of the following is a nontrivial solution of the problem

$$x'' + x = 0, \qquad x\left(\frac{\pi}{2}\right) = x'(0) = 0?$$

(a) $4 \cos t$ **(b)** $4 \cos t + 3 \sin t$ **(c)** $5 \sin 2t$ **(d)** $3 \sin t$

Answer true *or* false.

III. Nontrivial solutions exist for all values of p for the problem

$$x'' + p^2x = 0, \qquad x(0) = x(\pi) = 0.$$

Answers to Self-Quiz

I. b, c **II.** a **III.** False (not when p is not an integer)

PROBLEMS 11.1

1. What eigenfunctions and eigenvalues result in Example 1 if the boundary condition $x(1) = 0$ is replaced by $x(T) = 0$, where T is some fixed nonzero constant?
2. Consider the boundary value problem
$$\frac{d^2x}{dt^2} - kx(t) = 0, \qquad x(0) = x(1) = 0,$$
where k is any real number. What are the eigenvalues and eigenfunctions of this problem?
3. Answer Problem 2 given that the boundary condition $x(1) = 0$ is replaced by $x(T) = 0$ for some fixed nonzero constant T.
4. Answer Problem 2 given that the boundary condition $x(0) = 0$ is replaced by $x(-1) = 0$, all other conditions remaining the same.
5. Find the eigenfunctions and eigenvalues of the boundary value problem
$$x''(t) + k^2x(t) = 0, \qquad x(-T) = x(T) = 0,$$
where T is a fixed nonzero constant.
6. Find the eigenvalues and eigenfunctions of
$$x''(t) + k^2x(t) = 0, \qquad x(0) = x'(1) = 0.$$
7. Answer Problem 6 if the condition $x'(1) = 0$ is replaced by $x'(T) = 0$ for some fixed nonzero constant T.
8. Answer Problem 6 if the conditions $x(0) = x'(1) = 0$ are replaced by $x'(0) = x(1) = 0$.
9. Answer Problem 6 if the conditions $x(0) = x'(1) = 0$ are replaced by $x'(0) = x'(1) = 0$.
10. Find the eigenvalues and eigenvectors of the boundary value problem
$$(x^2y')' + \lambda y = 0, \qquad y(1) = y'(e) = 0.$$
11. Answer Problem 10 if the boundary conditions are replaced by $y'(1) = y(e) = 0$.

11.2 ORTHOGONAL SETS OF FUNCTIONS

Let $f_1(x)$ and $f_2(x)$ be two real-valued functions defined on an interval $a \le x \le b$, and suppose that the **inner product*** of f_1 and f_2, given by

The inner product of two functions

$$(f_1, f_2) = \int_a^b f_1(x) f_2(x)\, dx \tag{1}$$

---

* For the benefit of those readers familiar with the concept of the orthogonality of vectors in Euclidean n-space, we note that Equation **(1)** is a generalization of the usual inner product

$$(\mathbf{f}_1, \mathbf{f}_2) = f_{11}f_{21} + f_{12}f_{22} + \cdots + f_{1n}f_{2n} = \sum_{j=1}^{n} f_{1j}f_{2j}.$$

exists. We say that the functions f_1 and f_2 are **orthogonal on the interval** $a \le x \le b$ if integral **(1)** vanishes; that is, if $(f_1, f_2) = 0$.* A set of real-valued functions $f_1(x)$, $f_2(x), f_3(x), \ldots$ defined on $a \le x \le b$ is called an **orthogonal set of functions on the interval** $a \le x \le b$ if each integral

$$(f_n, f_m) = \int_a^b f_n(x) f_m(x)\, dx$$

exists and $(f_n, f_m) = 0$ whenever $n \ne m$. The nonnegative square root of the integral

$$(f_n, f_n) = \int_a^b f_n^2(x)\, dx$$

is called the **norm** of the function $f_n(x)$ and is often denoted by $\| f_n \|$ (see Problem 10):

The norm of a function

$$\| f_n \| = \sqrt{(f_n, f_n)} = \left[\int_a^b f_n^2(x)\, dx \right]^{1/2}. \tag{2}$$

An orthogonal set of functions $\{f_n(x)\}$ is called an **orthonormal set of functions on the interval** $a \le x \le b$ if $\| f_n \| = 1$ for all n.† Thus, if $\{f_n(x)\}$ is an orthonormal set of functions,

$$(f_n, f_m) = \int_a^b f_n(x) f_m(x)\, dx = \begin{cases} 0, & \text{whenever } n \ne m, \\ 1, & \text{whenever } n = m. \end{cases} \tag{3}$$

Given any orthogonal set of functions $\{f_n(x)\}$ whose norms are nonzero, it is always possible to construct an orthonormal set of functions $\{g_n(x)\}$ by defining

$$g_n(x) = \frac{f_n(x)}{\| f_n \|}, \qquad a \le x \le b.$$

Then

$$(g_n, g_m) = \int_a^b \frac{f_n(x)\ f_m(x)}{\| f_n \|\ \| f_m \|}\, dx = \frac{1}{\| f_n \| \| f_m \|}(f_n, f_m), \tag{4}$$

which is zero whenever $m \ne n$ since $\{f_n\}$ is an orthogonal set of functions. If $m = n$, then **(4)** yields

$$\| g_n \|^2 = (g_n, g_n) = \frac{1}{\| f_n \|^2}(f_n, f_n) = 1,$$

so that $\| g_n \| = 1$ for all n.

---

* Two vectors $\mathbf{f}_1$ and $\mathbf{f}_2$ are orthogonal in Euclidean n-space if $(\mathbf{f}_1, \mathbf{f}_2) = 0$. Also the **norm** of a vector is its Euclidean length

$$\| \mathbf{f}_1 \| = \sqrt{(\mathbf{f}_1, \mathbf{f}_1)} = \left(\sum_{j=1}^{n} f_{1j}^2 \right)^{1/2}.$$

† This corresponds to taking an orthonormal basis in Euclidean n-space, such as the **unit vectors** $\mathbf{e}_1 = (1, 0, \ldots, 0)$, $\mathbf{e}_2 = (0, 1, 0, \ldots, 0), \ldots, \mathbf{e}_n = (0, \ldots, 0, 1)$.

EXAMPLE 1 ▶ **An orthogonal set of functions**

The functions $f_n(x) = \sin nx$, $n = 1, 2, 3, \ldots$, are an orthogonal set of functions on the interval $-\pi \le x \le \pi$. To see this, we use the trigonometric identities

$$\cos(m - n)x = \cos mx \cos nx + \sin mx \sin nx,$$
$$\cos(m + n)x = \cos mx \cos nx - \sin mx \sin nx,$$

which imply (after subtracting and dividing by 2) that

$$\sin nx \sin mx = \frac{1}{2}[\cos(m - n)x - \cos(m + n)x].$$

Then

$$\begin{aligned}(f_n, f_m) &= \int_{-\pi}^{\pi} \sin nx \sin mx \, dx \\ &= \frac{1}{2}\int_{-\pi}^{\pi} [\cos(m - n)x - \cos(m + n)x] \, dx \\ &= \frac{1}{2}\left[\frac{\sin(m - n)x}{m - n} - \frac{\sin(m - n)x}{m + n}\right]\Bigg|_{-\pi}^{\pi} = 0, \end{aligned} \tag{5}$$

whenever $m \ne n$. If $m = n$, then $\cos(m - n)x = 1$ in the second integral of **(5)**, so

$$\begin{aligned}(f_n, f_n) &= \frac{1}{2}\int_{-\pi}^{\pi} (1 - \cos 2nx) \, dx \\ &= \frac{1}{2}\left(x - \frac{\sin 2nx}{2n}\right)\Bigg|_{-\pi}^{\pi} = \pi.\end{aligned}$$

Hence $\| f_n \| = \sqrt{\pi}$, and we may construct the orthonormal set of functions

$$g_n(x) = \frac{\sin nx}{\sqrt{\pi}}, \qquad n = 1, 2, \ldots . \blacktriangleleft$$

EXAMPLE 2 ▶ **An orthonormal set of functions**

The functions

$$\frac{1}{\sqrt{2\pi}}, \quad \frac{\cos x}{\sqrt{\pi}}, \quad \frac{\sin x}{\sqrt{\pi}}, \quad \frac{\cos 2x}{\sqrt{\pi}}, \quad \frac{\sin 2x}{\sqrt{\pi}}, \ldots$$

form an orthonormal set of functions on the internal $-\pi \le x \le \pi$. We have already demonstrated the orthonormality of the $\sin nx/\sqrt{\pi}$ terms, and clearly

$$\int_{-\pi}^{\pi} \frac{\cos nx}{\sqrt{\pi}} \frac{\cos mx}{\sqrt{\pi}} \, dx = \frac{1}{2\pi}\int_{-\pi}^{\pi} [\cos(m - n)x + \cos(m + n)x] \, dx = 0$$

whenever $m \ne n$, and

$$\begin{aligned}\int_{-\pi}^{\pi} \frac{\cos nx}{\sqrt{\pi}} \frac{\cos nx}{\sqrt{\pi}} \, dx &= \frac{1}{2\pi}\int_{-\pi}^{\pi} (1 + \cos 2nx) \, dx \\ &= \frac{1}{2\pi}\left(x + \frac{\sin 2nx}{2n}\right)\Bigg|_{-\pi}^{\pi} = 1.\end{aligned}$$

The identities

$$\sin(m + n)x = \sin mx \cos nx + \cos mx \sin nx,$$
$$\sin(m - n)x = \sin mx \cos nx - \cos mx \sin nx,$$

imply (after addition and division by 2) that

$$\int_{-\pi}^{\pi} \frac{\sin mx}{\sqrt{\pi}} \frac{\cos nx}{\sqrt{\pi}}\, dx = \frac{1}{2\pi}\int_{-\pi}^{\pi} [\sin(m-n)x + \sin(m+n)x]\, dx$$

$$= \frac{-1}{2\pi}\left[\frac{\cos(m-n)x}{m-n} + \frac{\cos(m+n)x}{m+n}\right]\Bigg|_{-\pi}^{\pi} = 0$$

for all m and n, since $\sin(m-n)x = 0$ if $m = n$, and $\cos(-\theta) = \cos\theta$. Finally, the integrals

$$\int_{-\pi}^{\pi} \frac{\sin mx}{\sqrt{\pi}} \frac{1}{\sqrt{2\pi}}\, dx = \frac{-1}{\pi\sqrt{2}} \frac{\cos mx}{m}\Bigg|_{-\pi}^{\pi} = 0,$$

$$\int_{-\pi}^{\pi} \frac{\cos mx}{\sqrt{\pi}} \frac{1}{\sqrt{2\pi}}\, dx = \frac{1}{\pi\sqrt{2}} \frac{\sin mx}{m}\Bigg|_{-\pi}^{\pi} = 0,$$

$$\int_{-\pi}^{\pi} \left(\frac{1}{\sqrt{2\pi}}\right)^2 dx = \frac{1}{2\pi}x\Bigg|_{-\pi}^{\pi} = 1$$

complete the verification of the orthonormality. ◄

Some important sets of functions $\{f_n(x)\}$ that occur in physical applications are not orthogonal but have the property that for some nontrivial nonnegative function $w(x)$ the integral

$$(f_n, f_m)_w = \int_a^b w(x) f_n(x) f_m(x)\, dx = 0 \tag{6}$$

whenever $n \neq m$. If this happens, the set $\{f_n(x)\}$ is said to be **orthogonal with respect to the weight function** $w(x)$ **on the interval** $a \le x \le b$. The **weighted norm** of f_n is defined as

$$\| f_n \|_w = \left[\int_a^b w(x) f_n(x)^2\, dx\right]^{1/2} \tag{7}$$

and $\{f_n(x)\}$ is said to be **orthonormal with respect to the weight function** $w(x)$, if, in addition to being orthogonal with respect to $w(x)$, it satisfies the condition $\| f_n \|_w = 1$ for all n.

EXAMPLE 3 ► **The Laguerre polynomials**

Let $y_n(x)$ be a polynomial solution of the **Laguerre equation,***

$$(xe^{-x}y')' + ne^{-x}y = 0. \tag{8}$$

[Equation **(8)** is obtained by multiplying Equation **(ii)**, p. 246, by e^{-x}.] Then y_n is called a **Laguerre polynomial.** Consider the set of Laguerre polynomials $\{y_n(x)\}$, $n = 1, 2, 3, \ldots$, and suppose $m \neq n$. Then we have the equations

$$(xe^{-x}y_n')' + ne^{-x}y_n = 0,$$
$$(xe^{-x}y_m')' + me^{-x}y_m = 0.$$

We multiply the first equation by y_m and the second by $-y_n$ and add them to obtain

$$y_m(xe^{-x}y_n')' - y_n(xe^{-x}y_m')' = e^{-x}(m-n)y_ny_m. \tag{9}$$

* Edmond Laguerre (1834–1886), a French mathematician, made many contributions to geometry and the theory of infinite series.

Integrating both sides of **(9)** on the interval $0 \le x < \infty$, we have

$$\int_0^\infty [y_m(xe^{-x}y_n')' - y_n(xe^{-x}y_m')']\,dx = (m-n)\int_0^\infty e^{-x}y_n y_m\,dx. \tag{10}$$

But by the product formula for differentiation,

$$\begin{aligned}[xe^{-x}(y_m y_n' - y_n y_m')]' &= (xe^{-x}y_n')'y_m + xe^{-x}y_n' y_m' - (xe^{-x}y_m')'y_n - xe^{-x}y_m' y_n' \\ &= y_m(xe^{-x}y_n')' - y_n(xe^{-x}y_m')'.\end{aligned}$$

Thus Equation **(10)** becomes

$$(m-n)\int_0^\infty e^{-x}y_n y_m\,dx = xe^{-x}(y_m y_n' - y_n y_m')\Big|_0^\infty = 0, \qquad m \ne n,$$

because $f(x) = x(y_m y_n' - y_n y_m')$ is a polynomial (since each term is a polynomial) and the quotient $f(x)/e^x$ tends to zero as $x \to \infty$. This may be verified by applying L'Hôpital's rule as many times as the degree of f to the quotient $f(x)/e^x$:

$$\lim_{x\to\infty}\frac{f(x)}{e^x} = \lim_{x\to\infty}\frac{f'(x)}{e^x} = \ldots = \lim_{x\to\infty}\frac{c}{e^x} = 0.$$

Therefore the set of functions $\{y_n\}$, $n = 1, 2, 3, \ldots$ is orthogonal with respect to the weight function e^{-x} on $0 \le x < \infty$. ◀

SELF-QUIZ

Answer true *or* false.

I. The functions $f_n(x) = \sin nx$, $n = 1, 2, 3, \ldots$ are an orthogonal set of functions on the interval $0 \le x \le \pi$.

II. The functions $f_n(x) = \cos nx$, $n = 1, 2, 3, \ldots$ are an orthogonal set of functions on the interval $0 \le x \le \pi$.

III. The functions $f_1(x) = 1/\sqrt{2}$ and $f_2(x) = x$ are orthonormal on the interval $-1 \le x \le 1$.

Answers to Self-Quiz

I. True **II.** True **III.** False

PROBLEMS 11.2

In Problems 1–8 show that each given set of functions is orthogonal on the given interval and determine the corresponding orthonormal set.

1. $\{\cos nx\}$, $n = 0, 1, 2, \ldots,$ $\quad 0 \le x \le 2\pi$

2. $\left\{\sin \dfrac{n\pi x}{T}\right\}$, $n = 1, 2, 3, \ldots,$ $\quad -T \le x \le T$

3. $\left\{\cos \dfrac{2n\pi x}{T}\right\}$, $n = 0, 1, 2, \ldots,$ $\quad 0 \le x \le T$

4. $\{\sin 2nx\}$, $n = 1, 2, 3, \ldots,$ $\quad 0 \le x \le \pi$

5. $\{\cos 2nx\}$, $n = 0, 1, 2, \ldots,$ $\quad 0 \le x \le \pi$

6. $\{\sin 3nx\}$, $n = 1, 2, 3, \ldots,$ $\quad -\pi \le x \le \pi$

7. $\{\cos 3nx\}$, $n = 0, 1, 2, \ldots,$ $\quad |x| \le \pi$

8. $\{\sin 2nx, \cos 2nx\}$, $n = 1, 2, 3, \ldots,$ $\quad |x| \le \pi$

9. Hermite polynomials.* The functions

$$H_0 = 1, \qquad H_n(x) = (-1)^n e^{x^2}\frac{d^n}{dx^n}e^{-x^2},$$

$$n = 1, 2, 3, \ldots,$$

* Charles Hermite (1822–1901), a French mathematician, was known for his work in number theory.

are called Hermite polynomials (see Problem 5.5.9). Prove that

(a) $H_1(x) = 2x, \quad H_2(x) = 4x^2 - 2,$
$H_3(x) = 8x^3 - 12x;$

(b) the Hermite polynomials satisfy the relation

$$H_{n+1}(x) = 2xH_n(x) - H_n'(x);$$

(c) $H_n(x)$ is a solution of the **Hermite equation:**

$$y'' - 2xy' + 2ny = 0;$$

(d) the set of functions $\{H_n(x)\}$ is orthogonal with respect to the weight function $\exp(-x^2)$ on the interval $-\infty < x < \infty$.

10. The norm of a function $f(x)$ on an interval $a \le x \le b$ is defined in Equation **(2)** by the integral

$$\|f\| = \left(\int_a^b f^2(x)\,dx\right)^{1/2}.$$

This concept is a generalization of the notion of the distance of a point (x, y) in the plane from the origin:

$$d(x, y) = \sqrt{x^2 + y^2}.$$

Assume that f and g are continuous functions on $a \le x \le b$. Prove that the norm satisfies the following three properties:

(a) $\|f\| \ge 0$, where the equality holds if and only if $f(x) \equiv 0$;

(b) $\|af\| = |a| \cdot \|f\|$ for any constant a;

(c) $\|f + g\| \le \|f\| + \|g\|$ (triangle inequality).

[*Hint:* (a) Use the fact that if $f(x_0) \ne 0$, then by continuity $f^2(x) \ge \epsilon > 0$ for all x in some interval $x_0 - \delta \le x \le x_0 + \delta$, where $\delta > 0$. (c) Show that for any two real numbers α and β

$$0 \le \|\alpha f - \beta g\|^2 = \alpha^2\|f\|^2 - 2\alpha\beta(f, g) + \beta^2\|g\|^2.$$

Then let $\alpha = \|g\|$ and $\beta = \|f\|$, showing that

$$(f, g) \le \|f\| \cdot \|g\|.$$

Finally, use the last inequality to show that

$$\|f + g\|^2 \le (\|f\| + \|g\|)^2.]$$

11.3 FOURIER SERIES*

An infinite series of the form

$$\frac{a_0}{2} + \sum_{n=1}^{\infty} (a_n \cos nx + b_n \sin nx), \tag{1}$$

where a_n and b_n are constants, is generally referred to as a **trigonometric** or **Fourier series.** Note that Equation **(11.1.19)** is a special case of this expression with $a_n = 0$ for all n.

If series **(1)** converges to a periodic function $F(x)$ on the interval $[-\pi, \pi]$, then it converges for all real x, since

$$F(x) = \frac{a_0}{2} + \sum_{n=1}^{\infty} (a_n \cos nx + b_n \sin nx)$$

is a **periodic function of period** 2π:

$$F(x + 2\pi) = F(x) \quad \text{for all } x. \tag{2}$$

*Named after the French physicist Jean Baptiste Joseph Fourier (1768–1830). See the biography at end of this section.

This observation is obvious, since the functions $\cos nx$ and $\sin nx$ are periodic of period 2π. The constants a_n and b_n are called the **Fourier coefficients** of $F(x)$ and are given by the **Euler formulas:**

Euler formulas

$$a_n = \frac{1}{\pi}\int_{-\pi}^{\pi} F(x)\cos nx\,dx, \qquad b_n = \frac{1}{\pi}\int_{-\pi}^{\pi} F(x)\sin nx\,dx. \tag{3}$$

To see why **(3)** is true, observe that

$$\frac{1}{\pi}\int_{-\pi}^{\pi} F(x)\cos kx\,dx = \frac{1}{\pi}\int_{-\pi}^{\pi}\left(\frac{a_0}{2} + \sum_{n=1}^{\infty} a_n\cos nx + b_n \sin nx\right)\cos kx\,dx;$$

if term-by-term integration is permissible,* we have

$$\begin{aligned}\frac{1}{\pi}\int_{-\pi}^{\pi} F(x)\cos kx\,dx = \frac{a_0}{2\pi}\int_{-\pi}^{\pi}\cos kx\,dx \;&+ \sum_{n=1}^{\infty}\left(\frac{a_n}{\pi}\int_{-\pi}^{\pi}\cos nx\cos kx\,dx\right.\\ &\left.+\frac{b_n}{\pi}\int_{-\pi}^{\pi}\sin nx\cos kx\,dx\right).\end{aligned} \tag{4}$$

However, as we saw in Example 11.2.2, the functions

$$\frac{1}{\sqrt{2\pi}},\quad \frac{\cos x}{\sqrt{\pi}},\quad \frac{\sin x}{\sqrt{\pi}},\quad \frac{\cos 2x}{\sqrt{\pi}},\quad \frac{\sin 2x}{\sqrt{\pi}},\ \ldots$$

form an orthonormal set of functions on the interval $-\pi \le x \le \pi$. Hence all of the integrals in (4) are zero except

$$\frac{a_k}{\pi}\int_{-\pi}^{\pi}\cos^2 kx\,dx = a_k, \quad \text{if } k \ne 0, \quad \text{or} \quad \frac{a_0}{2\pi}\int_{-\pi}^{\pi} 1\,dx = a_0, \quad \text{if } k = 0.$$

Thus the first of the Euler formulas has been verified. The second Euler formula is proved similarly.

Remark

There is nothing special about the period of $F(x)$. Suppose that instead of having period 2π, the function F has period $2T$:

$$F(x + 2T) = F(x).$$

Then define the function

$$G(x) = F\left(\frac{xT}{\pi}\right) = F(y),$$

where $y = \frac{xT}{\pi}$, and note that

$$G(x + 2\pi) = F\left(\frac{xT}{\pi} + 2T\right) = F\left(\frac{xT}{\pi}\right) = G(x).$$

---

* Most advanced calculus textbooks contain proofs that term-by-term integration is permissible if series **(1)** converges *uniformly* on $a \le x \le b$. Uniform convergence is discussed in Appendix 3.

If there is a Fourier series **(1)** that converges to $G(x)$, then

$$G(x) = \frac{a_0}{2} + \sum_{n=1}^{\infty} (a_n \cos nx + b_n \sin nx) \tag{5}$$

on the interval $-\pi \leq x \leq \pi$; the substitution $x = \frac{\pi y}{T}$ can be used in **(5)** to obtain a Fourier series for F:

$$F(y) = G\left(\frac{\pi y}{T}\right) = \frac{a_0}{2} + \sum_{n=1}^{\infty} \left(a_n \cos \frac{n\pi y}{T} + b_n \sin \frac{n\pi y}{T}\right). \tag{6}$$

The change of variables above can also be used to show the orthogonality of the functions

$$1, \quad \cos \frac{\pi y}{T}, \quad \sin \frac{\pi y}{T}, \quad \cos \frac{2\pi y}{T}, \quad \sin \frac{2\pi y}{T}, \ldots$$

over the interval $|y| \leq T$.

So far we have concentrated only on properties of the Fourier series **(1).** We have not discussed how to go about determining the Fourier series for a given periodic function $F(x)$. The task is not difficult: we merely compute the coefficients a_n and b_n using the Euler formulas **(3)** and substitute these values in **(1).** The result is called the **Fourier series corresponding to** $F(x)$. The remaining question is whether the Fourier series so obtained actually converges to $F(x)$. If the Fourier series does converge to $F(x)$, we call it a **representation** of $F(x)$.

The class of periodic functions $F(x)$ that can be represented by Fourier series is very large, so large in fact, that it originally aroused a big controversy. The following theorem gives sufficient conditions for almost all conceivable practical applications.

THEOREM 1 ♦ **Fourier convergence theorem**

Let $F(x)$ be a periodic function with period $2T$ and such that $F(x)$ and $F'(x)$ are piecewise continuous* on the interval $-T \leq x \leq T$. Then $F(x)$ has a Fourier series

$$F(x) = \frac{a_0}{2} + \sum_{n=1}^{\infty} \left(a_n \cos \frac{n\pi x}{T} + b_n \sin \frac{n\pi x}{T}\right), \tag{7}$$

whose coefficients are given by the **Euler formulas**

$$a_n = \frac{1}{T} \int_{-T}^{T} F(x) \cos \frac{n\pi x}{T}\, dx, \qquad n = 0, 1, 2, \ldots, \tag{8}$$

$$b_n = \frac{1}{T} \int_{-T}^{T} F(x) \sin \frac{n\pi x}{T}\, dx, \qquad n = 1, 2, \ldots. \tag{9}$$

The Fourier series **(7)** converges to $F(x)$ at all points where F is continuous, and to $\frac{F(x+0) + F(x-0)}{2}$ at all points of jump discontinuity of F.† ◄

Remark

Note that $[\frac{F(x+0) + F(x-0)}{2}]$ is the average of the right- and left-hand limits at the point x. At any point of continuity, both of these values coincide, so **(7)** converges to $[\frac{F(x+0) + F(x-0)}{2}]$ for all x in the interval $-T \leq x \leq T$.

---

* See Section 6.1.

† $F(x + 0) = \lim_{h \to 0+} F(x + h)$; $F(x - 0) = \lim_{h \to 0+} F(x - h)$.

The conditions given in this theorem guarantee the existence and convergence of a Fourier series for the function $F(x)$, but they are *not* the most general of such conditions. Moreover, the convergence of a Fourier series to a function $F(x)$ does not imply that F satisfies the conditions given in this theorem. In summary, the conditions given in this theorem are neither necessary nor the most general sufficient conditions. Furthermore, even with these limitations, the proof of the theorem as stated is too complicated to be presented here.*

Although the theory of Fourier series is complicated, the application of these series is very easy. It should be clear from Theorem 1 that Fourier series apply to a much wider class of functions than Taylor series do, since discontinuous functions cannot have a Taylor series representation. It is also useful to consider some functions for which Theorem 1 *does not* apply.

EXAMPLE 1 ▶ **A function for which the Fourier convergence theorem does not apply**

The functions $F_1(x) = \frac{1}{x}$ and

$$F_2(x) = (-1)^n, \qquad \frac{1}{n+1} < |x| \le \frac{1}{n}, \qquad n = 1, 2, 3, \ldots, \tag{10}$$

do not satisfy the hypotheses of Theorem 1 in the interval $-1 \le x \le 1$, since neither function is piecewise continuous. The function F_1 has an infinite jump discontinuity at $x = 0$, whereas F_2 has an infinite number of discontinuities in $-1 \le x \le 1$ (see Figure 11.3). ◀

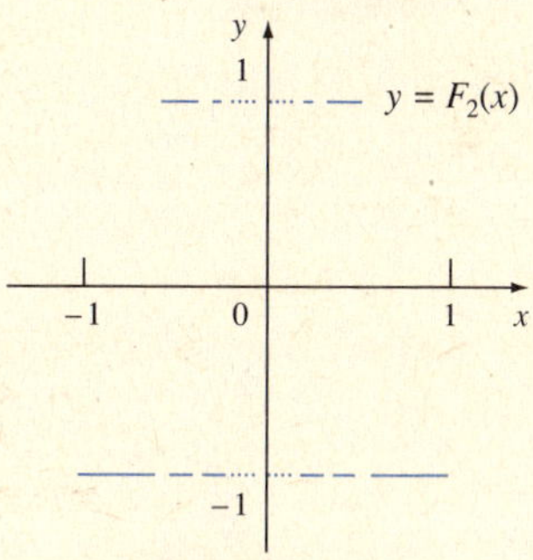

FIGURE 11.3
$F_2(x) = (-1)^n$ for $\frac{1}{n+1} < |x| \le \frac{1}{n}$

Let us now illustrate the procedure involved in obtaining the Fourier series of a function by some examples.

EXAMPLE 2 ▶ **The Fourier series of a periodic function**

Find the Fourier series of the periodic function (see Figure 11.4)

$$f(x) = (-1)^n k, \qquad n < x < n + 1.$$

Solution Functions of this type occur as off-on controls in mechanical systems. Observe that $f(x)$ has period $2T = 2$. By Equation **(8)** with $T = 1$,

$$a_n = \int_{-1}^{1} f(x) \cos n\pi x \, dx = -k \int_{-1}^{0} \cos n\pi x \, dx + k \int_{0}^{1} \cos n\pi x \, dx.$$

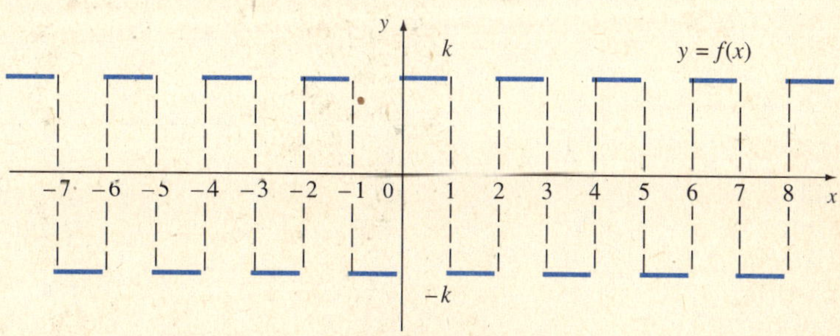

FIGURE 11.4
A periodic function

---

* Proofs of this theorem can be found in most books on advanced calculus or complex variables. See, for example, W. Kaplan, *Advanced Calculus* (Reading, Mass.: Addison-Wesley, 1973), p. 484; or W. R. Derrick, *Complex Analysis and Applications*, 2d ed. (Belmont, Calif.: Wadsworth, 1984), p. 267.

If $n = 0$, we obtain $a_0 = -k + k = 0$; and if $n \neq 0$,

$$a_n = \frac{-k}{n\pi} \sin n\pi x \Big|_{-1}^{0} + \frac{k}{n\pi} \sin n\pi x \Big|_{0}^{1} = 0.$$

Similarly, since $\cos(-t) = \cos t$,

$$\begin{aligned} b_n &= -k \int_{-1}^{0} \sin n\pi x \, dx + k \int_{0}^{1} \sin n\pi x \, dx \\ &= \frac{k}{n\pi} \cos n\pi x \Big|_{-1}^{0} - \frac{k}{n\pi} \cos n\pi x \Big|_{0}^{1} = \frac{2k}{n\pi}(1 - \cos n\pi), \end{aligned}$$

which is zero for even n and equals $\frac{4k}{n\pi}$ for odd n. Thus

$$f(x) = \frac{4k}{\pi}\left(\sin \pi x + \frac{1}{3} \sin 3\pi x + \frac{1}{5} \sin 5\pi x + \ldots\right). \quad \textbf{(11)}$$

The graphs of $f(x)$ and the first three partial sums of Equation **(11)** are shown in Figure 11.5.

As a bonus for our work above, note that if we set $x = \frac{1}{2}$ in **(11)**, we obtain the series

$$k = \frac{4k}{\pi}\left(1 - \frac{1}{3} + \frac{1}{5} - \frac{1}{7} + \ldots\right),$$

or

$$\frac{\pi}{4} = 1 - \frac{1}{3} + \frac{1}{5} - \frac{1}{7} + \ldots$$

since $\sin \frac{\pi}{2} = 1$, $\sin \frac{3\pi}{2} = -1$ and so on. This is a famous result that Leibniz obtained by means of a complicated geometrical construction.

As a final note, observe that the function $f(x)$ in this example satisfies the condition $f(-x) = -f(x)$. Functions having this property are said to be **odd**

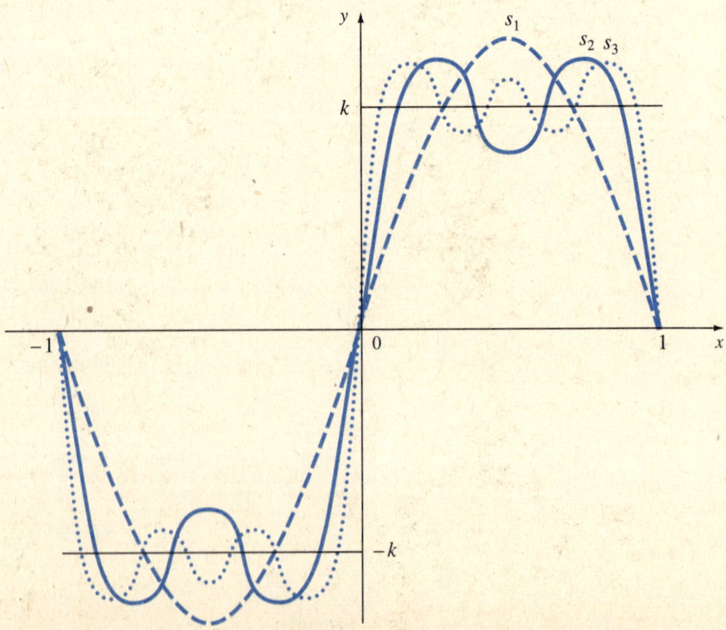

FIGURE 11.5
First three partial sums: dashed line = first term in **(11)**, solid line = first and second terms in **(11)**, dotted line = first three terms in **(11)**

functions. In particular, observe that all functions of the form $\sin n\pi x$ are odd functions. ◀

EXAMPLE 3 ▶ **The Fourier series of a sawtooth function**
Consider the sawtooth function

$$f(x) = \begin{cases} x + 1, & -1 \le x \le 0, \\ -x + 1, & 0 \le x \le 1, \end{cases} \qquad f(x + 2) = f(x),$$

shown in Figure 11.6(a). Again $T = 1$ and

$$a_n = \int_{-1}^{0} (x + 1) \cos n\pi x \, dx + \int_{0}^{1} (-x + 1) \cos n\pi x \, dx.$$

If $n = 0$, we have

$$a_0 = \left(\frac{x^2}{2} + x\right)\bigg|_{-1}^{0} + \left(x - \frac{x^2}{2}\right)\bigg|_{0}^{1} = 1,$$

and integrating by parts for $n \neq 0$, we obtain

$$\begin{aligned} a_n &= \frac{(x+1)}{n\pi} \sin n\pi x\bigg|_{-1}^{0} - \frac{1}{n\pi}\int_{-1}^{0} \sin n\pi x \, dx \\ &\quad + \frac{(1-x)}{n\pi} \sin n\pi x\bigg|_{0}^{1} + \frac{1}{n\pi}\int_{0}^{1} \sin n\pi x \, dx \\ &= \frac{1}{(n\pi)^2} \cos n\pi x\bigg|_{-1}^{0} - \frac{1}{(n\pi)^2} \cos n\pi x\bigg|_{0}^{1} = \frac{2}{(n\pi)^2}(1 - \cos n\pi). \end{aligned}$$

Therefore, a_n is zero for even $n \neq 0$ and equals $4/(n\pi)^2$ for odd n. Similarly,

$$\begin{aligned} b_n &= \frac{-(x+1)}{n\pi} \cos n\pi x\bigg|_{-1}^{0} + \frac{1}{n\pi}\int_{-1}^{0} \cos n\pi x \, dx \\ &\quad + \frac{(x-1)}{n\pi} \cos n\pi x\bigg|_{0}^{1} - \frac{1}{n\pi}\int_{0}^{1} \cos n\pi x \, dx \\ &= \frac{-1}{n\pi} + \frac{1}{(n\pi)^2} \sin n\pi x\bigg|_{-1}^{0} + \frac{1}{n\pi} - \frac{1}{(n\pi)^2} \sin n\pi x\bigg|_{0}^{1} = 0. \end{aligned}$$

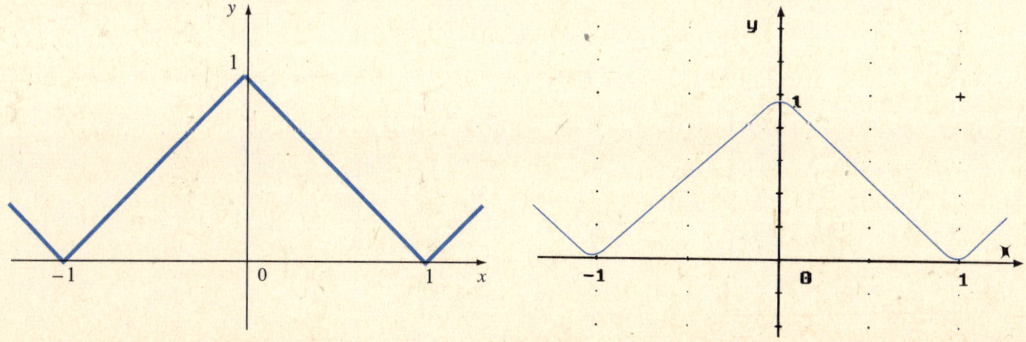

(a) graph of $f(x)$

(b) first four terms of (12)

FIGURE 11.6

Thus

$$f(x) = \frac{1}{2} + \frac{4}{\pi^2}\left(\cos \pi x + \frac{1}{3^2}\cos 3\pi x + \frac{1}{5^2}\cos 5\pi x + \frac{1}{7^2}\cos 7\pi x + \dots\right). \quad \textbf{(12)}$$

A computer-drawn sketch of the first four terms in **(12)** is given in Figure 11.6(b).

Incidentally, note that if we set $x = 1$, then

$$0 = \frac{1}{2} - \frac{4}{\pi^2}\left(1 + \frac{1}{3^2} + \frac{1}{5^2} + \frac{1}{7^2} + \dots\right),$$

or

$$\frac{\pi^2}{8} = 1 + \frac{1}{3^2} + \frac{1}{5^2} + \frac{1}{7^2} + \dots.$$

We observe that the term-by-term derivative of **(12)** yields

$$f'(x) = \frac{-4}{\pi}\left(\sin \pi x + \frac{1}{3}\sin 3\pi x + \frac{1}{5}\sin 5\pi x + \dots\right),$$

which is the same series as **(11)** for $k = -1$. An explanation for this fact is found by noting that $f'(x) = (-1)^n(-1)$ for $n < x < n + 1$.

Finally, note that this function satisfies the condition $f(-x) = f(x)$. All such functions are called **even** functions. In particular, $\cos n\pi x$ is an even function for every integer n. ◀

EXAMPLE 4 ▶ **The Fourier series of a clipped sine wave**

Consider the clipped sine wave

$$f(x) = \begin{cases} k \sin Tx, & 0 < x < \dfrac{\pi}{T}, \\ 0, & \dfrac{-\pi}{T} < x < 0, \end{cases} \qquad f\left(x + \frac{2\pi}{T}\right) = f(x),$$

which is obtained by passing a sinusoidal voltage $k \sin Tx$ through a half-wave rectifier (see Figure 11.7(a)). Here the period is $\frac{2\pi}{T}$, so

$$a_n = \frac{T}{\pi}\int_0^{\pi/T} k \sin Tx \cos nTx \, dx, \qquad a_0 = \frac{2k}{\pi}.$$

Using the identity

$$2 \sin Tx \cos nTx = \sin(n + 1)Tx - \sin(n - 1)Tx$$

as in Example 11.2.2, we have

$$\begin{aligned} a_n &= \frac{kT}{2\pi}\left[\frac{\cos(n-1)Tx}{(n-1)T} - \frac{\cos(n+1)Tx}{(n+1)T}\right]\Bigg|_0^{\pi/T} \\ &= \frac{k}{2\pi}\left[\frac{\cos(n-1)\pi - 1}{n-1} - \frac{\cos(n+1)\pi - 1}{n+1}\right], \qquad \text{for } n \neq 1, \end{aligned}$$

and

$$a_1 = \frac{kT}{2\pi}\int_0^{\pi/T} \sin 2Tx \, dx = \frac{kT}{2\pi}\left(\frac{1 - \cos 2\pi}{2T}\right) = 0.$$

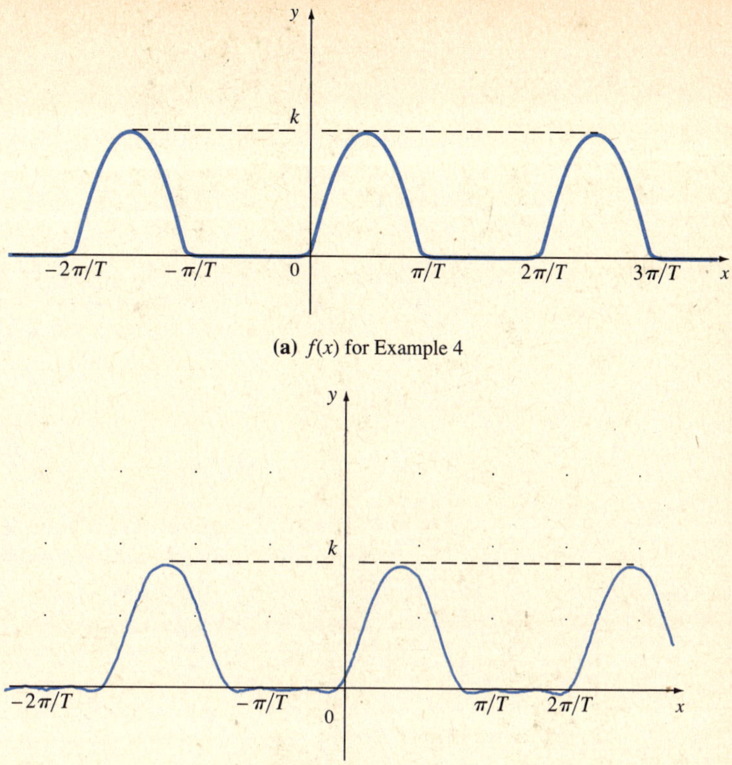

(a) $f(x)$ for Example 4

(b) first four terms of **(13)**

FIGURE 11.7

When n is odd, $a_n = 0$, and when n is even,

$$a_n = \frac{k}{\pi}\left(\frac{1}{n+1} - \frac{1}{n-1}\right) = \frac{-2k}{\pi(n^2-1)}.$$

Similarly, for $n \neq 1$,

$$\begin{aligned} b_n &= \frac{T}{\pi}\int_0^{\pi/T} k \sin Tx \sin nTx \, dx \\ &= \frac{kT}{2\pi}\int_0^{\pi/T} [\cos(n-1)Tx - \cos(n+1)Tx] \, dx \\ &= \frac{k}{2\pi}\left[\frac{\sin(n-1)Tx}{n-1} - \frac{\sin(n+1)Tx}{n+1}\right]\Bigg|_0^{\pi/T} = 0; \end{aligned}$$

and if $n = 1$, by Formula 100 of Appendix 1,

$$b_1 = \frac{k}{2\pi}(Tx - \sin Tx \cos Tx)\Big|_0^{\pi/T} = \frac{k}{2}.$$

Thus, since only b_1 and the evenly subscripted terms $a_{2n} = -2k/\pi\,(4n^2 - 1)$ are nonzero,

$$f(x) = \frac{k}{\pi} + \frac{k}{2}\sin Tx - \frac{2k}{\pi}\sum_{n=1}^{\infty}\frac{\cos 2nTx}{4n^2-1}. \qquad \textbf{(13)}$$

Incidentally, if we set $x = \frac{\pi}{2T}$, another intriguing identity results:

$$k = \frac{k}{\pi} + \frac{k}{2} - \frac{2k}{\pi}\sum_{n=1}^{\infty}\frac{\cos n\pi}{4n^2-1},$$

or

$$\frac{\pi}{2} - 1 = 2\left(\frac{1}{1\cdot 3} - \frac{1}{3\cdot 5} + \frac{1}{5\cdot 7} - \frac{1}{7\cdot 9} + \ldots\right).$$

Note that in this example the Fourier series **(13)** contains both sine and cosine terms. A computer-drawn sketch of the first four terms in **(13)** is given in Figure 11.7(b). ◀

Although we have concentrated in this section on trigonometric series, it is important to notice that other orthogonal sets of functions often provide useful series expansions. Let $\{f_n(x)\}$ be an orthogonal set of functions on an interval $a \le x \le b$ and suppose that the function $F(x)$ can be represented in terms of the functions $f_n(x)$ by a convergent series

$$F(x) = \sum_{n=1}^{\infty} c_n f_n(x) = c_1 f_1(x) + c_2 f_2(x) + \ldots. \tag{14}$$

Series **(14)** is called a **generalized Fourier series of** $F(x)$ and the coefficients $c_1, c_2, \ldots$ are called the **generalized Fourier coefficients of** $F(x)$ **with respect to the orthogonal set of functions** $\{f_n(x)\}$. It is easy to determine the constants $c_1, c_2, \ldots$ by the following procedure. Fixing m, we multiply both sides of **(14)** by $f_m(x)$ and integrate the result over the interval $a \le x \le b$:

$$(F, f_m) = \int_a^b F(x) f_m(x)\, dx = \sum_{n=1}^{\infty} c_n \int_a^b f_n(x) f_m(x)\, dx = \sum_{n=1}^{\infty} c_n (f_n, f_m).$$

Note that we have again assumed term-by-term integration to be permissible —an assumption that sometimes is not valid. Since $(f_n, f_m) = 0$ whenever $n \ne m$, we have

$$(F, f_m) = c_m \| f_m \|^2,$$

so that

Generalized Fourier coefficients

$$c_m = \frac{(F, f_m)}{\| f_m \|^2} = \frac{1}{\| f_m \|^2}\int_a^b F(x) f_m(x)\, dx, \qquad m = 1, 2, \ldots. \tag{15}$$

This procedure extends to orthogonal sets of functions $\{f_n(x)\}$ **with respect to a weight function** $w(x)$ on an interval $a \le x \le b$. If $F(x)$ can be represented by a generalized Fourier series in these weighted orthogonal functions,

$$F(x) = \sum_{n=1}^{\infty} c_n f_n(x), \tag{16}$$

then the general Fourier coefficients c_k can be determined by multiplying both sides of Equation **(16)** by $w(x) f_k(x)$ and integrating over the interval $a \le x \le b$:

$$\int_a^b w(x) F(x) f_k(x)\, dx = \sum_{n=1}^{\infty} c_n \int_a^b w(x) f_n(x) f_k(x)\, dx = c_k \| f_k \|_w^2,$$

or

$$c_k = \frac{(F, f_k)_w}{\| f_k \|_w^2}. \tag{17}$$

JEAN BAPTISTE JOSEPH FOURIER (1768–1830)

Source: Stock Montage, Inc.

One of the greatest mathematicians of the nineteenth century was the Frenchman Jean Baptiste Joseph Fourier.

Fourier was born in Auxerre in 1768 and died in Paris in 1830. The son of a tailor, he was orphaned at the age of eight and educated in a military school conducted by the Benedictines, where he was given a lectureship in mathematics before his twenty-first birthday. He assisted in the promotion of the French Revolution and was rewarded by a chair at the École Polytechnique. He resigned from this position so that he could accompany Napoleon on the Egyptian expedition. In 1798 he was appointed governor of Lower Egypt. Following the British victories and the capitulation of the French in 1801, Fourier returned to France and was made prefect of Grenoble. It was while at Grenoble that he started his experiments on heat.

In 1807, Fourier presented a paper before the French Academy of Sciences that initiated a new and highly fruitful chapter in the history of mathematics. The paper dealt with the practical problem of the flow of heat in metallic rods, plates, and solid bodies. In the course of the presentation of the paper, Fourier made the startling claim that any function, defined in a finite closed interval by an arbitrarily drawn graph, can be resolved into a sum of sine and cosine functions. To be more explicit, he claimed that *any* function whatever, no matter how capriciously it is defined in the interval $(-\pi, \pi)$, can be represented in that interval by

$$\frac{a_0}{2} + \sum_{n=1}^{\infty} (a_n \cos nx + b_n \sin nx),$$

where the a's and the b's are suitable real numbers. Such a series is known as a **trigonometric series,** and it was not new to the mathematicians of the time. Indeed, a number of more or less well behaved functions had been shown to be representable by such a series. But Fourier claimed that *any* function defined in $(-\pi, \pi)$ can be so represented. The scientists at the Academy were very skeptical of Fourier's claim, and the paper, which was judged by Lagrange, Laplace, and Legendre was rejected. However, to encourage Fourier to develop his ideas more carefully, the French Academy made the problem of heat propagation the subject of a grand prize to be awarded in 1812. Fourier submitted a revised paper in 1811, which was judged by a group containing, among others, the former three judges, and the paper won the prize, though it was criticized for lack of rigor and so was not recommended for publication in the Academy's *Mémoires.*

Resentful, Fourier continued his researches on heat, and, in 1822, after a move to Paris in 1816, he published one of the great classics of mathematics, his *Théorie analytique de la chaleur* (Analytical Theory of Heat). Two years after the publication of his great work, Fourier became secretary of the French Academy, and, in that capacity, was able to have his 1811 paper published in its original form in the Academy's *Mémoires.*

Though it has been shown that Fourier's claim that *any* function can be represented by a trigonometric series (or **Fourier series,** as they are commonly called today) is too extravagant, the class of functions so representable is very broad indeed. The Fourier series have proved to be highly valuable in such fields of study as acoustics, optics, electrodynamics, thermodynamics, and many others.

Lord Kelvin (William Thomson, 1824–1907) claimed that his whole career in mathematical physics was influenced by Fourier's work on heat, and Clerk

Maxwell (1831–1879) pronounced Fourier's treatise "a great mathematical poem."

A sad story is told about Fourier and his interest in heat. It seems that from his experience in Egypt, and maybe his work on heat, he became convinced that desert heat is the ideal condition for good health. He accordingly clothed himself in many layers of garments and lived in rooms of unbearably high temperature. It has been said by some that this obsession with heat hastened his death, by heart disease, so that he died, thoroughly cooked, in his sixty-third year.

Perhaps Fourier's most quoted sentence (it appeared in his early work on the mathematical theory of heat) is: "The deep study of nature is the most fruitful source of mathematical discovery."

SELF-QUIZ

Answer Questions I–V true *or* false.

I. Every function is equal to its Fourier series.

II. If $F(x + 0) = \frac{1}{2}$ and $F(x - 0) = 1$, then the Fourier series converges to $\frac{3}{4}$ at x.

III. The function $\cos 2x$ is an odd function.

IV. The function $\sin 3x$ is an odd function.

V. In Example 3 of Section 11.2, we showed that the Laguerre polynomials $\{y_n(x)\}$ are orthogonal with respect to the weight function e^{-x} on $(0, \infty)$. Then

$$\sum_{n=1}^{\infty} c_n y_n(x)$$

is a generalized Fourier series.

Answers to Self-Quiz

I. False **II.** True **III.** False **IV.** True **V.** True

PROBLEMS 11.3

1. Find the smallest positive period of the functions $\cos 2x$, $\sin \pi x$, $\cos(\frac{2\pi n x}{T})$, $\sin 2k\pi x$.

2. Show that a constant function is periodic with any period $2T > 0$.

3. Suppose that $f(x)$ has period $2T$. What is the period of $f(\frac{ax}{b})$?

4. Prove that a convergent infinite series of functions of period T is periodic of period T.

In Problems 5–13 find the Fourier series of each function $f(x)$ of period 2π, where one period is defined, and accurately plot the first three partial sums

$$\frac{a_0}{2} + \sum_{n=1}^{k} (a_n \cos nx + b_n \sin nx), \quad k = 1, 2, 3$$

5. $f(x) = x, |x| < \pi$

6. $f(x) = \begin{cases} 0, & -\pi < x < 0 \\ 1, & 0 < x < \pi \end{cases}$

7. $f(x) = x^2, |x| < \pi$

8. $f(x) = \begin{cases} 0, & -\pi < x < 0 \\ x, & 0 < x < \pi \end{cases}$

9. $f(x) = |x|, |x| < \pi$

10. $f(x) = \begin{cases} x, & -\pi < x < 0 \\ x - \pi, & 0 < x < \pi \end{cases}$

11. $f(x) = \begin{cases} -1, & -\pi < x < -1 \\ x, & -1 < x < 1 \\ 1, & 1 < x < \pi \end{cases}$

12. $f(x) = \begin{cases} \pi + x, & -\pi < x < 0 \\ \pi - x, & 0 < x < \pi \end{cases}$

13. $f(x) = e^x, |x| < \pi$

In Problems 14–21 find the Fourier series of each function $f(x)$ of period T, where one of the periods is defined.

14. $f(x) = x, |x| < 1, T = 2$

15. $f(x) = x, 0 < x < 2, T = 2$

16. $f(x) = x, 0 < x < 3, T = 3$

17. $f(x) = x^2, |x| < 1, T = 2$

18. $f(x) = x^2, 0 < x < 2, T = 2$

19. $f(x) = \begin{cases} 0 & 0 < x < 1, \\ 1, & 1 < x < 2, \quad T = 2 \end{cases}$

20 $f(x) = \begin{cases} 0 & 0 < x < 1, \\ x - 1, & 1 < x < 2, \quad T = 2 \end{cases}$

21. $f(x) = \begin{cases} x, & 0 < x < 1, \\ 1, & 1 < x < 2, \quad T = 2 \end{cases}$

22. Find the Fourier series of the periodic function of period 2π

$$f(x) = \frac{x^2}{4}, \qquad |x| < \pi,$$

and use this series to verify the identities:

$$\frac{\pi^2}{6} = 1 + \frac{1}{2^2} + \frac{1}{3^2} + \frac{1}{4^2} + \frac{1}{5^2} + \ldots,$$

$$\frac{\pi^2}{12} = 1 - \frac{1}{2^2} + \frac{1}{3^2} - \frac{1}{4^2} + \frac{1}{5^2} - \ldots,$$

$$\frac{\pi^2}{8} = 1 + \frac{1}{3^2} + \frac{1}{5^2} + \frac{1}{7^2} + \frac{1}{9^2} + \ldots.$$

11.4 HALF-RANGE EXPANSIONS

In Section 11.3 we defined the concepts of odd and even functions. A function $f(x)$ is **odd** if it satisfies the condition

$$f(-x) = -f(x) \tag{1}$$

and **even** if

$$f(-x) = f(x). \tag{2}$$

As we saw, the function $\sin n\pi x$ is odd, whereas $\cos n\pi x$ is even for $n = 1, 2, 3, \ldots$.

Knowing that a function $f(x)$ is even or odd can help us avoid unnecessary work in computing the Fourier coefficients of $f(x)$. This claim is based on the following facts.

THEOREM 1 ♦ **Facts about even and odd functions**

(a) The product of two even or two odd functions is even.

(b) The product of an even and an odd function is odd

(c) If $g(x)$ is an odd function, then

$$\int_{-T}^{0} g(x)\,dx = -\int_{0}^{T} g(x)\,dx \tag{3}$$

and, for every $T > 0$,

$$\int_{-T}^{T} g(x)\,dx = 0. \tag{4}$$

(d) If $g(x)$ is an even function, then

$$\int_{-T}^{0} g(x)\,dx = \int_{0}^{T} g(x)\,dx$$

and

$$\int_{-T}^{T} g(x)\,dx = 2\int_{0}^{T} g(x)\,dx. \tag{5}$$

Proof **(a)** Let $f(x)$ and $g(x)$ be even functions. By **(2)**,

$$f(-x)g(-x) = f(x)g(x),$$

indicating that their product is even. Now suppose that they are both odd functions. By **(1)**,

$$f(-x)g(-x) = [-f(x)][-g(x)] = f(x)g(x),$$

so again their product is even.

(b) If $f(x)$ is even and $g(x)$ is odd, then

$$f(-x)g(-x) = f(x)[-g(x)] = -f(x)g(x),$$

so the product is odd.

(c) Let $g(x)$ be an odd function and let $x = -t$, so that $dx = -dt$. Then

$$\int_{-T}^{0} g(x)\,dx = -\int_{T}^{0} g(-t)\,dt = \int_{0}^{T} g(-t)\,dt = -\int_{0}^{T} g(t)\,dt,$$

so we have

$$\int_{-T}^{T} g(x)\,dx = \int_{-T}^{0} g(x)\,dx + \int_{0}^{T} g(x)\,dx = 0.$$

(d)

$$\int_{-T}^{0} g(x)\,dx = -\int_{T}^{0} g(-t)\,dt = \int_{0}^{T} g(t)\,dt. \quad \blacklozenge$$

What precisely is the effect of **(4)** and **(5)** in computing the Fourier coefficients of an even or odd periodic function $F(x)$? If $F(x)$ is even and of period $2T$, then $F(x)\sin(\frac{n\pi x}{T})$ is odd and of period $2T$ by Theorem 1(b). Hence, by **(4)**,

$$\int_{-T}^{T} F(x)\sin\frac{n\pi x}{T}\,dx = 0,$$

and all the Fourier coefficients b_n are zero when $F(x)$ is even. Similarly, if $F(x)$ is odd and of period $2T$, then $F(x)\cos(\frac{n\pi x}{T})$ is odd and, by **(4)**,

$$\int_{-T}^{T} F(x)\cos\frac{n\pi x}{T}\,dx = 0,$$

indicating that all the Fourier coefficients a_n are zero. Thus we have the following result.

THEOREM 2 ♦ **Fourier sine and cosine theorem**

Let $F(x)$ be a periodic function of period $2T$ that has a Fourier series. If $F(x)$ is an even function, then all the Fourier coefficients b_n are zero, whereas if $F(x)$ is an odd function, then the Fourier coefficients a_n are zero. ♦

Remark

The Fourier series of an even function,

Fourier cosine series

$$F(x) = \frac{a_0}{2} + \sum_{n=1}^{\infty} a_n \cos\frac{n\pi x}{T}, \tag{6}$$

is called a **Fourier cosine series,** whereas that of an odd function,

Fourier sine series

$$F(x) = \sum_{n=1}^{\infty} b_n \sin \frac{n\pi x}{T}, \tag{7}$$

is said to be a **Fourier sine series.** Observe that Theorem 2 implies that for even and odd functions, we need to calculate only half as many coefficients as are generally required.

EXAMPLE 1 ▶ **A Fourier cosine series**

Find the Fourier series of the periodic function of period 2π given by

$$F(x) = |x|, \qquad |x| \le \pi.$$

Solution Since $F(-x) = |-x| = |x| = F(x)$ is even, we need only calculate a Fourier cosine series. By Equation **(5),**

$$a_n = \frac{1}{\pi}\int_{-\pi}^{\pi} |x| \cos nx\, dx = \frac{2}{\pi}\int_0^{\pi} x \cos nx\, dx.$$

Integrating by parts, we have

$$\begin{aligned} a_n &= \frac{2}{\pi}\left(\frac{x \sin nx}{n}\bigg|_0^{\pi} - \frac{1}{n}\int_0^{\pi} \sin nx\, dx \right) \\ &= \frac{2 \cos nx}{\pi n^2}\bigg|_0^{\pi} \\ &= \frac{2}{\pi n^2}[(-1)^n - 1], \qquad \text{for } n \ge 1. \end{aligned}$$

Also, by Equation **(5),**

$$a_0 = \frac{1}{\pi}\int_{-\pi}^{\pi} |x|\, dx = \frac{2}{\pi}\int_0^{\pi} x\, dx = \frac{x^2}{\pi}\bigg|_0^{\pi} = \pi,$$

so

$$F(x) = \frac{\pi}{2} + \sum_{n=1}^{\infty} \frac{2}{\pi n^2}[(-1)^n - 1] \cos nx,$$

or

$$|x| = \frac{\pi}{2} - \frac{4}{\pi}\sum_{k=0}^{\infty} \frac{\cos(2k+1)x}{(2k+1)^2}, \qquad |x| \le \pi. \tag{8}$$

In particular, setting $x = 0$ in Equation **(8)** we have

$$0 = \frac{\pi}{2} - \frac{4}{\pi}\sum_{k=0}^{\infty} \frac{1}{(2k+1)^2},$$

$$\frac{\pi^2}{8} = \sum_{k=0}^{\infty} \frac{1}{(2k+1)^2} = 1 + \frac{1}{3^2} + \frac{1}{5^2} + \cdots. \tag{9}$$ ◀

EXAMPLE 2 ▶ **A Fourier sine series**

What is the Fourier series of the periodic function

$$F(x) = x, \qquad |x| < T?$$

Solution Since $F(-x) = -x = -F(x)$ is odd, all the coefficients $a_n = 0$. By Theorem 1(a) and Equation **(5)**,

$$b_n = \frac{1}{T}\int_{-T}^{T} x \sin\frac{n\pi x}{T}\,dx = \frac{2}{T}\int_{0}^{T} x \sin\frac{n\pi x}{T}\,dx.$$

Integrating by parts, we have

$$b_n = \frac{2}{T}\left[\frac{-Tx\cos\left(\frac{n\pi x}{T}\right)}{n\pi}\Bigg|_0^T + \frac{T}{n\pi}\int_0^T \cos\frac{n\pi x}{T}\,dx\right]$$

$$= \frac{2}{T}\left[\frac{-(-1)^n T^2}{n\pi} + \left(\frac{T}{n\pi}\right)^2 \sin\frac{n\pi x}{T}\Bigg|_0^T\right]$$

$$= \frac{-2(-1)^n T}{n\pi}.$$

Hence

$$x = \frac{-2T}{\pi}\sum_{n=1}^{\infty}\frac{(-1)^n}{n}\sin\frac{n\pi x}{T}, \qquad |x| < T. \tag{10}$$

In particular, setting $x = \frac{T}{2}$ in **(10)** we have

$$\frac{T}{2} = \frac{-2T}{\pi}\sum_{n=1}^{\infty}\frac{(-1)^n}{n}\sin\frac{n\pi}{2},$$

or

$$\frac{\pi}{4} = 1 - \frac{1}{3} + \frac{1}{5} - \frac{1}{7} + \ldots. \tag{11}$$ ◀

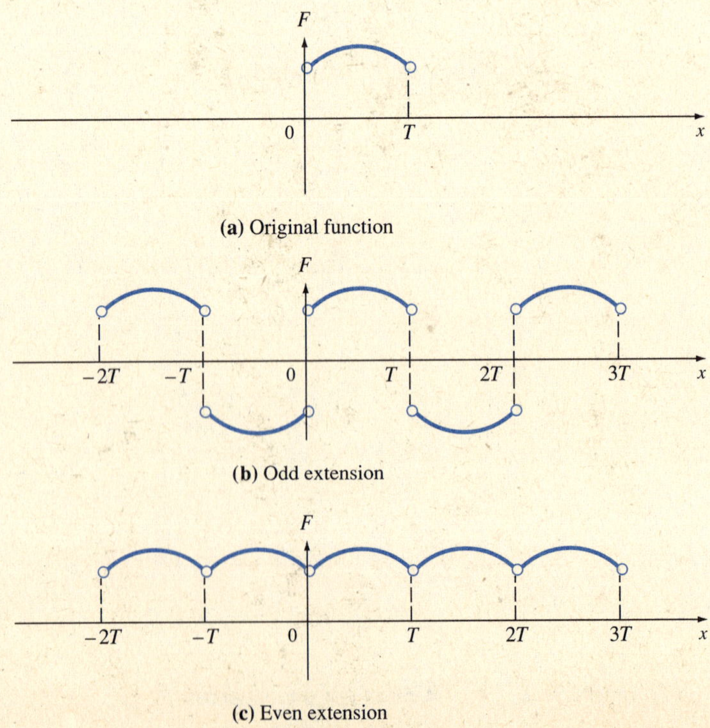

FIGURE 11.8
Odd and even extensions

In many problems of physics and engineering there is a practical need to apply a Fourier series to a nonperiodic function $F(x)$ that is defined only on the interval $0 < x < T$. Because of physical or mathematical considerations, it may be permissible to *extend* $F(x)$ over the interval $-T < x < T$, making it periodic of period $2T$. Figure 11.8 illustrates two such extensions. The **odd extension** in Figure 11.8(b) has a Fourier *sine* series; the **even extension** in Figure 11.8(c) has a Fourier *cosine* series.

EXAMPLE 3 ▶ Odd and even extensions of a function

Let $F(x) = x$ on $0 \leq x \leq \pi$. If we extend F as an even function of period 2π, we obtain the graph in Figure 11.9(c). Hence, for the even extension, $F_e(x) = |x|$ on $|x| \leq \pi$, so the result of Example 1 holds and

$$F_e(x) = |x| = \frac{\pi}{2} - \frac{4}{\pi}\sum_{k=0}^{\infty}\frac{\cos(2k+1)x}{(2k+1)^2}, \qquad |x| \leq \pi.$$

If we extend F as an odd function of period 2π [Figure 11.9(b)], then $F_0(x) = x$ on $|x| < \pi$ and the result (with $T = \pi$) of Example 2 applies:

$$F_0(x) = x = -2\sum_{n=1}^{\infty}\frac{(-1)^n}{n}\sin nx, \qquad |x| < \pi.$$

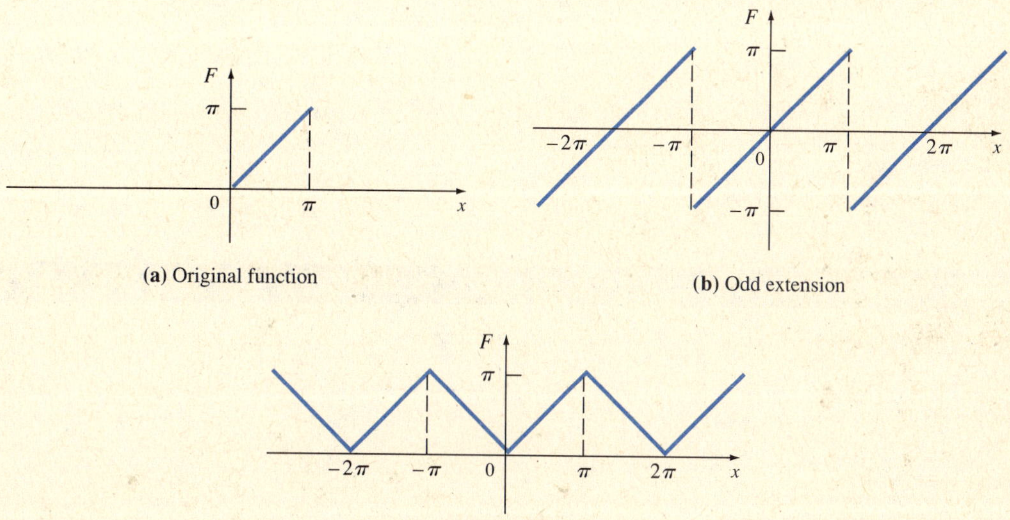

(c) Even extension

FIGURE 11.9
Extensions for Example 3 ◀

Since the Fourier series is based on a function $F(x)$ defined on only half of the interval of periodicity, the Fourier series for either an even or an odd extension is called a **half-range expansion.** Using **(5)** we obtain the following result.

THEOREM 3 ♦ Half-range expansion theorem

Let $F(x)$ be a function defined on the interval $0 \leq x \leq T$. Then an **even half-range expansion** of $F(x)$ is given by

$$F(x) = \frac{a_0}{2} + \sum_{n=1}^{\infty} a_n \cos\frac{n\pi x}{T},$$

where

$$a_n = \frac{2}{T}\int_0^T F(x)\cos\frac{n\pi x}{T}\,dx, \qquad n = 0, 1, 2, \ldots,$$

and an **odd half-range expansion** is given by

$$F(x) = \sum_{n=1}^{\infty} b_n \sin\frac{n\pi x}{T},$$

where

$$b_n = \frac{2}{T}\int_0^T F(x)\sin\frac{n\pi x}{T}\,dx, \qquad n = 1, 2, \ldots. \quad \blacklozenge$$

Examples of practical situations in which half-range expansions arise will be encountered in Sections 12.5–12.8.

SELF-QUIZ

Answer Questions I–IV true *or* false.

I. The product of three odd functions is odd.

II. If $F(x)$ is an odd function, then its Fourier series will be a Fourier sine series.

III. The function $F(x) = x^2$, $|x| \le \pi$, has a Fourier sine series.

IV. The odd half-range expansion of the function $F(x) = \cos x$, on $0 < x < \pi$, is a Fourier sine series.

V. The odd half-range expansion of the function in Question IV converges to ______ as $x \to 0$.
(a) 1 **(b)** -1 **(c)** 0 **(d)** $\frac{1}{2}$ **(e)** $-\frac{1}{2}$

Answers to Self-Quiz

I. True **II.** True **III.** False **IV.** True **V.** c

PROBLEMS 11.4

In Problems 1–10 determine whether each function is even, odd, or neither.

1. x^2 **2.** x^3 **3.** $x \sin x$ **4.** e^x

5. $x + x^3$ **6.** $x + x^2$ **7.** $\ln x$ **8.** $\sin x + \cos x$

9. $F(x) = \begin{cases} x, & -T < x \le 0, \\ 0, & 0 < x < T, \end{cases} \quad F(x + 2T) = F(x)$

10. $F(x) = \begin{cases} x^2, & -T < x < 0, \\ -x^2, & 0 < x < T, \end{cases} \quad F(x + 2T) = F(x)$

In Problems 11–20 find the Fourier series of the given functions, assuming that they are periodic. Pay particular attention to whether each function is even or odd.

11. $F(x) = x^2$, $|x| \le \pi$

12. $F(x) = x^3$, $|x| < \pi$

13. $F(x) = |x^3|$, $|x| \le \pi$

14. $F(x) = |\sin x|$, $|x| \le \pi$

15. $F(x) = \begin{cases} x + T, & -T < x < 0 \\ -x + T, & 0 < x < T \end{cases}$

16. $F(x) = \begin{cases} -x - T, & -T < x < 0 \\ x - T, & 0 < x < T \end{cases}$

17. $F(x) = \begin{cases} -x - T, & -T < x < 0 \\ -x + T, & 0 < x < T \end{cases}$

18. $F(x) = \begin{cases} x + T, & -T < x < 0 \\ x - T, & 0 < x < T \end{cases}$

19. $F(x) = \begin{cases} (x + 1)^2, & -1 < x < 0 \\ (x - 1)^2, & 0 < x < 1 \end{cases}$

20. $F(x) = \begin{cases} (x + 1)^2, & -1 < x < 0 \\ -(x - 1)^2, & 0 < x < 1 \end{cases}$

Extend each of the functions in Problems 21–30 as both an even and an odd periodic function. Sketch the resulting graphs and compute their corresponding Fourier cosine or sine series.

21. $F(x) = k, 0 < x < 1$

22. $F(x) = x^2, 0 < x < 1$

23. $F(x) = x^3, 0 < x < 1$

24. $F(x) = e^x, 0 < x < 1$

25. $F(x) = \begin{cases} 1, & 0 < x < 1 \\ 0, & 1 < x < 2 \end{cases}$

26. $F(x) = \begin{cases} 1, & 0 < x < 1 \\ \frac{1}{2}, & 1 < x < 2 \end{cases}$

27. $F(x) = \begin{cases} 1, & 0 < x < 1 \\ -x + 2, & 1 < x < 2 \end{cases}$

28. $F(x) = \begin{cases} x, & 0 < x < 1 \\ 1, & 1 < x < 2 \end{cases}$

29. $F(x) = \begin{cases} x^2, & 0 < x < 1 \\ 1, & 1 < x < 2 \end{cases}$

30. $F(x) = \begin{cases} x^2, & 0 < x < \frac{1}{2} \\ 1, & \frac{1}{2} < x < 1 \end{cases}$

31. Using the formula $e^{ix} = \cos x + i \sin x$, show that

(a) $\cos nx = \frac{1}{2}(e^{inx} + e^{-inx}), \sin nx = \frac{1}{2i}(e^{inx} - e^{-inx});$

(b) the Fourier series

$$F(x) = \frac{a_0}{2} + \sum_{n=1}^{\infty} (a_n \cos nx + b_n \sin nx)$$

can be written in the **complex form**

$$F(x) = \sum_{n=-\infty}^{\infty} c_n e^{inx};$$

(c) the **complex Fourier coefficients** c_n in (b) are given by

$$c_n = \frac{1}{2\pi}\int_{-\pi}^{\pi} F(x)e^{-inx}\,dx, \qquad n = 0, \pm 1, \pm 2, \ldots.$$

32. Show that the complex Fourier coefficients of an even function are real.

33. Show that the complex Fourier coefficients of an odd function are pure imaginary.

11.5 STURM-LIOUVILLE PROBLEMS

Examples 11.1.1 and 11.1.2, as well as several of the homogeneous boundary value problems we will encounter in Chapter 12, belong to a wide class of problems whose eigenfunctions and eigenvalues have particularly nice properties. We develop a small part of the fascinating theory of this class of problems in this and the next section.

Consider the differential equation

Sturm-Liouville equation

$$[r(x)y']' + [p(x) + \lambda q(x)]y = 0, \qquad (1)$$

where $r(x)$, $r'(x)$, $p(x)$, and $q(x)$ are continuous functions on some interval $a \le x \le b$ and λ is a real parameter. Many of the differential equations of applied mathematics, such as Legendre's equation and Bessel's equation, can be written in the form **(1)**. Suppose that we impose the boundary conditions

Boundary conditions for Sturm-Liouville problem

$$\begin{aligned} a_1 y(a) - a_2 y'(a) &= 0, \\ b_1 y(b) - b_2 y'(b) &= 0, \end{aligned} \qquad (2)$$

at the endpoints of the interval and require that at least one coefficient in each Equation in **(2)** be nonzero. Equation **(1),** together with the boundary conditions **(2),** is known as a **Sturm-Liouville problem.***

Observe, as in Example 11.1.1, that any Sturm-Liouville problem has the trivial solution $y \equiv 0$. For certain values of the parameter λ, nontrivial solutions of the Sturm-Liouville problem may exist. All such nontrivial solutions are called **eigenfunctions**† of the problem, and the corresponding values of λ that yield these solutions are called **eigenvalues** of the problem.

EXAMPLE 1 ▶ Finding eigenvalues and eigenvectors of a Sturm-Liouville problem

Find the eigenvalues and eigenfunctions of the Sturm-Liouville problem

$$y'' + \lambda y = 0, \qquad y(0) = y'(1) = 0. \qquad (3)$$

Solution Setting $r(x) \equiv q(x) \equiv 1$ and $p(x) \equiv 0$, we see that Equation **(3)** has the form of Equation **(1),** and letting $a = 0$, $b = 1$, $a_1 = -b_2 = 1$, and $a_2 = b_1 = 0$, we have the right type of boundary conditions. Thus Equation **(3)** is a Sturm-Liouville problem.

The roots of the characteristic equation are $\pm\sqrt{-\lambda}$, so we obtain the general solutions

$$y(x) = \begin{cases} c_1 e^{\sqrt{-\lambda}x} + c_2 e^{-\sqrt{-\lambda}x}, & \text{if } \lambda < 0, \\ c_1 + c_2 x, & \text{if } \lambda = 0, \\ c_1 \cos\sqrt{\lambda}\,x + c_2 \sin\sqrt{\lambda}\,x, & \text{if } \lambda > 0. \end{cases}$$

Thus we need to examine three cases:

Case 1

If $\lambda < 0$, the boundary conditions $y(0) = 0$ and $y'(1) = 0$ yield, after a short computation, the homogeneous system of equations

$$\begin{aligned} c_1 + c_2 &= 0, \\ \sqrt{-\lambda}\, e^{\sqrt{-\lambda}} c_1 - \sqrt{-\lambda}\, e^{-\sqrt{-\lambda}} c_2 &= 0. \end{aligned}$$

It is easy to verify that the only solution to this system is $c_1 = c_2 = 0$, which indicates that problem **(3)** has only trivial solutions if $\lambda < 0$.

Case 2

If $\lambda = 0$, the boundary condition $y(0) = 0$ implies that $c_1 = 0$, and since $y' \equiv c_2$, the second condition forces c_2 to be zero. Again **(3)** has only a trivial solution.

* The Swiss mathematician J. C. F. Sturm (1803–1855) made many significant contributions to the theory of differential equations. Joseph Liouville (1809–1882), a French mathematician, was noted for his work in complex analysis.

† Any constant multiple of an eigenfunction is also an eigenfunction.

Case 3

If $\lambda > 0$, the condition $y(0) = 0$ implies that $c_1 = 0$, so the solution to the problem must have the form

$$y(x) = c_2 \sin \sqrt{\lambda}\, x. \tag{4}$$

Differentiating Equation **(4)**, we have $y'(x) = c_2\sqrt{\lambda} \cos \sqrt{\lambda}\, x$, and setting $x = 1$, we obtain the equation $c_2\sqrt{\lambda} \cos \sqrt{\lambda} = 0$. Hence in this case we have nontrival solutions whenever $\cos \sqrt{\lambda}$ is zero. This occurs whenever the constant $\sqrt{\lambda}$ is one of the numbers

$$\frac{\pi}{2}, \frac{3\pi}{2}, \frac{5\pi}{2}, \frac{7\pi}{2}, \ldots;$$

that is, whenever λ takes on one of the values.

$$\frac{\pi^2}{4}, \frac{9\pi^2}{4}, \frac{25\pi^2}{4}, \frac{49\pi^2}{4}, \ldots. \tag{5}$$

The values **(5)** are the eigenvalues of **(3)**, and the eigenfunctions corresponding to the eigenvalues $\lambda_k = (2k - 1)^2\pi^2/4$, $k = 1, 2, 3, \ldots$ have the form

$$y_k(x) = A \sin \frac{(2k - 1)\pi x}{2}, \qquad A \neq 0. \quad ◀$$

EXAMPLE 2 ▶ Finding eigenvalues and eigenfunctions

Consider the Sturm-Liouville problem

$$(xy')' + \frac{\lambda}{x} y = 0, \qquad y(1) = y(e) = 0. \tag{6}$$

If we write $(xy')'$ as $xy'' + y'$ and multiply the differential equation on both sides by x, we obtain the Euler equation considered in Section 3.7. We therefore assume that $y = x^r$ is a solution of **(6)** and find, after a short calculation, that

$$(rx^r)' + \lambda x^{r-1} = (r^2 + \lambda)x^{r-1} = 0.$$

Since x does not vanish on $1 \leq x \leq e$, it follows that $r = \pm\sqrt{-\lambda}$. Again we must deal with three cases:

Case 1

If $\lambda < 0$, we have a general solution of the form

$$y(x) = c_1 x^{\sqrt{-\lambda}} + c_2 x^{-\sqrt{-\lambda}}.$$

Using the boundary conditions, we obtain the homogeneous equations

$$c_1 + c_2 = 0,$$
$$e^{\sqrt{-\lambda}} c_1 + e^{-\sqrt{-\lambda}} c_2 = 0.$$

Setting $c_2 = -c_1$ in the second equation, we obtain

$$2c_1 \sinh \sqrt{-\lambda} = c_1(e^{\sqrt{-\lambda}} - e^{-\sqrt{-\lambda}}) = 0,$$

and since $\sinh \sqrt{-\lambda} \neq 0$ if $\lambda \neq 0$, it follows that $c_1 = c_2 = 0$. Thus **(6)** has only trivial solutions when $\lambda < 0$.

Case 2

If $\lambda = 0$, Equation **(6)** reduces to the equation $(xy')' = 0$. Integrating both sides, we obtain $xy' = c_1$, so $y' = c_1/x$ and $y(x) = c_1 \ln x + c_2$. The condition $y(1) = 0$ implies that $c_2 = 0$. Then $y(e) = c_1 \ln e = c_1 = 0$. Again we have only trivial solutions.

Case 3

If $\lambda > 0$, we use the identity $e^{i\theta} = \cos\theta + i \sin\theta$ (see Appendix 5) to write

$$y_1(x) = x^{\sqrt{-\lambda}} = (e^{\ln x})^{i\sqrt{\lambda}} = \cos(\sqrt{\lambda} \ln x) + i \sin(\sqrt{\lambda} \ln x),$$
$$y_2(x) = x^{-\sqrt{-\lambda}} = (e^{\ln x})^{-i\sqrt{\lambda}} = \cos(\sqrt{\lambda} \ln x) - i \sin(\sqrt{\lambda} \ln x).$$

Letting $y_1^* = (y_1 + y_2)/2$ and $y_2^* = (y_1 - y_2)/2i$ (see Section 3.4), we may write the general solution of **(6)** in the form

$$y(x) = c_1 \cos(\sqrt{\lambda} \ln x) + c_2 \sin(\sqrt{\lambda} \ln x), \qquad \lambda > 0.$$

Setting $x = 1$, we find that the condition $y(1) = 0$ implies that $c_1 = 0$, so

$$y(x) = c_2 \sin(\sqrt{\lambda} \ln x).$$

Finally, setting $x = e$, we have $0 = y(e) = c_2 \sin \sqrt{\lambda}$. Therefore **(6)** has a nontrivial solution whenever $\sqrt{\lambda} = k\pi$, $k = 1, 2, 3, \ldots$. Thus the numbers $\lambda_k = k^2\pi^2$, $k = 1, 2, 3, \ldots$ are all eigenvalues of **(6)**, and the eigenfunctions corresponding to λ_k have the form

$$y_k(x) = A \sin(k\pi \ln x).$$

Note that it is not necessary to write this solution in the form $A \sin(k\pi \ln |x|)$, because we are interested only in values of x in the interval $1 \le x \le e$. Thus $x > 0$ and $\ln |x| = \ln x$. ◀

EXAMPLE 3 ▶ A more complicated boundary condition

Consider the Sturm-Liouville problem

$$y'' + \lambda y = 0, \qquad y(0) + y'(0) = 0, \qquad y(1) = 0.$$

Then, as in Example 1,

$$y(x) = \begin{cases} c_1 e^{\sqrt{-\lambda}x} + c_2 e^{-\sqrt{-\lambda}x}, & \text{if } \lambda < 0, \\ c_1 + c_2 x, & \text{if } \lambda = 0, \\ c_1 \cos \sqrt{\lambda}\, x + c_2 \sin \sqrt{\lambda}\, x, & \text{if } \lambda > 0. \end{cases}$$

It is easy to verify that the boundary conditions imply that $c_1 = c_2 = 0$ if $\lambda < 0$. If $\lambda = 0$, we have

$$y(0) = c_1, \qquad y'(0) = c_2, \qquad y(1) = c_1 + c_2,$$

and so both boundary conditions imply that $c_1 + c_2 = 0$. We therefore obtain the eigenfunctions

$$y_0(x) = c_1(1 - x).$$

If $\lambda > 0$, then

$$y(0) = c_1, \qquad y'(0) = \sqrt{\lambda}\, c_2, \qquad y(1) = c_1 \cos \sqrt{\lambda} + c_2 \sin \sqrt{\lambda}.$$

The boundary conditions imply that

$$c_1 + \sqrt{\lambda}\, c_2 = 0,$$
$$c_1 \cos \sqrt{\lambda} + c_2 \sin \sqrt{\lambda} = 0.$$

Setting $c_1 = -\sqrt{\lambda}\, c_2$, we find that the second equation becomes

$$-c_2 \sqrt{\lambda} \cos \sqrt{\lambda} + c_2 \sin \sqrt{\lambda} = 0,$$

or, dividing by $c_2 \cos \sqrt{\lambda}$ and simplifying,

$$\tan \sqrt{\lambda} = \sqrt{\lambda}.$$

We can easily see that there are an infinite number of real eigenvalues by plotting the curves $y = \tan x$ and $y = x$. The eigenvalues are the squares of the x-values of the points of intersection. From Figure 11.10 it seems clear that the square roots of the eigenvalues get closer and closer to the vertical asymptotes of $\tan x$; that is,

$$\sqrt{\lambda_n} \approx \left(\frac{2n+1}{2}\right)\pi \quad \text{or} \quad \lambda_n \approx \frac{(2n+1)^2}{4}\pi^2.$$

The eigenfunctions are

$$y_n(x) = A(\sin \sqrt{\lambda_n}\, x - \sqrt{\lambda_n} \cos \sqrt{\lambda_n}\, x), \qquad n \geq 1, \quad A \neq 0.$$ ◀

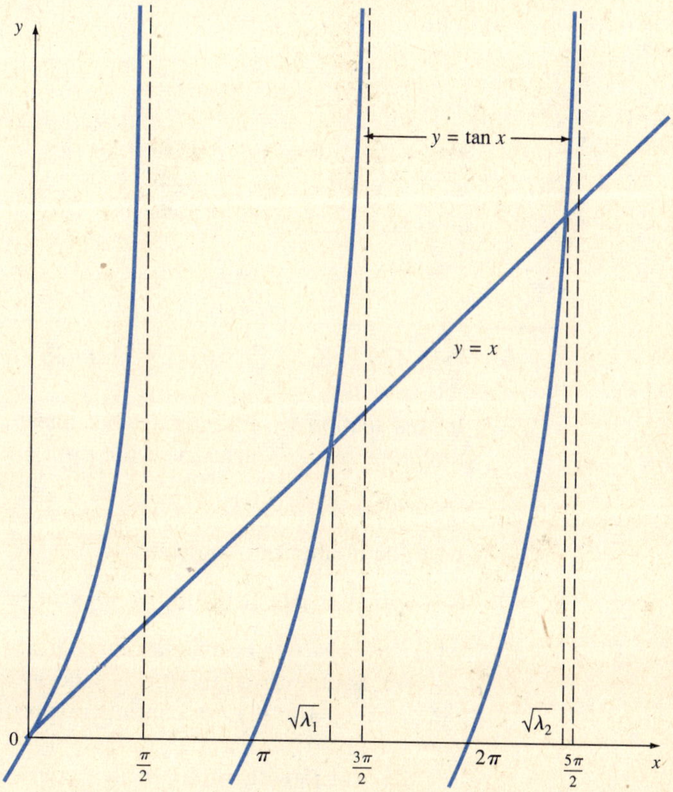

FIGURE 11.10
Graphs of $y = \tan x$ and $y = x$

SELF-QUIZ

Answer Questions I and II true or false.

I. The equation $y'' + \lambda y = 0$ with the conditions $y(0) = y(1)$, $y'(0) = y'(1)$, is a Sturm-Liouville problem.

II. Any constant multiple of an eigenfunction is also an eigenfunction.

III. Which of the following are eigenfunctions of the Sturm-Liouville problem

$$y'' + \lambda y = 0, \qquad y(0) = y(1) = 0?$$

(a) $\sin n\pi x$ **(b)** $\cos n\pi x$ **(c)** $\sin \frac{(2n-1)\pi x}{2}$

(d) $\cos \frac{(2n-1)\pi x}{2}$

IV. Which of the following are eigenfunctions of the Sturm-Liouville problem

$$y'' + \lambda y = 0, \qquad y'(0) = y(1) = 0?$$

(a) $\sin n\pi x$ **(b)** $\cos n\pi x$ **(c)** $\sin \frac{(2n-1)\pi x}{2}$

(d) $\cos \frac{(2n-1)\pi x}{2}$

Answers to Self-Quiz

I. False **II.** True **III.** a **IV.** d

PROBLEMS 11.5

Find the eigenvalues and corresponding eigenfunctions of the given Sturm-Liouville problems.

1. $y'' + \lambda y = 0,\ y(0) = y(\pi) = 0$

2. $y'' + \lambda y = 0,\ y(0) = y'(\pi) = 0$

3. $y'' + \lambda y = 0,\ y\left(\frac{-\pi}{2}\right) = y\left(\frac{\pi}{2}\right) = 0$

4. $y'' + \lambda y = 0,\ y'(0) = y'(\pi) = 0$

5. $(xy')' + \frac{\lambda}{x} y = 0,\ y(1) = y(e^2) = 0$

6. $(x^2y')' + \lambda y = 0,\ y(1) = y(e) = 0$

7. $(x^2y')' + \frac{\lambda^2}{x^2} y = 0,\ y\left(\frac{1}{2}\right) = y(1) = 0$

[*Hint*: Try $y_1 = \sin(\frac{\lambda}{x})$.]

8. $\left(\frac{1}{2x} y'\right)' - 2x\lambda y = 0,\ y(1) = y(\sqrt{2}) = 0$

11.6 THE ORTHOGONALITY THEOREM

In this section we investigate four basic properties of the eigenvalues and eigenfunctions of the Sturm-Liouville problem mentioned in the last section:

$$[r(x)y']' + [p(x) + \lambda q(x)]y = 0 \tag{1}$$

with the boundary conditions

$$a_1 y(a) - a_2 y'(a) = 0, \qquad b_1 y(b) - b_2 y'(b) = 0, \tag{2}$$

where at least one coefficient in each equation in **(2)** is nonzero.

We assume throughout this section that the real-valued functions $r(x)$, $r'(x)$, $p(x)$, and $q(x)$ in **(1)** are continuous on the interval $a \le x \le b$ and that $r(x)$ and $q(x)$ are positive on $a < x < b$.

Sturm-Liouville problems with these restrictions have the following four basic properties.

Properties of Sturm-Liouville problems

1. They have infinitely many eigenvalues.
2. The eigenvalues are real (so we need not worry about complex eigenvalues).
3. Each eigenvalue has a single linearly independent eigenfunction (simplifying our task of finding eigenfunctions). That is, the geometric multiplicity of each eigenvalue is 1.
4. Eigenfunctions corresponding to different eigenvalues are orthogonal with respect to the weight function $q(x)$ (allowing us to construct generalized Fourier series in terms of these eigenfunctions).

Not all of the restrictions are required for each of these properties, but all four hold when the restrictions apply. It will be clear in our proofs which restrictions must hold.

We prove only part (4). Parts (2) and (3) are left as exercises (see Problems 8 and 11); part (1) is beyond the scope of this book.

We begin by showing that the eigenfunctions of this problem are orthogonal with respect the weight function $q(x)$.

THEOREM 1 ♦ **Orthogonality theorem**

Let the real-valued functions $r(x)$, $r'(x)$, $p(x)$, and $q(x)$ of Equation **(1)** be continuous on the interval $a \le x \le b$, and let $y_m(x)$ and $y_n(x)$ be eigenfunctions corresponding to distinct eigenvalues λ_m and λ_n of the Sturm-Liouville problem **(1)**, **(2)**. Then y_m and y_n are orthogonal with respect to the weight function $q(x)$.

Proof The functions y_m and y_n satisfy the equations

$$(ry_m')' + (p + \lambda_m q)y_m = 0,$$
$$(ry_n')' + (p + \lambda_n q)y_n = 0.$$

Multiplying the first equation by y_n and the second by $-y_m$ and adding the resulting equations together yields

$$y_n(ry_m')' - y_m(ry_n')' = (\lambda_n - \lambda_m)qy_my_n.$$

Integrating all terms from a to b and the first two terms by parts, we obtain

$$\left(y_nry_m'\Big|_a^b - \int_a^b ry_m'y_n'\,dx\right) - \left(y_mry_n'\Big|_a^b - \int_a^b ry_n'y_m'\,dx\right)$$
$$= (\lambda_n - \lambda_m)\int_a^b qy_my_n\,dx. \quad \textbf{(3)}$$

Canceling the two identical integrals on the left-hand side, we obtain

$$r(b)[y_n(b)y_m'(b) - y_m(b)y_n'(b)] - r(a)[y_n(a)y_m'(a) - y_m(a)y_n'(a)]$$
$$= (\lambda_n - \lambda_m)\int_a^b qy_my_n\,dx. \quad \textbf{(4)}$$

By hypothesis, at least one of the constants a_1, a_2 in **(2)** is nonzero [see Equation **(2)**], so we have either

$$y(a) = \frac{a_2}{a_1}y'(a) \quad \text{or} \quad y'(a) = \frac{a_1}{a_2}y(a). \quad \textbf{(5)}$$

If the first of these equations is permissible, then it is satisfied by both $y_m(a)$ and $y_n(a)$, so the second term in brackets in **(4)** becomes

$$y_n(a)y_m'(a) - y_m(a)y_n'(a) = \frac{a_2}{a_1}y_n'(a)y_m'(a) - \frac{a_2}{a_1}y_m'(a)y_n'(a) = 0.$$

Similarly, if the second equation in **(5)** is permissible, then the second term in brackets is zero. In the same way, the second boundary condition in **(2)** causes the first term in brackets in **(4)** to be zero. Thus

$$(\lambda_n - \lambda_m)\int_a^b q y_m y_n \, dx = 0,$$

and since $\lambda_n \neq \lambda_m$, the proof is complete. ♦

Remark

Observe that if $r(a) = 0$, we do not need the first boundary condition in **(2)** to prove the orthogonality theorem. Similarly, the second boundary condition is not required if $r(b) = 0$. Finally, if $r(a) = r(b)$, we can also obtain the conclusion of the orthogonality theorem by assuming the two-point conditions

$$y(a) = y(b), \qquad y'(a) = y'(b) \tag{6}$$

instead of (2).* The proof of this fact is obvious, since the quantities in brackets in **(4)** are now identical, so that the left-hand side of **(4)** is zero. Furthermore, *if the functions $r(x)$, $p(x)$, and $q(x)$ are periodic with period $(b - a)$, then any eigenfunction of this problem is also periodic with period $(b - a)$* (see Problem 6).

EXAMPLE 1 ▶ **Applying the orthogonality theorem**

Consider the Sturm-Liouville problem

$$y'' + \lambda y = 0, \qquad y(0) = y'(1) = 0.$$

As we saw in Example 11.5.1, the eigenfunctions of this problem have the form

$$\sin\frac{\pi}{2}x,\ \sin\frac{3\pi}{2}x,\ \sin\frac{5\pi}{2}x,\ \ldots,\ \sin\frac{(2k-1)}{2}\pi x,\ \ldots.$$

Theorem 1 implies that these functions are orthogonal, a fact that we established directly in Section 11.3 (p. 538). ◀

EXAMPLE 2 ▶ **The Legendre polynomials are orthogonal**

Using the orthogonality theorem, it is easy to verify that the Legendre polynomials form an orthogonal set of functions. Since $[(1 - x^2)y']' = (1 - x^2)y'' - 2xy'$, we can write Legendre's equation (see Section 5.5) in the Sturm-Liouville form

$$[(1 - x^2)y']' + \lambda y = 0, \qquad \lambda = n(n + 1).$$

Since $r(x) = 1 - x^2$ vanishes at $x = \pm 1$, no boundary conditions are needed for the theorem to apply; and since $q(x) \equiv 1$, we immediately have

$$(P_m, P_n) = \int_{-1}^{1} P_m(x)P_n(x)\,dx = 0, \qquad \text{if } m \neq n.$$ ◀

* Note that property (3) fails if Equations **(2)** are replaced by the mixed two-point conditions **(6)** (see Problem 10).

The endpoints a and b in the orthogonality theorem need not be finite provided the improper integrals in **(3)** all exist. We illustrate this situation with the following example.

EXAMPLE 3 ▶ **The Laguerre polynomials are orthogonal**

The Laguerre equation

$$(xe^{-x}y')' + ne^{-x}y = 0$$

is of the Sturm-Liouville type, and $r(x) = xe^{-x}$ vanishes at $x = 0$. By L'Hôpital's rule,

$$\begin{aligned}\lim_{x\to\infty} r(x) &= \lim_{x\to\infty} \frac{x}{e^x} \\ &= \lim_{x\to\infty} \frac{1}{e^x} = 0.\end{aligned}$$

Thus for the Laguerre polynomials $\{y_n(x)\}$ no boundary conditions are required to apply the orthogonality theorem, and the Laguerre polynomials $\{y_n(x)\}$ form an orthogonal set of functions with respect to the weight function $q(x) = e^{-x}$ (compare this proof to that in Example 11.2.3):

$$\int_0^\infty e^{-x}y_m(x)y_n(x)\,dx = 0 \qquad \text{if } m \neq n. \quad ◀$$

In every example we have seen so far, the eigenvalues of the given Sturm-Liouville problem are real, not because the examples were carefully chosen, but because the situation holds for every Sturm-Liouville problem. The proof of this fact, stated below, is very similar to the proof of Theorem 1 and is therefore left as an exercise (see Problem 8).

THEOREM 2 ♦ **Real eigenvalue theorem**

The eigenvalues of the Sturm-Liouville problem **(1), (2)** are all real. ♦

Finally, we have observed that in every example there are an infinite number of eigenvalues, and that corresponding to each eigenvalue there is only one linearly independent eigenfunction; that is, if y_1 and y_2 are two eigenfuctions corresponding to the eigenvalue λ, then there is a constant c such that $y_1 = cy_2$. When the latter condition holds, the eigenvalue is said to be **simple.** These results are summarized in the theorem below, whose proof, however, is beyond the scope of this test.* (The simplicity of the eigenvalues can easily be proved; see Problem 11.)

THEOREM 3 ♦ **Simple eigenvalue theorem**

The eigenvalues of the Sturm-Liouville problem **(1), (2)** are simple. Moreover, there are an infinite number of them, which can be arranged in an increasing order

$$\lambda_1 < \lambda_2 < \lambda_3 < \ldots < \lambda_k < \ldots,$$

where λ_k tends to ∞ as k tends to ∞. ♦

All three theorems of this section are needed for the solution of nonhomogeneous boundary value problems in Section 11.7.

---

*See, for example, the excellent book by R. Courant and D. Hilbert, *Methods of Mathematical Physics* (New York: Wiley, 1953), vol. I, chap. 6.

SELF-QUIZ

Answer true *or* false.

I. The equation $((\sin \pi x)y')' + \lambda y = 0$ satisfies the Orthogonality theorem on the interval $[0, 1]$.

II. An eigenvalue of a Sturm-Liouville problem may have two linearly independent eigenfunctions.

Answers to Self-Quiz

I. True **II.** False

PROBLEMS 11.6

In Problems 1–5 verify the implications of Theorems 1, 2, and 3 for each given Sturm-Liouville problem.

1. $y'' + \lambda y = 0,\ y(0) = y(\pi) = 0$

2. $y'' + \lambda y = 0,\ y(0) = y'(\pi) = 0$

3. $y'' + \lambda y = 0,\ y\left(-\frac{\pi}{2}\right) = y\left(\frac{\pi}{2}\right) = 0$

4. $(xy')' + \left(\frac{\lambda}{x}\right)y = 0,\ y(1) = y(e^2) = 0$

5. $(x^2y')' + \left(\frac{\lambda}{x^2}\right)y = 0,\ y\left(\frac{1}{2}\right) = y(1) = 0$

6. Prove that if $r(x)$, $p(x)$, and $q(x)$ are continuous periodic functions of period $(b - a)$, then all eigenfunctions of the boundary value problem **(1)**, **(6)** are periodic with period $(b - a)$.

7. **(a)** Show that the Hermite equation [see Problem 11.2.9(c)] can be written in the form

$$(e^{-x^2}y')' + 2ne^{-x^2}y = 0.$$

(b) Using the orthogonality theorem, prove that the Hermite polynomials are orthogonal with respect to the weight function $\exp(-x^2)$ on $-\infty < x < \infty$.

8. Prove that the eigenvalues of the Sturm-Liouville problem **(1)**, **(2)** are all real by carrying out the following steps:

(a) Show that if $\lambda = \alpha + i\beta$ is an eigenvalue with the corresponding eigenfunction $y_\lambda(x) = u(x) + iv(x)$, then $\bar{\lambda} = \alpha - i\beta$ is an eigenvalue with the corresponding eigenfunction $\bar{y}_\lambda(x) = u(x) - iv(x)$.

(b) Using the same proof as that of Theorem 1, show that

$$(\lambda - \bar{\lambda}) \int_a^b y_\lambda(x)\bar{y}_\lambda(x)q(x)\,dx = 0.$$

(c) Use the result of part (b) of this exercise and the fact that $y_\lambda\bar{y}_\lambda = u^2 + v^2$ to show that $\beta = 0$.

9. The boundary value problem

$$y'' - \frac{\lambda}{x}y' + \frac{\lambda}{x^2}y = 0, \qquad y(1) = y(2) = 0,$$

is not a Sturm-Liouville problem. Show that none of the eigenvalues is real.

10. Consider the boundary value problem

$$y'' + \lambda y = 0, \qquad y(-1) = y(1),$$
$$y'(-1) = y'(1).$$

(a) Explain why it is not a Sturm-Liouville problem.

(b) Calculate the eigenvalues of this problem.

(c) Show that corresponding to each nonzero eigenvalue there are two linearly independent eigenfunctions; that is, show that the eigenvalues are not simple.

11. Prove that the eigenvalues of a Sturm-Liouville problem are simple. [*Hint:* Assume that y_1 and y_2 are two eigenfunctions corresponding to the eigenvalue λ. Use the boundary conditions **(2)** to show that the Wronskian $W(y_1, y_2)(x) = 0$ for $a \le x \le b$.]

11.7 NONHOMOGENEOUS BOUNDARY VALUE PROBLEMS

In Example 11.1.3 (p. 528) we considered a nonhomogeneous boundary value problem given by Equation **(11.1.12).** At that time we stated that the solution of (11.1.12), for $\omega \neq n\pi$, could be expressed by the Fourier sine series (11.1.20). In this section we justify that statement and unite the seemingly unrelated concepts of Fourier series and Sturm-Liouville problems. We discover that the solutions of certain nonhomogeneous boundary value problems can be obtained as generalized Fourier series in the eigenfunctions of the associated Sturm-Liouville problem.

This method of **eigenfunction expansions** is one of the main techniques used in solving nonhomogeneous boundary value problems. Furthermore, it provides the motivation for the very powerful method of **Green's functions** that we discuss at the end of this section, which is extremely useful for solving such problems.

Suppose that we wish to solve the *nonhomogeneous* differential equation

$$[r(x)w']' + p(x)w = f(x) \tag{1}$$

on $a \leq x \leq b$ with the boundary conditions

$$a_1 w(a) - a_2 w'(a) = 0, \qquad b_1 w(b) - b_2 w'(b) = 0. \tag{2}$$

We first consider a different boundary value problem,

$$[r(x)y']' + [p(x) + \lambda q(x)]y = 0, \tag{3}$$

with identical boundary conditions,

$$a_1 y(a) - a_2 y'(a) = 0, \qquad b_1 y(b) - b_2 y'(b) = 0, \tag{4}$$

and assume that its eigenvalues and eigenfunctions are known. We are free to choose $q(x)$ in any way we wish; however, the choice is limited, since we must be able to determine the eigenfunctions of **(3), (4).** Furthermore, different choices of $q(x)$ yield different sets of eigenfunctions, and some sets may be easier to work with than others. We assume that $r(x)$, $r'(x)$, $p(x)$, and $q(x)$ are continuous and that r and q are positive on $a \leq x \leq b$. By Theorem 11.6.3 and the orthogonality theorem, we know that the problem **(3), (4)** has an infinite orthogonal set $\{y_k\}$ of eigenfunctions with respect to the weight function $q(x)$.

The object now is to express the solution $w(x)$ of **(1), (2)** (if one exists) as a generalized Fourier series [see Equation **(11.3.16)**] of the form

$$w(x) = \sum_{n=1}^{\infty} c_n y_n, \tag{5}$$

where the functions y_n are eigenfunctions of **(3), (4).** Of course, since we are proceeding formally, when the process is complete we will have to check that our result is indeed a solution. As we saw in Section 11.3, the problem of determining which functions can be represented in this way is not easy. Although the proof is beyond the scope of this book, *any piecewise continuous function $w(x)$ with piecewise continuous derivative $w'(x)$ has a representation* **(5)** *at all points of continuity, provided that a and b are finite.** (At points of discontinuity, the generalized Fourier series converges to $\frac{w(x+0)+w(x-0)}{2}$.)

*Much more general theorems are known. See, for example, E. A. Coddington and N. Levinson, *Theory of Ordinary Differential Equations* (New York: McGraw-Hill, 1955), p. 199.

To obtain the formal representation **(5)**, we proceed in a manner similar to the proof of the orthogonality theorem. Multiplying **(1)** by y and **(3)** by $-w$, we obtain the equation

$$y(rw')' - w(ry')' = fy + \lambda qyw.$$

Integrating both sides from a to b and the left-hand side by parts, we have

$$\left(yrw'\Big|_a^b - \int_a^b rw'y'\,dx\right) - \left(wry'\Big|_a^b - \int_a^b ry'w'\,dx\right) = \int_a^b (fy + \lambda qyw)\,dx.$$

The integrals on the left-hand side cancel out, yielding the equation

$$r(b)[y(b)w'(b) - w(b)\,y'(b)] - r(a)[y(a)w'(a) - w(a)y'(a)] = \int_a^b (fy + \lambda qyw)\,dx. \tag{6}$$

Letting $\lambda = \lambda_k$ and $y = y_k$ we see, as in the proof of the orthogonality theorem, that the left-hand side of Equation **(6)** is zero. Thus

$$\lambda_k \int_a^b qwy_k\,dx = -\int_a^b fy_k\,dx,$$

and since, by Equation **(11.3.17)**,

$$c_k = \frac{(w, y_k)_q}{\|y_k\|_q^2} = \frac{\displaystyle\int_a^b qwy_k\,dx}{\displaystyle\int_a^b qy_k^2\,dx},$$

we finally obtain

Coefficients of a generalized Fourier series representation of the solution to a Sturm-Liouville problem

$$c_k = -\frac{\displaystyle\int_a^b fy_k\,dx}{\lambda_k \displaystyle\int_a^b qy_k^2\,dx}. \tag{7}$$

In particular, if the eigenfunctions y_k have been chosen so that they are orthonormal with respect to q, then the denominator in the right-hand side of **(7)** reduces to just the eigenvalue λ_k, and we have the representation

Series solution of a nonhomogeneous Sturm-Liouville problem

$$w(x) = -\sum_{n=1}^{\infty}\left(\lambda_n^{-1}\int_a^b fy_n\,dx\right)y_n(x) \tag{8}$$

for the solution of the boundary value problem **(1)**, **(2)**. We now apply this result to justify the solution of Example 11.1.3 [see Equation **(11.1.20)**].

EXAMPLE 1 ▶ A forced spring

Suppose that we again consider the forced vibrations of a spring-mass system, with spring constant k, attached to an external periodic force $K \sin \omega t$. For simplicity, we ignore the damping term in Equation **(4.3.1)** and merely consider the nonhomogeneous equation

$$x'' + \frac{k}{m}x = \frac{K}{m} \sin \omega t. \tag{9}$$

We again suppose that the object is at its equilibrium position initially and when $t = 1$ second, and we seek a solution to the nonhomogeneous boundary value problem.

To apply the eigenfunction expansion **(8)**, we consider the Sturm-Liouville problem

$$y'' + (\lambda + 1)\frac{k}{m}y = 0, \qquad y(0) = y(1) = 0. \tag{10}$$

The eigenvalues of **(10)** are obtained by setting $\sqrt{(\lambda + 1)k/m}$ equal to a multiple of π; hence

$$\lambda_n = \frac{mn^2\pi^2}{k} - 1, \qquad n = 1, 2, 3, \ldots .$$

The corresponding eigenfunctions have the form

$$y_n = A \sin n\pi t, \qquad n = 1, 2, 3, \ldots ,$$

and are orthonormal with respect to $q(x) = k/m$ if $A = \sqrt{2m/k}$. By Equation **(7)**, the Fourier coefficients c_n are given by

$$c_n = -\sqrt{\frac{2m}{k}} \cdot \frac{K}{m\lambda_n} \int_0^1 \sin \omega t \sin n\pi t \, dt.$$

Using Formula 103 of Appendix 1 or the identity

$$\sin(\omega \pm n\pi)t = \sin \omega t \cos n\pi t \pm \cos \omega t \sin n\pi t,$$

we have

$$\begin{aligned} c_n &= -\sqrt{\frac{2m}{k}} \cdot \frac{K}{2m\lambda_n}\left[\frac{\sin(\omega - n\pi)t}{\omega - n\pi} - \frac{\sin(\omega + n\pi)t}{\omega + n\pi}\right]\Bigg|_0^1 \\ &= -\sqrt{\frac{2m}{k}} \cdot \frac{K}{m\lambda_n} \sin \omega \cos n\pi \frac{n\pi}{\omega^2 - n^2\pi^2}, \qquad \omega \neq n\pi. \end{aligned}$$

If $\omega = n\pi$, then

$$c_n = -\sqrt{\frac{2m}{k}} \cdot \frac{K}{2m\lambda_n},$$

and all the other Fourier coefficients are zero by the orthogonality of the eigenfunctions. Thus the solution $x(t)$ must be stated for two cases:

(a) If $\omega \neq n\pi$, $n = 1, 2, 3, \ldots$, then by **(8)** and the identity $\cos n\pi = (-1)^n$,

$$x(t) = -\frac{2K\pi \sin \omega}{k} \sum_{n=1}^{\infty} \frac{(-1)^n n}{\lambda_n(\omega^2 - n^2\pi^2)} \sin n\pi t.$$

It is easy to show that this solution is identical to Equation **(11.1.20)**.

(b) Suppose that $\omega = n\pi$ for some positive integer n. Then using **(8)** and the value of λ_n, we obtain the particular solution

$$x_p(t) = -\frac{K}{k\lambda_n}\sin n\pi t = \frac{K \sin n\pi t}{k - mn^2\pi^2}.$$

This solution is valid provided that the spring constant $k \neq mn^2\pi^2$.

The general solution is now obtained as in Case 1 of Example 11.1.3. When $k = mn^2\pi^2$, there is no solution. Observe that in this case the eigenvalue,

$$\lambda_n = \frac{mn^2\pi^2}{k} - 1,$$

is zero. ◄

The following result is beyond the scope of this book, but it provides the necessary information to decide whether or not there is a solution in the case above.

THEOREM 1* ♦ **Existence theorem**

Let one of the eigenvalues λ_k of the problem in **(3), (4)** be zero. Then the nonhomogeneous boundary value problem in **(1), (2)** has a solution if and only if

$$\int_a^b f(t)y_k(t)\,dt = 0. \quad ♦$$

EXAMPLE 2 ► **Solving a nonhomogeneous boundary value problem**

Consider the boundary value problem

$$(xw')' + \frac{w}{x} = \frac{1}{x}, \qquad w(1) = w(e) = 0. \tag{11}$$

To apply the method of eigenfunction expansions **(8)** we let $p(x) = q(x) = f(x) = \frac{1}{x}$ and consider the Sturm-Liouville problem

$$(xy')' + \frac{(1 + \lambda)}{x}y = 0, \qquad y(1) = y(e) = 0. \tag{12}$$

By Example 11.5.2, we know that the eigenvalues $\lambda^* = 1 + \lambda$ of **(12)** are given by

$$1 + \lambda_k = \lambda_k^* = k^2\pi^2, \qquad k = 1, 2, 3, \ldots,$$

so $\lambda_k = k^2\pi^2 - 1$ and the eigenfunctions are of the form

$$y_k = A \sin(k\pi \ln x).$$

Using the substitution $u = k\pi \ln x$, $du = (\frac{k\pi}{x})\,dx$, we have

$$\int_1^e \frac{1}{x}\sin^2(k\pi \ln x)\,dx = \frac{1}{k\pi}\int_0^{k\pi} \sin^2 u\,du = \left.\frac{u - \cos u \sin u}{2k\pi}\right|_0^{k\pi} = \frac{1}{2}.$$

Letting $A = \sqrt{2}$, we obtain an orthonormal set of eigenfunctions. By **(7),** the coefficients c_k of the generalized Fourier series are given by

$$c_k = -\lambda_k^{-1}\int_1^e \frac{\sqrt{2}}{x}\sin(k\pi \ln x)\,dx = \left.\frac{\sqrt{2}\,\lambda_k^{-1}}{k\pi}\cos(k\pi \ln x)\right|_1^e$$

---

*See, for example, E. A. Coddington and N. Levinson, *Theory of Ordinary Differential Equations* (New York: McGraw-Hill, 1955), p. 294.

$$= \frac{\sqrt{2}\,(\cos k\pi - 1)}{k\pi\lambda_k} = \begin{cases} 0, & k \text{ even}, \\ \dfrac{-2\sqrt{2}}{k\pi(k^2\pi^2 - 1)}, & k \text{ odd}. \end{cases}$$

Hence the solution of problem **(11)** is given by the series

$$w(x) = -4\sum_{n=0}^{\infty} \frac{\sin[(2n+1)\pi \ln x]}{[(2n+1)\pi]^3 - [(2n+1)\pi]}.$$

A number of further examples of the use of eigenfunction expansions will be found in Chapter 12, where they are used in the solution of certain partial differential equations.

The methods we have presented in this chapter are not the only techniques that can be used to solve boundary value problems. A very powerful procedure, attributed to George Green (1793–1841), involves the representation of the solution of boundary value problem **(1)**, **(2)** in the form

$$w(x) = \int_a^b K(x, t) f(t)\, dt. \tag{13}$$

Actually, in light of the eigenfunction expansion **(8)**, a representation of the form **(13)** is not surprising, for if we use t as the variable of integration and interchange the sum and integral in **(8)**, we obtain the expression

$$\int_a^b \left[-\sum_{n=1}^{\infty} \frac{y_n(t) y_n(x)}{\lambda_n} \right] f(t)\, dt.$$

Hence the **Green's function** $K(x, t)$ may be presumed to equal

$$K(x, t) = -\sum_{n=1}^{\infty} \frac{y_n(t) y_n(x)}{\lambda_n}. \tag{14}$$

All the assumptions we made above can be justified.* Problems 8–12 indicate another method of obtaining the Green's function $K(x, t)$.

SELF-QUIZ

I. The solution to the nonhomogeneous boundary value problem

$$y'' + 2y = \sin x, \quad y(0) = y(\pi) = 0,$$

can be expressed as a generalized Fourier series in terms of the eigenfunctions of the Sturm-Liouville problem ________

(a) $y'' + 2y = 0$, $y(0) = y(\pi) = 0$
(b) $y'' + \lambda y = 0$, $y(0) = y(\pi) = 0$
(c) $y'' + [2 + \lambda]\, y = \sin x$, $y(0) = y'(\pi) = 0$
(d) $y'' + [2 + \lambda]\, y = 0$, $y(0) = y'(\pi) = 0$

Answer to Self-Quiz

I. b

PROBLEMS 11.7

In Problems 1–5 use the method of eigenfunction expansions to solve the given nonhomogeneous boundary value problems.

1. $y'' + y = x$, $y(0) = y(\pi) = 0$

2. $y'' + 2y = \cos x$, $y(0) = y(1) = 0$

3. $y'' + 4y = x^2$, $y(0) = y'(1) = 0$

4. $(xy')' + \dfrac{1}{x} y = \ln x$, $y'(1) = y'(e^{2\pi}) = 0$

---

*See, for example, E. A. Coddington and N. Levinson, *Theory of Ordinary Differential Equations* (New York: McGraw-Hill, 1955).

5. $(xy')' + \frac{3}{x}y = \frac{1}{x}\sin(\ln x)$, $y(1) = y(2) = 0$

6. Consider the nonhomogeneous problem

$$y'' + \lambda y = f(x),\ y(0) = y(1) = 0.$$

Show that if $f(x)$ is continuous, then there is a unique solution to this problem if and only if λ is *not* an eigenvalue of the associated homogeneous equation $y'' + \lambda y = 0$, $y(0) = y(1) = 0$

7. Prove the claim in Problem 6 for

$$y'' + \lambda y = f(x), \quad y(0) = y(1), \quad y'(0) = y'(1).$$

Consider the differential equation

$$(rw')' + qw = 0.$$

It is easy to find a nontrivial solution w_1 of this equation with $w_1(a) = 0$, and a nontrivial solution w_2 with $w_2(b) = 0$. Define the function

$$k(x, t) = \begin{cases} w_1(x)w_2(t), & a < x < t, \\ w_1(t)w_2(x), & t < x < b. \end{cases}$$

8. Ignoring any difficulties at $t = x$, show that

$$\frac{d}{dx}\int_a^b k(x,t)f(t)\,dt = \int_a^b \frac{\partial}{\partial x}k(x,t)f(t)\,dt.$$

9. Show that

$$\frac{d}{dx}\left[r(x)\frac{d}{dx}\int_a^b k(x,t)f(t)\,dt\right] = r(x)f(x)W(x) - q(x)\int_a^b k(x,t)f(t)\,dt,$$

where W is the Wronskian of the functions w_1 and w_2.

10. Using the result of Problem 9, conclude that

$$w_3(x) = \int_a^b k(x,t)f(t)\,dt$$

satisfies the equation

$$[r(x)w_3']' + q(x)w_3 = f(x)r(x)W(x).$$

11. Show that

$$\frac{d}{dx}(r(x)W(x)) = 0,$$

and conclude that $r(x)W(x) = C$, a constant.

12. Prove that

$$w(x) = C^{-1}w_3(x)$$

is a solution of the boundary value problem

$$(r(x)w')' + q(x)w = f(x), \qquad w(0) = w(b) = 0.$$

Hence conclude that the Green's function is given by $K(x,t) = C^{-1}k(x,t)$.

13. Using the ideas of Problems 8–12, find the Green's function for the boundary value problem

$$w''(x) - w(x) = f(x), \qquad w(0) = w(1) = 0.$$

14. Find the Green's function for the boundary value problem

$$w''(x) + k^2w(x) = f(x), \qquad w(0) = w(a) = 0.$$

15. Generalize the procedure used in Problems 8–12 to obtain the Green's function for the boundary value problem

$$(rw')' + qw = f, \qquad w(0) = w'(1) = 0.$$

Apply the generalized procedure of Problem 15 to Problems 16–18.

16. $w''(x) = f(x)$, $w(0) = w'(1) = 0$

17. $w''(x) + 4w(x) = f(x)$, $w(0) = w'(1) = 0$

18. $(xw')' = f(x)$, $w(1) = w'(e) = 0$

11.8 MEAN-SQUARE APPROXIMATION (OPTIONAL)

We have seen that if $F(x)$ is periodic of period $2T$ and if $F(x)$ can be represented by a trigonometric series, then

$$F(x) = \frac{a_0}{2} + \sum_{n=1}^{\infty}\left(a_n\cos\frac{n\pi x}{T} + b_n\sin\frac{n\pi x}{T}\right), \tag{1}$$

where the coefficients $a_0, a_1, b_1, a_2, b_2, \ldots$ are given by the Euler formulas (11.3.8) and (11.3.9).

In a practical situation no one adds up the infinite number of terms in **(1).** Rather, it is necessary to stop at some number N. We then have an **approximation** of $F(x)$:

$$F(x) \approx \frac{a_0}{2} + \sum_{n=1}^{N} \left(a_n \cos \frac{n\pi x}{T} + b_n \sin \frac{n\pi x}{T} \right) = P_N(x). \tag{2}$$

trigonometric polynomial

The function $P_N(x)$ is called a **trigonometric polynomial.** A question that naturally arises is

Is P_N the "best" approximation of the form **(2)** for $F(x)$?

In order to answer the question, we have to decide how to measure the "goodness" of an approximation. Here are four commonly used measures: Let lub denote the *least upper bound* of a set of numbers. (According to the completeness axiom, every bounded set of numbers has a least upper bound.)

$$\epsilon_a = \text{absolute error} = \operatorname*{lub}_{|x| \le T} |F(x) - P_N(x)|, \tag{3}$$

$$\epsilon_{as} = \text{absolute square error} = \operatorname*{lub}_{|x| \le T} [F(x) - P_N(x)]^2, \tag{4}$$

$$\epsilon_m = \text{mean error} = \int_{-T}^{T} |F(x) - P_N(x)|\, dx, \tag{5}$$

$$\epsilon_{ms} = \text{mean-square error} = \int_{-T}^{T} [F(x) - P_N(x)]^2\, dx. \tag{6}$$

The first two measure error at each point x in $[-T, T]$ and take the least upper bound of all these values. The sequence $\{P_N(x)\}$ converges to $F(x)$ at each point x if the term **(3)** or **(4)** approaches 0 as $N \to \infty$.

If we require ϵ_a (or ϵ_{as}) to be small, we shall require also that $\epsilon_m/2T$ (or $\epsilon_{ms}/2T$) be small, because

$$\epsilon_m = \int_{-T}^{T} |F(x) - P_N(x)|\, dx \le \epsilon_a \int_{-T}^{T} dx = 2\epsilon_a T. \tag{7}$$

In many situations it may not matter that $|F(x) - P_N(x)|$ [or $(F(x) - P_N(x))^2$] be large on small sets of x-values. Then mean error or mean-square error is used. If either of the terms **(5)** or **(6)** approaches zero as $N \to \infty$, then we say that $P_N(x)$ **converges in the mean** to $F(x)$.

convergence in mean

When dealing with Fourier series it is much easier to use mean-square error as our measure. With this measure P_N is the best approximation.

We prove this result in a more general setting. Before going further, reread the material on generalized Fourier series that begins on p. 544. Let $f_1(x), f_2(x), \ldots, f_N(x), \ldots$ be an orthonormal set of functions on the interval $[a, b]$ (see p. 532). That is,

$$\int_a^b f_n(x) f_m(x)\, dx = \begin{cases} 0, & \text{if } n \ne m, \\ 1, & \text{if } n = m. \end{cases} \tag{8}$$

Let $F(x)$ be a continuous function on $[a, b]$. We seek the polynomial

$$P_N(x) = a_1 f_1(x) + a_2 f_2(x) + \cdots + a_N f_N(x)$$

that best approximates $F(x)$ in the mean-square sense. That is, we seek coefficients $a_1, a_2, \ldots, a_N$ such that

$$\epsilon_{ms} = \int_a^b [F(x) - (a_1 f_1(x) + a_2 f_2(x) + \cdots + a_N f_N(x))]^2\, dx \tag{9}$$

is a minimum. We expand the integrand in **(9)**:

$$\epsilon_{ms} = \underbrace{\int_a^b [F(x)]^2\,dx}_{①} - \underbrace{2\int_a^b [a_1 f_1(x) + a_2 f_2(x) + \cdots + a_N f_N(x)]F(x)\,dx}_{②}$$

$$+ \underbrace{\int_a^b [a_1 f_1(x) + a_2 f_2(x) + \cdots + a_N f_N(x)]^2\,dx.}_{③} \qquad \textbf{(10)}$$

We consider each of the three integrals in **(10)**:

① $$\int_a^b [F(x)]^2\,dx = \|F\|^2.$$

② $$\int_a^b a_m f_m(x)F(x)\,dx = a_m \int_a^b f_m(x)F(x)\,dx = a_m c_m$$

Formula (15) on p.544 with $\|f_m\|^2 = 1$ ↓

$= a_m$ times the mth generalized Fourier coefficient of $F(x)$ with respect to the orthonormal set $\{f_n(x)\}$.

Thus

$$2\int_a^b [a_1 f_1(x) + \cdots + a_N f_N(x)]F(x)\,dx = 2[a_1c_1 + a_2c_2 + \cdots + a_Nc_N].$$

③ There are two kinds of terms when we expand the integrand:

(a) $$\int_a^b a_m^2 f_m^2(x)\,dx = a_m^2 \int_a^b f_m^2(x) \overset{\text{from (8)}}{=} a_m^2,$$

(b) $$\int_a^b 2a_k a_m f_k(x) f_m(x)\,dx = 2a_k a_m \int_a^b f_k(x) f_m(x)\,dx \overset{\text{from (8)}}{=} 0, \qquad k \neq m.$$

Thus

$$\epsilon_{ms} = \|F\|^2 + \sum_{m=1}^{N} (a_m^2 - 2a_m c_m).$$

But, completing the square, we find that

$$a_m^2 - 2a_m c_m = (a_m - c_m)^2 - c_m^2,$$

so

$$\epsilon_{ms} = \|F\|^2 - \sum_{m=1}^{N} c_m^2 + \sum_{m=1}^{N} (a_m - c_m)^2. \qquad \textbf{(11)}$$

Equation **(11)** enables us to prove a number of interesting results.

THEOREM 1 ♦ Fourier coefficients are best in the mean-square sense

Let $\{f_n(x)\}$ be an orthonormal set of functions in the interval $[a, b]$ and suppose that $F(x)$ is continuous on $[a, b]$. Define

$$c_m = \int_a^b F(x) f_m(x)\,dx. \qquad \textbf{(12)}$$

Let $P_N(x) = c_1 f_1(x) + c_2 f_2(x) + \cdots + c_N f_N(x)$. If $a_1, a_2, \ldots, a_N$ is another set of constants, let

$$q_N(x) = a_1 f_1(x) + a_2 f_2(x) + \cdots + a_n f_N(x). \tag{13}$$

Then

$$\int_a^b [F(x) - P_N(x)]^2\, dx \le \int_a^b [F(x) - q_N(x)]^2\, dx.$$

That is, the **Fourier polynomial** $P_N(x)$ gives the least mean-square error among all functions that can be written in the form **(13).**

Proof In **(11),** $\|F\|^2$ and $\Sigma_{m=1}^N c_m^2$ are constant. The last term, $\Sigma_{m=1}^N (a_m - c_m)^2$, is nonnegative and is minimized when $a_m = c_m$, for $m = 1, 2, \ldots, N$. In this case $\epsilon_{ms} = \|F\|^2 - \Sigma_{m=1}^N c_m^2$. ♦

THEOREM 2 ♦ **Bessel's inequality**

Under the assumptions of Theorem 1, the series $\Sigma_{m=1}^\infty c_m^2$ converges and

$$\sum_{k=1}^{\infty} c_m^2 \le \int_a^b [F(x)]^2\, dx. \tag{14}$$

Proof Let $a_m = c_m$ in **(11).** Then

$$0 \le \int_a^b [F(x) - P_N(x)]^2\, dx = \|F\|^2 - \sum_{m=1}^{N} c_m^2$$

or

$$\sum_{m=1}^{N} c_m^2 \le \|F\|^2 = \int_a^b [F(x)]^2\, dx.$$

Let $N \to \infty$ to obtain (14). ♦

Least squares polynomial approximation

In Section 5.5 we discussed **Legendre's differential equation**

$$(1 - x^2)y'' - 2xy' + p(p + 1)y = 0. \tag{15}$$

We showed (see p. 241) that if p is a nonnegative integer, then **(15)** has at least one solution that is a polynomial of degree p. This solution is called a **Legendre polynomial,** denoted by $P_p(x)$. We saw on p. 241 that

$$P_p(x) = \sum_{k=0}^{M} \frac{(-1)^k(2p - 2k)!}{2^p k!\,(p - k)!\,(p - 2k)!} x^{p-2k}, \qquad p = 0, 1, 2, \ldots, \tag{16}$$

where M is the largest integer not greater than $\frac{p}{2}$. In particular, $P_0(x) = 1$, $P_1(x) = x$, $P_2(x) = \frac{1}{2}(3x^2 - 1)$, $P_3(x) = \frac{1}{2}(5x^3 - 3x)$, and $P_4(x) = \frac{1}{8}(35x^4 - 30x^2 + 3)$. In Problems 16–18 you are asked to show that $\{P_n(x)\}$ is an orthogonal set on the interval $[-1, 1]$. (Also see Example 11.6.2.) Let

Equation (15) on p. 544
↓

$$c_m = \frac{1}{\|P_m\|^2} \int_{-1}^{1} F(x)P_m(x)\, dx. \tag{17}$$

Then, by Theorem 1, the Nth degree polynomial that best approximates $F(x)$ over $[-1, 1]$ in the mean-square sense is given by

$$q_N(x) = c_1 P_1(x) + c_2 P_2(x) + \cdots + c_N P_N(x). \tag{18}$$

least squares polynomial approximation The polynomial $q_N(x)$ is called the **least squares polynomial approximation** to $F(x)$ on the interval $[-1, 1]$.

EXAMPLE 1 ▶ **The least squares approximation of e^x**

Find the least squares approximation to e^x on $[-1, 1]$ by a straight line, that is, by a polynomial of degree 1.

Solution $P_0 = 1$ and $\|P_0\|^2 = \int_{-1}^{1} 1\,dx = 2$; $P_1(x) = x$ and $\|P_1\|^2 = \int_{-1}^{1} x^2\,dx = \frac{2}{3}$. Then

$$c_0 = \frac{1}{2}\int_{-1}^{1} e^x\,dx = \frac{e^1 - e^{-1}}{2} = \sinh 1,$$

$$c_1 = \frac{3}{2}\int_{-1}^{1} xe^x\,dx = \frac{3}{2}(x-1)e^x\Big|_{-1}^{1} = \frac{3}{e}.$$

Thus

$$q_1(x) = (\sinh 1)\,P_0(x) + \frac{3}{e}P_1(x) = \sinh 1 + \frac{3x}{e}.$$

Note that $q_1(x)$ is *not* the first two terms of the Maclaurin expansion of $e^x(= 1 + x + x^2/2! + x^3/3! + \ldots)$. ◀

PROBLEMS 11.8

1. Find constants a_0, a_1, and a_2 such that $\int_{-\pi}^{\pi} [1 - P_2(x)]^2\,dx$ is a minimum, where $P_2(x) = a_0/2 + a_1 \cos x + a_2 \sin 2x$.
2. Find constants a_1, a_2, and a_3 such that $\int_0^{\pi} [1 - P_3(x)]^2\,dx$ is a minimum, where $P_3(x) = a_1 \sin x + a_2 \sin 2x + a_3 \sin 3x$.
3. Find constants a_1, a_2, and a_3 such that $\int_{-1}^{1} [x - P_3(x)]^2\,dx$ is a minimum, where $P_3(x) = a_1 \sin \pi x + a_2 \sin 2\pi x + a_3 \sin 3\pi x$.
4. Find constants a_0, a_1, a_2, and a_3 such that
$$\int_0^1 [x - P_3(x)]^2\,dx$$
is a minimum, where $P_3(x) = a_0/2 + a_1 \cos \pi x + a_2 \cos 2\pi x + a_3 \cos 3\pi x$.
5. Find a polynomial $P_3(x) = \frac{a_0}{2} + a_1 \cos \pi x + a_2 \cos 2\pi x + a_3 \cos 3\pi x$ that best approximates x^2 in the mean-square sense in the interval $[-1, 1]$.
6. Let $\{f_n(x)\}$ denote the functions $\{1, \sin x, \cos x, \sin 2x, \cos 2x, \ldots\}$. Let $F(x)$ be continuous in $[-\pi, \pi]$ and suppose that a_n and b_n are given by formulas **(11.3.8)** and **(11.3.9)** on p. 538. Prove that
$$\frac{a_0^2}{2} + \sum_{n=1}^{\infty} a_n^2 + \sum_{n=1}^{\infty} b_n^2 \le \frac{1}{\pi}\int_{-\pi}^{\pi} [F(x)]^2\,dx.$$

In Problems 7–9 represent each polynomial as a linear combination of Legendre polynomials.

7. x^4
8. $x^5 - x + 3$
9. $9x^3 - 8x^2 + 7x - 6$

In Problems 10–14 find a quadratic polynomial that best approximates the given function in the least squares sense on the interval $[-1, 1]$.

10. e^x
11. $\sin x$
12. $\cos x$
13. $\sinh x$
14. $\dfrac{1}{1 + x^2}$
15. Given that $p(x)$ is a polynomial of degree n (≥ 1) and
$$\int_{-1}^{1} x^k p(x)\,dx = 0 \qquad \text{for } k = 0, 1, 2, \ldots, n-1,$$
show that $p(x) = cP_n(x)$ for some constant c.
16. Use Rodrigues' formula [Equation **(5.5.30)** on p. 241] to show that
$$\int_{-1}^{1} P_n(x)P_m(x)\,dx = \frac{1}{2^n n!}\int_{-1}^{1} P_m(x)\frac{d^n}{dx^n}[(x^2 - 1)^n]\,dx.$$

17. **(a)** Use the result of Problem 16 to show that

$$\int_{-1}^{1} P_n(x)P_m(x)\,dx = \frac{P_m(x)}{2^n n!}\frac{d^{n-1}}{dx^{n-1}}[(x^2-1)^n]\Bigg|_{-1}^{1} - \frac{1}{2^n n!}\int_{-1}^{1} P_m'(x)\frac{d^{n-1}}{dx^{n-1}}[(x^2-1)^n]\,dx.$$

(b) Show that the first term above is zero.

18. **(a)** Use the result of Problem 17 and repeated integration by parts to show that

$$\int_{-1}^{1} P_n(x)P_m(x)\,dx = \frac{(-1)^n}{2^n n!}\int_{-1}^{1} P_m^{(n)}(x)(x^2-1)^n\,dx.$$

(b) Explain why $\int_{-1}^{1} P_n(x)P_m(x)\,dx = 0$ if $n \neq m$.

19. Use the result of Problem 18 with $m = n$ to show that

$$\|P_n\|^2 = \frac{2}{2n+1}.$$

[*Hint:* Make the substitution $x = \sin\theta$ and use entry 218 in Appendix 1.]

11.9 USING SYMBOLIC MANIPULATORS FOR FOURIER SERIES

This section will show you how MATHEMATICA, MAPLE, DERIVE, MATLAB, or MATHCAD can be used to find the Fourier series of a periodic function $f(x)$. We will use the command for definite integrals:

INT($f(x)$, x, x_0, x_1) = {the definite integral of $f(x)$ with repect to x over the interval $[x_0, x_1]$: $\int_{x_0}^{x_1} f(x)\,dx$}

in calculating the Fourier series of $f(x)$.

We shall assume that the function $f(x)$ has period $2L$ and has been defined on the interval $[-L, L]$. Recall from Section 11.3 that the Fourier series of $f(x)$ is given by

$$\frac{a_0}{2} + \sum_{n=1}^{\infty}\left\{a_n\cos\frac{n\pi x}{L} + b_n\sin\frac{n\pi x}{L}\right\} \tag{1}$$

where

$$\left.\begin{matrix} a_n \\ b_n \end{matrix}\right\} = \frac{1}{L}\int_{-L}^{L} f(x)\left\{\begin{matrix} \cos\dfrac{n\pi x}{L} \\ \sin\dfrac{n\pi x}{L} \end{matrix}\right\}dx. \tag{2}$$

Thus, the algorithm for computing the partial sum of the first N terms of the Fourier series of $f(x)$ is readily apparent: we need to calculate

$$\frac{1}{2L}\int_{-L}^{L} f(x)\,dx + \frac{1}{L}\sum_{n=1}^{N}\left\{\cos\frac{n\pi x}{L}\int_{-L}^{L} f(x)\cos\frac{n\pi x}{L}\,dx + \sin\frac{n\pi x}{L}\int_{-L}^{L} f(x)\sin\frac{n\pi x}{L}\,dx\right\}. \tag{3}$$

or

```
FOURIER(f,x,L,N):=(.5*INT(f,x,-L,L)+SUM(cos(n*pi*x/L)*
     INT(f*cos(n*pi*x/L),x,-L,L)+sin(n*pi*x/L)*
     INT(f*sin(n*pi*x/L),x,-L,L),n,1,N))/L
```

EXAMPLE 1 ▶

Find the first five terms of the Fourier series of the function $f(x) = x^2$ on the interval $[-1, 1]$ which has period 2.

Solution We would give the command FOURIER (x^2, x, 1, 5) and it would return

$$\frac{1}{3} - \frac{4\cos\pi x}{\pi^2} + \frac{\cos 2\pi x}{\pi^2} - \frac{4\cos 3\pi x}{9\pi^2} + \frac{\cos 4\pi x}{4\pi^2} - \frac{4\cos 5\pi x}{25\pi^2}.$$

From this we might infer that the Fourier series of $f(x)$ is

$$\frac{1}{3} + \sum_{n=1}^{\infty} \frac{4(-1)^n \cos n\pi x}{n^2\pi^2}.$$ ◀

EXAMPLE 2 ▶

Find the first six terms of the function below with period 3:

$$f(x) = \begin{cases} 2, & -1.5 < x < 0, \\ x - 1, & 0 < x < 1.5. \end{cases}$$

Solution We need to write the function $f(x)$ as a single term. This can be done by using Heaviside functions:

$$H(x - a) = \begin{cases} 0, & \text{if } x < a, \\ 1, & \text{if } x > a. \end{cases}$$

In this case we write

f(x) = 2(1−H(x)) + (x − 1) ∗ H(x) = 2 + (x−3)∗H(x).

Thus,

FOURIER(2 + (x−3)*H(x), x, 1.5, 6)

which yields

$$\frac{7}{8} - \frac{3}{\pi^2}\left(\cos\frac{2\pi x}{3} + \frac{1}{3^2}\cos 2\pi x + \frac{1}{5^2}\cos\frac{10\pi x}{3}\right)$$
$$- \frac{3}{2\pi}\left(3\sin\frac{2\pi x}{3} + \frac{1}{2}\sin\frac{4\pi x}{3} + \sin 2\pi x \right.$$
$$\left. + \frac{1}{4}\sin\frac{8\pi x}{3} + \frac{3}{5}\sin\frac{10\pi x}{3} + \frac{1}{6}\sin 4\pi x\right).$$

This suggests that the Fourier series for $f(x)$ is

$$\frac{7}{8} - \frac{3}{\pi^2}\sum_{n=1}^{\infty}\frac{1}{(2n-1)^2}\cos\frac{(4n-2)\pi x}{3}$$
$$- \frac{3}{2\pi}\left(3\sum_{n=1}^{\infty}\frac{1}{2n-1}\sin\frac{(4n-2)\pi x}{3} + \sum_{n=1}^{\infty}\frac{1}{2n}\sin\frac{4n\pi x}{3}\right).$$ ◀

PROBLEMS 11.9

Find the first N terms of the Fourier series for the following functions defined on the given interval using MATHEMATICA , MAPLE, DERIVE, MATLAB, OR MATHCAD.

1. $f(x) = x, [-1, 1], N = 6$
2. $f(x) = x, [-2, 2], N = 6$
3. $f(x) = |x|, [-2, 2], N = 7$
4. $f(x) = |x|, [-3, 3], N = 6$
5. $f(x) = x^2 + x, [-2, 2], N = 6$
6. $f(x) = x^3 + x, [-2, 2], N = 6$

7. $f(x) = \begin{cases} 3, & -5 < x < 0, \\ x-2, & 0 < x < 5, \end{cases}$ $N = 5$

8. $f(x) = \begin{cases} 3, & -5 < x < 1, \\ x-2, & 1 < x < 5, \end{cases}$ $N = 5$

9. $f(x) = \begin{cases} x-2, & -5 < x < 0, \\ 3, & 0 < x < 5, \end{cases}$ $N = 5$

10. $f(x) = \begin{cases} 3, & -2.5 < x < 1, \\ x-2, & 1 < x < 2.5, \end{cases}$ $N = 5$

SUMMARY OUTLINE OF CHAPTER 11

Boundary value problem p. 525

Eigenfunction p. 526

Eigenvalue p. 526

Orthogonal
set of functions on an interval p. 532
with respect to a weight function on an interval p. 534

Orthonormal
set of functions on an interval p. 532
with respect to a weight function on an interval p. 534

Norm p. 532
weighted p. 534

Laguerre's equation and polynomials p. 534

Hermite's equation and polynomials pp. 535, 536

Fourier series or representation pp. 536, 538
Fourier coefficients p. 537
Euler formulas: $a_n = \frac{1}{\pi}\int_{-\pi}^{\pi} F(x)\cos nx\,dx$, $b_n = \frac{1}{\pi}\int_{-\pi}^{\pi} F(x)\sin nx\,dx$ p. 538
Generalized Fourier series p. 544
Generalized Fourier coefficients p. 544
Complex Fourier series p. 553
Complex Fourier coefficients $c_n = \frac{1}{2\pi}\int_{-\pi}^{\pi} F(x)\,e^{-inx}\,dx$ p. 553

Odd and Even functions p. 547
Fourier cosine series p. 548
Fourier sine series p. 549
Odd and even extensions p. 551
Odd and even half-range expansions pp. 551, 552

Sturm-Liouville problem p. 554
Eigenfunctions and eigenvalues p. 554
properties of pp. 559, 561, 564

Orthogonality theorem: If y_m and y_n are eigenfunctions corresponding to different eigenvalues λ_m and λ_n of the Sturm-Liouville problem

$$[r(x)y']' + [p(x) + \lambda q(x)]y = 0, \quad a_1 y(a) - a_2 y'(a) = 0,\ b_1 y(b) - b_2 y'(b) = 0,$$

then y_m and y_n are orthogonal with respect to $q(x)$, that is

$$\int_a^b q(x)\, y_m(x) y_n(x)\,dx = 0.$$ p. 559

Eigenfunction expansion p. 563

Green's function pp. 563, 567

Series solution of a nonhomogeneous Sturm-Liouville problem p. 564

Convergence in mean p. 569

Mean-square error: $\int_{-T}^{T} (F(x) - P_N(x))^2\,dx$ p. 569

Least squares polynomial approximation p. 572

REVIEW EXERCISES FOR CHAPTER 11

In Exercises 1–4 show that the given set of functions is orthogonal on the given interval and determine the corresponding othonormal set.

1. $\{\sin nx\}$, $n = 1, 2, 3, \ldots$, $0 \le x \le \pi$
2. $\{\cos n\pi x\}$, $n = 0, 1, 2, \ldots$, $0 \le x \le 2$
3. $\{\sin 2n\pi x, \cos 2n\pi x\}$, $n = 1, 2, \ldots$, $0 \le x \le 1$
4. $\{\sin n\pi x\}$, $n = 1, 2, 3, \ldots$, $0 \le x \le 1$

Expand each of the functions in Exercises 5–8 in a Fourier series. Examine each series at all the points of discontinuity.

5. $f(x) = x \sin x$, $0 < x < 2\pi$
6. $f(x) = \sqrt{1 - \cos x}$, $|x| < \pi$
7. $f(x) = |\sin x|$, $|x| < \pi$
8. $f(x) = \begin{cases} x, & 0 < x < \pi \\ 0, & \pi < x < 2\pi \end{cases}$

Find Fourier sine and cosine series for the functions in Exercises 9–12 by extending the given function on the interval [0, a] to an odd or even function on [−a, a].

9. $f(x) = x^2$, $0 < x < \pi$
10. $f(x) = e^x$, $0 < x < \pi$
11. $f(x) = x$, $0 < x < 1$
12. $f(x) = 2-x$, $0 < x < 2$
13. Find a Fourier sine series for $f(x) = x^2 - 2$ in $0 < x < 2$ and use this series to obtain a series for π^3.

In Exercises 14–19 find the values or approximate values of the eigenvalues and corresponding eigenfunctions of the given Sturm-Liouville problems.

14. $y'' - \lambda y = 0$, $y'(0) = y(1) = 0$
15. $y'' + \lambda y = 0$, $y'(0) = y(\pi) = 0$
16. $y'' - \lambda y = 0$, $y(0) + y'(0) = 0$, $y(1) = 0$
17. $y'' + \lambda y = 0$, $y(0) = 0$, $y(1) + y'(1) = 0$
18. $y'' + (-9 + \lambda)y = 0$, $y'(0) = y'(1) = 0$
19. $y'' + (-9 + \lambda)y = 0$, $y(0) = y(1) + y'(1) = 0$
20. Show that the generalized Laguerre equation

$$xy'' + (a + 1 - x)y' + ny = 0$$

is of Sturm-Liouville type, and determine the weight function for which the resulting polynomials are orthogonal.

12 PARTIAL DIFFERENTIAL EQUATIONS

As was indicated in Section 1.2, a **partial differential equation** is an equation involving a function of two or more variables and some of its partial derivatives. Thus the crucial difference between partial and ordinary differential equations is the number of independent variables involved in the equation.

In this chapter we present a brief introduction to partial differential equations. Since the general theory is much too difficult to be presented here, we merely examine a few simple examples. Because these examples have a number of important practical applications, even this cursory treatment is of significant value.

12.1 FIRST-ORDER LINEAR PARTIAL DIFFERENTIAL EQUATIONS

In this section we are concerned with first-order linear partial differential equations in two independent variables, that is, equations of the form

$$a(x, y)\frac{\partial z}{\partial x} + b(x, y)\frac{\partial z}{\partial y} + c(x, y)z = f(x, y), \tag{1}$$

where a, b, and c are functions of x and y. Equation **(1)** is **linear** since it does not involve any nonlinear functions of the dependent variable z and its partial derivatives; that is, Equation **(1)** is of **first degree** in z and its partial derivatives.

When a, b and c are constants, that is, when the coefficients of Equation **(1)** are constant, the partial differential equation is easy to solve by rotating the axes. Let

$$\tan\alpha = \frac{b}{a} \tag{2}$$

and set

$$\begin{aligned} u &= x\cos\alpha + y\sin\alpha, \\ v &= -x\sin\alpha + y\cos\alpha. \end{aligned} \tag{3}$$

Using **(3)** to change the independent variables x and y in Equation **(1),** we obtain a new differential equation in u and v that is an ordinary differential equation, because the partial derivative with respect to v is absent. We illustrate this with an example.

EXAMPLE 1 ▶ **Using rotation of axes to solve a partial differential equation**
Consider the partial differential equation

$$\frac{\partial z}{\partial x} + \frac{\partial z}{\partial y} - z = 0. \tag{4}$$

Since $a = b = 1$, it follows from **(2)** that $\tan \alpha = 1$ and $\alpha = \frac{\pi}{4}$ radians, so $\cos \alpha = \sin \alpha = 1/\sqrt{2}$. Thus, by Equation **(3),**

$$u = \frac{x + y}{\sqrt{2}}, \qquad v = \frac{-x + y}{\sqrt{2}}, \tag{5}$$

and we have (using the chain rule)

$$\begin{aligned} \frac{\partial z}{\partial x} &= \frac{\partial z}{\partial u}\frac{\partial u}{\partial x} + \frac{\partial z}{\partial v}\frac{\partial v}{\partial x} = \frac{1}{\sqrt{2}}\left(\frac{\partial z}{\partial u} - \frac{\partial z}{\partial v}\right), \\ \frac{\partial z}{\partial y} &= \frac{\partial z}{\partial u}\frac{\partial u}{\partial y} + \frac{\partial z}{\partial v}\frac{\partial v}{\partial y} = \frac{1}{\sqrt{2}}\left(\frac{\partial z}{\partial u} + \frac{\partial z}{\partial v}\right). \end{aligned} \tag{6}$$

Substituting the equations in **(6)** into Equation **(4),** we get

$$\sqrt{2}\,\frac{\partial z}{\partial u} - z = 0. \tag{7}$$

Although Equation **(7)** is still a partial differential equation in the independent variables u and v, we treat the variable v as if it were a parameter and solve the first-order linear differential equation

$$\sqrt{2}\,z' - z = 0, \quad \text{where} \quad z' = \frac{dz}{du}.$$

Since $(1/z)\,dz = (1/\sqrt{2})\,du$, we get

$$\ln|z| = \frac{u}{\sqrt{2}} + g(v),$$

where the last term is an arbitrary differentiable function of the parameter v (which is zero whenever we take its partial derivative with respect to u). Simplifying, we have

$$z = e^{u/\sqrt{2}}G(v), \qquad G(v) = e^{g(v)}$$

Replacing u and v by the substitutions in **(5),** we get

$$z = e^{(x+y)/2}G\left(\frac{y - x}{\sqrt{2}}\right). \tag{8}$$

That this is, indeed, a solution of Equation **(4)** for *any* differentiable function G is easily checked by performing the operations indicated in Equation **(4)**.

Up to now we have stressed the similarity between first-order ordinary and partial differential equations. We encounter a substantial difference when we consider initial-value problems. The solution of an ordinary differential equation may be completely determined by prescribing a value for it at a single point. This is generally not true for partial differential equations. In order to obtain a *unique* solution, it is usually necessary to prescribe values for the solution on an entire *line*. A suitable initial condition is

$$z(x, 0) = z_0(x)$$

for all x, where z_0 is a differentiable function. A solution satisfying this initial condition is then easily obtained by setting $y = 0$ in Equation **(8).** For example, suppose we are given the initial condition

$$z(x, 0) = \frac{x}{2} + e^x. \tag{9}$$

Setting $y = 0$ in Equation **(8)** and substituting the initial condition for the left-hand side of Equation **(8),** we have

$$\frac{x}{2} + e^x = e^{x/2}G\left(\frac{-x}{\sqrt{2}}\right),$$

or

$$G\left(\frac{-x}{\sqrt{2}}\right) = \frac{x}{2}e^{-x/2} + e^{x/2}.$$

Setting $w = -x/\sqrt{2}$, we have $x = -\sqrt{2}\,w$ and we obtain the exact form of the function G:

$$G(w) = \frac{-w}{\sqrt{2}}e^{w/\sqrt{2}} + e^{-w/\sqrt{2}}.$$

Replacing this function for G in Equation **(8)** yields

$$\begin{aligned} z &= e^{(x+y)/2}\left[\left(\frac{x-y}{2}\right)e^{(y-x)/2} + e^{(x-y)/2}\right] \\ &= \left(\frac{x-y}{2}\right)e^y + e^x, \end{aligned}$$

which is the solution to the initial-value problem given by Equations **(4)** and **(9).** ◀

Cauchy problem

A problem like that in Example 1, one in which we require that the solution satisfy a line of initial conditions, is called a **Cauchy problem** or an **initial-value problem.**

As Example 1 demonstrates, the solution of a Cauchy problem can be visualized as a surface $z = z(x, y)$ in three-dimensional Euclidean space. Thus we can use three-dimensional analytic geometry to increase our understanding of the solution.

integral surface

We call a surface $z = z(x, y)$ that is the solution of a Cauchy problem an **integral surface.** This name is used because "integrations" are required in solving (partial) differential equations. If we write the equation of an integral surface in the form

$$F(x, y, z) = z(x, y) - z = 0, \tag{10}$$

we can find the tangent plane to the integral surface by taking the total differential

$$z_x\,dx + z_y\,dy - dz = 0.$$

The vector $\mathbf{N} = z_x\mathbf{i} + z_y\mathbf{j} - \mathbf{k}$ is normal to the integral surface **(10)** at any point. Rewriting **(1)** in the form

$$a(x, y)z_x + b(x, y)z_y = g(x, y, z), \tag{11}$$

where $g(x, y, z) = f(x, y) - c(x, y)z$, we see that the vector

$$\mathbf{V} = a\mathbf{i} + b\mathbf{j} + g\mathbf{k} \tag{12}$$

is orthogonal to **N** since

$$\mathbf{V} \cdot \mathbf{N} = az_x + bz_y - g = 0.$$

Hence **V** is tangent to the integral surface and lies in the tangent plane at every point (see Figure 12.1). Thus the first-order partial differential equation provides the geometric *requirement* that any integral surface through a given point be tangent to the vector **V**. This means that if we begin at some point of the initial condition (which must belong to the integral surface) and move in the direction of the *known* tangent vector **V**, we move along a curve lying entirely on the integral surface

characteristic $F(x, y, z) = 0$. This curve is called a **characteristic.** The integral surface can often be described in terms of these characteristics.

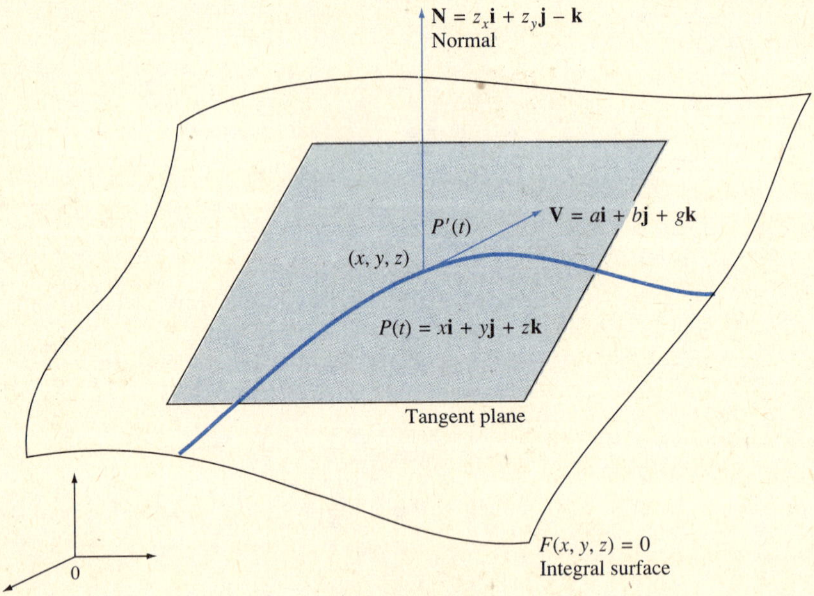

FIGURE 12.1
A characteristic on an integral surface

Points on a characteristic curve can be described parametrically by an expression of the form

$$\mathbf{P}(t) = x(t)\mathbf{i} + y(t)\mathbf{j} + z(t)\mathbf{k}. \tag{13}$$

If we differentiate **(13)** with respect to t, we obtain a tangent vector. This vector must belong to the tangent plane of the integral surface at the given point. Furthermore, it must be proportional to **V**, since the characteristic is obtained by moving in the **V** direction. Thus the coordinates of $\mathbf{P}'(t)$ and **V** are proportional; that is,

$$\frac{\dfrac{dx}{dt}}{a(x, y)} = \frac{\dfrac{dy}{dt}}{b(x, y)} = \frac{\dfrac{dz}{dt}}{g(x, y, z)},$$

or

$$\frac{dx}{a} = \frac{dy}{b} = \frac{dz}{g}. \tag{14}$$

Equations **(14)** provide a pair of first-order ordinary differential equations:

$$\frac{dy}{dx} = \frac{b(x, y)}{a(x, y)}, \qquad \frac{dz}{dx} = \frac{g(x, y, z)}{a(x, y)},$$

or

$$\frac{dx}{dy} = \frac{a(x, y)}{b(x, y)}, \qquad \frac{dz}{dy} = \frac{g(x, y, z)}{b(x, y)}. \tag{15}$$

The solutions of these equations determine what we call the **characteristics** of the partial differential equation. The following example illustrates the use of characteristics.

EXAMPLE 2 ▶ **Using characteristics to solve a partial differential equation**

Consider the Cauchy problem of Example 1:

$$\frac{\partial z}{\partial x} + \frac{\partial z}{\partial y} = z, \qquad z(x, 0) = \frac{x}{2} + e^x. \tag{16}$$

Since $a = b = 1$, and $g = z$, by Equations **(14)** we have

$$\frac{dx}{1} = \frac{dy}{1} = \frac{dz}{z},$$

and the system of differential equations **(15)** is

$$\frac{dx}{dy} = 1, \qquad \frac{dz}{dy} = z, \tag{17}$$

with solutions

$$x = y + c_1, \qquad z = c_2 e^y. \tag{18}$$

Setting $y = 0$ in **(18)**, we have $x = c_1$ and $z(x, 0) = c_2$. In the xz-plane we have

$$c_2 = z(x, 0) \overset{\text{initial condition}}{=} \frac{x}{2} + e^x = \frac{c_1}{2} + e^{c_1}. \tag{19}$$

Equation **(19)** determines the relation between the parameters c_1 and c_2. Rewriting **(18)** as

$$c_1 = x - y, \qquad c_2 = ze^{-y},$$

we have, by Equation **(19)**,

$$ze^{-y} = c_2 = \frac{c_1}{2} + e^{c_1} = \frac{x - y}{2} + e^{x-y},$$

or

$$z = \left(\frac{x - y}{2}\right)e^y + e^x.$$

Notice that this is the solution to the Cauchy problem that was obtained in Example 1. ◀

In the next example we show that the method of characteristics applies to Cauchy problems with variable coefficients.

EXAMPLE 3 ▶ **Solving an initial-value problem by characteristics**

Solve the initial-value problem

$$z_x + yz_y = 2xz, \qquad z(0, y) = y - y^2. \tag{20}$$

Solution In Equation **(14)** we have $a = 1$, $b(x, y) = y$, and $g(x, y, z) = 2xz$. Thus the characteristics are determined by the ordinary differential equations

$$\frac{dx}{1} = \frac{dy}{y} = \frac{dz}{2xz},$$

or

$$dx = \frac{dy}{y}, \qquad 2x\,dx = \frac{dz}{z}.$$

These differential equations have the solutions

$$x + c_1 = \ln|y|, \qquad x^2 + c_2 = \ln|z|,$$

or

$$y = c_3 e^x, \qquad z = c_4 e^{x^2}. \tag{21}$$

Setting $x = 0$, we have, from **(20)** and **(21)**, $y = c_3$ and

$$c_4 = z(0, y) = y - y^2 = c_3 - c_3^2. \tag{22}$$

Using **(21)**, we have

$$c_3 = ye^{-x}, \qquad c_4 = ze^{-x^2}. \tag{23}$$

Combining **(22)** and **(23)**, we obtain

$$ze^{-x^2} = c_4 = c_3 - c_3^2 = ye^{-x} - (ye^{-x})^2,$$

or

$$z = ye^{x^2-x} - y^2 e^{x^2-2x}. \tag{24}$$

To check that **(24)** is the solution to **(20)**, observe that $z(0, y) = y - y^2$,

$$z_x = (2x - 1)ye^{x^2-x} - (2x - 2)y^2 e^{x^2-2x},$$
$$yz_y = ye^{x^2-x} - 2y^2 e^{x^2-2x},$$

and

$$2xz = 2xye^{x^2-x} - 2xy^2 e^{x^2-2x}.$$ ◀

Solution of a Cauchy problem

To solve the Cauchy problem

partial differential equation: $a(x, y)\dfrac{\partial z}{\partial x} + b(x, y)\dfrac{\partial z}{\partial y} = g(x, y, z)$,

initial condition: $z(x, 0) = z_0(x)$,

by the method of characteristics, perform the following four steps:

Step 1 Write the differential equations

$$\frac{dx}{a(x, y)} = \frac{dy}{b(x, y)} = \frac{dz}{g(x, y, z)}.$$

Step 2 Solve (if possible) the first differential equation

$$\frac{dx}{dy} = \frac{a(x, y)}{b(x, y)}$$

for x in terms of y, obtaining $x = x(y, c_1)$. Substitute this expression (if necessary) for x in the second differential equation

$$\frac{dz}{dy} = \frac{g(x, y, z)}{b(x, y)}$$

and solve for z in terms of y, obtaining $z = z(y, c_2)$.

Step 3 Set $y = 0$ in the two solutions in step 2 and substitute the resulting expressions into the initial condition $z = z_0(x)$, obtaining an equation involving only c_1 and c_2:

$$z(0, c_2) = z_0(x(0, c_1)).$$

Step 4 Use the solutions in step 2 to eliminate the constants c_1 and c_2 in the final equation in step 3. The resulting equation in x, y, and z is the solution of the Cauchy problem.

SELF-QUIZ

I. Which of the following is *not* a linear partial differential equation?

(a) $\frac{\partial z}{\partial x} = 3\frac{\partial z}{\partial y}$

(b) $x^2\frac{\partial z}{\partial x} + xy^3\frac{\partial z}{\partial y} = \sin(x^2 - y^2)$

(c) $\frac{\partial z}{\partial x}\frac{\partial z}{\partial y} = 1$ **(d)** $\frac{1}{x}\frac{\partial z}{\partial x} - \cos y\frac{\partial z}{\partial y} = \frac{3z}{xy}$

II. The equation $4\frac{\partial z}{\partial x} - 3\frac{\partial z}{\partial y} + 5z = 0$ can be solved by rotating the axes through the angle α where $\tan\alpha =$ ________.

(a) $-\frac{3}{4}$ **(b)** $\frac{3}{4}$ **(c)** $\frac{4}{5}$

(d) $-\frac{3}{5}$ **(e)** $\frac{4}{3}$ **(f)** $-\frac{4}{3}$

Answer Questions III and IV true or false.

III. The problem $\frac{\partial z}{\partial x} + 2\frac{\partial z}{\partial y} + z = 0$, $z(0, 0) = 1$, is a Cauchy problem.

IV. The problem $\frac{\partial z}{\partial x} + 2\frac{\partial z}{\partial y} + z = 0$, $z(x, 0) = \sin x$, is a Cauchy problem.

V. Which set of equations helps determine the characteristic for the Cauchy problem

$$3\frac{\partial z}{\partial x} + 2y\frac{\partial z}{\partial y} = 5yz, \qquad z(x, 0) = x^2?$$

(a) $\frac{dx}{3} = \frac{dy}{2} = \frac{dz}{5y}$ **(b)** $\frac{dx}{3} = \frac{dy}{2y} = \frac{dz}{5yz}$

(c) $\frac{dx}{3} = \frac{dy}{2y} = \frac{dz}{5y}$ **(d)** $\frac{dx}{3} = \frac{dy}{2y} = \frac{dz}{5}$

Answers to Self-Quiz

I. c **II.** a **III.** False **IV.** True **V.** b

PROBLEMS 12.1

In Problems 1–8 use the method of rotation of axes (as in Example 1) to solve the given Cauchy problem.

1. $\frac{\partial z}{\partial y} - \frac{\partial z}{\partial y} + \sqrt{2}\,z = 0$, $z(x, 0) = x$

2. $\frac{\partial z}{\partial x} - \frac{\partial z}{\partial y} - \sqrt{2}\,z = 0$, $z(x, 0) = x^2$

3. $\frac{\partial z}{\partial x} + \frac{\partial z}{\partial y} + \sqrt{2}\,z = 0$, $z(x, 0) = x + e^x$

4. $\frac{\partial z}{\partial x} + \frac{\partial z}{\partial y} - \sqrt{2}\,z = 0$, $z(0, y) = e^y - y$

5. $3\frac{\partial z}{\partial x} - 4\frac{\partial z}{\partial y} + 2z = 7$, $z(x, 0) = e^x$

6. $\frac{\partial z}{\partial x} + \frac{\partial z}{\partial y} + z = 2$, $z(x, 0) = \sin x$

7. $\frac{\partial z}{\partial x} + \frac{\partial z}{\partial y} - z = e^x$, $z(x, 0) = 0$

8. $\frac{\partial z}{\partial x} - \frac{\partial z}{\partial y} + z = y$, $z(x, 0) = x^2$

In Problems 9–22 use the method of characteristics to solve the given Cauchy problem.

9. Problem 1

10. Problem 2

11. Problem 3

12. Problem 4

13. Problem 5

14. Problem 6

15. Problem 7

16. Problem 8

17. $xz_x + z_y = 1$, $z(1, y) = e^{-y}$

18. $xz_x + yz_y = z$, $z(1, y) = y^2$

19. $z_x + yz_y = z$, $z(x, 1) = xe^{-x}$

20. $xz_x + y^{-1}z_y = 1$, $z(x, 0) = 5 - x$

21. $z_x + \dfrac{z_y}{2y} - z = 2$, $z(x, 0) = \sin x - 2$

22. $xz_x + z_y - xz = x$, $z(e, y) = y - 2$

23. Adapt the method of characteristics to the Cauchy problem

$$z_x + z_y = z, \quad z(y, y) = e^y.$$

Discuss your result.

The following two problems show that Cauchy problems may have no solutions or infinitely many solutions.

24. Consider the Cauchy problem

$$z_x + z_y = 1, \quad z(y, y) = e^y.$$

Show that it has no solution.

25. Show that the Cauchy problem

$$z_x + z_y = 1, \qquad z(y, y) = y + 7$$

has infinitely many solutions.

26. Let f be a real-valued function on $\mathbb{R}^2$ that is **homogeneous of degree** n; that is,

$$f(tx, ty) = t^n f(x, y) \tag{i}$$

for all t real and (x, y) in $\mathbb{R}^2$. Show that $z = f(x, y)$ satisfies the partial differential equation

$$xz_x + yz_y = nz.$$

[*Hint:* Differentiate Equation **(i)** with respect to t and then set $t = 1$.]

12.2 INITIAL-VALUE PROBLEMS FOR QUASI-LINEAR FIRST-ORDER EQUATIONS

In Section 12.1 we developed the method of characteristics and applied it to linear first-order partial differential equations. In this section we show that this method applies to an even wider class of Cauchy problems. We also discuss some difficulties that arise when using the technique.

A first-order partial differential equation of the form

quasi-linear equation

$$a(x, y, z)\frac{\partial z}{\partial x} + b(x, y, z)\frac{\partial z}{\partial y} = g(x, y, z) \tag{1}$$

is called **quasi-linear.** Note that since a and b may involve the dependent variable z, quasi-linear equations need not be linear in z and its first partial derivatives. The set of quasi-linear partial differential equations of the first order includes the linear first-order equations as a subset. We now apply the method of characteristics to quasi-linear equations.

EXAMPLE 1 ▶ **Solving a quasi-linear equation by characteristics**

Solve the quasi-linear Cauchy problem

$$zz_x + yz_y = x, \qquad z(0, y) = \frac{1}{y}. \tag{2}$$

Solution Proceeding as in Section 12.1, we have

$$\frac{dx}{z} = \frac{dy}{y} = \frac{dz}{x}.$$

Then $x\,dx = z\,dz$, so $z^2 = x^2 + c_1$ and $z = \sqrt{x^2 + c_1}$. Also

$$\frac{dy}{y} = \frac{dx}{z} = \frac{dx}{\sqrt{x^2 + c_1}}.$$

Using Formula 67 in Appendix 1 we obtain

$$\begin{aligned}\ln|y| &= c_2 + \ln|x + \sqrt{x^2 + c_1}| \\ &= c_2 + \ln|x + z|,\end{aligned}$$

or

$$y = c_3(x + z).$$

Setting $x = 0$, we have

$$c_3 = \frac{y}{z(0, y)} \overset{\downarrow}{=} \frac{1}{z^2(0, y)} = \frac{1}{c_1}.$$

$z(0, y) = \frac{1}{y}$ **is given**

Therefore

$$\frac{y}{x + z} = c_3 = \frac{1}{c_1} = \frac{1}{z^2 - x^2},$$

or

$$x + z = y(z^2 - x^2). \tag{3}$$

It is easy to check that **(3)** satisfies the initial condition. To see that it satisfies the quasi-linear equation, we differentiate implicitly first with respect to x and then with respect to y:

$$\begin{aligned}1 + z_x &= 2yzz_x - 2xy, \\ z_y &= (z^2 - x^2) + 2yzz_y.\end{aligned}$$

Thus

$$z_x = \frac{1 + 2xy}{2yz - 1} \quad \text{and} \quad z_y = \frac{z^2 - x^2}{1 - 2yz}.$$

Then, using **(3)**,

$$zz_x + yz_y = \frac{z + 2xyz - y(z^2 - x^2)}{2yz - 1}$$

Equation (3)

$$\overset{\downarrow}{=} \frac{z + 2xyz - (x + z)}{2yz - 1} = x. \quad ◀$$

So far we have concentrated on situations where no difficulties are encountered in using the method of characteristics. The next examples provide a sample of various types of technical difficulties that may arise.

EXAMPLE 2 ▶ **Finding a general solution**

Find the general solution of the quasi-linear partial differential equation

$$zz_x + z_y = 0. \tag{4}$$

Solution If we follow the technique of Example 1 formally, we obtain the system of ordinary differential equations

$$\frac{dx}{z} = \frac{dy}{1} = \frac{dz}{0}.$$

Division by zero is not allowed. However, if we look again at the development leading to Equation **(12.1.14),** we see that all that is required is that the coordinates of the tangent $\mathbf{P}'(t)$ be proportional to those of $\mathbf{V}$. This is equivalent to the system

$$\frac{dx}{z} = \frac{dy}{1} \quad \text{and} \quad dz = 0. \tag{5}$$

Hence $z = c_2$ and $\frac{x}{z} = y + c_1$ (since z is constant).

If initial conditions for z are prescribed in the form $z(x, 0) = f(x)$, with f an arbitrary differentiable function, then setting $y = 0$, we obtain

$$c_2 = z(x, 0) = f(x) \overset{\substack{\text{set } y = 0 \text{ in} \\ x = z(y + c_1) \\ \downarrow}}{=} f(c_1 z) = f(c_1 c_2).$$

Hence

$$z = f\left(\left[\frac{x}{z} - y\right]z\right) = f(x - yz)$$

is the general solution of any such Cauchy problem. Observe that

$$z_x = f'(x - yz)(1 - yz_x), \qquad z_y = f'(x - yz)(-z - yz_y);$$

therefore

$$z_x = \frac{f'}{1 + yf'}, \qquad z_y = \frac{-zf'}{1 + yf'},$$

from which **(4)** follows immediately. ◀

EXAMPLE 3 ▶ **The effect of initial conditions**

Consider the partial differential equation

$$z_x + 2z_y = 0.$$

Here

$$\frac{dx}{1} = \frac{dy}{2}, \quad dz = 0,$$

so that

$$2\,dx = dy, \quad dz = 0,$$

and

$$2x + c_1 = y, \qquad z = c_2.$$

The constants c_1 and c_2 are arbitrary, so the characteristics are given by all horizontal lines (i.e., lines in a plane parallel to the xy-plane) parallel to $y = 2x$. Assume that $c_2 = f(c_1)$; then

$$z = f(y - 2x) \tag{6}$$

is the general solution of the partial differential equation. We now consider what effect various initial conditions have on Equation **(6).**

(a) If $z(x, x) = e^x$ is the initial condition, then setting $y = x$ in **(6)** we have

$$f(-x) = e^x \quad \text{or} \quad f(x) = e^{-x}.$$

Thus the solution of the corresponding Cauchy problem is $z = e^{2x-y}$.

(b) If $z(x, 2x) = e^x$ is the initial condition, then when we set $y = 2x$ in **(6)** we obtain $e^x = f(0)$. This is impossible, since x is variable. The problem here is that the initial condition is defined on a characteristic ($y = 2x$) on which *z must be constant,* because of the partial differential equation. Thus we have *no* solution in this case.

(c) If $z(x, 2x) = k$ (a constant), setting $y = 2x$ we have $f(0) = k$. In this case, *any* differentiable function satisfying $f(0) = k$ is a solution: For example, if $f(t) = at^n + k$, then $z = a(y - 2x)^n + k$ is a solution to the Cauchy problem for every number a and integer $n > 0$.

It is clear from (b) and (c) above that if the initial condition is given along a characteristic, we cannot guarantee the existence or uniqueness of a solution. It can be shown* that the Cauchy problem **(1)** with initial condition $z(x_0(t), y_0(t)) = z_0(t)$, $\alpha \le t \le \beta$, has a unique solution in a disk centered at $(x_0(t_0), y_0(t_0))$, where t_0 is the initial value of t, provided

$$a(x_0(t), y_0(t), z_0(t))\frac{dy_0}{dt} \neq b(x_0(t), y_0(t), z_0(t))\frac{dx_0}{dt}. \quad ◀$$

EXAMPLE 4 ▶ **Fluid flow**

A partial differential equation that arises in the one-dimensional motion of an ideal gas or fluid, because of the conservation of mass, is

$$\rho_t + (\rho v)_x = 0, \tag{7}$$

where ρ, v, x, and t represent density, velocity, displacement, and time in the flow of the gas or fluid. For certain (**isentropic**†) flows, we are interested in solutions where the density is a function of the velocity; that is, where $\rho = f(v)$. In this case, **(7)** becomes

$$f'(v)v_t + [f(v) + vf'(v)]v_x = 0,$$

or

$$\left[v + \frac{f(v)}{f'(v)}\right]v_x + v_t = 0. \tag{8}$$

For simplicity, assume that $f(v) = a + bv$. Then we get

$$\left(\frac{a}{b} + 2v\right)v_x + v_t = 0. \tag{9}$$

The characteristics for **(9)** are determined by the system

$$\frac{dx}{\left(\frac{a}{b}\right) + 2v} = \frac{dt}{1}, \qquad dv = 0.$$

---

*See G. C. Zachmanoglou and D. W. Thoe, *Introduction to Partial Differential Equations with Applications* (Baltimore: Williams & Wilkens, 1976).

† **Isentropic** means having constant entropy. **Entropy** (for a gas) is a scalar field that is proportional to the potential temperature.

Hence v is constant on each characteristic, and since

$$\frac{dx}{dt} = \frac{a}{b} + 2v,$$

$\frac{dx}{dt}$ is also constant on characteristics:

$$x = \left(2c_2 + \frac{a}{b}\right)t + c_1, \qquad v = c_2.$$

Thus the characteristics are straight lines that are parallel only in each plane $v = c_2$. Since c_1 and c_2 are arbitrary, we can assume that $c_2 = g(c_1)$, so

$$v = g\left(x - \left(2v + \frac{a}{b}\right)t\right) \tag{10}$$

is the general solution to **(9)**.

Now suppose the initial condition is

$$v(x, 0) = \begin{cases} \epsilon(1 + \cos x), & |x| < \pi, \\ 0, & |x| > \pi. \end{cases}$$

Then $g(x) = v(x, 0)$, and

$$v(x, t) = \begin{cases} \epsilon\left[1 + \cos\left(x - \left(2v + \frac{a}{b}\right)t\right)\right], & \text{if} \left|x - \left(2v + \frac{a}{b}\right)t\right| < \pi, \\ 0, & \text{if} \left|x - \left(2v + \frac{a}{b}\right)t\right| > \pi. \end{cases}$$

If we look at Figure 12.2 we see that the initial velocity wave propagates along the characteristics $x = (\frac{at}{b}) \pm \pi$ for $v = 0$, but that the *shape* of the wave changes, since the maximum of 2ϵ is located above the line $x = (4\epsilon + \frac{a}{b})t$ in the base plane. More will be said about this example in Section 12.3.

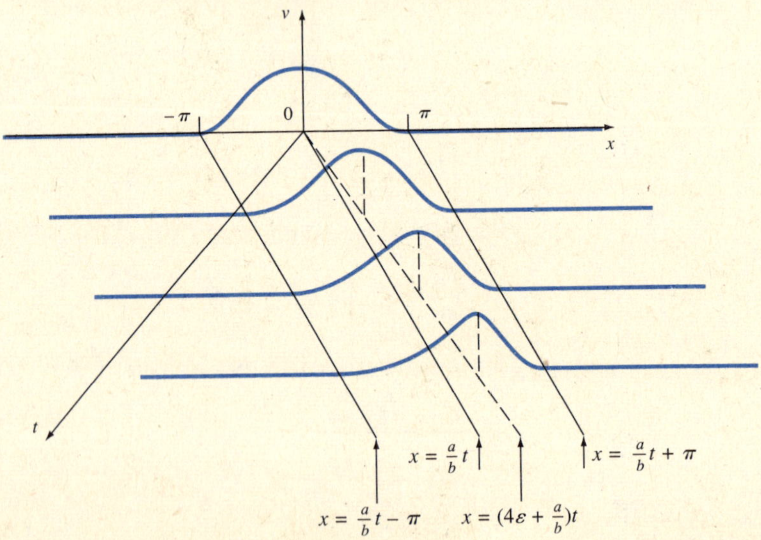

FIGURE 12.2
Wave propagation

SELF-QUIZ

I. Which of the following partial differential equations are quasi-linear?

(a) $z\dfrac{\partial z}{\partial x} + \dfrac{1}{z}\dfrac{\partial z}{\partial y} = xyz$

(b) $x^2\dfrac{\partial z}{\partial x} - \sin xy\dfrac{\partial z}{\partial y} = e^{x-2y}$

(c) $\dfrac{\partial z}{\partial x}\dfrac{\partial z}{\partial y} = x^2 + y^2$

(d) $\dfrac{\ln z}{e^x}\dfrac{\partial z}{\partial x} - \dfrac{3xe^z}{y^2}\dfrac{\partial z}{\partial y} = \dfrac{xz - z^4 + z^3}{y}$

II. If $xz_x + 2zz_y = y^2$, then ________.

(a) $xdx = 2zdy = y^2\,dz$ **(b)** $\dfrac{dx}{2z} = \dfrac{dy}{x} = \dfrac{dz}{y^2}$

(c) $\dfrac{dx}{x} = \dfrac{dy}{2y} = \dfrac{dz}{y^2}$ **(d)** $\dfrac{dx}{x} = \dfrac{dy}{2z} = \dfrac{dz}{y^2}$

III. If $x\,z_x + z\,z_y = 2yz$, then the characteristics are ________.

(a) $\ln|x| = \dfrac{y}{z} + c$ **(b)** $z = y^2 + c$

(c) $y^2 = \ln|z| + c$ **(d)** $x = k\left|\dfrac{y-c}{y+c}\right|^{1/2c}$

Answers to Self-Quiz

I. a, b (this is also linear), d **II.** d **III.** b, d

PROBLEMS 12.2

In Problems 1-20 find either the general solution or, if initial conditions are given, the solution to the Cauchy problem.

1. $z_x + z_y = 0$
2. $z_x - z_y = 0$
3. $xz_x + z_y = 0$
4. $zz_x - z_y = 0$
5. $xz_x + zz_y = 1$
6. $zz_x + z_y = z^2$
7. $z_x + z^2z_y = 0,\ z(x, 0) = \cos x$
8. $z_x + zz_y = 0,\ z(0, y) = \cosh y$
9. $x^2z_x + y^2z_y = z^2,\ z(x, 2x) = 1$
10. $xz_x = yz_y,\ z(x, x) = x^2$
11. $zz_x + yz_y = x,\ z(x, 1) = 2x$
12. $(y + z)z_x + yz_y = 0$
13. $zz_x + z_y = 0,\ z(x, 0) = x$
14. $z^2z_x + z_y = 0$
15. $xz_x + yz_y = z^2$
16. $(1 + x)z_x - yz_y = 1 - z^2$
17. $xz_x + yz_y = 1 - z^2,\ z(x, 1) = e^x$
18. $zz_x + zz_y = z^2$
19. $yz_x + zz_y = xy$
20. $zz_x + zyz_y = y$
21. The quasi-linear equation
$$\left(1 + \frac{3}{2}z\right)z_x + z_y = 0$$
arises in the study of gravity water waves in shallow water.* Find a general solution.
22. Traffic flow on a highway can be modeled as a one-dimensional flow along the x-axis (as the highway). Let $\rho(x, t)$ be the density (cars per unit length) at the point x of the highway at time t, and let $q(x, t)$ be the rate at which cars pass the point x per unit time at time t.
    (a) Assuming that there are no sources or sinks of cars at any point of the highway and that the functions ρ and q are sufficiently smooth, show that ρ and q satisfy the conservation equation
    $$\rho_t + q_x = 0.$$
    (b) Assume that the flow rate depends only on the density, $q = f(\rho)$, for some smooth function f. What quasi-linear equation is obtained?
    (c) Find the general solution of the equation in part (b) if $f(\rho) = c\rho(1 - \frac{\rho}{k})$.

---

* Robert L. Street, *Partial Differential Equations* (Monterey, Calif.: Brooks/Cole, 1973).

12.3 APPLICATIONS TO THE THEORY OF SHOCKS IN GAS DYNAMICS (OPTIONAL)

In Example 12.2.4 we showed that, under certain conditions, conservation laws arising in many physical applications give rise to quasi-linear Cauchy problems. In this section we study the one-dimensional, time-dependent flow of a compressible fluid under constant pressure p. Let u, ρ, and E represent the fluid velocity, density, and internal energy per unit volume. The basic equations of gas dynamics* are

$$\begin{aligned} u_t + uu_x &= 0, \\ \rho_t + (\rho u)_x &= 0, \\ E_t + (Eu)_x + pu_x &= 0. \end{aligned} \tag{1}$$

The substitution $s = E + p$ allows us to replace the third equation in **(1)** by

$$s_t + (su)_x = 0, \tag{2}$$

since p is constant. Note that **(2)** has the same form as the second equation in **(1)**. Finally, to convert this into a Cauchy problem, we assume that initial conditions

$$\begin{aligned} u(x, 0) &= f(x), \\ \rho(x, 0) &= g(x), \\ s(x, 0) &= h(x), \end{aligned} \tag{3}$$

where f, g, and h have sufficient smoothness for what is yet to come.

If we solve the first equation in **(1)**, which is independent of ρ and s, we can use the solution u to solve the remaining *linear* partial differential equations.

As in Section 12.2, the characteristics of the first equation in **(1)** are given by the system of differential equations

$$\frac{dt}{1} = \frac{dx}{u} \quad \text{and} \quad du = 0.$$

Here $u = c_2$ and $ut = x + c_1$ (since u is constant), in three-dimensional txu-space. Applying the first set of initial conditions in **(3)**, we have (with $t = 0$)

$$c_2 = u(x, 0) = f(x) = f(-c_1).$$

Hence we obtain the general solution

$$u = f(x - tu). \tag{4}$$

To check that this is indeed a solution, observe that, by the chain rule,

$$u_t = (-u - tu_t)f'(x - tu) \quad \text{or} \quad (1 + tf')u_t = -uf', \tag{5}$$

$$u_x = (1 - tu_x)f'(x - tu) \quad \text{or} \quad (1 + tf')u_x = f'. \tag{6}$$

Thus the first equation in **(1)** is satisfied if $1 + tf' \neq 0$. Observe that

$$1 + tf'(x - tu) > 0 \tag{7}$$

for very small values of t.

We may rewrite **(5)** and **(6)** as

$$u_t = \frac{-uf'(x - tu)}{1 + tf'(x - tu)}, \quad u_x = \frac{f'(x - tu)}{1 + tf'(x - tu)}. \tag{8}$$

* See, for example, S. I. Pai, *Magnetogas-Dynamics and Plasma Dynamics* (Englewood Cliffs, N.J.: Prentice-Hall, 1962).

shocks

Notice that u_t and u_x tend to ∞ if $1 + tf'$ tends to zero. When $1 + tf'$ becomes zero, the solution u develops a discontinuity known as a **shock.** Shocks are a common phenomenon in fluid and gas dynamics.

EXAMPLE 1 ▶ **A shock**

Consider the general solution **(12.2.10)** of Example 12.2.4:

$$v = g\left(x - \left(2v + \frac{a}{b}\right)t\right).$$

In that example we let the initial condition be

$$v(x, 0) = \begin{cases} \epsilon(1 + \cos x), & |x| < \pi, \\ 0, & |x| > \pi, \end{cases}$$

so that

$$v(x, t) = \begin{cases} \epsilon(1 + \cos \eta), & |\eta| < \pi, \\ 0, & |\eta| > \pi, \end{cases}$$

where $\eta = x - (2v + \frac{a}{b})t$. Hence

$$v_x = -\epsilon(1 - 2tv_x) \sin \eta,$$

or

$$v_x = \frac{-\epsilon \sin \eta}{1 - 2t\epsilon \sin \eta}. \tag{9}$$

Since ϵ is constant, the shock first appears for $\eta = \frac{\pi}{2}$ ($\sin \eta = 1$) when $t = \frac{1}{2\epsilon}$. Hence the waves shown in Figure 12.2 occur only as long as $t < \frac{1}{2\epsilon}$.

To understand why shocks occur, consider Equation **(4).** Select any fixed number x_0 and calculate $u_0 = f(x_0)$. The points lying on the intersection of the two surfaces

$$x - tu_0 = x_0 \quad \text{and} \quad u = u_0 \tag{10}$$

satisfy Equation **(4),** since

$$u = u_0 = f(x_0) = f(x - tu_0).$$

Thus the straight line defined by **(10)** lies on the solution surface. In particular, this means that along the line

$$x - tu_0 = x_0 \tag{11}$$

in the tx-plane, which passes through the point $(x_0, 0)$, the solution u to the Cauchy problem is constant and equal to $u_0 = f(x_0)$.

Now suppose we pick a different number x_1, calculate $u_1 = f(x_1) \neq u_0$, and consider the line

$$x - tu_1 = x_1. \tag{12}$$

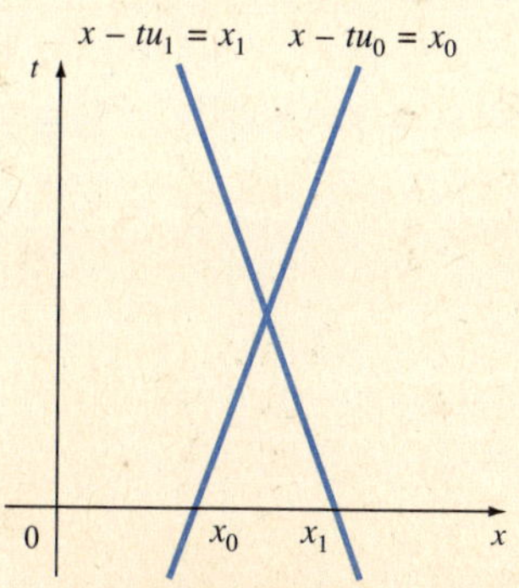

FIGURE 12.3
Shock occurs where straight lines intersect

As long as lines **(11)** and **(12),** for arbitrary choices of x_0 and x_1, do not intersect in the half plane $t > 0$, the solution exists for all $t > 0$. But if two of these lines intersect in $t > 0$ (see Figure 12.3), then at the point of intersection we have a difficulty: by Equation **(11)** the solution u must equal u_0, whereas Equation **(12)** requires that it equal $u_1 \neq u_0$. Thus the solution u cannot exist at the point of intersection given by

$$x = x_1 + tu_1 = x_0 + tu_0,$$

or

$$t = \frac{x_1 - x_0}{u_0 - u_1}, \qquad x = \frac{u_0 x_1 - x_0 u_1}{u_0 - u_1}. \tag{13}$$

In particular, the solution does not exist for

$$t \geq \frac{x_1 - x_0}{u_0 - u_1}. \quad ◀$$

Although the existence of shocks presents a technical difficulty, there are ways of handling this obstacle by generalizing the meaning of a solution of a partial differential equation.* These concepts are beyond the scope of this book.

Returning to the original Cauchy problem of Equations **(1)–(3),** we write the second equation in **(1)** as

$$\rho_t + u\rho_x + \rho u_x = 0. \tag{14}$$

Since

$$u_x = \frac{f'(x - tu)}{1 + tf'(x - tu)}, \qquad u = f(x - tu),$$

we try a solution for ρ of the form

$$\rho = \frac{G(x - tu)}{1 + tf'(x - tu)} \tag{15}$$

so that all terms in **(14)** have a denominator of $(1 + tf')^2$. Then

$$\rho_t = \frac{(1 + tf')G'(-u - tu_t) - G(f' + tf''(-u - tu_t))}{(1 + tf')^2},$$

$$u\rho_x = u\left[\frac{(1 + tf')G'(1 - tu_x) - Gtf''(1 - tu_x)}{(1 + tf')^2}\right],$$

$$\rho u_x = \frac{Gf'}{(1 + tf')^2}.$$

Substituting in Equation **(14)** we find that

$$\rho_t + u\rho_x + \rho u_x = \frac{-tG'(1 + tf')(u_t + uu_x) + t^2 Gf''(u_t + uu_x)}{(1 + tf')^2} = 0,$$

since $u_t + uu_x = 0$ by **(1).** Setting $t = 0$ in Equation **(15),** we get

$$\rho(x, 0) = G(x),$$

implying, by the second initial condition, that $G = g$ and that the general solution of the density equation is

$$\rho = \frac{g(x - tu)}{1 + tf'(x - tu)}.$$

Similarly, since **(2)** is of the same form as the second equation in **(1),** we get

$$s = \frac{h(x - tu)}{1 + tf'(x - tu)},$$

*See the survey article by P. D. Lax, "The Formation and Decay of Shock Waves," *American Mathematical Monthly* 79 (1972): 227–241.

from which it follows that

$$E = \frac{h(x - tu)}{1 + tf'(x - tu)} - p.$$

PROBLEMS 12.3

1. Solve the Cauchy problem

$$uu_x + u_t = 0, \qquad u(x, 0) = x.$$

Do shocks occur in this problem for $t \geq 0$?

2. Repeat Problem 1 for the Cauchy problem

$$u^2u_x + u_t = 0, \qquad u(x, 0) = x.$$

[*Hint:* Use the Maclaurin series of $\sqrt{1 + a}$.]

3. Repeat Problem 1 for

$$uu_x + u_t = 0, \qquad u(x, 0) = x^2.$$

4. For the Cauchy problem of Equations **(1)–(3),** show that if $f(x)$ is nondecreasing, the shocks never occur for $t \geq 0$.

5. For the Cauchy problem of Equations **(1)–(3),** show that if $f(x)$ is strictly decreasing over any interval, then a shock eventually develops at some positive t. Derive a formula for the time and location of the shock.

6. Discuss whether shocks occur for the Cauchy problem

$$u_t + F(u)u_x = 0, \qquad u(x, 0) = f(x).$$

7. Discuss whether shocks occur for the Cauchy problem

$$u_t + (F(u))_x = 0, \qquad u(x, 0) = f(x).$$

12.4 CLASSIFICATION OF LINEAR SECOND-ORDER EQUATIONS

In this section we are concerned only with second-order linear partial differential equations of two independent variables with constant coefficients, that is, equations of the form

$$a\frac{\partial^2 w}{\partial x^2}(x, y) + 2b\frac{\partial^2 w}{\partial x\, \partial y}(x, y) + c\frac{\partial^2 w}{\partial y^2}(x, y) + k\frac{\partial w}{\partial x}(x, y) + m\frac{\partial w}{\partial y}(x, y) + nw(x, y) = f(x, y), \tag{1}$$

where a, b, c, k, m, and n are constants (the number 2 in front of the coefficient b simplifies later computations). Since we want to consider second-order equations, we assume that at least one of the coefficients, a, b, or c is nonzero. If so, Equation **(1)** is called a **second-order** equation. The procedure for solving second-order equations is much more complicated than that for first-order equations. Fortunately, as we show in this section, the general second-order equation **(1)** can be transformed into one of four standard simpler forms:

$$\frac{\partial^2 z}{\partial x^2} + \frac{\partial^2 z}{\partial y^2} + \lambda z = G(x, y) \qquad \text{(elliptic)}$$

$$\frac{\partial^2 z}{\partial x^2} - \frac{\partial^2 z}{\partial y^2} + \lambda z = G(x, y) \qquad \text{(hyperbolic)}$$

$$\frac{\partial^2 z}{\partial x^2} + \lambda\frac{\partial z}{\partial y} = G(x, y) \qquad \text{(parabolic)}$$

$$\frac{\partial^2 z}{\partial x} + \lambda z = G(x, y) \qquad \text{(degenerate)}$$

λ constant.

In addition to being easier to solve, these standard forms arise naturally in a number of important applications. In Sections 12.5 and 12.6 we will study a particular hyperbolic equation, called the **wave equation,** that arises in modeling small transverse vibrations in a tightly stretched string. In Sections 12.7 and 12.8 we will consider a parabolic equation, called the **heat equation,** that is used to analyze heat conduction in thin round bars. In Sections 12.8 and 12.9 we will study a particular case of an elliptic equation, called **Laplace's equation,** that is used in determining steady-state flows (of fluids, heat, etc.). The degenerate equation has already been considered in Chapter 4, when we looked at spring-mass vibrations.

Initially, it is convenient to simplify Equation **(1)** by another rotation of axes. Using the substitution

$$\begin{aligned} u &= x \cos \alpha + y \sin \alpha, \\ v &= -x \sin \alpha + y \cos \alpha, \end{aligned} \tag{2}$$

with $\tan 2\alpha = \frac{2b}{a-c}$, we can eliminate the mixed second partial $\partial^2 w/\partial u\, \partial v$ and obtain an equation of the form

$$\begin{aligned} A\frac{\partial^2 w}{\partial u^2} + C\frac{\partial^2 w}{\partial v^2} + K\frac{\partial w}{\partial u} + M\frac{\partial w}{\partial v} + nw \\ = f(u\cos\alpha - v\sin\alpha,\, u\sin\alpha + v\cos\alpha) \\ = F(u, v), \end{aligned} \tag{3}$$

where

$$\begin{aligned} A &= a\cos^2\alpha + 2b\sin\alpha\cos\alpha + c\sin^2\alpha, \\ C &= a\sin^2\alpha - 2b\sin\alpha\cos\alpha + c\cos^2\alpha, \\ K &= k\cos\alpha + m\sin\alpha, \\ M &= m\cos\alpha - k\sin\alpha. \end{aligned}$$

To see why this happens, observe that

$$\begin{aligned} \frac{\partial w}{\partial x} &= \frac{\partial w}{\partial u}\frac{\partial u}{\partial x} + \frac{\partial w}{\partial v}\frac{\partial v}{\partial x} = \cos\alpha\frac{\partial w}{\partial u} - \sin\alpha\frac{\partial w}{\partial v}, \\ \frac{\partial w}{\partial y} &= \frac{\partial w}{\partial u}\frac{\partial u}{\partial y} + \frac{\partial w}{\partial v}\frac{\partial v}{\partial y} = \sin\alpha\frac{\partial w}{\partial u} + \cos\alpha\frac{\partial w}{\partial v}, \end{aligned}$$

and

$$\begin{aligned} \frac{\partial^2 w}{\partial x^2} &= \frac{\partial}{\partial u}\left(\frac{\partial w}{\partial x}\right)\frac{\partial u}{\partial x} + \frac{\partial}{\partial v}\left(\frac{\partial w}{\partial x}\right)\frac{\partial v}{\partial x} \\ &= \cos\alpha\left(\cos\alpha\frac{\partial^2 w}{\partial u^2} - \sin\alpha\frac{\partial^2 w}{\partial u\,\partial v}\right) - \sin\alpha\left(\cos\alpha\frac{\partial^2 w}{\partial u\,\partial v} - \sin\alpha\frac{\partial^2 w}{\partial v^2}\right) \\ &= \cos^2\alpha\frac{\partial^2 w}{\partial u^2} - 2\sin\alpha\cos\alpha\frac{\partial^2 w}{\partial u\,\partial v} + \sin^2\alpha\frac{\partial^2 w}{\partial v^2}. \end{aligned}$$

Similarly,

$$\frac{\partial^2 w}{\partial x\,\partial y} = \sin\alpha\cos\alpha\frac{\partial^2 w}{\partial u^2} + (\cos^2\alpha - \sin^2\alpha)\frac{\partial^2 w}{\partial u\,\partial v} - \sin\alpha\cos\alpha\frac{\partial^2 w}{\partial v^2},$$

and

$$\frac{\partial^2 w}{\partial y^2} = \sin^2\alpha\frac{\partial^2 w}{\partial u^2} + 2\sin\alpha\cos\alpha\frac{\partial^2 w}{\partial u\,\partial v} + \cos^2\alpha\frac{\partial^2 w}{\partial v^2}.$$

Then

$$a\frac{\partial^2 w}{\partial x^2} + 2b\frac{\partial^2 w}{\partial x\,\partial y} + c\frac{\partial^2 w}{\partial y^2} = A\frac{\partial^2 w}{\partial u^2} + 2B\frac{\partial^2 w}{\partial u\,\partial v} + C\frac{\partial^2 w}{\partial v^2},$$

where

$$B = (c - a)\sin\alpha\cos\alpha + b(\cos^2\alpha - \sin^2\alpha).$$

By the double angle formulas of trigonometry,

$$B = \frac{(c-a)}{2}\sin 2\alpha + b\cos 2\alpha = b\cos 2\alpha\left[\frac{(c-a)}{2b}\tan 2\alpha + 1\right].$$

Since

$$\tan 2\alpha = \frac{2b}{a-c},$$

it follows that $B = 0$. An easy but messy calculation (see Problem 21) shows that

$$b^2 - ac = B^2 - AC = -AC. \tag{4}$$

elliptic, hyperbolic, and parabolic equations

By analogy with analytic geometry, we call the partial differential equation **(1)** an **elliptic** equation if $b^2 - ac < 0$, that is, if A and C have the same sign. If A and C have opposite signs or, equivalently, if $b^2 - ac > 0$, we say that Equation **(1)** is a **hyperbolic** equation. Finally, if $b^2 - ac = 0$, Equation **(1)** is called a **parabolic** equation. In summary, Equation **(1)** is called

$$\left.\begin{array}{r}\text{hyperbolic}\\ \text{parabolic}\\ \text{elliptic}\end{array}\right\}\quad \text{if } b^2 - ac\begin{cases} > 0,\\ = 0,\\ < 0.\end{cases} \tag{5}$$

EXAMPLE 1 ▶ **Classifying an equation**

Classify the second-order linear partial differential equation

$$\frac{\partial^2 w}{\partial x^2} + 2\frac{\partial^2 w}{\partial x\,\partial y} + 4\frac{\partial^2 w}{\partial y^2} + 7\frac{\partial w}{\partial x} + w = 0. \tag{6}$$

Solution Here $a = b = 1$ and $c = 4$, so $b^2 - ac = -3$. Thus the partial differential equation **(6)** is elliptic. ◀

Once we have eliminated the mixed second partial term, we can, for suitable values of α and β, use the substitution

$$w = e^{\alpha u + \beta v} z$$

to eliminate *at least one* of the first partials in **(3).** Note that by the product rule of differentiation,

$$\frac{\partial w}{\partial u} = \alpha w + e^{\alpha u + \beta v}\frac{\partial z}{\partial u}$$

and

$$\frac{\partial w}{\partial v} = \beta w + e^{\alpha u + \beta v}\frac{\partial z}{\partial v},$$

so that

$$\frac{\partial^2 w}{\partial u^2} = \alpha^2 w + 2\alpha e^{\alpha u+\beta v}\frac{\partial z}{\partial u} + e^{\alpha u+\beta v}\frac{\partial^2 z}{\partial u^2}$$

and

$$\frac{\partial^2 w}{\partial v^2} = \beta^2 w + 2\beta e^{\alpha u+\beta v}\frac{\partial z}{\partial v} + e^{\alpha u+\beta v}\frac{\partial^2 z}{\partial v^2}.$$

Equation **(3)** becomes

$$A\frac{\partial^2 z}{\partial u^2} + C\frac{\partial^2 z}{\partial v^2} + (2\alpha A + K)\frac{\partial z}{\partial u} + (2\beta C + M)\frac{\partial z}{\partial v} + (\alpha^2 A + \beta^2 C + \alpha K + \beta M + n)z = e^{-(\alpha u+\beta v)}F(u, v). \tag{7}$$

If A and $C \neq 0$, then choosing $\alpha = -\frac{K}{2A}$ and $\beta = -\frac{M}{2C}$ allows us to eliminate the first partial terms, yielding the partial differential equation

$$A\frac{\partial^2 z}{\partial u^2} + C\frac{\partial^2 z}{\partial v^2} + \left(n - \frac{K^2}{4A} - \frac{M^2}{4C}\right)z = e^{-(\alpha u+\beta v)}F(u, v). \tag{8}$$

If either A or C is zero, we cannot eliminate one of the first partial terms.

A change of scale in the independent variables provides a final simplification: If A and $C \neq 0$, set $u = \sqrt{|A|}\,x$ and $v = \sqrt{|C|}\,y$, so that Equation **(8)** becomes

$$\frac{A}{|A|}\frac{\partial^2 z}{\partial x^2} + \frac{C}{|C|}\frac{\partial^2 z}{\partial y^2} + \left(n - \frac{K^2}{4A} - \frac{M^2}{4C}\right)z = G(x, y). \tag{9}$$

From the calculations above, it follows that Equation **(1)** can be transformed into one of the following four standard forms.

Standard forms for a second-order, linear partial differential equation with constant coefficients

$$\frac{\partial^2 z}{\partial x^2} + \frac{\partial^2 z}{\partial y^2} + \lambda z = G(x, y) \quad \text{(elliptic)} \tag{10}$$

$$\frac{\partial^2 z}{\partial x^2} - \frac{\partial^2 z}{\partial y^2} + \lambda z = G(x, y) \quad \text{(hyperbolic)} \tag{11}$$

$$\frac{\partial^2 z}{\partial x^2} + \lambda\frac{\partial z}{\partial y} = G(x, y) \quad \text{(parabolic)} \tag{12}$$

$$\frac{\partial^2 z}{\partial x^2} + \lambda z = G(x, y) \quad \text{(degenerate)} \tag{13}$$

λ constant.

Remark

Equation **(10)** occurs when A and C in **(9)** are nonzero and have the same sign. Equation **(11)** occurs when A and C are nonzero and have opposite signs. Equation **(12)** arises when $C = 0$ and $AM \neq 0$, by choosing $\alpha = -\frac{K}{2A}$ and $\beta = (K^2/4A - n)/M$; here $\lambda = \frac{M}{A}$. Equation **(13)** occurs when $C = M = 0$; here $\lambda = n - K^2/4A$.

The degenerate case may be treated as an ordinary differential equation with parameter v, in a manner very similar to Example 12.1.1. For this reason, we do not consider this case any further. In the next five sections we consider examples of elliptic, hyperbolic, and parabolic equations, showing how each arises in practice and indicating what steps can be taken to obtain a solution.

EXAMPLE 2 ▶ **Finding the standard form of an equation**

To transform the equation

$$\frac{\partial^2 w}{\partial x^2} + 4\frac{\partial^2 w}{\partial x\, \partial y} + \frac{\partial^2 w}{\partial y^2} + \frac{\partial w}{\partial x} = 0$$

into standard form, observe that $b^2 - ac = 3$, since $a = c = 1$ and $b = 2$, and therefore that the equation is hyperbolic. Since $\tan 2\alpha$ is infinite, we have $\alpha = 45° = \frac{\pi}{4}$. Substituting this value into **(3)**, we obtain

$$3\frac{\partial^2 w}{\partial u^2} - \frac{\partial^2 w}{\partial v^2} + \frac{1}{\sqrt{2}}\frac{\partial w}{\partial u} - \frac{1}{\sqrt{2}}\frac{\partial w}{\partial v} = 0.$$

Setting

$$w = z \exp\left[-\frac{\sqrt{2}}{12}(u + 3v)\right]$$

(or $\alpha = -K/2A = -1/6\sqrt{2} = -\sqrt{2}/12$ and $\beta = -M/2C = -1/2\sqrt{2} = -\sqrt{2}/4$), we obtain from Equation **(8)**

$$3\frac{\partial^2 z}{\partial u^2} - \frac{\partial^2 z}{\partial v^2} + \frac{z}{12} = 0.$$

Finally, letting $u = \sqrt{3}\, x$ and $v = y$, we get the hyperbolic equation

$$\frac{\partial^2 z}{\partial x^2} - \frac{\partial^2 z}{\partial y^2} + \frac{z}{12} = 0. \quad ◀$$

SELF-QUIZ

I. The partial differential equation $4\partial^2 w/\partial x^2 + 6\,\partial^2 w/\partial x \partial y + 5\,\partial^2 w/\partial y^2 + 3\partial w/\partial x + 4w = x^2 + y^2$, is ______.
(a) elliptic **(b)** hyperbolic **(c)** parabolic **(d)** degenerate

II. The partial differential equation $4\partial^2 w/\partial x^2 + 6\,\partial^2 w/\partial x\, \partial y + 9\partial^2 w/\partial y^2 + 3\partial w/\partial x + 4w = x^2 + y^2$, is ______.
(a) elliptic **(b)** hyperbolic **(c)** parabolic **(d)** degenerate

III. The partial differential equation $4\partial^2 w/\partial x^2 + 6\,\partial^2 w/\partial x\, \partial y + 10\,\partial^2 w/\partial y^2 + 3\,\partial w/\partial x + 4w = x^2 + y^2$, is ______.
(a) elliptic **(b)** hyperbolic **(c)** parabolic **(d)** degenerate

IV. The partial differential equation $\partial^2 w/\partial y^2 + 4w = x^2 + y^2$, is ______.
(a) elliptic **(b)** hyperbolic **(c)** parabolic **(d)** degenerate

Answers to Self-Quiz

I. b **II.** c **III.** a **IV.** d

PROBLEMS 12.4

In Problems 1–10 classify the given equation as elliptic, parabolic, hyperbolic, or degenerate second-order partial differential equations.

1. $\dfrac{\partial^2 w}{\partial x^2} - \dfrac{\partial^2 w}{\partial y^2} = 0$

2. $\dfrac{\partial^2 w}{\partial x^2} + \dfrac{\partial^2 w}{\partial y^2} + w = 0$

3. $w_{xx} + 2w_{xy} + w_{yy} + w_x + w_y = 0$

4. $w_{xx} + w_{xy} + 4w_{yy} + 5w_x = 2$

5. $w_{xx} + 2w_{xy} + 2w_{yy} = 0$

6. $w_{xx} + 2w_{xy} + w_{yy} + w = 0$

7. $w_{xx} + 2w_{xy} + 4w_{yy} + 5w = 0$

8. $w_{xx} + 2w_{xy} - w_{yy} + w_y = 0$

9. $3w_{xx} - 5w_{xy} + 2w_{yy} = \sin(xy)$

10. $7w_{xy} + 5w_{yy} + 8w = \dfrac{x}{y}$

In Problems 11–20 transform the given equation into standard form.

11. $4\dfrac{\partial^2 w}{\partial x^2} + 3\dfrac{\partial^2 w}{\partial y^2} - w = 0$

12. $7\dfrac{\partial^2 w}{\partial x^2} + 4\dfrac{\partial^2 w}{\partial y^2} + 3w = 2$

13. $\dfrac{\partial^2 w}{\partial x^2} - \dfrac{\partial^2 w}{\partial y^2} + 2\dfrac{\partial w}{\partial x} - 5\dfrac{\partial w}{\partial y} + 7w = 0$

14. $\dfrac{\partial^2 w}{\partial x^2} + 2\dfrac{\partial^2 w}{\partial x\,\partial y} + 2\dfrac{\partial^2 w}{\partial y^2} = 0$

15. $\dfrac{\partial^2 w}{\partial x^2} + 2\dfrac{\partial^2 w}{\partial x\,\partial y} + \dfrac{\partial^2 w}{\partial y^2} + w = 0$

16. $\dfrac{\partial^2 w}{\partial x^2} + 2\dfrac{\partial^2 w}{\partial x\,\partial y} - \dfrac{\partial^2 w}{\partial y^2} = 0$

17. $\dfrac{\partial^2 w}{\partial x^2} + 2\dfrac{\partial^2 w}{\partial x\,\partial y} + 4\dfrac{\partial^2 w}{\partial y^2} + 5w = 0$

18. $\dfrac{\partial^2 w}{\partial x^2} + 2\dfrac{\partial^2 w}{\partial x\,\partial y} + \dfrac{\partial^2 w}{\partial y^2} + \dfrac{\partial w}{\partial y} = 0$

19. $\dfrac{\partial^2 w}{\partial x^2} - 4\dfrac{\partial^2 w}{\partial x\,\partial y} + \dfrac{\partial^2 w}{\partial y^2} + \dfrac{\partial w}{\partial y} = 0$

20. $w_{xx} - 4w_{xy} + 2w_{yy} + w_x - 3w_y + w = 0$

21. Obtain Equation **(4)** by a direct calculation.

The classification of a partial differential equation can be done even if the coefficients in **(1)** *are not constants. In such a case we divide the plane into the regions where the* **discriminant**

$$\Delta = b^2 - ac$$

is everywhere positive, everywhere negative, or everywhere zero. For example

$$w_{xx} + xw_{yy} = 0$$

has discriminant $\Delta = -x$, *so the partial differential equation is hyperbolic in the region* $\{(x, y)\colon x < 0\}$, *elliptic in* $\{(x, y)\colon x > 0\}$, *and parabolic in* $\{(x, y)\colon x = 0\}$. *In Problems 22–27 describe the regions in the plane where the given equation is elliptic, parabolic, or hyperbolic.*

22. $w_{xx} - xw_{yy} = 0$

23. $w_{xx} - xyw_{yy} = 0$

24. $x^2w_{xx} + w_{yy} = xy$

25. $w_{xx} + (x^2 - y^2)w_{yy} = 0$

26. $w_{xx} + 2xw_{xy} + w_{yy} + 5w = 0$

27. $y\dfrac{\partial^2 w}{\partial x^2} - 2\dfrac{\partial^2 w}{\partial x\,\partial y} + e^x\dfrac{\partial^2 w}{\partial y^2} + y^2\dfrac{\partial w}{\partial y} - w = 0$

12.5 THE VIBRATING STRING: d'ALEMBERT'S METHOD

As our first physical application, consider a string that is tightly stretched between two fixed points 0 and L on the x-axis (see Figure 12.4). Suppose that the string is pulled back vertically a distance that is very small compared to the length L and released at time $t = 0$, causing it to vibrate. Our problem is to determine the displacement $y(x, t)$ of the point on the string that is x units away from the end 0, at any time t.

To avoid making our equation too complicated, we make two simplifying assumptions:

1. Our "ideal" string has uniform mass m per unit length and offers no resistance to bending.
2. The tension T in the string is so large that the gravitational force on the string may be neglected.

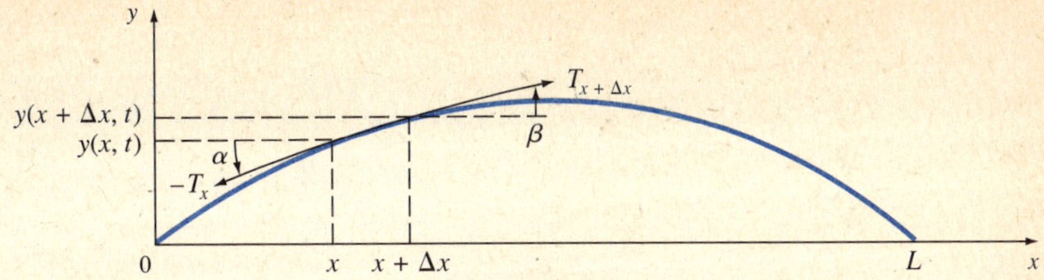

FIGURE 12.4
A vibrating string

Consider a small segment of length Δx. By Newton's second law of motion, the total force acting on this piece of string is equal to the mass of the piece of string multiplied by its acceleration:

$$F = (m\,\Delta x)\frac{\partial^2 y}{\partial t^2}. \tag{1}$$

We assume in this equation that the string is moving only in the xy-plane and that each particle in the string moves only vertically.

Let T_x and $T_{x+\Delta x}$ be the tension vectors at the endpoints of the given segment. These forces are applied tangentially (see Figure 12.4) since the string offers no resistance to bending. Since there is no motion in the x-direction, the x-components of the tension vectors must coincide:

$$T_{x+\Delta x}\cos\beta = T_x\cos\alpha \equiv T. \tag{2}$$

Thus T is constant, since x and Δx are arbitrary. Similarly, the difference in the y-components of the tension vectors must equal the total force acting on the string. Then by Equation **(1)**,

$$T_{x+\Delta x}\sin\beta - T_x\sin\alpha = m\,\Delta x\,\frac{\partial^2 y}{\partial t^2}. \tag{3}$$

Dividing each term in Equation **(3)** by the corresponding term in **(2)**, we have

$$\frac{T_{x+\Delta x}\sin\beta}{T_{x+\Delta x}\cos\beta} - \frac{T_x\sin\alpha}{T_x\cos\alpha} = \frac{m}{T}\,\Delta x\,\frac{\partial^2 y}{\partial t^2},$$

or

$$\tan\beta - \tan\alpha = \frac{m}{T}\,\Delta x\,\frac{\partial^2 y}{\partial t^2}. \tag{4}$$

Since

$$\tan\alpha = \left.\frac{\partial y}{\partial x}\right|_x \quad \text{and} \quad \tan\beta = \left.\frac{\partial y}{\partial x}\right|_{x+\Delta x},$$

we may rewrite **(4)** in the form

$$\frac{1}{\Delta x}\left[\left.\frac{\partial y}{\partial x}\right|_{x+\Delta x} - \left.\frac{\partial y}{\partial x}\right|_x\right] = \frac{m}{T}\frac{\partial^2 y}{\partial t^2}.$$

Letting Δx tend to zero, we obtain in the limit the equation

$$\frac{\partial^2 y}{\partial x^2} = \frac{m}{T}\frac{\partial^2 y}{\partial t^2}.$$

Since $\frac{m}{T}$ is positive, it is clear that this is a hyperbolic partial differential equation.

The equation, written in the form

The wave equation

$$\frac{\partial^2 y}{\partial t^2} = c^2 \frac{\partial^2 y}{\partial x^2}, \qquad c^2 = \frac{T}{m}, \tag{5}$$

is often called the one-dimensional **wave equation,** where the constant c^2 indicates that the coefficient is positive.

As yet we have not made use of the fact that the string is fixed at its endpoints. We may write these boundary conditions as

$$y(0, t) = y(L, t) = 0, \qquad t \geq 0. \tag{6}$$

In addition, we have not taken into account the initial distortion of the string and the fact that it was at rest when released. These initial conditions can be written as

$$y(x, 0) = f(x), \qquad 0 \leq x \leq L, \tag{7}$$

$$\left.\frac{\partial y}{\partial t}\right|_{t=0} = 0, \tag{8}$$

where $f(x)$ is a function depicting the original distortion (see Examples 1 and 2, following), and $(\frac{\partial y}{\partial t})(x, t)$ is the velocity of the point x units away from the origin at time t.

The direct method we use to solve this problem is due to d'Alembert.* Since it is only rarely possible to apply this technique, we develop two more applicable methods in this chapter.

We begin by defining the variables

$$u = x + ct, \qquad v = x - ct, \tag{9}$$

and transforming the wave equation **(5)** into one involving variables u and v. By the chain rule, we have

$$\frac{\partial y}{\partial t} = \frac{\partial y}{\partial u}\frac{\partial u}{\partial t} + \frac{\partial y}{\partial v}\frac{\partial v}{\partial t} = c\left(\frac{\partial y}{\partial u} - \frac{\partial y}{\partial v}\right), \tag{10}$$

where we used **(9)** to obtain the partial derivatives $\frac{\partial u}{\partial t} = c = -\frac{\partial v}{\partial t}$. Taking the partial derivative with respect to t of **(10),** we obtain

$$\begin{aligned}\frac{\partial^2 y}{\partial t^2} &= c\frac{\partial}{\partial t}\left(\frac{\partial y}{\partial u} - \frac{\partial y}{\partial v}\right) = c\left[\frac{\partial}{\partial u}\left(\frac{\partial y}{\partial u} - \frac{\partial y}{\partial v}\right)\frac{\partial u}{\partial t} + \frac{\partial}{\partial v}\left(\frac{\partial y}{\partial u} - \frac{\partial y}{\partial v}\right)\frac{\partial v}{\partial t}\right] \\ &= c^2\left[\left(\frac{\partial^2 y}{\partial u^2} - \frac{\partial^2 y}{\partial u\, \partial v}\right) - \left(\frac{\partial^2 y}{\partial v\, \partial u} - \frac{\partial^2 y}{\partial v^2}\right)\right].\end{aligned}$$

Since the mixed second partials are equal, we finally have

$$\frac{\partial^2 y}{\partial t^2} = c^2\left(\frac{\partial^2 y}{\partial u^2} - 2\frac{\partial^2 y}{\partial u\, \partial v} + \frac{\partial^2 y}{\partial v^2}\right).$$

Similarly, we find that

$$\frac{\partial^2 y}{\partial x^2} = \left(\frac{\partial^2 y}{\partial u^2} + 2\frac{\partial^2 y}{\partial u\, \partial v} + \frac{\partial^2 y}{\partial v^2}\right).$$

---

* The French mathematician Jean Le Rond d'Alembert (1717–1883) is known for his contributions in mechanics.

Substituting these results into **(5)** and canceling like terms yields the equation

$$\frac{\partial^2 y}{\partial u\, \partial v} = 0, \tag{11}$$

since $4c^2 \neq 0$. It is now easy to solve Equation **(11)** by performing two successive integrations. Integrating with respect to u, we obtain

$$\frac{\partial y}{\partial v} = g'(v), \tag{12}$$

where g' is an unknown function in v. From **(12)** it is evident that g' is the partial derivative, with respect to v, of the solution y—which is the reason for using the prime in our notation. Integrating **(12)** with respect to v yields

$$y = g(v) + h(u), \tag{13}$$

where h is an unknown function in u. Substituting **(9)** into **(13)** yields **d'Alembert's solution,**

$$y(x, t) = g(x - ct) + h(x + ct), \tag{14}$$

of the wave equation **(5).**

Since h is arbitrary, if we ignore **(7)** we see that one possible solution of the wave equation is $g(x - ct)$. It we set $t = 0$, the initial distortion of the string is given by $y(x, 0) = g(x)$. At a fixed time later, its distortion is $y(x, t) = g(x - ct)$. Thus the shape of the string can be obtained from the graph of the function $g(x)$ by moving an interval of length L to the right with velocity c. For this reason, $g(x - ct)$ is called a **traveling wave.** Similarly, $h(x + ct)$ is a traveling wave that moves to the left with velocity c. Hence d'Alembert's solution is a superposition of two traveling waves, one moving to the right, the other to the left, with the same velocity c.

The functions g and h can be determined from the initial conditions. Setting $t = 0$, we find that

$$f(x) = y(x, 0) = g(x) + h(x). \tag{15}$$

Noting that g and h are functions of one variable, we may use the chain rule to differentiate **(14)** with respect to t. Setting $t = 0$, we obtain

$$0 = \left.\frac{\partial y}{\partial t}\right|_{t=0} = -cg'(x - ct) + ch'(x + ct)|_{t=0} = c[h'(x) - g'(x)].$$

From the last equation we known that $h' = g'$, so

$$h(x) = g(x) + k,$$

where k is a constant. Replacing this function for h in **(15)** yields $f(x) = 2g(x) + k$ or $g = \frac{f-k}{2}$. Hence $h = \frac{f+k}{2}$, so the solution of the problem finally becomes

$$y(x, t) = \frac{1}{2}[f(x - ct) + f(x + ct)]. \tag{16}$$

Note that since the boundary conditions **(6)** must be satisfied by Equation **(16),** setting $x = 0$ in **(16)** yields

$$0 = \frac{1}{2}[f(-ct) + f(ct)],$$

or

$$f(-ct) = -f(ct).$$

Thus $f(x)$ is odd. This result may seem curious, since $f(x)$ has been defined only on the interval $0 \le x \le L$. However, it causes no difficulty, since we may extend f to the interval $-L \le x \le L$ by defining

$$f(-x) = -f(x) \quad \text{for} \quad -L \le x \le 0.$$

This extension is always possible as $f(0) = 0$. Similarly, setting $x = L$ in Equation **(16)** and using the fact that f is odd, we find that

$$\begin{aligned} f(L + ct) &= -f(L - ct) \\ &= f(ct - L), \end{aligned}$$

so that f has period $2L$. Again, it is clear that we can extend $f(x)$ to the entire real line by letting

$$f(x + 2L) = f(x),$$

for all x. This extension of $f(x)$ to the entire real line seems very reasonable if we return to the notion of traveling waves. It is clear, from Equation **(16),** that we must immediately leave the interval $0 \le x \le L$ when $t > 0$. Now suppose we move $f(x)$ gradually to the right. Since $y(0, t) = y(L, t) = 0$ for all t, any positive contribution at 0 or L from the traveling wave $f(x - ct)$ must be offset by an equal negative contribution from $f(x + ct)$ [which is obtained by moving $f(x)$ the same distance to the left], and vice versa. In effect, this procedure amounts to a reflection of wave forms at the boundary.

By Theorem 11.4.2, the function $f(x)$ has a Fourier sine series—a fact we will refer to in the next section. (The continuity of f is guaranteed by the fact that the string is unbroken. We assume that f' is piecewise continuous, since that is what it is in most practical situations, such as in the example below.)

EXAMPLE 1 ▶ **A vibrating string with piecewise linear distortion**

Suppose that we pluck the string at its center (see Figure 12.5), a distance $y(\frac{L}{2}, 0) = y_0$ meters (y_0 is assumed to be small). We can assume that the distortion consists of two straight lines from $(0, 0)$ to $(\frac{L}{2}, y_0)$ and from $(\frac{L}{2}, y_0)$ to $(L, 0)$. Hence

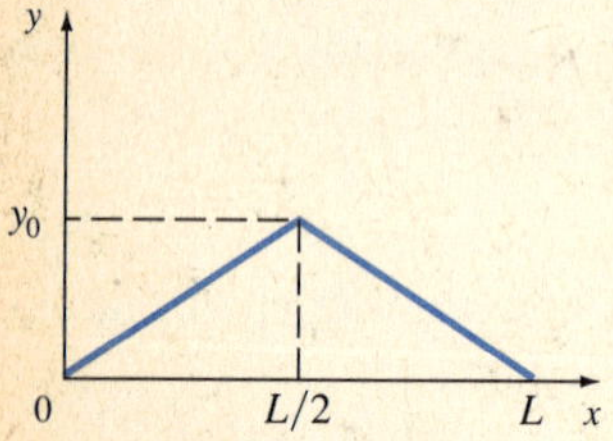

FIGURE 12.5
A string initially distorted into two straight lines

$$f(x) = \begin{cases} \dfrac{2y_0}{L}x, & 0 \le x \le \dfrac{L}{2}, \\ \dfrac{2y_0}{L}(L - x), & \dfrac{L}{2} \le x \le L, \end{cases}$$

which has the piecewise continuous derivative

$$f'(x) = \begin{cases} \dfrac{2y_0}{L}, & 0 \le x < \dfrac{L}{2}, \\ \dfrac{-2y_0}{L}, & \dfrac{L}{2} < x \le L. \end{cases}$$

If the string is 4 meters long, has a total mass of 0.08 kilogram, and is subject to a constant tension of 2 newtons (= kg-m/s^2), we obtain the solution

$$y(x, t) = \frac{1}{2}[f(x - 10t) + f(x + 10t)],$$

since

$$c^2 = \frac{T}{m} = \frac{2}{\frac{0.08}{4}} = 100 \text{ (meters/second)}^2.$$

To discover the exact location of a point 1 meter from the end 0 after exactly 1 second, we set $x = t = 1$. Then

$$y(1, 1) = \frac{1}{2}[f(-9) + f(11)],$$

and since f has period $2L = 8$ and is odd, we get

$$y(1; 1) = \frac{1}{2}[f(-1) + f(3)] = \frac{1}{2}[f(3) - f(1)]$$

$$= \frac{1}{2}\left(\frac{2y_0}{L} - \frac{2y_0}{L}\right) = 0. \blacktriangleleft$$

EXAMPLE 2 ▶ **A string distorted sinusoidally**

Suppse that $f(x) = y_0 \sin(\frac{\pi x}{L})$. By Equation **(16)**,

$$y(x, t) = \frac{y_0}{2}\left[\sin\frac{\pi}{L}(x - ct) + \sin\frac{\pi}{L}(x + ct)\right],$$

and by the addition formulas of trigonometry, this reduces to

$$y(x, t) = y_0 \sin\frac{\pi x}{L} \cos\frac{\pi c t}{L}.$$

Thus, if we fix our attention on the point x_0 units from the left, we see that it oscillates with a period $\frac{2L}{c}$ and with an amplitude $y_0 \sin(\pi x_0/L)$. ◀

SELF-QUIZ

I. One solution to the wave equation $\partial^2 y/\partial t^2 = 4\partial^2 y/\partial x^2$, $y(x, 0) = \sin x$, $\frac{\partial y}{\partial t}(x, 0) = 0$, is $y(x, t) =$ ________.

(a) $\sin(x - 2t)$ **(b)** $\frac{1}{2}\sin(x - 2t)$

(c) $\frac{1}{2}\sin(x + 2t)$ **(d)** $\frac{1}{2}[\sin(x - 2t) + \sin(x + 2t)]$

Answer true *or* false.

II. A solution to the problem $\partial^2 y/\partial t^2 = 4\,\partial^2 y/\partial x^2$, $y(x, 0) = \cos x$, $y(0, t) = y(\frac{\pi}{2}) = 0$, $\frac{\partial y}{\partial t}(x, 0) = 0$, is $y(x, t) = \frac{1}{2}[\cos(x - 2t) + \cos(x + 2t)]$.

Answers to Self-Quiz

I. d **II.** False ($y(0, t) = \cos 2t \neq 0$)

PROBLEMS 12.5

In Problems 1–8 use d'Alembert's solution **(16)** *to find the displacement $y(x, t)$ for each given function $f(x)$, point x, and time t. Assume that $c = 5$ and $L = 6$ meters.*

1. $f(x) = 0.01 \sin\left(\frac{\pi x}{L}\right)$, $x = 2$, $t = 1$

2. $f(x) = 0.1 \sin\left(\frac{\pi x}{L}\right)$, $x = 3$, $t = 2$

3. $f(x) = 0.01 \sin\left(\frac{2\pi x}{L}\right)$, $x = 3$, $t = 1$

4. $f(x) = 0.1 \sin\left(\frac{2\pi x}{L}\right)$, $x = 2$, $t = 2$

5. $f(x) = \begin{cases} 0.1x, & 0 \le x \le 3, \\ 0.1(6 - x), & 3 \le x \le 6 \end{cases}$ $x = 2$, $t = 2$

6. $f(x) = \begin{cases} 0.1x, & 0 \le x \le 2 \\ 0.2(3 - x), & 2 \le x \le 4, \\ 0.1(x - 6), & 4 \le x \le 6 \end{cases}$ $x = 3$, $t = 5$

7. $f(x) = \begin{cases} 0.1x, & 0 \le x \le 2 \\ 0.2, & 2 \le x \le 4, \\ 0.1(6 - x), & 4 \le x \le 6 \end{cases}$ $x = 3$, $t = 5$

8. $f(x) = \begin{cases} 0.01x^2, & 0 \le x \le 3, \\ 0.01(6 - x)^2, & 3 \le x \le 6 \end{cases}$ $x = 2$, $t = 2$

9. Assume instead of the initial conditions **(7)** and **(8)**, that a string is initially in its equilibrium position and is set vibrating by each of its points x being given an initial velocity $f'(x)$, where f is a differentiable function. Prove that the solution is given by $y(x, t) = \frac{1}{2c}[f(x + ct) - f(x - ct)]$.

Using the result of Problem 9 and setting $c = 5$ and $L = 6$ meters, do Problems 10–15 for the given functions $f'(x)$ and values x and t.

10. $f'(x) = 0.01 \sin\left(\frac{\pi x}{L}\right)$, $x = 2$ meters, $t = 3$ seconds

11. $f'(x) = 0.1 \sin\left(\frac{2\pi x}{L}\right)$, $x = 1$ meter, $t = 1$ second

12. $f'(x) = 3(6x - x^2)$, $x = 3$ meters, $t = 0.01$ second

13. $f'(x) = 0.1 \sin^2\left(\frac{\pi x}{L}\right)$, $x = 1$ meter, $t = 0.1$ second

14. $f'(x) = \begin{cases} 0.01x, & 0 \le x \le 3, \quad x = 2, \quad t = 1 \\ 0.01(6 - x), & 3 \le x \le 6 \end{cases}$

15. $f'(x) = \begin{cases} 0.01(3 - x), & 0 \le x \le 3, \quad x = 2, \quad t = 1 \\ 0.01(x - 3), & 3 \le x \le 6 \end{cases}$

16. Suppose that we subject a vibrating string to both an initial displacement $y(x, 0) = f(x)$, $0 \le x \le L$, and an initial velocity $y_t(x, 0) = g'(x)$, $0 \le x \le L$. Show that the solution of this initial-value problem is

$$y(x, t) = \frac{1}{2}[f(x - ct) + f(x + ct)] + \frac{1}{2c}[g(x + ct) - g(x - ct)].$$

17. Use the result of Problem 16 to solve the initial-value problem

$$y_{tt} = y_{xx}, \quad |x| < \infty, \quad t > 0$$

$$y(x, 0) = \sin \frac{\pi x}{2} = y_t(x, 0).$$

18. Suppose that we subject a vibrating string to a damping force that is proportional at each instance to the velocity at each point. Show that the resulting hyperbolic differential equation is of the form

$$\frac{\partial^2 y}{\partial t^2} - k\frac{\partial y}{\partial t} = c^2\frac{\partial^2 y}{\partial x^2}.$$

19. The distortionless solution of the transmission line problem of a telegraph cable satisfies the partial differential equation

$$v_{xx} = \alpha v_{tt} + \beta v_t + \gamma v, \quad 0 < x < \infty, \quad t > 0,$$

where $\alpha = LC$, $\beta = RC + GL$, $\gamma = RG$, and v, L, R, C, and G are the voltage, inductance, resistance, capacitance, and conductance between the line and ground. What is the form of the solution?

12.6 SEPARATION OF VARIABLES: THE WAVE EQUATION

One of the simplest techniques for solving ordinary differential equations consists of separating the variables. Although we now have *two* independent variables, we can nevertheless adapt the technique to all partial differential equations of the forms **(12.4.10)–(12.4.13)** whenever the function $F(u, v) \equiv 0$. The method is best explained by means of an example.

Suppose that we again consider the problem of a vibrating string given by the wave equation

$$\frac{\partial^2 y}{\partial t^2} = c^2\frac{\partial^2 y}{\partial x^2}, \qquad c^2 = \frac{T}{m}, \tag{1}$$

with boundary conditions

$$y(0, t) = y(L, t) = 0, \qquad t \ge 0, \tag{2}$$

and initial conditions

$$y(x, 0) = f(x), \qquad \left.\frac{\partial y}{\partial t}\right|_{t=0} = 0, \qquad 0 \le x \le L. \tag{3}$$

We begin by seeking a solution of the form

$$y(x, t) = X(x)T(t), \tag{4}$$

where X is a function of x alone and T is a function involving only the independent variable t. Of course there is no guarantee that such a solution exists. In fact, as we will see later, separation is often not possible. Nevertheless, we can try it out and see if it works. Using primes to denote differentiation, we observe that

$$\frac{\partial^2 y}{\partial t^2} = \frac{\partial^2}{\partial t^2}[XT] = XT'' \quad \text{and} \quad \frac{\partial^2 y}{\partial x^2} = X''T,$$

so that Equation **(1)** becomes

$$XT'' = c^2X''T. \tag{5}$$

It is now possible to separate the variables and write Equation **(5)** as

$$\frac{T''}{T} = c^2\frac{X''}{X}. \tag{6}$$

Initially we had no assurance that it would be possible to separate the variables as we have done. Should such a separation prove impossible, this method would not apply and we would have to turn to other techniques.

Now suppose we fix t and allow x to vary. Since the left-hand side of Equation **(6)** is constant, the right-hand side must be constant for all values of x. Consequently, the left-hand side is also constant *regardless* of the value of t, and we have

$$\frac{T''}{T} = c^2\frac{X''}{X} = k,$$

where k is constant. From this set of equalities we obtain two ordinary differential equations

$$T'' - kT = 0, \tag{7}$$

$$X'' - \frac{k}{c^2}X = 0. \tag{8}$$

Thus the problem of solving the partial differential Equation **(1)** has been reduced to solving two ordinary differential equations, **(7)** and **(8)**. The constant k is arbitrary, but its value must be the same for both equations. Using the characteristic equation (see Chapter 3) $\lambda^2 - k = 0$, we can easily obtain the general solution

$$T(t) = \begin{cases} a_1e^{\sqrt{k}\,t} + a_2e^{-\sqrt{k}\,t}, & \text{if } k > 0, \\ a_1t + a_2, & \text{if } k = 0, \\ a_1 \cos \sqrt{-k}\,t + a_2 \sin \sqrt{-k}\,t, & \text{if } k < 0. \end{cases}$$

Similarly, we also have the general solution

$$X(x) = \begin{cases} b_1e^{\sqrt{k}\,x/c} + b_2e^{-\sqrt{k}\,x/c}, & \text{if } k > 0, \\ b_1x + b_2, & \text{if } k = 0, \\ b_1 \cos \dfrac{\sqrt{-k}\,x}{c} + b_2 \sin \dfrac{\sqrt{-k}\,x}{c}, & \text{if } k < 0. \end{cases}$$

Since our solution must satisfy the boundary conditions **(2)**, if a solution of the form **(4)** exists, we must have the equations

$$X(0)T(t) = X(L)T(t) = 0 \qquad \text{for all } t.$$

If $T(t) \equiv 0$, then $y \equiv 0$ and the string is constantly at rest, which is impossible if $f(x) \not\equiv 0$. For $f \not\equiv 0$, T must be nonzero for at least one value of t, which implies that $X(0) = X(L) = 0$. Thus Equation **(8)** becomes a Sturm-Liouville problem with these boundary conditions.

Setting $x = 0$ and $x = L$ in the general solution yields, for $k > 0$, the homogeneous system of equations

$$\begin{aligned} X(0) &= b_1 + b_2 = 0, \\ X(L) &= b_1 e^{\sqrt{k}L/c} + b_2 e^{-\sqrt{k}L/c} = 0. \end{aligned}$$

This system has a nonzero solution if and only if the determinant of its coefficients is zero. But

$$\begin{vmatrix} 1 & 1 \\ e^{\sqrt{k}L/c} & e^{-\sqrt{k}L/c} \end{vmatrix} = e^{-\sqrt{k}L/c} - e^{\sqrt{k}L/c} = -2 \sinh \frac{\sqrt{k}\,L}{c} \not\equiv 0,$$

so that $X(x) \equiv 0$ for $k > 0$. Then $y \equiv 0$, which is impossible if $f(x) \not\equiv 0$. For $k = 0$, the condition $X(0) = 0$ implies that $b_2 = 0$, while $X(L) = 0$ requires that $b_1 = 0$, again leading to the zero solution. So we are left with the remaining possibility that k is negative, and we set $k = -r^2$. This yields the general solution

$$X(x) = b_1 \cos \frac{rx}{c} + b_2 \sin \frac{rx}{c}.$$

Letting $x = 0$ and $x = L$, we have

$$X(0) = b_1 = 0 \quad \text{and} \quad X(L) = b_2 \sin \frac{rL}{c} = 0.$$

To prevent our again having the zero solution, we must choose an r such that $\frac{rL}{c}$ is a positive multiple of π:

$$\frac{rL}{c} = n\pi \quad \text{or} \quad r = \frac{n\pi c}{L}.$$

Thus we obtain an infinite set of solutions,

$$X_n(x) = b_2 \sin \frac{n\pi x}{L}, \tag{9}$$

each associated with the choice

$$k = -\frac{n^2\pi^2c^2}{L^2}. \tag{10}$$

Hence the corresponding solutions for T are

$$T_n(t) = a_1 \cos \frac{n\pi ct}{L} + a_2 \sin \frac{n\pi ct}{L}.$$

Finally, we obtain an infinite set of solutions,

$$y_n(x, t) = X_n(x)T_n(t) = \left(B_n \cos \frac{n\pi ct}{L} + B_n^* \sin \frac{n\pi ct}{L}\right) \sin \frac{n\pi x}{L}, \tag{11}$$

where B_n and B_n^* are constants, that satisfy the partial differential Equation **(1)** and the boundary conditions **(2)**. In the language of Chapter 11 we see that we have found the *eigenvalues* **(10)** and the corresponding *eigenfunctions* **(11)** of the boundary value problem **(1)**, **(2)**.

We now try to select a solution that satisfies the initial conditions **(3).** It is extremely likely that no one solution **(11)** will satisfy **(3).** However, any finite sum of solutions **(11)** is again a solution of the boundary value problem (by the *principle of superposition*), so we can try to find a finite linear combination of these solutions that satisfies the initial conditions **(3).** Generally, even this will fail, so as a last resort we try an infinite series of the solutions **(11):**

$$y(x, t) = \sum_{n=1}^{\infty} y_n(x, t) = \sum_{n=1}^{\infty} \left(B_n \cos \frac{n\pi ct}{L} + B_n^* \sin \frac{n\pi ct}{L} \right) \sin \frac{n\pi x}{L}. \tag{12}$$

There is no guarantee at this point that such a series converges, but since its form is very similar to that of the Fourier series **(11.3.7),** there is every reason to be optimistic. Furthermore, if **(12)** converges and the series obtained by the formal term-by-term partial differentiations $\partial/\partial t$, $\partial^2/\partial t^2$, $\partial/\partial x$, and $\partial^2/\partial x^2$ all converge uniformly, then **(12)** is again a solution of the boundary value problem **(1), (2).** If, in addition, we can also satisfy the initial conditions **(3),** then we have solved the vibrating string problem.

Setting $t = 0$ in Equation **(12)** and using the first of Equations **(3),** we have

$$y(x, 0) = \sum_{n=1}^{\infty} B_n \sin \frac{n\pi x}{L} = f(x). \tag{13}$$

We again assume that $f(x)$ and $f'(x)$ are piecewise continuous (see Section 12.5). The infinite series **(13)** is a Fourier sine series, which requires that the function $f(x)$ be odd $[f(-x) = -f(x)]$, and periodic with period $2L$. As in Section 12.5, we can extend the function $f(x)$, defined on the interval $0 \le x \le L$, in such a way that it is odd and periodic with period $2L$. Since $f(0) = 0$, we let $f(-x) = -f(x)$ on $-L \le x \le 0$ and require that $f(x + 2L) = f(x)$ for all real numbers x. Then the coefficients B_n are given by the Euler formula

$$\begin{aligned} B_n &= \frac{1}{L} \int_{-L}^{L} f(x) \sin \frac{n\pi x}{L}\, dx \\ &= \frac{2}{L} \int_{0}^{L} f(x) \sin \frac{n\pi x}{L}\, dx, \qquad n = 1, 2, 3, \ldots . \end{aligned} \tag{14}$$

The last equality holds because $f(x)$ and $\sin(\frac{n\pi x}{L})$ are odd and their product is even [see Equation **(11.4.5),** p. 547]. Taking the partial derivative of **(12)** with respect to t and using the second of the initial conditions **(3),** we have

$$\frac{\partial y}{\partial t} = \frac{\pi c}{L} \sum_{n=1}^{\infty} \left(nB_n^* \cos \frac{n\pi ct}{L} - nB_n \sin \frac{n\pi ct}{L} \right) \sin \frac{n\pi x}{L} \bigg|_{t=0} = 0,$$

or

$$\sum_{n=1}^{\infty} nB_n^* \sin \frac{n\pi x}{L} = 0.$$

Again we obtain an expression of the form **(13)** for B_n^*, but this time the function involved is zero. Hence all the coefficients B_n^* vanish. Thus solution **(12)** reduces to

$$y(x, t) = \sum_{n=1}^{\infty} B_n \cos \frac{n\pi ct}{L} \sin \frac{n\pi x}{L}, \tag{15}$$

where the coefficients B_n are given by **(14).** It is possible to write series **(15)** in closed form by using the trigonometric identity

$$2 \sin A \cos B = \sin(A + B) + \sin(A - B).$$

Then

$$y(x, t) = \frac{1}{2}\sum_{n=1}^{\infty} B_n\left[\sin\frac{n\pi}{L}(x + ct) + \sin\frac{n\pi}{L}(x - ct)\right]$$
$$= \frac{1}{2}\left[\sum_{n=1}^{\infty} B_n \sin\frac{n\pi}{L}(x + ct) + \sum_{n=1}^{\infty} B_n \sin\frac{n\pi}{L}(x - ct)\right],$$

and the series can be evaluated by substituting $x + ct$ and $x - ct$, respectively, for the variable x in **(13)**. Hence

$$y(x, t) = \frac{1}{2}[f(x + ct) + f(x - ct)],$$

which is precisely the same result as we obtained directly in Section 12.5.

EXAMPLE 1 ▶ **Using equation (15) to solve a wave equation problem**

Let a vibrating string have length π and let $c^2 = 1$. Suppose that the initial velocity is zero and the initial distortion $f(x) = x(\pi - x)$. Instead of using d'Alembert's solution, we here obtain an expansion of the form **(15)** by calculating the coefficients B_n.

From **(14)** we have

$$B_n = \frac{2}{\pi}\int_0^{\pi} x(\pi - x)\sin nx\,dx = \frac{2}{\pi}\int_0^{\pi}(\pi x - x^2)\sin nx\,dx.$$

We integrate by parts twice:

$$B_n = \frac{2}{\pi}\left[-(\pi x - x^2)\frac{\cos nx}{n}\Big|_0^{\pi} + \frac{1}{n}\int_0^{\pi}(\pi - 2x)\cos nx\,dx\right]$$
$$= \frac{2}{n\pi}\left[(\pi - 2x)\frac{\sin nx}{n}\Big|_0^{\pi} + \frac{2}{n}\int_0^{\pi}\sin nx\,dx\right]$$
$$= \frac{4}{n^2\pi}\left(-\frac{\cos nx}{n}\Big|_0^{\pi}\right) = \frac{4}{n^3\pi}[1 - (-1)^n].$$

Hence the solution is

$$y(x, t) = \frac{4}{\pi}\sum_{n=1}^{\infty}\frac{[1 - (-1)^n]}{n^3}\cos nt\sin nx$$
$$= \frac{8}{\pi}\left(\cos t\sin x + \frac{1}{3^3}\cos 3t\sin 3x + \frac{1}{5^3}\cos 5t\sin 5x + \ldots\right)$$
$$= \frac{8}{\pi}\sum_{n=1}^{\infty}\frac{\cos(2n + 1)t\sin(2n + 1)x}{(2n + 1)^3}.$$ ◀

SELF-QUIZ

Answer Question I true or false.

I. The problem given by the partial differential equation $\partial^2 y/\partial t^2 = 4\,\partial^2 y/\partial x^2$, and the conditions $y(0, t) = y(\pi, t) = 0$, $y(x, 0) = \cos x$, $\frac{\partial y}{\partial t}(x, 0) = 0$, can be solved by the method of separation of variables.

II. The partial differential equation $\partial^2 y/\partial t^2 = 9\,\partial^2 y/\partial x^2$ has solutions of the form $y(x, t) = \sum_{n=1}^{\infty}(B_n \cos\frac{n\pi ct}{L} + B_n^* \sin\frac{n\pi ct}{L})\sin\frac{n\pi x}{L}$, for $c =$ ______.
(a) 9 **(b)** −9 **(c)** 3 **(d)** −3

Answers to Self-Quiz

I. True (but $y(x, 0) = \cos x$ must be extended as an odd function on the interval $[-\pi, \pi]$, that is $y(x, 0) = -\cos x$ for $-\pi \le x < 0$) **II.** c

PROBLEMS 12.6

In Problems 1–6 use the method of separation of variables to find the displacement $y(x, t)$ in the wave equation for each initial condition $y(x, 0) = f(x)$.

1. $f(x) = 0.1 \sin \dfrac{\pi x}{L}, 0 \le x \le L$

2. $f(x) = 0.01 \sin \dfrac{2\pi x}{L}, 0 \le x \le L$

3. $f(x) = 0.1x(L - x), 0 \le x \le L$

4. $f(x) = 0.1x^2(L - x), 0 \le x \le L$

5. $f(x) = \begin{cases} 0.1x, & 0 \le x \le 3 \\ 0.1(6 - x), & 3 \le x \le 6 \end{cases}$

6. $f(x) = \begin{cases} 0.1x, & 0 \le x \le 2 \\ 0.2, & 2 \le x \le 4 \\ 0.1(6 - x), & 4 \le x \le 6 \end{cases}$

In Problems 7–14 reduce the given equations to pairs of ordinary differential equations by the method of separation of variables.

7. $\dfrac{\partial^2 y}{\partial x^2} + \dfrac{\partial y}{\partial t} = 0$

8. $\dfrac{\partial^2 y}{\partial x^2} + x\dfrac{\partial y}{\partial t} = 0$

9. $\dfrac{\partial^2 y}{\partial x\, \partial t} + \dfrac{\partial y}{\partial t} = 0$

10. $z_{xx} = y^2 z_{yy}$

11. $z_{yy} + xyz_x = 0$

12. $z_{xx} - z_{yy} = z$

13. $u_{rr} + \dfrac{1}{r}u_r + \dfrac{1}{r^2}u_{\theta\theta} = 0$, where $u = u(r, \theta)$

14. $u_{rr} + \dfrac{1}{r}u_r + \dfrac{1}{r^2}u_{\theta\theta} + u_{zz} = 0$, where $u = u(r, \theta, z)$

For Problems 15–20 do the following:

(a) *Find every solution of the given equation by the technique of separation of variables.*

(b) *Find the nontrivial solutions if we require the solutions to satisfy the boundary conditions*

$$y(0, t) = y(\pi, t) = 0.$$

(c) *Using the nontrivial solutions in part (b), indicate whether it would be possible to find a solution that also satisfies the initial condition*

$$y(x, 0) = f(x), \qquad 0 < x < \pi,$$

for $f(x) = x(\pi - x)$.

15. $\dfrac{\partial^2 y}{\partial x^2} + \dfrac{1}{x^2}\dfrac{\partial^2 y}{\partial t^2} = 0$

16. $\dfrac{\partial^2 y}{\partial x\, \partial t} = y$

17. $\dfrac{\partial^2 y}{\partial x^2} + \dfrac{\partial^2 y}{\partial t^2} = 0$

18. $\dfrac{\partial^2 y}{\partial x^2} - 2\dfrac{\partial y}{\partial x} = -\dfrac{\partial y}{\partial t}$

19. $\dfrac{\partial^2 y}{\partial t^2} = \dfrac{\partial^2 y}{\partial x^2} + y$

20. $\dfrac{\partial^2 y}{\partial t^2} = x^2\dfrac{\partial^2 y}{\partial x^2}$

12.7 HEAT FLOW AND THE HEAT EQUATION

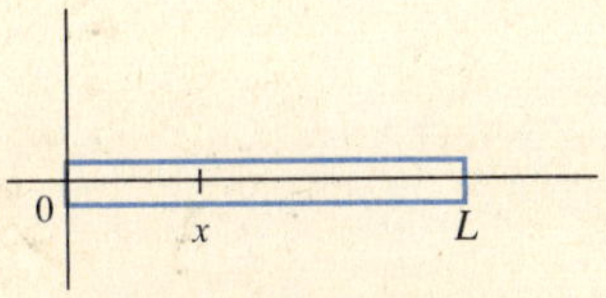

FIGURE 12.6
A rod of length L

Consider a cylindrical rod of length L and radius R (so that the cross-sectional area is $A = \pi R^2$) composed of a uniform heat-conducting material. We assume that heat can enter and leave the rod only through its ends; that is, the lateral surface of the rod is completely insulated. Let x measure the distance along the rod (see Figure 12.6) and let $T(x, t)$ denote the absolute temperature at time t at a point x units along the rod (we assume that the temperature is uniform across any cross-sectional area of the rod). Let ρ be the **density** of the rod (mass per unit volume), which is assumed to be constant. The **specific heat** c of the rod is defined as the amount of heat (joules, BTUs) that must be supplied to raise the temperature of one unit mass (kilograms, slugs) of the rod by one degree (Celsius, Fahrenheit).

Consider a section of the rod between x and $x + dx$. Since mass = volume × density, the mass between these two points is $\rho A\, dx$. In order to change the temperature of the rod between these two points from 0 to $T(x, t)$ degrees, we must supply $T(x, t) \cdot c\rho A \cdot dx$ units of heat. Thus, between any two points x_0 and x_1, the heat energy contained in the rod at time t is

$$Q(t) = \int_{x_0}^{x_1} T(x, t) c\rho A \, dx. \tag{1}$$

By the law of conservation of energy, if there are no heat sources within the rod, the heat energy in any part of the rod can increase or decrease only because of a lateral flow of heat through the boundaries of that part of the rod. This **heat flux,** written as $Q_F(x, t)$, is the quantity of heat energy per unit time passing through a unit area in the cross section x units from the left-hand end in a positive (rightward) direction. Thus, to find the rate of change of the heat energy between the x_0 and x_1 cross sections, we need only take the difference

$$\frac{dQ}{dt} = AQ_F(x_0, t) - AQ_F(x_1, t), \tag{2}$$

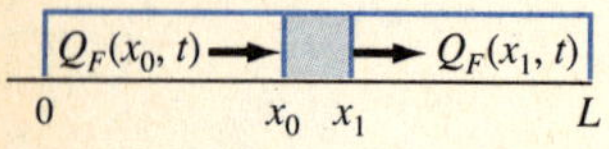

FIGURE 12.7
Heat flux

where the first term on the right-hand side is the heat energy flowing in and the second is that flowing out (see Figure 12.7).

We now assume that the heat flux at any point is proportional to the **temperature gradient** $\frac{\partial T(x,t)}{\partial x}$ at that point. This assumption has been shown by experiment to be reasonable in many situations. The constant of proportionality is called the **thermal conductivity** of the rod and is denoted by κ. Since heat flows in the direction of decreasing temperature, we have

$$Q_F(x, t) = -\kappa \frac{\partial T}{\partial x}(x, t),$$

where $\kappa > 0$. Since no heat sources are in this part of the rod, we have

$$\begin{aligned} AQ_F(x_0, t) - AQ_F(x_1, t) &= A\left[\kappa \frac{\partial T}{\partial x}(x_1, t) - \kappa \frac{\partial T}{\partial x}(x_0, t)\right] \\ &= \int_{x_0}^{x_1} \frac{\partial}{\partial x}\left[\kappa \frac{\partial T}{\partial x}(x, t)\right] A \, dx. \end{aligned} \tag{3}$$

Substituting the right-hand side of **(3)** for the right-hand side of **(2)** and differentiating **(1),** we have

$$\int_{x_0}^{x_1} \frac{\partial}{\partial x}\left(\kappa \frac{\partial T}{\partial x}\right) A \, dx = \frac{dQ}{dt} = \int_{x_0}^{x_1} c\rho \frac{\partial T}{\partial t} A \, dx,$$

or

$$\int_{x_0}^{x_1} \left[c\rho \frac{\partial T}{\partial x} - \frac{\partial}{\partial x}\left(\kappa \frac{\partial T}{\partial x}\right)\right] A \, dx = 0. \tag{4}$$

Equation **(4)** must hold in any interval $x_0 \leq x \leq x_1$, so if we assume that the functions involved are all continuous, the expression in brackets must be zero. To see this, note that if it were positive (negative) at some point, then by continuity it would be positive (negative) on some interval, in which case Equation **(4)** would not hold. Thus, for all x in the interval $[0, L]$,

$$c\rho \frac{\partial T}{\partial t} = \frac{\partial}{\partial x}\left(\kappa \frac{\partial T}{\partial x}\right), \tag{5}$$

and if we assume that κ is constant,

The heat equation

$$\frac{\partial T}{\partial t} = \delta \frac{\partial^2 T}{\partial x^2}, \qquad (6)$$

where $\delta = \kappa/c\rho$. Equation **(6)** is called the **heat (conduction) equation** and is clearly parabolic. The constant δ is positive and measures the **diffusivity** of the material of the rod. To completely specify our problem, we select the case in which the boundary conditions are

$$T(0, t) = T(L, t) = 0, \qquad t \geq 0, \qquad (7)$$

and the initial condition is

$$T(x, 0) = f(x), \qquad 0 \leq x \leq L, \qquad (8)$$

where f is a given function. In physical terms, we are keeping the ends of the rod at zero temperature and letting $f(x)$ denote the initial temperature at any point x of the rod. It is easy to modify the problem by selecting other, possibly time-dependent, boundary conditions (see Problems 7–9).

We begin to solve the heat equation by assuming the existence of a solution to the problem of the form

$$T(x, t) = X(x)\mathcal{T}(t).$$

Substituting this equation into **(6)** yields

$$X\mathcal{T}' = \delta X''\mathcal{T},$$

or

$$\frac{\mathcal{T}'}{\delta \mathcal{T}} = \frac{X''}{X}. \qquad (9)$$

The function on the left-hand side depends only on t, so it is constant for fixed t and arbitrary x. On the other hand, the function on the right side of (9) depends only on x and is constant for fixed x and arbitrary t. The only way these situations can hold simultaneously is for each function to be constant, say, k. Then we obtain the pair of ordinary differential equations

$$\mathcal{T}' - k\delta\mathcal{T} = 0, \qquad (10)$$

$$X'' - kX = 0. \qquad (11)$$

The boundary conditions **(7)** may be written as

$$X(0)\mathcal{T}(t) = X(L)\mathcal{T}(t) = 0,$$

implying that $X(0) = X(L) = 0$ unless the rod has zero initial temperature at every point. If we ignore this uninteresting case, we note that the boundary value problem **(11)**, $X(0) = X(L) = 0$, is almost identical to the boundary value problem for X in the case of the vibrating string [see Equation **(12.6.8)**]. Paralleling the development of Section 12.6, we see that the only nonzero solutions for the boundary value problem **(11)** with $X(0) = X(L) = 0$ arise for negative values of k. Setting $k = -r^2$, we find that k is an eigenvalue of the problem only if r is a multiple of $\frac{\pi}{L}$, and the eigenfunctions corresponding to $-(n\pi/L)^2$ have the form

$$X_n(x) = A \sin \frac{n\pi x}{L}.$$

Setting $k = -(n\pi/L)^2$ in Equation **(10)**, we see that this first-order equation has the general solution

$$\mathcal{T}_n(t) = Be^{-(n\pi/L)^2 \delta t}.$$

Hence we consider an infinite series of the form

$$T(x, t) = \sum_{n=1}^{\infty} X_n(x)\mathcal{T}_n(t) = \sum_{n=1}^{\infty} B_n \sin \frac{n\pi x}{L} e^{-(n\pi/L)^2 \delta t}. \tag{12}$$

Setting $t = 0$ and making use of the initial condition **(8)**, we have

$$T(x, 0) = \sum_{n=1}^{\infty} B_n \sin \frac{n\pi x}{L} = f(x).$$

Thus, in order that separation of variables work, $f(x)$ must be representable as a Fourier sine series, implying that f must be extended as an odd, piecewise continuous function of period $2L$ with piecewise continuous derivatives. When we make this extension, the coefficients B_n are given by

$$\begin{aligned} B_n &= \frac{1}{L}\int_{-L}^{L} f(x) \sin \frac{n\pi x}{L}\, dx \\ &= \frac{2}{L}\int_{0}^{L} f(x) \sin \frac{n\pi x}{L}\, dx, \qquad n = 1, 2, 3, \ldots . \end{aligned} \tag{13}$$

The solution **(12)** is completely determined and the series must converge, since $T(x, 0)$ converges and the exponential factors are less than 1 for all $t > 0$. Observe that the exponential factors in **(12)** cause $T(x, t)$ to approach zero as t tends to infinity.

EXAMPLE 1 ▶ **Using Equations (12), (13) to solve a heat flow problem**

Suppose that the rod has length π, $\delta = 1$, and initial temperature $f(x) = \sin 2x$. By the orthogonality of the set of functions $\{\sin nx\}$ (see Example 11.2.1), the coefficients B_n are all zero for $n \neq 2$. For $n = 2$, we have, from Equation **(13)** and from Example 11.2.1,

$$B_2 = \frac{1}{\pi}\int_{-\pi}^{\pi} \sin^2 2x\, dx = 1.$$

Thus the solution of this problem is given by

$$T(x, t) = (\sin 2x)e^{-4t}. \quad ◀$$

EXAMPLE 2 ▶ **Extending the function *f(x)***

Let the rod have length 1, $\delta = 1$, and initial temperature $f(x) = x(1 - x^2)$. Observe that $f(x)$ is an odd function for all $|x| \leq 1$, since

$$f(-x) = -x[1 - (-x)^2] = -x(1 - x^2) = -f(x).$$

Extend f to the reals $\mathbb{R}$ periodically by letting $f(x + 2) = f(x)$. Then

$$B_n = 2\int_0^1 f(x) \sin n\pi x\, dx = 2\int_0^1 (x - x^3) \sin n\pi x\, dx.$$

We integrate by parts three times to obtain (the details should now be familiar)

$$B_n = \frac{12(-1)^{n+1}}{n^3\pi^3}.$$

Hence, by **(12)**, the solution to this initial-value problem is

$$T(x, t) = \frac{12}{\pi^3} \sum_{n=1}^{\infty} \frac{(-1)^{n+1} \sin n\pi x \, e^{-n^2\pi^2 t}}{n^3}. \blacktriangleleft$$

SELF-QUIZ

I. Consider the heat equation $\partial T/\partial t = \delta(\partial^2 T/\partial x^2)$, with conditions $T(0, t) = T(L, t) = 0$, $T(x, 0) = f(x)$, where $L = \pi$, $\delta = 1$, and $f(x) = \sin 9x$. Then a solution to the problem is $T(x, t) =$ ______.

(a) $e^{-9t} \sin 3x$ **(b)** $e^{-9t} \sin 9x$ **(c)** $e^{-3t} \sin 9x$
(d) $e^{-81t} \sin 9x$

Answer Questions II and III true *or* false.

II. In question I, suppose that $f(x) = x$. Then the problem has a solution of the form $T(x, t) = \sum_{n=1}^{\infty} B_n e^{-n^2 t} \sin nx$.

III. In question I, suppose that $\delta = 2$ and $f(x) = e^x$. Then the problem has a solution of the form $T(x, t) = \sum_{n=1}^{\infty} B_n e^{-n^2 t} \sin nx$.

Answers to Self-Quiz

I. d **II.** True **III.** False

PROBLEMS 12.7

In Problems 1–6 find the temperature $T(x, t)$ in an insulated rod π units long whose ends are kept at 0°C, where the initial temperature is given by

1. $f(x) = x(\pi - x)$

2. $f(x) = x(\pi^2 - x^2)$

3. $f(x) = x^2(\pi - x)$

4. $f(x) = x \cos \dfrac{x}{2}$

5. $f(x) = \begin{cases} x, & 0 \le x \le \dfrac{\pi}{2} \\ 0, & \dfrac{\pi}{2} < x < \pi \end{cases}$

6. $f(x) = \begin{cases} 0, & 0 \le x < \dfrac{\pi}{3} \\ 1, & \dfrac{\pi}{3} \le x < \dfrac{2\pi}{3} \\ 0, & \dfrac{2\pi}{3} \le x \le \pi \end{cases}$

7. Suppose that the ends of the rod are kept at different constant temperatures:

$$T(0, t) = T_1, \qquad T(L, t) = T_2.$$

Find the temperature at x as t tends to ∞; that is, find the **steady-state** temperature. [*Hint:* The steady-state temperature is *not* time dependent.]

8. Denote the steady-state temperature by $T_s(x)$ and define the **transient** temperature in the rod by

$$T_t(x, t) = T(x, t) - T_s(x).$$

Verify that $T_t(x, t)$ is given by Equation **(12)**. This shows that we need only superimpose the steady-state temperature on the solution **(12)** to obtain the solution for nonzero boundary conditions.

9. Suppose that the ends of the rod are kept at the temperatures

$$T(0, t) = T_1 e^{-c^2\delta t}, \qquad T(L, t) = T_2 e^{-c^2\delta t}.$$

For what initial conditions $f(x)$ can a solution be found by the separation of variables? What happens if cL is a multiple of π?

10. Consider the nonhomogeneous heat equation

$$\frac{\partial u}{\partial t} - k\frac{\partial^2 u}{\partial x^2} = A(x, t).$$

where $A(x, t)$ is given by

$$A(x, t) = \sum_{n=1}^{\infty} a_n(t) \sin \frac{n\pi x}{L}$$

for known functions $a_n(t)$. Suppose that initial and boundary conditions are given by

$$u(0, t) = u(L, t) = 0, \qquad t \ge 0,$$
$$u(x, 0) = f(x), \qquad 0 \le x \le L,$$

with $f(x)$ known. Find a Fourier series for the solution $u(x, t)$. [*Hint:* Assume that B_n in Equation **(12)** is a function of t.]

12.8 TWO-DIMENSIONAL HEAT FLOW AND LAPLACE'S EQUATION

Consider a rectangular sheet of heat-conducting material of uniform thickness θ, density ρ, specific heat c, and thermal conductivity κ. We may suppose that the set of points (x, y) with $0 \le x \le L$ and $0 \le y \le M$ is a face of the sheet. We assume that the faces of the sheet are insulated and that heat can enter and leave only through the edges of the sheet.

Select any interior point (x, y) on the face and consider a rectangular region $ABCD$ whose corner coordinates are given in Figure 12.8. Under previous assumptions, it is reasonable to assume the heat flow in each plane to be identical, so that the flow is two-dimensional. In the discussion below we assume that the distances Δx and Δy are very small.

The rate of change of heat energy in the sheet $ABCD$ at any time t is approximately (see Section 12.7).

$$c\rho\theta\, \Delta x\, \Delta y \frac{\partial T}{\partial t}, \tag{1}$$

where $T = T(x, y, t)$ is the temperature at any point (x, y) at time t. Since Δx and Δy are small, we assume that the heat flux along each edge is constant. Then the heat energy passing through $ABCD$ in the vertical direction is

$$\theta\, \Delta x[Q_F(x, y, t) - Q_F(x, y + \Delta y, t)], \tag{2}$$

where Q_F is the heat flux. In the horizontal direction the heat energy is

$$\theta\, \Delta y[Q_F(x, y, t) - Q_F(x + \Delta x, y, t)]. \tag{3}$$

In Section 12.7 we mentioned that it is reasonable to assume that the heat flux at any point is proportional to the temperature gradient at that point. Adding **(2)** and **(3)**, we get another approximation of the rate of change of heat energy in the sheet $ABCD$:

$$\kappa\theta\, \Delta y\left[\frac{\partial T}{\partial x}(x + \Delta x, y, t) - \frac{\partial T}{\partial x}(x, y, t)\right] + \kappa\theta\, \Delta x\left[\frac{\partial T}{\partial y}(x, y + \Delta y, t) - \frac{\partial T}{\partial y}(x, y, t)\right]. \tag{4}$$

Equating **(4)** and **(1)** and dividing by $\Delta x\, \Delta y$ yields

$$c\rho\theta \frac{\partial T}{\partial t} = \kappa\theta \frac{\left[\dfrac{\partial T}{\partial x}(x + \Delta x, y, t) - \dfrac{\partial T}{\partial x}(x, y, t)\right]}{\Delta x} + \kappa\theta \frac{\left[\dfrac{\partial T}{\partial y}(x, y + \Delta y, t) - \dfrac{\partial T}{\partial y}(x, y, t)\right]}{\Delta y}.$$

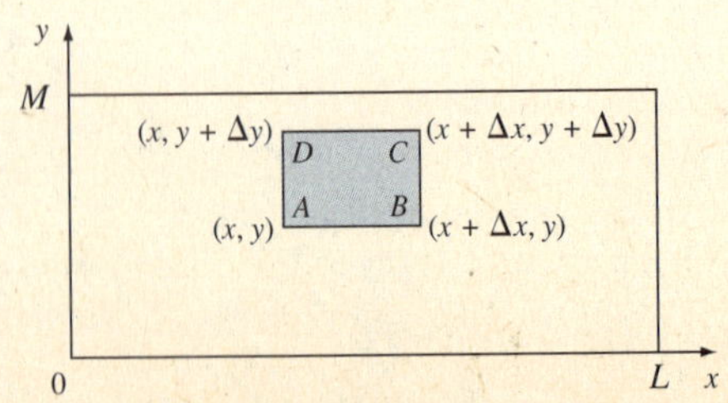

FIGURE 12.8
Heat flow in a sheet of material

Passing to the limit as Δx and Δy both approach zero, we obtain

$$c\rho\theta \frac{\partial T}{\partial t} = \kappa\theta\left[\frac{\partial^2 T}{\partial x^2} + \frac{\partial^2 T}{\partial y^2}\right],$$

which may be written in the form

$$\frac{\partial^2 T}{\partial x^2} + \frac{\partial^2 T}{\partial y^2} = \frac{1}{\delta}\frac{\partial T}{\partial t}, \tag{5}$$

where $\delta = \kappa/c\rho$ is the **diffusivity.** Equation **(5)** is called the **two-dimensional heat equation.**

If the edges of the sheet are kept at constant temperatures, and an extremely long time is allowed to pass, the temperature at any given point stabilizes and remains almost constant for all larger values of t; that is, steady-state conditions are attained. The temperature T at a point on the sheet then depends only on its position and not on time. Thus $\frac{\partial T}{\partial t}$ is zero throughout the sheet and Equation **(5)** becomes

Laplace's equation

$$\frac{\partial^2 T}{\partial x^2} + \frac{\partial^2 T}{\partial y^2} = 0. \tag{6}$$

This elliptic partial differential equation, commonly referred to as **Laplace's equation,** arises in many other problems of applied mathematics. We now give an example of the solution of Laplace's equation by the method of separation of variables.

EXAMPLE 1 ▶ **Solving Laplace's equation**

We begin by studying the steady-state heat flow problem in the rectangular sheet of size $L \times M$ in which Laplace's equation arose. Suppose that the upper horizontal edge is kept at 100°C, while the other three edges are kept at 0°C. We may write these boundary conditions for $T(x, y)$ as follows:

$$T(0, y) = T(L, y) = 0, \qquad 0 < y < M, \tag{7}$$

$$T(x, 0) = 0,\ T(x, M) = 100, \qquad 0 < x < L. \tag{8}$$

Since we are seeking the steady-state solution, time is not a factor in this problem, and therefore no initial condition is needed.

Setting $T(x, y) = X(x)Y(y)$, we find that Equation **(6)** becomes

$$X''Y + XY'' = 0,$$

which may be written in separated form as

$$\frac{X''}{X} = -\frac{Y''}{Y}. \tag{9}$$

The left-hand side of Equation **(9)** is constant for fixed x and arbitrary y. Hence Y''/Y must be constant. Letting k be this constant, we obtain the pair of ordinary differential equations

$$X'' - kX = 0,$$

$$Y'' + kY = 0. \tag{10}$$

From **(7)** we have

$$X(0)Y(y) = X(L)Y(y) = 0,$$

and $Y(y) \not\equiv 0$ [since otherwise $T(x, y) \equiv 0$, contradicting the second equation of **(8)**], so that $X(0) = X(L) = 0$. Thus again we have the situation encountered in Section 12.7, and the eigenvalues are all negative. Setting $k = -r^2$, we find that rL must be a multiple of π in order that we have a nonzero solution of the boundary value problem for X. Hence $r = \frac{n\pi}{L}$, and the functions

$$X_n(x) = A \sin \frac{n\pi x}{L}, \qquad n = 1, 2, 3, \ldots, \tag{11}$$

are the eigenfunctions of the problem. Setting $k = -(n\pi/L)^2$ in the second equation of **(10)** yields the general solution

$$Y_n(y) = B_1 e^{n\pi y/L} + B_2 e^{-n\pi y/L}.$$

The first condition of **(8)** implies that $Y(0) = 0$; thus $B_1 = -B_2$, and we can rewrite Y_n as

$$Y_n(y) = B \sinh \frac{n\pi y}{L}. \tag{12}$$

As in Section 12.7, to enlarge the class of possible solutions, we consider an infinite sum of products of the terms **(11)** and **(12)**:

$$T(x, y) = \sum_{n=1}^{\infty} c_n \sin \frac{n\pi x}{L} \sinh \frac{n\pi y}{L}. \tag{13}$$

Evaluating Equation **(13)** at any point (x, M), we obtain, by **(8)**,

$$100 = \sum_{n=1}^{\infty} \left(C_n \sinh \frac{n\pi M}{L} \right) \sin \frac{n\pi x}{L}, \qquad 0 < x < L,$$

which again is a Fourier sine series, requiring that we extend the boundary condition $T(x, M)$ to the interval $-L \le x \le L$ as an odd piecewise continuous periodic function of period $2L$ with a piecewise continuous derivative. Clearly the function

$$f(x) = \begin{cases} 100, & 0 < x < L, \\ -100, & -L < x < 0, \end{cases}$$

with $f(x + 2L) = f(x)$ satisfies these conditions. Hence $c_n \sinh(\frac{n\pi M}{L})$ must equal the nth Fourier (sine) coefficient of $f(x)$:

$$c_n \sinh \frac{n\pi M}{L} = \frac{1}{L} \int_{-L}^{L} f(x) \sin \frac{n\pi x}{L}\, dx = \frac{200}{L} \int_{0}^{L} \sin \frac{n\pi x}{L}\, dx$$

$$= \frac{200}{n\pi}(1 - \cos n\pi) = \frac{200}{n\pi}[1 - (-1)^n].$$

Hence $c_{2k} = 0$, while

$$c_{2k+1} = \frac{400}{\pi(2k + 1) \sinh\left[(2k + 1)\frac{\pi M}{L}\right]},$$

and the solution is given by

$$T(x, y) = \frac{400}{\pi}\left[\frac{\sin\left(\frac{\pi x}{L}\right)\sinh\left(\frac{\pi y}{L}\right)}{\sinh\left(\frac{\pi M}{L}\right)} + \frac{\sin\left(\frac{3\pi x}{L}\right)\sinh\left(\frac{3\pi y}{L}\right)}{3\sinh\left(\frac{3\pi M}{L}\right)} + \dots\right]$$

$$= \frac{400}{\pi}\sum_{n=1}^{\infty}\frac{\sin\left[\frac{(2n-1)\pi x}{L}\right]\sinh\left[\frac{(2n-1)\pi y}{L}\right]}{(2n-1)\sinh\left[\frac{(2n-1)\pi M}{L}\right]}.$$

We may use this formula to compute the steady-state temperature at any point in the rectangle. For example,

$$T\left(\frac{L}{2}, \frac{M}{2}\right) = \frac{400}{\pi}\sum_{n=1}^{\infty}\frac{\sin\left[\frac{(2n-1)\pi}{2}\right]\sinh\left[\frac{(2n-1)\pi M}{2L}\right]}{(2n-1)\sinh\left[\frac{(2n-1)\pi M}{L}\right]}.$$

Since

$$\frac{\sinh\left(\frac{A}{2}\right)}{\sinh A} = \frac{e^{A/2} - e^{-A/2}}{e^{A} - e^{-A}} = \frac{1}{e^{A/2} + e^{-A/2}} = \frac{1}{2\cosh\left(\frac{A}{2}\right)},$$

we obtain

$$T\left(\frac{L}{2}, \frac{M}{2}\right) = \frac{200}{\pi}\left[\frac{1}{\cosh\left(\frac{\pi M}{2L}\right)} - \frac{1}{3\cosh\left(\frac{3\pi M}{2L}\right)} + \frac{1}{5\cosh\left(\frac{5\pi M}{2L}\right)} - \dots\right],$$

which, for given L and M, can be approximated to any desired degree of accuracy. ◀

SELF-QUIZ

I. Which of the following are solutions of Laplace's equation $\Delta T = 0$?
(a) $T(x, y) = x^2 - y^2$
(b) $T(x, y) = 3x + 12y$
(c) $T(x, y) = e^x \cos y$
(d) $T(x, y) = 2xy$

Answer Questions II–IV true *or* false.

II. $T(x, y) = ax + by$ is a solution of Laplace's equation for all real numbers a and b.

III. $T(x, y) = axy$ is a solution of Laplace's equation for every real number a.

IV. $T(x, y) = ax^2 - by^2$ is a solution of Laplace's equation for all real numbers a and b.

Answers to Self-Quiz

I. All of them **II.** True **III.** True **IV.** False (True only when $a = b$.)

PROBLEMS 12.8

1. A square plate with sides of length L has both faces insulated. The upper horizontal edge is kept at 50°C, while all the other edges are at 0°C. Find the steady-state temperatures at the points $(\frac{L}{2}, \frac{L}{2})$ and $(\frac{L}{4}, \frac{L}{4})$.
2. If in Problem 1 the temperatures along the edge $y = L$ are given by
$$T(x, L) = x(L - x),$$
and all the other conditions remain the same, what is $T(x, y)$?
3. If in Problem 1 the temperatures along the upper horizontal edge are
$$T(x, L) = 100 \sin\left(\frac{\pi x}{L}\right),$$
with all the other conditions remaining uncharged, what is $T(x, y)$?
4. Suppose that the temperature along the edge $y = L$ in Problem 2 is changed to
$$T(x, L) = x(L^2 - x^2),$$
with all the other conditions remaining the same. Find the temperature at the points $(\frac{L}{4}, \frac{L}{4})$ and $(\frac{L}{4}, \frac{3L}{4})$.

Assume in Problems 5–8 that all other previous conditions are unchanged.

5. Repeat Problem 4 with $T(x, L) = x^2(L - x)$.
6. Repeat Problem 2 with $T(x, L) = x \cos \frac{\pi x}{2L}$.
7. Repeat Problem 2 with $T(x, L) = 10x \sin \frac{\pi x}{L}$.
8. Repeat Problem 4 with $T(x, L) = 100 \sin^2(\frac{\pi x}{L})$.
9. A rectangular plate of length $2L$ and height L has both faces insulated. The upper edge is kept at 50°C and the lower edge at 0°C, while the vertical edges have temperature
$$T(-L, y) = T(L, y) = 50 \sin \frac{\pi y}{2L}, \qquad 0 \leq y \leq L.$$
Find the temperature at any point in the plate.
10. Repeat Problem 9 with
$$T(-L, y) = T(L, y) = \frac{50y}{L}, \qquad 0 \leq y \leq L,$$
all other conditions remaining unchanged.
11. Repeat Problem 10 with the upper edge now changed to
$$T(x, L) = 50\left(\frac{x}{L}\right)^2,$$
all other conditions remaining unchanged.
12. Extend the development in this section to a solid cube of side L.
   - **(a)** What is the three-dimensional heat equation for such a situation?
   - **(b)** What is the steady-state equation for part (a)?
   - **(c)** Assume that all faces, except $z = L$, of the cube are kept at 0°C, while the $z = L$ face is kept at 100°C. What is the steady-state solution in the form of a Fourier series?

12.9 LAPLACE'S EQUATION IN POLAR AND SPHERICAL COORDINATES

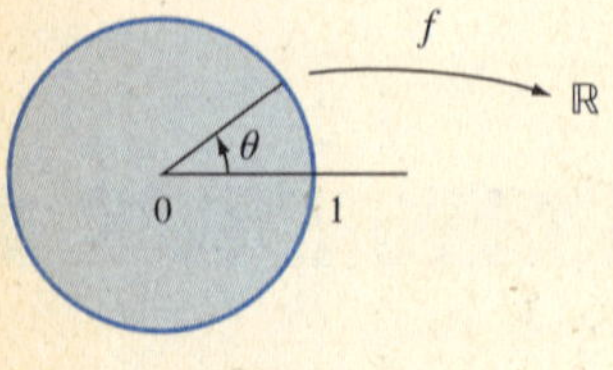

FIGURE 12.9
Boundary condition $f(\theta)$ on the disk

We now consider the steady-state heat flow problem in the disk D of radius 1 centered at the origin. It should be clear from the nature of the situation that the problem will be simplified if we can use polar coordinates (r, θ) instead of rectangular coordinates (x, y). Assume that the boundary of the disk is kept at the temperature

$$T(1, \theta) = f(\theta), \qquad 0 \leq \theta \leq 2\pi, \tag{1}$$

where the function $f(\theta)$ is periodic with period 2π and has a continuous derivative (see Figure 12.9). Now we need only write Laplace's equation, $\partial^2 T/\partial x^2 + \partial^2 T/\partial y^2 = 0$, in polar coordinates. Recall the transformation formulas

$$r = \sqrt{x^2 + y^2}, \qquad x = r \cos \theta,$$

$$\theta = \tan^{-1} \frac{y}{x}, \qquad y = r \sin \theta.$$

By the chain rule, we have

$$\frac{\partial T}{\partial x} = \frac{\partial T}{\partial r}\frac{\partial r}{\partial x} + \frac{\partial T}{\partial \theta}\frac{\partial \theta}{\partial x} = \frac{\partial T}{\partial r}\cos\theta - \frac{\partial T}{\partial \theta}\frac{\sin\theta}{r},$$

since

$$\frac{\partial r}{\partial x} = \frac{\partial}{\partial x}\sqrt{x^2 + y^2} = \frac{x}{\sqrt{x^2 + y^2}} = \frac{x}{r} = \cos\theta,$$

and

$$\frac{\partial \theta}{\partial x} = \frac{\dfrac{-y}{x^2}}{1 + \left(\dfrac{y}{x}\right)^2} = \frac{-y}{r^2} = \frac{-\sin\theta}{r}.$$

Then

$$\begin{aligned}\frac{\partial^2 T}{\partial x^2} &= \frac{\partial}{\partial r}\left(\frac{\partial T}{\partial r}\frac{\partial r}{\partial x} + \frac{\partial T}{\partial \theta}\frac{\partial \theta}{\partial x}\right)\frac{\partial r}{\partial x} + \frac{\partial}{\partial \theta}\left(\frac{\partial T}{\partial r}\frac{\partial r}{\partial x} + \frac{\partial T}{\partial \theta}\frac{\partial \theta}{\partial x}\right)\frac{\partial \theta}{\partial x}\\ &= \frac{\partial^2 T}{\partial r^2}\cos^2\theta - \frac{\partial^2 T}{\partial r\,\partial \theta}\frac{\sin\theta\cos\theta}{r} + \frac{\partial T}{\partial \theta}\frac{\sin\theta\cos\theta}{r^2}\\ &\quad - \frac{\partial^2 T}{\partial \theta\,\partial r}\frac{\sin\theta\cos\theta}{r} + \frac{\partial T}{\partial r}\frac{\sin^2\theta}{r} + \frac{\partial^2 T}{\partial \theta^2}\frac{\sin^2\theta}{r^2} + \frac{\partial T}{\partial \theta}\frac{\cos\theta\sin\theta}{r^2}.\end{aligned}$$

Similarly,

$$\begin{aligned}\frac{\partial^2 T}{\partial y^2} &= \frac{\partial}{\partial r}\left(\frac{\partial T}{\partial r}\sin\theta + \frac{\partial T}{\partial \theta}\frac{\cos\theta}{r}\right)\frac{\partial r}{\partial y} + \frac{\partial}{\partial \theta}\left(\frac{\partial T}{\partial r}\sin\theta + \frac{\partial T}{\partial \theta}\frac{\cos\theta}{r}\right)\frac{\partial \theta}{\partial y}\\ &= \frac{\partial^2 T}{\partial r^2}\sin^2\theta + \frac{\partial^2 T}{\partial r\,\partial \theta}\frac{\sin\theta\cos\theta}{r} - \frac{\partial T}{\partial \theta}\frac{\cos\theta\sin\theta}{r^2}\\ &\quad + \frac{\partial^2 T}{\partial \theta\,\partial r}\frac{\sin\theta\cos\theta}{r} + \frac{\partial T}{\partial r}\frac{\cos^2\theta}{r} + \frac{\partial^2 T}{\partial \theta^2}\frac{\cos^2\theta}{r^2} - \frac{\partial T}{\partial \theta}\frac{\sin\theta\cos\theta}{r^2}.\end{aligned}$$

Adding the two equations, we obtain

Laplace's equation in polar coordinates

$$\frac{\partial^2 T}{\partial x^2} + \frac{\partial^2 T}{\partial y^2} = \frac{\partial^2 T}{\partial r^2} + \frac{1}{r^2}\frac{\partial^2 T}{\partial \theta^2} + \frac{1}{r}\frac{\partial T}{\partial r} = 0. \qquad \textbf{(2)}$$

Now we let $T(r, \theta) = R(r)\Theta(\theta)$ and transform **(2)** into

$$R''\Theta + \frac{1}{r^2}R\Theta'' + \frac{1}{r}R'\Theta = 0,$$

or

$$\left(R'' + \frac{1}{r}R'\right)\Theta = -\frac{R}{r^2}\Theta''.$$

Separating the variables, we have

$$\frac{r^2R'' + rR'}{R} = -\frac{\Theta''}{\Theta} = k,$$

since each side of the first equality is a function of only one variable. Hence we obtain two ordinary differential equations:

$$\Theta'' + k\Theta = 0, \tag{3}$$

and

$$r^2R'' + rR' - kR = 0. \tag{4}$$

Since $f(\theta) = T(1, \theta) = R(1)\Theta(\theta)$, it is clear that $\Theta(\theta)$ is periodic with period 2π. We assume that f is not identically zero [see Problem 12 for the case where $f(\theta) \equiv 0$], so $R(1) \neq 0$. Observe that $\Theta(0) = \Theta(2\pi)$ and $\Theta'(0) = \Theta'(2\pi)$. By the remark following the orthogonality theorem (p. 559) and Problem 11.6.6, the boundary value problem in Θ has periodic eigenfunctions of period 2π. Thus the eigenvalues must be $k = n^2$, with the corresponding eigenfunctions

$$\Theta_n(\theta) = A_1 \cos n\theta + A_2 \sin n\theta, \qquad n = 0, 1, 2, 3, \ldots.$$

Equation **(4)** is an Euler equation (see Section 3.7) and can be solved by letting $R = r^\lambda$. Letting $k = n^2$, we see that **(4)** becomes

$$r^\lambda[\lambda(\lambda - 1) + \lambda - n^2] = r^\lambda(\lambda^2 - n^2) = 0,$$

and the roots of the equation $\lambda^2 - n^2 = 0$ are $\lambda = \pm n$. Hence the general solution of Equation **(4)** with $k = n^2$ is

$$R_0(r) = B_1 + B_2 \ln r; \qquad \text{for } n = 0, \text{ and}$$
$$R_n(r) = B_1 r^n + B_2 r^{-n}, \qquad n = 1, 2, 3, \ldots.$$

Since we want R_n to exist for all values in the range $0 \le r \le 1$, B_2 must be zero and $R_n(r) = Br^n$, $n = 0, 1, 2, 3, \ldots$. Thus we seek a solution in the form of the infinite series

$$T(r, \theta) = \frac{A_0}{2} + \sum_{n=1}^{\infty} r^n(A_n \cos n\theta + B_n \sin n\theta). \tag{5}$$

To evaluate the constants A_n, B_n, we note that

$$f(\theta) = T(1, \theta) = \frac{A_0}{2} + \sum_{n=1}^{\infty} A_n \cos n\theta + B_n \sin n\theta. \tag{6}$$

Hence **(6)** is the Fourier series of f, and the coefficients A_n and B_n are the Fourier coefficients of f (now over the range $|\theta| < \pi$) given by the Euler formulas

$$A_n = \frac{1}{\pi}\int_{-\pi}^{\pi} f(\phi) \cos n\phi \, d\phi, \qquad B_n = \frac{1}{\pi}\int_{-\pi}^{\pi} f(\phi) \sin n\phi \, d\phi.$$

Since **(6)** converges and $r \le 1$, series **(5)** also converges. For example, if $f(\theta) = \cos\theta$, all the coefficients except A_1 are zero, and $A_1 = 1$. Hence the temperature is given by

$$T(r, \theta) = r \cos\theta$$

at all points in the unit disk.

It is interesting to note what occurs if we replace the coefficients A_n and B_n in **(5)** by the Euler formulas and interchange the sums and integrals:

$$\begin{aligned} T(r, \theta) &= \frac{1}{2\pi}\int_{-\pi}^{\pi} f(\phi)\, d\phi \\ &\quad + \sum_{n=1}^{\infty} r^n\left[\frac{\cos n\theta}{\pi}\int_{-\pi}^{\pi} f(\phi)\cos n\phi \, d\phi + \frac{\sin n\theta}{\pi}\int_{-\pi}^{\pi} f(\phi) \sin n\phi \, d\phi\right] \\ &= \frac{1}{\pi}\int_{-\pi}^{\pi} f(\phi)\left[\frac{1}{2} + \sum_{n=1}^{\infty} r^n(\cos n\theta \cos n\phi + \sin n\theta \sin n\phi)\right] d\phi. \end{aligned}$$

Since

$$\cos n(\theta - \phi) = \cos n\theta \cos n\phi + \sin n\theta \sin n\phi,$$

we obtain

$$T(r, \theta) = \frac{1}{\pi}\int_{-\pi}^{\pi} f(\phi)\left[\frac{1}{2} + \sum_{n=1}^{\infty} r^n \cos n(\theta - \phi)\right] d\phi.$$

But

$$\cos n(\theta - \phi) = \frac{e^{in(\theta-\phi)} + e^{-in(\theta-\phi)}}{2},$$

so

$$\sum_{n=1}^{\infty} r^n \cos n(\theta - \phi) = \frac{1}{2}\sum_{n=1}^{\infty} r^n[e^{in(\theta-\phi)} + e^{-in(\theta-\phi)}]$$
$$= \frac{1}{2}\left\{\sum_{n=1}^{\infty} [re^{i(\theta-\phi)}]^n + \sum_{n=1}^{\infty} [re^{-i(\theta-\phi)}]^n\right\}.$$

The complex number $e^{it} = \cos t + i \sin t$ can be represented as a vector of length $\sqrt{(\cos t)^2 + (\sin t)^2} = 1$ (see Appendix 5), so $re^{i(\theta-\phi)}$ and $re^{i(\phi-\theta)}$ both have length equal to $r(< 1)$. Using the geometric series, we have

$$\sum_{n=1}^{\infty} r^n \cos n(\theta - \phi) = \frac{1}{2}\left[\frac{re^{i(\theta-\phi)}}{1 - re^{i(\theta-\phi)}} + \frac{re^{-i(\theta-\phi)}}{1 - re^{-i(\theta-\phi)}}\right]$$
$$= \frac{1}{2}\left\{\frac{r[e^{i(\theta-\phi)} + e^{-i(\theta-\phi)}] - 2r^2}{r^2 - r[e^{i(\theta-\phi)} + e^{-i(\theta-\phi)}] + 1}\right\}$$
$$= \frac{r\cos(\theta - \phi) - r^2}{r^2 - 2r\cos(\theta - \phi) + 1}.$$

Thus

$$\frac{1}{2} + \sum_{n=1}^{\infty} r^n \cos n(\theta - \phi) = \frac{1}{2} + \frac{r\cos(\theta - \phi) - r^2}{r^2 + 1 - 2r\cos(\theta - \phi)}$$
$$= \frac{1 - r^2}{2[r^2 + 1 - 2r\cos(\theta - \phi)]}$$

and

The Poisson integral formula

$$T(r, \theta) = \frac{1 - r^2}{2\pi}\int_{-\pi}^{\pi} \frac{f(\phi)}{r^2 + 1 - 2r\cos(\theta - \phi)}\, d\phi. \quad (7)$$

Equation **(7)**, the **Poisson integral formula***, is valid for all values $r < 1$ (see Problem 9). It indicates that the temperature at any interior point (r, θ) of the unit disk may be obtained by integrating the boundary temperatures according to formula **(7)**. In particular, if $r = 0$, then the temperature at the center of the disk is

$$T(0, \theta) = \frac{1}{2\pi}\int_{-\pi}^{\pi} f(\phi)\, d\phi;$$

that is, **the temperature at the center is the integral average of the boundary temperatures.** This fact, often called the **mean value theorem,** holds for all functions that satisfy Laplace's equation on the unit disk $r \leq 1$.

* Named in honor of Siméon Denis Poisson (1781–1840), a French mathematician and physicist.

EXAMPLE 1 ▶ **Using cylindrical coordinates**

In this example we consider the three-dimensional Laplace equation

$$\frac{\partial^2 T}{\partial x^2} + \frac{\partial^2 T}{\partial y^2} + \frac{\partial^2 T}{\partial z^2} = 0 \tag{8}$$

associated with the steady-state heat flow problem in a solid.

Suppose that the solid is a rectangular cylinder of height π and radius 1 (see Figure 12.10). Since such a solid is easily described in cylindrical coordinates, it is convenient to obtain Laplace's equation **(8)** in terms of the coordinates (r, θ, z), where

$$x = r\cos\theta, \qquad y = r\sin\theta, \qquad z = z.$$

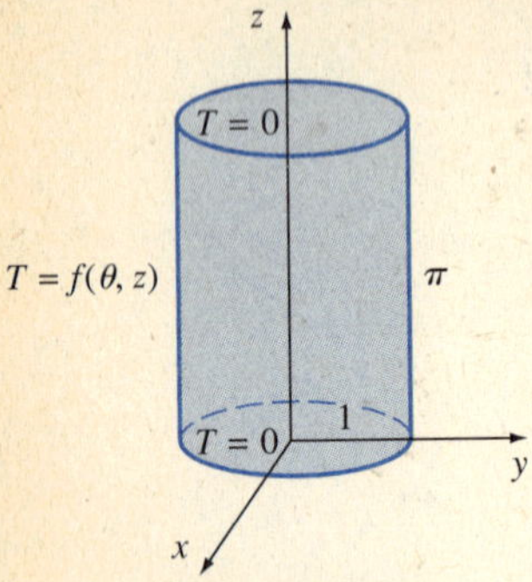

FIGURE 12.10
Boundary conditions on a cylinder

Obviously we are simply writing the xy-coordinates in polar form. Thus the calculations we did earlier still apply, and Laplace's equation becomes [from Equation **(2)**]

Laplace's equation in cylindrical coordinates

$$\frac{\partial^2 T}{\partial r^2} + \frac{1}{r}\frac{\partial T}{\partial r} + \frac{1}{r^2}\frac{\partial^2 T}{\partial \theta^2} + \frac{\partial^2 T}{\partial z^2} = 0. \tag{9}$$

We now apply the method of separation of variables to the function T and, accordingly, set $T(r, \theta, z) = R(r)\Theta(\theta)Z(z)$. Equation **(9)** becomes

$$R''\Theta Z + \frac{1}{r}R'\Theta Z + \frac{1}{r^2}R\Theta''Z + R\Theta Z'' = 0,$$

or

$$[(R'' + r^{-1}R')\Theta + r^{-2}R\Theta'']Z = -R\Theta Z''.$$

Separating out the z-variables, we have

$$\frac{R'' + r^{-1}R'}{R} + r^{-2}\frac{\Theta''}{\Theta} = -\frac{Z''}{Z} = k_1,$$

since the left-hand side of the equation is a function of r and θ, whereas the right-hand side is a function of z alone. Thus

$$Z'' + k_1 Z = 0,$$

and

$$\frac{R'' + r^{-1}R'}{R} = k_1 - r^{-2}\frac{\Theta''}{\Theta}.$$

Multiplying both sides of the last equation by r^2, we have

$$\frac{r^2R'' + rR'}{R} = k_1 r^2 - \frac{\Theta''}{\Theta},$$

or

$$\frac{r^2R'' + rR'}{R} - k_1 r^2 = -\frac{\Theta''}{\Theta} = k_2,$$

since each side is a function of only one variable. Hence we also obtain the ordinary differential equations

$$\Theta'' + k_2\Theta = 0$$

and

$$r^2R'' + rR' - (k_2 + k_1r^2)R = 0. \tag{10}$$

If we now assume the boundary conditions

$$T(r, \theta, 0) = T(r, \theta, \pi) = 0, \qquad T(1, \theta, z) = f(\theta, z),$$

where f is a continuously differentiable function in θ and z, we can translate the conditions into terms of R, Θ, and Z:

$$R(r)\Theta(\theta)Z(0) = R(r)\Theta(\theta)Z(\pi) = 0,$$
$$R(1)\Theta(\theta)Z(z) = f(\theta, z).$$

Since R and Θ are not constantly zero [assuming $f(\theta, z) \not\equiv 0$], it follows that $Z(0) = Z(\pi) = 0$. Also, since f is continuous, it must be periodic with period 2π in θ (since f is the temperature on the surface of the cylinder). Thus $\Theta(\theta)$ is periodic with period 2π, so $\Theta(0) = \Theta(2\pi)$ and $\Theta'(0) = \Theta'(2\pi)$. Finally, we know that $R(1) \neq 0$ and that $R(0)$ is finite. As in **(3)**, the eigenvalues of the boundary value problem

$$\Theta'' + k_2\Theta = 0, \qquad \Theta(0) = \Theta(2\pi), \qquad \Theta'(0) = \Theta'(2\pi),$$

are $k_2 = m^2$, $m = 0, 1, 2, \ldots$, with the corresponding eigenfunctions

$$\Theta_m(\theta) = A_1 \cos m\theta + A_2 \sin m\theta.$$

Similarly, the boundary value problem

$$Z'' + k_1 Z = 0, \qquad Z(0) = Z(\pi) = 0,$$

has the eigenvalues $k_1 = n^2$, $n = 1, 2, 3, \ldots$, and the corresponding eigenfunctions

$$Z_n(z) = B \sin nz.$$

Setting $k_1 = n^2$ and $k_2 = m^2$ in Equation **(10)**, we obtain

$$r^2R'' + rR' - (n^2r^2 + m^2)R = 0, \tag{11}$$

which may be transformed into Bessel's equation by the substitution $\rho = inr$. To verify this fact, we note that

$$\frac{dR}{d\rho} = \frac{dR}{dr}\frac{dr}{d\rho} = \frac{1}{in}R'(r),$$

and

$$\frac{d^2R}{d\rho^2} = \frac{d}{d\rho}\left(\frac{1}{in}R'\right) = \frac{1}{(in)^2}R''(r) = -\frac{1}{n^2}R''(r).$$

Substituting these values into Equation **(11)**, we obtain

$$-n^2r^2\frac{d^2R}{d\rho^2} + inr\frac{dR}{d\rho} - (n^2r^2 + m^2)R = 0,$$

or

$$\rho^2\frac{d^2R}{d\rho^2} + \rho\frac{dR}{d\rho} + (\rho^2 - m^2)R = 0. \tag{12}$$

By Theorem 5.5.2 (p. 237), Equation **(12)** has the general solution

$$R(\rho) = C_1 J_m(\rho) + C_2 Y_m(\rho), \qquad \rho \neq 0.$$

And since $R(0)$ is finite, the constant $C_2 = 0$, since the Bessel function of the second kind, Y_m, involves a $\ln \rho$ term that approaches $-\infty$ as ρ approaches zero. We are, therefore, led to the double series

$$T(r, \theta, z) = \sum_{m=0}^{\infty} \sum_{n=1}^{\infty} (a_{mn} \cos m\theta + b_{mn} \sin m\theta) \sin nz\, J_m(inr).$$

The coefficients a_{mn} and b_{mn} can be determined by setting $r = 1$ and requiring (with $|\theta| < \pi$) that

$$f(\theta, z) = \sum_{m=0}^{\infty} \left[\left(\sum_{n=1}^{\infty} a_{mn} J_m(in) \sin nz \right) \cos m\theta + \left(\sum_{n=1}^{\infty} b_{mn} J_m(in) \sin nz \right) \sin m\theta \right]. \quad \textbf{(13)}$$

If we define

$$a_m(z) = \sum_{n=1}^{\infty} a_{mn} J_m(in) \sin nz$$

and

$$b_m(z) = \sum_{n=1}^{\infty} b_{mn} J_m(in) \sin nz, \quad \textbf{(14)}$$

we see that **(13)** becomes

$$f(\theta, z) = \sum_{m=0}^{\infty} a_m(z) \cos m\theta + b_m(z) \sin m\theta,$$

which is a Fourier series in θ, where z is an arbitrary parameter. By the Euler formulas, it follows that

$$a_m(z) = \frac{1}{\pi} \int_{-\pi}^{\pi} f(\theta, z) \cos m\theta \, d\theta, \qquad b_m(z) = \frac{1}{\pi} \int_{-\pi}^{\pi} f(\theta, z) \sin m\theta \, d\theta.$$

If the functions $a_m(z)$ and $b_m(z)$ can be calculated, we can then use the Fourier sine series **(14)** in z to calculate the coefficients a_{mn} and b_{mn}:

$$a_{mn} = \frac{2}{\pi J_m(in)} \int_0^{\pi} a_m(z) \sin nz \, dz,$$

$$b_{mn} = \frac{2}{\pi J_m(in)} \int_0^{\pi} b_m(z) \sin nz \, dz. \quad ◀$$

Although the formal procedure delineated above is straightforward, the calculations involved are, to say the least, very tedious. In Problem 8 the reader is asked to develop a similar procedure that leads to the use of series of Legendre polynomials in place of Bessel functions.

SELF-QUIZ

I. The two-dimensional form of Laplace's equation in polar coordinates is ________.

(a) $\dfrac{\partial^2 T}{\partial x^2} + \dfrac{\partial^2 T}{\partial y^2} = 0$ **(b)** $\dfrac{\partial^2 T}{\partial r^2} + \dfrac{\partial^2 T}{\partial \theta^2} = 0$

(c) $\dfrac{\partial^2 T}{\partial r^2} + \dfrac{1}{r^2}\dfrac{\partial^2 T}{\partial \theta^2} + \dfrac{1}{r}\dfrac{\partial T}{\partial r} = 0$

(d) $\dfrac{\partial^2 T}{\partial r^2} + \dfrac{1}{r}\dfrac{\partial^2 T}{\partial \theta^2} + \dfrac{1}{r^2}\dfrac{\partial T}{\partial r} = 0$

Answer true *or* false.

II. The Poisson integral formula

$$T(r, \theta) = \frac{1 - r^2}{2\pi} \int_{-\pi}^{\pi} \frac{f(\phi)}{r^2 + 1 - 2r\cos(\theta - \phi)} \, d\phi$$

is valid whenever $r \neq 1$.

Answers to Self-Quiz

I. c **II.** False (only for $r < 1$)

PROBLEMS 12.9

1. Consider the steady-state heat flow problem on the unit disk D [see Equation **(2)**] and suppose that the boundary condition **(1)** is given by $f(\theta) = |\theta - \pi|$, $0 \le \theta \le 2\pi$. Find the Fourier series solution **(5)** of this problem.

2. Repeat Problem 1 with $f(\theta) = \cos 2\theta$.

3. Find the Poisson integral solution **(7)** of Problem 1. What is the temperature at the origin? What is the temperature at the point $(r, \theta) = (\frac{1}{2}, 0)$?

4. Find the Poisson integral solution for Problem 2. What is the temperature at the origin?

5. Consider the steady-state heat flow problem on the unit disk and suppose that $f(\theta) = \sin^2\theta$. Find the temperature at the point $r = \frac{1}{2}$, $\theta = 0°$ by
(a) the eigenfunction expansion method;
(b) the Poisson integral formula.

6. Repeat Problem 5 with $f(\theta) = \sin\theta - \cos\theta$.

7. Repeat Problem 5 with $f(\theta) = \theta(2\pi - \theta)$, $0 \le \theta \le 2\pi$.

8. **(a)** Show that the three-dimensional Laplace equation **(8)** may be written in spherical coordinates r, θ, ϕ ($x = r\cos\theta\sin\phi$, $y = r\sin\theta\sin\phi$, $z = r\cos\phi$) in the form

$$\frac{\partial^2 T}{\partial r^2} + \frac{2}{r}\frac{\partial T}{\partial r} + \frac{1}{r^2}\frac{\partial^2 T}{\partial \phi^2} + \frac{\cot\phi}{r^2}\frac{\partial T}{\partial \phi} + \frac{1}{r^2\sin^2\phi}\frac{\partial^2 T}{\partial \theta^2} = 0. \qquad \text{(i)}$$

(b) Separate the variables in Equation **(i)** to obtain the ordinary differential equations

$$\Theta'' + k_1\Theta = 0,$$

$$R'' + \left(\frac{2}{r}\right)R' + \left(\frac{k_2}{r^2}\right)R = 0,$$

$$[(\sin\phi)\Phi']' - \left(\frac{k_2\sin^2\phi + k_1}{\sin\phi}\right)\Phi = 0.$$

(c) Assume the boundary condition $T(1, \theta, \phi) = f(\theta, \phi)$ for a sphere of radius 1, where f is continuously differentiable in θ and ϕ and periodic with period 2π in θ. Find the eigenfunctions of the three boundary value problems thus defined. [*Hint:* Use Legendre polynomials.]

(d) Obtain the eigenfunction expansion for $T(r, \theta, \phi)$.

9. Justify the steps involved in obtaining the Poisson integral formula **(7)** from the solution **(5)** of the steady-state heat equation in the disk $r \le 1$. [*Hint:* Show that series **(5)** converges uniformly in the disk $r \le r_0 < 1$.]

10. Generalize the problem in Figure 12.9 to obtain the steady-state temperature in a disk of radius R centered at the origin, and prove that the corresponding Poisson integral formula is

$$T(r, \theta) = \frac{R^2 - r^2}{2\pi}\int_{-\pi}^{\pi} \frac{f(\phi)\, d\phi}{R^2 + r^2 - 2rR\cos(\theta - \phi)},$$
$$r < R,$$

where the continuously differentiable function $f(\phi)$ describes the boundary temperature in the direction $0 \le \phi < 2\pi$.

11. Let D be any bounded polygon in the xy-plane, and let (x_0, y_0) be a point interior to D. Let u be a real-valued function defined on D that has continuous derivatives of at least order two.

(a) Show that if u attains its maximal value M at (x_0, y_0), then

$$\left.\frac{\partial^2 u}{\partial x^2} + \frac{\partial^2 u}{\partial y^2}\right|_{(x_0, y_0)} \le 0.$$

(b) Show that if

$$\frac{\partial^2 u}{\partial x^2} + \frac{\partial^2 u}{\partial y^2} > 0$$

at all interior points of D, then u does not attain its maximal value at any interior point of D.

(c) Suppose that

$$\frac{\partial^2 u}{\partial x^2} + \frac{\partial^2 u}{\partial y^2} \ge 0 \qquad \text{(ii)}$$

at all interior points of D. Show that the function

$$w(x, y) = u(x, y) + \epsilon[(x - x_0)^2 + (y - y_0)^2]$$

does not attain its maximal value in D at (x_0, y_0), for any value $\epsilon > 0$.

(d) Conclude that any function satisfying Equation **(ii)** at all interior points of D must attain its maximal value on the boundary of D. (This is called the **maximum principle.**)

12. Use the maximum principle obtained in Problem 11 to show that any function u that satisfies Laplace's equation attains both its maximum and its minimum on the boundary of D. (Physically, this means that, in a steady-state heat flow problem, the maximum and minimum temperatures occur on the boundary of the region.)

12.10 LAPLACE TRANSFORM METHODS FOR PARTIAL DIFFERENTIAL EQUATIONS

The one-dimensional heat equation

$$\frac{\partial T}{\partial t} = \delta \frac{\partial^2 T}{\partial x^2}$$

is sometimes called the **thermal diffusion equation,** since it describes the flow of heat along a uniform rod. The same equation may also be used to describe one-dimensional fluid flow, since we can replace temperature by mass (or concentration) in the considerations of Equation **(12.7.6).** By doing this, we can use the heat equation to study such diverse problems as the dispersal of pollutants in the atmosphere or the flow of a chemical across a membrane. In some cases, the fluid flow is aided (or retarded) by other forces; for example, the motion of a pollutant emitted from a smokestack involves both a mixing **(diffusion)** of the pollutant in the air and the transport **(convection)** of the pollutant by air currents. The change in the concentration T in the interval from x_0 to x_1, due to convection, is

$$v[T(x_0, t) - T(x_1, t)],$$

where v is the velocity of the current. Adding this term to Equation **(12.7.3)** (p. 610), we obtain

$$\frac{dQ}{dt} = \int_{x_0}^{x_1} \frac{\partial}{\partial x}\left(\kappa \frac{\partial T}{\partial x} - vT\right) A\, dx.$$

Differentiating Equation **(12.7.1),** we finally have, by considerations similar to those in Section 12.7,

The diffusion equation

$$c\rho \frac{\partial T}{\partial t} = \frac{\partial}{\partial x}\left(\kappa \frac{\partial T}{\partial x}\right) - v\frac{\partial T}{\partial x}. \qquad \textbf{(1)}$$

Equation **(1)** is generally called the **diffusion equation.** If κ is zero, we call **(1)** a **pure convection;** if $v = 0$, a **pure diffusion.**

The method of separation of variables is again very useful in solving diffusion equations. However, in some situations the boundary and initial conditions are such that it is not possible to obtain a solution by this technique. Consider the following example of a pure diffusion.

EXAMPLE 1 ▶ **When separation of variables fails**

Suppose that κ is constant and $v = 0$ in Equation **(1)**. Then **(1)** reduces to the thermal diffusion equation

$$\frac{\partial T}{\partial t} = \delta \frac{\partial^2 T}{\partial x^2}. \tag{2}$$

Assume that we are given the initial condition

$$T(x, 0) = 0, \qquad 0 \leq x < \infty, \tag{3}$$

and the boundary conditions

$$\lim_{x \to +\infty} T(x, t) = \lim_{x \to +\infty} \frac{\partial T}{\partial x}(x, t) = 0, \tag{4}$$

$$T(0, t) = T_0(\neq 0), \qquad 0 < t. \tag{5}$$

These conditions could be used to describe the heat flow in an infinitely long rod when a constant source of heat is applied at $x = 0$, or to describe the concentration of a pollutant at varying distances from a smokestack, in the absence of winds, with the stack constantly emitting pollutants, for all time $t > 0$. If we let $T(x, t) = X(x)\mathcal{T}(t)$, we have

$$X\mathcal{T}' = \delta X''\mathcal{T} \quad \text{or} \quad \frac{\mathcal{T}'}{\delta\mathcal{T}} = \frac{X''}{X} = k,$$

from which we obtain the two differential equations

$$\mathcal{T}' = k\delta\mathcal{T} \quad \text{and} \quad X'' - kX = 0. \tag{6}$$

The first differential equation in **(6)** has the general solution $\mathcal{T}(t) = ce^{k\delta t}$. Using the initial condition **(3)**, we observe that

$$0 = T(x, 0) = X(x)\mathcal{T}(0) = cX(x).$$

Hence either $c = 0$, implying that $\mathcal{T} \equiv 0$, or $X \equiv 0$. In either case $T \equiv 0$, contradicting the second boundary condition **(5)**. Clearly, the method of separation of variables fails for this problem. It is therefore apparent that other techniques must be developed to treat problems of this type. ◀

The effectiveness of Laplace transform methods for ordinary differential equations suggests that they might be useful in this context. The rest of this section is devoted to a discussion of how transformation techniques may be applied to partial differential equations. It should be noted, however, that the effectiveness of Laplace transforms is subject to several limitations. The equation *must* be linear and should have constant coefficients, and there must be appropriate initial conditions. Even when these conditions are met, there is no guarantee of success; and even if a solution can be obtained, it may be easier to obtain by other methods.

To apply Laplace transforms to a partial differential equation, we must begin by considering the ranges of the independent variables. We illustrate here the entire procedure as it relates to Example 1. In this problem, the independent variables x and t may both assume all values in the range 0 to ∞. This property is very desirable, since the definition of the Laplace transform of a function requires its integration over this range [see Equation **(6.1.1)**]. At this point both variables look promising. We recall, however, that the differentiation property (Theorem 6.2.1) of Laplace transforms requires an additional initial condition for each order of the derivative involved. In this case, **(2)** contains first derivatives in t and second derivatives in x. Since $(\frac{\partial T}{\partial x})(0, t)$ is not known, there are not enough data for the x variable. Thus we treat x as a parameter and consider **(2)** as an ordinary differential equation with t as the only independent variable. Let

$$\mathcal{L}\{T(x, t)\} = \int_0^\infty e^{-st}T(x, t)\, dt.$$

Then $\mathcal{L}\{T\}$ is a function of s and the parameter x. Using the differentiation property of Laplace transforms and the initial condition **(3)**, we may transform **(2)** into the ordinary differential equation

$$s\mathcal{L}\{T\} - 0 = \mathcal{L}\left\{\frac{\partial T}{\partial t}\right\} = \delta\mathcal{L}\left\{\frac{\partial^2 T}{\partial x^2}\right\} = \delta\frac{\partial}{\partial x^2}\mathcal{L}\{T\}. \tag{7}$$

Note that we have interchanged the operations of taking the Laplace transform and differentiating with respect to x. This exchange may not be valid, but the objection can be ignored if the method succeeds in producing a solution to the problem. Hence it is essential that we verify any result obtained by this method.

It is important now to interpret the boundary conditions **(4)** and **(5)** for $\mathcal{L}(T)$ whenever possible. Note that **(5)** yields

$$\mathcal{L}\{T\}(0, s) = \int_0^\infty e^{-st}T(0, t)\, dt = \frac{T_0}{s}. \tag{8}$$

Also, after warning that the indicated operations may not always be valid, we obtain

$$\begin{aligned}\lim_{x\to+\infty} \mathcal{L}\{T\}(x, s) &= \lim_{x\to+\infty}\int_0^\infty e^{-st}T(x, t)\, dt \\ &= \int_0^\infty e^{-st}\left[\lim_{x\to+\infty} T(x, t)\right] dt = 0.\end{aligned} \tag{9}$$

Now we treat s as a parameter and x as the independent variable. Setting $z = \mathcal{L}\{T\}$, we may rewrite **(7)** as

$$\frac{d^2z}{dx^2} = \frac{s}{\delta}z, \tag{10}$$

where s is fixed. The characteristic equation for the differential equation **(10)** has the roots $\pm\sqrt{s/\delta}$, so z has the general solution

$$z(x) = c_1e^{\sqrt{s/\delta}x} + c_2e^{-\sqrt{s/\delta}\,x}.$$

Since $s > 0$, the coefficient c_1 must be zero, because the term $e^{\sqrt{s/\delta}x}$ tends to infinity as x approaches $+\infty$, violating the identity **(9)**. Setting $x = 0$, we find that $c_2 = T_0/s$ by **(8)**, so we finally have

$$\mathcal{L}\{T\} = z(x) = \frac{T_0}{s}e^{-\sqrt{s/\delta}x}. \tag{11}$$

Again we treat x as a parameter and seek the inverse Laplace transform of **(11)**. By Formula 43 of Appendix 2, with $r = x/\sqrt{\delta}$, we see that

$$T(x, t) = T_0\left(1 - \frac{2}{\sqrt{\pi}}\int_0^{x/2\sqrt{\delta t}} e^{-u^2}\, du\right). \tag{12}$$

To check that **(12)** is indeed the solution of Example 1, we note from Formula 216 of Appendix 1 that

$$\int_0^\infty e^{-u^2}\,du = \frac{\sqrt{\pi}}{2}. \tag{13}$$

The integral in **(12)** tends to $\sqrt{\pi}/2$ if t approaches zero or if x approaches $+\infty$. Thus **(3)** and the first limit of **(4)** hold. If $x = 0$, we obviously obtain **(5).** Differentiating **(12)** with respect to x, we have

$$\frac{\partial T}{\partial x} = -\frac{T_0}{\sqrt{\pi\delta t}}e^{-(x^2/4\delta t)},$$

which vanishes as x approaches $+\infty$, and differentiating again, we obtain

$$\delta\frac{\partial^2 T}{\partial x^2} = \frac{T_0 x}{2\sqrt{\pi\delta}}\,\frac{e^{-(x^2/4\delta t)}}{t^{3/2}} = \frac{\partial T}{\partial t}.$$

Thus in this case the Laplace transform method does yield a solution.

EXAMPLE 2 ▶ **Using Laplace transforms to solve a partial differential equation**

We now consider the vibrating string problem of Section 12.5 given by the wave equation

$$\frac{\partial^2 y}{\partial t^2} = c^2\frac{\partial^2 y}{\partial x^2}, \tag{14}$$

with the boundary condition

$$y(0, t) = y(L, t) = 0 \tag{15}$$

and initial conditions

$$y(x, 0) = f(x), \qquad \left.\frac{\partial y}{\partial t}\right|_{(x,\,0)} = 0. \tag{16}$$

Since $0 \le x \le L$, we select t as our independent variable for the Laplace transform. By the differentiation property [see Equation **(6.2.4)**], we transform **(14)** into

$$s^2\mathcal{L}\{y\} - sf(x) - 0 = c^2\frac{\partial^2}{\partial x^2}\mathcal{L}\{y\}.$$

Thus we have the nonhomogeneous second-order ordinary differential equation

$$c^2Y'' - s^2Y = -sf(x), \tag{17}$$

where $Y(x) = \mathcal{L}\{y\}$ and s is a parameter. By **(15),** we have $Y(0) = Y(L) = 0$. Our problem thus reduces to a nonhomogeneous boundary value problem. It may then be solved by the methods of Section 11.7. To simplify matters, we assume that $f(x) = y_0 \sin(\frac{\pi x}{L})$ and so avoid having to use eigenfunction expansions to find the solution Y. A particular solution of Equation **(17)** is of the form

$$Y_p = A\cos\frac{\pi x}{L} + B\sin\frac{\pi x}{L}.$$

Hence

$$-A\left(\frac{c^2\pi^2}{L^2} + s^2\right)\cos\frac{\pi x}{L} - B\left(\frac{c^2\pi^2}{L^2} + s^2\right)\sin\frac{\pi x}{L} = -sy_0\sin\frac{\pi x}{L},$$

implying that $A = 0$ and $B = sy_0/[s^2 + (c^2\pi^2/L^2)]$. Thus the general solution of Equation **(17)** is

$$Y = ae^{sx/c} + be^{-sx/c} + \frac{sy_0}{s^2 + \left(\frac{c^2\pi^2}{L^2}\right)} \sin\frac{\pi x}{L}. \tag{18}$$

Setting $x = 0$ and $x = L$ in **(18),** we obtain the homogeneous simultaneous equations

$$\begin{aligned} a \quad\; + b \quad\; &= 0, \\ ae^{sL/c} + be^{-sL/c} &= 0. \end{aligned} \tag{19}$$

Since the determinant of the coefficients on the left-hand side of **(19)** is nonzero, $a = b = 0$ and **(18)** reduces to

$$\mathcal{L}\{y\} = Y = \frac{sy_0}{s^2 + \left(\frac{c^2\pi^2}{L^2}\right)} \sin\frac{\pi x}{L}.$$

Solving for y, we obtain

$$y = y_0 \sin\frac{\pi x}{L} \cos\frac{\pi ct}{L},$$

which agrees with the solution of Example 12.5.2, which was obtained by d'Alembert's method. ◄

SELF-QUIZ

I. If we solve the problem $\partial^2 y/\partial t^2 = 4\,\partial^2 y/\partial x^2$, $y(0, t) = y(1, t) = 0$, $y(x, 0) = \cos x$, $\partial y/\partial t\,(x, 0) = 0$, and if Y denotes the Laplace transform of the solution, then Y satisfies the differential equation ______.

(a) $4Y'' - s^2Y = -s\cos x$

(b) $2Y'' - s^2Y = -s\cos x$

(c) $4Y'' - s^2Y = -s\left(\frac{s}{s^2+1}\right)$

(d) $2Y'' - s^2Y = -s\left(\frac{s}{s^2+1}\right)$

$\left[\textit{Hint}: \mathcal{L}\{\cos t\} = \frac{s}{s^2+1}.\right]$

Answer to Self-Quiz

I. a

PROBLEMS 12.10

In Problems 1–6 obtain the solution of the given boundary and initial condition problem by the method of Laplace transforms. Be sure to verify your solution.

1. $\frac{\partial^2 y}{\partial x^2} = 4\frac{\partial^2 y}{\partial t^2}$, $t > 0$, $x > 0$

$y(x, 0) = 0$, $\left.\frac{\partial y}{\partial t}\right|_{(x,0)} = 1$, $x \geq 0$

$y(0, t) = t$, $\lim_{x\to\infty} y(x, t)$ is finite

2. $\frac{\partial^2 y}{\partial x^2} = 16\frac{\partial^2 y}{\partial t^2}$, $t > 0$, $x > 0$

$y(x, 0) = 0$, $\left.\frac{\partial y}{\partial t}\right|_{(x,0)} = 1$, $x \geq 0$

$y(0, t) = \sin t$, $\lim_{x\to\infty} y(x, t)$ is finite

3. $\frac{\partial T}{\partial t} = \delta\frac{\partial^2 T}{\partial x^2}$, $t > 0$, $x > 0$

$T(x, 0) = 0$, $0 \leq x < \infty$, $T(0, t) = 1$, $t > 0$

$\lim_{x\to\infty} T(x, t) = \lim_{x\to\infty} \frac{\partial T}{\partial x}(x, t) = 0$, $t > 0$

4. $\frac{\partial T}{\partial t} = \delta\frac{\partial^2 T}{\partial x^2}$, $t > 0$, $x > 0$

$T(x, 0) = e^{-x}$, $x \geq 0$, $T(0, t) = T_0(\neq 0)$

$\lim_{x\to\infty} T(x, t) = \lim_{x\to\infty} \frac{\partial T}{\partial x}(x, t) = 0$, $t > 0$

5. $\frac{\partial T}{\partial t} = \delta \frac{\partial^2 T}{\partial x^2}, t > 0, x > 0$
$T(x, 0) = 0, x \geq 0,$
$T(0, t) = \sin \omega t$

$$\lim_{x\to\infty} T(x, t) = \lim_{x\to\infty} \frac{\partial T}{\partial x}(x, t) = 0, t > 0$$

6. $\frac{\partial T}{\partial t} = \delta \frac{\partial^2 T}{\partial x^2} + \mu \frac{\partial T}{\partial x}, t > 0, x > 0$
$T(x, 0) = 0, x \geq 0$
$T(0, t) = T_0(\neq 0)$

$$\lim_{x\to\infty} T(x, t) = \lim_{x\to\infty} \frac{\partial T}{\partial x}(x, t) = 0, t > 0$$

12.11 SOLVING PARTIAL DIFFERENTIAL EQUATIONS WITH A SYMBOLIC MANIPULATOR

We now demonstrate how MATHEMATICA, MAPLE, DERIVE, MATLAB, or MATHCAD may be used to solve partial differential equations. Since our techniques for solving partial differential equations involve replacing the partial differential equation by two associated ordinary differential equations, we have already developed all the scripts we need. Two examples illustrate how to use the scripts we developed in previous chapters.

EXAMPLE 1 ▶

Solve the equation

$$xyz_x + (x^2 + y^2)z_y = yz, \qquad z(x, 0) = xe^x.$$

Solution From Section 12.2 we have $dx/xy = dy/(x^2 + y^2) = dz/yz$ so that we obtain two differential equations

$$\frac{dy}{dx} = \frac{x^2 + y^2}{xy} \quad \text{and} \quad \frac{dz}{z} = \frac{dx}{x}.$$

The second equation is trivial: its solution is $z = c_2 x$. To solve the first we note that it is homogeneous of degree zero so we give the command

```
HOMOGENEOUS(x, y,(x^2+y^2)/(x*y), c)
```

and obtain the solution

$$\left(\frac{y}{x}\right)^2 = 2 \ln x + c_1.$$

Now give the commands

```
SUBST((y/x)^2=2*log(x)+c₁, x=t, y=0)
SUBST(z=c₂*x, x=t, z=t*exp(t))
```

obtaining

$$0 = 2 \log(t) + c_1 \quad \text{and} \quad \exp(t) = c_2 \quad \text{or} \quad c_2 = \exp(\exp(-c_1/2)).$$

Hence,

$$\frac{z}{x} = c_2 = e^{e^{-c_1/2}} = e^{e^{\ln(x)-(y^2/x^2)/2}} \quad \text{or} \quad z = x\, e^{x e^{-y^2/(2x^2)}}.$$

Checking that this answer is correct is easy: define the variable z as above and compute

```
x*y*DIF(z, x)+(x^2+y^2)*DIF(z, y) - y*z.
```

If the answer is zero, the expression for z is the correct solution. ◀

EXAMPLE 2 ▶ Solve the heat equation problem

$$y_t = y_{xx}, \qquad y(0, t) = y(L, t) = 0, \qquad y(x, 0) = x(L - x).$$

Solution Separating variables we obtain the two differential equations

$$T' + kT = 0, \qquad T(0) = 1,$$
$$X'' + kX = 0, \qquad X(0) = X(L) = 0.$$

The first has the solution $T(t) = \exp(-kt)$. The second has nontrivial solutions only when $\sqrt{k} = \frac{n\pi}{L}$. So the eigenvalues are $k = n^2\pi^2/L^2$, $n = 1, 2, \ldots$ and the associated eigenfunctions are $X_n(x) = \sin \frac{n\pi x}{L}$. Thus, the solution of the problem has the form

$$y(x, t) = \sum_{n=1}^{\infty} b_n \sin \frac{n\pi x}{L} e^{-n^2\pi^2 t/L^2}.$$

Setting $t = 0$, we need to find a Fourier sine series of the function $f(x) = x(L - x)$ extended to the interval $[-L, L]$. We must extend $f(x)$ as an odd function, so

$$f(x) = \begin{cases} x(L - x), & 0 < x < L, \\ x(L + x), & -L < x < 0, \end{cases}$$

or

$$f(x) = x(L + x)(1 - H(x)) + x(L - x)H(x) = x(L + x(1 - 2H(x))).$$

We shall calculate the first four terms hoping to discover the form of the coefficients b_n. Thus, we give the command

```
FOURIER(x*(L+x*(1 - 2*H(x))), x, L, 4)
```

and get

```
L^2*[(1/2*π))*sin(4*π*x/L)
    +(2*(4 - 9*π^2)/(27*π^3))*sin(3*π*x/L)
    +(1/π)*sin(2*π*x/L)+(2*(4 - π^2)/π^3)*sin(π*x/L)]
```

This suggests that the complete answer is

$$y(x, t) = \frac{8L^2}{\pi^3} \sum_{n=1}^{\infty} \frac{1}{(2n - 1)^3} \sin \frac{(2n - 1)\pi x}{L} e^{-(2n-1)^2\pi^2 t/L^2}$$
$$+ \frac{2L^2}{\pi} \sum_{n=1}^{\infty} \frac{(-1)^n}{n} \sin \frac{n\pi x}{L} e^{-n^2\pi^2 t/L^2}.$$ ◀

PROBLEMS 12.11

Solve each of the following problems using MATHEMATICA, MAPLE, DERIVE, MATLAB, *or* MATHCAD.

1. $xyz_x + (x^2 - y^2)z_y = xy$, $z(x, 0) = x^2$

2. $xyz_x + (x^2 - xy)z_y = yz$, $z(0, y) = e^y$

3. $zz_x + 2zz_y = y^2$, $z(x, 1) = e^x$

4. $xz_x + zz_y = 2yz$, $z(1, y) = y^2$

5. Solve the heat equation

$$y_t = 4y_{xx}, \qquad y(0, t) = y(2, t) = 0,$$
$$y(x, 0) = 1 - \cos \pi x.$$

6. Solve the vibrating string problem

$$y_{tt} = 4y_{xx}, \qquad y(0, t) = y(1, t) = 0,$$
$$y(x, 0) = \cos x, \qquad y_t(x, 0) = 0.$$

SUMMARY OUTLINE OF CHAPTER 12

Partial differential equation (PDE): an equation involving a function of two or more variables and its partial derivatives p. 577
Linear: $a(x, y)z_x + b(x, y)z_y + c(x, y)z = f(x, y)$ p. 577

Rotation of Axes Method: set $\tan\alpha = b/a$ and let $u = x\cos\alpha + y\sin\alpha$, $v = -x\sin\alpha + y\cos\alpha$ p. 577

Cauchy problem: first degree PDE with line of initial conditions p. 579

Integral surface: solution to a Cauchy problem p. 579

Characteristic: a curve on the integral surface p. 580

Method of characteristics: given $az_x + bz_y = g$, then $\frac{dx}{a} = \frac{dy}{b} = \frac{dz}{g}$ pp. 580, 581

Quasi-linear equation: $a(x, y, z)z_x + b(x, y, z,)z_y = g(x, y, z)$ p. 584

Shock: discontinuity in the solution p. 591

Second-order PDE: $aw_{xx} + 2bw_{xy} + cw_{yy} + kw_x + mw_y + nw = f$ pp. 593, 595

if $b^2 - ac < 0$, then the equation is elliptic

if $b^2 - ac = 0$, then the equation is parabolic

if $b^2 - ac > 0$, then the equation is hyperbolic

Standard forms of second-order equations: pp. 594–596
Rotation of Axes: $\tan 2\alpha = \frac{2b}{a - c}$

Degenerate equation p. 596

Wave equation: $y_{tt} = c^2 y_{xx}$ p. 600
d'Alembert's method: $y(x, t) = \frac{1}{2}[f(x + ct) + f(x - ct)]$ satisfies wave equation with conditions $y(0, t) = y(L, t) = 0$, $y(x, 0) = f(x)$, $y_t(x, 0) = 0$ p. 601

Traveling wave: two-dimensional functions of the form $f(x \pm ct)$ p. 601

Separation of variables: Substitute $y(x, t) = X(x)T(t)$ into problem then separate variables and solve resulting ODE's. Obtain an eigenfunction expansion, set $t = 0$, and solve Fourier series for $f(x)$. May need to expand $f(x)$ to an odd (or even) function. pp. 605–607

Heat equation: $T_t = \delta T_{xx}$ (here δ is diffusivity) p. 611
Two-dimensional: $T_t = \delta[T_{xx} + T_{yy}] = \delta\,\Delta T$ p. 615

Laplace's equation: $\Delta T = 0$ p. 615
In polar coordinates: $T_{rr} + r^{-2}T_{\theta\theta} + r^{-1}T_r = 0$ p. 619
Poisson integral formula, p. 621
In cylindrical coordinates: $T_{rr} + r^{-2}T_{\theta\theta} + r^{-1}T_r + T_{zz} = 0$ p. 622

Maximum principle, p. 625

Diffusion equation: $c\rho\, T_t = (\kappa T_x)_x - vT_x$ p. 626
Pure convection if $\kappa = 0$
Pure diffusion if $v = 0$

REVIEW EXERCISES FOR CHAPTER 12

In Exercises 1–4 find the solution $z(x, y)$ of each partial differential equation that satisfies the given condition.

1. $z_x + z_y - z = e^y$, $z(x, 0) = x$

2. $z_x + z_y - 2z = 0$, $z(x, 0) = e^x$

3. $z_x + z_y - z = 1 - x$, $z(x, 0) = x + 1$

4. $z_x - z_y + z = 0$, $z(x, 0) = \sin x$

In Exercises 5–8 find the temperature in an insulated rod a units long whose ends are kept at 0° C, where the initial temperature is given by

5. $f(x) = x(a - x)$

6. $f(x) = x^2(a^2 - x^2)$

7. $f(x) = \begin{cases} x, & 0 \le x \le \frac{a}{2}, \\ a - x, & \frac{a}{2} \le x \le a \end{cases}$

8. $f(x) = \begin{cases} x^2, & 0 \le x \le \frac{a}{2}, \\ 0, & \frac{a}{2} < x \le a \end{cases}$

Find solutions of the equations in Exercises 9–12 by separation of variables.

9. $\dfrac{\partial^2 y}{\partial x^2} = \dfrac{t^2}{x^2}\dfrac{\partial^2 y}{\partial t^2}$

10. $\dfrac{\partial^2 y}{\partial x^2} + \dfrac{\partial^2 y}{\partial x\,\partial t} = \dfrac{\partial y}{\partial t}$

11. $\dfrac{\partial^2 y}{\partial t^2} = \dfrac{\partial^2 y}{\partial x^2} - y$

12. $\dfrac{\partial^2 y}{\partial x^2} + \dfrac{\partial y}{\partial x} = \dfrac{\partial y}{\partial t}$

13. A tightly stretched string fixed at $x = 0$ and $x = a$ is initially displaced to the position $y(x, 0) = 2\sin^3(\frac{\pi x}{a})$. At time $t = 0$, it is released from rest from this position. Find the displacement of any point on the string at any time t.

14. Repeat Exercise 13 if the initial displacement is $y(x, 0) = a^2x - x^3$.

15. The vibrating string in Exercise 13 is subjected to a damping force that is proportional to the velocity at each point and instant in time.
(a) Find the differential equation that the damped string satisfies.
(b) Solve the equation by separation of variables.
(c) If the initial displacement is $y(x, 0) = f(x)$, $0 \le x \le a$, express the solution as an infinite series.

16. A rectangular plate is bounded by the lines $x = 0$, $y = 0$, $x = 100$, $y = 50$. Its surfaces are insulated and the temperature along the upper edge is given by $T(x, 50) = x(100 - x)$, while the other edges are kept at 0°C. Find the steady-state temperature.

APPENDIX 1 Integral Tables*

Standard forms

1. $\int a\,dx = ax + C$

2. $\int af(x)\,dx = a\int f(x)\,dx + C$

3. $\int u\,dv = uv - \int v\,du$ (integration by parts)

4. $\int u^n\,du = \dfrac{u^{n+1}}{n+1} + C, \quad n \neq -1$

5. $\int \dfrac{du}{u} = \ln u \quad \text{if } u > 0 \quad \text{or} \quad \ln(-u) \quad \text{if } u < 0$

$= \ln|u| + C$

6. $\int e^u\,du = e^u + C$

7. $\int a^u\,du = \int e^{u\ln a}\,du$

$= \dfrac{e^{u\ln a}}{\ln a} = \dfrac{a^u}{\ln a} + C, \quad a > 0, a \neq 1$

8. $\int \sin u\,du = -\cos u + C$

9. $\int \cos u\,du = \sin u + C$

10. $\int \tan u\,du = \ln|\sec u| = -\ln|\cos u| + C$

11. $\int \cot u\,du = \ln|\sin u| + C$

12. $\int \sec u\,du = \ln|\sec u + \tan u|$

$= \ln \tan\left|\dfrac{u}{2} + \dfrac{\pi}{4}\right| + C$

13. $\int \csc u\,du = \ln|\csc u - \cot u| = \ln\left|\tan\dfrac{u}{2}\right| + C$

14. $\int \sec^2 u\,du = \tan u + C$

15. $\int \csc^2 u\,du = -\cot u + C$

16. $\int \sec u \tan u\,du = \sec u + C$

17. $\int \csc u \cot u\,du = -\csc u + C$

18. $\int \dfrac{du}{u^2 + a^2} = \dfrac{1}{a}\tan^{-1}\dfrac{u}{a} + C$

19. $\int \dfrac{du}{u^2 - a^2} = \dfrac{1}{2a}\ln\left|\dfrac{u-a}{u+a}\right| + C$

$= -\dfrac{1}{a}\coth^{-1}\dfrac{u}{a} + C, \quad u^2 > a^2$

20. $\int \dfrac{du}{a^2 - u^2} = \dfrac{1}{2a}\ln\left|\dfrac{a+u}{a-u}\right| + C$

$= \dfrac{1}{a}\tanh^{-1}\dfrac{u}{a} + C, \quad u^2 < a^2$

21. $\int \dfrac{du}{\sqrt{a^2 - u^2}} = \sin^{-1}\dfrac{u}{|a|} + C$

22. $\int \dfrac{du}{\sqrt{u^2 + a^2}} = \ln(u + \sqrt{u^2 + a^2}) + C$

*All angles are measured in radians.

23. $\displaystyle\int \frac{du}{\sqrt{u^2 - a^2}} = \ln|u + \sqrt{u^2 - a^2}| + C$

24. $\displaystyle\int \frac{du}{u\sqrt{u^2 - a^2}} = \frac{1}{|a|}\sec^{-1}\left|\frac{u}{a}\right| + C$

25. $\displaystyle\int \frac{du}{u\sqrt{u^2 + a^2}} = -\frac{1}{a}\ln\left|\frac{a + \sqrt{u^2 + a^2}}{u}\right| + C$

26. $\displaystyle\int \frac{du}{u\sqrt{a^2 - u^2}} = -\frac{1}{a}\ln\left|\frac{a + \sqrt{a^2 - u^2}}{u}\right| + C$

Integrals involving $au + b$

27. $\displaystyle\int \frac{du}{au + b} = \frac{1}{a}\ln|au + b| + C$

28. $\displaystyle\int \frac{u\,du}{au + b} = \frac{u}{a} - \frac{b}{a^2}\ln|au + b| + C$

29. $\displaystyle\int \frac{u^2\,du}{au + b} = \frac{(au + b)^2}{2a^3} - \frac{2b(au + b)}{a^3} + \frac{b^2}{a^3}\ln|au + b| + C$

30. $\displaystyle\int \frac{du}{u(au + b)} = \frac{1}{b}\ln\left|\frac{u}{au + b}\right| + C$

31. $\displaystyle\int \frac{du}{u^2(au + b)} = -\frac{1}{bu} + \frac{a}{b^2}\ln\left|\frac{au + b}{u}\right| + C$

32. $\displaystyle\int \frac{du}{(au + b)^2} = \frac{-1}{a(au + b)} + C$

33. $\displaystyle\int \frac{du}{(au + b)^2} = \frac{b}{a^2(au + b)} + \frac{1}{a^2}\ln|au + b| + C$

34. $\displaystyle\int \frac{du}{u(au + b)^2} = \frac{1}{b(au + b)} + \frac{1}{b^2}\ln\left|\frac{u}{au + b}\right| + C$

35. $\displaystyle\int (au + b)^n\,du = \frac{(au + b)^{n-1}}{(n + 1)a} + C, \quad n \neq -1$

36. $\displaystyle\int u(au + b)^n\,du = \frac{(au + b)^{n+2}}{(n + 2)a^2} - \frac{b(au + b)^{n-1}}{(n + 1)a^2} + C, \quad n \neq -1, -2$

37. $$\int u^m(au + b)^n\,du = \begin{cases} \dfrac{u^{m+1}(au + b)^n}{m + n + 1} + \dfrac{nb}{m + n + 1}\displaystyle\int u^m(au + b)^{n-1}\,du \\[2ex] \dfrac{u^m(au+b)^{n+1}}{(m + n + 1)a} - \dfrac{mb}{(m + n + 1)a}\displaystyle\int u^{m-1}(au + b)^n\,du \\[2ex] \dfrac{-u^{m+1}(au + b)^{n+1}}{(n + 1)b} + \dfrac{m + n + 2}{(n + 1)b}\displaystyle\int u^m(au + b)^{n+1}\,du \end{cases}$$

Integrals involving $\sqrt{au + b}$

38. $\displaystyle\int \frac{du}{\sqrt{au + b}} = \frac{2\sqrt{au + b}}{a} + C$

39. $\displaystyle\int \frac{u\,du}{\sqrt{au + b}} = \frac{2(au - 2b)}{3a^2}\sqrt{au + b} + C$

40. $$\int \frac{du}{u\sqrt{au + b}} = \begin{cases} \dfrac{1}{\sqrt{b}}\ln\left|\dfrac{\sqrt{au + b} - \sqrt{b}}{\sqrt{au + b} + \sqrt{b}}\right| + C, & b > 0 \\[2ex] \dfrac{2}{\sqrt{-b}}\tan^{-1}\sqrt{\dfrac{au + b}{-b}} + C, & b < 0 \end{cases}$$

41. $\displaystyle\int \sqrt{au + b}\,du = \frac{2\sqrt{(au + b)^3}}{3a} + C$

42. $\displaystyle\int u\sqrt{au + b}\,du = \frac{2(3au - 2b)}{15a^2}\sqrt{(au + b)^3} + C$

43. $\displaystyle\int \frac{\sqrt{au + b}}{u}\,du = 2\sqrt{au + b} + b\int \frac{du}{u\sqrt{au + b}}$ (See 40.)

Integrals involving $u^2 + a^2$

44. $\displaystyle\int \frac{du}{u^2 + a^2} = \frac{1}{a}\tan^{-1}\frac{u}{a} + C$

45. $\displaystyle\int \frac{u\,du}{u^2 + a^2} = \frac{1}{2}\ln(u^2 + a^2) + C$

46. $\displaystyle\int \frac{u^2\,du}{u^2 + a^2} = u - a\tan^{-1}\frac{u}{a} + C$

47. $\displaystyle\int \frac{du}{u(u^2 + a^2)} = \frac{1}{2a^2}\ln\left(\frac{u^2}{u^2 + a^2}\right) + C$

48. $\displaystyle\int \frac{du}{u^2(u^2+a^2)} = -\frac{1}{a^2u} - \frac{1}{a^3}\tan^{-1}\frac{u}{a} + C$

49. $\displaystyle\int \frac{du}{(u^2+a^2)^n} = \frac{u}{2(n-1)a^2(u^2+a^2)^{n-1}} + \frac{2n-3}{(2n-2)a^2}\int \frac{du}{(u^2+a^2)^{n-1}}$

50. $\displaystyle\int \frac{u\,du}{(u^2+a^2)^n} = \frac{-1}{2(n-1)(u^2+a^2)^{n-1}} + C, \quad n \neq 1$

51. $\displaystyle\int \frac{du}{u(u^2+a^2)^n} = \frac{1}{2(n-1)a^2(u^2+a^2)^{n-1}} + \frac{1}{a^2}\int \frac{du}{u(u^2+a^2)^{n-1}}, \quad n \neq 1$

Integrals involving $u^2 - a^2$, $u^2 > a^2$

52. $\displaystyle\int \frac{du}{u^2-a^2} = \frac{1}{2a}\ln\left|\frac{u-a}{u+a}\right| + C$

53. $\displaystyle\int \frac{u\,du}{u^2-a^2} = \frac{1}{2}\ln(u^2-a^2) + C$

54. $\displaystyle\int \frac{u^2\,du}{u^2-a^2} = u + \frac{a}{2}\ln\left|\frac{u-a}{u+a}\right| + C$

55. $\displaystyle\int \frac{du}{u(u^2-a^2)} = \frac{1}{2a^2}\ln\left|\frac{u^2-a^2}{u^2}\right| + C$

56. $\displaystyle\int \frac{du}{u^2(u^2-a^2)} = \frac{1}{a^2u} + \frac{1}{2a^3}\ln\left|\frac{u-a}{u+a}\right| + C$

57. $\displaystyle\int \frac{du}{(u^2-a^2)^2} = \frac{-u}{2a^2(u^2-a^2)} - \frac{1}{4a^3}\ln\left|\frac{u-a}{u+a}\right| + C$

58. $\displaystyle\int \frac{du}{(u^2-a^2)^n} = \frac{-u}{2(n-1)a^2(u^2-a^2)^{n-1}} - \frac{2n-3}{(2n-2)a^2}\int \frac{du}{(u^2-a^2)^{n-1}}, \quad n \neq 1$

59. $\displaystyle\int \frac{u\,du}{(u^2-a^2)^n} = \frac{-1}{2(n-1)(u^2-a^2)^{n-1}} + C, \quad n \neq 1$

60. $\displaystyle\int \frac{du}{u(u^2-a^2)^n} = \frac{-1}{2(n-1)a^2(u^2-a^2)^{n-1}} - \frac{1}{a^2}\int \frac{du}{u(u^2-a^2)^{n-1}}, \quad n \neq 1$

Integrals involving $a^2 - u^2$, $u^2 < a^2$

61. $\displaystyle\int \frac{du}{a^2-u^2} = \frac{1}{2a}\ln\left|\frac{a+u}{a-u}\right| + C = \frac{1}{a}\tanh^{-1}\frac{u}{a} + C$

62. $\displaystyle\int \frac{u\,du}{a^2-u^2} = -\frac{1}{2}\ln|a^2-u^2| + C$

63. $\displaystyle\int \frac{u^2\,du}{a^2-u^2} = -u + \frac{a}{2}\ln\left|\frac{a+u}{a-u}\right| + C$

64. $\displaystyle\int \frac{du}{u(a^2-u^2)} = \frac{1}{2a^2}\ln\left|\frac{u^2}{a^2-u^2}\right| + C$

65. $\displaystyle\int \frac{du}{(a^2-u^2)^2} = \frac{u}{2a^2(a^2-u^2)} + \frac{1}{4a^3}\ln\left|\frac{a+u}{a-u}\right| + C$

66. $\displaystyle\int \frac{u\,du}{(a^2-u^2)^2} = \frac{1}{2(a^2-u^2)} + C$

Integrals involving $\sqrt{u^2 + a^2}$

67. $\displaystyle\int \frac{du}{\sqrt{u^2+a^2}} = \ln(u + \sqrt{u^2+a^2}) + C = \sinh^{-1}\frac{u}{|a|} + C$

68. $\displaystyle\int \frac{u\,du}{\sqrt{u^2+a^2}} = \sqrt{u^2+a^2} + C$

69. $\displaystyle\int \frac{u^2\,du}{\sqrt{u^2+a^2}} = \frac{u\sqrt{u^2+a^2}}{2} - \frac{a^2}{2}\ln(u + \sqrt{u^2+a^2}) + C$

70. $\displaystyle\int \frac{du}{u\sqrt{u^2+a^2}} = -\frac{1}{a}\ln\left|\frac{a+\sqrt{u^2+a^2}}{u}\right| + C$

71. $\int \sqrt{u^2 + a^2}\, du = \frac{u\sqrt{u^2 + a^2}}{2} + \frac{a^2}{2} \ln(u + \sqrt{u^2 + a^2}) + C$

72. $\int u\sqrt{u^2 + a^2}\, du = \frac{(u^2 + a^2)^{3/2}}{3} + C$

73. $\int u^2\sqrt{u^2 + a^2}\, du = \frac{u(u^2 + a^2)^{3/2}}{4} - \frac{a^2 u\sqrt{u^2 + a^2}}{8} - \frac{a^4}{8} \ln(u + \sqrt{u^2 + a^2}) + C$

74. $\int \frac{\sqrt{u^2 + a^2}}{u}\, du = \sqrt{u^2 + a^2} - a \ln\left|\frac{a + \sqrt{u^2 + a^2}}{u}\right| + C$

75. $\int \frac{\sqrt{u^2 + a^2}}{u^2}\, du = -\frac{\sqrt{u^2 + a^2}}{u} + \ln(u + \sqrt{u^2 + a^2}) + C$

Integrals involving $\sqrt{u^2 - a^2}$

76. $\int \frac{du}{\sqrt{u^2 - a^2}} = \ln|u + \sqrt{u^2 - a^2}| + C$

77. $\int \frac{u\, du}{\sqrt{u^2 - a^2}} = \sqrt{u^2 - a^2} + C$

78. $\int \frac{u^2\, du}{\sqrt{u^2 - a^2}} = \frac{u\sqrt{u^2 - a^2}}{2} + \frac{a^2}{2} \ln|u + \sqrt{u^2 - a^2}| + C$

79. $\int \frac{du}{u\sqrt{u^2 - a^2}} = \frac{1}{|a|} \sec^{-1}\left|\frac{u}{a}\right| + C$

80. $\int \sqrt{u^2 - a^2}\, du = \frac{u\sqrt{u^2 - a^2}}{2} - \frac{a^2}{2} \ln|u + \sqrt{u^2 - a^2}| + C$

81. $\int u\sqrt{u^2 - a^2}\, du = \frac{(u^2 - a^2)^{3/2}}{3} + C$

82. $\int u^2\sqrt{u^2 - a^2}\, du = \frac{u(u^2 - a^2)^{3/2}}{4} + \frac{a^2 u\sqrt{u^2 - a^2}}{8} - \frac{a^4}{8} \ln|u + \sqrt{u^2 - a^2}| + C$

83. $\int \frac{\sqrt{u^2 - a^2}}{u}\, du = \sqrt{u^2 - a^2} - |a| \sec^{-1}\left|\frac{u}{a}\right| + C$

84. $\int \frac{\sqrt{u^2 - a^2}}{u^2}\, du = -\frac{\sqrt{u^2 - a^2}}{u} + \ln|u + \sqrt{u^2 - a^2}| + C$

85. $\int \frac{du}{(u^2 - a^2)^{3/2}} = -\frac{u}{a^2\sqrt{u^2 - a^2}} + C$

Integrals involving $\sqrt{a^2 - u^2}$

86. $\int \frac{du}{\sqrt{a^2 - u^2}} = \sin^{-1} \frac{u}{|a|} + C$

87. $\int \frac{u\, du}{\sqrt{a^2 - u^2}} = -\sqrt{a^2 - u^2} + C$

88. $\int \frac{u^2\, du}{\sqrt{a^2 - u^2}} = -\frac{u\sqrt{a^2 - u^2}}{2} + \frac{a^2}{2} \sin^{-1} \frac{u}{|a|} + C$

89. $\int \frac{du}{u\sqrt{a^2 - u^2}} = -\frac{1}{a} \ln\left|\frac{a + \sqrt{a^2 - u^2}}{u}\right| + C$

90. $\int \frac{du}{u^2\sqrt{a^2 - u^2}} = -\frac{\sqrt{a^2 - u^2}}{a^2 u} + C$

91. $\int \sqrt{a^2 - u^2}\, du = \frac{u\sqrt{a^2 - u^2}}{2} + \frac{a^2}{2} \sin^{-1} \frac{u}{|a|} + C$

92. $\int u\sqrt{a^2 - u^2}\, du = -\frac{(a^2 - u^2)^{3/2}}{3} + C$

93. $\int u^2\sqrt{a^2 - u^2}\, du = -\frac{u(a^2 - u^2)^{3/2}}{4} + \frac{a^2 u\sqrt{a^2 - u^2}}{8} + \frac{a^4}{8} \sin^{-1} \frac{u}{|a|} + C$

94. $\int \frac{\sqrt{a^2 - u^2}}{u}\, du = \sqrt{a^2 - u^2} - a \ln\left|\frac{a + \sqrt{a^2 - u^2}}{u}\right| + C$

95. $\int \frac{\sqrt{a^2 - u^2}}{u^2}\, du = -\frac{\sqrt{a^2 - u^2}}{u} - \sin^{-1} \frac{u}{|a|} + C$

Integrals involving the trigonometric functions

96. $\int \sin au\, du = -\frac{\cos au}{a} + C$

97. $\int u \sin au\, du = \frac{\sin au}{a^2} - \frac{u \cos au}{a} + C$

98. $\int u^2 \sin au\, du = \frac{2u}{a^2} \sin au + \left(\frac{2}{a^3} - \frac{u^2}{a}\right)\cos au + C$

99. $\int \frac{du}{\sin au} = \frac{1}{a} \ln|\csc au - \cot au|$
$= \frac{1}{a} \ln\left|\tan \frac{au}{2}\right| + C$

100. $\int \sin^2 au\, du = \frac{u}{2} - \frac{\sin 2au}{4a} + C$

101. $\int u \sin^2 au\, du = \frac{u^2}{4} - \frac{u \sin 2au}{4a} - \frac{\cos 2au}{8a^2} + C$

102. $\int \frac{du}{\sin^2 au} = -\frac{1}{a} \cot au + C$

103. $\int \sin pu \sin qu\, du = \frac{\sin(p - q)u}{2(p - q)} - \frac{\sin(p + q)u}{2(p + q)} + C, \quad p \neq \pm q$

104. $\int \frac{du}{1 - \sin au} = \frac{1}{a} \tan\left(\frac{\pi}{4} + \frac{au}{2}\right) + C$

105. $\int \frac{u\, du}{1 - \sin au} = \frac{u}{a} \tan\left(\frac{\pi}{4} + \frac{au}{2}\right) + \frac{2}{a^2} \ln\left|\sin\left(\frac{\pi}{4} - \frac{au}{2}\right)\right| + C$

106. $\int \frac{du}{1 + \sin au} = -\frac{1}{a} \tan\left(\frac{\pi}{4} - \frac{au}{2}\right) + C$

107. $\int \frac{du}{p + q \sin au} = \begin{cases} \frac{2}{a\sqrt{p^2 - q^2}} \tan^{-1}\left[\sqrt{\frac{p + q}{p - q}} \tan\left(\frac{1}{2}au + \frac{\pi}{4}\right)\right] \\ \frac{1}{a\sqrt{q^2 - p^2}} \ln\left|\frac{p \tan \frac{1}{2}au + q - \sqrt{q^2 - p^2}}{p \tan \frac{1}{2}au + q + \sqrt{q^2 - p^2}}\right| \end{cases} + C, \quad |p| < |q|$

108. $\int u^m \sin au\, du = -\frac{u^m \cos au}{a} + \frac{mu^{m-1} \sin au}{a^2} - \frac{m(m - 1)}{a^2} \int u^{m-2} \sin au\, du$

109. $\int \sin^n au\, du = -\frac{\sin^{n-1} au \cos au}{an} + \frac{n - 1}{n} \int \sin^{n-2} au\, du$

110. $\int \frac{du}{\sin^n au} = \frac{-\cos au}{a(n - 1)\sin^{n-1} au} + \frac{n - 2}{n - 1} \int \frac{du}{\sin^{n-2} au}, \quad n \neq 1$

111. $\int \cos au\, du = \frac{\sin au}{a} + C$

112. $\int u \cos au\, du = \frac{\cos au}{a^2} + \frac{u \sin au}{a} + C$

113. $\int u^2 \cos au\, du = \frac{2u}{a^2} \cos au + \left(\frac{u^2}{a} - \frac{2}{a^3}\right)\sin au + C$

114. $\int \frac{du}{\cos au} = \frac{1}{a} \ln|\sec au + \tan au| = \frac{1}{a} \ln\left|\tan\left(\frac{\pi}{4} + \frac{au}{2}\right)\right| + C$

115. $\int \cos^2 au\, du = \frac{u}{2} + \frac{\sin 2au}{4a} + C$

116. $\int u \cos^2 au\, du = \frac{u^2}{4} + \frac{u \sin 2au}{4a} + \frac{\cos 2au}{8a^2} + C$

117. $\int \frac{du}{\cos^2 au} = \frac{\tan au}{a} + C$

118. $\int \cos qu \cos pu\, du = \frac{\sin(q - p)u}{2(q - p)} + \frac{\sin(q + p)u}{2(q + p)} + C, \quad q \neq \pm p$

119. $$\int \frac{du}{p + q\cos au} = \begin{cases} \dfrac{2}{a\sqrt{p^2 - q^2}} \tan^{-1}\left[\sqrt{\dfrac{p-q}{p+q}} \tan \frac{1}{2}au\right] + C, & |p| > |q| \\[2ex] \dfrac{1}{a\sqrt{q^2 - p^2}} \ln\left[\dfrac{\tan \frac{1}{2}au + \sqrt{\dfrac{q+p}{q-p}}}{\tan \frac{1}{2}au - \sqrt{\dfrac{q+p}{q-p}}}\right] + C, & |p| < |q| \end{cases}$$

120. $$\int u^m \cos au\, du = \frac{u^m \sin au}{a} + \frac{mu^{m-1}}{a^2}\cos au - \frac{m(m-1)}{a^2}\int u^{m-2}\cos au\, du$$

121. $$\int \cos^n au\, du = \frac{\sin au \cos^{n-1} au}{an} + \frac{n-1}{n}\int \cos^{n-2} au\, du$$

122. $$\int \frac{du}{\cos^n au} = \frac{\sin au}{a(n-1)\cos^{n-1} au} + \frac{n-2}{n-1}\int \frac{du}{\cos^{n-2} au}, \quad n \neq 1$$

123. $$\int \sin au \cos au\, du = \frac{\sin^2 au}{2a} + C$$

124. $$\int \sin pu \cos qu\, du = -\frac{\cos(p-q)u}{2(p-q)} - \frac{\cos(p+q)u}{2(p+q)} + C, \quad p \neq \pm q$$

125. $$\int \sin^n au \cos au\, du = \frac{\sin^{n+1} au}{(n+1)a} + C, \quad n \neq -1$$

126. $$\int \cos^n au \sin au\, du = -\frac{\cos^{n+1} au}{(n+1)a} + C, \quad n \neq -1$$

127. $$\int \sin^2 au \cos^2 au\, du = \frac{u}{8} - \frac{\sin 4au}{32a} + C$$

128. $$\int \frac{du}{\sin au \cos au} = \frac{1}{a}\ln|\tan au| + C$$

129. $$\int \frac{du}{\cos au(1 \pm \sin au)} = \mp\frac{1}{2a(1 \pm \sin au)} + \frac{1}{2a}\ln\left|\tan\left(\frac{au}{2} + \frac{\pi}{4}\right)\right| + C$$

130. $$\int \frac{du}{\sin au(1 \pm \cos au)} = \pm\frac{1}{2a(1 \pm \cos au)} + \frac{1}{2a}\ln\left|\tan\frac{au}{2}\right| + C$$

131. $$\int \frac{du}{\sin au \pm \cos au)} = \frac{1}{a\sqrt{2}}\ln\left|\tan\left(\frac{au}{2} \pm \frac{\pi}{8}\right)\right| + C$$

132. $$\int \frac{\sin au\, du}{\sin au \pm \cos au} = \frac{u}{2} \mp \frac{1}{2a}\ln|\sin au \pm \cos au| + C$$

133. $$\int \frac{\cos au\, du}{\sin au \pm \cos au} = \pm\left[\frac{u}{2} \pm \frac{1}{2a}\ln|\sin au \pm \cos au|\right] + C$$

134. $$\int \frac{\sin au\, du}{p + q\cos au} = -\frac{1}{aq}\ln|p + q\cos au| + C$$

135. $$\int \frac{\cos au\, du}{p + q\sin au} = \frac{1}{aq}\ln|p + q\sin au| + C$$

136. $$\int \sin^m au \cos^n au\, du = \begin{cases} -\dfrac{\sin^{m-1} au \cos^{n+1} au}{a(m+n)} + \dfrac{m-1}{m+n}\displaystyle\int \sin^{m-2} au \cos^n au\, du, & m \neq -n \\[2ex] \dfrac{\sin^{m+1} au \cos^{n-1} au}{a(m+n)} + \dfrac{n-1}{m+n}\displaystyle\int \sin^m au \cos^{n-2} au\, du, & m \neq -n \end{cases}$$

137. $$\int \tan au\, du = -\frac{1}{a}\ln|\cos au| = \frac{1}{a}\ln|\sec au| + C$$

138. $$\int \tan^2 au\, du = \frac{\tan au}{a} - u + C$$

139. $$\int \tan^n au \sec^2 au\, du = \frac{\tan^{n+1} au}{(n+1)a} + C, \; n \neq -1$$

140. $$\int \tan^n au\, du = \frac{\tan^{n-1} au}{(n-1)a} - \int \tan^{n-2} au\, du + C, \quad n \neq 1$$

141. $\int \cot au\, du = \frac{1}{a} \ln|\sin au| + C$

142. $\int \cot^2 au\, du = -\frac{\cot au}{a} - u + C$

143. $\int \cot^n au \csc^2 au\, du = -\frac{\cot^{n+1} au}{(n+1)a} + C, \quad n \neq -1$

144. $\int \cot^n au\, du = -\frac{\cot^{n-1} au}{(n-1)a} - \int \cot^{n-2} au\, du, \quad n \neq 1$

145. $\int \sec au\, du = \frac{1}{a} \ln|\sec au + \tan au| = \frac{1}{a} \ln\left|\tan\left(\frac{au}{2} + \frac{\pi}{4}\right)\right| + C$

146. $\int \sec^2 au\, du = \frac{\tan au}{a} + C$

147. $\int \sec^3 au\, du = \frac{\sec au \tan au}{2a} + \frac{1}{2a} \ln|\sec au + \tan au| + C$

148. $\int \sec^n au \tan au\, du = \frac{\sec^n au}{na} + C$

149. $\int \sec^n au\, du = \frac{\sec^{n-2} au \tan au}{a(n-1)} + \frac{n-2}{n-1} \int \sec^{n-2} au\, du, \quad n \neq 1$

150. $\int \csc au\, du = \frac{1}{a} \ln|\csc au - \cot au| = \frac{1}{a} \ln\left|\tan \frac{au}{2}\right| + C$

151. $\int \csc^2 au\, du = -\frac{\cot au}{a} + C$

152. $\int \csc^n au \cot au\, du = -\frac{\csc^n au}{na} + C$

153. $\int \csc^n au\, du = -\frac{\csc^{n-2} au \cot au}{a(n-1)} + \frac{n-2}{n-1} \int \csc^{n-2} au\, du, \quad n \neq 1$

Integrals involving inverse trigonometric functions

154. $\int \sin^{-1} \frac{u}{a}\, du = u \sin^{-1} \frac{u}{a} + \sqrt{a^2 - u^2} + C$

155. $\int u \sin^{-1} \frac{u}{a}\, du = \left(\frac{u^2}{2} - \frac{a^2}{4}\right)\sin^{-1} \frac{u}{a} + \frac{u\sqrt{a^2 - u^2}}{4} + C$

156. $\int \cos^{-1} \frac{u}{a}\, du = u \cos^{-1} \frac{u}{a} - \sqrt{a^2 - u^2} + C$

157. $\int u \cos^{-1} \frac{u}{a}\, du = \left(\frac{u^2}{2} - \frac{a^2}{4}\right)\cos^{-1} \frac{u}{a} - \frac{u\sqrt{a^2 - u^2}}{4} + C$

158. $\int \tan^{-1} \frac{u}{a}\, du = u \tan^{-1} \frac{u}{a} - \frac{a}{2} \ln(u^2 + a^2) + C$

159. $\int u \tan^{-1} \frac{u}{a}\, du = \frac{1}{2}(u^2 + a^2)\tan^{-1} \frac{u}{a} - \frac{au}{2} + C$

160. $\int u^m \sin^{-1} \frac{u}{a}\, du = \frac{u^{m+1}}{m+1} \sin^{-1} \frac{u}{a} - \frac{1}{m+1} \int \frac{u^{m+1}}{\sqrt{a^2 - u^2}}\, du$

161. $\int u^m \cos^{-1} \frac{u}{a}\, du = \frac{u^{m+1}}{m+1} \cos^{-1} \frac{u}{a} + \frac{1}{m+1} \int \frac{u^{m+1}}{\sqrt{a^2 - u^2}}\, du$

162. $\int u^m \tan^{-1} \frac{u}{a}\, du = \frac{u^{m+1}}{m+1} \tan^{-1} \frac{u}{a} - \frac{a}{m+1} \int \frac{u^{m+1}}{\sqrt{u^2 - a^2}}\, du$

Integrals involving e^{au}

163. $\int e^{au}\, du = \frac{e^{au}}{a} + C$

164. $\int ue^{au}\, du = \frac{e^{au}}{a}\left(u - \frac{1}{a}\right) + C$

165. $\int u^2 e^{au}\, du = \frac{e^{au}}{a}\left(u^2 - \frac{2u}{a} + \frac{2}{a^2}\right) + C$

166. $\int u^n e^{au}\,du = \frac{u^n e^{au}}{a} - \frac{n}{a}\int u^{n-1}e^{au}\,du$

$= \frac{e^{au}}{a}\left[u^n - \frac{nu^{n-1}}{a} + \frac{n(n-1)u^{n-2}}{a^2} - \cdots + \frac{(-1)^n n!}{a^n}\right]$, if n is a positive integer

167. $\int \frac{du}{p + qe^{au}} = \frac{u}{p} - \frac{1}{ap}\ln|p + qe^{au}| + C$

168. $\int e^{au}\sin bu\,du = \frac{e^{au}(a\sin bu - b\cos bu)}{a^2 + b^2} + C$

169. $\int e^{au}\cos bu\,du = \frac{e^{au}(a\cos bu + b\sin bu)}{a^2 + b^2} + C$

170. $\int ue^{au}\sin bu\,du = \frac{ue^{au}(a\sin bu - b\cos bu)}{a^2 + b^2} - \frac{e^{au}[(a^2 - b^2)\sin bu - 2ab\cos bu]}{(a^2 + b^2)^2} + C$

171. $\int ue^{au}\cos bu\,du = \frac{ue^{au}(a\cos bu + b\sin bu)}{a^2 + b^2} - \frac{e^{au}[(a^2 - b^2)\cos bu + 2ab\sin bu]}{(a^2 + b^2)^2} + C$

172. $\int e^{au}\sin^n bu\,du = \frac{e^{au}\sin^{n-1} bu}{a^2 + n^2b^2}(a\sin bu - nb\cos bu) + \frac{n(n-1)b^2}{a^2 + n^2b^2}\int e^{au}\sin^{n-2} bu\,du$

173. $\int e^{au}\cos^n bu\,du = \frac{e^{au}\cos^{n-1} bu}{a^2 + n^2b^2}(a\cos bu + nb\sin bu) + \frac{n(n-1)b^2}{a^2 + n^2b^2}\int e^{au}\cos^{n-2} bu\,du$

Integrals involving ln *u*

174. $\int \ln u\,du = u\ln u - u + C$

175. $\int u\ln u\,du = \frac{u^2}{2}\left(\ln u - \frac{1}{2}\right) + C$

176. $\int u^m\ln u\,du = \frac{u^{m+1}}{m+1}\left(\ln u - \frac{1}{m+1}\right), \quad m \neq -1$

177. $\int \frac{\ln u}{u}\,du = \frac{1}{2}\ln^2 u + C$

178. $\int \frac{\ln^n u\,du}{u} = \frac{\ln^{n+1} u}{n+1} + C, \quad n \neq -1$

179. $\int \frac{du}{u\ln u} = \ln|\ln u| + C$

180. $\int \ln^n u\,du = u\ln^n u - n\int \ln^{n-1} u\,du + C, \quad n \neq -1$

181. $\int u^m\ln^n u\,du = \frac{u^{m+1}\ln^n u}{m+1} - \frac{n}{m+1}\int u^m\ln^{n-1} u\,du + C, \quad m, n \neq -1$

182. $\int \ln(u^2 + a^2)\,du = u\ln(u^2 + a^2) - 2u + 2a\tan^{-1}\frac{u}{a} + C$

183. $\int \ln|u^2 - a^2|\,du = u\ln|u^2 - a^2| - 2u + a\ln\left|\frac{u+a}{u-a}\right| + C$

Integrals involving hyperbolic functions

184. $\int \sinh au\,du = \frac{\cosh au}{a} + C$

185. $\int u\sinh au\,du = \frac{u\cosh au}{a} - \frac{\sinh au}{a^2} + C$

186. $\int \cosh au\,du = \frac{\sinh au}{a} + C$

187. $\int u\cosh au\,du = \frac{u\sinh au}{a} - \frac{\cosh au}{a^2} + C$

188. $\int \cosh^2 au\,du = \frac{u}{2} + \frac{\sinh au\cosh au}{2a} + C$

189. $\int \sinh^2 au\,du = \frac{\sinh au\cosh au}{2a} - \frac{u}{2} + C$

190. $\int \sinh^n au\,du = \frac{\sinh^{n-1} au\cosh au}{an} - \frac{n-1}{n}\int \sinh^{n-2} au\,du$

191. $\int \cosh^n au\,du = \frac{\cosh^{n-1} au\sinh au}{an} + \frac{n-1}{n}\int \cosh^{n-2} au\,du$

192. $\displaystyle\int \sinh au \cosh au\, du = \frac{\sinh^2 au}{2a} + C$

193. $\displaystyle\int \sinh pu \cosh qu\, du = \frac{\cosh(p+q)u}{2(p+q)} + \frac{\cosh(p-q)u}{2(p-q)} + C$

194. $\displaystyle\int \tanh au\, du = \frac{1}{a}\ln\cosh au + C$

195. $\displaystyle\int \tanh^2 au\, du = u - \frac{\tanh au}{a} + C$

196. $\displaystyle\int \tanh^n au\, du = \frac{-\tanh^{n-1} au}{a(n-1)} + \int \tanh^{n-2} au\, du$

197. $\displaystyle\int \coth au\, du = \frac{1}{a}\ln\left|\sinh au\right| + C$

198. $\displaystyle\int \coth^2 au\, du = u - \frac{\coth au}{a} + C$

199. $\displaystyle\int \operatorname{sech} au\, du = \frac{2}{a}\tan^{-1} e^{au} + C$

200. $\displaystyle\int \operatorname{sech}^2 au\, du = \frac{\tanh au}{a} + C$

201. $\displaystyle\int \operatorname{sech}^n au\, du = \frac{\operatorname{sech}^{n-2} au \tanh au}{a(n-1)} + \frac{n-2}{n-1}\int \operatorname{sech}^{n-2} au\, du$

202. $\displaystyle\int \operatorname{csch} au\, du = \frac{1}{a}\ln\left|\tanh\frac{au}{2}\right| + C$

203. $\displaystyle\int \operatorname{csch}^2 au\, du = -\frac{\coth au}{a} + C$

204. $\displaystyle\int \operatorname{sech} u \tanh u\, du = -\operatorname{sech} u + C$

205. $\displaystyle\int \operatorname{csch} u \coth u\, du = -\operatorname{csch} u + C$

Some definite integrals

Unless otherwise stated, all letters stand for positive numbers.

206. $\displaystyle\int_0^\infty \frac{dx}{x^2 + a^2} = \frac{\pi}{2a}$

207. $\displaystyle\int_0^\infty \frac{x^{p-1}}{1+x}\, dx = \frac{\pi}{\sin p\pi}$

208. $\displaystyle\int_0^a \frac{dx}{\sqrt{a^2 - x^2}} = \frac{\pi}{2}$

209. $\displaystyle\int_0^a \sqrt{a^2 - x^2}\, dx = \frac{\pi a^2}{4}$

210. $\displaystyle\int_0^\pi \sin mx \sin nx\, dx = \begin{cases} 0, & \text{if } m, n \text{ integers and } m \neq n \\ \dfrac{\pi}{2}, & \text{if } m, n \text{ integers and } m = n \end{cases}$

211. $\displaystyle\int_0^\pi \cos mx \cos nx\, dx = \begin{cases} 0, & \text{if } m, n \text{ integers and } m \neq n \\ \dfrac{\pi}{2}, & \text{if } m, n \text{ integers and } m = n \end{cases}$

212. $\displaystyle\int_0^\pi \sin mx \cos nx\, dx = \begin{cases} 0, & \text{if } m, n \text{ integers and } m + n \text{ is even} \\ \dfrac{2m}{(m^2 - n^2)}, & \text{if } m, n \text{ integers and } m + n \text{ is odd} \end{cases}$

213. $\displaystyle\int_0^{\pi/2} \sin^2 x\, dx = \int_0^{\pi/2} \cos^2 x\, dx = \frac{\pi}{4}$

214. $\displaystyle\int_0^\infty e^{-ax}\cos bx\, dx = \frac{a}{a^2 + b^2}$

215. $\displaystyle\int_0^\infty e^{-ax}\sin bx\, dx = \frac{b}{a^2 + b^2}$

216. $\displaystyle\int_0^\infty e^{-a^2x^2}\, dx = \frac{\sqrt{\pi}}{2a}$

217. $\displaystyle\int_0^{\pi/2} \sin^{2m} x\, dx = \int_0^{\pi/2} \cos^{2m} x\, dx = \frac{1\cdot 3\cdot 5\cdot \cdots \cdot (2m-1)}{2\cdot 4\cdot 6\cdot \cdots \cdot 2m}\,\frac{\pi}{2}, \quad m = 1, 2, 3, \ldots$

218. $\displaystyle\int_0^{\pi/2} \sin^{2m+1} x\, dx = \int_0^{\pi/2} \cos^{2m+1} x\, dx = \frac{2\cdot 4\cdot 6\cdot \cdots \cdot 2m}{1\cdot 3\cdot 5\cdot \cdots \cdot (2m+1)}, \quad m = 1, 2, 3, \ldots$

219. $\displaystyle\int_0^\infty \frac{e^{-x}}{\sqrt{x}}\, dx = \sqrt{\pi}$

220. $\displaystyle\int_0^1 x^m (\ln x)^n\, dx = \frac{(-1)^n n!}{(m+1)^{n+1}}$

APPENDIX 2

TABLE OF LAPLACE TRANSFORMS*

$F(s) = \mathscr{L}\{f(t)\}$	$f(t)$
1. $\dfrac{1}{s^r}, \quad r > 0$	$\dfrac{t^{r-1}}{\Gamma(r)}, \quad \Gamma(n+1) = n!$
2. $\dfrac{1}{(s-a)^r}, \quad r > 0$	$\dfrac{t^{r-1}e^{at}}{\Gamma(r)}$
3. $\dfrac{1}{(s-a)(s-b)}, \quad a \neq b$	$\dfrac{e^{at} - e^{bt}}{a - b}$
4. $\dfrac{s}{(s-a)(s-b)}, \quad a \neq b$	$\dfrac{ae^{at} - be^{bt}}{a - b}$
5. $\dfrac{1}{s^2 + k^2}$	$\dfrac{1}{k} \sin kt$
6. $\dfrac{s}{s^2 + k^2}$	$\cos kt$
7. $\dfrac{1}{s^2 - k^2}$	$\dfrac{1}{k} \sinh kt$
8. $\dfrac{s}{s^2 - k^2}$	$\cosh kt$
9. $\dfrac{1}{(s-a)^2 + k^2}$	$\dfrac{1}{k} e^{at} \sin kt$
10. $\dfrac{s-a}{(s-a)^2 + k^2}$	$e^{at} \cos kt$
11. $\dfrac{1}{(s-a)^2 - k^2}$	$\dfrac{1}{k} e^{at} \sinh kt$
12. $\dfrac{s-a}{(s-a)^2 - k^2}$	$e^{at} \cosh kt$
13. $\dfrac{1}{s(s^2 + k^2)}$	$\dfrac{1}{k^2}(1 - \cos kt)$
14. $\dfrac{1}{s^2(s^2 + k^2)}$	$\dfrac{1}{k^3}(kt - \sin kt)$
15. $\dfrac{1}{(s^2 + k^2)^2}$	$\dfrac{1}{2k^3}(\sin kt - kt \cos kt)$

*For a more extensive list of Laplace transforms and their inverses, see A. Erdelyi et al., *Tables of Integral Transforms*, 2 vols. (New York: McGraw-Hill, 1954).

$F(s) = \mathscr{L}\{f(t)\}$	$f(t)$
16. $\dfrac{s}{(s^2 + k^2)^2}$	$\dfrac{t}{2k}\sin kt$
17. $\dfrac{s^2}{(s^2 + k^2)^2}$	$\dfrac{1}{2k}(\sin kt + kt\cos kt)$
18. $\dfrac{1}{(s^2 + a^2)(s^2 + b^2)}, \quad a^2 \neq b^2$	$\dfrac{a\sin bt - b\sin at}{ab(a^2 - b^2)}$
19. $\dfrac{s}{(s^2 + a^2)(s^2 + b^2)}, \quad a^2 \neq b^2$	$\dfrac{\cos bt - \cos at}{a^2 - b^2}$
20. $\dfrac{1}{s^4 - k^4}$	$\dfrac{1}{2k^3}(\sinh kt - \sin kt)$
21. $\dfrac{s}{s^4 - k^4}$	$\dfrac{1}{2k^2}(\cosh kt - \cos kt)$
22. $\dfrac{1}{s^4 + 4k^4}$	$\dfrac{1}{4k^3}(\sin kt\cosh kt - \cos kt\sinh kt)$
23. $\dfrac{s}{s^4 + 4k^4}$	$\dfrac{1}{2k^2}\sin kt\sinh kt$
24. $\sqrt{s-a} - \sqrt{s-b}$	$\dfrac{e^{bt} - e^{at}}{2\sqrt{\pi t^3}}$
25. $\dfrac{s}{(s-a)^{3/2}}$	$\dfrac{e^{at}}{\sqrt{\pi t}}(1 + 2at)$
26. $\dfrac{1}{\sqrt{s+a}\sqrt{s+b}}$	$e^{-(a+b)t/2}I_0\left(\dfrac{a-b}{2}t\right)$
27. $\dfrac{(\sqrt{s^2+k^2} - s)^r}{\sqrt{s^2+k^2}}, \quad r > -1$	$k^rJ_r(kt)$
28. $\dfrac{1}{(s^2+k^2)^r}, \quad r > 0$	$\dfrac{\sqrt{\pi}}{\Gamma(r)}\left(\dfrac{t}{2k}\right)^{r-1/2}J_{r-1/2}(kt)$
29. $(\sqrt{s^2+k^2} - s)^r, \quad r > 0$	$\dfrac{rk^r}{t}J_r(kt)$
30. $\dfrac{(s - \sqrt{s^2-k^2})^r}{\sqrt{s^2-k^2}}, \quad r > -1$	$k^rI_r(kt)$
31. $\dfrac{1}{(s^2-k^2)^r}, \quad r > 0$	$\dfrac{\sqrt{\pi}}{\Gamma(r)}\left(\dfrac{t}{2k}\right)^{r-1/2}I_{r-1/2}(kt)$
32. $\dfrac{e^{-k/s}}{s^r}, \quad r > 0$	$\left(\dfrac{t}{k}\right)^{(r-1)/2}J_{r-1}(2\sqrt{kt})$
33. $\dfrac{e^{-k/s}}{\sqrt{s}}$	$\dfrac{1}{\sqrt{\pi t}}\cos 2\sqrt{kt}$
34. $\dfrac{e^{k/s}}{s^r}, \quad r > 0$	$\left(\dfrac{t}{k}\right)^{(r-1)/2}I_{r-1}(2\sqrt{kt})$
35. $\dfrac{e^{k/s}}{\sqrt{s}}$	$\dfrac{1}{\sqrt{\pi t}}\cosh 2\sqrt{kt}$
36. $\dfrac{1}{s}\ln s$	$-\ln t - \gamma, \quad \gamma \approx 0.5772$
37. $\ln\dfrac{s-a}{s-b}$	$\dfrac{e^{bt} - e^{at}}{t}$

$F(s) = \mathcal{L}\{f(t)\}$	$f(t)$
38. $\ln\left(1 + \dfrac{k^2}{s^2}\right)$	$\dfrac{2}{t}(1 - \cos kt)$
39. $\ln\left(1 - \dfrac{k^2}{s^2}\right)$	$\dfrac{2}{t}(1 - \cosh kt)$
40. $\arctan\left(\dfrac{k}{s}\right)$	$\dfrac{\sin kt}{t}$
41. $\dfrac{1}{s}\arctan\left(\dfrac{k}{s}\right)$	$\operatorname{Si}(kt) = \displaystyle\int_0^{kt} \frac{\sin u}{u}\,du$
42. $e^{-r\sqrt{s}}, \quad r > 0$	$\dfrac{r}{2\sqrt{\pi t^3}} \exp\left(-\dfrac{r^2}{4t}\right)$
43. $\dfrac{e^{-r\sqrt{s}}}{s}, \quad r \geq 0$	$1 - \operatorname{erf}\left(\dfrac{r}{2\sqrt{t}}\right) = 1 - \dfrac{2}{\sqrt{\pi}} \displaystyle\int_0^{r/2\sqrt{t}} e^{-u^2}\,du$
44. $e^{r^2s^2}(1 - \operatorname{erf}(rs)), \quad r > 0$	$\dfrac{1}{r\sqrt{\pi}} \exp\left(-\dfrac{t^2}{4r^2}\right)$
45. $\dfrac{1}{s}e^{r^2s^2}(1 - \operatorname{erf}(rs)), \quad r > 0$	$\operatorname{erf}\left(\dfrac{t}{2r}\right)$
46. $\operatorname{erf}\left(\dfrac{r}{\sqrt{s}}\right)$	$\dfrac{1}{\pi t}\sin(2k\sqrt{t})$
47. $e^{rs}(1 - \operatorname{erf}\sqrt{rs}), \quad r > 0$	$\dfrac{\sqrt{r}}{\pi\sqrt{t}\,(t + r)}$
48. $\dfrac{1}{\sqrt{s}}e^{rs}(1 - \operatorname{erf}\sqrt{rs}), \quad r > 0$	$\dfrac{1}{\sqrt{\pi(t + r)}}$

3 APPENDIX

THE EXISTENCE AND UNIQUENESS OF SOLUTIONS

In this appendix we prove a number of results that guarantee the existence of unique solutions to first-order initial-value problems and first-order systems of differential equations. We begin with the first-order initial-value problem

$$x'(t) = f(t, x(t)), \qquad x(t_0) = x_0, \tag{1}$$

where t_0 and x_0 are real numbers. Equation **(1)** includes all the first-order equations we have discussed in this book. For example, for the linear nonhomogeneous equation $x' + a(t)x = b(t)$,

$$f(t, x) = -a(t)x + b(t).$$

We here show that if $f(t, x)$ and $(\frac{\partial f}{\partial x})(t, x)$ are continuous in some region containing the point (t_0, x_0), then there is an interval (containing t_0) on which a unique solution of the initial-value problem **(1)** exists.

Before continuing with our discussion, we advise the reader that in this appendix we use theoretical tools from calculus that have not been widely used earlier in the text. In particular, we need the following facts about continuity and convergence of functions. They are discussed in most intermediate and advanced calculus texts:*

(a) Let $f(t, x)$ be a continuous function of the two variables t and x and let the closed, bounded region D be defined by

$$D = \{(t, x) : a \le t \le b, c \le x \le d\},$$

where a, b, c, and d are finite real numbers. Then $f(t, x)$ is bounded for (t, x) in D; that is, there is a number $M > 0$ such that $|f(t, x)| \le M$ for every pair (t, x) in D.

(b) Let $f(x)$ be continuous on the closed interval $a \le x \le b$ and differentiable on the open interval $a < x < b$. Then the **mean value theorem** of differential calculus states that there is a number ξ between a and b $(a < \xi < b)$ such that

$$f(b) - f(a) = f'(\xi)(b - a).$$

* See, for example, R. C. Buck. *Advanced Calculus,* 3d ed., (New York: McGraw-Hill, 1978).

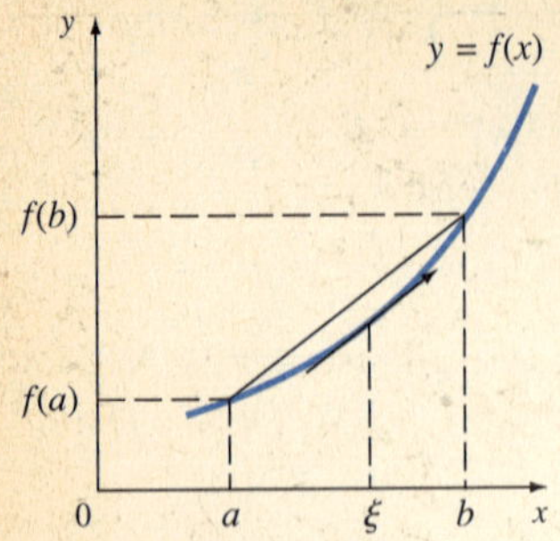

FIGURE A3.1

This equation can be written as

Mean value theorem

$$\frac{f(b) - f(a)}{b - a} = f'(\xi),$$

which says, geometrically, that the slope of the tangent to the curve $y = f(x)$ at the point ξ between a and b is equal to the slope of the secant line passing through the points $(a, f(a))$ and $(b, f(b))$ (see Figure A3.1).

(c) Let $\{x_n(t)\}$ be a sequence of functions. Then $x_n(t)$ is said to **converge uniformly** to a (limit) function $x(t)$ on the interval $a \le t \le b$ if for every real number $\epsilon > 0$, there exists an integer $N > 0$ such that whenever $n \ge N$, we have

$$|x_n(t) - x(t)| < \epsilon,$$

for every t, $a \le t \le b$.

(d) If the functions $\{x_n(t)\}$ of statement (c) are continuous on the interval $a \le t \le b$, then the limit function $x(t)$ is also continuous there. This fact is often stated as "the uniform limit of continuous functions is continuous."

(e) Let $f(t, x)$ be a continuous function in the variable x and suppose that $x_n(t)$ converges to $x(t)$ uniformly as $n \to \infty$. Then

$$\lim_{n\to\infty} f(t, x_n(t)) = f(t, x(t)).$$

(f) Let $f(t)$ be an integrable function on the interval $a \le t \le b$. Then

$$\left|\int_a^b f(t)\,dt\right| \le \int_a^b |f(t)|\,dt,$$

and if $|f(t)| \le M$, then

$$\int_a^b |f(t)|\,dt \le M \int_a^b dt = M(b - a).$$

(g) Let $\{x_n(t)\}$ be a sequence of functions with $|x_n(t)| \le M_n$ for $a \le t \le b$. Then, if $\Sigma_{n=0}^{\infty} |M_n| < \infty$ (that is, if $\Sigma_{n=0}^{\infty} M_n$ converges absolutely), then $\Sigma_{n=0}^{\infty} x_n(t)$ converges uniformly on the interval $a \le t \le b$ to a unique limit function $x(t)$. This is often called the **Weierstrass *M*-test** for uniform convergence.

(h) Let $\{x_n(t)\}$ converge uniformly to $x(t)$ on the interval $a \le t \le b$ and let $f(t, x)$ be a continuous function of t and x in the region D defined in statement (a). Then

$$\lim_{n\to\infty} \int_a^b f(s, x_n(s))\,ds = \int_a^b \lim_{n\to\infty} f(s, x_n(s))\,ds = \int_a^b f(s, x(s))\,ds.$$

The first result we need was proved in Section 2.8 (see p. 86).

THEOREM 1 ♦ **Converting an initial-value problem to an integral equation**

Let $f(t, x)$ be continuous for all values t and x. Then the initial-value problem **(1)** is equivalent to the integral equation

$$x(t) = x_0 + \int_{t_0}^{t} f(s, x(s))\,ds \qquad \textbf{(2)}$$

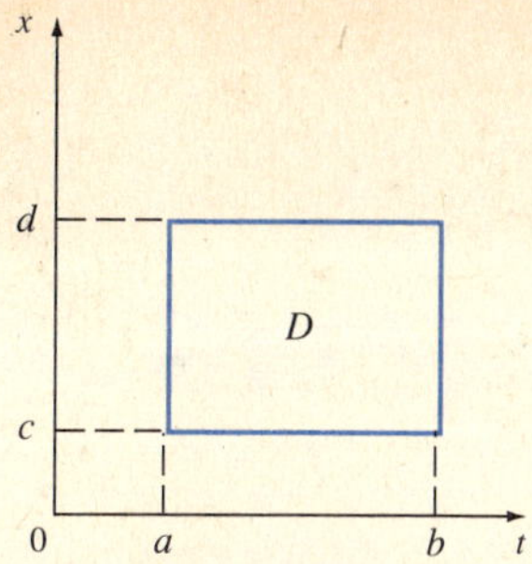

FIGURE A3.2

in the sense that $x(t)$ is a solution of the initial-value problem **(1)** if and only if $x(t)$ is a solution of Equation **(2).** ♦

Let D denote the rectangular region in the tx-plane defined by

$$D\colon a \le t \le b,\ c \le x \le d, \tag{3}$$

where $-\infty < a < b < +\infty$ and $-\infty < c < d < +\infty$ (see Figure A3.2). We say that the function $f(t, x)$ is **Lipschitz continuous** in x over D if there exists a constant k, $0 < k < \infty$, such that

Lipschitz condition

$$|f(t, x_1) - f(t, x_2)| \le k|x_1 - x_2|, \tag{4}$$

whenever (t, x_1) and (t, x_2) belong to D. The constant k is called a **Lipschitz constant.** Clearly, according to inequality **(4),** every Lipschitz continuous function is continuous in x for each fixed t. However, *not every continuous function is Lipschitz continuous.*

EXAMPLE 1 ▶ **A function that is not Lipschitz continuous at a point**

Let $f(t, x) = \sqrt{x}$ on the set $0 \le t \le 1$, $0 \le x \le 1$. Then $f(t, x)$ is certainly continuous on this region. But

$$|f(t, x) - f(t, 0)| = |\sqrt{x} - 0| = \frac{1}{\sqrt{x}}|x - 0|,$$

for all $0 < x < 1$, and $x^{-1/2}$ tends to infinity as x approaches zero. Thus no finite Lipschitz constant can be found to satisfy Equation **(4).** ◀

However, Lipschitz continuity is not a rare occurrence, as shown by the following theorem.

THEOREM 2 ♦ **A condition for Lipschitz continuity**

Let $f(t, x)$ and $(\frac{\partial f}{\partial x})(t, x)$ be continuous on D. Then $f(t, x)$ is Lipschitz continuous in x over D.

Proof Let (t, x_1) and (t, x_2) be points in D. For fixed t, $(\frac{\partial f}{\partial x})(t, x)$ is a function of x, and so we may apply the mean value theorem of differential calculus [statement (b)] to obtain

$$|f(t, x_1) - f(t, x_2)| = \left|\frac{\partial f}{\partial x}(t, \xi)\right||x_1 - x_2|,$$

where $x_1 < \xi < x_2$. But since $\frac{\partial f}{\partial x}$ is continuous in D, it is bounded there [according to statement (a)]. Hence there is a constant k, $0 < k < \infty$, such that

$$\left|\frac{\partial f}{\partial x}(t, x)\right| \le k,$$

for all (t, x) in D. ♦

EXAMPLE 2 ▶ **A Lipschitz continuous function**

If $f(t, x) = tx^2$ on $0 \le t \le 1$, $0 \le x \le 1$, then

$$\left|\frac{\partial f}{\partial x}\right| = |2tx| \le 2,$$

so that

$$|f(t, x_1) - f(t, x_2)| \leq 2|x_1 - x_2|. \blacktriangleleft$$

In Section 2.8 (p. 86) we defined the

Picard iterations

$$\begin{aligned} x_0(t) &= x_0, \\ x_1(t) &= x_0 + \int_{t_0}^{t} f(s, x_0(s))\, ds, \\ x_2(t) &= x_0 + \int_{t_0}^{t} f(s, x_1(s))\, ds, \\ &\vdots \\ x_n(t) &= x_0 + \int_{t_0}^{t} f(s, x_{n-1}(s))\, ds. \end{aligned} \tag{5}$$

We here show that under certain conditions the Picard iterations defined by **(5)** converge uniformly to a solution of Equation **(2)**. In Example 2.8.1 (p. 86) we showed that the Picard iterations converge to the unique solution of the initial-value problem

$$x'(t) = x(t), \qquad x(0) = 1. \tag{6}$$

We now state and prove the main result of this appendix.

THEOREM 3 ♦ Existence theorem

Let $f(t, x)$ be Lipschitz continuous in x with the Lipschitz constant k on the region D of all points (t, x) satisfying the inequalities

$$|t - t_0| \leq a, \qquad |x - x_0| \leq b.$$

(See Figure A3.3.) Then there exists a number $\delta > 0$ with the property that the initial-value problem

$$x' = f(t, x), \qquad x(t_0) = x_0,$$

has a solution $x = x(t)$ on the interval $|t - t_0| \leq \delta$.

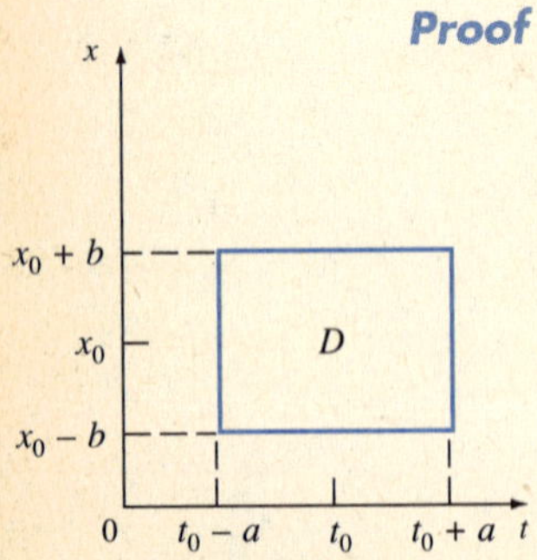

FIGURE A3.3

Proof The proof of this theorem is complicated and is done in several stages. However, the basic idea is simple: We need only justify that the Picard iterations converge uniformly and yield, in the limit, the solution of the integral equation **(2)**.

Since f is continuous on D, it is bounded there [statement (a)] and we may begin by letting M be a finite upper bound for $|f(t, x)|$ on D. We then define

$$\delta = \min\left\{a, \frac{b}{M}\right\}. \tag{7}$$

1. We first show that the iterations $\{x_n(t)\}$ are continuous and satisfy the inequality

$$|x_n(t) - x_0| \leq b. \tag{8}$$

Inequality **(8)** is necessary in order that $f(t, x_n(t))$ be defined for $n = 0, 1, 2, \ldots$. To show the continuity of $x_n(t)$, we first note that $x_0(t) = x_0$ is continuous (a constant function is always continuous). Then

$$x_1(t) = x_0 + \int_{t_0}^{t} f(t, x_0(s))\, ds.$$

But $f(t, x_0)$ is continuous [since $f(t, x)$ is continuous in t and x], and the integral of a continous function is continuous. Thus $x_1(t)$ is continuous. In a similar fashion, we can show that

$$x_2(t) = x_0 + \int_{t_0}^{t} f(t, x_1(s))\, ds$$

is continuous, and so on for $n = 3, 4, \ldots$.

Obviously inequality **(8)** holds when $n = 0$, because $x_0(t) = x_0$. For $n \neq 0$, we use definition **(5)** and Equation **(7)** to obtain

$$\begin{aligned}|x_n(t) - x_0| &= \left|\int_{t_0}^{t} f(s, x_{n-1}(s))\, ds\right| \leq \left|\int_{t_0}^{t} |f(s, x_{n-1}(s))|\, ds\right| \\ &\leq M\left|\int_{t_0}^{t} ds\right| = M|t - t_0| \leq M\delta \leq b.\end{aligned}$$

These inequalities follow from statement (f). Note that the last inequality helps explain the choice of δ in Equation **(7)**.

2. Next, we show by induction that

$$|x_n(t) - x_{n-1}(t)| \leq Mk^{n-1}\frac{|t - t_0|^n}{n!} \leq \frac{Mk^{n-1}\delta^n}{n!}. \qquad \textbf{(9)}$$

If $n = 1$, we obtain

$$\begin{aligned}|x_1(t) - x_0(t)| &\leq \left|\int_{t_0}^{t} f(s, x_0(s))\, ds\right| \leq M\left|\int_{t_0}^{t} ds\right| \\ &= M|t - t_0| \leq M\delta.\end{aligned}$$

Thus the result is true for $n = 1$.

We assume that the result is true for $n = m$ and prove that it holds for $n = m + 1$; that is, we assume that

$$|x_m(t) - x_{m-1}(t)| \leq \frac{Mk^{m-1}|t - t_0|^m}{m!} \leq \frac{Mk^{m-1}\delta^m}{m!}.$$

Then, since $f(t, x)$ is Lipschitz continuous in x over D,

$$\begin{aligned}|x_{m+1}(t) - x_m(t)| &= \left|\int_{t_0}^{t} f(s, x_m(s))\, ds - \int_{t_0}^{t} f(s, x_{m-1}(s))\, ds\right| \\ &\leq \left|\int_{t_0}^{t} |f(s, x_m(s)) - f(s, x_{m-1}(s))|\, ds\right| \\ &\leq k\left|\int_{t_0}^{t} |x_m(s) - x_{m-1}(s)|\, ds\right| \\ &\leq \frac{Mk^m}{m!}\left|\int_{t_0}^{t} (s - t_0)^m\, ds\right|^* \\ &= \frac{Mk^m|t - t_0|^{m+1}}{(m + 1)!} \leq \frac{Mk^m\delta^{m+1}}{(m + 1)!},\end{aligned}$$

which is what we wanted to show.

---

* This inequality follows from the induction assumption that inequality **(9)** holds for $n = m$.

3. We now show that $x_n(t)$ converges uniformly to a limit function $x(t)$ on the interval $|t - t_0| \leq \delta$. By statement (d), this shows that $x(t)$ is continuous.

We first note that

$$x_n(t) - x_0(t) = x_n(t) - x_{n-1}(t) + x_{n-1}(t) - x_{n-2}(t) + \cdots + x_1(t) - x_0(t)$$
$$= \sum_{m=1}^{n} [x_m(t) - x_{m-1}(t)]. \quad \textbf{(10)}$$

But by inequality **(9)**,

$$|x_m(t) - x_{m-1}(t)| \leq \frac{Mk^{m-1}\delta^m}{m!} = \frac{M}{k}\frac{k^m\delta^m}{m!},$$

so

$$\sum_{m=1}^{\infty} |x_m(t) - x_{m-1}(t)| \leq \frac{M}{k}\sum_{m=1}^{\infty}\frac{(k\delta)^m}{m!} = \frac{M}{k}(e^{k\delta} - 1),$$

since

$$e^{k\delta} = \sum_{m=0}^{\infty}\frac{(k\delta)^m}{m!} = 1 + \sum_{m=1}^{\infty}\frac{(k\delta)^m}{m!}.$$

By the Weierstrass M-test [statement (g)], we conclude that the series

$$\sum_{m=1}^{\infty} [x_m(t) - x_{m-1}(t)]$$

converges absolutely and uniformly on $|t - t_0| \leq \delta$ to a unique limit function $y(t)$. But

$$y(t) = \lim_{n\to\infty}\sum_{m=1}^{n} [x_m(t) - x_{m-1}(t)]$$
$$= \lim_{n\to\infty} [x_n(t) - x_0(t)] = \lim_{n\to\infty} x_n(t) - x_0(t),$$

or

$$\lim_{n\to\infty} x_n(t) = y(t) + x_0(t).$$

We denote the right-hand side of this equation by $x(t)$. Thus the limit of the Picard iterations $x_n(t)$ exists and the convergence $x_n(t) \to x(t)$ is uniform for all t in the interval $|t - t_0| \leq \delta$.

4. It remains to be shown that $x(t)$ is a solution to Equation **(2)** for $|t - t_0| < \delta$. Since $f(t, x)$ is a continuous function of x and $x_n(t) \to x(t)$ as $n \to \infty$, we have, by statement (e),

$$\lim_{n\to\infty} f(t, x_n(t)) = f(t, x(t)).$$

Hence, by Equation **(5)**,

$$x(t) = \lim_{n\to\infty} x_{n+1}(t) = x_0 + \lim_{n\to\infty}\int_{t_0}^{t} f(s, x_n(s))\, ds$$
$$= x_0 + \int_{t_0}^{t} \lim_{n\to\infty} f(s, x_n(s))\, ds = x_0 + \int_{t_0}^{t} f(s, x(s))\, ds.$$

The step in which we interchange the limit and integral is justified by statement (h). Thus $x(t)$ solves Equation **(2)**, and therefore it solves the initial-value problem **(1)**. ◆

It turns out that the solution obtained in Theorem 3 is unique. Before proving this, however, we derive a simple version of a very useful result known as **Gronwall's inequality.**

THEOREM 4 ♦ **Gronwall's inequality**

Let $x(t)$ be a continuous nonnegative function and suppose that

$$x(t) \le A + B\left|\int_{t_0}^{t} x(s)\,ds\right|, \tag{11}$$

where A and B are positive constants, for all values of t such that $|t - t_0| \le \delta$. Then

$$x(t) \le Ae^{B|t-t_0|} \tag{12}$$

for all t in the interval $|t - t_0| \le \delta$.

Proof We prove this result for $t_0 \le t \le t_0 + \delta$. The proof for $t_0 - \delta \le t \le t_0$ is similar (see Problem 14). Since $x(t) \ge 0$ and $t > t_0$,

$$\left|\int_{t_0}^{t} x(s)\,ds\right| = \int_{t_0}^{t} x(s)\,ds$$

and we define

$$y(t) = B\int_{t_0}^{t} x(s)\,ds.$$

Then

$$y'(t) = Bx(t) \le B\left[A + B\int_{t_0}^{t} x(s)\,ds\right] = AB + By,$$

or

$$y'(t) - By(t) \le AB. \tag{13}$$

We note that

$$\frac{d}{dt}[y(t)e^{-B(t-t_0)}] = e^{-B(t-t_0)}[y'(t) - By(t)].$$

Therefore, multiplying both sides of Equation **(13)** by the integrating factor $e^{-B(t-t_0)}$ (which is greater than zero), we have

$$\frac{d}{dt}[y(t)e^{-B(t-t_0)}] \le ABe^{-B(t-t_0)}.$$

An integration of both sides of the inequality from t_0 to t yields

$$y(s)e^{-B(s-t_0)}\Big|_{t_0}^{t} \le AB\int_{t_0}^{t} e^{-B(s-t_0)}\,ds = -Ae^{-B(s-t_0)}\Big|_{t_0}^{t}.$$

But $y(t_0) = 0$, so

$$y(t)e^{-B(t-t_0)} \le A(1 - e^{-B(t-t_0)}),$$

from which, after multiplying both sides by $e^{B(t-t_0)}$, we obtain

$$y(t) \le A[e^{B(t-t_0)} - 1].$$

Then by Equation **(11),**

$$x(t) \le A + y(t) \le Ae^{B(t-t_0)}. \quad ♦$$

THEOREM 5 ♦ Uniqueness theorem
Let the conditions of Theorem 3 (existence theorem) hold. Then $x(t) = \lim_{n\to\infty} x_n(t)$ is the only continuous solution of the initial-value problem **(1)** in $|t - t_0| \le \delta$.

Proof Let $x(t)$ and $y(t)$ be two continuous solutions of Equation **(2)** in the interval $|t - t_0| \le \delta$ and suppose that $(t, y(t))$ belongs to the region D for all t in that interval.* Define $v(t) = |x(t) - y(t)|$. Then $v(t) \ge 0$ and $v(t)$ is continuous. Since $f(t, x)$ is Lipschitz continuous in x over D, by Theorem 1

$$\begin{aligned} v(t) &= \left|\left[x_0 + \int_{t_0}^{t} f(s, x(s))\, ds\right] - \left[x_0 + \int_{t_0}^{t} f(s, y(s))\, ds\right]\right| \\ &\le k\left|\int_{t_0}^{t} |x(s) - y(s)|\, ds\right| = k\left|\int_{t_0}^{t} v(s)\, ds\right| \\ &\le \epsilon + k\left|\int_{t_0}^{t} v(s)\, ds\right| \end{aligned}$$

for every $\epsilon > 0$. By Gronwall's inequality, we have

$$v(t) \le \epsilon e^{k|t-t_0|}.$$

But $\epsilon > 0$ can be chosen arbitrarily close to zero, so that $v(t) \le 0$. Since $v(t) \ge 0$, it follows that $v(t) \equiv 0$, implying that $x(t)$ and $y(t)$ are identical. Hence the limit of the Picard iterations is the only continuous solution. ♦

THEOREM 6 ♦ Local existence-uniqueness theorem
Let $f(t, x)$ and $(\frac{\partial f}{\partial x})$ be continuous on D. Then there exists a constant $\delta > 0$ such that the Picard iterations $\{x_n(t)\}$ converge to a unique continuous solution of the initial-value problem **(1)** on $|t - t_0| \le \delta$.

Proof This theorem follows directly from Theorem 2 and the existence and uniqueness theorems. ♦

We note that Theorems 3, 5, and 6 are *local* results. By this we mean that unique solutions are guaranteed to exist only "near" the initial point (t_0, x_0).

EXAMPLE 3 ▶ An initial-value problem that has a unique solution
Let

$$x'(t) = x^2(t), \qquad x(1) = 2.$$

Without solving this equation, we can show that there is a unique solution in some interval $|t - t_0| = |t - 1| \le \delta$. Let $a = b = 1$. Then $|f(t, x)| = x^2 \le 9$ $(= M)$ for all $|x - x_0| \le 1$, $x_0 = x(1) = 2$. Therefore, $\delta = \min\{a, \frac{b}{M}\} = \frac{1}{9}$, and Theorem 6 guarantees the existence of a unique solution on the interval $|t - 1| \le \frac{1}{9}$. The solution of this initial-value problem is easily found by a separation of variables to be $x(t) = \frac{2}{3 - 2t}$. This solution exists as long as $t \ne \frac{3}{2}$. Starting at $t_0 = 1$, we see that the maximum interval of existence is $|t - t_0| < \frac{1}{2}$. Hence the value $\delta = \frac{1}{9}$ is not the best possible. ◀

EXAMPLE 4 ▶ An initial-value problem with multiple solutions
Consider the initial-value problem

$$x' = \sqrt{x}, \qquad x(0) = 0.$$

*Note that without this assumption, the function $f(t, y(t))$ may not even be defined at points where $(t, y(t))$ is not in D.

As we saw in Example 1, $f(t, x) = \sqrt{x}$ does *not* satisfy a Lipschitz condition in any region containing the point (0, 0). By a separation of variables, it is easy to calculate the solution

$$x(t) = \left(\frac{t}{2}\right)^2.$$

However $y(t) \equiv 0$ is also a solution. Hence, without a Lipschitz condition the solution to an initial-value problem (if one exists) may fail to be unique. ◀

The last two examples illustrate the local nature of our existence-uniqueness theorem. We now show that it is possible to extend our local results to systems and derive *global* existence-uniqueness results for certain linear differential equations; that is, we show that a unique solution exists for every real number t.

Consider the initial-value system

$$\begin{aligned} x_1' &= f_1(t, x_1, x_2, \ldots, x_n), & x_1(t_0) &= x_{10}, \\ x_2' &= f_2(t, x_1, x_2, \ldots, x_n), & x_2(t_0) &= x_{20}, \\ &\ \vdots & &\ \vdots \\ x_n' &= f_n(t, x_1, x_2, \ldots, x_n), & x_n(t_0) &= x_{n0}. \end{aligned} \tag{14}$$

Suppose we write the vectors

$$\mathbf{x} = \begin{pmatrix} x_1 \\ x_2 \\ \vdots \\ x_n \end{pmatrix}, \qquad \mathbf{f}(t, \mathbf{x}) = \begin{pmatrix} f_1(t, x_1, \ldots, x_n) \\ f_2(t, x_1, \ldots, x_n) \\ \vdots \\ f_n(t, x_1, \ldots, x_n) \end{pmatrix}, \qquad \mathbf{x}_0 = \begin{pmatrix} x_{10} \\ x_{20} \\ \vdots \\ x_{n0} \end{pmatrix}.$$

Then we may write **(14)** in the compact form

$$\mathbf{x}' = \mathbf{f}(t, \mathbf{x}), \qquad \mathbf{x}(t_0) = \mathbf{x}_0. \tag{15}$$

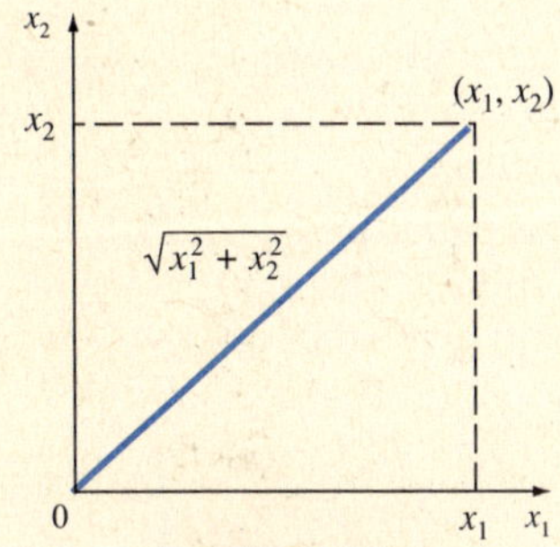

FIGURE A3.4

In order to generalize the results of Theorems 3 and 5, it is necessary to generalize the notion of the absolute value of a number to the absolute value or *norm* of a vector $\mathbf{x}$ or matrix A. If $\mathbf{x} = (x_1, x_2)$ is a two-vector, then the distance from (x_1, x_2) to the origin is given by the Pythagorean theorem (see Figure A3.4):

$$\|\mathbf{x}\| = \sqrt{(x_1^2 + x_2^2)}. \tag{16}$$

Therefore, it is natural to define the length or **norm** of an n-vector $\mathbf{x} = (x_1, x_2, \ldots, x_n)$ by

Norm of a vector

$$\|\mathbf{x}\| = \sqrt{(x_1^2 + x_2^2 + \cdots + x_n^2)}. \tag{17}$$

If A is an $n \times n$ matrix, there are several ways to define its norm. For our purposes, the simplest choice for the norm of A is

Matrix norm

$$\|A\| = \sum_{i=1}^{n} \sum_{j=1}^{n} |a_{ij}|; \tag{18}$$

that is, the norm of A is the sum of the absolute values of the components of A.

Using Equations **(17)** and **(18),** we may prove that

(a) $\|\mathbf{x}\| \geq 0$, where the equality holds only if $\mathbf{x}$ is the zero vector; **(19)**

(b) $\|\alpha\mathbf{x}\| = |\alpha|\|\mathbf{x}\|$ when α is a scalar; **(20)**

(c) $\|\mathbf{x} + \mathbf{y}\| \leq \|\mathbf{x}\| + \|\mathbf{y}\|$. **(21)**

Moreover, if A is an $n \times n$ matrix, it is not difficult to show (see Problems 17–20) that

$$\|A\mathbf{x}\| \leq \|A\| \|\mathbf{x}\|. \tag{22}$$

Now using this notation and a norm $\|\cdot\|$ for every absolute value $|\cdot|$ that appeared in the proofs of Theorems 3 and 5, we obtain the existence and uniqueness of a local vector solution $\mathbf{x}(t)$ of the initial-value problem **(15).** The verification of these details is left as an exercise. In fact, all the steps of the proof of Theorem 3 are as before with $\|\cdot\|$ replacing $|\cdot|$. The notions of continuity and uniform convergence apply to the component functions of $\mathbf{x}$ and $\mathbf{f}$. We have the following result.

THEOREM 7 ♦ **Existence-uniqueness theorem for systems**

Let D denote the region [in $(n + 1)$-dimensional space, one dimension for t and n dimensions for the vector $\mathbf{x}$]

$$|t - t_0| \leq a, \qquad \|\mathbf{x} - \mathbf{x}_0\| \leq b, \tag{23}$$

and suppose that $\mathbf{f}(t, \mathbf{x})$ satisfies the Lipschitz condition

$$\|\mathbf{f}(t, \mathbf{x}_1) - \mathbf{f}(t, \mathbf{x}_2)\| \leq k\|\mathbf{x}_1 - \mathbf{x}_2\|, \tag{24}$$

whenever the pairs (t, $\mathbf{x}_1$,) and (t, $\mathbf{x}_2$) belong to D, where k is a positive constant. Then there is a constant $\delta > 0$ such that there exists a unique continuous vector solution $\mathbf{x}(t)$ of system **(15)** in the interval $|t - t_0| \leq \delta$. ♦

Before leaving this theorem, we should briefly discuss condition **(24).** It is easy to see that **(24)** is implied by the inequalities

$$|f_i(t, x_{11}, \ldots, x_{1n}) - f_i(t, x_{21}, \ldots, x_{2n})| \leq k_i \sum_{j=1}^{n} |x_{1j} - x_{2j}|, \tag{25}$$

for $i = 1, \ldots, n$ with $k = n\sqrt{\Sigma_{i=1}^{n} k_i^2}$. This fact follows from the double inequality

$$\frac{1}{n}\sum_{j=1}^{n} |x_j| \leq \|x\| \leq \sum_{j=1}^{n} |x_j|, \tag{26}$$

which is an immediate consequence of the definition of $\|\mathbf{x}\|$ (see Problem 21).

Another condition that implies inequality **(24)** is

$$|f_i(t, x_{11}, \ldots, x_{1n}) - f_i(t, x_{21}, \ldots, x_{2n})| \leq k_2 \max_j |x_{1j} - x_{2j}|, \tag{27}$$

for $i = 1, 2, \ldots, n$. The two inequalities **(25)** and **(27)** are useful, since it is often very difficult to verify inequality **(24)** directly.

Finally, if the partial derivatives $\partial f_i/\partial x_j$, $i, j = 1, 2, \ldots, n$, are continuous in D, then they are bounded on D and conditions **(25)** and **(27)** both follow from the mean value theorem of differential calculus.

Global existence and uniqueness for a linear system

Consider the system

$$\mathbf{x}' = A(t)\mathbf{x} + \mathbf{f}(t), \qquad \mathbf{x}(t_0) = \mathbf{x}_0, \tag{28}$$

where t_0 is a point in the interval $\alpha \leq t \leq \beta$ and $A(t)$ is an $n \times n$ matrix.

THEOREM 8 ♦ **Global existence-uniqueness theorem**

Let $A(t)$ and $\mathbf{f}(t)$ be a continuous matrix and vector function, respectively, on the interval $a \le t \le \beta$. Then there exists a unique vector function $\mathbf{x}(t)$ that is a solution of Equation **(28)** on the entire interval $\alpha \le t \le \beta$.

Remark

We proved this theorem for the scalar case ($n = 1$) in Section 2.8 (see p. 87).

Proof We define the Picard iterations

$$\begin{aligned} \mathbf{x}_0(t) &= \mathbf{x}_0, \\ \mathbf{x}_1(t) &= \mathbf{x}_0 + \int_{t_0}^{t} [A(s)\mathbf{x}_0(s) + \mathbf{f}(s)]\, ds, \\ &\vdots \\ \mathbf{x}_{n+1}(t) &= \mathbf{x}_0 + \int_{t_0}^{t} [A(s)\mathbf{x}_n(s) + \mathbf{f}(s)]\, ds, \\ &\vdots \end{aligned} \tag{29}$$

Clearly these iterations are continuous on $\alpha \le t \le \beta$ for $n = 1, 2, \ldots$. Since $A(t)$ is a continuous matrix function, we have

$$\sup_{\alpha \le t \le \beta} \| A(t) \| = \sup_{\alpha \le t \le \beta} \sum_{i=1}^{n} \sum_{j=1}^{n} |a_{ij}(t)| = k < \infty.$$

Let

$$M = \sup_{\alpha \le t \le \beta} \| A(t)\mathbf{x}_0 + \mathbf{f}(t) \|,$$

which is finite [according to statement (a)] since $A(t)$ and $\mathbf{f}(t)$ are continuous. Observe that we are unable to define the value δ of Equation **(7),** since the domain of the vector function $\mathbf{x}(t)$ is *unrestricted.* Thus a slight modification of the proof of Theorem 3 is necessary:

(a) Note that for any vectors $\mathbf{x}_1$ and $\mathbf{x}_2$ and for $\alpha \le t \le \beta$, we have

$$\begin{aligned} \| [A(t)\mathbf{x}_1 + \mathbf{f}(t)] - [A(t)\mathbf{x}_2 + \mathbf{f}(t)] \| &\le \| A(t) \| \, \|\mathbf{x}_1 - \mathbf{x}_2\| \\ &\le k \| \mathbf{x}_1 - \mathbf{x}_2 \|. \end{aligned}$$

Thus the vector function $A(t)\mathbf{x} + \mathbf{f}$ satisfies a Lipschitz condition for any $\mathbf{x}$ and $\alpha \le t \le \beta$.

(b) By induction, we now show that the Picard iterations **(29)** satisfy the inequality

$$\| \mathbf{x}_n(t) - \mathbf{x}_{n-1}(t) \| \le Mk^{n-1} \frac{|t - t_0|^n}{n!} \le Mk^{n-1} \frac{(\beta - \alpha)^n}{n!}. \tag{30}$$

We prove the first inequality. The second follows easily. First, if $n = 1$, then

$$\begin{aligned} \| \mathbf{x}_1(t) - \mathbf{x}_0(t) \| &= \left\| \int_{t_0}^{t} [A(s)\mathbf{x}_0 + \mathbf{f}(s)]\, ds \right\| \\ &\le \left| \int_{t_0}^{t} \| A(s)\mathbf{x}_0 + \mathbf{f}(s) \| \, ds \right| \\ &\le M \left| \int_{t_0}^{t} ds \right| = M|t - t_0| \le M(\beta - \alpha).* \end{aligned}$$

* The first of these inequalities is called the **Cauchy-Schwarz inequality.** Its proof may be found in any advanced calculus text. See, for example, R. Buck, *Advanced Calculus,* 3d ed., (New York: McGraw-Hill, 1978).

Assume that the result holds for $n = m$. Then by part (a),

$$\begin{aligned}\| \mathbf{x}_{m+1}(t) - \mathbf{x}_m(t) \| &\le \left| \int_{t_0}^{t} \| [A(s)\mathbf{x}_m(s) + \mathbf{f}(s)] - [A(s)\mathbf{x}_{m-1}(s) + \mathbf{f}(s)] \| \, ds \right| \\ &\le k \left| \int_{t_0}^{t} \| \mathbf{x}_m(s) - \mathbf{x}_{m-1}(s) \| \, ds \right| \le k \left| \int_{t_0}^{t} Mk^{m-1} \frac{|s - t_0|^m}{m!} \, ds \right| \\ &= Mk^m \frac{|t - t_0|^{m+1}}{(m+1)!} \le Mk^m \frac{(\beta - \alpha)^{m+1}}{(m+1)!},\end{aligned}$$

which is the desired result.

The major difference between this proof and the proof of Theorem 3 is that, in the linear case, the bound M can be defined independently of x, whereas in Theorem 3 we must restrict x as well as t. This difference enables us to prove the global result that solutions exist and are unique on the entire interval $\alpha \le t \le \beta$.

With estimate **(30)** [which is identical to estimate **(9)**], the proof proceeds exactly as those for Theorems 3 and 5. ♦

COROLLARY ♦ **Extension of the solution to the real axis**

Let $A(t)$ and $\mathbf{f}(t)$ be continuous on the interval $-\infty < t < \infty$. Then there exists a unique continuous solution $\mathbf{x}(t)$ of **(28)** defined for *all* t on $-\infty < t < \infty$.

Proof Suppose $|t_0| \le n$. We define $\mathbf{x}_n(t)$ to be the unique solution of **(28)** on the interval $|t| \le n$ which is guaranteed by Theorem 8. We observe that $\mathbf{x}_n(t)$ coincides with $\mathbf{x}_{n+k}(t)$ on the interval $|t| \le n$ for $k = 1, 2, 3, \ldots$, since both are solutions of **(28)** and Theorem 8 requires any solution on $|t| \le n$ to be unique. Thus the vector function

$$\mathbf{x}(t) = \lim_{n\to\infty} \mathbf{x}_n(t)$$

is defined for all real values t, since $|t| \le n$ for sufficiently large n. Clearly $\mathbf{x}(t)$ is a solution of **(28)** on $-\infty < t < \infty$, and such a solution must be unique since it is uniquely defined on each finite interval containing t_0. ♦

PROBLEMS APPENDIX 3

For each initial-value problem of Problems 1–10 determine whether a unique solution can be guaranteed. If so, let $a = b = 1$ if possible, and find the number δ as given by Equation **(7)**. *When possible, solve the equation and find a better value for δ, as in Example 3.*

1. $x' = x^3$, $x(2) = 5$

2. $x' = x^3$, $x(1) = 2$

3. $x' = \dfrac{x}{t - x}$, $x(0) = 1$

4. $x' = x^{1/3}$, $x(1) = 0$

5. $x' = \sin x$, $x(1) = \dfrac{\pi}{2}$

6. $x' = \sqrt{x(x-1)}$, $x(1) = 2$

7. $x' = \ln|\sin x|$, $x\left(\dfrac{\pi}{2}\right) = 1$

8. $x' = \sqrt{x(x-1)}$, $x(2) = 3$

9. $x' = |x|$, $x(0) = 1$

10. $x' = tx$, $x(5) = 10$

11. Compute a Lipschitz constant for each of the following functions on the indicated region D:

(a) $f(t, x) = te^{-2x/t}$, $t > 0$, $x > 0$

(b) $f(t, x) = \sin tx$, $|t| \le 1$, $|x| \le 2$

(c) $f(t, x) = e^{-t^2}x^2 \sin \dfrac{1}{x}$, all t, $-1 \le x < 1$, $x \ne 0$

[Part (c) shows that a Lipschitz constant may exist even when $\frac{\partial f}{\partial x}$ is not bounded in D.]

(d) $f(t, x) = (t^2x^3)^{3/2}$, $|t| \le 2$, $|x| \le 3$

12. Consider the initial-value problem

$$x' = x^2, \qquad x(1) = 3$$

Show that the Picard iterations converge to the unique solution of this problem.

13. Construct the sequence $\{x_n(t)\}$ of Picard iterations for the initial-value problem

$$x' = -x, \quad x(0) = 3,$$

and show that it converges to the unique solution $x(t) = 3e^{-t}$.

14. Prove Gronwall's inequality (Theorem 4) for $t_0 - \delta \leq t \leq t_0$. [*Hint:* For $t < t_0$, $y(t) \leq 0$.]

15. Let $v(t)$ be a positive function that satisfies the inequality

$$v(t) \leq A + \int_{t_0}^{t} r(s)v(s)\, ds,$$

where $A \geq 0$, $r(t)$ is a continuous, positive function, and $t \geq t_0$. Prove that

$$v(t) \leq A \exp\left[\int_{t_0}^{t} r(s)\, ds\right],$$

for all $t \geq t_0$. This is a general form of Gronwall's inequality. [*Hint:* Define $y(t) \equiv \int_{t_0}^{t} r(s)v(s)\, ds$ and show that $y'(t) = r(t)v(t) \leq r(t)[A + y(t)]$. Finish the proof by following the steps of the proof of Theorem 4, using the integrating factor $\exp[-\int_{t_0}^{t} r(s)\, ds]$.]

16. Consider the initial-value problem

$$x'' = f(t, x), \qquad x(0) = x_0, \qquad x'(0) = x_1, \quad \textbf{(i)}$$

where f is defined on the rectangle D: $|t| \leq a$, $|x - x_0| \leq b$. Prove under appropriate hypotheses that if a solution exists, then it must be unique. [*Hint:* Let $x(t)$ and $y(t)$ be continuous solutions of **(i)**. Verify by differentiation that

$$x(t) = x_0 + x_1 t - \int_0^t (t - s) f(s, x(s))\, ds,$$

$$y(t) = x_0 + x_1 t - \int_0^t (t - s) f(s, y(s))\, ds,$$

in some interval $|t| \leq \delta$, $\delta > 0$. Then subtract these two expressions, use an appropriate Lipschitz condition, and apply the Gronwall inequality of Problem 15.]

17. Prove inequality **(19)**.

18. Prove Equation **(20)**.

19. Prove inequality **(21)**. [*Hint:* (a) First show that the **inner product**

$$\mathbf{x} \cdot \mathbf{y} = x_1 y_1 + x_2 y_2 + \cdots + x_n y_n$$

satisfies the condition $|\mathbf{x} \cdot \mathbf{y}| \leq \|\mathbf{x}\|\,\|\mathbf{y}\|$, where $\mathbf{x} = (x_1, x_2, \ldots, x_n)$ and $\mathbf{y} = (y_1, y_2, \ldots, y_n)$. (b) Show that $\mathbf{x} \cdot \mathbf{x} = \|\mathbf{x}\|^2$. (c) Apply (b) to $\|\mathbf{x} + \mathbf{y}\|^2$ and use the inequality in (a) to verify **(21)**.]

20. Prove inequality **(22)**.

21. Prove inequality **(26)**.

22. Using inequality **(26)**, show that inequalities **(24)**, **(25)**, and **(27)** are all equivalent.

23. Show that if the partial derivatives $\partial f_i/\partial x_j$ are continuous on the region D defined by **(23)**, then there exists a constant $k_i > 0$ such that inequality **(24)** holds for $i = 1, \ldots, n$.

24. Prove Theorem 7 by carrying out the following steps:

(a) Show that system **(15)** is equivalent to the integral equation

$$\mathbf{x}(t) = \mathbf{x}_0 + \int_{t_0}^{t} \mathbf{f}(s, \mathbf{x}(s))\, ds. \quad \textbf{(ii)}$$

(b) Define the vector-valued Picard iterates $\{\mathbf{x}_n\}$ as in Equations **(5)**.

(c) Prove that

$$M = \sup_{(t,\,x)\in d} \|\mathbf{f}(t, \mathbf{x})\|$$

is finite.

(d) Prove that if $\delta = \min\{a, \frac{b}{M}\}$, then the Picard iterates defined in (b) satisfy the condition $\|\mathbf{x}_n(t) - \mathbf{x}_0\| \leq b$, for $|t - t_0| \leq a$.

(e) Show by induction that

$$\|\mathbf{x}_n(t) - \mathbf{x}_{n-1}(t)\| \leq Mk^{n-1}\frac{|t - t_0|^n}{n!} \leq Mk^{n-1}\frac{\delta^n}{n!}.$$

(f) Prove that the series

$$\sum_{m=1}^{\infty} \|\mathbf{x}_m(t) - \mathbf{x}_{m-1}(t)\|$$

converges uniformly for $|t - t_0| < \delta$ and conclude that $\mathbf{x}_n(t)$ converges uniformly to a vector function $\mathbf{x}(t)$ as n tends to ∞.

(g) Show that the $\mathbf{x}(t)$ defined in part (f) is a continuous solution to **(ii)** on the interval $|t - t_0| \leq \delta$.

(h) Use Gronwall's inequality to show that this solution is unique.

25. Consider the system

$$x_1' = t x_1 x_2 + t^2, \qquad x_1(0) = 1,$$
$$x_2' = x_1^2 + x_2^2 + t, \qquad x_2(0) = 2.$$

(a) Write the system in form **(15)**.

(b) Find a Lipschitz constant for $\mathbf{f}$ in the region

$$D\colon |t| \leq 1, \quad (x_1 - 1)^2 + (x_2 - 2)^2 \leq 1.$$

(c) Find $\delta = \min\{a, \frac{b}{m}\}$ as in Problem 24.

26. Follow the steps of Problem 25 for the system

$$x_1' = x_2^2 + 1, \qquad x_1(1) = 0,$$
$$x_2' = x_1^2 + t, \qquad x_2(1) = 1,$$

in the region D: $|t - 1| \leq 3$, $x_1^2 + (x_2 - 1)^2 \leq 4$.

27. Follow the steps of Problem 25 for the system

$$x_1' = x_1^2 x_2,$$
$$x_2' = x_3 + t, \qquad \mathbf{x}(0) = \mathbf{c},$$
$$x_3' = x_3^2,$$

in the region D: $|t| \leq a$, $\|\mathbf{x} - \mathbf{c}\| \leq b$.

28. Under the assumption of Theorem 7, consider the systems

$$\mathbf{x}' = \mathbf{f}(t, \mathbf{x}), \qquad \mathbf{x}(t_0) = \mathbf{x}_{10},$$
$$\mathbf{x}' = \mathbf{f}(t, \mathbf{x}), \qquad \mathbf{x}(t_0) = \mathbf{x}_{20},$$

Let $\mathbf{x}_1$ and $\mathbf{x}_2$ be solutions of these two systems, respectively. Prove that

$$\|\mathbf{x}_1(t) - \mathbf{x}_2(t)\| \leq \|\mathbf{x}_{10} - \mathbf{x}_{20}\| e^{k\delta}.$$

This shows that the solutions of Theorem 7 vary continuously with the initial vector $\mathbf{x}(t_0)$.

29. Consider the system

$$\mathbf{x}' = \mathbf{f}(t, \mathbf{x}). \qquad \textbf{(iii)}$$

Let D denote the region $a \leq t \leq b$, $\mathbf{x} \in R^n$, and assume that

$$\|\mathbf{f}(t, \mathbf{x}_1) - \mathbf{f}(t, \mathbf{x}_2)\| \leq k\|\mathbf{x}_1 - \mathbf{x}_2\|,$$

for any pair of points $(t, \mathbf{x}_1)$ and $t, \mathbf{x}_2)$ in D. Prove that for any pair $(t_0, \mathbf{x}_0)$ in D there exists a unique solution $\mathbf{x}(t)$ of Equation **(iii)** defined on the entire interval $a \leq t \leq b$, which satisfies $\mathbf{x}(t_0) = \mathbf{x}_0$. [*Hint:* Proceed as in the proof of Theorem 8. This is a more general global existence-uniqueness result.]

APPENDIX 4 DETERMINANTS

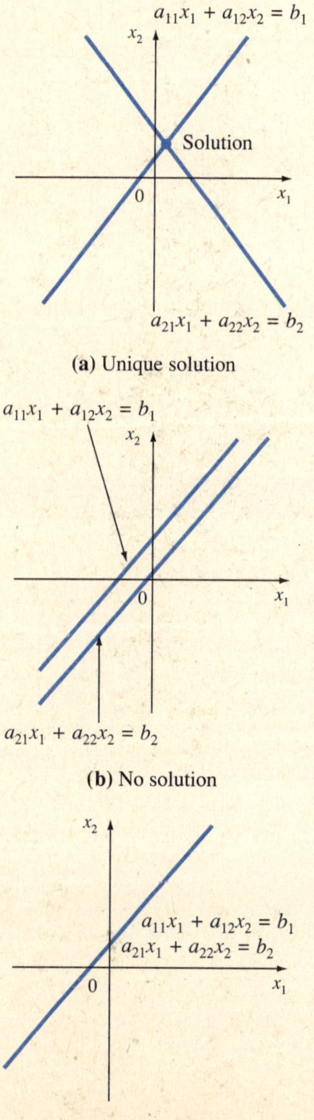

(a) Unique solution

(b) No solution

(c) Infinitely many solutions

FIGURE A4.1

In several parts of this book we make use of determinants. In this appendix we show how determinants arise and discuss their uses.

We begin by considering the system of two linear equations in two unknowns:

$$\begin{aligned} a_{11}x_1 + a_{12}x_2 &= b_1, \\ a_{21}x_1 + a_{22}x_2 &= b_2. \end{aligned} \tag{1}$$

For simplicity we assume that the constants a_{11}, a_{12}, a_{21}, and a_{22} are all nonzero (otherwise, the system can be solved directly). To solve system **(1)**, we reduce it to one equation in one unknown. To accomplish this, we multiply the first equation by a_{22} and the second by a_{12} to obtain

$$\begin{aligned} a_{11}a_{22}x_1 + a_{22}a_{12}x_2 &= a_{22}b_1, \\ a_{12}a_{21}x_1 + a_{22}a_{12}x_2 &= a_{12}b_2. \end{aligned} \tag{2}$$

Subtracting the second equation from the first, we have

$$(a_{11}a_{22} - a_{12}a_{21})x_1 = a_{22}b_1 - a_{12}b_2. \tag{3}$$

Now we define the quantity

Determinant

$$D = a_{11}a_{22} - a_{12}a_{21}. \tag{4}$$

If $D \neq 0$, then Equation **(3)** yields

$$x_1 = \frac{a_{22}b_1 - a_{12}b_2}{D}, \tag{5}$$

and x_2 may be obtained by substituting this value of x_1 into either of the equations of **(1)**. *Thus, if $D \neq 0$, system (1) has a unique solution.*

On the other hand, suppose that $D = 0$. Then $a_{11}a_{22} = a_{12}a_{21}$, and Equation **(3)** leads to the equation

$$0 = a_{22}b_1 - a_{12}b_2.$$

Either this equation is true or it is false. If it is false, that is, if $a_{22}b_1 - a_{12}b_2 \neq 0$, then system **(1)** has *no* solution. If the equation is true, that is, $a_{22}b_1 - a_{12}b_2 = 0$, then the second equation of **(2)** is a multiple of the first and **(1)** consists essentially of only one equation. In this case we may choose x_1 arbitrarily and calculate the corresponding value of x_2, which means that there are an *infinite* number of solutions. In sum, we have shown that *if $D = 0$, then system (1) has either no solution or an infinite number of solutions.*

These facts are easily visualized geometrically by noting that **(1)** consists of the equations of two straight lines. A solution of the system is a point of intersection of the two lines. If the slopes are different, then $D \neq 0$ and the two lines intersect at a single point, which is the unique solution. It is easy to show (see Problem 21) that $D = 0$ if and only if the slopes of the two lines are the same. If $D = 0$, either we have two parallel lines and no solution, since the lines never intersect, or both equations yield the same line and every point on this line is a solution. These results are illustrated in Figure A4.1.

EXAMPLE 1 ▶ **Determinants of systems of equations**

Consider the following systems of equations:

$$\text{(i)}\ \begin{aligned} 2x_1 + 3x_2 &= 12 \\ x_1 + x_2 &= 5 \end{aligned} \qquad \text{(ii)}\ \begin{aligned} x_1 + 3x_2 &= 3 \\ 3x_1 + 9x_2 &= 8 \end{aligned} \qquad \text{(iii)}\ \begin{aligned} x_1 + 3x_2 &= 3 \\ 3x_1 + 9x_2 &= 9 \end{aligned}$$

In system (i), $D = 2 \cdot 1 - 3 \cdot 1 = -1 \neq 0$, so there is a unique solution, which is easily found to be $x_1 = 3, x_2 = 2$. In system (ii), $D = 1 \cdot 9 - 3 \cdot 3 = 0$. Multiplying the first equation by 3 and then subtracting this from the second equation, we obtain the equation $0 = -1$, which is impossible. Thus there is no solution. In (iii), $D = 1 \cdot 9 - 3 \cdot 3 = 0$. But now the second equation is simply three times the first equation. If x_2 is arbitrary, then $x_1 = 3 - 3x_2$, and there are an infinite number of solutions. ◀

Returning again to system **(1)**, we define the **determinant of the system** as

$$D = a_{11}a_{22} - a_{12}a_{21}. \tag{6}$$

For convenience of notation, we denote the determinant by writing the coefficients of the system in a square array:

Determinant of a 2 × 2 system

$$D = \begin{vmatrix} a_{11} & a_{12} \\ a_{21} & a_{22} \end{vmatrix} = a_{11}a_{22} - a_{12}a_{21}. \tag{7}$$

Therefore, a 2×2 determinant is the product of the two components in the upper-left-to-lower-right diagonal minus the product of the other two components.

We have proved the following theorem.

THEOREM 1 ♦ **Unique solution when the determinant is nonzero**

For the 2×2 system **(1)**, there is a unique solution if and only if the determinant D is not equal to zero. If $D = 0$, then there is either no solution or an infinite number of solutions. ♦

Let us now consider the general system of n equations in n unknowns,

$$\begin{aligned} a_{11}x_1 + a_{12}x_2 + \cdots + a_{1n}x_n &= b_1, \\ a_{21}x_1 + a_{22}x_2 + \cdots + a_{2n}x_n &= b_2, \\ \vdots \qquad \vdots \qquad\quad \vdots \quad &\ \ \vdots \\ a_{n1}x_1 + a_{n2}x_2 + \cdots + a_{nn}x_n &= b_n, \end{aligned} \tag{8}$$

and define the determinant of such a system in order to obtain a theorem like the one above for $n \times n$ systems. We begin by defining the determinant of a 3×3 system.

Determinant of a 3 × 3 system

$$D = \begin{vmatrix} a_{11} & a_{12} & a_{13} \\ a_{21} & a_{22} & a_{23} \\ a_{31} & a_{32} & a_{33} \end{vmatrix} = a_{11}\begin{vmatrix} a_{22} & a_{23} \\ a_{32} & a_{33} \end{vmatrix} - a_{12}\begin{vmatrix} a_{21} & a_{23} \\ a_{31} & a_{33} \end{vmatrix} + a_{13}\begin{vmatrix} a_{21} & a_{22} \\ a_{31} & a_{32} \end{vmatrix}. \qquad (9)$$

We see that to calculate a 3×3 determinant, it is necessary to calculate three 2×2 determinants.

EXAMPLE 2 ▶ Determinant of a 3 × 3 system

$$\begin{vmatrix} 3 & 5 & 2 \\ 4 & 2 & 3 \\ -1 & 2 & 4 \end{vmatrix} = 3\begin{vmatrix} 2 & 3 \\ 2 & 4 \end{vmatrix} - 5\begin{vmatrix} 4 & 3 \\ -1 & 4 \end{vmatrix} + 2\begin{vmatrix} 4 & 2 \\ -1 & 2 \end{vmatrix}$$

$$= 3 \cdot 2 - 5 \cdot 19 + 2 \cdot 10 = -69. \quad ◀$$

The general definition of the determinant of the $n \times n$ system of Equations **(8)** is simply an extension of this procedure:

Determinant of $n \times n$ system

$$D = \begin{vmatrix} a_{11} & a_{12} & \cdots & a_{1n} \\ a_{21} & a_{22} & \cdots & a_{2n} \\ \vdots & \vdots & & \vdots \\ a_{n1} & a_{n2} & \cdots & a_{nn} \end{vmatrix} = a_{11}A_{11} - a_{12}A_{12} + \cdots + (-1)^{n+1}a_{1n}A_{1n}, \qquad ($$

where A_{1j} is the $(n-1) \times (n-1)$ determinant obtained by crossing out the first row and jth column of the original $n \times n$ determinant. Thus an $n \times n$ determinant can be obtained by calculating n $(n-1) \times (n-1)$ determinants. Note that in definition **(10)** the signs alternate. The signs of the n^2 $(n-1) \times (n-1)$ determinants can easily be illustrated by the following schematic diagram:

$$\begin{vmatrix} + & - & + & - & + & - & \cdots \\ - & + & - & + & - & + & \cdots \\ + & - & + & - & + & - & \cdots \\ - & + & - & + & - & + & \cdots \\ + & - & + & - & + & - & \cdots \\ \vdots & \vdots & \vdots & \vdots & \vdots & \vdots & \ddots \end{vmatrix}$$

EXAMPLE 3 ▶ Determinant of a 4 × 4 system

$$\begin{vmatrix} 1 & 3 & 5 & 2 \\ 0 & -1 & 3 & 4 \\ 2 & 1 & 9 & 6 \\ 3 & 2 & 4 & 8 \end{vmatrix} = 1\begin{vmatrix} -1 & 3 & 4 \\ 1 & 9 & 6 \\ 2 & 4 & 8 \end{vmatrix} - 3\begin{vmatrix} 0 & 3 & 4 \\ 2 & 9 & 6 \\ 3 & 4 & 8 \end{vmatrix} + 5\begin{vmatrix} 0 & -1 & 4 \\ 2 & 1 & 6 \\ 3 & 2 & 8 \end{vmatrix} - 2\begin{vmatrix} 0 & -1 & 3 \\ 2 & 1 & 9 \\ 3 & 2 & 4 \end{vmatrix}$$

$$= 1(-92) - 3(-70) + 5(2) - 2(-16) = 160.$$

(The values in parentheses are obtained by calculating the four 3×3 determinants.) ◀

The reason for considering determinants of systems of n equations in n unknowns is that Theorem 1 also holds for these systems (this fact is not proven here).*

THEOREM 2 ♦ **Uniqueness of solutions**

For system **(8)** there is a unique solution if and only if the determinant D, defined by **(10),** is not zero. If $D = 0$, then there is either no solution or an infinite number of solutions. ♦

It is clear that calculating determinants by formula **(10)** can be extremely tedious, especially if $n \geq 5$. For that reason techniques are available for greatly simplifying these calculations. Some of these techniques are described in the theorems below. The proofs of these theorems can be found in the text cited in the footnote.

We begin with the result that states that the determinant can be obtained by expanding in any row.

THEOREM 3 ♦ **Expansion by cofactors using rows**

For any i, $i = 1, 2, \ldots, n$,

$$D = \begin{vmatrix} a_{11} & a_{12} & \cdots & a_{1n} \\ a_{21} & a_{22} & \cdots & a_{2n} \\ \vdots & \vdots & & \vdots \\ a_{n1} & a_{n2} & \cdots & a_{nn} \end{vmatrix} = (-1)^{i+1} a_{i1} A_{i1} + (-1)^{i+2} a_{i2} A_{i2} + \ldots + (-1)^{i+n} a_{in} A_{in},$$

where A_{ij} is the $(n-1) \times (n-1)$ determinant obtained by crossing off the ith row and jth column of D. Notice that the signs in the expansion of a determinant alternate. ♦

EXAMPLE 4 ▶ **Finding the determinant by expanding by cofactors**

Calculate

$$\begin{vmatrix} 3 & 5 & 2 \\ 4 & 2 & 3 \\ -1 & 2 & 4 \end{vmatrix}$$

by expanding in the second row (see Example 2).

Solution

$$\begin{vmatrix} 3 & 5 & 2 \\ 4 & 2 & 3 \\ -1 & 2 & 4 \end{vmatrix} = (-1)^{2+1}(4)\begin{vmatrix} 5 & 2 \\ 2 & 4 \end{vmatrix} + (-1)^{2+2}(2)\begin{vmatrix} 3 & 2 \\ -1 & 4 \end{vmatrix} + (-1)^{2+3}(3)\begin{vmatrix} 3 & 5 \\ -1 & 2 \end{vmatrix}$$

$$= -4(16) + 2(14) - 3(11) = -69.$$

We remark that we can also get the same result by expanding in the third row of D. ◀

---

* For a proof, see S. I. Grossman, *Elementary Linear Algebra,* 5th ed., (Philadelphia: Harcourt Brace, 1995), Chapter 2.

THEOREM 4 ♦ **Expansion by cofactors using columns**

For any j, $j = 1, 2, \ldots, n$,

$$D = \begin{vmatrix} a_{11} & a_{12} & \cdots & a_{1n} \\ a_{21} & a_{22} & \cdots & a_{2n} \\ \vdots & \vdots & & \vdots \\ a_{n1} & a_{n2} & \cdots & a_{nn} \end{vmatrix} = (-1)^{1+j}a_{1j}A_{1j} + (-1)^{2+j}a_{2j}A_{2j} + \cdots + (-1)^{n+j}a_{nj}A_{nj},$$

where A_{ij} is as defined in Theorem 3. ♦

EXAMPLE 5 ▶ **Expanding using columns**

Calculate

$$D = \begin{vmatrix} 3 & 5 & 2 \\ 4 & 2 & 3 \\ -1 & 2 & 4 \end{vmatrix}$$

by expanding in the third column.

Solution

$$\begin{vmatrix} 3 & 5 & 2 \\ 4 & 2 & 3 \\ -1 & 2 & 4 \end{vmatrix} = (-1)^{1+3}(2)\begin{vmatrix} 4 & 2 \\ -1 & 2 \end{vmatrix} + (-1)^{2+3}(3)\begin{vmatrix} 3 & 5 \\ -1 & 2 \end{vmatrix} + (-1)^{3+3}(4)\begin{vmatrix} 3 & 5 \\ 4 & 2 \end{vmatrix}$$

$$= 2(10) - 3(11) + 4(-14) = -69.$$ ◀

THEOREM 5 ♦ **Facts about determinants**

Let the determinant be

$$D = \begin{vmatrix} a_{11} & a_{12} & \cdots & a_{1n} \\ a_{21} & a_{22} & \cdots & a_{2n} \\ \vdots & \vdots & & \vdots \\ a_{n1} & a_{n2} & \cdots & a_{nn} \end{vmatrix}.$$

(i) If any row or column of D is a zero vector, then $D = 0$.

(ii) If any row (column) is a multiple of any other row (column), then $D = 0$.

(iii) Interchanging any two rows (columns) of D has the effect of multiplying D by -1.

(iv) Multiplying a row (column) of D by a constant α has the effect of multiplying D by α.

(v) If any row (column) of D is multiplied by a constant and added to a different row (column) of D, then the determinant D is unchanged. ♦

EXAMPLE 6 ▶ **Finding a determinant**

Calculate

$$D = \begin{vmatrix} 2 & 1 & 4 & 3 \\ 3 & 1 & -2 & -1 \\ 14 & -2 & 0 & 6 \\ 6 & 2 & -4 & -2 \end{vmatrix}.$$

Solution $D = 0$ according to (ii) since the fourth row is twice the second row. This can easily be verified. ◀

EXAMPLE 7 ▶ Finding a determinant

Calculate

$$D = \begin{vmatrix} 1 & 14 & 3 \\ 2 & 28 & -2 \\ 0 & -42 & 1 \end{vmatrix}.$$

Solution According to (iv), we may divide the second column by 14, which has the effect of dividing D by 14. Then

$$\frac{D}{14} = \begin{vmatrix} 1 & 1 & 3 \\ 2 & 2 & -2 \\ 0 & -3 & 1 \end{vmatrix},$$

or

$$D = 14\begin{vmatrix} 1 & 1 & 3 \\ 2 & 2 & -2 \\ 0 & -3 & 1 \end{vmatrix} = 14(-24) = -336. \blacktriangleleft$$

EXAMPLE 8 ▶ Finding a determinant

Calculate

$$D = \begin{vmatrix} 0 & -42 & 1 \\ 2 & 28 & -2 \\ 1 & 14 & 3 \end{vmatrix}.$$

Solution D is obtained from the determinant of Example 7 by interchanging the first and third rows. Hence, according to (iii), $D = -(-336) = 336$. ◀

The results in Theorem 5 can be used to simplify the calculation of determinants.

EXAMPLE 9 ▶ Finding a determinant

Calculate

$$D = \begin{vmatrix} 1 & 3 & 5 & 2 \\ 0 & -1 & 3 & 4 \\ 2 & 1 & 9 & 6 \\ 3 & 2 & 4 & 8 \end{vmatrix}.$$

Solution This determinant was calculated in Example 3. The idea is to use Theorem 5 to make the evaluation of the determinant almost trivial. We begin by multiplying the first row by -2 and adding it to the third row. By (v), this manipulation leaves the determinant unchanged:

$$D = \begin{vmatrix} 1 & 3 & 5 & 2 \\ 0 & -1 & 3 & 4 \\ 2+(-2)1 & 1+(-2)3 & 9+(-2)5 & 6+(-2)2 \\ 3 & 2 & 4 & 8 \end{vmatrix}$$

$$= \begin{vmatrix} 1 & 3 & 5 & 2 \\ 0 & -1 & 3 & 4 \\ 0 & -5 & -1 & 2 \\ 3 & 2 & 4 & 8 \end{vmatrix}.$$

Now we multiply the first row by -3 and add it to the fourth row:

$$D = \begin{vmatrix} 1 & 3 & 5 & 2 \\ 0 & -1 & 3 & 4 \\ 0 & -5 & -1 & 2 \\ 0 & -7 & -11 & 2 \end{vmatrix}.$$

We now expand D by its first column:

$$D = 1\begin{vmatrix} -1 & 3 & 4 \\ -5 & -1 & 2 \\ -7 & -11 & 2 \end{vmatrix} - 0\begin{vmatrix} 3 & 5 & 2 \\ -5 & -1 & 2 \\ -7 & -11 & 2 \end{vmatrix} + 0\begin{vmatrix} 3 & 5 & 2 \\ -1 & 3 & 4 \\ -7 & -11 & 2 \end{vmatrix} + 0\begin{vmatrix} 3 & 5 & 2 \\ -1 & 3 & 4 \\ -5 & -1 & 2 \end{vmatrix}$$
$$= \begin{vmatrix} -1 & 3 & 4 \\ -5 & -1 & 2 \\ -7 & -11 & 2 \end{vmatrix},$$

which is a 3×3 determinant. We can calculate it by expansion or we can reduce further. By Theorem 5, parts (iv) and (v),

$$\begin{vmatrix} -1 & 3 & 4 \\ -5 & -1 & 2 \\ -7 & -11 & 2 \end{vmatrix} = -\begin{vmatrix} 1 & -3 & -4 \\ -5 & -1 & 2 \\ -7 & -11 & 2 \end{vmatrix} = -\begin{vmatrix} 1 & -3 & -24 \\ 0 & -16 & -18 \\ 0 & -32 & -26 \end{vmatrix}$$
$$= -\begin{vmatrix} -16 & -18 \\ -32 & -26 \end{vmatrix}$$
$$= -[(-16)(-26) - (-18)(-32)] = 160.$$

In the second step we multiplied the first row by 5 and added it to the second and multiplied the first row by 7 and added it to the third. Note how we were able to reduce the problem to the calculation of a single 2×2 determinant. ◀

There is one further useful result about determinants.

THEOREM 6 ♦ Sum of determinants

Let

$$D = \begin{vmatrix} a_{11} & a_{12} & \cdots & a_{1n} \\ a_{21} & a_{22} & \cdots & a_{2n} \\ \vdots & \vdots & & \vdots \\ a_{i1} + b_{i1} & a_{i2} + b_{i2} & \cdots & a_{in} + b_{in} \\ \vdots & \vdots & & \vdots \\ a_{n1} & a_{n2} & & a_{nn} \end{vmatrix}.$$

Then

$$D = \begin{vmatrix} a_{11} & a_{12} & \cdots & a_{1n} \\ a_{21} & a_{22} & \cdots & a_{2n} \\ \vdots & \vdots & & \vdots \\ a_{i1} & a_{i2} & \cdots & a_{in} \\ \vdots & \vdots & & \vdots \\ a_{n1} & a_{n2} & \cdots & a_{nn} \end{vmatrix} + \begin{vmatrix} a_{11} & a_{12} & \cdots & a_{1n} \\ a_{21} & a_{22} & \cdots & a_{2n} \\ \vdots & \vdots & & \vdots \\ b_{i1} & b_{i2} & \cdots & b_{in} \\ \vdots & \vdots & & \vdots \\ a_{n1} & a_{n2} & \cdots & a_{nn} \end{vmatrix}. \quad (11)$$ ♦

EXAMPLE 10 ▶ **Find a sum of determinants**

To illustrate Theorem 6, we note that

$$\begin{vmatrix} 2 & 1 & 4 \\ 3+5 & 2-3 & 1+2 \\ 0 & -4 & 2 \end{vmatrix} = \begin{vmatrix} 2 & 1 & 4 \\ 3 & 2 & 1 \\ 0 & -4 & 2 \end{vmatrix} + \begin{vmatrix} 2 & 1 & 4 \\ 5 & -3 & 2 \\ 0 & -4 & 2 \end{vmatrix}$$
$$= -38 - 86 = -124. \quad ◀$$

We conclude this appendix by showing how determinants can be used to obtain the unique solution (if one exists) of system **(8)** of n equations in n unknowns. We define the determinants

$$D_1 = \begin{vmatrix} b_1 & a_{12} & \cdots & a_{1n} \\ b_2 & a_{22} & \cdots & a_{2n} \\ \vdots & \vdots & & \vdots \\ b_n & a_{n2} & \cdots & a_{nn} \end{vmatrix},$$

$$D_2 = \begin{vmatrix} a_{11} & b_1 & a_{13} & \cdots & a_{1n} \\ a_{21} & b_2 & a_{23} & \cdots & a_{2n} \\ \vdots & \vdots & \vdots & & \vdots \\ a_{n1} & b_n & a_{n3} & \cdots & a_{nn} \end{vmatrix}, \ldots,$$

$$D_k = \begin{vmatrix} a_{11} & a_{12} & \cdots & a_{1,k-1} & b_1 & a_{1,k+1} & \cdots & a_{1n} \\ a_{21} & a_{22} & \cdots & a_{2,k-1} & b_2 & a_{2,k+1} & \cdots & a_{2n} \\ \vdots & \vdots & & \vdots & \vdots & \vdots & & \vdots \\ a_{n1} & a_{n2} & \cdots & a_{n,k-1} & b_n & a_{n,k+1} & \cdots & a_{nn} \end{vmatrix}, \ldots, \tag{12}$$

$$D_n = \begin{vmatrix} a_{11} & a_{12} & \cdots & a_{1,n-1} & b_1 \\ a_{21} & a_{22} & \cdots & a_{2,n-1} & b_2 \\ \vdots & \vdots & & \vdots & \vdots \\ a_{n1} & a_{n2} & \cdots & a_{n,n-1} & b_n \end{vmatrix},$$

obtained by replacing the kth column of D by the column

$$\begin{pmatrix} b_1 \\ b_2 \\ \vdots \\ b_n \end{pmatrix}.$$

Then we have the following theorem.

THEOREM 7 ♦ **Cramer's rule**

Let D and D_k, $k = 1, 2, \ldots, n$, be given as in **(10)** and **(12)**. If $D \neq 0$, then the unique solution to system **(8)** is given by the values

$$x_1 = \frac{D_1}{D}, \quad x_2 = \frac{D_2}{D}, \ldots, x_n = \frac{D_n}{D}. \quad ♦ \tag{13}$$

EXAMPLE 11 ▶ **Using Cramer's rule**

$$\begin{aligned} 2x_1 + 4x_2 - x_3 &= -5, \\ -4x_1 + 3x_2 + 5x_3 &= 14, \\ 6x_1 - 3x_2 - 2x_3 &= 5. \end{aligned}$$

We have

$$D = \begin{vmatrix} 2 & 4 & -1 \\ -4 & 3 & 5 \\ 6 & -3 & -2 \end{vmatrix} = 112, \qquad D_1 = \begin{vmatrix} -5 & 4 & -1 \\ 14 & 3 & 5 \\ 5 & -3 & -2 \end{vmatrix} = 224,$$

$$D_2 = \begin{vmatrix} 2 & -5 & -1 \\ -4 & 14 & 5 \\ 6 & 5 & -2 \end{vmatrix} = -112, \qquad D_3 = \begin{vmatrix} 2 & 4 & -5 \\ -4 & 3 & 14 \\ 6 & -3 & 5 \end{vmatrix} = 560.$$

Therefore

$$x_1 = \frac{D_1}{D} = 2, \qquad x_2 = \frac{D_2}{D} = -1, \qquad x_3 = \frac{D_3}{D} = 5.$$ ◀

Note

As a general rule, avoid Cramer's rule if $n > 3$. There is too much work involved.

PROBLEMS APPENDIX 4

For each of the 2×2 systems in Problems 1–8 calculate the determinant D. If $D \neq 0$, find the unique solution. If $D = 0$, determine whether there is no solution or an infinite number of solutions.

1. $\begin{aligned} 2x_1 + 4x_2 &= 6 \\ x_1 + x_2 &= 3 \end{aligned}$

2. $\begin{aligned} 2x_1 + 4x_2 &= 6 \\ x_1 + 2x_2 &= 5 \end{aligned}$

3. $\begin{aligned} 2x_1 + 4x_2 &= 6 \\ x_1 + 2x_2 &= 3 \end{aligned}$

4. $\begin{aligned} 6x_1 - 3x_2 &= 3 \\ -2x_1 + x_2 &= -1 \end{aligned}$

5. $\begin{aligned} 6x_1 - 3x_2 &= 3 \\ -2x_1 + x_2 &= 1 \end{aligned}$

6. $\begin{aligned} 6x_1 - 3x_2 &= 3 \\ -2x_1 + 2x_2 &= -1 \end{aligned}$

7. $\begin{aligned} 2x_1 + 5x_2 &= 0 \\ 3x_1 - 7x_2 &= 0 \end{aligned}$

8. $\begin{aligned} 2x_1 - 3x_2 &= 0 \\ -4x_1 + 6x_2 &= 0 \end{aligned}$

In Problems 9–20 calculate the determinant.

9. $\begin{vmatrix} 1 & 2 & 3 \\ 6 & -1 & 4 \\ 2 & 0 & 6 \end{vmatrix}$

10. $\begin{vmatrix} 4 & -1 & 0 \\ 2 & 1 & 7 \\ -2 & 3 & 4 \end{vmatrix}$

11. $\begin{vmatrix} 7 & 2 & 3 \\ 0 & 4 & 1 \\ 0 & 0 & 5 \end{vmatrix}$

12. $\begin{vmatrix} 1 & -1 & 4 \\ 3 & -2 & 1 \\ 5 & 1 & 7 \end{vmatrix}$

13. $\begin{vmatrix} 4 & 2 & 7 \\ 1 & 5 & 3 \\ -1 & 1 & 4 \end{vmatrix}$

14. $\begin{vmatrix} -1 & 0 & 4 \\ 7 & 3 & 2 \\ 4 & 1 & 5 \end{vmatrix}$

15. $\begin{vmatrix} 1 & 4 & 7 & 2 \\ 0 & 5 & 8 & 1 \\ 0 & 0 & -3 & 4 \\ 0 & 0 & 0 & 8 \end{vmatrix}$

16. $\begin{vmatrix} a_1 & a_2 & a_3 & a_4 \\ 0 & b_1 & b_2 & b_3 \\ 0 & 0 & c_1 & c_2 \\ 0 & 0 & 0 & d_1 \end{vmatrix}$

17. $\begin{vmatrix} 2 & 1 & 3 & 4 \\ 3 & -2 & 5 & 1 \\ 4 & 0 & 4 & 5 \\ 2 & 1 & 7 & -4 \end{vmatrix}$

18. $\begin{vmatrix} 1 & 3 & -1 & 7 \\ -2 & 5 & 2 & 8 \\ -3 & 7 & 3 & 3 \\ 5 & 0 & -5 & 11 \end{vmatrix}$

19. $\begin{vmatrix} 2 & 3 & 1 & 4 \\ 2 & 2 & 4 & 6 \\ 3 & -1 & -2 & 4 \\ 4 & 2 & -3 & -5 \end{vmatrix}$

20. $\begin{vmatrix} 1 & 0 & 2 & 3 & 1 \\ 0 & 4 & -1 & -2 & 3 \\ 2 & 1 & 0 & -1 & 1 \\ -3 & 2 & 2 & 0 & 5 \\ 0 & 3 & 6 & 1 & -3 \end{vmatrix}$

21. Show that two lines in system **(1)** have the same slope if and only if the determinant of the system is zero.

In Problems 22–28 solve the system by using Cramer's rule.

22.
$$\begin{aligned} 3x_1 - x_2 &= 13 \\ -4x_1 + 6x_2 &= -8 \end{aligned}$$

23.
$$\begin{aligned} 2x_1 + 6x_2 + 3x_3 &= 9 \\ -3x_1 - 17x_2 - x_3 &= 4 \\ 4x_1 + 3x_2 + x_3 &= -7 \end{aligned}$$

24.
$$\begin{aligned} 2x_1 \phantom{{}- 2x_2} + x_3 &= 0 \\ 3x_1 - 2x_2 + 2x_3 &= -4 \\ 4x_1 - 5x_2 \phantom{{}+ 2x_3} &= 3 \end{aligned}$$

25.
$$\begin{aligned} x_1 - 8x_2 - x_3 &= -1 \\ -x_1 + 4x_2 + x_3 &= 3 \\ 3x_1 - 2x_2 + 6x_3 &= 5 \end{aligned}$$

26.
$$\begin{aligned} x_1 + 2x_2 - x_3 - 4x_4 &= 1 \\ -x_1 \phantom{{}+ 2x_2} + 2x_3 + 6x_4 &= 5 \\ -4x_2 - 2x_3 - 8x_4 &= -8 \\ 3x_1 - 2x_2 \phantom{{}- 2x_3} + 5x_4 &= 3 \end{aligned}$$

27.
$$\begin{aligned} 2x_1 + 5x_2 - 3x_3 + 2x_4 &= 3 \\ -x_1 - 3x_2 + 2x_3 - x_4 &= -1 \\ -3x_1 + 4x_2 + 8x_3 - 2x_4 &= 4 \\ 6x_1 - x_2 - 6x_3 + 4x_4 &= 2 \end{aligned}$$

28.
$$\begin{aligned} x_1 - 2x_2 + x_3 &= 0 \\ 2x_1 - x_2 - 3x_3 &= 0 \\ 5x_1 + 7x_2 - 8x_3 &= 0 \end{aligned}$$

APPENDIX 5

COMPLEX NUMBERS

In algebra we encounter the problem of finding the roots of the polynomial

$$\lambda^2 + a\lambda + b = 0. \tag{1}$$

To find the roots, we use the quadratic formula to obtain

$$\lambda = \frac{-a \pm \sqrt{a^2 - 4b}}{2}. \tag{2}$$

If $a^2 - 4b > 0$, there are two real roots. If $a^2 - 4b = 0$, we obtain the single root (of multiplicity 2) $\lambda = -\frac{a}{2}$. To deal with the case $a^2 - 4b < 0$, we introduce the **imaginary number***

Imaginary number

$$i = \sqrt{-1}. \tag{3}$$

Then for $a^2 - 4b < 0$,

$$\sqrt{a^2 - 4b} = \sqrt{(4b - a^2)(-1)} = \sqrt{4b - a^2}\,\sqrt{-1} = \sqrt{4b - a^2}\, i,$$

and the two roots of **(1)** are given by

$$\lambda_1 = -\frac{a}{2} + \frac{\sqrt{4b - a^2}}{2} i \quad \text{and} \quad \lambda_2 = -\frac{a}{2} - \frac{\sqrt{4b - a^2}}{2} i.$$

---

* You should not be troubled by the term "imaginary." It is just a name. The British mathematician Alfred North Whitehead, in the chapter on imaginary numbers in his *Introduction to Mathematics,* wrote:

> At this point it may be useful to observe that a certain type of intellect is always worrying itself and others by discussion as to the applicability of technical terms. Are the incommensurable numbers properly called numbers? Are the positive and negative numbers really numbers? Are the imaginary numbers imaginary, and are they numbers?—are types of such futile questions. Now, it cannot be too clearly understood that, in science, technical terms are names arbitrarily assigned, like Christian names to children. There can be no question of the names being right or wrong. They may be judicious or injudicious; for they can sometimes be so arranged as to be easy to remember, or so as to suggest relevant and important ideas. But the essential principle involved was quite clearly enunciated in Wonderland to Alice by Humpty Dumpty, when he told her, apropos of his use of words, 'I pay them extra and make them mean what I like'. So we will not bother as to whether imaginary numbers are imaginary, or as to whether they are numbers, but will take the phrase as the arbitrary name of a certain mathematical idea, which we will now endeavour to make plain.

EXAMPLE 1 ▶ **Finding roots**

Find the roots of the quadratic equation $\lambda^2 + 2\lambda + 5 = 0$.

Solution We have $a = 2$, $b = 5$, and $a^2 - 4b = -16$. Thus $\sqrt{a^2 - 4b} = \sqrt{-16} = \sqrt{16}\sqrt{-1} = 4i$, and the roots are

$$\lambda_1 = \frac{-2 + 4i}{2} = -1 + 2i \quad \text{and} \quad \lambda_2 = -1 - 2i. \quad ◀$$

Definition Complex number

A **complex number** is a number of the form

$$z = \alpha + i\beta, \tag{4}$$

where α and β are real numbers. α is called the **real part** of z and is denoted by Re z. β is called the **imaginary part** of z and is denoted by Im z. Representation **(4)** is sometimes called the **Cartesian form** of the complex number z.

Remark

If $\beta = 0$ in Equation **(4),** then $z = \alpha$ is a real number. In this context we can regard the set of real numbers as a subset of the set of complex numbers.

EXAMPLE 2 ▶ **Real and imaginary parts**

In Example 1, Re $\lambda_1 = -1$ and Im $\lambda_1 = 2$. ◀

We can add and multiply complex numbers by using the standard rules of algebra.

EXAMPLE 3 ▶ **Adding, subtracting, and multiplying complex numbers**

Let $z = 2 + 3i$ and $w = 5 - 4i$. Calculate (a) $z + w$, (b) $3w - 5z$, and (c) zw.

Solution

(a) $z + w = (2 + 3i) + (5 - 4i) = (2 + 5) + (3 - 4)i = 7 - i$.

(b) $3w = 3(5 - 4i) = 15 - 12i$, $5z = 10 + 15i$, and
$3w - 5z = (15 - 12i) - (10 + 15i) = (15 - 10) + i(-12 - 15)$
$= 5 - 27i$.

(c) $zw = (2 + 3i)(5 - 4i) = (2)(5) + 2(-4i) + (3i)(5) + (3i)(-4i)$
$= 10 - 8i + 15i - 12i^2 = 10 + 7i + 12 = 22 + 7i$. Here we use the fact that $i^2 = -1$. ◀

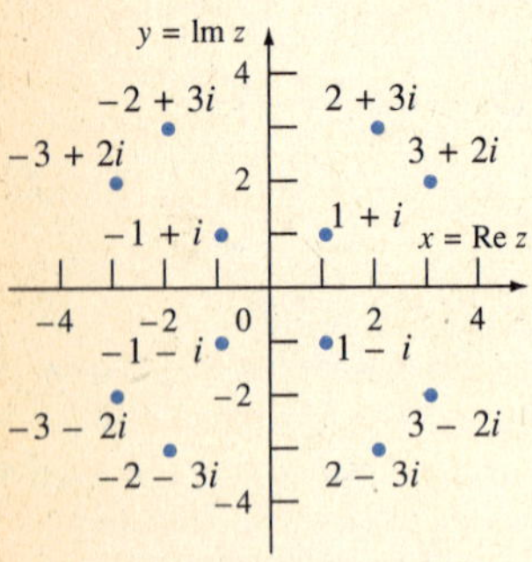

FIGURE A5.1

We can plot a complex number z in the xy-plane by plotting Re z along the x-axis and Im z along the y-axis. Thus each complex number can be thought of as a point in the xy-plane. With this representation the xy-plane is called the **complex plane.** Some representative points are plotted in Figure A5.1.

If $z = \alpha + i\beta$, then we define the **conjugate** of z, denoted $\bar{z}$, by

$$\bar{z} = \alpha - i\beta. \tag{5}$$

Figure A5.2 depicts a representative value of z and $\bar{z}$.

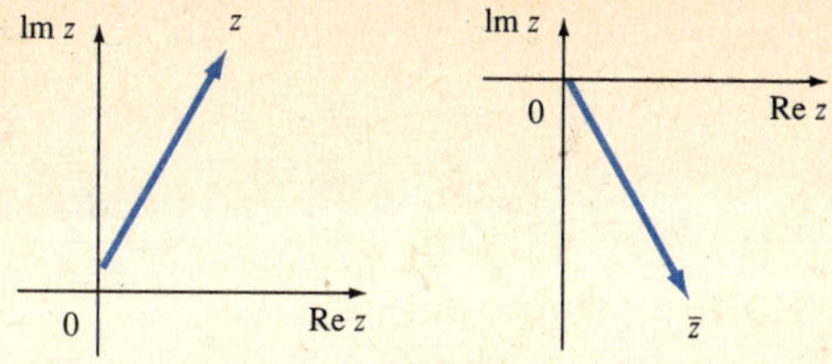

(a) (b)

FIGURE A5.2

EXAMPLE 4 ▶ **Conjugate of a complex number**

Compute the conjugates of (a) $1 + i$, (b) $3 - 4i$, (c) $-7 + 5i$, and (d) -3.

Solution

(a) $\overline{1 + i} = 1 - i$. **(b)** $\overline{3 - 4i} = 3 + 4i$. **(c)** $\overline{-7 + 5i} = -7 - 5i$.
(d) $\overline{-3} = -3$. ◀

It is not difficult to show (see Problem 35) that

$$\bar{z} = z, \qquad \text{if and only if } z \text{ is real.} \tag{6}$$

If $z = \beta i$ with β real, then z is said to be **pure imaginary.** We can then show (see Problem 36) that

$$\bar{z} = -z, \qquad \text{if and only if } z \text{ is pure imaginary.} \tag{7}$$

Let $p_n(x) = a_0 + a_1x + a_2x^2 + \cdots + a_nx^n$ be a polynomial with real coefficients. Then it can be shown (see Problem 41) that the complex roots of the equation $p_n(x) = 0$ occur in complex conjugate pairs; that is, if z is a root of $p_n(x) = 0$, then so is $\bar{z}$. We saw this fact illustrated in Example 1 in the case in which $n = 2$.

For $z = \alpha + i\beta$ we define the **magnitude** of z, denoted $|z|$, by

$$|z| = \sqrt{a^2 + \beta^2}, \tag{8}$$

and we define the **argument** of z, denoted by arg z, as the angle θ between the line $0z$ and the positive x-axis. From Figure A5.3 we see that $r = |z|$ is the distance from z to the origin, and, if $-\frac{\pi}{2} < \theta < \frac{\pi}{2}$,

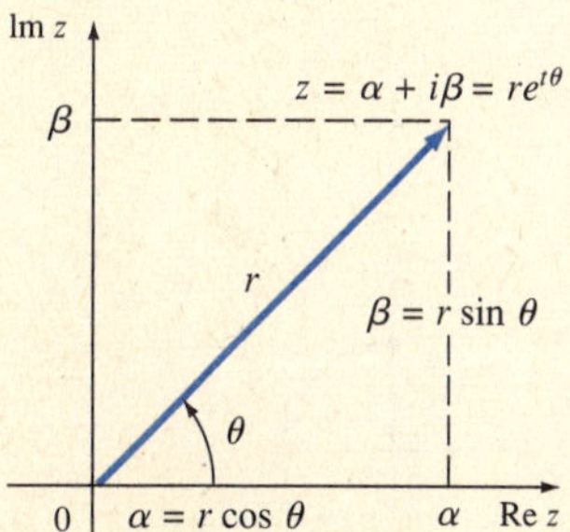

FIGURE A5.3

$$\theta = \arg z = \tan^{-1} \frac{\beta}{\alpha}. \tag{9}$$

By convention, we always choose values of arg z that lie in the interval

$$-\pi < \theta \leq \pi. \tag{10}$$

From Figure A5.4 we see that

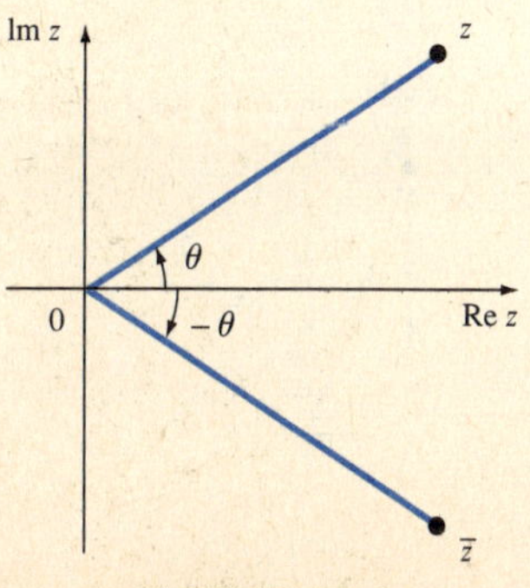

FIGURE A5.4

$$|\bar{z}| = |z| \tag{11}$$

and

$$\arg \bar{z} = -\arg z. \tag{12}$$

We can use $|z|$ and arg z to describe what is often a more convenient way to represent complex numbers. From Figure A5.3 it is evident that, if $z = \alpha + i\beta$, $r = |z|$, and $\theta = \arg z$, then

$$\alpha = r \cos \theta \quad \text{and} \quad \beta = r \sin \theta. \tag{13}$$

We see at the end of this appendix that

$$e^{i\theta} = \cos\theta + i\sin\theta. \tag{14}$$

Since $\cos(-\theta) = \cos\theta$ and $\sin(-\theta) = -\sin\theta$, we also have

$$e^{-i\theta} = \cos(-\theta) + i\sin(-\theta) = \cos\theta - i\sin\theta. \tag{14'}$$

Formula **(14)** is called the **Euler formula.** Using the Euler formula and Equations **(13),** we have

$$z = \alpha + i\beta = r\cos\theta + ir\sin\theta = r(\cos\theta + i\sin\theta),$$

or

$$z = re^{i\theta}. \tag{15}$$

Representation **(15)** is called the **polar form** of the complex number z.

EXAMPLE 5 ▶ **Finding the polar form of complex numbers**

Determine the polar forms of the following complex numbers: (a) 1, (b) -1, (c) i, (d) $1 + i$, (e) $-1 - \sqrt{3}\,i$, and (f) $-2 + 7i$.

Solution The six points are plotted in Figure A5.5.

(a) From Figure A5.5(a) it is clear that $\arg 1 = 0$. Since $\operatorname{Re} 1 = 1$, we see that, in polar form,

$$1 = 1e^{i0} = 1e^0 = e^0.$$

(b) Since $\arg(-1) = \pi$ [Figure A5.5(b)] and $|-1| = 1$, we have

$$-1 = 1e^{\pi i} = e^{i\pi}.$$

(c) From Figure A5.5(c) we see that $\arg i = \frac{\pi}{2}$. Since $|i| = \sqrt{0^2 + 1^2} = 1$, it follows that

$$i = e^{i\pi/2}.$$

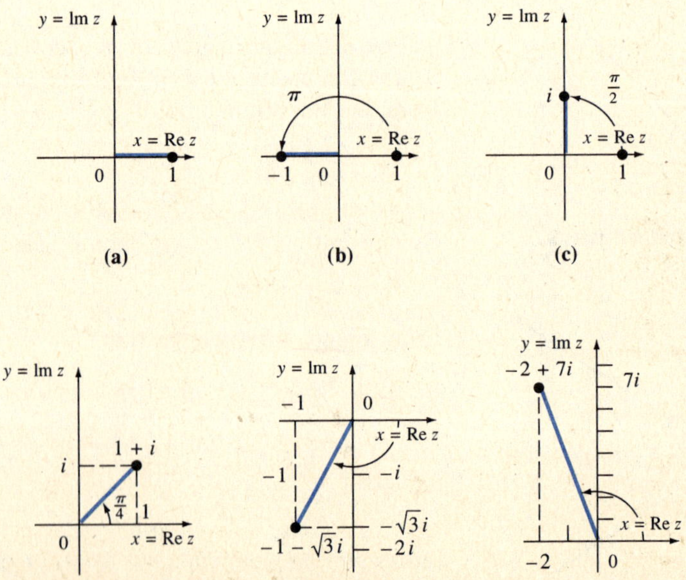

(a) (b) (c)

(d) (e) (f)

FIGURE A5.5

(d) $\arg(1 + i) = \tan^{-1}(\frac{1}{1}) = \frac{\pi}{4}$, and $|1 + i| = \sqrt{1^2 + 1^2} = \sqrt{2}$, so

$$1 + i = \sqrt{2}\, e^{i\pi/4}.$$

(e) Here $\tan^{-1}(\frac{\beta}{\alpha}) = \tan^{-1}\sqrt{3} = \frac{\pi}{3}$. However, $\arg z$ is in the third quadrant, so

$$\arg z = (\tfrac{\pi}{3}) - \pi = -\tfrac{2\pi}{3}.$$

Also, $|-1 - \sqrt{3}\, i| = \sqrt{1^2 + (\sqrt{3})^2} = \sqrt{1 + 3} = 2$, so

$$-1 - \sqrt{3}\, i = 2e^{-2\pi i/3}.$$

(f) To compute this complex number, we need a calculator. A calculator indicates that

$$\tan^{-1}(-\tfrac{7}{2}) = \tan^{-1}(-3.5) \approx -1.2925.$$

But $\tan^{-1} x$ is defined as a number in the interval $(-\frac{\pi}{2}, \frac{\pi}{2})$. Since from Figure A5.5(f) θ is in the second quadrant, we see that $\arg z = \tan^{-1}(-3.5) + \pi \approx 1.8491$. Next, we see that

$$|-2 + 7i| = \sqrt{(-2)^2 + 7^2} = \sqrt{53}.$$

Hence

$$-2 + 7i \approx \sqrt{53}\, e^{1.8491i.}$$ ◀

EXAMPLE 6 ▶ **Converting from polar to Cartesian form**

Convert the following complex numbers from polar to Cartesian form:
(a) $2e^{i\pi/3}$, **(b)** $4e^{3\pi i/2}$.

Solution

(a) $e^{i\pi/3} = \cos\left(\frac{\pi}{3}\right) + i\sin\left(\frac{\pi}{3}\right) = \frac{1}{2} + \left(\frac{\sqrt{3}}{2}\right)i$. Thus $2e^{i\pi/3} = 1 + \sqrt{3}\, i$.

(b) $e^{3\pi i/2} = \cos\left(\frac{3\pi}{2}\right) + i\sin\left(\frac{3\pi}{2}\right) = 0 + i(-1) = -i$. Thus $4e^{3\pi i/2} = -4i$.

◀

If $\theta = \arg z$, then by Equation **(12)**, $\arg \bar{z} = -\theta$. Thus, since $|\bar{z}| = |z|$, we have the following:

$$\text{If } z = re^{i\theta}, \quad \text{then } \bar{z} = re^{-i\theta}. \qquad \textbf{(16)}$$

Suppose we write a complex number in its polar form $z = re^{i\theta}$. Then

$$z^n = (re^{i\theta})^n = r^n(e^{i\theta})^n = r^n e^{in\theta} = r^n(\cos n\theta + i\sin n\theta). \qquad \textbf{(17)}$$

Formula **(17)** is useful for a variety of computations. In particular, when $r = |z| = 1$, we obtain the

De Moivre formula*

$$(\cos\theta + i\sin\theta)^n = \cos n\theta + i\sin n\theta. \qquad \textbf{(18)}$$

*Abraham De Moivre (1667–1754) was a French mathematician well known for his work in probability theory, infinite series, and trigonometry. He was so highly regarded that Newton often told those who came to him with questions on mathematics, "Go to M. De Moivre; he knows these things better than I do."

EXAMPLE 7 ▶ **Using De Moivre's formula**

Compute $(1+i)^5$.

Solution In Example 5(d) we showed that $1+i=\sqrt{2}\,e^{\pi i/4}$. Then

$$(1+i)^5=(\sqrt{2}\,e^{\pi i/4})^5=(\sqrt{2})^5e^{5\pi i/4}=4\sqrt{2}\left(\cos\frac{5\pi}{4}+i\sin\frac{5\pi}{4}\right)$$
$$=4\sqrt{2}\left(-\frac{1}{\sqrt{2}}-\frac{1}{\sqrt{2}}i\right)=-4-4i.$$

This can be checked by direct calculation. If the direct calculation seems no more difficult, then try to compute $(1+i)^{20}$ directly. Proceeding as above, we obtain

$$(1+i)^{20}=(\sqrt{2})^{20}e^{20\pi i/4}=2^{10}(\cos 5\pi+i\sin 5\pi)$$
$$=2^{10}(-1+0)=-1024.$$ ◀

Proof of Euler's formula

We now show that

$$e^{i\theta}=\cos\theta+i+\sin\theta \tag{19}$$

by using power series. We have

$$e^x=1+x+\frac{x^2}{2!}+\frac{x^3}{3!}+\ldots,\text{*} \tag{20}$$

$$\sin x=x-\frac{x^3}{3!}+\frac{x^5}{5!}-\ldots, \tag{21}$$

$$\cos x=1-\frac{x^2}{2!}+\frac{x^4}{4!}-\ldots. \tag{22}$$

Then

$$e^{i\theta}=1+(i\theta)+\frac{(i\theta)^2}{2!}+\frac{(i\theta)^3}{3!}+\frac{(i\theta)^4}{4!}+\frac{(i\theta)^5}{5!}+\ldots. \tag{23}$$

Now $i^2=-1$, $i^3=-i$, $i^4=1$, $i^5=i$, and so on. Thus **(23)** can be written

$$e^{i\theta}=1+i\theta-\frac{\theta^2}{2!}-\frac{i\theta^3}{3!}+\frac{\theta^4}{4!}+\frac{i\theta^5}{5!}-\ldots$$
$$=\left(1-\frac{\theta^2}{2!}+\frac{\theta^4}{4!}-\ldots\right)+i\left(\theta-\frac{\theta^3}{3!}+\frac{\theta^5}{5!}-\ldots\right)$$
$$=\cos\theta+i\sin\theta.$$

This completes the proof. ♦

PROBLEMS APPENDIX 5

In Problems 1–5 perform the indicated operation.

1. $(2-3i)+(7-4i)$
2. $3(4+i)-5(-3+6i)$
3. $(1+i)(1-i)$
4. $(2-3i)(4+7i)$
5. $(-3+2i)(7+3i)$

*Although we do not prove it here, these series expansions are also valid when x is a complex number.

In Problems 6–15 convert the complex number to its polar form.

6. $5i$ **7.** $5 + 5i$ **8.** $-2 - 2i$ **9.** $3 - 3i$ **10.** $2 + 2\sqrt{3}i$
11. $3\sqrt{3} + 3i$ **12.** $1 - \sqrt{3}\,i$ **13.** $4\sqrt{3} - 4i$ **14.** $-6\sqrt{3} - 6i$ **15.** $-1 - \sqrt{3}\,i$

In Problems 16–25 convert from polar to Cartesian form.

16. $e^{3\pi i}$ **17.** $2e^{-7\pi i}$ **18.** $\frac{1}{2}e^{3\pi i/4}$ **19.** $\frac{1}{2}e^{-3\pi i/4}$ **20.** $6e^{\pi i/6}$
21. $4e^{5\pi i/6}$ **22.** $4e^{-5\pi i/6}$ **23.** $3e^{-2\pi i/3}$ **24.** $\sqrt{3}\,e^{23\pi i/4}$ **25.** e^{i}

In Problems 26–34 compute the conjugate of the given number.

26. $3 - 4i$ **27.** $4 + 6i$ **28.** $-3 + 8i$
29. $-7i$ **30.** 16 **31.** $2e^{\pi i/7}$
32. $4e^{3\pi i/5}$ **33.** $3e^{-4\pi i/11}$ **34.** $e^{0.012i}$

35. Show that $z = \alpha + i\beta$ is real if and only if $z = \bar{z}$. [*Hint*: If $z = \bar{z}$, show that $\beta = 0$.]

36. Show that $z = \alpha + i\beta$ is pure imaginary if and only if $z = -\bar{z}$. [*Hint*: If $z = -\bar{z}$, show that $\alpha = 0$.]

37. For any complex number z, show that $z\bar{z} = |z|^2$.

38. Show that the circle of radius 1 centered at the origin (the **unit circle**) is the set points in the complex plane that satisfy $|z| = 1$.

39. For any complex number z_0 and real number a, describe $\{z: |z - z_0| = a\}$.

40. Describe $\{z: |z - z_0| \le a\}$, where z_0 and a are as in Problem 39.

41. Let $p(\lambda) = \lambda^n + a_{n-1}\lambda^{n-1} + a_{n-2}\lambda^{n-2} + \cdots + a_1\lambda + a_0$ with $a_0, a_1, \ldots, a_{n-1}$ real numbers. Show that if $p(z) = 0$, then $p(\bar{z}) = 0$; that is, *that the roots of polynomials with real coefficients occur in complex conjugate pairs.*

42. Derive expressions for $\cos 4\theta$ and $\sin 4\theta$ by comparing the De Moivre formula and the expansion of $(\cos\theta + i\sin\theta)^4$.

43. Prove the De Moivre formula by mathematical induction. [*Hint*: Recall the trigonometric identities $\cos(x + y) = \cos x \cos y - \sin x \sin y$ and $\sin(x + y) = \sin x \cos y + \cos x \sin y$.]

APPENDIX 6 SUMMARY LIST OF COMMANDS

General commands

`ITERATE(f(v),v,v₀,n)` = {do n iterations of the function $v_{k+1} = f(v_k)$ beginning at v_0 and save the values $v_0, v_1, \ldots, v_n$ in a vector. The symbol v can be either a variable or a vector of variables.}

`SUBST(f(x,y)[=c],x=x₀,y=y₀)` = {substitute $x = x_0$ and $y = y_0$ in the function $f(x, y)$ [or in the equation $f(x, y) = c$]}

`SOLVE(f(x,y)=c,y)` = {solve the equation $f(x, y) = c$ for the variable y in terms of the variable x}

`SOLVE(EQ(c)=0,c)` = {solves the system of equations $\mathbf{EQ(c)=0}$ for the vector of unknowns $\mathbf{c}$}.

`IF(x,f₊,f₀,f₋)` = if $x > 0$ return f_+, if $x = 0$ return f_0, else return f_-}.

`LIM(f(x),x,x₀)` = {limit of $f(x)$ as $x \to x_0$}

`DIF(f(x),x)` = {differentiate the function $f(x)$ with respect to x}

`INT(f(x),x)` = {integrate the function $f(x)$ with respect to x. The result will be the antiderivative with constant equal to zero}

`INT(f(x),x,x₀,x₁)` = {the definite integral of $f(x)$ with respect to x over the interval $[x_0, x_1]$: $\int_{x_0}^{x_1} f(x)\,dx$}

`SUM(f(n),n,0,N)` = {the partial sum $\Sigma_{n=0}^{N} f(n)$}

`EIGENVALUES(A)` = {finds n eigenvalues for the matrix $\mathbf{A}$}

`EIGENVECTOR(A,λ)` = {finds an eigenvector $\mathbf{v}$ corresponding to the eigenvalue λ satisfying $\mathbf{Av} = \lambda\mathbf{v}$ and $\|\mathbf{v}\| = 1$}

Special commands for differential equations

First order

```
SEPARABLE(x,y,p,q,c):=SOLVE(INT(1/q,y)=INT(p,x)+c,y)
LINEAR(x,y,p,q,c):=
            SOLVE(y*exp(INT(p,x)=INT(q*exp(INT(p,x)),x)+c,y)
BERNOULLI(x,y,p,q,k,c):=
           LINEAR(x,y^(1-k),(1-k)*p,(1-k)*q,c)
HOMOGENEOUS(x,y,r,c):=
      SUBST(SEPARABLE(x,z,1/x,SUBST(r-y,x=1,y=z),c),x=x,z=y/x)
EXACT(x,y,p,q,c):=SOLVE(INT(p,x)+INT(q-DIF(INT(p,x),y),y)=c,y)
IV(x,y,x0,y0,GS,c):=SOLVE(SUBST(GS,x=x0,y0),c)
```

Second order

```
WRNSK(x,u,v):= u*DIF(v,x)-v*DIF(u,x)
LIN2(x,p,q):=IF(p*p-4*q,exp((-p+sqrt(p*p-4*q))*x/2),
              exp(-p*x/2),exp(-p*x/2)*cos(sqrt(4*q-p*p)*x/2))
LIN2G1(x,u,p):=(u*INT(exp(-p*x)/u^2,x))
LIN2G2(x,u,v,r):=c1*u+c2*v-u*INT(v*r/WRNSK(x,u,v),x)
              +v*INT(u*r/WRNSK(x,u,v),x)
LINEAR2(x,y,p,q,r):=
              (y=LIN2G2(x,LIN2(x,p,q),LIN2G1(x,LIN2(x,p,q),p),r)
IV2(x,y,GS,x0,y0,dy0,c1,c2):=SOLVE([SUBST(GS=y,x=x0,y=y0),
              SUBST(DIF(GS,x)=dy,x=x0,dy=dy0)],[c1,c2])
SYS2G1(t,a,b,c,d,f,g):=LIN2G2(t,LIN2(t,-a-d,a*d-b*c),
   LIN2G1(t,-a-d,a*d-b*c,-a-d),b*g-d*f+DIF(f,t))
SYS2G2(t,a,b,c,d,f,g):=(DIF(SYS2G1(t,a,b,c,d,f,g),t)
              -a*SYS2G1(t,a,b,c,d,f,g)-f)/b
SYS2G(t,x,y,a,b,c,d,f,g):= ([x=SYS2G1(t,a,b,c,d,f,g),
              y=SYS2G2(t,a,b,c,d,f,g)])
LINSYS2(t,x,y,a,b,c,d,f,g):=IF(b,SYS2G(t,x,y,a,b,c,d,f,g),
              IF(c,SYS2G(t,y,x,d,c,b,a,g,f),"uncoupled:use first
              order methods",SYS2G(t,y,x,d,c,b,a,g,f)),
              SYS2G(t,x,y,a,b,c,d,f,g))
IVSYS2(t,[x=GS_1,y = GS_2],t0,x0,y0,c_1,c_2):=
          SOLVE(SUBST([x=GS_1,y=GS_2],t=t0,x=x0,y=y0),[c_1,c_2])
JACOBIAN([f,g],[x,y],[x_0,y_0]):= EIGENVALUES(SUBST([
      [DIF(f,x),DIF(f,y)],[DIF(g,x),DIF(g,y)]],x=x_0,y=y_0))
```

Numerical methods

Euler

```
EULER(x,y,f(x,y),x_0,y_0,h,n):=
              ITERATE(v+h*[1,SUBST(f,[x,y]=v)],v,[x_0,y_0],n)
```

Improved Euler

```
IE(v,f,u,h):= u + h*[1,LIM(f,v,u)]
IEULER (v,f,v_0,h,n):=ITERATE((u+IE(f,v,u,h)+h*
              [1,LIM(f,v,IE(f,v,u,h)]/2,u,v_0,n)
```

Runge-Kutta

```
RK3(p,v,u,c1,c2,c3):=(c1+LIM(p,v,u+c3)+2*(c2+c3))/6
RK2(p,v,u,c1,c2):= RK3(p,v,u,c1,c2,LIM(p,v,u+c2/2))
RK1(p,v,u,c1):=RK2(p,v,u,c1,LIM(p,v,u+c1/2))
RK0(p,v,u):=RK1(p,v,u,LIM(p,v,u))
RK(v,f,v_0,h,n):=ITERATE(u+RK0(h*[1,f],v,u),u,v_0,n)
```

Power series

```
TAYLOR1(x,y,r(x,y),x_0,y_0,n):=
              ITERATE([v_1+1,[1,r] • GRAD(v_2),v_3+SUBST(v_2,x=x_0,y=y_0)*
              (x-x_0)^v_1/v_1!],v,[1,r,y_0],n)
```

```
TAYLOR2(x,y,z,p(x),q(x),x₀,y₀,z₀,n):=
        ITERATE([v₁+1,[1,z,−q*y−p*z]•GRAD(v₂),v₃+
        SUBST(v₂,x=x₀,y=y₀,z=z₀)*(x−x₀)^v₁/v₁!],v,[1,z,y₀],n)
INDICIAL(x,m(x),p(x),q(x)):=
      SOLVE(r*(r−1)+r*LIM(x*p/m,x,0)+LIM(x^2*q/m,x,0)=0,r)
FROBENIUS(x,r,m,p,q,N):=
      SOLVE(x*x*m*DIF(DIF(Z(x,N),x),x)+(x^2*p+2*r*x*m)*
      DIF(Z(x,N),x)+(m*r^2−m*r+x*p*r+x^2*q)*Z(x,N)=0,c)
```

Other commands

```
LAPLACE(t,f(t)):=
     LIM(INT(f*exp(−s*t),t),t,∞)−LIM(INT(f*exp(−s*t),t),t,0)
FOURIER(f,x,L,N):= (1/2)*INT(f,x,−L,L)+SUM(cos(n*pi*x/L)*
         INT(f*cos(n*pi*x/L),x,−L,L)+sin(n*pi*x/L)*
         INT(f*sin(n*pi*x/L),x,−L,L),n,1,N))/L
```

ANSWERS TO ODD-NUMBERED PROBLEMS

CHAPTER 1

PROBLEMS 1.1 *(page 13)*

1. 55.1 minutes **3.** 38.7 minutes **5.** 62,500 after 20 days; 156,250 after 30 days
7. $250{,}000(1.06)^{10} \approx 447{,}712$ in 1980; $250{,}000(1.06)^{30} \approx 1{,}435{,}873$ in 2000

9. (a) $150(0.16)^2 \approx 3.84°\text{C}$ **(b)** $-\dfrac{\ln 15}{\ln 0.16} \approx 1.48$ minutes

11. $5580\dfrac{\ln 0.7}{\ln 0.5} \approx 2871$ years **13. (a)** $25(0.6)^{2.4} \approx 7.34$ kg **(b)** $\dfrac{10 \ln 0.02}{\ln 0.6} \approx 76.6$ hours

15. $\dfrac{\ln 0.5}{(-\alpha)} \approx 4.62 \times 10^6$ years **17.** $t = \dfrac{\ln (8/4.845)}{0.019} \approx 26.4$ years or A.D. 2012 **19.** approx. 4,032,650,000

21. (a) 19,057,000 approx. **(b)** Between A.D. 2003 and 2004 **23.** $\beta = \left(\dfrac{1}{1500}\right) \ln \left(\dfrac{845.6}{1013.25}\right) \approx -1.205812 \times 10^{-4}$

(a) 625.526 mbar **(b)** 303.416 mbar **(c)** 429.033 mbar **(d)** 348.632 mbar **(e)** $-\left(\dfrac{1}{\beta}\right) \ln 1013.25 \approx 57.4$ km

PROBLEMS 1.2 *(page 19)*

1. First **3.** Third **5.** Second **7.** Third **9.** Initial-value **11.** Initial-value **13.** Boundary value
19. $y_1 = e^{2x} \cos x$, $y_2 = e^{2x} \sin x$ **25.** $\phi''(-1) = -3$, $\phi'''(-1) = 2$ **27.** All points except those on the line $x + y = 0$
29. All points in the horizontal strip $-1 < y < 1$ **31.** Valid for all $x < 1$

PROBLEMS 1.3 *(page 24)*

1.

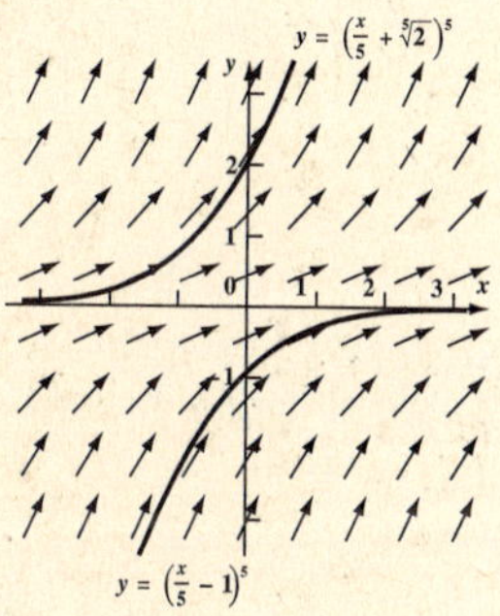

3.

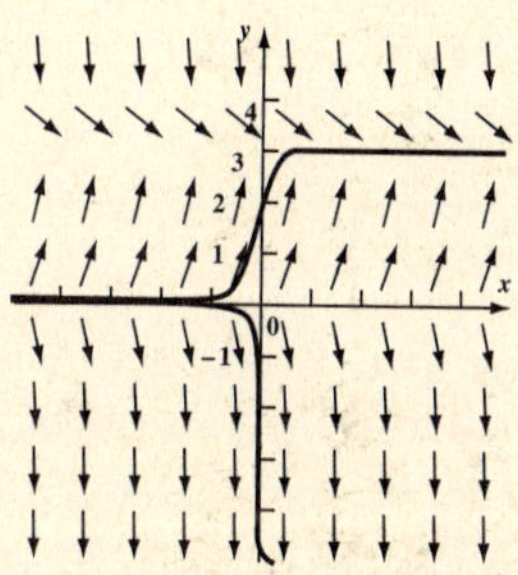

5.

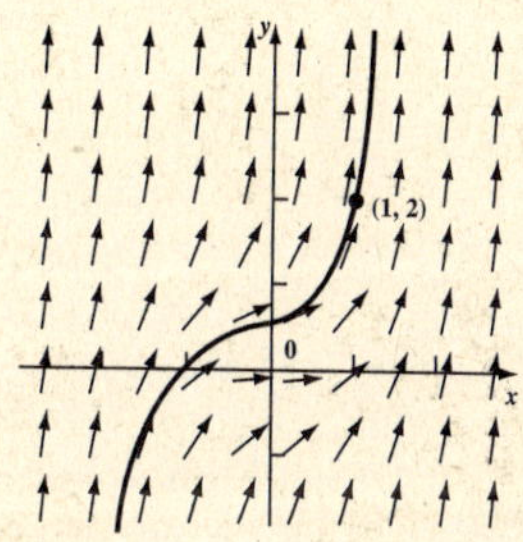

7.

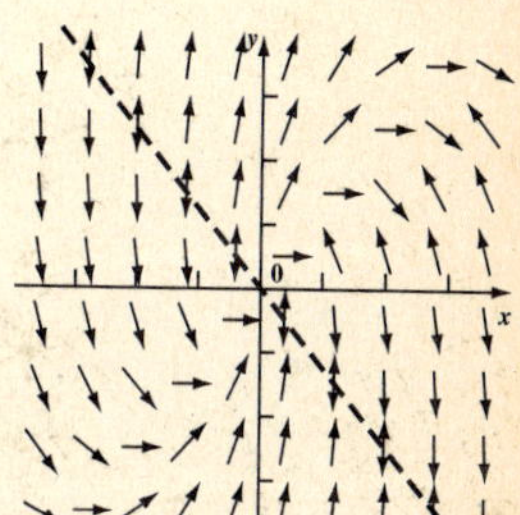

9. (a)

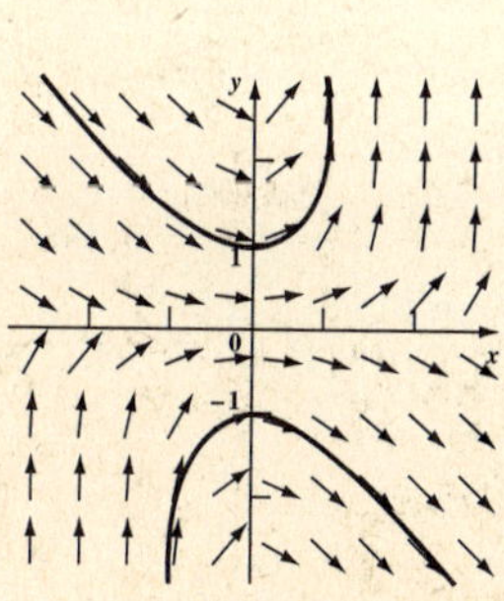

(b) It is the constant solution $y = 0$. **(c)** Their graphs are symmetric with respect to the origin.

11. 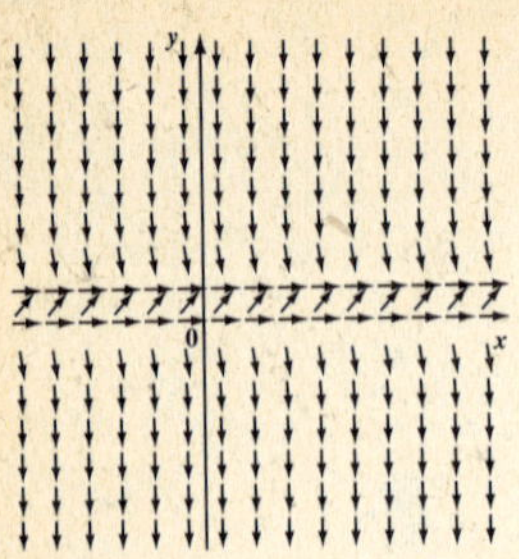

13. 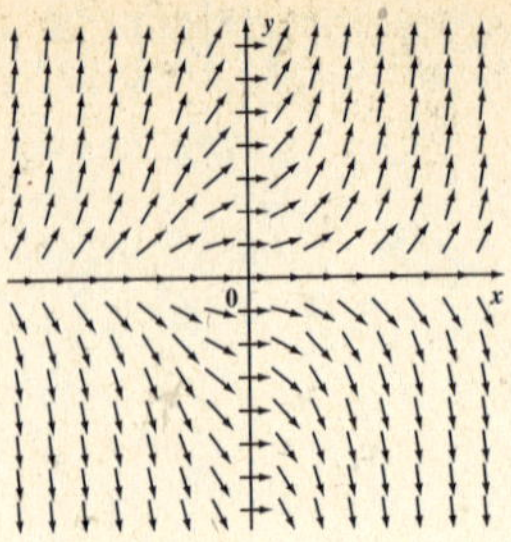

15. 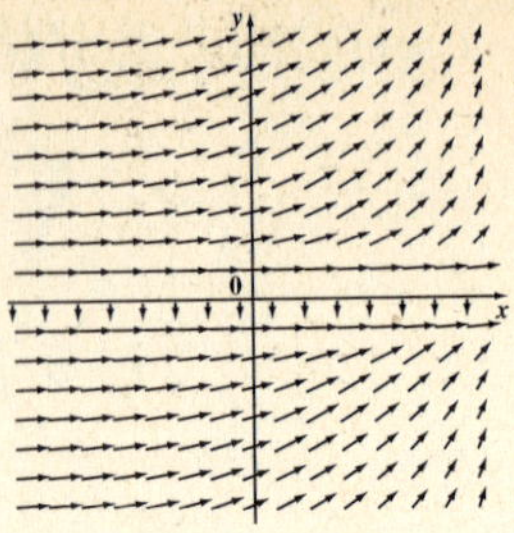

PROBLEMS 1.4 *(page 30)*

1. 2.98 **3.** 0.71 **5.** 8.31 **7.** 0.34 **9.** 156.45 **11.** $y_1 = 1.02, y_2 = 1.04, y_3 = 1.07, y_4 = 1.09, y_5 = 1.12$

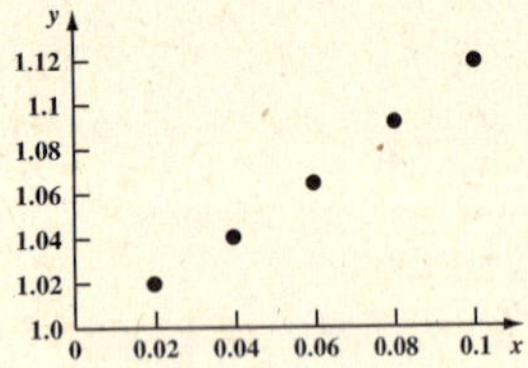

13. $y_1 = 0.40, y_2 = 0.68, y_3 = 0.79, y_4 = 0.95, y_5 = 1.20$

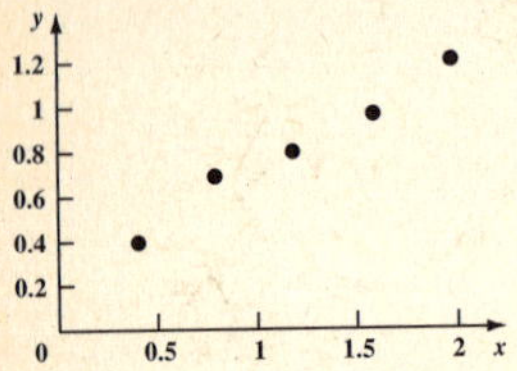

15. $y_1 = 1.00, y_2 = 1.56, y_3 = 2.06, y_4 = 1.28, y_5 = 0.67, y_6 = 1.05, y_7 = 0.28, y_8 = 1.07$

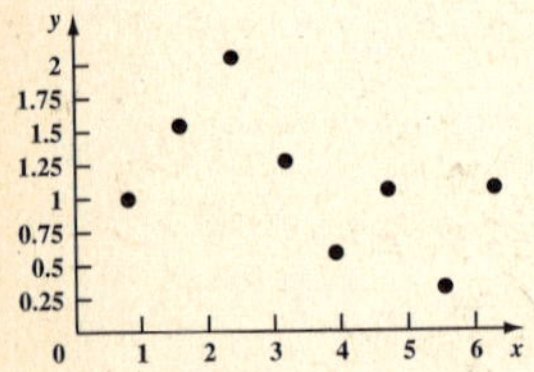

17. $y_1 = 1.10, y_2 = 1.21, y_3 = 1.33, y_4 = 1.46, y_5 = 1.60,$
$y_6 = 1.75, y_7 = 1.91, y_8 = 2.10, y_9 = 2.29, y_{10} = 2.50$

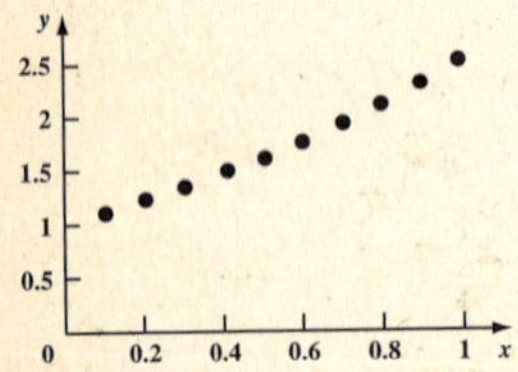

19. $y_1 = 1.55, y_2 = 1.19, y_3 = 0.91, y_4 = 0.69, y_5 = 0.52$

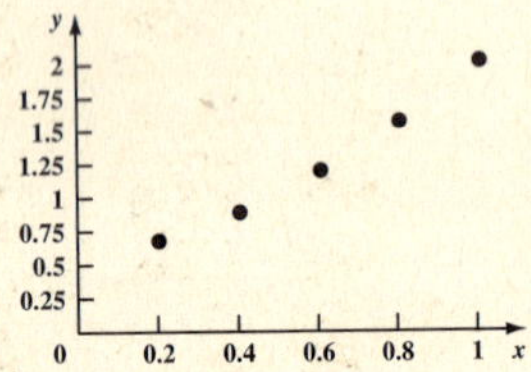

PROBLEMS 1.5 *(page 32)*

1. {{0, 1}, {0.02, 1.02}, {0.04, 1.04164}, {0.06, 1.06511}, {0.08, 1.09064}, {0.1, 1.11849}, {0.12, 1.14898}, {0.14, 1.18248}, {0.16, 1.21946}, {0.18, 1.26049}, {0.2, 1.30627}, {0.22, 1.35767}, {0.24, 1.41583}, {0.26, 1.48222}, {0.28, 1.55877}, {0.3, 1.64813}, {0.32, 1.75396}, {0.34, 1.88157}, {0.36, 2.03886}, {0.38, 2.2383}, {0.4, 2.50066}}

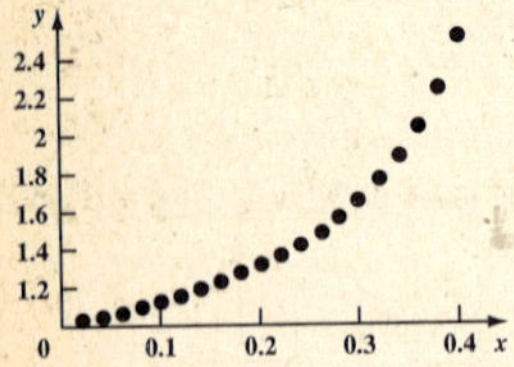

3. {{0, 1}, {0.05, 1.05}, {0.1, 1.10256}, {0.15, 1.15791}, {0.2, 1.21629}, {0.25, 1.27792}, {0.3, 1.34303}, {0.35, 1.41184}, {0.4, 1.48457}, {0.45, 1.56144}, {0.5, 1.64269}, {0.55, 1.72855}, {0.6, 1.81924}, {0.65, 1.91503}, {0.7, 2.01614}, {0.75, 2.12285}, {0.8, 2.23543}, {0.85, 2.35414}, {0.9, 2.47928}, {0.95, 2.61116}, {1.0, 2.75009}}

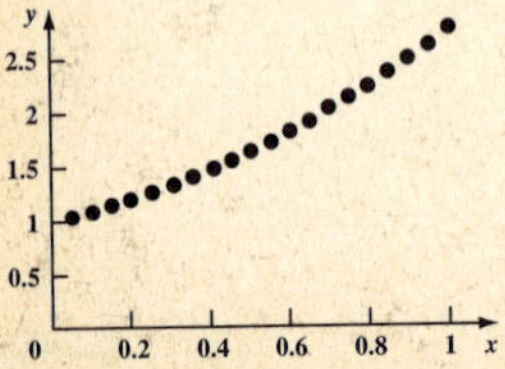

5. {{1, 2}, {1.1, 2.05556}, {1.2, 2.11131}, {1.3, 2.16725}, {1.4, 2.22332}, {1.5, 2.27953}, {1.6, 2.33583}, {1.7, 2.39223}, {1.8, 2.44869}, {1.9, 2.50522}, {2.0, 2.56179}, {2.1, 2.6184}, {2.2, 2.67504}, {2.3, 2.73171}, {2.4, 2.78838}, {2.5, 2.84507}, {2.6, 2.90176}, {2.7, 2.95844}, {2.8, 3.01512}, {2.9, 3.0718}, {3.0, 3.12846}}

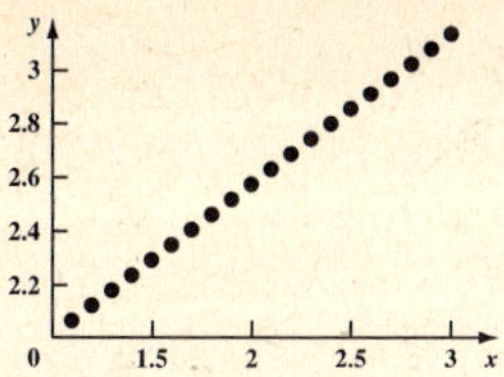

REVIEW EXERCISES FOR CHAPTER 1 *(page 33)*

1. $y = \frac{3}{2}x^2 + C$ **3.** $x = -3e^{t/2}$ **5.** $P(5) = 10{,}000(1.15)^5 \approx 20{,}114$; $P(10) = 10{,}000(1.15)^{10} \approx 40.456$

7. **(a)** $T(20) = 23 + 102\left(\frac{57}{102}\right)^2 \approx 54.9°\text{C}$ **(b)** $t = \frac{10\ln(2/102)}{\ln(57/102)} \approx 67.6$ minutes **9.** $h = \frac{\ln 0.5}{\ln 0.8} \approx 3.11$ weeks

11.

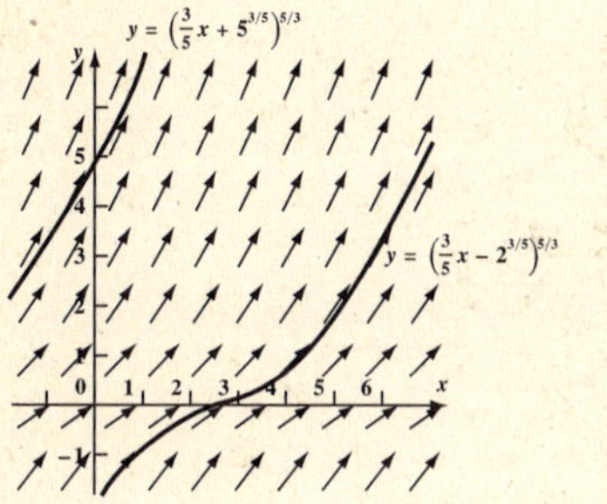

13. $y(3) \approx 6.076$ **15.** $y(3) \approx 5.445$ **17.** $y(3) \approx 0.625$

19. **(a)** $y(1) \approx 25.32$ **(c)** $(1 - x)^{-1/2}$ is not defined at $x = 1$

CHAPTER 2

PROBLEMS 2.1 *(page 46)*

1. $y = -\ln\left(c - \frac{x^2}{2}\right)$ **3.** $P = (Q + C)^2$ **5.** $x = c\,e^{(1/2)\,e^{t^2}}$ **7.** $y = c\,e^{-(e^{-t})}$ **9.** $y = \frac{1}{2}\ln(x^2 + 1) + c$

11. $z = \tan\left(\frac{r^3}{3} + c\right)$ **13.** $P = ce^{\sin Q - \cos Q}$ **15.** $s = e^{(t^3/3) - 2t}$ **17.** $y = c\cos x - 3$

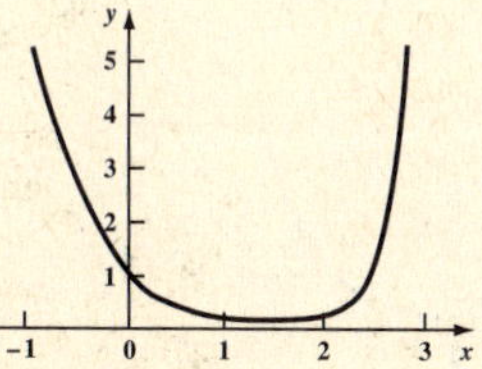

19. $y = \frac{3x}{4x - 3}$ **21.** $y = \pm\sqrt{2e^x + c}$ **23.** $x = \ln\left(1 - \frac{1 - e}{e^t}\right)$ **25.** $x = 2t + 1$

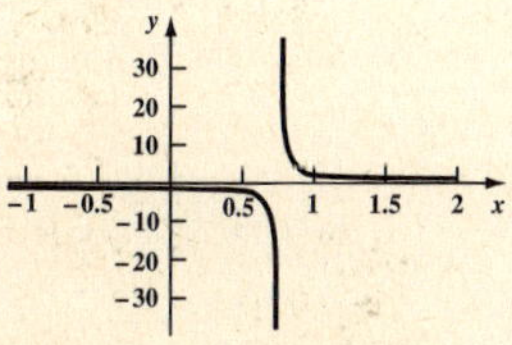

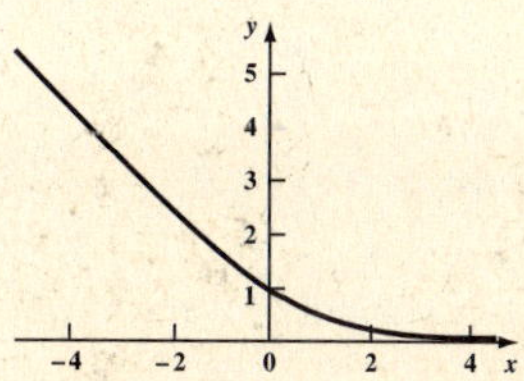

27. $x = x_0 + v_0(\cos\theta)t,\quad y = y_0 + v_0(\sin\theta)t - \frac{1}{2}gt^2$ **29.** $t = 3\,\frac{\ln 10}{\ln 2} \approx 9.97$ hr **31.** Yes **33.** 12: 19: 31.7 PM

PROBLEMS 2.2 *(page 52)*

1. Show $(x \sin C + \sqrt{1 - x^2} \cos C) = \sqrt{1 - y^2}$ **3.** $y = \dfrac{7^4}{(x+1)^3}$ **5.** $y^2 = \dfrac{4}{(1 - ce^{-8t^2})}$

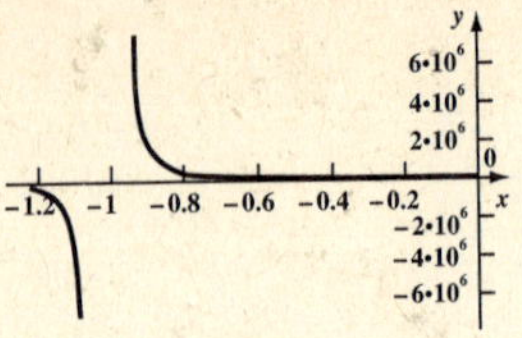

7. $y^2 = x^2 - 2x + c$ **9.** $x = 4 \sec^2\left(c + \dfrac{t^2}{2}\right)$ **11.** $z = \sec\left(\ln|x| + \dfrac{\pi}{3}\right)$ **13.** $y = 2 \sin[\ln(x + \sqrt{x^2 - 1}) + c]$

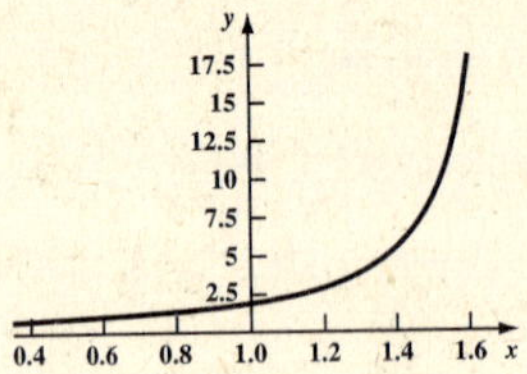

15. $y = 3 \sin(c - \sin^{-1}(x))$ or $y = c\sqrt{1 - x^2} \pm |x|\sqrt{1 - c^2}$ **17.** $y^2 = 4 \sec^2(x^2 + c)$

19. $t = \dfrac{\pi d}{r}\left[\dfrac{\sqrt{i}}{2(k\pi)^{3/2}} \ln\left|\dfrac{\sqrt{i} + \sqrt{k\pi}\,u}{\sqrt{i} - \sqrt{k\pi}\,u}\right| - \dfrac{u}{k\pi}\right]$ where $u = \left(\dfrac{3rv}{\pi d}\right)^{1/3}$; to prevent overflow, $i = k\pi u^2$

21. Revolve $y = kx^4$ about the y-axis to get the cistern's shape.

PROBLEMS 2.3 *(page 61)*

1. $x = ce^{3t}$ **3.** $x = 1$ **5.** $y = ce^{7x} - \dfrac{x}{7} - \dfrac{1}{49}$ **7.** $z = ce^{-5x} + \dfrac{x}{5} + \dfrac{4}{25}$ **9.** $y = \dfrac{5x - 4 - 26e^{5(x-1)}}{25}$

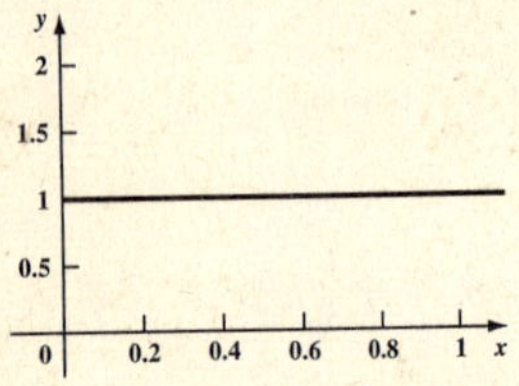

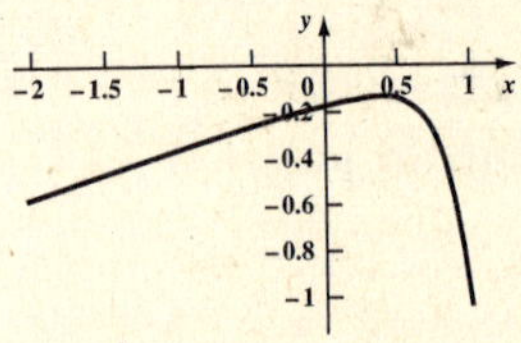

11. $y = ce^x - x^2 - 2x - 4$ **13.** $x = 2e^{t^2/2} - 1$ **15.** $x = y^y(1 + ce^{-y})$ **17.** $y = x^4 + 3x^3$

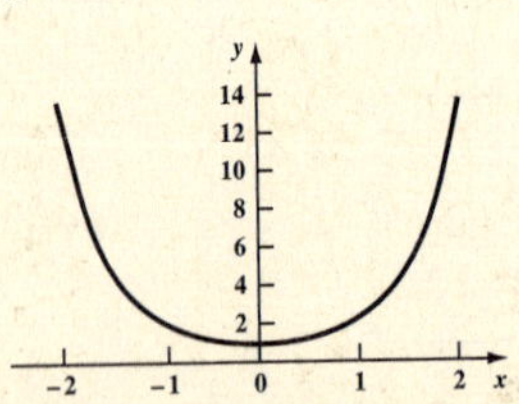

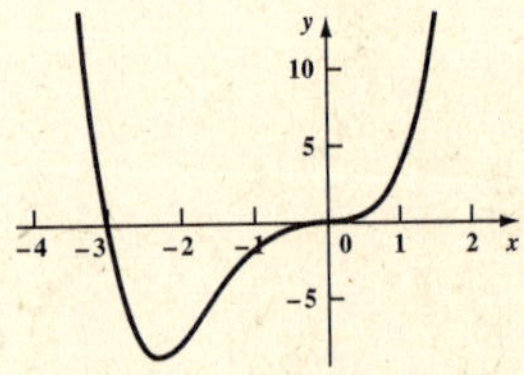

19. $x = ce^{2t} + \dfrac{t^3 e^{2t}}{3}$ **21.** $s = \left(c + \dfrac{u^2}{2}\right)e^{-u} + 1$ **23.** $x = e^{-y}(c + y)$ **25.** $x = \dfrac{1}{k} + ce^{-kt}$ **27.** $y = B + ce^{-kt}$

29. $P = \dfrac{\beta}{\delta + (\beta/P_0 - \delta)\, e^{-\beta t}}$ **31. (a)** $\dfrac{dN}{dt} = rN\left(1 - \dfrac{N}{K}\right) - h = -\dfrac{r}{K}\left(N^2 - KN + \dfrac{hK}{r}\right)$

(b) $N = \dfrac{1}{2}K + \dfrac{1}{2}D \tanh\left[\dfrac{1}{2}D\dfrac{r}{K}t + c\right]$ where $D = \sqrt{K^2 - \dfrac{4Kh}{r}}$ and $c = \tanh^{-1}\left[2N_0 - \dfrac{K}{D}\right]$

33. Graph $y = rN\left(\dfrac{1 - N}{K}\right)$ and observe maximum $\dfrac{rK}{4}$ occurs at $N = \dfrac{K}{2}$ **35.** The object satisfies $m\dot{v} = -mg + kv^2$. Thus $\lim_{t\to\infty} v = -\sqrt{\dfrac{mg}{k}}$. **37.** $y = \pm\sqrt{\dfrac{x}{6 + ce^{-x}}}$ **39.** $y = \dfrac{1}{x\sqrt{2 - x^2}}$, $0 < x < \sqrt{2}$

41. $y = \dfrac{1}{x^3\left(\dfrac{1}{2} - \ln x\right)}$, $0 < x < \sqrt{e}$

43. $y = \left[\int g(x)e^{\int f(x)\,dx}\,dx + k\right]e^{-\int f(x)\,dx}$ **45.** **(b)** $y = -ce^{-3x} + ke^{-2x}$ **47.** $y = (cx + k)e^{-ax}$

49. $y = \begin{cases} e^{-x}, & \text{if } 0 \le x \le 1 \\ e^{-1}, & \text{if } x > 1 \end{cases}$ **51.** See the reference.

PROBLEMS 2.4 *(page 68)*

1. $x^2y + y = c$ **3.** $xy(x + y) = c$ **5.** $x^2 - 2y\sin x = \dfrac{\pi^2}{4} - 2$ **7.** $e^{xy} + 4xy^3 - y^2 + 3 = 0$ **9.** $x^2y + xe^y = c$

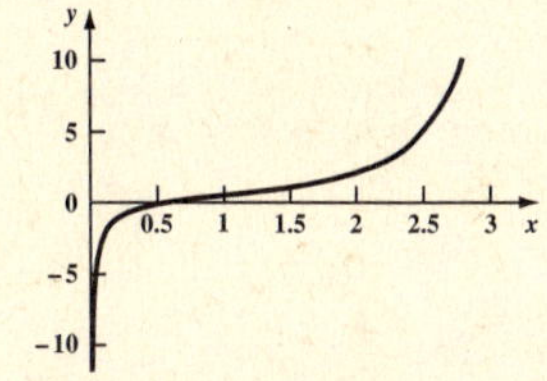

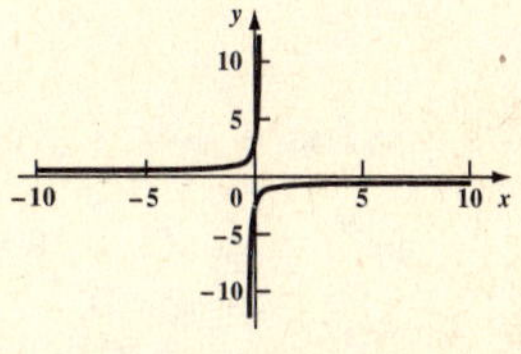

11. $\tan^{-1}\dfrac{y}{x} + \ln\left|\dfrac{x}{y}\right| = c$ **13.** $x^4y^3 + \ln\left|\dfrac{x}{y}\right| = e^3 - 1$ **15.** $\ln|y| + \left(\dfrac{x}{y}\right) = c$ **17.** $xy^{3/2}e^{-1/y} = c$

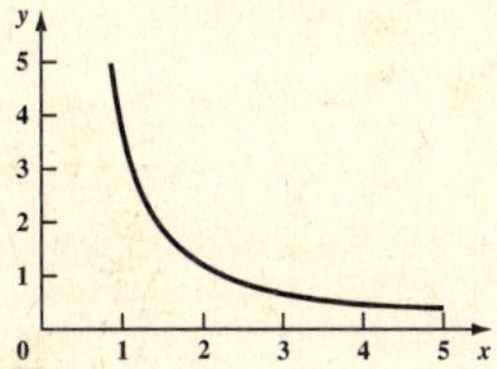

19. $x^{2/3}(x^2 + 4y^2) = c$ **21.** $x^3 + 3x^2y^2 - 3xy + y^3 = c$ **23.** $x^3\sin y + x^4y^2 = c$ **25.** $y^2(x^2 + y^2 + 2) = c$
27. $x^2y = c$

PROBLEMS 2.5 *(page 75)*

1. $y = \dfrac{1}{4}x(c + \operatorname{sgn} x \ln|x|)^2$; $y \equiv 0$ is also a solution **3.** $y = -x\ln(1 - \ln x)$ **5.** $y = \dfrac{x}{(1 - \ln x)}$

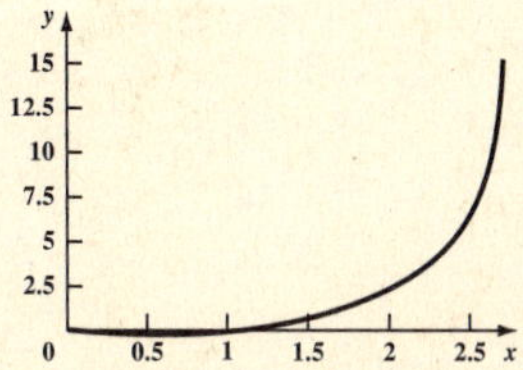

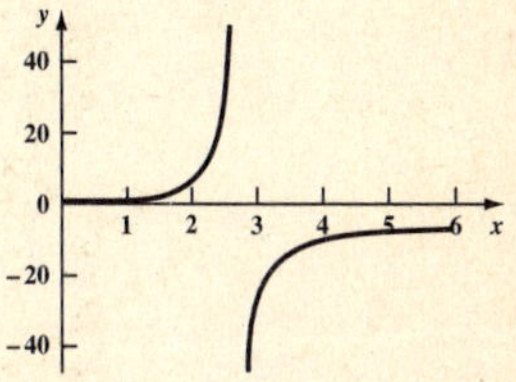

7. $y = \dfrac{1}{2}\left(x + \sqrt{x^2 + \dfrac{8}{x}}\right)$ **9.** $y = \dfrac{(c^2x^2 - 1)}{2c}$ **11.** $y = \tan(2x + c) - \dfrac{1}{2}x$ **13.** $3x + c = x + y + \ln(x + y - 1)$

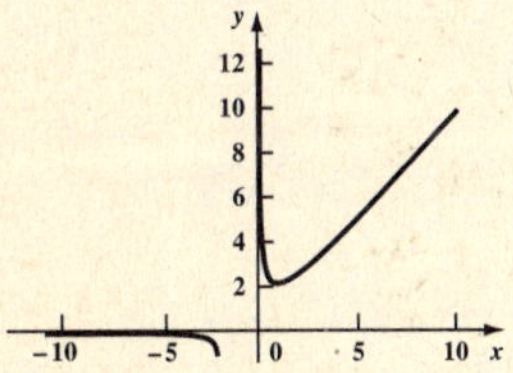

15. $y = \tan^{-1}(x + c) - x$ **17.** $y = \dfrac{\pm 1}{x\sqrt{c - 2x}}$ **19.** $y = \dfrac{-\ln|1 + e^{-1} - x|}{x}$ **21.** $y = e^{ce^x} - x$

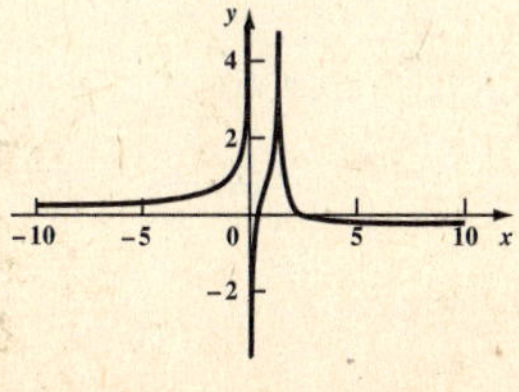

23. Solve the system $\begin{array}{l} ah + bk = -m \\ ch + dk = -n \end{array}$ simultaneously for h and k. **25.** $y = 2x \pm \sqrt{5\left(x + \frac{2}{5}\right)^2 + c}$

27. $(x + 1)e^{(1-t)/(x+1)} = c$ **29.** $y = 5 - x \pm \sqrt{4(x - 2)^2 + c}$ **31.** $xe^{tx} = ct$, by setting $v = tx$

33. General solution: $y = cx + c^3$; singular solution: $y = \pm\frac{2}{3}x\sqrt{-\frac{1}{3}x}$

35. General solution: $y = cx + \frac{1}{4}c^4$; singular solution: $y = -\frac{3}{4}x^{4/3}$

PROBLEMS 2.6 *(page 80)*

1. 2000-year-old wood has $\left(\frac{1}{2}\right)^{20/57} \approx 0.784$ the C^{14} concentration of freshly cut wood. **3.** $I(t) = \dfrac{1 + N}{1 + Ne^{-k(1+N)t}}$

5. $t = 25 \ln 2 \approx 17.3$ min **7.** $47\frac{43}{91}$ g/liter **9.** $p(t) = 400 + \frac{2}{\pi} - \frac{1}{3}t + \frac{2}{\pi}\sin\frac{\pi}{12}(t - 6)$

11. $x(t) = \dfrac{A}{k} + \dfrac{B}{k^2 + \omega^2}(k \sin \omega t - \omega \cos \omega t) + \left(x_0 + \dfrac{B\omega}{\omega^2 + k^2} - \dfrac{A}{k}\right)e^{-kt}$

PROBLEMS 2.7 *(page 84)*

1. $I = \frac{6}{5}(1 - e^{-10t})$ **3.** $I = 2(1 - e^{-25t})$ **5.** $I = \frac{1}{20}(e^t - e^{-t})$ **7.** $Q = \frac{1}{680}[3 \sin 60t + 5 \cos 60t - 5e^{-100t}]$

9. $Q = \frac{1}{100}(1 + 99e^{-100t})$ **11.** $1000\frac{dQ}{dt} + 10^6 Q = 0$; $Q(0) = 10$, $Q = 10e^{-1000t}$ **13.** $Q(60) = \frac{1}{2000}[1 - e^{-10^5(600+3600)}]$

15. $I(t) = \dfrac{E_0 C}{1 + (\omega RC)^2}\left[RC\omega^2 \cos \omega t - \omega \sin \omega t + \dfrac{1}{RC}e^{-t/RC}\right]$; $Q(t) = \dfrac{E_0 C}{1 + (\omega RC)^2}[(\cos \omega t - e^{-t/RC}) + \omega RC \sin \omega t]$

17. $I_{\text{transient}}(0) = \dfrac{E_0}{R[1 + (\omega RC)^2]} \approx \dfrac{E_0}{R}$ for very small R

19. $I(t) = \begin{cases} \dfrac{3}{50}(1 - e^{-50t}), & 0 \le t \le 10 \\ \left[\dfrac{e^{500}}{(70)^2} - \dfrac{3}{50}\right]e^{-50t} + \dfrac{7}{100} - \dfrac{e^{10-t}}{98}, & t \ge 10 \end{cases}$

PROBLEMS 2.8 *(page 88)*

1. $y = e^{-x}$ **3.** $y = e^x - 1 - x$

PROBLEMS 2.9 *(page 93)*

1. $y_n = y_0 + 2\left(1 - \frac{1}{2^n}\right)$ **3.** $y_n = (-1)^n y_0$ **5.** $y_n = n + 1$ **7.** $y_n = \frac{5^{n+1} - 1}{2}$ **9.** $y_n = e^{-n^2+n}\left(y_0 + \frac{e^n - 1}{e - 1}\right)$

11. $P_n = \frac{2}{3} + \frac{1}{3}\left(-\frac{1}{2}\right)^n$ **13.** $P_n = \frac{P_0}{n!}$, $1 = P_0\left(1 + \frac{1}{1!} + \frac{1}{2!} + \ldots\right) = P_0 e$

15. $x_k = \dfrac{n - (k - 1)}{k} \cdot \dfrac{n - (k - 2)}{k} \cdot \ldots \cdot \dfrac{n - 1}{2} \cdot x_1$

PROBLEMS 2.10 *(page 99)*

1. $y = (1 - \ln|x|)^{-1}$ **3.** $y = 2e^{(x^2-1)/2}$ **5.** $y = e^{\sin x}\left[\int x^2 e^{-\sin x}\,dx + c\right]$ **7.** $y = \dfrac{(-x + \sqrt{3x^2 + 4})}{2}$

9. Divide by x^4y^3. Then the solution is $\frac{y}{x^3} + \frac{3}{x} + \frac{x}{y^2} = c$.

REVIEW EXERCISES FOR CHAPTER 2 *(page 101)*

1. $y = \dfrac{1}{1 - \ln|x|}$ **3.** $y = \sin(\ln|x| + c)$ **5.** $y = \dfrac{2}{c - x^2}$ **7.** $y = \sqrt{2 \sin x}$ **9.** $y = cx + \frac{1}{c}$ and $y = 2\sqrt{x}$

11. $y = e^{(x^2-1)/2}$ **13.** $y = ce^{-\cos x} - 1$ **15.** $y = \dfrac{x + c}{\sqrt{1 + x^2}}$ **17.** $y = x^2(1 + \ln|x|)$

19. $y = \begin{cases} e^{x^2/2}\left(c - \int_x^0 e^{-t^2/2}\,dt\right), & x \le 0 \\ ce^{x^2/2}, & x > 0 \end{cases}$ **21.** $y = \frac{1}{2}(\sqrt{3x^2+4} - x)$ **23.** $y = \frac{(x+c)^2 - 4}{4x}$

25. $y = \frac{1}{1 - x + ce^{-x}}$ **27.** $xy + c = e^y \tan x$ **29.** $\frac{y}{x^3} + \frac{3}{x} + \frac{x}{y^2} = c$

31. $x = \frac{1 - e^{-bt}}{1001}$ where $b = 1.001 \times 10^{-6}$; $x(36{,}000) \approx 3.536 \times 10^{-5}$; $x(360{,}000) \approx 3.023 \times 10^{-4}$; $x(31{,}536{,}000) \approx 9.99 \times 10^{-4}$

33. Approximate by assuming that pollutants are dumped continuously at the rate $\frac{1}{24}$ m³/sec; $x \approx \frac{(1 - e^{-bt})}{24{,}024}$; maximum concentration is $\frac{1}{24{,}024}$ **35.** 1.18 hr **37.** $P + \exp\frac{1}{b}[a - e^{-b(t+c)}]$; $\lim_{t\to\infty} P(t) = e^{a/b}$ **39.** 3.125 ft.

41. $6 - \cosh\left(\sqrt{\frac{g}{6}}\,t\right)$ as long as this is nonnegative **45.** $y = e^x - 1$ **47.** $y = x$

CHAPTER 3

PROBLEMS 3.1 *(page 119)*

1. Linear, homogeneous, variable coefficients **3.** Nonlinear **5.** Linear, nonhomogeneous, constant coefficients **7.** Linear, nonhomogeneous, variable coefficients **9.** Linear, homogeneous, constant coefficients **11.** Independent **13.** Independent **15.** Independent **17.** $W = -x^{-4}$ **19.** $W = \sin x \tan^2 x$
21. $W = (m-n)p(p-1) + (p-m)n(n-1) + (n-p)m(m-1)$ **23.** Independent **25.** $y = c_1 + c_2x$
27. $y = c_1 + c_2x^2$ **29.** $y = c_1 \sin 3x + c_2 \cos 3x$ **31.** $y = c_1e^{3x} + c_2e^{-5x}$ **33.** $y = (c_1 + c_2x)e^{3x}$
35. $y = e^{-x}(c_1 \cos x + c_2 \sin x)$ **37.** $y = c_1 \sin 3x + c_2 \cos 3x + \frac{4}{9}x$ **39.** $y = c_1e^{3x} + c_2e^{-5x} - x^2 - \frac{4}{15}x - \frac{8}{225}$
41. At that point $y_1(x_1) = y_1'(x_1) = 0$. But $y \equiv 0$ also satisfies that initial condition
43. **(b)** No; $a(x) = -\frac{3}{x}$ is not continuous on $|x| < 1$. **45.** $y_3 = 2y_1 - \frac{1}{2}y_2$ **47.** $W(y_1, y_2)(0) = 0$, so $y_2 = xy_1$.
49. $W = -\frac{2}{x}$

PROBLEMS 3.2 *(page 123)*

1. $y_2(x) = xe^x$ **3.** $y_2(x) = e^{-2x}$ **5.** $y_2(x) = e^{-3x}$ **7.** $y_2(x) = e^{-x/2} \sin \frac{3\sqrt{3}}{2}x$ **9.** $y_2(x) = e^{10x}$ **11.** $y_2(x) = xe^{-x/2}$
13. $y_2(x) = x^{-2}$ **15.** $y_2(x) = x \cos x$ **17.** $y_2(x) = (7x + 1)e^{-4x}$
19. $y_2(x) = e^x \int x^n e^{-x}\,dx = -(x^n + nx^{n-1} + n(n-1)x^{n-2} + \cdots + n!)$ **21.** $y_2 = 1 + \frac{x}{\sqrt{3}} \tan^{-1}\left(\frac{x}{\sqrt{3}}\right)$
23. $y_2 = [2x - \sin 2x]$ **25.** $y_2 = x - \frac{8}{3}\left[\cot^3 x + \cot^2 x \csc x - \frac{1}{2}\csc x\right]$ **27.** $y_2 = ce^{x^2} \operatorname{erf}(x) = ce^{x^2} \int e^{-x^2}\,dx$
29. $y_2(x) = \frac{\cos x}{\sqrt{x}}$

PROBLEMS 3.3 *(page 126)*

1. $y = c_1e^{2x} + c_2e^{-2x}$ **3.** $y = c_1e^x + c_2e^{2x}$ **5.** $x = \left(1 + \frac{9}{2}t\right)e^{-5t/2}$ **7.** $x = -\frac{1}{5}(e^{3t} + 4e^{-2t})$ **9.** $y = c_1 + c_2e^{5x}$

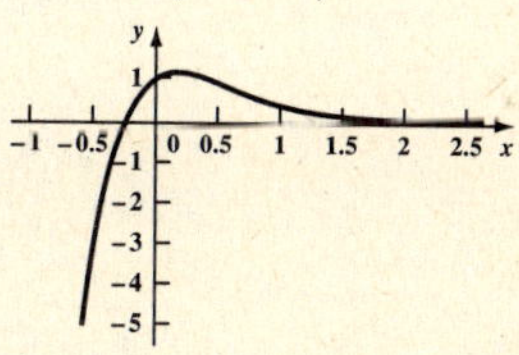

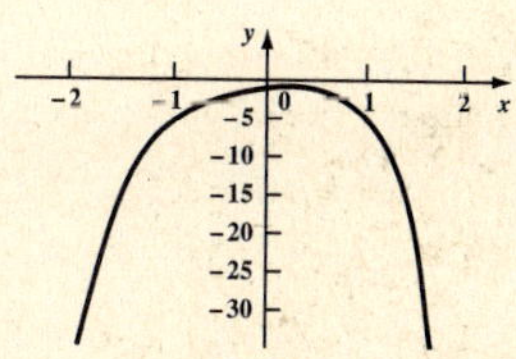

11. $y = (c_1 + c_2x)e^{-\pi x}$ **13.** $z = c_1e^{-5x} + c_2e^{3x}$ **15.** $y = (1 + 2x)e^{4x}$ **17.** $y = c_1e^{\sqrt{2}x} + c_2e^{-\sqrt{2}x}$

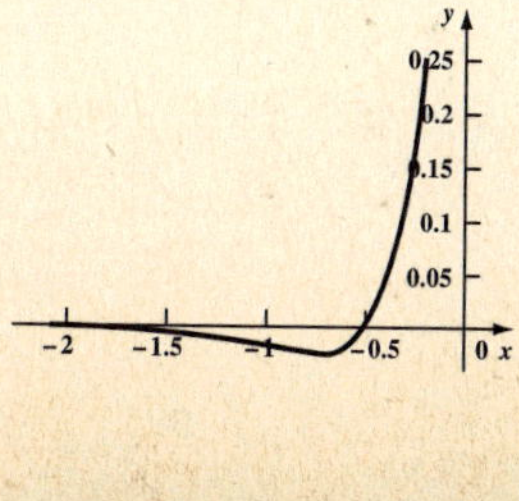

19. $y = e^{\sqrt{5}x} + 2e^{-\sqrt{5}x}$

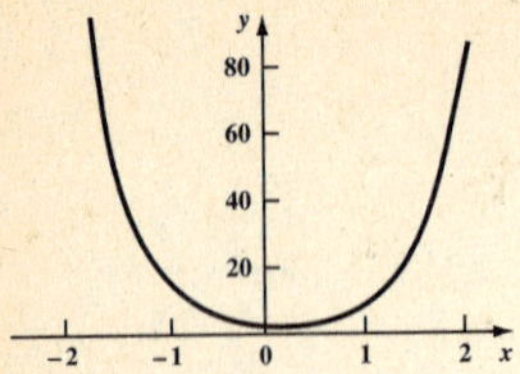

21. $y = \frac{1}{3}(4 - e^{-3x})$

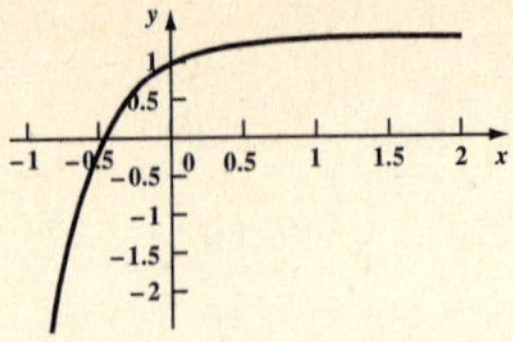

23. $y = c_1e^{-2x} + c_2e^{-8x}$

25. $y = e^{4x} - 2e^{3x}$; Fig. c.

27. $y = e^{4x} - 2e^{-3x}$; Fig. a.

29. $y = 5e^{4x} - 6e^{3x}$; Fig. f.

31. $\dfrac{c_1z_1' + c_2z_2'}{c_1z_1 + c_2z_2} = \dfrac{z_1' + (c_2/c_1)z_2'}{z_1 + (c_2/c_1)z_2}$

33. $y = 1 - \dfrac{4}{e^{2x} + 2}$

35. $y = 1 + \dfrac{1}{c + x}$

37. $y = 2 - \dfrac{3c}{e^{3x} + c}$

39. **(b)** $\phi_n(x) = e^{\lambda_1 x}\left(\dfrac{e^{hx} - 1}{h}\right) \to xe^{\lambda_1 x}$

PROBLEMS 3.4 *(page 132)*

1. $y = e^{-x}(c_1 \cos x + c_2 \sin x)$

3. $x = e^{-t/2}\left(c_1 \cos \dfrac{3\sqrt{3}}{2}t + c_2 \sin \dfrac{3\sqrt{3}}{2}t\right)$

5. $x = -\dfrac{3}{2}\cos 2\theta + \sin 2\theta = \dfrac{\sqrt{13}}{2}\sin(2\theta - 0.9828)$

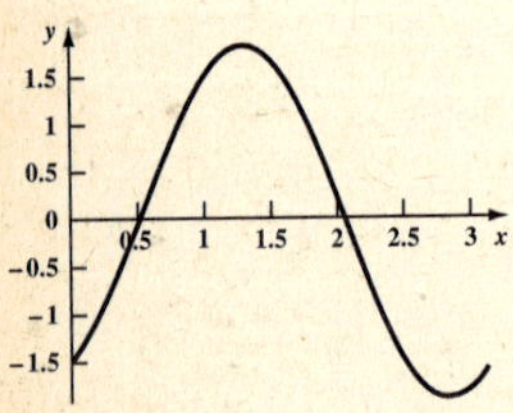

7. $y = \sin \dfrac{x}{2} + 2\cos \dfrac{x}{2} = \sqrt{5}\sin\left(\dfrac{x}{2} + 1.1071\right)$

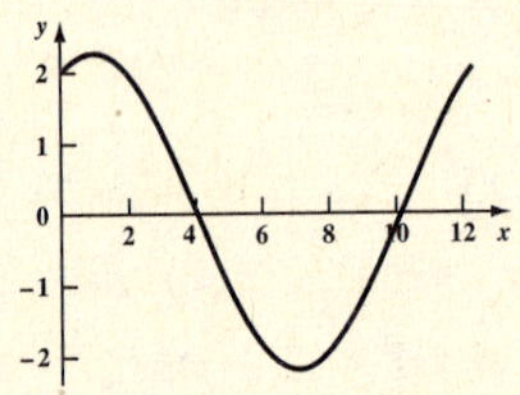

9. $y = e^{-x}(c_1 \cos 2x + c_2 \sin 2x)$

11. $y = e^{-x}(\sin x - \cos x) = \sqrt{2}\,e^{-x}\sin(x - 0.7854)$

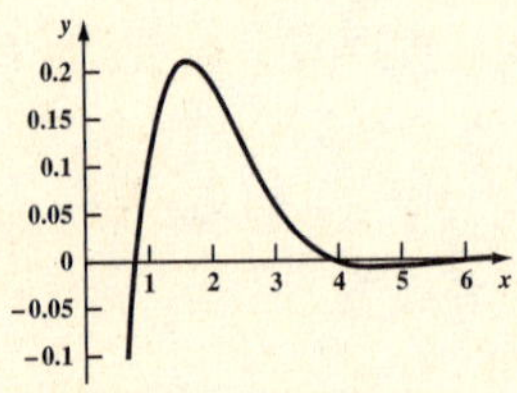

13. $\sqrt{13}\cos(t - 0.9828) = \sqrt{13}\sin(t + 0.588)$

15. $\sqrt{13}\cos(t - 4.1244) = \sqrt{13}\sin(t + 3.73)$

17. $5\cos(2\theta - 0.6435) = 5\sin(2\theta + 0.9273)$

19. $13\cos\left(\dfrac{\alpha}{2} - 1.966\right) = 13\sin\left(\dfrac{\alpha}{2} - 0.3948\right)$

21. $\dfrac{\sqrt{5}}{4}\cos(4\beta + 0.4636) = \dfrac{\sqrt{5}}{4}\sin(4\beta + 2.0344)$

23. $\sqrt{58}\cos\left(\dfrac{\pi x}{2} - 3.5465\right) = \sqrt{58}\sin\left(\dfrac{\pi x}{2} + 4.0375\right)$

25. See Fig. b, $y = e^x(\cos x - 3\sin x) = \sqrt{10}\,e^x\cos(x + 1.25)$

27. See Fig. a, $y = \cos x - 2\sin x = \sqrt{5}\cos(x + 1.107)$

29. See Fig. c, $y = e^x\left(\cos 2x - \dfrac{3}{2}\sin 2x\right) = \dfrac{\sqrt{13}}{2}e^x\sin(2x + 2.5536)$

31. $y = c_1 \sin x$ is zero at $x = \pi$.

PROBLEMS 3.5 *(page 139)*

1. $y = c_1e^{2x} + c_2e^x + 5$

3. $y = \sin x + \cos 2x + \dfrac{1}{2}\sin 2x$

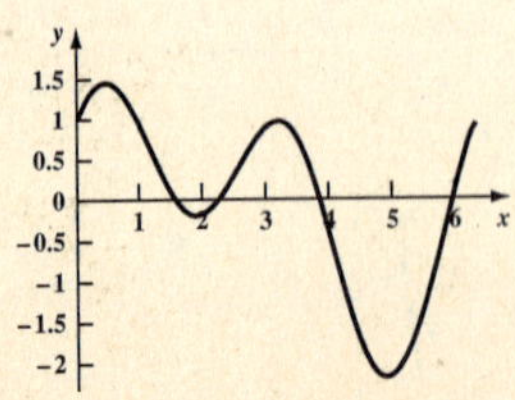

5. $y = c_1e^{2x} + c_2e^x + 3e^{3x}$

7. $y = (c_1 + c_2x - 2x^2)e^x$

9. $y = c_1e^{5x} + c_2e^{2x} + 10x + 7$ **11.** $y = 1 + 2e^x - \ln(\cos x)$

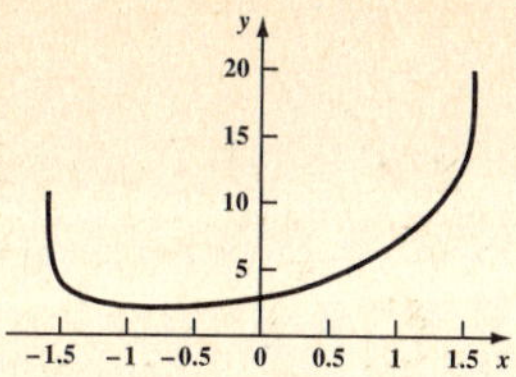

13. $y = c_1 \cos 2x + \left(c_2 + \frac{x}{2}\right) \sin 2x + \frac{1}{4} \cos 2x \ln(\cos 2x)$

15. $y = e^x(c_1 + c_2x - \ln|x - 1|)$ **17.** $y = e^x\left(\frac{7}{e} - \frac{1}{2} + \ln x\right) + \left(2 + \frac{e}{2}\right)e^{1-x}$

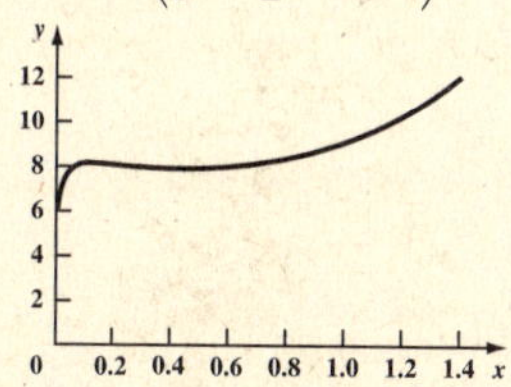

19. $y = \frac{e^{2(x-1)}}{6}(9 - 5x) - e^{2x}(x - 1) + e^{2x}(x + 1) \ln\left|\frac{x + 1}{2}\right|$ **21.** $y = \left(c_1 + \frac{x}{3}\right)e^{3x} + c_2e^{-3x} - \frac{\ln(e^{6x} + 1)\,\text{sech}(3x)}{9}$

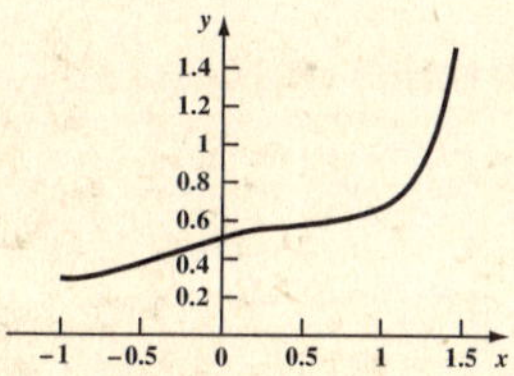

23. $y = \frac{e^x}{2}[c_1 + c_2e^{2x} + e^x + (1 - e^{2x}) \ln(1 + e^{-x})]$ **25.** $y_p = \frac{1}{2}\left[-1 - x \ln x + \left(x - \frac{1}{x}\right) \ln(1 + x)\right]$

27. $y' = \int_0^x f(t) \cos \omega(x - t)\, dt$ and differentiate again.

PROBLEMS 3.6 *(page 146)*

1. $y = c_1 \sin 2x + c_2 \cos 2x + \sin x$ **3.** $y = c_1e^x + c_2e^{2x} + 3e^{3x}$ **5.** $y = e^x(c_1 + c_2x - 2x^2)$

7. $y = 3e^{5x} - 10e^{2x} + 10x + 7$ **9.** $y = c_1 + c_2e^{-x} + \frac{x^4}{4} - \frac{4}{3}x^3 + 4x^2 - 8x$ **11.** $y = e^{2x}(c_1 \cos x + c_2 \sin x) + xe^{2x} \sin x$

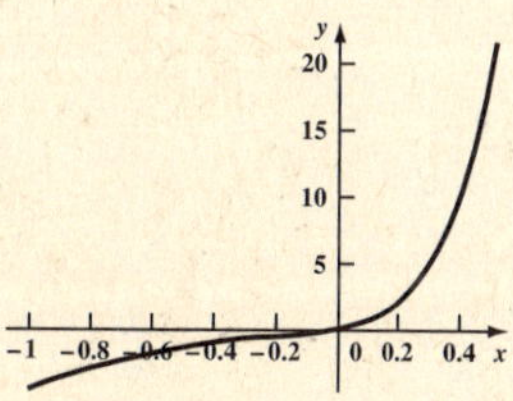

13. $y = e^{-3x}(c_1 + c_2x + 5x^2)$ **15.** $y = (c_1 + c_2x)e^{-4x} + \frac{(4x - 1)e^{4x}}{128}$

17. $y = \frac{4111}{4096}\, e^{-4x}(1 + 2x) + \frac{5}{4096}\, e^{4x}(64x^3 - 48x^2 + 18x - 3)$ **19.** $y = e^{-3x}\left(c_1 \sin 2x + \left(c_2 - \frac{x}{4}\right) \cos 2x\right)$

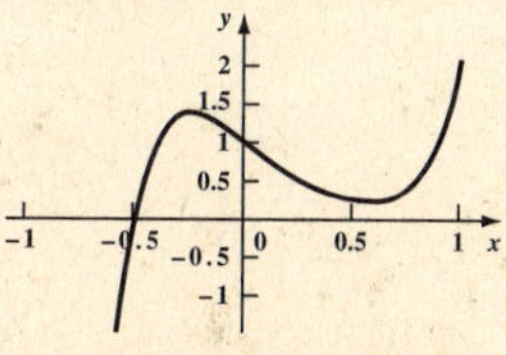

21. $y = e^{-3x}\left[\left(c_1 + \frac{5x^2}{16}\right)\sin 2x + \left(c_2 + \frac{5}{32}x - \frac{1}{12}x^3\right)\cos 2x\right]$ **23.** $y = c_1e^{2x} + c_2e^x - \frac{1}{5}(\cos x + 2\sin x)$

25. $y = \frac{148}{243}\cos 3x + \frac{27x^4 - 63x^2 + 95}{243}$ **27.** $y = \left(\frac{x^2}{4} + 1\right)\cos 3x - \frac{x}{12}\sin 3x$

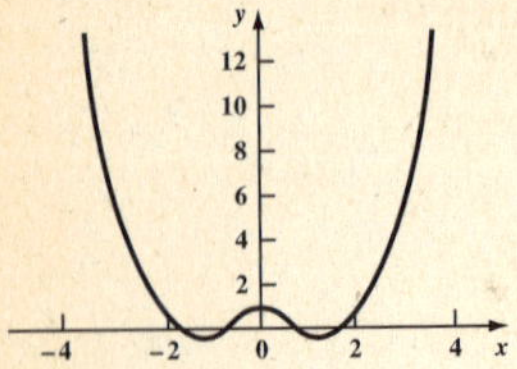

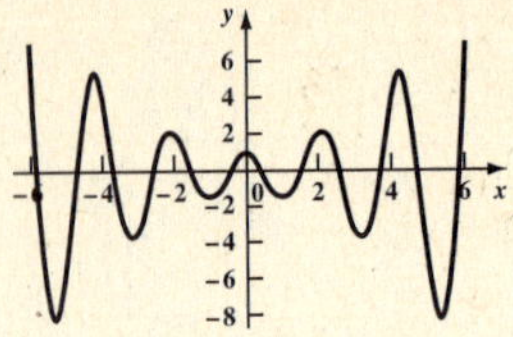

29. $y = \frac{103e^{-2x}}{128} + \frac{e^{2x}}{384}(32x^3 + 24x^2 + 84x + 459)$

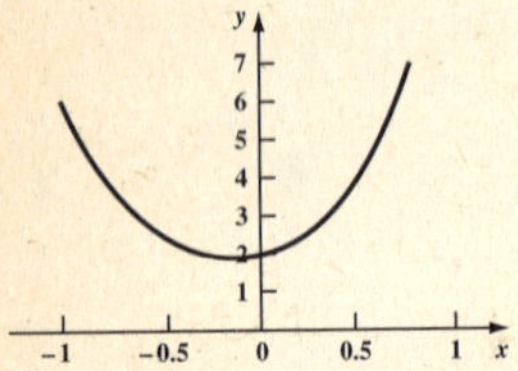

31. $y = c_1e^{-x} + c_2e^{-3x} + \left(\frac{4x^2 + 4x - 5}{32}\sin 2x\right)e^{-x} - \left(\frac{4x^2 - 8x - 1}{32}\cos 2x\right)e^{-x}$

33. $y = c_1e^{2x} + c_2e^{-x} - \frac{e^x}{2} - \frac{3}{2}$ **35.** $y = c_1e^{3x} + c_2e^{-x} + \frac{20}{27} - \frac{7}{9}x + \frac{1}{3}x^2 - \frac{1}{4}e^x$

37. $y = (c_1 + c_2x)e^{-2x} + \left(\frac{1}{9}x - \frac{2}{27}\right)e^x + \frac{3}{25}\sin x - \frac{4}{25}\cos x$

39. Let $y = c_0 + c_1x + \cdots + c_nx^n$ and solve for $c_n, c_{n-1}, \ldots, c_1, c_0$.

41. $y = \left(-\frac{1}{6}x^3 - \frac{1}{4}x^2 + \frac{1}{4}x + c_1\right)\cos x + \left(\frac{1}{4}x^2 + \frac{1}{4}x + c_2\right)\sin x$ **43.** $y = \left(\frac{1}{12}x^4 + c_1x + c_2\right)e^x$

45. (a) $y = \begin{cases} a\sin x + b\cos x + x, & 0 \le x \le 1 \\ a\sin x + b\cos x + 1, & x \ge 1 \end{cases}$ **(b)** $y = \begin{cases} x, & 0 \le x \le 1 \\ 1, & x \ge 1 \end{cases}$

PROBLEMS 3.7 *(page 151)*

1. $y = c_1x + c_2x^{-1}$ **3.** $y = x[c_1\cos(\ln x) + c_2\sin(\ln x)]$ **5.** $y = x^{3/2} - x^{1/2}$

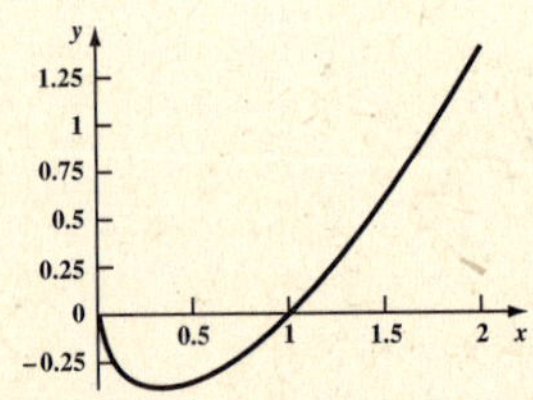

7. $y = c_1x + c_2x^3$ **9.** $y = x^{-2}[c_1\cos(\ln x) + c_2\sin(\ln x)]$ **11.** $y = c_1x^3 + c_2x^{-4}$ **13.** $y = c_1x^3 + c_2x^{-5} - \frac{1}{16x}$

15. $y = x^3(c_1 + c_2\ln x + \frac{1}{2}\ln^2 x)$ **17.** $y = c_1\cos(\ln x) + c_2\sin(\ln x) + 10$ **19.** $y = c_1x^{-5} + c_2x^{-1} + \frac{x}{12}$

21. $y = c_1x^2 + c_2x^{-1} + \frac{1}{4} - \frac{1}{2}\ln x$ **23.** $y = c_1 + c_2x^{-3}$

PROBLEMS 3.8 *(page 156)*

1. $y = (c_1 + c_2x)\cos x + (c_3 + c_4x)\sin x$ **3.** $y = (1 + x)e^x$ **5.** $y = 1 + e^{3x} + e^{-3x}$

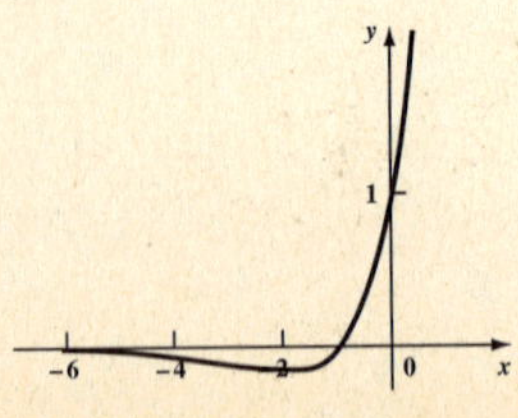

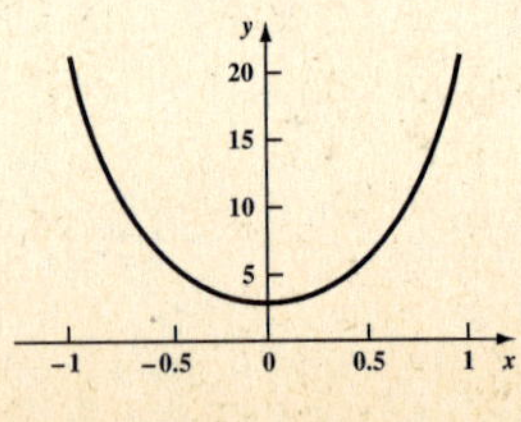

7. $y = c_1 + c_2x + c_3x^2 + c_4x^3$ **9.** $y = c_1e^x + c_2e^{-x} + c_3e^{2x} + c_4e^{-2x}$ **11.** $y = 1 - x + e^{2x} - e^{-2x}$

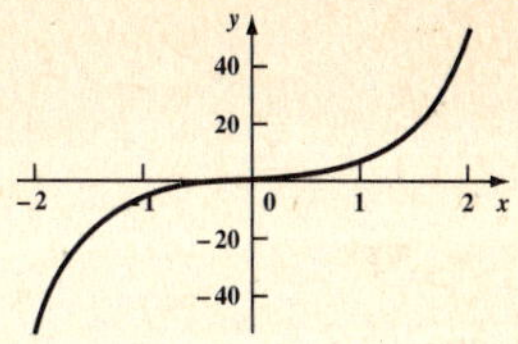

13. $y = c_1e^x + c_2 \cos x + c_3 \sin x$ **15.** $y = c_1e^{3x} + e^{-3x/2}\left(c_2 \cos \dfrac{3\sqrt{3}}{2}x + c_3 \sin \dfrac{3\sqrt{3}}{2}x\right)$ **17.** If $c_1y_1 + c_2y_2 + c_3y_3 \equiv 0$, then we also must have $c_1y_1' + c_2y_2' + c_3y_3' \equiv 0$, $c_1y_1'' + c_2y_2'' + c_3y_3'' \equiv 0$. In order for all three of these identities to hold at x_0, it must be that $c_1 = c_2 = c_3 = 0$. **19. (a)** $y_1(v')'' + (3y_1' + ay_1)(v')' + (3y_1'' + 2ay_1' + by_1)v' = 0$ **(c)** $v' = \dfrac{W(y_1, y_2)}{y_1^2} \times \displaystyle\int \frac{y_1e^{-\int a(x)\,dx}}{W^2(y_1, y_2)}\,dx$ **21.** The proof is similar to that of Theorem 3.1.3. **23.** The proof is similar to that of Theorem 3.1.5.

25. $y = c_1e^x + c_2xe^x + c_3e^{-x} + \dfrac{1}{4}x^2e^x$ **27.** $y = c_1e^{2x} + c_2e^{-3x} + c_3e^{4x} + \dfrac{1}{24}x + \dfrac{41}{288}$

29. (b) $y_p = \dfrac{1}{10}e^{2x}\displaystyle\int e^{-3x} \tan x\, dx + \frac{1}{10}e^{-x}[(3 \cos x - \sin x) \ln|\sec x + \tan x| + 1]$ **31.** $y = c_1x^{-1} + (c_2 + c_3 \ln|x|)x$

33. $y = c_1x^{-1} + c_2 \cos(\ln|x|) + c_3 \sin(\ln|x|)$

PROBLEMS 3.9 *(page 163)*

1. $y_n = c_14^n + c_2(-1)^n$ **3.** $y_n = 3\left(-\dfrac{1}{3}\right)^n - 2\left(-\dfrac{1}{2}\right)^n$ **5.** $y_n = (c_1 + c_2n)(-1)^n$ **7.** $y_n = 2^{n/2}\left(\cos \dfrac{n\pi}{4} + \sin \dfrac{n\pi}{4}\right)$

9. $y_n = \dfrac{2^n}{\sqrt{3}} \sin \dfrac{n\pi}{3}$ **11.** $P_n = \sqrt{5}\left[\left(\dfrac{1}{2} + \dfrac{1}{2\sqrt{5}}\right)^n - \left(\dfrac{1}{2} - \dfrac{1}{2\sqrt{5}}\right)^n\right]$; $P_{10} \approx 0.088$ **13.** If there are at least 125,000 pairs in any one year, there is no danger extinction. **15. (a)** \$4; \$18 **(b)** Even if B has unlimited resources, the probability of B bankrupting A is only $\dfrac{8}{27}$.

PROBLEMS 3.10 *(page 168)*

1. $y = c_1e^{7x} + c_2e^{3x}$ **3.** $y = e^{9x/2}\left[c_1 \sin \dfrac{\sqrt{3}x}{2} + c_2 \cos \dfrac{\sqrt{3}x}{2}\right]$ **5.** $y = 13e^{4x} - 10e^{5x}$

7. $y = c_1 \sin 2x + c_2 \cos 2x - \dfrac{1}{4} \cos 2x \ln\left(\dfrac{1 + \sin 2x}{\cos 2x}\right)$ **9.** $y = \dfrac{e^{3x}(72x^2 - 60x + 19)}{1728} + \dfrac{95}{64}e^{-x} - \dfrac{107e^{-3x}}{216}$

REVIEW EXERCISES FOR CHAPTER 3 *(page 172)*

1. $y_2 = \cos 2x$ **3.** $y_2 = x^{-2}$ **5.** $y_2 = -2 + x \ln\left|\dfrac{1 + x}{1 - x}\right|$ **7.** $y = 13e^{4x} - 10e^{5x}$ **9.** $y = \dfrac{2}{\sqrt{7}}e^{(3/2)x} \sin \dfrac{\sqrt{7}}{2}x$

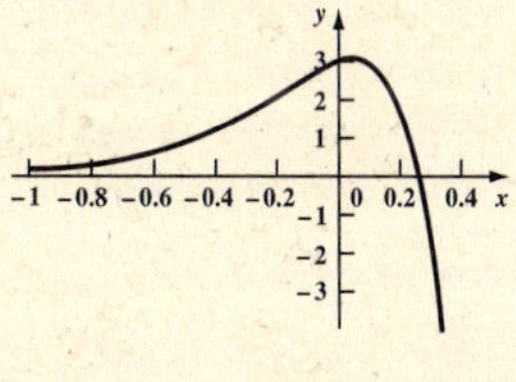

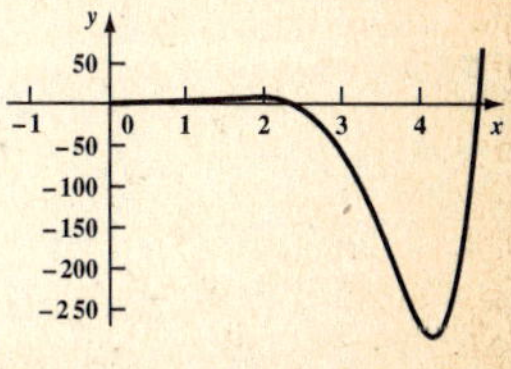

11. $y = (c_1 + c_2x)e^{-x/2}$ **13.** $y = e^x(c_1 \cos \sqrt{6}\,x + c_2 \sin \sqrt{6}\,x)$ **15.** $y = c_1e^x + c_2e^{2x} + c_3e^{3x}$

17. $y = -3e^x + x^2 + 4x + 5$ **19.** $y = e^x(c_1 + c_2x + x^{-1})$ **21.** $y = e^{-3x}\left(c_1 + c_2x + \dfrac{x^2}{2}\right)$

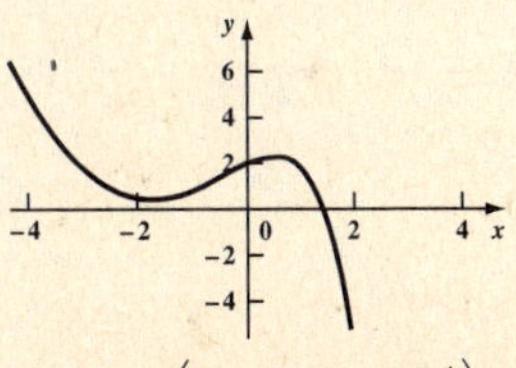

23. $y = e^{-3x}\left(c_1 + c_2x + \dfrac{x^4}{12}\right)$ **25.** $y = e^{-x}(c_1 \sin x + c_2 \cos x) + \dfrac{1}{2}(x - 1)^2$

27. $y = e^{-x}(c_1 \sin x + c_2 \cos x) - \frac{1}{10}(\sin 2x + 2\cos 2x)$ **29.** $y = (c_1 + x)e^{3x} + c_2 e^{2x}$ **31.** $y = (c_1 + c_2 \ln x)x^{-2}$ **33.** $y = (c_1 + c_2 x)e^{-2x} + c_3 e^{3x}$ **35.** $y'' = 0$ at an inflection point so $ay_1' + by_1 = 0 = ay_2' + by_2$. If $a \neq 0$ then $y_1' = by_1/a$, $y_2' = by_2/a$ so that the Wronskian is zero at the inflection point, contradicting their independence. **37.** Assume y_2 is nonzero in $[x_1, x_2]$ consecutive zeros of y_1. By Rolle's theorem $\frac{y_1}{y_2}$ has zero slope someplace in $[x_1, x_2]$. This implies the Wronskian is zero.

CHAPTER 4

PROBLEMS 4.1 *(page 177)*

1. $x = \cos 10t$ **3.** $x = 3\cos t + 4\sin t$ **5.** $x = \frac{3}{5}\sin 5t$ **7.** $v = -0.2\sqrt{g}$ m/sec

9. $\sqrt{\frac{8\pi}{g}}$ seconds; $x(t) = \left(1 - \frac{2}{\pi}\right)\cos\left(\sqrt{\frac{\pi g}{2}}\,t\right)m$ **11.** $\theta = \frac{1}{2}\cos\sqrt{2g}\,t,\ f = \frac{\sqrt{2g}}{2\pi} \approx 1.28$ Hz

13. $f = \frac{\sqrt{96g}}{2\pi} \approx 8.85$ Hz **15.** 844 sec = 14.1 min **17.** $x'' + \frac{T_0 g}{L\omega}x = 0,\ 2\pi\sqrt{\frac{L\omega}{T_0 g}}$

PROBLEMS 4.2 *(page 181)*

1. $x(t) = (1 + 10t)e^{-10t}$

3. $x(t) = e^{-\sqrt{5}t/2}\left[3\cosh\frac{1}{2}t + (8 + 3\sqrt{5})\sinh\frac{1}{2}t\right] = \left(\frac{3\sqrt{5} + 11}{2}\right)e^{[(1-\sqrt{5})t/2]} - \left(\frac{3\sqrt{5} + 5}{2}\right)e^{[(-1-\sqrt{5})t/2]}$

5. $x(t) = e^{-4t}\sin 3t$ **7.** $k = \frac{25}{4}\left(1 + \frac{\pi^2}{10000}\right)$

In Problems 9–13, $p = \sqrt{\frac{g}{96}} \approx 0.579$

9. $x(t) = e^{-3pt}\left(\frac{1}{6}\cos\sqrt{91}\,pt - \frac{1 - \frac{1}{2}p}{\sqrt{91}\,p}\sin\sqrt{91}\,pt\right)$ **11.** $x(t) = e^{-10pt}\left[\frac{1}{6} - \left(1 - \frac{5}{3}p\right)t\right]$

13. $x(t) = e^{-3pt/\sqrt{2}}\left(-\frac{24}{25}\cos\frac{1}{2}\sqrt{182}\,pt - \frac{72}{25\sqrt{91}}\sin\frac{1}{2}\sqrt{182}\,pt\right)$ **15.** No

PROBLEMS 4.3 *(page 188)*

1. $x(t) = \frac{2001}{2000}(1 + 10t)e^{-10t} - \frac{1}{2000}\cos 10t$

3. $x(t) = \frac{1}{50}\left\{e^{-\sqrt{5}t/2}\left[(150 + \sqrt{5})\cosh\frac{1}{2}t + (405 + 150\sqrt{5})\sinh\frac{1}{2}t\right] - \sqrt{5}\cos t\right\}$

$= \frac{1}{100}[(555 + 151\sqrt{5})e^{(1-\sqrt{5})t/2} - (255 + 149\sqrt{5})e^{(-1-\sqrt{5})t/2} - 2\sqrt{5}\cos t]$

5. $x(t) = \frac{1}{104}[e^{-4t}(3\cos 3t + 106\sin 3t) - 3\cos 3t + 2\sin 3t]$

7. $x(t) = -0.1419e^{-2t} + 0.0029e^{-19.62t} - 0.1390\cos 2t + 0.1133\sin 2t$

9. Use $c^2 = \frac{144(240)}{g(4\pi^2 + 0.36)}$ and solve the system

$$\left(\frac{240}{100}g - 144\right)A + \frac{12}{100}cgB = \frac{g}{5}$$

$$\frac{-12}{100}cgA + \left(\frac{240}{100}g - 144\right)B = 0.$$

11. (b) Let $A = \frac{F_0}{k - m\omega^2}$. With $x(0) = x'(0) = 0$, $x = -\frac{A\omega}{\omega_0}\sin\omega_0 t + A\sin\omega t$. If $\omega \approx \omega_0$, $x \approx A(\sin\omega t - \sin\omega_0 t) = 2A\sin\frac{1}{2}(\omega - \omega_0)t\cos\frac{1}{2}(\omega + \omega_0)t$, the product of a slowly varying component and a rapidly varying one.

PROBLEMS 4.4 *(page 192)*

1. $I(t) = 6(e^{-5t} - e^{-20t})$; $Q(t) = \dfrac{1}{10}(9 - 12e^{-5t} + 3e^{-20t})$. **3.** $I_{\text{steady state}} = \cos t + 2\sin t$

5. $I_{\text{steady state}} = \dfrac{1}{13}(70\sin 10t - 90\cos 10t)$ **7.** $I_{\text{transient}} = e^{-t}(-\cos t - 3\sin t)$

9. $I_{\text{transient}} = \dfrac{40}{3}e^{-5t} - \dfrac{250}{39}e^{-2t}$ **11.** $I_{\text{transient}} = \dfrac{e^{-t}}{481}\left(-245\cos 7t - \dfrac{255}{7}\sin 7t\right)$

13. $I_{\text{transient}} = \dfrac{e^{-600t}}{2320}[-153\sin 800t - 96\cos 800t]$; $I_{\text{steady state}} = \dfrac{1}{580}[24\cos 600t + 27\sin 600t]$

15.
$$Q = \begin{cases} c_1\cos\dfrac{t}{\sqrt{LC}} + c_2\sin\dfrac{t}{\sqrt{LC}} + \left(\dfrac{CE_0}{1 - CL\omega^2}\right)\sin\omega t, & \omega \neq \dfrac{1}{\sqrt{LC}}, \\ \left(c_1 - \dfrac{E_0 t}{2\omega L}\right)\cos\omega t + c_2\sin\omega t, & \omega = \dfrac{1}{\sqrt{LC}} \end{cases}$$

17. **(b)** $\omega = \dfrac{\sqrt{(4L/C) - R^2}}{2L}$ **19.** **(b)** No ω produces resonance.

PROBLEMS 4.5 *(page 193)*

1. $x = e^{-10t}\left(\dfrac{201}{200} + \dfrac{201t}{20}\right) - \dfrac{\cos 10t}{200}$ **3.** $e^{(1-\sqrt{5})t/2}\left(\dfrac{8}{\sqrt{5}} + 6\right) - e^{(-1-\sqrt{5})t/2}\left(\dfrac{7}{\sqrt{5}} + 3\right) - \dfrac{1}{\sqrt{5}}\cos t$

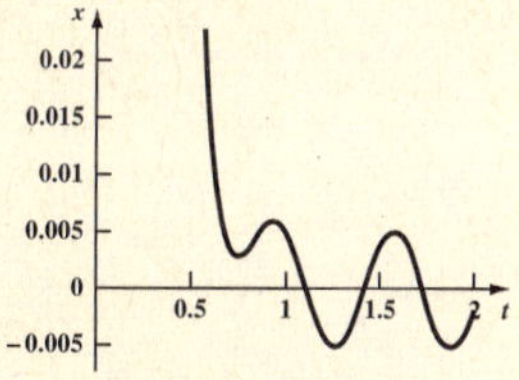

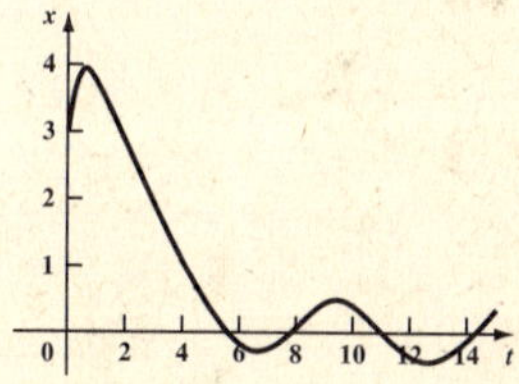

5. $x = \dfrac{1}{104}[e^{-4t}(3\cos 3t + 106\sin 3t) - 3\cos 3t + 2\sin 3t]$

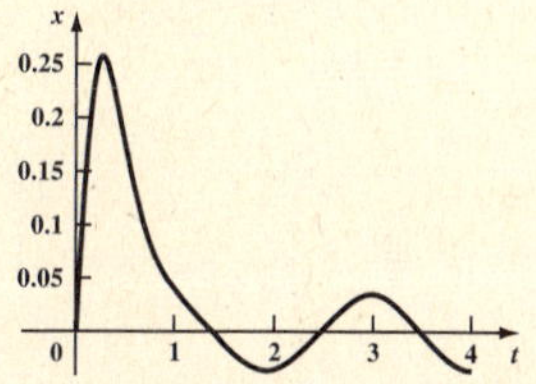

REVIEW EXERCISES FOR CHAPTER 4 *(page 194)*

1. **(a)** $x = \cos\sqrt{50}\,t$ **(b)** $x = e^{-5t}(\cos 5t + \sin 5t)$ **(c)** $x = \dfrac{1}{1000}[1000\cos\sqrt{50}\,t + \sqrt{2}\sin\sqrt{50}\,t - \sin 10t]$

(d) $x = \dfrac{1}{2500}[e^{-5t}(2501\cos 5t + 2502\sin 5t) - \cos 10t - \dfrac{1}{2}\sin 10t]$ **3.** **(a)** $x = 3\cos\dfrac{2\sqrt{10}}{5}t + \sqrt{10}\sin\dfrac{2\sqrt{10}}{5}t$

(b) $x = e^{-t/\sqrt{5}}\left(3\cos\sqrt{\dfrac{7}{5}}\,t + \dfrac{3 + 4\sqrt{5}}{\sqrt{7}}\sin\sqrt{\dfrac{7}{5}}\,t\right)$ **(c)** $x = 3\cos\dfrac{2\sqrt{10}}{5}t + \dfrac{59\sqrt{10}}{60}\sin\dfrac{2\sqrt{10}}{5}t + \dfrac{1}{15}\sin t$

(d) $x = \dfrac{1}{145}\left[e^{-t/\sqrt{5}}\left((435 + 2\sqrt{5})\cos\sqrt{\dfrac{7}{5}}\,t + \left(579\sqrt{\dfrac{5}{7}} + \dfrac{435}{\sqrt{7}}\right)\sin\sqrt{\dfrac{7}{5}}\,t\right) - 2\sqrt{5}\cos t + 3\sin t\right]$

5. **(a)** $x = \dfrac{6}{5}\sin\dfrac{5}{2}t$ **(b)** $x = \dfrac{2}{7}\sqrt{21}\,e^{-t}\sin\dfrac{\sqrt{21}}{2}t$ **(c)** $x = \dfrac{72}{55}\sin\dfrac{5}{2}t - \dfrac{1}{11}\sin 3t$

(d) $x = \dfrac{1}{697}\left[e^{-t}\left(24\cos\dfrac{\sqrt{21}}{2}t + \dfrac{4296}{\sqrt{21}}\sin\dfrac{\sqrt{21}}{2}t\right) - 24\cos 3t - 11\sin 3t\right]$

7. $Q_{ss} = \frac{1}{10}$, $Q_{tran} = -\frac{1}{10}e^{-25t/2}\left[\cos\frac{25\sqrt{31}}{2}t + \frac{1}{\sqrt{31}}\sin\frac{25\sqrt{31}}{2}t\right]$ **9.** $I_{ss} = \frac{50}{14841}[120\cos 5t + 21\sin 5t]$

11. $I_{ss} = 4\cos 5t + 2\sin 5t$ **13.** $x = c_1\cos\sqrt{2g}\,t + c_2\sin\sqrt{2g}\,t + \frac{g}{10(g-32)}\cos 8t$

15. $x = e^{-gt/2000}(c_1\cos rt + c_2\sin rt) - \frac{13g^2}{8000D}\cos\frac{65}{8}t + \left(\frac{2g^2 - \frac{4225g}{64}}{5D}\right)\sin\frac{65t}{8}$ where $r^2 = 2g - \frac{g^2}{4}\times 10^6$ and

$D = \left(2g - \frac{4225}{64}\right)^2 + \left(\frac{65g}{8000}\right)^2.$

CHAPTER 5

PROBLEMS 5.1 *(page 205)*

1. $e^x = e\left[1 + (x-1) + \frac{(x-1)^2}{2!} + \frac{(x-1)^3}{3!} + \ldots\right]$

3. $\cos x = \frac{\sqrt{2}}{2}\left[1 + \left(x - \frac{\pi}{4}\right) - \frac{\left(x - \frac{\pi}{4}\right)^2}{2!} - \frac{\left(x - \frac{\pi}{4}\right)^3}{3!} + \ldots\right]$

5. $e^{bx} = e^{-b}\left[1 + b(x+1) + \frac{b^2}{2!}(x+1)^2 + \frac{b^3}{3!}(x+1)^3 + \ldots\right]$ **7.** $x^2e^{-x^2} = \sum_{n=0}^{\infty}\frac{(-1)^n x^{2n+2}}{n!}$

13. (a) Integrate term-by-term **(b)** Substitute x^2 for x and integrate. **(c)** Differentiate term-by-term **17.** $y = 2e^x - 1$

19. $y = \frac{9}{4}e^{2(x-1)} - \frac{2x^2 + 2x + 1}{4}$ **21.** $y = \frac{x}{2} + \frac{3}{4}\sin 2x$ **23.** $y = 6e^{-x} + (x^3 - 3x^2 + 6x - 2)$

25. $y = c_0(1 + x\tan^{-1}x) + c_1x$ **27.** $y = xe^x$ **29.** $y = 1 - 2x^2$ **31.** $y = x + \sin x$ **33.** $y = \frac{1}{1-x}$

35. $y = e^{x^2} - \frac{x}{4}$ **37.** $y = c_0\left[1 + \sum_{n=1}^{\infty}\frac{1\cdot 4\cdot\cdots\cdot(3n-2)}{(3n)!}x^{3n}\right] + c_1\left[x + \sum_{n=1}^{\infty}\frac{2\cdot 5\cdot\cdots\cdot(3n-1)}{(3n+1)!}x^{3n+1}\right]$

39. $y = c_1x + c_0\left[1 - x^2 - \sum_{n=2}^{\infty}\frac{1\cdot 3\cdot 5\cdot\cdots\cdot(2n-3)}{(2n)!}(2x^2)^n\right]$

41. $y = c_0\left[1 + 2(x-1)^2 + \frac{1}{2}\sum_{n=3}^{\infty}(n+1)(1-x)^n\right] + c_1\left[(x-1) - \frac{1}{2}(x-1)^2 - \frac{1}{4}\sum_{n=3}^{\infty}(n+1)(1-x)^n\right]$

43. $y = c_0(1 - 3x^2) + c_1\left[x - \frac{2}{3}x^3 - 2\sum_{n=2}^{\infty}\frac{2^n[3\cdot 12\cdot 25\cdot\cdots\cdot(n+1)(2n-3)]}{(2n+1)!}x^{2n+1}\right]$

PROBLMS 5.2 *(page 215)*

1. $y_1 = x$; $y_2 = 1 - \frac{x^2}{2!} - \frac{x^4}{4!} - \frac{3x^6}{6!} - \frac{3\cdot 5}{8!}x^8 - \ldots$

3. $y_1 = 1 + \frac{x^3}{2} + \frac{1}{2!\,(5\cdot 2)}x^6 + \frac{1}{3!\,(8\cdot 5\cdot 2)}x^9 + \ldots = 1 + \sum_{n=1}^{\infty}\frac{x^{3n}}{n!\,[2\cdot 5\cdot\cdots\cdot(3n-1)]}$

$y_2 = x + \frac{1}{4}x^4 + \frac{1}{2!\,(7\cdot 4)}x^7 + \frac{1}{3!\,(10\cdot 7\cdot 4)}x^{10} + \ldots = \sum_{n=0}^{\infty}\frac{x^{3n+1}}{n!\,[1\cdot 4\cdot\cdots\cdot(3n+1)]}$

5. $y_1 = (x-2)e^x$; $y_2 = (x-2)e^{x-2}\int\frac{e^{(x-2)^2/2}}{(x-2)^2}\,dx$ **7.** $y_1 = e^{-x^3/3}$; $y_2 = e^{-x^3/3}\int e^{x^3/3}\,dx$ **9.** $y_1 = x$, $y_2 = 1 + x\tan^{-1}x$

11. $y_1 = 1 - x$, $y_2 = \frac{1}{1-x}$ **13.** $y_1 = x + \frac{8}{3}x^3$, $y_2 = 1 + 9x^2 + \frac{15}{2}x^4 - \frac{7}{2}x^6 + \frac{27}{8}x^8 - \ldots$

15. $y = c_1x + c_0\left(1 - \frac{1}{2}x^2 - \frac{1}{4!}x^4 - \frac{3}{6!}x^6 - \ldots\right) + \sin x = c_1x - c_0x\int\frac{e^{x^2/2}}{x^2}\,dx + \sin x$

17. $y = 1 + \frac{x^4}{4!} + \frac{1\cdot 5x^8}{8!} + \frac{1\cdot 5\cdot 9}{12!}x^{12} + \ldots$ **19.** $y = e^{x^2/2}$ **21.** $y = 1 + x + \frac{x^2}{2!} + \frac{2}{3!}x^3 + \frac{4}{4!}x^4 + \ldots$

23. $y = x - \frac{x^3}{3!} + \frac{4x^5}{5!} - \frac{19x^7}{7!} + \ldots$ **25.** $y = e^x(c_1 + c_2 \ln|x|)$

27. $y = c_1 x + c_2 x \int \frac{e^{-x}}{x}\,dx = c_1 x + c_2\left(x \ln|x| + \sum_{n=1}^{\infty} \frac{(-1)^n x^{n+1}}{(n!)n}\right)$

29. $y = c_1 x^2 + c_2 x^2 \int \frac{e^{-x}}{x^3}\,dx = c_1 x^2 + c_2\left(-\frac{1}{2} + x + \frac{1}{2}x^2 \ln|x| + \sum_{n=3}^{\infty} \frac{(-1)^n x^n}{n!(n-2)}\right)$ **31.** $y = c_1(x^2 + 2x + 3) + \frac{c_2 x^4}{(1-x)^2}$

33. No; the solutions $y = ce^{-1/x}$ and $y = ce^{-1/2x^2}$ do not have power series about $x = 0$.

PROBLEMS 5.3 *(page 223)*

1. $y = c_0 \cos\sqrt{x} + c_1 \sin\sqrt{x}$ if $x > 0$; $c_0 \cosh\sqrt{-x} + c_1 \sinh\sqrt{-x}$ if $x < 0$

3. $y = c_0\left[1 + \sum_{n=1}^{\infty} \frac{(-1)^n x^{2n}}{2^n n!\, 3 \cdot 7 \cdots (4n-1)}\right] + c_1\sqrt{x}\left[1 + \sum_{n=1}^{\infty} \frac{(-1)^n x^{2n}}{2^n n!\, 5 \cdot 9 \cdots (4n+1)}\right]$

5. $y = c_0\left[1 + \sum_{n=1}^{\infty} \frac{x^n}{n!\, 1 \cdot 4 \cdot 7 \cdots (3n-2)}\right] + c_1 x^{2/3}\left[1 + \sum_{n=1}^{\infty} \frac{x^n}{n!\, 5 \cdot 8 \cdots (3n+2)}\right]$

7. $y = c_0\left[1 + \sum_{n=1}^{\infty} \frac{(-2x)^n}{n!\, 3 \cdot 5 \cdots (2n-1)}\right] + c_1\sqrt{x}\left[1 + \sum_{n=1}^{\infty} \frac{(-2x)^n}{n!\, 3 \cdot 5 \cdots (2n+1)}\right]$

9. $y = c_0 x^{1/3}\left[1 - \frac{x}{18} + \frac{x^2}{405} - \frac{7x^3}{87480} + \ldots\right] + c_1 x^{2/3}\left[1 - \frac{x}{18} + \frac{5x^2}{2268} - \ldots\right]$

11. $y = c_0\left[1 + 2x + \frac{x^2}{3}\right] + c_1\sqrt{x}\left[1 + \frac{x}{2} + \frac{x^2}{2^2 2!\, 5} - \ldots\right]$

13. $y = c_0(1 - 3x + 2x^2 + \cdots) + c_1\sqrt{x}\left(1 - \frac{7}{6}x + \frac{21}{40}x^2 + \ldots\right)$

15. $y = \frac{c_0}{x}\left[1 - x - \frac{x^2}{2!} - \frac{x^3}{3 \cdot 3!} - \frac{x^4}{3 \cdot 5 \cdot 4!} - \cdots\right] + c_1\sqrt{x}\left[1 + \frac{x}{5} + \frac{x^2}{5 \cdot 7 \cdot 2!} + \frac{x^3}{5 \cdot 7 \cdot 9 \cdot 3!} + \ldots\right]$

17. $y = c_0\left[1 - \frac{x^2}{2} + \frac{x^4}{40} - \frac{x^6}{2160} + \cdots\right] + c_1 x^{3/2}\left[1 - \frac{x^2}{14} + \frac{x^4}{616} - \ldots\right]$

19. $y = c_0 x^{1/3}\left[1 - \frac{3x}{2} + \frac{3^2 x^2}{20} - \frac{3^2 x^3}{160} + \cdots\right] + c_1 x^{2/3}\left[1 - \frac{3}{4}x + \frac{3^2 x^2}{56} - \frac{3^2 x^3}{560} + \ldots\right]$

21. $y = c\left[x - \frac{x^2}{2} + \frac{x^3}{12} - \frac{x^4}{48} + \ldots\right]$; use Section 5.2 to find other solutions.

23. **(b)** $y_2 = xe^{-1/x}$

PROBLEMS 5.4 *(page 232)*

1. $y = c_0 x + \frac{c_1}{x^2}$ **3.** $y = c_0\frac{\sin x^2}{x^2} + c_1\frac{\cos x^2}{x^2}$ **5.** $y = c_0\frac{\sin x}{x} + c_1\frac{\cos x}{x}$

7. $y = c_0 x^{1-\sqrt{2}}\left[1 + \frac{(2-\sqrt{2})}{2\sqrt{2}-1}x + \frac{(3-\sqrt{2})x^2}{2^{3/2}(2\sqrt{2}-1)} + \cdots\right] + c_1 x^{1+\sqrt{2}}\left[1 - \frac{2+\sqrt{2}}{2\sqrt{2}+1}x + \frac{(3+\sqrt{2})x^2}{2^{3/2}(2\sqrt{2}+1)} + \ldots\right]$

9. $y = c_0\left[1 - \frac{x}{2} + \frac{x^2}{12} - \frac{x^3}{144} + \ldots\right]$; use reduction of order. **11.** $y = c_0 e^x + c_1 e^x \ln|x|$ **13.** $y = \sqrt{|x|}\, e^x(c_0 + c_1 \ln|x|)$

15. $y = \frac{c_0}{1-x} + c_1\frac{\ln|x|}{1-x}$ **17.** $y = (c_0 + c_1 \ln|x|)\left(1 + \frac{x^2}{2^2} + \frac{x^4}{(2 \cdot 4)^2} + \cdots\right) - c_1\left(\frac{x^2}{4} + \frac{3x^4}{8 \cdot 16} + \ldots\right)$

19. $y = c_0\frac{\sinh x}{x^3} + c_1\frac{\cosh x}{x^3}$ **21.** $y = c_0 x + c_1\left(x \ln|x| + \sum_{n=1}^{\infty} \frac{(-1)^n}{n!\, n} x^{n+1}\right)$ **23.** $y = \sqrt{|x|}\,(c_0 + c_1 \ln|x|)$

PROBLEMS 5.5 *(page 244)*

13. Use Equation (17). **15.** Set $z = \sqrt{x}$, then $y = AJ_p(\sqrt{x}) + BY_p(\sqrt{x})$. **17.** Set $z = xu$, then $y = x[AJ_1(x) + BY_1(x)]$.

19. Set $y = ux^{-k}$, then $y = x^{-k}(AJ_k(x) + BY_k(x))$. **21.** Set $y = \sqrt{x}\,u$, $z = kx^3/3$, then $y = \sqrt{x}\,(AJ_{1/6}(kx^3/3) + BY_{1/6}(kx^3/3))$.

29. b. $u^2\frac{d^2x}{du^2} + \left(u^2 + \frac{4k_1}{ma^2}\right)x = 0$ **c.** Substitute $x = \sqrt{uz}$ getting $u^2 z'' + uz' + \left(u^2 - \left(\frac{1}{4} - \frac{4k_1}{ma^2}\right)\right)z = 0$.

31. $P_5 = \dfrac{63x^5 - 70x^3 + 15x}{8}$,

$P_6 = \dfrac{231x^6 - 315x^4 + 105x^2 - 5}{16}$,

$P_7 = \dfrac{429x^7 - 693x^5 + 315x^3 - 35x}{16}$,

$P_8 = \dfrac{429(15x^8 - 28x^6) + 630(11x^4 - 2x^2) + 35}{128}$

43. $H_0(x) = 1, H_1(x) = 2x, H_2(x) = 4x^2 - 2, H_3 = 8x^3 - 12x, H_4(x) = 16x^4 - 48x^2 + 12$

PROBLEMS 5.7 *(page 255)*

1. $y = 2x$ **3.** $y = x - \dfrac{2x^3}{3} + \dfrac{4x^5}{15} - \dfrac{8x^7}{105} + \dfrac{16x^9}{945}$ **5.** $y = 1 + \dfrac{x^4}{12} + \dfrac{x^8}{672}$

7. $y = c_1x\left[1 - \dfrac{x}{1!\,2!} + \dfrac{x^2}{2!\,3!} - \dfrac{x^3}{3!\,4!} + \dfrac{x^4}{4!\,5!} - \dfrac{x^5}{5!\,6!} + \dfrac{x^6}{6!\,7!} - \dfrac{x^7}{7!\,8!} + \dots\right]$; use reduction of order.

9. $y = c_0x^2\left[1 - \dfrac{x}{4} + \dfrac{x^2}{40} - \dfrac{x^3}{720} + \dots\right]$; use reduction of order.

REVIEW EXERCISES FOR CHAPTER 5 *(page 257)*

7. $y = x^2 + e^x$ **9.** $y = x(\sin x + 1)$ **11.** $y = e^{-x^2}$ **13.** $y = c_0e^x + c_1\sqrt{x}$ **15.** $y = \dfrac{c_0}{x} + \dfrac{c_1}{1 - x}$

17. $y = c_1\dfrac{e^{x/2}}{\sqrt{x}} + c_2x\left(1 + \sum_{n=1}^{\infty} \dfrac{x^n}{5 \cdot 7 \cdots (2n + 3)}\right)$ **19.** $y = \dfrac{c_0}{x}\left[1 - \sum_{n=1}^{\infty} \dfrac{(-1)^n x^{2n}}{(2n)(2n - 2)!}\right] + 3c_3x^2\left[\sum_{n=0}^{\infty} \dfrac{(-1)^n x^{2n}}{(2n + 3)(2n + 1)!}\right]$

23. $x^2I_p'' + xI_p' - (x^2 + p^2)I_p = i^{-p}[(ix)^2J_p'' + (ix)J_p' + [(ix)^2 - p^2]J_p] = 0$

25. $e^{xt/2}e^{x/2t} = \sum_{n=-\infty}^{\infty} A_n t^n$; where $A_n = \begin{cases} J_n(x) & n \ge 0 \\ (-1)^n J_{-n}(x) & n < 0 \end{cases}$ and let $t^n = e^{i\theta n}$.

29. $c_{n+2} = \dfrac{n^2 - p^2}{(n + 2)(n + 1)}c_n$; so terminates if p is an integer. **31.** $y = P_n(\cos x)$

33. $J_{1/4}\left(\dfrac{1}{2}\right) = 0.799432$

$J_2(2.5) = 3.56847$

$J_{1.3}(2.4) = 1.54874$

CHAPTER 6

PROBLEMS 6.1 *(page 269)*

1. $\dfrac{5}{s^2} + \dfrac{2}{s}, s > 0$ **3.** $\dfrac{18}{s^3} - \dfrac{7}{s}, s > 0$ **5.** $\dfrac{2}{s^3} + \dfrac{8}{s^2} - \dfrac{16}{s}, s > 0$ **7.** $\dfrac{3}{4s^4} + \dfrac{1}{2s^3} + \dfrac{1}{2s^2} + \dfrac{1}{s}, s > 0$ **9.** $\dfrac{a}{s^2} + \dfrac{b}{s}, s > 0$

11. $\dfrac{e^2}{s - 5}, s > 5$ **13.** $\dfrac{1}{s - \frac{1}{2}}, s > \dfrac{1}{2}$ **15.** $\dfrac{e^{-1/2}}{s + 1}, s > -1$ **17.** $\dfrac{3}{s^2 + 9}, s > 0$ **19.** $\dfrac{s}{s^2 + 49}, s > 0$

21. $\dfrac{5 \cos 2}{s^2 + 25} + \dfrac{s \sin 2}{s^2 + 25}, s > 0$ **23.** $\dfrac{s \cos b}{s^2 + a^2} - \dfrac{a \sin b}{s^2 + a^2}, s > 0$ **25.** $\dfrac{s}{s^2 - \frac{1}{4}}, s > \dfrac{1}{2}$

27. $\dfrac{s \cosh 2}{s^2 - 25} - \dfrac{5 \sinh 2}{s^2 - 25}, s > 5$ **29.** $\dfrac{a \cosh b}{s^2 - a^2} + \dfrac{s \sinh b}{s^2 - a^2}, s > |a|$ **31.** $\dfrac{1}{(s - 1)^2}, s > 1$ **33.** $\dfrac{6}{(s + 1)^4} - \dfrac{1}{s + 1}, s > -1$

35. $\dfrac{1}{(s - 1)^2 + 1}, s > 1$ **37.** $\dfrac{s - 4}{(s - 4)^2 + 4}, s > 4$ **39.** $\dfrac{s + 2}{(s + 1)^2 + 1}, s > -1$ **43.** $9t^2 + 7$ **45.** $\cos t + \sin t$

47. $\cosh \sqrt{2}\, t - \sqrt{2} \sinh \sqrt{2}\, t$ **49.** te^t **51.** $\dfrac{3}{\sqrt{5}}e^{-2t} \sin \sqrt{5}\, t$ **53.** $2e^{-t} \cos \sqrt{7}\, t - \dfrac{3}{\sqrt{7}}e^{-t} \sin\sqrt{7}\, t$

55. $ce^{-at} \cos\sqrt{b - a^2}\, t + \dfrac{d - ac}{\sqrt{b - a^2}}e^{-at} \sin \sqrt{b - a^2}\, t$ **63.** $\dfrac{2abs}{(s^2 - a^2 - b^2)^2 - 4a^2b^2}$ **65.** $\dfrac{s(s^2 - a^2 + b^2)}{(s^2 + a^2 + b^2)^2 - 4a^2s^2}$

67. $\dfrac{a(s^2 - a^2 - b^2)}{(s^2 + a^2 + b^2)^2 - 4a^2s^2}$ **71.** $\dfrac{6}{(s + 1)^4}$ **73.** $\dfrac{2s^3 - 54s}{(s^2 + 9)^3}$ **75.** $\dfrac{(s - a)^2 - b^2}{[(s - a)^2 + b^2]^2}$ **77.** $\dfrac{3(s + 1)^2 + 3}{[(s + 1)^2 - 1]^2}$

81. **(a)** $\sqrt{\dfrac{\pi}{s}}$ **(b)** $\dfrac{\sqrt{\pi}}{2s^{3/2}}$ **(c)** $\dfrac{15\sqrt{\pi}}{8s^{7/2}}$

PROBLEMS 6.2 *(page 280)*

1. $\cos t$ **3.** $A\cosh at + \frac{B}{a}\sinh at$ **5.** $e^{-t}(\cos 2t + \sin 2t)$ **7.** $\frac{1}{3} - e^{t} + \frac{5}{3}e^{3t}$ **9.** $-\frac{t}{9} + \frac{23}{27}e^{3t} + \frac{4}{27}e^{-3t}$ **11.** e^{-t}

13. $\sinh t$ **15.** $\left(1 + \frac{t}{2k}\right)\sin kt$ **17.** $\left(a + \frac{1}{2a^2}\right)\sin at + \left(a - \frac{t}{2a}\right)\cos at$

19. $\frac{5}{4}\cosh t - \frac{1}{4}(\cos t + t\sin t)$ **21.** $\frac{s^2 + 2a^2}{s(s^2 + 4a^2)}$ **23.** $\frac{2a(3s^2 - a^2)}{(s^2 + a^2)^3}$ **25.** $\frac{8 + 6s^2}{s^2(s^2 + 4)^2}$ **27.** $\frac{2a^2(4a^2 + 3s^2)}{s^2(s^2 + 4a^2)^2}$

29. $\frac{16}{s^2 + 16}\left[\frac{1}{s^3} + \frac{1}{s(s^2 + 16)} + \frac{4s}{(s^2 + 16)^2}\right]$ **33.** $\frac{\pi}{2} - \tan^{-1}s = \tan^{-1}\left(\frac{1}{s}\right)$ **35.** $\frac{\pi}{2} - \tan^{-1}\left(\frac{s}{3}\right) = \tan^{-1}\left(\frac{3}{s}\right)$

37. $\tan^{-1}\left(\frac{k}{s}\right)$ **39.** $\frac{1}{2}\ln\frac{s^2 - a^2}{s^2}$ **41.** $\frac{1}{2s}\ln\frac{s^2 - a^2}{s^2}$ **43.** $\frac{e^{s^2/4}}{s}\left(1 - \text{erf}\left(\frac{s}{2}\right)\right)$ **45.** $\frac{2}{t}(1 - \cos at)$ **47.** $\frac{\sin t}{t}$

49. **(a)** $Y' - s^2Y = -1$ **(b)** $Y(s) = \left[Y(0) - \int_0^s e^{-u^3/3}du\right]e^{s^3/3}$

PROBLEMS 6.3 *(page 293)*

9. $\frac{b}{s(e^{as} + 1)}$ **11.** $\frac{b(1 - e^{-as})}{s(1 + e^{-2as})}$ **13.** $\frac{1}{s^2(1 + e^{-as})}$ **15.** $\frac{1}{(s^2 + 1)(1 - e^{-\pi s})}$ **17.** $\frac{1 + se^{-4\pi s}}{s^2 + 1}$

19. $\sin(t - \pi)H(t - \pi)$ **21.** $\cos t(1 + H(t - \pi))$ **23.** $\frac{1}{4!}[t^4 + (t - 4)^4H(t - 4)]$

25. $f(t) = 1 + 2H(t - 1) + 2H(t - 7);\ \mathcal{L}\{f\} = \frac{1}{s}[1 + 2e^{-s} + 2e^{-7s}]$ **27.** $\frac{1}{2}\int_0^t (t - u)^3\sin u\,du = \frac{t^3}{2} - 3t + 3\sin t$

29. $\frac{1}{a^2}(1 - \cos at)$ **31.** $\frac{1}{2}(t - 3)^2H(t - 3)$ **33.** $\frac{3!}{s^4(s^2 + 1)}$ **35.** $\frac{3!\,5!}{s^{10}}$

37. $y = t - \sin t - H(t - \pi)(t + \sin t + \pi\cos t)$

39. $y = \frac{1}{10}\left[1 - e^{-t}(\cos 3t + \frac{1}{3}\sin 3t)\right] - \frac{1}{10}\left[1 - e^{-(t-1)}(\cos 3(t - 1) + \frac{1}{3}\sin 3(t - 1))\right]H(t - 1)$

41. $y = \frac{1}{10} + \frac{9}{10}e^{-t}\cos 3t + \frac{3}{10}e^{-t}\sin 3t - \frac{1}{10}H(t - 1)[1 - e^{-(t-1)}\cos 3(t - 1) - \frac{1}{3}e^{-(t-1)}\sin 3(t - 1)]$

43. $\frac{1}{13}\left[1 - e^{-2t}\left(\cos 3t + \frac{2}{3}\sin 3t\right)\right] - \frac{1}{13}\left[1 - e^{-2(t-\pi)}\left(\cos 3(t - \pi) + \frac{2}{3}\sin 3(t - \pi)\right)\right]H(t - \pi)$ **45.** $e^{-t}(1 - t)^2$

PROBLEMS 6.4 *(page 299)*

1. $y = 2(1 - e^{3-t})H(t - 3)$ **3.** $y = 2e^{-2t} - \frac{1}{4}(7e^{8-2t} - 2t + 1)H(t - 4)$ **5.** $y = \frac{1}{2}e^{-2t}[|t - 1| - |t - 2| + 7]$

7. $y = [1 - \cos(t - 1)]H(t - 1)$

9. $y = te^{-t} + (1 - te^{1-t})H(t - 1) + (1 - (t - 1)e^{2-t})H(t - 2) + (1 - (t - 2)e^{3-t}))H(t - 3)$

11. $y = \frac{H(t - 3)}{18}[\sin 3t - 3(t - 3)\cos 3t - \sin 9\cos(3t - 9)] + \cos 3t - \frac{1}{3}\sin 3t$

13. $y = e^{2t}\sin 3t + \frac{1}{3}e^{2(t-1)}\sin 3(t - 1)H(t - 1)$ **15.** $y = \cos t + \sin(t - 2\pi)H(t - 2\pi)$

17. $y = e^{-3t}[(1 - \cos(t - 1))H(t - 1) + e^6\sin(t - 2)H(t - 2) + \cos t + 3\sin t]$

19. $y = e^{-t}\left[2\cosh 2t + \frac{1}{2}\sinh 2t + \frac{1}{2}\sinh 2(t - 1)H(t - 1)\right] - \frac{1}{16}[e^{t-16} - e^{-3t} - 4(t - 4)e^{-3t}]H(t - 4)$

21. $\mathcal{L}\{y\} = \sum_{n=0}^{\infty}\frac{e^{-2\pi ns}}{s^2 + 1}$, so $y = \sum_{n=0}^{\infty}\sin(t - 2\pi n)H(t - 2\pi n)$, which grows without bound as $t \to \infty$.

PROBLEMS 6.5 *(page 304)*

1. $I(t) = \frac{10}{\sqrt{6}}e^{-20(t-30)}\sin\sqrt{600}\,(t - 30)H(t - 30)$ **3.** $I(t) = 1 - \cos 4t + H(t - 5)[\cos 4(t - 5) - 1]$

5. $I(t) = 1 - \cos\sqrt{10}\,t - H(t - 2)[1 - \cos\sqrt{10}\,(t - 2)] + 2H(t - 4)[1 - \cos\sqrt{10}\,(t - 4)]$
$+ \frac{60}{\sqrt{10}}H(t - 4)\sin\sqrt{10}\,(t - 4)$

7. (a) $x(t) = \frac{t}{2} - \frac{\sin 2t}{4} - \frac{1}{4}H\left(t - \frac{\pi}{2}\right)[2t - \pi\cos(2t - \pi) - \sin(2t - \pi)]$

(b) $x(t) = \frac{t}{2} - \frac{\sin 2t}{4} + \cos 2t - \frac{1}{4}H\left(t - \frac{\pi}{2}\right)[2t - \pi\cos(2t - \pi) - \sin(2t - \pi)]$

PROBLEMS 6.6 *(page 305)*

1. $\frac{5}{s^2 - 6s + 34}$ **3.** $\frac{6(s^4 + 150s^2 + 625)}{(s+5)^4(s-5)^4}$ **5.** $\frac{1}{50} + \frac{\sqrt{5\pi}\, s}{20{,}000}\left[(s^2 + 30)e^{s^2/20}\,\mathrm{erf}\left(\frac{\sqrt{5}\, s}{10}\right) + i(30 - s^2)e^{-s^2/20}\,\mathrm{erf}\left(\frac{\sqrt{5}\, is}{10}\right)\right]$

REVIEW EXERCISES FOR CHAPTER 6 *(page 305)*

1. $\frac{3}{s^2} - \frac{2}{s}$ **3.** $\frac{1}{e(s-2)}$ **5.** $\frac{1}{(s+1)^2}$ **7.** $\frac{3\cosh 4 - s(\sinh 4)}{s^2 - 9}$ **9.** $\frac{(s-1)^2 - 1}{[(s-1)^2 + 1]^2}$ **11.** $\frac{s^2 + 2}{s(s^2 + 4)}$ **13.** $\frac{2s(s^2 - 12)}{(s^2 + 4)^3}$

15. e^{-3s} **17.** $\frac{1 + se^{-2\pi s}}{s^2 + 1}$ **19.** -14 **21.** te^{2t} **23.** $e^{-2t}\sin t$ **25.** $2e^{2t} - e^t$ **27.** $-2e^{-t} - 5te^{-t} + \cos t + 4\sin t$

29. $\frac{2}{t}(1 - \cos 2t)$

31. $\cos t - \cos\left(t - \frac{\pi}{2}\right)H\left(t - \frac{\pi}{2}\right) = \begin{cases} \cos t, & t < \frac{\pi}{2} \\ \cos t - \sin t, & t > \frac{\pi}{2} \end{cases}$

33. $y = \cos t + 3\sin t$ **35.** $y = 5e^{2t} - 3e^{3t}$ **37.** $y = \frac{e^{2t}}{9}(3t - 4) + e^t - \frac{5}{9}e^{-t}$

39. $y = \frac{1}{2}[1 - e^{-(t-3)}(\cos(t-3) - \sin(t-3))]H(t-3)$ **41.** $y = te^{2t} - (t-1)e^{2(t-1)}H(t-1)$ **43.** $\frac{4!\,7!}{s^{13}}$

45. $\frac{19!}{s^{20}(s-17)}$ **47. (a)** $\int_0^t \sin(t-u)u^2\,du = t^2 + 2\cos t - 2$; **(b)** $\int_0^t \frac{1}{2}\sin 2u\,du = \frac{1}{4}(1 - \cos 2t)$

49. $I(t) = \frac{1}{676}[12\cos t + 5\sin t - (12 + 65t)e^{-5t}]$ **51.** $Q(t) = \frac{1}{125}[-4\cos 10t - 3\sin 10t + (4 + 50t)e^{-5t}]$

CHAPTER 7

PROBLEMS 7.1 *(page 313)*

1. $x_1' = x_2$, $x_2' = -3x_1 - 2x_2$ **3.** $x_1' = x_2$, $x_2' = x_3$, $x_3' = x_1^3 - x_2^2 + x_3 + t$ **5.** $x_1' = x_2$, $x_2' = x_3$, $x_3' = -x_1x_3 + x_1^4x_2 + \sin t$ **7.** $x_1' = x_2$, $x_2' = x_3$, $x_3' = x_1 - 4x_2 + 3x_3$

9. (a) $x_1 = c_1e^t$, $x_2 = c_2e^t$

PROBLEMS 7.2 *(page 318)*

1. (b) $W = e^{-6t}$ **(c)** $([c_1 + c_2(1-t)]e^{-3t}, (-c_1 + c_2t)e^{-3t})$ **3. (d)** System is not defined at $t = 0$.

PROBLEMS 7.3 *(page 328)*

1. $x = c_1e^{-t} + c_2e^{4t}$, $y = -c_1e^{-t} + \frac{3}{2}c_2e^{4t}$ **3.** $x = (c_1 + c_2t)e^{-3t}$, $y = -(c_1 + c_2 + c_2t)e^{-3t}$

5. $x = (c_1 + c_2t)e^{10t}$, $y = -(2c_1 + c_2 + 2c_2t)e^{10t}$ **7.** $x = c_1 + c_2e^{2t} - \frac{1}{4}(t^2 + 9t)$, $y = -c_1 - \frac{1}{3}c_2e^{2t} + \frac{1}{4}(t^2 + 7t - 3)$

9. $x = c_1e^{(1+\sqrt{3})t} + c_2e^{(1-\sqrt{3})t}$, $y = \frac{\sqrt{3}}{2}[c_1e^{(1+\sqrt{3})t} - c_2e^{(1-\sqrt{3})t}] - \frac{1}{4}\sin 2(c_1e^{(1+\sqrt{3})t} + c_2e^{(1-\sqrt{3})t})$

11. $x_1 = 3c_1e^{-t} + 4c_2e^{-2t}$, $x_2 = -4c_1e^{-t} - 5c_2e^{-2t} - c_3e^{2t}$, $x_3 = -2c_1e^{-t} - 7c_2e^{-2t} + c_3e^{2t}$ **13.** $(3e^{3t}, -2e^{3t})$, $(2e^{10t}, e^{10t})$

15. (e^{2t}, e^{2t}), $(te^{2t}, (1+t)e^{2t})$ **17.** $\left(e^{3t}\cos 3t, \frac{1}{2}e^{3t}(3\sin 3t + \cos 3t)\right)$; $\left(e^{3t}\sin 3t, \frac{1}{2}e^{3t}(\sin 3t - 3\cos 3t)\right)$

19. $\left(\frac{1}{4}te^{2t}, -\frac{11}{4}e^{2t}\right)$ **21.** $(-\sin t, -\cos t - 2\sin t)$ **23.** $\left(-t + 3e^t, \frac{1}{2} - t + 3e^t\right)$

25. $\dfrac{dx}{dt} = -\dfrac{3x}{100+t} + \dfrac{2y}{100-t}, \quad x(0) = 100$

$\dfrac{dy}{dt} = \dfrac{3x}{100+t} - \dfrac{4y}{100-t}, \quad y(0) = 0$

27. $y_{\max} = \dfrac{500}{\sqrt{3}}[(2+\sqrt{3})^{(1-\sqrt{3})/2} - (2+\sqrt{3})^{(-1-\sqrt{3})/2}] = 500\sqrt{\dfrac{2}{3}}(2+\sqrt{3})^{-\sqrt{3}/2},\ t_{\max} = \dfrac{25}{\sqrt{3}}\ln(2+\sqrt{3})$ min.

29. $x_1 = c_1\cos\sqrt{10}\,t + c_2\sin\sqrt{10}\,t + c_3\cos t + c_4\sin t,\ x_2 = -\dfrac{1}{4}c_1\cos\sqrt{10}\,t - \dfrac{1}{4}c_2\sin\sqrt{10}\,t + 2c_3\cos t + 2c_4\sin t$

31. The concentration of H_2S approaches $\dfrac{\gamma}{\alpha}$; the concentration of SO_2 approaches $\dfrac{\gamma+\delta}{\beta}$.

39. $x = e^{2t}\left[\dfrac{t}{4} - \dfrac{c_2}{2}\cos 2t + c_1\sin 2t\right]$

$y = e^{2t}\left[\dfrac{-11}{4} + 2c_1\cos 2t + c_2\sin 2t\right]$

41. $x = e^t(c_1\sin t - c_2\cos t) + \sin t - \cos t,\ y = e^t((c_1 - 2c_2)\cos t + (2c_1 + c_2)\sin t) - \cos t + 2\sin t$

PROBLEMS 7.4 *(page 333)*

1. $x = \cosh t,\ y = \sinh t$ **3.** $x = e^{3t}(2\cos 3t + 2\sin 3t),\ y = e^{3t}(-2\cos 3t + 4\sin 3t)$

5. $x = \dfrac{3}{2}\cos t + \dfrac{1}{2}\sin t + \dfrac{7}{2}e^t - 5e^{-t} - 4t,\ y = \cos t + \dfrac{7}{2}e^t - \dfrac{5}{2}e^{-t} - 1 - 3t$

7. $x = \dfrac{e^{3t}}{4(13)^3}(6887\cos 2t + 2637\sin 2t) + \dfrac{e^t}{4} - \dfrac{5t^2}{13} - \dfrac{34t}{(13)^2} - \dfrac{74}{(13)^3},\ y = \dfrac{e^{3t}}{4(13)^3}(-9524\cos 2t + 4250\sin 2t) + \dfrac{4}{13}t^2 + \dfrac{48t}{(13)^2} + \dfrac{184}{(13)^3}$ **9.** $x = \cos t,\ y = -\cos t - \sin t$ **11.** $x = \cos t - \sin t,\ y \equiv 1,\ z = \sin t + \cos t$

13. $x(t) = \dfrac{u}{2\omega}(2\omega t\cos^2\beta + \sin^2\beta\sin 2\omega t) + \dfrac{v}{\omega}\sin\beta\sin^2\omega t$

$- \dfrac{w}{2\omega}\sin\beta\cos\beta(2\omega t - \sin 2\omega t) + \dfrac{g}{2\omega^2}\sin\beta\cos\beta(\omega^2t^2 - \sin^2\omega t);$

$y(t) = -\dfrac{u}{\omega}\sin\beta\sin^2\omega t + \dfrac{v}{2\omega}\sin 2\omega t$

$- \dfrac{w}{\omega}\cos\beta\sin^2\omega t + \dfrac{g}{4\omega^2}\cos\beta(2\omega t - \sin 2\omega t);$

$z(t) = -\dfrac{u}{2\omega}\sin\beta\cos\beta(2\omega t - \sin 2\omega t) + \dfrac{v}{\omega}\cos\beta\sin^2\omega t$

$+ \dfrac{w}{2\omega}(2\,\omega t\sin^2\beta + \sin 2\omega t\cos^2\beta)$

$- \dfrac{g}{2\omega^2}(\omega^2t^2\sin^2\beta + \cos^2\beta\sin^2\omega t).$

PROBLEMS 7.5 *(page 338)*

1. $(I_L, I_R) = (1 - 1.025e^{-0.05},\ 1 - 1.05e^{-0.05})$

3. $(100(k_1e^{\lambda_1 t} + k_2e^{\lambda_2 t}) + 1,\ -8(k_1\lambda_1e^{\lambda_1 t} + k_2\lambda_2e^{\lambda_2 t}) + 1)$ where $\left.\begin{matrix}k_1\\k_2\end{matrix}\right\} = \dfrac{\mp 3\sqrt{2} - 4}{800},\ \left.\begin{matrix}\lambda_1\\\lambda_2\end{matrix}\right\} = -50 \pm 25\sqrt{2}$

5. The general solution $(I_L, I_R) = (I_L, I_R)_h + (I_L, I_R)_p$ is given by $(I_L, I_R)_p = (A\sin\omega t + B\cos\omega t,\ C\sin\omega t + D\cos\omega t)$, where $\omega = 60\pi,\ \Delta = \omega^2 + 2500,\ A = \left(\dfrac{2500}{\Delta}\right)^2,\ B = \dfrac{-25\omega(\omega^2 + 7500)}{\Delta^2},\ C = \dfrac{-2500(\omega^2 - 2500)}{\Delta^2},\ D = -250{,}000\dfrac{\omega}{\Delta^2}$, and

(a) $(I_L, I_R)_h = \left(\dfrac{e^{-50t}}{2}\left[k_1 + k_2\left(t + \dfrac{1}{25}\right)\right],\ e^{-50t}\left[k_1 + k_2\left(t + \dfrac{1}{25}\right)\right]\right).$

(b) $(I_L, I_R)_h = (e^{-50t}[k_1\cos 50\sqrt{3}\,t + k_2\sin 50\sqrt{3}\,t],\ e^{-50t}[k_3\cos 50\sqrt{3}\,t + k_4\sin 50\sqrt{3}\,t])$

(c) $(I_L, I_R)_h = (e^{-50t}(k_1e^{25\sqrt{2}t} + k_2e^{-25\sqrt{2}t}),\ e^{-50t}[(2 - \sqrt{2})\,k_1e^{25\sqrt{2}t} + (2 + \sqrt{2})k_2e^{-25\sqrt{2}t}])$

7. $I_1 = \dfrac{1}{20}(8\sin t - 6\cos t + 5e^{-t} + e^{-3t});\ I_2 = \dfrac{1}{20}(2\sin t - 4\cos t + 5e^{-t} - e^{-3t})$ **9.** $I = 50t$

11. $E_k = A\dfrac{\sinh a(n-k)}{\sinh an}\cos\omega t$ **13.** Natural frequencies are $\omega_N = \left[2\sqrt{CL}\sin\dfrac{N\pi}{2(n+1)}\right]^{-1}$. Cut-off frequency is $\omega_n \approx \dfrac{1}{2\sqrt{CL}}$.

15. See the reference.

PROBLEMS 7.6 *(page 344)*

1. When $t = \sqrt{5}\ln\left(\frac{3+\sqrt{5}}{3-\sqrt{5}}\right)$, $y_{\max} = \frac{40}{3-\sqrt{5}}\left(\frac{3+\sqrt{5}}{3-\sqrt{5}}\right)^{(-5-3\sqrt{5})/10}$

3. When $t = \frac{150}{\sqrt{3}}\ln\frac{7+\sqrt{13}}{7-\sqrt{13}}$, $y_{\max} = \frac{3000}{7-\sqrt{13}}\left(\frac{7-\sqrt{13}}{7+\sqrt{13}}\right)^{(7+\sqrt{13})/(2\sqrt{13})}$

5. Let $k = AB - \frac{(a+b)^2}{4}$. If $k > 0$, $x = e^{(a-b)t/2}\left[x_0\cos\sqrt{k}t + \left(\frac{1}{2}x_0(a+b) - Ay_0\right)\frac{\sin\sqrt{k}t}{\sqrt{k}}\right]$,

$y = e^{(a-b)t/2}\left[y_0\cos\sqrt{k}t - \left(\frac{1}{2}y_0(a+b) - Bx_0\right)\frac{\sin\sqrt{k}t}{\sqrt{k}}\right]$. If $k < 0$, replace cos, sin, $\sqrt{k}$ with cosh, sinh, $\sqrt{-k}$.

7. Let $D = a_{11}a_{22} - a_{12}a_{21} \neq 0$, λ_1 and λ_2 be the roots of $\lambda^2 - (a_{11} + a_{22})\lambda + D = 0$. Then $x = c_1e^{\lambda_1 t} + c_2e^{\lambda_2 t} + \frac{(b_2a_{12} - b_1a_{22})}{D}$,

$y = \frac{c_1(\lambda_1 - a_{11})e^{\lambda_1 t}}{a_{12}} + \frac{c_2(\lambda_2 - a_{11})e^{\lambda_2 t}}{a_{12}} + \frac{(b_1a_{21} - b_2a_{11})}{D}$.

9. Let $D = a_1b_2 - a_2b_1 \neq 0$, $\omega = \frac{\pi}{4}$, λ_1 and λ_2 be the roots of $\lambda^2 + (a_1 + b_2)\lambda + D = 0$, and A_1, A_2, A_3, A_4 be the solution of the system $a_1A_1 + \omega A_2 - a_2A_3 = -a$, $-\omega A_1 + a_1A_2 - a_2A_4 = 0$, $-b_1A_1 + b_2A_3 + \omega A_4 = 0$, $b_1A_2 + \omega A_3 - b_2A_4 = 0$.

$x = c_1e^{\lambda_1 t} + c_2e^{\lambda_2 t} + \frac{ab_2}{D} + A_1\cos\omega t + A_2\sin\omega t$, $y = \frac{(c_1\lambda_1 + a_1)e^{\lambda_1 t}}{a_2} + \frac{(c_2\lambda_2 + a_2)e^{\lambda_2 t}}{a_2} + \frac{ab_1}{D} + A_3\cos\omega t + A_4\sin\omega t$.

PROBLEMS 7.7 *(page 346)*

1. If $\tan\theta = \frac{B}{A}$, then $A\cos\omega t + B\sin\omega t = \sqrt{A^2 + B^2}\cos(\omega t - \theta)$.

3. $x_1 = \left(\frac{1}{2}x_{10} + \frac{1}{4}x_{20}\right)\cos t + \left(\frac{1}{2}x_{30} + \frac{1}{4}x_{40}\right)\sin t + \left(\frac{1}{2}x_{10} - \frac{1}{4}x_{20}\right)\cos\sqrt{3}t + \left(\frac{1}{2}x_{30} - \frac{1}{4}x_{40}\right)\frac{\sin\sqrt{3}t}{\sqrt{3}}$

$x_2 = \left(x_{10} + \frac{1}{2}x_{20}\right)\cos t + \left(x_{30} + \frac{1}{2}x_{40}\right)\sin t - \left(x_{10} - \frac{1}{2}x_{20}\right)\cos\sqrt{3}t - \left(x_{30} - \frac{1}{2}x_{40}\right)\frac{\sin\sqrt{3}t}{\sqrt{3}}$

5. $\omega_n = \frac{n^2\pi^2}{64}\sqrt{\frac{4.32\times 10^9}{\frac{24(8)^4}{\pi g}}} \approx 325n^2$ rev/sec. The lowest frequency is 325 rev/sec.

7. $l_1(m_1 + m_2)\theta'' + m_2l_2\phi''\cos(\theta - \phi) + m_2l_2\phi'^2\sin(\theta - \phi) + g(m_1 + m_2)\sin\theta = 0$, $l_1\theta''\cos(\theta - \phi) + l_2\phi'' - l_1\theta'^2\sin(\theta - \phi) + g\sin\phi = 0$ **9.** The characteristic equation is $m_1l_1l_2\lambda^4 + (m_1 + m_2)(l_1 + l_2)g\lambda^2 + (m_1 + m_2)g^2 = 0$. The discriminant as a quadratic in λ^2 is positive, so the roots are pure imaginary, and the solution is a superposition of two simple harmonics.

PROBLEMS 7.8 *(page 352)*

1. Yes; $W = 907$ **3.** No epidemic **5.** Yes; $W = 22{,}994$ **7.** From **(11)** $x(t) = \frac{\beta}{\alpha}$ at $y_{\max}$. Substitute this into **(13).** Then use $N \geq x(0)$ and a power series for the logarithm.

PROBLEMS 7.9 *(page 355)*

1. $x = e^{-3t}(c_1 + c_2t)$, $y = -e^{-3t}(c_1 + c_2(t+1))$ **3.** $x = -\frac{1}{\sqrt{2}}\sin\sqrt{2}t - \cos\sqrt{2}t$, $y = \frac{3}{\sqrt{2}}\sin\sqrt{2}t$

5. $x = c_1e^{\sqrt{13}s} + c_2e^{-\sqrt{13}s} - \frac{e^s}{4} - \frac{2(s+1)}{13}$, $y = c_1\left(\frac{\sqrt{13}-1}{3}\right)e^{\sqrt{13}s} - c_2\left(\frac{\sqrt{13}+1}{3}\right)e^{-\sqrt{13}s} - \frac{8s}{13}$

7. See answer to Problem 7.3.39. **9.** See answer to Problem 7.3.41 with $u = x$, $v = y$, and $s = t$.

REVIEW EXERCISES FOR CHAPTER 7 *(page 359)*

1. $x_1' = x_2$, $x_2' = x_3$, $x_3' = 6x_3 - 2x_2 + 5x_1$ **3.** $x_1' = x_2$, $x_2' = x_3$, $x_3' = \frac{(\ln t - x_1x_3)}{x_2}$

5. $x = c_1e^{3t} + c_2(1+t)e^{3t}$, $y = (c_1 + c_2t)e^{3t}$ **7.** $x = c_1e^{5t} + c_2e^{-t}$, $y = 2c_1e^{5t} - c_2e^{-t}$

9. $x = c_1e^{-t} + 3e^{-2t}$, $y = (c_2 - 2c_1t)e^{-t} + 12e^{-2t}$

11. $x = \frac{e^{-t}}{4}[(2 + 3\sqrt{2})e^{2\sqrt{2}t} + (2 - 3\sqrt{2})e^{-2\sqrt{2}t}]$; $y = \frac{e^{-t}}{4}[(4 - \sqrt{2})e^{2\sqrt{2}t} + (4 + \sqrt{2})e^{-2\sqrt{2}t}]$

13. $x = \frac{8e^t}{7} - 5t^2 + 50t - 240 + \frac{e^{-5t/2}}{7}\left[1679 \cosh \frac{\sqrt{21}\,t}{2} + 371\sqrt{21} \sinh\left(\frac{\sqrt{21}\,t}{2}\right)\right]$, $y = \frac{3e^t}{7} - 2(t^2 - 11t + 53)$ $+ \frac{e^{-5t/2}}{7}\left[732 \cosh \frac{\sqrt{21}\,t}{2} + 166\sqrt{21} \sinh \frac{\sqrt{21}\,t}{2}\right]$ **15. (a)** $\frac{dE}{dx} = -IR, \frac{dI}{dx} = -EG$ **(b)** $E = E_0 e^{-\sqrt{RG}\,x}, I = \sqrt{\frac{G}{R}}\, E_0 e^{-\sqrt{RG}\,x}$
17. (a) $x = 30(1 - e^{-9t/75}), y = 30(1 + e^{-9t/75})$ **(b)** y decreases steadily to 30 lb, x increases steadily to 30 lb.
(c) There is no maximum. **19.** $y = \frac{bd}{(a + c)(a + d)}[a + ce^{-(a+c)t}]$

CHAPTER 8

PROBLEMS 8.1 *(page 363)*

1. 0.33333333×10^0 **3.** -0.35×10^{-4} **5.** 0.77777777×10^0 **7.** 0.77272727×10^1 **9.** -0.18833333×10^2
11. 0.23705963×10^9 **13.** 0.83742×10^{-20} **15.** $\epsilon_a = 0.1; \epsilon_r = 0.0002$ **17.** $\epsilon_a = 0.005; \epsilon_r = 0.04$
19. $\epsilon_a = 0.333\cdots \times 10^{-2}; \epsilon_r = 0.571428 \times 10^{-3}$ **21.** $\epsilon_a = 1, \epsilon_r = 0.14191 \times 10^{-4}$

PROBLEMS 8.2 *(page 370)*

1. $y = 2e^x - x - 1, y(1) = 2(e - 1) \approx 3.4366$; **(a)** $y_E = 2.98$ **(b)** $y_{IE} = 3.405$
3. $y = \sqrt{2x^2 + 1} - x; y(1) = \sqrt{3} - 1 \approx 0.73205$ **(a)** $y_E = 0.71$ **(b)** $y_{IE} = 0.73207$
5. $y = \sinh\left(\frac{1}{2}x^2 - \frac{1}{2}\right); y(3) \approx 27.2899$ **(a)** $y_E = 8.31$ **(b)** $y_{IE} = 21.671$ **7.** $y = (x^3 + x^{-2})^{-1/2}, y(2) = \frac{2}{\sqrt{33}} \approx 0.3481553$;
(a) $y_E = 0.343$ **(b)** $y_{IE} = 0.34939$ **9.** $y = 2e^{e^x - 1}, y(2) = 2e^{e^2 - 1} \approx 1190.59$; **(a)** $y_E = 1.56$ **(b)** $y_{tE} = 781.56$.

In 11–19 answers are given for the improved Euler method. Part (a) is given in the answers for Section 1.3.

11. $y_1 = 1.02, y_2 = 1.04, y_3 = 1.07, y_4 = 1.09, y_5 = 1.12$ **13.** $y_1 = 0.34, y_2 = 0.56, y_3 = 0.76, y_4 = 0.98, y_5 = 1.34$
15. $y_1 = 1.28, y_2 = 1.65, y_3 = 1.46, y_4 = 1.09, y_5 = 0.87, y_6 = 0.79, y_7 = 0.93, y_8 = 0.95$
17. $y_1 = 1.10, y_2 = 1.22, y_3 = 1.35, y_4 = 1.48, y_5 = 1.63, y_6 = 1.79, y_7 = 1.97, y_8 = 2.16, y_9 = 2.37, y_{10} = 2.60$
19. $y_1 = 1.60, y_2 = 1.27, y_3 = 1.00, y_4 = 0.80, y_5 = 0.64$ **21.** No method will provide a correct answer since the solution to the differential equation is the hyperbola $x^2 - 2xy - y^2 = 4$ or $y = \sqrt{2x^2 - 4} - x$, which is not defined if $x = 1$.
23. The solution to the differential equation is $y = [x^2(2 - x^2)]^{-1/2}$, which is not defined at $x = 3$.
25. Using $h = 0.2, y(1) = 3.1700$; using $h = 0.1, y(1) = 3.1586$; exact value is 3.1548548.
27. Use the fact that $y'' = \frac{d}{dx}(f(x, y)) = f_x + f_y y'$ to show $e_{1E} = \frac{h^2}{3!} y_0'''$. **29.** $y(2) \approx 0.321324$ **31.** $y(2) \approx 5.80427$

PROBLEMS 8.3 *(page 380)*

1. $y = 2e^x - x - 1; y(1) = 2(e - 1) \approx 3.4365637; y_{RK} = 3.436502$
3. $y = \sqrt{2x^2 + 1} - x; y(1) = \sqrt{3} - 1 \approx 0.732050807; y_{RK} = 0.73205044$
5. $y = \sinh\left(\frac{1}{2}x^2 - \frac{1}{2}\right); y(3) \approx 27.2899; y_{RK} = 27.0275$ **7.** $y = (x^3 + x^{-2})^{-1/2}; y(2) = \frac{2}{\sqrt{33}} \approx 0.3481553; y_{RK} = 0.348161$
9. $y = 2e^{e^x - 1}, y(2) = 2e^{e^2 - 1} \approx 1190.59; y_{RK} = 1164.76$ **11.** $y_1 = 1.02, y_2 = 1.04, y_3 = 1.07, y_4 = 1.10, y_5 = 1.12$.
13. $y_1 = 0.38, y_2 = 0.61, y_3 = 0.78, y_4 = 0.98, y_5 = 1.35$. **15.** $y_1 = 1.33, y_2 = 1.91, y_3 = 1.61, y_4 = 1.16, y_5 = 0.92$, $y_6 = 0.79, y_7 = 0.54, y_8 = 0.77$. **17.** $y_1 = 1.11, y_2 = 1.22, y_3 = 1.35, y_4 = 1.48, y_5 = 1.63, y_6 = 1.79, y_7 = 1.97, y_8 = 2.16$, $y_9 = 2.37, y_{10} = 2.60$. **19.** $y_1 = 1.59, y_2 = 1.26, y_2 = 1.00, y_4 = 0.79, y_5 = 0.64$. **21.** See answer to Problem 8.2.21.
23. See answer to Problem 8.2.23. **25.** $a = \frac{1}{4}, b = 0, c = \frac{3}{4}, d = 0$ **27.** $n = p = \frac{2}{3}$
29. Because the determinant of the resulting 4×4 system is zero. There are an infinite number of solutions; no.

PROBLEMS 8.4 *(page 385)*

1. $y = 2e^x - x - 1, y(1) = 2(e - 1) \approx 3.43656$; **(a)** $y = 3.43761$ **(b)** $y = 3.43022$ **3.** $y = \sqrt{2x^2 + 1} - x$, $y(1) = \sqrt{3} - 1 \approx 0.73205$; **(a)** $y = 0.74146$ **(b)** $y = 0.74566$ **5.** $y = \sinh\left(\frac{1}{2}x^2 - \frac{1}{2}\right), y(3) = \frac{(e^4 - e^{-4})}{2} \approx 27.28992$;
(a) $y = 26.19221$ **(b)** $y = 26.55498$ **7.** $y = (x^3 + x^{-2})^{-1/2}, y(2) = \frac{2}{\sqrt{33}} \approx 0.34816$; **(a)** $y = 0.34840$ **(b)** $y = 0.34465$
9. $y = 2e^{e^x - 1}, y(2) = 2e^{e^2 - 1} \approx 1190.58883$; **(a)** $y = 1082.39224$ **(b)** $y = 1086.87818$ **19.** $y = 3.41076$

PROBLEMS 8.5 *(page 389)*

1. (a) $|e_n| \le 0.859h$ **(b)** $h = 0.1, |e_n| \le 0.086; h = 0.2, |e_n| \le 0.172$ **3.** $|e_n| \le 6.06h$; **(a)** 1,212,000 **(b)** 12,120,000

PROBLEMS 8.7 *(page 398)*

1. Exact solution: $x = e^{3t}(\cos 3t - \sin 3t)$, $y = e^{3t}(2\cos 3t + \sin 3t)$, $x(1) = -22.719$, $y(1) = -36.935$
(a) $x = -17.43$, $y = 7.76$ **(b)** $x = -30.09$, $y = -31.31$ **(c)** $x = -22.907$, $y = -37.412$ **3.** Exact solution: $x = \left(\frac{2}{5}\right)(e^{4t} - e^{-t})$, $y = \left(\frac{1}{5}\right)(3e^{4t} + 2e^{-t})$, $x(1) = 21.692$, $y(1) = 32.906$ **(a)** $x = 7.43$, $y = 11.47$ **(b)** $x = 16.98$, $y = 25.84$ **(c)** $x = 21.54$, $y = 32.68$

5. Exact solution: $x = \left(\frac{1}{27}\right)(16e^{3t} + 9t^2 + 6t + 11)$, $y = \left(\frac{1}{27}\right)(16e^{3t} - 18t^2 - 21t - 16)$, $x(1) = 12.866$, $y(1) = 9.866$
(a) $x = 7.13$, $y = 4.43$ **(b)** $x = 11.58$, $y = 8.61$ **(c)** $x = 12.84$, $y = 9.84$

7. Exact solution: $x = \left(\frac{33}{8}\right)(e^{2t} - 1) - \left(\frac{1}{4}\right)(t^2 + 9t)$, $y = \left(\frac{1}{8}\right)(27 - 11e^{2t}) + \left(\frac{1}{4}\right)(t^2 + 7t)$, $x(1) = 23.855$, $y(1) = -4.785$
(a) $x = 15.61$, $y = -2.07$ **(b)** $x = 22.67$, $y = -4.39$ **(c)** $x = 23.85$, $y = -4.78$ **9.** $y(1) = 1.09$ **11.** Maximum satisfies $y(x) = \sqrt{2(y'(0) - x)}$. Exact solution can be obtained by transforming the Riccati equation $y' + \frac{y^2}{2} = y'(0) - x$ and using Example 5.4.4. Numerically with $h = 0.1$ use $y'(0) = 0.504$ and get $y_{\max} \approx 0.125$.

PROBLEMS 8.8 *(page 402)*

1. $y_{IE}(1) \approx 3.40542$, $y_{RK}(1) \approx 3.4365$ **3.** $y_{IE}(0.4) \approx 3.08882$, $y_{RK}(0.4) \approx 3.14092$ **5.** $y_{IE}(1) \approx 2.82235$, $y_{RK}(1) \approx 2.82343$ **7.** $y_{IE}(1) \approx 0.732051$, $y_{RK}(1) \approx 0.724813$ **9.** $x_{RK}(1) \approx -22.7286$, $y_{RK}(1) \approx -36.9746$
11. $y_{IE}(1) \approx -0.70923$, $y_{RK}(1) \approx -1.82224$ but solution is $y^2 + 2xy - x^2 = -4$. When $x = 1$, $y^2 + 2y + 3 = 0$ has roots $y = -1 \pm i\sqrt{2}$.

REVIEW EXERCISES FOR CHAPTER 8 *(page 403)*

1. $y^2 = 2(e^x + 1)$, $y(3) = \sqrt{2(e^3 + 1)} \approx 6.4939$; **(a)** $y_{IE} = 6.56$ **(b)** $y_{RK} = 6.4942$ **3.** $y = x + \sqrt{1 + x^2}$, $y(3) = 3 + \sqrt{10} \approx 6.1623$; **(a)** $y_{IE} = 5.96$ **(b)** $y_{RK} = 6.1605$ **5.** $y^{-2} = \frac{(2x - 1 + 3e^{-2x})}{2}$, $y(3) = 0.6320$; **(a)** $y_{IE} = 0.6356$ **(b)** $y_{RK} = 0.6329$ **7.** The exact solution is $y^2(y + 1) = x^2$. If we graph this curve and begin moving to the right from the initial point, we cannot proceed *on that branch* beyond $x = \frac{2}{3\sqrt{3}}$. **9.** Solution $y = \tan x$ becomes infinite at $x = \frac{\pi}{2}$, so no numerical method will yield $y(2)$. **11.** $y = \cos(\ln|x|) + \sin(\ln|x|)$, $y(2) = \cos(\ln 2) + \sin(\ln 2) \approx 1.4082$; **(a)** $y_E = 1.42163$ **(b)** $y_{RK} = 1.40820$ **(c)** $y_{PC} = 1.39531$ **13.** $x = \left(\frac{2 - \sqrt{2}}{4}\right)e^{(3+\sqrt{8})t} + \left(\frac{2 + \sqrt{2}}{4}\right)e^{(3-\sqrt{8})t}$, $y = \frac{\sqrt{2}}{4}e^{(3+\sqrt{8})t} - \frac{\sqrt{2}}{4}e^{(3-\sqrt{8})t}$, $x(2) \approx 16912.88$, $y(2) \approx 40827.91$. **(a)** $x_E = 333.57$, $y_E = 801.94$ **(b)** $x_{RK} = 15783.44$, $y_{RK} = 38101.20$ **(c)** $x_{PC} = 5090.404$, $y_{PC} = 12285.920$

CHAPTER 9

PROBLEMS 9.1 *(page 418)*

1. $\begin{pmatrix} 3 & 9 \\ 6 & 15 \\ -3 & 6 \end{pmatrix}$ **3.** $\begin{pmatrix} 2 & 2 \\ -2 & -1 \\ 6 & -1 \end{pmatrix}$ **5.** $\begin{pmatrix} 0 & 0 \\ 0 & 0 \\ 0 & 0 \end{pmatrix}$ **7.** $\begin{pmatrix} -2 & 4 \\ 7 & 15 \\ -15 & 10 \end{pmatrix}$ **9.** $\begin{pmatrix} 4 & 10 \\ 17 & 22 \\ -9 & 1 \end{pmatrix}$ **11.** $\begin{pmatrix} 0 & 6 \\ 5 & 14 \\ -9 & 9 \end{pmatrix}$

13. $\begin{pmatrix} 1 & -5 & 0 \\ -3 & 4 & -5 \\ -14 & 13 & -1 \end{pmatrix}$ **15.** $\begin{pmatrix} 1 & 1 & 5 \\ 9 & 5 & 10 \\ 7 & -7 & 3 \end{pmatrix}$ **17.** $\begin{pmatrix} -1 & -1 & -1 \\ -3 & -3 & -10 \\ -7 & 3 & 5 \end{pmatrix}$ **19.** $\begin{pmatrix} -1 & -1 & -5 \\ -9 & -5 & -10 \\ -7 & 7 & -3 \end{pmatrix}$ **21.** $\begin{pmatrix} 8 & 20 \\ -4 & 11 \end{pmatrix}$

23. $\begin{pmatrix} -3 & -3 \\ 1 & 3 \end{pmatrix}$ **25.** $\begin{pmatrix} 13 & 35 & 18 \\ 20 & 26 & 20 \end{pmatrix}$ **27.** $\begin{pmatrix} 19 & -17 & 34 \\ 8 & -12 & 20 \\ -8 & -11 & 7 \end{pmatrix}$ **29.** $\begin{pmatrix} 18 & 15 & 35 \\ 9 & 21 & 13 \\ 10 & 9 & 9 \end{pmatrix}$ **31.** (7, 16) **33.** $\begin{pmatrix} 3 & -2 & 1 \\ 4 & 0 & 6 \\ 5 & 1 & 9 \end{pmatrix}$

35. $\begin{pmatrix} 0 & -8 \\ 32 & 32 \end{pmatrix}$ **37.** $\begin{pmatrix} 11 & 38 \\ 57 & 106 \end{pmatrix}$ **39.** $\begin{pmatrix} \frac{1}{2} & -\frac{1}{2} \\ -\frac{1}{4} & \frac{3}{4} \end{pmatrix}$ **41.** $\begin{pmatrix} 0 & 1 \\ 1 & 0 \end{pmatrix}$ **43.** $\begin{pmatrix} \frac{1}{3} & -\frac{1}{4} & -\frac{1}{6} \\ 0 & \frac{1}{4} & \frac{1}{2} \\ 0 & \frac{1}{4} & -\frac{1}{2} \end{pmatrix}$ **45.** $\begin{pmatrix} 0 & 1 & -1 \\ 2 & -2 & -1 \\ -1 & 1 & 1 \end{pmatrix}$

47. Not invertible **49. (a)** $\begin{pmatrix} i & 2 \\ 1 & -i \end{pmatrix}$ **(b)** $\begin{pmatrix} \frac{1}{2}+\frac{1}{2}i & 0 \\ 0 & \frac{1}{2}-\frac{1}{2}i \end{pmatrix}$ **(c)** $\frac{1}{2}\begin{pmatrix} 1+i & 1+i & -i \\ -1-i & -1+i & 1 \\ -1+i & 1+i & 1 \end{pmatrix}$ **51.** $\begin{pmatrix} \frac{1}{2} & 0 & 0 \\ 0 & \frac{1}{3} & 0 \\ 0 & 0 & \frac{1}{4} \end{pmatrix}$

53. $\begin{pmatrix} \frac{1}{a_{11}} & 0 & \cdots & 0 \\ 0 & \frac{1}{a_{22}} & \cdots & 0 \\ & & \vdots & \\ 0 & 0 & \cdots & \frac{1}{a_{nn}} \end{pmatrix}$ **55.** $A_{1,2}(2)$ yields a row of zeros. **57.** $\mathbf{x}'(t) = (1, \cos t)$; $\int \mathbf{x}(t)\,dt = \left(\frac{1}{2}t^2, -\cos t\right) + \mathbf{c}$

59. $\mathbf{y}'(t) = \begin{pmatrix} e^t \\ -\sin t \\ \sec^2 t \end{pmatrix}$; $\int \mathbf{y}(t)\,dt = \begin{pmatrix} e^t \\ \sin t \\ -\ln|\cos t| \end{pmatrix} + \mathbf{c}$

PROBLEMS 9.2 *(page 424)*

1. $\mathbf{x}' = \begin{pmatrix} 2 & 3 \\ 4 & -6 \end{pmatrix}\mathbf{x}$ **3.** $\mathbf{x}' = \begin{pmatrix} 0 & 1 & 0 \\ 0 & 0 & 1 \\ 1 & -4t & 2 \end{pmatrix}\mathbf{x} + \begin{pmatrix} 0 \\ 0 \\ \sin t \end{pmatrix}$ **5.** $\mathbf{x}' = \begin{pmatrix} 2t & -3t^2 & \sin t \\ 2 & 0 & -4 \\ 0 & 17 & 4t \end{pmatrix}\mathbf{x} + \begin{pmatrix} 0 \\ -\sin t \\ e^t \end{pmatrix}$

PROBLEMS 9.3 *(page 435)*

1. Fundamental **3.** Fundamental **5.** Not fundamental **7.** $C = \begin{pmatrix} 1 & -2 \\ -1 & 1 \end{pmatrix}$ **9.** $C = \begin{pmatrix} 1 & -1 & 1 \\ 0 & 1 & 1 \\ 1 & 0 & 0 \end{pmatrix}$

13. $\Psi(t) = e^{6t}\begin{pmatrix} \cos 2t - \sin 2t & \frac{1}{2}\sin 2t \\ -4\sin 2t & \cos 2t + \sin 2t \end{pmatrix}$ **15.** $\Psi(t) = \frac{1}{6}\begin{pmatrix} -e^{-t}+9e^t-2e^{2t} & -2e^{-t}+2e^{2t} & 7e^{-t}-9e^t+2e^{2t} \\ 6e^t-6e^{2t} & 6e^{2t} & -6e^t+6e^{2t} \\ -e^{-t}+3e^t-2e^{2t} & -2e^{-t}+2e^{2t} & 7e^{-t}-3e^t+2e^{2t} \end{pmatrix}$

17. (a) $e^t\begin{pmatrix} 1-2t \\ 2-2t \end{pmatrix}$ **(b)** $e^t\begin{pmatrix} -10t-2 \\ -10t+3 \end{pmatrix}$ **(c)** $e^{t-1}\begin{pmatrix} 2-2t \\ 3-2t \end{pmatrix}$ **(d)** $e^{t+1}\begin{pmatrix} 4+2t \\ 3+2t \end{pmatrix}$ **(e)** $e^{t-3}\begin{pmatrix} 3 \\ 3 \end{pmatrix}$ **(f)** $e^{t-a}\begin{pmatrix} 2(c-b)(a-t)+b \\ 2(c-b)(a-t)+c \end{pmatrix}$

19. (a) $\mathbf{x}' = A(t)\mathbf{x}$, where $A(t) = \begin{pmatrix} 0 & 1 \\ -b(t) & -a(t) \end{pmatrix}$ **21.** $\phi_2(t) = e^{t^2}\int e^{-t^2}\,dt$

PROBLEMS 9.4 *(page 449)*

In the following answers, each eigenvalue λ is followed by linearly independent eigenvectors corresponding to λ.

1. $\lambda_1 = -4$, $\begin{pmatrix} 1 \\ 1 \end{pmatrix}$; $\lambda = 3$, $\begin{pmatrix} 2 \\ -5 \end{pmatrix}$ **3.** $\lambda_1 = i$, $\begin{pmatrix} 2+i \\ 5 \end{pmatrix}$; $\lambda_2 = -i$, $\begin{pmatrix} 2-i \\ 5 \end{pmatrix}$ **5.** $\lambda_1 = -3$ (alg. mult. 2), $\begin{pmatrix} 1 \\ 0 \end{pmatrix}$

7. $\lambda_1 = 0$, $\begin{pmatrix} 1 \\ 1 \\ 1 \end{pmatrix}$; $\lambda_2 = 1$, $\begin{pmatrix} -1 \\ 0 \\ 1 \end{pmatrix}$; $\lambda_3 = 3$, $\begin{pmatrix} 1 \\ -2 \\ 1 \end{pmatrix}$ **9.** $\lambda_1 = 1$ (alg. mult. 2), $\begin{pmatrix} 1 \\ 0 \\ -2 \end{pmatrix}\begin{pmatrix} 0 \\ 1 \\ -2 \end{pmatrix}$; $\lambda_2 = 10$, $\begin{pmatrix} 2 \\ 2 \\ 1 \end{pmatrix}$

11. $\lambda_1 = 1$ (alg. mult. 3), $\begin{pmatrix} 1 \\ 1 \\ 1 \end{pmatrix}$ **13.** $\lambda_1 = -1$, $\begin{pmatrix} 0 \\ -1 \\ 1 \end{pmatrix}$; $\lambda_2 = i$, $\begin{pmatrix} 1+i \\ 1 \\ 1 \end{pmatrix}$; $\lambda_3 = -i$, $\begin{pmatrix} 1-i \\ 1 \\ 1 \end{pmatrix}$

15. $\lambda_1 = 1$, $\begin{pmatrix} 4 \\ 1 \\ -3 \end{pmatrix}$, $\lambda_2 = 2$ (alg. mult. 2), $\begin{pmatrix} 3 \\ 1 \\ -2 \end{pmatrix}$

17. $\lambda_1 = a$ (alg. mult. 4), $\begin{pmatrix} 1 \\ 0 \\ 0 \\ 0 \end{pmatrix}, \begin{pmatrix} 0 \\ 1 \\ 0 \\ 0 \end{pmatrix}, \begin{pmatrix} 0 \\ 0 \\ 1 \\ 0 \end{pmatrix}, \begin{pmatrix} 0 \\ 0 \\ 0 \\ 1 \end{pmatrix}$ **19.** $\lambda_1 = a$ (alg. mult. 4), $\begin{pmatrix} 1 \\ 0 \\ 0 \\ 0 \end{pmatrix}, \begin{pmatrix} 0 \\ 0 \\ 0 \\ 1 \end{pmatrix}$

PROBLEMS 9.5 *(page 460)*

1. $\frac{1}{7}\begin{pmatrix} 5e^{-4t}+2e^{3t} & 2e^{-4t}-2e^{3t} \\ 5e^{-4t}-5e^{3t} & 2e^{-4t}+5e^{3t} \end{pmatrix}$ **3.** $\begin{pmatrix} 2\sin t+\cos t & -\sin t \\ 5\sin t & -2\sin t+\cos t \end{pmatrix}$ **5.** $e^{-t}\begin{pmatrix} 1+4t & -2t \\ 8t & -4t+1 \end{pmatrix}$

7. $e^{-t}\begin{pmatrix} \frac{3}{4}+t \\ 1+2t \end{pmatrix}$ (see Problem 5) **9.** $e^{-t}\begin{pmatrix} 2-2t \\ 5-4t \end{pmatrix}$ (see Problem 5) **11.** $\frac{1}{6}\begin{pmatrix} 2+3e^t+e^{3t} & 2-2e^{3t} & 2-3e^t+e^{3t} \\ 2-2e^{3t} & 2+4e^{3t} & 2-2e^{3t} \\ 2-3e^t+e^{3t} & 2-2e^{3t} & 2+3e^t+e^{3t} \end{pmatrix}$;

$\frac{1}{6}\begin{pmatrix} 4-6e^t+8e^{3t} \\ 4-16e^{3t} \\ 4+6e^t+8e^{3t} \end{pmatrix}$ **13.** $\begin{pmatrix} 4e^t-3e^{2t}+6te^{2t} & -12e^t+12e^{2t}-6te^{2t} & 6te^{2t} \\ e^t-e^{2t}+2te^{2t} & -3e^t+4e^{2t}-2te^{2t} & 2te^{2t} \\ -3e^t+3e^{2t}-4te^{2t} & 9e^t-9e^{2t}+4te^{2t} & -4te^{2t}+e^{2t} \end{pmatrix}$

15. $\frac{1}{9}\begin{pmatrix} 5e^t+4e^{10t} & -4e^t+4e^{10t} & -2e^t+2e^{10t} \\ -4e^t+4e^{10t} & 5e^t+4e^{10t} & -2e^t+2e^{10t} \\ -2e^t+2e^{10t} & -2e^t+2e^{10t} & 8e^t+e^{10t} \end{pmatrix}$ **17.** $\frac{1}{2}\begin{pmatrix} e^{2t}+e^{6t} & -e^{2t}+e^{4t} & 0 & -e^{4t}+e^{6t} \\ -e^{2t}+e^{6t} & e^{2t}+e^{4t} & 0 & -e^{4t}+e^{6t} \\ e^{2t}-e^{6t} & -e^{2t}+e^{4t} & 2e^{2t} & 2e^{2t}-e^{4t}-e^{6t} \\ -e^{2t}+e^{6t} & e^{2t}-e^{4t} & 0 & e^{4t}+e^{6t} \end{pmatrix}$

19. $\frac{1}{6}\begin{pmatrix} 4+2e^{3t} \\ 4-4e^{3t} \\ 4+2e^{3t} \end{pmatrix}$ (see Problem 11) **21.** $\begin{pmatrix} -4e^t+3e^{2t}+6te^{2t} \\ -e^t+e^{2t}+2te^{2t} \\ 3e^t-e^{2t}-4te^{2t} \end{pmatrix}$ (see Problem 13)

PROBLEMS 9.6 *(page 466)*

1. $\boldsymbol{\phi}_p(t)=\begin{pmatrix} \frac{1}{3}e^{2t} \\ \frac{1}{2}e^t-\frac{2}{3}e^{2t} \end{pmatrix}$ **3.** $\boldsymbol{\phi}(t)=\begin{pmatrix} -t\cos t \\ -t\sin t-2t\cos t+\cos t+\sin t \end{pmatrix}$

5. $\boldsymbol{\phi}(t)=\begin{pmatrix} \frac{3}{2}te^t+\frac{1}{3}te^{2t}+\frac{5}{2}e^t-\frac{115}{72}e^{-t}-\frac{7}{9}e^{2t}-\frac{1}{8}e^{3t} \\ te^t+te^{2t}+\frac{5}{2}e^t-2e^{2t}+\frac{1}{2}e^{3t} \\ \frac{1}{2}te^t+\frac{1}{3}te^{2t}+e^t-\frac{115}{72}e^{-t}-\frac{7}{9}e^{2t}+\frac{3}{8}e^{3t} \end{pmatrix}$ **7.** $\boldsymbol{\phi}_p(t)=\begin{pmatrix} t \\ t^2 \\ t^4 \end{pmatrix}$

9. $\boldsymbol{\phi}_p(t)=\begin{pmatrix} 5\sin t\ln|\csc t+\cot t| \\ 1+(2\sin t-\cos t)\ln|\csc t+\cot t| \end{pmatrix}$ **11.** $\Phi(t)=\begin{pmatrix} \frac{1}{t} & \frac{\ln t}{t} \\ \frac{-1}{t^2} & \frac{(1-\ln t)}{t^2} \end{pmatrix}$

$\Phi^{-1}(t)=\begin{pmatrix} t(1-\ln t) & -t^2\ln t \\ t & t^2 \end{pmatrix}$

PROBLEMS 9.7 *(page 474)*

1. $\begin{pmatrix} e^t & 0 \\ 0 & e^{3t} \end{pmatrix}$ **3.** See answer to Problem 9.5.1. **5.** See answer to Problem 9.5.3. **7.** See answer to Problem 9.5.11.

9. See answer to Problem 9.5.15 **11.** $\begin{pmatrix} e^{4t} & te^{4t} \\ 0 & e^{4t} \end{pmatrix}$ **13.** $\begin{pmatrix} e^t & 2te^t \\ 0 & e^t \end{pmatrix}$ **15.** See answer to Problem 9.5.5.

17. See answer to Problem 9.5.11. **19.** See answer to Problem 9.5.21.

PROBLEMS 9.8 *(page 477)*

1. $\frac{1}{\sqrt{2}}\begin{pmatrix} 1 & 1 \\ 1 & -1 \end{pmatrix}\begin{pmatrix} e^{2t} & 0 \\ 0 & e^{4t} \end{pmatrix}\begin{pmatrix} k_1 \\ k_2 \end{pmatrix}$ **3.** $\begin{pmatrix} \frac{\sqrt{10}}{\sqrt{35}} & \frac{\sqrt{10}}{\sqrt{14}} \\ \frac{-5}{\sqrt{35}} & \frac{2}{\sqrt{14}} \end{pmatrix}\begin{pmatrix} e^t & 0 \\ 0 & e^{8t} \end{pmatrix}\begin{pmatrix} k_1 \\ k_2 \end{pmatrix}$ **5.** $\begin{pmatrix} \sqrt{\frac{6}{15}} & \sqrt{\frac{6}{10}} \\ \frac{-3}{\sqrt{15}} & \frac{2}{\sqrt{10}} \end{pmatrix}\begin{pmatrix} e^{2t} & 0 \\ 0 & e^{7t} \end{pmatrix}\begin{pmatrix} k_1 \\ k_2 \end{pmatrix}$

7. $\begin{pmatrix} \frac{1}{\sqrt{2}} & \frac{1}{\sqrt{3}} & \frac{1}{\sqrt{6}} \\ \frac{-1}{\sqrt{2}} & \frac{1}{\sqrt{3}} & \frac{1}{\sqrt{6}} \\ 0 & \frac{-1}{\sqrt{3}} & \frac{2}{\sqrt{6}} \end{pmatrix} \begin{pmatrix} e^{3t} & 0 & 0 \\ 0 & e^{4t} & 0 \\ 0 & 0 & e^{7t} \end{pmatrix} \begin{pmatrix} k_1 \\ k_2 \\ k_3 \end{pmatrix}$ **9.** $\begin{pmatrix} \frac{1}{\sqrt{2}} & \frac{1}{\sqrt{2}} & \frac{1}{\sqrt{3}} \\ \frac{-1}{\sqrt{2}} & 0 & \frac{1}{\sqrt{3}} \\ 0 & \frac{-1}{\sqrt{2}} & \frac{1}{\sqrt{3}} \end{pmatrix} \begin{pmatrix} e^{3t} & te^{3t} & 0 \\ 0 & e^{3t} & 0 \\ 0 & 0 & e^{6t} \end{pmatrix} \begin{pmatrix} k_1 \\ k_2 \\ k_3 \end{pmatrix}$

REVIEW EXERCISES FOR CHAPTER 9 *(page 481)*

1. $\begin{pmatrix} x_1 \\ x_2 \end{pmatrix}' = \begin{pmatrix} 3 & -4 \\ -2 & 7 \end{pmatrix} \begin{pmatrix} x_1 \\ x_2 \end{pmatrix}$ **3.** $\begin{pmatrix} x_1 \\ x_2 \end{pmatrix}' = \begin{pmatrix} 1 & 1 \\ -3 & 2 \end{pmatrix} \begin{pmatrix} x_1 \\ x_2 \end{pmatrix} + \begin{pmatrix} e^t \\ e^{2t} \end{pmatrix}$ **5. (a)** $\frac{1}{5}\begin{pmatrix} -2e^t + 12e^{6t} \\ 3e^t + 12e^{6t} \end{pmatrix}$ **(b)** $\frac{1}{5}\begin{pmatrix} -2e^t - 3e^{6t} \\ 3e^t - 3e^{6t} \end{pmatrix}$
(c) $\begin{pmatrix} 0 \\ 0 \end{pmatrix}$ **(d)** $\frac{1}{5}\begin{pmatrix} 18e^t + 17e^{6t} \\ -27e^t + 17e^{6t} \end{pmatrix}$ **(e)** $\frac{1}{5}\begin{pmatrix} 2(a-b)e^t + (3a+2b)e^{6t} \\ -3(a-b)e^t + (3a+2b)e^{6t} \end{pmatrix}$ **7.** $\lambda_1 = 2$(alg. mult. 2),$\begin{pmatrix} 1 \\ 0 \end{pmatrix}$

9. $\lambda_1 = 1, \begin{pmatrix} 1 \\ 0 \\ -1 \end{pmatrix}, \lambda_2 = 2, \begin{pmatrix} -1 \\ 1 \\ 1 \end{pmatrix}, \lambda_3 = -1, \begin{pmatrix} -1 \\ -2 \\ 7 \end{pmatrix}$ **11.** $\lambda_1 = -2$(alg. mult. 3), $\begin{pmatrix} 1 \\ 0 \\ 0 \end{pmatrix}$ **13.** $\frac{1}{3}\begin{pmatrix} 2e^{2t} + e^{5t} & e^{2t} - e^{5t} \\ 2e^{2t} - 2e^{5t} & e^{2t} + 2e^{5t} \end{pmatrix}$

15. $e^{-2t}\begin{pmatrix} -5t^2 + t + 1 & \frac{5}{2}t^2 - 18t & -\frac{5}{2}t^2 - 7t \\ -3t^2 + t & \frac{3}{2}t^2 - 11t + 1 & -\frac{3}{2}t^2 - 4t \\ 7t^2 - t & -\frac{7}{2}t^2 + 25t & \frac{7}{2}t^2 + 10t + 1 \end{pmatrix}$ **17.** $\boldsymbol{\phi}(t) = e^{2t}\begin{pmatrix} 3\cos 2t + \frac{19}{8}\sin 2t + \frac{1}{4}t \\ -6\sin 2t + \frac{19}{4}\cos 2t - \frac{11}{4} \end{pmatrix}$

CHAPTER 10

PROBLEMS 10.1 *(page 489)*

1. (a) $x' = x$ **(b)** $x = 0, x = 1$ **(c)** $x(t) = \dfrac{x(0)e^t}{x(0)e^t + 1 - x(0)}$ **3. (a)** $x' = 2x$ **(b)** $x = 0, x = -\dfrac{2}{3}$

(c) $x(t) = \dfrac{2x(0)e^{2t}}{2 + 3x(0)(1 - e^{2t})}$ **5. (a)** $x' = 2x$ **(b)** $x = 0, x = 1, x = 2$
(c) $\dfrac{x^2 - 2x}{(x-1)^2} = ce^{2t}$ where $c = \dfrac{x(0)^2 - 2x(0)}{(x(0) - 1)^2}$ **7. (a)** The orbits are the circles $x^2 + y^2 = a^2 + b^2$.
9. The orbits are $x = \exp\left[\frac{1}{2}(\ln^2 y - 2t_0 \ln y)\right]$.

PROBLEMS 10.2 *(page 500)*

1. Saddle point **3.** Unstable focus **5.** Stable focus **7.** Unstable focus **15. (a)** $x = c_1, y = c_1 + c_2e^t$
(c) (c, c) is a critical point for any real number c.

PROBLEMS 10.3 *(page 512)*

1. $(0, 0)$ is a saddle point (unstable). **3.** $(0, 0)$ is unstable. **5.** $(0, 0)$ is an unstable node.
7. $(0, 0)$ is the only critical point and is an unstable node or focus. **9.** $(0, 0)$ is an unstable node or focus, $\left(0, \frac{1}{3}\right)$ and $(1, 0)$ are saddle points, and $\left(\frac{8}{7}, \frac{1}{7}\right)$ is a stable focus. **11.** $(0, 0)$, $(3, -9)$, and $(-2, -4)$ are saddle points; $(1, -1)$ is a stable focus and $(-1, -1)$ is an unstable focus. **13.** $(0, 0)$ is the only critical point and is a stable focus. **15.**

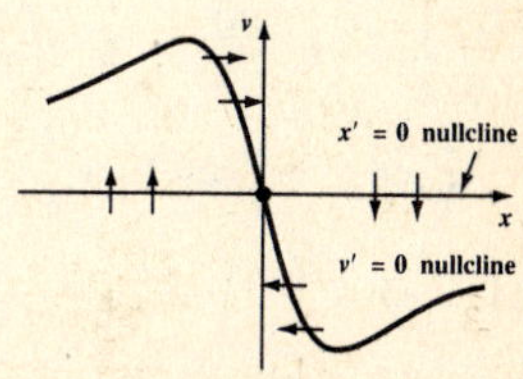

PROBLEMS 10.4 *(page 518)*

1. Let $V(x, y) = x^2 + 2y^2$. **3.** Let $V(x, y) = \frac{y^2}{2} + \omega^2(1 - \cos x)$. **5. (a)** Unstable **(b)** Asymptotically stable **(c)** Stable

7. Let $V(x, y) = \frac{y^2}{2} + \int_0^x g(x)\,dx$. Then $\dot{V} = -yf(x, y) < 0$.

PROBLEMS 10.5 *(page 523)*

1. Unstable node at $(0, 0)$; saddle at $(1, -1)$ **3.** Saddle at $(3, 9)$

5. Saddle at $(0, 0)$;
if $a > 4d > 0$ unstable node
$0 < a < 4d$ unstable focus
at $\left(\frac{d}{c}, \frac{ad}{bc}\right)$

7. $x' = a\left(\frac{x^3}{3} - x\right) + y$, $y' = -x$; equilibrium point is at origin. Find roots of $\lambda^2 + a\lambda + 1 = 0$.

REVIEW EXERCISES FOR CHAPTER 10 *(page 524)*

1. (a) $x' = x$ **(b)** $x \equiv 0, x \equiv -1$ **(c)** $x(t) = [(x(0)^{-1} + 1)e^{-t} - 1]^{-1}$ **(d)** Solutions become infinite as $t \to \ln \frac{x(0) + 1}{x(0)}$.

3. (a) $x' = x$ **(b)** $x \equiv 0, x \equiv 1, x \equiv -1$ **(c)** $x(t) = \sqrt{\frac{1}{1 - ke^t}}$ where $k = \frac{x^2(0) - 1}{x^2(0)}$ **(d)** Solutions become infinite as $t \to \ln \frac{x^2(0)}{x^2(0) - 1}$. **5.** Stable node **7.** Stable node (star-shaped) **9.** $(0, 0)$ is a saddle point (unstable).

11. $(0, 0)$ is a saddle point; $\left(\frac{1}{\sqrt{2}}, 0\right)$ and $\left(-\frac{1}{\sqrt{2}}, 0\right)$ are unstable (star-shaped) nodes. **13.** $(0, 0)$ is a saddle point; $(3, 4)$ is a saddle point; $\left(0, \frac{5}{2}\right)$ is a stable node; $(1, 0)$ is an unstable node.

CHAPTER 11

PROBLEMS 11.1 *(page 531)*

1. eigenvalues: $\left(\frac{n\pi}{T}\right)^2$; eigenfunctions: $A \sin \frac{n\pi t}{T}$, $n = 1, 2, 3, \ldots$ **3.** eigenvalues: $-\left(\frac{n\pi}{T}\right)^2$; eigenfunctions: $A \sin \frac{n\pi t}{T}$, $n = 1, 2, 3, \ldots$ **5.** eigenvalues: $\left(\frac{n\pi}{2T}\right)^2$; eigenfunctions: $A \sin \frac{n\pi(t + T)}{2T}$, $n = 1, 2, 3, \ldots$ **7.** eigenvalues: $\left[\left(n + \frac{1}{2}\right)\frac{\pi}{T}\right]^2$; eigenfunctions: $A \sin\left(n + \frac{1}{2}\right)\frac{\pi t}{T}$, $n = 0, 1, 2, 3, \ldots$ **9.** eigenvalues: $(n\pi)^2$; eigenfunctions: $A \cos n\pi t$, $n = 0, 1, 2, 3, \ldots$

11. eigenvalues: all values of λ such that if $\lambda > \frac{1}{4}$ and $\beta = \frac{\sqrt{4\lambda - 1}}{2}$, then β satisfies $2\beta + \tan \beta = 0$; eigenfunctions: $Ax^{-1/2}[2\beta \cos(\beta \ln |x|) + \sin(\beta \ln |x|)]$.

PROBLEMS 11.2 *(page 535)*

1. $\frac{1}{\sqrt{2\pi}}, \frac{\cos x}{\sqrt{\pi}}, \frac{\cos 2x}{\sqrt{\pi}}, \ldots$ **3.** $\frac{1}{\sqrt{T}}, \sqrt{\frac{2}{T}} \cos \frac{2\pi x}{T}, \sqrt{\frac{2}{T}} \cos \frac{4\pi x}{T}, \ldots$ **5.** $\sqrt{\frac{1}{\pi}}, \sqrt{\frac{2}{\pi}} \cos 2x, \sqrt{\frac{2}{\pi}} \cos 4x, \ldots$

7. $\frac{1}{\sqrt{2\pi}}, \frac{1}{\sqrt{\pi}} \cos 3x, \frac{1}{\sqrt{\pi}} \cos 6x, \frac{1}{\sqrt{\pi}} \cos 9x, \ldots$

PROBLEMS 11.3 *(page 546)*

1. $\pi, 2, \frac{T}{n}, \frac{1}{k}$ **3.** $\frac{2bT}{a}$ **5.** $2\left(\sin x - \frac{1}{2} \sin 2x + \frac{1}{3} \sin 3x - \frac{1}{4} \sin 4x + \ldots\right)$

7. $\frac{\pi^2}{3} - 4\left(\cos x - \frac{1}{4} \cos 2x + \frac{1}{9} \cos 3x - \frac{1}{16} \cos 4x + \ldots\right)$ **9.** $\frac{\pi}{2} - \frac{4}{\pi}\left[\cos x + \frac{1}{3^2} \cos 3x + \frac{1}{5^2} \cos 5x + \ldots\right]$

11. $\frac{2}{\pi}[(1 + \sin(\pi - 1)) \sin x - \frac{1}{2}\left(1 + \frac{1}{2} \sin 2(\pi - 1)\right) \sin 2x + \frac{1}{3}\left(1 + \frac{1}{3} \sin 3(\pi - 1)\right) \sin 3x - \ldots$

13. $\frac{\sinh \pi}{\pi}\left[1 + 2\sum_{n=1}^{\infty} \frac{(-1)^n}{1 + n^2} \cos nx - 2\sum_{n=1}^{\infty} \frac{n(-1)^n}{1 + n^2} \sin nx\right]$

15. $1 - \frac{2}{\pi}\sum_{n=1}^{\infty}\frac{\sin n\pi x}{n}$ **17.** $\frac{1}{3} + \frac{4}{\pi^2}\sum_{n=1}^{\infty}\frac{(-1)^n}{n^2}\cos n\pi x$ **19.** $\frac{1}{2} - \frac{2}{\pi}\left(\sin \pi x + \frac{1}{3}\sin 3\pi x + \frac{1}{5}\sin 5\pi x + \dots\right)$

21. $\frac{3}{4} - \frac{1}{\pi}\sum_{n=1}^{\infty}\frac{1}{n}\sin n\pi x - \frac{2}{\pi^2}\left(\cos \pi x + \frac{1}{3^2}\cos 3\pi x + \frac{1}{5^2}\cos 5\pi x + \dots\right)$

PROBLEMS 11.4 *(page 552)*

1. Even **3.** Even **5.** Odd **7.** Neither **9.** Neither

11. $\frac{\pi^2}{3} - 4\left(\cos x - \frac{1}{4}\cos 2x + \frac{1}{9}\cos 3x - \frac{1}{16}\cos 4x + \dots\right)$

13. $\frac{\pi^3}{4} - 6\pi\left(\cos x - \frac{\cos 2x}{4} + \frac{\cos 3x}{9} - \frac{\cos 4x}{16} + \dots\right) + \frac{24}{\pi}\left(\cos x + \frac{\cos 3x}{3^4} + \frac{\cos 5x}{5^4} + \dots\right)$

15. $\frac{T}{2} + \frac{4T}{\pi^2}\left(\cos\frac{\pi x}{T} + \frac{1}{9}\cos\frac{3\pi x}{T} + \frac{1}{25}\cos\frac{5\pi x}{T} + \dots\right)$ **17.** $\frac{2T}{\pi}\left(\sin\frac{\pi x}{T} + \frac{1}{2}\sin\frac{2\pi x}{T} + \frac{1}{3}\sin\frac{3\pi x}{T} + \frac{1}{4}\sin\frac{4\pi x}{T} + \dots\right)$

19. $\frac{1}{3} + \frac{4}{\pi^2}(\cos \pi x + \frac{1}{4}\cos 2\pi x + \frac{1}{9}\cos 3\pi x + \frac{1}{16}\cos 4\pi x + \dots)$

21. even: k odd: $\frac{4k}{\pi}\left(\sin \pi x + \frac{1}{3}\sin 3\pi x + \frac{1}{5}\sin 5\pi x + \dots\right)$

23. even: $\frac{1}{4} - \frac{6}{\pi^2}\left(\cos \pi x - \frac{1}{4}\cos 2\pi x + \frac{1}{9}\cos 3\pi x - \frac{1}{16}\cos 4\pi x + \dots\right) + \frac{24}{\pi^4}\left(\cos \pi x + \frac{\cos 3\pi x}{3^4} + \frac{\cos 5\pi x}{5^4} + \dots\right)$

odd: $\frac{2}{\pi}\left[\sin \pi x - \frac{1}{2}\sin 2\pi x + \frac{1}{3}\sin 3\pi x - \frac{1}{4}\sin 4\pi x + \cdots\right] - \frac{12}{\pi^3}\left[\sin \pi x - \frac{\sin 2\pi x}{2^3} + \frac{\sin 3\pi x}{3^3} - \frac{\sin 4\pi x}{4^3} + \dots\right]$

25. even: $\frac{1}{2} + \frac{2}{\pi}\left(\cos\frac{\pi x}{2} - \frac{1}{3}\cos\frac{3\pi x}{2} + \frac{1}{5}\cos\frac{5\pi x}{2} - \frac{1}{7}\cos\frac{7\pi x}{2} + \dots\right)$

odd: $\frac{2}{\pi}\left(\sin\frac{\pi x}{2} + \sin \pi x + \frac{1}{3}\sin\frac{3\pi x}{2} + \frac{1}{5}\sin\frac{5\pi x}{2} + \frac{1}{3}\sin 3\pi x + \frac{1}{7}\sin\frac{7\pi x}{2} + \frac{1}{9}\sin\frac{9\pi x}{2} + \frac{1}{5}\sin 5\pi x + \dots\right)$

27. even: $\frac{3}{4} + \frac{4}{\pi^2}\left(\cos\frac{\pi x}{2} - \frac{2}{2^2}\cos \pi x + \frac{1}{3^2}\cos\frac{3\pi x}{2} + \frac{1}{5^2}\cos\frac{5\pi x}{2} - \frac{2}{6^2}\cos 3\pi x + \frac{1}{7^2}\cos\frac{7\pi x}{2} + \frac{1}{9^2}\cos\frac{9\pi x}{2}\right.$
$\left. - \frac{2}{10^2}\cos 5\pi x + \dots\right)$ odd: $\frac{2}{\pi}\left(\sin\frac{\pi x}{2} + \frac{1}{2}\sin \pi x + \frac{1}{3}\sin\frac{3\pi x}{2} + \frac{1}{4}\sin 2\pi x + \frac{1}{5}\sin\frac{5\pi x}{2} + \dots\right)$
$+ \frac{4}{\pi^2}\left(\sin\frac{\pi x}{2} - \frac{1}{3^2}\sin\frac{3\pi x}{2} + \frac{1}{5^2}\sin\frac{5\pi x}{2} - \dots\right)$

29. even: $\frac{2}{3} - \frac{16}{\pi^3}\left(\cos\frac{\pi x}{2} - \frac{1}{3^3}\cos\frac{3\pi x}{2} + \frac{1}{5^3}\cos\frac{5\pi x}{2} - \dots\right) - \frac{8}{\pi^2}\left(\frac{1}{2^2}\cos \pi x - \frac{1}{4^2}\cos 2\pi x + \frac{1}{6^2}\cos 3\pi x - \dots\right)$

odd: $\frac{2}{\pi}\left(\sin\frac{\pi x}{2} - \frac{1}{2}\sin \pi x + \frac{1}{3}\sin\frac{3\pi x}{2} - \frac{1}{4}\sin 2\pi x + \frac{1}{5}\sin\frac{5\pi x}{2} - \dots\right) + \frac{8}{\pi^2}\left(\sin\frac{\pi x}{2} - \frac{1}{3^2}\sin\frac{3\pi x}{2} + \frac{1}{5^2}\sin\frac{5\pi x}{2} - \dots\right)$
$- \frac{16}{\pi^3}\left(\sin\frac{\pi x}{2} + \frac{2}{2^3}\sin \pi x + \frac{1}{3^3}\sin\frac{3\pi x}{2} + \frac{1}{5^3}\sin\frac{5\pi x}{2} + \frac{2}{6^3}\sin 3\pi x + \frac{1}{7^3}\sin\frac{7\pi x}{2} + \frac{1}{9^3}\sin\frac{9\pi x}{2} + \dots\right)$

PROBLEMS 11.5 *(page 558)*

1. $\lambda_k = k^2, y_k = A\sin kx, k = 1, 2, 3, \dots$ **3.** $\lambda_k = k^2, y_k = A\sin k\left(x + \frac{\pi}{2}\right); k = 1, 2, 3, \dots$

5. $\lambda_k = \frac{k^2\pi^2}{4}, y_k = A\sin\left(\frac{k\pi\ln x}{2}\right), k = 1, 2, 3, \dots$ **7.** $\lambda_k = k^2\pi^2, y_k = A\sin\frac{k\pi}{x}, k = 1, 2, 3, \dots$

PROBLEMS 11.6 *(page 562)*

1. $\lambda_k = k^2$ are real and simple, $y_k = A\sin kx, k = 1, 2, 3, \dots,$ are orthogonal.

3. $\lambda_k = k^2$ are real and simple, $y_k = A\sin k\left(x + \frac{\pi}{2}\right), k = 1, 2, 3, \dots,$ are orthogonal. **5.** $\lambda_k = k^2\pi^2$ are real and simple, $y_k = A\sin\frac{k\pi}{x}, k = 1, 2, 3, \dots,$ are orthogonal with respect to $\frac{1}{x^2}$ on $\frac{1}{2} \le x \le 1$.

9. The eigenvalues of this Euler equation are $1 \pm \frac{2k\pi}{\ln 2}i, k = 1, 2, 3, \dots$.

PROBLEMS 11.7 *(page 567)*

1. $\sum_{k=1}^{\infty} \frac{2(-1)^k}{k(k^2-1)} \sin kx$ **3.** $4\sum_{k=1}^{\infty} \frac{\left[1+(-1)^k\left(\frac{2k-1}{2}\pi\right)\right]}{\left(\frac{2k-1}{2}\pi\right)^3\left[\left(\frac{2k-1}{4}\pi\right)^2-1\right]} \sin\left(\frac{2k-1}{2}\right)\pi x$

5. $2(\ln 2)^2 \sin(\ln 2)\pi \sum_{k=1}^{\infty} \frac{(-1)^k k}{[3(\ln 2)^2-(k\pi)^2][(\ln 2)^2-(k\pi)^2]} \sin\left(\frac{k\pi}{\ln 2}\ln x\right)$ **13.** $K(x,t)=\begin{cases} -\frac{\sinh(1-t)\sinh x}{\sinh(1)}, & x<t, \\ -\frac{\sinh(1-x)\sinh t}{\sinh(1)}, & t<x \end{cases}$

17. $K(x,t)=\begin{cases} -\frac{\sin 2x\cos 2(1-t)}{2\cos 2}, & x<t, \\ -\frac{\cos 2(1-x)\sin 2t)}{2\cos 2}, & x<t \end{cases}$

PROBLEMS 11.8 *(page 572)*

1. $a_0=2,\ \ a_1=a_2=0$ **3.** $a_1=\frac{2}{\pi},\ \ a_2=\frac{-1}{\pi},\ \ a_3=\frac{2}{(3\pi)}$ **5.** $a_0=\frac{2}{3},\ \ a_n=\frac{4(-1)^n}{n^2\pi^2},\ \ n=1,2,3$

7. $\frac{1}{5}P_0+\frac{4}{7}P_2+\frac{8}{35}P_4$ **9.** $\frac{18}{5}P_3-\frac{16}{3}P_2+\frac{62}{5}P_1-\frac{26}{3}P_0$ **11.** $\sin x\approx 3(\sin 1-\cos 1)x$ **13.** $\sinh x\approx 3e^{-1}x$

15. $\int_{-1}^{1} x^k p(x)\,dx=0$, for $k=0,1,\ldots,n-1$ implies $p(x)$ is orthogonal to $P_0, P_1, \ldots, P_{n-1}$. Since deg $p=n$, p must be a multiple of P_n.

17. (a) Integrate by parts with $u=P_m$ and $v=\frac{d^{n-1}}{dx^{n-1}}[(x^2-1)^n]$. **(b)** $x^2=1$ is a factor of v.

19. The nth degree term of P_n has the coefficient $\frac{(2n)!}{2^n(n!)^2}$. Thus $P_n^{(n)}=\frac{(2n)!}{2^n n!}$ so that $\|P_n\|^2=\frac{(2n)!}{2^{2n}(n!)^2}\int_{-1}^{1}(1-x^2)^n\,dx$.

Substituting $x=\sin\theta$, we get by entry 218, $\|P_n\|^2=\frac{(2n)!}{2^{2n}(n!)^2}\int_{-\pi/2}^{\pi/2}\cos^{2n+1}\theta\,d\theta=\frac{2}{2n+1}$.

PROBLEMS 11.9 *(page 574)*

1. $\frac{1}{\pi}[2\sin\pi x-\sin 2\pi x+2\sin 3\pi x-\sin 4\pi x+2\sin 5\pi x-\sin 6\pi x]$

3. $1-\frac{8}{\pi^2}\left[\cos\frac{\pi x}{2}+\frac{1}{3^2}\cos\frac{3\pi x}{2}+\frac{1}{5^2}\cos\frac{5\pi x}{2}+\frac{1}{7^2}\cos\frac{7\pi x}{2}\right]$

5. $\frac{4}{3}+\frac{1}{\pi}\left[4\sin\frac{\pi x}{2}-2\sin\pi x+4\sin\frac{3\pi x}{2}-\sin 2\pi x+4\sin\frac{5\pi x}{2}-2\sin 3\pi x\right]-\frac{1}{\pi^2}\left[16\cos\frac{\pi x}{2}-4\cos\pi x+\right.$ $\left.16\cos\frac{3\pi x}{2}-\cos 2\pi x+16\cos\frac{5\pi x}{2}-4\cos 3\pi x\right]$

7. $\frac{7}{4}-\frac{5}{\pi}\left[\sin\frac{\pi x}{5}+\frac{1}{2}\sin\frac{2\pi x}{5}+\frac{1}{3}\sin\frac{3\pi x}{5}+\frac{1}{4}\sin\frac{4\pi x}{5}+\frac{1}{5}\sin\pi x\right]-\frac{10}{\pi^2}\left[\cos\frac{\pi x}{5}+\frac{1}{3^2}\cos\frac{3\pi x}{5}+\frac{1}{5^2}\cos\pi x\right]$

9. $\frac{-3}{4}+\frac{15}{\pi}\left[\sin\frac{\pi x}{5}-\frac{1}{6}\sin\frac{2\pi x}{5}+\frac{1}{3}\sin\frac{3\pi x}{5}-\frac{1}{12}\sin\frac{4\pi x}{5}+\frac{1}{5}\sin\pi x\right]+\frac{10}{\pi^2}\left[\cos\frac{\pi x}{5}+\frac{1}{3^2}\cos\frac{3\pi x}{5}+\frac{1}{5^2}\cos\pi x\right]$

REVIEW EXERCISES FOR CHAPTER 11 *(page 576)*

1. $\left\{\sqrt{\frac{2}{\pi}}\sin nx\right\}$ **3.** $\{\sqrt{2}\sin 2n\pi x, \sqrt{2}\cos 2n\pi x\}$ **5.** $-1+\pi\sin x-\frac{1}{2}\cos x+2\sum_{n=2}^{\infty}\frac{\cos nx}{n^2-1}$

7. $\frac{2}{\pi}\left[1-2\sum_{n=1}^{\infty}\frac{\cos 2nx}{4n^2-1}\right]$ **9.** $2\pi\sum_{n=1}^{\infty}\left\{\left(\frac{1}{2n-1}-\frac{4}{(2n-1)^{2n-1}\pi^2}\right)\sin(2n-1)x-\frac{1}{2^n}\sin 2nx\right);\ \frac{\pi^2}{3}+4\sum_{n=1}^{\infty}\frac{(-1)}{n^2}\cos nx$

11. $\frac{2}{\pi}\sum_{n=1}^{\infty}\frac{(-1)^{n-1}}{n}\sin n\pi x;\ \frac{-4}{\pi^2}\sum_{n=1}^{\infty}\frac{1}{(2n-1)^2}\cos(2n-1)\pi x$

13. $\frac{-32}{\pi^2}\left(\sin\frac{\pi x}{2}+\frac{\pi^2}{8}\sin\frac{2\pi x}{2}+\frac{1}{3^3}\sin\frac{3\pi x}{2}+\frac{\pi^2}{16}\sin\frac{4\pi x}{2}+\ldots\right);\ \pi^3=32\left(1-\frac{1}{3^3}+\frac{1}{5^3}-\frac{1}{7^3}+\ldots\right)$

15. $\lambda_n = \frac{(2n-1)^2}{4}$, $y_n(x) = \cos\frac{(2n-1)x}{2}$, $n = 1, 2, 3, \ldots$

17. $y_n(x) = \sin\sqrt{\lambda_n}x$, where λ_n are the roots of $\tan k = -k$. For large n, $\sqrt{\lambda_n} \approx \frac{(2n-1)\pi}{2}$. **19.** Same as Exercise 17.

CHAPTER 12

PROBLEMS 12.1 *(page 583)*

1. $z(x, y) = (x + y)e^{\sqrt{2}y}$ **3.** $z(x, y) = (x - y)e^{-\sqrt{2}y} + e^{x-(1+\sqrt{2})y}$ **5.** $z(x, y) = \frac{7}{2} + e^{x+(5/4)y} - \frac{7}{2}e^{y/2}$ **7.** $z(x, y) = ye^x$

9. $z(x, y) = (x + y)e^{\sqrt{2}y}$ **11.** $z(x, y) = (x - y)e^{-\sqrt{2}y} + e^{x-(1+\sqrt{2})y}$ **13.** $z(x, y) = \frac{7}{2} + e^{x+(5/4)y} - \frac{7}{2}e^{y/2}$ **15.** $z(x, y) = ye^x$

17. $z(x, y) = \ln x + xe^{-y}$ **19.** $z(x, y) = y^2(x - \ln y)e^{-x}$ **21.** $z(x, y) = \sin(x - y^2)e^{y^2} - 2$

23. $dx = dy = \frac{dz}{z}$ so $x + c_1 = y$ and $z = c_2e^x$. If $c_2 = f(c_1)$, then $ze^{-x} = f(y - x)$ or $z(x, y) = e^xf(y - x)$. Setting $x = y$, we get $e^y = z(y, y) = e^yf(0)$, implying that $f(0) = 1$. Hence, any such function f will do. For example, $z(x, y) = e^x(y - x + 1)$ is a solution. **25.** $z - x = f(x - y)$ so $f(0) = 7$. Hence $z = x + f(x - y)$ for any differentiable function satisfying $f(0) = 7$ will do. For example $z(x, y) = (n + 1)x - ny + 7$, for all n.

PROBLEMS 12.2 *(page 589)*

1. $z(x, y) = f(x - y)$ (for any differentiable function f) **3.** $z(x, y) = f(xe^{-y})$ **5.** $z^2 - 2y = f(xe^{-z})$

7. $z = \cos\left(x - \frac{y}{z^2}\right)$ **9.** $z = \left(1 - \frac{1}{x} + \frac{2}{y}\right)^{-1}$ **11.** $z = y\sqrt{3(z^2 - x^2)} - x$ **13.** $z = \frac{x}{1+y}$ **15.** $\frac{1}{z} = f\left(\frac{x}{y}\right) - \ln y$

17. $z = \frac{y^2(1 + e^{x/y}) - (1 - e^{x/y})}{y^2(1 + e^{x/y}) + (1 - e^{x/y})}$ **19.** $y^2 - 2xz + \frac{2}{3}x^3 = f\left(\frac{x^2}{2} - z\right)$ **21.** $z = f\left(x - \left(1 + \frac{3}{2}z\right)y\right)$

PROBLEMS 12.3 *(page 593)*

1. No; $u = \frac{x}{1+t}$ **3.** $u = \frac{1 + 2tx - \sqrt{1 + 4tx}}{2t^2} = \frac{1 + 2tx - (1 + 2tx - 2t^2x^2 + \ldots)}{2t^2} \to x^2$ as $t \to 0$ so no shocks develop. **5.** The shock (x_s, t_s) lies on the line $x - tf(x_1) = x_1$ where $f'(x - tu) = f'(x_1)$. Hence $t_s = -\frac{1}{f'(x_1)}$ and $x_s = x_1 - \frac{f(x_1)}{f'(x_1)}$. **7.** $u = f(x - F'(u)t)$ so $u_x = \frac{f'}{(1 + tf'F'')}$ and shocks develop whenever $1 + tf'(x - F'(u)t)F''(u) = 0$.

PROBLEMS 12.4 *(page 597)*

1. Hyperbolic **3.** Parabolic **5.** Elliptic **7.** Elliptic **9.** Elliptic **11.** $\frac{\partial^2 z}{\partial x^2} + \frac{\partial^2 z}{\partial y^2} - z = 0$

13. $\frac{\partial^2 z}{\partial x^2} - \frac{\partial^2 z}{\partial y^2} + \frac{57}{4}z = 0$ **15.** $\frac{\partial^2 z}{\partial x^2} + \frac{z}{2} = 0$ **17.** $\frac{\partial^2 z}{\partial x^2} + \frac{\partial^2 z}{\partial y^2} + 5z = 0$ **19.** $\frac{\partial^2 z}{\partial x^2} - \frac{\partial^2 z}{\partial y^2} - \frac{1}{12}z = 0$

21. Multiply $-AC$ to get $\left[b^2 - \frac{(a-c)^2}{4}\right]\sin^2 2\alpha + b(a - c)\sin 2\alpha\cos 2\alpha - ac$. Then substitute $a - c = \frac{2b}{\tan 2\alpha}$ to obtain $b^2 - ac$. **23.** Elliptic in second and fourth quadrants, hyperbolic in first and third quadrants, and parabolic on axes. **25.** Elliptic for $|y| < |x|$, hyperbolic for $|y| > |x|$, and parabolic for $|y| = |x|$. **27.** Elliptic for $y > e^{-x}$, hyperbolic for $y < e^{-x}$, and parabolic for $y = e^{-x}$.

PROBLEMS 12.5 *(page 603)*

1. -0.0075 m **3.** 0.0 m **5.** 0.1 m **7.** 0.2 m **9.** By **(14)** $y(x, t) = h(x + ct) + g(x - ct)$. Since $y(x, 0) = 0$ we get $h(x) = -g(x)$. Thus $f'(x) = y_t(x, 0) = c[h'(x + ct) + h'(x - ct)]|_{t=0} = 2ch'(x)$. Thus $h(x) = \frac{1}{2c}[f(x) + k]$ from which the result follows. Furthermore, since $y(0, t) = 0$, it follows that f is even with period $2L$. **11.** $-\frac{0.045}{\pi}$ m **13.** $\frac{0.03}{\pi}\left[\frac{\pi}{6} - \frac{1}{4}\right]$ m

15. 0.002 m **17.** $y(x, t) = \sin\frac{\pi x}{2}\left[\cos\frac{\pi t}{2} + \frac{2}{\pi}\sin\frac{\pi t}{2}\right]$ **19.** If $RC = LG$, let $v(x, t) = u(x, t)e^{-Rt/L}$. Then u satisfies the wave equation $u_{tt} = \left(\frac{1}{LC}\right)u_{xx}$, so that v has the form $v(x, t) = e^{-Rt/L}\left[g\left(x - \frac{1}{\sqrt{LC}}t\right) + h\left(x + \frac{1}{\sqrt{LC}}t\right)\right]$. A much more complicated solution involving integral equations exists when $RC \neq LG$ (see Zachmanoglou and Thoe, *Introduction to Partial Differential Equations*, Williams & Wilkins, Baltimore, MD (1976), p. 375).

PROBLEMS 12.6 (page 609)

1. $y(x, t) = 0.1 \cos \frac{\pi c t}{L} \sin \frac{\pi x}{L}$ **3.** $y(x, t) = \frac{0.8L^2}{\pi^3} \sum_{n=0}^{\infty} \frac{1}{(2n+1)^3} \cos \frac{(2n+1)\pi c t}{L} \sin \frac{(2n+1)\pi x}{L}$

5. $y(x, t) = \frac{2.4}{\pi^2} \sum_{n=1}^{\infty} \frac{1}{n^2} \sin\left(\frac{n\pi}{2}\right) \cos\left(\frac{n\pi c t}{6}\right) \sin\left(\frac{n\pi x}{6}\right)$ **7.** $T' = kT$ and $X'' + kX = 0$

9. $X' + X = 0$ or $T' = 0$; hence $y(x, t) = c_1 t e^{-x} + c_2$ **11.** $Y'' - kyY = 0$ and $xX' + kX = 0$

13. $\Theta'' + k\Theta = 0$ and $r^2R'' + rR' - kR = 0$ **15. (a)** $y(x, t) = \sqrt{x}(b_1 e^{\sqrt{k}t} + b_2 e^{-\sqrt{k}t})(c_1 x^{\sqrt{1-4k}/2} + c_2 x^{-\sqrt{1-4k}/2})$, all k

(b) $y(x, t) = c_1\sqrt{x}\left(x^{\sqrt{1-4k}/2} - \frac{\pi^{\sqrt{1-4k}}}{x^{\sqrt{1-4k}/2}}\right)(b_1 e^{\sqrt{k}t} + b_2 e^{-\sqrt{k}t})$, $k > 0$

17. (a) $y(x, t) = (b_1 e^{\sqrt{k}t} + b_2 e^{-\sqrt{k}t})(c_1 \cos \sqrt{k}\,x + c_2 \sin \sqrt{k}\,x)$, all k

(b) $c_1 = 0$ and $k = 1^2, 2^2, 3^2, \ldots$, so $y(x, t) = \sin nx(b_1 e^{nt} + b_2 e^{-nt})$ **(c)** Use a Fourier sine series.

19. (a) $y(x, t) = (b_1 e^{\sqrt{k}x} + b_2 e^{-\sqrt{k}x})(c_1 e^{\sqrt{k-1}t} + c_2 e^{-\sqrt{k-1}t})$

(b) Let $k < 0$, then $b_1 = 0$ and $k = -1^2, -2^2, \ldots$ so that $y(x, t) = \sin nx(c_1 \cos \sqrt{n^2+1}\,t + c_2 \sin \sqrt{n^2+1}\,t)$.

(c) Use a Fourier sine series.

PROBLEMS 12.7 (page 613)

1. $T(x, t) = \frac{4}{\pi} \sum_{n=1}^{\infty} \frac{[1 - (-1)^n]}{n^3} (\sin nx) e^{-n^2 \delta t}$ **3.** $T(x, t) = 4 \sum_{n=1}^{\infty} \frac{[2(-1)^{n+1} - 1]}{n^3} (\sin nx) e^{-n^2 \delta t}$

5. $T(x, t) = \frac{2}{\pi} \sum_{n=1}^{\infty} \frac{1}{n^2} \left[\sin \frac{n\pi}{2} - \frac{n\pi}{2} \cos \frac{n\pi}{2}\right] (\sin nx) e^{-n^2 \delta t}$

7. $\frac{[(L - x)T_1 + xT_2]}{L}$ **9.** $f(x) = \frac{c_0^2}{T_1} \cos cx + c_0 c_2 \sin cx$, where $c_2 = \frac{c_0(1 - \cos cL)}{T \sin cL}$. If cL is a multiple of π, then c_2 is arbitrary.

PROBLEMS 12.8 (page 618)

1. $T(x, y) = \frac{100}{\pi} \sum_{n=1}^{\infty} \frac{[1 - (-1)^n]}{n \sinh n\pi} \sin \frac{n\pi x}{L} \sinh \frac{n\pi y}{L}$; evaluate at $\left(\frac{L}{2}, \frac{L}{2}\right)$ and $\left(\frac{L}{4}, \frac{L}{4}\right)$. **3.** $T(x, y) = \frac{100}{\sinh \pi} \sin \frac{\pi x}{L} \sinh \frac{\pi y}{L}$

5. $T(x, y) = \frac{-4L^3}{\pi^3} \sum_{n=1}^{\infty} \frac{[1 + 2(-1)^n]}{n^3 \sinh n\pi} \sin \frac{n\pi x}{L} \sinh \frac{n\pi y}{L}$; evaluate at $\left(\frac{L}{4}, \frac{L}{4}\right)$ and $\left(\frac{L}{4}, \frac{3L}{4}\right)$.

7. $T(x, y) = \frac{-160L}{\pi^2} \sum_{n=1}^{\infty} \frac{n \sin \frac{2n\pi x}{L} \sinh \frac{2n\pi y}{L}}{(4n^2 - 1)^2 \sinh (2n\pi)}$ **9.** The solution is the superposition of the solutions T_1 and T_2 of two problems: first where $T_1 = 50°$ on the top edge and 0° on the other edges; second where $T_2 = 50 \sin \frac{\pi y}{2L}°$ on the side edges and 0° on the horizontal edges. Thus

$$T(x, y) = T_1(x, y) + T_2(x, y) = \frac{200}{\pi} \sum_{n=0}^{\infty} \frac{(-1)^n}{(2n+1) \sinh \frac{(2n+1)\pi}{2}} \cos\left(\frac{2n+1}{2}\right)\frac{\pi x}{L} \sinh\left(\frac{2n+1}{2}\right)\frac{\pi y}{L}$$
$$- \frac{400}{\pi} \sum_{n=1}^{\infty} \frac{(-1)^n n}{(4n^2 - 1) \cosh n\pi} \cosh \frac{n\pi x}{L} \sin \frac{n\pi y}{L}.$$

11.
$$T(x, y) = 200 \sum_{n=0}^{\infty} \frac{(-1)^n}{\frac{\sinh(2n+1)\pi}{2}} \left[\frac{1}{(2n+1)\pi} - \left(\frac{2}{(2n+1)\pi}\right)^3\right] \cos\left(\frac{2n+1}{2}\right)\frac{\pi x}{L} \sinh\left(\frac{2n+1}{2}\right)\frac{\pi y}{L}$$
$$- \frac{100}{\pi} \sum_{n=1}^{\infty} \frac{(-1)^n}{n \cosh n\pi} \cosh \frac{n\pi x}{L} \sin \frac{n\pi y}{L}$$

PROBLEMS 12.9 (page 625)

1. $T(r, \theta) = 2 \sum_{n=1}^{\infty} \frac{r^n(-1)^n}{n} \sin n\theta$ **3.** $T(r, \theta) = \frac{1 - r^2}{2\pi} \int_{-\pi}^{\pi} \frac{(\pi - \phi)\, d\phi}{r^2 + 1 - 2r \cos(\theta - \phi)} \Bigg|_{(\frac{1}{2}, 0)} = \frac{3}{8\pi} \int_{-\pi}^{\pi} \frac{(\pi - \phi)\, d\phi}{\frac{5}{4} - \cos \phi}$

5. (a) $A_0 = 1$, $A_2 = -\frac{1}{2}$; all other coefficients are zero, so that $T(r, \theta) = \frac{1}{2}(1 - r^2 \cos 2\theta)$. Hence $T\left(\frac{1}{2}, 0\right) = \frac{3°}{8}$.

(b) $T\left(\frac{1}{2}, 0\right) = \frac{3}{8\pi} \int_{-\pi}^{\pi} \frac{\sin^2 \phi}{\frac{5}{4} - \cos \phi}\, d\phi = \frac{3°}{8}$. (The last equality can be proved using the calculus of residues.)

7. (a) $T(r,\theta) = -\dfrac{\pi^2}{3} - 4\sum_{n=1}^{\infty}(-r)^n\left(\dfrac{\cos n\theta}{n^2} + \dfrac{\pi \sin n\theta}{n}\right)$ **(b)** $T(r,\theta) = \dfrac{1-r^2}{2\pi}\displaystyle\int_{-\pi}^{\pi}\dfrac{\phi(2\pi-\phi)\,d\phi}{r^2+1-2r\cos(\theta-\phi)}$

9. Integrating by parts, we get $|a_n| \le \dfrac{T}{n^2\pi^2}\left|\displaystyle\int_{-T}^{T} F''(x)\cos\dfrac{n\pi x}{T}\,dx\right| < \dfrac{2MT^2}{n^2\pi^2}$; compare with $\sum \dfrac{1}{n^2}$.

11. (a) If u attains a maximal value at (x_0, y_0), the cross sections of the surface with the planes $x = x_0$ and $y = y_0$ have slope zero at (x_0, y_0) and are concave down. Thus $u_{xx} \le 0$ and $u_{yy} \le 0$ at (x_0, y_0). **(c)** $w_{xx} + w_{yy} = u_{xx} + u_{yy} + 4\epsilon > 0$

PROBLEMS 12.10 *(page 630)*

1. $y(x,t) = t$ **3.** $T(x,t) = 1 - \operatorname{erf}\left(\dfrac{x}{2\sqrt{\delta t}}\right)$ **5.** $T(x,t) = \displaystyle\int_0^t \sin\omega(t-u)\,\dfrac{x}{2\sqrt{\pi\delta u^3}}\,e^{-x^2/4\delta u}\,du$

PROBLEMS 12.11 *(page 632)*

1. $z = x + \sqrt{x^2(x^2-2y^2)} - \sqrt[4]{x^2(x^2-2y^2)}$ **3.** $z^2 = \dfrac{y^3-1}{3} + e^{2x-y+1}$

5. $y(x,t) = \dfrac{16}{\pi}\left[\dfrac{1}{3}\sin\left(\dfrac{\pi t}{2}\right)e^{-\pi^2 t} - \sum_{n=1}^{\infty}\dfrac{1}{(4n^2-1)(2n+3)}\sin\dfrac{(2n+1)\pi x}{2}\,e^{-(2n+1)^2\pi^2 t}\right]$

REVIEW EXERCISES FOR CHAPTER 12 *(page 633)*

1. $z(x,y) = xe^y$ **3.** $z(x,y) = x + e^y$ **5.** $T(x,t) = \dfrac{4a^2}{\pi^3}\sum_{n=1}^{\infty}\dfrac{[1-(-1)^n]}{n^3}\sin\dfrac{n\pi x}{a}\,e^{-(n\pi/a)^2\delta t}$

7. $T(x,t) = \dfrac{4a}{\pi^2}\sum_{n=0}^{\infty}\dfrac{(-1)^n}{(2n+1)^2}\sin\dfrac{(2n+1)\pi x}{a}\,e^{-[(2n+1)\pi/a]^2\delta t}$

9. $y = X(x)T(t)$ where the general solutions to X and T are

$$X(x) = \begin{cases} c_1x^{\lambda_1} + c_2x^{\lambda_2}, & k > -\frac{1}{4} \\ c_1\sqrt{x} + c_2\sqrt{x}\ln x, & k = -\frac{1}{4} \\ \sqrt{x}\,[c_1\cos(\ln|x|^{\beta}) + c_2\sin(\ln|x|^{\beta})], & k < -\frac{1}{4}, \end{cases}$$

$$T(t) = \begin{cases} b_1t^{\lambda_1} + b_2t^{\lambda_2}, & k > -\frac{1}{4} \\ b_1\sqrt{t} + b_2\sqrt{t}\ln t, & k = -\frac{1}{4} \\ \sqrt{t}\,[b_1\cos(\ln|t|^{\beta}) + b_2\sin(\ln|t|^{\beta})], & k < -\frac{1}{4}, \end{cases}$$

where $\left.\begin{matrix}\lambda_1\\ \lambda_2\end{matrix}\right\} = \dfrac{1 \pm \sqrt{1+4k}}{2}$, respectively, and $\beta = \dfrac{\sqrt{-(1+4k)}}{2}$.

11. $y = X(x)T(t)$ where the general solutions to X and T are

$$X(x) = \begin{cases} c_1e^{\sqrt{k+1}\,x} + c_2e^{-\sqrt{k+1}\,x}, & k > -1 \\ c_1 + c_2x, & k = -1 \\ c_1\cos\sqrt{-(k+1)}\,x + c_2\sin\sqrt{-(k+1)}\,x, & k < -1 \end{cases}$$

$$T(t) = \begin{cases} b_1e^{\sqrt{k}\,t} + b_2e^{-\sqrt{k}\,t}, & k > 0 \\ b_1 + b_2t, & k = 0 \\ b_1\cos\sqrt{-k}\,t + b_2\sin\sqrt{-k}\,t, & k < 0. \end{cases}$$

13. $y(x,t) = \dfrac{1}{2}\left(3\sin\dfrac{\pi x}{a}\cos\dfrac{c\pi t}{a} - \sin\dfrac{3\pi x}{a}\cos\dfrac{3c\pi t}{a}\right)$ **15. (a)** $\dfrac{\partial^2 y}{\partial t^2} = c^2\dfrac{\partial^2 y}{\partial x^2} - 2b\dfrac{\partial y}{\partial t}$, where b is a constant of proportionality; **(b)** $y(x,t) = \dfrac{e^{-bt}}{4}\left[3\left(\cos k_1t + \dfrac{b}{k_1}\sin k_1t\right)\sin\dfrac{\pi x}{a} - \left(\cos k_3t + \dfrac{b}{k_3}\sin k_3t\right)\sin\dfrac{3\pi x}{a}\right]$;
(c) $y(x,t) = e^{-bt}\sum_{n=1}^{\infty}A_n\left(\cos k_nt + \dfrac{b}{k_n}\sin k_nt\right)\sin\dfrac{n\pi x}{a}$, where $k_n = \dfrac{\sqrt{n^2\pi^2c^2 - b^2a^2}}{a}$ and $A_n = \dfrac{2}{a}\displaystyle\int_0^a f(x)\sin\dfrac{n\pi x}{a}\,dx$.

APPENDICES

PROBLEMS APPENDIX 3 *(page A24)*

1. Yes; $\delta = \dfrac{1}{216}\left(\text{better: } \delta = \dfrac{1}{50}\right)$ **3.** Yes, but $a = b = 1$ is not possible. You need $a + b < 1$. If $a = b = \dfrac{1}{3}$, then $\delta = \dfrac{1}{4}$.

5. Yes; $\delta = 1$ **7.** Yes; but you need $b < 1$. If $b = \dfrac{1}{2}$, then $\delta = 1$.

9. Yes; $\delta = \dfrac{1}{2}$ (better: $\delta = +\infty$; there is a unique solution defined for $-\infty < t < \infty$) **11.** **(a)** $k = 1$ **(b)** $k = 1$

(c) $k = 3\left(\text{a smaller constant is } \sup_{|x|\le 1}\left|2x\sin\dfrac{1}{x} - \cos\dfrac{1}{x}\right|\right)$; **(d)** $972\sqrt{3} \approx 1683$ **13.** $x_n(t) = 3\displaystyle\sum_{k=0}^{n}\frac{(t-3)^k}{k!}$

25. **(a)** $\begin{pmatrix} x_1 \\ x_2 \end{pmatrix}' = \begin{pmatrix} tx_1x_2 + t^2 \\ x_1^2 + x_2^2 + t \end{pmatrix}$; $x(0) = \begin{pmatrix} 1 \\ 2 \end{pmatrix}$ **(b)** $k = 10$ **(c)** $\delta = \dfrac{1}{7\sqrt{5}} \approx 0.064$

27. **(a)** $\begin{pmatrix} x_1 \\ x_2 \\ x_3 \end{pmatrix}' = \begin{pmatrix} x_1^2x_2 \\ x_3 + t \\ x_3^2 \end{pmatrix}$; $x(0) = c$ **(b)** $k \le \displaystyle\max_{\substack{1\le i\le 3 \\ (t,x)\in D}}\left|\frac{\partial f}{\partial x_i}\right| \le \max\{1 + a, 2(b + |c|)^2\}$

(c) $\delta = \min\left\{a, \dfrac{b}{M}\right\}$ where $M \le ((b + |c|)^6 + (b + |c| + a)^2 + (b + |c|)^4)$

PROBLEMS APPENDIX 4 *(page A35)*

1. $D = -2$; $x_1 = 3, x_2 = 0$ **3.** $D = 0$; infinite number—the entire line $x_1 + 2x_2 = 3$ **5.** $D = 0$; no solution
7. $D = -29$; $x_1 = x_2 = 0$ **9.** -56 **11.** 140 **13.** 96 **15.** -120 **17.** -132 **19.** -398
23. $x_1 = -3, x_2 = 0, x_3 = 5$ **25.** $x_1 = -\dfrac{26}{9}, x_2 = -\dfrac{1}{2}, x_3 = \dfrac{19}{9}$ **27.** $x_1 = -2, x_2 = 0, x_3 = 1, x_4 = 5$

PROBLEMS APPENDIX 5 *(page A42)*

1. $9 - 7i$ **3.** 2 **5.** $-27 + 5i$ **7.** $5\sqrt{2}\,e^{\pi i/4}$ **9.** $3\sqrt{2}\,e^{-\pi i/4}$ **11.** $6e^{\pi i/6}$ **13.** $8e^{-\pi i/6}$ **15.** $2e^{-2\pi i/3}$
17. -2 **19.** $-\left(\dfrac{\sqrt{2}}{4}\right) - \left(\dfrac{\sqrt{2}}{4}\right)i$ **21.** $-2\sqrt{3} + 2i$ **23.** $-\dfrac{3}{2} - \left(\dfrac{3\sqrt{3}}{2}\right)i$ **25.** $\cos 1 + i\sin 1 \approx 0.54 + 0.84i$
27. $4 - 6i$ **29.** $7i$ **31.** $2e^{-\pi i/7}$ **33.** $3e^{4\pi i/11}$
39. Circle of radius a centered at z_0 if $a > 0$, single point z_0 if $a = 0$, empty set if $a < 0$

INDEX